W9-CLP-918

Man, settlement and urbanism

Man, settlement and urbanism

---

edited by
PETER J. UCKO,
RUTH TRINGHAM
and
G. W. DIMBLEBY

Proceedings of a meeting of the
Research Seminar in Archaeology and Related Subjects
held at the Institute of Archaeology, London University

Duckworth

First published in 1972 by
Gerald Duckworth and Co. Ltd.
43 Gloucester Crescent, NW1

ISBN 0 7156 0589 5

Made and printed in Great Britain by
The Garden City Press Limited, Hertfordshire, SG6 1JS, England

CONTENTS

The plates are between pp. 948 and 949.

Preface

On 5th, 6th and 7th December 1970 a meeting of the Research Seminar in Archaeology and Related Subjects, was held at the Institute of Archaeology, London University, to discuss Settlement Patterns and Urbanization. This meeting—like its predecessor on the Domestication and Exploitation of Plants and Animals, which also resulted in a book—was called to enable researchers in different disciplines to meet in an attempt to bridge the gaps between them by personal contact and discussion. As shown by the title of the meeting, we envisaged discussion on a very broad front indeed, with the third day devoted specifically to urban forms of settlement. The response to the announcement of the meeting was overwhelming and we had reluctantly to decline a considerable number of suggested contributions. But we were able to provide close circuit television relay to the lecture theatre of the Institute of Archaeology where, we understand, a large number of people entered into lively discussion on their own.

The range and scope of papers included in this volume reflect the unique importance of the subject and, we believe, the essential nature of inter-disciplinary approaches to such subjects. Both the concept of non-urban settlement and the nature of urbanism itself are discussed. Mobility and settled life are considered in detail as well as such factors as demarcation and defence, population and disease. Finally, there are several studies of specific communities and particular periods of antiquity. Although in such a meeting as this there is obviously no chance for more than partial coverage of a particular geographical area or chronological period, we believe that the papers collected here on Ancient Egypt, for example, represent the most complete coverage available on the subject.

As is the practice with these seminars the papers were all circulated beforehand and the book consists of these papers revised by their authors (sometimes very considerably) in the light of the discussions which took place. Some of the contributors were unable at the last moment to attend the seminar (see the list of participants) and some of their papers were not discussed in detail at the meeting. One (Professor V. N. Masson's) has been included here in an English translation which unfortunately we have been unable to check with the author. The important job of chairing discussion sessions was undertaken by Mr. P. Burnham, Dr. D. Clarke, Mr. P. Fowler, Dr. J. Garlick, Professor J. Hawkes, Professor I. Lewis, Dr. G. Lienhardt,

Dr. I. Longworth, Professor A. Momigliano and Mr. K. Painter, to all of whom we offer our grateful thanks. As before (see *The Domestication and Exploitation of Plants and Animals*) Professor Stuart Piggott undertook the superhuman task of contributing a concluding paper, and we have again placed this paper, revised but in its original form, as the general conclusion to the book.

We have already stressed the inter-disciplinary nature of these meetings. Some of the participants were worried by the breadth of approach adopted. We ourselves make no apologies for this; we have received confirmation from a wide range of participants since the meeting of the positive (and in some cases, practical) co-operation which has resulted from contact between specialists from different disciplines. Many of the participants are seasoned "conference-goers"; the number who have written to us to say how stimulating they found the meeting has greatly encouraged us. We still derive immense pleasure from reading in one letter the following:

> Your symposium was disconcerting to a number of old and comfortable patterns of thought, and I suppose that many of us will require some months to get all our ducks lined up in a row again . . . a symposium that was truly one of the most educational—and enjoyable—that I have ever encountered.

The meeting itself was organized into a number of discussion sessions under different chairmen (see above) but these sessions are not reflected exactly in the organization of this volume. The division and grouping of the papers published here are inevitably arbitrary and dependent on a personal assessment of the quality and implications of urbanism. We have tried various combinations of papers and have settled on groupings which, we hope, emphasize important variables and at the same time allow ease of reference to particular geographical areas or ancient cultures. As we are aware that there cannot be one ideal way of organizing this volume, it is worth recording the following headings and order of the discussion sessions at the meeting itself: concepts of non-urban settlement; methods of investigation; mobility; factors influencing non-urban settlement; demarcation and defence; factors influencing agricultural settlements; concepts of urbanization; relationship between peripheral and urban settlement; population and settlement; factors influencing urbanization; urban organization. At the meeting, as well as the major groupings, we drew attention to sections of other papers relevant to that particular discussion session. It is interesting to record the success of this, to us, novel method of organization. In more than three-quarters of cases either the author of these sections contributed to the discussion or the papers were directly referred to by the main participants of that particular session.

After the seminar meeting, we received from Professor H. S. Smith an "additional note for publication *only* at discretion of editors".

This is a personal statement but it is one with which we sympathize and which we consider of intrinsic importance. It is also the kind of reaction to Research Seminar meetings which affords us immense encouragement. We are therefore pleased to publish Professor Smith's note exactly as it reached us:

> A consideration of the papers and discussions at this seminar and its predecessor confirms the necessity for archaeologists, historians, anthropologists, geographers and natural scientists concerned with aspects of the human environment to work together if we are to obtain a true picture of the way in which man and his society have evolved. At present, the modes of interpretation and analysis used in these various disciplines are inevitably largely dictated by the material with which they work. Prehistoric archaeologists have, it is true, adopted an interdisciplinary approach with some fruitful results; but because of the strict limitations on the types of material physically preserved from prehistoric times, interpretations of early human activity are largely materialistic, mechanistic and evolutionary in character. Social anthropologists, dealing with material gathered from a single society in a single environment at a single time, tend to analyse it in terms mainly of social and institutional relationships, though some emphasize the importance of environmental and historical aspects for which their evidence is defective. Because of the complexity of most of the societies with which they deal, the vast scale of the records and remains known to exist, and the countervailing inadequacy of these to provide a complete cultural picture at any one moment or a complete development through time, historians and historical archaeologists tend to be less analytical in approach, more catholic and less explicit in their assumptions. Often the sheer mass of material and the difficulties of physical, financial and political circumstances have led to failure to attain objectives postulated many years ago; for instance, no single Dynastic Egyptian town has been stratigraphically excavated over its whole area. The consciousness that such basic tasks are still not achieved, and that a wealth of archaeological and documentary information already recovered has still not been adequately published or studied, often tends to inhibit historical archaeologists from questing for new types of information and new conceptual and analytical frameworks for their treatment. But the combination of a long, detailed and well-preserved archaeological record in a relatively stable environment with a large body of social and historical information from written documents, from scenes preserved on standing moments and from works of art, still renders the great civilizations of the ancient Near East among the most apt of topics for any attempt to weld together the evolutionary, environmental, mechanistic, sociological, economic, historical and other approaches to the study of man.
>
> For the success of such studies it seems to me essential that scholars in various disciplines should not merely combine in the interpretation of

evidence already culled, as for the most part at present, but join in the planning and the execution of original field work; for the importance of the answers obtained depends necessarily upon the quality and relevance of the questions asked, and the adequacy of the techniques evolved to provide evidence to answer them. It is certainly expecting too much of an archaeologist dealing with a complex historic civilization to expect him to have the theoretical and conceptual knowledge of all the necessary disciplines to enable him to plan field experiments and evolve field techniques that will provide evidence adequate to answer all the questions that specialists consider relevant. Indeed most of us are inevitably prisoners of our own education, and in practice, under the heavy pressures of most field situations, are unlikely even to ask ourselves questions outside our own conceptual framework (as the past history of historical archaeology demonstrates). In my view there ought ideally to emerge from this seminar groups of scholars of varied training and approach working together to try to ensure that at a future date, instead of considering piecemeal evidence from a variety of regions, dates, and social contexts, we have before us integrated studies of particular societies in particular environments at particular dates, which will furnish a true basis for the comparative work which alone can lead to sound generalization. Such studies demand a sort of co-operative endeavour and self-abnegation rare in the annals of scholarship and are difficult in practice to achieve. If, however, the various branches of the study of man are eventually to allow an intellectually satisfying understanding of his nature and effect in the world, such co-operation will be required.

My advocacy here of co-operation and combination between disciplines normally distinguished as natural sciences, social sciences, and arts, does not imply any belief that the study of man can become a single scientific discipline in which all information can be quantified, all hypotheses tested, and all theoretical structures evaluated. On the contrary, as a large part of the life of man is in the mind, and his thoughts and beliefs have in history often dictated his action, I believe that human imagination will always be necessary for the understanding of the human species' evolution and history. For me, as for the ancient Greeks, Clio will ever be a Muse, and the story of humanity a work of human art. It is indeed in artistic quality and clarity that so much that we write today of man falls short and compares ill with the work of some of our forebears. But factual evidence must be the basis; and the recovery of fact is a scientific operation. Time and circumstance have obliterated, mercifully, as some think, the majority of past facts; if we are to elicit truth from what remains, then all available intellectual tools must be used at source in concert.

As for *Domestication*, we were extremely fortunate in having the services of Miss Susan Elizabeth Johnson, who not only undertook all the secretarial work essential to the meeting but gave invaluable assistance in the preparation of the book and compiled the Indexes.

Once again, we would like to conclude the Preface by stressing the informal and happy atmosphere in which this meeting took place. We are glad to see that this important feature of our meetings has come through the print of *Domestication* and communicated itself to some reviewers. We very much hope that the same will be true of the present volume. Last time we were fortunate enough to be able to conclude informally with two poems. This meeting produced no poetic masterpieces. We conclude instead with a splendid set of "instructions to editors" we received for one of the most interesting of the papers included in this volume:

> This is too long, it is bad, it is irrelevant. Tear it up, throw it away, burn it. . . . If, in the unlikely event of your accepting it, you feel the need to change the words, or the meaning, or the title, please go ahead. But *do not change* the punctuation.

PETER J. UCKO
RUTH TRINGHAM
G. W. DIMBLEBY
London 1971

*List of participants**

(Excluding those in overflow television room)

R. McC. Adams (The Oriental Institute, University of Chicago)
J. Alexander (Department of Extra-Mural Studies, University of London)
A. M. Ali (Selwyn College, Cambridge and Department of Archaeology, University of Khartoum)
W. Allan (School of Agriculture, Cambridge)
B. Allchin* (Department of Archaeology and Anthropology, Cambridge)
G. H. A. Bankes (Department of Prehistoric Archaeology, Institute of Archaeology, London)
B. W. Blouet (Department of Geography, University of Nebraska)
S. Bökönyi (Magyar Nemžeti Múzeum, Budapest)
R. Boyd (Paediatric Department, University College Hospital, London)
K. Branigan (Department of Classics, University of Bristol)
W. Bray (Department of Prehistoric Archaeology, Institute of Archaeology, London)
D. R. Brothwell (Sub-department of Anthropology, British Museum—Natural History—London)
P. Burnham (Department of Anthropology, University College London)
J. D. Clark (Department of Anthropology, University of California, Berkeley)
D. L. Clarke (Peterhouse, Cambridge)
B. A. L. Cranstone (Department of Ethnography, British Museum, London)
H. E. W. Crawford (Cambridge)
G. Danişman (Department of West Asiatic Archaeology, Institute of Archaeology, London and Ankara)
D. A. Davidson (Department of Geography, The University, Sheffield)
K. C. Day* (Peabody Museum, Harvard University)
G. W. Dimbleby (Department of Human Environment, Institute of Archaeology, London)
D. M. Dixon (Department of Egyptology, University College London)
M. Douglas (Department of Anthropology, University College London)
C. Doumas (Santorini Museum, Greece)
K. V. Flannery (Museum of Anthropology, University of Michigan, Ann Arbor)

* Authors who were unable to attend.

A. Forge (Department of Anthropology, London School of Economics)
P. J. Fowler (Department of Extra-Mural Studies, University of Bristol)
D. H. French (British Institute of Archaeology at Ankara)
J. P. Garlick (Duckworth Laboratory of Physical Anthropology, University of Cambridge)
P. Gathercole (University Museum of Archaeology and Ethnology, Cambridge)
I. C. Glover (Department of Prehistoric Archaeology, Institute of Archaeology, London)
W. F. Grimes (Institute of Archaeology, London)
D. Grove* (Shankland, Cox and Associates, London)
J. R. Harlan (Department of Agronomy, University of Illinois)
D. R. Harris (Department of Geography, University College London)
M. W. C. Hassall (Department of Archaeology of the Roman Provinces, Institute of Archaeology, London)
C. F. C. Hawkes (Institute of Archaeology, Oxford)
J. G. Hawkes (Department of Botany, University of Birmingham)
C. B. Haycraft (Gerald Duckworth and Co. Ltd., London)
E. S. Higgs* (Department of Archaeology and Anthropology, Cambridge)
H. W. M. Hodges (Department of Conservation, Institute of Archaeology, London)
S. C. Humphreys (Departments of Ancient History and Anthropology, University College London)
J. G. Hurst (Inspectorate of Ancient Monuments, Department of the Environment, London)
G. Ll. Isaac* (Department of Anthropology, University of California, Berkeley)
M. R. Jarman (Department of Archaeology and Anthropology, Cambridge)
P. A. Jewell (Department of Zoology, University College London)
G. A. Johnson (Museum of Anthropology, University of Michigan, Ann Arbor)
S. E. Johnson (Institute of Archaeology, London)
P. K. Kaberry (Department of Anthropology, University College London)
B. J. Kemp (Faculty of Oriental Studies, Cambridge)
R. Layton (Department of Anthropology, University College London)
R. B. Lee (Department of Anthropology, Rutgers University, New Jersey)
I. M. Lewis (Department of Anthropology, London School of Economics)
G. Lienhardt (Institute of Anthropology, Oxford)
I. M. Longworth (Department of British and Medieval Antiquities, British Museum, London)
R. S. MacNeish (Peabody Foundation for Archaeology, Massachusetts)
R. D. Martin (Department of Anthropology, University College London)
V. N. Masson* (Institute of Archaeology, Soviet Academy of Sciences, Leningrad)

A. McNicoll (Magdalen College, Oxford)
J. Mellaart (Department of Western Asiatic Archaeology, Institute of Archaeology, London)
J. Middleton* (Department of Anthropology, New York University)
A. Momigliano (Department of Ancient History, University College London)
P. Mortensen (Institute of Prehistory and Ethnography, University of Aarhus, Denmark)
P. Morton-Williams (Department of Anthropology, University College London)
H. J. Nissen (The Oriental Institute, Chicago)
D. Oates (Department of Western Asiatic Archaeology, Institute of Archaeology, London)
J. Oates (Girton College, Cambridge)
D. O'Connor (Egyptian Section, University Museum; Department of Oriental Studies, University of Pennsylvania)
K. Painter (Department of British and Medieval Antiquities, British Museum, London)
P. J. Parr (Department of Western Asiatic Archaeology, Institute of Archaeology, London)
I. Pavlů* (Institute of Archaeology, Prague)
S. Piggott (Department of Archaeology, University of Edinburgh)
J. N. Postgate (Department of Near and Middle East, School of Oriental and African Studies, London)
P. Pratt (Department of Conservation, Institute of Archaeology, London)
W. G. L. Randles (Ecole Pratique des Hautes Etudes (VI[e] Section), Paris)
J. D. Ray (Department of Egyptian Antiquities, British Museum, London)
C. Renfrew (Department of Ancient History, University of Sheffield)
B. Reynolds (National Museum of Man, Ottawa)
V. Reynolds (Department of Sociology, University of Bristol)
A. Ritchie (Department of Archaeology, University of Edinburgh)
M. Roaf (Department of Western Asiatic Archaeology, Institute of Archaeology, London)
D. J. Robinson (Department of Geography, University College London)
A. Rosenfeld (Department of British and Medieval Antiquities, British Museum, London)
I. Rouse (Department of Anthropology, Yale University, Connecticut)
M. J. Rowlands (Department of Prehistoric Archaeology, Institute of Archaeology, London and Department of Extra-Mural Studies, University of London)
F. W. Schwerdtfeger (School of Environmental Studies, University College London)
G. de G. Sieveking (Department of British and Medieval Antiquities, British Museum, London)
P. Simonsen (Tromsö Museum, Norway)

R. Singer* (Department of Anatomy, University of Chicago, Illinois)
H. S. Smith (Department of Egyptology, University College London)
M. G. Smith (Department of Anthropology, University College London)
P. E. L. Smith (Department of Anthropology, University of Montreal)
B. Soudský* (Institute of Archaeology, Prague)
D. Sturdy (Institute of Archaeology and School of Slavonic and East European Studies, London)
C. C. Taylor (Royal Commission on Historical Monuments, England)
B. G. Trigger (Department of Anthropology, McGill University, Montreal)
R. E. Tringham (Department of Anthropology, University College London)
P. J. Ucko (Department of Anthropology, University College London)
E. Uphill (Department of Extra-Mural Studies, University of London)
C. Vita-Finzi (Department of Geography, University College London)
J. B. Ward-Perkins* (The British School at Rome)
W. Watson (Percival David Foundation of Chinese Art, School of Oriental and African Studies, London)
M. Weaver (Department of Architecture and Conservation, Middle Eastern Technical University, Ankara)
W. A. West (Department of Law, University of Reading)
P. Wheatley (Department of Geography, University College London)
A. Whitty (Ford, Newman and Whitty, Chartered Architects, Eastbourne, Sussex)
F. Willett (Department of Anthropology, Northwestern University, Evanston)
J. C. Woodburn (Department of Anthropology, London School of Economics)
J. J. Wymer (Department of Anatomy, University of Chicago, Illinois)
T. C. Young (West Asian Department, Royal Ontario Museum)

RUTH TRINGHAM

Introduction: Settlement patterns and urbanization

Settlement archaeology

The concept of "settlement archaeology" and the "settlement pattern" was first systematically applied to archaeological evidence by Gordon Willey[1]. As he was the first to admit, however, "settlement archaeology" does *not* comprise a "self-contained approach to prehistory . . . a 'new archaeology' " but is a "new approach" *within* archaeology, which must nevertheless begin with the same factual analytical data[2].

The concept of "the settlement" and the use of the term in descriptions of prehistoric cultures was by no means new in 1953. It had been used since the nineteenth century to distinguish archaeological sites which contained habitation debris (referred to as "settlement sites") from non-occupation sites (including burial sites, "ritual sites", etc.). Archaeologists, at least in the Old World, have continued to use the word "settlement" with this connotation, as a purely locatory term[3]. "Distribution of settlements" in Old World archaeology refers to the geographical distribution of artifacts and other tangible remains in locations where there is evidence that domestic activities were carried out. Such a usage of the term "settlement" may be at the root of many of the misunderstandings between Old World and New World (since 1953) archaeologists on the subject of settlement, settlement patterns and settlement archaeology, to which Rouse also refers[4].

The "settlement archaeology" of the New World archaeologists is essentially *not* the analysis of archaeological settlement sites. It has aptly been defined as "the study of social relationships using archaeological data"[5]. Elsewhere it has been defined as the study of archaeological data in terms of the effect of "the natural environment, the level of technology on which the builders operated, and various institutions of social interaction and control with the cultures maintained"[6]. It is, therefore, a study which provides one of the strongest links between archaeology and ethnography.

The use of structural or socio-cultural models rather than historical models in the interpretation of archaeological data has generally been discouraged or, at least, looked on with disapproval by Old World prehistorians, who regard it, for example, as "beyond the bounds of archaeological inference"[7]. V. G. Childe was an exception in that his ultimate aim was the socio-cultural, preferably sociological, explanation of prehistoric evidence: "as the peculiarities of the component types are determined by convention rather than by function, the culture must correspond to a social group which sanctifies the distinctive conventions and carries the social tradition"[8]. His interpretations, however, were made with greatest care and caution at the level of the archaeological culture. In referring to data at the level of the archaeological site or locality (its size and population, external relationships, internal organization and activities), he tended to draw more loosely on sociological and geographical terms, which he interchanged quite freely with the archaeological term "site". Even though he was familiar with ethnographic analogy, Childe used the sociological term "community" and the geographical terms "settlement", "village", and "hamlet" with little reference to their sociological, ethnographic, or geographical significance[9]. It is as if they were interchanged with the dry archaeological word "site" or "locality" in order to give the study greater applicability to the modern context and therefore enhance its readability.

Many other Old World archaeologists use these sociological and geographical terms, but, like Childe, without defining them. Thus although they may use the same terminology as the New World studies of "settlement archaeology", they are not writing about the same subject. In fact they are asking a completely different set of questions. For example, in the sentence, "we have . . . a well-established pattern of peasant communities, with villages and small towns of mud-walled and mud-brick buildings forming permanent settlements. . ."[10], is the writer interchanging these terms for the sake of readability and variety, or does he really have evidence, in the tangible remains of social organization, of peasants forming a community? Does he mean that each community formed a settlement, in which case what is the difference between peasant community and permanent settlement? What does he mean by settlement? What does he mean by community?

Community and settlement

New World studies and discussions on settlement archaeology frequently revolve round a definition of the terms "settlement" and "community" and round the question whether any correlation can be made between the two entities[11]. The term "settlement" clearly refers to more than a *locus* of domestic activity, but there is still disagreement about its meaning in social terms, its validity and usefulness. Chang took the bold step of defining the "archaeological settlement" as "the local context wherein the community is presumed to have resided and to have gone about its daily business"[12]. Thus the settlement is "the physical locale or cluster of locales where the members of a community lived, ensured their subsistence, and pursued their social functions in a delineable time-period"[13]. In other words, the archaeological settlement "must substitute for community"[14]. Chang's definitions and conclusions have been severely criticised by Rouse who nevertheless agrees with two of his assumptions. In stating that a settlement is "all those space/time components in which a particular community carried out its various activities"[15], Rouse would accept, first, that an *archaeological* settlement, which comprises the material, tangible remains of social activity, *can* be correlated with the sociological concept of "the community"[16]. Secondly, he would agree that the "community" is a universal concept and one which has been well defined by social anthropologists and sociologists. Willey, writing in the same volume, was rather more cautious but agreed that, in general, "there is a good correspondence between site or settlement unit and the social unit, the community. It is a task for archaeological research, however, to demonstrate the relationship by amplifying or correcting its conclusions in order to increase the probabilities of such correspondences"[17].

The term "community" and its definition have been the source of lively discussion[18]. A definition which has some acceptance among anthropologists and sociologists is Murdock's: "the maximal group of persons who normally reside in face-to-face association"[19]. If the term "settlement" is defined sociologically as "an aggregate of people living in the same locality"[20], it is clear that even among modern small-scale societies, it is not always possible to equate the community with the settlement. In certain cases, face-to-face association occurs at several settlements, in which the material culture and activities may be quite different, as with most mobile hunters and gatherers[21]. In other contexts, particularly those in which high densities of large populations are involved, as in urban centres, there may be several communities in one settlement[22], though in smaller sedentary agricultural settlements it is more usual to find that the settlement does correspond to the maximal group residing in face-to-face association[23]. Ethnographic evidence, therefore, should warn against blindly equating archaeological settlements with communities on the basis of a common material culture or one aspect of it.[24] Such a correlation is in fact a very complex problem, since it immediately involves the interpretation of the tangible remains of the settlement in sociological terms. On the basis of the present evidence, it seems doubtful that any correlation between archaeological settlement and the sociological concept of community *can* be made. But in any case it. would seem extremely unwise to proceed to answer such a complex set of questions without first examining the tangible remains themselves—the location, size, spacing, activities and material culture of the settlements and the interaction of their environmental, economic and technological determinants[25]. In this way settlement archaeology is probably the most direct route to the reconstruction of pre-historic activities. The social determinants of settlements should by no means comprise an isolated, or even a primary, aim of study.

Gordon Willey envisaged the development of "settlement archaeology" on a broad front of unified purpose[26], but both Willey and Trigger have remarked that there is frequently a large discrepancy between the original concept of the method and its actual application[27]. Quite early in the history of settlement archaeology, Chang suggested that the "settlement pattern" could be studied as a separate entity from the "community pattern"[28]; frequently the zonal, regional, geographical distribution of settlements (the macrosettlement pattern) and the ecological and economic aspects of their patterning[29] have been studied without reference to individual settlements or small clusters of activity *loci* (microsettlement pattern or community pattern) and the reconstruction of their socio-political activities and *vice versa*[30]. However, to separate the study of the ecological-economic from the sociological determinants of settlement would seem to be inviting a most unrealistic picture not only of each settlement situation, but also of broader macrosettlement patterns and the nature of cultural development in general. The set of ecological factors cannot be studied in isolation from the products of human activities (the tangible remains of the community—the archaeological settlement); and the social factors cannot be isolated from the landscape of which the society was a part. The factors—ecological, economic, technological, socio-cultural—which cause variation in settlements and settlement patterns are to a great extent interdependent and interrelated.

Levels of analysis

Trigger has suggested that these interdependent factors should be studied at three levels in order to understand more fully the composition of the settlement as a whole and the activities of its inhabitants[31]. His three suggested levels comprise: (*a*) the individual building or structure; (*b*) the settlement layout, which he equates with community layout; (*c*) settlement distribution or the spatial relationships between different communities on a zonal scale. In the third level of analysis, it may theoretically be possible to distinguish, by examining the social forces involved in the spatial relationships of the different residential units, whether the units comprised one or several communities, i.e. whether each unit consisted of one community of several settlements, and the degree of political or social interaction between the settlements. The spatial relationships between settlements could possibly be studied more informatively at two levels rather than one: first, at the level of the smaller geographical region of one ecological zone (possibly comprising a number of micro-ecological zones)[32]; and secondly at the level of the larger geographical region comprising a number of ecological zones and a number of settlement networks (the macrosettlement pattern)[33].

At the level of the *individual building*, the analytical components include size, form, method of construction, internal organization including the demarcation of internal divisions, the activities which take place in and immediately around the building i.e. its function and the external appearance and decoration of its superstructure. The second level of analysis—the "settlement site" of the Old World archaeologists, the *locus* of activity, the residential location, the "core area"—comprises much more than the conglomeration of individual buildings. Additional analytical components include, for example, geographical location, total number and density of inhabitants per building, internal organization of the buildings and space, and the activities within and immediately around the settlement and their differential distribution within this area, as well as the external demarcation of the settlement. The form and content of the components at both these levels of analysis depend on factors of natural environment, technology and subsistence economy which have been described in detail by Cranstone and Hodges[34]. These factors may restrict, but may not determine the possible form and function of buildings and settlements. In the settlements of many modern small-scale societies buildings with a large range of shapes and sizes constructed by a number of different methods may occur at the same location. These may all have the same function, but it is equally possible for buildings of identical shape and size to enclose entirely different social and domestic activities[35].

In many societies, the socio-cultural factors may be very important in determining the internal organization of residential areas, i.e. the distribution of activities in the various buildings and spaces, and the spatial relationships of the individual buildings. For the archaeologist, whose data comprise the ground-plan only of structures, frequently without the original floor surface, the reconstruction of social and domestic activities in the buildings and intervening spaces is at best speculative, and at worst impossible, so that the effect of the socio-cultural factors cannot be assessed. This is not to say that archaeologists have not attempted to reconstruct the function and social organization of prehistoric structures and settlements. Frequently such reconstructions have assumed that all the structural features of a site, even pits, were lived in[36]. Less common is the attempt to interpret buildings with "abnormal" internal arrangement as "shrines"[37] or larger than normal buildings as "clubhouses"[38], or long buildings as the dwelling-houses of extended families[39].

Recent studies in prehistory have attempted by the use of ethnographic analogies to reconstruct the social activities in prehistoric settlements by considering a range of possible interpretations of archaeological features in social terms, on the basis of cross-cultural comparison with similar features in modern and historical small-scale societies[40]. In this way it is possible to see, for example, the range of social forces which cause one house in a village to be built larger than the rest. The examination of the effect of the environmental, economic and technological factors on the composition of the archaeological settlement will enable part of this range to be eliminated, but never the whole, so that a one-to-one correlation between an ethnographic observation and an archaeological feature is virtually impossible.

At the third and fourth levels of analysis, comprising a number of "settlement sites" in one or more ecological zones, the spatial, ecological, social and political relationships may be studied, on the one hand between the various "settlements" of a single "community" (as with hunters and gatherers), or, on the other hand, between various "settlement-communities" (as with sedentary agriculturalists). The analysis of the settlement pattern comprises: (*a*) the location of the settlements and their relationship to the ecology, the available natural resources, the relative ease of communication; (*b*) the overall density and distribution of the settlements; (*c*) the exploitation area and the subsistence economy of the settlements; (*d*) the distribution of specialized activity *loci*—burial sites, butchering sites etc; (*e*) the population movement among the various *loci*, the degree of seasonal and non-seasonal mobility; (*f*) the exchange of commodities between settlements by short- and long-range trade, and the distribution of markets; (*g*) the social relationships between the settlements (community ties, kinship ties etc.); (*h*) the political relationships between the settlements including political domination and hierarchy, and centralization of political, religious and military activities; warfare, defence; (*i*) the differential exploitation of the same gross ecological zone, for example pastoral/agricultural symbiosis, and rural/urban dichotomy[41].

The environmental and economic factors are clearly very important in determining the form of the settlement pattern. A study of settlements at this level, however, is also particularly important for an understanding of the development of social and political hierarchies and settlement networks, and the emergence of urban centres[42]. Again, it is clear that these grosser levels of analysis will be more meaningful if the evidence is studied in the same detail and with equal care at the level of the individual settlement and structure. As Trigger has pointed out, each level may be studied independently[43], but the study becomes more informative when the data are analysed at all three (or four) levels.

Settlements and settlement patterns are not, and never have been, fixed situations. Their interrelated determinants and components are in a state of continual change—change in their ecological situation, their technology and material culture, their economy, their social organization and their political relationships. The change may be so minute as to be indiscernible either to the living population, or in the archaeological data. Or the change may be obvious and swift, in which case it may be clear to modern observers or felt by the community in question, and may be reflected in an equally clear modification of the settlement or settlement pattern, as for example when a sudden threat of attack is answered by the construction of defensive fortifications[44]. More frequently, however, the change is slow and not directly visible to those taking part in the process. Such a process, however, especially in its later stages, may be clearly reflected in the archaeological data, for example in the emergence of sedentary agricultural settlements in the Near East and Europe. The archaeologist has the advantage over the ethnographer in that his evidence of change, even if it is not a continuous record or very full in some respects, can provide information over a very long period of time, sometimes over thousands of years. The archaeological evidence can give the fullest information on the effect of changing natural environment and economies. It is particularly lacking, without spoken or written records, on social and political activity. This lack becomes sadly obvious in the studies of the so-called "emergence of urban centres", in which the process from non-urban to urban centres is still obscure[45].

Urban and non-urban settlements

In present-day urban studies there is great disagreement on the definitions, significance, and validity of terms such as "urban", "urbanism", and the "process of urbanization". Wheatley has indicated the very limited criteria, mostly morphological, on which many of the urbanists have based their classifications, and has stressed the need for much greater attention to the "urban context" than to the urban centre itself[46].

There is no reason why the same environmental, economic, technological socio-politico-cultural factors which affect the settlements of small-scale societies should not also affect the large-scale settlements of more complex societies; nor is there any reason why such large-scale settlements should not be analysed informatively at the levels suggested by Trigger. Such analyses of "urban centres", however, are extremely rare. One of the difficulties in distinguishing non-urban from urban settlements, particularly in the recognition of early stages in the "process of urbanization", is undoubtedly caused by the fact that there is no sharp dividing line in the development from non-urban to urban settlements but rather a continuum. At each end of the continuum (the modern city and the pre-urban village), the various determining factors clearly affect the composition of the settlements in very different ways[47]. Thus the characteristics of modern towns, for example political and social centralization, may frequently be recognized in non-urban situations but on a much smaller scale. To a certain extent it would seem that the difference between urban and non-urban settlements is a quantitative rather than a qualitative one—a difference in the degree of social complexity, so that a multitude of "communities", or interaction networks, can exist in one settlement; a difference in the degree of political complexity, so that one settlement can dominate a network of other settlements, not necessarily by any physical compulsion but by the centralization of all kinds of activities; a difference in the degree of economic complexity, so that the inhabitants of the urban settlement need no longer be self-sufficient but may be fed and supported by the surrounding network of settlements; thus their immediate exploitation area cannot support them and yet they do not need to move their settlement to save themselves from starvation; and a difference in the degree of technological complexity, so that centralized organization of activity, producers, power and the exploitation of resources from various ecological zones make available forms of machinery which would otherwise be impossible.

At the far end of the continuum, however, is the modern city. The differences between this phenomenon and the non-urban village are clearly qualitative ones. Complexity in every aspect of settlement activity produces in aggregate a settlement of qualitatively different composition. These are settlements in which the communities are fragmented and the concept of "community" is lost in the cult of the individual. The individual is urged to make his independent choice[48]; social relationships of the kind studied by social anthropologists in small-scale societies are at a minimum among the inhabitants of the modern city. And yet the individual is dependent and cannot survive without the interaction of his fellow metropolids. Certainly this is an overpainted gloomy picture of the future of modern cities and their inhabitants, but it is a picture which many town-planners hold in their minds when they give their panic-stricken cry to reverse the irreversible process, to revert back along the continuum to urban or non-urban settlements of lower population density, greater self-sufficiency, and greater "community spirit".

Notes

1 Willey, G. R. (1953). Prehistoric settlement patterns in the Viru Valley. *Bull. Bureau of Amer. Ethnology*, **155**.
2 Willey, G. R. (ed.) (1956). Prehistoric settlement patterns in the New World, *Viking Fund Publications in Anthropology*, **23,** p.1; Willey, G. R. (1968). Settlement archaeology: an appraisal, *in* Chang, K. C. (ed.) *Settlement Archaeology*, New Haven. p. 208; Trigger, B. (1967). Settlement archaeology—its goals and promise, *Amer. Antiq.*, **32,** p. 149.
3 Viz. Piggott, S. (1965). *Ancient Europe*. Edinburgh, p. 40: "in the famous site of Jericho, a settlement and probable shrine of these people formed the first occupation of the place".
4 Rouse, I., this volume, pp. 95-106.
5 Trigger, B. (1967). *op. cit.* p. 153.
6 Willey, G. R. (1953). *op. cit.* p. 1.
7 Hawkes, C. F. C. (1954). Archaeological theory and method: some suggestions from the Old World, *Amer. Anthrop.* **56,** pp. 161-2.
8 Childe, V. G. (1942). *Social Evolution*. London. p. 18.
9 Childe, V. G. (1942). *op-cit.* p. 47; Childe, V. G. (1951). *What Happened in History*. London. p. 87.

10 Piggott, S. (1965). *op. cit.* p. 41.
11 See especially: Chang, K. C. (1958). Study of the Neolithic social grouping: examples from the New World, *Amer. Anthrop.*, **60,** pp. 208-35; Chang, K. C. (1968). Towards a science of prehistoric society, *in* Chang, K. C. (ed.) *op.cit.* pp. 1-9; Rouse, I. (1968). Prehistoric typology, and the study of society, *in* Chang, K. C. (ed.) *op. cit.* p. 23; Rouse, I., this volume; Willey, G. R. (1968). *op. cit.* pp. 211–12.
12 Chang, K. C. (1968). *op. cit.* p. 3.
13 Chang, K. C. (1968). *ibid.*
14 Chang, K. C. (1968). *ibid.*
15 Rouse, I. *op. cit.* p. 23.
16 Rouse repeats his ideas of 1968, but rather more cautiously in his paper in this volume.
17 Willey, G. R. (1968). *op. cit.* p. 216, note 1.
18 See, for example, the views expressed by M. G. Smith and R. Layton in this volume.
19 Murdock, G. P. (1949). *Social Structure.* New York. p. 79.
20 Layton, R., this volume, p 337.
21 See for example the studies of the Bushmen (R. Lee) and Hadza (J. Woodburn) settlements in this volume.
22 Smith, M. G., this volume.
23 As studied by Chang, K. C. (1958). *op. cit.*, as well as several papers in this volume (e.g. A. Forge), but not for example in the Kwandu settlements (B. Reynolds, this volume).
24 e.g. using pottery as the main criterion, Sears, W. (1968). The state and settlement patterns in the New World, *in* Chang, K. C. (ed.) *op. cit.* p. 135. Sears, as well as Chang, may have had a more legitimate basis for the assumption that archaeological settlement should be correlated with community, since they were both discussing exclusively sedentary agricultural settlements.
25 Trigger, B. (1968a). The determinants of settlement patterns, *in* Chang, K. C. (ed.) *op. cit.* pp. 60-1; Trigger, B. (1968b). *Beyond History: the Methods of Prehistory.* New York. p. 21.
26 Willey, G. R. (1953). *op. cit.* p. 1.
27 Trigger, B. (1968a). *op. cit.* p. 54; Willey, G. R. (1968). *op. cit.* p. 215.
28 Chang, K. C. (1958). *op. cit.* p. 299.
29 See for example MacNeish, R., Jarman, M. *et al.*, Bankes, G. etc. in this volume.
30 In spite of the outcries from Old World archaeologists at the seminar, this volume contains remarkably few leanings towards this kind of isolation of the two kinds of settlement study.
31 Trigger, B. (1967). *op. cit.* p. 151; Trigger, B. (1968a). *op. cit.* p. 59.
32 e.g. Soudský and Pavlů, Davidson, D., Mortensen, P., French, D., MacNeish R., this volume.
33 Renfrew, C., Doumas, Ch., Gathercole, P., Clark, J. D., Masson, V. etc., this volume.
34 Cranstone, B., this volume, pp. 487-503; Hodges, H., this volume, pp. 523-30.
35 Warnings from ethnographic studies for the interpretation of the function of prehistoric buildings may be found in Douglas, M., this volume, Schwerdtfeger, F. W., this volume, and David, N. (1971). The Fulani compound and the archaeologist. *World Archaeology.* 1, IV: 3.
36 Viz. the many widespread references in Old World archaeology to "pit-dwellings" even though there are remarkably few ethnographic references to such constructions. In spite of Childe's criticism of the interpretation of Linear Pottery pits as dwellings (Childe, V. G. (1949) Neolithic houses—types in temperate Europe, *Proc. Prehist. Soc.* **15,** 77P86), this interpretation of pits in prehistoric settlements is still frequently used,

37 Mellaart, J. (1965). *The Earliest Civilizations of the Near East*. London. pp. 89–98.
38 Soudský, B., this volume.
39 Childe, V. G. (1957). *The Dawn of European Civilization*. London. pp. 105–10; Piggott, S. (1965). *op. cit.* p. 52.
40 e.g. Ucko, P. (1969). Ethnography and archaeological interpretation of funerary remains, *World Archaeology*, **II: 2,** pp. 262–80; David, N. (1971). *ibid.*; Rowlands, M. (1970). The archaeological interpretation of metallurgy, *Research Seminar in Archaeology and Related Subjects* (Mimeo). London. Also e.g. bibliography in Chang, K. C. (1967). Major aspects of the interrelationship of archaeology and ethnography. *Current Anthropology*, **VIII: 3.**
41 Many of these aspects of settlement patterns are dealt with by papers in this volume: ecology—Davidson, D., Woodburn, J., Allan, W., population density—Smith, P., Young, C., Renfrew, C., Forge, A; economy —Harris, D., Bray, W., Jarman, M. *et al.*, Allan, W.; mobility of population—Ali, A., Clark, J. D.; Glover, I., Mortensen, P., Simonsen, P., Woodburn, J., Lee, R., Harris, D., MacNeish, R., Reynolds, B. special activity *loci*—Clark, J. D. social relationships—Layton, R.; political hierarchies and settlement networks—Adams, R., Johnson, G., Ward-Perkins, J., Humphreys, S., Bray, W.; defence—Rowlands, M., Tringham, R.; rural/urban dichotomy—Humphreys, S.; Ward-Perkins, J.
42 See Johnson, G., Adams, R. and Young C., this volume.
43 Trigger, B. (1967). *op. cit.* p. 151.
44 Rowlands, M., this volume.
45 See Adams, R., Alexander, J., Wheatley, P., Blouet, B., Hassall, M., Oates, D., Masson, V., Young, C., Robinson, D., this volume.
46 Wheatley, P., this volume, p. 601 and p. 623.
47 Trigger, B., this volume, suggests the factors which affect the growth of urban centres.
48 Grove, D., this volume, puts forward the various views and visions of modern town-planners.

Part 1: Non-urban settlement

Section I: Concepts, in theory and practice

BRIAN W. BLOUET

Factors influencing the evolution of settlement patterns[1]

This essay is an attempt to study the impact of various economic changes upon the settlement pattern of a hypothetical region. The proposed stages of development will not necessarily "fit" any particular area, nor do they take account of certain cultural forces which may override economic matters; this is an attempt to isolate and focus upon a group of economic processes which may give rise to alterations in settlement patterns.

Although adjustment is always imperfect there is a strong link between settlement patterns and economic activity. If this basic assumption can be accepted, then it follows that as an area undergoes economic development the settlement pattern of that area will be liable to undergo structural alteration.

An historical examination of settlements in the Western world suggests that they have evolved through the following phases although no set time span is suggested for the development of any phase[2]. The growth phases are assumed to develop on a plain of uniform relief and agricultural potential[3].

Phase One: limited function settlements

On a uniform plain, *sensu latu*, inhabited by subsistence farmers, after an initial distribution which contained random elements, settlements would tend to be equally spread to make the most efficient use of resources. The competitive processes which would help to bring about this distribution have recently been carefully analysed by Hudson[4]. The term efficient distribution is used here specifically in relation to agricultural exploitation of the land. Historically, settlement distributions have frequently had to respond to other major pressures besides the need to promote the efficient utilization of land. For instance, the need to provide for the defence of inhabitants has been a common additional, or sometimes, paramount pressure. In short, the word efficient means efficient in relation to some function or group of functions. In this study we are specifically interested in a group of economic functions. If we add additional functions (i.e. alter the controls of the experiment) we will change the form of the theoretically most efficient distribution. In sum, we would start a different experiment with different pre-conditions. The patterns generated by the new pre-conditions will influence later phases of development.

Phase Two: the emergence of villages (Fig. 1)

Assuming that the farmers on the plain acquire the ability to produce surpluses then natural variation will ensure that some farmers have a surplus of certain products while others have surpluses of different commodities. If it is desired to exchange these surpluses with nearby farmers or for goods or services in other areas then the phase one settlement pattern will not facilitate exchange in the most efficient manner. An inhabitant in a group of settlements desiring to exchange goods would find a large number of journeys involved if he wished to deal with other inhabitants in the group. Exchanges would best be carried out by using a centrally placed settlement for the purpose. In practice exchanges require a degree of management in order to bring parties together and eventually the centrally placed settlement will house inhabitants who specialize in these tasks and associated functions. Exchanges create wealth surpluses and there will also be opportunities for specialist craftsmen of a limited number of types to offer goods and services at the central settlement. At this point we may refer to the central settlement as a village. In order to minimize journeys the villages will tend to be equally spaced. Inhabitants of the plain may decide that the disadvantages of having to make longer journeys to their farmland are outweighed by the social and economic advantages of living in a village. Economic advantages would include those benefits derived from co-operative agricultural effort. In this event a tendency could develop for the outlying settlements to decline in number[5]. However, it must be recognized that there are a number of paths by which villages may appear in a landscape. The above example simply suggests some economic forces which may give rise to agglomeration tendencies within a population. Historically villages have frequently been planted as the result of a colonization and thus enter a landscape as part of the organization of a group which has migrated. The Anglo-Saxon settlement of England is an example of this type of village planting. Many of the earliest villages must have emerged as populations acquired agricultural skills and gradually established permanent settlements in the regions exploited for farming purposes. No doubt such emergent settlement patterns contained random elements but the forces of competition identified by Hudson would tend to produce, eventually, an equal spacing of villages[6].

It is extremely difficult to isolate the factors responsible for the development of villages in a landscape but the argument here is that although such settlements may develop by several routes they will tend to be equally spaced. Thus the problems associated with generating the villages does not invalidate the subsequent analysis.

PHASE II

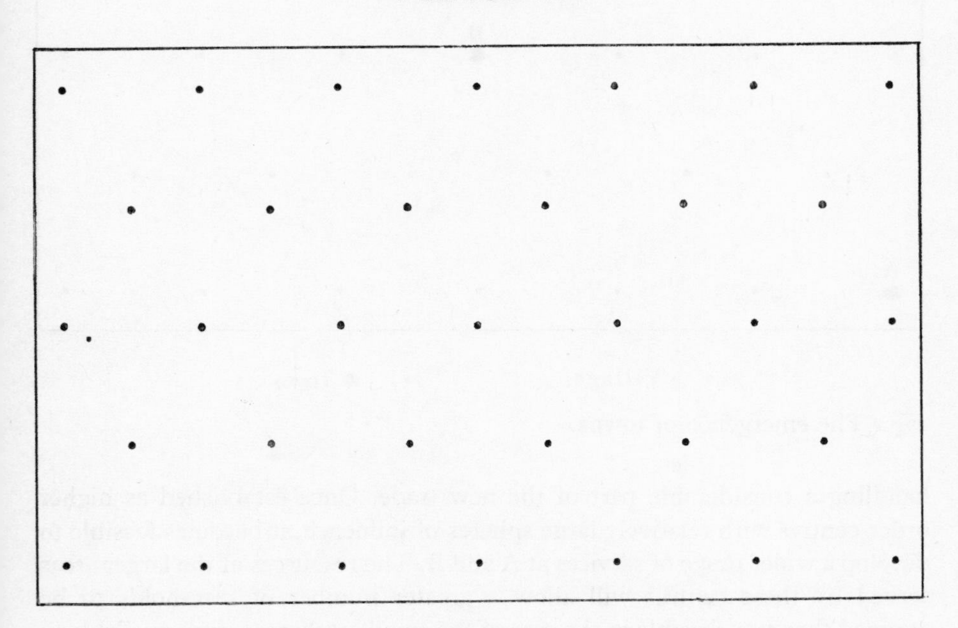

Fig 1 The emergence of villages.

Phase Three: the emergence of towns (Fig. 2)

Assuming that economic development on the plain continues it is likely to be characterized by a marked increase of inter-regional trade[7].

If we assume that an important source of goods lies to the south-west of the plain and that these goods break bulk on entering the region, then Village A has locational advantages which fit it to act as a trade centre. A has potential connectivity with the largest number of settlements on the plain[8]. The degree of connectivity, assuming a shortest distance path, declines along a line from Settlement A to Settlement C. However, the fact that Village B enjoys the highest connectivity for intra-regional purposes, will probably lead to it

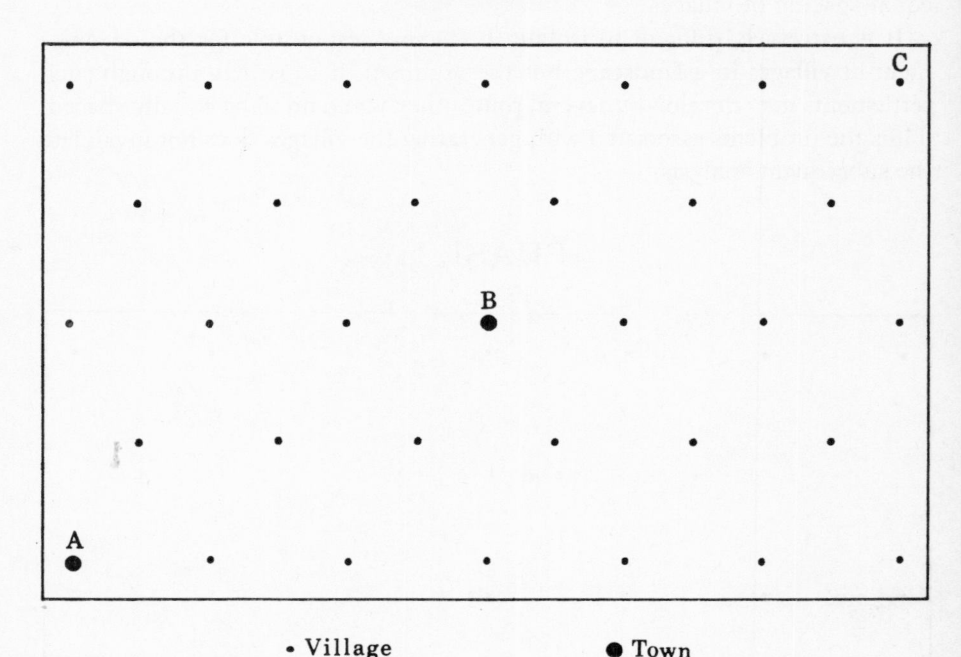

Fig 2 The emergence of towns.

handling a considerable part of the new trade. Once established as higher order centres with relatively large spheres of influence, it becomes feasible to develop a wider range of services at A and B. The resources of the larger areas served by these centres will allow a greater number of thresholds to be attained than was feasible in the case of the small catchment areas available to a village. With the establishment of this greater range of services, we may refer to A and B as towns. These urban settlements would display the characteristics which are commonly associated with the term pre-industrial.

The existence of the towns has a widespread influence in that they become in themselves important markets. The existence of these new markets may trigger numerous complementary economic developments. Patterns of agricultural land-use may alter markedly for there are now opportunities for greater specialization of production, particularly in areas close to the towns[9]. The towns offer a variety of employment and the possibility of marketing a much increased range of goods. The towns resulted from the greater circulation of energy in the economic system and once in existence, they provide opportunities for the generation of yet more energy. Economic activity will tend to multiply most rapidly at these new nodes in the settlement pattern.

The emergence of the towns and the associated economic changes may have important effects upon the other elements in the settlement pattern. Previously the region had an economy which was very largely subsistence orientated. Once a greater range of goods is produced for sale it becomes necessary to have access to service centres which offer the necessary commercial facilities.

The smaller, more isolated settlements will not be able to provide such services. There will be an increasing tendency for small badly located villages to decline in importance[10]. The emergence of the towns in part of the development of a more efficient economic system, to lie at the peripheries of this system will carry disadvantages[11].

Due to differential economic development, with the centres A and B as poles of economic growth which are developing their own locational momentum and attraction as markets, suppliers, and transportation centres, the plain becomes less and less uniform in an economic sense.

Phase Four: the establishment of centres of manufacturing (Fig. 3)

While the towns and villages on the plain will contain craft industries such production will be characterized by the small scale of the production units. Historically the next phase has been the introduction of new technologies (an industrial revolution) which allow goods to be manufactured on a relatively large scale by mass production methods. Such methods were dependent upon the existence of plentiful sources of power and raw materials which are frequently bulky. Today the transmission of energy and the movement of raw materials presents us with few problems, in relation to existing transportation techniques. In the earliest phases of the industrial revolution this was not the case and industry tended to grow up largely on the energy and raw material sources.

Substances like coal are usually localized in their distribution and this fact would have a profound influence upon the evolution of settlement patterns. Let us assume that a coalfield lies beneath the northern half of the plain and dips to the north. Initially coal will tend to be extracted where it is nearest to the surface and most easily mined. In these terms the villages along the line XY will all have an equal opportunity to develop as mining centres. However the villages nearest to B, with its financial resources, entrepreneurial skills and market potential, should develop most rapidly[12]. Simultaneously new settlements might also grow up around B at mines sunk close to the town. Heavy manufacturing industries would tend to be located at the raw material sources whilst production concerned with lighter consumer goods and labour intensive processes would be more influenced by market and labour considerations. B would be the site for much of this activity.

The new mining and manufacturing centres would rapidly become the major zones of wealth creation upon the plain and would attract strong migratory flows towards them.

Town A, although not well situated initially to become a manufacturing centre, will probably be instrumental financially in the development of the coalfield. The existence of new sources of wealth within the catchment area of A would increase the importance of the town as a transportation centre. With the improvement of communications (the first major railway would be built from B to A)[13] it would easily become possible to set up manufacturing at A.

The shallow part of the coalfield would be rapidly worked out and new pits established to the north. However, improvements in transportation techniques would render it unnecessary for manufacturing units to move to the pit heads. The existence of raw materials has ceased to be a decisive location factor for it is becoming progressively easier to move the raw materials to the manufacturing plant, the labour, and the market. The locational momentum of A and B already apparent in phase three is becoming increasingly dominant.

Until this point we have been dealing with a fairly distinct hierarchy of settlements: isolated dwellings, hamlets, villages, and towns. By the end of phase four the categories are much less distinct. A whole new range of mining and manufacturing settlements possessing some urban functions is developing. The structure of the settlement pattern is altering from a hierarchy to a continuum.

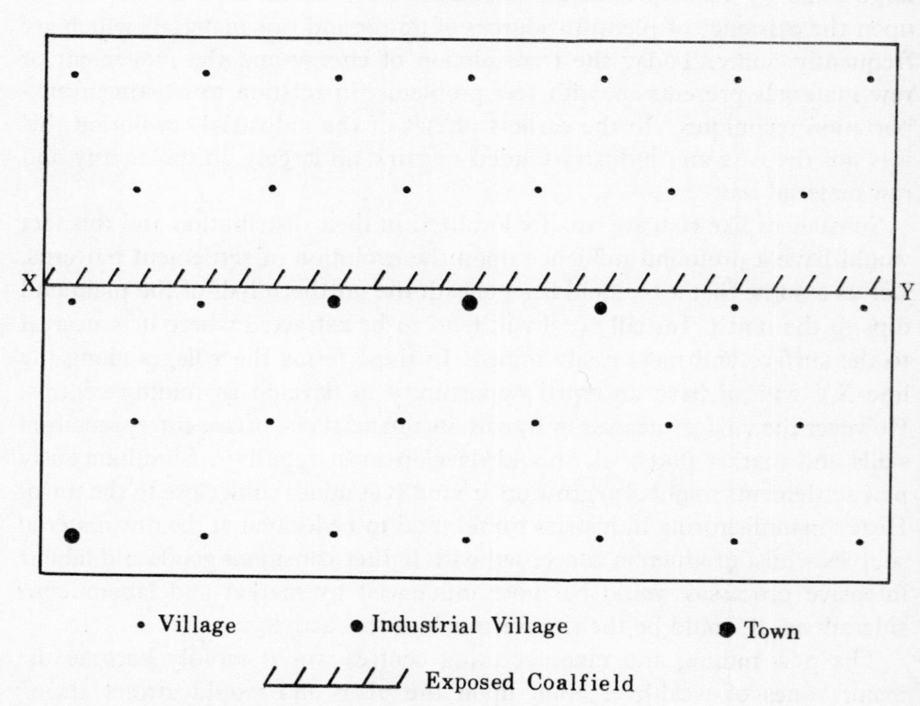

Fig 3 The establishment of centres of manufacturing.

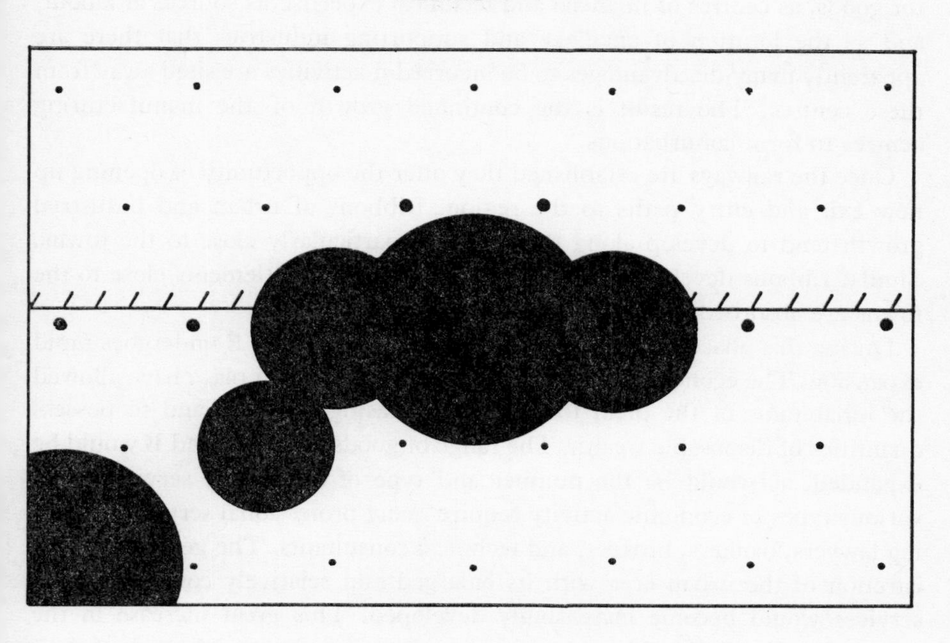

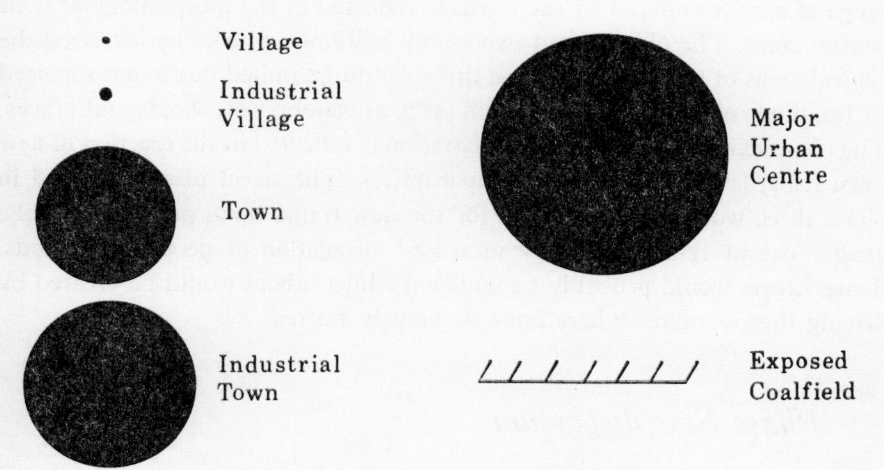

Fig 4 The growth of conurbations.

Phase Five: the growth of conurbations (Fig. 4)

In this phase the major urban centres have become so attractive as markets for goods, as centres of financial and technical expertise, as sources of labour, and as the location of ancillary and supporting industries that there are apparently many disadvantages to be incurred if activities are sited away from these centres. The result is the continued growth of the manufacturing centres to form conurbations.

Once the railways are established they offer the opportunity of opening up new exit and entry paths to the region. Ribbons of urban and industrial growth tend to develop along the railways, particularly close to the towns. Similar ribbons develop along the roads and smaller settlements close to the towns are absorbed into the urban area.

During this phase the range of services offered in A and B undergoes rapid expansion. The economic developments which have taken place have allowed the inhabitants of the plain to become increasingly mobile and to possess quantities of disposable wealth. The range of goods sold in A and B would be expanded, as would be the number and type of shops and services. The various types of economic activity require many professional services including lawyers, bankers, brokers, and technical consultants. The general administration of the urban area with its enlarged and relatively complex public services would become increasingly developed. This great increase in the range of services offered by the towns is reflected in the morphology of their central areas. The old one and two storey buildings which characterized the central areas of the towns in phase three would be pulled down and replaced by buildings of several storeys which serve as warehouses, shops, and offices. The complexity of the urban administration is reflected in the erection of new town halls, law courts, and municipal offices. The street plan developed in phase three would be inadequate for the new traffic flows generated by the greater use of vehicles and the increased circulation of people and goods. Some streets would probably be widened whilst others would be created by driving thoroughfares where none previously existed.

Phase Six: dispersion

Phase six is not yet visible in a fully developed form. In this phase land, labour[14] and congestion costs become high and this combined with developments in communications which permit a high degree of contact over relatively wide areas, allows and encourages dispersion to take place to urban satellites set away from the conurbations[15].

This new mobility leads to a questioning of locational values[16] and towns whose advantages spring from phase four conditions may find they face stagnation or decline.

The correspondence of settlement phase and stage of economic development

It has been suggested in the preceding pages that there may be a strong correlation between the structure of settlement patterns and the type of economy they serve. Settlement patterns and the economy probably form a system. The system is imperfect in that residual elements from earlier phases are not automatically eliminated when conditions change.

It is interesting to speculate how far it is possible to relate the settlement phases suggested here to W. W. Rostow's stages of economic growth.

Settlement Phase	*Stage of Economic Growth* (W. W. Rostow)[17]
Phases 1 and 2—hamlets and villages	Traditional society
Phase 3—the emergence of towns	Preconditions
Phase 4—the development of manufacturing centres	Take off into self-sustaining growth
Phase 5—growth of conurbations	Drive towards the mature economy
Phase 6—dispersion	High mass consumption

Naturally the correspondences have not to be pushed too rigidly for the settlements themselves have flexibility of function; a village may exist on the same site for hundreds of years while an economy is developing and during this time the settlement may undergo alteration of function.

The size and distribution of population in the settlement phases

The process of settlement evolution is characterized by

(i) an increasing population, and
(ii) an increasingly high proportion of the population living within urban areas.

In phases one and two an extremely low percentage of the population could be described as urban, if any. It is possible that some tribal centre exists to which allegiance is given and the inhabitants of this centre may or may not qualify to be called urban. Such centres would be relatively few and the percentage of urban population would not exceed 5%.

During phase three with the emergence of towns the urban population will increase in number but it is unlikely to constitute more than 10–20% of the total population. With the emergence of centres of manufacturing industry the urban population increases steeply and probably represents about 50% of the total by the time the phase is established. An obvious corollary of the growth of population and the increase in urbanization is the expansion of the built-up area.

Changes in the hierarchy of settlements

As the economy of an area develops there is a tendency for the units of population to become increasingly large and for the range of services required by individuals and businesses to become greater. A corollary of this statement is that relatively small settlements carry a built-in tendency to decline for they are found to be lacking when placed in competition with larger places. This tendency may be counterbalanced by added wealth available in the system which may allow settlements to survive even where they are not of optimum size and location.

In the modern economic setting the propensity of farms to get larger and the farmhouses fewer and for central places to get bigger and further apart is well known. Scott, in a recent study, was able to conclude that "the lower the order of central places, the higher is the proportion of declining centres within it"[18].

These trends are equally well marked at earlier phases in development. For instance in 1940 72% of the communities in Mexico were hamlets with a hundred inhabitants or less. Ninety-four per cent of all the inhabited places had under 500 persons living in them[19]. During the past three decades as a higher level of economic development has been attained so larger centres have emerged and many of the smaller central places have declined in population numbers or have stagnated. All this has taken place against a general trend of sharply rising population numbers.

Examples of similar trends emerging in Western Europe in the seventeenth, eighteenth and nineteenth centuries could be cited. In prehistory, analogous structural alterations took place and Hole and Flannery have recently shown that important changes in south-western Iran in the size and spacing of settlements developed as the productivity of land was increased[20].

Conclusion

The landscapes of Christaller and Lösch have been given to us in a developed form. There has been very little discussion of how the elements in their patterns have been derived historically[21]. One of the purposes of the present paper is to expand the discussion. This in no way is to be construed as a criticism of the authors mentioned for their work was formulated with a different set of objectives.

The proposed scheme will not serve as a model of how settlement patterns will evolve in newly developing areas. The scheme is derived from Western European experience. That experience, particularly with regard to the industrial revolution, has made available a considerable technology to less developed areas. Once populations have access to transport technology it may be unnecessary to develop early industries at the fuel and raw material sources. It is possible for some industries to locate at the market at a much earlier stage than was possible in Western Europe. This fact, incidentally, is a contributory cause of the emergence of strongly primate cities in certain developing areas.

Because so much more technical knowledge is available the less developed countries are not going to follow an exactly similar sequence. This study, however, by focusing attention on the processes which give rise to changes in the structure of settlement patterns, may help in the understanding of evolutionary processes in less developed areas. In such areas, all the stages sometimes appear to be taking place together. In some Latin American countries rural settlement is nucleating, new market centres are emerging and large conurbations developing at the same time. Although the sequence and time scale differs from Western European experience, the processes involved appear to be much the same.

Notes

1 In addition to members of the seminar the author wishes to thank colleagues in the Department of Geography at the University of Sheffield who contributed ideas to this study. In particular Miss J. M. Wilkes and L. J. Wood gave freely of their time. The Research Council of the University of Nebraska contributed generously to the costs of preparation and presentation of the paper.

2 The development of Anglo-American settlements requires a modified scheme in that growth started at a later period with a different set of pre-conditions. The modified scheme would start at phase three as a trans-oceanic colonial off-shoot. See Whebell, C. F. J. (1969). Corridors: a theory of urban systems, *Ann. Assoc. Amer. Geogr.*, **59** (1), pp. 1–26.

3 It is frequently pointed out that such preconditions render a model unrealistic. Such criticisms are based upon a misunderstanding of the functions of models. The assumptions "are a necessary part of my argument, allowing me to establish the operation of a certain factor, a factor whose operation we see but dimly in reality, where it is in incessant conflict with others of its kind". Thunen, J. H. V. (1842). *Von Thunen's Isolated State*. London. trans. by C. M. Wartenberg (1966).

4 Hudson, J. C. (1969). A location theory for rural settlement, *Ann. Assoc. Amer. Geogr.*, **59** (2), pp. 365–81. The general principle involved has been stated by Zelinsky in the following form: Within "the nexus of economic interchange, under rules that vary from culture to culture, people *tend* to sort themselves out areally to attain the greatest efficiency, *i.e.* minimum cost and maximum return, in performing their roles as producers and as consumers of goods and services", Zelinsky, W. (1966). *Prologue to Population Geography*. New Jersey. p. 35. This statement

can be accepted in relation to areas of relatively limited economic development, although there are numerous exceptions. However, if the statement is allowed to stand unmodified in relation to industrial regions it comes close to perpetuating the idea of Adam Smith that in economic matters a guiding but unseen hand worked within economic structures to promote the most efficient of all possible worlds. This is just not so, unfortunately, otherwise many industrial regions would not be facing the problems of congestion and pollution on such a vast scale. The individual decision-making which is the competitive basis of efficient distribution can only function effectively in straightforward situations. Once the environment becomes complex, imperfect knowledge may produce poor decisions. After all, problems of congestion are merely one result of numerous individuals and corporations all opting for the same "right" decision.

5 Numerous examples of this type of rural settlement concentration can be gathered from the developing world. In Cuba, for instance, the peasants have moved "from their dispersed *bohios* to small agricultural villages, usually of less than 1,000 inhabitants. During the last 30 years, hundreds of such villages have sprung up in the Cuban countryside, especially in Oriente". West, R. C. and Augelli, J. P. (1966). *Middle America.* New Jersey, p. 125. Similar examples could be drawn from other Latin American countries.

6 It is not the intention at this stage to attempt to be precise about the relative size and spacing of settlements. The major purpose is to identify the forces which lead to the emergence of the elements in the pattern. A useful starting point for a study in size and spacing is Rashevsky, N. (1968). *Looking at History through Mathematics.* Cambridge. Chapter 16 contains "Some speculations on the formation of early villages".

7 This statement is open to contest for not everyone would agree with Alfred Marshall's statement that "the causes which determine the economic progress of nations [regions] belong to the study of international [interregional] trade". Yet it is beginning to appear that the first *networks* of towns emerged from precisely this cause. In a recent survey of the Aegean area in the Bronze Age, Renfrew has found a striking correspondence between metallurgical development, the growth of trade, and the emergence of something approximating a network of towns. Renfrew, C. (1969). Trade and Culture Process in European Prehistory, *Curr. Anthrop.*, **10** (2–3), p. 159. A good review of urban origins in part of Western Europe is to be found in Nicholas, D. M. (1969). Medieval Urban Origins in Northern Continental Europe: State of Research and some Tentative Conclusions, *Studies in Medieval and Renaissance History*, **6.** It is to be noted that the first sentence of this section does not preclude the existence of a complex of socio-economic changes giving rise to the economic development referred to.

8 The problem becomes more complex if we assume several entry and exit paths for goods but historically the number of paths has normally been limited at this phase.

9 The situation is now susceptible to a von Thunen *ideal-state* type of analysis.

10 While there is a general awareness of the fact that small communities are liable to decline in a modern economic setting, it is not widely recognized that relatively small settlements have, historically, been prone to similar forces. The operation of economic forces in this context is easily masked by more spectacular processes, for instance warfare, which are more liable to catch the pen of the eloquent chronicler.

11 This process of town emergence in response to changes in the structure of an economy can be illustrated by numerous examples from the less developed world. Mabogunje has recently demonstrated that alterations

in the economy of Nigeria had a marked impact upon the urban hierarchy of the country. Mabogunje, A. L. (1968). *Urbanization in Nigeria*. London. Soja has analysed the spread of a money economy in Kenya and shown how "a whole cluster of events occurs as an area modernizes". One of the most important events in the cluster is the emergence of urban centres. Soja, E. W. (1968). *The Geography of Modernization in Kenya*. Syracuse. See also, Morgan, W. T. W. (1969). Urbanization in Kenya, *Inst. Brit. Geogr., Trans.*, **46,** pp. 167–78. In Latin America Preston has indicated how the introduction of a more widespread money economy has resulted in the emergence of new market centres, Preston, D. A. (1969). The Revolutionary Landscape of Highland Bolivia, *Geogr. J.*, **135** (1), pp. 1–16; Preston, D. A. (1970). New Towns—A major Change in the Rural Settlement Pattern in Highland Bolivia, *J. Latin Amer. Studies*, **2** (1), pp. 1–27.

12 Wilson has worked out the development of a hypothetical coalfield. Wilson, M. G. A. (1968). Changing patterns of pit location on the New South Wales Coalfields, *Ann. Assoc. Amer. Geogr.*, **58** (1), pp. 79–81.

13 It is to be noted that the assumption here is that the industrial revolution develops on the plain. If the new technology is imported into the region then this will alter the manner in which the coalfield is exploited and initially the relative positions of A and B.

14 "In the past 30 years one of the most important changes in the field of industrial location can be described as the growing role of labour problems and the diminishing role of transport costs". *Criteria for Location of Industrial Plants*. (1967). U.N. Publ. New York, p. 5.

15 Lampard, E. E. (1955). The history of cities in the economically advanced areas, *Economic Development and Cultural Change*, **3.** He describes the factors which help to cause dispersion and notes that they have been operating in the New York area since the end of the nineteenth century.

16 Friedmann, J. (1968). An information model of urbanization, *Urban Affairs Quarterly*, **4** (2). He outlines some possible changes in locational values.

17 Rostow, W. W. (1961). *The Process of Economic Growth*. Cambridge. In a recent paper Whebell, C. F. J. (1969) *op. cit.* p. 6, f.n. 22, has expressed the view that the "institutional changes, with which Rostow is mainly concerned, are not the same thing as distributional changes, which can both reflect institutional changes . . . and also be a cause of them. . . ." It is the basis of the present article that the economy of an area and its settlement patterns form part of a system, however open, and that changes in the one are reflected in the structure of the other. Many economic historians have difficulty accepting Rostow's ideas relating to the stages of economic growth. The reservations, however, are usually concerned with rates of capital accumulation, the problem of explaining the mechanisms of economic development, and the length of the stages.

18 Scott, P. (1968). Trade Center, Population Change, Civilization, and Trade Area Economy Type, *Rural Sociology*, **33,** p. 428.

19 Tannenbaum, F. (1965). *Mexico*. London, p. 10.

20 Hole, F. and Flannery, K. V. (1968). The prehistory of south-western Iran: a preliminary report, *Proc. Prehist. Soc.*, **33.**

21 Since this paper was prepared originally a text by James E. Vance, Jr., 1970. *The Merchant's World: The Geography of Wholesaling*, Englewood Cliffs, has appeared. The book contains a sequence of settlement development somewhat similar to the one outlined here and also some pertinent observations on the need to view central places in a historical perspective, (see pp. 138–66).

DONALD A. DAVIDSON

Terrain adjustment and prehistoric communities

One contribution which geomorphology can make to archaeology is the recognition of terrain types which were of significance to prehistoric communities. Many studies have recognized the importance of variation of terrain in itself[1] or terrain within ecological associations[2]. The terrain of an area can be defined as the sum of the form of the land, the soil and the water attributes. The aim of this short paper is to focus attention on the methodology of terrain classification with respect to archaeology and the resultant implications in the interpretation of such spatial variation in terrain. Emphasis must be given to the recognition of terrain elements which were of behavioural significance; this theme will be illustrated by reference to the Plain of Drama, Macedonia.

Consideration must first be given to the physical environmental system which includes terrain as an integral sub-system. The physical environment of an area at a point in time can be considered to consist of a set of elements. These elements would describe fully the terrain, flora, fauna and climate. In Fig. 1a the physical environment (A_1) is represented by a set of elements which are shown as dots. These elements must be viewed as inter-related within an ecosystem. The physical environment as perceived by a group of people living within this area can be visualized to be a subset of A_1 and is represented in Fig. 1a by set P[3]. The perceived environment contains only the environmental elements which are included in decision-making processes. This perceived environment or psycho-milieu[4] consists of images or ideas derived from the physical environment. The actual behavioural environment which elicits a behavioural response is a subset of the elements within the perceived environment (P). The behavioural environment (B_1) can be defined to consist of elements e_1 to e_6 inclusive. These significant elements of B_1 are essential parts of the man-environment system and necessitate a behavioural response. Kirk[5] has distinguished between the phenomenal environment and the behavioural environment; the former includes natural phenomena, perhaps altered or created by man while the latter consists of behaviourally significant patterns or structures (*gestalten*).

The cultural and technological development of a community has the important corollary that the behavioural environment will also change, a theme which has been very well illustrated by Bennett[6]. Brookfield[7] has been able to present a comprehensive review of the processes of environmental perception and by use of a system approach he was able to describe succinctly the constantly changing relationship between the perceived environment and the overall system. This theme can be exemplified by reference to Fig. 1.

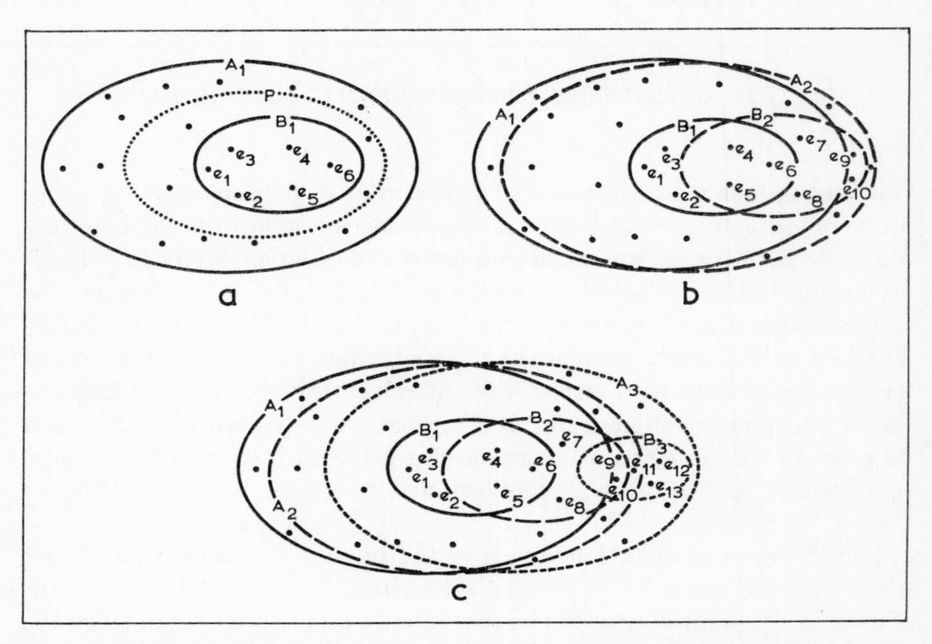

Fig 1 The development over three periods of the relationships between behavioural environments (B_1, B_2 and B_3) and physical environments (A_1, A_2 and A_3).

At a second point in time (Fig. 1b) the physical environment (A_2) can be considered to consist of another set of elements, many of which occurred in A_1. The elements not common to A_1 and A_2 are a result of human interference and changes within the environmental system. Similarly, the behavioural environment has also changed; the second behavioural environment (B_2) includes elements e_4, e_5 and e_6 common to both B_1 and B_2, but elements e_7, e_8, e_9 and e_{10} which do not occur in B_1 are included in B_2. A third point in time can be envisaged (Fig. 1c) when a behavioural environment (B_3) consisting of elements e_9, e_{10}, e_{11}, e_{12} and e_{13}, within a new physical environment (A_3) has no element common with the first behavioural environment (B_1).

These general considerations have important implications for the geomorphologist who is involved in an archaeological project. His precise problem is to recognize elements of the terrain which were included within the behavioural environment of communities in the past. The geomorphologist must first make a comprehensive inventory of the terrain. Some type of geomorphological map is required to provide such an inventory. Fig. 2 is a simplified geomorphological map, which also shows the distribution of known prehistoric sites, of part of the Plain of Drama. This map places emphasis on the major variations in terrain type as well as on the distribution of springs and on the drainage pattern. Such an inventory must be followed by a reconstruction of the terrain during the settlement phases. It has been suggested[8] that during the period of known prehistoric settlement in this area springs were more numerous towards the base of alluvial fans and also along the edge of the alluvial strips flanking the streams. The lower alluvial fans were still in active formation and the alluvium of the Angitis river was subjected to several phases of aggradation and incision.

A terrain inventory together with a reconstruction of the former elements of the terrain provide the first step in an environmental study; in order for the second step, which is concerned with the evaluation of the behavioural attributes of the terrain, to be taken, all the relevant elements must be included in this comprehensive inventory.

The recognition of the terrain elements which were of behavioural significance to a people in the past is the most difficult part of an environmental study. For example, from a visual examination of the relationship between the distribution of prehistoric sites and certain attributes of the terrain which are shown on Fig. 2, five possible significant attributes can be suggested. The association of at least a few sites with springs and finer textured alluvial land could be suggested whilst the absence of sites on the steep foothills, the stony upper fans and the central part of the lowland limestone area could suggest that these terrain types inhibited permanent settlement. If, for the purpose of this discussion, the period of prehistoric occupation on the Plain is divided into three as in Fig. 1, then it can be suggested that these terrain characteristics are included in a set composed of the union of sets B_1, B_2 and B_3. The significance of terrain in each phase must be initially assessed by reference to the man-environment system. It would be wrong at this stage to emphasize the significance of any one terrain characteristic on settlement on the Plain; during the 2700 years of occupation at the tell near the village of Sitagroi the relative importance of terrain characteristics would have drastically altered.

It is important to note the feedback effect of behavioural adjustment at an early phase on a later phase. For example, in the Plain of Drama, at a first hypothetical phase when hunting and collecting would still have been dominant, localization of site may have been influenced by proximity to a spring. At a later phase, when cultivation of crops became important, the change in perception of terrain would have been in part conditioned by settlement location which was established at an earlier phase. Perception of terrain is not only dependent upon the character of the terrain, but also upon distance from settlement sites. Brookfield noted[9] that he had in an earlier study[10]

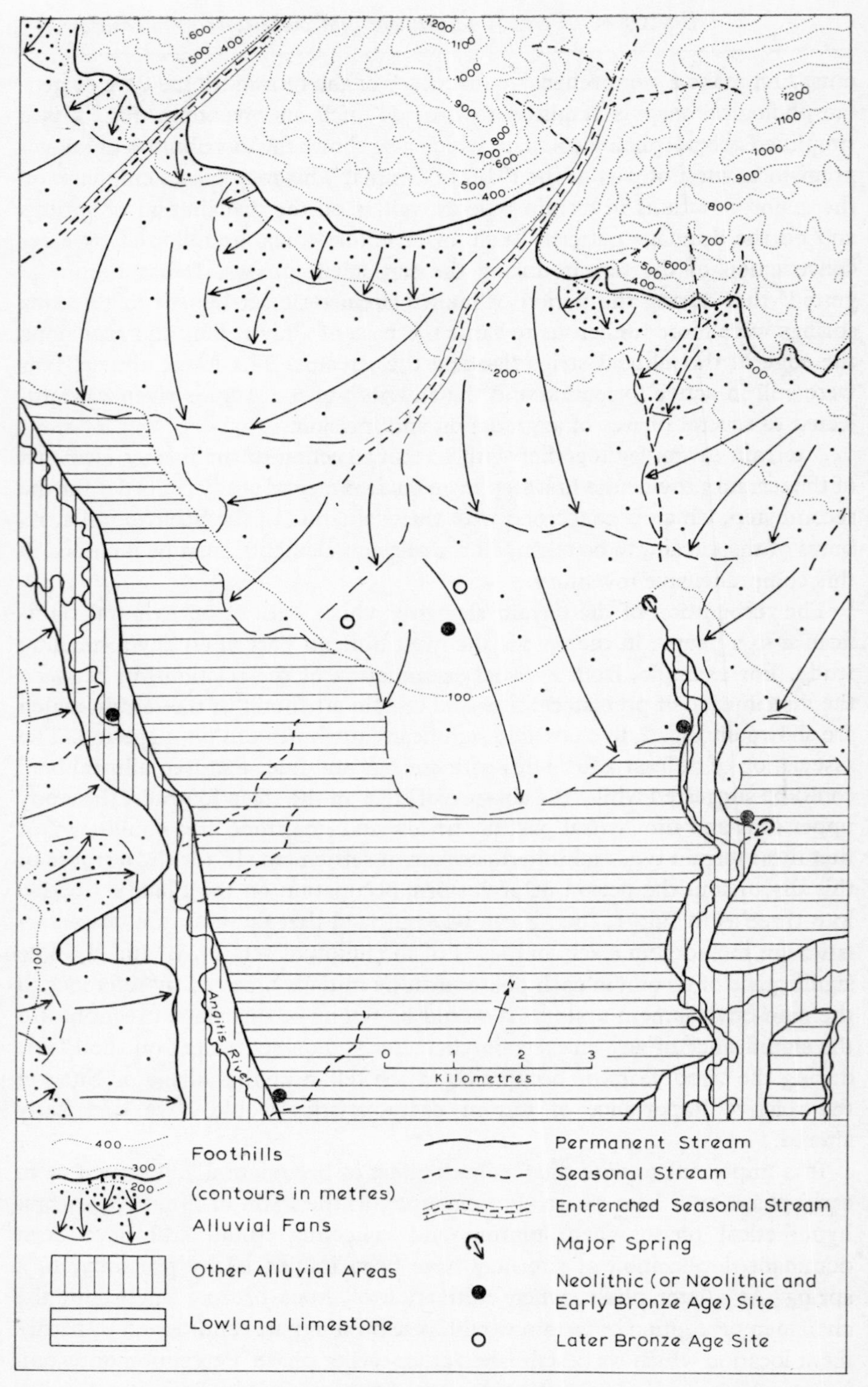

Fig 2 Geomorphology and prehistoric settlement of part of the Plain of Drama.

failed to realise that land evaluation depended as much on proximity to site and political conditions as it did on terrain characteristics. There is thus the suggestion that certain terrain factors may be of lesser importance than expected because the feedback effect of earlier terrain factors may have in part restricted re-evaluation of the terrain at later phases. In other words, not only must the terrain be considered within each phasal man-environment system, but the inheritance of terrain adjustment from former systems must be incorporated within subsequent systems. Carson[11] has stressed that perception of a given environment depends upon the other environments which have been experienced. The important corollary is that the continuity of a settlement pattern over a long period does not indicate a similar constancy in a behavioural environment.

These considerations emphasize that simple explanations of adjustment by prehistoric communities must not be sought. The difficulty in interpreting the significance of certain elements of the terrain has been exemplified by reference to part of the Plain of Drama. During the earliest Neolithic the agricultural attraction of the alluvial land was probably not realized; instead localization beside a water supply and access to hunting, herding and collecting areas were more likely important. The riverine zone may have been of aquatic resource value during this phase whilst during later phases, this alluvial area may have become attractive for cultivation. Perhaps a suitable conclusion which can be drawn is that prehistoric settlement evolved in the areas where a good water supply was coincident with light alluvial land[12].

The important point which emerges is the dynamic relationship between man and terrain. Kirk[13] wrote that "as in Historical Geography we are concerned with the behaviour of human groups in relation to environment it behoves us to reconstruct the environment not only as it was at various dates but as it was observed and thought to be, for it is in this behavioural environment that physical features acquire values and potentialities which attract or repel human action". Although Kirk did not differentiate between the perceived and behavioural environments, his argument is of marked relevance to environmental archaeology. The physical environment must not be viewed as a static backcloth upon which man is superimposed, but instead the constantly changing behavioural environment must be extracted from the physical environment and interpreted within the general man-environment system[14]. There is the danger in site catchment analysis[15] that the definition of land types results in a static appraisal of resource evaluation. It is argued that a cognitive behavioural approach to environmental archaeology provides a fuller comprehension of the processes of adjustment[16] to terrain by prehistoric communities[17].

Notes

1 Flannery, K. V., Kirkby, A. V. T., Kirkby, M. J. and Williams, A. W. (1967). Farming systems and political growth in Ancient Oaxaca, *Science, N.Y.*, **158,** pp. 445–54; Butzer, K. W. and Hansen, C. L. (1968). *Desert and River in Nubia.* Madison; Vita-Finzi, C. (1969). Geological opportunism, *in* Ucko, P. J. and Dimbleby, G. W. (eds.). *The Domestication and Exploitation of Plants and Animals.* London.

2 Coe, M. D. and Flannery, K. V. (1964). Micro environments and Mesoamerican pre-history, *Science, N.Y.*, **143,** pp. 650–54; Hole, F., Flannery, K. V. and Neely, J. A. (1969). Pre-history and human ecology of the Deh Luran Plain, *Memoirs of the Museum of Anthropology, Univ. of Michigan, 1.*

3 Sonnenfeld, J. (1968). Geography, perception and the behavioural environment, a paper presented at Dallas AAAS, 27th December 1968, in a symposium on *The Use of Space by Animals and Man;* quoted by Saarinen, T. F. (1969). *Perception of Environment.* Commission on College Geography, Resource Paper No. 5, Washington.

4 Sprout, H. and Sprout, M. (1965). *The Ecological Perspective on Human Affairs.* Princeton.

5 Kirk, W. (1951). Historical geography and the concept of the behavioural environment, *Indian Geogr. J., Silver Jubilee Vol.*, pp. 152–60; Kirk, W. (1963). Problems in geography, *Geogr.*, **48,** pp. 357–71.

6 Bennett, J. W. (1944). The interaction of culture and environment in the smaller societies, *Amer. Anthrop.*, **46,** pp. 461–78.

7 Brookfield, H. C. (1969). On the environment as perceived, *Progress in Geogr.*, **1,** pp. 51–80.

8 Davidson, D. A. (in press). Geomorphology and prehistoric settlement of the Plain of Drama, *Rev. Géom. dyn.*

9 Brookfield, H. C. (1969). *op. cit.* pp. 51–80.

10 Brookfield, H. C. and Brown, P. (1963). *Struggle for land: agriculture and group territories among the Chimbu of the New Guinea Highlands.* Melbourne.

11 Carson, D. H. (1965). The interactions of man and his environment, *in Environmental Evaluations.* Publication of Architectural Research Laboratory of the Schools Environments Research Project, Michigan Univ.

12 Davidson, D. A. (in press). *ibid.*

13 Kirk, W. (1951) *op. cit.* p. 159.

14 For a recent review of perception see Wood, L. J. (1970). Perception studies in geography, *Trans. Inst. Br. Geogr.*, **50,** pp. 129–42.

15 Jarman, M. R., Vita-Finzi, C. and Higgs, E. S., Site Catchment analysis in archaeology, this volume, pp. 61-6.

16 The distinction between *adjustment* and *adaptation* is ignored in Sonnenfeld, J. (1966). Variable values in space and landscape: an enquiry into the nature of environmental necessity, *J. Social Issues*, **22,** pp. 71–82.

17 My thanks are due to Mr. G. M. Lewis, Dr. A. C. Renfrew and to Dr. J. M. Renfrew for their helpful comments, and to Miss S. Ottewell for drawing Fig. 2. The base map for Fig. 2 was provided by the Greek Army Geographical Service.

KENT V. FLANNERY

The origins of the village as a settlement type in Mesoamerica and the Near East: A comparative study

Introduction

The village—one of the most widespread settlement types in the world today—seems to have been unknown during the first two million years of the human career. The first steps toward truly sedentary life may have been taken by hunter-gatherers in Europe during the Late Pleistocene, but the archaeological evidence is still ambiguous and the trend seems to have broken down at the end of the Würm glaciation. Unmistakeable villages appeared, apparently independently, in several different parts of the world after the close of the Pleistocene epoch. By 7500 B.C. in the Near East, by 2500 B.C. in the Andes, by 1500 B.C. in Mesoamerica, villages were not only widespread but architecturally diversified. In some the houses were of wattle-and-daub, in others of mud or mud-brick, in still others of stone masonry, with or without mortar.

Here is a situation made to order for comparative studies, and its potential did not escape the eyes of the multilinear evolutionists who have contributed so much to anthropological theory over the last two decades[1]. Struck by the apparent contemporaneity of early villages and the Neolithic revolution, the evolutionists came up with a by-now familiar reconstruction: "once agriculture had freed man from the eternal food quest he was able to give up his ceaseless wandering and settle in villages where he perfected pottery making, loom weaving, and all the hallmarks of sedentary life".

Archaeological discoveries over the last ten years have not been kind to this reconstruction. From the Near East came the discovery of fully sedentary communities dating to 8000 B.C., yet lacking all evidence of domestic animals or phenotypically domestic cereals[2]. From Mesoamerica came the discovery of prehistoric groups who cultivated four or five species of plants by 5000 B.C., yet were still nomadic—and remained so for the next 3500 years in spite of substantial increases in the number and variety of cultivars[3].

Gradually it dawned on both archaeologist and evolutionist that neither had been sufficiently multilinear: perhaps agriculture and village life had begun in these various parts of the world for reasons so different, and through processes so different, that it might not be possible to explain them all by means of the same model[4]. Our Western eyes, grown somewhat heavy-lidded during the long search for cross-cultural regularities, had confused three variables which are not necessarily interdependent. These were agriculture (which does not require either sedentary life or villages); sedentary life (which does not necessarily require agriculture, and may not take the form of a village); and villages (which need not require agriculture, nor require year-round sedentary life).

The problems outlined above are sufficiently discouraging to restrain me from presenting a general model to account for the origins of the village everywhere. Instead, I will compare and contrast two specific areas—Mesoamerica and the Near East—and concentrate on the early permanent settlements which arose there, largely ignoring the hundreds of other kinds of villages which have existed around the world. The model of archaeological process I will use is that recently proposed by Clarke[5], in which culture is regarded as a *system* which passes with time through a succession of *states*; the sequence of states exhibited, called the *trajectory*, depending upon the history of the system and on the influence of the *environment*, both physical and cultural, with which it interacts. It is my impression that Mesoamerican and Near Eastern cultures, starting with quite different histories, passed through somewhat different trajectories on their way to village society. Such convergence on a common state through different pathways implies something about the adaptive value of the village as a settlement type[6].

Differences in the history of the system

In Clarke's scheme, the current state of any culture is partly the product of its past states. In other words, if we go back to the period *before* agriculture or sedentary life in Mesoamerica and the Near East, it may be possible to find some pre-existing differences in settlement pattern which help explain the later differences between the two areas. The differences I will suggest must be regarded as tentative, since they are based on archaeological data which could be given an alternative interpretation. They depend mainly on the very different role played by the hunting of large ungulates in the two different regions.

In the Near East, the later part of the "food-gathering" era (20,000–10,000 B.C.) featured skilled and intensive hunting, with composite weapons, of herd animals like the wild goat, sheep, gazelle, onager, deer, and aurochs, supplemented by the collection of a wide range of molluscs, land snails, fish, crabs, turtles, partridges, and presumably also wild fruits and legumes (although remains of the latter are not preserved). For the Zagros Mountains, Hole and I[7] have interpreted the archaeological record as indicating that the basic unit of settlement was a "base camp" of perhaps 15–40 persons; from this base, hunting parties of 3–8 males made periodic trips to "transitory stations" or vantage points from which they stalked and eventually killed the herbivores which provided 99% of their meat supply. This game was sometimes cut up into portable sections at temporary "butchering stations", from whence it could be transported back to the base camp. For inland Syria, Binford and Binford[8] have suggested a similar pattern, with both men's "hunting camps" and women's "temporary work camps" for plant processing made at some distance from the base camp.

Such a pattern may be fairly typical of hunting groups in areas of abundant herd ungulates. In a recent comparative study, Wilmsen[9] has concluded that a modal band of this type might consist of an "optimally effective number of hunting males (6–8)", each associated with 1–3 adult females (and their offspring) who constitute a plant-collecting unit. As in the case of the surviving old world hunters and gatherers recently described in the Lee and DeVore symposium[10], such bands periodically formed loose aggregates with neighbouring bands at times of high resource availability. Lee and DeVore stress that the composition of such large base camps is not a formal set of related families, but a fluctuating aggregate of all ages and both sexes, with considerable coming and going; size of the camp varies more with the availability of water, game, and wild plant foods than with residence rules. Concepts of descent and territoriality are very weakly developed, and division of labour is along the lines of sex rather than families. According to Wilmsen, such a pattern may be common where the major resource is mobile or very unevenly distributed.

In the arid highlands of Mesoamerica, on the other hand, MacNeish[11] has interpreted the archaeological data as indicating a different pattern. Here the late food-gathering era (8000–5000 B.C.) featured intensive rainy-season harvests of localized wild plants, trapping of small game, and limited hunting of white-tailed deer—the only ungulate of any size, and one which did not form large herds. In the dry season, groups fragmented and dispersed into the smallest units consistent with survival, using widespread, stable, but less appetizing plant foods. MacNeish has distinguished two types of settlements: "macroband" camps, occupied for most of a season by 15–20 persons, and "microband" camps, occupied by 2–5 individuals for anything from a day or two to most of a season. Since even microband camps contain both men's and women's tools, MacNeish attributes them to the kinds of "family collecting groups" recorded by Steward[12] for the Indians of the arid Great Basin of the western United States. Presumably, therefore, the "macroband" camps resulted from the periodic coalescence of numerous scattered families (who normally foraged by themselves) during times of high food resource availablity in certain areas. In the study already mentioned, Wilmsen equates the maintenance of small, family collecting areas with widely distributed, stable resources of the kind used during the dry season "microband" phase.

The implications of these two pre-village settlement patterns—if future research proves them to be valid—are interesting. As one who has worked in both areas, my impression is that population densities were considerably higher in the Near East at this stage, and Near Eastern "base camps" larger and more nearly permanent than anything yet found in highland Mesoamerica. Indeed, some of the base camps in caves like Abu Halka, Ksâr 'Akil, Yafteh, and Shanidar may have been occupied repeatedly for the greater part of a year on a scale unknown in Mexico[13].

But perhaps more significant for what was to follow were the possible differences in the lines along which society segmented. Only in Mesoamerica is there any evidence that "the family", as such, was a basic unit of residence, food procurement, and storage, as at "microband" camps. In the Near East, the basic residential unit seems to have been an encamped band of flexible composition, crosscut by men's and women's work groups, not necessarily composed of siblings. This may have been due, at least in part, to the far greater importance of ungulate hunting in the Near East, which selected for modal aggregates of 6–8 co-operating meals. In Mesoamerica, on the other hand, a low density of big game coupled with great food scarcity during the late dry season selected for maximum dispersal into family collecting groups like those of the Paiute and Shoshone. Thus two ingredients which were later to characterize the village—family residential units and male/female work groups—already existed in the food-gathering era, but perhaps not in the same combination and configuration that we see in the later villages in Mesoamerica and the Near East.

Interactions with the physical environment

Additional similarities and differences between early sedentary communities in Mesoamerica and the Near East can be traced to the ways in which they interacted with the physical environment.

1. Geographer Philip Wagner[14] has made a useful distinction between two kinds of "human artefacts": implements and facilities. Implements are the human artefacts which transmit or move kinetic energy; facilities store up potential energy, or impede its transfer. More recently, Binford[15] has pointed out that one long-term trend discernible in human prehistory is an increasing reliance on facilities relative to implements. Especially striking is the increased use of immobile facilities such as storage pits, wells, check dams, canals, and fish weirs.

Such a trend can already be seen in the final stages of the Palaeolithic in the Near East (as, for example, in the digging of storage pits up to a metre in diameter in caves like Shanidar[16]), and it manifests itself clearly in the scores of plastered storage pits of the Natufian period, the mortars dug permanently into bedrock at El Wad and Aïn Mallaha, and the grain roasting pits of Tell

Mureybit[17]. Early Formative villages in Mesoamerica not only had hundreds of bell-shaped storage pits for maize, but also—in some areas—wells of a type used widely for irrigation over the arid parts of the New World[18]. Actual irrigation canals appeared at least as early as the Samarran period in the Near East[19] and the Middle Formative period in Mesoamerica[20].

2. Granting some of the parallels between Mesoamerica and the Near East in the development of storage, processing, and land-improvement facilities, we are still faced with the paradox: why did such a complex appear *before* cultivation in the Near East and take 3000 years to develop *after* cultivation started in Mesoamerica?

Recent field studies in Mexico by Kirkby[21] indicate that major reliance on maize as a staple food is not considered practical unless a return of 200–250 kg/ha can be expected. Such yields can be expected even on relatively marginal land with today's races, but would be difficult to obtain even on the best soils with the tiny races of early maize whose progress has been summarized by Mangelsdorf[22]. In fact, it is instructive to examine some of the innate differences between maize and the Near Eastern staples, wild wheat and barley.

Under optimum conditions, in the best parts of their ranges, wild wheat and barley form stands as dense as those of a cultivated field. For example, Zohary[23] estimates that mixed stands of wild emmer wheat and wild barley in eastern Galilee will produce 500–800 kg of grain/ha in rainy years. In his now-famous experiment in Turkey, Harlan[24] harvested a kilogramme of wild einkorn wheat in an hour and estimated that a family of four could harvest a metric ton in three weeks. Theoretically, fully sedentary life could be based on such yields. Nowhere, however, is it imagined that the wild ancestor of maize occurred in stands that dense; no one could have harvested enough to underwrite sedentary life. Indeed, only under the very best conditions could Mesoamerican farmers have produced 250 kg/ha even with the races of *domestic* maize known as late as 1500 B.C.[25].

Some of the favourable genetic changes taking place in the early domestication of wheat and barley—for example, the appearance of the tough rachis, the "naked" grain, and the six-row head—are apparently controlled by no more than one or two genes[26]; hence, they could take place in a generation or two. Moreover, wheat and barley are self-pollinating[27], so that favourable mutant strains can be rapidly stabilized. So far as is known, none of the comparable genetic "improvements" in maize are controlled by so few gene loci; the favourable characters are all polygenic and many may only have been stabilized after considerable crossing and back-crossing[28]. In addition, maize is wind-pollinated; each generation of early cultivated maize, no matter how "improved", was subject to contamination from the pollen of its wild ancestor, which in many cases grew as a weed alongside the domestic strain. Thus long periods were required to stabilize the more productive domestic races.

Interaction with the social environment

Most of the hunting-gathering groups discussed in the Lee and DeVore symposium have a rather weakly developed sense of territoriality. In part, this is due to the large size of the "home ranges" they utilize, the scattered (and sometimes mobile) nature of their resources, and the need for widespread sharing. Archaeological data from Mesoamerica and the Near East suggest that early farmers, on the other hand, concentrated their planting in marsh areas, seasonally-flooded bottomlands and outwash fans, or zones of permanently high subsurface water wherever these were available[29]. Maintaining "ownership" of these far smaller and more localized resource areas in the face of competition from neighbouring groups requires a kind of communal territoriality which, although far from rigid, exceeds that of the hunters described by Lee and DeVore.

Three methods of defending territory observed among aboriginal farmers have been (*a*) permanent settlement on the critical resource area, (*b*) development of a territorial gradient from the "core" to the "periphery" of their home range, and (*c*) an ideology of descent which stresses maintenance of land ownership through many generations, with continued participation of deceased ancestors in the affairs of the descent group. To cite only one example, Ford's recent study of a Tewa Pueblo in New Mexico[30] shows the home range to be divided into (1) an immense area of mountains and piedmont which is shared relatively amicably with neighbouring groups, and (2) a small area of river bottom land and fringing upland where the village is located, cultivation takes place, and ancestors are buried; the latter area would be defended, by violence if necessary, against intrusion by outsiders. Such "core" areas are critical, and their value is enhanced once they have been improved with "investments" like permanent storage, processing, resource-maintenance and water-control facilities.

Response to the physical and social environments: a summary

Tentatively, I suggest that the origins of "sedentary life" had more to do with the installation and maintenance of permanent facilities, and the establishment and maintenance of hereditary ownership of limited areas of high resource potential, than it did with agriculture *per se*. There can, however, be little doubt that the decision to make cereal grasses the focal point of the subsistence strategy increased the need for permanent storage and processing facilities. Such a decision could have been made before agriculture began in parts of the Near East; in Mesoamerica, even after several millennia of cultivation, maize was not yet sufficiently improved to serve as a major staple.

The placing of permanent, nucleated communities on or near localized areas of strategic resources probably changed group ideology from one of weak territoriality to the pattern of a small, defended core area versus a large, undefended periphery, further emphasized in concepts of descent. In this regard, we should perhaps not be surprised to find burials under house floors or in adjacent courtyards, as in Early Formative Mesoamerica[31]; repositories for multiple secondary burials, as in both regions[32]; or the saving of ancestor's skulls, as in the Near East[33]. In a world without written deeds, the presence of the ancestors frequently serves as a group's best evidence that the land has been theirs "since time began".

Early settlement types: the archaeological evidence

In the course of preparing this paper, I detected in the archaeological record what appeared to be two types of early sedentary communities: one, a compound or homestead of small circular houses, and the other, a true village of somewhat larger rectangular houses. My next move was to search the ethnographic literature on Mesoamerica, the Near East, and adjacent areas for examples of societies with similar communities. Examples were so many, and so varied, that I was forced to set up a series of *ideal types* against which the archaeological data could be tested. Each ideal type was an abstraction from various ethnographic examples, defined on the basis of structural features which I felt should be preserved archaeologically. I evaluated the ideal types on the basis of shared structural features with actual prehistoric communities, and will discuss below those that seemed to give the best "fit". I did not expect any one archaeological site to resemble precisely any one ethnographically-documented community; all I asked was that both appear to be examples of the same ideal type.

In setting up the ideal types, I have drawn on some recent cross-cultural studies of the relationship between man and his dwellings. Particularly helpful were papers by Naroll[34] and Cook and Heizer[35] on the allometric (loglog) relationship between population and floor area, and between population and village size; equally useful were the contributions of Robbins[36] and Whiting and Ayres[37] on the relationship of house shape to settlement type. At the risk of oversimplifying, their conclusions might be summarized as follows:

(*a*) On the average, roughly 10 m^2 of house floor space are available per individual in societies of Neolithic type[38].

(*b*) Circular dwellings tend to correlate (at a statistically significant level) with nomadic or semi-nomadic societies; rectangular dwellings tend to correlate with fully sedentary societies (although numerous exceptions occur)[39].

(*c*) Rectangular structures replace circular ones through time in many archaeological areas around the world[40] (although reversals of this trend occur).

(*d*) While circular structures may be easier to construct (and often portable), it is much easier to add units to rectangular structures[41].

1 The circular hut compound

The first type of permanent settlement I will discuss is the circular hut compound. I could find no surviving examples in the Near Eastern ethnographic literature, but such compounds are still widespread in nearby Africa. They are found among sedentary cultivators, herders, and people combining cereal cultivation with goat domestication (as did the Neolithic Near Easterners); the compound thus reflects a very flexible type of social grouping which can occur in a variety of subsistence systems. At the outset, I regarded all these African compounds as independent (convergent) examples of a settlement type once characteristic of the ancient Near East. Later, I found so many examples in the archaeological literature of Egypt (from the Nile delta to Badarian Hememieh) that I am no longer firmly convinced of their full independence. Until all the facts are in, I now feel we must hold out the possibility (intriguing, if remote) that the African compounds to be described below are simply the last surviving examples of a settlement type that may once, in early post-Pleistocene times, have stretched from the Levant across Sinai and up the Nile.

Today, excellent ground plans are available for compounds used by the Penhalonga and Inyanga of Rhodesia, the Nuba of the Sudan, the Massa (Fig. 1), Mousgoum and Matakam of Cameroun[42] and the Tiv of Nigeria[43]. In addition, Bohannan's monograph on Tiv settlements gives further insight into the social relationships of the occupants of a compound, who are usually patrilineal polygynous kinsmen. The typical compound consists of an elder—the compound head—his wives, minor children, unmarried adult daughters, his adult sons and their wives and children[44]. Among the Tiv, the average number of persons per compound is 17 (with a range from 2 to 80 persons); adult males average five per compound, adult females six.

From an archaeologist's viewpoint, the critical aspects of the compound are as follows:

(*a*) It is composed of circular huts, often with stone foundations, *each of which is designed to house only one, or at most two, persons*. In other words, no structure is occupied by "a family"; each man has his own hut, and each of his wives expects to have her own hut, especially after she has borne her first child[45]. Thus most huts have no more than 10 m^2 of floor space, confirming Naroll's observations[46]. A sample of thirty huts from three compounds in Cameroun, illustrated by Fraser[47], brings to light two interesting points. First, the typical residence hut is approximately 3 m in diameter and houses only one adult (a "man" or a "wife"), for an average area (πr^2) of 7 m^2 per adult; the largest huts in the sample had a diameter of 5 m, but were often shared by "a man and two cows" or "a wife and several goats", rather than two individuals. Second, out of the sample of thirty huts, no more than twenty can be construed as residences; the rest are kitchens, stables, or storage buildings. Among the Tiv, the largest huts are used only for the reception and entertaining of guests[48]. Thus *the number of adults occupying a compound may be somewhat less than the number of huts;* for an archaeologist to multiply the number of huts by some figure chosen to represent the "average family size" would give a grossly inflated estimate of the population.

(*b*) The huts are frequently (but not always) arranged in a circle or oval, surrounding a large cleared space where most of the work is carried on. In some compounds, the buildings may be clumped so that a man's and his wives' huts form a group separate from those of his brother and the latter's wives. Among the Massa, all the men's huts may be on one side of the circle while all the wives' huts are on the opposite side[49]. A small compound, with its cleared work area, could be only 30 m in diameter, while a large one could cover a hectare or more.

(*c*) Although the headman of a compound may have his own granary (and a man's senior wife may have her own, as well), by and large *the food storage is open and shared by all occupants of a compound.* Most granaries are huts by themselves, or storage units out in the open in the cleared area of the compound. Nor are the granaries of the headman or the senior wives truly private, for "in severe shortage what food is held openly must be shared equally by all within the compound"[50].

The communal nature of such a settlement is perhaps best shown by two Rhodesian compounds illustrated by Fraser[51]. In each, the individual huts, rarely more than 3–4 m in diameter, are joined together like cells in a honeycomb by a series of dry laid stonemasonry walls, the west wall of each hut being the east wall of the next. Within this self-contained unit the storage huts, separate and distinct from the sleeping huts, occur as smaller cells (perhaps 2 m in diameter) constructed as interstices in the very same walls: obviously shared by the occupants of more than one hut, and perhaps on occasion by the whole compound.

(*d*) The social organization of the compound still shares many characteristics with the hunting-gathering groups idealized in Wilmsen's model, and postulated for the Palaeolithic Near East. The typical compound is a group of 6–8 males, each associated with 1–3 women and their children. Generally, the people of a single compound form a basic labour group, and below this level, "full siblings tend to work together"[52]; the sexual division of labour is so basic that even when males and females do perform the same tasks, they structure the work

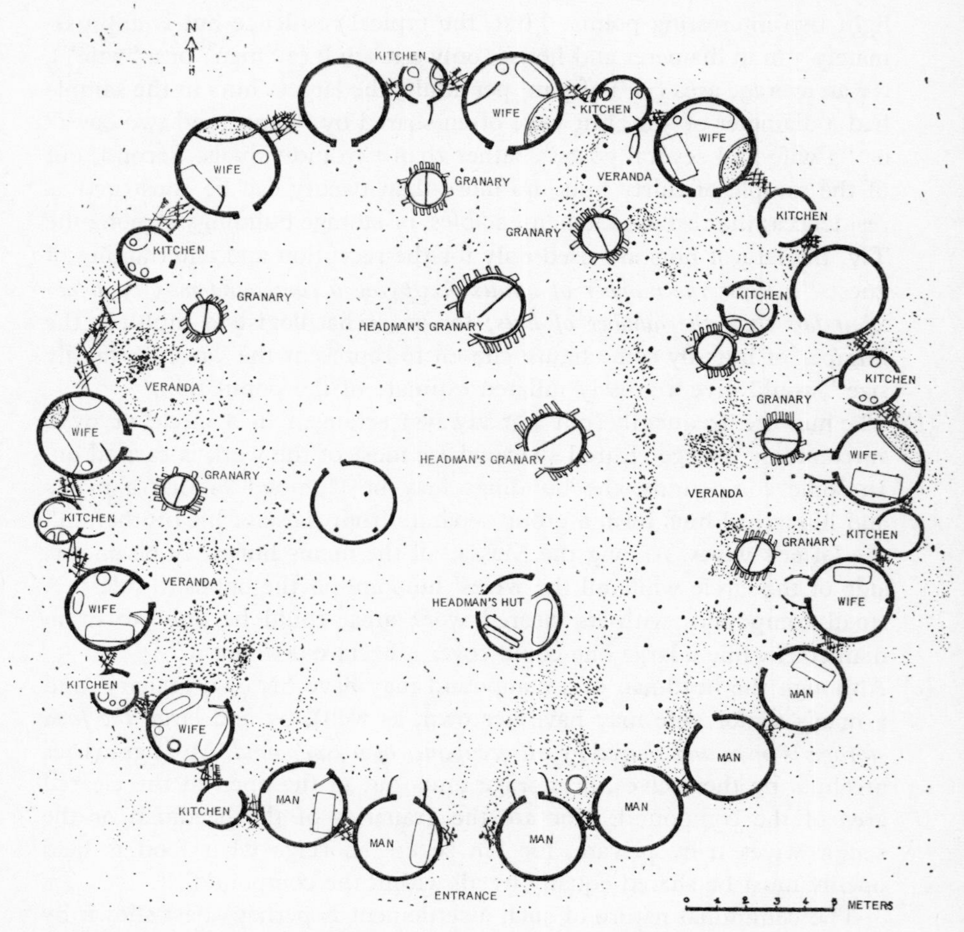

Fig 1 Plan of Massa circular hut compound, Cameroun, Africa.
After Fraser, D. (1968). *Village Planning in the Primitive World.* New York. Fig. 52.

differently. Almost the only point at which the "family" (as we think of it) comes into focus is during the fissioning of a compound through discontent.

The sort of compound described above may have been the earliest type of permanent community in the Near East. Between 9000 and 6500 B.C. the archaeological record shows abundant examples, from the Natufian and Pre-Pottery Neolithic cultures of the coastal Levant to Tell Mureybit on the upper Euphrates. Each of these compounds usually had between ten and fifty huts with stone foundations, most probably housing only one person, a few large enough for two to three. Storage units were usually out in the open and presumably shared; many activities seem to have been carried on in open areas between the huts. These compounds may reflect groups of related polygynous males, each accompanied by one to three females and their children. This could not be proved on the basis of presently published archaeological data, but I gather that (in the case of some sites, at least) there do exist unpublished data which could be used as an independent test of this hypothesis. I refer to the total inventory of tools from each hut, which should reflect differences in occupancy, and could be used to suggest where males and females probably lived. There are hints, for example, that such a test could be run at Nahal Oren (see below).

Cereals reported from these compounds are either phenotypically "wild" (Mureybit) or in the earliest stages of domestication (Beidha); with the possible exception of Beidha, domestic animals were not kept, though wild animal bones are abundantly represented. A few contemporary sites in the Zagros Mountains with circular "structures"—such as Tepe Asiab and Zawi Chemi Shanidar[53]—hint that the pattern may have extended farther to the east than I have suggested. But let us briefly examine a few of the sites in the Levant.

A. *Name:* Aïn Mallaha
Location: Near Lake Huleh, Israel
Probable date: 9000–8000 B.C. (Natufian)
Excavator: Jean Perrot[54]
Description: Settlement here covered about a fifth of a hectare (2000 m^2), and Perrot estimates that at any one time there were probably no more than fifty huts in use on the site. The huts were circular, semi-subterranean, and rock-lined, and Perrot's partial plan of Level I suggests that they may *possibly* have been arranged in a large circle around an open area (see Fig. 2); at least, those illustrated can be construed as forming an arc.

The measurable huts range in diameter from roughly 2·5 to 9 m, and fall into two groups. One group has 5–7 m^2 of floor space (one occupant), another 20–28 m^2 (two to three occupants, by Naroll's criteria). Shelter 51, an unusually large one (64 m^2) may, by analogy with the Tiv, have been for receiving visitors; one atypically well-made, lime-plastered hut (Habitation no. 1, 20 m^2) may have belonged to the compound head. Storage pits occurred either inside, or (more often) outside the huts.

B. *Name:* Nahal Oren (Level II)
Location: On terraced talus slope below rock shelter, Mt. Carmel range, Israel
Probable date: 7000 B.C. ? (Pre-Pottery Neolithic)
Excavators: Stekelis and Yizraely[55]
Description: Level II covered at least 500 m^2, and more huts may have been removed by highway construction. The thirteen reasonably well-preserved huts excavated by Stekelis and Yizraely ranged from 4 m^2 to

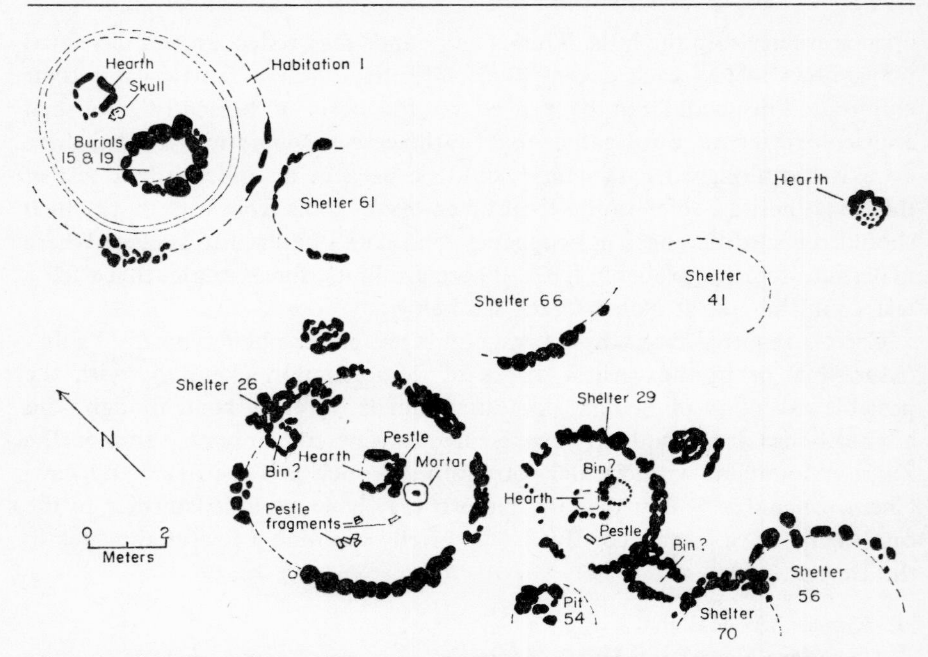

Fig 2 Partial plan of Level I at Aïn Mallaha, Israel (greatly simplified). After Perrot, J. (1966). Le gisement Natufien de Mallaha (Eynan), Israel, *L'Anthropologie*, **70** (5–6).

15 m^2 in extent, with half the specimens in the range of 6 to 9 m^2 (Fig. 3); most, by Naroll's criteria, would have housed no more than one individual. Particularly interesting are the differences in chipped stone tools between House 9 (14 m^2) and House 10 (4 m^2). Tools present in nine but absent in ten include awls, notched tools, retouched blades, bladelets, and microliths. House 10, on the other hand, has many more planes. Such differences are what one might expect in a compound where the larger huts were occupied by man-wife pairs, and the smaller huts by individual wives or single adult males, engaged in a narrower range of tasks.

C. *Name:* Beidha (Levels IV–VI)
Location: Near Petra, in the arid zone of southern Jordan
Probable date: 7000–6500 B.C. (Pre-Pottery Neolithic)
Excavator: Diana Kirkbride[56]
Description: During the period of Levels IV–VI, Beidha is perhaps the finest example available in the archaeological literature of a circular hut compound where, in Kirkbride's own words, the structures fit together "like cells in a honeycomb"[57] and storage units were built into the interstices between dwellings; Fig. 4 compares these levels at Beidha with two almost identical compounds from Rhodesia. Most huts in these

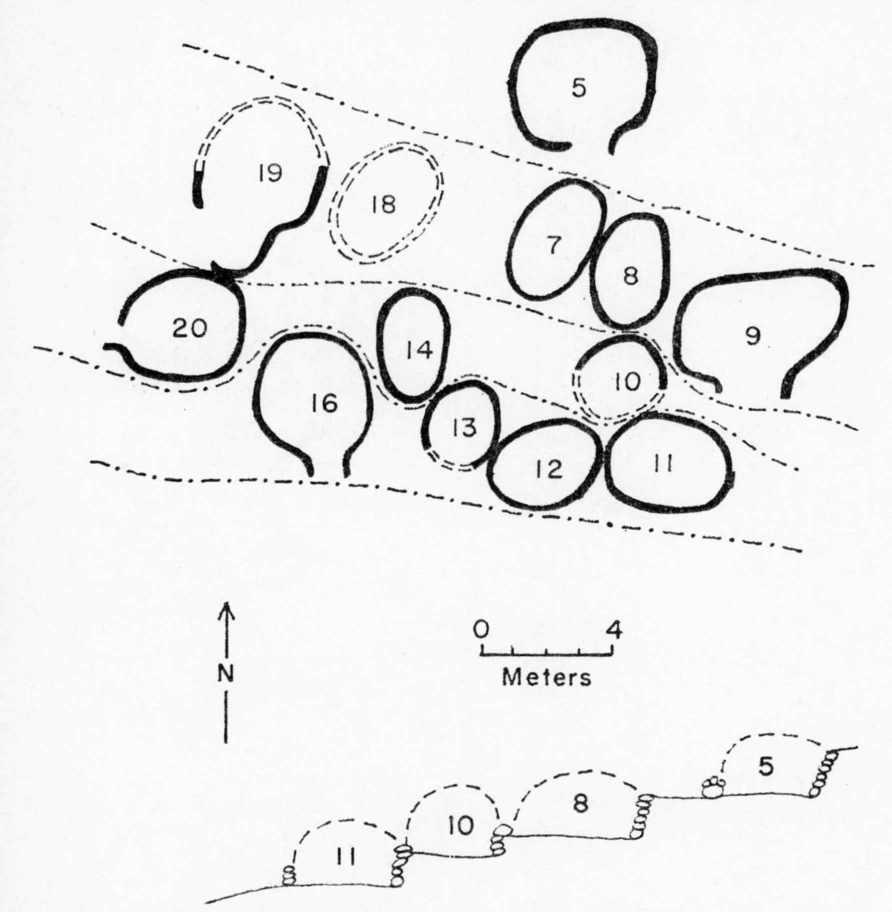

Fig 3 Hut foundations at Nahal Oren, Israel, Level II. *Top*, plan view; *Bottom*, cross-section through huts 5, 8, 10, 11, showing terraces.
After Stekelis, M. and Yizraely, T. (1963) Excavations at Nahal Oren, *Israel Explor. J.*, **13** (1).

levels at Beidha have 5–28 m² of floor space (with a modal area of 13 m²) and were arranged in clusters, probably around open work areas. Outside the compound (but at no great distance) was a series of unusual structures which may be a "shrine area"—a feature not unknown from the African compounds we have discussed, where patrilineal ancestors are revered for reasons already suggested above.

D. *Name:* Tell Mureybit (Levels I–VIII)
Location: On the Euphrates River, east of Aleppo, Syria
Probable date: 8200–8000 B.C.
Excavators: van Loon and Skinner[58]
Description: A limited exposure of the lowest levels at Mureybit produced

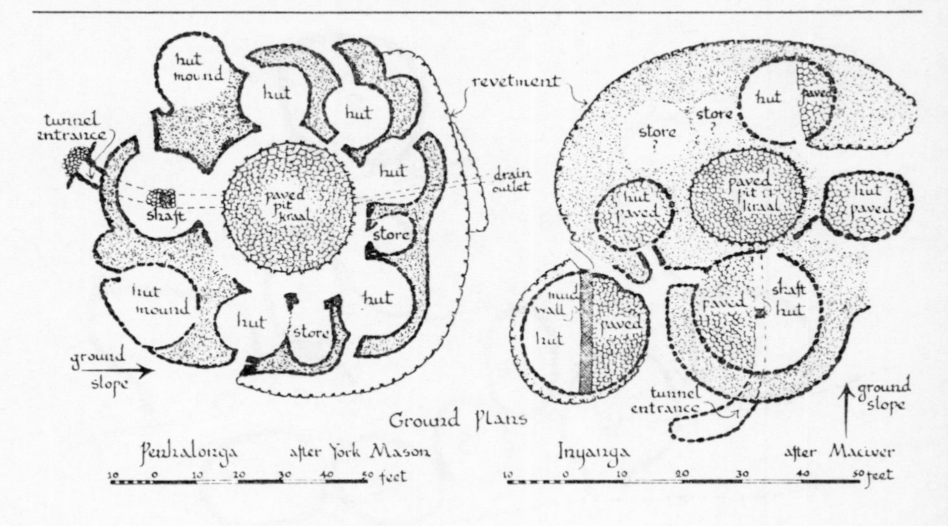

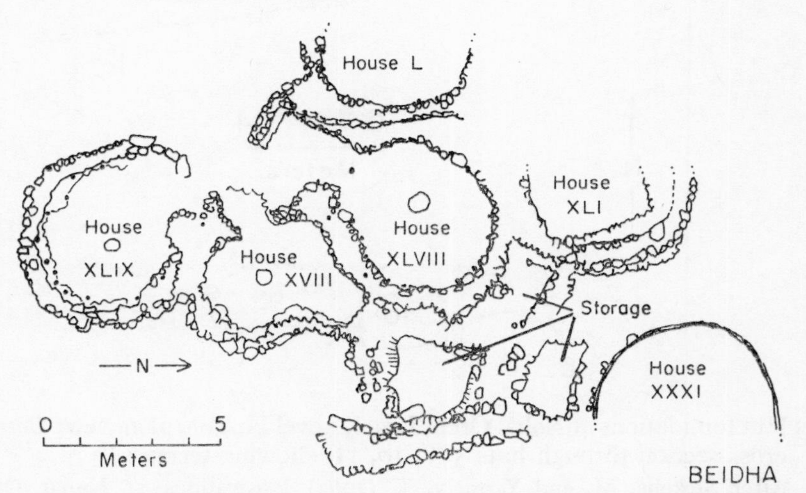

Fig 4 Top: ground plans of two present-day circular hut compounds from Rhodesia.

After Fraser, D. (1968). *Village Planning in the Primitive World.* New York.

Bottom: circular huts from lower levels at Pre-Pottery Neolithic Beidha, Jordan.

After Kirkbride, D. (1967). Beidha 1965: an interim report, *Palestine Explor. Quarterly*, **99**. Note, in both cases, how stone foundations fit together like cells in a honeycomb, with interstices used as storage units.

a few measureable circular huts with stone foundations, ranging from 2·7 to 4 m^2 in diameter, and with floor areas of between 5 and 14 m^2. Some 45 m south of the main excavation, the excavators also encountered the remnant of an oval structure which, if completed at the same rate of curvature, would have provided 112 m^2 of floor space—once again, suggesting a guest house or some other "public" building. Houses were apparently grouped into clusters at Mureybit, and there were open work areas, one of which (in Level XIV) was riddled with roasting pits, apparently for processing grain.

Nothing found so far in early Mesoamerica closely resembles the circular hut compounds of the early Near East. MacNeish has excavated a single example of what may be an "oval" structure from the late pre-pottery Abejas phase in Mexico, but this period remains one of the most poorly-known in all of Mesoamerican archaeology. I feel a few words must be said about this site because recent textbooks and secondary sources have begun to refer somewhat uncritically to "Abejas phase villages" in spite of the scanty and ambiguous nature of the data. I am extremely grateful to Dr. Richard S. MacNeish for allowing me to examine his unpublished manuscript on the following site.

Name: Site Ts–381
Location: In canyon, 2 km west of Chilac in the Tehuacán Valley, Puebla, Mexico
Probable date: 3000 B.C. ? (Abejas Phase)
Excavator: MacNeish[59]
Description: As reconstructed by MacNeish, this was an oval shelter 3·9 m by 5·3 m in extent, with its base excavated 60 cm into sterile clay. The flat "floor" of the excavated base was only 2 by 3 m (6 m^2), or equivalent to the smallest of the Nahal Oren shelters. A central ridge pole was supported by two uprights only 15 cm in diameter, and accompanied by "leaners" coming in at an angle from the sides of the shelter; the roofing and sides were presumably of brush and twigs. No evidence of daub or plastered mud was found, and a dozen more test excavations at 10 m intervals failed to produce another structure on the site.

In spite of the fact that no further structures were found, MacNeish regards Ts–381 as a "hamlet". He does so in part because of another Abejas phase site, Ts–388, located 4·5 km west of Chilac and thus not far from Ts–381. At Ts–388, one can see several slabs of stone projecting up above the surface of the site at three different places, forming a line of rough ovals parallel to the course of an arroyo. MacNeish feels these slabs may indicate stone-lined pit houses, but unfortunately this could not be checked by excavation due to the fact that the terrace on which the site lay was under cultivation. The claims for "village life" in the Abejas phase which appear in recent textbooks thus depend on one unexcavated site and two-thirds of one excavated structure at another. On the basis of these fragmentary data, I do not feel I can justifiably assign Ts–381 to either of the two "ideal types" discussed in this paper. It is neither a compound nor a village by the definitions used here. Should it eventually prove to be, as MacNeish suggests, a linear alignment of pit houses, its strongest resemblance would probably be with some of the basket-maker sites in the south-western U.S., which represent still another settlement type. Let me close with three final observations on the structure at Ts–381.

(*a*) The structure appears to be (1) a single, isolated hut and (2) too small, in view of Naroll's findings, to house more than one person, or two at the most.

(*b*) It differs in several fundamental and perhaps significant ways from the houses of Early Formative villages in Mesoamerica, which typically have (1) rectangular shape, (2) four corner posts, and (3) cane walls with plastered mud (daub) over them. Lacking the latter, it seems little more substantial than the temporary shelters used by Hadza hunter-gatherers[60] or Kalahari Desert Bushmen[61] at their seasonal camps.

(*c*) MacNeish cites multiple ash lenses in the shelter as evidence for occupation over many seasons, but neither the plant remains nor the number of ash beds conclusively demonstrates year-round occupation; the site could as easily be a long-term planting or harvesting camp. In fact, this period in Mesoamerican prehistory is one in which I would expect a very wide range of site types (for planting, hunting, and wild plant gathering) with little standardization. It may be a long time before we find even two sites so similar that we can begin to define "settlement types" for the Abejas phase.

2 *The village of rectangular houses*

A second type of permanent settlement, widespread in Mesoamerica and the Near East, is the village of rectangular houses, often with patios or courtyards. These communities have a wide range of sizes and population figures. In Mesoamerica, several recent settlement pattern studies have referred to communities with populations between ten and 100 persons as " hamlets", reserving the term "village" for communities of 100 to 1000 persons[62]. This follows the work of Sanders, who originally defined the Mesoamerican village as "a nucleated community with populations running at least into the hundreds and in which at least 75% of the population derive at least 75% of their income from agriculture or some other extractive activity"[63]. In addition, both Blanton[64] and MacNeish[65] have suggested that hamlets may be further distinguished from villages by the absence of public (ceremonial or civic) architecture. While this will be a useful distinction if it turns out to be valid, I will not apply it here because (*a*) its proof requires more extensive excavation than most sites have received, and (*b*) there are already hints that even small hamlets may have some kinds of "public" buildings, like community shrines, or the "visitors' reception huts" of the Tiv. In this paper, I will regard hamlets and villages as the small and large ends of a size continuum, and their differences in public architecture as being more a matter of degree than kind.

These early villages share a number of features which contrast with the circular hut compound:

(*a*) They are composed of rectangular houses designed to accommodate *families*, rather than individuals. In our sample drawn from the archaeological record, no complete house is smaller than 15 m, and most are 25–35 m^2—large enough, in Naroll's terms, to shelter at least three to four persons. Where published, the household inventories include everything from spear points and arrowheads or microlithic projectile elements to grinding slabs, needles, and cooking pots; that is, both men's and women's tools. Probably the residential unit was a man, his wife, unmarried children and occasionally more distant relatives, single or widowed.

(*b*) Although there may also be a "headman's granary" or "village granary", each household has its own storage facilities, either inside the house (Near East) or immediately adjacent to it (Mesoamerica). The implication is that the individual household was the basic unit of production, and that (by analogy with today's villages in the same regions) these household storage units were not subject to the same kind of obligatory sharing as those of the compound. In addition, some early Near Eastern rectangular houses have courtyards which are walled off from the rest of the village; in this case, even work space was not necessarily shared.

(*c*) The rectangular ground plan of these houses made it easy to add or subtract rooms. Thus, as families grew or as nuclear families became extended families, new rooms could be added on; as children married and moved away, the unused rooms could be sealed off. Near Eastern site reports provide numerous examples of houses to which later additional units had been added, and of still other houses whose unneeded rooms had been bricked up.

The rectangular principle pervaded the building-up of larger multi-family units. In Early Mesoamerica (see below), houses might be grouped around three sides or four sides of a rectangular patio; these Early Formative groupings were probably ancestral to the standard-sized "module" of four buildings plus patio which formed the basic building unit of the later Classic cities[66]. The rectangular houses of the Neolithic Near East eventually were united into rectangular multi-house units, which by Samarran times had taken on the outward impression of the modern Arabic *bayt* or extended household[67].

(*d*) It is not necessary to go beyond the borders of Mesoamerica or the Near East to find ethnographically documented examples of such villages today. By examining these, we can find some of the social institutions which serve to hold households together beyond the bounds of simple kinship. These institutions, sometimes called *sodalities*[68], are so common that I will discuss only one—the *guelaguetza* of the Zapotec Indians of the Valley of Oaxaca, Mexico—which happens to occur in one of the areas of Mesoamerica where Early Formative rectangular-house villages are well represented.

Among the ancient Zapotec (and, to a lesser extent, even today), the *guelaguetza* had two aspects: it was a voluntary communal work group and a ceremonial organization at the same time. As Schmieder[69] has reconstructed the situation, the task of preparing new land for cultivation was carried out collectively by a volunteer group of adult males from various families (more recent ethnographic information might suggest that the work group was made up primarily of married men of middle age or younger). After the land had been cleared, "it was subdivided into equal lots and each family received one of these subdivisions"[70]. The other face of the *guelaguetza* was an organization in which some of the same men co-operated in sponsoring a village ceremony, dance (or today, the *fiesta* of a patron saint). Here the *guelaguetza* functioned much as the dance societies of the Pueblo Indians of the south-western United States[71], whose members earn participation by achieving a certain marital status and level of responsibility. As far back as 1500 to 1000 B.C. archaeological evidence suggests that such societies may have cut across families and lineages in Mesoamerica's early rectangular-house villages. Villagers at Tlatilco in the Valley of Mexico were often buried with pottery masks like those used by Pueblo or Iroquois dancers[72]; fragments of similar masks are found at least as far south as the Valley of Oaxaca. And throughout this whole area at the same time occur hundreds of figurines showing dancers with false faces, beads, rattles, feathers, and fantastic animal disguises.

Between 7000 and 5500 B.C. over much of the Near East, villages of rectangular houses like those described above replaced the circular hut compound. The villages themselves usually varied between one and four hectares, though rare larger examples occurred. Some differences in size and storage facilities can already be detected between houses at this time, implying differences in production, surplus, or accumulation of "wealth" between households which served as the units of production. Let us look at a few samples:

A. Name: Beidha (Levels I–III)
Location: (see above)
Probable date: c. 6500 B.C. (Pre-Pottery Neolithic)
Excavator: Kirkbride
Description: These upper levels at Beidha had literally dozens of rectangular houses averaging 5 × 6 m (approx. 30 m²). Stairways and strong buttresses suggest that the remains found may, in fact, be the lower

floors of two-storey buildings. According to Kirkbride, smaller rectangular houses occur on the perimeter of the site. The larger houses have enclosed courtyards; storage areas and specialized workshops occur within the lower stories of the buildings. Although not published in detail, some of the workshops suggest that certain households may have had differential access to exotic raw materials used in ornament manufacture.

B. *Name:* Tell Mureybit (Levels X–XVII)
Location: (see above)
Probable date: 7500 B.C.
Excavators: van Loon and Skinner
Description: These upper levels at Mureybit had rectangular houses built of stone blocks, but in most cases, due to limited exposures, only parts of houses were found. Structure 16 had one area 3·5 m on a side (12·5 m^2), divided into four small spaces which could be storage rooms (cf. Jarmo); a courtyard lay adjacent to them. Structure 19 was at least 5 × 5 m (25 m^2), and it is unlikely that all of this house was recovered. By this period, the village covered 3 hectares.

C. *Name:* Jarmo
Location: On low silt bluff near wadi, Chemchemal Valley, Iraq
Probable date: 6750–6000 B.C.
Excavator: Robert J. Braidwood[73]
Description: Jarmo may originally have covered 2 hectares. A house described by Braidwood as "representative" had approximately 30 m^2 under its roof, plus 3–4 m^2 of walled courtyard or work area. The roofed area was divided into seven rectangular spaces, four of which may have been storage rooms; thus the total "living" (sleeping?) area may have been only 15 m^2 (Fig. 5).

D. *Name:* Matarrah
Location: On steppe-piedmont south of Kirkuk, Iraq
Probable date: c. 5500 B.C.?
Excavators: Braidwood et al[74].
Description: Matarrah occupied roughly 2 hectares. The best preserved single house (Operation IX, Level 2) had about 24 m^2 of floor space divided into four rooms (largest 3·5 × 2·0 m, smallest 2·0 × 1·0 m) (Fig. 6B). It may have been one of the smaller houses, however. In Operation IX, Level 1, there were complexes of four to six rooms, some rooms being up to 5·0 × 3·5 m (17·5 m^2) (Fig. 6A). Some of the smallest rooms were presumably for storage, and other rooms had massive storage jars in and around the corners.

Why did the rectangular house village replace the circular hut compound in the Near East, and how was it accomplished? I should briefly mention

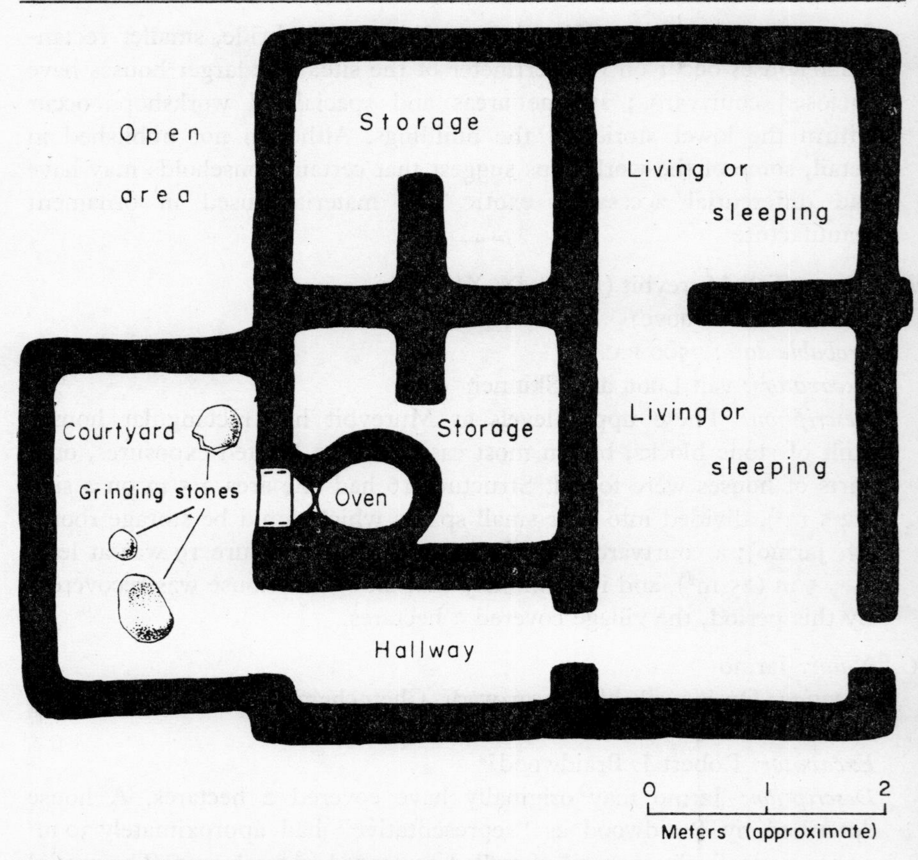

Fig 5 "Idealized" rectangular house from Jarmo, Iraq.
Reconstructed from data given in Braidwood, R. J. and Howe, B. (1960). Prehistoric investigations in Iraqi Kurdistan, *Studies in Ancient Oriental Civilization*, 31.

here a number of possibilities. First, I should stress a point frequently misunderstood: it is not the "circular" or "rectangular" shape of the house which is critical, but whether it is intended for a single individual or a family —in other words, whether the minimal unit of production and storage is a polygynous extended household or a (primarily) monogamous nuclear family. To be sure, the later rectangular-house villages had extended households, but they were organized into large multi-room residences with common courtyards. Among other things, these large multi-room residences may reflect the fact that certain kinds of farming and herding systems select for large families because there are so many subsistence tasks which can be performed by children.

Very revealing, from an archaeological point of view, are Schwerdtfeger's illustrations of the conversion of a Hausa circular hut compound into a series of multi-room rectangular houses in the space of one generation.

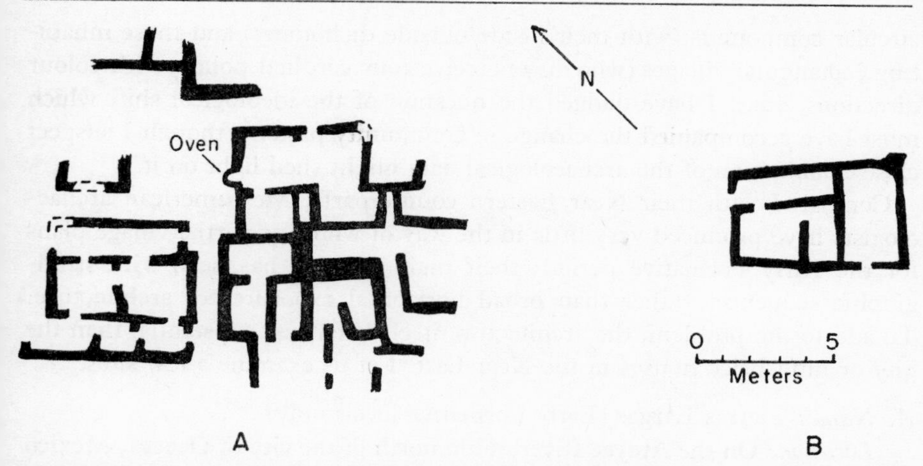

Fig 6 Rectangular houses at Tell Matarrah, Iraq. *A*, partial plan of Operation IX, Level 1. *B*, complete house from Operation IX, Level 2.
After Braidwood, R. J. et al. (1952). Matarrah, *J. Near East. Stud.*, **XI** (1).

Schwerdtfeger's data[75] make it seem unlikely that all the huts in any prehistoric compound were occupied simultaneously—further reducing archaeological population estimates—and clearly show that the entire transition from circular to rectangular could take place during the deposition of a single "building level" in an archaeological site. (Previous generations of Near Eastern archaeologists would almost certainly have interpreted such a rapid transition as the result of an invasion by new people!) Finally, Schwerdtfeger's analysis showed that one underlying cause for the transition was *higher income*, which encouraged the division of the extended communal household into smaller, nuclear family units, which would previously not have been self-sufficient.

Was the trend reversible? Joan Oates suggests[76] that it was, citing the unpublished results of the Russian expedition to Yarim Tepe in northern Iraq, where Late Neolithic rectangular houses gave way to small clusters of circular structures. Morever, she raises the important question of the degree to which different house styles might be related to different ethnic groups. It seems inescapable that more than one ethnic group was involved in the Neolithic Near East, but we are still left with the problem of explaining the differences in social segmentation implied by the two patterns. Moreover, there were probably good reasons why some communities reverted to the smaller compound—dispersal into smaller units, as Forge points out[77], is one of the commonest means of defence.

Finally, before passing to an examination of Early Mesoamerican rectangular-house villages, I should admit to avoiding a still thornier problem which may face future Near Eastern archaeologists. Lévi-Strauss and others have suggested[78] real ideological differences between the occupants of circular compounds (with their inside/outside dichotomy) and those inhabiting rectangular villages (who may perceive four cardinal points, with colour directions, etc.). I have dodged the question of the ideological shift which must have accompanied the change in community pattern, though I suspect close examination of the archaeological data might shed light on it.

Compared with their Near Eastern counterparts, Mesoamerican archaeologists have produced very little in the way of whole or partial village plans for the Early Formative period; their main concern has been with stratigraphic sequences rather than broad horizontal exposures of architecture. To add to the problem, the architecture itself is far less substantial than the *tauf* or mud-brick houses in the Near East. Let us examine a few sites.

A. Name: Tierras Largas (Early Formative levels only)
Location: On the Atoyac River, 5 km north of the city of Oaxaca, Mexico
Probable date: 1400–900 B.C.
Excavator: Marcus Winter[79]
Description: Because of its small size and shallowness, and Winter's extensive excavation by means of a rigorous sampling programme, we can reconstruct the settlement at Tierras Largas with some confidence. At 1300 B.C. it was a hamlet of some six to twelve households, occupying 0·5 ha; by 900 B.C., it was a hamlet or small village of twelve to twenty-two households, occupying 1·5 ha. During both these periods, the unit of settlement was a wattle-and-daub house 6–7 m long by 4–5 m wide (24–35 m^2), presumably occupied by a family of roughly two to four persons. Each house was accompanied by two to six large, bell-shaped storage pits which, from their location, do not seem to have been shared with other households. In some instances, there seems to be a larger unit composed of a rectangular patio or open area some 30 m wide, with houses on each of three sides, the fourth side being left open (see Figs. 7–8).

B. Name: San José Mogote (Early Formative levels only)
Location: On the Atoyac River, 15 km north of the city of Oaxaca, Mexico
Probable date: 1400–900 B.C.
Excavators: Flannery et al.[80]
Description: This site, contemporary with Tierras Largas, was perhaps the largest village of its period in the region. At 1300 B.C. it was a village of at least fifteen to thirty households covering 1–2 ha. In the central part of this village was an area of 200–300 m^2 of lime-plastered "public buildings" of undetermined function (accounting for one-fifth of the total village area). By 900 B.C., San José Mogote covered 20 ha and must have had in the order of 240–320 households. Perhaps a dozen houses of this period have been recovered, and suggest a pattern similar to that at Tierras Largas: wattle-and-daub houses of 24–35 m^2 with

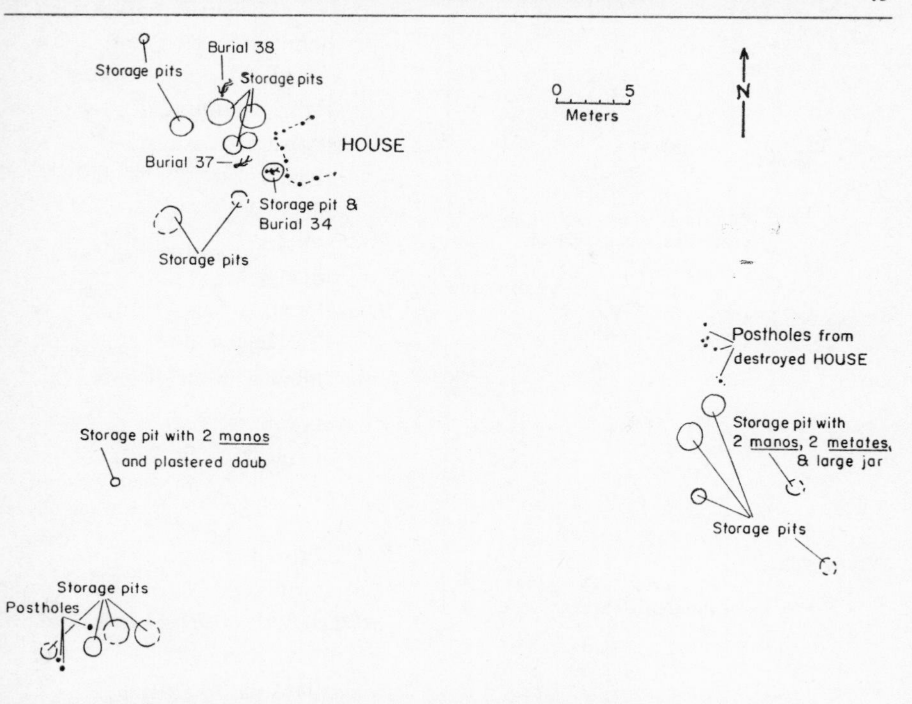

Fig 7 Partial plan of Tierras Largas, Oaxaca, Mexico, at 1300 B.C.
Revised, on the basis of more refined chronological data, from Winter, M.C. (1970), Excavations at Tierras Largas (Atzompa Oaxaca) *in* Flannery, K. V. (ed.) *Preliminary Archaeological Investigations in the Valley of Oaxaca, Mexico. Ann Arbor.*

clusters of storage pits, separated from other houses by courts or patios 20–30 m across.

In addition, tool scatters which could be plotted on the house floors suggest that houses were "conceptually divided" into men's and women's work areas, as are present-day Indian households in highland Chiapas[81]. In every case where extensive floor scatters could be mapped, "women's tools" (two-hand grinding stones, deer bone maize huskers, lime-crusted cooking pots, etc.) were in the northern half (to the right as one entered the door). To the south (on the left as one entered the door), and especially in the south-east corner, were the flint-chipping areas with cores, scrapers, knives, burins, drills, and presumed "men's tools". Moreover, houses differed considerably in the amounts of luxury materials (magnetite, *Spondylus* and pearl oyster shell, or imported ceramics) found on the floor. The differences found did not seem to be solely a question of differentially represented tool kits (as between huts 9 and 10 at Nahal Oren), but implied differences in production, surplus, or the accumulation of "wealth" between households.

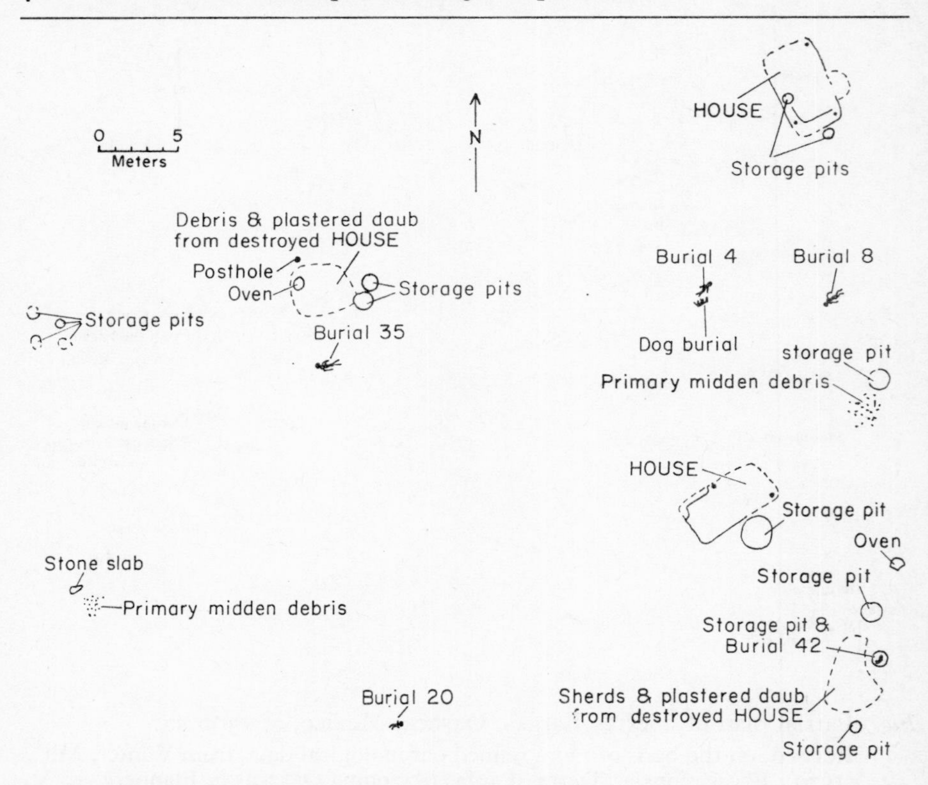

Fig 8 Partial plan of Tierras Largas, Oaxaca, Mexico, at 900 B.C.
Revised, on the basis of more refined chronological data, from Winter, M.C. (1970), Excavations at Tierras Largas (Atzompa, Oaxaca) *in* Flannery, K. V. (ed.) *Preliminary Archaeological Investigations in the Valley of Oaxaca, Mexico. Ann Arbor.*

Although published village plans are hard to come by, a search through the Mesoamerican literature does produce additional evidence for the features mentioned above. San Lorenzo in the Gulf Coast lowlands, perhaps the largest Early Formative site in Mesoamerica, also has houses arranged around three-sided patios[82]. Moyotzingo in the Valley of Puebla also shows clusters of bell-shaped pits associated with houses[83]. Several other early sites, from Tampico to Guatemala, have yielded posthole patterns or burned daub fragments with square corners which indicate that houses were rectangular.

Mesoamerica seems never to have gone through a circular hut compound stage: its earliest sedentary communities are already of village or hamlet type. In part, this may be because—in contrast to the Near East—*the family was already established as the basic unit of procurement and storage long before agriculture began,* in the microband camps of the hunting-gathering era.

Advantages of the village over the compound

Why should Mesoamerica and the Near East, starting with different past histories and following quite different trajectories, both have stabilized at a state of rectangular-house villages? As the suspicious reader will already have guessed, the reason lies not in the village itself, but in the systems of social and political organization of which it is merely the architectural manifestation. The organization of society into large, nucleated settlements composed of related households (with their own landholdings and storage facilities), cross-cut by sodalities composed of family heads (co-operating in specific acts of ceremony and labour), gives the village certain strengths not shared by the compound.

Schmieder[84] drew an important contrast between the Aztec and Zapotec (who lived in nucleated villages) and their less powerful Mixe Indian neighbours (who lived in scattered farmsteads or tiny compounds near their fields). According to Schmieder, one selective force helping to maintain nuclearity was the *guelaguetza* land clearance system, followed by division of the land among co-operating families. "This resulted in a dispersion of the land owned by one family which . . . kept the inhabitants from moving out of the village and building up isolated farmsteads on their increasingly distant and scattered fields . . . field dispersion resulted in the growth of larger, more compact settlements, in which differentiation of activities became possible. Crafts, art, and science developed and were maintained by the mass of the population which nevertheless remained agricultural"[85]. Among the Mixe, on the other hand, "each family cleared for itself the land it needed for cultivation, and erected a hut in the midst of its fields. Numerous isolated farmsteads sprang up in the forest. The distance from neighbour to neighbour was often great. Every man was confronted by all the tasks of everyday life. There was no differentiation of activities, and, in spite of the proximity of more highly developed tribes, no higher civilization was obtained"[86].

A more important point, however, is the difference between the two types of settlement in terms of potential for growth and intensification of production. Among unstratified societies, both compounds and villages suffer from the same weakness—the inability to grow beyond a certain size because of their weak polity. There is a widespread belief among archaeologists that village size is primarily a function of agricultural potential; ethnographic data suggests, on the contrary, that what limits village size may be an inability to keep the peace when the community has grown too large. Among the Tiv, compounds fission when disputes grow too great to resolve, as for example after the death of a compound head. It is extremely unlikely that a Tiv compound will grow to include eighty persons except under exceptional circumstances (and in fact, the average number of persons is under twenty). Consider, also, the problem of those compounds whose huts form a circle; each expansion creates an increase in the diameter of the circle and produces real problems of spacing and social distance. However, growth problems are by no means limited to compounds. Among agricultural tribesmen of the Amazon basin, Carneiro[87] maintains that the "ease and frequency of village fissioning for reasons not related to subsistence" is far more crucial than any limitations imposed by slash-and-burn cultivation, since "the political mechanisms for holding a growing community together in the face of increasingly strong divisive forces are all but lacking".

This weakness of political organization is responsible for still another characteristic, shared by unstratified compound and village dwellers alike: the difficulty of intensifying production. In this case, there is a widespread belief among archaeologists that primitive agriculturalists are limited primarily by their technology; ethnographic data suggests, on the contrary, that a far more important limitation is the lack of real authority. One hundred years ago, social philosophers (including Marx and Engels!) were pointing out that society's production was limited more by the *organization* of labour than by the *means* of labour. In a forthcoming book, Sahlins[88] has put together a convincing body of ethnographic data to indicate that all unstratified societies are probably underproductive, and that the challenge of intensification lies not in new farming systems but in "getting people to work more, or more people to work".

With respect to intensification of production, however, compounds come off more poorly than villages. For the compound is a real commune in which, as we have seen, storage facilities are shared openly, and there is no reward for intensification. Indeed, the reverse is often true: among Bohannan's Tiv, for example, the compound head prevents profit-taking by anyone wanting to plant more yams than he needs, or so many that his own planting and harvesting would prevent him from participation in communal labour for the compound[89]. Disparities of land, wealth, and material goods between members of a compound are almost non-existent, and there are social institutions which prevent them from arising.

However, when we shift to societies where the individual household is the basic production unit and the sharing of storage more selective (such as those typically occupying rectangular-house villages), the opportunities for intensification greatly increase. To be sure, there are village societies where the strictures on personal gain are as rigorous as those of the Tiv, or where "levelling mechanisms" (such as sponsored *fiestas*) spread the benefit of any accumulated surplus through the whole community. On the other hand, there are many—and here the "Big Man" societies of Melanesia come to mind—where great fluctuations of wealth and property between households occur even without social stratification[90]. Here a family can, in fact, plant more yams than it needs, work harder, store its own harvest, and share the "profit" with no one, should it choose. More than likely, however, it will have been spurred to such intensive production by the continual demands of a kinsman of great charisma, whose lavish feasting and gift-giving are one source of his prestige—and, by association, theirs. The situation is only intensified in "chiefdoms" or truly stratified societies, where the demands on a household can be backed up by real authority, not just persuasion. Such societies also have the power and authority to hold together large villages which would surely fission if organized along the lines of the compound[91]. Hence, we should not be surprised to find much greater variation in exotic raw materials, luxury goods, and storage facilities between households in rectangular-house villages than in circular hut compounds (see above). However, archaeologists should take care to distinguish this variation from the "functional" differences in tool kits to be expected between one-person and two-person huts in a compound (once again, see above).

Finally, we should not close without pointing out that the nucleated village constitutes a defensible unit on a scale not seen among farmsteads or compounds. It is still too early to evaluate the role that raiding (and its prevention) played in the evolution of larger and more nucleated villages, but the defensive walls at pre-pottery Neolithic Jericho[92] and later Samarran sites[93], as well as the defensible locations chosen by some Formative Mesoamerican sites[94], suggest that it may have been an important consideration. Two classic studies by Vayda[95] and Sahlins[96] make it clear that the tribesman's inherent tendencies to fission and disperse can be quickly overcome when the need for mutual defence arises. These studies show that expansion and raiding between primitive cultivators may take place even when no land shortage exists[97]; and in Sahlins' analysis, our old friends the Tiv play a prominent role.

There is also a suggestion, in the studies mentioned above, that once large nucleated village societies had arisen, no neighbouring compound or farmstead-based society could really have been secure. Their only alternatives were to nucleate in their own defence, to be absorbed into village society, or to retreat continually before the expanding village. This "retreat" might, in some regions, take the form of still more dispersal into tiny scattered farmsteads, which is a form of defence in itself—or a reversion into the compound pattern, as suggested above. We should search the archaeological record for evidence that this happened, for it may be that the success of village society was a prerequisite for further political evolution. The subsequent stages of cultural development in Mesoamerica and the Near East—the chiefdoms, stratified societies, and early states which followed—were based on great intensification of production, great disparities of wealth and prestige. So far as we can tell from the ethnographic record, these are precisely the developments most effectively prevented by the kinds of societies occupying circular hut compounds[98].

Notes

1 See for example Steward, J. H. (1955). *Theory of Culture Change: the Methodology of Multilinear Evolution*. Urbana. Chapter 11.
2 van Loon, M. (1968). The Oriental Institute excavations at Mureybit, Syria: preliminary report on the 1965 campaign, *J. Near East. Stud.*, **27** (4), pp. 265–90.
3 MacNeish, R. S. (1964). Ancient Mesoamerican civilization, *Science*, **143**, pp. 531–7.
4 For a criticism of multilinear evolution, see Renfrew, C. (1969). Trade and culture process in European pre-history, *Curr. Anthrop.*, **10**, pp. 151–69.
5 Clarke, D. L. (1968). *Analytical Archaeology*. London. See also Doran, J. (1970). Systems theory, computer simulations, and archaeology, *World Archaeol.*, **1**, p. 291.
6 For a recent discussion see Lamberg-Karlovsky, C. C. (1970). Operations problems in archaeology, *in* Fischer, A. (ed.), Current directions in Anthropology, *Bull. Amer. Anthrop. Assoc.*, **3**, p. 113, with references.
7 Hole, F. and Flannery, K. V. (1967). The prehistory of south-western Iran: a preliminary report, *Proc. Prehist. Soc.*, **33**, pp. 147–206.
8 Binford, L. R. and Binford, S. R. (1966). A preliminary analysis of functional variability in the Mousterian of Levallois facies, *Amer. Anthrop.*, **68**, pp. 238–95.
9 Wilmsen, E. (1970). Late Pleistocene human ecology, *Abstracts, 1st Annual Meeting Amer. Quaternary Assoc.*, p. 153.
10 Lee, R. B. and DeVore, I. (eds.) (1968). *Man the Hunter*. Chicago. (especially Chapter 1).
11 MacNeish, R. S. (1964). *op. cit.*; MacNeish, R. S. (In press). *Prehistoric Settlement Patterns in the Tehuacán Valley*. Austin.
12 Steward, J. H. (1955). *op cit.* Chapter 6.
13 For descriptions of Abu Halka and Ksâr 'Akil, see Howell, F. C. (1959). Upper Pleistocene stratigraphy and early man in the Levant, *Proc. Amer. Phil. Soc.*, **103** (1), pp. 1–65, with references. For Yafteh Cave, see Hole, F. and Flannery, K. V. (1967) *op. cit.* For Shanidar, see Solecki, R. S. (1963). Prehistory in Shanidar Valley, northern Iraq, *Science*, **139**, pp. 179–93, with references.
14 Wagner, P. (1960). *The Human Use of the Earth*. New York, Chapter 6.
15 Binford, L. R. (1965). Unpublished comments at seminar, *The Social Organization of Prehistoric Communities*, 64th Annual Meetings, *Amer. Anthrop. Assoc.*, Denver, November 1965.
16 Solecki, R. S. (1963) *op. cit.*; Solecki, R. S. (1955). Shanidar Cave, a Palaeolithic site in northern Iraq, *Ann. Report of Smithsonian Inst. for* 1954, (Pub. 4190), Fig. 6, p. 406.
17 See for example Perrot, J. (1966). Le gisement Natufien de Mallaha (Eynan), Israel, *L'Anthrop.*, **70**, pp. 437–84; Garrod, D. A. E. and Bate, D. M. A. (1937). *The Stone Age of Mt. Carmel: Excavations at the Wadi el-Mughara*, I. Oxford; van Loon, M. (1968). *ibid.*
18 Flannery, K. V. (ed.) (1970). *Preliminary Archaeological Investigations in the Valley of Oaxaca, Mexico*, 1966–1969 (mimeographed). Ann Arbor.
19 Oates, J. (1969). Choga Mami, 1967–68: a preliminary report, *Iraq*, **31**, pp. 115–52.
20 Fowler, M. L. (1969). A pre-classic water distribution system in Amalucan, Mexico, *Archaeology*, **22**, pp. 208–15. See also Neely, J. A. (1967). Organización hidráulica y sistemas de irrigación prehistóricos en el Valle de Oaxaca, *Bol. Inst. Nac. de Antrop. e Hist.*, **27**, pp. 15–17; Neely, J. A. (1970). Terrace and water control systems in the Valley of Oaxaca

region: a preliminary report *in* Flannery, K. V. (ed.) (1970). *op. cit.* pp. 83–6.

21 Kirkby, A. V. (1970). *Modern and Prehistoric Land and Water Use in the Valley of Oaxaca, Mexico.* (Unpublished doctoral dissertation, Johns Hopkins Univ.)

22 Mangelsdorf, P. C., MacNeish, R. S. and Galinat, W. C. (1967). Prehistoric wild and cultivated maize, *in* Byers, D. S. (ed.) *Prehistory of the Tehuacán Valley.* I. Austin.

23 Zohary, D. (1969). The progenitors of wheat and barley in relation to domestication and agricultural dispersal in the Old World, *in* Ucko, P. J. and Dimbleby, G. W. (1969). *The Domestication and Exploitation of Plants and Animals.* London, p. 56.

24 Harlan, J. R. (1967). A wild wheat harvest in Turkey, *Archaeology*, **20** (3), pp. 197–201.

25 Kirkby, A. V. (1970). *ibid.* Maize of that period was approximately one-fifth the size of today's races.

26 Stubbe, H. (1959). Considerations on the genetical and evolutionary aspects of some mutants of Hordeum, Glycine, Lycopersicon, and Antirrhinum, *Cold Spring Harbor Symposium on Quantitative Biology*, **24**, pp. 31–40; Helbaek, H. (1960). Ecological effects of irrigation in ancient Mesopotamia, *Iraq*, **22**, pp. 186–96.

27 Zohary, D. (1969). *op. cit.* p. 63.

28 Mangelsdorf, P. C., MacNeish, R. S., and Galinat, W. C. (1967). *ibid.*, with references.

29 Flannery, K. V. (1969). Origins and ecological effects of early domestication in Iran and the Near East, *in* Ucko, P. J. and Dimbleby, G. W. (1969). *op. cit.* p. 81, with references.

30 Ford, R. I. (1968). *An Ecological Analysis Involving the Population of San Juan Pueblo, New Mexico.* (Unpublished doctoral dissertation, University of Michigan.)

31 Flannery, K. V. (ed.) (1970) *ibid.*

32 cf. Perrot, J. (1966) *op cit.*; Pina Chán, R. (1955). *Las Culturas Preclásicas de la Cuenca de México.* Mexico City; Cornwall, I. W. (1956). The pre-Pottery Neolithic burials, Jericho, *Palestine Explor. Quart.*, **88**, pp. 110–124.

33 Kenyon, K. (1967). Jericho, *Archaeology*, **20**, pp. 268–75; Cornwall, I. W. (1956) *op. cit.*

34 Naroll, R. (1962). Floor area and settlement population, *Amer. Antiq.*, **27** (4), pp. 587–89.

35 Cook, S. F. and Heizer, R. F. (1968). Relationships among houses, settlement areas, and population in aboriginal California, *in* Chang, K. C. (ed.) *Settlement Archaeology.* Palo Alto, pp. 79–116.

36 Robbins, M. C. (1966). House types and settlement patterns: an application of ethnology to archaeological interpretation, *Minnesota Archaeologist*, **28** (1), pp. 3–26.

37 Whiting, J. W. M. and Ayres, B. (1968). Inferences from the shape of dwellings, *in* Chang, K. C. (ed.) *op. cit.*

38 Naroll, R. (1962) *op. cit.*

39 Robbins, M. C. (1966). *ibid.*; Whiting, J. W. M. and Ayres, B. (1968). *ibid.*

40 Robbins, M. C. (1966). *ibid.*

41 Robbins, M. C. (1966) *op. cit.* p. 7, quoting Ascher, R. (personal communication).

42 Fraser, D. (1968). *Village Planning in the Primitive World.* New York. Figs. 48–54. See also Rapoport, A. (1969). *House Form and Culture.* Englewood Cliffs, pp. 56–7.

43 Bohannan, P. (1954). *Tiv Farm and Settlement.* London.

44 Bohannan, P. (1954) *op. cit.* p. 4.
45 Bohannan, P. (1954) *op. cit.* p. 3.
46 Naroll, R. (1962) *op. cit.*
47 Fraser, D. (1968) *op. cit.* Figs. 52–4.
48 Bohannan, P. (1954) *op. cit.* p. 3.
49 Fraser, D. (1968) *op. cit.* Fig. 52.
50 Bohannan, P. (1954) *op. cit.* p. 30.
51 Fraser, D. (1968) *op. cit.* Fig. 50.
52 Bohannan, P. (1954) *op. cit.* p. 23.
53 Braidwood, R. J. (1960). Seeking the world's first farmers in Persian Kurdistan: a full-scale investigation of prehistoric sites near Kermanshah, *Ill. London News*, October, **22**, pp. 695–97; Solecki, R. L. (1964), Zawi Chemi Shanidar, a post-Pleistocene village site in northern Iraq, *Reports, VI Int. Cong. on Quaternary*, **4**, pp. 405–12.
54 Perrot, J. (1966) *ibid.*
55 Stekelis, M. and Yizraely, T. (1963). Excavations at Nahal Oren: preliminary report, *Israel Explor. J.*, **13** (1), pp. 1–12.
56 Kirkbride, D. (1966). Five seasons at the pre-pottery Neolithic village of Beidha in Jordan, *Palestine Explor. Quart.*, **98** (1), pp. 8–61; Kirkbride, D. (1967). Beidha 1965: an interim report, *Palestine Explor. Quart.*, **99**, pp. 5–13; Kirkbride, D. (1968). Beidha 1967: an interim report, *Palestine Explor. Quart.*, **100**, pp. 91–6.
57 van Loon, M. (1968). *ibid.*
58 van Loon, M. (1968). *ibid.*
59 MacNeish, R. S. (In press). Excavations at site Ts–381, *in Prehistory of the Tehuacán Valley* (future volume). Austin.
60 Woodburn, J., this volume, p. 194.
61 Marshall, L. K. (1965). The !Kung Bushmen of the Kalahari Desert, *in* Gibbs, J. (ed.) *Peoples of Africa.* New York.
62 See for example, Blanton, R. (n.d.). *Prehispanic Settlement Patterns of the Ixtapalapa Peninsula, Valley of Mexico.* (Unpublished doctoral dissertation, University of Michigan.)
63 Sanders, W. T. (1956). The central Mexican symbiotic region: a study in prehistoric settlement patterns, *in* Willey, G. R. (ed.) Prehistoric settlement patterns in the New World, *Viking Fund Pubs. in Anth.*, **23**, p. 117.
64 Blanton, R. (n.d.). *ibid.*
65 MacNeish, R. S. (in press). *Prehistoric Settlement Patterns in the Tehuacán Valley.* Austin; MacNeish, R. S. (1969). Comments at symposium, *The Origins of the Village*, held at annual meetings of Soc. Amer. Archaeol., Milwaukee, May, 1969.
66 Millon, R. (1967). Teotihuacán, *Sci. Amer.*, **216** (6), pp. 38–48; Millon, R. (1970). Personal communication; Millon, R. (1964). Teotihuacán, *Interim Topographic and Archaeological*, Map, 1.
67 Oates, J. (1969). *ibid.*
68 Service, E. R. (1962). *Primitive Social Organization.* New York, Chapter 4.
69 Schmieder, O. (1930). The settlements of the Tzapotec and Mije Indians, state of Oaxaca, Mexico, *Univ. California Pubs. in Geography*, **4**, Berkeley.
70 Schmieder, O. (1930). *op. cit.* p. 15.
71 Ortiz, A. (1969). *The Tewa World.* Chicago; Ford, R. I. (1968). *ibid.*
72 Coe, M. D. (1965). *The Jaguar's Children: Pre-classic Central Mexico.* New York. Figs. 161–9.
73 Braidwood, R. J. and Howe, B. (1960). Prehistoric investigations in Iraqi Kurdistan, *Studies in Ancient Oriental Civilization.* Chicago, 31.
74 Braidwood, R. J., Braidwood, L., Smith, J. G. and Leslie, C. (1952).

Matarrah: a southern variant of the Hassunan assemblage, excavated in 1948, *J. Near East. Stud.*, **11** (1), pp. 1–75.
75 Schwerdtfeger, F., this volume, p. 553.
76 Oates, J., personal communication.
77 Forge, A., this volume, p. 369.
78 Lévi-Strauss, C. (1963). *Structural Anthropology*, New York.
79 Winter, M. C. (1970). Excavations at Tierras Largas (Atzompa, Oaxaca): a preliminary report, *in* Flannery, K. V. (ed.) *op. cit.* pp. 61–9.
80 Flannery, K. V. (ed.) (1970). *op. cit.* pp. 38–58.
81 Vogt, E. Z. (1969). *Zinacantan: a Maya Community in the Highlands of Chiapas*, Cambridge (Mass.). p. 83.
82 Coe, M. D. (1967). *Map of San Lorenzo, an Olmec Site in Veracruz, Mexico*, New Haven.
83 Aufdermauer, J. (1970). Excavaciones en dos sitios preclásicos de Moyotzingo, Puebla, *Communicaciones, Fundación Alemana para la Investigación Científica*, **1**, pp. 9–24. Similar pits accompany houses at the nearby, later site of San Francisco Acatepec; see Walter, H. (1970). Informe preliminar sobre una excavación realizada en el sitio Preclásico de San Francisco Acatepec, Puebla, *Comunicaciones, F.A.I.C.*, **1**, pp. 25–36.
84 Schmieder, O. (1930). *ibid.*
85 Schmieder, O. (1930). *op. cit.* p. 76.
86 Schmieder, O. (1930). *op. cit.* p. 77.
87 Carneiro, R. L. (1961). Slash-and-burn cultivation among the Kuikuru and its implications for cultural development in the Amazon basin, *Antropologica*, (2).
88 Sahlins, M. D. (In press). *Essays in Stone Age Economics*. New York.
89 Bohannan, P. (1954). *op. cit.* p. 15.
90 Oliver, D. (1955). *A Solomon Island Society*. Cambridge (Mass.); Sahlins, M. D. (1968). *Tribesmen*, Chapter 5. Englewood Cliffs; Reay, M. (1959). *The Kuma: Freedom and Conformity in the New Guinea Highlands*. Melbourne.
91 Service, E. R. (1962). *op. cit.* Chapter 5.
92 Kenyon, K. (1967). *ibid.*
93 Oates, J. (1969). *ibid.* El-Wailly, F. and Abu al-Soof, B. (1956). The excavations at Tell es-Sawwan: first preliminary report (1964), *Sumer*, **21** (1–2), pp. 17–32.
94 Varner, D. M., Drennan R., Kuttruff, L. C. and Plog, S. (n.d.). Unpublished settlement pattern data from the Valley of Oaxaca, Mexico, collected 1970.
95 Vayda, A. P. (1961). Expansion and warfare among swidden agriculturalists, *Amer. Anthropologist*, **63**, pp. 346–58.
96 Sahlins, M. D. (1961). The segmentary lineage: an organization of predatory expansion. *Amer. Anthropologist*, **63**, pp. 322–45.
97 Sahlins, M. D. (1961). *op. cit.* p. 341, with references.
98 An earlier vesion of this paper was given constructive criticism by Richard I. Ford, Richard S. MacNeish, Joan Oates, Marshall D. Sahlins and Marcus C. Winter. I would like to thank them for numerous ideas and improvements, without implying that they necessarily believe a word I have said. Much of the Mesoamerican research mentioned here (including the excavation of Tierras Largas and San José Mogote) was supported by National Science Foundation Grant GS–2121.

PETER GATHERCOLE

The study of settlement patterns in Polynesia[1]

While patterns of settlement have been of interest to ethnologists of Polynesia for at least the past half-century, the study of settlement archaeology, in the sense employed by Trigger[2], has become important in the area only during the last decade. Most archaeologists in Polynesia would accept his definition of its scope as "the study of social relationships using archaeological data"[3], and would regard settlement patterns as an interpretative aspect of what Kennedy has termed "the spatial organization of a human group, which is held to reflect economy, social organization and the resources of the physical environment"[4]. In practice, archaeologists have tended to follow Groube's stress on the need to disentangle from the total recognized field evidence "the spatial distribution of archaeological sites *at any one moment in time*"[5] (my emphasis). This particular approach has derived mainly from the history of anthropological and archaeological research in Polynesia, and is widely seen as a basis for tackling the more profound and much more complex question of the nature of prehistoric social relationships referred to by Trigger. Thus, so far the conclusions which can be safely drawn for Polynesia are very limited—and these apply to only parts of Polynesia, at that. But they raise some exciting hypotheses, as well as a number of interesting methodological problems.

Prior to the development of modern archaeological research in tropical Polynesia after 1950, opinion on prehistoric settlement patterns was based largely on the nineteenth and early twentieth century ethnological evidence, with the inference made that this generally reflected a situation prior to European contact in the late eighteenth or early nineteenth centuries. Insofar as it was thought possible to generalize from this information, it was often considered that settlement on the high volcanic islands was primarily coastal and sedentary, with ancillary movement inland, of a temporary or more permanent character, when this was necessary and practicable. On atolls, of course, it was hardly feasible to make an inland/coastal distinction, which indeed was not always meaningful on high but small islands. During the last ten years, it has become necessary to examine this generalization in detail on particular islands, not least because fieldwork has demonstrated that settlement in East Polynesia goes back to A.D. 300 (Marquesas), and in West Polynesia to the second millennium B.C. (Tonga)[6].

The work by Green and his associates in the Society Islands provides an instructive example[7]. For a number of reasons connected with the apparent suitability and accessibility of the sites, Green chose to examine the 'Opunohu valley on the island of Mo'orea, approximately 10 miles west of Tahiti. Surveys indentified "more than 500 separate structures which could be functionally distributed among four main groups: agricultural terraces, residence areas, religious structures and specialized sites, with each group then broken down into various structural types on the basis of specific features. . . Moreover, it was the numerical relationship between the various groups of structures and the special distributions of the various types within each group, and within the valley itself, that proved to have a major role in making various sociological interpretations"[8]. The 'Opunohu valley was appropriate for more than archaeological analysis in that it is referred to in both historical records and indigenous traditions, and does not appear to have been in much contact with Europeans prior to its abandonment between 1800–10. On the basis of this variety of evidence, it was possible to state that settlement began in the thirteenth century A.D., with a "climax" in the late eighteenth century. Some of the structures could be identified sociologically. Community houses, religious structures and residences of "ranking members of the society" were distinguished from "more common residences of the general population", as well as archery and assembly platforms[9]. Green also argued that nearly all the structural types examined in the valley are to be found in the Windward Island group of the Society Islands, and that, because the settlement was not involved in the attempts at political consolidation found in Tahitian society at the time of Contact, "it may also represent a typological step preceding the development of that coastal society"[10].

This work is interesting not only for its intrinsic value, but also because it illustrates nicely the positive aspects of current settlement pattern research and underscores some of its problems. Previously in the study of Tahitian stone remains, attention had been focused on religious structures. Now it was possible to see them in a broader and yet more precise archaeological and historical context, aided to a degree by the existence of an outside control, Garanger's survey of the Tautira valley in Tahiti[11]. Most significant, however, was the use of ethnohistorical and traditional evidence to establish the pattern of settlement and its significance at the time of European contact, and slightly later. This pattern can now be used as a cultural and chronological base line, against which to measure the nature and extent of post-European change elsewhere in the Group, and as a starting point for assessing modifications during the full prehistoric period. Obviously there are problems deriving primarily from the extent to which evidence from periods other than that of the eighteenth and early nineteenth centuries can be made amenable to a comparable though not necessarily identical approach. But the investigation has clearly established a useful frame of reference for subsequent work throughout the Society Islands, and perhaps elsewhere in Central Polynesia.

A rather similar approach has recently been adopted by Davidson[12] to the settlement patterns of Western Samoa. She is concerned to clarify the concept of "aboriginal Samoa" which has figured in studies by certain anthropologists on the nature and possible history of non-unilinear descent groups as land-holding units. She maintains that at present archaeology can do little to support any particular interpretation in this important discussion. On the other hand, it is now possible to define more precisely the immediate effects of European influence on local settlement patterns and so give some reality to the character of late prehistoric site distribution. Archaeological work since 1965 suggests that prehistoric settlement was spread over both inland and coastal regions in the western islands, not, as previously suggested, concentrated in areas less than one mile from the coast. The latter distribution was a feature of post-missionary settlement after 1830–40 and is still true today[13]. Superficially, this may seem a rather limited conclusion, but it is important to appreciate some of the nuances of the problem involved, revealed by reference to the ethnographic evidence. The fundamental political unit was and is the *nu'u*, which Davidson defines as "a cluster of titles, which together are considered a localized political entity, together with the land over which the various title holders are recognized to have *pule* (authority)"[14]. In theory, this institution could be expressed in a variety of settlement forms—as nucleated or dispersed units, for instance, or as various combinations of these. Thus, although dispersal was usual in prehistoric Samoa, Davidson stresses that nucleation could occur around the dwelling of a chief of high rank. This suggestion of the possibility of marked social stratification is in line with the existence of only a few large mounds, which might be interpreted as sites for chiefs' houses—although they may also have housed gods[15].

Although Western Samoa has been fairly intensively investigated in comparison with the rest of tropical Polynesia, it is not yet possible to say much of value concerning prehistoric population size, land pressure and warfare, which are clearly crucial for any consideration of the dynamics of prehistoric social change. It is also relevant to note that archaeologically oriented settlement pattern studies have been undertaken in Mangareva, the Marquesas, the Cook Islands and Hawaii[16]. Although the results are still preliminary, it is worth drawing attention to the relative lateness of inland settlement in the Marquesas (Hane: after A.D. 1600) and the range of settlement patterns so far revealed in several ecological regions in Hawaii.

The study of settlement patterns is well developed in some areas of New Zealand (culturally part of East Polynesia), which has had a longer history of excavation than anywhere else in Oceania. For many years, the picture of Maori society at the time of Contact was also perceived in terms of ethnological records of varying quality. Groube has argued that, generally speaking, these accounts projected back to beyond the time of Cook, situations which were typical primarily of the nineteenth century, when Maori society had been profoundly modified by European influence[17]. Thus, for domestic structures, he maintained that there was no evidence either in the ethnohistoric sources, or from the restricted archaeological remains available, for the existence at Contact of village settlements, with the meeting house and other dwellings arranged around the *marae* (open gathering space), nor for the *pa* (fortified site) with permanent occupation. Such nucleation was a post-Contact feature, although in the Bay of Islands there were probably elements of this in late prehistoric times associated with particularly productive *kumara* cultivation, a qualification subsequently demonstrated by Kennedy for its south east area[18]. It would seem, however, that the most typical settlement form in the immediate pre-Contact period, at least, was one of small kin groups, living in units of three or four houses and "forming a cluster or compound with associated cooking, dumping and latrine sites"[19]. The way of life appears to have been largely semi-nomadic, well adapted to small but varied resources zones (reflecting the country's ecological range), and to seasonal movement within defined areas, presumably recognized as tribal or sub-tribal territory. The *pa* appears to have been occupied only at times of danger[20]. Its overwhelming concentration in the North Island reflects the continual density of population there in contrast to the South.

Groube has also proposed a re-interpretation of the evidence concerning the earlier Archaic or Moa-hunter period, at least for some areas. Previously it was considered by archaeologists that this was the period *par excellence* of mobile groups dependent to a considerable degree on the hunting of moa and some other animals, both on the coast and inland. Groube suggested that, at many sites, the richness of the avifauna allowed for the establishment of more permanent settlements, with a diversification of associated economic activity not seen again until post-Contact times[21]. Recent work by Higham and his colleagues in Southern New Zealand suggests that in this area, at least, seasonal movement was normal throughout prehistory[22].

In his most recent review of the evidence, Green has extended his survey of 1967 by combining and contrasting the archaeological data from East Polynesia with the largely ethnographic material from West Polynesia[23]. He sees a much greater range of settlement patterns in the East, particularly in Hawaii and New Zealand, where the regional environments are so varied. In the West, however, "given the diversity among various technological items of culture, one of the most striking results is the high degree of similarity in the internal patterning among its settlements". This, he regards, in part, as a reflection of "similar ideological patterns of social, political and religious behaviour". In general, he concludes that "the most common internal patterning in Polynesia ranges from dispersed to loosely-clustered, though largely unplanned settlements which are organized in relation to the localized kin-group ownership of land. Thus, it was only at the communal or political level, . . . that any tendency towards nucleation appeared and only occasionally was this clustering transformed into permanantly-occupied and formally-planned villages"[24].

Such revaluations naturally bring in their wake numerous elaborations and modifications. It is interesting to note that forms of settlement pattern have sometimes been used to mark stages of possible social evolution in several Polynesian societies, notably the Marquesas and New Zealand[25]. These have been useful studies because, at least, they suggested chronological settings in which archaeological data might be plausibly interpreted in socio-economic terms, although much of the discussion has naturally been somewhat theoretical. The task now is to extend and deepen the settlement pattern approach, where, among other things, such hypotheses can be more rigorously tested.

Notes

1 I am grateful to the following for their comments on an earlier version of this review: Miss J. M. Davidson (Auckland Institute and Museum), Dr. R. C. Green and Dr. A. L. Kaeppler (Bernice P. Bishop Museum, Honolulu) and Miss J. Kennedy (University of Hawaii).
2 Trigger, B. G. (1967). Settlement archaeology—its goals and promise, *Amer. Antiq.*, **32.**
3 Trigger, B. G. (1967). *op. cit.* p. 151.
4 Kennedy, J. (1970). Settlement and subsistence in Northland, New Zealand, *N.Z. Archaeol. Ass. Newsl.*, **13,** p. 28.
5 Groube, L. M. (1965). Settlement patterns in New Zealand prehistory, *Anthrop. Dept., Univ. Otago, Occ. Pap. Archaeol.*, **1,** p. 2.
6 Marquesas: Sinoto, Y. H. (1970). An archaeologically based assessment of the Marquesas Islands as a dispersal center in East Polynesia, *in* Green, R. C. and Kelly, M. (eds) Studies in oceanic culture history, vol. 1, *Pacific Anthrop. Rec.*, **11** (Bernice P. Bishop Museum). p. 113; Tonga: Groube, L. M. (1971). Tonga, Lapita pottery and Polynesian origins, *J. Polynes. Soc.* **80.**
7 Green, R. C., Green, K., Rappaport, R. A., Rappaport, A. and Davidson, J. (1967). Archeology on the Island of Mo'orea, French Polynesia, *Anthrop. Pap. Amer. Mus.*, **51** (2); Green, R. C. (1967). Settlement patterns: four case studies from Polynesia, *in* Solheim, W. G. (ed.) *Archaeology at the Eleventh Pacific Science Congress.* (Soc. Sci. Res. Inst., Univ. Hawaii). pp. 112–15.
8 Green, R. C. (1967). *op. cit.* p. 112.
9 Green, R. C. (1967). *op. cit.* p. 114.
10 Green, R. C. (1967). *ibid.*
11 Garanger, J. (1964). Recherches archéologiques dans le district de Tuatira (Tahiti, Polynésie Française), rapport préliminaire, *J. Soc. Océanist.*, **20,** pp. 5–21.

12 Davidson, J. M. (1969). Settlement patterns in Samoa before 1840, *J. Polynes. Soc.*, **78**.
13 Davidson, J. M. (1969). *op. cit.* pp. 48–55.
14 Davidson, J. M. (1969). *op. cit.* p. 55.
15 Davidson, J. M. (1969). *op. cit.* pp. 67–72.
16 Green, R. C. (1967). *op. cit.* pp. 115–9 (Mangareva); Bellwood, P. (1969). Archaeology on Rarotonga and Aitutaki, Cook Islands: a preliminary report, *J. Polynes. Soc.*, **78**; Green, R. C. (1970). Settlement pattern archaeology in Polynesia, *in* Green, R. C. and Kelly, M. (eds) *op. cit.* pp. 30–1 (Marquesas and Hawaii).
17 Groube, L. M. (1965). *op. cit.*; Groube, L. M. (1967). Models in prehistory: a consideration of the New Zealand evidence, *Archaeol. Phys. Anthrop. Oceania*, **2**, pp. 17–19.
18 Groube, L. M. (1965). *op. cit.* esp. pp. 35–57; Kennedy, J. (1969). Settlement in the South East Bay of Islands, 1772: a study in text-aided field archaeology, *Studies in Prehistoric Anthropology*, **3** (Univ. Otago).
19 Groube, L. M. (1965). *op. cit.* p. 43.
20 Groube, L. M. (1965). *op. cit.* p. 52; for a recent study of the classification and distribution of *pa*, cf. Groube, L. M. (1970). The origin and development of earthwork fortifications in the Pacific, *in* Green, R. C. and Kelly, M. (eds) *op. cit.*
21 Groube, L. M. (1965). *op. cit.* pp. 15–21; Groube, L. M. (1967). *op. cit.* pp. 15–7.
22 Higham, C. F. W. (1970). The role of economic prehistory in the interpretation of the settlement of Oceania, *in* Green, R. C. and Kelly, M. (eds) *op. cit.*
23 Green, R. C. (1970). *op.cit.*
24 Green, R. C. (1970). *op. cit.* p. 32. He adopts Charlton's terminology of *external* and *internal* patterns to relate his discussion to other settlement pattern literature (cf. Charlton, T. H. (1969). Review of Chang, K. C. (ed) *Settlement Archaeology*, *Amer. Anthrop.*, **71**.
25 Suggs, R. C. (1961). The archeology of Nuku Hiva, Marquesas Islands, French Polynesia, *Anthrop. Pap. Amer. Mus.*, **49** (1); Sinoto, Y. H. (1966). A tentative prehistoric cultural sequence in the Northern Marquesas Islands, French Polynesia, *J. Polynes. Soc.*, **75**; Green, R. C. (1963). *A Review of the Prehistoric Sequence in the Auckland Province.* (Auckland Archaeol. Soc.); Groube, L. M. (1967). *op. cit.*

M. R. JARMAN, C. VITA-FINZI,
and E. S. HIGGS

Site catchment analysis in archaeology

Few excavations are now carried out without some attempt, however cursory, to determine the environmental conditions that prevailed during occupation of the site. But whether or not it is the archaeologist who does the work, the degree of integration achieved between the environmental and the archaeological aspects of the investigation often leaves much to be desired. Even when it is sought to illustrate the interaction between "cultural" and "natural" factors by interdisciplinary study, a coherent set of results is unlikely to emerge since each of the contributing specialists submits the data to the concepts and objectives of his own discipline. Classifications which have been grafted from one field of study on to another are a common source of confusion in archaeology today. Such experiments have a long history and can no longer be justified on the grounds that, however minimal their current value, they have a promising future.

One reason for the difficulty encountered in fusing the evidence of sites with that of their settings is the reliance by environmental scientists on a zonal approach. While this has undoubted didactic and other virtues and contributes to the process of classification that must precede explanation, it is questionable to what degree environmental or ecological zones meet the needs of archaeology, by which we mean the study of past human behaviour. The definition of zones can permit useful generalizations about otherwise intractable phenomena, but it may be more informative to formulate economic or agricultural zones, which relate to aspects of human behaviour, than to insert units of economic exploitation into climatic and vegetational zones, defined by the relative percentage of tree species or other criteria not immediately relevant to any form of human behaviour. Sites frequently occupy positions which are not typical of their general zonal setting: in a desert, oases will be preferred because of the presence of water; in a flood-plain, settlement may have been attracted to the higher ground, and in a forest to the more open areas. Man sought the optimum and hence the atypical situation to meet his needs. Even if sites should prove after study to lie in an environmentally typical situation, it cannot be safely assumed at the outset that their presence there is due to the factors which determine the environmental type. One means of ensuring that the environmental attributes investigated relate primarily to the site, its occupation and its occupants, whatever its overall zonal situation may be, is by the intensive study of the small area immediately adjacent to it. This is a practical and satisfactory procedure from the archaeological, ecological and economic points of view.

Site catchments

The functions of sites are varied. We shall confine ourselves to home base sites[1], whether they were occupied all the year round or only at certain seasons.

A site of this nature will be composed of an accumulation of deposits brought there by many different agencies; the catchment area of the different agencies will vary greatly. Finds in river gravels may originate hundreds of miles upstream; rodent bones brought into a cave by owls will have come from their hunting territory, an area no greater than 12 ha within 2 km of the site; material brought by hunters will commonly come from within a radius of 10 km. Hence every assemblage at a site may represent a diversity of environments each of which will react differently to changes in external factors. This difficulty tends to pass unremarked in the interpretation of pollen diagrams and faunal assemblages.

In what follows we are using "territory" in an economic sense, to refer to the area exploitable from a particular site, and no reference is intended to the concept of territory as a defended entity. Territories exploited from archaeological sites will vary according to a number of factors, among which the most obvious are the nature of the economy, the means of transport available, the population pressure, and the character of the terrain. Nonetheless the range of possible behaviour is both in theory and practice limited by the capacity of a human population to exploit an area effectively, and is further reduced once the concept of the optimum is introduced. The Italian peasant of the Mezzogiorno will exploit plots of land several hours' walk away from his home if need be, but he will not do this if equally acceptable land is available at his back door. The complications introduced by insecurity, the incidence of diseases such as malaria, religious and social customs and the like could be allowed to dissuade us from undertaking an analysis of human exploitation territories. Fortunately the nature of archaeological evidence enables the assumption to be made that overall long-term trends underlie minor oscillations and discrepancies in the record and that they can be observed and studied by reference to optimum situations.

Studies of modern agricultural[2] and hunting and gathering[3] economies have shown that the territory exploited from a site tends to lie within certain well-defined limits. Other things being equal the further the land is from the site the less likely it is to be exploited from it. In the former case the costs of exploitation rise to oppressive heights at a distance of 4–5 km from the settlement. Among the !Kung Bushmen the site exploitation territory lies within a radius of about 10 km; beyond this, exploitation becomes uneconomic and the home base has to be moved. Clearly, distance has to be qualified in terms of local topography; the operative factor is the time and effort involved in travelling the distance, rather than the absolute distance itself. For the purposes of preliminary study we have adopted the distance covered in two hours' walking as the critical threshold for hunting and gathering economies, and in one hour's walking for agricultural exploitation. These figures approximate to the geographical limits proposed by Lee and Chisholm, and seem reasonable in that they would permit an effective working day after discounting travelling time. In a flat or relatively uniform environment, territories will tend to be circular; where relief is pronounced they will be distorted.

On the assumption that at least some limitations on exploitation operate now as they did in the past, the present-day situation has been taken as a known platform from which to work. For our preliminary assessment of the productive potential of site territories we have measured the percentage of the total area within the catchment boundary that is occupied by arable land, rough grazing, marsh and so on (Fig 1). More refined land capability classifications are available, but they are suitable neither for rapid field survey nor for the comparative study of diverse catchments.

The extent and influence of environmental changes during occupation of the site has been estimated. For instance, where it is clear that alluvial material now supporting arable crops had been deposited since prehistoric occupation[4], it is eliminated from the total. Such figures can help to establish the function and nature of individual sites. Clearly some site locations are not chosen primarily because of their land potential. Coastal sites such as Romanelli were deprived of half their optimal territory by the sea (Fig. 2), although some increase of land would be expected during the periods of low sea-level that accompanied the last glaciation. Where the archaeological record does not show that sea food formed an important element in the economy coastal occupation was probably seasonal. Sites whose characteristics invite the label "urban" and which by implication were occupied throughout the year, such as Jericho, have been found by catchment analysis not to have had the resources available to support the population estimated on other grounds[5], and here again a seasonal influx seems the most likely explanation.

Site catchment analysis has also been found useful in comparing the locations of sites. Where several contemporaneous sites are available it is instructive to see whether their territories have resources or properties in common, or indeed whether any sites or groups of sites can be considered economically complementary to each other. Sites which are inexplicable in

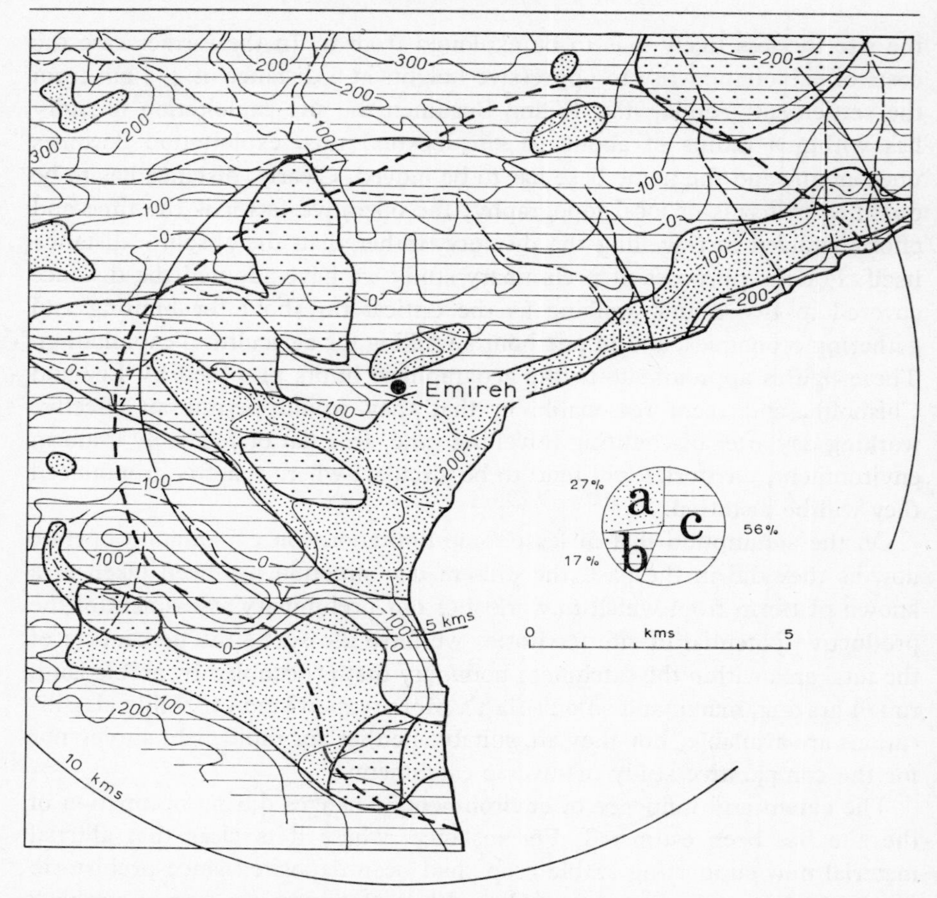

Fig 1 Site-catchment map of Emireh (Palestine). *a*: arable; *b*: good grazing/potentially arable; *c*: rough grazing.

isolation can acquire verisimilitude when compared with other sites. By this means it has been shown that two sites 30 miles apart were complementary to each other for a hunter-gatherer economy, providing suitable winter and summer territories. The method has also demonstrated that two different artefactually defined cultures may represent two different economies rather than two different human groups.

In Palestine the exploitation territories of Nahal Oren, El Wad, Kebarah, and Iraq El Baroud include a very low proportion of potentially arable land (cumulatively 20% or less); the rough grazing quotient is variable, but if it is compounded with marsh it exceeds 60% in all cases[6]. In spite of the fact that these sites were occupied by what are thought to have been both hunter-gatherers and agriculturalists it seems unlikely that any form of arable agriculture played a significant part in the economy of any of them. Thus catchment analysis furnishes us with a ready check for the economic conclu-

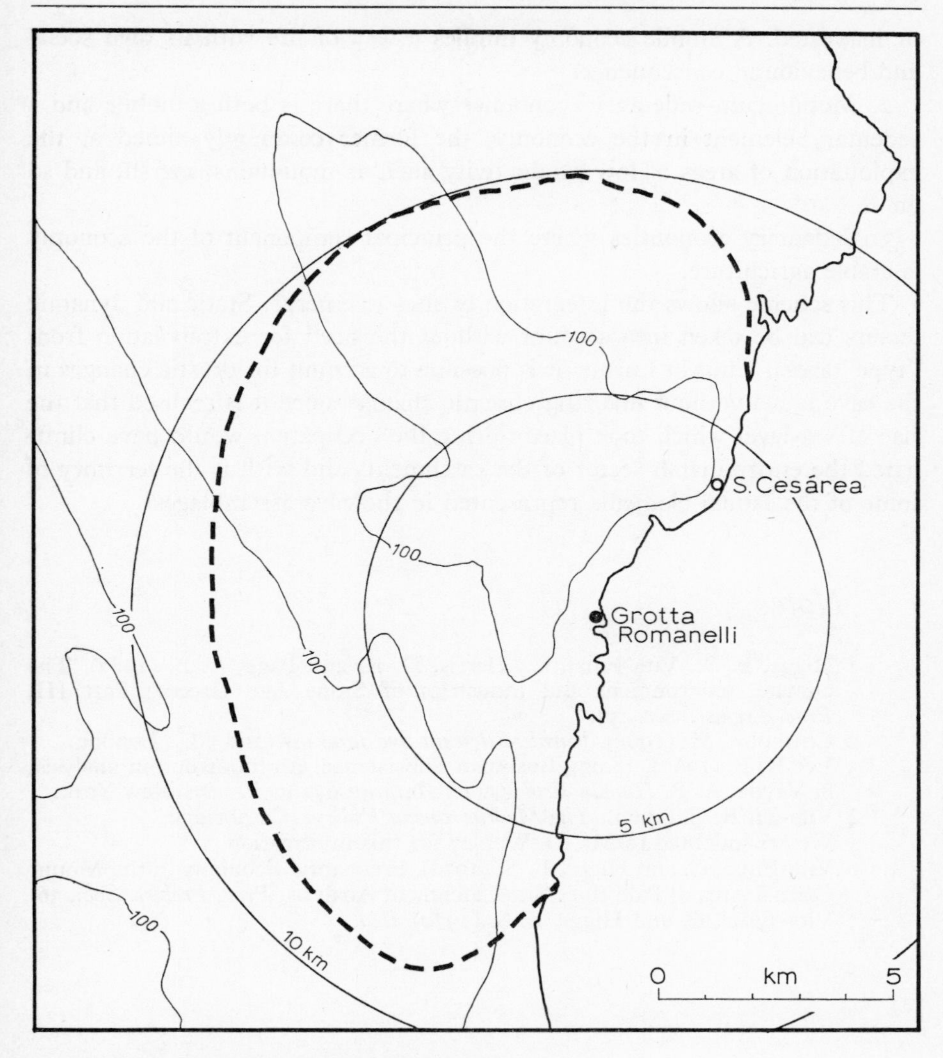

Fig 2 The coastal site of Grotta Romanelli

sions arrived at by excavation. The close integration of territories that has been revealed in Palestine also suggests that at certain times in prehistory the area was fully exploited by human groups and that the level of the population was approaching the limits of that which could be maintained by the available technology.

Our findings from catchment analysis have also led us to classify the economies practised from sites into three groups.

1. Mobile economies which include hunter-gatherers, pastoralists, etc. In terms of human behaviour this classification seems more useful than one based on the manner in which plant and animal species were killed, tended

or harvested. A mobile economy implies a way of life with its own social and behavioural consequence.

2. Mobile-cum-sedentary economies where there is both a mobile and a sedentary element in the economy, the former commonly aimed at the exploitation of areas of low productivity such as mountains, marsh, and so on.

3. Sedentary economies where the principal component of the economy is arable agriculture.

This scheme allows the integration of sites to emerge. Static and dynamic factors can be taken into account without the need for extrapolation from "type" areas. Thus at Carmel it is possible to account for drastic changes in the cave fauna without invoking climatic changes once it is realised that the rise of sea-level which took place during the occupation would have eliminated the entire marsh sector of the catchments and with it the territory of some of the faunal elements represented in the cave assemblages[7].

Notes

1 Higgs, E. S., Vita-Finzi, C., Harris, D. R. and Fagg, A. E. (1967). The climate, environment and industries of Stone Age Greece: Part III, *Proc. Prehist. Soc.*, **33**.
2 Chisholm, M. (1968). *Rural settlement and land use* (2nd ed.). London.
3 Lee R. B. (1967). !Kung Bushman Subsistence. An input-output analysis, *in* Vayda, A. P. *Human Ecology: an Anthropological Reader*. New York.
4 Vita-Finzi, C. (1969). *The Mediterranean Valleys*. Cambridge.
5 We are indebted to Mr. D. Webley for this information.
6 Vita-Finzi, C. and Higgs, E. S. (1970). Prehistoric Economy in the Mount Carmel Area of Palestine: Site Catchment Analysis, *Proc. Prehist. Soc.*, **36**.
7 Vita-Finzi, C. and Higgs, E. S. (1970). *ibid*.

RICHARD S. MACNEISH

The evolution of community patterns in the Tehuacán Valley of Mexico and speculations about the cultural processes

From 1960 to 1965 archaeological investigations greatly supplemented by inter-disciplinary studies were undertaken in the Tehuacán Valley of south-central Mexico by the R. S. Peabody Foundation[1]. From the study of the materials from twelve stratified sites with about 138 components (floors) as well as the artefacts from the 457 sites found in survey it has been possible to define a long sequence of cultural phases[2]. Although much of this material is in print I believe it is best to summarize briefly the ecology of the Tehuacán Valley before describing the cultural sequence and the changes in settlement pattern (Fig. 1). These will be discussed in terms of the ecological or micro-environmental zones that have been defined in Volume I of the Tehuacán reports plus the further topographic sub-divisions we have added[3].

Sub-area 1 was termed The Western Valley Floor of the Tehuacán and Zapotitlan Valleys. It was characterized by travertine-impregnated salty soils with a predominance of short grass vegetation, except along the flanks of hills or terraces and the main waterways (such as the Zapotitlan, Rio Salado, Calapilla and Xiquila) where gallery forest occurred. As a whole, it is not an inviting area, but under certain conditions at least parts of it can be exploited relatively successfully. One sub-division of Sub-area 1, termed the Western Travertine Terraces, has some alluvial soil which during the rainy season or with irrigation can be used for agriculture. The rainy season also brings some game. Further, in these slopes and terraces there are deposits of chert and other materials with conoidal fractures which can be quarried. A second sub-division, the Humid River Bottoms, can be used for agriculture and plant collecting at any time of the year, and it contains abundant water-fowl and game during the winter dry seasons. The Valley Centre Steppe region just west of Rio Salado, a third sub-division, does have some game and edible seeds in the wet season and can be turned to agriculture if canal irrigation is introduced. The other two sub-divisions of this zone are mainly differentiated on the basis of topographical features and are the hills and hill flanks within this western zone. During the rainy season, these do have

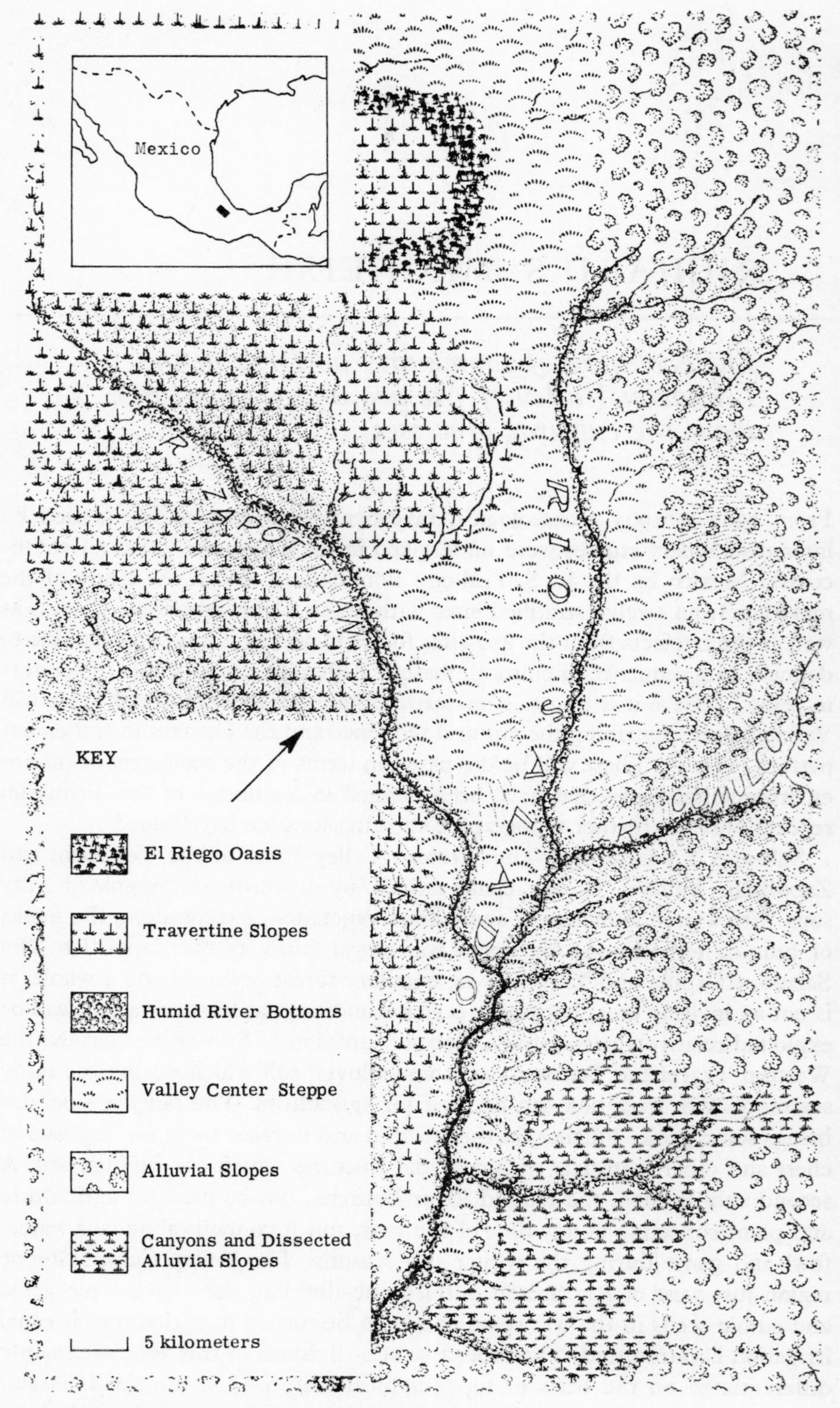

Fig 1 Micro-environments of the Tehuacán Valley

some game and can be dry farmed, but, more importantly, they can provide a good look-out for game, are defensible in a military sense and can be a home base for farming some of the surrounding sub-division. Finally, it must be mentioned that in all these sub-divisions there are seeps of saline deposits that can be worked to obtain salt.

Sub-area 2 may be termed the Western Travertine Slopes of the Tehuacán Valley and the North-eastern Travertine Slopes and Canyons of the Zapotitlan Valley. This is an even more inhospitable zone with a Sonora desert-like vegetation, limited amounts of game, few springs and little rain, even in the summer rainy season, and shallow soils covering limestone or travertine beds. Perhaps, the most inviting area of this region is a zone or sub-division that in reality is transitional between our Zones 1 and 2. This we are calling the El Riego Oasis. This sub-zone is a lush patch with permanent springs rich in plants and animals. However, it covers a relatively small area and could accommodate but a small population. The second sub-zone, the Travertine Slopes, is a poor region under any conditions, but it does have salt and some game, and edible plants do appear in some profusion during the rainy seasons. Much the same may be said of the Canyons eroded into the Travertine Slopes, but here at least there are deeper alluvial soils. The other sub-divisions, involving the hills and hill flanks, are equally unprofitable for human exploitation, but they do have occasional deposits of onyx, salt, flint and blocks of stone suitable for construction. Again, these elevated sections offer strategic military positions.

Now to examine the third major sub-area, the Alluvial (Eastern) Slopes of the Tehuacán Valley. The region, in terms of vegetation, is characterized by thorn forest, but even this is far from uniform. At the edge of the Rio Salado, and along some of the south-westerly flowing arroyos, there is a more lush gallery forest vegetation, and as one ascends the flanks of the mountains, the thorn forest gradually includes more and more oaks. Then higher still there is an oak-pine forest, with the summits of some of the higher peaks having a zone that has features of tropical rain forest. The statistical data on rainfall is also deceiving, for it shows that most of this region has the same limited rainfall (under 800 m) as the rest of the valley, where it occurs only in the rainy season. However, rainfall is much greater and of longer duration in the mountains, and some of this flows down into the arroyos of this eastern zone. The hill-top and flanks make up the other sub-zones of this sub-area, and it must be added that besides being defensible, they often overlook arroyos and valleys that could be conveniently worked from habitations situated on them. Generally speaking, this is a habitable sub-area during all but the driest parts of the dry season.

The final ecological sub-area, which might well be considered a sub-zone of sub-area 3, consists of the Canyon Bottoms and Dissected Alluvial Eastern Slopes in the south-eastern portion of the Tehuacán Valley. It has the same advantage of plants, animals and water as sub-area 3, but animals do hide here during the dry season and the area can be dammed to provide water throughout the year.

We can now examine the various cultural phases, touching briefly on their subsistence, technology and other features and emphasizing the types of settlements and the types of community patterns[4].

The earliest components from 10,000 to 6700 B.C. have been classified into the Ajuereado phase. The sustenance of these earliest peoples seems to have been mainly meat (perhaps over 70%) with a lesser amount from vegetal foods[5]. In terms of subsistence activities, hunting was of prime importance; drives, ambushing with lance and dart-stalking techniques were used in all seasons. In the winter this was probably their only means of subsistence. In other periods of the year the dominant activity was supplemented by the collecting of pods in the spring, the collecting of seeds in the spring and summer, the collecting of fruits in the autumn and cutting of leaves (opuntia and agave) in all these seasons[6].

Evidence of their industrial activities mainly indicated by their stone tools. Generally these were chipped by percussion with a hard stone from flakes derived from blocky cores, usually without prepared striking platforms, and finished by pressure retouching. Diagnostic tool types include Lerma and Abasolo points, flake and slab choppers, crude blades as well as well-made end-scrapers, gravers, and side-scraper knives. The latter hints that there also was a hide-working industry. Chipped spokeshaves, also present, probably indicate a wood-working industry[7].

Four open sites (Ts-380, Ts-383, Ts-372 and Ts-457), two cave sites (Tc-39 and Tc-391) and six floors (XXIII–XVIII) in Coxcatlan Cave and one floor in El Riego Cave yielded artefacts that allowed these thirteen components to be classified as of the Ajuereado phase. Three other sites (Ts-367, Ts-204 and Tr-40) also had Lerma points on their surface, indicating possible Ajuereado occupations. In terms of micro-environments, Ajuereado sites appear in all zones, except the canyons and dissected south-eastern Alluvial Slope zone. Tc-391 and Zones XXIII and XXIV of Coxcatlan Cave lie on the Eastern Alluvial Slopes. The four earlier Coxcatlan occupations from zoological and geological evidence seem to have been Valley Centre Steppes, as were Tr-40 and Ts-457; Ts-367 and Ts-368 were in the Humid River Bottoms, while Ts-380 was on a Western Travertine Terrace above the Rio Zapotitlan Bottomland. Two sites were in the El Riego Oasis area, one on the Travertine Slopes and one in a canyon in the Travertine Slopes. Thus, there was little correlation between Ajuereado sites and any particular micro-environment. Further, when one examines the floral and faunal remains from the eight excavated components where preservation occurred, one finds that all are of a single brief season and that little correlation exists between any particular season and any particular micro-environment. One camp in the El Riego Oasis is spring and the other winter, one occupation in the Alluvial Slopes in Coxcatlan Cave is spring and the other summer, while the four earliest occupations of Coxcatlan Cave then in the steppes were all wet season occupations. In terms of site size, all components covered less than 100 m^2 and all excavated components had no more than two fireplaces. I therefore consider all components to have been occupied by less than three families and have classified all sites as microband camps. When one looks at their distribution over the valley one can discern no clusters of sites, so there is no evidence of territoriality. From all the above data I have classified the community pattern of Ajuereado as *Nomadic Microbands*—that is, groups of families who hunted game in all seasons without regard to any well-regulated subsistence scheduling or well-defined territories (Fig. 2).

The following El Riego phase, 5000–7000 B.C., saw some changes in the subsistence pattern, although 54% of their food was still meat, 40% from vegetal remains and 0–6% from agricultural produce[8]. The major difference was, however, in the scheduling of their subsistence activities. In the winter these people seemed to obtain most of their food by hunting—lance-ambushing or dart-stalking—and supplemented this activity with leaf-cutting and trapping. In the spring their predominant activity was seed-collecting and pod-picking and it was supplemented by hunting and leaf-cutting. The summer season was much the same except the activities were now supplemented by barranca horticulture (of mixta squash, amaranths, and chile). Autumn saw fruit-picking becoming the predominant activity which in turn was supplemented by hunting, leaf-cutting, and hydro-horticulture (avocado cultivation)[9].

Industrial activities also changed, although our most abundant evidence still comes from chipped stone tools. Flakes and crude blades were now struck from cores, usually with prepared platforms, by hard and soft (antler) hammers and then retouched. Tools so made included such types as Flacco, El Riego, Trinidad and Hidalgo points, gouges, crude blades, etc. At this time, however, many tools were made from the cores themselves for instance bifacial chopper and scraper-plane types by percussion chipping with soft hammers. A new ground-stone industry occurred and we now find mortars, pestles, milling stones, and mullers. We also find wooden tools, the weaving of non-interlocking stitch baskets and knotted and knotless nets, twining of mats, string-making, and bone awl- and hammer-manufacturing and bead-making[10]. Some group burials occurred and there is some evidence of cremation and possibly human sacrifice[11]. This evidently was the beginning of the ritual activities that became so prevalent in Mesoamerica.

Their community and settlement pattern is very different from that of the previous Ajuereado phase. First of all, there were eleven macrobands, that is, sites covering between 50 and 500 m^2 and with three or more fire-places (but no evidence of architectural features). Seven of the nine that could positively be identified as to season were from the wet summer months, the other occupations on the Alluvial Slopes being in winter and one

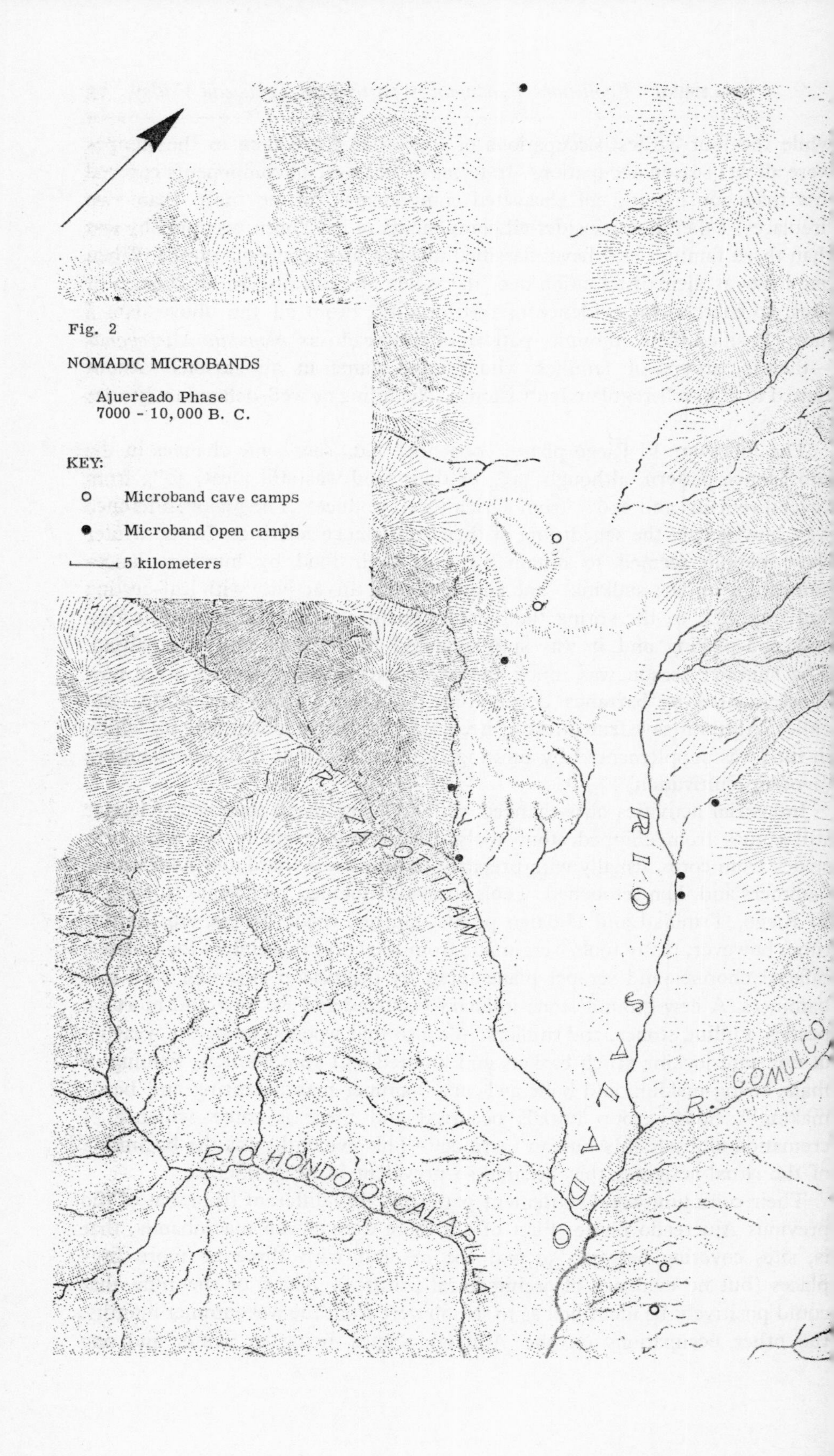
Fig. 2
NOMADIC MICROBANDS
Ajuereado Phase
7000 - 10,000 B. C.
KEY:
Microband cave camps
Microband open camps
5 kilometers
R. ZAPOTITLAN
RIO SALADO
R. COMULCO
RIO HONDO O CALAPILLA

in the River Bottoms being in the spring. Second, all these macroband sites occurred in only two micro-environs, five in the Humid Valley Bottoms and the other six in the Eastern Alluvial Slopes. There were twenty-eight microband sites or occupations at spring sites in the South-eastern Canyons and one in the Alluvial Slopes, as well as two spring-summer occupations in this same zone, one in the oasis zone and one in the steppes. Microband summer sites were almost non-existent, there being but a single occupation from Purron Cave in the Dissected Canyons in the south-eastern portion of the valley, one in the Travertine Canyons and another from nearby Coxcatlan Cave. There were three winter microband camps in the Humid River Bottoms, three at Coxcatlan Cave, and one each in the Oasis, Travertine Slopes, and Dissected Canyons respectively. Of summer autumn camps one was in the Dissected Canyons and one in the Alluvial Slopes, while a single autumn microband camp occurred in the Alluvial Slopes. In summer, macroband camps usually occurred in one or two zones, the Alluvial Slopes and River Bottoms, while microband camps of the winter occurred in similar zones, plus the travertine area, but spring and summer microband camps were in many zones and autumn microband camps were confined to the Alluvial Slopes or the dissected part of it.

When one examines the distribution of the sites one finds three or four concentrations of micro-macro-band camps. One concentration is near the south-eastern part of the valley with six macroband occupations and nineteen microband camps. Another to the central north-west with three microbands and three macroband camps, a third area in the north with four microband camps and two macroband camps, and a possible final group near Ajalpan in the centre of the valley with three possible microband camps. I believe these concentrations represent three- or four-band territories (Fig. 3).

The community pattern here represented might be called *Seasonal micro-macrobands* with territoriality and scheduled subsistence system. Here a number of family groups or microbands would gather together into macrobands in various spots in the River Bottoms or more lush spots in Alluvial Slopes within their own territories in the summer when food was plentiful. Then with the coming of autumn they would break up into microbands to hunt and collect fruits in the Alluvial Slopes or Dissected Canyons or Steppes in various parts of their band territories. Then with drier winter months, the microband would move again to hunt in the Oasis area, the Travertine Canyons, Humid River Bottoms, the Alluvial Slopes or Dissected Canyons or anywhere else where they could find game within their territories. As spring came on and pods and seeds became available, the family groups would move into any of the various zones of their territories where these foods were available, and then the cycle would start over again with the coming of the wet summer.

Coxcatlan, from 5000 to 3400 B.C., saw some changes from El Riego in the proportions of foodstuffs, with wild plants composing 52% of their

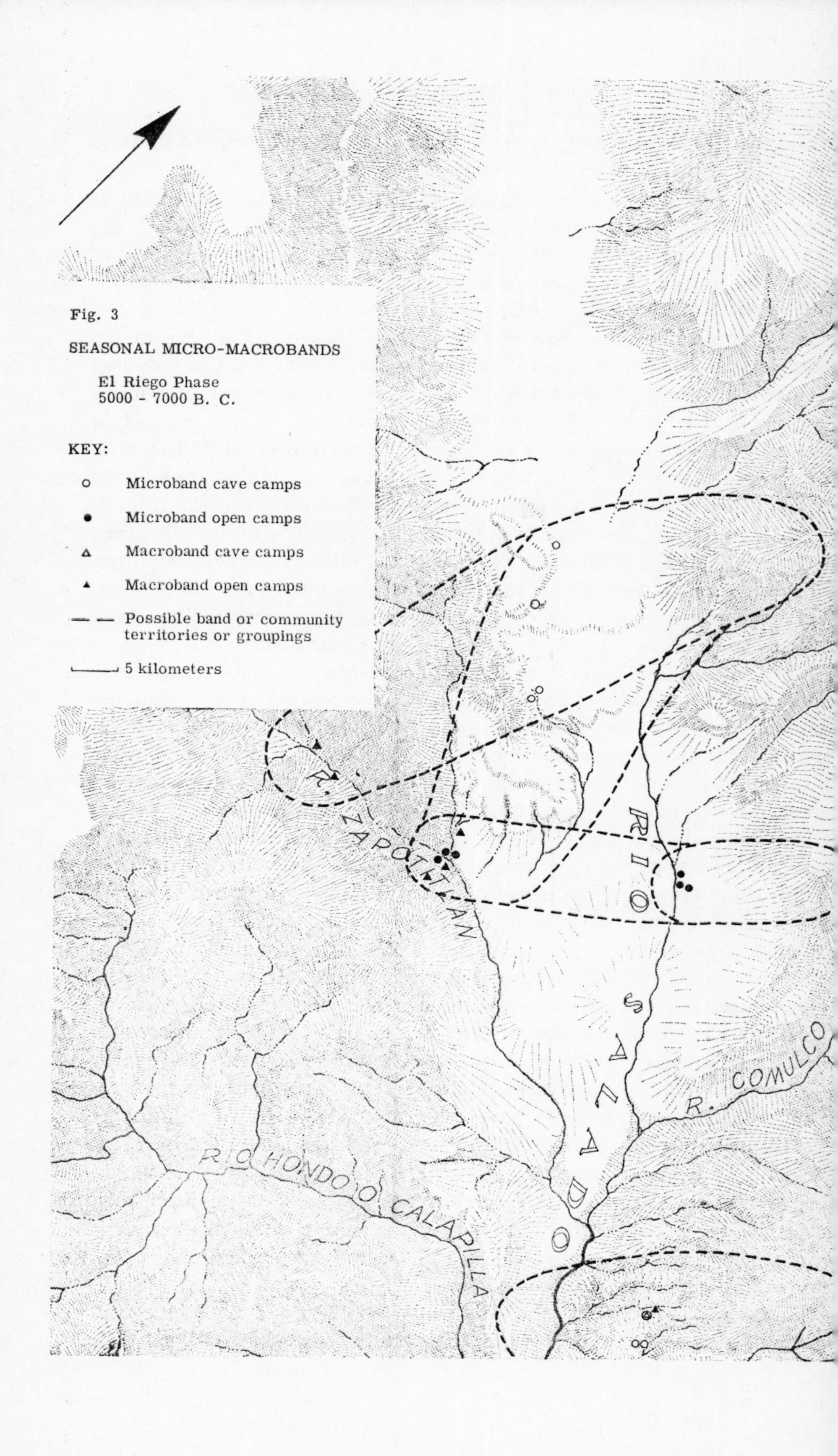

Fig. 3
SEASONAL MICRO-MACROBANDS
El Riego Phase
5000 - 7000 B. C.
KEY:
Microband cave camps
Microband open camps
Macroband cave camps
Macroband open camps
Possible band or community territories or groupings
5 kilometers
R. ZAPOTITLAN
RIO SALADO
R. COMULCO
RIO HONDO O CALAPILLA

diet, meat 34% and agricultural produce 14%, but the subsistence activities and their scheduling was much the same[12]. The autumn and winter activities were about the same, and in the spring seed-collecting was still predominant, but now supplemented by leaf-cutting, pod-picking, hunting, root-digging, and barranca horticulture (amaranth). The summer season was slightly different with seed-collecting and barranca horticulture of a wide variety of plants such as corn, beans, zapotes, gourds, etc., being dominant, with only other minor supplemental subsistence activities[13].

Techniques of chipping flint were little different but there was more fine pressure retouching and some blades were struck from cylindrical cores by indirect percussion. There was, however, a host of new chipped stone types such as Tilapa, Coxcatlan, Almagre and Garyito points, fine blades, end-scrapers, etc. The ground-stone industry continued with manos, metates and anvil stones being new types. More wooden tools occurred, including digging sticks and atlatl dart parts, and bone needles were added to the bone tool complex. The weaving and string-making industry had changed little but interlocking and split stitch baskets appeared[14]. Complex burials continued to occur.

The following Coxcatlan phase had a settlement and community pattern similar to that of El Riego, but there are some noticeable differences. First of all, less Coxcatlan components are known, there being only nine macroband and eleven microband occupations. Seemingly this would indicate a smaller population, but in fact it does not, for the Coxcatlan macroband camps are two or three times larger than those of El Riego and moreover macroband as well as microband camps were occupied for longer periods of time. In fact, all the Coxcatlan macroband camps were occupied for two or more seasons. The two in Dissected Canyons were occupied for the spring and summer as was one of those from the Humid River Bottoms, while one from Coxcatlan Cave (Zone XI) was occupied winter-spring and summer and another (Zone XIII) in summer-autumn and winter. Two of the other four macroband camps in the River Bottoms were occupied in the autumn and winter and the other two in the winter and spring seasons. This contrasts with El Riego that had only one camp which lasted two seasons.

The microband camps, while roughly the same size as those of El Riego and occupying many of the same ecological niches, are different in that seven of the eleven were occupied for two seasons while only three of twenty-eight in El Riego were occupied this length of time. Spring-summer camps occurred as follows: one in the Canyons in the Travertine Slopes, two in the Humid River Bottoms, and one in Purron Cave in the Dissected Alluvial Slopes. A summer-autumn occupation occurred in Zone XII of Coxcatlan in the Alluvial Slopes, and Zone F of San Marcos Cave was a summer microband occupation in the Canyons in the Travertine Slopes. One winter-spring micro-band camp was in the Oasis and one in the Dissected Canyons. Three winter camps occurred in the Travertine Slopes, Humid River Bottoms and Alluvial Slopes respectively.

Like El Riego, Coxcatlan sites cluster into three to five groups—one near Coxcatlan Cave in the southern portion of the valley, one near Tehuacán in the north end of the valley, one to the north-west along the Zapotitlan arroyo, one in the middle reaches of the Zapotitlan River and perhaps one in the centre near Ajalpan which also might have connected with the Zapotitlan cluster or perhaps all the latter three were connected. The areas, I believe, represent band territories (Fig. 4).

In terms of community patterns, Coxcatlan should be classified as *Seasonal macro-microbands* with territoriality and a scheduled subsistence system that included incipient agriculture. In their calendar-round it appears that in the spring, groups of microbands usually coalesced to form macrobands in favoured spots in the Humid River Bottoms, Dissected Canyons or Alluvial Slopes to gather wild seeds and pod foods as well as hunt. While collecting in the spring they probably planted gourds, squash, corn and beans that allowed them to continue to live as macrobands in the same spot through the summer. Less often, microbands went through the same process in the same areas in the same season without joining together into macrobands, but this occurred mainly in early Coxcatlan times. Macrobands at Coxcatlan Cave in Zone XIII did not occupy it until summer, but their summer crops allowed them to stay there through the autumn and winter dry season; one microband camp in the Humid Bottoms lasted until the autumn while another was only occupied in the summer. Generally speaking, however, autumn saw people move out of their spring-summer abodes, perhaps to new zones with fruits (avocado, zapotes, chupandilla, cosahuico, etc.) and more game. Two of these macroband camps in the Humid River Bottom to the west lasted through the winter while a macroband camp in the Alluvial Slopes had a similar tenure. Most winter camps were, however, microband camps, occurring in all possible zones, perhaps because game had to be sought in widely dispersed areas. A few of these camps in the Zapotitlan Humid River Bottoms that started as microbands may have gradually grown into macrobands with the coming of spring, which, of course, started their cycle of activities once again. Thus, although cultivation, plant domestication and horticulture began in Coxcatlan, it affected their community pattern and way of life but little—it changed it quantitatively rather than qualitatively.

Abejas, 3400 to 2300 B.C., saw the rise of agricultural produce (25%) at the expense of both meat food (25%) and wild plant foods (50%)[15]. The scheduling of the subsistence activities was like that of Coxcatlan, but in the summer and spring the barranca horticulture increased. These increased activities may have at this time produced surpluses that could be stored and then eaten in the leaner autumn and winter months[16].

Chipping techniques were much the same as those of Coxcatlan, although there were more fine blades and cores. New types of chipped stone artefacts,

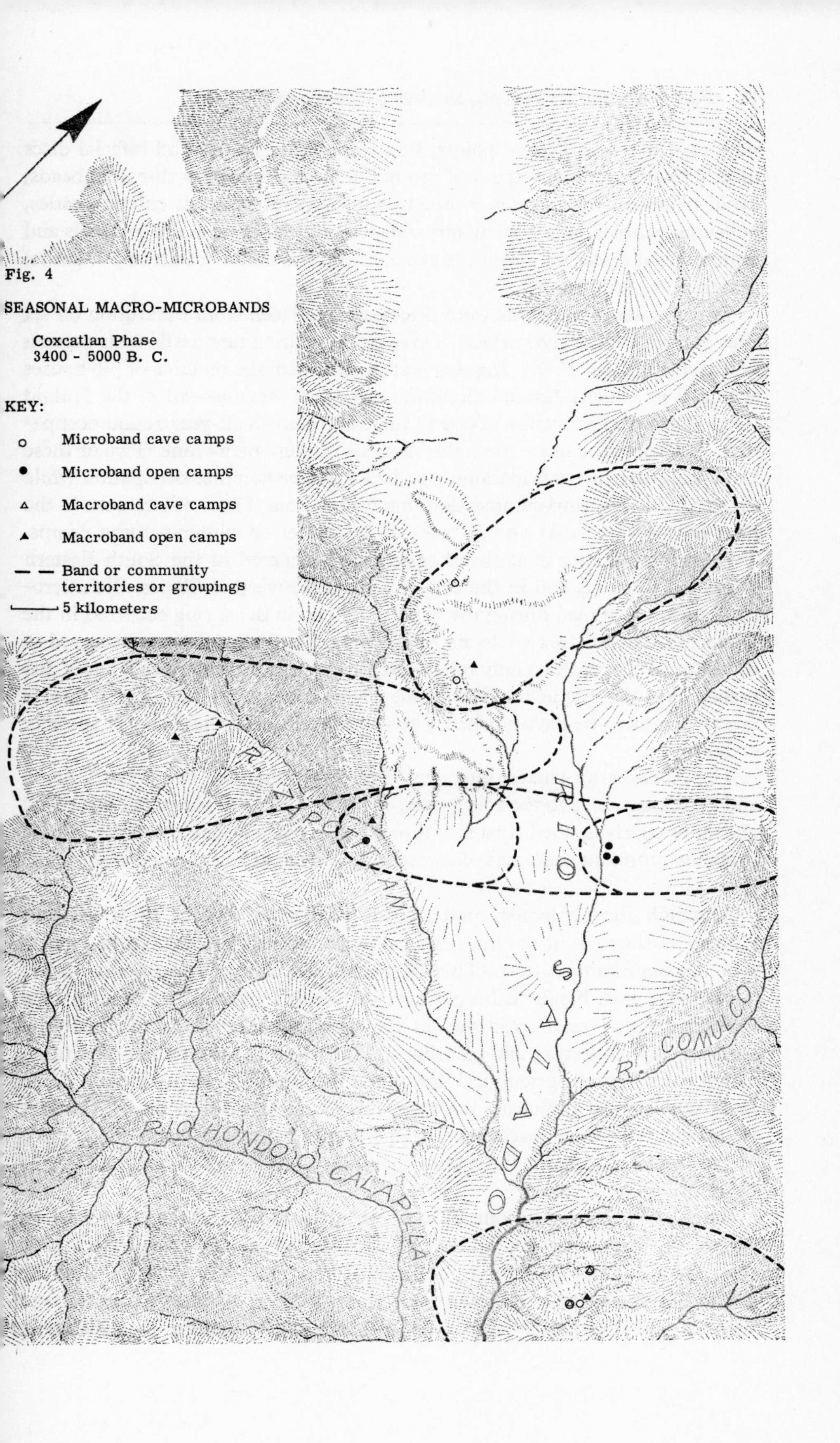
Fig. 4
SEASONAL MACRO-MICROBANDS
Coxcatlan Phase
3400 - 5000 B. C.
KEY:
Microband cave camps
Microband open camps
Macroband cave camps
Macroband open camps
Band or community territories or groupings
5 kilometers
R. ZAPOTITLAN
RIO SALADO
R. COMULCO
RIO HONDO O CALAPILLA

such as Catan and Pelona points, fine rectangular blades and bifacial discs together with more new types of ground-stone tools such as discoidal beads, ovoid plano-convex metates, rubbed pebbles, paint palettes, cuboid pestles, polishing pebbles, spherical manos and stone bowls. Bone and wood tools and types of cord remained about the same, as did the weaving industry[17]. Burial practices are little known.

Twenty-two components were discovered, thirteen from excavation on six sites and nine from the surface. Three of these are a new settlement pattern type termed *Linear Valley Hamlets* that is, sites with the remains of pit-houses arranged in a linear fashion along the terrace of streams—all in the Humid River Bottoms. Excavation of one of these revealed an all-year-round occupation. Macroband camps were also numerous, there being nine. Two of these had spring-summer occupations, one had a winter-summer occupation while the other had a spring-summer-autumn occupation. Three others were in the river bottoms or flanks and all seem to be winter or winter-autumn camps. Two spring-summer macroband camps also occurred in the South-Eastern Dissected Canyons and in the canyons in the Travertine Slopes. Six microband camps occupied during the winter and one in the spring occurred in the Dissected Canyon area while a spring-summer and winter camp occurred in the Alluvial Slopes. The only other microband camp came from a badly disturbed cave deposit in the Travertine Slope zone, and since it contained a mortar fragment we have guessed it was occupied during the spring or summer or both.

The Abejas sites cluster into two or three groups; one in Purron Canyon and Coxcatlan Cave to the south-east, one near Chilac to the north-west, and one more poorly defined cluster of three sites in the centre of the valley. These probably represent some sort of communities or some sort of territorial groups (Fig. 5).

Although the settlement pattern was similar to that of El Riego and Coxcatlan, the presence of linear waterway hamlets suggests that it had a new type of community pattern. This I have termed *Central-based Bands*. Here, apparently, most bands had a permanent base at a hamlet, but some groups banded together as macrobands in the spring and stayed throughout the summer, first collecting and planting and then later in the summer or autumn eating their limited agricultural produce. Their agricultural production in the main was still not sufficient, however, for all microbands to live all the year round either in their macroband camps or the hamlets, so with the coming of the leaner months in the winter they had to move and to exploit (mainly by hunting) other zones.

Some of them moved out as microbands into the less favoured Alluvial Slopes or South-Eastern Dissected Canyons while others moved out as macrobands into the more lush Humid River lands. Nevertheless, with the coming of spring they returned once again to their hamlets or macroband base camps. Obviously, this type of community pattern is transitional

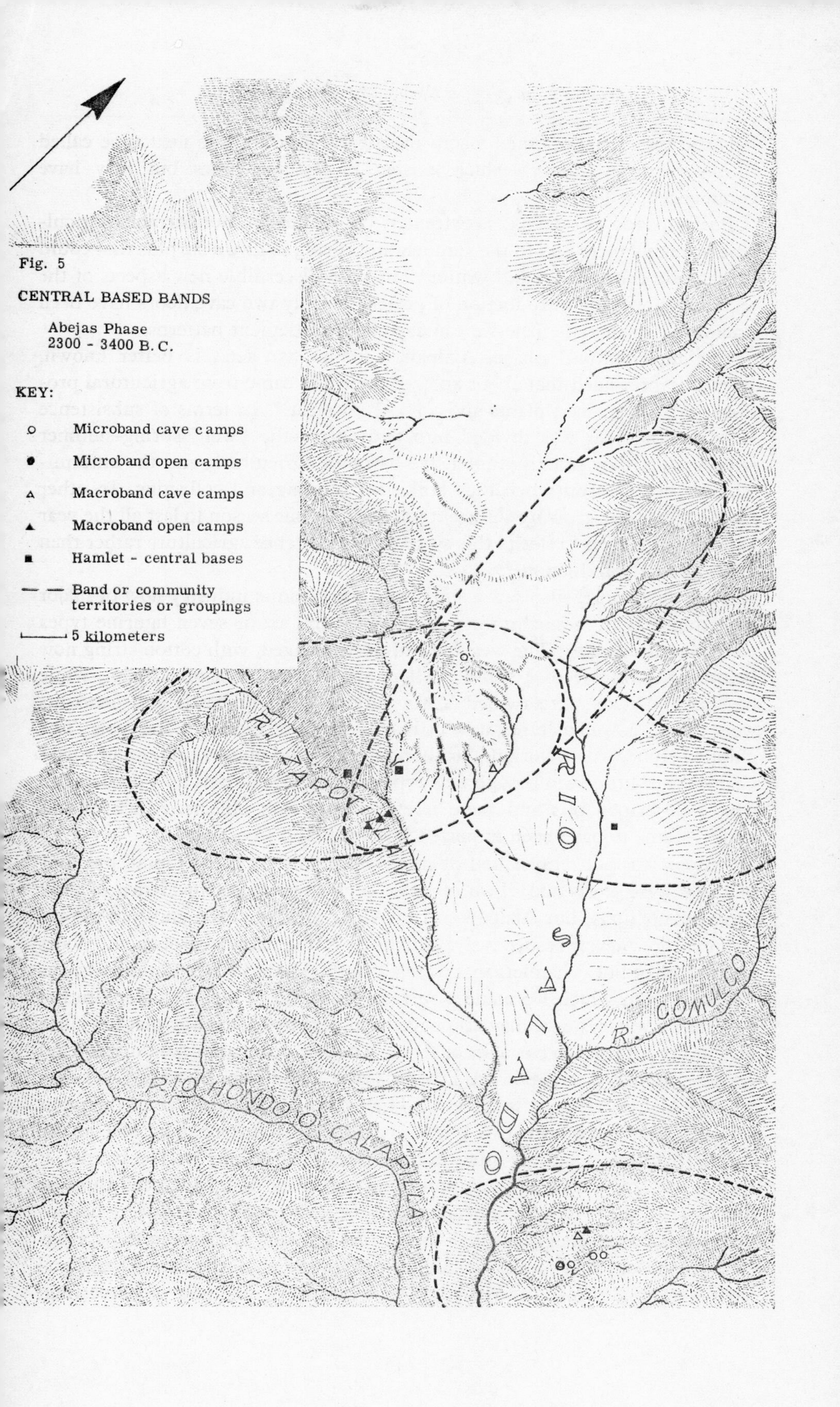
Fig. 5
CENTRAL BASED BANDS
Abejas Phase
2300 - 3400 B.C.
KEY:
Microband cave c amps
Microband open camps
Macroband cave camps
Macroband open camps
Hamlet - central bases
Band or community territories or groupings
5 kilometers
R. ZAPOTITLAN
RIO SALADO
R. COMULCO
RIO HONDO O CALAPILLA

between the earlier seasonal macro-microband type and the next type called *Semi-permanent Hamlets* which existed in Ajalpan times but may have commenced in Purron.

Purron, 2300 to 1500 B.C. is extremely poorly known, but barranca agriculture probably increased in the summer and spring to give a surplus that could be eaten in the autumn and winter. The only discernible new aspects of the technology was the introduction of pottery[18]. Only two cave floors have been uncovered so there is little we can say about settlement patterns.

The next cultural phase, Ajalpan, 1500 to 850 B.C., is better known. Excavation revealed that about 40% of their diet came from agricutural produce, 31% from wild plants and 29% from meat[19]. In terms of subsistence activities the year was divided into two parts—the "wet" spring-summer season when they practised intensive barranca agriculture, and the autumn-winter "dry" season when they still went hunting and collecting. In other words, they were growing sufficient food in only one season to last all the year and for this reason I refer to this system as subsistence agriculture rather than effective food-production[20].

Their technology also saw a shift, with the ceramic industry now of major importance. Ten monochrome pottery types and six or seven figurine types were manufactured[21]. The weaving industry changed, with cotton string now being made by use of a spindle whorl and cloth woven on a loom, while baskets and mats were twilled. Also, various kinds of beads and pendants were made, as well as other new utilitarian types of ground stone tools. The chipped stone, wood-working, leather-working, and bone-working industries changed but little even though new types occurred[22]. Burials now occurred in deep bell-shaped pits and there is evidence of cult objects. Much of this "cult" seems to have been imported from the Gulf Coast area.

Ajalpan was not represented by many sites, only six being found, but four of these were excavated. Two of these cave sites yielded evidence of only three occupations, but Ajalpan and Coatepec yielded seven and four superimposed floor levels respectively. All eleven of the excavated open site floors seem to be of linear hamlets that were along the Humid River Bottoms as was Ts-204B from which we made a surface collection. Ts-4, a hamlet site not now near a water source may also have been next to a waterway that has since disappeared, for it too has a linear form. Another occupation was macroband spring-summer occupation in Purron Cave while San Marcos Cave also yielded Ajalpan remains in Zone C. The latter, on the evidence of the plant remains, may have been two brief occupations by microbands, one in the spring (planting ?) and the other in the autumn (harvesting ?). The hamlet, it seems, was now the basic community unit and there may have been three or four in the valley; they were occupied all the year round and all were dependent upon barranca agriculture—that is, the growing of crops in the spring-summer seasons that produced sufficient to last them all the year. Occasional trips were made from the village to cave camps for planting or harvesting,

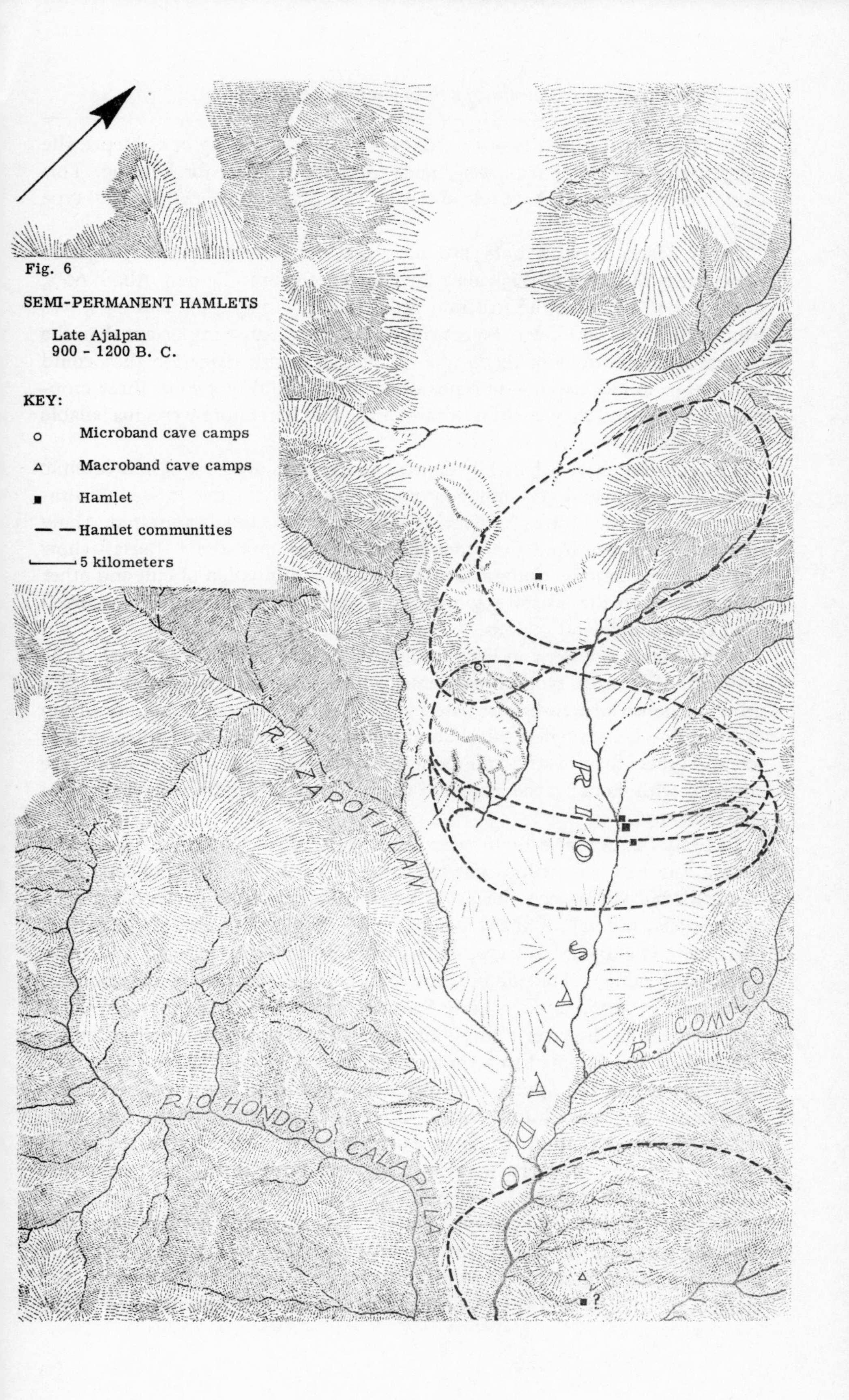
Fig. 6
SEMI-PERMANENT HAMLETS
Late Ajalpan
900 - 1200 B. C.
KEY:
Microband cave camps
Macroband cave camps
Hamlet
Hamlet communities
5 kilometers
R. ZAPOTITLAN
RIO SALADO
R. COMULCO
RIO HONDO O CALAPILLA

while floors F1, G1 and H of the Ajalpan site and Zone K3 of Coatepec site suggest that in winter trips were made from the hamlets for hunting. This community pattern I have classified as the *Semi-permanent Hamlet* type (Fig. 6).

Santa Maria, from 850 to 150 B.C., saw the introduction of irrigation agriculture, the basis in this valley for effective food production. About 60% of their food was from agricultural products while only 15% and 25% was from wild plants and meat respectively[23]. However, more important than the change in proportions of foods was the fact that with irrigation they could grow crops in any season—in other words there could be two or three croppings a year—and they could now farm areas that heretofore were unavailable in drier seasons[24].

The cultural phase is basically defined on the basis of a host of new ceramic types with white and grey monochrome surfaces, distinctive incised decoration, and new bowl forms[25]. Weaving improved and sandals appeared. Many new artefact types from the other industries also appeared[26]. Burials show status differentiations. Ceramics, also, reveal a continuation of cult and other influences from the coastal regions, but some trade seems to have occurred with various highland regions.

Santa Maria saw both an increase in number of components as well as the addition of a new type of settlement-village. A village is defined as a group of house structures associated with a central plaza or mound or group of mounds—i.e. a single central administrative or ceremonial centre. Architecture other than wattle-and-daub now appears and some mounds are surfaced with stone; stone steps are used and some buildings have dry laid slab masonry.

In Early Santa Maria there were ten sites, three of which were stratified, having fourteen occupations giving a total of twenty-one components. Late Santa Maria had sixteen sites, four of which were stratified with eighteen occupations, a total of thirty components. In terms of the whole period, there were twenty-five villages, twenty hamlets, three macroband and three microband camps. The picture, however, becomes clear when one examines these sites by sub-phase. In Early Santa Maria twelve of the fourteen linear hamlets were in the river bottom and the other two were in Purron Canyon which then was dammed to form an artificial waterway. The spring-autumn macrobands and the spring macrobands were also in Purron Canyon and obviously were agricultural camps. Of the villages, three of the four were also in the river bottom. Further, except for the village in the steppe, all villages had one or more hamlets associated with them, and they were clustered into five or six community groups. This community pattern is termed the *Nuclear Village* type (Fig. 7). The agricultural system was still basically barranca wet season agriculture, but the increasing use of irrigation made it possible now for one village to be established in the steppes and clusters of hamlets and camps oriented to a nuclear village to occur in the

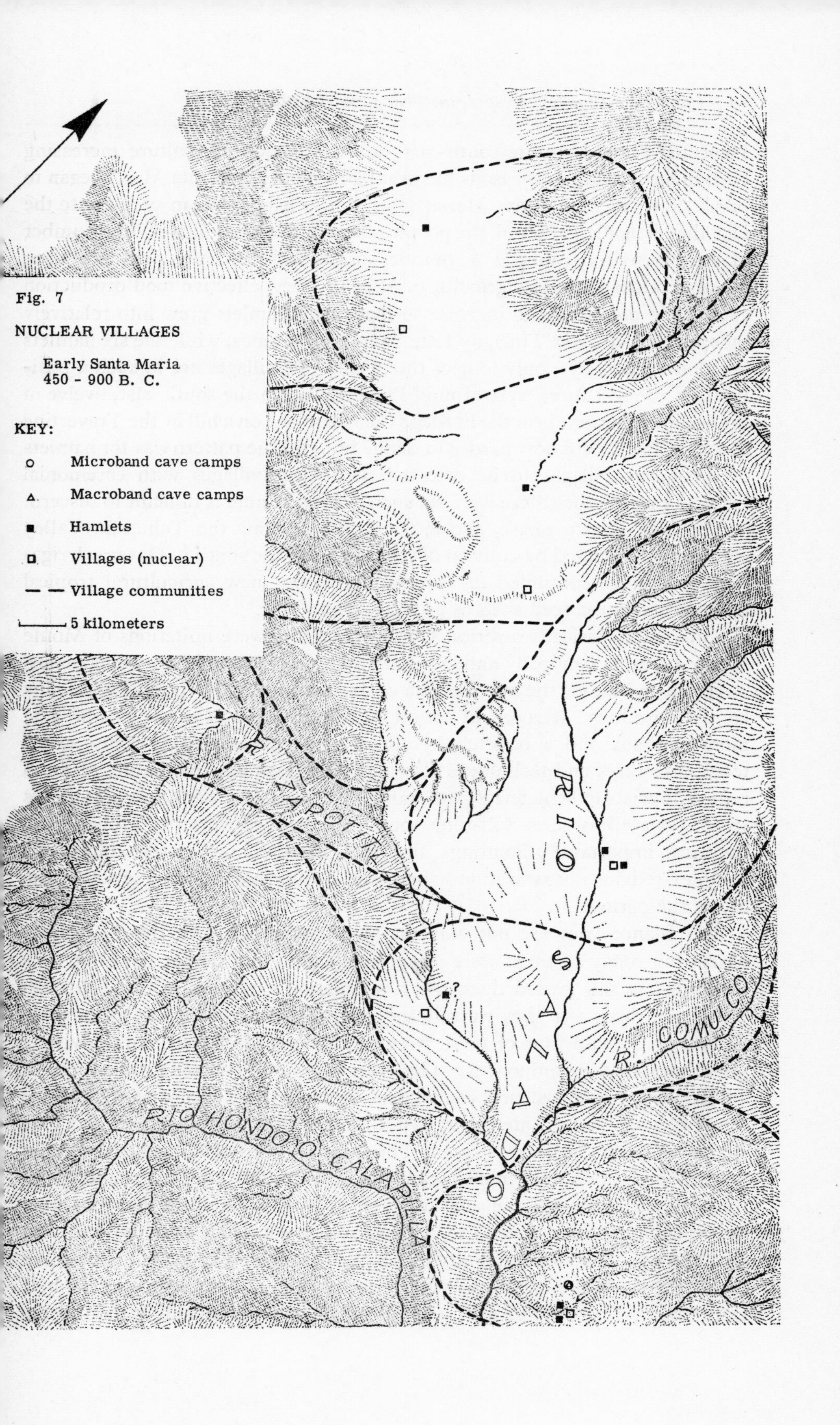
Fig. 7
NUCLEAR VILLAGES
Early Santa Maria
450 - 900 B. C.
KEY:
Microband cave camps
Macroband cave camps
Hamlets
Villages (nuclear)
Village communities
5 kilometers
R. ZAPOTITLAN
RIO SALADO
R. COMULCO
RIO HONDO O CALAPILLA

Dissected Canyons to the south-east. With irrigation agriculture increasing both in the wet and dry seasons the neat pattern of Early Santa Maria began to break down in Late Santa Maria times. This change was in part due to the fact that irrigation allowed the people effective food production in a number of micro-environs during a number of seasons and this had not been possible under barranca agriculture. Also, this new effective food production caused the population to increase so that many hamlets grew into relatively independent villages. Thus, by Late Santa Maria times, while the six hamlets are along waterways, only four of the twenty-one villages are in such a position. Of the rest three were around Purron Dam in the south-east, twelve in the Steppe Zone, one near the El Riego Oasis, and one on a hill in the Travertine Slopes. Clusters are now harder to define but still the pattern was for hamlets or secondary villages to be oriented to nuclear villages with ceremonial features but whether there were six such groups or nine is difficult to discern.

The Palo Blanco phase, 150 B.C. to A.D. 700, saw the Tehuacán Valley becoming dominated by cultural developments to the south in Oaxaca. Irrigation agriculture continued to improve and a few new agricultural tropical plants were introduced such as guava, peanuts, etc.[27].

Pottery was basically a series of grey wares that were imitations of Monte Alban types[28]. Chipped- and ground-stone tools were more finely manufactured, and we see the introduction of the bow and arrow (points) and many lapidary products. Weaving of cotton on various types of looms also became more complex and a bark cloth industry was introduced[29]. Many architectural features and techniques also appeared. Further, the finding of tomb burials, burials in cemeteries, and individual burials hint at a complex status system as well as some full-time specialists, with the priestly group being the most important[30]. Paintings, figurines, incense burners, etc. indicate a complex religious organization which seems to have dominated much of the life of the period.

Palo Blanco sites are not only far more numerous than those of Santa Maria but many of them were much larger, and one cannot help but conclude that in this period there was a population explosion in the Tehuacán Valley. In total, 151 sites were found with materials belonging to the Palo Blanco phase, and four of these were stratified with eleven floors, giving a grand total of 158 components. The older types of settlement still appeared, with twenty-two microband camps, sixteen macroband camps as well as two linear hamlets along waterways, two linear hamlets in the steppes and nine steppe or river bottom villages. The hamlets were about the same but house foundations were more inclined to be made of slab construction with frequent use of mortar. Villages also were slightly different in that they were not so definitely linear in form. Houses and truncated pyramids were now faced mainly with stone masonry rather than earth, and a few had talus-talud construction. Also, none of the villages were really nuclear centres of a whole community. There were, also, some new hamlet types including four on the

Travertine Slopes and three on the Alluvial Slopes or on the sides of hills. These usually had their house structures dispersed roughly along contour lines and may have belonged to groups of people who were involved in dry (slash-and-burn) farming. The other two types of hamlets were on hilltops and had a roughly concentric arrangement of masonry house structures. Two of the twenty-five seem to be associated with the salt industry. Three new village types also occur, all with the more complex architectural features mentioned previously. Twenty-eight of the villages were on hilltops with houses surrounding the plaza or mound areas; seven were strung out along hill flanks, one was on Travertine Slopes and four on the valley flanks. There were also eight salt industrial sites without habitations—a new feature. The other completely new settlement type comprised towns, of which there were twenty-eight. Towns were distinguished from villages in being large and by having two or three central plaza areas surrounded by pyramids, ball courts and houses. Twenty of the towns occurred on hilltops. Twelve of these we consider to be nuclear hilltop towns as they have their two or three plazas in juxtaposition, often terraced, with talus-talud construction and sometimes I-shaped ball courts, and surrounded by well-cut masonry structures (often one room) that in turn are surrounded by many poorer structures of not-so-well-cut masonry or of wattle-and-daub. There also were six similar nuclear towns in the steppes. Less common were similar dispersed towns on hilltops with their plazas widely separated, often being on different spurs of the hilltops. There were eight of these while two dispersed towns occurred on the steppes.

In terms of community pattern, these nuclear towns seem to have been the centre of communities and in Early Palo Blanco times there were eight to ten of them. These communities were usually dispersed along a river valley, whereas a non-nucleated town or towns were on a hill or rarely in flats. Nuclear or dispersed villages and hamlets that are on hilltops as well as elsewhere are all oriented to a centre nuclear town, usually on a hilltop. These nuclear towns were probably the administrative and ceremonial centres for the settlements oriented to them and the whole community was further held together by the use of the water for irrigation from a common drainage system. This type of community I have called *Advanced Nuclear Centres* or sometimes *Dioceses* (Fig. 8).

The final phase, Venta Salda, A.D. 700 to 1520, sees the valley fall under the influence of the Cholula or Mixteca-Puebla area to the north and west[31]. Irrigation agriculture continued to expand and improve[32]. Further, there are hints that their foodstuffs (such as varieties of beans and fruits) were imported and exchanged for exported products (cotton textiles and salt). In other words, there was a trade and commerce, hand-in-hand with which is evidence of militarism and the rise of "city states".

Much of the technology now seemed to be in the hands of full-time specialists. Pottery was made in moulds, the weaving of textiles reached a

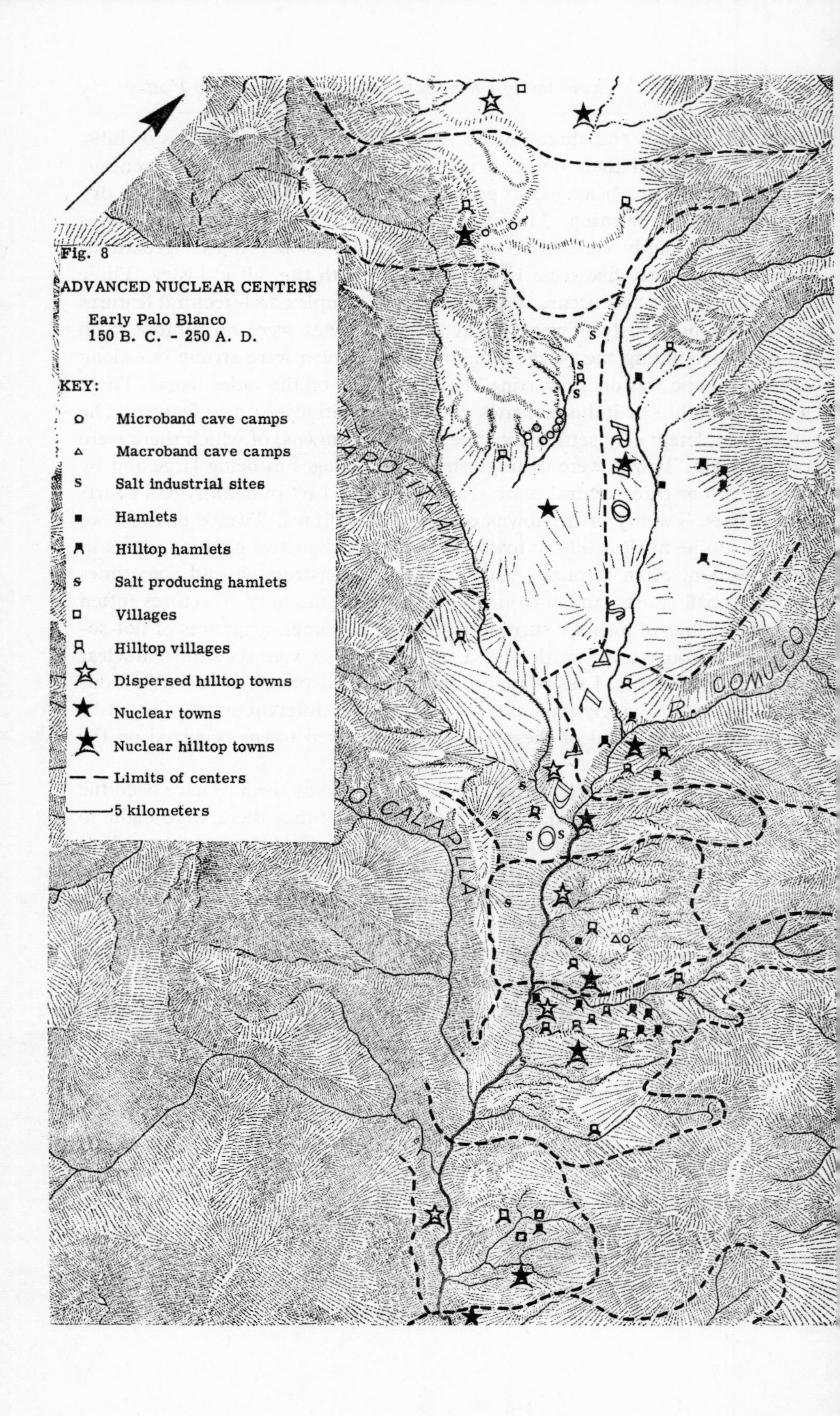
Fig. 8
ADVANCED NUCLEAR CENTERS
Early Palo Blanco
150 B. C. - 250 A. D.
KEY:
Microband cave camps
Macroband cave camps
Salt industrial sites
Hamlets
Hilltop hamlets
Salt producing hamlets
Villages
Hilltop villages
Dispersed hilltop towns
Nuclear towns
Nuclear hilltop towns
Limits of centers
5 kilometers
ZAPOTITLAN
R. COMULCO
RIO SALADO
RIO CALAPILLA

high degree of skill and there are hints that certain villages, hamlets, or barrios specialized in these industries[33]. Other industries, to judge by the many artefact types, may have been equally specialized[34]. Monumental architecture, long canals, and aqueducts also occurred in this period, and there is good evidence that codices were being written in this valley[35].

Ethnohistorical evidence, partly confirmed by archaeological data, reveal that there was a class-structured society with two main divisions—the *principales* (nobility) and *macehuales* (commoners). At the top of the *principales* were members of the despotic ruling lineages who were "living gods" and rulers of city-states. Under them were the nobles and governors and below, the specialists in religion, the military and craftsmen. Finally there were the peasants. Ethnohistorical evidence also reveals there was a complex religious institution under the aegis of the "living god" rulers[36].

One hundred and seventy-nine components of the final phase exist, only six of which come from stratified excavated sites. Most of the older types continue to exist, with seven microband camps, fifteen macroband camps, twenty-one salt industrial sites, eighteen salt hamlets, thirty-two hamlets (one in the steppes, three on rivers, eighteen on hills and ten on hill flanks), twenty-six villages (ten on hills, seven on valley flanks, five in the steppes or on slopes and four near waterways) and sixteen towns (two nucleated ones on hilltops, three in the valley, eight dispersed on hilltops and three on the hill flanks). There are, however, a number of new significant types and these include five fortified hamlets or guardhouses, a salt village, four fortified villages or garrisons, six fortified dispersed towns and two nucleated ones on hilltops as well as five cities. We distinguished cities from towns not only by their greater size, but by the fact that they had three or more plazas usually with one bigger than the others, and that within the cities there were major barrios or areas for full-time specialists and non-agricultural workers not peasants, indicating that the majority population was of full-time specialists not farmers. The five cities, all with a modern town equivalent are all slightly different, one being a fortified hilltop city (Zapotitlan), another a fortified city in the steppes (Venta Salada), another being a valley fortified flank and hilltop city (Tehuacán) while two are valley flank cities (Coxcatlan and Teotitlan del Camino) with no discernible fortifications.

In all these later settlement pattern types we see new features of architecture such as fortification, monumental staircases, cylindrical columns, cruciform tombs, complexes of nested pyramids and terraces, ball courts with two parallel walls, and multi-room as well as single-room masonry houses with plastered covering; and many other features. In fact, the architectural innovations are almost as distinctive as the settlement patterns themselves.

Distribution of sites in the valley reveal clusters of various types of sites around the five cities as well as two or three of the fortified nuclear towns (Acatepec, Zapotitlan and Chilac). Ethno-historical data indicate that these (and perhaps others) were the capitals of small city-states incorporated into a large political entity called the Senorio del Teotitlan. This final type of community is termed the *Primitive City State* (Fig. 9). This last state has all the characteristics of civilization as defined by Childe[37].

In conclusion I would like to speculate about *why* these changes in community pattern occurred in the Tehuacán Valley. Was the sequence of causal factors that operated in this small region of Mexico also in operation in other nuclear areas or at least parts of other nuclear areas where primary or pristine civilizations began?

The initial stimulus for the change from *Nomadic Macrobands* to *Seasonal Macro-microbands* seems to have been a slow process that began about a thousand years before the transition actually took place—roughly during the late Ajuereado phase—say about 9600 B.C. Previous to this time people had specialized in various kinds of hunting activities, but as they developed these techniques they acquired, as a by-product, knowledge of the food potential of their micro-environments at various seasons. Also, as a supplement to their meat diet they had experimented in seed-collection, pod- and fruit-picking and leaf-cutting subsistence activities. This accumulation of subsistence and scheduling knowledge coincided with the waning of the Pleistocene, with its changes in climate and rainfall that resulted in greater diversity in seasons, a drier and warmer climate, diminution of grassland steppes and water-holes, as well as an expansion of the thorn forests on the Alluvial Slopes and an expansion of the cactus forests on the Travertine Slopes. The result of these changes brought about the extinction of the various "herd" animals such as horse, antelope, jackrabbit, as well as other Pleistocene fauna (mammoth, mastodon, etc.). Thus, changes in the ecosystem became interconnected with change in the scheduled subsistence system to form a larger closed system—or an encompassing feedback system. For example, changes in micro-environments may have caused hunting to be undertaken in different ways in different zones at different times. These in turn further diminished the fauna and led to different kinds of hunting in different seasons as well as new subsistence activities in certain seasons as hunting became unprofitable in these zones. This, of course, would lead to other changes in technology with lance points developing into atlatl dart points and paint stones developing into mortar and milling stones, skin scrapers evolving into plant pulpers, etc. Further, it meant that peoples moved in certain well-regulated regional cycles so that bands became tied to territories and gradually, as they found abundant foodstuffs at certain spots in the wet season, small groups coalesced into larger groups (macrobands) for brief periods. Changes in the social system may also have taken place in response to the new way of life and increased population concentration. Is this hypothesis about how and why Nomadic Microbands changed to Seasonal Macro-microbands in the Tehuacán Valley perhaps applicable to changes from "Early Hunter to Incipient Agriculturists" in South America[38] from the "Big game hunting tradition" to the "Archaic and Desert traditions" in North America[39] and

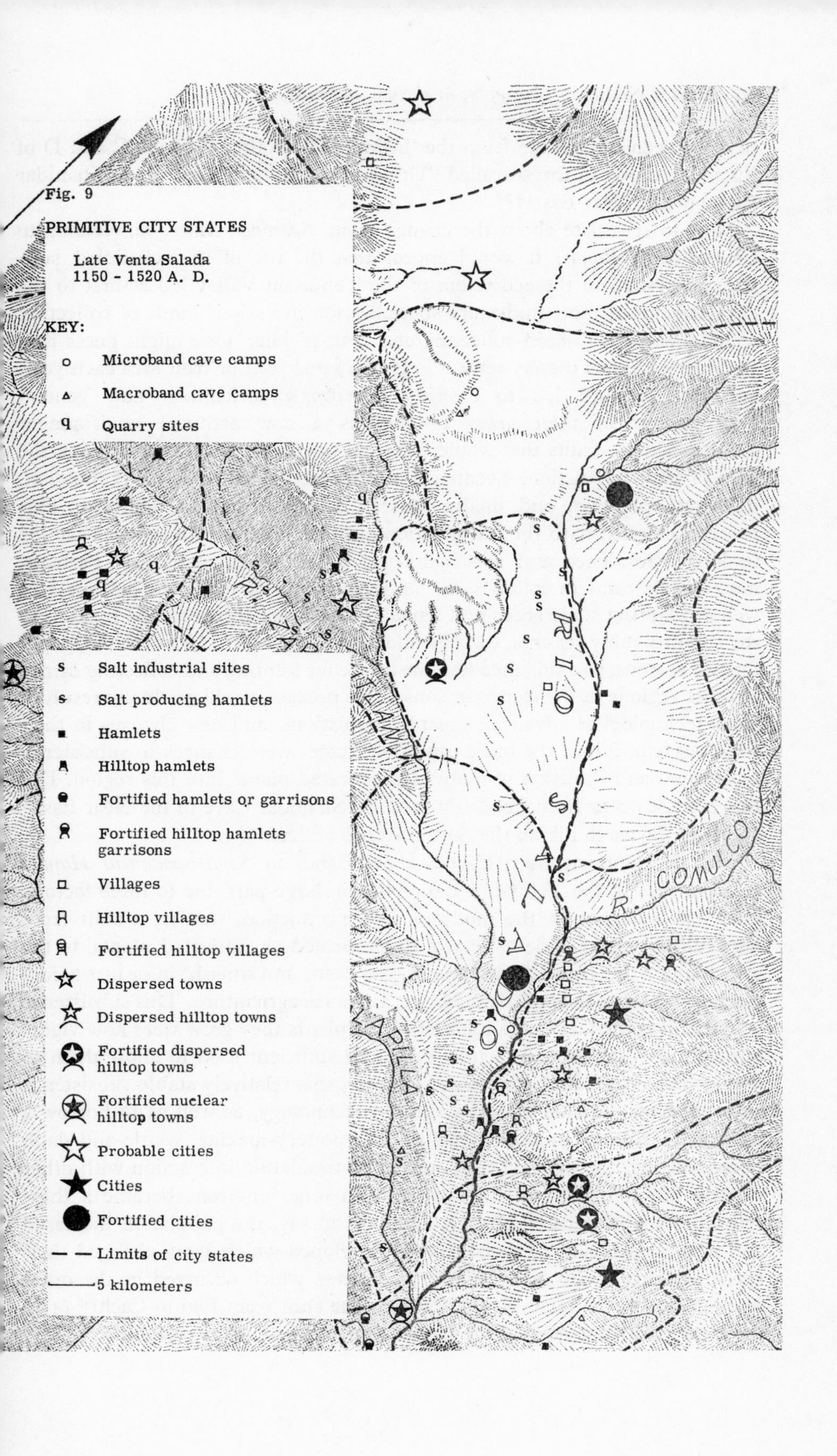

Fig. 9
PRIMITIVE CITY STATES
Late Venta Salada
1150 - 1520 A. D.
KEY:
Microband cave camps
Macroband cave camps
Quarry sites
Salt industrial sites
Salt producing hamlets
Hamlets
Hilltop hamlets
Fortified hamlets or garrisons
Fortified hilltop hamlets garrisons
Villages
Hilltop villages
Fortified hilltop villages
Dispersed towns
Dispersed hilltop towns
Fortified dispersed hilltop towns
Fortified nuclear hilltop towns
Probable cities
Cities
Fortified cities
Limits of city states
5 kilometers
RIO SALADO
R. COMULCO
R. ZAPOTITLAN
ZAPILLA

perhaps even the change from the "Middle Palaeolithic" (Zone C and D of Shanidar Cave) to the so-called "Upper Palaeolithic" (Zone B2 of Shanidar Cave) of the Near East[40]?

Now to speculate about the change from *Seasonal Macro-microbands* to *Central-based Bands*, it would appear that the use of the scheduled subsistence system in the ecosystem of the Tehuacán Valley led at first to the development of more and more specialization in various kinds of collecting techniques such as seed-collecting and fruit-picking. One might guess that an early stage was merely returning to the same seed or fruit area each year, and this in turn led to clearing (weeding), enrichment and general improvement of their areas as well as a new artificial environment for the seeds or fruits that would have led to various genetic changes in the seed or fruit population. Eventually this may have led to actual cultivation, deliberate planting, and finally to seed selection and horticulture. This process, coupled with the introduction of other domesticates, must have led to longer residence and re-scheduling of macroband activities. Perhaps, macrobands came to collect some seeds in the spring along the barrancas as well as planting some seeds and fruits. These latter then reached fruition in the summer rainy seasons, thereby allowing the bands to be in the same spot for two seasons, and this then became a base for hunting and collecting camps in other regions in the leaner seasons. This process would gradually result in some technological advances, greater populations, and new changes in their social system. Thus, the basic causative factors were changes in subsistence techniques and diffusion of many domesticated plants into this region. Did this process occur from Zones B2 to B1 in Shanidar Cave in the Near East[41] and from Jaywa to Piki in the central Andes of Peru[42]?

The development from *Central-based Bands* to *Semi-permanent Hamlet* communities in the Tehuacán Valley was in large part due to three factors. One was a change in the mode of food production, that is, a shift from horticulture with a wide variety of domesticated or cultivated plants, to the growing of a few plants (corn, and maybe beans and squash) in or just before the rainy seasons in the barrancas, i.e. barranca agriculture. This subsistence activity, plus the fact that a number of the plants they grew were now vigorous hybrids—the second factor—produced sufficient food to allow them to live in hamlets all the year round. Further, this relatively stable subsistence allowed them to make improvements in technology, as well as to borrow a number of cultural improvements such as pottery-making, wattle-and-daub house-making, the figurine cult, etc. This stimulating interaction with other slightly different cultural developments in other environs became a third crucial factor in cultural change. Needless to say, the population increased and new mechanisms of social control developed which again changed their social system. Is this not the sort of process which occurred in the other nuclear area in the New World, Peru, in the shift from Piki to Cachi[43] or in the Near East from Zarzi of the Upper Palaeolithic to Early Neolithic[44]?

Although process and the causal factors for the previous developments have probably been oversimplified, the change from *Semi-permanent Hamlet* communities to simple *Nuclear Village* communities was probably even more complex and thus more difficult to understand. One fact, however, does seem apparent at least for the Tehuacán Valley, and that is that the causal factors were not economic or subsistence factors but something still more basic. The great population increases in Late Ajalpan as well as the influx or development of a new ceremonial-religious complex lead to the conclusion that these were the major causal factors. In this way, for the period of Early to Middle and Late Formative in Tehuacán (and perhaps all of Mesoamerica as well as Peru[45]), the increased population pressures apparently resulted in the development and use of new subsistence techniques (such as irrigation, etc.)[46]. Moreover, the development or adoption of the new ceremonial-religious complex seems to have resulted in a new type of social system that reorganized the technology as well as subsistence system and furthered stimulation and change by much greater outside contacts with other developing groups. Is not this the sort of process that was occurring in the Near East in the Neolithic in pre-Uruk times?

Changes from the *Simple Nuclear Village* to *Advanced Nuclear communities* were even more complicated and difficult to understand. However, the studies at Tehuacán strongly indicate that a major causal factor was the development of water control and various kinds of irrigation agriculture. Obviously, it was not the only factor, but in our sequence it does occur just before the greatest population increase, the final major concentrations of population, many other technological advances and great changes in the social system. Is this not true in the Near East, China, and Peru? In fact, is it not about time to re-examine and re-investigate the whole role of water control and the hydraulic society concept rather than throw the baby out with the canal water just because all parts of Wittfogel's theories do not fit all the facts[47]? I suspect that interaction between various regional developments was also very important but our knowledge of this part of the Tehuacán sequence is so incomplete that it is difficult to discern the nature of these interactions.

The final development from *Advanced Nuclear Centres* to *Primitive City States* is obviously the most complex and obscure. But for the Tehuacán Valley, the expansion of water control systems and the rise of a commercial economy with various systems of exchange involving imports and exports seems very important. The latter is a stimulating interaction between different regional developments *par excellence*. Are some of these factors not also crucial in the development of primary or pristine civilizations in other nuclear areas?

In conclusion then, it would appear that the evolution from savagery to civilization, at least in the Tehuacán Valley of Mexico, was due to multicausal factors which occurred in a very definite sequence. Is this true of other developments of primary civilizations in other nuclear areas and is the sequence of causal factors the same? Although the facts are not all available for Tehuacán, let alone the other areas, the hypotheses put forward here should be tested by further investigation. I hope these hypotheses will stimulate constructive discussion that perhaps will lead eventually to some generalizations about how and why primary or pristine civilizations arose in nuclear areas.

Notes

1 MacNeish, R. S. (1964). Ancient Mesoamerican civilization, *Science*, **143**, pp. 531–7; Byers, D. S. (ed.), (1967). *Environment and Subsistence, The Prehistory of the Tehuacán Valley*, vol. 2, Texas.
2 MacNeish, R. S., Nelken, A. and Johnson, I. (1967). The non-ceramic artefacts *in* Byers, D. S. (ed.) *op. cit*. vol. 2, pp. 231–4; MacNeish, R. S., Peterson, F. A. and Flannery, K. V. (1970). Ceramics *in The Pre-history of the Tehuacán Valley*, vol. 3, Texas. pp. 238–66.
3 Flannery, K. V. (1967). The vertebrate fauna and hunting patterns *in* Byers, D. S. (ed.) *op. cit*. vol. 1, pp. 132–78.
4 MacNeish, R. S. (1970). Social implications of changes in population and settlement pattern of the 12,000 years of pre-history in the Tehuacán Valley of Mexico *in* Deprez, P. (ed.) *Population and Economics*, Manitoba.
5 MacNeish, R. S. (1967). A summary of the subsistence *in* Byers, D. S. (ed.) *op. cit*. vol. 1, pp. 290–309.
6 MacNeish, R. S. (in press). The scheduling factor in the rise of effective food production in the Tehuacán Valley *in* Lothrop, D. (ed.). Volume in honour of J. MacGregor.
7 MacNeish, R. S., Nelken, A. and Johnson, I. (1967). *op. cit*. p. 231.
8 MacNeish, R. S. (1967). *op. cit*. pp. 290–309.
9 MacNeish, R. S. (In press). *ibid*.
10 MacNeish, R. S., Nelken, A. and Johnson, I. (1967). *op. cit*. pp. 231–2.
11 Anderson, J. E. (1967). The human skeleton, *in* Byers, D. S. (ed.) *op. cit*. vol. 1, pp. 94–7.
12 MacNeish, R. S. (1967). *op. cit*. pp. 290–309.
13 MacNeish, R. S. (In press). *ibid*.
14 MacNeish, R. S., Nelken, A. and Johnson, I. (1967). *op. cit*. p. 232.
15 MacNeish, R. S. (1967). *op. cit*. pp. 290–309.
16 MacNeish, R. S. (In press). *ibid*.
17 MacNeish, R. S., Nelken, A. and Johnson, I. (1967). *op. cit*. pp. 232–3.
18 MacNeish, R. S., Peterson, F. A. and Flannery, K. V. (1970). *op. cit*. pp. 21–6, 238–9.
19 MacNeish, R. S. (1967). *op. cit*. pp. 290–309.
20 MacNeish, R. S. (In press). *ibid*.
21 MacNeish, R. S., Peterson, F. A. and Flannery, K. V. (1970). *op cit*. pp. 26–58, 239–41.
22 MacNeish, R. S., Nelken, A. and Johnson, I. (1967). *op. cit*. p. 233.
23 MacNeish, R. S. (1967). *op. cit*. pp. 290–309.
24 MacNeish, R. S. (In press). *ibid*.
25 MacNeish, R. S., Peterson, F. A. and Flannery, K. V. (1970). *op. cit*. pp. 58–145, 241–6.
26 MacNeish, R. S., Nelken, A. and Johnson, I. (1967). *op. cit*. p. 233.
27 MacNeish, R. S. (1967). *op. cit*. 290–309.

28 MacNeish, R. S., Peterson, F. A. and Flannery, K. V. (1970). *op. cit.* pp. 145–76, 246–56.
29 MacNeish, R. S., Nelken, A. and Johnson, I. (1967). *op. cit.* pp. 233–4.
30 MacNeish, R. S. (1970). *op. cit.* pp. 246–7.
31 MacNeish, R. S., Peterson, F. A. and Flannery, K. V. (1970). *op. cit.* pp. 267–306.
32 MacNeish, R. S. (1967). *op. cit.* pp. 290–309.
33 MacNeish, R. S., Peterson, F. A. and Flannery, K. V. (1970). *op. cit.* pp. 177–238.
34 MacNeish, R. S., Nelken, A. and Johnson, I. (1967). *op. cit.*
35 Chadwick, R. and MacNeish, R. S. (1967). Borgia Codex and the Venta Salada Phase *in* Byers, D. S. (ed.) *op. cit.* vol. 1, pp. 114–31.
36 MacNeish, R. S. (1970). *op. cit.* pp. 246–9.
37 Childe, V. G. (1946). *What Happened in History*. New York.
38 Lanning, E. P. (1967). *Peru before the Incas. New Jersey.*
39 Willey, G. (1966). *An Introduction to American Archaeology*, vol. 1.
40 Solecki, R. S. (1964). Shanidar Cave, a Late Pleistocene site in northern Iraq, *Report of Sixth International Quaternary Congress of Lodz.*
41 Solecki, R. S. (1964). *ibid.*
42 MacNeish, R. S. (1969). *First Annual Report of Ayacucho Archaeological-Botanical Project*, R. S. Peabody Foundation for Archaeology.
43 MacNeish, R. S. (1969). *ibid.*
44 Braidwood, R. J. (1967). *Prehistoric Men*. Chicago. pp. 81–112.
45 MacNeish, R. S. (1969). *ibid.*
46 Boserup, E. (1965). The inter-relations between population trends and agricultural methods, *United Nations World Population Conference*, Belgrade, WPC/WP/318.
47 Wittfogel, K. A. (1964). *Oriental Despotism*. Yale.

IRVING ROUSE

Settlement patterns in archaeology[1]

Introduction

One aim of archaeological research is to reconstruct the lives of extinct peoples from their remains. In the present paper, it will be assumed that the archaeologist reconstructs a people's ecology, culture, and social structure, viewing each of them as a separate system. A people's ecological system may be said to consist of its adaptation to the local environment and its utilization of the resources available there. Its cultural system comprises the activities which it carries out during the course of its daily life. Its social system, finally, is composed of the institutions or social groups into which it is organized for the purpose of carrying out its various activities.

The reconstruction of ecological, cultural, and social systems has often been likened to detective work. Just as a detective reconstructs a crime by searching for clues and interpreting them in terms of his knowledge of crimes, so also the archaeologist reconstructs a people's life by searching for archaeological evidence and interpreting that evidence in terms of his knowledge of ecological, cultural, and social systems. His knowledge is derived primarily from geography in the case of ecological systems, from cultural anthropology in the case of cultural systems, and from social anthropology in the case of social systems. Hence, we may say that he combines archaeological evidence with geographical and anthropological knowledge in order to reconstruct the three kinds of systems.

The geographical and anthropological knowledge includes both patterns and models of behaviour. The patterns have been observed among contemporary peoples and may also be documented for past peoples[2]. The models are abstract patterns which have been worked out theoretically by specialists in geography and anthropology[3].

The present paper is concerned with one kind of observed pattern, the settlement pattern, which is singled out because it has proved to be a key to the reconstruction of the three kinds of systems. The paper will discuss the nature of settlement patterns, their survival among archaeological remains, and their use in reconstructing ecological, cultural, and social systems. In addition, an attempt will be made to define civilization and urbanization in terms of settlement patterns.

The concept of settlement pattern and its application to archaeology

By a settlement pattern is meant the manner in which a people's cultural activities and social institutions are distributed over the landscape[4]. Such a pattern embodies all three kinds of systems, cultural, social, and ecological, and provides a record of the relationships among them.

The student of a contemporary settlement pattern will determine the range of a people's activities, e.g. from food production to art and religion, and will plot the distribution of these activities. He will also determine the range of a people's institutions, e.g. from residential communities to work groups and religious sects, and will plot the distribution of these institutions, showing how they participated in the various activities. He will thus produce a network of different places, which may be called loci, and a record of the activities which were carried out in each locus, together with the institutions which operated there. This network of loci and the record of its activities and institutions comprise the settlement pattern.

To the best of my knowledge, the concept of settlement pattern was first applied to archaeological remains in the United States, as a by-product of the close relationship between ethnographers and archaeologists in the American academic system. During World War II and immediately thereafter, Julian H. Steward, an ethnographer, and Gordon R. Willey, an archaeologist, were colleagues at the Bureau of American Ethnology in Washington, D.C. Steward had done research on the semi-nomadic Indian bands of the Great Basin in the western United States, had followed the bands from place to place, and had observed them depositing their remains in different sites. He and Willey often discussed ways of increasing the amount of ecological, cultural, and social information that can be inferred from a people's remains. Steward advised Willey to adopt the approach he had used in studying the Indian bands of the Great Basin. He suggested that, instead of focusing upon the remains left by a people in a single site, Willey make a study of the distribution of the people's remains from site to site. In other words, he advised Willey to approach the people's remains in terms of its settlement pattern.

Willey first employed the new approach in a study of the archaeology of Virú Valley, Peru, which attracted much attention among Americanists[5]. He subsequently organized a conference in which we who specialized in other parts of the New World applied the approach to our own areas[6]. Archaeologists in both Old and New Worlds have also been attracted to the study of settlement patterns by geographic research on the subject[7], and in

America it has become the cornerstone of a new sub-discipline known as settlement archaeology[8].

Survival of settlement patterns as archaeological remains

It is important to note that archaeologists never recover an entire settlement pattern, as it is observed by ethnographers and geographers. They find only the traces of it which have survived as archaeological remains. These take the form of various types of sites, in each of which the people under study carried out one or more of its activities. If only a single activity was performed at a site, as in the case of an isolated burial mound, we may refer to that site as an *activity locus*. If several different activities took place at the site, as is normally the case where people lived, the site has to be divided, insofar as possible, into a number of activity loci[9]. Hence, an activity locus may be defined as any spot where a minimum number of activities took place.

An archaeologist who wishes to recover the surviving traces of a settlement pattern must first locate the entire range of a people's activity loci and then collect remains from representative loci. We may refer to the remains obtained from each locus as an *activity assemblage*.

What the archaeologists obtains, then, is a network of activity loci, some of which consist of whole sites and others of parts of sites, together with an activity assemblage from each locus. We shall call this combination of network and assemblages a *remnant settlement pattern* in order to distinguish it from the full settlement pattern which is observed by the ethnographer or geographer. A remnant settlement pattern is to its full pattern as a fossil skeleton is to its living counterpart. The archaeologist bases his reconstruction of the full pattern on its remnant in the same way that the palaeontologist bases his reconstruction of an animal on its fossil remains.

The present paper is concerned only with remnant settlement patterns, not with their full counterparts. To go into the nature of full settlement patterns would be to encroach upon the subject matter of other papers in this volume[10].

Remnant settlement patterns in the strategy of archaeology

In an attempt to clarify the concept of remnant settlement pattern, let us see how it fits into the strategy of archaeology. There are two ways of approaching the excavation of archaeological remains. One is to recover all possible kinds of remains in an effort to learn the nature of the local archaeology. This approach may be termed *analytic*, since it focuses on the component parts of the remains, such as food materials and artefacts, and identifies or classifies them in order to determine what they are like. The second approach is to work selectively by recovering only the kinds of remains needed to learn about the local people. This approach may be called *synthetic*, because it brings together the pertinent bits of archaeological evidence with facts drawn from other sources and uses them to synthesize a picture of peoples, their lives, and times[11].

The concept of a remnant settlement pattern is not analytic. It should not be equated with concepts like site and artefact, which are designed to express the nature of remains. Neither should it be confused with analytic usages of the term *settlement*, such as those of Chang[12] and Piggott[13]. Chang applies the term to that part of a site which has yielded homogeneous remains, i.e. to the unit which other archaeologists call a component[14]. Piggott uses the term for the arrangement and ground plan of houses within a component. Instead of signifying features of remains like these, the phrase *remnant settlement pattern* refers to the people who produced the remains. It is thus synthetic rather than analytic.

As I have noted elsewhere[15], the synthesizer needs to answer four main questions about the people: who were they, where and when did they live, what were they like, and how and why did they get that way? To answer the first question, he selects a number of typical cultural assemblages and groups them into classes on the basis of their similarities and differences. He assumes that each class is indicative of a people and defines that people by formulating a complex of traits which is diagnostic of the class and which can also be used to identify the people as additional sites are dug. The class, the people, and the diagnostic complex are named after a site or a trait, e.g. Aurignacian or Bandkeramik[16].

In answering the question where and when the peoples lived, the synthesizer searches for stratigraphies, in which assemblages or other remains of several peoples lie one above another. He uses the stratigraphies in combination with other, less reliable evidence to construct a chronology showing the distributions of peoples in both space and time. He also collects radiocarbon samples from which to estimate calendric values for the temporal dimension of the chronology[17].

It is in answering the third question, "What were the people like?", that the concept of remnant settlement pattern is used. like the concepts of cultural assemblage, diagnostic complex, and stratigraphy, it is a heuristic device which enables the investigator to select the particular kind of data he needs in order to answer a question about the peoples he is studying. As we have seen, it is means of obtaining the data needed to reconstruct the people's ecological, cultural and social systems.

The comparable heuristic device used in answering the question "How and why?" is a pattern of change[18]. This refers to the changes in the people's systems which one is able to observe by comparing the successive layers of a stratigraphy or adjacent units in a chronology. Patterns of change answer the *how* of the question. Having reconstructed them, the synthesizer may proceed to explain *why* the changes took place by postulating processes such as migration, convergence, or adaptation to the environment[19].

The remnant settlement pattern, then, is a device for finding out about a people's ecological, cultural, and social systems in answer to the question "What?". Before using this device, the synthesizer will have only a partial knowledge of the people's systems, consisting of the diagnostic complex which he uses to identify the people, and the traits which mark its chronological system. The concept of remnant settlement patterns helps him to flesh out this skeleton by reconstructing other parts of the people's systems.

Activity assemblages, which form parts of remnant settlement patterns and are used to answer the question "What?", should not be confused with cultural assemblages, which are used to distinguish peoples in answer to the question "Who?". If a community carried out only a single activity at a site, as in the case of a separate burial mound, the activity assemblage from that site will also be a cultural assemblage. More commonly, though, the community will have performed several activities in the same site, such as cooking and tool making, and will therefore have deposited several different activity assemblages, in which case all of them, taken together, will constitute a cultural assemblage. Activity assemblages and the loci in which they occur may therefore be distributed within as well as among sites, and the investigator must plot both kinds of distribution in order fully to work out the remnant settlement pattern.

The network of loci and activity assemblages which comprise a remnant settlement pattern may be said to parallel the structure of a site and its cultural assemblages. The dual procedure of tracing activity loci among and within sites and recovering an activity assemblage from each locus is comparable to the procedure of demarcating the components of a site and excavating a cultural assemblage from each component.

Kinds of remnant settlement patterns

Three main kinds of remnant settlement patterns are to be found archaeologically, as follows:

1. *Dispersed pattern.* A people's activity loci may be widely distributed over the landscape, in which case the archaeologist will have to dig a number of sites in order to obtain the total range of that people's activity assemblages. This kind of a remnant pattern results when the residential communities, work groups, or other institutions into which a people were divided have moved from place to place, whether haphazardly or on a regular schedule. The groups may have been obtaining seasonally available game and plant

foods, if they lived by gathering, or they may have sought new pastures, if they relied upon herding.

2. *Concentrated pattern.* At the other extreme, each of a people's sites may contain virtually the same set of activity loci. This will be true if a people's residential communities and other institutional groups concentrated their activities in one spot, as simple farming communities tend to do, regardless of whether they have a sedentary or shifting pattern of residence.

3. *Co-patterns.* Some remnant patterns are partial, and have to be combined in order to obtain the total range of a people's activities. This condition is a result of the process of becoming civilized. As the process took place, simple farming communities gradually ceased to be uniform in culture. Some of the communities continued to maintain their previous peasant way of life, while others began to specialize in activities which were not, strictly speaking, essential for the survival of the people, such as religion among the Sumerians, burial among the Egyptians, and trade among the Minoans[20]. The surviving peasant communities may be termed *sustaining*, since they supported the people as a whole, and the specialized communities, *professional*, since they were composed of individuals whom we would consider professionals if they were alive today—priests, merchants, scribes, artisans, etc. The sustaining and professional communities each had a different settlement pattern, which I would suggest calling a co-pattern. Taken together, all of a people's sustaining or professional communities may then be termed a co-people, and each co-people may be said to possess a distinct co-culture. The combination of sustaining and professional co-cultures constitutes a *civilization* (Table 1)[21].

As a civilized people developed, its sustaining and professional activities were often transferred from residential communities to other kinds of institutions, such as farmsteads and factories. This sometimes blurred the distinction between the co-peoples, because they tended to live interspersed among each other. Among ourselves, however, the sustaining institutions are still predominantly rural and the professional institutions urban.

The terms *co-pattern*, *co-people*, and *co-culture* are used because neither sustaining nor professional institutions are fully self-sufficient. Neither can exist without the other; the two must be combined in order to obtain the total range of activities one normally finds in a people's culture. In my opinion, this is the essential difference between civilized and non-civilized peoples.

	Simple patterns	*Sustaining and professional co-patterns*
Cultural group	People	Sustaining and professional co-peoples = a civilized people
Diagnostics of the group	Complex	Complex
Cultural system	Culture	Sustaining and professional co-cultures = a civilization

Table 1 Implications of settlement patterns and co-patterns

Non-civilized peoples can also have symbiotic relationships, as when farming and herding peoples live side by side and exchange food products. However, such peoples differ in two ways from co-peoples who share the same civilization. First, each non-civilized people has its own diagnostic complex and, second, each has a separate settlement pattern, covering the total range of potential activities. By contrast, each pair of co-peoples has the same diagnostic complex, which may be considered indicative of their joint civilization; and, because of the high degree of specialization, neither co-people performs the total range of the civilization's activities (Table 1).

Reconstruction of ecological systems

Few people live in completely uniform environments. Instead, each has access to several micro-environments, differing somewhat in climate, topography, drainage, rocks, soils, flora, and/or fauna. A people tends to exploit several micro-environments, moving from one to another at different seasons of the year or else visiting each when in the need of one of its resources. In the earlier stages of human development, it did so haphazardly, but in the later stages, has tended to follow a regular schedule[23].

The remnant settlement pattern can be used to study the manner in which a people exploits its various micro-environments. As we have seen, it is a record of the manner in which a people's activities were distributed over the landscape. The archaeologist finds out by studying its activity assemblages which activities, if any, were carried on in each micro-environment, and thereby obtains an idea of the extent to which the people utilized that micro-environment[24].

By studying the relationships of activity assemblages to micro-environments, the archaeologist can also obtain clues as to a people's adjustment to its micro-environments. He may discover, for example, which kind of environment the people preferred to live in, and which they considered most suitable for other kinds of activities.

Reconstruction of cultural systems

We have seen that a cultural system (alternatively called a culture, a pattern of culture, or a civilization) consists of a series of activities which a people undertakes in order to maintain itself. Some of its activities result in procurement of materials from the environment[25], others result in the manufacture of artefacts[26], and still others involve the use of materials and artefacts in the people's daily life[27], but it is the activities themselves which are the central part of any cultural system.

When we study a people's activities, we find that they exhibit regularities which can be ascribed to the operation of custom and hence may be conceptualized as cultural norms. These norms are a way of expressing the essential nature of the activities and of the resultant materials and artefacts. When we speak of a people's activities, we mean its normal and customary ways of doing things, as opposed to variations and deviations from the norms.

There have been various attempts to portray the total range of a people's activities, both by ethnographers[28] and archaeologists[29]. One such attempt is given in Table 2[30]. Though over-simple, it will give an idea of the range of activities for which an archaeologist should look when recovering a settlement pattern. He will have to proceed differently depending upon the kind of settlement pattern with which he is dealing.

If the remnant settlement pattern is dispersed, it will be necessary to begin with a survey of sites, designed to identify the loci in which different activities were performed. Then the investigator must attempt to distinguish a network of loci including as many as possible of the activities listed in Table 2, and must recover assemblages from the loci[31]. If the network and assemblages

Ethnic systems	*Categories of subsystems*	*Subsystems*
Culture or civilization	Subsistence activities	Tool making Clothing Shelter Transportation Recreation Education
	Intellectual activities	Burial Religion Aesthetics Commerce Science Medicine Philosophy
Social structure	Residential institutions	Communities: bands, villages, towns, cities, etc. Divisions of communities: households, wards, suburbs, etc. Groups of communities: tribes, states, etc.
	Activity institutions	Work groups: hunting parties, workshops, pilgrimage groups, etc. Sustaining and professional enterprises: farms, estates, churches, factories, markets, etc. Professional associations: trade unions, guilds, scientific institutions, etc.
	Relational institutions	Interaction groups: political parties, committees, police, etc. Descent groups: families, clans, moieties, etc. Rank groups: age grades, classes, castes, etc.

Table 2 Cultural and social systems and sub-systems

appear to be typical of the people, he can use them to reconstruct the people's culture.

When studying a concentrated settlement pattern, the archaeologist should instead begin his search for its loci at a typical dwelling site. He should attempt to distinguish areas in that site which were used for different purposes, such as house sites, workshops, and graves. Each of these areas constitutes an activity locus. Many will be self-evident, but others will be discoverable only by statistical analysis of the distribution of the remains, designed to reveal concentrations of artefacts usable in a particular activity[32]. This will require extensive excavation. It is desirable to sample all parts of the site in an effort to obtain evidences of all the activities carried out there. The operation should subsequently be repeated at several other dwelling sites in a search for activities which the people's communities normally performed where they lived but of which there did not happen to be traces at the first site.

After learning about the activities performed at dwelling sites, the archaeologist should compare his results with a list of the people's potential activities, such as that in Table 2. When doing so, he may find that the pattern he has reconstructed is incomplete, in that it does not include certain activities which the people may be expected to have undertaken. He must assume that the missing activities were performed outside the people's dwelling places and must search elsewhere for the loci of these activities. He may find the missing loci grouped together in a single site or he may find each locus by itself, as when a burial mound was constructed some distance away from a place of residence and a cave was used for religious rituals. The search for the missing loci should continue until the entire range of potential activities has been covered.

An archaeologist who is studying the settlement co-patterns of a civilized people should follow a similar procedure, but rarely does so. He normally begins his search for activity loci in a site of the professional co-people, to which he is attracted by the wealth of its remains or by the opportunity of studying the rise of civilizations, if he is interested in the question "How and why?". All too often, he confines his research to that site and others like it, becoming so preoccupied with the professional co-pattern that he completely overlooks the sustaining co-pattern[33].

This is a mistake, if it is desired to provide a satisfactory answer to the question "What?". Knowledge of the sustaining co-pattern is just as important for that purpose as knowledge of the professional co-pattern. Let us suppose, for example, that the investigator has excavated a city site without finding any evidence that the inhabitants cultivated the surrounding terrain. He should then search for village sites in which the producers of the city dwellers' food might have lived. Let us further suppose that he finds the village sites but that they contain no evidences of burial or religion. He should then review the evidences for burial and religion in the city site. If it seems reasonable that the sustaining co-people went to the city for those purposes, he need search no further. If not, he will have to look elsewhere, possibly in caves, for the remnants of the sustaining co-people's burial and religion.

Reconstruction of social systems

We may define a social system (or structure) as a set of institutions into which a society is organized. Two words in this definition require clarification, *institution* and *society*. The term *institution* signifies any individual or group of individuals who form a socially recognized category. Some institutions are based upon common residence, others are organized to carry out particular activities, and still others are an expression of social relationships (Table 2). In all three cases, we may say that each institution is held together by the behaviour of interaction among its members, including both physical and mental behaviour. There are numerous regularities of behaviour, which we may conceptualize as social norms. The triad of institutions, interaction among the members of each institution, and social norms parallels the triad of materials and artefacts, activities, and cultural norms abstracted from the activities.

Just as a people consists of all individuals who perform similar activities, so a society may be said to consist of all individuals who belong to similar institutions. A people and a society may coincide, but not necessarily so. For example, western and eastern Europeans currently share the same civilization and hence may be said to constitute a single people, but they belong to separate societies, the one in the west being characterized by capitalist institutions and the one in the east by communist institutions.

Since peoples and societies may not coincide, as in this instance, the archaeologist must keep an open mind about the number of sets of institutions (social structures) he is going to have to reconstruct from a single settlement pattern. By definition, a settlement pattern can only have been laid down by a single cultural group (people), but the members of that group may have belonged to several different social groups (societies), as is the case today among the nationalist and communist Chinese, the West and East Germans, and the two Koreas and Vietnams.

An archaeologist working with historic peoples should have no difficulty in distinguishing instances such as these, since he will find clues to them in his documentary evidence. The prehistoric archaeologist faces a more difficult problem. He must reconstruct the social system or systems solely from archaeological remains.

In reconstructing residential institutions (Table 2), the prehistorian should logically focus on residential sites, i.e. on the sites of camps, villages, or cities, depending on the degree of social development of the people under study. In other words, he should restrict himself to that part of the remnant settlement pattern which is to be found in the site components where the people lived.

We may refer to this part of the settlement pattern as a *remnant residential pattern*—or co-pattern, if the people were civilized[34]. The investigator should compare the remnant residential pattern or co-pattern he finds in his sites with those present among historic and contemporary people; through this comparison he will be able to infer the residential institutions of the prehistoric people.

In reconstructing activity institutions (Table 2), on the other hand, the entire settlement pattern must be used. One should focus on each locus and attempt to infer the institution which operated there and which produced the activity assemblage to be found there. In a particular locus, for example, one may find structures and/or artefacts made for the purpose of government, such as meeting halls and symbols of political authority. One may then infer the institution which might have produced such paraphernalia by comparing them with the paraphernalia of historic and contemporary institutions.

Reconstruction of relational institutions (Table 2) is the most difficult of all, since settlement patterns are not pertinent. One can only hope to infer such institutions from the nature and content of the houses in which the members lived or the graves in which they were buried. If, for example, certain individuals lived in houses of appreciably larger size and were buried with appreciably more grave-goods than other individuals in a society, it is reasonable to assume that they belonged to different rank groups, though the nature of the groups may be difficult, if not impossible, to determine unless one is able to project historically known groups back into prehistory, as is done in studying the American Indian[35].

Settlement patterns, civilization, and urbanization

The process of becoming civilized is often equated with the process of becoming urbanized, notably by Childe in his concept of the urban revolution[36]. Childe, however, subsequently recognized that a people can be civilized without also being urbanized[37].

The fact is that civilization and urbanization are independent variables. The professional and sustaining co-peoples who share civilization may both be urban, both rural, or one may be urban and the other rural. For example, the early Sumerian co-peoples of Mesopotamia appear to have both lived in cities, the professional co-people in the city centres and the sustaining co-people in the peripheries and in suburbs[38]. The classic Maya co-peoples of Mesoamerica were both rural, the professional co-people living in ceremonial centres and the sustaining co-people in satellite villages[39]. Among the classic Greeks and Romans, on the contrary, the professional co-people lived in cities and most of the sustaining co-people, in the country[40].

This variability is understandable when one realizes that civilization and urbanization are different kinds of processes. Civilization refers to the development of a people's activities, and hence is cultural. Urbanization, on the contrary, refers to the development of an institution, the city, and hence is social. It is not surprising, therefore, that the two may develop differently.

Archaeologists sometimes differ as to whether a given people was civilized and/or urbanized. For example, Coe[41] considers the Olmec to be the first civilized people in the New World, while Sanders[42] argues that they were not civilized. Such disagreements may arise, as in this case, because the authors are relying upon inferences concerning the nature of a people, which are uncertain, instead of using archaeological evidence to reach an empirically based conclusion.

The concepts of remnant settlement and residential patterns offer an empirical basis for distinguishing civilized and urban peoples respectively. The presence of sustaining and professional co-patterns can be used to identify civilizations, and the existence of a sufficiently large and concentrated residential pattern, to identify urbanization. It would only be necessary to reach agreement as to how large and concentrated a remnant residential pattern ought to be for the purpose.

Notes

1 This paper is extracted from Rouse, I. (In press). *Introduction to Prehistory: a Systematic Approach.*

2 Benedict, R. (1934). *Patterns of Culture*. Boston.

3 Chorley, R. J. and Haggett, P. (eds.) (1967). *Models in Geography*. London.

4 Trigger, B. G. (1968). The determinants of settlement patterns, *in* Chang, K. C. (ed.) *Settlement Archaeology*. Palo Alto, p. 55.

5 Willey, G. R. (1953). Prehistoric settlement patterns in Virú Valley, Peru, *Bull. Bureau Amer. Ethnology*, 155.

6 Willey, G. R., editor (1956). Prehistoric settlement patterns in the New World, *Viking Fund Publ. in Anthrop.*, 23.

7 Trigger, B. G. personal communication.

8 Trigger, B. G. (1967). Settlement archaeology: its goals and promises, *Amer. Antiq.*, **32**, pp. 149–60.

9 Binford, L. R. (1964). A consideration of archaeological research design, *Amer. Antiq.*, **29**, p. 424.

10 See, for example, Glover, I. C. this volume, pp. 157–64,

11 In Rouse, I. (In press). *op. cit.* the term *archaeology* is limited to the analytic approach and the synthetic approach is called *pre-history*.

12 Chang, K. C. (1967). *Rethinking Archaeology*. New York. pp. 38–56.

13 Piggott, S. (1965). *Ancient Europe from the Beginnings of Agriculture to Classical Antiquity*. Edinburgh. Fig. 84, p. 151.

14 Willey, G. R. and Phillips, P. (1962). *Method and Theory in American Archaeology*. Chicago. pp. 21–2.

15 Rouse, I. (1965). The place of "peoples" in prehistoric research, *J. Roy. Anthrop. Inst.*, **95**, pp. 1–15; see also Rouse, I. (In press). *op. cit.*

16 McKern, W. C. (1939). The mid-western taxonomic method as an aid to archaeological culture study, *Amer. Antiq.*, **4**, pp. 301–13; Childe, V. G. (1956). *Piecing together the Past: the Interpretation of Archaeological Data.* London. pp. 15, 173.

17 Clark, J. G. D. (1960). *Archaeology and Society*. London and New York. pp. 132–68.
18 Caldwell, J. R. (1958). Trend and tradition in the pre-history of the eastern United States, *Mem. of Amer. Anthrop. Assoc.*, **88,** pp. 1–2; Barth, F. (1967). On the study of social change. *Amer. Anthrop.*, **69,** p. 665.
19 Kroeber, A. L. (1963). *Anthropology: Cultural Patterns and Processes*. New York.
20 Rouse, I. (In press). *op. cit.*
21 This table is adapted from Rouse, I. (In press). *op. cit.* Table 5.
22 Coe, M. D. and Flannery, K. V. (1964). Micro-environments and Meso-american pre-history, *Science*, **143,** pp. 650–4.
23 MacNeish, R. S. this volume, pp. 67–93.
24 Flannery, K. V. (1968). Archaeological systems theory and early Meso-america, *in* Meggers, B. J. (ed.) *Anthropological Archaeology in the Americas*. Washington, pp. 67–87.
25 Flannery, K. V. (1968). *op. cit.*
26 Rouse, I. (In press). *op. cit.* Fig. 5.1; Clarke, D. L. (1968). *Analytical Archaeology*. London. Fig. 19.
27 Clarke, D. L. (1968). *op. cit.*
28 Anonymous (1952). *Notes and Queries on Anthropology*. 6th ed. London; Murdock, G. P. *et al.* (1950). *Outline of Cultural Materials*. 3rd rev. ed. New Haven.
29 Clarke, J. G. D. (1960). *op. cit.* Fig. 25; Clarke, D. L. (1968). *op. cit.* Fig. 13.
30 This table is adapted from Rouse, I. (In press). *op. cit.* Table 5.1.
31 MacNeish, R. S. (1962). *Second Annual Report of the Tehuacán Archaeo-logical-Botanical Project*. Andover, Mass.
32 Tugby, D. J. (1965). Archaeological objectives and statistical methods: a frontier in archaeology, *Amer. Antiq.*, 31, p. 11.
33 The specialists in Mesoamerican civilisations have only recently begun to excavate the sites of the sustaining as well as the professional co-peoples, e.g. Willey, G. R. (1966). *An Introduction to American Archaeology I*. Englewood Cliffs. To my knowledge, none of the prehistoric civilisations of the Old World has been studied this way, except for Shang in China: Chang, K. C. (1968). *The Archaeology of Ancient China*, 2nd ed. New Haven.
34 MacNeish, R. S. this volume, p. 70, prefers to call it a community pattern.
35 e.g. Longacre, W. A. (1964). Archaeology as anthropology: a case study, *Science*, **144,** pp. 1454–5.
36 Childe, V. G. (1936). *Man Makes Himself*. London.
37 Childe, V. G. (1950). The urban revolution, *Town Planning Review*, **21,** pp. 3–17.
38 Frankfort, H. (1951). *The Birth of Civilization in the Near East*. London. pp. 57–8; Oppenheim, A. L. (1964). *Ancient Mesopotamia: Portrait of a Dead Civilization*. Chicago. pp. 109–17.
39 Willey, G. R. (1966), *op. cit.* pp. 117–24.
40 Frankfort, H. (1951), *op. cit.*
41 Coe, M. D. (1968). *America's First Civilization: Discovering the Olmec*. New York.
42 Sanders, W. T. and B. J. Price (1968). *Mesoamerica: the Evolution of a Civilization*. New York, pp. 115–34.

C. C. TAYLOR

The study of settlement patterns in pre-Saxon Britain

Geographers have given many of their techniques and ideas to archaeologists, on the whole greatly to the advantage of the latter. Amongst these techniques the reconstruction and study of settlement patterns is one that is often undertaken. Yet this writer doubts if such work is valid, and believes that the recovery of the pattern of settlement of pre-Saxon societies in Britain is something that archaeologists cannot achieve.

The basic requirements for the successful study of settlement patterns in any period are that the recoverable pattern must be reasonably complete in a given area, and that there must be a fairly accurate idea of the form, size, purpose and organization of most of the settlements in the same area. This is something that can be easily achieved for the last two or three centuries and it can sometimes be done as far back as the Late Saxon period[1]. But when one moves into periods when archaeological material and techniques are the only form of evidence it becomes impossible to ascertain either the true pattern of settlement or to recover the types of settlement with any certainty. This is not the fault of archaeologists. It is the nature of the evidence which though useful for the understanding of many aspects of the past is, and always will be, defective for the study of settlement patterns.

It should be obvious that we can never hope to have anything like the total number of settlements of any period before the tenth century A.D. in any area large or small except in rare cases. The one exception in Britain is that of the fenlands of eastern England where the existence of a near complete Romano-British pattern of settlement is recoverable in outline[2]. Elsewhere the destruction by successive peoples of the habitations of their predecessors results in the obliteration in archaeological terms of the total settlement pattern. The evidence for prehistoric and Roman settlement in this country falls into two zones which are perhaps more important than the old concept of Lowland and Highland Zones, evolved by the late Sir Cyril Fox. These are the Zone of Survival and the Zone of Destruction. The Zone of Survival comprises small and often upland areas where later peoples never lived, and only rarely cultivated. In this zone prehistoric and Roman settlement sites have remained reasonably intact, at least into the present century. The

Zone of Destruction comprises the greater part of the country where later people have largely or partly destroyed the evidence. In this zone we can and do recover part of the pattern of settlement by various means, and in the future we shall recover much more. But we shall never obtain anything like the complete pattern. This inability to recover the total pattern is made worse by the fact that we have no idea how incomplete our evidence is. If we knew that any given pattern was 75% or even 50% complete it would help, but we never do. The recoverable number of settlements may be only 1% of the original total, or it may be as much as 90%. It is very difficult to draw valid conclusions about patterns of settlement from such evidence.

An actual example of the problem is the remarkable work done by P. J. Fowler over many years on the Fyfield and Overton Downs of Central Wiltshire[3]. There, a large area of open downland, apparently untouched by post-Roman activity, is dotted with prehistoric and Roman remains. But the downland is not untouched. It has medieval and later ploughing over parts of it, there is a medieval farmstead, and medieval and later ponds, tracks, etc. exist, all of which have destroyed earlier remains. Even when the extant sites are excavated they continually reveal parts of earlier sites of which no trace exists on the ground. Thus when some Roman fields were excavated an Iron Age settlement was discovered which the fields had obliterated, while the settlement itself had cut through a Beaker cemetery whose settlement is unknown. Short of excavating the entire downland, an impossible task, there is no way of recovering the true pattern of settlement in the area. Even if this were possible these downlands cannot be considered in isolation at any period. The lower slopes and adjacent valley bottoms inhabited by people from medieval times onwards were also occupied by prehistoric and Roman people as the incomplete records of chance finds, excavation and air-photography show only too well. Indeed there is evidence that in the chalklands of Wessex during the later prehistoric and Roman periods the valley bottoms were more intensively settled and cultivated than the uplands. It is only the lack of large-scale human activity on these uplands since the end of the Roman period that has preserved the evidence of earlier occupation on what has perhaps always been marginal land. This has resulted in the quite wrong impression that most people in Wessex during the pre-Saxon period lived on higher ground[4].

In mountainous areas too the same applies. The distribution of prehistoric and Roman sites in, for example, Caernarvonshire is interesting in this respect[5]. The higher mountains are devoid of remains except for major defensive sites. The rich lowlands along the coast are also largely empty, except for a scatter of small finds and a few important Roman military sites. Between these two areas, on the lower slopes of the mountains are many settlements, often of considerable complexity. This pattern is meaningless in terms of prehistoric and Roman occupation and is only explainable in terms of later destruction. On the higher ground nothing is found because prehistoric and Roman people could not or did not want to live there. But on the lower slopes nothing is found because it has been completely destroyed by later occupation. All we have left to study are the mutilated edges of a cloth which was once a complete settlement pattern.

The same is true of other areas in a reverse way. Air-photography of the river gravels of Britain has revealed complex evidence for prehistoric and Roman occupation often on such a scale that it is impossible to interpret without massive excavation[6]. Yet it would be wrong to use the evidence of river gravel sites alone in an area to reconstruct settlement patterns. The gravel areas are usually relatively narrow strips of land along the rivers. But what of the areas immediately beyond? The soils there are often not the type which reveal soil or crop marks from the air and the lack of surface archaeological material is not evidence of absence of occupation. One of the most complex sites in the Peterborough region has not produced a single sherd of pottery of any date[7]. Yet it would be misleading to ignore these areas merely because we have no evidence. In some places, such as the upper Nene valley, where the valley-side soils do reveal crop marks many prehistoric and Roman sites are known.

These problems have always been with archaeologists though they are often not sufficiently appreciated. However recent developments have added new difficulties. The main one is the sheer bulk of new archaeological evidence which is now becoming available to us. Well-organized local archaeological groups are discovering large numbers of new occupation sites on every type of soil and in all positions. Work such as the watching of motorway constructions, natural gas and oil pipelines and urban and rural developments is producing a multitude of new archaeological sites. Systematic air-photography of selected areas, over long periods of time, in all light, crop and weather conditions produces a constant stream of new sites from both areas already known for their archaeological wealth and those previously empty of recorded occupation. This new material, so large that some archaeologists have suggested that we give up trying to assess its actual meaning for the time being, indicates that prehistoric and Romano-British people in Britain lived almost everywhere and were seemingly not controlled by their environment to the extent we often suppose. In addition this evidence is still extremely uneven in its distribution, and dependent upon the interest and activities of local archaeologists, on crop and soil conditions and on purely chance discoveries resulting from modern developments. An example of this may be seen in the concentration of archaeological sites and finds around Cambridge[8]. This is not, as is often suggested, because of the importance of Cambridge as a river crossing at the edge of the fens, but merely the combination of large-scale nineteenth century building development, gravel and phosphate digging, with the existence of a flourishing contemporary archaeological society whose members took the trouble to record the sites and finds. The second development which adds to the problem is the rate of destruction. Not only do we have to take into account the vast amount of evidence which has been destroyed in the last 1500 years, but today we are living through a time of the greatest destruction of the remains of the past which has ever been seen. In spite of all the careful work of discovery and recording noted above only a small proportion of the evidence is being found. The rest is irrecoverably lost. The present rate of destruction is now so fast that many of us who are professionally engaged in recording sites believe that there will be little work to do by 2000 A.D. These facts pose serious problems for most aspects of archaeological study in this country but they supply a completely inadequate basis for the reconstruction of settlement patterns even in distributional terms.

When we come to try to assess the size, function and relationships of the component parts of the pattern of settlement, the problems are even greater. The amount of excavation required to ascertain these facts, even in a small area, is almost beyond comprehension and certainly beyond our resources. The present method of excavating a specific type of settlement and then assuming that others of the same date and general form have a similar purpose, population, etc., is extremely dangerous. Also it is now clear that there was as much variety in size, form and organization of settlements in the later prehistoric and Roman periods as there is today. The different types of Iron Age and Roman settlements now known from Wessex are remarkable. Everything from single farmsteads through hamlets and villages to towns and near urban fortresses can be shown to have existed[9]. The problems involved in ascertaining by archaeological means how these varied settlements actually function in terms of the pattern of settlement are enormous.

Thus the study of settlement patterns in pre-Saxon Britain seems to this writer to be impossible either now, or in the foreseeable future. The concept of erecting models for the study of settlement patterns, so favoured by geographers and by some archaeologists, would also seem to be fraught with difficulties, for whatever model is proposed, there is no way, nor will there ever be a way, of testing it.

Notes

1 Taylor, C. C. (1967). Whiteparish, a study of the development of a forest edge parish, *Wilts. Archaeol. Mag.*, **62,** pp. 79–102.
2 Phillips, C. W. (ed.) (1970). *The Fenland in Roman Times.* Roy. Geogr. Soc., Research Series No. 5.
3 Fowler, P. J. (1962–7). The Archaeology of Fyfield and Overton Downs, *Wilts. Archaeol. Mag.*, **57,** pp. 98–115; **58,** pp. 342–50; **62,** pp. 16–33; Fowler, P. J. (1969). Fyfield Down, *Curr. Archaeol.*, **16.**
4 Taylor, C. C. (1970). *The Making of the English Landscape; Dorset.* London, pp. 25–39.
5 R.C.A.M. (Wales). (1956–1964). *Caernarvonshire*, I–III. London.

6 R.C.H.M. (Eng.). (1960). *A Matter of Time*. London. Figs. 6–10.
7 R.C.H.M. (Eng.). (1969). *Peterborough New Town. A Survey of the Antiquities*. London. Fig. 7.
8 Fox, C. (1923). *The Archaeology of the Cambridge Region*. Cambridge.
9 Bowen, H. C. and Fowler, P. J. (1960). Romano-British Settlements in Dorset and Wilts *in* Thomas, A. C. (ed.). *Rural Settlement in Roman Britain*. C.B.A. Research Report, 7. London.

Section 2: The influence of mobility on non-urban settlement

BRIDGET ALLCHIN

Hunters or pastoral nomads? Late Stone Age settlements in Western and Central India

The complex of stone industries variously described as Microlithic, Mesolithic, and Late Stone Age are widely distributed throughout the Indian sub-continent. There are marked regional differences both in the choice of available raw materials and in technique.

Certain among the remaining body of Late Stone Age industries can be associated with reasonable certainty with particular ways of life. The coastal industries from around Bombay, for example, which frequently occur at sites on islands and headlands, can be fairly confidently ascribed to fishing people, perhaps to the ancestors of the present day fishing communities of the same localities. The Late Stone Age industry found in rock shelters and at open air factory sites in the uplands of south-eastern Ceylon can be attributed to people with a mixed hunting and gathering economy, closely comparable to that of the Vedda tribes of recent historic times. Temporal, regional and cultural sub-divisions of the whole complex of blade industries have been discussed at some length elsewhere[1].

The regions with which I propose to deal in this short paper are the western half of the hills of Central India, and the plain of Gujarat, with certain extensions into western Rajasthan. The Central Indian hills stretch in a broad belt across India from the borders of Gujarat on the west to West Bengal on the east, effectively dividing the plains of the Indus and the Ganges river systems from the plateau of peninsular India. In contrast to the Ganges and all the major rivers of peninsular India which flow to the east, the Narmada (Narbada) and Tapi (Tapti) flow west across the plain of Gujarat into the Gulf of Cambay. Thus there is a clear structural relationship between the Central Indian hills and the Gujarat plain. The plain extends along the eastern shore of the Gulf of Cambay, where its maximum width is about a hundred miles, and northward to meet the Rajasthan Desert. Today it is one of the most fertile and productive regions of India. Much of Central India on the other hand is rocky and thinly populated. Here live many of the so called tribal people of India who still maintain, or until recently maintained, a variety of archaic forms of social and economic life, many of which have been put on record. The whole complex of these people and their varied culture are of the greatest interest to the prehistorian in attempting to understand past cultures, but at present none of them can be taken as representing in its entirety any particular cultural group from the archaeological record. It is possible that relationships between past and present cultures may emerge as the result of further research.

Western Central India has long been known for its proliferation of microlithic sites. These include vast numbers of surface sites, ranging from factories covering several acres to spots where a passing hunter appears to have made a few tools and gone on his way. Due to the climate and the hard rocky country stone tools tend to remain on the surface. Late Stone Age tools are also found in rock shelters which are fairly numerous in certain localities. Here there is a fairly obvious but ill-defined relationship between the stone industries and a characteristic range of rock paintings. Some rock shelters contain a certain depth of occupation deposit, but only a few have been systematically excavated—we shall return shortly to consider one of these. Much random collecting has gone on during the past century, and the general tendency has been to assume that this was the culture of hunters and gatherers. Recently the quantity of tools and the standard of craftsmanship have caused a number of Indian archaeologists to have reservations about this.

Late Stone Age sites are also widely distributed in Gujarat and Rajasthan, notably in northern Gujarat and western Rajasthan—on the edge of the desert—where they are associated with ancient sand dunes and sources of water of limited duration following the monsoon. The sites, several of which have been excavated, appear to be temporary but regular camping places. There are no rock shelters owing to the nature of the country, and for the same reason stone tools do not tend to remain on the surface here, for this is an alluvial plain constantly affected by deposition and erosion, both by water and wind.

At Langhnaj, in northern Gujarat, the first of the sand dune sites to be excavated in the 1940's[2], only the bones of wild animals were found. These correspond to the fauna of remote parts of Western India today, and appear to be the food débris of hunting people[3] who camped intermittently on the dunes over a long period of time. The continuity of occupation and the general character of the industry appear to have survived two minor climatic fluctuations, and to extend from a period contemporary with, or perhaps considerably antedating, the Harappan civilization of adjacent regions forward into the Iron Age. Two sites of a similar nature, recently excavated in Rajasthan seem to show comparable results[4]. As yet only preliminary reports of the excavations are available, and opinions appear to differ as to whether the animal bones include those of domestic species.

The most positive evidence on this question comes from the excavation of Adamgarh rock shelter, near Hoshangabad, in Central India[5]. Here together with some thousands of microlithic blade tools characteristic of the Late Stone Age of Central India were quantities of bones of domestic species —humped cattle (*Bos indicus*), water buffalo, goat, sheep, pig and dog—in an occupation deposit approximately one metre in depth.

It would appear that a systematic study of the animal bones from a Late Stone Age site should indicate to what extent its occupants were hunters or pastoralists at any period of its occupation. However, if we turn to modern pastoral peoples we find that such a simple distinction based upon material remains is not borne out by analogy with their practices. Such residual groups of hunters and gatherers who still survive, or accounts of whom were put on record during the last century and a half in Central India, are clearly too much influenced and restricted in their movements by outside pressures to provide a meaningful analogy, except in most general terms. Pastoral people, however, still have an important economic function in the rural society of India generally, particularly in the dry regions we are discussing. Their presence alongside settled communities in Rajasthan has already been noted by Leshnik[6] and others, as being a significant indicator of possible relationships in the past. Throughout India there are communities or castes who specialize in the management of flocks and herds, either on behalf of villages to which they are attached, or as independent owners or agents who herd them in remote or uncultivated areas and bring their young stock to fairs and markets at certain times of year to sell to farmers on the plains as draught oxen or milking cows. Today such pastoral people operate in a kind of symbiotic relationship with settled agriculturalists with whose yearly routine their movements are closely integrated.

During two seasons' archaeological field work in Gujarat in 1967–8 and 1969 I had the opportunity to watch one of the major groups of pastoralists in the region, the Bhārvāds, and although these observations are incomplete they seem to me to be of interest in considering past cultures and their inter-relationships. The Bhārvāds keep large herds of cattle and sheep, together with a certain number of donkeys, camels, dogs, etc. During the monsoon (June–September approx.) they graze on the salt marshes around the top of the Gulf of Cambay, and in northern Kathiawar along the edge of the Rann of Cutch, following the available supplies of grass and fresh water. As these become exhausted they move eastwards to the agricultural lands of Gujarat, across which they move along long established routes. They have regular camping places to which they return each year, different groups sometimes using the same site at intervals of a few days. Local farmers welcome their arrival, and are said to pay the herdsmen a small sum of money to pen their animals overnight on fields which need to be manured before cultivation for the next season's crops. They sell curds and clarified butter in the villages and small towns near which they camp, and also lambs for slaughter. Today these are bought in large numbers by dealers from Bombay who take them to the city by train. I did not see them selling young cattle, but this definitely takes place at certain stages of their journey. After a leisurely progress across the plain they move into the forested hills of Central India where they spend the winter. It is said that in the increasingly dry conditions of late winter and spring they herd their animals near perennial rivers such as the Narmada and its tributaries. The herds consist of young animals and breeding stock, but no excessively old or ailing animals are to be seen. There can be little doubt that as occasion demands these are disposed of to the leather-working communities of towns and villages, who in turn discreetly get rid of the bones. At none of the camping places of the Bhārvāds, which we saw on the plain of Gujarat, were cattle bones or the bones of any other animal to be seen. It seems unlikely that young cattle are ever slaughtered for food, as they are such a valuable source of income alive, but there is little doubt that they kill lambs for food from time to time. How far the absence of bones at camping places is in deference to the feelings of orthodox Hindu villagers, and how far these practices are modified when the Bhārvāds are in remote forests or marshes where they have no one else to consider are questions which deserve further careful study. But as far as the camp sites of the densely populated plain are concerned, all the future archaeologist could glean from them would be a few fragments of pottery, and broken or discarded objects from the rather limited range that the Bhārvāds carry with them—jewellery, knives and simple cooking equipment. Their real wealth is in their flocks and herds, the heavy gold ornaments of the women, and, it is said, in their bank accounts.

I shall not try to draw a close analogy between the Bhārvāds and the Late Stone Age inhabitants of Western India or any other cultural group, but simply put these observations forward in the hope that they may provide food for thought.

Notes

1 Allchin, B. (1966). *The Stone Tipped Arrow.* London; Allchin, F. R. and B. (1968). *The Birth of Indian Civilization.* London.

2 Zeuner, F. E. (1952). The microlithic industry of Langhnaj, Gujarat, *Man*, **182**, pp. 1–3.

3 Sankalia, H. D. (1965). Excavations at Langhnaj: 1944–63, Part I, *Deccan College Building Centenary and Jubilee Series.* Poona; Clutton-Brock, J. (1965). *in* Sankalia, H. D. *op. cit.* Part II.

4 Leshnik, L. S. (1968). Prehistoric Exploration in North Gujarat and parts of Rajasthan, *East and West*, N.S., **18**, pp. 3–4; Misra, V. N. (1968). Excavation at Tilwara, Dist., Barmer, *Indian Archaeology* 1967–68, *a review*, **48**, pp. 39–42; Misra, V. N. (1968). Excavation at Bagor, Dist., Bhilwara, *Indian Archaeology* 1967–68, *a review*, **49**.

5 Joshi, R. V. (1968). Late Mesolithic Culture in Central India, *in La Préhistoire, Problèmes et Tendances*. Paris. pp. 245–54.

6 Leshnik, L. S. (1968). *op. cit.*

S. BÖKÖNYI

Zoological evidence for seasonal or permanent occupation of prehistoric settlements

A perennial problem of archaeology is how to determine whether a settlement was occupied permanently (all the year round) or seasonally (only during certain parts of the year). In the temperate zone of Europe and in south-west Asia this question arises especially in connection with the settlements of the Early Neolithic and those of the Late Migration Period. The former is the period of the beginning of food production, and the latter is the period of the stabilization of peoples of the Migration Period. In both cases the permanency or seasonality of the settlements is a key problem.

In attempting to solve this problem zoology can give some help to the archaeologist through the study and interpretation of animal bone samples from settlements. Attempts have been made in this field on New World bone samples[1], but no systematic evaluation of fauna has yet been obtained from Old World settlements. For this reason in this paper I would like to discuss the biological bases of these investigations, and to demonstrate their limits, difficulties and dangers. I also intend to give a short account of the results of the analyses of fauna associated with certain settlements in the temperate zone of Europe and south-west Asia in order to elucidate the nature of the occupation of these sites.

The problem can be approached from two directions: from the fish and avifauna on the one hand, and from the mammals, especially the domesticated species on the other. Concerning the fish and avifauna, the starting point is the known seasonal migration of certain species. The aim of these migrations is either to wander to the south before the cold of the winter and then back again to the north with the spring (birds), or to complete certain phases of a particular process of propagation (fishes, among which certain species, e.g. salmon spend the greater part of their life in the sea and enter the rivers at spawning time, and eels who do the opposite).

According to the palaeontological data these migrations have not changed in character in the last 12,000–15,000 years. This is particularly true of birds, whose seasonal migrations have been exhaustively studied for a long time, and are now well established. In most cases, ornithologists know exactly where each of the migrating species spends the different seasons of the year. In this way, if their remains occur in a given archaeological site they provide good evidence for the occupation of the settlement in a certain season.

The migrations of fish, particularly those of south-west Asia, are not as well known as are those of the birds. For example, in the Early Neolithic site of Tepe Sarab (West Iran, Kermanshah Valley, *c.* 5000 B.C.) remains of two individuals of a tiny fish species called *Strongylura strongylura van Hasselt* have been found. Handbooks mention that this species lives in the Persian Gulf (and in the Indian Ocean) and occasionally penetrates into the Tigris and Euphrates, and also their tributaries. The problem is, however, that Iraqi ichthyologists do not know—or at least do not mention—in which season of the year this happens[2].

Nevertheless, the occurrence of migrating fish and birds can give valuable information about the seasons in which the settlements were in use. Their great disadvantage is that their bones are generally very poorly represented in prehistoric sites of temperate Europe and south-west Asia. There are a few exceptions in which fish bones represent a very high proportion of the animal remains[3], though the archaeological age of these is still a problem, e.g. some settlements of the Iron Gates Gorge of the Danube which precede the earliest Pottery Neolithic, and certain Neolithic settlements of the southern Ukraine[4]; in addition one or two settlements of the Körös culture occurring in a swampy environment in southern Hungary have a large number of bones of waterfowl[5].

The situation regarding mammals is much better since the bones of mammals represent the overwhelming majority of animal remains of settlements. Conclusions can easily be drawn from the presence or absence of certain age groups (which can be determined on the grounds of the set of teeth, the sutures of the skull and the terminals of the long bones) as to the occupation of the settlements in the various seasons of the year.

Wild mammals bear their young in a strictly determined season of the year, in which there is the best chance for the young to survive and mature. The young of modern improved breeds of domestic animals are born all the year round since Man can secure the conditions necessary for their breeding. This is not the case with primitive domestic animals who produce young only once a year just like wild mammals. In this respect there are very good data from Iran where sheep and goats, for example, lamb between January and March even in modern times. Since the animal breeding practised by the peasants of the Middle East and most parts of the Balkans represents roughly the medieval stage of animal breeding in Europe, one is justified in supposing that a similar situation prevailed throughout Europe and south-west Asia in prehistoric times and also in the Migration Period. In central Europe—because of its colder climate in comparison to the Middle East—lambing may have been a little later, between February and April, and calving in April-May.

In general the season for producing young is the spring. Each species has its strictly determined period (which can indicate slight geographical differences, as we have seen) when not only the climatic but also the food conditions are the best for a given species. And since there has been no drastic climatic changes in a given region since the Pleistocene one can suppose that the time of giving birth for wild and domestic mammals has remained unchanged too.

Table 1 shows the periods of producing young for the most common (in the bone samples of settlement) wild mammal species—among them are the wild forms of the most important European domestic animals (except cattle)[6].

species	*periods of producing young*
wolf—*Canis lupus* L.	March-April
fox—*Vulpes vulpes* L.	April-May
wild cat—*Felis silvestris* Schreb.	April-May
tarpan (wild horse)—*Equus ferus* Boddaert	April-May
wild swine—*Sus scrofa* L.	April-May
red deer—*Cervus elaphus* L.	May-June
roe deer—*Capreolus capreolus* L.	May-June
moufflon—*Ovis musimon* Pall.	March-April
bezoar goat—*Capra aegagrus* Erxl.	April-May

Table 1 Periods of producing young.

Knowing the period of producing young and having determined, on the bases of the dentition, the sutures and the ends of long bones, how old the animal was when it was killed, one can also determine—within a time range of one or two months—the season of the year when the animal was killed. And the fact that animals were killed on the site at a certain period is good evidence that the settlement was occupied in that season.

For this purpose remains of small domestic and wild mammals (dog, pig, sheep, goat; wolf, fox, wild cat, wild swine, roe deer, moufflon and bezoar goat) are the best indicators. In these species the eruption and changing of the various teeth occur at shorter intervals, and the different phases of the whole process takes place within shorter time-spans than in large species. In these species the entire dentition develops within two to two and a half years, whereas the large species sometimes need four and a half years. The remains of young animals (under one year) are particularly suitable, since in later years the time-spans between changes become increasingly longer.

It is unnecessary to enumerate all characteristics that can be used in age determination, since they are described in detail in various handbooks. For general information, however, Table 2 contains the data of the most important phases in the development of the dentition of the most common domestic and wild mammals.

Using Tables 1 and 2 together it is relatively easy to determine from the animal remains in which seasons of the year a given settlement was occupied.

eruptions	*dog*	*pig*	*sheep*	*goat*	*cattle*	*horse*	*roe deer*	*red deer*
I_1 occurs	1 m	1–4 w	1 w	wb	wb	1 w	1–4 w	wb
I_2 occurs	1 m	6–10 w	2 w	wb	wb	3–8 w	1–4 w	wb
I_3 occurs	1 m	6–10 w	3 w	wb	wb	5–9 m	1–4 w	wb
C occurs	1 m	wb	4 w	3 m	1–14 d	2–6 m	1–4 w	2–4 m
I_1 is shed	4–6 m	11–14 m	17–18 m	14–16 m	18 m	2½ y	5–7 m	15 m
I_2 is shed	4–6 m	16–20 m	24–25 m	19–22 m	30 m	3½ y	9–11 m	15–16 m
I_3 is shed	4–6 m	6–9 m	32–35 m	21–26 m	33 m	4½ y	12 m	18–20 m
C is shed	5–6 m	6–9 m	42–45 m	29–36 m	45–54 m	4 y	13–14 m	21 m
P_1 occurs	5–6 w	3–6 m	1–4 w	3 m	1–14 d	wb	wb	2–4 m
P_2 occurs	5–6 w	8–30 d	1–4 w	3 m	1–14 d	wb	wb	2–4 m
P_3 occurs	5–6 w	3–5 w	1–4 w	3 m	1–14 d	wb	wb	2–4 m
P_4 occurs	5–6 w	4–7 w	—	—	—	—	—	—
P_1 is shed	—	—	24 m	17–20 m	28 m	2½ y	12–15 m	27–28 m
P_2 is shed	16 m	14–16 m	24 m	17–20 m	28 m	2½ y	12–15 m	27–28 m
P_3 is shed	16 m	14–16 m	24 m	17–20 m	34 m	3½ y	12–15 m	27–28 m
P_4 is shed	16 m	14–16 m	—	—	—	—	—	—
M_1 occurs	4–6 m	5–6 m	3 m	3 m	5–6 m	1 y	5–7 m	4–5 m
M_2 occurs	5–6 m	10–13 m	9 m	8–9 m	18 m	2 y	6–10 m	11–12 m
M_3 occurs	6–7 m	17–22 m	18 m	18–24 m	28 m	4 y	12–17 m	24–28 m

d = day; w = week; m = month; y = year; wb = with birth.

Table 2 Phases of dentition (based on Duerst[a] and Habermehl[b])

(*a*) Duerst, U. J. (1926). Vergleichende Untersuchungsmethoden am Skelett bei Säugern *in* Abderhalden, O. *Handbuch der biologischen Arbeitsmethoden*, 7, pp. 169–210.
(*b*) Habermehl, K.-H. (1961). *Die Altersbestimmung bei Haustieren, Pelztieren und beim jagdbaren Wild*. Berlin-Hamburg.

On the other hand it is not so easy to prove that a settlement was not occupied in certain other seasons.

It would be simple to say that the absence of bones of a certain age group proves that man did not live on the site in the corresponding season of the year. However, if the excavation of a settlement yields only a few bones, the evidence cannot be considered to be at all conclusive. Only if one is in possession of a large bone sample can the complete absence of certain age groups be significant evidence in this respect.

It is impossible, merely on the basis of the relative frequencies of the age groups, to determine whether the whole population lived in a given settlement all the year round or not. One need only think of transhumance, a special form of migrating pastoralism, in which a large part of the inhabitants spend the whole year in the settlement, and only the herdsmen migrate with the herds in specific seasons. One would expect this to show up in the bone sample in that certain age groups would occur in higher, and others in lower proportions. Unfortunately this is not so, since the killing period (or dying period; in most of the cases one can not determine whether the animal remains found in a site are from animals which have been killed or have died) for animals depends also on many other factors: for example:

1 Local taste. Man generally killed young animals rather than old ones, but not too young since these would supply him with too little meat.

2 Fodder situation. In prehistoric times many young animals, especially the superfluous males, were killed in the autumn because of the winter fodder shortage, and their meat was preserved by smoking or other method for the winter (Early Neolithic sites of south-west Asia and the Balkans).
3 Animal diseases very often increased in the same season each year, causing the death of animals of a particular group. Sometimes also the mortality rate of the new-born animals in the spring could greatly increase the ratio of the animals killed (or which had died) (Tepe Sarab in Iran, and Early Neolithic sites of the Balkans).

In the past some authorities used to claim that the presence of shed antlers of cervids was evidence for the occupation of a site in winter time. Of the Old World deers, red deer sheds its antlers between February and May[7], and roe deer between November and January[8]. However, since antler has always been a valuable raw material, man may have used not only the antlers of the deer which he killed, but also the shed ones which he collected. In this way shed antlers can be present in a settlement in any season of the year. On the other hand, antlers found in a site with a skull or skull fragment may originate from the middle of the summer to May of the next year, a very large time span.

Finally I should like to give some example of information which animal remains have provided about the nature of the occupation of some important prehistoric sites.

1 *Érd* (*Hungary*). Palaeolithic site. One part of the site was used in the spring, the other in the summer, or at least its occupation extended also into the summer[10].
2 *Tepe Asiab* (*Iran*). Pre-pottery Neolithic, *c.* 8000 B.C. On the basis of the distribution of the age groups of the few animal remains, whose age was well determinable, the site was certainly used between February and April, and occasionally between August and April.
3 *Tepe Sarab* (*Iran*). Early Neolithic, *c.* 5000 B.C. The age distribution of the very large mammal bone sample demonstrates a perennial occupation.
4 *Sitagroi-Photolivos* (*Greece*). Phase I, Middle Neolithic, *c.* 4600–4100 B.C. Perennial occupation.
5 *Anzabegovo* (*Yugoslavia*). Early and Middle Neolithic. Perennial occupation.
6 *Vlassac* (*Yugoslavia*). Before the Pottery Neolithic. On the basis of the small mammal bone sample of well-identifiable age, the site was certainly occupied between January and June, between August and November and probably also at some time during the winter.
7 *Lepenski Vir* (*Yugoslavia*). Phase I, before the Pottery Neolithic: there are no mammal remains of well-determined age. The red deer skulls with ripe antlers show that the settlement was occupied at some time between August and May. Phase III, Early Neolithic (Starčevo culture): the settlement was certainly in use between December and May, and from July to October, and

probably at some time during the winter, and, therefore, apparently all the year round.

8 *Obre* (*Yugoslavia*). Late Neolithic (Butmir culture). On the basis of the age distribution of the mammal bones the site was used all the year round.

9 *Ludaž-Budšak* (*Yugoslavia*). Early Neolithic (Körös culture). Its very rich avifauna remains included species which live in the Carpathian Basin from the spring to the autumn, and others which live only in the winter, proving all the year round occupation. The age distribution of small domestic species (pig, sheep, goat) also proves perennial occupation.

10 *Maroslele-Pana* (*Hungary*). Early Neolithic (Körös culture). Avifauna and small domestic mammals identical to those of Ludaš.

11 *Röszke-Ludvár* (*Hungary*). Early Neolithic (Körös culture). The same as for Ludaš and Maroslele.

12 *Deszk-Olajkút* (*Hungary*). Early Neolithic (Körös culture). The same as for Ludaš, Maroslele and Röszke.

Notes

1 Heizer, R. F. (1960). Physical analysis of habitation residues, ***Viking Fund** Publ. in Anthrop.*, **28**, pp. 93–142.

2 Khalaf, K. T. (1961). *The Marine and Freshwater Fishes of Iraq*. Baghdad. p. 59; Mahdi, N. (n.d.). *Fishes of Iraq*. Baghdad. p. 44; in addition, Iraqi fishermen could not give me any information about this, which is not at all surprising, since the species has no practical importance for them.

3 Bökönyi, S. (1970). The vertebrate fauna of Lepenski Vir, *Science*, **167**, pp. 1702–4.

4 Pidopličko, I. G. (1956). *Materials for the Study of the Fossil Fauna of the Ukrainian SSR*. Kiev. pp. 16, 26.

5 Bökönyi, S. (1964). The vertebrate fauna of the Neolithic settlement at Maroslele-Pana, *Arch. Ért.*, **91**, pp. 87–93.

6 Haltenorth, Th. (1957). Gebissformel- und Lebensdaten-Tabelle, *in* van den Brinck, F. H. *Die Säugetiere Europas*. Hamburg-Berlin. pp. 193–213.

7 Szederjei, A. (1960). *Red Deer*. Budapest. p. 6.

8 Kretzoi, M. (1968). Etude paléontologique, *in* Gábori-Csánk, V. *La station paléolithique moyen d'Érd—Hongrie*. Budapest. pp. 92–94.

J. DESMOND CLARK

Mobility and settlement patterns in sub-Saharan Africa: a comparison of late prehistoric hunter-gatherers and early agricultural occupation units

Any attempt to make comparisons between the settlement patterns and behaviour of "Later Stone Age" populations and those of early farming communities in Africa south of the Sahara is immediately faced with a number of questions and problems that the available evidence can do little to elucidate satisfactorily.

First, even with the increased amount of field survey and excavations being undertaken in many parts of the sub-continent today, there are very few, if any, settlements that have been completely excavated and for the most part work has been concentrated on recovering a chronological and stratified sequence, generally from caves and rock shelters. Even though, in some cases, approximate measurements of the area of an occupation mound in the open may be given, these and the limited extent of the excavations, can provide only a very imperfect estimate of the amount of ground occupied at any one time, of the number of dwelling units and other structures, of the dispersal and range of activity units or of population density.

It is generally assumed that the larger the occupation area the more numerous the population. However, while this may, in general, be true it takes no account of any differential use of various parts of the site through time or of the rate of accumulation of the occupation layers relative to the materials—food waste, artefacts, etc.—of which they are composed. Occupation middens at open sites are subject to natural and artificial dispersal but caves and rock shelters impose limits on the available space and so upon the size of the group that could occupy them. If, therefore, such sites were occupied, as indeed they were, by agriculturalists, then their doing so implies a dispersed population pattern which could be expected to reflect seasonal scattering. If such occupation were permanent then it would be the exception rather that the rule and, in general, their main places of settlement would more probably be in those areas most favourable for stock-raising or cultivation. While some shift in the centres of population can be seen in, for example, Rhodesia as between "Late Stone Age" hunter-gatherers, early and late Iron Age[1], it is not yet possible to demonstrate this as between the "pre-Neolithic" and "Neolithic" anywhere south of the Sahara.

The second problem is the rather small number of sites at which it has been possible to discern at all certainly the nature of the economy and any such determinations are mostly based on the differentiation between domestic and wild mammalian species and much more rarely upon the identification of plant remains. Bone is preserved on the majority of East African sites but in the West African forests, the Congo basin and large parts of the central African savanna, climatic conditions usually, though not invariably, preclude the survival of bone. Industrial units relating to an economy based on domestic animals and plants will more generally, in the "Neolithic" south of the Sahara, be determined by the introduction of new forms into what is otherwise a traditional tool-kit, rather than by the complete replacement of this tool-kit by another. Such a situation is interpreted as showing that diffusion and not ethnic change was mainly responsible for the introduction of the new economy.

In southern Africa, the change from hunting and gathering to agriculture that took place in the first few centuries A.D. can be more directly related to ethnic movements and closely associated with the diffusion of metallurgical techniques, new plants and animals and the spread of Bantu. These early Iron Age settlements are in marked contrast to those of the "Later Stone Age", Khoisan, hunting groups and in both the earlier and later Iron Age patterns more intensive regional exploitation can be seen. The staple food source in the higher rainfall regions was provided mainly by plants and, in the drier regions by livestock—supplemented always by hunting which probably for long continued to be the chief source of meat. In some regions this economic intensification is seen by changes in the pollen spectrum, as in the Victoria Basin[2], by the presence of a widespread buried charcoal horizon believed to be evidence of intentional burning and clearing of forest and bush for hunting and cultivation, as in the Congo and Barotseland[3], and by rapid colluvial filling of stream channels due to soil erosion after the vegetation cover had been stripped off[4].

The degree of mobility, regular seasonal transhumance and the relative size of settlements are primarily determined, during prehistoric times as today, by ecological factors. The concept of favourable, marginal and unfavourable areas for human settlement depends upon subsistence needs and which environments could best provide these has been well demonstrated for southern Africa by Lee[5]. Both geomorphological[6] and palaeobotanical[7] evidence as well as cultural distributions[7] show how changes in climate and vegetation patterns affected the population spread into the Congo basin during the later Pleistocene. These and similar data must have continent-wide application.

Settlements of hunters in the most favourable areas—grassland savanna with permanent surface water and a large biomass provided by one or two species of large game animals[9]—can be expected to produce some of the largest semi-permanent groupings possible at the hunter-gatherer level. Waterside sites concentrating on the exploitation of aquatic mammals, fish and shellfish, both near freshwater lakes and rivers and in coastal habitats, also provide evidence of large groupings, judging from the dimensions of the occupation areas and the quantity of accumulated midden[10]. Where there was also present an abundance of terrestrial fauna and plant foods, such sites may well have been near-permanent settlements. One group of Rer Magno fishermen encountered on the Somalia coast in 1946 had remained at one living site—a rock shelter—for ten years until the fishing became poor when they moved to another further down the coast[11].

Savanna woodland with a high biomass of the large Ethiopian fauna and a regular seasonal supply of plant foods, as in south central Africa or the West African Sudan belt, will, similarly, have supported larger groupings as well as an overall greater density of population than do the marginal or what today would be called the unfavourable regions of low rainfall, with scarce surface water and, comparatively speaking, less abundant sources of animal protein, such as in the Kalahari.

It would seem that lowland, humid forest should be considered a marginal habitat for prehistoric African hunters since it was only occupied comparatively late in the Pleistocene and large gregarious populations of antelopes and other large animals are usually absent. It might, therefore, be expected that at the food-gathering level social groupings there would be small. Turnbull's[12] Mbuti evidence, however, suggests that it is social rather than environmental factors that make for comparatively large camps (7–30 nuclear families) among those that practise communal hunting with nets and much smaller groupings among those who depend on bow and arrow hunting by individuals. Clearly, therefore, lowland forest must be classed as a favourable environment. With the development of plant domestication and cultivation, however, this type of habitat would support, in fact would necessitate, groupings of much larger size for co-operative clearing of forest for cultivation. The success of the various "Later Stone Age" populations was dependent upon the seasonal availability of resources from several different micro-environments as well as on the ingenuity with which these were exploited.

A considerable amount of data is available to archaeologists from ethnographic sources at both the hunter-gatherer and the village farming level. This should, however, be used only with caution and if it is employed as comparative material to aid interpretation of archaeological situations, then it is necessary to demonstrate similar ecological and economic contexts. Thus, Nemadi behaviour in the Mauretanian desert[13]; !Kung Bushmen behaviour in the central Kalahari[14]; that of Pygmy groups in the Ituri forests of the Congo[15]; of Nderobo in the montane forests of the Eastern Rift[16]; of Bon and Ribi in the dry steppe and bush country of southern

Somalia[17]; of the Hadza in a somewhat similar habitat in the Lake Eyasi Rift[18]; or of the OvaTjimba in the stony desert of South West Africa[19]; all have much of general technological and socio-economic interest to contribute to the understanding of the archaeological record. Ethnography can, however, provide at best only approximations that need to be tested against the specific archaeological data.

More reliable comparisons can probably be made between early Iron Age populations and some of the patterns of village farming seen, for example, among Bantu speakers or West African Negroes. Such comparisons can perhaps also be made between Neolithic herders of the southern Sahara and Fulani or historic Hottentot nomadic pastoralists. If greater reliability can be attached to such evidence it is because it is more closely related in time to the archaeological situations with which it is being compared and not necessarily because of any continuity of tradition that may be preserved. Rather is it the result of a lengthy adaptation of broadly comparable economies within habitats, changed to some extent certainly, but still closely similar to those of prehistoric times.

If, however, the environment is the main limiting agent controlling group size and regional population density, it is various social factors such as interaction, territoriality and antagonisms that appear to keep the optimum size well below the carrying capacity of the habitat[20]. Scudder[21] has shown that while hunter-gatherers (!Kung Bushmen and Hadza) make use of only a very small number of their plant resources and generally move when these are eaten out, the agricultural Gwembe Tonga continue to support higher density village settlements of between a hundred and more than four hundred individuals[22] by the use during times of food shortage of a wide range of wild "famine foods". These (131 wild species are known to be used) are necessary because of the frequent failure of their crops; and, moreover, some of these foods are poisonous if eaten without long and laborious preparation. Such situations suggest that this knowledge is of some antiquity and, therefore, of significance for interpreting archaeological patterns among farming communities over a large part of the drier African savanna regions. Of course, in the past in the Gwembe and in south central Africa generally, greater emphasis must have been given to hunting as a means of combating famine similar to the practice described by Turnbull[23] among the Ik in the southern Sudan who resort to hunting about once in every four or five years when drought produces a total crop failure.

For the early Neolithic farming sites south of the Sahara the importance of wild relishes cannot usually be substantiated though it has been shown to be the case for later Neolithic groups in southern Mauretania[24]. Today, however, such foods still form a very important item of regular diet and an essential "stop gap" in time of scarcity among many modern African peoples. A large number of wild leaf relishes, fruits, nuts and tubers is in use among the agricultural tribes of Malawi[25].

Similarly, the practice of hunting as a major source of the meat consumed by many agriculturalists today can be traced back into the early Iron Age[26]. Where cattle and other stock were regarded as wealth, as among the Ila[27], the Fulani[28] or the Karamojong[29], animals are usually slaughtered only on ceremonial occasions, so leading to a lower percentage of bones of domesticated animals in the midden deposits. This may also have been the case among the Neolithic pastoralists of the central and northern Sahara[30] if the apparent scarcity of domestic species in the occupation sites is any indication. At the early Iron Age Kalomo Culture site of Isamu Pate in Zambia (*c.* A.D. 750–1200) the number of bones of domestic animals gradually increases through time. In the lower levels, wild species make up about 80–90%, while in the upper levels these proportions are nearly reversed[31]. At the later (mid-thirteenth to early sixteenth century A.D.) site of Ingombe Ilede the ratio of domestic to wild species is probably the same[32] while at the last century Tonga site of Behrens Farm domestic cattle make up 98·1% of the bone waste[33].

It may, therefore, be expected that gathered foods and wild animal species continued to form a significant part of the diet of all early food-producers in sub-Saharan Africa. With time these traditional food sources were supplanted and later largely superseded by the locally developed strains of domestic plants and cereal grasses and, where they were not precluded by tsetse fly, by live-stock mostly of non-African origin. Agricultural settlements supported by the various environments depended upon the natural productivity of the soil and the manner of land use. Most, if not all of the early farming communities are believed to have used *swidden* agriculture like those of historic times and evidence for forest clearance first becomes apparent in the first to second millennia B.C.[34]. The degree of permanency of settlements is determined by the extent to which the natural vegetation can regenerate. In the forest zone, settlements may usually be expected to be large and permanent while those in dry, marginal savanna or steppe are more likely to be small and mobile or to give place to pastoralism.

It is now necessary to examine the archaeological evidence for the main ecological zones.

West Africa and the southern Sahara

From the Nile valley and Nubia comes evidence for intensive exploitation of cereal grasses after *c.* 15,000 B.C. Settlements seem to have varied in size from 20 m^2 to 6300 m^2 (approx) and there is also evidence for conflict and competition between populations using traditionally distinct tool-kits[35]. Such evidence has not yet been observed for other parts of the continent and it may have been induced by the special restrictions on movement in the Nile area imposed by the surrounding desert. There is still no satisfactory evidence for domestication in north-east Africa before the beginning of the fifth millennium B.C. for livestock[36] and about the middle of the fourth millennium[37] for cereals.

The dates from the Sahara suggest that pastoral Neolithic peoples were widely dispersed there between the mid-fourth and the mid-third millennium B.C. with indications that cattle may have been introduced into the north-central parts as early as the beginning of the fifth millennium[38]. If the Tenerian settlement localities at Adrar Bous in northern Niger are in any way typical of the Neolithic pattern in the Sahara then it would seem that this was one of seasonal settlements of cattle herders occupying topographically restricted areas of pasture, often adjacent to lakes and swamps. One such midden area at Adrar Bous covered approximately 25,000 m². Stone circles and semi-circles mark the bases of features—probably dwellings and, perhaps, storage structures—but differences in the amount of disturbance these have undergone suggest that the whole area was never occupied all at one time. The structures were clearly light and so easily transportable so that it appears likely that the Tenerian population practised regular transhumance and such sites may have been strictly seasonal encampments set up anew each year as among the pastoralists of the Sudanese belt[39] or the Horn[40] today.

Dates for the Tenerian are still being received, but, from those so far available, it appears to have existed for some 1300 years, perhaps longer (3800–2490 B.C.). Using Cook's and Heizer's estimate[41] of 10 m² of living space per person and suggesting that perhaps one quarter of an open area was occupied at any one time by small composite groupings, we obtain an approximate population of about 625 persons. On the other hand Neolithic open sites at Meniet[42] north of the Hoggar, with dispersed occupation débris, covered 325,000 m² and 175,000 m² respectively. Comparable ethnographic data for Fulani and Somali suggest that such computations of population would be considerable overestimates although certainly Adrar Bous, because of the extensive lake and swamp area, was a place on which a number of extended family units were wont to converge.

Munsen's work[43] at Dar Tichitt in Mauretania shows that by *c.* 1200 B.C. the lakes had practically disappeared and the defended, stone-walled settlements of the succeeding Chebka Phase (*c.* 1150 B.C.), situated on the scarp and of which more than fifty are known, contained populations of between 500 and 1000 persons as well as cattle and goats. A number of rock shelters and caves with Neolithic occupation of this time in the Hoggar and Tassili can probably be interpreted as the camping places, again probably seasonal, of a number of families. That at Amekni[44] with a surface area of approximately 54 m² gives an estimate of five resident individuals.

Hunting-fishing communities occupied Adrar Bous immediately prior to the Tenerian and the observed size of these camps was markedly different from that of the Tenerian. The occupation débris from two of them covered, each, approximately 3600 m² which, using the same method of computation, gives a population of about ninety individuals—not necessarily an overestimate if the lake was an important focus for seasonal fishing activities. Indeed, when compared with the figure of ten individuals suggested for the restricted 100 m^2 area covered by the lower occupation horizon at the Ishango hunting-fishing camp on Lake Edward (*c.* 6500–6000 B.C.)[45] and that of 162 persons for the occupation mound of early Khartoum (approximately 6500 m^2)[46], it would seem that the figure for the fishing camp at Adrar Bous might be an underestimate. The comparable figure for the Khartoum Neolithic mound at Es Shaheinab[47] (approximately 5600 m^2) is 140 individuals and, although a small domestic goat is present, the great bulk of the food waste shows that this was a seasonal settlement for fishing, fowling, and hunting[48]. If these figures have any validity—and there is, as yet, no means of knowing this—they suggest that Neolithic pastoral units may have been larger than those living under the earlier hunting-fishing economy.

While there is much evidence from the rock art of stock herding, in particular cattle, in many parts of the Sahara from the Nile to the Atlantic Ocean, there is as yet no certain evidence before *c.* 1100 B.C. for grain cultivation. At this date there is domestic *Pennisetum* with Dar Tichitt sites of the Chebka Phase and possible evidence for domestic sorghum occurs there by *c.* 800 B.C.[49]. The many grindstones at some Saharan sites may equally well have been employed for grinding the wild *Panicum* and other grasses that abound in certain restricted localities even today and are still regularly used.

The wide distribution of certain traits and artefacts—"Wavy Line" and "Rocked Stamp" designs on pottery, harpoons, Shaheinab- and *hache à gorge*-type axes and adzes, amazonite beads, etc.—indicates a situation of regular interchange that must have permitted the rapid spread of new techniques and equipment within the Saharan culture area. It is to be expected, therefore, that knowledge of herding and cultivation must have reached the hunting-gathering populations of the West African Sudan belt and forest during the fourth millennium B.C., if not earlier. The failure of the Saharan technology to spread south of the sixteenth parallel is surely due, therefore, not to lack of opportunity, but to the inappropriateness of the stock-raising economy in the higher rainfall and tsetse zones south of the desert as well as to the great richness of the wild plant and animal resources there. That contact did take place is shown by the Ghanaian sites at Kintampo[50] and Ntereso[51] where Saharan elements occur together with traditional forest/savanna forms. South of the desert the stone technology and the pottery are in general quite different. In addition the early importance attached there to wild grain and plant foods and to the development of local plant domesticates rather than animal ones, can be seen in the later emergence of various regionally distributed domesticated species—dry rice, bullrush millet, sorghum, *Digitaria*, yams and minor species of fruits, roots, tubers, peas and beans[52].

The known settlements of pre-Neolithic hunting communities south of the sixteenth parallel are mostly situated in caves and rock shelters which probably give a false impression of the size of the social unit. The commonest artefacts are microlithic tranchets and deep crescents and it seems probable that a proportion of these formed the heads of arrows. If this is so, then it suggests that individual hunting methods rather than co-operative techniques were the more usual ones employed and so the camps may have been consistently small and composed of two or three nuclear families. Davies'[53] maps, showing the distribution of microliths with and without Neolithic or Iron Age associations, show a proliferation of sites in the forest and along the rivers from the pre-Neolithic and fewer but more dispersed sites later. The implication that this may indicate fewer but larger groupings in the forest zone after the adoption of agriculture needs to be tested.

The pre-Neolithic industries of the forest and savanna zones are best seen in the lower levels at Kintampo and Buobini[54] in Ghana; in the lower levels at Rop[55], Iwo Eleru[56] in Nigeria and in Yengema cave in Sierra Leone[57], Most of these comprise a high proportion of microlithic tools and at Yengema, certainly, a large number of splintered artefacts (*outils esquillés*). In the Buobini horizon at Kintampo abundant remains of *Celtis* seeds and other fruits were found and animal bone, all apparently from wild species. About 1450 B.C. the Buobini industry was replaced by the Kintampo industry and a displacement of population is inferred since the associated pottery is completely different. Polished stone axes now become common and other new forms appear in association with a sequence of clay floors separated by layers of ashy midden.

Davies'[58] excavations at Ntereso in northern Ghana throw new light on the Kintampo Neolithic. This was a fishing settlement close to the White Volta and there occur here Saharan arrowhead forms, bone harpoons and fishhooks, various other "exotic" elements of northern type and small clay models of animals. The site was first occupied about the thirteenth century B.C. and the occupation lasted, it is believed, for about 100 years. It covered a comparatively restricted area, approximately 750 m^2 which suggests a confined village settlement of, perhaps, about seventy-five individuals. Plans of one rectangular and one possibly circular dwelling were found, the walls and roof having been formed of spaced poles and probably mud daub since many fragments of this material lay on the floors[59]. Two thermoluminescence dates for the introduction of pottery and ground stone tools at Yengema cave[60] suggest that Neolithic technology and so, perhaps, a planting economy may have reached the forest zone before the end of the third millennium B.C. The introduction of pottery and ground stone into the forest/savanna zones of Nigeria probably started as early as 3000 B.C. as shown by dates from rock shelters at Iwo Eleru and Afikpo[61]. Rock shelter sites at Rop and Mejiro cave, Old Oyo[62] with pre-Neolithic microlithic technology have only small living areas (possibly 20 m^2 and 35 m^2, respectively) and can only have served as living quarters for one, perhaps two, nuclear families. In both cases the main microlithic layers are shallow. With the exceptions of Old Oyo the material is incompletely published, however.

Little can be said concerning Neolithic open settlement patterns in the southern savanna and forest belts of West Africa since insufficient excavation has been carried out. The tool-kits at Iwo Eleru (Nigeria) and Bosumpra (southern Ghana) rock shelters probably show the range of artefacts in the local industries and numbers of stone axes and "hoes" have been found at open sites, mostly, it would seem, to the west of the Volta[63]. Since ground stone axes and pottery are found in buried horizons underlying what were believed to be stands of primary forest[64] it can be inferred that the forest populations of the eastern part of West Africa were now probably organized into larger groupings for co-operative forest clearance and for yam cultivation, while those in the western part were probably similarly organized for rice culture. For the efficient clearing of evergreen forest by slash-and-burn methods a large and permanent labour force is essential and it may be expected that village settlements may have comprised several hundred members well before the middle of the first millennium B.C. when metallurgy appears in Nigeria. No systematic investigation of any pre-metal-working open settlement site has yet been undertaken in the high rainfall zone, however.

North of the forest in the northern savanna grasslands and Sahel there is evidence for fairly large settlements of stock owners who were probably also cultivators of grain crops as well as fishers and hunters. The occupation mound or "tell" over 10·5 m high at Daima in Borku in the annually inundated heavy clay lands south of Lake Chad, was occupied from the sixth century B.C. into the earlier part of the second millennium A.D. (i.e. 1500–2000 years of occupation). Today this area supports many cattle and two kinds of locally domesticated guinea corn are extensively grown. The socio-economic pattern in this *firki* region may not have been too different from that of the Nuer of the Upper Nile—"Nuer are forced into villages for protection against floods and mosquitoes and to engage in horticulture and are forced out of villages into camps by drought and barrenness of vegetation and to engage in fishing"[65].

From the beginning Daima was a permanent settlement of houses with clay floors and wooden walls. Equipment comprised numbers of grindstones, ground-stone celts, barbed harpoons and other bone tools, distinctive red, burnished pottery, animal figurines, etc. Iron began to replace stone about the fifth or sixth century A.D.[66]. The population possessed domestic cattle and goats but most of their animal protein appears on preliminary examination to have been derived from fish and wild bovids[67].

Numbers of such mounds also exist in the east and north of the Chad basin[67] and it would seem that this is most likely one of the nuclear areas for domestication south of the Sahara. Not only is this region favourable for the concentration of large holdings of livestock but it now appears most likely to have been here that the domestic strains of bullrush millet, sorghum and *Digitaria* were first developed. The area of the Daima tell is not yet available but a somewhat similar situation of purely artificial mound villages is found also in seasonally inundated regions along the middle Niger and its tributaries. Two such long occupation mounds exist at Karkarichinkat on the lower Tilemsi, one of which covers approximately 30,000 m^2 though both rise only 3–4 m above the plain[69]. Other mounds with Neolithic occupation have recently been reported from Upper Volta[70].

So far as West Africa is concerned, therefore, the situation, on the available archaeological evidence, suggests the gradual replacement after *c.* 3000 B.C. of small social groupings of hunter-gatherers occupying seasonal sites by large village settlements of 100 to 300 individuals living by herding and cereal cultivation and having a Neolithic technology in the Sahel and northern Sudan belts. Although it cannot yet be proved by excavation, it is most probable that the populations exploiting the yams and dry rice staples of the eastern and western regions after 3000 B.C. were also communities of several hundred individuals occupying large, virtually permanent villages. It has been suggested by Livingstone that the spread of agriculture and the formation of large settlements here were responsible for the spread of the sickle cell gene.

In some regions, for example northern Ghana, actual population replacement appears to have been responsible for introducing agricultural economy. In others—most of the eastern half of West Africa, for example—the continuing microlithic tool-kit suggests instead that ethnic replacement did not accompany this economic change which was brought about by culture contact and diffusion.

East Africa

Little is known, as yet, about the late prehistoric, pre-Axumite populations of Ethiopia, but four permanent village sites are known from the vicinity of Jebel Khorkan, Agordat, on the Eritrean plateau. They have been dated provisionally and by extrapolation to the mid-second millennium B.C.; they yielded a material culture suggesting an economy based on cereal cultivation and stock-raising[71].

If the region of "Punt" visited by the ships of Queen Hatshepsut (1496 B.C.) can be identified with some part of the northern Eritrean coast, as seems most probable, the inhabitants grew cereal crops and possessed two breeds of cattle[72]. Recent survey in the Khashm el Girba area on the Atbara close to the Sudanese/Ethiopian border, has revealed the presence of three distinct groups of early ceramic settlements. One group with dwellings spreading over an area of some 300 m^2 produced "dotted wavy line" pottery similar to that associated with the Khartoum Neolithic. Another group of four sites consisted of large village "tells" 3–5 m high, one with an area of more than 96,000 m^2. A charcoal sample from one of them gave a date of 2460 B.C. A third grou represented by a single site, situated away from the river, yielded microliths associated with yet a third distinctive pottery ware; the occupation area covers approximately 17,500 m^2 and dates to 1100 B.C.[73]. Thus, more than one distinctive pattern of land-use, probably having an agricultural basis, can be expected in the Upper Nile and Ethiopia between the third and first millennia B.C. It was, presumably, sometime between the third and the second millennia B.C. also that the local domesticates *teff* and *ensete* were developed and wheat and barley cultivation spread from North Africa to the Ethiopian plateau.

The evidence of the rock art[74] indicates that the more arid regions of Danakil and Somalia were occupied by pastoral nomads and, though as yet their first appearance there cannot be satisfactorily dated, they were certainly present by the middle of the first millennium B.C. when the pre-Axumite settlers were established on the Eritrean plateau[75] and, quite possibly, as much as a millennium or more before that. The long-horned cattle, art styles and lithic assemblages, especially those with hollow-based arrowheads of eastern Sahara type, suggest a north-western origin for these herding populations and an emphasis on hunting. A date around the beginning of the second millennium for their first appearance might indicate that these movements were directly related to the onset of the main desiccation in the Sahara as was also, it is believed, the appearance of the predominantly pastoral C-Group population in Nubia[76].

Although numbers of ground-stone axes and what may be hoes have been found at certain sites in south-western Ethiopia, nothing is known concerning the patterns of settlement with which they must be associated.

The Central African Republic and Cameroun can be expected to provide evidence of settled agricultural communities with a Neolithic tool-kit but, as yet, only initial surveys have been carried out. However, one extensive Neolithic occupation site is recorded from Batalimo, close to the junction of the Lobaye and Ubangui with pottery and much worked stone; it has not yet been dated but systematic excavation may be expected to give important information for understanding the initial penetration of the Congo basin by northern cultivators[77].

Pastoral peoples are present in East Africa by the beginning of the first millennium B.C. In association are varieties of what have been termed the Stone Bowl Cultures that are generally described as Neolithic and pottery is now known from the Nakuru basin as early as 1400 B.C.[77]. Several different groups with distinctive pottery wares are sometimes chronologically differentiated as has been shown to be the case also in West Africa.

Unfortunately most of the sites that have been investigated are either in caves and rock shelters or are burial cairns and, while they may provide data on diet and domesticates, contribute little to knowledge of settlement patterns.

The stone platters and bowls with accompanying pestles are a possible, though not a certain, indication that some form of cultivation was practised but, as yet, all evidence of a vegetable staple is also lacking. The regular use of caves for burial, as the Njoro River Cave (970 ± 80 B.C.) where seventy-eight individuals were buried[79], suggests reasonably permanent occupation while the finely pierced and polished beads of hard and semi-precious stones again point to an economy more complex than that of hunting and gathering. The association of an engraved *Lagenaria* gourd here and also perhaps in Mound C[80] at Ilkek also suggests a community practising cultivation.

Domestic cattle are associated with stone bowls, pottery and ground-stone axes at an open settlement site on Prospect Farm, Elementeita, dated to 740 ± 80 B.C.; the extent of the settlement is not known. Another open site with cattle and probably sheep/goat commonly represented and also with a small percentage of wild species, is situated at Narosura near Narok on the Mau Escarpment. Although full details of the site await publication by the excavator, K. Odner, it is of importance to note that there is evidence of daga plaster from structures suggesting a permanent village settlement with dates that range between 850 and 200 B.C.[81]. The Hyrax Hill, Nakuru, Neolithic occupation and cemetery with eighteen burials on the living site is also, though no domestic stock was reported, strongly suggestive of a regularly occupied settlement of pastoralists and marginal cultivators, the centre of the site being formed by a roughly cobbled area yielding much pottery and stone artefacts of evolved Kenya Capsian type[82].

Caves and rock shelters were also occupied by stock-owning groups and cattle and possibly sheep, together with bones of wild animals, were present at Tunnel Rock in western Kenya[83] associated with an industry of Wilton or Kenya Capsian affinities[84].

Although wild species, including fish and shellfish, made up most of the food waste in six rock shelters on the northern shore of the Kavirondo Gulf of Lake Victoria in Uganda, cattle were present at three of them; sheep or goats at one and possibly two others and domestic cats also at two of them. A date of *c.* 365 ± 185 B.C. is suggested, and again the industry is a microlithic one associated with some pottery[85]. Three of these shelters had floor areas of 20 m^2 or less and the others varied between 50 m^2 and 100 m^2 suggesting seasonal hunting/fishing occupation by small local groups of one or more nuclear family units. Like the western Kenya assemblages, these can be interpreted most probably as representing a dispersed population of agriculturalists who moved temporarily to the lake shore from permanent settlements further inland. Alternatively, but less likely, this kind of assemblage could be interpreted as that of predominantly hunting-gathering groups who had acquired stock and pottery by diffusion through contact with agriculturalists[86]. Regular transhumance of the kind suggested was practised by the inland settlements of Tonga in the Gwembe who dispersed to occupy single family sites along the Zambezi during the dry season[87]. Similar seasonal movements were made to the coast by tribal groups in Natal and southern Mozambique[88].

An open camp site at Long's Drift (Gr Ji 1), Enderit River, Nakuru, in the

Kenya Highlands recently excavated by C. M. Nelson, covers an area of approximately 400–450 m². Large quantities of butchered wild animal bones and possible domestic stock were also associated with a very rich microlithic industry of "Wilton A" type, some pottery, stone bowl fragments and a ground-stone axe. This dates to after 1590 ± 120 B.C., the age determined for the base of the alluvial sequence in which the site is situated, and so probably belongs in the first half of the first millennium B.C.[89]. Until the fauna has been analysed the site might be interpreted as either a camp of hunter-gatherers or as a seasonal hunting camp of an agricultural group.

To date the only burial cairn from which cattle bones are reported is that from MacDonald's Farm, Nakuru[90]. As yet there is no direct information from cattle bones to show the type to which the Kenya breed belonged but a rock shelter on Mount Elgon contains paintings of long-horned, humpless animals similar to those depicted at sites in the Harrar region of Ethiopia[91].

The persistence of the Kenya Capsian tradition in the percussion flaked obsidian industry suggests that the introduction of agriculture was due most probably to diffusion rather than to ethnic movement. The physical type, moreover, shows, it would seem, no marked difference from that of the older Afro-Mediterranean hunting population that, in the Rift area, was organized into groups of specialist hunters and fishers occupying rock shelters and open sites along the high shorelines of the lakes within the Gregory Rift where they lived by exploiting several rich micro-environments[92].

The pattern that is emerging suggests that the economy of the earliest agricultural peoples to occupy the highlands of the East African Rift and the Victoria Basin from Elgon in the north to Ngorongoro in the south, was based primarily on pastoralism but with cultivation playing, through time, an increasingly important role. It may be expected that the populations were organized in medium-sized settlements that practised regular transhumance for hunting and pasturage and obtained, at first, a major part of their meat from wild animals. In fact, it was a situation very similar to that now found in the Gwembe or which the survivors of Portuguese wrecks in the sixteenth and seventeenth centuries encountered among the Bantu-speakers living along the south-east African coast who moved regularly between the interior and the coast[93].

If these early East African agricultural settlements were in any way comparable to the later iron-using settlements at Hyrax Hill, North-East Village, then they may have comprised groups of between 100 and 200 individuals. The sixteenth or seventeenth century A.D. North-East Village consisted of thirteen sunken cattle-enclosures with associated dwelling areas (each possibly housing one extended family), middens and walling. It covered an area of approximately 1830 m² giving an estimate of 183 persons on the basis of 10 m² for an individual.

The sixteenth century A.D. enclosure site at Lanet covers approximately 40,500 m² and if it was indeed used as a place of settlement and not simply as a stock kraal or temporary defensive work, as seems more probable, it could have housed a pastoral population of 4000 over short periods. Longer occupation by any such improbably large grouping would have produced more than the thin scatter of occupation débris within the enclosure and greater disturbance outside as well as completely stripping the grazing for miles around. More probably the adjacent fifty to sixty hollows represent sunken cattle-enclosures and homesteads of the Sirikwa Hole type excavated at Hyrax Hill, North-East Village[94]. If this were so and if all were occupied at one time which, on the evidence seen by the writer in the Iraqiw area in northern Tanzania seems unlikely, the population, on an estimate of four individuals per family, would have been between 200 and 240 persons.

Probably the best comparisons are with present-day pastoral Cushites and "Nilo-Hamites" such as the Somali[95] and the Karamajong[96] or with groups such as the Nilotic Jie[97] where cattle are of major importance but most of the staple food is supplied by cultivation. The pattern of settlement among the nomadic pastoralists is one of groups of several nuclear families, between three and seven on an average, but sometimes more, where there is need for security as, for example, among the Murille. Among the Jie the pattern is one of larger permanent settlements (sometimes reaching 130 persons per square mile) together with separate mobile camps where the stock is kept.

Southern Africa

In southern Africa the earliest agricultural settlements appear all to date to the Early Iron Age in the first half of the first millennium A.D. Whether or not the pastoral Hottentots, who apparently did not cultivate, obtained their cattle and sheep before this is not known but seems unlikely. The economic advantage of the metal-using technology in the drier savanna regions cannot as yet be precisely determined but it was probably considerable. The settlement areas of some of these earlier sites in Zambia and Rhodesia show communities to have been small and dispersed rather than large and concentrated as, for example, on the Tonga plateau in southern Zambia where Fagan[98] using the formula developed by Cook and Treganza[99] has computed populations of 90–200 for the mound site of Isamu Pate (A.D. 725–1200) and about 132 for the same type of site at Kalundu (*c.* A.D. 790–1080). The Kalambo Group settlement (A.D. 550–870) at Kalambo Falls stretches over some 3000 m² and suggests a larger grouping, though it is unlikely that the area was fully occupied all at one time.

Mason[100] has analysed Late Iron Age settlement patterns on air photographs in the southern Transvaal and northern Natal and distinguishes five types of settlement ranging from dense clustering of walled enclosures to isolated circle patterns and believes that these plan forms directly reflect social behaviour. Settlements are sited mostly on higher ground and length of occupation depended upon the availability of water and grazing for cattle and other stock and successful defence of the site and resources. Shallow depth of deposit suggests no long continuous period of occupation but one such site had an estimated population of 16,000 when it was occupied in A.D. 1820!

The situation at Early Iron Age sites appears to be rather different. The sites are generally but not always smaller and the midden accumulation, as with the Zambia mounds, is often greater, pointing to more regular or more continuous occupation. At Isamu Pate and Kalundu the depth was approximately 2·5 m and a maximum of 2 m is recorded for the 200 m diameter mound at Kapwirimbwe near Lusaka[101]. But the K2 mound at Bambandyanalo in the Limpopo valley which is over 180 m in diameter has occupation débris up to 6 m deep and the settlement appears to have been fairly continuously occupied, the plan consisting of a central cattle kraal with houses surrounding it[102]. More generally, however, in south central Africa it would seem that the disturbed and built-up occupation layer reaches a depth of around 1·0–1·10 m, as at Kangila[103] and Dambwa[104] or the Kalambo Group settlement at the Kalambo Falls. The same is the case at the Basanga and Mwanamaipa mounds in the Kafue valley near Namwala. Fagan[105] suggests that it represents the normal accumulation that can be expected in the life of the average village settlement of the first millennium A.D. The dates for Dambwa suggest a duration of 260 years[106] and for Isamu Pate of 475 years[107].

Caves and shelters were also extensively occupied in southern Africa during the Iron Age but in general could have provided temporary quarters for only small units[108]. The accumulated deposit appears to be about half that of the open settlements, i.e. about 0·30–0·60 m but is sometimes deeper.

By contrast there are less settlement data available for "Later Stone Age" sites. Those investigated are generally situated in caves and rock shelters and the rate of build up of deposit was clearly dependent upon several variables for which as yet little control exists. Nachikufu yielded a maximum of 3·60 m of occupation waste covering more than 900 years; while the 1·60 m at Leopard's Hill accumulated in nearly 1400 years[109]. At some South African south coast caves as much as 9 m of shell midden accumulated in some 900 years[110]. Usually, however, Late Stone Age occupation layers are appreciably thinner and the available area for habitation is restricted and provided living space for only small units. At Riversmead in the Orange Free State some 10 m^2 were available with a shallow deposit of between 0·96 and 0·46 m[111]; at Zaayfontein the available space is approximately 90 m^2 and the deposit attains a depth of between 0·60 and 0·90 m[112]; at Glen Eliot shelter the area is 43 m^2 and the depth 0·25–0·80 m[113].

These shelters are generally typical of a large number of such occupation sites in southern Africa and can have provided temporary accommodation for one or two mobile family units at best. A similar but probably more recent situation is that in the "Great Elephant Shelter", Ameib, in the arid Erongo

Mountains area of South West Africa. Here two occupations on two separate and consecutive occasions can be clearly seen, represented in the one instance by four "living hollows" and in the other by three. In the case of the latter, the brushwood screens, bedding materials and hearth areas were still in place though the screens had collapsed. The maximum diameter of each "living hollow" was approximately 3 m and they are approximately 35·5 cm deep; the area they occupy with the floor space in front is approximately 65 m², giving an estimate of six to seven occupants[114].

At the rock-engraving site of Vosberg, Goodwin[115] records a group of some nine living areas, again clearly a temporary, wet-season camp sixty miles south of the nearest permanent water in the Orange River. Another but larger wet-season camp of this kind is situated at Half Way House Kopje about 50 km north of the Vaal River between Barkly West and Kimberley where at one time between 100 and 200 circular living areas could be seen round the foot and slopes of the hill. An open site, in the Upper Zambezi valley west of Livingstone covered an area of approximately 100 m² and was a single period occupation possibly during dry-season convergence on the river by a nuclear family unit[116]. At Kalambo Falls occupation waste from a microlithic assemblage dated to *c.* 1900 B.C. covered an area of 85 m² round a single hearth and possibly represents a single family dwelling area; the environment is one of woodland savanna.

The Late Stone Age camp of hunter-gatherers at Gwisho Hot Springs, Zambia is interpreted[117] as a near-permanent settlement forming a low mound of approximately 2·75 m of occupation deposit over a 700 m² area. The excavation yielded excellent evidence of the diet of the population including large numbers of fruits and seeds (80% from five species) and much bone waste. Buffalo is represented 334 times, lechwe 225, zebra 270 and warthog 278, as well as eleven other species including rhinoceros 120 times, elephant 43 and hippopotamus 43 together with a large quantity of fishbone predominantly from barbel. The site appears to have been occupied sporadically over 1000 years (2835–1710 B.C.)[118] and may have been the wet-season base of several nuclear families that moved in the dry-season to exploit more intensively the fishing along the Kafue River and the lagoons, as do the Batwa today.

Gwisho is a concentrated occupation and unlikely to have accommodated a large group at any one time but the dispersal and number of stone artefacts at some of the Late Pleistocene and Holocene Tshitolian camps in the Congo Basin suggest that these may have accommodated more temporary but larger units, e.g. Cauma and Mbalambala in Angola[119] and Dinga, Kwango[120] where a high concentration of artefacts occurred within an area of approximately 800 m², suggesting groupings of eighty or more individuals exploiting the plant and animal reserves of the forest as well as those of the rivers and streams. Comparability may be suggested with the resident populations of the Mbuti net hunters but the precise limits of these Tshitolian camps have still not been determined.

I am indebted to G. Ll. Isaac for the following data on the relationship between settlement areas and the size of the resident population which correlates data from R. Lee on the Central Kalahari Bushmen. Bushman camps range between ten individuals occupying an area of 130 m^2; twenty-eight occupying approximately 817 m^2 and thirty-four occupying 464 m^2. The average is around thirty persons in a camp area of between 371 m^2 and 464 m^2. These figures probably compare favourably with prehistoric hunting-gathering groups in a similar habitat but they are almost certainly underestimates for richer, more favourable habitats.

In conclusion, therefore, it would seem that there is probably almost as much variability in the form of early and late agricultural settlements in sub-Saharan Africa as there appears to be among hunting-gathering settlements. The chief factors determining settlement size and resident population as well as the amount of seasonal movement, then as now, is the richness of the natural resources of the territory exploited by the group and the time taken for these to be replenished. In the most favourable circumstances the camps of hunter-gatherers may have been nearly permanent and have housed populations of over 100 persons but in most cases group size is unlikely to have exceeded thirty-five individuals for any length of time.

The large size of the occupation area, the kinds of housing and the often considerable height of the Neolithic mounds in the savanna grasslands of West Africa and the Sudan support the view that these are permanent villages, the population of which subsisted on stock-raising, with cultivation being of variable importance. The same can be said of sites with domestic animals in the East African highlands though the settlement areas here are neither so extensive nor have such a depth of accumulated deposit. In southern Africa occupation sites of this kind appear only with the coming of the Iron Age agriculturalists.

In the regions adjacent to the Sahara in Ethiopia and also in southern Africa with the coming of the Iron Age, there is some indication that ethnic movement and replacement were responsible for the introduction of agriculture, food-producing technology and permanent settlement. Elsewhere, however, both in the southern savanna and forest of West Africa and in the East African highlands an appreciation of the advantages of an economy based on animal husbandry and cultivation was probably the result of diffusion and readjustment by the indigenous populations who still gave considerable emphasis to hunting and the gathering of wild plants.

With very few exceptions hunting continued to be a major activity and source of meat and, in localities of uncertain rains and frequent crop failure, wild vegetable foods would have played a very important role in maintaining the population level. Similarly, fishing was of importance in certain localities, as in the Lake Chad basin and the southern Sudan. Some of these populations almost certainly practised regular transhumance for grazing and water or for hunting and/or fishing at certain seasons as, for example, in the Lake

Victoria and Nakuru Basins. At the same time, however, clear evidence exists for the late persistence of groups of hunter-gatherers up to comparatively recent times, one of the best archaeological examples being at the Nakapapula rock shelter in northern Zambia[121]. Some were symbiotically adjusted to sharing territory and resources with agriculturalists as do most Bush, Pygmy and Nderobo groups today, but others, unable to preserve a balanced relationship[122], eventually succumbed for one reason or another and disappeared[123].

Notes

1 Summers, R. (1960). Environment and culture in Southern Rhodesia: a study in the "Personality" of a land locked country, *Proc. Amer. Philos. Soc.*, **104** (3).

2 Kendall, R. L. and Livingstone, D. A. (1967). Palaeoecological studies on the East African Plateau. Paper delivered at the VIth Pan-African Congress on Prehistory, Dakar.

3 Clark, J. D. (1968). Further palaeo-anthropological studies in Northern Lunda, *Museu do Dundo, Lisbon, Pub. Cult.*, **78.**

4 Clark, J. D. (1968). *op. cit.* pp. 146–7.

5 Lee, R. B. (1963). The population ecology of man in the Early Upper Pleistocene of southern Africa, *Proc. Prehist. Soc.*, **29.**

6 Ploey, J. de (1965). Position géomorphologique, genèse et chronologie de certains dépôts superficiels au Congo occidental, *Quaternaria*, **7.**

7 Bakker, E. M. van Zinderen and Clark, J. D. (1962). Pleistocene climates and cultures in north-western Angola, *Nature*, **196** (4855).

8 Clark, J. D. (1967). *Atlas of African Prehistory*. Chicago.

9 Bourlière, F. (1963). Observations on the ecology of some large African mammals, *in* Howell, F. C. and Bourlière, F. (eds.) African ecology and human evolution, *Viking Fund Publ. in Anthrop.*, **36.**

10 Arkell, A. J. (1949). *Early Khartoum.* Oxford; Heinzelin, J. de (1957). Les fouilles d'Ishango, *Inst. des Parcs. Nats. du Congo Belge*, **2**; Louw, J. T. (1960). Prehistory of the Matjes River Rock Shelter, *Mem. no.* **1** *Nat. Mus. Bloemfontein.*

11 Clark, J. D. (1954). *The Prehistoric Cultures of the Horn of Africa.* Cambridge. pp. 286–7 and Frontispiece.

12 Turnbull, C. M. (1968). The importance of flux in two hunting societies, *in* Lee, R. B. and DeVore, I. (eds.). *Man the Hunter*. Chicago. pp. 132–7.

13 Grabus, J. (1951–2). Contribution à l'étude des Ne'madi chasseurs archaïques au Djouf, *Bull. der Schweizerischen Gesellschaft f. Anthrop. und Ethnologie*, **28.**

14 Lee, R. B. (1968). What hunters do for a living or how to make out on scarce resources, *in* Lee, R. B. and DeVore, I. (eds.). *op. cit.* pp. 30–48; Silberbauer, G. B. (1965). *Bushman Survey*. Gaberones.

15 Turnbull, C. M. (1961). *The Forest People.* New York.

16 Huntingford, G. W. B. (1929). Modern hunters: some account of the Kamelilo-Kapchepkendi Dorabo (Oriek) of Kenya Colony, *J. Roy. Anthrop. Inst.*, **69.**

17 Puccioni, N. (1919). Studi sui materiali antropologici ed etnografici raccolte della Missione Stefanini-Paolo nella Somalia Italiana meridionale, *Archivo per l'Antrop. e l'Etnol.*, **49**; personal observation.

18 Woodburn, J. (1968). An introduction to Hadza ecology and stability

and flexibility in Hadza residential groupings, *in* Lee, R. B. and DeVore, I. (eds.). *op. cit.* pp. 49–55 and 103–10.
19 MacCalman, H. R. and Grobbelaar, B. J. (1965). Preliminary report of the stone-working OvaTjimba groups in the northern Kaokoveld of South West Africa, *Cimbebasia*, **13**.
20 Deevey, E. (1968). Pleistocene family planning, *in* Lee, R. B. and DeVore, I. (eds.) *op. cit.* pp. 248–9.
21 Scudder, T.—The transition from gathering and hunting to consumption agriculture in central Africa. Paper presented at Univ. of Calif. Berkeley Archaeological Facility Seminar, 1969. In Press.
22 Scudder, T. (1962). *The Ecology of the Gwembe Tonga.* Manchester. p. 139.
23 Turnbull, C. M. (1968). *op. cit.* pp. 132–7.
24 Munsen, P. J. (1968). Recent archaeological research in the Dhar Tichitt region of south-central Mauretania, *West African Newsletter*, **10.** Ibadan, Nigeria, pp. 6–13 and personal communication.
25 Williamson, J. (1955). *Useful Plants of Nyasaland.* Zomba.
26 Fagan, B. M. (1967). *Iron Age Cultures in Zambia* (*Kalomo and Kangila*), I. London, pp. 83–85.
27 Smith, E. W. and Dale, A. (1920). *The Ila Speaking Peoples of Northern Rhodesia.* London.
28 Dupire, M. (1962). Peuls nomades: Etude descriptive des Wodaabe du Sahel Nigérien, *Trav. et Mem. Inst. d'Ethnol.*, **64**.
29 Deshler, W. W. (1965). Native cattle keeping in eastern Africa, *in* Leeds, A. and Vayda, A. P. (eds). *Man, Culture and Animals.* Washington.
30 Camps, G. (1969). Amekni: Néolithique ancien du Hoggar, *Mem.* 10. *Centre de Recherches Anthrop. Préhist. et Ethnog.* Paris. Personal observation, Adrar Bous, Ténéré.
31 Fagan, B. M. (1967). *op. cit.* p. 218.
32 Fagan, B. M., Phillipson, D. W. and Daniels, S. G. H. (1969). *Iron Age Cultures of Zambia, II, Dambwa, Ingombe Ilede and the Tonga.* London, p. 87.
33 Fagan, B. M. (1967). *op. cit.* p. 199.
34 Kendall, R. L. and Livingstone, D. A. (1967). *ibid.*; Clark, J. D. (1968). *op. cit.*
35 Wendorf, F. (1968). *The Prehistory of Nubia*, II. Dallas. pp. 595, 808, 941–953.
36 McBurney, C. B. M. (1967). *The Haua Fteah* (*Cyrenaica*) *and the Stone Age of the South-East Mediterranean.* Cambridge. pp. 271–323.
37 Caton-Thompson, G. and Gardner, E. W. (1934). *The Desert Fayum.* London; Wendorf, F. *et al.* (In Press). Problems of dating the Late Palaeolithic in Egypt, *Polish Acad. Sci.*
38 Mori, F. (1965). *Tadrart Acacus: Arte rupestre e culture del Sahara preistorico.* Torino. pp. 234–40.
39 Stenning, D. J. (1965). The Pastoral Fulani of northern Nigeria, *in* Gibbs, J. L. (ed.) *Peoples of Africa.* New York. pp. 361–402.
40 Lewis, I M. (1965). The northern pastoral Somali of the Horn, *in* Gibbs, J. L. (ed.) *op. cit.*
41 Cook, S. F. and Heizer, R. F. (1965). The quantitative approach to the relation between population and settlement size, *Report of the Univ. of California Archaeol. Survey*, **64**.
42 Hugot, H. (1963). Recherches préhistoriques dans l'Ahagger nord-occidental, 1950–57. Mem. 1. *Centre Recherches Anthrop. Préhist. et Ethnog*, Paris.
43 Munsen, P. J. (1968). *op. cit.* p. 9.
44 Camps, G. (1969). *op. cit.* p. 14.
45 Heinzelin, J. de (1957). *op. cit.* pp. 7–8.

46 Arkell, A. J. (1949). *op. cit.* pl. 4.
47 Arkell, A. J. (1953). *Shaheinab.* Oxford. pl. 2.
48 Arkell, A. J. (1953). *op. cit.* pp. 10–18.
49 Munsen, P. J. (1948). *op. cit.*
50 Flight, C. (1967). The prehistoric sequence in the Kintampo area of Ghana. *Paper presented at VIth Pan-African Congress on Prehistory*, Dakar.
51 Davies, O. (1966). The invasion of Ghana from the Sahara in the Early Iron Age, *in Actas del V Congresso Panafricano de Prehistoria y de estudio del Cuaternario*, II. Santa Cruz de Tenerife.
52 Portères, R. (1962). Berceaux agricoles primaires sur le continent africain, *J. Afr. Hist.*, **3** (2).; Schnell, R. (1957). *Plantes alimentaires et vie agricole de l'Afrique noire*. Paris.
53 Davies, O. (1967). *West Africa before the Europeans*. London. figs. 51 a and b.
54 Flight, C. (1967). *ibid.*
55 Fagg, B. E. B. (1944). Preliminary report on a microlithic industry at Rop rock shelter, northern Nigeria, *Proc. Prehist. Soc.*, **10**.; Eyo, E. (1965). 1964 excavations at Rop rock shelter, *West African Archaeol. Newsletter*, **3**.
56 Shaw, T. (1969). Archaeology in Nigeria, *Antiquity*, **43** (171).
57 Coon, C. S. (1968). *Yengema cave report*. Philadelphia Mus. Monographs.
58 Davies, O. (1966). *op. cit.*
59 Davies, O. (1967). Timber-construction and wood-carving in West Africa in the second millennium B.C., *Man*, **2** (1).
60 Coon, C. S. (1968). *ibid.*
61 Shaw, T. (1969). *op. cit.* p. 191.
62 Willett, F. (1962). The microlithic industry from Old Oyo, western Nigeria, *in Actes du IVe Congres Panafricain de Préhistoire et de l'Etude du Quaternaire*, 2. Tervuren. pp. 361–72; Eyo, E. (1965). *op. cit.* p. 7.
63 Davies, O. (1967). *op. cit.* p. 198.
64 Allison, P. A. (1962). Historical inferences to be drawn from the effect of human settlement on the vegetation of Africa, *J. Afr. Hist.*, **3** (2), p. 244.
65 Evans-Pritchard, E. E. (1940). *The Nuer*. Oxford. p. 63.
66 Connah, G. (1967). Progress report on archaeological work in Bornu, 1964–66, with particular reference to the excavations at Daima Mound, *Northern History Research Scheme, 2nd Interim Report*, Zaria. pp. 17–31; Connah, G. (1968). Radiocarbon dates for Benin City and further dates for Daima, N.E. Nigeria, *J. Hist. Soc. Nigeria*, **4** (2), p. 317.
67 Fagan, B. M. Personal communication.
68 Lebeuf, J.-P. (1962). *Archéologie tchadienne*. Paris.
69 Mauny, R. (1955). Les gisements Néolithiques de Karkarichinkat (Tilemsi, Soudan français), *in* Balout, L. (ed.) *Actes du IIe Congrès Panafricain de Préhistoire*. Paris.
70 B. Wai-Ogosu. Personal communication.
71 Arkell, A. J. (1954). Four occupation sites at Agordat, *Kush*, **2**.
72 Naville, E. (1898). *The Temple of Deir el Bahari*, Part III. Pub. Egypt. Exploration Fund, London.
73 Shiner, J. C. and Wendorf, F. Personal communication.
74 Bailloud, G. (1959). La Préhistoire de l'Ethiopie, *Cahiers de l'Afrique et l'Asie*, **5**; Graziosi, P. (1964). New discoveries of rock paintings in Ethiopia, *Antiquity*, **38**, pp. 91–98, 150–151, 187–90.
75 Anfray, F. (1968). *Aspects de l'Archéologie Ethiopienne*, **9** (3).
76 Arkell, A. J. (1961). *A history of the Sudan*. London. pp. 46–54.
77 De Bayle des Hermens, R. (1969). Résultats d'ensemble des Missions de Recherches Préhistoriques effectuées en 1966, 1967 et 1968 au République Centrafricaine, *Bull. Soc. Roy. Belge Anthrop. Prehist.*, **80**.

78 Nelson, C. M. Personal communication.
79 Leakey, M. D. (1945). Report on the excavations at Hyrax Hill, Nakuru, Kenya Colony, 1937–38, *Trans. Roy. Soc. S. Africa*, **30** (4).
80 Brown, J. (1966). The excavation of a group of burial mounds at Ilkek, near Gilgil, Kenya, *Azania*, **1**.
81 Nelson, C. M. and Merrick, H. V. Personal communication.
82 Leakey, M. D. (1945). *op. cit.* p. 279.
83 Sutton, J. E. G. (1966). The archaeology and early peoples of the Highlands of Kenya and northern Tanzania, *Azania*, **1**, p. 41.
84 Posnansky, M. (1967). The Iron Age in East Africa, *in* Bishop, W. W. and Clark, J. D. (eds.) *Background to Evolution in Africa*. Chicago.
85 Gabel, C. (1969). Six rock shelters on the northern Kavirondo shore of Lake Victoria, *Afr. Hist. Studies*, **2** (2), p. 240.
86 Gabel, C. (1969). *op. cit.* p. 253.
87 Scudder, T. (1962). *op. cit.* pp. 140–7.
88 Schoute-Vanneck, C. A. and Walsh, R. C. (1960). Two Early Iron Age settlements near Durban, *S. Afr. J. Sci.*, **56** (4).
89 Isaac, G. Ll. and Nelson, C. M. Personal communication.
90 Brown, J. (1966). *op. cit.* pp. 72–3.
91 Wright, R. (1961). A painted rock shelter on Mount Elgon, Kenya, *Proc. Prehist. Soc.*, **27**.
92 Leakey, L. S. B. (1931). *The Stone Age Cultures of Kenya Colony*. Cambridge. pp. 91–187; Leakey, L. S. B. (1942). The Naivasha fossil skull and skeleton, *J. East Afr. Nat. Hist. Soc.*, **16** (4 and 5).
93 Boxer, C. R. (1969). *The Tragic History of the Sea*, 1589–1622. Cambridge.
94 Posnansky, M. (1967). Excavations at Lanet, Kenya, 1957, *Azania*, **2**; but see also Sutton, J. E. G. (1966). *Azania*, **1**, pp. 37–58 for suggestion that "Sirikwa Holes" are cattle kraals.
95 Lewis, I. M. (1965). *ibid.*
96 Deshler, W. W. (1965). *ibid.*
97 Gulliver, P. H. (1965). The Jie of Uganda, *in* Gibbs, J. L. (1965). *op. cit.*
98 Fagan, B. M. (1967). *op. cit.* pp. 57–8.
99 Cook, S. F. and Treganza, A. E. (1950). The quantitative investigation of Indian mounds, *Univ. of Calif. Amer. Archaeol. Ethn. Pub.*, **40** (5).
100 Mason, R. J. (1968). Transvaal and Natal Iron Age settlement revealed by aerial photography and excavation, *African Studies*, **25** (4).
101 Phillipson, D. W. (1968). The Early Iron Age site at Kapwirimbwe, Lusaka, *Azania*, **3**.
102 Fagan, B. M. (1964). The Greefswald sequence: Bambandyanalo and Mapungubwe, *J. Afr. Hist.*, **5** (3), p. 338.
103 Fagan, B. M. (1967). *op. cit.* p. 130.
104 Fagan, B. M. *et al.* (1969). *op. cit.* pp. 6–7.
105 Fagan, B. M. Personal communication.
106 Fagan, B. M. *et al.* (1969). *op. cit.* p. 42.
107 Fagan, B. M. (1967). *op. cit.* p. 41.
108 Mason, R. J. (1962). *Prehistory of the Transvaal*. Johannesburg. pp. 301–330; Phillipson, D. W. (1969). The prehistoric sequence at Nakapapula rock shelter, Zambia, *Proc. Prehist. Soc.*, **35**; Robinson, K. R. and Sandelowsky, B. (1968). The Iron Age of northern Malawi: recent work, *Azania*, **3**; Robinson, K. R. (1963). Further excavations in the Iron Age deposits at the Tunnel site, Gokomere Hill, Southern Rhodesia, *S. Afr. Archaeol. Bull.*, **18** (72).
109 Miller, S. F. (1969). The Nachikufan industries of the Late Stone Age in Zambia. Doctoral dissertation, Dept. of Anthropology, Univ. of Calif., Berkeley.
110 Louw, J. T. (1960). *ibid.*

111 Sampson, G. and Sampson, M. (n.d.). Riversmead Shelter: excavations and analysis, *Mem. no.* 3, *National Museum, Bloemfontein.*
112 Sampson, C. G. (1967). Excavations at Zaayfontein Shelter, Norvalspont, northern Cape. *Res. Nat. Mus. Bloemfontein*, **2** (4).
113 Sampson, C. G. (1967). Excavations at Glen Eliot Shelter, Colesberg District, northern Cape, *Res. Nat. Mus. Bloemfontein*, **2** (5 and 6).
114 Clark, J. D. and Walton, J. (1962). A Late Stone Age site in the Erongo Mountains, South West Africa, *Proc. Prehist. Soc.*, **28.**
115 Goodwin, A. J. H. (1936). Vosberg: its petroglyphs, *Ann. S. Afr. Mus.*, **24**; Clark, J. D. (1959). *The Prehistory of Southern Africa.* Harmondsworth. p. 281.
116 Inskeep, R. R. (1959). A Late Stone Age camping site in the Upper Zambezi Valley, *S. Afr. Archaeol. Bull.*, **14** (55).
117 Fagan, B. M. Personal communication.
118 Fagan, B. M. and Noten, F. L. Van (1966). Wooden implements from Late Stone Age sites at Gwisho Hot-Springs, Lochinvar, Zambia, *Proc. Prehist. Soc.*, **32.**
119 Clark, J. D. (1968). *op. cit.* pp. 137–8, 156.
120 Bequaert, M. (1953). Fouilles à Dinga (Congo Belge), *in Actes du Congrès Panafricain de Préhistoire, Algiers* (1952). Paris. p. 348.
121 See Phillipson, D. W. (1969). *ibid.*
122 Clark, J. D. (1950). A note on the pre-Bantu inhabitants of Northern Rhodesia and Nyasaland, *S. Afr. J. Sci.*, **47** (3); Ellenberger, V. (1953). *La fin tragique des Bushmen.* Paris.
123 The writer wishes to record his grateful thanks to Dr. G. Ll. Isaac for his unpublished data on the relationship between settlement area and resident population among the Bushmen and North American groups, as also, together with C. M. Nelson and H. V. Merrick, for information regarding the Long's Drift excavation and other Kenya sites; to Dr. B. M. Fagan for details of diet and settlement area for the Gwisho B. site, Zambia and to Drs. J. C. Shiner and F. Wendorf for the information concerning the ceramic sequence in the Khashm el Girba region, Sudan.

J. DESMOND CLARK

Palaeolithic butchery practices

Meat and the various activities associated with ensuring a regular supply of it are probably the most significant factors involving group social structure among existing hunting populations. The most important sources of meat are mammalian but, with rare exceptions, meat forms only a comparatively small part (about 25%) of the diet of hunter-gatherers and the bulk of their food resources are vegetable[1].

Bones, representing food waste, when found in primary context in association with evidence of the activities of prehistoric man, constitute one of the main sources for reconstructing the economy; indeed, for the Palaeolithic almost the *only* source, in the many regions where vegetable remains have not survived. As a result, early man's propensity for eating meat is often over-emphasized and the stone artefacts associated with food bones are, perhaps too readily, taken to represent equipment connected with hunting and butchering.

Butchering practices can be shown to vary depending often on the species and the size of the animal, but in most cases detailed analysis of the bone waste at Palaeolithic occupation sites is still lacking, rendering any precise comparisons difficult or impossible. The few studies available suggest that it is possible to distinguish several different synchronic patternings of bone with human occupation sites—butchery and kill sites, hunting camps and settlement sites. It is of interest that each of these appears to show very little change through time from the beginning to the end of the Palaeolithic. However, at some Last Glacial sites there is more evidence of the use of bone for various purposes—tools, fuel, dwelling construction, fences, and the by-product of cold storage of meat for food, etc[2].

It cannot, however, be assumed that all split and fractured bone on an archaeological site has been broken by man. Scavengers habitually visit human settlements and camping places and are responsible for fracture and dispersal of the food débris. It is, therefore, necessary to be able to distinguish damage and dispersal due to this and other natural agencies if butchering and food preparation techniques are to be precisely identified[3].

With the Oldowan industry, as at DKI and FLKI sites at the Olduvai Gorge in Tanzania, the scatter pattern and concentrations of the bones, more than the fractures themselves, together with the stone artefacts, provide the best evidence for butchery techniques. Choppers and chopping tools, polyhedrals and unmodified flakes are associated with concentrations of highly comminuted bone while larger bones, little altered, occur in a peripheral situation[4].

Contrary to general belief there is no evidence for a small game economy in the Lower Pleistocene followed, in the Middle Pleistocene, by one where large animals were exploited. A full range of large, medium and small mammal species is represented at the Oldowan sites in Bed I and lower Bed II[5] but it is probable that remains of large animals betokening a greater poundage of meat per animal are more common during the Middle Pleistocene. The site of BKII in upper Bed II at Olduvai provides the first evidence for the driving and butchering of a number of animals (*Pelorovis* and *Libytherium*)[6], and the evidence is even clearer at the Spanish sites (Torralba and Ambrona)[7] where numbers of *Elephas antiquus* and other game were driven and butchered and some use was made of bone tools. Small game are well represented at some Olduvai sites and in possible hominid faeces also, where whole rodents seem to have been chewed up[8]; also at the Mindel Age site of Verteszöllos in Hungary[9] and at the Rissian Upper Acheulean site of Lazaret[10]; fish are present on Oldowan, Acheulean and Perigordian/Magdalenian sites; butchered game is found on site at the Lower and Middle and off site at the Upper Palaeolithic settlements[11]. In Africa, by contrast, certain sites, in particular those with many Acheulean bifaces, have very little bone associated.

It can be shown that Upper Palaeolithic populations made use of one or two mammal species, as at Solutre[12], Hamburgian and Ahrensbergian sites[13], at Vogelherd[14] and at the Moravian[15] and southern Russian Gravettian sites[16]. The same can also be seen to occur with the Mousterian in southern Russia[17] and at Middle Stone Age sites in sub-Saharan Africa[18]. Comparing the African evidence with that from Europe, the range of medium-sized mammal species was perhaps wider in Africa during the Mousterian[19] and Evolved Acheulean[20], no doubt because of the greater variety of species in the Ethiopian fauna.

Butchery techniques are rarely described in other than general terms, with some notable exceptions. At single context butchery sites of large animals, artefacts consist of cutting and chopping equipment[21]. At Upper Palaeolithic camps, projectile points as well as cutting and flensing tools predominate, while settlement sites contain the full range of equipment for the culture stage represented and the bone will be broken into smaller pieces and often burnt as at the Gravettian camps in south Russia[22]. Skinning practices, removing the foot bones and sometimes the tail bones with the skins, can be seen at Lazaret (Upper Acheulean)[23], at Kostenki[24] and at North American sites[25].

Bone waste on Palaeolithic human settlement sites often shows the following characteristics:

(1) The skulls will generally have been broken up into a number of pieces and the fragments widely dispersed, presumably in the process of removing the brain. Mandibles usually have the rami and symphyses removed and the lower margin of the body of the mandible is also generally broken away.

(2) The shafts of long bones, toe bones, etc. containing marrow are fractured, first by removal of the articular ends using a chopper or crushing implement, leaving a depressed or comminuted fracture near the end on one or other face of the bone. Cannon or other bones not containing marrow are rarely broken. Shafts may be broken by twisting the ends in opposite directions producing what has been described as a spiral fracture. The splintered fragments and articulated joints often lie together as discarded suggesting the little meat on them did not warrant further processing and that the bone fragments were held together by sinew.

(3) Vertebrae and rib bones, pelvis, shoulder blades and other bones were more rarely deliberately broken but may show evidence of cutting and chopping as at Choukoutien.

Some bones and antlers also show evidence of percussion fracture and cutting from the use of chopper or pounder and a small percentage show intentional flaking to produce a bone tool, as do those from the Developed Oldowan sites in Bed II at Olduvai[26], from Vallonet[27], from the *Sinanthropus* layers at Choukoutien[28] or from Torralba and Ambrona[29]. A general description of fracture techniques, initially it must be presumed for butchery, are given by Breuil and Lantier[30], by Dart for the Australopithecine site of Makapansgat[31], by Martin for the Mousterian from La Quina and by Semenov for the Upper Palaeolithic Russian sites[32], as also by Kitching for the Creswellian and Mousterian from Pin Hole Cave[33].

It has been generally believed that the selected nature of the surviving parts in bone concentrations also pointed to a human origin for the remains[34]. However, the recent work of Brain on Hottentot bone refuse from butchered goats[35] and of Simons and Isaac on bones in carnivore lairs[36] has shown that there is selective survival of bone fragments in all these cases. Generally speaking, however, where culture exists there is little difficulty in distinguishing the agency of the bone accumulation.

Small and medium-sized animals were usually carried to the camp, just as is done today, but generally only a small proportion of the bones of the skeleton of any large animal are present on the site. Butchering of single large animals away from camp may be carried out with minimal disturbance of the articulated skeleton, the meat only being cut off and the bones left as with the *Elephas reckii* in Bed I or the *Dinotherium* in the base of Bed II at Olduvai[37]. Alternatively, a part only may be left behind and the remainder removed to camp—for example, the hippopotamus carcase at Isimila[38]. Or again, the remains may be disarticulated and broken as with the hippopotamus at Olorgesailie[39], the Mwanganda elephant[40] or with certain New World mammoth kills, for example at Locality 1, Murray Springs[41], Lehner[42], Naco[43] or Santa Isabel Iztapan[44]. At such sites in Africa and Europe the associated artefacts generally consist of choppers, hammers and flake or blade knives and scrapers[45] with, at the New World sites, the addition of projectile points, perhaps used as knives as well as being the weapon with which the kill may have been made[46].

At the Upper Palaeolithic sites in the loess of Moravia and southern Russia the bones of mammoth were regularly carried back to the settlement and separated into piles according to their particular uses—for house or fence construction, as fuel, artefacts or, of course, food[47].

The most complete knowledge of Upper Palaeolithic butchery practices, however, comes from Palaeo-Indian kill sites in the New World. A single animal is generally associated with a small number of fluted points; sometimes, as at Lehner in Arizona where nine immature mammoth and other animal remains occur, choppers and side scrapers are also associated[48]. The remains are usually disarticulated and lie within or adjacent to gullies or swampy basins. They are also displaced though not usually dispersed. The meat seems to have been cut off and either eaten at the site or removed to the settlement for consumption. At the kill and butchery site of Murray Springs in Arizona, the disarticulated bones of five mammoth and two bison lie partly in the stream channel. Here, associated with a hearth and the disarticulated remains of an adult mammoth are three flaking concentrations, the results of resharpening knives and flensing tools most of which had, however, been removed from the site after butchering was complete[49].

After about 8000 B.C. bison replaced mammoth as the main large mammal hunted by the Palaeo-Indians and these were driven over cliffs or into gullies. At Bonfire Cave, Texas, some 120 animals are represented in the lower of two layers of bison bones (*c.* 8280 B.C.), the remains of three separate drives. The carcasses were dismembered by being cut into large joints and, with similar joints from different animals, piled together for further processing, techniques of dismemberment being clearly demonstrated. Twenty-one implements (mostly projectile points and scrapers) and seventeen unworked flakes were associated[50]. Basically the same techniques and the sequence of stages in butchering as well as the method of killing 193 bison are brilliantly recorded at the Olsen-Chubbock bison kill site in Colorado[51].

Clearly, more sophisticated techniques of analysis and classification of fracture types will permit of differentiation between the work of humans and that of other agencies and one of the best ways of producing such knowledge, as well as for interpreting the different identifiable patternings, is from ethnographic and zoological data and by experiment. Because so little change in the nature of the bone waste is recognizable over some two million years, it is believed the ethnographic and zoological evidence has particular relevance. The work of White and others[52] on the butchering techniques used for bison and other animals in the northern Great Plains of America in late prehistoric times shows clearly how little these differ from those of the historic Indians in that region as also from those of the Palaeo-Indian bison hunters. In Africa, Woodburn has described[53] the equipment and methods used in the butchery practices of the hunting-gathering Hadza of the Lake Eyasi Rift and the resulting bone fracture. The use of stone in skinning and cutting up a springbok has been described recently by MacCalman and Grobbelaar[54] for the only hunting-gathering groups still known to be using stone in the African continent and for western Australia Gould[55] gives an account of butchering, transporting and preparing game by the Ngatatjara. Valuable comparative material can also be obtained from experiment as, for example, in the splitting of elephant bone[56], in experiments in skinning, dismembering and removing meat and in breaking bovid bone carried out by Dr. L. S. B. Leakey, Dr. B. M. Fagan and the writer. All this evidence is most relevant for understanding butchery practices and food processing by Palaeolithic man[57].

Notes

1 Lee, R. B. (1968). What hunters do for a living or How to make out on scarce resources, *in* Lee, R. B. and DeVore, I. (eds.) *Man the Hunter*. Chicago. pp. 30–43; Woodburn, J. (1968). An introduction to Hadza ecology, *in* Lee, R. B. and DeVore, I. (eds.) *op. cit.* pp. 49–55.

2 Klein, R. G. (1969). *Man and culture in the Late Pleistocene: a case study*. San Francisco. pp. 75–230; Klima, B. (1962). The first ground plan of an Upper Palaeolithic loess settlement in middle Europe and its meaning, *in* Braidwood, R. J. and Willey, G. R. (eds.) *Courses towards urban life*. V.F.P.A.

3 Leakey, M. D. (in press). *Olduvai Gorge: excavations in Beds I and II, 1960–1963*. Cambridge. Reference to grooving and toothmarks by carnivores on bones on living floors, p. 246. However, depressed fractures on the frontlets of *Parmularis* in Bed I at Olduvai suggest mode of killing by hominids, p. 262. Reference to bones, of which 24% had been gnawed by porcupine showing a preference for post-cranial bones: Mason, R. J., Dart, R. A. and Kitching, J. W. (1958). Bone tools at the Kalkbank Middle Stone Age site and the Makapansgat Australopithecine locality, central Transvaal, *S. Afr. Archaeol. Bull.*, **13** (51); Brain, C. K. (1970). New finds at the Swartkrans Australopithecine site, *Nature*, **225** (5238): Reference work of leopards.

4 Leakey, M. D. (in press). *op. cit.* pp. 21–4; 49–50; figs. 7 and 24.

5 Leakey, M. D. (in press). *op. cit.* pp. 249–57.

6 Leakey, L. S. B. (1958). Recent discoveries at Olduvai Gorge, Tanganyika, *Nature*, **181**.

7 Howell, F. C. (1966). Observations on the earlier phases of the European Lower Palaeolithic, *Amer. Anthrop.*, **68** (2).

8 Leakey, M. D. (in press). *op. cit.* pp. 67; 259.

9 Vertes, L. (1965). Upper Biharian (Intermindel) pebble industry occupation site in western Hungary, *Curr. Anthrop.*, **6** (1).

10 de Lumley, H. (ed.) (1969). Une cabane Acheuléenne dans la grotte du Lazaret (Nice), *Mem. Préhist. Française*, *no.* 7. Paris. Small game pp.

75–94; fish p. 111; bone tools pp. 164–5; possible human coprolites pp. 121–2.

11 Leakey, M. D. (in press). *op. cit.* pp. 251–3; de Lumley-Woodyear, H. (1969). A Palaeolithic camp at Nice, *Sci. Amer.*, **220** (5); Posnansky, M. (1962). Recent Palaeolithic discoveries in Uganda, *in* Mortelmans, G. and Nenquin, J. (eds.) *Actes du 4e Congrès panafricain de Préhistoire et de l'étude du Quaternaire*. Tervuren. pp. 210–211; Clark, J. G. D. (1952). *Prehistoric Europe: the economic basis*. London. pp. 22–33.

12 Arcelin, A. (1890). Les nouvelles fouilles de Solutré, *L'Anthropologie*, **1**; Combier, J. (1967). *Le Paléolithique de l'Ardeche*. Bordeaux.

13 Clark, J. G. D. (1952). *op. cit.*

14 Kurtén, B. (1968). *Pleistocene mammals of Europe*. Chicago. pp. 270–4.

15 Klima, B. (1962). *ibid.*

16 Klein, R. G. (1969). *op. cit.* pp. 216–7, 222.

17 Klein, R. G. (1969). *op. cit.* p. 225.

18 Clark, J. D. (1970). *The Prehistory of Africa*. London. pp. 138–41.

19 McBurney, C. B. M. and Hey, R. W. (1955). *Prehistory and Pleistocene geology in Cyrenaican Libya*. Cambridge. pp. 154–6.

20 Inskeep, R. R. and Hendy, B. (1966). An interesting association of bones from the Elandsfontein fossil site, *in* Cuscoy, L. D. (ed.) *Actas del V Congresso panafricano de Preistoria y de Estudio del Cuaternario*, **II**. Santa Cruz de Tenerife. pp. 109–24.

21 Clark, J. D. and Haynes, C. V., Jr. (1970). An elephant butchery site at Mwanganda's village, Karonga, Malawi and its relevance for Palaeolithic archaeology, *World Archaeol.*, **1** (3), pp. 390–411.

22 Klein, R. G. (1969). *op. cit.*

23 Pillard, B. (1969). Les données palethnographiques apportées par la faune découverte sur le sol de la cabane du Lazaret, *in* de Lumley, H. (ed.) *op. cit.* pp. 177–81.

24 Klein, R. G. (1969). *op. cit.* pp. 140–3.

25 Dibble, D. S. and Lorrain, D. (1968). *Bonfire shelter: a stratified bison kill site, Val Verde County, Texas*. Texas. pp. 9–138.

26 Leakey, M. D. (in press). *op. cit.* pp. 235–47.

27 de Lumley-Woodyear, H. (1969). Le Paléolithique inférieur et moyen du midi mediterrannéen dans son cadre géologique, 1: Ligurie-Provence, *C.N.R.S. Ve supplement "Gallia Préhistoire"*. p. 106.

28 Breuil, H. (1939). Bone and antler industry of the Choukoutien *Sinanthropus* site, *Palaeontologia Sinica*, N.S:D, **6**. Pekin.

29 Biberson, P. and Aguirre, E. (1965). Expériences de taille d'outils préhistoriques dans des os d'éléphant, *Quaternaria*, **7**.

30 Breuil, H. and Lantier, R. (1965). *The men of the Old Stone Age*. London. pp. 38–43.

31 Dart, R. A. (1957). The Osteodontokeratic culture of *Australopithecus prometheus*, *Transvaal Museum Mem.*, **10**. Pretoria.

32 Martin, M. (1963). Comment vivait l'homme de La Quina à l'époque Moustérienne, *Préhistoire*, **5**, pp. 7–23; Semenov, S. A. (1964). *Prehistoric technology*. London. pp. 144–7.

33 Kitching, J. W. (1963). *Bone, tooth and horn tools of Palaeolithic man: an account of the Osteodontokeratic discoveries in Pin Hole cave, Derbyshire*. Manchester.

34 Dart, R. A. (1957). *ibid.*; and Tobias, P. V. (1967). Cultural hominization among the East African Pleistocene Hominids, *Proc. Prehist. Soc.*, **33**.

35 Brain, C. K. (1967). Hottentot food remains and their bearing on the interpretation of fossil bone assemblages, *Scientific Papers of the Namib Desert Research Station*, **22**.

36 Simons, J. W. (1966). The presence of leopard and a study of the food débris in the leopard lairs of the Mount Suswa caves, Kenya, *Bull. Cave*

Exploration Group of East Africa, **1**, pp. 51–9; Isaac, G. Ll. (1967). Towards the interpretation of occupation débris: some experiments and observations, *Kroeber Anthrop. Soc. Papers*, **37**, pp. 31–57; Bones collected from the mouths of hyaena dens by H. Kruuk in the Serengeti, northern Tanzania and now at Berkeley, show longitudinal splitting, spiral fracture and enlargement of the gap between the condyles of long bones, all characteristics claimed as clearly of hominid origin.

37 Leakey, M. D. (in press). *op. cit.* pp. 64, 85–6, fig. 32.

38 Howell, F. C., Cole, G. H. and Kleindienst, M. R. (1962). Isimila, an Acheulian occupation site in the Iringa Highlands, Southern Highlands Province, Tanganyika, *in* Mortelmans, G. and Nenquin, J. (eds.) *op. cit.* sec. III, pp. 43–105.

39 Isaac, G. Ll. (1968). Traces of Pleistocene hunters: an East African example, *in* Lee, R. B. and DeVore, I. (eds.) *op. cit.* p. 259.

40 Clark, J. D. and Haynes, C. V. (1970). *ibid.*

41 Hemmings, E. T. (1968). Preliminary archaeological report on the Murray Springs Clovis site, Arizona. Unpublished paper presented at the 33rd Annual Meeting of the Society for American Archaeology (Santa Fé).

42 Haury, E. W., Sayles, E. B. and Wasley, W. W. (1959). The Lehner mammoth site, south-western Arizona, *Amer. Antiq.*, **25** (1).

43 Haury, E. W. (1953). Artefacts with mammoth remains, Naco, Arizona, *Amer. Antiq.*, **19** (1).

44 Arroyo de Anda, L. A. and Maldonado-Koerdell, M. (1953). Association of artefacts with mammoth in the valley of Mexico, *Amer. Antiq.*, **18** (4).

45 Leakey, M. D. (In press). *op. cit.* pp. 64–6, 85–6, 199–222; Movius, H. L. (1950). A wooden spear of Third Interglacial Age from Lower Saxony, *S.W. J. Anthrop.*, **6** (2); Vertes, L. (1966). The Upper Palaeolithic site on Mt. Henye at Bodrogkereszur, *Acta Archaeol. Hungaricae*, **18**.

46 Hemmings, E. T. (1968). *ibid.*; Haury, E. W., Sayles, E. B. and Wasley, W. W. (1959). *ibid.*; Haury, E. W. (1953). *ibid.*, Arroyo de Anda, L. A. and Maldonado-Koerdell, M. (1953), *ibid.*

47 Klein, R. G. (1969). *op. cit.*; Klima, B. (1962). *op. cit.* Comparative studies are required for bone waste from settlement sites before and after the regular use of fire and the cooking of meat; different hearth patterns appear to be associated with different butchery and cooking techniques with the Upper Palaeolithic in the Dordogne. (See Movius, H. L. (1966). The hearths of the Upper Perigordian and Aurignacian horizons at the Abri Pataud, Les Eyzies (Dordogne) and their possible significance, *Amer. Anthrop.*, **68** (2).)

48 Haury, E. W., Sayles, E. B. and Wasley, W. W. (1959). *ibid.*

49 Hemmings, E. T. (1968). *ibid.*

50 Dibble, D. S. and Lorrain, D. (1968). *op. cit.*

51 Wheat, J. B. (1967). A Palaeo-Indian bison kill, *Sci. Amer.*, *216* (*1*), pp. 44–61.

52 White, T. E. (1952). Observations on the butchery technique of some aboriginal peoples: 1, *Amer. Antiq.*, **17** (4); White, T. E. (1953). Observations on the butchery technique of some aboriginal peoples: 2, *Amer. Antiq.*, **19** (2); White, T. E. (1954). Observations on the butchery technique of some aboriginal peoples: 3, 4, 5, 6, *Amer. Antiq.*, **19** (3); White, T. E. (1955). Observations on the butchery technique of some aboriginal peoples: 7, 8, 9, *Amer. Antiq.*, **21** (2); Kehoe, T. F. and Kehoe, A. B. (1960). Observations on the butchery technique at a prehistoric bison kill in Montana, *Amer. Antiq.*, **25** (3); Wood, W. R. (1962). Notes on the bison bone from Paul Brave, Huff and Demery sites (Oahe Reservoir), *Plains Anthropologist*, 7 (17).

53 Unpublished note on the extraction of marrow from the bones of game animals by the Hadza (1961) written to accompany a collection of fractured bone and photographic record now at the University of California, Berkeley.

54 MacCalman, H. R. and Grobbelaar, B. J. (1965). Preliminary report of the stone-working OvaTjimba groups in the northern Kaokoveld of South West Africa, *Cimbebasia*, **13**.

55 Gould, R. A. (1967). Notes on hunting, gathering and sharing among the Ngatatjara and their neighbours in the west Australian desert, *Kroeber Anthrop. Soc. Papers*, **36**.

56 Biberson, P. and Aguirre, E. (1965). *ibid.*

57 The writer wishes to express his most grateful thanks to Dr. Mary D. Leakey for permission to quote from her forthcoming volume on the archaeology of Beds I and II at the Olduvai Gorge.

I. C. GLOVER

Settlements and mobility among the hunter-gatherers of south-east Asia

In this paper I will discuss some aspects of the ethnography of recently surviving hunter-gatherer societies of tropical south-east Asia which are relevant to problems faced by archaeologists working in that area.

It has been said that tropical forests generally offer an unfavourable hunting environment, since game is mostly to be found in the tree canopy or it is aquatic; and that tropical forest hunting culture is less an adaptation to the forest than to a riverine environment in the forest[1]. On the other hand, tropical rain forest is regarded as the most generalized, productive and stable of the major terrestrial ecosystems[2], and it seems probable that some of the tropical forest collectors of south-east Asia have survived because the equatorial forest is more stable and resistant to man-induced changes than monsoon or temperate forms.

The region considered lies between about 90° to 150° E and from 20° N to 10° S; that is roughly from Assam and East Bengal to the Philippine Islands and from south China to Indonesia and New Guinea. Extensive mountain ranges are found only in Assam, Burma and south China, elsewhere climatic variation depends more on the length of the dry season, which increases with the distance from the equator. The Malayan peninsula, Sumatra and Borneo have an equatorial climate with little seasonal variation while Burma, Thailand and Indo-China to the north, and Java, the Lesser Sunda Islands and southern Papua have more pronounced wet and dry seasons. Seasonal variation in rainfall is the dominating factor controlling vegetation and human settlement. The areas of equatorial climate are still largely covered with rain forest and are sparsely inhabited. In these regions are the surviving hunter-gatherer groups. Human settlement based on intensive agriculture is still mostly confined to areas of greater seasonality, where the original forest has been cleared[3].

South-east Asia, on account of its great variety of locally domesticated food plants, has long been regarded by plant geographers as a probable independent and early centre of plant domestication[4], an opinion which gained some support as a result of excavations at Spirit Cave, Thailand[5].

However, archaeological investigation of the region has been neither systematic nor thorough and many areas are quite unexplored. Only in parts of Malaysia, Thailand, Vietnam and western Indonesia, has a broad outline of prehistoric cultural sequence for the Late Pleistocene been built up, with a slowly developing tradition, the Hoabinhian, of flaked pebble tools on which edge grinding appears towards the end of the Pleistocene and to which pottery is added by at least 6000 B.C.[6]. A seemingly intrusive, and more developed tradition appears by 3000–2000 B.C. and this is often associated with the arrival of Mongoloid peoples from the north bringing a settled way of life, with a developed agriculture based on cereals, root crops and cultivated fruits[7].

Archaeological sites so far investigated in south-east Asia dating from the Late Pleistocene to Early Holocene are nearly all coastal shell middens or caves situated on or close to major river systems[8]. The evidence for the Hoabinhian culture which has been obtained from these sites tends to support the argument that tropical forest hunting culture is an adaptation to riverine (and coastal) environments rather than to the forest itself. How much this is due to the ease of access to, and the visibility of, these sites is problematical. The surviving hunters and collectors of the forests in south-east Asia, as elsewhere, are today found in remote and difficult places since the more favoured and accessible areas have long been appropriated by people with more advanced and aggressive cultures. Some of those that do survive, however, illustrate a degree of adaptation to the forest that would be difficult to reconstruct from the archaeological materials now known.

A recent survey of surviving hunter-gatherer peoples included only brief notes on this area[9]. Nevertheless in many parts of south-east Asia there are groups which until recently were entirely, or largely, dependent on collecting wild foods, and some, their numbers unknown, continue this life today. Best known are the people of the Andaman Islands[10], the Negritos, Senoi and Malayan Aborigines of Malaya and South Thailand[11], the Kubus, Orang Batin and other small groups of central and east Sumatra[12], possibly the Toala of south-west Sulawesi[13], the Punan and Penan of Borneo[14], and some groups in north-east Thailand, variously called Yumbri, Mrabri, Khon Pa and Phi Tong Luang[15]. The correct identification of these forest groups is difficult[16]; many names in the literature are reference terms, often derogatory, given by nearby villagers, such as *sakai* and *kubu*, and others mean no more than "jungle dwellers" in various local languages, e.g. *khon pa*, *orang utan* and *toala*. The hunter-gatherers of south-east Asia belong to a variety of racial groups although there is little agreement on how to classify them. The Andaman Islanders, Semang of Malaya and south Thailand, together with the Aeta of the Philippines are usually called Negrito; the Senoi, Kubu and Toala have, though the term is not accepted today, been described as Veddoid, with the implication that they are relicts of a pre-Mongoloid, non-Negrito population of south-east Asia, while the groups of the Nam Wa valley in north-east Thailand, the Aboriginal Malay (or Jakun), Punan and Penan are largely or completely Mongoloid. Of these peoples, the Negritos of the Andamans, the Kubus of Sumatra between Djambi and Palembang, and the nomads of the Nam Wa valley depend most on hunting and collecting, although all of these people obtain some artefacts, especially iron, and sometimes food from surrounding agricultural populations in exchange for forest products, such as matting, rattan, camphor and beeswax.

Information on these groups which is of particular interest relates to patterns of residence and movement, group size, utilisation of local and wild foods, especially their use of the various tubers and fruit trees which were the staples of the early agricultural peoples of south-east Asia. Chang has recently discussed these crops in detail and conveniently listed them[17].

All the groups under discussion were to some extent nomadic, wandering over extensive territorial ranges in search of wild pigs and smaller game, edible tubers and seasonal fruits. The degree to which these people were sedentary seems to have depended on whether they had access to the coast, although the evidence relating to this is often conflicting. In the Andaman Islands Radcliffe-Brown and Man noted that "forest dwellers are less nomadic than coastal dwellers"[18]. On the other hand, on Little Andaman, Cipriani found possession of a coastal range enabled bands of five to ten families (perhaps twenty to forty people) to live almost permanently in one location in well-built, circular, communal huts, roofed with pandanus matting and furnished with separate raised sleeping platforms for each family[19]. As recently as 1963–4, of the twenty-four communal huts on Little Andaman, fifteen were on the coast and only nine inland, most of those situated near tidal creeks[20]. Communal huts of the Onge on Little Andaman are said to have been regularly swept and surrounded with circles of food refuse. After many rebuildings a hut perched on a mound of shell and bone and organic débris. Cipriani believes that the shell mounds of Great Andaman developed in this way although communal huts had been abandoned there by the middle of the nineteenth century.

In the Andamans local groups are said to have defined and named, though not exclusive, territorial ranges, and the inland tribes, Man's "Eremetaga", did not appear to have much access to the coast[21]. Brown noted that within each territory a number of recognized camps were used repeatedly and sometimes for long periods[22]. A single group might build three sorts of huts for different purposes, of which the circular, communal hut, where it was still built, was the largest and most permanent, serving as a base camp. For brief visits to the forest in the drier season a hunter might build no more than a rectangular lean-to, with a roof of palm or pandanus leaves angled from the ground to two forked sticks. A family, away from the communal hut for more than a few days, would build a lean-to shelter in which four posts may be used to raise the roof further from the ground allowing room for a sleeping frame about eighteen inches high resting on four stakes. A group of families might build a number of such huts around an oval dancing ground, with a communal hearth at one end. In Great Andaman such a village consisting of ten to fifteen structures seems to have replaced the circular hut of Little Andaman by the time of European settlement[23].

The temporary hunter's lean-to of the Andamans is found among most of the other forest groups of south-east Asia. Brandt describes a Negrito camp in south Thailand where the construction and arrangement of houses in an irregular circle facing inwards is close to the Andaman style, except that large bamboos, absent from the Andamans, were used to construct the sleeping platforms[24]. Hagen illustrates such a hut built by the Kubu on the Merung River, Sumatra[25]. The Semang built a greater variety of hut forms than the Andaman Islanders; long oval-shaped communal shelters[26], occasional tree huts for protection against tigers, and small store houses on posts[27]. Caves and rock shelters were also used in which sleeping platforms were built with fires between them and the cave mouth closed by a light fence of branches and rattan[28]. However, the lack of permanent resources of the sea forced the surviving Semang to move frequently and their communal houses were neither so large nor so well built as in the Andamans. Semang bands were said to stay in a single camp for no more than seven to ten days unless they were acculturated to Malay ways[29].

Local groups among the Semang and in the Andamans consisted of five to seven families (twenty to thirty people), one of the "magic numbers" of hunter-gatherer demography, but these figures, as well as those relating to total populations and carrying capacity are not to be relied on because of the effect of disease introduced in the early years of European contact. Within a single camp only the simplest of social divisions have been noted. Evans, for instance, found that in some Semang camps, as in the Andamans, married people, bachelors, widows and unmarried girls occupied separate huts or even areas[30].

A valuable description of a forest camp of the Mrabri, at Dai Thong in Thailand, has been given by Velder, although the wary occupants had moved out before his arrival in January 1963[31]. The camp, in thick bamboo forest, stretched in an irregular line for 160 m across the slope of the hillside. It contained six family sleeping places each with between three and eight beds and had room for forty-five people, although only thirty-one beds were in current use. Some of the sleeping places were as much as 60 m from their neighbours. The only structures were wind-breaks of palm branches placed at the north side of some of the beds, which themselves comprised either bundles of palm leaves or strips of bark lying directly in the ground. Between many beds were small fires to give warmth at night. No permanent artefacts were seen by Velder in the camp; bamboo pipes served for water containers and for cooking, while baskets and mats contained surplus food and any valuable objects obtained in trade from Lao and Thai villagers. Velder suggested that such a large group of Mrabri came together only in the dry season and smaller groups of nine to fifteen people roamed the forest at other seasons, but this is not consistent with Thai accounts which reported that groups were larger and more sedentary in the wet season[32]. Bernatzik, the only anthropologist to spend any time with these "spirits of the yellow leaves", has described six bands ranging in size from three to eleven persons (two to four male adults) and pointed out that frequent shifts of camp (every three to four days) were occasioned more by the desire for social contacts among small family groups, which foraged independently, than from a shortage of food in the vicinity of the camp[33].

Bernatzik stresses that these people whom, he says, called themselves "Yumbri", kept to the dense bamboo forest on the steep slopes of the upper reaches of the rivers. Not only were these areas rich in edible tubers and tree fruits, but here they were comparatively safe from tigers and elephants. For the last reason they avoided waterholes frequented by game and obtained water from the stems of large bamboos. In the wet season the palm leaf wind-break would be leant on two forked sticks and waterproofed with wild banana leaves. A platform of split bamboo angled against the slope kept the sleeper off the damp ground. In addition to small hearths for personal warmth a large fire would be built outside the line of screens to frighten off tigers. Rather more permanent structures are mentioned by other visitors to the area although they are based on second-hand accounts[34]. But even these seem to consist of no more than two lean-to shelters, of the type already described, built close and facing each other with a narrow gap at the ridge to allow smoke to escape. Almost identical huts are described by Evans for some of the Malayan Semang, and by Brown in the Andamans. The latter regarded them as a stage in the evolution of the communal hut[35].

How many of these wandering forest groups still exist in Thailand and Laos is not known. Statements by anthropologists visiting the area for short periods indicate that they are few and their numbers declining. But the forest people are exceptionally shy of unnecessary contacts with strangers, even their trade with Thai and Lao villagers is by silent exchange where this is possible. Haemoglobin analysis of blood samples collected from the Mrabri by Flatz suggested to him that their breeding group is more than the few hundred suggested by Bernatzik and others[36].

The Kubu are the least well known of the four forest peoples. They own dogs, as do all the other groups discussed, obtain iron and cloth by silent barter for jungle products such as beeswax, dammar, gutta percha and rattan, and speak the languages of the neighbouring villagers. Three to five lean-to huts form a camp which is abandoned after a few weeks of occupation. No mention is made of larger and more permanent camps, except among the "tame" Kubu, and the lack of seasonality in the Sumatran climate may account for this. On the other hand, ethnographers appear to have had only fleeting contact with these elusive people, some of whom still live in the forests north of Palembang[37].

In the ethnographic sources which have been mentioned, there are indications that the degree of seasonality in the climate affects the mobility as well as size of collecting groups. In the Andamans, and perhaps in Thailand, the wetter months are spent in larger groups which move less often. In the Andaman Islands local bands have several rather than a single base camp, and where shellfish are available in quantities, these camps are easily recognizable. In northern Thailand the Mrabri and allied groups appear to have developed a continuing way of life in dense bamboo forests on the steep slopes of the headwaters of major rivers. They are not dependent on river or coastal resources, and their settlements would be very difficult to recognize archaeologically. In the equatorial forests, however, where small land game and edible tubers are less common, access to rivers or coasts may be essential. Despite the uncertain accuracy of the demographic data there seems to be agreement that local bands of twenty to thirty people are the norm, that much larger groups are virtually never seen, that such bands, which are not based on any single rule of association, have known territories but do not exercise exclusive rights over a carefully defined stretch of territory. Ownership of resources within a general area may be more important than the territory itself.

Notes

1 Lee, R. B. and DeVore, I. (1968). Problems in the study of hunters and gatherers, *in* Lee, R. B. and DeVore, I. (eds.) *Man the Hunter*. Chicago. p. 6.

2 Harris, D. R. (1969). Agricultural systems, ecosystems and the origins of agriculture, *in* Ucko, P. J. and Dimbleby, G. W. (eds.) *The domestication and exploitation of plants and animals*. London, p. 4.

3 Richards, P. W. (1964). *The tropical rain forest*. Cambridge. p. 12.

4 Vavilov, N. I. (1951). *The origin, variation, immunity and breeding of cultivated plants*. New York. pp. 20–48; Sauer, C. O. (1952). *Agricultural origins and dispersals*. New York; Harris, D. R. (1967). New light on plant domestication and the origin of agriculture: a review, *Geogr. Rev.*, **57**, pp. 90–107; Harris, D. R. (1969). *op. cit.*

5 Gorman, C. F. (1969a). Hoabinhian: a pebble-tool complex with early plant association in south-east Asia, *Science*, **163**, pp. 671–3; Gorman, C. F. (1969b). Hoabinhian transformations in early south-east Asia: a cultural-chronological sequence *c.* 10,000 B.C. to 5500 B.C. (Background data paper presented at the 68th Annual General Meeting of the American Anthropological Association, November 1969, New Orleans (unpublished).

6 Gorman, C. F. (1969b). *op. cit.*

7 Tweedie, M. W. F. (1957). *Prehistoric Malaya*. Singapore. pp. 20–2; Sørensen, P. (1963). North-South. Indications of a prehistoric migration into Thailand, *East and West* (N.S.), **14**, pp. 211–7.

8 Matthews, J. (1961). *A checklist of Hoabinhian sites excavated in Malaya 1860–1939*. Singapore; Matthews, J. (1964). *The Hoabinhian in south-east Asia and elsewhere*. Ph.D. thesis, Canberra, ANU (unpublished).

9 Murdock, G. P. (1968). The current status of the world's hunting and gathering peoples, *in* Lee, R. B. and DeVore, I. (eds.) *op. cit.* p. 17.
10 Mouat, F. J. (1863). *Adventures and researches among the Andaman Islanders*. London; Portman, M. V. (1899). *A history of our relations with the Andamanese*. 2 vols. Calcutta; Brown A. R. (1922). *The Andaman Islanders*. Cambridge; Man, E. H. (1932). *On the aboriginal inhabitants of the Andaman Islands*. London; Sen, P. (1962). *Land and Peoples of the Andamans*. Calcutta; Bose, S. (1964). Economy of Onge of Little Andaman, *Man in India*, **44**, pp. 298–310; Cipriani, L. (1966). *The Andaman Islands*. London.
11 Skeat, W. W. and Blagden, C. O. (1906). *The Pagan Races of the Malay Peninsula*. 2 vols. London; Schebesta, P. (1927). *Among the Forest Dwarfs of Malaya*. London; Evans, I. H. N. (1937). *Negritos of Malaya*. Cambridge; Williams-Hunt, P. D. R. (1952). *An Introduction to the Malayan Aborigines*. Kuala Lumpur; Brandt, J. H. (1965). The south-east Asian Negrito, *J. Siam Soc.*, **53**, pp. 27–44; Dentan, R. K. (1968). *The Semai: a non-violent people of Malaya*. New York.
12 Forbes, H. O. (1885). On the Kubus of Sumatra, *J. Anthrop. Inst.*, **14**, pp. 121–7; Hagen, B. (1908). *Die Orang Kubu auf Sumatra*. Frankfurt; Schebesta, P. (1928). *Orang-Utan: bei den Urwaldmenschen Malayas und Sumatras*. Leipzig. pp. 218–64.
13 Sarasin, P. and F. (1905). *Reisen in Celebes*. 2 vols. Wiesbaden.
14 Hose, C. and McDougall, W. (1912). *The Pagan Tribes of Borneo*, 2. London. pp. 177f; Needham, W. R. (1954). Penan and Punan, *J. Malay Branch Roy. Asiatic Soc.*, **27**, pp. 73–83.
15 Bernatzik, H. A. and E. (1958). *The Spirits of the Yellow Leaves*. London; Kraisri, N. and Hartland-Swan, J. (1962). Expedition to the "Khon Pa", *J. Siam Soc.*, **50**, pp. 165–86; Boeles, J. J. (1963). A second expedition to the Mrabri ("Khon Pa") of North Thailand, *J. Siam Soc.*, **51**, pp. 133–60; Velder, C. (1963). A description of the Mrabri camp, *J. Siam Soc.*, **51**, pp. 185–8; Velder, C. (1964). Die Geister der gelben Blätter—ein Urvolk Thailands ?, *Zeits. f. Ethnologie*, **89**, pp. 10–23.
16 See, for instance, the controversy over the existence of the Punan of Borneo in Needham, W. R. (1954). *op. cit.*, and a number of expeditions in search of Bernatzik's "Spirits of the Yellow Leaves" were unable to discover if they were the same people found by earlier and later visitors to the Nam Wa Valley, Boeles, J. J. (1963). *op. cit.*
17 Chang, K. C. (1970). The beginnings of agriculture in the Far East, *Antiquity*, **44**, pp. 178–85.
18 Brown, A. R. (1922). *op. cit.* p. 30; Man, E. H. (1932). *op. cit.* p. 36.
19 Cipriani, L. (1966). *op. cit.* pp. 56–61.
20 Bose, S. (1964). *op. cit.* pp. 305–6.
21 Man, E. H. (1932). *op. cit.* pp. 30–6.
22 Brown, A. R. (1922). *op. cit.* p. 29.
23 Brown, A. R. (1922). *op. cit.* pp. 409–14.
24 Brandt, J. H. (1965). *op. cit.* pp. 37–8.
25 Hagen, B. (1908). *op. cit.* Fig. 16.
26 Skeat, W. W. and Blagden, C. O. (1908). *op. cit.* 1, pp. 169–73.
27 Evans, I. H. N. (1937). *op. cit.* pp. 49–50.
28 Skeat, W. W. and Blagden, C. O. (1908). *op. cit.* 1, p. 173; Wray, L. (1905). Further notes on the cave dwellers of Perak, *J. Federated Malay States Museum*, **1**, pp. 13–5.
29 Williams-Hunt, P. D. R. (1952). *op. cit.* p. 45.
30 Evans, I. H. N. (1937). *op. cit.* p. 51; Man, E. H. (1932). *op. cit.* pp. 39–40; Brown, A. R. (1922). *op. cit.* p. 35.
31 Velder, C. (1963). *op. cit.*
32 Kraisri, N. and Hartland-Swan, J. (1962). *op. cit.* pp. 172–3.

33 Various observations in Bernatzik, A. H. and E. (1958). *op. cit.*
34 Kraisri, N. and Hartland-Swan, J. (1962). *op. cit.*
35 Evans, I. H. N. (1937). *op. cit.* pp. 51–2; Brown, A. R. (1922). *op. cit.* pp. 413–4.
36 Flatz, G. (1963). The Mrabri, *J. Siam Soc.*, **51**, pp. 161–78.
37 According to information I was given in Djakarta in 1969.

GLYNN LL. ISAAC

Comparative studies of Pleistocene site locations in East Africa

The organization of activities around a localized home base is a distinctive and crucial feature of human behaviour, and in consequence, hominids have surely attached importance to the specific features of potential camp sites through most of the Pleistocene. Factors considered presumably included then as now, comfort, safety and a view over the terrain, as well as economic factors such as proximity to water, gatherable food and hunting grounds. Thus, studies of the location of sites may amongst other things, enable us to augment the biased evidence of tools and food refuse with regard to diet and subsistence activity.

The East African Rift Valley currently provides opportunities for compiling a regional history of patterns of site location preference that spans more than 2 million years[1]. It is at least possible that such a long continuous record will prove of unique value both in reconstructing the earliest phases of hominid activity and in understanding some of the long-term evolutionary transformations that brought human behaviour to the levels of complexity which elsewhere made possible the radical reorganization of economy and society that were involved in "farming" and "urbanization".

One can recognize two scales in the study of Pleistocene settlement patterns. On the one hand, site distribution data can be considered in relation to gross geographic features and ecological zonation; the *Atlas of African Prehistory* compiled by Clark[2] provides an admirable source book for such studies. On the other hand, investigation can be carried out by treating minutiae of the character and placement of sites within a single natural physiographic division such as a lake basin or river valley. Studies of this latter kind are possible even for remote periods of Pleistocene time, but depend on detailed stratigraphic work and careful palaeogeographic reconstruction.

This contribution is concerned with research of the second kind. However, since such investigations are just getting under way in Africa, it is only possible to offer information regarding a few examples in order that some of the problems and the possible implications can be discussed. The data have been

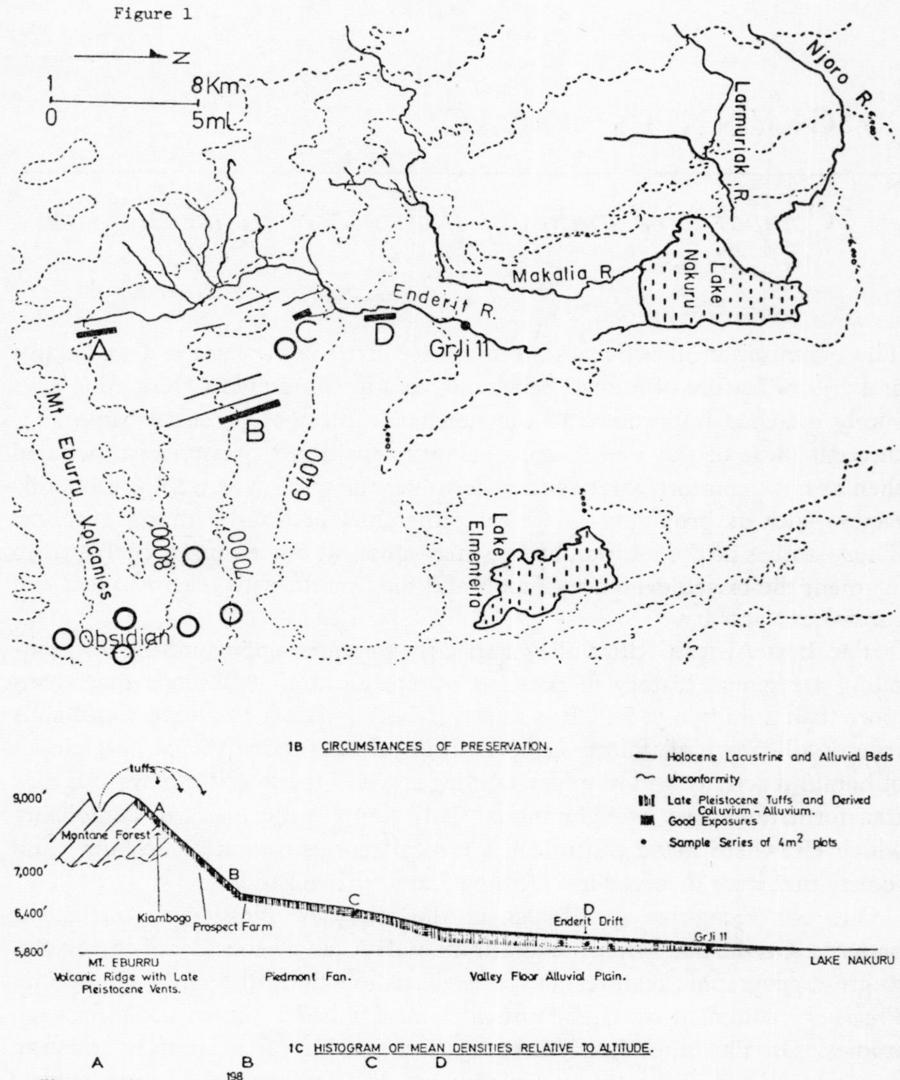
Figure 1
N
1 8Km
0 5ml
Njoro R.
Larmuriak R.
Lake Nakuru
Makalia R.
Enderit R.
A
B
C
D
GrJi 11
Mt. Eburru Volcanics
6400
8000
7000
Lake Elmenteita
Obsidian
1B CIRCUMSTANCES OF PRESERVATION.
Holocene Lacustrine and Alluvial Beds
Unconformity
Late Pleistocene Tuffs and Derived Colluvium - Alluvium.
Good Exposures
Sample Series of 4m2 plots
tuffs
9,000'
7,000'
6,400'
5,800'
Montane Forest
Kiambogo
Prospect Farm
Enderit Drift
MT. EBURRU
Volcanic Ridge with Late Pleistocene Vents.
Piedmont Fan.
Valley Floor Alluvial Plain.
LAKE NAKURU

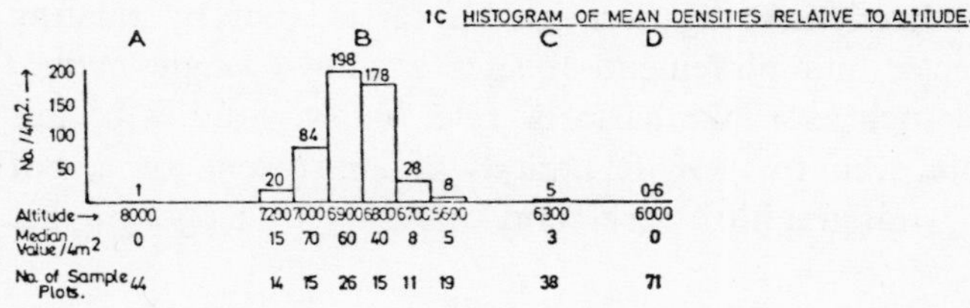
1C HISTOGRAM OF MEAN DENSITIES RELATIVE TO ALTITUDE.
A
B
C
D
No. /4m2
200
150
100
50
1 20 84 198 178 28 8 5 0·6
Altitude → 8000 7200 7000 6900 6800 6700 6600 6300 6000
Median Value/4m2 0 15 70 60 40 8 5 3 0
No. of Sample Plots. 44 14 15 26 15 11 19 38 71

Fig 1 An obsidian density survey in the southern portion of the Nakuru basin. (0°20′ – 0°40′ S, 36°05′ – 36°20′).

A. Contour map showing the sample localities from which the diagrammatic section B has been drawn (heavy lines). Other sample transects are shown with thin lines. Known obsidian outcrops are marked 'o'.

B. Diagrammatic profile from Mt. Eburru to Lake Nakuru showing the stratified Late Pleistocene tuffs with inter-stratified palaeosols and the laterally equivalent colluvium/alluvium on the valley floor. Stratigraphy shows that the lake was at least as small and shallow as it is now. These strata are bracketed between C-14 dates of $>$ 21,000 and $>$ 12,000. Evidence from elsewhere in East Africa indicate colder conditions during part of this period and drier conditions. J. L. Richardson is conducting palaeo-limnological studies in order to determine details of palaeo-environment. Excavations at site GrJi ll on the valley floor showed it to be associated with a sand-filled channel.

C. Histogram showing MEAN obsidian counts per 4 m^2 plot along a composite transect. Obsidian densities on erosion exposures were sampled largely by throwing a hammer about on exposures and recording densities in a 2 × 2 m square wherever it fell. The number of plots and the median counts are shown for each locality included in the histogram. On Mt. Eburru the highest densities occur along the ridge crests. The histogram shows data for the most densely littered ridge, but in spite of between ridge variation *all* of the 4 ridges sampled showed similar patterning with the maximum densities around 7000 ft altitude. The maximum for all ridges is at least 10× greater than any locality mean on the valley floor. The high densities on the ridges are presumed to result from the lag concentration of numerous overlapping occupations catters. The results of Miss B. Anthony's excavations in the area tend to confirm this.

Potential lines of explanation for the localized high densities of obsidian include the following:

1. Abundance of raw material nearby.
2. Localized high frequency of camp occupation owing to preferences with regard to (*a*) topographic situation, and/or
   (*b*) vegetational zonation, and/or
   (*c*) climatic zonation, and/or
   (*d*) access to a variety of differing ecotypes.

Vegetation patterns in the area today have been extensively modified by burning and grazing but the 7000 ft density maximum occurs not far below the lower limit of montane forest vegetation—the boundary being marked by a very pleasant fringe of parkland with table-top acacias grading down to leleshwa bush scrub. Sites at this altitude also enjoy a very wide view of the landscape. The upshot of this is that the maximum densities occur close to the junction of two ecotypes and that this, plus a range of other factors, may well have led to frequent occupation of camps at this zone. Relative proximity (2–3 miles) to raw material may have helped these camps to become especially obvious.

drawn from work in four sedimentary formations along the Gregory Rift Valley. The strata range in age through the full span of Pleistocene time.

Differential preservation poses problems in connection with studies of Early Pleistocene distribution patterns, and this applies to features at both macro- and micro-scales. What survives for study is a highly biased sample of what originally existed over the whole landscape. Preservation is effectively restricted to areas of stable sedimentation, which in turn are ordinarily associated only with the floors of valleys. It is only under rare circumstances, such as may prevail in volcanic terrain or in the loess areas of Eurasia that whole landscapes were covered by mantles of sediments. Knowledge of the existence of such rare conditions of preservation in the Nakuru lake basin, Kenya, led to its choice in 1969 as a region for detailed interdisciplinary study of palaeo-environments and settlement patterns[3]. The research is still in progress, but certain striking features of the Late Pleistocene archaeological distribution have been worked out and a brief account of these is presented as the first example. Because of the unusual comprehensiveness of topographic representation it can be used in part as a yardstick by which the data for the other three basins can be judged.

Example 1:

The Nakuru Lake Basin in the Late Pleistocene (*c.* 30,000–13,000 C-14 years B.P.) (Fig. 1).

A continuous mantle of stratified tuffs and reworked tuffaceous colluvium and alluvium extends from the top of Mount Eburru (approximately 9000 ft) down to the more or less flat plains north of the modern Lake Nakuru (approximately 5800 ft). The mantle contains obsidian artefacts of Middle Stone Age or "Kenya Stillbay" character. Locally these are superabundant, while elsewhere there may only be a very sparse scatter. A systematic sampling programme enabled us to define aspects of the density pattern and to explore possible inferences regarding land use and settlement patterns.

Maximum obsidian artefact density is exposed within a zone extending along the north face of Mt. Eburru at altitudes ranging only 100 or 200 ft (40–60 m) on either side of the 7000 ft (2140 m) contour. At this altitude very considerable areas are littered with impressive quantities of artefacts.

The less extensive exposures on the alluvial plains show general densities lower by at least an order of magnitude than those on the mountain. Only one site concentration was found (GrJi ll); and this was situated in association with a sandy stream channel—a point of great interest in relation to features of the earlier settlement patterns discussed below.

The 7000 ft zone of maximum artefact density coincides approximately with the lower limits of the montane forest, and it is of interest to note that the same altitudinal zone and ecotone is said to be preferred for settlements belonging to the modern Wa'Ndorobo, who practise hunting and collecting on the escarpment west of the Nakuru basin[4].

Further details of this study and some discussion of interpretation are given in the explanation of Fig. 1.

Example 2:

The Olorgesailie Basin in the Middle Pleistocene (?400,000 K/Ar years B.P.)[5].

The map in Fig. 2 illustrates the main features of the evidence. It is immediately clear that the sample is restricted to the basin floor and thus covers only one half of the topographic range represented in the Nakuru basin. The fact that more sites have been located on the alluvial and lacustrine flats at Olorgesailie than at Nakuru may be due to more extensive exposure of this sedimentary facies at Olorgesailie and the available samples do not allow us to distinguish density differences. Clearly at Olorgesailie there are no means of assessing the relative importance of upland and lowland site locations.

The data indicate a very specific pattern of preferences for site locations within the basin floor environment (table in Fig. 2). Camp places were habitually chosen along the course of abandoned or seasonal channels with sandy stream beds. At least 75% of sites for which a distance could be determined were more than 4 km away from the contemporary lake shores. No sites have been found that were definitely very close to the margin of lake-waters or swamps.

Example 3:

The Natron Basin. The Early Middle Pleistocene (?1·5 m K/Ar years B.P.)[6].

Good exposures enable artefact distribution to be observed over a wide area as illustrated in Fig. 3. The western margin of the basin consists of well-watered hilly country and as at Olorgesailie no archaeological study of Middle Pleistocene archaeological distribution was possible in the uplands. However, transects have been traced from the hills into the drier centre of the basin where a somewhat salty lake exists. There was a thin scatter of artefacts, and though densities are very much lower than in either the Olorgesailie or Nakuru basins, one hominid jaw and two Early Acheulean sites were found. It was not feasible to collect quantitative data on such very low densities, but the frequency of artefacts was noticeably higher at the eastern extremity of the exposed transect and the richest site, RHS, was found there. This may indicate that the focus of land-use in this basin was in the hilly catchment area with only intermittent activity out on the deltaic flats and lake flood plains.

The two Natron sites conform to the Olorgesailie site preference patterns and both were located on a sandy sub-stratum in or adjacent to stream channels.

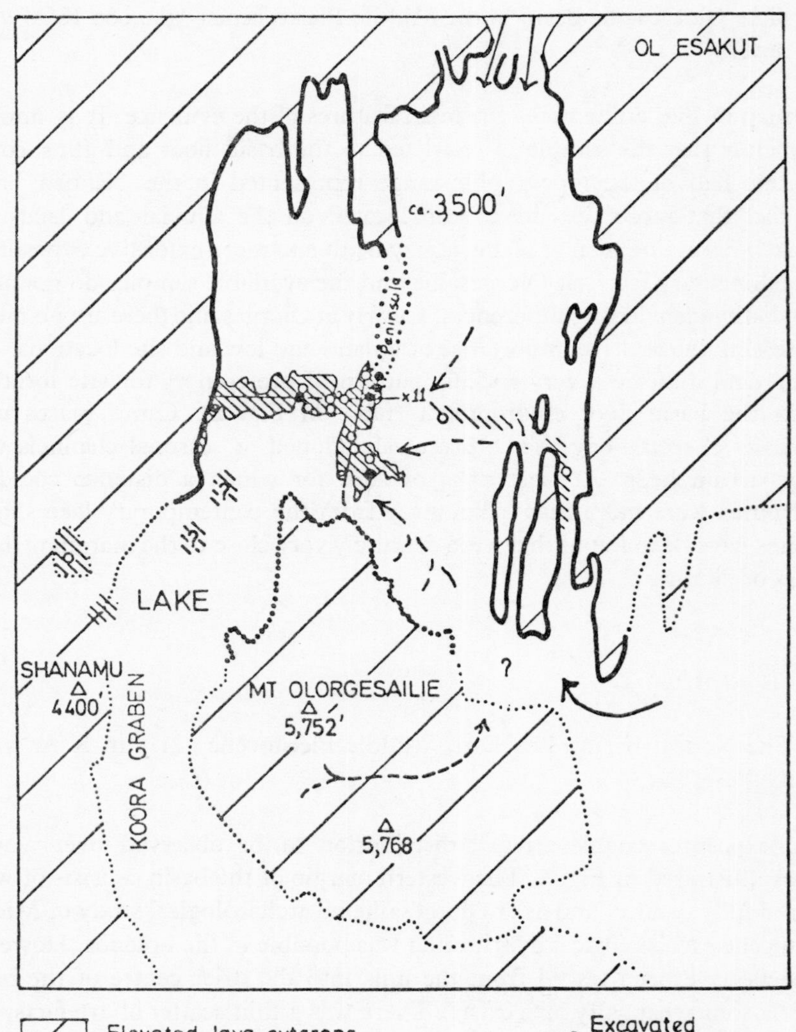

Example 4:

Part of the Lake Rudolf Basin in the Early Lower Pleistocene (?2·61 ± 0·26 m K/Ar years B.P.)[7].

Fig. 4 provides a sketch map of the small area hitherto studied in detail. Lower Pleistocene artefacts were recovered by excavation from two localities: FxJi 1 and FxJj 3; but other material was located along outcrops as shown. Artefacts, which consist largely of unretouched flakes, plus a few core-choppers, were found to be distributed at low densities all along the course of one or more meandering stoneless stream channels that had silted with stream-borne tuffs and which had in some localities been filled by fine aeolian tuff after abandonment. Tool making and using may have been effectively restricted to the stream courses, since no artefacts could be found on laterally equivalent surfaces or beds. The principal site, KBS or FxJj 1, appears to be a low intensity "home base" site with a thin scatter of stone (approximately 1·15 per m^2). At one other site, FxJj 3, there appears from surface evidence to be a significant association between flakes and parts of a hippopotamus carcass. Further excavation will be required to verify this.

Available stratigraphic data indicate that these channels meander across a flood plain of low relief and were fairly close to the margin of the contemporary lake, which was fresh or at any rate only very weakly brackish. The area was remote from highlands or hard rock outcrops by distances that were probably in excess of 15 km.

Fig 2 A reconstruction of the palaeogeography of the Olorgesailie basin in the Middle Pleistocene.

The available sample series of sites are mainly located out on the basin floor, with the densest cluster (marked xll) occurring where the drainage channels from the eastern half of the basin passed between Mt. Olorgesailie and a rocky peninsula, which was also at times a popular camp locale.

The lowest portion of the basin is in the Koora graben. During the times represented by most of the sites, lake waters were confined to that arm of the basin and were thus remote from the focus of occupation.

The table summarizes details of site locations.

Relation to major topographic feature	Excavated	Surface observations	Total %
Foot of bounding slopes	1	3	4–13%
On rocky peninsula	1	–	1– 3%
Flats near peninsula	4	–	4–13%
'Alluvial' flats	4	–	4–13%
Associated with sandy stream channels	10	7	17–51%*
	—		—
*Includes all the most prolific sites			30

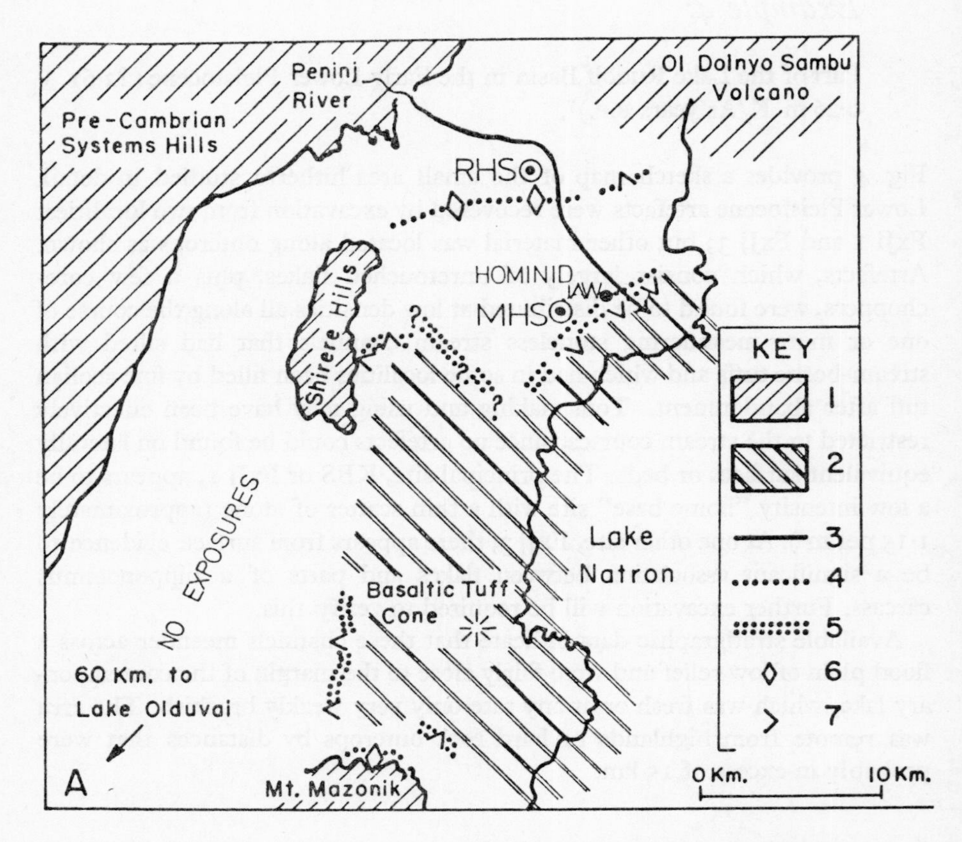

Fig 3 Reconstruction of the palaeogeography of the Natron basin during Humbu Formation times (early Middle Pleistocene).

1. Uplands bounding the basin.
2. Depression ± permanently flooded.
3. Shoreline of the modern lake.
4. Palaeo-shoreline at a time of high water level.
5. Ditto at low water level.
6. Locality at which a shoreline facies change has been observed.
7. Indicates direction in which lake shorelines which cannot be observed are known to lie.

Terrestrial fossils and artefacts are largely confined to exposures in the northern half of the basin with the most dense scatters of artefacts around the foot of the hills to the west. RHS and MHS are the two Acheulean site concentrations.

The proto Lake Natron appears like its modern successor to have been saline and alkaline, except at the time of maximum extent, when it was dilute. The stratum containing the two Acheulean camp sites can be shown to have been a time of low lake level.

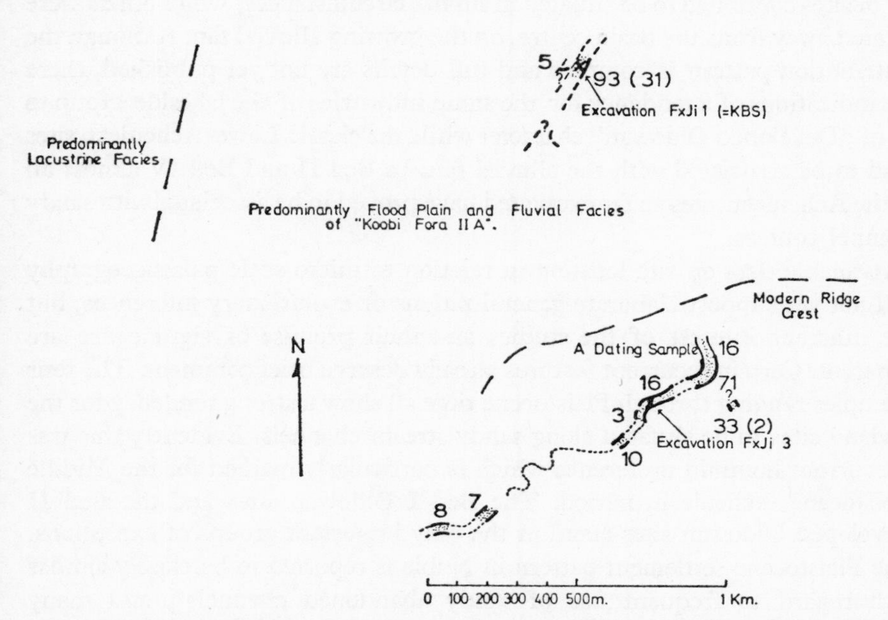

Fig 4 Sketch map showing the reaches of palaeo-channels which it was possible to map within the stratum Koobi Fora II A. (Based on triangulations by Isaac, Barthelme and Mudoga.)

The channels were filled first with stream-borne tuff, and later by aeolian tuff. A large number of K/Ar age determinations on pumices and crystals produced dates in the range 2·25–4·62, with 2·61±0·26 as the best estimate. Technically this is an excellent mean date. Samples from other tuffs are needed to check for possible inherent distortions of true age.

A single channel system can be traced on the south side of the outcrop ridge, but a ridge crest divides the area mapped and precludes the tracing of channel courses in the central portion of the map.

The tuff outcrops are generally so situated that the majority of stones collected on them must be derived from them by erosion. The figures on the map show numbers of stones recorded from each outcrop. At both outcrops where excavation was carried out, artefacts were found stratified within the consolidated tuffaceous siltstones (numbers shown in parentheses).

This is a preliminary report on work that is to be continued, and details are subject to revision.

Discussion

In addition to these examples, research of a similar nature and of special importance has been done by R. L. Hay and M. D. Leakey at Olduvai Gorge, where the sample area is similarly restricted to the floor of a lake basin below volcanic highlands[8]. The Oldowan sites of Beds I and Lower II (approximately 1·9–1·7 million K/Ar years) are confined to flood plain flats adjacent to a salt lake which also had fresh water swamps around it. In Upper Bed II times, one set of sites continued to be situated in similar circumstances, while others were located away from the basin centre, on the growing alluvial fan. Although the distribution pattern is complex and full details are not yet published, there are indications of a tendency for the stone industries of the lakeside group to be of "Developed Oldowan" character while the classic Lower Acheulean sites tend to be associated with the alluvial fan. In Bed II and Bed IV almost all of the Acheulean sites so far excavated have proved to be associated with sandy channel courses.

Available data on site location in relation to micro-scale palaeogeography will not yet support elaborate generalizations or evolutionary inferences, but the inherent interest of the studies and their promise of significance are apparent. Certain recurrent features already deserve brief comment. The four examples ranging through Pleistocene time all show a strong tendency for the lowland sites to be situated along sandy stream channels. Evidently this was a recurrent hominid preference which is particularly marked for the Middle Pleistocene, Acheulean, period. The Bed I Oldowan sites and the Bed II Developed Oldowan sites stand as the only important groups of exceptions. The Pleistocene settlement pattern in Nubia is reported to be closely similar with regard to frequent use of sandy abandoned channels[9], and many Pleistocene sites in other regions appear to reflect similar tastes (e.g. Kalambo Falls, Latamne, Casablanca S.T.I.C., Melka Kontouré, etc.). This kind of site location may also help to account for the numerous localized concentrations of Acheulean tools that have been recovered from European gravel deposits. Behavioural and ecological reasons for this pattern of preference were probably complex and varied. In Africa such ephemeral courses often carry strips of riverine bush, which would have provided shade and gatherable fruits. Also "sand between the toes" as Christopher Robin well knew, provides pleasant sensations[10].

The Nakuru basin study, sounds a warning that although long term preservation of archaeological evidence occurs mainly in low-lying parts of the landscape, the greatest occupation intensities may often have been elsewhere. The fact that stratified settlement pattern samples are liable to bias in this way does not however preclude valuable inferences on economy and activity being based on observed, recurrent features.

Extrapolation from the palaeogeographic data would suggest that the sites dealt with in the examples were all located in proximity to a variety of micro-environments: grassland, swamps, fruit-bearing riverine bush, and in some cases also montane forest. The data could be made to fit with any dietary hypothesis[11] but perhaps in truth it is the *environmental diversity* that should be stressed. This view of Early Pleistocene settlement pattern would be consistent with observations on the broad range of subsistence activity amongst recent non-agricultural peoples of the tropics.

The discussion has been confined to matters pertaining to the location of sites. There are also some vague implications in available data of a trend towards rising densities of artefacts on sites through Pleistocene time, and possible contrasts exist between Pleistocene and Holocene hunter-gatherer sites. For instance, several Late Stone Age sites exist where the density of artefacts and bone refuse is far beyond anything that has yet been observed on any African Pleistocene site: for example the "Prolonged Drift" Kenya Wilton site, GrJi 1, in the Nakuru basin, or the Gwisho Spring site in Zambia[12]. It seems probable that these higher densities may result from increased tendencies to sedentary life, and as such deserve further comparative study. The characteristics of all of the Pleistocene sites known seem consistent with interpretations involving mobility and recurrent shifting of the home base.

As stated at the outset, the "home base" behaviour pattern is one of the fundamental features that distinguish the life of man from that of other primates. The institution of the home bases as an integral part of human socio-economic organization must itself have had evolutionary beginnings. As indicated, home base camp sites were well established 1·8 million years ago at Olduvai[13]; perhaps further studies of the distribution of archaeological material in still earlier strata, such as those of Omo and Koobi Fora, will reveal stages with lower intensities in the localization of food sharing and tool using activities. It may eventually prove possible to discuss not only diversity of land use patterns in space and time, and the evolution of settlement patterns, but also the very processes by which any phenomenon that could be called a "settlement", emerged.

Notes

1 See Isaac, G. L. (1969). Studies of early culture in East Africa, *World Archaeol.*, **1**, pp. 1–28, for discussion of the chronology and nomenclature used here.
2 Clark, J. D. (1967). *Atlas of African Prehistory*. Chicago.
3 Isaac, G. L. with Richardson, J. L. (unpublished research proposal 1968). "Prehistory and Later Quaternary Environments in the Naivasha and Nakuru basins, East Africa." Research funding by the National Science Foundation (GS-2344) is gratefully acknowledged.
4 Personal communication from R. Blackburn.
5 Previous reports on the Olorgesailie basin include the following: Leakey, L. S. B. (1952). The Olorgesailie prehistoric site, *in* Leakey, L. S. B. (ed.) *Proc. First Pan African Congress of Prehistory*. Oxford. p. 209; Isaac, G. L. (1966). New evidence from Olorgesailie relating to the character of Acheulean occupation sites, *in* Cuscoy, L. D. (ed.). *Actas del V Congreso Panafricano de Prehistoria y de Estudio del Cuaternario*, II. Tenerife. pp. 135–145; Isaac, G. L. (1968). Traces of Pleistocene Hunters: an East African example, *in* Lee, R. B. and DeVore, I. (eds.) *Man the Hunter*. Chicago, pp. 253–61.
6 Isaac, G. L. (1967). The stratigraphy of the Peninj Group—Early Middle Pleistocene formations west of Lake Natron, Tanzania, *in* Bishop, W. W.

and Clark, J. D. (eds.) *Background to Evolution in Africa*. Chicago. pp. 229–57.

7 Leakey, R. E. F., Behrensmeyer, A. K., Fitch, F. J., Miller, J. A. and Leakey, M. D. (1970). New hominid remains and early artefacts from Northern Kenya, *Nature*, **226,** pp. 223–30. The expedition is supported by the National Geographic Society. The author's participation was made possible by a grant from the Wenner Gren Foundation.

8 Personal communications and Hay, R. L. (1967). Hominid-bearing deposits of Olduvai Gorge, in *Time and Stratigraphy in the Evolution of Man*. Washington, D.C. pp. 30–42; and Leakey, M. D. (1967). Preliminary summary of the cultural material from Beds I and II, Olduvai Gorge, Tanzania, *in* Bishop, W. W. and Clark, J. D. (eds.) *Background to Evolution in Africa*. Chicago. pp. 417–42.

9 Wendorf, F. (eds.) (1968). *The Prehistory of Nubia*. 2 vols. Dallas; and personal communication by J. de Heinzelin de Braucourt, who also gave advice in the field at Koobi Fora.

10 Milne, A. A. (1924). *When We Were Very Young*. London and New York.

11 Jolly, C. (1970). The seed-eaters: a new model of hominid differentiation based on a baboon analogy, *Man*, **5,** (1), pp. 5–26.

12 See Isaac, G. L. *et al.* (1970 unpublished). Preliminary report on the work of the University of California Archaeological Research Group in Kenya 1969–1970, Berkeley, mimeographed, and J. D. Clark's contribution to this seminar.

13 Discussed in Isaac, G. L. (1969). *op. cit.* p. 11.

R. B. LEE

Work effort, group structure and land-use in contemporary hunter-gatherers[1]

Recent developments in the study of the social organization of hunter-gatherers may have a bearing on problems of prehistory. The rigid territorial model of hunter social groups, epitomized by Radcliffe-Brown's Australian horde, has been superseded by a more flexible grouping frequently changing in size and composition. These flexible groupings, observed in many contemporary hunter-gatherer societies, are not simply a product of depopulation and breakdown brought on by contact. Many of these groupings in the Americas, Australia and Africa show considerable time depth. Furthermore the flexible group can be shown to possess certain adaptive advantages over the more rigid patrilocal form and these features would have been no less advantageous during the pre-contact period.

This paper presents an explanatory model of hunter-gatherer group structure and land-use, based on the three fundamental variables, space, population and energy. The model builds upon the major contributions to hunter-gatherer research of Steward. As the founder of cultural ecology, Steward was the first to combine social systems and environmental variables in a meaningful analytical framework[2]. In contrast to the simplistic view of hunter local groups found in the writings of Radcliffe-Brown[3], Steward saw the hunters as living in a variety of kinds of social organization with a diversity of economic situations ranging from loose- to tight-knit social groups, and ranging from primary dependence on meat in parts of the Arctic to a primary dependence on vegetable foods in tropical areas[4]. The form of economic and social organization, Steward argued, was determined by the particular environmental circumstances to which the group was adapted. Steward proposed a three-part typology of hunter-gatherer bands: the composite, the patrilineal, and the family bands, each of which had its geographic representatives, and the occurrence of each could be explained by reference to significant features of the resource base.

Recent field-work has shown the patrilineal band to be empirically rare as a form of hunter social organization. In its place many field-workers have reported a flexible group structure with local groups assuming a variety of genealogical forms and changing in size and composition with the seasons and even from week to week with a continual reshuffling of sub-groups[5]. The groupings observed by the ethnographers were more or less encompassed by Steward's *composite* band type. Among the advantages noted for this form over the more rigidly territorial patrilocal form were, for example, the ability to adjust group size to resources, the levelling out of demographic variance in sex ratio and family size, and the ability to resolve conflicts by group splitting[6]. These widespread advantages cast doubt on Service's proposition that "the composite band was obviously a product of near destruction of aboriginal bands after contact with civilization[7].

Despite these adaptive advantages, the composite band presents a not altogether satisfying picture. If nothing else the patrilineal band model had possessed a certain elegance in its formulation. In the model, the society is structured by the operation of a small number of fundamental jural rules: territorial ownership by males, band exogamy and viripatrilocal post-marital residence. Similarly, the land-use pattern was very neat: a mosaic of territories arranged in a honeycomb pattern each containing its land-owning group. In comparison the flexible living groups of recent hunters seems confused and disorderly. It was a society in which everyone could live wherever, and with whom, he or she pleased.

Another puzzling element in the understanding of hunter-gatherer society concerned their population dynamics. Hunters live in smaller groups and at lower population densities than any other level of society. In fact it is possible to speak of a Pleistocene "carrying capacity" which DeVore and I set at one person per square mile, a level that probably remained in force throughout human history until it was broken for the first time at the origin of agriculture[8]. How this low level of population density was maintained for so long a period is the crucial question of hunter-gatherer demography. Since food supply appears to be abundant in modern hunters[9], the constant threat of starvation has probably been overestimated as a means of keeping hunting populations in check. But if food supply is not the limiting factor, then what is? This question becomes even more puzzling since it appears that some observed hunter population densities have become stabilized at a mere fraction of the numbers that could be supported by the food supply. How are these low levels achieved and maintained?[10]

The answers to such questions were not to be found by simply collecting more ethnographic data and analysing these data by conventional social anthropological means. What was required was the development of novel approaches for the analysis of group structure, land-use, and population dynamics. And such an approach required the bringing to bear of quantitative ecological, social and demographic data collected over a period of years[11].

The basis of my approach has been to build a model that places a human hunting and gathering population into an ecosystem and watches what happens. The model starts with three components: a population, a space with food resources consumed by the population, and the work that the population performs in order to maintain itself within that space.

I have adopted an explicitly cultural materialist research strategy; in this perspective social organization and ideology are seen as outcomes of the interaction of techno-environmental and demographic variables[12]. The advantages of the cultural materialist research strategy over conventional methodology in social anthropology are: first, it uses units of measurement that can be precisely defined such as spatial units and units of time, and most important, units of energy, such as calories; second it deals with societies as integrated systems, and allows the analyst to examine the relationship between the variables in the system, and to observe how a change in one variable affects the other components of the system. Also I have found it analytically necessary clearly to distinguish between the activities performed by individuals and groups, that can be measured and confirmed by a community of observers, and the individual culture bearer's perceptions of these activities. One of the shortcomings of earlier work on the hunters was that *etic* and *emic* categories were interwoven in the analysis. This is not to say that one approach is more valuable than another—both have a role to play—but only that they must be distinguished analytically. Finally, the materialist research strategy places the study of human society firmly within the context of the natural sciences. By our choice of units of measurement we are enabled to treat natural systems and human systems within a single analytical framework. This latter feature is particularly valuable and necessary if one of our purposes is to illuminate the evolution of human behaviour.

The model of hunter-gatherer adaptation starts with the three variables: numbers, resources, and work; later additional variables such as group size, seasonal variation, conflict and even cognitive variables will be incorporated.

Let us consider each of the primary variables and how each may vary:

The *population* of people occupies the space and its size is subject to increase through birth and immigration and to decrease by deaths and emigration. Also by weighing the people as we have, one can add the concept of biomass.

The *space* contains a patterned distribution of resources of a given density including vegetable and animal foods perceived by the people to be edible; the amount and types of food may vary seasonally and on a longer term basis; for example, with a cycle of drought years and high rainfall. The inter-section of population and space gives one, of course, population density.

Work is the energy—expressed in calories—that the population performs in order to maintain itself within its space.

These are the variables. The constant or the given in this simple model is the techno-economic system: the tools, knowledge and organization necessary to making a living in that environment.

When a research worker enters into a field situation he finds an ongoing system, and his task is to unravel all the relationships between the components of that system.

Every population has to behave in a sufficiently business-like manner to ensure its own survival. Since the group we are studying is surviving it must be that the work level at which it is performing is sufficient for maintaining the population at its current numbers. We can confirm this point empirically by several simple measures such as taking the weights and skinfold measurements and certain biochemical tests on the individuals in the population. If they are maintaining weight, fitness, and fat and muscle mass, we are reasonably sure that intake is adequate. At this stage we may calculate energy figures for the total society in *x* millions of calories consumed per annum against *y* millions of calories expended in subsistence work effort. And we calculate from the total work effort the work effort *per capita* for the society in hours or calories[13].

But totals tells us little about what would happen if there were a change in the parameters. In order to get a handle on that problem we have to observe what happens when a change occurs in variables one and two. If numbers of people increase, due to a rise in births, and food resources remain the same, then work effort will have to increase to feed the greater population, not only total work, but work *per capita* as well. Or if food resources decrease, as in a drought year, but the numbers to be fed remain the same, then work effort *per capita* will have to increase in order to maintain adequate caloric levels. The reason for this *per capita* increase is that work effort is partly a function of distance travelled between the home base and the food supply. As I have discussed elsewhere[14], the longer a group stays in one place the farther the workers have to walk each week in order to reach the food supply. These relationships between numbers, work, and resources can be precisely plotted and reduced to mathematical terms.

The key point I want to make is the central position that *work* plays in the model. Work effort is the intervening variable between population and food supply. And work is the major way that individuals and groups of men respond to changes in the parameters. If food supply decreases, humans do not respond by dying off like fruit flies, they respond by working harder.

Nothing has been said so far about the organization of this population into groups. A great deal of scholarly effort has been directed to the study of ideal group size for hunters, and a figure of twenty-five has been called the magic number of hunter group size[15]. However, given the tremendous variations in the size and composition of individual groups from season to season, I feel this approach is not as fruitful as the approach that starts from work.

Let us assign a strictly arbitrary mean group size of say thirty, just to be different. With a given age-sex composition we note that there will be a level of work effort appropriate to that particular group size—work effort measured in kilometres walked in order to find food. This distance will increase through time as are foods eaten out over an increasing radius. In this illustration, let us say that the mean work effort is two hours per adult per day. Now double the group size to sixty, while holding resources at a constant, and we see that the mean work effort will have to rise accordingly, let us say to three hours. For an even larger group double the group size again, to 120 and the mean work effort will again rise, this time to four and a half hours per adult per day. The point is that each one of these group sizes is quite feasible given the food resources, it is simply a matter that the larger the group, the more work the individuals in it will have to do.

There is a corollary to this relation between work and group size. Resources may be denser in one season than in another so that it may be possible to move from a scarce season group of thirty to an abundant season group of 120 with no increase in the mean daily work effort—at least in the short run. These relationships—work effort against group size and work effort against density of resources—can also be expressed mathematically.

Now we can ask: what determines whether populations of hunters will arrange themselves into groups of thirty, sixty, or 120, or twenty-five, fifty, or 100, and for how long will such groups remain together? The "principle of least effort" would specify that people always arrange themselves into the smallest groups since doing so would always keep work effort at a minimum. Field observations, however, indicate radical departures from this picture. Among the Bushmen large aggregations of 100 persons and over are observed in which individuals put out a considerably larger work effort than would be required if they split up and dispersed into smaller groups of say, twenty to thirty. At other times the Bushmen *were* observed to divide up into smaller living groups and here the work effort was observed to be modest.

These data indicate that the principle of least effort alone is not sufficient to explain the observed living groups of the Bushmen and other hunters. The clue to a possible answer comes from a classic paper by Mauss and Beuchat on the seasonal life of the Eskimo[16]. They distinguished two phases in the annual cycle of the Central Eskimo—the winter phase, when large groups of 100–150 gathered for sealing through the breathing holes and for ceremonials, and a summer phase when the large groups split up into small domestic groups of fifteen to thirty, for fishing and hunting caribou. They called the large group phase, "the public life" and the small group phase, "the private life."

Such a division of the year seems to be characteristic of most hunter-gatherer groups including the Bushmen. In the winter dry season, when water points are limited, the !Kung would aggregate in large groups for a period of more intense social life. This was an exciting time of the year, the time when the men's initiation ceremonies were performed, a time of frequent trance-dancing and curing, and a time for catching up on the news from distant camps, trading, and marriage brokering[17].

This was a period when individuals had a much higher social velocity than at other times of the year. But this intense social life had its disadvantages:

first, because the group was so large, people had to work a lot harder to bring in food and, second, fights were much more likely to break out in larger camps than in smaller camps. I heard about initiation camps disbanding because of disagreement on procedural matters and the groups fanning out back to their home localities. And in eighteen case histories of homicides—for which I have data on the size of the camp in which the killing took place—fifteen killings occurred in larger camps of forty to 150 people and only three in camps of less than forty people.

In short, living in large groups was a mixed blessing. It offered the people a more intense social life but it also meant harder work and a higher frequency of conflict. But, like many other hunter-gatherers, the Bushmen sought both kinds of social existence, the intensity and excitement of a larger grouping and its attendant risks, and the domestic tranquillity and leisure time of smaller groupings. The small group was easier to support, but the larger group was possible to maintain if people were strongly motivated to stay together. Work effort was the sliding scale between group size and resources.

There is another aspect to work. It may be useful as a tie-in between the objective material conditions of life, or *etic* situation, and the individual culture bearer's perception of his role in society and the forces placed upon him, or *emic* situation. How are group size and work effort perceived and reacted to by the individual members of the group? In a large gathering of the kind we have been discussing, an increasingly wider and wider radius of food resources are eaten out with each succeeding week that the group stays together. Accordingly, the work effort required of individuals rises with each week. As work increases, the individual perceives his "costs" of staying together as increasing. He has to work harder than he did last week, or accept a substandard diet or, if he slacks off, come in for the criticism of his fellows.

People get on each other's nerves and as the situation tightens, this leads to heightened irritability and a lowered threshold of conflict. With food getting harder to provide it becomes more important that everyone does his fair share of the work. In this atmosphere even small disagreements may erupt into conflict and this conflict usually results in one or both parties splitting off to seek greener pastures. Hunters say "to hell with it". This is not to say that Bushmen only argue when food is scarce or when the group is large. They argue all the time about lots of things. What I am suggesting is that arguments in large, harder working groups are more likely to lead to a split than in smaller groups.

In contrast to agricultural and urban peoples, hunters have a great deal of latitude to vote with their feet, to walk out of an unpleasant situation. And they do so, not when the food supply is exhausted, but well before that point when only their patience is exhausted.

This mobility has a profound ecological adaptive significance. Fear and avoidance of conflict has the effect of keeping people apart. This perception of the threat of conflict functions to maintain group size and population density at a much lower level than could be supported by the food resources, if the population could be organized to use those resources more efficiently. By stabilizing numbers at a lower level through a behavioural spacing mechanism, the population is buffered against a wide range of variation in the abundance of food resources. Thus conflict, far from being a causal factor that is opposed to an ecological explanation, may best be seen as an ecological variable, a usage that is consistent with contemporary thinking in animal behaviour and animal ecology[18].

The model of Bushmen land-use and social organization presented here may be of use in illuminating Steward's typology of hunter-gatherer bands—particularly the composite band and family band types[19]. About forty-five years ago Bantu-speaking Herero pastoralists began to move into the Dobe Area. They set up their cattle posts at large waterholes and the Bushmen came in and built camps near them in order to drink some of the surplus milk. At /ai/ai for example, where I spent fourteen months in 1968–69 there were at various times 147 Bushmen resident, along with seventy Herero and their 500 head of cattle. Bushmen subsistence was eclectic: they drank some milk and made frequent short hunting and gathering trips to the hinterland. For a long time I was not able to figure out what was going on at /ai/ai, but it is now clear that the Bushmen have achieved there what amounts to a permanent "public life" situation. Where formerly they spent eight to ten months of the year dispersed in their home localities and a few months together in the large groups for dancing and trading, today they have reversed this ratio and can spend most of the year together dancing and trading and just spend a few months split up into smaller groups[20].

The question becomes, how are they able to enjoy the public life without experiencing its disadvantages? Two reasons are suggested. First they get a substantial food input from the Herero—milk, meat and, at times, agricultural products—and this reduces the level of work effort by comparison with what would be required if the same size group were dependent strictly on wild foods. But the second and also important reason is that the Herero provide a legal umbrella under which this large number of feisty Bushmen can live together. Arguments and fights are always breaking out at /ai/ai, but when they do, someone yells "stop everything" and runs to get a Herero, who immediately comes over to break up the fight and adjudicate the dispute. The outside mediator thus becomes a crucial element in stretching the duration of the Bushmen public life beyond the few weeks or months of former years. Even though the food supplies sufficient for such a large group are present, the added element of someone to maintain order is also necessary[21].

In this perspective Steward's family bands and composite bands may actually be different seasonal manifestations of the same social system. The composite band of several hundred persons observed among the northern Athapaskans may be an expression of the public life phase made permanent as the population stabilizes around mission stations or Hudson's Bay posts. And the family bands of the Great Basin Shoshone may have been a stabilization in the private or domestic life phase. The ethnographies note that the Athapaskans split up for part of each year into smaller groups and the Shoshone came together for part of each year into much larger groups than the extended family-sized units that gave them their name[22].

The model presented here offers partial answers to the questions of how hunter-gatherers are organized with reference to food resources and how their organization is affected by culture contact. I have also explored the relation between population aggregation and political development. What happens when more Bushmen get together in one place may have bearing on problems of political evolution.

The model also has something to say about the factors that keep population densities of hunter-gatherers at adaptively low levels. Threat of conflict, it is argued, contributes to keeping people apart, so that the Bushmen's *perception* of the appropriate number of people that can safely live together in one place falls far short of the number that could be supported by the food supply.

The possible utility of this model for prehistory may be briefly discussed. It has been customary to regard hunter-gatherer land-use in terms of a mosaic of territories with a group of a given size encapsulated in each one. The present model visualizes group size as a continuum along the dimension of concentration-dispersion. Any specific group size postulated in this framework is likely to be an ephemeral event in the continual reshaping of groups into larger and smaller aggregates.

On the other hand if the view presented here has validity then it may be useful for prehistorians to consider their settlement pattern data in terms of concentration and dispersion. Sites indicating large population size and sites indicating small population size may represent seasonal manifestations of the same society.

Instead of postulating mean group size of twenty-five, or fifty, or 100 for prehistoric populations it may be analytically more useful to think in temporal terms of the amount of *time* members spend in groups of various sizes.

Notes

1 Data for this paper were collected during three years of field work among the !Kung Bushmen of the Republic of Botswana (1963–1969). I wish to thank the National Science Foundation (U.S.), the National Institute of Mental Health (U.S.), and the Wenner-Gren Foundation for Anthropological Research for financial support of the research.

2 Harris, M. (1968). *The Rise of Anthropological Theory*. New York. p. 654.

3 Radcliffe-Brown, A. R. (1931). Social organization of Australian tribes, *Oceania Monographs*, 1, p. 35.

4 Steward, J. H. (1936). The economic and social basis of primitive bands *in* Lowie, R. H. (ed.) *Essays in Anthropology Presented to A. L. Kroeber*. Berkeley. pp. 331–45; Steward, J. H. (1955). *Theory of Culture Change: the Methodology of Multilinear Evolution*. Urbana.

5 Helm, J. (1965). Bilaterality in the socio-territorial organization of the Arctic drainage Dene, *Ethnology*, **4,** pp. 361–85; Hiatt, L. R. (1962). Local organization among the Australian Aborigines, *Oceania*, **32,** pp. 267–86; Marshall, L. (1960). !Kung Bushman bands, *Africa*, **30,** pp. 325–55; Turnbull, C. M. (1965). *Wayward Servants: the Two Worlds of the African Pygmies*. New York; Woodburn, J. C. (1968). Stability and flexibility in Hadza residential groupings, *in* Lee, R. B. and DeVore, I. (eds.) *Man the Hunter*. Chicago. pp. 103–10.

6 Anderson, J. N. (1968). Comments on the analysis of group composition, *in* Lee, R. B. and DeVore, I. (eds.) *op. cit.* pp. 153–5; Turnbull, C. M. (1968). The importance of flux in two hunting societies, *in* Lee, R. B. and DeVore, I. (eds.) *op. cit.* pp. 132–7.

7 Service, E. R. (1962). *Primitive Social Organization: an Evolutionary Perspective*. New York. p. 108.

8 Lee, R. B. and DeVore, I. (1968). Problems in the study of hunters and gatherers, *in* Lee, R. B. and Devore, I. (eds.) *op. cit.* p. 11.

9 Lee, R. B. (1968). What hunters do for a living, or How to make out on scarce resources, *in* Lee, R. B. and DeVore, I. (eds.) *op. cit.* pp. 30–48.

10 Birdsell, J. B. (1957). Some population problems involving Pleistocene man, *Cold Springs Harbor Symposia in Quantitative Biology*, **22,** pp. 47–69.

11 Birdsell, J. B. (1968). Some predictions for the Pleistocene based on equilibrium systems among recent hunter-gatherers, *in* Lee, R. B. and DeVore, I. (eds.) *op. cit.* pp. 229–40.

12 Harris, M. (1968). *op. cit.* pp. 1–7.

13 Harris, M. (1970). Energy and the factors of production. Unpublished, Columbia University.

14 Lee, R. B. (1969). !Kung Bushman subsistence: an input-output analysis, *in* Vayda, A. P. (ed.) *Environment and Cultural Behaviour*. New York. pp. 47–79. A mathematical formulation of the above is found in: Thomas, H. C. (n.d.). Population dynamics of primitive societies. Unpublished, Harvard University.

15 Birdsell (1968). *op. cit.* pp. 234–6.

16 Mauss, M. and Beuchat, H. (1904–5). Essai sur les variations saisonnières des sociétés Eskimos, *L'Année Sociologique*, neuvième année, pp. 39–132.

17 Lee, R. B. (In Press). The intensification of social life among the !Kung Bushmen, *in* Spooner, B. J. (ed.) *Population Growth: Anthropological Implications*. Cambridge.

18 Wynne-Edwards, V. C. (1962). *Animal Dispersion in Relation to Social Behaviour*. Edinburgh.

19 Steward, J. H. (1955). *op. cit.*

20 Dunning notes a remarkably similar development among the Ojibwa of Pekangekum. Dunning, R. W. (1960). *Social and Economic Change among the Northern Ojibwa*. Toronto. p.85.

21 See also Netting, R. McC. (In press). Sacred power and centralization; some notes on political adaptation in Africa, *in* Spooner, B. J. (ed.). *op. cit.*

22 Steward, J. H. (1938). Basin-plateau socio-political groups, *Bureau Amer. Ethnology*, Bull. **120**; Helm, J. (1961). The Lynx Point people: the dynamics of a northern Athapaskan band, *Bull. Nat. Museum Canada*, **176**.

POVL SIMONSEN

The transition from food-gathering to pastoralism in North Scandinavia and its impact on settlement patterns

On the basis of a series of excavations carried out in the last two decades in the easternmost parts of North Norway, close to the Soviet border, it has been possible to obtain detailed information about the culture, the type of settlement and the method of house construction in that area in the later parts of the Late Stone Age, i.e. in periods contemporary with the Bronze Age and the Early Iron Age farther south[1]. The Bronze Age does not exist in North Scandinavia, the Iron Age culture was introduced *c.* A.D. 200–300, until that time the Stone Age culture survived. The term Late Stone Age therefore covers the whole period 3000 B.C.–A.D. 300, and may be divided into four periods: (1) 3000–2300 B.C., (2) 2300–1800 B.C., (3) 1800–1200/1000 B.C., and (4) 1200/1000 B.C.–A.D. 300[2]. The evidence of settlement patterns and houses is very scanty for period (1), better for period (2), and extensive for periods (3)–(4). But data on annual migration cycles, means of subsistence, and methods of obtaining food, are more or less absent before the beginning of period (3). Thus the following discussion is based on evidence from the periods (3)–(4), *c.* 1500 B.C.–0[3]. This evidence takes the form of a series of settlement sites with house remains, containing rich material, including well-preserved bone implements, and refuse heaps with well-preserved organic material. All the settlements are situated along the old sea coast, which is now elevated to 15–30 m above the modern sea level. The faunal remains show that these coastal sites were seasonally inhabited, viz. in the winter, from October to May, and that they were re-occupied for many years, possibly centuries in succession. The surface burials, solid houses, and the lack of pottery production in the settlements also point to the same season[4].

The determination of the seasons of occupation is based on the following evidence:

(*a*) *Fish:* the most abundant species is cod, although coalfish and haddock are also common. The cod season in this district lasts from February to April. On the other hand, salmon, which abounds in summer and autumn, is completely absent.

(*b*) *Seals:* there a large number of bones of juvenile animals, aged two to four months, which indicates occupation from March to May.

(*c*) *Birds:* there are no bones of young birds, which means that the sites cannot have been occupied after mid-June. Among the bird bones there are a large number of Arctic species which winter in the Varanger fjord from October–April, but there is not a single bone of species which live there between May and August.

All this evidence indicates that the sites were occupied from November to early May (assuming that the climate then was similar to that of the present-day, as is suggested by the geological evidence).

The earliest houses were round or square, light turf huts with a fireplace in the centre; later the shape became more rectangular, with normally two fireplaces, situated on the long axis. In the last millennium B.C. a very solid, large, long ovoid house type was introduced, with two or three fireplaces, sleeping platforms, and from the middle of the lower long side an entrance-corridor of 6–8 m, with doors at both ends. These houses were half sunk in the ground and covered with very heavy turf roofs. They closely resemble the West Greenlander's winter house in near modern times. During 2000 years the interior area of the houses increased from approximately 15–20 m^2 to a maximum of 70 m^2. This increase in association with the increase from one to two or three fireplaces, suggests the possibility of a parallel social development. Does this represent the development of a society of nuclear families into a society of extended families or united family groups? In the seventeenth century A.D. we have the same transition in Greenland, resulting from changing economic conditions, i.e. a change in the hunting and fishing activities. Was the same thing happening two and a half millennia earlier in North Scandinavia?[5]

Turning to the fishing settlements as a whole, the small huts of the earlier periods were crowded in clusters of thirty to forty each, while in the later periods there were only ten to fifteen big houses on each site, indicating a relatively stable number of inhabitants in each "group". Just as among the Eskimos of North Alaska, the houses were arranged in two rows separated by a kind of main street, the backs of the houses of the lower row facing the sea and near to the high tidemark. The doors of the houses faced the street, not the sea, in the early periods, but with the introduction of the entrance-corridor all the houses had to face the sea, because of the drainage of the terrain.

The material found in the refuse heaps on both sides of the outer end of the house corridor indicate that the two most important means of obtaining food were fishing cod and hunting seals. Fishing took place from boats with the help of lines with bone-hooks and stone-sinkers. Very heavy sinkers and the bones of deep-sea fish prove that fishing lines of more than 100 m must have been used. The boats may have been skin-boats of the umiak type, but certain evidence points to the existence also of boats made of hollowed-out tree-trunks. Seals and small whales were harpooned from boats. It is possible that the group inhabiting each house may have been the same as the crew of one boat with a kind of chieftainship similar to the Eskimo "umialiq", or boat-leader[6].

There is far less evidence of activities in the other seasons. On the outer coast and the smaller islands we have found a few settlement sites, where small groups collected eggs, hunted sea-birds, and looked for stranded whales. The animal bones in the refuse heaps show that these places were inhabited in the spring. There are a large number of bones of young birds and of birds belonging to migratory species visiting this area in the summer. In addition, coalfish outnumber cod. No remains of houses apart from a few tent circles have been found at these sites. The small, thin layers of black earth indicate that the sites were only inhabited for short visits and that they were not regularly re-occupied in successive years.

Summer settlements have been found in the interior, on good fishing and hunting grounds near big rivers running into the Arctic Ocean from the wide plains of North Scandinavia. The sites have thick layers of cultural remains. This, however, is not the occupation débris from a series of successive years, but results from much longer periods, although each layer is only of a moderate thickness. The activities at these sites did not result in the accumulation of so much refuse each season. These settlements, however, were occupied or re-occupied for a much longer period than the coastal sites. This has an explanation in the fact that on the coast the post-glacial movement of the land forced people to change their dwelling places to a lower position after some generations. Along the rivers, however, this lowering of the water level was not apparent, and it was possible to live on the same spot for thousands of years. That these river sites were summer settlements, inhabited from June to September, is shown by the houses, which were light structures of birch bark or wattle plastered with clay. Many of them have no fireplaces at all, others have one placed outside the door in the open. The fireplaces were apparently only for cooking, not for warming the house. Unfortunately we have no preserved faunal remains from the inland settlements, but from the implements it is clear that the main occupations were river fishing and hunting. As in the spring settlements, each settlement was occupied by only three to six, or at most ten families. The sea in special fishing seasons can support more people in one place than the land can at any season. Thus each village "group" divided into five to six smaller groups for the summer half of the year[7].

It is impossible to tell whether there were special autumn settlement sites. We can assume, however, that the hunt for wild reindeer must have been of importance, and this is most easily carried out in September and October during the big reindeer migrations from the coast inland. In Finnmark thousands of fall pits for reindeer have been found, and several stone fences for another method of reindeer hunting. None of these, however, can be firmly dated earlier than the eleventh century A.D. However, it is possible that many of them date in fact from the Stone Age. In the winter settlements hardly any reindeer bones or antlers have been found in the refuse heaps, but most of the bone and antler implements were made of reindeer. Thus there must have been a "reindeer season", which was not in the winter. The spring and summer settlements are not situated locations for reindeer hunting. It is assumed, therefore, that there were autumn settlements for hunting reindeer and picking berries on the plains, but no bases of this kind have yet been found.

During the Late Stone Age of North Norway the population lived a semi-nomadic life, migrating along a fixed route every year, living in three to four seasonal sites, and subsisting by hunting and fishing. The winter sites were coastal; the summer sites were inland. This is the natural cycle for a population which depends on a coast which is never frozen. In contrast to this, in northern Russia where the sea freezes every winter for several months, the population lived inland in winter, and along the coast in summer. Inland, near the large lakes Onega, Ladoga, Inari, etc., and near the inner parts of the Baltic Sea, the situation was the same as in northern Russia. Their inland sea (or lake) froze in the winter, so they lived there in the summer, and had their winter settlements in the deep forests of the taiga zone. The result for these people with little or no contact with the sea, was hazardous. In times of stress, as a result of such factors as climatic deterioration or over-population, the inland tribes migrated to the coast and settled there. This has happened several times from *c.* 2200 B.C. to modern times, and parallels to it can be found in every part of the Arctic[8].

As described above, there is a sharp cultural and social boundary line just east of Murmansk on the Kola Peninsula, which for thousands of years has marked the limits of the Gulf Stream, and therefore the limits of the ice-free sea. At *c.* A.D. 200–300 this cultural borderline disappeared. In eastern Finnmark and near Murmansk the people changed their annual cycles to conform with the tribes further east. The faunal remains from refuse heaps along the coast have a distinct summer character, the coastal settlements frequently have no traces of solid structures, and the burials are sunk deep in the ground and must have been dug in a season when the soil was not frozen. It is certain that people dwelt along the sea-coast in summer (salmon predominating among the fish-bones), but the evidence is not detailed enough to indicate whether they lived there for the whole year or had a migration cycle including inland winter places. There is little evidence of the settlement pattern associated with this transition. The village group continued to exist. The interior area of individual houses was reduced, so that it is possible that there was a return to the nuclear family after at least several generations. The Lapps of the district in historical time always lived in nuclear families in small huts[9].

We know that at this period, in the transition from Stone Age to Iron Age, this area received many influences—perhaps also a minor immigration—from the east, partly from as far away as central Siberia. I think, however, that an immigration from an area with summer coastal sites would not have been a sufficient stimulus to change the patterns of behaviour to such an extent. The natural environment itself with its hunting and fishing opportunities, its climate and sea-currents was unaltered and, under hard and marginal living conditions, would have strongly affected every population. Immigrants could not have altered the environment, but would have had to acclimatize themselves to it. There could only have been one exception to this rule: that the immigrants—or the cultural influence—brought with them an innovation in daily life which altered the relationship between man and nature, making him independent of nature or opening up to him a new ecological niche.

In central Siberia, whence the influences ultimately originated, the domestication of reindeer had taken place about 300 years earlier. Could reindeer breeding have been introduced to North Scandinavia about A.D. 200? And could that be the reason for the new annual cycle? The answer to the first question is that domesticated reindeer were apparently unknown in North Scandinavia in the Late Stone Age. On the other hand we know of domesticated reindeer in Scandinavia in the ninth century A.D. This way of life, therefore, must have been introduced before this date. The answer to the second question is that every type of reindeer nomadism, even the most primitive combination of it with hunting and fishing, makes it necessary for the reindeer owner to follow the herd the whole year, not like the huntsman, who only crosses the path of the reindeer once or twice a year. Thus the reindeer owner has to be at the coast in summer, and inland in winter, since this is the annual cycle of the reindeer herds. Definite proof that reindeer breeding was introduced to Finnmark in the second or third century A.D., alongside the transition from stone to iron technology, new shapes in the pottery, new patterns in bone ornamentations, and a new type of fishing hook, can only be obtained by an anatomical analysis of the reindeer bones from the settlement sites dating to before and after this period. Unfortunately there is very little difference between the bones of tame and wild reindeer; the only sure trait is the deformation of the antlers on the gelding rein-ox. If this deformation is common, castration has been practised, and domestication has taken place. There has not yet been any zoological investigation for this purpose, nor can any be undertaken until new and more extensive material has been excavated.

Notes

1 Gjessing, G. (1942). *Yngre steinalder i Nord-Norge*. Oslo.
2 Simonsen, P. (1958). Recent research on East Finnmark's Stone Age, *Rivista di Scienze Preistoriche*, **13**, pp. 131–50.

3 Simonsen, P. (1961–3). *Varangerfunnene II–III*. Tromsø.
4 Simonsen, P. (1965). Settlement and occupation in the Younger Stone Age, *Hunting and Fishing*. Luleå. pp. 397–406.
5 Petersen, R. (In press). On the variations in settlement pattern and hunting occupations in three Greenlandic districts, *Proc. of the Luleå/Tromsø symposium*, 1969.
6 Olsen, H. (1967). *Varangerfunnene IV*. Tromsø.
7 Simonsen, P. (1961–3). *op. cit.*; Simonsen, P. (1957). Bopladserne ved Noatun i Pasvikdalen, *Finska Fornminnesföreningens Tidskrift*, **58**, pp. 233–67.
8 Simonsen, P. (In press). The cultural concept in the Arctic Stone Age, *Proc. of the Luleå/Tromsø symposium*, 1969.
9 Solberg, O. (1909). *Eisenzeitfunde in Ostfinnmarken*. Oslo; Gurina, N. N. (1953). Arkheologičeskiye issledovaniya na severo-zapade Evropeiskoi casti SSSR, *Kratkiye Soobščeniye Instituta Istorii Materialnoi Kulturi*, **49**; Gurina, N. N. (1956). Vestiges de l'age du metal ancien sur le rivage nord de la presqu'île de Kola, *Paléolithique et Neolithique de l'URSS*, Paris; Simonsen, P. (1967). Relations between the Lapps and Scandinavians in early times, *Lapps and Norsemen in Olden Times*, Oslo.

JAMES WOODBURN

Ecology, nomadic movement and the composition of the local group among hunters and gatherers: an East African example and its implications

Although in this paper I write specifically with reference to the Hadza, hunters and gatherers of northern Tanzania[1], my intention is to illuminate some issues which seem to me to be important in the study by both anthropologists and archaeologists of the local organization of nomadic[2] hunting and gathering societies of the present, the recent past and even, with appropriate qualification, of the pre-Neolithic period. My main concern is with ecology and residential alignment, with the extent to which some of the more important characteristics of residential groupings are and are not co-ordinated with, and dependent on, the availability of local wild products—game animals, wild plant foods, the raw material needed for artefacts and so on. I also look briefly at the way in which residential arrangements are associated with some other aspects of the economy and of the social organization more generally.

In earlier publications on Hadza ecology, social organization and material culture[3], and in an ethnographic film about Hadza hunting and gathering[4], I have shown how the 400 or so Hadza who live by hunting and gathering in some 1000 square miles of bush to the east of Lake Eyasi in northern Tanzania obtain their sustenance with relative ease from a wide variety of wild animals and plants and from the honey of wild bees; they satisfy their nutritional needs very adequately without much effort, much forethought, much equipment or much organization. No rights over land and its ungarnered resources are held either by individuals or by groups; any Hadza lives, hunts and gathers anywhere he chooses without any restriction[5].

At any one time these 400 or so Hadza may be living in camps containing very varied numbers of people—from a single person to almost 100 people. The average camp contains about eighteen adults. But the camp is not a fixed unit, either in place, in numbers or in composition. Camps move every few weeks and sometimes much more often. At the time members of a camp all move, they may go together to a new site; they may split up and form camps at two or more new sites; they may go as a body to join some existing camp; or they may divide, some joining an existing camp and others building a camp at new site. Even while people are living together in a camp at a particular site, the composition of the camp changes: some people move in and some move out. In spite of this flexibility there are, needless to say, consistencies in the composition of camps which will be discussed below.

Camp sites and their layouts

Hadza have no base camps and it is unusual to return to huts which have been previously occupied. Certain places are particularly favoured as camping grounds and Hadza may return to them more than once in the course of a year. But usually they choose a fresh site when they return (unless the old site is a rock shelter or has some other intrinsic attraction) even though this may be only 100 yds or less from where they were previously camping. One of the most popular camping grounds in the country is Ugulu hill and there within a circle with a diameter of about a mile it used to be possible to find the remains of some twenty camps which had all been occupied within the previous few years.

For most of the year the Hadza live in simple grass huts put up by the women in an hour or two. In the wet season these are well constructed and are covered with a thick layer of grass to keep out the rain. When really heavy rains occur, some of the Hadza choose to move into rock shelters of which there are large numbers in their country. Perhaps the most important criterion influencing the choice of a particular shelter is the degree of protection which the rock provides from rain and wind. Caves, however, are almost never occupied; the Hadza say they are uninhabitable because of the vermin they contain and certainly on one occasion when some Hadza and I spent a night in a cave at Ketl'ako we were attacked in the night by bugs.

During the dry season, when for six months almost no rain falls, the Hadza often construct purely nominal shelters with a rough framework of branches among which possessions can be stored but with little or no grass covering. Sometimes a camp is built in the middle of a hollowed-out bush (*Cordia gharaf Ehrenb.*, which bears prolific quantities of edible berries, is particularly favoured) and even nominal huts are dispensed with. The branches of the bush provide places where possessions may be hung out of reach of white ants and some shade in the heat of the day.

Unobtrusive sites off the paths used by game and wandering strangers are favoured. There is nothing remarkable in the failure of many of the travellers who have walked through Hadza country to see a single Hadza. Even in 1960 one could spend weeks in their country without encountering them. The population density of very roughly one person to 2 square miles, is comparatively high for a hunting and gathering tribe, but since each camp is usually only about 20 or 30 yds across and is, for most of the year, some miles away from the nearest neighbouring camp, a casual visitor might well imagine the country to be quite uninhabited.

The layout of huts within a camp and the direction that their entrances face is determined in part by the physical characteristics of the site. The direction of the prevailing wind, the amount of flat ground, the presence of trees and rocks—all these and other factors are taken into account by the Hadza when they set up their huts within a camp. But the constraints of the particular bonds of kinship and affinity which they value and which provide routes for important exchanges of goods and services, are equally important. This point can be illustrated quite simply by reference to a dry-season camp of seventeen huts at Ugulu in October 1959. The camp was no more than 30 yds across and was about four miles from the nearest neighbouring camp. The layout of the camp and the relationships of some of its members are shown in Fig. 1.

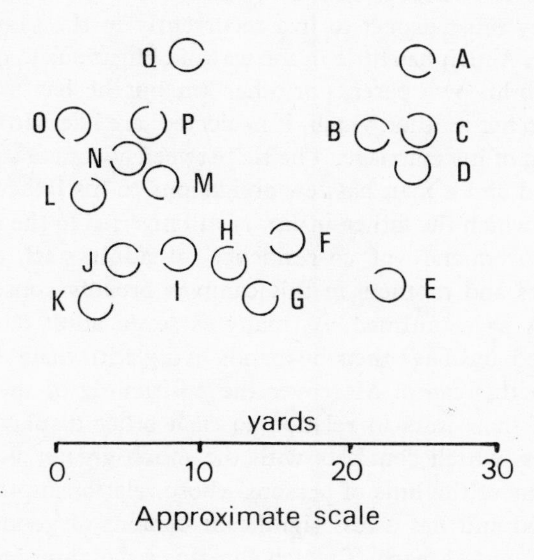

Mother & Father
Son—None
Father—Son
Hut Q—Hut F
Mother—Son
Hut J—Hut K
Mother & Father
—Daughter
Hut A—Hut B
Hut A—Hut C
Hut A—Hut D

Father — Daughter
Hut E—Hut N
Mother—Daughter
Hut J—Hut F
Hut J—Hut H
Hut J—Hut I
Hut Q—Hut O

Fig 1 The residential arrangement of parents and their adult offspring in a camp at Ugulu in October 1959

Of the seventeen huts in the camp, twelve were occupied by a married couple. Five of these married couples shared their huts with their small children up to about six years old. The five huts which did not contain a married couple were each occupied by persons exclusively of the same sex from among the older children, the unmarried adults and the married adults whose spouses were not living in the camp. Sometimes a mother and her adult daughter will share a hut if both have no husband, but otherwise parents and their adult children do not normally live in the same hut. In the camp shown in Fig. 1 there were no instances of parents and their adult children living in the same hut but many instances of parents and their adult children living in the same camp. These instances are listed in Fig. 1.

I have discussed elsewhere[6] the way in which the Hadza, who have a bilateral kinship system, stress the importance of the frequent co-residence of a man with his wife in the camp of her mother. He and his wife may, and often do, live elsewhere from time to time but in general, if the marriage is to persist, they must expect to live recurrently in the same camp as the woman's mother. A man has little in the way of obligations to give goods or to render services to his own parents or other kin but he has important duties to his wife and to her mother which, if neglected, are likely in the end to lead to the breakdown of his marriage. The tie between a woman and her father is not highly valued and a man has few obligations to his father-in-law except in those cases in which the father-in-law is still married to the mother-in-law.

The relative frequency of co-residence of adult sons and daughters with their fathers and mothers in this camp is broadly consistent with the evaluation that I have outlined. As many as seven adult daughters (six of whom are married and have their husbands living with them) are living with their mothers in this camp. Moreover the positioning of their huts and of the entrances of their huts in relation to each other displays a reasonably consistent pattern which contrasts with the much greater flexibility in the mutual alignments of the huts of persons whose relationship to each other is less highly valued and has a less significant content of goods and services. The huts of B, C and D, each of which contains a daughter and a son-in-law of the old couple who live in hut A, are grouped together between 5 and 10 yds away from hut A. Their entrances do not face towards the entrance of hut A, but do face towards each other. Hut O is at a similar distance and similarly faces away from hut Q, in which the mother/mother-in-law is living, and towards the huts of various other Hadza who are not bound to the occupants of hut O by important obligations. The same applies to the positioning of huts H and F in relation to their mother/mother-in-law in hut J. But the woman in hut J has another adult daughter in hut I whose alignment contrasts with the pattern I have described. In this case the huts are close together and face directly towards each other: it is no coincidence that in this case the daughter is not married (although it happens that she is pregnant) and is living in the hut on her own.

Although the pattern is in no sense a rigid one and variations do occur, the same general principles apply in other camps and at other times. The hut of a married couple is normally placed in relation to the hut of the wife's mother, in such a way that she is not directly facing them and is neither too close nor too far away. What underlies this patterning? As I have said, a man has important obligations towards his mother-in-law and she also has some obligations to him. They also have an avoidance relationship of which the residential alignment is one aspect. A man may not talk directly to his mother-in-law, nor approach close to her, nor enter her hut. He may not utter her name. As is well known, avoidance relationships occur widely where both associative and divergent interests are present in the same relationship; the actual and notional segregation of the persons in the relationship undoubtedly reduces the potentiality for confusion and for conflict.

I have earlier stressed the great flexibility of Hadza camp composition and have described this in detail elsewhere[7]. The flexibility of Hadza residential groups is certainly much greater than in the great majority of small-scale societies and is consistent with the relatively small number of what might be described as load-bearing relationships in this society. Outside the man–wife–wife's mother relationship, individual Hadza are not closely bound to *specific* other individuals through kinship, political or other bonds which are load-bearing, that is relationships in which the participants have a clearly defined, systematic set of obligations to perform services or give property to each other. Individual Hadza have some obligations to all the other Hadza who happen to be in the same camp, and other obligations to all the other adults of the same sex who happen to be in the same camp. They do not have important differentiated property obligations to specific individuals within these groupings except in the man–wife–wife's mother triad.

I think that it is likely that these findings are applicable to other societies. Where individuals within a residential unit have, by custom, a load-bearing relationship with each other, then it is probable that their domestic arrangements will be patterned in relation to each other. They may be residentially associated or residentially divided from each other, but if they have control over the placing of their dwellings they will not build without reference to each other.

Camp size

The wide variations that occur in the numbers of people who live in a single camp do not seem to be closely related to fluctuations in the available food resources which can be exploited by Hadza techniques. The numbers of people actually found, while of course they fall within the limits defined by environmental necessities, are not closely patterned by these necessities. I here examine the question of the minimal and maximal numbers of people who find it convenient to live together and the reasons for these effective limits.

An individual on his own can subsist and live entirely by his own efforts for very long periods. Hunting is not essential: roots and berries may provide adequate sustenance. One old man, probably nearing eighty years old, frail and often ill, went off to a place some miles away from his camp to find soft stone to make a stone pipe. He did not return to his camp, nor did he arrive at any other camp and he was widely assumed to have died. For about three months nothing was heard about him but then he suddenly arrived back at the camp where his wife was living. During this time he had lived on his own and kept himself fed adequately although he is far too old to be able to hunt effectively. Another man, a murderer and mentally unstable, lived on his own for a period of years. During this time he occasionally visited camps to obtain metal for arrow-heads and other articles he was unable to obtain himself. But for food he relied exclusively on his own efforts at hunting and gathering. So the acquisition of sufficient food to keep alive and healthy is relatively easy and requires no co-operative effort.

But women are never found living on their own and indeed it would be rare for a woman to go on a journey from one camp to another by herself. The reason is that a woman has no weapons to defend herself against dangerous animals—particularly rhinoceros—or, much more important, against men, Hadza or non-Hadza. Hadza men place a high value on sexual intercourse and a woman encountered on her own in the bush would be in serious danger of being raped. To meet a man of a neighbouring tribe may be even more dangerous. The Tatoga, for example, acquire great status by killing either a person or a lion. The person must be of another tribe but may be male or female. The right hand is cut off to take back as evidence[8]. For a Tatoga to kill a member of an administered tribe is dangerous: the administration are likely to hear of the crime and to hang the murderer. But to kill a Hadza is relatively safe. The Hadza do not normally report murders although a number of them have died by Tatoga spears in the past few decades. Men never venture unarmed into the bush on their own. Women have no weapons with which to defend themselves and justifiable fear keeps them from travelling or living on their own.

A man and his wife often travel together to visit other camps. But they do not normally go to spend nights away from other Hadza except for one purpose: the collecting of honey. Honey is the most valuable of all products used for barter with neighbouring tribes and a woman, who depends on her husband to supply her with trade goods, accompanies her husband and helps him to obtain the product in exchange for which trade goods can so readily be acquired. Honey-gathering is also the only food-gathering pursuit in which husband and wife normally co-operate. For a husband and wife to live in a camp on their own by ordinary Hadza food-gathering methods would mean separating, as a wife does not accompany her husband while he is hunting.

To separate would leave her alone and in danger. Even to remain with the husband is not too safe, for a woman relies on the assistance of other Hadza, especially her close female kin, to protect her against her husband if he attempts to beat her. In the smallest camps one almost never finds less than two adult women. If there are two women they are able to go together to gather food and to assist each other if danger threatens.

What then are the advantages and disadvantages of joining with other Hadza to form large camps? Perhaps the main advantage lies in the fact that where there are more men, more large animals are killed. Although among the Hadza hunting is a solitary affair, the meat of every large animal is shared widely among the members of a camp, as it is among hunters and gatherers almost everywhere whether or not their hunting methods involve systematic co-operation. For a Hadza hunter to fail to share a large animal with other members of the camp in which he is living is to invite violent retribution and is believed in addition to attract serious supernatural penalties[9]. The more hunters there are in a camp, the more frequent in general will be the opportunities to share in eating meat which the Hadza value far more highly than vegetable foods. But living in a large camp does not only increase the frequency of hunting kills: it also reduces the unpredictability. Hunting large animals with bow and arrow is inevitably unpredictable; where the number of hunters is greater, the intervals between kills are not only smaller but are more equal in length. However, these advantages of aggregation apply only to camps of moderate size: it is rare to find a camp containing more than about fifty adults and even a camp of this size would be unlikely to remain together for long. If numbers become as large as this, the amount of meat each individual obtains from, say, a zebra or similar sized animal, tends to drop below an acceptable quantity. The lack of an acceptable amount of meat per capita is particularly evident at the eating of the sacred *epeme* meat by all the initiated men of the camp as a group. The men consider that this communal meal should satisfy their hunger and if it does not do so, there will be much dissatisfaction and the camp is likely to split. The more successful hunters in particular will be tempted to depart.

Lee has presented[10] a cogent argument which suggests that in a given environment individuals in a hunting and gathering camp of above-average size for the area have to work harder to obtain a given quantity of food than individuals in a camp of below-average size exploiting identical resources. Although obviously this is true in many circumstances, I do not believe it to be generally true even in those societies in which there is hardly any systematic co-operation in obtaining food. The more frequent access to the meat of large animals in Hadza camps of average or above-average size reduces work *per capita* in that preservation of meat by drying—which is tedious, time-consuming and renders the meat less palatable—is unnecessary when meat is obtained regularly and when the amount is not so great that it cannot be consumed before it becomes too rotten for Hadza taste. With vegetable resources, Lee's argument will tend to apply when the resources are scarce and widely dispersed but when people are exploiting vegetable foods which occur in such abundance that their use of them does not diminish the quantity available in the area at a much greater rate than new supplies ripen, then members of camps of moderate size will be at no disadvantage as compared with members of smaller camps. Current evidence suggests to me that most hunters and gatherers of which we have knowledge live for at least part of the year in conditions in which members of moderately large camps suffer no economic penalty.

There is never any question of the Hadza (and this probably applies equally well to all other hunting and gathering peoples) being forced to join together as members of large camps because at certain seasons water or other vital resources are concentrated in a small number of localities. One water-hole, or any other localized resource, may be simultaneously exploited by people from different camps. Two or more Hadza camps are often found within a mile of each other and using a single source of water; the vegetable resources the camps exploit will overlap substantially and men will hunt over much the same area. To aggregate into a single camp would be ecologically entirely practicable but could nevertheless create difficulties for members of the two camps. To live together imposes the obligation to share the meat of large animals and I have already pointed out how, if the unit is too large, this obligation creates dissatisfaction. People living in two camps within the same locality are members of residentially segregated meat-sharing groups. Members of two neighbouring camps will reluctantly share meat to some extent with each other if they are unable to conceal the fact that they have made a kill but sharing is always very much less than if they were living in a single camp.

To live in a moderately large camp, has other attractions. Social activities in the camp tend to be more intensive than in small camps and at times men are so preoccupied with these activities that hunting suffers and the advantages of improved access to supplies of meat may cease to apply. In large camps gambling is generally continuous during the hours of daylight. The larger the number of men who participate, the greater the number of valuables there are to be won and the smaller the chance of play being stopped by one individual accumulating all the valuables in the camp. The monthly sacred dance, to which great importance is attached by the Hadza, can also be held much more effectively in a camp where there are plenty of people.

Joint interests in sharing meat and the pleasures of gambling are not enough to bind people together in regular co-residence. The relatively few people who regularly associate with each other in a series of camps are linked to each other by the ties of kinship and marriage that I have, in part, already described. These ties and the sentiments and "property" obligations associated with them assist in the solution of conflict between members of a camp. The larger a camp becomes, the more distant are the ties linking some of the individual members of the camp and the greater the chance that conflict, if it breaks out, will be difficult to solve.

Conflicts which cannot be settled lead immediately to the break-up of the camp unit. Even the threat or possibility of conflict is sufficient to cause people to move away. The danger of conflict is greater, then, in large camps and in these same camps the quantity of meat per head from each kill may also give rise to dissatisfaction. Together these two factors are important and tend to restrict the maximum practicable camp size to about fifty adults, and even camps of this size are essentially unstable and do not maintain themselves as a single unit for long. The relative advantages of aggregation and of dispersal apply rather differently at different times of the year and will be discussed elsewhere; all that needs to be stressed here is that differences in camp size both in a given season and between seasons are a product of multiple causes and are not neatly co-ordinated with variations in the overall amount of food that is locally available.

The movement of camps from one site to another

Camps move (i.e. everyone in the camp leaves the site) on average about once each fortnight. I found that my presence tended to stabilize camps in one place for rather longer than would ordinarily have happened and this estimate is therefore based on camp movements which took place when I was not living in the immediate vicinity. Malangasu, an elderly man who rarely moves away from a site unless all the members of the camp are moving, moved camp seven times during a period of two months (December 1959–February 1960) and twenty-eight times during a period of ten months (September 1960–July 1961). In the dry season (June–November inclusive) Malangasu moves rather less often than in the wet season and may at times spend well over a month at one site; these figures therefore suggest more frequent moves than would be the case if I had a full record of his movements over the whole of the annual cycle. He moved, on average, about every ten days in the wet season and perhaps about half as often in the dry season. Other data suggest similar frequency of movement for other camps in the Sipunga region and probably in the other regions of Hadza country.

Malangasu's twenty-eight moves over a period of ten months all took place within a relatively small part of Hadza country. Five of these moves were of less than a mile and the average for all twenty-eight moves is under five miles. The longest move was less than ten miles. Hadza camps do, at times, move much farther than this, sometimes as much as twenty or thirty miles, but this occurs comparatively rarely.

To discover how Hadza decide when to move to a new site and how they select a new site is a matter of some difficulty. Camps have no acknowledged leaders to take responsibility for making the necessary decisions. It is a matter for all the members of the camp (except for small children who are directly dependent on their parents) to decide individually. Often they come to different decisions and the camp then splits into two or more groups.

The initiative for taking decisions about movement is in the hands of the men. Women do not normally discuss a decision with the men: they either accept it or reject it. Rejection may not even be verbal: the women may show their objection by continuing with their ordinary day-to-day activities and making no preparations for a move. Men have no alternative but to accept a veto of this type.

Among men informal discussions take place about the desirability of moving to various possible sites but no attempt is made to secure general agreement. A move is not delayed if objection is made to a proposed site. The man or men who object have to make up their minds whether it is more to their advantage to go with the other people to the site they have chosen, or to stay in the present place or to move to some different site. Often there is considerable doubt about whether everybody is going to the same place until they actually set off. No planning is necessary before a move and sometimes people set off little more than half an hour after the subject has been raised. Occasionally, if danger threatens, a camp may be abandoned in a few minutes.

The Hadza do not look upon camp movements as tedious or difficult and indeed for people who recognize no rights over objects which are immovable or difficult to move and whose environment is suitable for exploitation by moving, nomadic movements may involve no real problems. The weight of possessions which have to be moved is small enough to be carried with ease over the short distances necessary. And the building of huts at the new site demands only an hour or two's labour by the women. For the Hadza to move is so easy that even a minor reason, such as a vague feeling of dissatisfaction, is quite sufficient to cause them to leave a site.

What are the factors which encourage people to move camp? In the literature on nomadic hunting and gathering tribes great stress is commonly laid on the shortage of food and water and the way in which nomadic movements improve access to food and water resources. For the Hadza at least, and I would imagine more generally among hunters and gatherers, too much emphasis on this type of interpretation is misleading. Movements occur for many other reasons as well as those connected with access to food and water.

In this connection I refer again to Malangasu's twenty-eight moves in ten months of which five were of less than a mile. It is very difficult to imagine how travelling as short a distance as this can make very much difference to access to food. The women of the camp commonly go out three or four miles or more to obtain wild berries and roots; the men may go out further than this if they intend to hunt. Camp movement of this order hardly affects the ease or the success of food-gathering activities. Again, camps often move when food supplies in the neighbourhood are still plentiful and the area to which the camp moves may provide no better access. Sometimes the departure of one camp from a particular site is immediately followed by the arrival of another camp in the same area. This would hardly suggest that access to food and water is the decisive factor.

I have collected some quite detailed quantitative material on the amount of food gathered daily and on the time and effort used. Although the figures are not yet fully worked out, it seems clear that the time and effort spent in obtaining food do not, in general, show any straightforward, progressive increase the longer people stay in a camp. There seems to be no question of a normal pattern of orderly and systematic exploitation of an area until gradually its food resources are depleted and then of movement at this point rather than earlier or later. Far more often the available foods are exploited casually and without much system and then movement occurs well before even reasonably convenient resources are exhausted.

This is not to say that food and water are unimportant. Camps are certainly sometimes moved because food supplies have run short in a particular place and quite often because water sources are exhausted.

Members of a Hadza camp may live on a very meagre supply of water in a hollow in a baobab tree or in a puddle under a rock. When the water is finished and none remains elsewhere within reasonable distance of the camp, a move is regarded as essential. The possibility of obtaining liquid solely from the blood and stomach contents of the animals they hunt and from succulent roots as the G/wi bushmen do for nine months of the year[11] is to a Hadza quite inconceivable.

Availability of some desirable food may stimulate movement to exploit it. If a large animal, such as a giraffe, eland or zebra, is killed some distance from camp, very often people move to eat the animal to avoid having to carry the meat for a long distance. Out of Malangasu's twenty-eight camps, four were described to me as being of this type: twice people of the camp moved to eat giraffes shot by members of the camp and twice they moved to eat elephants shot by European hunters.

On two other occasions Malangasu camped close to a government tsetse clearance camp from which maize flour could be obtained. This is far preferred to ordinary bush vegetable produce. Other advantages are also present: for example, the foreigners working in clearance camps are often keen to trade with the Hadza. Sometimes also neighbours' harvests attract Hadza camps to the area to feast off cereal obtained by begging.

Some camps move, then, because of the failure of their supplies and others because movement will provide an obvious increment of some desirable foodstuff even though existing supplies may be adequate for subsistence. More often than not a move does improve access to food supplies and Hadza regard this as an important aspect of movement but it is misleading to interpret either the frequency or the distance of moves as though these derive in general from the necessities of the food quest.

The desire for products other than food sometimes influences movements.

When trade goods are desired the camp may move close to a neighbouring tribe. When stone is needed for making stone pipes, a camp may be set up close to the area where the appropriate type of stone can be found.

Other environmental factors play a part. Heavy rainfall which penetrates the grass covering of the huts or floods the floor may lead to movement to a rock shelter or to a sheltered site where plenty of grass is available for thatching. After a camp has remained some time in one place insect pests—lice, mosquitos, white ants, etc.—multiply and become troublesome, while the surroundings of the camp become dirty with human excreta and camp débris. The Hadza are less disturbed by these conditions than might be expected, but they do play some part in encouraging frequent movements, especially in the wet season when insects are at their worst and dirt is more obtrusive.

Camps invariably move when a death occurs. Even when there is a miscarriage or a child is stillborn, a move will often be made although this is not always the case. Hadza tend to regard illnesses as linked with the localities in which they occur; if illness breaks out in a camp, the Hadza describe the place as having become hot (*petene*) and rapidly move away. Even a whole region may be thought of as hot and two families whose children had been ill (some had died) while they were living in Tli'ika, gave this as their reason for moving to Sipunga. Other dangers which threaten can be averted by movement. A nightmare of some impending disaster is said to lead to movement to avert the threat. The sound of a bird which the Hadza call the *kurupuko* (a type of woodpecker) is another indication of danger which may, in Hadza theory at least, provoke movement to a new site. Simply to live for a long time in one place is itself thought to be dangerous to the health. Mahela became seriously ill with what appeared to be pneumonia at Ke/he. The universal opinion of the members of the camp was that his illness arose from their long residence—about two months—at this site. Good health is seen as an attribute of regular nomadic movement and failure to move frequently enough leads to illness.

People are constantly moving in and out of camp. The huts of those who have left may remain empty or they may be occupied by newcomers. But as I have already described, the arrangement of huts within a camp is ideally patterned in accordance with the social relations existing between at least some of the individual members of the camp. While this pattern is relatively flexible (much less precise in its form than, for example, the Hei-//om Bushman camp as described by Fourie[12]) individuals who arrive in the camp after it has already been set up may be difficult to accommodate appropriately. Often they move into vacated huts which are far from the kinsmen and affines with whom they should be aligning themselves. Rocks and trees and existing arrangements of huts within the camp may preclude building in any appropriate spot. Camps which have been established in one place for some time may occasionally move a distance of only a few hundred yards so that the huts within the camp may be rearranged.

Movement not only permits the rearrangement of huts in the camp, but also in itself provides a means of validating or, alternatively, of relinquishing social relationships between individuals. For, every time individuals move from one camp to another, they are associating themselves publicly with certain kin and affines and dissociating themselves from others. If a camp divides and a man's mother goes in one direction with one group of people and his wife's mother in another direction with another group of people, he is faced with a direct choice about whom he should accompany. If a relationship is to be maintained, it must be constantly validated by co-residence, for separation may be in itself a denial of relationship. Each time a camp moves and is rebuilt at a new site the alignment of the huts sets out the evaluation by the members of their relationships with each other.

This applies especially to the marital tie. A man is married only so long as he lives regularly with his wife. Once he leaves her for longer than a few weeks "his house has died" (*tan//iaakwa /ets'akoma*) and he has no further rights over her or his children by her. She may marry again as soon as she likes. If her husband returns before she remarries, she may accept or reject him as she chooses.

Hadza believe in the existence of witchcraft, but those who are thought to practise it are not Hadza but neighbouring tribes. The absence of accusations against other Hadza may relate to the ease with which any man or woman may terminate the social relationships in which he or she is involved without endangering his or her vital interests. No important property has to be divided and the actual break itself may be made to coincide with an ordinary camp movement. Enduring hostility of the sort that is in other societies so widely characteristic of the relationship between persons accused of witchcraft and their accusers or supposed victims does not normally arise when the conflicting persons can segregate themselves from each other with such ease.

Nomadic movement therefore may be said to provide a means for the regular positive affirmation of multiple enduring social ties and for the easy segregation of conflicting individuals from each other; the value which Hadza place on movement as such, on movement more frequent than can be explained on ecological grounds alone, is related to its importance for every individual Hadza as a means through which his or her social ties may be manipulated without strain. I would expect that similar considerations are relevant in the interpretation of nomadic movement in other societies, whether of hunters and gatherers or of pastoral nomads, where residential entities change through time not just in size but in composition and where nomadic movement is accomplished without too substantial difficulties. The very widespread enthusiasm for nomadic movement among both hunters and gatherers and pastoral nomads suggests that its benefits for the people concerned are far greater than the immediately obvious ecological ones.

Notes

1 I would like to thank the following bodies which provided funds for field research: The Royal Society; the World Health Organization; the Wenner-Gren Foundation for Anthropological Research; the East African Institute of Social Research; the Goldsmiths' Company; the Smuts Memorial Fund; the Sir Bartle Frere Fund; the Mary Euphrasia Mosley Fund.

2 My use of the term *nomadic* was queried at the Conference. I use the term which by origin of course relates to the seasonal movements of pastoral peoples with their flocks, to refer to the residential movements of non-sedentary hunters and gatherers. It is of course, well known that the residential movements of contemporary hunters and gatherers typically resemble those of pastoral nomads in that they are systematic and seasonally recurrent and it seems appropriate to use this term rather than a less specific and rather misleading term like *migratory*.

3 Woodburn, J. C. (1968a). An introduction to Hadza ecology, *in* Lee, R. B. and DeVore, I. (eds.) *Man the Hunter*. Chicago; Woodburn, J. C. (1968b). Stability and flexibility in Hadza residential groupings, *in* Lee, R. B. and DeVore, I. (eds.) *op. cit.*; Woodburn, J. C. (1970). *Hunters and Gatherers: the Material Culture of the Nomadic Hadza*. London. The material in the present paper is drawn directly from Woodburn, J. C. (1964). *The Social Organization of the Hadza of North Tanganyika*. Doctoral Dissertation, University of Cambridge.

4 Woodburn, J. C. and Hudson, S. (1966). *The Hadza: the Food Quest of an East African Hunting and Gathering Tribe*. (16 mm film: prints are available from Dr. Woodburn and on hire from the Royal Anthropological Institute Film Library at the British Film Institute, or from Concord Films Council).

5 I describe the Hadza as they were in 1958–60 during the period when I carried out most of my field research. In 1964 a government settlement scheme was put into effect which radically altered their way of life.

6 Woodburn, J. C. (1964). *op. cit.*; Woodburn, J. C. (1968b). *op. cit.*

7 Woodburn, J. C. (1964). *op. cit.*; Woodburn, J. C. (1968b). *op. cit.*

8 This is how the Hadza describe the institution: their account is not necessarily accurate.

9 Woodburn, J. C. (1964). *op. cit.* p. 294f.

10 Lee, R. B., this volume. pp. 179-81; Lee, R. B. (in press). The intensification of social life among the !Kung Bushmen, *in* Spooner, B. (ed.) *Population Growth: Anthropological Implications*. Philadelphia.

11 Silberbauer, G. (1960). *First Interim Report of the Bushmen Survey*. (Submitted to the Bechuanaland Protectorate Government.) pp. 11–12; Thomas, E. M. (1959). *The Harmless People*. London. pp. 14, 48–9, 105.

12 Fourie, L. (1925–6). Preliminary notes on certain customs of the Hei-//om Bushmen, *J. S.W. Africa Sci. Soc.*, **1**.

J. J. WYMER and R. SINGER

Middle Stone Age occupational settlements on the Tzitzikama coast, eastern Cape Province, South Africa[1]

The Middle Stone Age (M.S.A.) in South Africa is a term applicable to the stone industries known to be mainly more recent than the hand-axe industries of that country, but generally preceding the semi-microlithic or other industries of the Later Stone Age (L.S.A.). Assessments of absolute date for the M.S.A. have varied considerably but there is little doubt now that they cover most of the Late Pleistocene period. A distinctive semi-microlithic industry, first recognized at Howieson's Poort in the eastern Cape, is generally considered to intervene between the M.S.A. and L.S.A., but in some areas it has been shown that the Howieson's Poort industry is contemporary with the later phases of the M.S.A. This transitional period was termed for convenience the "second intermediate period"[2] but this term is now thought to be inappropriate[3]. Variations in the M.S.A. industries in different parts of South Africa have produced a terminology based on such local variants, i.e. Mossel Bay variant, Pietersburg variant, etc., but these terms are also thought to be misleading and it is considered more rational to differentiate the various M.S.A. industries from a consideration of their chronological phases[4].

M.S.A. industries are known from open sites, sealed or on the present surface, inland and coastal rock shelters and caves, and coastal middens. Recent investigations by the authors of a complex of caves and rock shelters on the Tzitzikama coast have shown that settlement in this particular area was more static and intensive than suspected. The sites investigated were close to the mouth of the Klasies River, about 40 km west of Cape St. Francis, the nearest town being Humansdorp. Excavation was concentrated on one particular complex of shelters and caves, but observations and trial excavations were made at three other caves about 1 and 2 km to the east.

Settlement at Klasies River mouth began at the time of the 6–8 m sea, for the earliest hearths and spreads of occupational material interdigitate with marine beach sand. From this time, occupation of the site appears to have been continuous in the vicinity of the cave mouth and adjacent rock shelters.

A vast midden of occupational refuse, soil, sand and rock fragments built up to an eventual height of 22 m from the original 6–8 m beach level, and there is nothing in the stratigraphy to suggest any significant break in occupation throughout the entire sequence: no sterile layers, no obviously weathered soils, no truncations or unconformities. It is unlikely that seasonal occupation would leave any different stratigraphical record to annual, unbroken occupation, but it is difficult to interpret the evidence other than as continuous occupation in the sense of no major periods when it was not occupied. 22 m of deposits, with considerable lateral spread, implies a very long period of time. A series of over thirty radiocarbon dates is of little help here as they are almost all minimum dates in the 30–40,000 years' range. The time span involved, based on the dating of the 6–8 m sea, and radiocarbon dates from the Howieson's Poort levels, and in the M.S.A. stratified above them, is likely to be in the order of 50,000 years. The continuous occupation of one site for such a lengthy period is contrary to the expected activities of a Palaeolithic community of hunter-fisher-food gatherers.

The M.S.A. stone industries throughout the vertical sequence of 22 m (excluding the Howieson's Poort levels) show no major developments which might reflect changes in activity. There is a slight reduction in flake-blade size and a little more use of finer-grained non-quartzitic rocks in the more recent phases, but other factors than activity patterns could account for this. The Howieson's Poort levels, in marked contrast, contain such a different type of stone industry, that it would not be surprising if this indicated both changes in patterns of activity and settlement. In respect of the latter it is perhaps significant that the Howieson's Poort levels are only about 1·5 m thick and are wedged between thick M.S.A. deposits. The type of occupational deposit also presents a considerable contrast: the M.S.A. levels abound with spreads of ash with intervening or mixed layers of food refuse and discarded artefacts in a matrix of sand or clay, whereas the Howieson's Poort levels are composed of numerous individual hearths divided by black, carbonaceous soil and little sand. Assessments have yet to be made on the nature of the food refuse throughout the sequence and it will be interesting to compare that from the Howieson's Poort with the M.S.A. There is no immediately obvious difference, but the faunal and molluscan remains in the former are so much more fragmentary that it may not be discerned until complete analysis has been done. However, it is clear that throughout the whole sequence food was abundant. The animals hunted include fifteen species of antelope, the extinct African buffalo, elephant, hippopotamus and horse, implying open veld, forest and vleis all within convenient hunting distance. Shellfish and marine birds occur prolifically in all the layers, as do seal remains. There is some whale and dolphin, but fish is surprisingly rare.

Insufficient work has been done on the other three major caves nearby to show whether there were similar long, unbroken sequences of occupation, but M.S.A. material belonging to an early phase has been found in one and is presumably contemporary with one of the early phases in the main complex investigated. There is other evidence of contemporary occupation about 14 km east along the coast at Oysterbaai, where low middens occur on raised beaches. A few kilometres inland from Klasies River mouth large quantities of M.S.A. quartzite artefacts abound on the deflated dune sands. Much further to the west there are M.S.A. cave sites at Plettenberg Bay and Mossel Bay. Others have been reported in between.

The implication is that there was relatively intensive occupation along the Tzitzikama coast in the M.S.A. period, and continuous settlement at Klasies River mouth at least. For hunter and food-gathering communities to live close to each other and apparently for generations to continue doing so for so great a period of time suggests unique conditions. The food remains certainly indicate a highly favourable environment for such communities and the explanation is probably that there was an adequate faunal biomass to enable large groups to live without serious competition. Other factors were also favourable: fresh water was no problem as numerous springs occur along the cliff face, apart from the small rivers which flow all the year; there was unlimited raw material for the stone industries available in the form of beach cobbles.

If it is correct to interpret the evidence from Klasies River mouth in this way, i.e. that highly favourable environmental conditions were so successfully exploited that continuous occupation at the same site was possible for tens of thousands of generations, it is significant to note that:

(i) There was little, if any, change in the pattern of M.S.A. settlement over a period of possibly 50,000 years (excluding the intrusive Howieson's Poort occupation).

(ii) No grindstones were found to indicate the exploitation of vegetable foods, in spite of the evidence for the use of grindstones in the M.S.A. at the Cave of Hearths[5]. However, the latter site is some 300 km inland and this could well account for the difference.

(iii) There were no structures to indicate any more elaborate activities at any stage of the sequence.

(iv) Although red ochre crayons were found throughout the sequence, there was no indication of any artistic activity in the form of wall paintings, or engravings or carvings. However, a nearby cave produced painted stones in L.S.A. midden deposits radiocarbon dated to 335 ± 105 years B.C.

The conclusion is that, in spite of permanent settlement under favourable conditions, there was no apparent cultural advance over a span of probably more than 50,000 years, let alone any rise to food production or urbanization.

Notes

1 The studies on which this paper is based were supported by grants from the U.S. Public Health Service (GM10113), the National Science Foundation, the Wenner-Gren Foundation for Anthropological Research, the Boise Fund and the University of Chicago.

2 Clark, J. D. (ed.) (1957). *Proc. 3rd. Pan-Afr. Congr. Prehist.* 1955. London. p. xxxiii.

3 Bishop, W. W. and Clark, J. D. (eds.) (1967). *The background to Evolution in Africa.* Papers presented at a symposium at Burg Wartenstein, Austria, in 1965. Chicago.

4 Sampson, C. G. (1968). *The Middle Stone Age Industries of the Orange River Scheme Area.* Nat. Mus. Bloemfontein. Memoir No. 4.

5 Mason, R. J. (1962). *Prehistory of the Transvaal.* Johannesburg.

Section 3: The influence of ecology and agriculture on non-urban settlement

WILLIAM ALLAN

Ecology, techniques and settlement patterns

Human ecology and settlement patterns are determined not only by environments but by the techniques available for their exploitation. Hunting and food-gathering economies probably require, on a global average, about 26 km² of land for the maintenance of an individual[1]. The most extensive system of cultivation practised on the poorest soils with no implement but the axe and only finger millet as a major crop, can support forty times as many people[2]. Until the advent of agriculture man was a comparatively rare creature, whose permanent settlements were probably restricted to unusually favourable food-gathering sites and fishing communities.

The nucleus sites

The earliest agricultural settlements show two features that are significant: their small size and the ecology of the sites with which they were associated. In general, the proved site areas are such as to suggest groups of twenty to fifty households. The Tell es-Sultan fortified establishment in the Dead Sea Valley, which is one of the early agricultural settlements,[3] is exceptional in size. It appears to have covered more than 4 ha and may have housed a population of 2000, but a settlement of this size is not inconsistent with the extent and exceptional potential of the cultivation site. It would be hard to find a more probable setting for a genesis of agriculture than the Oasis of Jericho. Here, in the natural "greenhouse" of the Rift, 656 ft below sea level, are "permanent cultivation"[4] soils still fertile after millennia of use, an excellent water source and extensive seepage areas. In such a setting elementary irrigation would be technically simple, a matter merely of guiding the flow of water to extend the area of moist soils. Water spreading of this sort may even have preceded the evolution of cultivars.

Many of the sites where agriculture arose at an early date, or independently, are in valleys and depressions, flood plains, fans, deltas and pediments with "permanent" soils periodically refreshed by alluvial and colluvial deposits. They had admirable climates, abundant sunshine and ample water provided by running streams, springs, seepage sites and lakes. Fish, game birds and game were to be had for the taking. Such were, for example, Sialk, Fayum, the Oasis of Jericho and, on the other side of the world, the Valley of Mexico. Such, too was the narrow belt of rich black soils between the Elburz mountains and the Caspian: a "prehistoric Côte d'Azur", as Whyte[5] has called it, "playground of the Mesolithic idle rich". The same sort of paradisean environment is described in the oldest record of all, the Eden legend. In the garden where man was settled "to cultivate and take care of it", "a flood was rising from the earth and watering all the surface of the soil". Could there be a clearer description of an extensive seepage area?

An ideal pabulum for the growth of agriculture would be found in such highly favoured places, in seepage sites on permanent soils. These would be the best collecting grounds for wild cereals. Surviving food-gathering peoples take a special interest in particularly good patches of food plants: they even loosen the soil with digging sticks and clear weeds from around the desirable plants. Men of early Holocene times probably did the same, and from this it is no very long step to cultivation. The moist land would be weeded, perfunctorily at first and then more thoroughly as the effects became obvious. At a later stage the wild crop may have been thinned, leaving the strongest plants with the plumpest grain. Man weeded, and he probably applied a simple plant selection technique, before he sowed. Perhaps sowing, with seed from the patches of selected plants, came with the idea of guiding the flow of water—the beginning of water management. Only on some such supposition can one explain the early association of agriculture with irrigation.

These processes may have been aided by cultural adaptations during the Pleistocene-Holocene transition. Climatic change appears to have induced regional specialization and a tendency towards intensive exploitation of limited environmental niches—the beginning of land selection. It is highly conceivable that there would be pressure on the more productive food-gathering sites, accelerated by natural increase in such favourable environments. They may have been early centres of population explosion. Population pressure would provide the motive, the compulsion, to exploit the environment as fully as possible—first by guiding the flow of water, then by sowing. Even with these developments, there must have been very strict limits to the carrying capacities of the nucleus sites, set by their small extent and the low genetic potential of the first crop plants.

There is no need to postulate a prehistoric science of plant breeding to explain the origin of relatively high-yielding cultivars. Nature was the plant breeder, and Neolithic man took what he was given. By initiating cultivation he released and amplified forces of genetic change far beyond his understanding; but he aided the process by deliberate selection of the emergent plant types most useful to him.

The cereal-fallow rotation

The new food-producing economy could not have had a profound effect on human numbers and settlement patterns until agriculture spread from the nucleus sites on to the variable soils of the rain-fed lands. This must have posed new problems to the early innovators. Soils vary greatly in their natural capacity to sustain cultivation. Some lose their fertility when cultivated for a year or two and regain it only after a long rest under regenerating vegetation. Others can be maintained in fertility at acceptable yield levels under systems of alternate crop and fallow periods of short duration. The most fertile and durable soils of all, a very small fraction of the earth's mantle, can be cultivated at relatively high yield levels for long periods, or even indefinitely, with little or no break. We know how to improve land, to *make* good soils—at a cost—but the early farmers did not have these resources. They were dependent on the natural fertility and regenerative capacity of the soils they worked, as many of the world's cultivators still are.

In the Near East area of agricultural origins the problems of rain-fed cropping would not be too difficult, once suitable cultivars had become available. "Nucleus sites" often lie close to or within extensive areas of "semi-permanent cultivation" soils which can be worked on a basis of alternate crop and fallow. They include alluvials and young soils from alluvium, Red and Brown Mediterranean soils and better variants of the Brown Forest soils and Brown Steppe soils of the region. Although fertile, they are much less productive than the perennially moist permanent soils. Some years ago I measured the yield of Durum wheat on a still highly fertile though long-cultivated seepage site in southern Anatolia. The output of grain was nearly twelve times as great as that from neighbouring crops grown on Red Mediterranean soils under rain-fed conditions.

It seems probable that there would be no wide spread of cultivation beyond the nucleus sites before the evolution of cultivars capable of giving acceptable yields on less fertile and durable soils without seepage. This raises the question, what is an "acceptable" yield? If we are to answer this question and others involving quantitative terms, we must make certain assumptions. Archaeology does not provide us with basic data on yields of early cultivars or on systems of field management, seed rates, consumption and inputs of land and labour. The best we can do is to make tentative guesses by analogy with existing or recent communities of hand cultivators not very much better equipped than the Neolithic farmer.

The area that can be maintained in cultivation with no implements but the small axe and hand hoe is strictly limited by the labour power of the exploiting group. There is, of course, very considerable variation with the nature of soil and vegetation, but for the types we have in mind the limit is probably little more than half a hectare per head[6], and the minimum acceptable yield, for a single-season cereal culture, would be about 550 kg per hectare[7]. We might tentatively assume a general yield level of 600 kg per hectare and an annual grain requirement of the order of 300 kg per head of population, with some supplement from other activities, such as livestock, food-gathering, hunting and fishing. Fallowing would have been the only means of fertility maintenance available to the early cultivators and we might assume the fallow to have been a simple break of one to four years under stubble and weed growth, following one year of cropping.

The land requirement of such a system in this environment would be one and a half hectares of *cultivable* land per head of population: that is, land capable of giving a sustained yield at the acceptable level under the system and with the cultivars available. Land varies very greatly in cultivability, whatever the definition of the term. Any large land area is a complex which may contain much or little cultivable soil, or none.

The example can be carried further only by applying the hypothetical concepts to actual land areas for which adequate data are available. For this purpose, we may take the Turkish provinces of Antalya, Burdur and Isparta, a total area of 37,000 km^2, which was the subject of intensive study over a period of four years[8]. This region contains four major ecological zones with the following salient features:

(i) The Western Plateau: an area of closed drainage basins and large depression plains, set in a matrix of broken, "uncultivable" land. Extensive alluvial and hydromorphic soils of the plains and basins are excellent for wheat and barley.

(ii) The Lakes zone: comprising the drainages of three large lakes on the Anatolian plateau. The zone consists mainly of rough and locally mountainous country in which lithosols and bare rock outcrops are prominent. Cultivation is restricted to alluvial plains radiating from the lakes, and a limited range of foothill, plateau and valley soils.

(iii) The Coastal zone: a strip of the Mediterranean coast extending inland to the foothills of the Taurus mountains. Cultivable soils include those of extensive flood plains, severely limited for cereal cultivation by annual flooding, *terra rossa* and travertine types and mixed soils of outlet fans and terraces.

(iv) The Mountain zone: part of the main massif of the Taurus system. The terrain is rugged in the extreme, with great areas of exposed rock surface and unweathered talus cones. Soil development is slight owing to the high rate of natural erosion and cultivation is restricted to small and scattered areas of colluvial soils on the lower slopes and mountain shelves, and to deposits in the steep and narrow valleys.

On the basis of the hypothetical parameters, this region as a whole might have been able to support an overall population density of about twelve per km², under a simple system of grain production at low yield levels such as we have postulated (Table 1). There would, however, be great differences in the carrying capacities of the ecological zones, ranging from twenty-four per km², in the very favourable conditions of the Western Plateau, to sixteen in the Lakes zone, eleven in the Coastal zone with its excellent soils but severe physical limitations, and less than two per km² in the Mountain zone[9].

	R	*Zone 1*	2	3	4	*Region*
Areas						
(*a*) Total km²		10,000	5500	6500	15,000	37,000
(*b*) Cultivable km² [a]						
Class I	1 : 1	580	300	370	80	1330
II	1 : 2	1510	390	190	155	2245
III	1 : 3	960	410	360	105	1835
IV	1 : 4	970	360	150	100	1580
R = ratio of cultivation and fallow						
(*c*) Total cultivable km²		4020	1460	1070	440	6990
(*d*) Cultivable percentage		40·2	26·5	16·5	2·9	18·9
Hypothetical carrying capacities						
(*e*) Persons, total		242,500	91,000	73,500	27,500	437,500
(*f*) Persons, per km²		24·3	16·5	11·3	1·8	11·8

[a] Based on unpublished data from a government (*Toprak Su*) soil survey carried out in connection with the Antalya region pre-investment surveys.

Table 1 Hypothetical Critical Population Densities for a simple system of cereal-fallow grain production in four ecological zones of south-west Anatolia

In general terms, the carrying capacity of early agriculture in the Near East region may have been some 200 times as great as that of the former exploitive economies. But the ceiling on population growth had not been removed: it had been raised to a new level. Growth could proceed to the new limit unchecked by famine, except temporarily in disastrous seasons. Such checks would be fairly frequent, but if we assume a growth rate such as would double the population once in every century, the limit would be reached within a millennium.

No doubt, the process took much longer. There was no "revolution" in terms of time but rather a gradual absorption of communities at the level of intensified food-collecting within an expanding matrix of the food-producing economy. The introduction of a cultivar to a new environment would probably be followed by a period of "patch" cultivation such as is still practised by some wandering peoples, including the Yoruk of Anatolia. These nomadic pastoralists plant small grain fields by their winter camps in the lowlands, and on mountain slopes and ledges during the movement to the uplands. The fields are left uncared for and the meagre harvest is taken as the nomads return along their seasonal routes. This may have been an important means of dissemination of the early cultivars. They may have been spread not only by pastoralists but by traders and other travellers, and even by migrant hunters and food-gatherers—as the South American crops were spread from the coast throughout Africa in comparatively recent times.

As reliance on agriculture increased, the phase of "patch" cultivation would probably be followed by one of temporary or "shifting" cultivation, with larger areas in crop and reversion of the land to waste for indefinite periods. This extensive form of land-use can still be found where populations are extremely sparse and there is a plenitude of land, as in parts of the Congo Basin and over much of the Kalahari Sands region of west-central Africa.

Eventually, population increase and a high degree of reliance on cultivation would impose a need to conserve land resources and bring about, by empirical processes, the establishment of fixed and balanced cereal-fallow rotation systems. Such systems are permanent and can sustain permanent settlement so long as the "critical population density" is not exceeded. When this happens the systems enter a phase of decay, with falling yields and increasing incidence of famine, unless the excess population moves elsewhere. African folklore and the unwritten histories of the tribes are full of stories of such movements in the recent past, "when the land for cultivation became too small for the people"[10].

The archaeological evidence from south-west Asia points to a fairly wide diffusion of cultivation by 5000 B.C.[11] The later diffusion of cultivars and cultivation techniques into western Europe resulted in the emergence of locally adapted systems of land-use based on varying but generally short periods of fallow, and this device remained a primary basis of fertility maintenance until the Agrarian Revolution of the eighteenth century. With few exceptions, the "traditional" systems of rain-fed agriculture still in use over much of the world are based on fallow.

Surviving and recent hand-cultivation systems

In the tropical regions of the world there are great areas of Ferrallitic, Ferruginous Tropical and other soils of low agricultural value in the natural state. Their transient fertility is rapidly exhausted under cultivation and regained only after a long or very long period of fallow, which may grow to bush, secondary forest or regenerating woodland before the next cycle of cultivation. This accounts for the prevalence of so-called "shifting cultivation" or "slash-and-burn" in Africa and elsewhere in the tropics. In fact, these systems are often ingenious adaptations to unfavourable environments, based on a remarkably complete knowledge of local ecology and soil potential. Cultivation periods commonly vary from one to six years and rest periods from six to twenty years or more, in various combinations determined by the capability of the land.

In the case of a simple long-fallow system, we may think of the production unit, the "farm", as a fairly large area only a fraction of which—commonly one-quarter to one-eighth—is in cultivation at any one time. The remainder is in various stages of regeneration, and the cropped area is rotated over the whole in such a way as to maintain the fallow-cultivation balance and preserve the fertility of the soil. These two parts are, however, held under different forms of land rights, and this facilitates the breakdown of the systems when the Critical Population Density is exceeded[12]. In the context of traditional subsistence agriculture, the C.P.D. is the maximum population density an environment can support permanently, under a given system of land-use, without damage to the land. The term "carrying capacity" has a similar connotation. We may say that an area of 100 km^2 has a carrying capacity of 1200 people for a given system, and that C.P.D. is twelve per km^2.

There are few simple systems of land-use. Even where the ecology imposes a high degree of reliance on long-fallow cultivation, environmental niches are utilized for more intensive or even permanent cultivation; but these are subsidiary sources of food supply, and the general carrying capacity of these environments is low. In the great region of central and eastern Africa with a modified Sudanian climate, the Critical Population Density for such systems commonly ranges from 1·5–10 per km^2, if we include the peculiar fire-cultivation systems of the Congo-Zambesi watershed[13]. A similar range of Critical Population Densities (4·1–10·6 per km^2) has been demonstrated for normal long-fallow ("shifting cultivation") systems in certain areas of India with monsoon climates of marked wet and dry seasons[14] (see Table 2).

In regions of equatorial and subequatorial climates, where a more or less continuous sequence of crops is possible, the requirement of land per head of population is less and cultivable percentages are generally higher. Consequently, Critical Population Densities for the long-fallow systems rise to about nineteen per km^2 even on poorer variants, and to as much as forty in the more productive environments where such relatively high-yielding crops as plantains, yams and cocoyams provide the bulk of the diet. There are, however, other factors which inhibit population growth in the equatorial forest regions, including the high labour input required for continuous clearing even of secondary forest growth, an enervating climate, and the prevalence of disease. Over much of the diminishing but still vast forest zone of Africa present population densities are far below the critical levels, in contrast with the Savanna regions.

"Semi-permanent" lands, containing strong and fertile soils which can be maintained in cultivation with fallows equal to or little longer than the cultivation periods, have, of course, still higher carrying capacities. Under the Modified Sudanian climatic regime, with cassava as a major crop, Critical Population Densities may range from thirty-five to fifty per km². The figure rises with improving climate and may exceed ninety-five per km² in the most favourable regions of double rainfall with two effective cropping seasons in the year.

	C.P.D. per km²
Long-fallow cultivation environments	
Africa: modified Sudanian climate	
"Fire-cultivation" systems	1·5–6·0
Normal long-fallow systems	3·0–10·0
India: monsoon climate	
"Shifting cultivation"	4·1–10·6
Sarawak: hill rice cultivation	8·0–9·5
Africa: equatorial and sub-equatorial climates	
Normal long-fallow systems	19·5–40·0
Semi-permanent cultivation environments	
Africa: modified Sudanian climate	
Short-fallow systems	35·0–50·0
Africa: double rainfall climates	
Short-fallow systems	38·6–96·5
Permanent cultivation environments	
Africa: Plantain and banana cultures	230·0–350·0
Madagascar and south-east Asia: wet rice cultivation	100·00–1000·0

Table 2 Estimated Critical Population Densities for current and recent hand-cultivation systems in various environments

Very high carrying capacities are attained on Eutrophic Brown Soils under unusually favourable rainfall conditions, allowing of permanent cropping and permanent crops. These are young or rejuvenated soils developed on volcanic material, alluvial deposits or basic crystalline rocks, under warm and humid conditions which favour rapid soil formation. They are of exceptionally high agricultural value, but they comprise only a very small proportion of the soil mantle, and where they do occur they are intensively exploited. In East Africa, such environments are generally utilized for intensive cultivation of permanent groves of plantains and bananas with a range of subsidiary crops. Critical Population Densities for these systems, which are practicable only in such highly favoured environmental niches, probably range from about 230 to 350 per km² or, exceptionally, even more (Table 2).

Wet rice cultivation

Still higher carrying capacities are attained in the wet rice cultures typical of south-eastern Asia and China. These cultures are of ancient but uncertain origin. The first historical record dates from 2800 B.C. when a Chinese Emperor established a ceremonial for the planting of rice, but the crop was probably grown in India before 3000 B.C.

Cultivation of wet rice is, obviously, restricted to environments of suitable soil, climate and water supply: coastal plains, deltas, river basins and flood plains with fertile soils of low permeability, where fresh water is available to submerge the land. In such environments cultivation is permanent, but "cultivability" is largely determined by the extent of land on which the water level can be maintained at the required depth during the growing season, without excessive flooding or drying out.

As a relatively simple example, we may consider the basic features of a system which is still practised on the coastal deltas and flood plains of south-eastern Madagascar. The only implement used is the *ngade*, a sort of combination spade, hoe and shovel on a long pole, but cattle are employed for the essential operation of stirring and trampling the soil of the small bunded paddy fields (*piétinage*). The general cultivable percentage of the land, under present conditions, is about sixteen, the remainder being too dry or, more usually, seasonally flooded to excessive depth. The normal yield, after deduction of the seed requirement, is 1290 kg of paddy per ha, and annual consumption averages 188 kg of paddy per head of population. On the basis of these figures[15], the Critical Population Density for the area, under the current system of land-use, is likely to be of the order of 110 per km^2. The present density in the most highly populated canton is 105 per km^2.

In some of the coastal deltas of Sarawak, where the cultivable percentage appears to be twenty-six, Critical Population Densities may be about 140 per km^2. The normal yield is 1385 kg of paddy per ha and the average annual consumption is 250 kg per head, for some is used in ritual observances and as poultry feed. In contrast, dry-land rice cultivation, which supports a considerable proportion of the population, gives a yield of about 750 kg per ha, and the long fallow of twelve to fifteen years required to maintain this yield reduces the Critical Population Density to 8–9·5 per km^2 [16].

Conditions in the great river deltas of south-east Asia allow of extraordinarily high population densities. Where the cultivable percentage of the land is in the range of thirty-five to seventy-five and double cropping of wet rice is practicable, Critical Densities of 500 to 1000 may be reached. Present densities are of this order in the Red River delta of North Vietnam. Here, canton populations vary from 500 to 1000 per km^2 and the overall average is 575. The almost incredible rural population density of 1700 per km^2 has been reported from one canton, but this probably represents a very considerable degree of over-population[17].

Such dense concentrations of people may be, and often are, in juxtaposition with land of low carrying capacity and correspondingly meagre population. The "intermediate" region west of the Red River ends abruptly at the edge of the delta and on this line of juncture the general population density changes from 575 to two per km^2 [18].

Forms of settlement

These differences in the carrying capacity of land determined the pattern of human settlement, as agriculture spread over the earth. They are still reflected in the distribution of population throughout that great part of the world in which agrarian economies remain at a near-subsistence level. Where people, or most of them, can live only by producing their own food, the limitations imposed by environmental ecology continue to operate.

While the population that can be fed is strictly limited by environment, the form of settlement is less rigidly constrained. There is a considerable range of choice.

In environments of low carrying capacity villages tend to be small communities of a few hundred people, so that the exploited area is limited and the crop lands do not move too far from the homesteads. Distance from the lands is not, however, an absolute constraint. The Tswana, for example, live in an environment which supports a general population of two per km^2, yet they have comparatively large towns housing thousands of inhabitants: Serowe and Kgwaketse both have populations of about 35,000. This density of habitation implies, of course, that most of the people must spend part of the year away from the town, at work in the distant crop lands or far afield with the cattle.

Relatively large and static communities are commonly associated with permanent and semi-permanent cultivation environments, such as the strip of fertile *chipya* soils along the east bank of the Luapula River in Zambia. Here, a "street" of villages extends for 115 miles, and some of the villages have been in the same place for three centuries or more. This is clearly an imposed pattern, set by river and lake on one hand and the parallel ribbon of fertile, durable soils on the other. In contrast with this concentration of habitation, the plantain and banana cultivators of East Africa, in their environment of high carrying capacity, have no villages. They live in small family homesteads beside or within the groves, from preference and for convenience. Permanent crops require constant attention and household refuse is applied to the land as fertiliser.

This pattern of dispersed homesteads is not uncommon in many parts of the world and in widely varying environments. The "nucleated" pattern, with a parent village from which segments have split off to form surrounding and more or less distant hamlets of a few families, is also common and unrelated to environment. It is found in Africa and the Near East, Asia and Madagascar. In India the rural scene is built around villages which are usually concentrated or nucleated, but in Travancore-Cochin and west Madras the dispersed homestead type prevails and it extends widely over north Iran and east and west Turkestan.

Permanent houses and villages do not necessarily indicate static populations, as Higgs and Jarman point out[19]. Transhumance continues to be practised where two or more separate ecosystems are exploited. In south-western Anatolia villages are quite frequently duplicated and the whole or a large part of the community moves between the two sites. The dominant movement is a winter-summer transhumance between the coastal lowlands and the uplands, over distances of 50 km or more. Commonly, land is held and cultivated both at the lowland and upland site, but in some cases the movement involves only the livestock and their herdsmen. Even the quite considerable towns of Kas and Kalkan are affected by this movement: their winter populations are double those of summer.

Forms of settlement are numerous, but not immutable. They change with circumstances. Richards, in her classic account of the fire-cultivating Bemba, points out that their villages had formerly been larger than they were at the time of her study. "They tended to group themselves under the protection of their chiefs in large stockaded villages"[20]. This grouping resulted in over-population of the readily exploitable area, and periodic movement of villages to new land. With the establishment of European rule and the prevention of war, the Bemba villages divided and sub-divided as often as possible within the limits allowed by the law of the new rulers[21]. A similar process of dispersion took place over much of Africa, but settlement patterns which can be explained only by former defence needs still survive[22]. The dispersal tendency may, however, be reversed by the attraction of modern amenities. In the Anatolian study area previously mentioned, the settlement pattern is strongly nucleated, as a result of repeated fission in the past; but the present tendency is towards increasing concentration of population in the larger centres, which offer such amenities as the cinema and a bus route, while the little villages and hamlets tend to remain static or decline.

Forms of settlement are determined more by human preferences, needs and traditions, than by environmental ecology. Some people are gregarious (the "long house" of Sarawak, Borneo and Papua, in which a whole village lives under a single roof, must represent something like the ultimate in human propinquity): others prefer some measure of domestic privacy. Neolithic people were probably much the same. There is a marked contrast between "the closed plan of the settlements of the Near East, with houses so concentrated that (as at Çatal Hüyük) there may be no space at all between them, and the open villages of individual houses which are found in eastern Europe"[23].

The first cities

In the fifth millennium B.C. there was a marked movement of population to the southern alluvial plain of Mesopotamia. It may be that these processes, operating over millennia, had culminated in balanced but static land-use systems and generated an outflow of population in excess of the critical limit. Other explanations have been offered: desiccation, for which there is no clear evidence at this time, and a movement of what has been called new groups of the depressed and the oppressed, the disgruntled and dissatisfied. Whatever their origins, these people of the southern plain achieved a mastery of their environment placing them within classic definitions of "civilized" life.

It may seem strange at first sight that this region of sun-baked desert soils, floods and marshlands, should be the growth point of the first civilization: yet the process would not have been possible in any other type of environment. Here the Tigris and Euphrates, depositing their alluvium on a bed of sedimentary rocks, have built up the only really large and fertile plain between the Nile and the Indus. They meander over this wide, flat expanse of their own creation and, as meandering rivers do, they have built up their beds so that the water surface is often above the level of the plain. Here were all the elements of the nucleus sites greatly magnified. The earliest settlers would find marshlands and lakes teeming with water birds, fish and other game. There would be seepage sites and limited but readily cultivable soils on the levees. In addition, and of far greater importance for the future, there were vast stretches of dry but potentially highly fertile alluvial soils commanded by the waters of the raised rivers. The basic techniques of water management had been learned in the nucleus sites: their expansion and development provided the key to the first civilizations. Large and reliable surpluses could be produced; capital could be generated.

The nucleus sites were too small and scattered to serve as growth points of civilizations, and in the transitional stages of rain-fed agriculture there would be little or no surplus. All fully-developed agricultural systems at the subsistence level do, however, produce a "normal" surplus, so long as they remain in balance with their environment. When dependence on agriculture is complete, or almost so, the exploiting group must cultivate an area large enough to ensure the food supply in a season of poor yield. Otherwise, the community would not survive for long. Consequently there is a surplus in the "normal" year, none in the poor year and shortage or famine in the unusually bad season[24]. This is as much as the hoe-cultivator can do on fallow-rotation soils under rain-fed conditions. The surpluses he produces may allow of a tenuous trade or support a chiefly hierarchy but they are far too unreliable to maintain complex societies with a high degree of occupational specialization.

The early civilizations of the Old World grew from the alluvial soils of the great river basins: the Tigris-Euphrates, Nile, Indus, Ganges-Bramaputra, Huang Ho and Yangtze. These are among the most persistently fertile and naturally productive soils in the world: they have an almost inexhaustible supply of plant nutrients brought down from the upper lands drained by the rivers. With good growth temperatures, abundant sunshine, suitable cultivars and an adequate water supply assured by irrigation, such soils can produce a surplus large and reliable enough to allow of capital accumulation, the growth of non-agricultural classes and the genesis of civilization. A hoe cultivator might well have been able to produce twice or three times his domestic needs, provided he had enough land.

Considering its relatively homogeneous nature and high cultivable percentage, and assuming a general yield level of about 1200 kg of grain per ha, the Mesopotamian plain may have been able to support densities of 300 per km^2 or more, with less than half of this population wholly engaged in agriculture. Herodotus suggests much higher yields. He says that wheat yielded two-hundred fold, but this seems most unlikely. The common wheat of the plain before 1000 B.C. could not have had a very high output potential: it was a form of emmer. Barley, which was very widely grown, may have yielded more than the wheat.

Mastery of these environments called for concerted action. This required the establishment of co-ordinating authorities controlling large populations and powerful enough to draft mass labour for the construction of irrigation and other public works. It also involved development planning on a large scale. The controlling authorities, kings, royal officials, nobles and priesthoods, also controlled the surplus of production: they accumulated wealth and power, so long as the surplus continued to increase. In the homogeneous environment of the great plain or river valley with its navigable river, the collection and distribution of surpluses would not be difficult. Food and labour could be turned to account in the construction of increasingly elaborate public and mortuary buildings, and the support of kingly, noble and official classes. The needs of those classes then had to be met by a growing multitude of servants, soldiers, artists, technicians and workers in many skills and trades. Such societies were anchored to and could not extend beyond the fertile, irrigable soils from which they grew; and they were subject to recurrent disaster by flood, famine and invasion. On the ill-drained plains of Mesopotamia, the most insidious enemy of all was salinization of the soil. New and more malleable civilizations were able to grow from the rain-fed lands, when the limitations of the hoe had been overcome.

Ox-power and iron

By the beginning of the third millennium many fundamental technical innovations had been made within the areas of the first civilizations; including the wheel, the breaking and harnessing of oxen for draught, and, some would add, the plough.

Representations of what have been described as ploughs appear in the pictographs of the Uruk period. It seems unlikely that these were really true ploughs in the sense of implements used for *primary* tillage. There is no evidence that they had metal shares or blades. If they had, the metal must have been copper or bronze. How long would such substances withstand the abrasion of ploughing?

The function of these "ploughs" may have been to assist in planting. Egyptian pictures of land preparation and planting in the third millennium show a labour gang breaking the land with hoes. After the gang comes a sower, broadcasting seed. He is followed by the "plough" which appears to be covering the seed and consolidating a seed-bed already prepared with the hoe. An ox-drawn wooden drag is still widely used for the same purpose. The famous "seed-drill" plough of Babylonia, depicted on a Kassite cylinder-seal of the second millennium, was almost certainly a planter. It was fitted with a tube through which seed could be dropped from a funnel. Such things may have been unusual novelties peculiar to the estates of temple gods and the god-kings and nobles on whose seals and monuments they are depicted. The ordinary cultivator had to make do with his wooden or stone-bladed hoe. There was no effective plough; certainly none that was widely available, until iron came into common use.

The great discovery of the second millennium B.C. was not iron itself; the metal had long been known as a rarity; but processes for producing it from such abundant ores as haematite and limonite and treating the furnace product so as to carburize and toughen it to the durability of mild steel. These discoveries were made, probably in eastern Anatolia, about 1400 B.C., and a few centuries later the processes were widely known.

With cheap and plentiful iron, the family farmer could have a simple but effective ox-drawn plough. Perhaps it was very like one that is still widely used: it could hardly have been simpler or more basic. This "modern" plough has a wooden stock shod with an iron blade, and a handle which is a backward projection of the stock. The draught pole projects from this piece and it carries a cross-beam, fixed with a wooden peg, to which two oxen are yoked. Stock, handle and pole may be made in one piece from a small tree suitably shaped. The iron blade can readily be attached to and removed from the stock: it looks like a large, broad spearhead. This implement is still in use from the Sudan to Sinkiang. It is the most common plough of Turkey, where more than two million farm families have no other primary equipment.

With an implement such as this, which is remarkably efficient on many soils, the dry-land farmer was free from the limitations of the hoe: ox-power and iron were his servants. He was probably as well equipped as many of his living successors, for he would have the drag, the threshing sled and iron tools, including axes and scythes. Inter-tillage would be done with the hoe, and field crops cut with the scythe or pulled by hand, as is the case today. He could plough his fallows. Judging by Homer's references to "the thrice ploughed fallow", he was doing so early in the first millennium B.C. Thus equipped, the farmer of thirty centuries ago could clear and break land which had been "uncultivable" for his predecessors. He could maintain three or four times as much land in cultivation and rival the productivity of the hoe-cultivator on the irrigated alluvium: not, of course, in terms of production per hectare but in output per unit of labour and total productivity. The Anatolian peasant of today, though little better equipped, can maintain in cultivation about 2 ha per head of the farm family, in addition to ploughed fallow, but he can do so only if he has enough land and is not limited, as he now very often is, by lack of winter feed for his oxen. The first plough farmers were probably much less limited by the availability of land and feed for livestock. They could produce a surplus large and reliable enough to permit of capital accumulation.

The civilizations of Greece and Rome were built on this surplus of plough farming and based on the cereal-fallow rotation and the Mediterranean triad—grain, vine and olive. Irrigation, within the limited areas of practicability, was subsidiary to production from the rain-fed lands.

Notes

1 Allan, W. (1965). *The African Husbandman*. Edinburgh and London. pp. 261–6.
2 Allan, W. (1965). *op. cit.* pp. 107–33.
3 Kenyon, K. (1956). Jericho and its setting in Near East History, *Antiquity*, **30** (120); Kenyon, K. (1957). *Digging up Jericho*. New York; Kenyon, K. (1959). Earliest Jericho, *Antiquity*, **33**; Zeuner, F. E. (1956). The Radiocarbon Age of Jericho, *Antiquity*, **30** (120); Ucko, P. J. and Dimbleby, G. W. (eds.) (1969). *The Domestication and Exploitation of Plants and Animals*. London.
4 For definition of this and other land classification terms, see Allan, W. (1965). *op. cit.*
5 Whyte, R. O. *Evolution of Land-use in south-western Asia*. UNESCO.
6 Allan, W. (1965). *op. cit.* pp. 48–65.
7 Allan, W. (1965). *op. cit.* pp. 94–6.
8 The material and data used in this estimate were obtained by the author in the course of this study. See *Preinvestment Surveys of the Antalya Region, Turkey*, Vol. II, UNDP(SF)/FAO, Rome, 1966.
9 The populations of the present day, still largely rural, are more than twice as great as those we have postulated, but their distribution still reflects the ecological differences in a marked degree.
10 Allan, W. (1965). *op. cit.* pp. 284–6.

11 See Ucko, P. J. and Dimbleby, G. W. (eds.) (1969). *op. cit.*
12 Allan, W. (1969). Land Tenure and Productivity, *in* Hutchinson, Sir J. (ed.) *Population and Food Supply*. Cambridge.
13 *Citemene* systems: see Allan, W. (1965). *op. cit.* Ch. 9.
14 Bose, S. (1967). Carrying Capacity of Land under Shifting Cultivation, *Asiatic Society, Monograph Series*, **12**. Calcutta.
15 Data from *Données de Bas* and *Comptes Régionaux*, Groupe Economique, Project d'Etudes de Farafangana, 1970 (unpublished).
16 Estimated from data given in Government publications and personal communications.
17 Dumont, R. (1957). *Types of Rural Economy*. London. p. 125.
18 Dumont, R. (1957). *op. cit.* p. 125.
19 Higgs, E. S. and Jarman, M. R. (1969). The origins of agriculture: a reconstruction, *Antiquity*, **43** (169).
20 Richards, A. (1939). *Land, Labour and Diet in Northern Rhodesia*. Oxford.
21 Richards, A. (1939). *ibid.*
22 Allan, W. (1965). *op. cit.* pp. 245–6.
23 Nandris, J. (1968). Lepenski Vir, *Sci. J.*, **4** (1).
24 See Allan, W. (1965). *op. cit.* Ch. 4.

CHRISTOS DOUMAS

Early Bronze Age settlement patterns in the Cyclades

The economy of Early Cycladic communities strongly influenced the character of settlement in these islands during the third millennium B.C. The situation, as well as the size of an E.B.A. Cycladic village, reflected the occupations of its inhabitants. The lack of large cultivable areas, the mountainous character of the islands and the rough surrounding sea resulted in the islanders developing a mixed economy consisting of farming, stock-raising and fishing. Such economies did not favour the formation of large self-sufficient settlements, but rather small communities, composed of a small number of possibly related families. The character of these villages explains, perhaps, the fact that not a single house in the settlement can be considered as belonging to a central authority; the "governing principle" of centrifugal planning, which characterizes Minoan architecture[1], is absolutely lacking in the Cyclades. Security and defence also influenced Early Cycladic settlements; thus the slopes of low hills were often chosen for the siting of villages for they offered security against probable floods, caused by heavy rains, and facilitated defensive vigilance. A carefully selected slope could also be easily protected by the construction of a defensive wall.

The Cycladic E.B.A. is sub-divided into three main periods characterized by the Pelos-Lakkoudhes Culture, the Keros-Syros Culture and the Phylakopi City I Culture respectively[2]. The aim of this paper is to outline the settlement patterns of these cultures, which cover the whole third millennium B.C. and perhaps part of the fourth.

Although the Late Neolithic in the Cyclades, as has been recently proved for the islet of Saliagos near Paros, is represented by well preserved architectural remains[3], no E.B.A. ruins can be ascribed to the Pelos-Lakkoudhes Culture. The absence of architectural remains dating to this earliest phase of the Early Cycladic from the settlement site of Phylakopi led Mackenzie to the conclusion that people during this phase lived in huts made of perishable material[4]. A large number of Early Cycladic excavated cemeteries belong to the Pelos-Lakkoudhes Culture; none contained more than fifty graves, the normal number being ten to fifteen single tombs in each burial ground. The location of the cemeteries mainly on sloping ground may be indicative that the settlements occupied similar areas nearby. Perhaps, these small cemeteries, or at least some of them, were used by nomadic families wandering from place to place in search of food for themselves and pasture for their animals. In this case, the absence of permanent architecture can be more easily explained, and Mackenzie's theory of perishable huts becomes plausible.

At an early stage of the second period of the Cycladic E.B.A., before the middle of the third millennium B.C., small rectangular houses appear scattered here and there not far from the cemetery areas. These isolated houses were built of small slab-like stones embedded in clay and consisted of one or two rooms. Although no such settlement has been fully excavated to date, judging from the fact that it was usual for multiple successive burials to be interred in each tomb, the size of the community seems to be greater than that which existed in the Pelos-Lakkoudhes Culture. An isolated house of rectangular plan has been found near the cemetery of Pyrgos in Paros under the ruins of a slightly later settlement (House B)[5]. A solitary house was also found under the fortified settlement of Panormos in Naxos[6], while similar houses have been uncovered near the cemeteries at Avdheli on Naxos[7], in Keros[8] and at Avyssos on Paros[9]. Settlements of this phase are found usually on sloping areas either close to the sea or at a considerable distance from it. The location of each settlement reflects the main occupation of its inhabitants. The principal activities of these people were farming and stock-raising in both the inland and coastal areas. Fishing was also carried out in the coastal sites.

During this early phase of the Keros-Syros Culture no special security measures, such as the construction of defensive walls around the settlement, were taken. In any case it would have been practically impossible to enclose within a wall all the houses of a settlement which tended to be scattered over a large area.

The need for fortifications arose only later during the main phase of the Keros-Syros Culture. Settlements of this period shared the following features:

(i) They were sited on hilltops.
(ii) These hills were close to the sea.
(iii) Very often, one slope of the hill was very steep.
(iv) The settlements were surrounded with a fortification wall, which was usually reinforced with bastions.
(v) Within this enclosure houses, built of small irregular stones embedded in clay, were crowded together, separated by corridor-like streets.
(vi) Curvilinear houses occurred parallel to the rectilinear ones.

Settlements of this kind are known from Kastri on Syros[10], Spedhos[11] and Panormos on Naxos[12], from the islet of Dhaskaleio near Keros[13], and from the top of Kynthos on Delos[14]. It seems that fishing became gradually the main occupation of the islanders and their settlements had, therefore, to be located in places providing safe anchorage for their small boats. It also seems that, besides fishing, other activities, such as trading and piracy, were developed by the islanders. The latter activity would have been a serious obstacle to seafarers from neighbouring lands, in particular from Crete. The Minoans by this time had started to develop their naval empire and would not easily have accepted the Aegean being controlled by Cycladic pirates. Perhaps the reason behind the Cretan attempt to clear the seas of the pirates was the creation of the fortified settlements in the Cyclades. The threat from the Minoan navy possibly forced the islanders to concentrate inside the walls, where every corner was exploited to house as many families as possible. The shape of the ground available thus dictated the plan of the house, and this is the reason why curvilinear buildings appeared in this period for the first time in the Cyclades.

Nevertheless, neither the steep slopes of the selected hill sides nor the structural fortifications were enough to resist the increasing naval power of Minoan Crete. It may be that Thucydides' statement that Minos cleared the seas and colonized the islands refers to this same period[15]. The Keros-Syros fortified settlements entirely disappeared during the final period of the Cycladic E.B.A. The two excavated sites of the Phylakopi City I Culture show a quite different settlement pattern. Phylakopi on Melos and Phrourion at Paroikia on Paros are coastal sites with no fortifications. The houses, built again of small irregular stones embedded in clay, are well spaced out within the settlement area. The Minoan presence is manifest at both sites through pottery imported from Crete[16]. In the struggle against Minoan Crete the Cyclades were defeated. They became the stepping stones for the expansion of Minoan influence throughout the Aegean. The islanders appeared to continue their trading activities in a climate of mutual acceptance. The settlements of the final period of the E.B.A. were no longer refuges for pirates and they became commercial centres and naval stations.

Notes

1 Hutchinson, R. W. (1962). *Prehistoric Crete*. Harmondsworth. p. 162.

2 Renfrew has successfully distinguished the existence of these three main Cultures in the Cycladic E.B.A. and introduced this terminology instead of the arbitrary distinction between Early Cycladic I, E.C. II and E.C. III. Here the term Pelos-Lakkoudhes Culture is used instead of the term Grotta-Pelos Culture for reasons to be given elsewhere (Doumas, C. (In press)).

3 Evans, J. D. and Renfrew, C. (1968). *Excavations at Saliagos near Antiparos*. London.

4 Atkinson, T. D. *et al.* (1904). Excavations at Phylakopi in Melos, *J. Hellenic Studies, Supp. paper no.* 4, p. 244.

5 Tsountas, Chr. (1898). Kykladika, *Archaeologikè Ephemeris*. p. 170.

6 Doumas, C. (1964). Archaiotikies kai mnimeia Kykladon, *Archaeologikon Deltion*, **19**, B, p. 411, Fig. 2.
7 Doumas, C. (1963). Archaiotikies kai mnimeia Kykladon, *Archaeologikon Deltion*, **18**, B, p. 279.
8 Doumas, C. (1964). *op. cit.* p. 410, Fig. 1.
9 Tsountas, Chr. (1898). *op. cit.* pp. 175–6.
10 Tsountas, Chr. (1898). *op. cit.* pp. 115–22; see also Bossert, E. M. (1967). Kastri auf Syros, *Archaeologikon Deltion*, **22**, A.
11 Stephanos, K. (1903). *Praktika tes Archaeologikes Etaireias*. p. 53.
12 Doumas, C. (1964). *op. cit.* n. 8, p. 411, Fig. 2.
13 Doumas, C. (1964). *op. cit.* p. 410.
14 Plassart, A. (1928). Les sanctuaires et les cultes du Mont Cynthe, *Délos*, **11**.
15 Thucydides, I, 4 and 8.
16 Hutchinson, R. W. (1962). *op. cit.* p. 111; Rubensohn, O. (1917). Die Prähistorischen und Frühgeschichtlichen Funde auf dem Burghügel von Paros, *Athenische Mitteilungen*, **42**, p. 46.

DAVID FRENCH

Settlement distribution in the Konya Plain, south central Turkey [1]

The evidence available for a complete study of the Konya Plain is seriously inadequate, particularly in the field of physical environment[2]. The purpose of this paper is, nevertheless, to construct a model which may *perhaps* have both theoretical and practical significance.

Elsewhere[3] the hypothesis has been put forward that, within the limits of the contemporary technological capacities, a ratio between the size and the number of settlement sites was maintained in the exploitation of the total available resources. In the model adopted, there is no implication that an "ideal" ratio was ever achieved or that the total potential was ever realized. The concept of an "ideal" (i.e. optimum) potential and exploitation is here a discardable construct. On the other hand, there is the explicit suggestion that societies (or communities) can go beyond an ideal ratio and achieve an imbalance which can lead to pressure on the available food resources; it is here that social factors, i.e. controls, would become relevant. It is intended here to extend yet further the application of the original hypothesis, since it has direct relevance to population size and, thus, to any study of the definition of settlement status. This paper concentrates on the known basic technology and the known available food resources (plants and animals), on the assumption or postulate that populations are supported from the total of the basic resources available to them. The distribution patterns[4] and (apparent size of settlement sites datable to the time-range, 6500–1000 B.C., can be investigated, and their significance interpreted, on the basis of this hypothesis.

The chronological framework, given below, concerns the expansion and development (i.e. in number and size) of settlement sites within the limits of existing resources and technology; it relies on the idea of *continual change* in the total potential; by which is meant the concept of a number of interacting and *changing* factors:

1 *chronology*, i.e. period differences, e.g. inherited experience[5];
2 *space*, i.e. the physical area occupied by the total resources[6];
3 *ecology*, i.e. the diversity of resources whether exploited or not;

4 *technology*, i.e. the means of exploitation;
5 *social structure*, i.e. the exercise of social controls;
6 *aggression/defence*, i.e. factors introduced by warfare.

In other words, the relation of man to his environment is always variable (and varying, i.e. according to region):

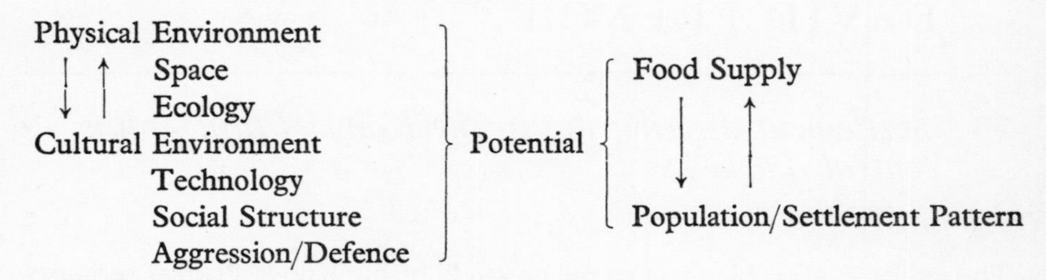

The changing pattern of man's exploitation of the existing resources may then be set down with reference to these six factors.

A chronological framework for sites in the Konya Basin might run as follows:

6500–6000 B.C. small agglomerations

Small sites; structures closely grouped, single-roomed, i.e. "cellular", built of pisé and/or mud-brick; perhaps permanent (i.e. year round) settlements; evidence from Can Hasan III of a wide diversity of animals and plants[7], but this need not imply an absence of "domestication"; *varied* obsidian/flint assemblage, implying several *functional* categories[8]; the total number of *known* sites: two, Can Hasan III and perhaps Çatal Hüyük I (i.e. the eastern mound); Can Hasan III is an excavated site, its size is probably not greater than 100 m in diameter. Site structurally defendable?

6000–5500 B.C. large agglomeration, "Çatal" phase

One major site, Çatal Hüyük I: structures closely grouped, single roomed; "cellular", built of mud-brick (up to 1·00 m long); perhaps, like Can Hasan III above, permanent; evidence of a diversity of plant and animal resources, perhaps both domesticated and wild[9]; *varied* obsidian/flint assemblage, close to Can Hasan III, but with some additions, e.g. "large daggers"[10]; total number of *known* sites: one, Çatal Hüyük I, excavated, stated dimensions approximately 450 by 275 m. Site structurally defendable?

5500–4000 B.C. small agglomerations

Several sites known, small but with impressive "cellular", mud-brick structures, closely grouped, single-roomed but two-storied, fairly large (up to 7–8 m across) and built of large mud-bricks (approximately 0·80 by 0·40 by 0·12 m); presumably permanent; *perhaps* specialized (i.e. concentration on particular species) exploitation of plant and animal resources, both wild and domesticate[11]; restricted range of stone tools; first *metal* tool[12]; total number of known sites: seven, perhaps eight; scattered, much smaller than Çatal Hüyük I; excavated sites: Can Hasan I (phases 5–2A) and Çatal Hüyük II (the western mound), exact extent of settlement not known. Site structurally defendable?

4000–3000 B.C. small agglomerations and farmsteads

A greater number of sites than earlier; mostly small and, when compared with earlier phases, showing easily recordable differences, e.g. in building features (narrow wall, i.e. small mud-brick, approximately 0·30 by 0·18 by 0·12 m) and in settlement plan (spreading, complex, free-standing structures); probably the exploitation of plant and animal resources remained unchanged (the evidence is limited)[13]; narrow range of stone tools; metal tools not recorded[14]; total number of known sites: seventeen, perhaps twelve more, scattered, none very large, i.e. much smaller than Çatal Hüyük I; one site excavated, Can Hasan I, exact extent of settlement not known. Defence systems (e.g. fortifications) not yet attested in Konya Plain.

3000–2000 B.C. "?cities", large and small agglomerations, farmsteads

Large number of sites, of all sizes; certainly structure plans, building features and settlement plans but probably also technological equipment and food resources represent a follow-on from the previous phase[15]; total number of known sites: 102, some (Samih, Emirler, Kerhane) very large, i.e. greater than 1 km in diameter, i.e. larger than Çatal Hüyük I; some very small, less than 100 m in diameter; only one excavated site, Karahüyük—Konya. Fortifications (i.e. structurally independent defences) not yet attested in Konya Plain.

2000–1000 B.C. "cities", large and small agglomerations

Few sites of varying size; very little known from excavation; technology and food resources unchanged[16]; total number of known sites: seven; two, Domuzboğazliyan and Karahüyük, very large, i.e. larger than Çatal Hüyük I; one excavated site Karahüyük—Konya. Earliest known fortifications (at Karahüyük—Konya).

Certain features of this chronological outline should perhaps be clarified and simplified:

1 Intentionally excluded is a discussion both of the results and effects of communal-effort in the size and plan of settlements, and of religious and social devices in the control and exploitation of resources and, therefore, in social grouping. The reasons are two:

(*a*) The regularity of the settlement plans at Çatal Hüyük I and Can Hasan I and III is deceptive; the original orientation determined the succeeding orientations. This point is stratigraphically demonstrable whereas the idea of "planning", i.e. deliberate lay-out of a group of structures, is not so easily proven.

(*b*) There is very real difficulty in using archaeological data for problems of a sociological or religious nature. This latter field has been left to others[17]. Nevertheless the possible significance and influence of prehistoric religion, although here omitted, must, on the evidence of written sources, be accepted; the presence of religious centres in Anatolia is attested for the second millennium B.C.

2 It is irrelevant that Çatal Hüyük I, for example, is simply a *big* site; the emphasis in the scheme outlined above has been put not on size *per se* but also on site distribution, technology and resources. The "urbanization" of Çatal Hüyük I, when assessed in relation to the stated hypothesis and not in terms of size alone or of apparent opulence (i.e. material richness), is better interpreted as the result of an efficient exploitation which has possibly reached an optimum or, indeed, a maximum. In the scheme suggested above, the size and uniqueness (there are no certain contemporary sites in the Konya Plain) of Çatal Hüyük I can be interpreted as the result of a full (? maximum) utilization of resources. In other words, more and similar sites were not viable because greater exploitation i.e. acquisition of more food, was not possible within the limits of the existing resources and the existing technology. Thus one big site can be the equivalent of (*exempli gratia*) six small sites but the amount of resources, i.e. food available, does not permit a greater number of sites (a higher

population ?) until there is an increase (1) in the food-production, perhaps by technological innovation, and/or (2) in the area of exploitation. Here the available data is very poor for ancient conditions; modern examples, however, from the Konya Plain support this generalization.

Why the concentration at Çatal Hüyük I? This is a question perhaps answerable by an analysis of its location[18]. Mellaart, for example, sees the site as a religious and trade centre[19] and at the same time sees it as "a super-nova amid the rather dim galaxy of contemporary peasant cultures"[20]; it may, he suggests, have "burnt itself out". Is there here a suggestion of the idea that Çatal Hüyük I collapsed because it had over-reached itself?, that it could not continue to be supported on the available resources? Çatal Hüyük I was preceded by a small site: it was succeeded by small sites. Large sites again emerge in the third millennium. Whether or not these sites, as opposed to Çatal Hüyük I, are urban (i.e. political) centres is discussed below.

The scheme is, therefore, the result of generalizations constructed on this basis, and, thus, in looking for distribution-patterns in the occupation and use of the Konya Plain, it is possible to put forward, as follows, a set of definitions and characterizations of the sites which are themselves architectural expressions of the exploitation of the environment. These architectural expressions, however, can function independently of technological development and can be adopted because of factors (outside the scope of this paper) of which the technological may, but need not, be one.

Farmstead

Single structure or complex; small; suited in some ways to agricultural exploitation by small population groups; something like the modern *çiftlik*.

Village

Small agglomeration of structures or buildings; the number and size of the agglomerations depending on the factors of space, technology, etc., cf. Can Hasan III, Çatal Hüyük II.

Town

Large agglomeration of structures; basically with the same economy as a small agglomeration; perhaps but not necessarily, the result of especially favourable natural conditions (i.e. natural resources, "basic elements")[21]; the modern Turkish *kasaba*, translated in the Oxford Dictionary as a "small town". Not necessarily politically structured or used as a political centre, e.g. Çatal Hüyük I.

City

Perhaps relying on the (traditional) natural resources but structured around a political system; probably a political centre; not necessarily defined simply on size, e.g. Karaman, Domuzboğazlıyan.

In these definitions there has been excluded, for the moment, a treatment of the effects of trade, i.e. outside resources, and of specialization; neither factor can be assessed on present evidence from the Konya Plain.

The emergence of political (and perhaps religious) centres, i.e. cities, in the third millennium is not certain. Three sites, each larger than 1 km in diameter (Samih, Emirler, Kerhane), occur close together in the northern part of the Konya Plain. Their closeness (less than 5 km distant in one instance) may *perhaps* suggest that they are not political centres of the kind we know from second millennium written sources although the unique and immense second millennium site, Domuzboğazlıyan, in the middle of the plain, could be the capital of a kingdom of the kind recorded by the Hittites. It is possible that the third millennium *did* see the emergence of political groupings, i.e. kingdoms with capitals, palaces and so on, if only for the reason that such kingdoms are recorded in the Assyrian texts at Kültepe. i.e. *c.* 2000 B.C. Certainly by that date (if not earlier) trade was a vigorous activity and thus a new factor had been introduced into *local* economies and, therefore, into social and political groupings.

Notes

1 Grateful acknowledgement is made, for their suggested improvements, to R. A. Watson, J. Bordaz, A. W. McNicoll and M. E. Weaver.

2 For the geomorphology of the Konya Plain: Ridder, N. A. de (1965). Sediments of the Konya Basin, central Anatolia, Turkey, *Palaeogeography, Palaeoclimatology, Palaeoecology* **1**; Driessen, P. M. and Meester, T. de (1969). *The Soils of the Çumra Area, Turkey*. Wageningen; Meester, T. de (1970). *Soils of the Great Konya Basin, Turkey*. Wageningen; Meester, T. de (1971). *Morphological Studies in the Great Konya Basin, Turkey*. Wageningen; Cohen H. R. and Erol, O. (1969). Aspects of the palaeogeography of central Anatolia, *Geogr. J.* **135**; Cohen, H. R. (1970). The palaeoecology of south central Anatolia at the end of the Pleistocene and the beginning of the Holocene, *Anat. Stud.* **20**. For vegetational history: Helbaek, H. (1964). First impressions of the Çatal Hüyük plant husbandry, *Anat. Stud.*, **14**; Birand, H. (1970). Die Verwüstung der Artemisia- Steppe bei Karapınar in Zentralantolien. *Vegetatio*, **20** (which includes earlier references). For faunal remains (from excavations): Perkins, D. and Daly, P. (1968). A hunter's village in Neolithic Turkey, *Sci. Amer.*, **219** (5). For mollusca: Ridder, N. A. de (1965) *op. cit.*; Meester, T. de (1971) *op. cit.*

3 French, D. H. (1970). Notes on site distribution in the Konya Plain, *Anat. Stud.*, **20**.

4 French, D. H. (1970). *op. cit.* maps, figs. 2–8.

5 cf. Davidson, D. A., this volume, p. 21.

6 cf. Jarman, M. R., Vita-Finzi, C. and Higgs, E. S., this volume, p. 62.

7 Can Hasan III plant remains (preliminary list) *inter alia: Triticum diccocum, T. aestivum* s.l., hulled barley, *Secale cereale* sp., *Vicia ervilia* and various small-seeded legumes, *Prunus* sp., *Vitis* sp., *Celtis* sp., various Gramineae. Can Hasan III faunal remains *inter alia:* equid, *Sus* sp., *Bos* sp., *Ovis/Capra, Capreolus* sp., *Cervus* sp., *Lepus* sp., *Vulpes* sp., *Erinaceus* sp.; fish; on this basis, Can Hasan III was probably not a single animal dependent community.

8 Can Hasan III tool assemblages: (1) stone (a) chipped: mostly obsidian, less flint; microtools, e.g. obliquely truncated bladelets (approximately 1·5 cm long), parallel-sided blades, unworked or with one- or two-edged retouch; unifacially worked leaf-shaped points (mostly in the range 3–5 cm length); tanged unifacially worked points including "arrowheads"; scrapers (including triangular-shaped and "thumb"), burins, etc. etc. (b) ground: stone-axes; also pounders and querns. (2) bone: spatulae (pierced or not) and points (including one needle), both of several types.

9 Çatal Hüyük I plant remains: Helbaek, H. (1964) *op. cit.*; Renfrew, J. (1969). The archaeological evidence for the domestication of plants: methods and problems, *in* Ucko, P. J. and Dimbleby, G. W. (eds.) *The Domestication and Exploitation of Plants and Animals.* London. Çatal Hüyük I faunal remains: Perkins, D. and Daly, P. (1968). *op. cit.* p. **101**; Mellaart, J. (1962). Excavations at Çatal Hüyük, 1961. *Anat. Stud.*, **12,** pp. 56–7.

10 Çatal Hüyük I tool assemblage: (1) stone (a) chipped: Bialor, P. A. (1962). The chipped stone industry of Çatal Hüyük, *Anat. Stud.*, **12**; Mellaart, J. (1964). Excavations at Çatal Hüyük 1963, *Anat. Stud.*, **14**. pp. 103–11. (b) ground: Mellaart, J. (1962). *op. cit.* p. 55; Mellaart, J. (1963). Excavations at Çatal Hüyük 1962, *Anat. Stud.*, **13**. pp. 46, 101, axes, "adzes", maces; also mortars, pounders, querns. (2) bone: Mellaart, J. (1962). *op. cit.* pp. 55–6; Mellaart, J. (1964). *op. cit.* pp. 100–3, awls and punches, needles, spoons, spatulae, scrapers *inter alia.* (3) clay: "sling-bullets", Mellaart, J. (1962). *op. cit.* p. 56.

11 Can Hasan I faunal remains: Perkins, D. and Daly, P. (1968). *op. cit.* p. 104; the list of identified species is diverse but certain species (e.g. sheep) are numerically dominant (S. Payne: personal communication.) A quantative/qualitative distinction can be made here. Can Hasan I plant remains: Renfrew, J. (1968). A note on the Neolithic grain from Can Hasan, *Anat. Stud.*, **18**; Renfrew, J. (1969). *op. cit.*

12 Can Hasan I tool assemblage: (1) stone (a) chipped: largely obsidian; dominantly parallel-sided blades (cf. Çatal Hüyük II: Mellaart, J. (1965). Çatal Hüyük West, *Anat. Stud.*, **15**, p. 136). (b) ground: some stone axes; pounders and querns (cf. Çatal Hüyük II, Mellaart, J. (1965). *ibid.*). (2) bone: points (plus holders); antler sleeves (also cf. Çatal Hüyük II: Mellaart, J. (1965). *ibid.* includes awls and spatulae). (3) clay: sling-bullets. (4) metal *copper* mace-head.

13 The faunal and floral remains have not been studied in detail; sheep, goat, cow, pig seem to dominate.

14 A preliminary study of the obsidian tools suggests a narrowing and less abundant production of stone tools; obsidian parallel-sided blades dominate; also present are bone points and antler sleeves. Metal objects are numerously found but none have been identified as tools; the problem may be one of sampling.

15 Our sole source of evidence is Karahüyük—Konya, which has not been published but a superficial examination of the faunal, floral and artefactual remains from the site *suggests* a follow-on from the previous phase. Metal objects (? bronze) are numerous including axes, pins and points. The preceding chipped stone assemblages seems to have disappeared, but bone tools (points, awls, etc.) are found.

16 See Note 14. Our sole source of evidence is Karahüyük—Konya. Structures, settlement plans, etc. are certainly unchanged; probably also, technological equipment and food resources remained the same. There is, however, the real problem of the introduction of iron. The following are known from Karahüyük: (1) Faunal remains: sheep, goat, cow, pig, horse. (2) Plant remains: not studied. (3) Tools: (a) stone: none recovered. (b) bone: decorative objects. (c) metal: bronze knives, axes, pins.

17 Mellaart, J. (1965). *Earliest Civilizations in the Near East*. London; Mellaart, J. (1967). *Çatal Hüyük*. London.

18 For this aspect see Cohen, H. R. (1970). *ibid.*

19 Mellaart, J. (1967). *op. cit.* p. 213.

20 Mellaart, J. (1965). *op. cit.* p. 77; Mellaart, J. (1967). *op. cit.* p. 177.

21 Chisholm, M. (1968). *Rural Settlement and Land-use*. London. pp. 102–103.

JACK R. HARLAN

Crops that extend the range of agricultural settlement

In crossing mountain passes in the Himalaya, the Karakoram and the Hindu Kush, I have been impressed by the importance of barley in the ecology of the highest villages. At the upper limits of agriculture, barley is essentially the *only* field crop grown, and it is evident that the very highest villages are possible only because of this crop. The extension of the range of agricultural settlement covers only a thin altitudinal zone because it is usual for the second or third village downslope to grow wheat, horse-bean, pease, vetch, and other cool-season crops originating in the Near East[1].

In like manner, barley can extend the range of settlement into the Arctic where it flourishes at higher latitudes than other cereals except rye. Along the northern fringes of the Sahara and the Near Eastern deserts, barley can be grown with less rainfall than any other cereal and makes possible a scattering of permanent or semi-permanent villages at the arid limits of agriculture.

In addition, barley can tolerate more saline soils than other cereals. Adams has assembled impressive evidence for a shift to a near monoculture of barley in southern Mesopotamia by the start of the second millennium B.C.[2] Both the archaeobotanical evidence and the cuneiform literature agree that wheat had virtually disappeared as a crop at Agade by 2300 B.C., at Ur III by *c.* 2100 B.C., at Isin/Larsa by *c.* 2000 B.C. and Old Babylonia by *c.* 1700 B.C. The shift to barley monoculture is attributed to salinization of the irrigated lands. At any rate, it is apparent that these city states were able to persist for some centuries only because of the barley crop.

The ecological amplitude of barley, then, is such that it can extend the range of agricultural settlement in several directions. This is primarily due to the fact that barley is an ephemeral, short-season crop. Early maturity makes it possible for it to fit into a short growing season at high altitudes and latitudes and it can exploit a short season due to low rainfall. While barley does have good physiological salt tolerance, early maturity is also important in evading salt damage in saline soils.

Pearl millet (*Pennisetum*) plays a similar role in the dry, summer-rainfall areas of Africa and India. Around the southern fringes of the Sahara and about the Kalahari and Indian deserts it is the only cereal that can be consistently grown near the limits of agriculture. It is especially well suited to sandy soils and can be grown in areas of 200 mm annual rainfall or even less. In Africa, it is especially appreciated as a cereal and the belt of near monoculture that stretches across Sudan, Chad, Niger to Senegal and Mauretania is, in some places, rather wide. Some thousands of village communities are able to exist at the arid fringes of agriculture because of pearl millet[3].

In many areas of the thornbush savanna of Africa, lack of water for man is even more limiting than lack of moisture for crops. There are vast plains that are very level and which have no surface water in the dry season although puddles and water holes are abundant in the rainy season. I have visited villages in Sudan, for example, in which people are resident the year around only because of the baobab tree (*Adansonia*). The boles of these trees are often very large and hollow and can be filled with water during the rainy season in quantities sufficient to support the village through the dry season. The baobab is cultivated by villagers not only for water storage, but the leaves and fruits are eaten and the bark used for fibre. In a sense, it is a crop that extends the range of agricultural settlement.

Agriculture may be limited by climates that are too wet as well as by climates that are too dry. Contrary to general opinion, agriculture in the wet tropics is never easy and demands a set of specialized and, sometimes, rather sophisticated practices. Low, swampy areas are among the most difficult of all. Wet-land crops *par excellence* are rice (*Oryza*) and taro (*Colocasia*), both of which can thrive under continuous shallow inundation. Indeed, the floating rices can produce a crop after flooding up to three metres deep. Both rice and taro can, of course, be grown on uplands as well. Other aroid tubers such as *Alocasia*, *Cyrtosperma*, *Xanthosoma*, and *Maranta* are also adapted to swampy conditions, but are less important on a world scale. In restricted regions of the South Pacific, the sago palm and related species (*Metroxylon*) are important in making settlement of low swampy regions possible[4].

The most critical feature of agriculture in the uplands of the wet tropics is the maintenance of soil fertility and structure. The traditional solution to the problem, worked out independently in south Asia, Africa, the South Pacific Islands, and tropical America, is "shifting cultivation", "slash-and-burn", or "bush fallow" rotations. As a general rule, these rotations are one to three years in crop and ten to twenty years in bush. No adequate substitute for the system has yet been found, and when population pressures force a reduction in the length of the bush fallow period, productivity of the whole region tends to sink. The balance is, therefore rather delicate and we must consider areas subject to traditional shifting cultivation to be on the fringe of agriculture[5]. Root and tree crops, together with upland rice, make the system possible. Root crops include the aroids mentioned above, plus yams (*Dioscorea*) and manioc (*Manihot*). The "tree" crops include oil palm (*Elaeis*), breadfruit (*Artocarpus*), bananas and plantains (*Musa*), and a variety of other fruits.

The lush exhuberance of tropical rainforest growth has deceived many into thinking that agriculture in the wet tropics is easy, and some have even thought that agriculture originated in such ecological zones. Actually, regions of tropical rainforest are, even today, rather sparsely settled and new solutions to their limitations must be found before they can be fully exploited.

Storage of foodstuffs is always a problem in the wet tropics. Materials are continuously subject to attack by fungi, bacteria, insects, rats, mice, and other creatures and spoilage is always considerable. Root and tree crops that can be harvested essentially all year round are helpful. A further defence is in root crops that are poisonous without special treatment. Almost all of the root crops have cultivars that fall into this category. Manioc has both poisonous (prussic acid) and non-poisonous forms, but the poisonous ones are generally preferred in the most difficult areas. The tuberous aroids have poisonous types (oxalic acid) as well, and the same pattern prevails. Some of the yams are also poisonous (alkaloids) without treatment. While processing to remove the poisonous principles is sometimes elaborate and time-consuming, the poisons serve an important function in the preservation of food resources under very difficult conditions[6].

The high islands of the South Seas usually provide fairly adequate agricultural resources, but the low, relatively dry coral atolls are decidedly marginal for agriculture. They would probably be entirely uninhabitable for man without the introduction of the coconut and pandanus. Even when planted, the resources are so meagre that the Polynesian inhabitants frequently resorted to a kind of nomadism, moving from atoll to atoll exploiting the scant harvest in rotation. During residence, they often cultured *Cyrtosperma* and taro by digging deep pits into the fresh water lenses in the centres of the islands[7]. The more desirable crops of Oceania such as bread-fruit, banana and sugar-cane are poorly adapted to these marginal conditions and the coconut is the mainstay for the inhabitants of the low atolls.

The filling of an empty ecological niche by an appropriate crop can have striking effects on the human population. A dramatic example is the introduction of the sweet potato (*Ipomoea*) to the highlands of New Guinea. The sweet potato is an American domesticate, probably introduced to New Guinea by the Portuguese after A.D. 1500. In some communities at 6000 ft elevation or higher, 80% or more of the food consumed is sweet potato and the pigs reared by the villagers for food are also fed on the sweet potato. Watson speaks of an "ipomoean revolution" in which a population explosion occurred in the New Guinea highlands[8]. Indeed, the present population in the sweet potato belt is the most dense on the island, reaching 200–250 per square mile in some valleys. Brookfield and White[9] tend to discount the "revolutionary" nature of the ipomoean impact, but admit that the sweet potato extended the range of agriculture to higher elevations and out of serious malaria hazards, thereby making a population expansion possible.

The chief advantage of the sweet potato over the more indigenous root crops such as taro, yams, and *Pueraria lobata* is that it is adapted to more temperate conditions and will thrive at cooler temperatures. The yams do not do well above 5500 ft in New Guinea and the taro and *Pueraria* are poorly adapted above 6000 ft elevation. As Bulmer and Bulmer[10] indicated, it is unlikely that the indigenous crops could have supported anything approaching present population densities at between 5000 and 6000 ft and could only have supported very sparse populations at higher altitudes. The sweet potato has extended the range of agricultural settlement to over 9000 ft elevation[11].

A similar, but less striking case can be made for the introduction of maize (*Zea mays* L.) into the hill country of India and Pakistan. Prior to the introduction of maize, the cereals available were wheat and barley for high altitudes and rice and sorghum for the lower valleys. There seems to have been a belt at intermediate altitudes rather sparsely settled. With the introduction of maize, the intermediate zone filled up rapidly; the forests were cut, the hillsides were terraced, and maize became a dominant cereal of the middle zone. It is not that agriculture was impossible in these regions, but the crops available were not especially appropriate and there was a niche that maize could fill efficiently.

The replacement of less efficient crops by more efficient ones can have profound effects on population patterns. The introducion of the potato into Europe had a profound impact[12]. It did not extend the *range* of agriculture, but helped to convert some areas of very modest food production to much higher levels of productivity, thereby affecting settlement patterns.

No doubt, other examples might be cited, but perhaps I have given enough cases to indicate the general patterns. A large percentage of the land surface of the earth is considered non-arable. The climate may be too cold, too dry, or too wet, the growing season too short; the terrain may be too steep or the soils too wet, too dry or too infertile to sustain agricultural settlements. In the better agricultural lands, crops can be grown in great variety and large populations can be supported. As one moves towards marginal areas, fewer and fewer crops can be grown and the population becomes less dense. At the very fringes of agriculture, the system moves towards a monoculture of the crop most tolerant to the primary factors limiting production. There is nothing unexpected in this ecological pattern, but it might be useful to take these features into consideration with respect to settlement patterns.

Notes

1 Harlan, J. R. (1968). On the origin of barley, *in Barley: Origin, Botany, Culture, Winterhardiness, Genetics, Utilization, Pests*. USDA Agric. Handbook, **338**, pp. 9–31. Harlan, J. R. (In press). On the origin of barley: A second look, *Proc. Second International Barley Genetics Symposium*, Pullman. It is to be noted that people who *must* eat barley prefer

the naked sorts. The percentage of naked cultivars in Tibet and other Asian highlands is very high, and this preference may also explain why naked barley was relatively more important in the European Neolithic than it has been since.

2 Adams, R. M. (1958). Salinity and irrigation agriculture in antiquity, *Diyālā Basin Archaeological Project Progress Report*, 1 June 1957 to 1 June 1958. Univ. Chicago Mimeo, p. 104. Adams, R. M. (1965). *Land behind Baghdad*. Chicago.

3 Adrian, J. and Jacquot, R. (1964). *Le sorgho et les mils en alimentation humaine et animale*. Paris. Gast, M. and Adrian, J. Mils et sorgho en Ahaggar, étude ethnologique et nutritionelle. *Mémoires C.R.A.P.E.*, **4**. Pearl millet is not only a preferred cereal in the drier parts of Africa, but is one of the most nutritious of all cereals.

4 Barrau, J. (1958). Subsistence agriculture in Melanesia, *Bishop Museum Bul.*, **219**. Useful and enlightening ethnobotanical comments are contained in: Barrau, J. (1965). L'humide et le sec, an essay on ethnobotanical adaptation to contrastive environments in the Indo-Pacific area, *J. Polynesian Soc.*, **74**, pp. 329–346.

5 Literature on shifting cultivation is voluminous but not always enlightening. One might consult: Conklin, H. C. (1963). El estudio del cultivo de roza. The study of shifting cultivation, *Est. Monagr.*, II. Union Panamerican. Washington; Jurion, F. and Henry, J. (1967). De l'agriculture itinérante à l'agriculture intensifiée. *I.N.E.A.C.*

6 Barrau, J. (1958). *ibid.*

7 Barrau, J. (1961). Subsistence agriculture in Polynesia and Micronesia, *Bishop Museum Bul.*, **223**.

8 Watson, J. B. (1965). From hunting to horticulture in the New Guinea highlands, *Ethnol.*, **4**, pp. 295–309.

9 Brookfield, H. C. and White, J. P. (1968). Revolution or evolution in the pre-history of the New Guinea highlands: a seminar report, *Ethnol.*, **7**, pp. 43–52.

10 Bulmer, S. and Bulmer, R. (1964). The pre-history of the Australian New Guinea highlands, *in* J. B. Watson (ed.) *New Guinea: the Central Highlands*, *Amer. Anthrop.*, **66** (2) special publ., pp. 39–76.

11 Barrau, J. (1958). *ibid.*

12 Salaman, R. N. (1949). *The History and Social Influence of the Potato*. Cambridge.

DAVID R. HARRIS

Swidden systems and settlement

The question of what formal and functional relationships link agricultural systems and settlement patterns has long engaged the interest of anthropologists, geographers and other social scientists. An ecological approach to the analysis of agricultural systems, which recognizes them as distinctive types of man-modified ecosystems or *ecotypes*, has brought a new dimension to the study of traditional systems of palaeotechnic cultivation[1]. But when attention is focused on the linkage between ecotype and settlement it is necessary to consider other criteria in addition to those that are overtly ecological. For present purposes palaeotechnic agricultural systems will be classified primarily by reference to the temporal continuity of agricultural land-use because a relationship exists between the length of time a field or plot is cultivated and the degree of concentration and permanence of the associated settlement. By reference to this criterion agricultural systems can be ranged from those in which land is cultivated in staple crops for shorter periods than it is fallowed (long-term fallowing or swidden systems), through systems in which the period of cultivation of staple crops exceeds the fallow period (short-term fallowing systems), to systems of "continuous" or "permanent" cultivation[2]. The latter depend for the maintenance of agricultural productivity on soil fertilization which may be effected primarily either by direct soil enrichment ("edaphic" systems) or through the control of water supplies ("hydraulic" systems).

It is clear that continuous palaeotechnic systems of both edaphic and hydraulic type, such as fixed-plot horticulture in the Mediterranean region and wet-padi farming in south and east Asia, allow or even demand a high degree of concentration and permanence of settlement. They are normally associated with sedentary and socially stratified populations living in nucleated villages or towns, and very commonly with peasant societies in which a majority of rural cultivators produce surpluses which are transferred to a dominant ruling group[3]. Palaeotechnic short-term fallowing systems are also associated with permanent settlement in villages and towns and with peasant societies. Their most significant historical development took place in Atlantic Europe where a system of mixed grain-livestock farming evolved based on crop rotation, fallowing and the use of animal and other organic manures to maintain soil fertility. It is when we consider palaeotechnic long-term fallowing systems that we encounter the phenomenon of impermanent and often widely dispersed agricultural settlement. Indeed it is commonly assumed that swidden cultivation necessitates at least a semi-nomadic life on the part of the cultivators and the term "shifting cultivation" is sometimes erroneously interpreted as implying shifts of settlements rather than of fields; hence the adoption of the neutral and unambiguous term "swidden cultivation" which was first proposed by the Swedish anthropologist Izikovitz in 1951[4].

Over several millennia long-term fallowing or swidden systems have become adapted to widely diverse habitats and have incorporated a great variety of cultivated plants and techniques of cultivation, but they are universally characterized by (*a*) partial or complete clearance of the vegetation cover by cutting and burning, (*b*) the temporary cultivation of crops in the cleared area, and (*c*) the abandonment of the plot to fallow under regenerating vegetation for a longer period than the preceding phase of cultivation. In this paper assumptions about the relationship between settlement and swidden cultivation in its various manifestations will be examined and the question raised as to whether and in what circumstances swidden systems can sustain complex societies living in permanent villages or towns.

General attributes of swidden

It is difficult and often misleading to generalize about so geographically and historically diverse a mode of cultivation, but certain common denominators of swidden ecology and economy can be discerned which are relevant to the discussion. First, swidden is essentially a small-scale form of agriculture: cleared plots are seldom more than 1 ha (2·47 acres) in extent and are commonly only an acre or less in size. In the tropics today a plot may be planted "monoculturally", with one regionally preferred staple crop, usually either a seed-sown grain crop such as maize, millet, sorghum or rice or a vegetatively reproduced root crop such as manioc, sweet potato, yam or taro; or, alternatively, it may be planted "polyculturally" with a diverse assemblage of useful plants—trees, shrubs and herbs—grown together as one plant community which broadly simulates the wild vegetation it replaces[5].

A second common denominator of swidden cultivation is that it is a "land-extensive" and "labour-intensive" system. Because plots are only cultivated for short periods of time—perhaps one to three years on average—before being abandoned for longer periods, there is normally a considerable excess of fallow over cultivated land within the effective agricultural area of a given population. At the same time the processes of clearance, cultivation and harvesting involve intensive human effort in the use of hand tools such as axes, knives, hoes and digging-sticks, and—less typically—animal-drawn ploughs. It is estimated that a swidden cultivator expends on average between 500 and 1000 man-hours per year on agricultural tasks exclusive of food preparation and other subsistence activities and it has been suggested that this expenditure of time represents little or no reduction on the time devoted to the food quest by many non-agricultural hunters and gatherers[6]. Although it is a labour-intensive system swidden cultivation seldom demands concerted action by large groups of people: most agricultural tasks can be undertaken on a family basis and only exceptionally, as when a new swidden is first cleared, is there need for joint action on a larger scale involving the co-operation of members from several families.

Thirdly, and contrary to the common assumption that swidden is an inefficient method of cultivation, it can be shown that such systems are often highly productive. Western observers of contemporary swidden cultivators in the tropics have tended to judge the system from the standpoint of their own European tradition of fixed- and clean-field farming and to condemn it as unproductive and wasteful of forest resources. It is an unproductive system per unit area of land cultivated, but in terms of yields per unit of labour expended its productivity can equal or even exceed that of some types of permanent, fixed-field agriculture. Provided that no land shortage threatens the maintenance of an optimum cycle of cultivation and fallowing, swidden plots can yield as much as or more than comparable fields under continuous cultivation. For example, a comparison of data from Conklin and Gourou shows that the Hanunóo of the Philippines can grow as much rice per unit of labour on their swiddens as is produced on double-cropped padi fields in the Tonkin delta[7], and Leach has demonstrated that tribal peoples in northern Burma recognize that swidden cultivation in forested areas produces higher yields for less work than continuous cultivation of terraced fields[8]. Comparable examples may be drawn from the African and American tropics[9]. Most available comparisons are between single-crop cultivations, as of rice in south-east Asia and maize in Mexico, but in many areas swidden cultivators engage in polycultural planting and yields from a single plot can then reach still higher levels as different useful plants mature and are harvested successively through the year. Striking examples of this polycultural swidden pattern incorporating root as well as seed crops have been described by Conklin who notes that over forty different types of crop have been observed growing in one Hanunóo swidden at the same time[10], and by Montgomery who reported twenty different crops planted in one small plot in the Chimbu highlands of New Guinea[11].

Fourthly, and despite their tendency to produce high yields per unit of labour, swidden systems are nevertheless characteristically associated with low densities of population. Because fallow must substantially exceed cultivated land if there is to be adequate time before re-cultivation for soil fertility to be restored under a regenerating cover of vegetation, there is always need for a large amount of land per head of the population. It is this limitation on productivity per unit of area, rather than a limit to productivity per unit of labour, that restricts the capacity of swidden cultivation to support concentrated populations. Gross densities of population do in fact vary widely from one area of swidden cultivation to another, but they seldom exceed 150 per square mile and more usually they fall below 100 per square mile[12]. Densities as low as 1·5 per square mile are known—as for present maize cultivators in the Lake Peten area of Guatemala[13]—and among South American tropical forest cultivators densities of less than ten per square mile are usual. At the other extreme there are cases of swidden cultivators achieving population densities well in excess of 150 per square mile, most remarkably in the tropical forest zone of south-eastern Nigeria[14] and in the central highlands of New Guinea[15] where predominantly root-crop cultivators employing modified swidden methods live at densities of over 400 per square mile.

Lastly the characteristic size, distribution and stability of swidden settlements calls for comment. The usual unit of settlement is the village or hamlet which, in accordance with prevailing low population densities, seldom houses more than 200–250 people. In the South American tropical forest, for example, the population of villages occupied by swidden cultivators typically falls between fifty and 150[16], although it is probable that substantially larger villages existed in the past, especially along the major rivers. In areas where the characteristic unit of settlement is the hamlet rather than the village populations tend to fall below 100, as for example among the Isneg of the Philippines where there are on average eighty-five people to a hamlet[17].

As is to be expected, the spatial distribution of settlements normally relates to their average size in such a way that the larger the village units the greater the cultivable area that separates one village from another; conversely the smaller the units the more closely spaced they tend to be. Thus one end of the settlement spectrum is characterized by a relatively close scatter of hamlets, each surrounded by its swidden lands, whereas the other is typified by an open pattern of villages sited at wide intervals. These extremes may be exemplified on the one hand by the hamlets of the Isneg, located from two to three miles apart[18], and on the other by the villages of the pre-European Maori of New Zealand, which contained several hundred people each and were situated many miles apart[19]. However, the relationship between settlement size and spacing is not always simple and direct. Some swidden cultivators maintain a dual settlement system in which small hamlets occupied by extended family groups surround a centrally placed village which is the focus of supra-family activities. For example, the Akawaio Indians of the Guiana Highlands maintain villages which are usually occupied by some twenty to sixty people and are distinguished by the presence of a large, communally built house used for ceremonial and other social activities, around which are located a number of small "garden places" each occupied by an extended family of usually less than fifteen people[20].

In relating the size of settlements to their spacing it is of course misleading merely to consider linear distance for, as Vayda among others has pointed out[21], pheric distances (i.e. distances expressed in the time required to travel them) provide a more realistic means of measuring the intervals between communities, as also between a settlement and its swidden plots. However, whether linear distance only is considered or, more realistically, pheric distance, the common denominator of swidden settlement patterns is their dispersed character. A degree of nucleation is sometimes achieved in the form of villages of above-average size, such as those that are located at convenient sites on navigable waterways, but this is normally compensated by wider spacing of the villages themselves so that the spatial pattern remains one of dispersed settlement and scattered population.

The aspect of swidden settlement that has evoked most discussion is the question of its stability or permanence. As has already been noted, it is often assumed that shifting cultivation implies impermanence of settlement and it is sometimes asserted that long-term occupation of village sites is incompatible with the constant need to clear new swiddens. In some cases a chronic instability of settlement does apparently prevail, as for example among the Iban of Sarawak who prefer to cut new swiddens out of primary rather than secondary forest and who move their settlements frequently in order to do so[22]. But in this respect the Iban resemble pioneer cultivators in other parts of the world who employ swidden techniques as a convenient and productive means of land clearance and who tend to regard forest resources as either unlimited or expendable. This mode of swidden cultivation, involving progressive linear rather than cyclic shifts of clearings and frequent relocation of settlements, has been adopted widely in the past in "frontier" situations, as by the early European settlers in the forests of eastern North America, and presumably also by Neolithic cultivators in the forests of temperate Europe. It is still often adopted as the most effective pioneering technique where forested land is being opened up for agricultural use in the wake of new road construction or lumbering operations, as in many tropical forest areas today, for example the eastern lowlands of Guatemala[23]. The distinctiveness of this pioneering mode of cultivation, with its attendant instability of settlement, has not been adequately emphasized in the literature on swidden, and, as will be argued later, it may have been of greatest historical significance in temperate areas of seed-crop cultivation such as Atlantic Europe.

In contrast to this pioneering mode stands a more stable form of swidden cultivation that is associated with much greater permanence of settlement. It is characterized by fields shifting in cyclical rather than progressive linear patterns, and, provided population increase does not upset the equilibrium of the system by causing a reduction of the fallow period to a point where soils cannot recover their fertility and overall productivity declines, it does not necessitate the periodic relocation of village sites.

Carneiro has attempted to quantify the capacity of a swidden system to support a sedentary population permanently and has derived three formulas which—given numerical data on the area of arable land within practicable walking distance of the settlement, the area required to provide cultivated plant food for one individual for one year, the numbers of years a plot is cultivated and fallowed, and the total population of the community—make it possible to determine (*a*) how large a population can be supported permanently at one site, (*b*) the smallest area of cultivable land that will support a village of a given size at one site indefinitely, and (*c*) how long a community can remain in the same place before reduction of the fallow period and soil exhaustion obliges it to move (if it ever does)[24]. He works out the values for a Kuikuru village in the Upper Xingú region of Brazil and concludes, on the basis of the Kuikuru's present swidden system, (*a*) that some 2000 persons could live permanently where in fact only 145 do now, (*b*) that the present population could still remain sedentary indefinitely even if they used only 7% of the arable land that lies within an accessible radius of the village, and (*c*) that the present Kuikuru are sedentary by a wide margin since it would take nearly 400 years to plant and exhaust all the arable land available to them which would allow more than ample time for forest regeneration to take place, the usual interval before recultivation of a plot being twenty-five years. Carneiro is thus able to demonstrate that under present conditions there is nothing incompatible between the Kuikuru's system of swidden cultivation and permanent settlement and in fact his field enquiries showed that they had maintained their village in the same locale for the previous ninety years, moving its site only very short distances a number of times for supernatural rather than ecological reasons. Generalizing on the basis of his data for the Kuikuru, Carneiro further postulates that under "low average" conditions of swidden subsistence in the tropical forest environment it would still be possible for villages of about 500 people to remain sedentary indefinitely.

Carneiro's analysis of the relation between swidden and settlement stability provides a valuable corrective to the assumption that impermanence of settlement is inherent in all swidden systems and it brings into sharp focus the more stable alternative to the pioneering mode of swidden cultivation. But in claiming, on the grounds that the South American tropical forest is a typical area of swidden cultivation, that his results show "that for primitive peoples in general permanence of settlement is certainly compatible with slash-and-burn agriculture"[25], Carneiro ignores other significant ecological variables. Most important of these is probably the nature of the crop complex on which the swidden system is based. The Kuikuru, like many other South American forest cultivators, subsist primarily on manioc which is extremely easy to propagate, yields starch abundantly and makes low demands on soil fertility. Assumptions based on a system that is adjusted to the cultivation of this undemanding staple crop cannot with confidence be applied to other swidden systems based on different crop complexes, particularly those that depend primarily on seed-crop cultivation.

The degree of stability or permanence of swidden settlements thus ranges from ephemeral camps associated with agricultural pioneering to permanently established tropical forest villages, but because of the large amounts of land per head that swidden cultivation requires the settlements themselves tend to be widely scattered and population densities normally remain low.

Ecological factors

Having characterized the common denominators of swidden production, population and settlement we may now examine the ecological factors that limit the capacity of swidden systems to sustain large populations in concentrated patterns of settlement. Several of these factors have already been mentioned. Chief among them is the high demand for cultivable land which, because there must always be a substantial excess of fallow over cultivated plots if the system is not to break down, limits productivity per unit area and maintains a dispersed pattern of settlement. The need for land varies with the time interval necessary to allow a plot to recover its former fertility and this fallow period varies widely according to other ecological factors such as soil quality, climate and the crop complex. Fallow periods may be as short as two to six years, as among present-day Mayan- and Kekchi-speaking Indians of Guatemala[26], or as long as twenty-five or more years, as among the Kuikuru and Akawaio of Brazil and Guyana[27], but ultimately the need to maintain a high proportion of fallow to cultivated land sets limits to the concentration of settlement possible within a given system.

A second set of ecological variables that affects the population and settlement potential of swidden systems consists of those factors of the physical environment that condition agricultural production. They include locally variable factors that defy broad generalization, such as surface topography, soil fertility, microclimatic conditions and the presence or absence of weed grasses and insect pests. But they also include zonal variations in climate, soils and vegetation that have more general significance, particularly differences in the length of the growing season. Because swidden cultivation depends for its successful continuance on a sufficiently long fallow interval to allow the regeneration of a cover of wild vegetation and an associated build-up of soil structure and fertility to a level comparable with that prevailing before initial clearance, it could be argued that it will have its greatest potential for human subsistence in those areas of the tropics where the growing, season is not curtailed either by cold or by drought and where, therefore, natural regeneration proceeds most rapidly. This view would imply that the capacity for swidden to support population is greatest in the humid, equatorial tropics, but it fails to take account of other ecological factors that work in a contrary direction. Chief of these is the marked infertility below the surface horizon of most humid tropical soils and the fact that, once the nearly closed nutrient cycle that maintains the productivity of primary tropical rainforest is broken by clearance and cultivation, the restoration of soil fertility even in the surface horizon proceeds very slowly despite the rapid regeneration of secondary vegetation. Also significant is the fact that as dry seasons become shorter and less dependable towards the equator so the difficulty increases of thoroughly burning cleared vegetation; and because crop growth depends to a great extent on nutrients derived from the ash an inefficient burn usually results in poor yields. This factor is probably more significant where maize or other grains are the staple crop for, as has been argued elsewhere, the successful swidden cultivation of manioc or other root crops in the humid tropics may sometimes depend less on the burning process than on the provision of a mulch-like litter of organic débris to make nutrients available[28].

If the humid tropics do not provide optimum climatic and edaphic conditions for swidden cultivation neither do the cold and temperate zones of high- and mid-latitudes where the rate of natural regeneration is checked by low winter temperatures and also, in interior continental situations, by low and irregular rainfall. These limitations are to a considerable extent compensated in mid-latitude temperate areas by the widespread occurrence of more fertile and easily tilled soils, but, because plant nutrients characteristically occur more deeply through the soil profile than in the tropics, because traditional methods of tillage by digging-stick, hoe and even light plough disturb only the uppermost layers of the soil, and because cereals and other mid-latitude herbaceous seed crops tend to be relatively shallow-rooting, the greater fertility of temperate soils may well have had only a minor compensating effect on the yields of swidden crops.

Between the temperate zone and the humid tropics seasonal drought is the most critical factor limiting the rate of regeneration of vegetation on abandoned swidden plots. As the mid-latitude climatic deserts are approached and the dry season lengthens, swidden cultivation becomes less and less feasible, but where there is a short but regular dry season, in the intermediate tropical zone between the hot deserts and the humid rainforests, climatic conditions for swidden cultivation are at their optimum. Here (except at high altitudes) plant growth is never checked by low temperatures and the main climatic restraint on natural regeneration is the brief dry season that is in any case desired for the successful burning of cleared vegetation. Edaphic conditions vary widely in this intermediate zone but soil fertility tends to increase away from the humid rainforests.

Parallel with these major climatic and edaphic contrasts that condition the rate of regeneration and hence the length of the fallow period necessary for successful swidden cultivation are variations in vegetation type that also affect the productive potential of the system. For both ecological and technological reasons swidden is best adapted to forest ecosystems. Because the living vegetation cover of a plot represents a major potential source of nutrients for food production, which clearance and burning makes available to the crops chiefly in the form of ash, it follows that clearance of forest vegetation tends to provide a larger and richer supply of nutrients than clearance of shrubs or herbaceous vegetation. Indeed one of the ways in which swidden systems become ecologically maladapted is when too frequent clearance of regenerating secondary forest results in there being available for conversion into ash a progressively smaller woody component in the cleared vegetation. In the tropics one of the well known signs of a maladapted swidden system is when abandoned plots are colonized only by herbs and shrubs rather than by trees. In many tropical areas this downgrading process has led to the conversion of forests into grasslands dominated by species, such as the widespread perennial grass *Imperata cylindrica*, which are exceedingly difficult to eradicate and which if burned supply insufficient nutrients to allow productive swidden cultivation to continue.

Technologically, too, swidden cultivation is essentially a forest ecotype for the simple tools and techniques traditionally used for clearance and tillage—axes, knives of the *machete* type, ring-barking, fire, digging-sticks and, on a more restricted scale, hoes and light ploughs—are well suited to the disposal of woody vegetation and the shallow disturbance of the soil; but against fire-tolerant and tussock- or sward-forming grasses they are—with the exception of some hoes and ploughs—largely ineffective.

Viewed broadly therefore the factors of zonal climate, soils and vegetation suggest that the productive potential of swidden cultivation is greater in forested than in non-forested areas and that it reaches a maximum in the short dry-season forests of the intermediate tropical zone. But it is necessary to qualify these very broad generalizations by reference to a third set of ecological variables that directly affects swidden productivity: variation in the crop complexes themselves. The significance of this factor has been widely overlooked in discussions of swidden but, as has already been pointed out in relation to Carneiro's analysis, the nature of the crops raised must be taken into account when assessing the capacity of a swidden system to support a sedentary population.

The most significant distinction appears to be between systems that depend primarily either on seed crops or on roots and other vegetatively reproduced crops. This distinction corresponds, in the Asian and American tropics at least, if not also in Africa, to a marked cultural division between "seed-culture" and "vegeculture"; and its main ecological significance lies in the contrasted demands on soil fertility made by the two traditions of cultivation. When protein-rich seed crops are raised greater demands are made on the supply of nutrients in the ash, litter and soil than when starch-rich root crops—many of which have, under domestication, partially or completely lost their capacity to form seeds—are cultivated; and this difference is accentuated at harvest when the seed-crop cultivator removes from his field for consumption highly concentrated nutrients in the form of, say, beans or cereal grains, whereas the vegeculturalist removes a much smaller fraction of fertility in his harvest of tubers, corms or rhizomes. Assuming other ecological variables to be equal, therefore, we would expect seed-crop swidden systems to be less stable than vegecultural ones, in the sense that they would require longer average fallow periods for the restoration of fertility to pre-cultivation levels. It can also be inferred that they would, more readily than vegeculture systems, become ecologically maladapted under conditions of stress such as sustained population growth. Furthermore, if a system is monocultural in the sense that either seed or root crops are grown to the virtual exclusion of other types of crop then these differences between the two traditions of cultivation will be accentuated.

This ecological distinction has interesting implications for our understanding of the evolution of swidden systems and settlements. First it leads us to expect that swidden cultivation will achieve its most stable manifestation in areas of tropical forest where there is a vegecultural tradition of root-crop planting. The stability of settlements associated with the swidden cultivation of manioc in South America has already been commented upon and Spencer has described a similar vegecultural tradition based on taro and yams in south-east Asia[29]. Secondly it suggests that in those parts of the dry-season tropics where traditions of seed-crop swidden have evolved, as among the maize cultivators of the American tropics, the upland rice growers of the Asian tropics and the sorghum and millet farmers of tropical Africa, swidden settlements will tend to be less stable and more liable to progressive migration than those of the tropical vegeculturalists. In fact in both tropical America and tropical south-east Asia there is suggestive evidence for the instability of maize- and rice-dominated swidden systems and for the progressive intrusion of seed-culture into areas of vegeculture[30].

Thirdly the ecological argument implies that the tendency for seed-crop swidden systems to turn maladaptive may be still greater in the temperate zone where natural forest regeneration proceeds more slowly and fallow intervals need to be longer than in areas of equivalent soil fertility in the intermediate tropical zone. This view affords a fresh perspective on the colonization of temperate forested Europe by Neolithic cultivators. It lends support to the hypothesis, early advocated by V. Gordon Childe and Grahame Clark, that the rate of migration of Danubian and other Neolithic groups into and across Europe was related to the necessity of relocating their settlements at relatively frequent intervals, or in other words, that they were swidden cultivators in the pioneering mode. Evidence from both pollen analysis and macro-fossil finds indicates that Neolithic European agriculturalists depended on a seed-crop complex based on wheat, barley and other grains[31] which, in the absence of any fertilizer other than the ash derived from swidden burning and the haphazard droppings of free-ranging livestock, must have made heavy demands on soil fertility. Such a system would—save in areas of exceptional soil fertility—have had only a limited capacity to support population without becoming maladapted. Under pressure of population increase the fallow interval would have been reduced below the optimum and the need to relocate the village or hive off daughter settlements in unoccupied areas of forest would have become compelling. That this did probably occur has recently been demonstrated at the Neolithic site of Bylany (Linear Pottery culture) in Czechoslovakia[32]. Here excavation and inference suggests that about 30 ha of chernozem soil, developed over loessic parent material and cultivated in grains and legumes by swidden methods, supported some twenty-five families or a population of approximately 150 adults and children. There is evidence that the village site was relocated within the area at intervals of about fourteen years and also that part of the population periodically broke away and left.

In the Neolithic European context an additional factor that may have significantly checked the rate of forest regeneration and further curtailed the capacity of swidden to absorb an increasing population was the presence of domestic livestock, particularly cattle and goats. In the wild these animals browse the protein-rich buds and young growth of woody plants in preference to grazing herbaceous vegetation and it is probable that where domestic livestock were raised by Neolithic swidden cultivators they were allowed to browse and graze in abandoned plots as well as in the forest, thus retarding the regeneration of woody plants and accelerating the conversion of forest into grassland. Livestock would therefore have undermined the capacity of the swidden system alone to support population, but by providing additional food they may have increased the overall productivity of the local economy and hastened the transition from swidden to more intensive systems of agriculture which led ultimately to the emergence in Atlantic Europe of the mixed grain-livestock farming tradition.

From this examination of the ecological factors that limit the capacity of swidden systems to sustain large populations in concentrated patterns of settlement it is evident that, although in some situations particularly in the tropical forest *permanence* of settlement is possible, the high land demands of swidden cultivation ensure that—until the system gives way to some more land-intensive form of agriculture—population densities remain low and settlements widely dispersed. Where there are exceptions to the normal low densities of population they occur, as ecological inference leads us to expect, among tropical swidden cultivators who raise low-nutrient demanding root crops either as staples in largely monocultural systems or as important components of polycultural systems. Thus the rural population densities of over 400 per square mile that have for long characterized much of south-eastern Nigeria have been sustained largely by a vegecultural swidden system in which yams and manioc are dominant crops, and similar densities in the Chimbu area of New Guinea rest primarily on the cultivation of sweet potato[33]. Nevertheless, although swidden systems can in certain circumstances support relatively dense populations, there appears to be no convincing evidence that they can ever develop to a point where they alone sustain a complex, stratified society distributed in a concentrated pattern of settlement in large villages or towns.

It has so far been argued that under swidden ecological factors operate to maintain a dispersed pattern of settlement and that population densities normally remain low, but ecology alone provides an insufficient explanation for the apparent failure of swidden to sustain any advances to the higher cultural levels we associate with the word "civilization". In the last part of this paper some of the social factors that contribute to this failure will be summarily reviewed and in conclusion the "problem" of Maya civilization, reputedly based exclusively on swidden cultivation, will be briefly examined.

Social factors

In its pioneering mode, both in tropical and temperate areas, swidden cultivation is clearly so ecologically unstable a system that no appeal to social factors is necessary to explain its failure to support complex societies and concentrated patterns of settlement. It is the similar failure of its more stable mode, which is most closely associated with root-crop cultivation in tropical areas, that demands more than an ecological explanation. It is in such areas, as Carneiro has shown, that population remains at levels well below those which the system could support without ecological deterioration. Why tropical swidden populations should normally stabilize at levels below the ecologically permitted maximum is obscure, but the explanation probably relates to the difficulty of generating appropriate social controls to integrate larger populations[34].

The characteristic pattern of social organization among swidden cultivators is that of simple segmentary tribes living as decentralized autonomous communities in small dispersed settlements. The transition from this pattern to one of a dependent peasantry under centralized control is a critically difficult one which populations wholly dependent on swidden cultivation appear unable, or at least most unlikely, to make. When in modern times swidden cultivators have been forced to give up their autonomy and change their status to that of peasants working a permanent-field system of agriculture, as in parts of south-east Asia and Africa[35], they have strongly opposed the change and returned to swidden cultivation when opportunity allowed.

Other social factors reinforce the characteristic swidden pattern of decentralized autonomous communities and prevent the reorganization of simple kinship groups into conical clans or pyramidal chiefdoms that might permit the development of social stratification and centralized control. The dispersed pattern of settlement itself militates against the emergence of centralized authority, particularly in forested environments where, if navigable waterways are absent, communication between villages is usually slow and laborious. The fact that rights to land are normally held in common at the level of the family, hamlet, lineage or village also limits the chances of wealth and power becoming concentrated in the hands of an elite; as does the absence of specialization in agricultural production and craft industry. As most communities are self-sufficient there is little opportunity for local or regional trade to develop and for specialized merchant groups to emerge. Lastly, competition for land between swidden cultivators can readily lead to a state of chronic institutionalized warfare which eliminates any prospect of social integration at more than strictly local level[36]. Wolf has commented on the fragmenting effect of this social-ceremonial form of warfare which he distinguishes from instrumental or "true" war which is carried on for reasons of state and often results in social integration[37].

Swidden and civilization: the Maya "problem"

In view of all the ecological and social factors conspiring to limit the evolutionary potential of swidden it would seem right to conclude that when practised as the only form of agricultural subsistence it is incapable of sustaining "civilization" or indeed any level of socio-economic complexity above that of autonomous tribal communities living in small dispersed settlements. Before drawing this conclusion however the apparent exception of Maya civilization invites comment.

It is often suggested that the lowland Maya, who developed their complex "theocratic" civilization in south-eastern Mexico, Guatemala and adjacent British Honduras between 900 B.C. and A.D. 900, did so exclusively on the basis of the swidden cultivation of maize and associated seed crops. If so then they achieved what perhaps no other swidden society has achieved: namely to overcome all the ecological and social limitations inherent in the system and to generate mechanisms of centralized authority sufficiently powerful to control land rights and regulate shifts in cultivation over such an extensive area that swidden could continue for many centuries without getting out of equilibrium with the environment. Although it has been argued that the seed-crop swidden system of the Maya was quite capable of producing sufficient surplus to support a non-cultivating elite[38], and that "the development of a centralized and stratified society was by no means impossible under conditions of shifting cultivation"[39], these views undervalue the inhibiting ecological and social effects of swidden, particularly the inherent instability of a maize-dominated, high nutrient-demanding system of cultivation such as is assumed to have been the foundation of Maya subsistence. A more convincing explanation of the apparent paradox of Maya civilization is to suppose that it depended not only on seed-crop swidden but also on less demanding or more intensive forms of cultivation. Until recently little attention had been paid to this possibility, but in 1966 Bronson argued that root crops may have contributed significantly to Maya subsistence[40].

If low nutrient-demanding crops such as manioc and sweet potatoes were indeed important components in the crop complex of Maya swidden farmers then it becomes much easier to comprehend how the large populations involved in the construction and maintenance of numerous monumental ceremonial centres were supported without the system rapidly becoming maladapted. We have seen that the most stable mode and the highest population densities of swidden cultivation are associated with tropical vegecultural systems which yield carbohydrate heavily without overtaxing limited soil fertility. It may well be that the development of such a system in the lowland forests of Middle America was a necessary prelude to the rise of Maya civilization and it is tempting to speculate that a comparable development may have underlain the rise of other non-irrigation civilizations based largely or at first on swidden cultivation in tropical south-east Asia, Ceylon and West Africa[41].

Although the hypothesis of root-crop cultivation helps to explain how the large Maya population may have been supported it leaves unresolved the problem of how mechanisms of centralized control were generated to convert a population of autonomous cultivators into a dependent peasantry. This transition may be attributable to the operation of non-ecological factors, such as external pressure, but it is also possible to argue, by analogy with modern examples, that root-crop swiddens, which are normally cultivated for longer periods than seed-crop swiddens, may have served as foci around which population could have begun to concentrate while the shorter-term cultivation of maize continued in more peripheral areas. This process would have been accentuated if other crops amenable to continuous, intensive cultivation, particularly tropical tree crops, also contributed to Maya subsistence. The descendants of the Maya today make use of many tropical fruit trees, including the pejibaye palm (*Guilielma gasipaes* = *Bactris gasipaes*) and the ramon or bread-nut tree (*Brosimum alicastrum*) which yield abundant supplies of carbohydrate, and strong evidence has recently been put forward by Puleston which suggests that intensive "garden" cultivation of the ramon in residential areas may have contributed substantially to the Mayan food supply[42]. If this, and perhaps other forms of intensive cultivation[43], were practised by the Maya in addition to swidden then it becomes easier to envisage the emergence of a dependent peasantry and the paradox of a civilization based on seed-crop swidden is resolved.

Examination of the "problem" of Maya subsistence, therefore, lends support to the general argument advanced in this paper that an appreciation of ecological factors, particularly the nature of the crop complexes involved, is critical to an evaluation of the population and settlement potential of swidden systems. When the emphasis is on seed-crop cultivation the pioneering mode of swidden with associated low population and instability of settlement is the norm; and this tendency may be accentuated both in mid- and high-latitudes and when monoculture is practised. Conversely, when the emphasis is on root crops, swidden cultivation achieves its most stable manifestation and its potential for supporting relatively high populations in lasting settlements is greatest; although, in the absence of any form of intensive and continuous cultivation, social factors are still likely to prevent the emergence of loci of centralized power and authority and the evolution of a simple into a complex society.

Notes

1 For example, Geertz, C. (1963). *Agricultural Involution: the Process of Ecological Change in Indonesia.* Berkeley and Los Angeles. See also Harris, D. R. (1969). Agricultural systems, ecosystems and the origins of agriculture, *in* Ucko, P. J. and Dimbleby, G. W. (eds.). *The Domestication and Exploitation of Plants and Animals.* London. In this paper the term *ecotype* is used not in its original biological sense of a plant or animal race genetically adapted to a particular environment but in its anthropological sense to indicate a system of energy transfers by which a social group adapts to and exploits its environment.

2 For similar schemes, which classify agricultural systems primarily by reference to continuity of cultivation, see Boserup, E. (1965). *The Conditions of Agricultural Growth.* London. pp. 15–16 and Wolf, E. R. (1966). *Peasants.* Englewood Cliffs. pp. 20–21.

3 Wolf, E. R. (1966). *op. cit.* pp. 3–4.

4 Izikovitz, K. G. (1951). Lamet: hill peasants in French Indo-China, *Etnologiska Studier*, **17**, p. 7.

5 For an account of these contrasted systems of planting in one part of the South American tropics see Harris, D. R. (1971). The ecology of swidden cultivation in the upper Orinoco rainforest, Venezuela. *Geogr. Rev.*, **61**.

6 Sahlins, M. D. (1968). *Tribesmen.* Englewood Cliffs. p. 30.

7 Conklin, H. C. (1957). *Hanunóo Agriculture in the Philippines.* F.A.O., Rome; Gourou, P. (1956). The quality of land-use of tropical cultivators, *in* Thomas, W. L., Jr. (ed.). *Man's Role in Changing the Face of the Earth.* Chicago. pp. 342–3.

8 Leach, E. R. (1959). Some economic advantages of shifting cultivation, *Proc. 9th Pacific Sci. Congr.*, Bangkok, 7, pp. 64–6.

9 Gourou, P. (1956). *op. cit.* p. 345; Allan, W. (1965). *The African Husbandman.* London; Clark, C. and Haswell, M. (1967). *The Economics of Subsistence Agriculture.* 3rd ed. London; Carter, W. E. (1969). *New Lands and Old Traditions. Kekchi Cultivators in the Guatemalan Lowlands.* Gainesville. pp. 132–46; Lewis, O. (1951). *Life in a Mexican Village: Tepoztlán Restudied.* Urbana. p. 156.

10 Conklin, H. C. (1957). *op. cit.* p. 147.

11 Montgomery, D. E. (1960). Patrol of Upper Chimbu Census Division, Eastern Highlands, *Papua and New Guinea Agric. J.*, **13**.

12 Ooi Jin-Bee (1958). The distribution of present-day man in the tropics: historical and ecological perspective, *Proc. 9th Pacific Sci. Congr.*, Bangkok, **20**, p. 113; Dumond, D. E. (1961). Swidden agriculture and the rise of Maya civilization, *S.W.J. Anthrop.*, **17**, pp. 308–11.

13 Cowgill, U. M. (1962). An agricultural study of the southern Maya lowlands, *Amer. Anthrop.*, **64**, p. 278.

14 Exceedingly high rural population densities characterize much of south-eastern Nigeria where the traditional Ibo pattern of polycultural swidden cultivation with yams and other root crops as major staples persists in a modified form. For example in the 1952 census gross rural population densities of well over 400 per square mile were recorded over an extensive area between Onitsha and Calabar and also around Enugu and Nsukka: Udo, R. K. (1965). Disintegration of nucleated settlement in eastern Nigeria, *Geogr. Rev.*, **55**, p. 55. Still higher rural densities, which in certain areas such as the Abak and Uyo divisions of Calabar Province exceeded 1000 per square mile, were recorded in the 1963 census: Okonjo, C. (1968). A preliminary medium estimate of the 1962 mid-year population of Nigeria, *in* Caldwell, J. C. and Okonjo, C. (eds.). *The Population of Tropical Africa*. London. p. 88. Under pressure of increased population during and before the inter-censal period the traditional fallow interval of 10 to 15 years was progressively reduced to an average of 5 years and much land formerly cultivated in a long-term fallowing cycle was given over to short-term fallowing or to the continuous cultivation of staple crops with the use of household manure: Udo, R. K. (1965). *op. cit.* p. 58; F.A.O. (1966). *Agricultural Development in Nigeria 1965–1980*. Rome. pp. 11–14; Buchanan, K. M. and Pugh, J. C. (1955). *Land and People in Nigeria*. London. pp. 103–6; Morgan, W. B. (1955). Farming practice, settlement pattern and population density in south-eastern Nigeria, *Geogr. J.*, **121**. The Tiv of east-central Nigeria south of the Benue River afford another example of primarily root-crop swidden cultivators whose traditional long-term fallowing and crop rotation cycles have been shortened and altered in areas where rural population density has risen above 400 per square mile: Vermeer, D. E. (1970). Population pressure and crop rotational changes among the Tiv of Nigeria. *Ann. Assoc. Amer. Geogr.*, **60**.

15 Very high rural population densities also occur in the Chimbu highlands of New Guinea where, for example, the Dengla-Maguagu tribe was living in 1960 in an area of 9.3 square miles at a density of 421 per square mile and the much smaller Bindegu sub-clan of the Naregu tribe occupied 322 acres at a density of 524 per square mile. Brookfield, H. C. and Brown, P. (1963). *Struggle for Land: Agriculture and Group Territories among the Chimbu of the New Guinea Highlands*. Melbourne. pp. 105–22. The staple crops of the Chimbu tribes include sweet potato, bananas and sugar cane and the cultivation cycle commonly incorporates intervals of short fallow as well as long fallow periods. The better land is cultivated for periods of from 4 to 10 years, interspersed with short-fallow intervals of several months' to 3 years' duration, before being abandoned to long fallow for periods of from 5 to 20 years. On poorer land cultivation is for periods of from 1 to 3 years separated by long-fallow periods of from 20 to 50 or more years. Only on the most valued land does the period of cultivation probably equal the fallow period; Brookfield, H. C. and Brown, P. (1963). *op. cit.* pp. 46–54. Despite Brookfield and Brown's suggestion that Chimbu agriculture "is so far removed from the classic "swidden" type that the term is a misnomer" (*op. cit.* p. 162) it may be regarded as a swidden system—albeit an atypical one—in the sense that most land is cultivated for shorter periods than it is fallowed.

16 Steward, J. H. (1949). South American cultures: an integrative summary, *in* Steward, J. H. (ed.) *Handbook of South American Indians*, **5**: *The Comparative Ethnology of South American Indians*. Washington, D.C. p. 676.

17 Keesing, F. M. (1962). The Isneg: shifting cultivators of the northern Philippines, *S.W.J. Anthrop.*, **18**.

18 Keesing, F. M. (1962). *op. cit.* p. 4.

19 Cumberland, K. B. (1949). Aoteroa Maori: New Zealand about 1780, *Geogr. Rev.*, **39**; Vayda, A. P. (1960). *Maori Warfare*. Wellington. pp. 19–21.
20 Butt, A. J. (1970). Land-use and social organization of tropical forest peoples of the Guianas, *in* Garlick, J. P. and Keay, R. W. J. (eds.) *Human Ecology in the Tropics*. Oxford. pp. 26–42.
21 Vayda, A. P. (1961). Expansion and warfare among swidden agriculturalists, *Amer. Anthrop.*, **63**.
22 Freeman, J. D. (1955). Iban agriculture: a report on the shifting cultivation of hill rice by the Iban of Sarawak, *Colonial Research Studies*, **18**, p. 32.
23 Carter, W. E. (1969). *op. cit.*
24 Carneiro, R. L. (1960). Slash-and-burn agriculture: a closer look at its implications for settlement patterns, *in* Wallace, A. F. C. (ed.). *Men and Cultures*. Philadelphia.
25 Carneiro, R. L. (1960). *op. cit.* p. 233.
26 Cowgill, U. M. (1962). *op. cit.* p. 276; Reina, R. E. (1967). Milpas and milperos: implications for prehistoric times, *Amer. Anthrop.*, **69**; Carter, W. E. (1969). *op. cit.* pp. 124–8.
27 Carneiro, R. L. (1961). Slash-and-burn cultivation among the Kuikuru and its implications for cultural development in the Amazon Basin, *in* Wilbert, J. (ed.) *The Evolution of Horticulture Systems in Native South America: Causes and Consequences, Antropologica Supplement*, **2**. Caracas. p. 48; Butt, A. J. (1970). *op. cit.* p. 37.
28 Harris, D. R. (1971). *ibid.*
29 Spencer, J. E. (1966). Shifting cultivation in Southeastern Asia, *Univ. Calif. Publ. Geogr.*, **19**, pp. 110–22.
30 Harris, D. R. (1969). *op. cit.* pp. 9–14; Spencer, J. E. (1966). *ibid.*
31 Godwin, H. (1965). The beginnings of agriculture in North-West Europe, *in* Hutchinson, Sir Joseph (ed.). *Essays on Crop Plant Evolution*. Cambridge.
32 Soudský, B. and Pavlů, I., this volume, p. 325; Soudský, B. (1968). Criteria to distinguish cultural phases—methods employed in the excavation at Bylany. Paper presented to the Research Seminar on Archaeology and Related Subjects, Institute of Archaeology, London.
33 A food survey conducted among the Sinasina people in the south-east of the Chimbu area showed that sweet potato contributed 92% of the calories in the adult diet: Oomen, H. A. P. C. and Malcolm, S. H. (1958). Nutrition and the Papuan child, *S. Pacific Com. Tech. Paper*, **118**, quoted in Brookfield, H. C. and Brown, P. (1963). *op. cit.* p. 60.
34 Forge suggests that among Neolithic cultivators there may be a threshold in the size of the basic population unit—which he tentatively sets at about 150 or approximately thirty-five adult males—below which ties of kinship and affinity are capable of ordering social relationships and above which division of the population takes place into status and/or occupational groups of a higher order than those of the household and family: Forge, A., this volume, p. 371.
35 Gourou, P. (1956). *op. cit.* p. 345; Leach, E. R. (1954). *Political Systems of Highland Burma: a Study of Kachin Social Structure*. London. pp. 27–8.
36 Vayda, A. P. (1961). *ibid.*
37 Wolf, E. R. (1961). Concluding comments, *in* Wilbert, J. *op. cit.* pp. 111–6.
38 Cowgill, U. M. (1961). Soil fertility and the ancient Maya, *Trans. Connecticut Acad. Arts Sci.*, **42**.
39 Dumond, D. E. (1961). *op. cit.* p. 312.
40 Bronson, B. (1966). Roots and the subsistence of the ancient Maya, *S.W.J. Anthrop.*, **22**. Bronson's argument is challenged in Sanders, W. T. and Price, B. (1968). *Mesoamerica. The Evolution of a Civilization*. New York. pp. 92–3.

41 As Dumond points out (*op. cit.* p. 312) Pelzer cites suggestions that swidden cultivation was of fundamental importance to the civilizations of Angkor in Cambodia and Anuradapura in Ceylon. Pelzer, K. J. (1945). Pioneer settlement in the Asiatic tropics, *Amer. Geogr. Soc. Special Publ.*, **29**. New York. p. 20.

42 Puleston, D. E. (1968). New data from Tikal on Classic Maya subsistence. Paper presented to the Society for American Archaeology. Santa Fé.

43 For example, remains of ridged fields have recently been reported along the Rio Candelaria in Yucatan which may imply that intensive cultivation of wet-lands for root or other crops was practised in the Maya area. See Siemens, A. and Puleston, D. (1970). Prehistoric ridged fields and related features in Campeche, Mexico. Paper presented to the International Congress of Americanists, Lima; and for general discussions of such ridged fields see Parsons, J. J. and Denevan, W. M. (1967). Pre-Columbian ridged fields, *Sci. Amer.*, **217**, and Denevan, W. M. (1970). Aboriginal drained-field cultivation in the Americas, *Science*, **169**.

V. M. MASSON

*Prehistoric settlement patterns in Soviet central Asia**

Ancient settlements and their characteristic features, which change and develop over time and space, serve as a most important source for studying the history of society. Features of settlements associated with 1. ecology, 2. economy, 3. social organization, and 4. "ethnos" can be found correlated in various ways. Frequently, specific features which have arisen under the influence of ecology or economy are consolidated by tradition and transformed into ethnic features. When studying the initial character of a historic process, it is essential to examine features associated with changes in the economy and social organization. It is often possible to observe that elements in the structure of settlements which are associated with the socio-economic environment are sociologically synonymous, in spite of specific local characteristics, caused by ecological conditions. The basic constituent, planning and social unit of the settlement is the dwelling house, but the history of settlements is not synonymous with the history of dwellings. The character of a specific settlement is determined by the character of the organization of the houses as part of the settlement organism, and also by features specific to that organism (shrines, general meeting places, community storehouses, production areas). In this respect, settlements reflect the history of societies more fully than can houses by themselves.

The earliest types of settlement are the occupation sites of gatherers, associated with a "gathering" type of economy. In central Asia, the earliest of these are caves and rock shelters with a Late Mousterian stone industry. The most interesting of these were the sites whose habitation deposits showed long continuous occupation. Hole and Flannery have convincingly interpreted these caves and rock shelters as seasonal camps[1]. Generally the traces of occupation in these sites include hearths, an assemblage of stone tools of various types as well as a large number of bone remains. One of these cave settlements in central Asia is the site of Tešik-Taš, a cave 20 m wide and 21 m deep. Five stratified cultural layers are separated by sterile levels. In one of the layers, the burial of a boy was found[2]. In the valley

*See Preface, p. ix.

of the River Čatkal, north-east of Taškent, are two other very similar caves, Khodžikent 2 and Obirakhmat. The latter is 21 m wide and 8·5 m deep, and its 10 m of occupation deposits are divided into twenty-one layers[3]. I am not convinced that it is possible to distinguish between butchering stations and transitory stations on the basis of archaeological evidence[4]. Smaller and less commodious caves were used during the Mousterian period in central Asia as temporary shelters. Although separate hearths are sometimes found in these small caves, the amount of implements and waste from their manufacture is minimal. A cave of this type is that of Khodžikent 1 and the long narrow limestone cave of Aman-Kutan near Samarkand, which has an entrance 1·5 m wide and 0·9 m high[5]. Judging by the large number of animal bones, the latter had been used as a butchering station. In the Mousterian period in central Asia there also existed open temporary camps, situated on the banks of rivers. But at the present time these are represented only by accumulations of redeposited flint artefacts (Kairakkumi on the middle course of the Syr-Darya, and Kara-bura on the river Vakhš). Thus the dominance of ecological and economic factors is very clear in these early settlements. Man still did very little towards organizing his dwelling place. He preferred to utilize natural refuges. In addition, the choice of such natural refuges depended on, apart from factors of amenity, social factors such as being able to accommodate all members of the hunting group. It is possible that the group of large and small caves in the Taškent region corresponded to the seasonal migration of one such hunting group (the distance between the Obirakhmat and Khodžikent caves is about 25 km). The sterile levels between the cultural layers in the Tešik-Taš cave may also indicate some kind of seasonal cycle in the activities of the nomadic hunters. In addition, cave settlements seem to have been the centres or base-camps of hunting areas of separate groups, and their material indicates a relatively settled way of life.

The tradition of settling in caves was preserved until at least the Mesolithic and Early Neolithic periods as may be seen, for example, in the hunting and fishing settlements of the Caspian Sea area (Džebel[6], Dam-Dam-Češme[7]). Among the open camp-sites, it is interesting to note the site of Samarkand which dates to the Upper Palaeolithic period, since this would seem to have been a camp of steppe hunters[8]. At this site hearths were excavated, around which were concentrated artefacts and occupation débris. The character of the dwellings, however, is still obscure. Sites which were suitable for such seasonal nomadic camps were repeatedly occupied, which would explain why the finds are spread over such a large area, as at this site. Sometimes, as at the Mesolithic site of Oš-Khona in the eastern Pamirs, which was a temporary summer camp of mountain hunters[9], the occupation débris is spread over several hectares. It would seem that some process of development and improvement of dwelling conditions was taking place in these open camp-sites. In the settlement of Tutkaul, for example, which is situated in the mountain region by the river Vakhš, semi-subterranean dwellings have been excavated in the layers dating from the tenth to fifth millennium B.C., and paved areas of flat stones in the same layers indicate the existence of some sort of building with a firm floor[10].

The houses of the Neolithic Kelteminar culture, of hunting/fishing groups living in the northern region of central Asia in the fifth to third millennium B.C. have very distinctive and original features. These were oval in plan, constructed on a frame of wooden posts, with a wood and reed roof. The lower parts of the walls were packed with compact earth. According to another reconstruction these were large semi-subterranean dwellings. Their area is large, ranging from 320 m² (Džanbas)[11] to 360–380 m² (Kavat 7[12]). In the centre of the pit-dwelling was a large permanent hearth; on its periphery were numerous temporary fireplaces, as well as pits for domestic use. On the basis of Tolstov's calculations, 100–125 people could have lived in a house of this size, but this is possibly rather an exaggeration. The settlements in fact consisted of one huge house of this type. In addition, material of the Kelteminar culture is sometimes found scattered over an area of 1–3 ha. As at the site of Oš-Khona, this probably reflects regularly repeated temporary occupation of suitable sites. The actual character of the Kelteminar dwellings is closely affected by the natural conditions in the ancient delta of the Amu-darya with its "tugaina" vegetation providing an abundance of wood and reeds. It is difficult to judge for how long the Kelteminar houses were occupied. The location of the hearths for eating and cooking activities at the site of Kavat 7 next to the entrance of the house would seem to indicate that it was occupied also in the cold season of the year. Intensive fishing would have made a relatively settled way of life possible. The social factor may be seen clearly at this site. The whole group or community was joined together in one house with a collective hearth, so that the size of the group determined the size of the house. Sociologically, the Kelteminar house is comparable to a sizeable cave transferred to an open space. It cannot be excluded that this type of dwelling, which developed as a result of ecological, economic and social factors, was also an ethnic peculiarity of the Kelteminar tribes.

The change to agriculture and animal husbandry was a radical economic step in the history of society, justifiably referred to by a number of specialists as a "Neolithic revolution" or "food-production revolution". The changes are particularly clearly reflected in the form of the settlements. In arid conditions, where agriculture depended on some form of artificial irrigation, the pattern of settled life took on particular permanence. In these conditions, an important discovery was made in the techniques of house construction. This was the construction of houses of clay blocks, a method soon replaced by the use of sun-dried clay bricks of standard sizes. This discovery was responsible for the qualitatively new form of stable agricultural settlements in the arid zone. But these were only external changes. The most important new feature

Fig 1 Early Neolithic settlement of Džeitun, S. Turkmenia.

was that among the number of factors influencing the character of settlements, the socio-economic factor now became more and more dominant.

In southern Turkmenia, the evolution of the agricultural settlements has been studied in great detail[13]. A comparatively detailed classification of settlement types has been suggested by researchers working in Peru[14]; this work, although extremely interesting from the theoretical point of view, has only limited application to other archaeological sites with poorly preserved material. Thus, for example in central Asia, as well as in the whole of the Near East, "achorism" of certain settlements could have been determined only by the destruction of building remains outside the central "tell". It is certainly possible to distinguish between "settlements" and "large settlements" on the basis of number of inhabitants, the dividing line being about 1000 inhabitants (cf. Lanning's distinction between "village" and "town").

In central Asia the earliest agricultural settlements are the sites of the Neolithic Džeitun culture. The settlement of Džeitun itself[15] consists of thirty small houses of standard plan, which, judging from their size (16–30 m²) were intended for one paired nuclear family (Fig. 1). Judging by the distribution of finds of tools in the area of the settlement, each paired family carried out its own processing of skins, manufacture of chipped stone tools and woodworking. The disposition of the houses shows no deliberate planning or organization in the early phase of the culture; no central open area or central "clubhouse" or "shrine" can be distinguished. It seems probable that the population of Džeitun numbered about 150–180 people. In the middle phase of the Džeitun culture, a shrine was constructed in the centre of the settlement. This was identical in plan to the dwelling houses, but was twice as large in area (64 m²). A shrine of this type was discovered at Pessedžik-depe[16]. During the excavations of autumn 1969, it was found that the walls of this shrine were covered by geometric and zoomorphic polychrome painted designs. The unplanned disposition of the dwelling houses, however, continued. This feature is also characteristic of the late phase of the Džeitun culture, as shown by the excavations of Čagilli-depe, which comprises thirteen to fourteen one-roomed houses[17] (Fig. 2). The largest of these houses was situated in the centre of the settlement, and may have been used as a shrine and a general meeting-place. The influence of the social factor on the character of the settlements of the Džeitun culture is extremely significant. The importance of the paired family, as the basic component unit of society, was responsible for the appearance of small one-roomed houses. In addition, the necessity to unite these families into a community in order to ensure the continuation of the agrarian cycle is reflected in the size of the settlement. A symbol of this unification, which was basically for economic purposes, is the shrine, a house for meetings of the collective. The presence at Džeitun of two types of house (distinguished by the position of the hearths) originally led to the assumption that the Džeitun settlement was divided into two parts or phratries. With the completion of the excavations, however, it was established that such houses were randomly distributed. In addition, there is more or less definite evidence that groups of two or three houses were joined together with a communal yard for domestic activities. It is possible that this is an indication of the earliest stage in the development of extended-family communities. If we compare the planning of the Džeitun settlements with chronologically similar sites, such as Çatal Hüyük, detached houses which do not form agglomerated blocks, are clearly a distinctive feature of the Džeitun culture.

The peculiar characteristics which distinguish the planning of the Anau IA settlements is well known. This culture is characterized by strong influences from Iran, which may even have taken the form of an immigration of people from the Sialk region. The planning of contemporary settlements in Iran has not been studied. In southern Turkmenia, the settlements of Anau IA are small in size and are clearly divided by a street into two parts (Mondžukli-depe, Čakmakli-depe[18]) (Fig. 3). Each of these parts consists of several house-complexes, comprising living and domestic structures. These are not detached houses as in the Džeitun culture, but are blocks of agglomerated rooms. The

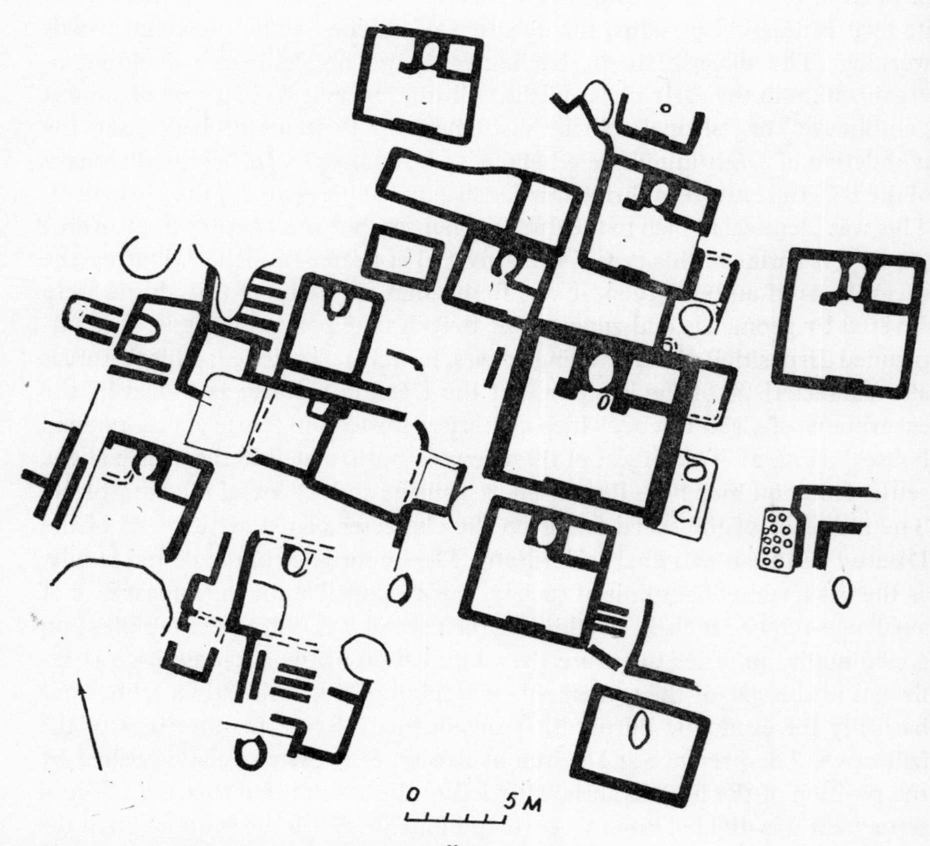

Fig 2 Early Neolithic settlement of Čagilli-depe.

plan of these settlements may reflect the division of their inhabitants into two phratries or separate communities, especially as each part contains a structure of identical plan whose walls were decorated with red paint and which have therefore been interpreted as shrines. Such planning, which has no analogies in later sites, may reflect ethnic and social features which are peculiar to this culture.

The settlements of the Namazga I period are in character more like those of the Džeitun culture, unrelated to the Čakmakli-depe traditions. The small settlement of Dašlidži-depe consists of small one-roomed houses with random distribution, and a large central house—presumably the communal meeting place[19] (Fig. 4). In addition large settlements also appear during this period,

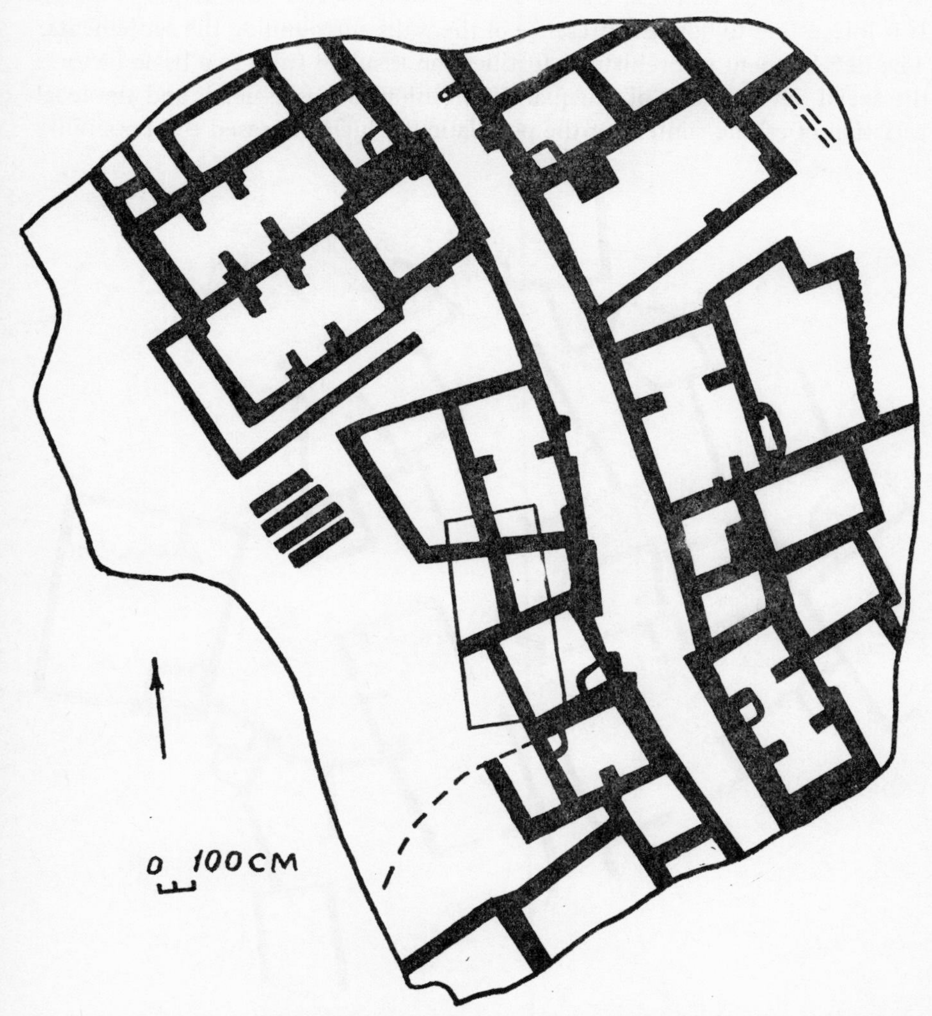

Fig 3 Plan of the settlement at Čakmakli-depe belonging to the Anau IA culture.

for example Kara-depe with an area of 15 ha and Namazga-depe, with as large a living area. Unfortunately, the planning of these large centres is still obscure. Their very presence testifies to the success of the economy and social organization, which enabled quite large numbers of people to be concentrated in one place. If we assume that the density of structures was the same as at Dašlidži-depe, then the population must have reached 2500 people.

In the fourth millennium B.C., in the settlements of the Namazga II period, there was a tendency to agglomerate the buildings into multi-roomed blocks[20]. Such settlements as Mullali-depe and Yalangač-depe comprise enclosed units and a large communal building—a shrine in the centre—and a surrounding wall of 0·5 m thickness. At Mullali-depe there lived a community of seventeen to twenty paired families, that is about ninety to 120 inhabitants (Fig. 5). It is interesting to note the presence of the walls surrounding the settlements. The development of prehistoric fortification resulted from two basic factors: the social development of unequal distribution of possessions, and the local variation of ethnic content of the population, which increased the possibility

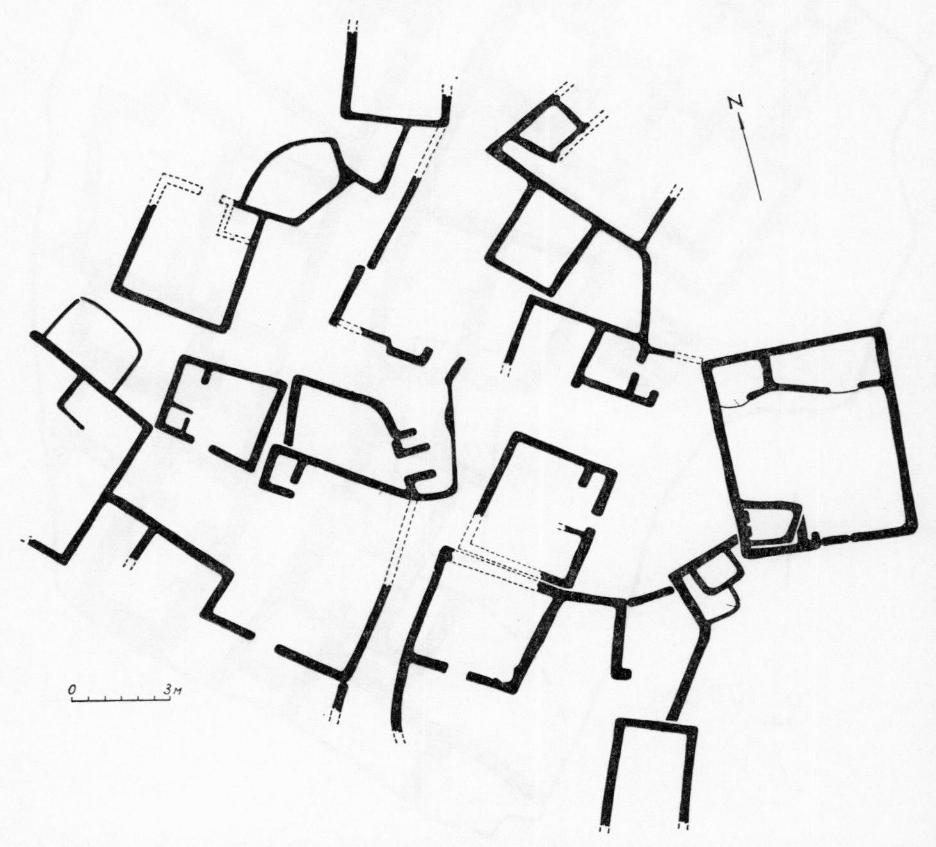

Fig 4 Plan of Chalcolithic settlement of Dašlidži-depe, S. Turkmenia, dating from the Namazga I culture.

of the occurrence of inter-tribal conflict. The latter factor, in my view, was important in determining the development of early fortifications at Jericho and Hacilar. It is interesting to note that we also find fortified settlements in the Geoksyur Oasis in southern Turkmenia, which formed a frontier region of the agricultural communities, bordering on the culture areas of hunters and fishermen[21].

Important changes occurred in the southern Turkmenian sites at the end of the fourth to the beginning of the third millennium B.C. The planning of these settlements is well known from excavations at Kara-depe[22], Geoksyur (Fig. 6) and Čong-depe[23]. In this period, these settlements consisted of multi-roomed blocks, divided by narrow streets. The basic unit of each house or block was the living room, about 10–25 m² in area, with a hearth for heating purposes in a hollow in the floor, and a narrow domestic room which was joined only to the living room by a passage. There were about four to eight such units in each block. In addition there was a communal yard for domestic activities and communal store-rooms. It is likely that such a house was inhabited by four to six paired families, who were united in an extended family group, comparable to the households of the Pueblo Indians[24]. It is not surprising that for the first time in southern Turkmenia, collective graves appear in this period. The number buried in these graves ranged from eight to twenty-three

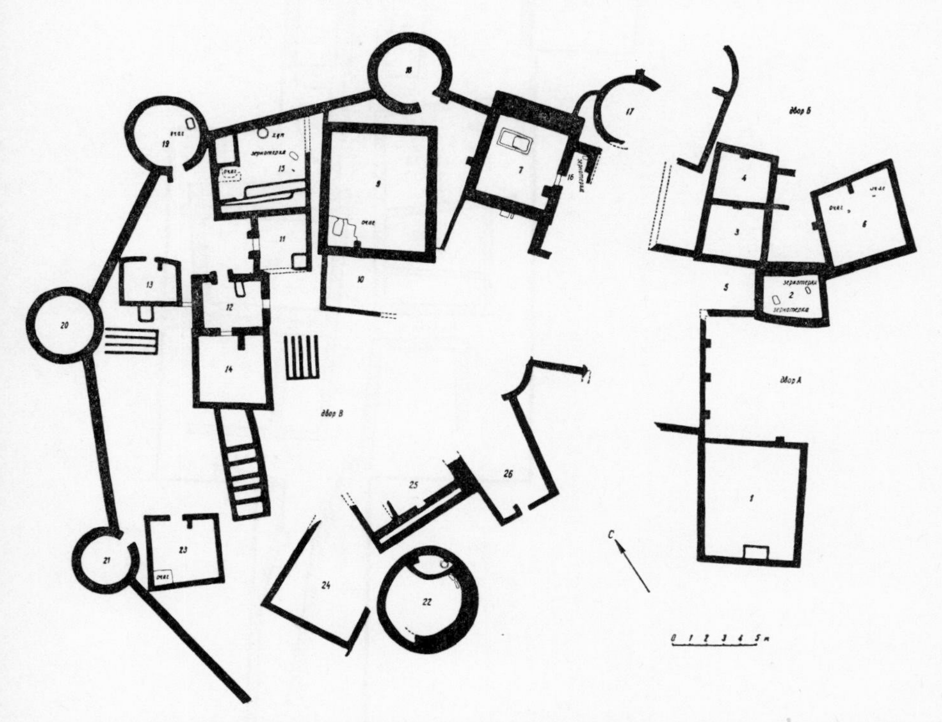

Fig 5 Plan of the Late Chalcolithic site of Mullali-depe, S. Turkmenia, which was surrounded by a wall.

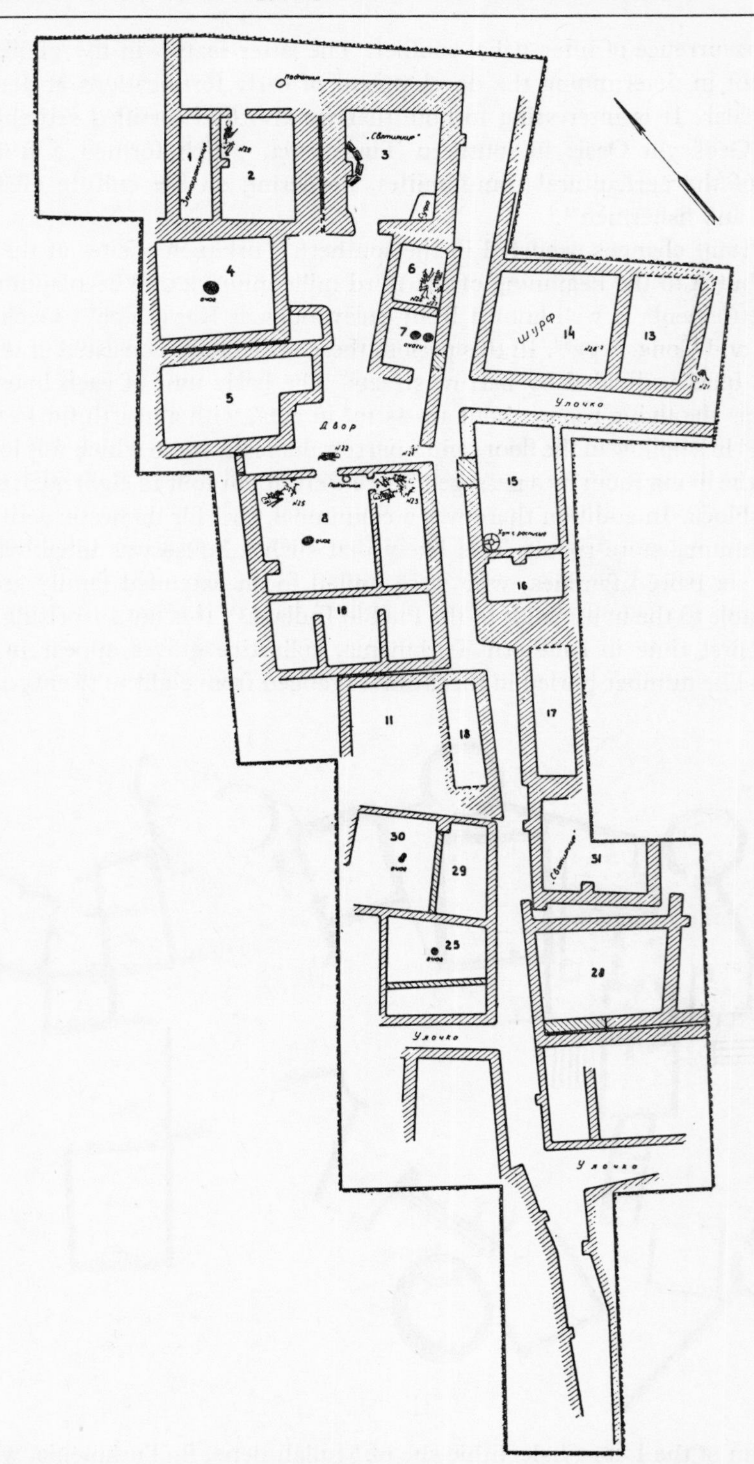

Fig 6 Plan of the Late Chalcolithic settlement at Geoksyur, S. Turkmenia.

people, who may very well have belonged to one such family group or household. The population of one large house would have varied with the number of families in each group, but probably ranged from twenty to forty-five people; at Kara-depe, judging from the extent of the Namazga III layer, there was a total of about twenty-five of these agglomerated blocks, which would correspond to a total population of the settlement of about 1000–1100. The Kara-depe houses are sociologically analogous to the Iroquois houses, although their external appearances were very different, due to the differences of the natural environment. Thus we can see here very clearly that the sociological factor (as seen in the development of the extended family group) was reflected in the whole character of the settlement and in the partial differentiation of two ethno-cultural regions of southern Turkmenia: a western region (Kara-depe ware), and an eastern region (Geoksyur ware). Secondary features, which are assumed to have been associated with ethnic peculiarities, include the distribution of circular sacrificial hearths in the east, at Geoksyur and Čong-depe, and their absence in the west. It is interesting to note the presence of a pottery kiln in one of the houses at Geoksyur. This may indicate the beginning of inter-group specialists. Judging by the grave-goods in the burial chambers at Geoksyur, several of the family groups had developed basket production. Features indicating unity of the whole settlement as a single complex are much less obvious. However, they do include a vast central open space, for example at Kara-depe, to which the narrow streets dividing the houses led.

In later periods in southern Turkmenia, where the ecological and ethnic factors remained constant, there is evidence of an ever-increasing influence of economic and social factors on the character of the settlement. I am referring to the process of urbanization, named by Childe relatively successfully "The Urban Revolution"[25]. The most convenient criteria of civilization as seen in archaeological evidence include such relatively superficial features as the presence of monumental architecture, literacy, and large centres of population of more than 5000 inhabitants[26]. More fundamental are such basic changes as 1. the distinction of crafts from agricultural activities, and the appearance not of isolated specialists, but of whole craft workshops; 2. the development of trade replacing the prehistoric exchange system; 3. social differentiation, developing into class stratification, which may be revealed in the distinction of secular from religious hierarchy as clearly reflected in the burial procedures. We may regard the urban revolution as a process whereby the economic and cultural bases of early class society were formed[27]. All these phenomena were also reflected in the development of settlements, where there were the same motivating forces, in spite of external variation. Thus it is well known that in Peru in the Moche period and in Mesoamerica in the Mayan period nucleated dwelling centres did not exist and this would seem an even more obvious element of urban civilization. However, monumental temple complexes were centres of economic, administrative and social leadership, in both regions, and were functionally analogous to the Mesopotamian cities. The concentration of the population in the one case in close proximity to these centres, and in the other case the dispersal of the population over a large area reflects the historical and ecological peculiarities, which may have been associated with the lesser importance of defence and the smaller density of the population in general (itself dependent on the character of the economy)[28]. It is possible that the replacement of the term "urbanization" by the term "civilizing process" would lead to the recognition of the striking and indisputable typological resemblances between the earliest civilizations of the New and Old Worlds.

In southern Turkmenia this process, still partially incomplete, took place at the end of the third to beginning of the second millennium B.C. The excavations of Altin-depe best reflect the effect of the urban revolution on the character of settlement[29] (Fig. 7). The settlement itself, like the preceding

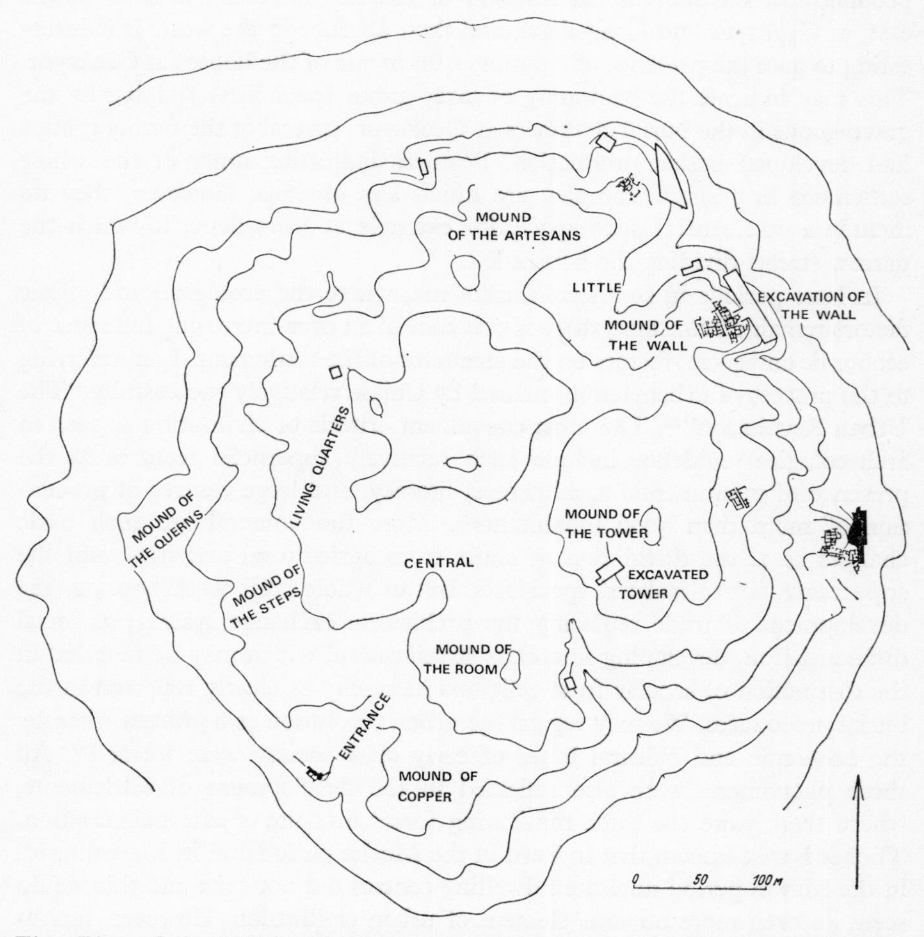

Fig 7 Plan of the Middle Bronze Age settlement at Altin-depe, S. Turkmenia.

Eneolithic settlements, consisted of multi-roomed blocks, separated by narrow winding streets. However, an examination of the interlinking of rooms by passages shows that these blocks break down into separate "apartments", consisting of two small living rooms and two smaller rooms for domestic uses. Sometimes such an "apartment" also had its own yard. It is possible that these were the dwelling units of individual families, which again became the main nucleus of society. At the same time they were united in blocks which would seem to reflect multi-family groups. In addition there were important collective burials, in which both men and women were buried (two graves were excavated, with fourteen people in each). The size of the settlement increased—the area of Altin-depe was 46 ha. In places, there were traces of a surrounding wall with primitive towers. But behind these quantitative changes are to be found qualitative changes. The separation of craftsmen from the rest of society was consolidated by their territorial isolation. The potters with their two-tiered kilns occupied special quarters of about 2 ha in area. In this period certain houses of a special type were constructed according to a strict geometrical plan and situated separately; for example the "house of the chief", excavated in 1969, which has an area of approximately 100 m^2. Without further excavation it is difficult to judge whether the social differentiation in architecture was accompanied by a territorial division of the "house of the chief". Those features which characterize the settlement as a single unit take on a special significance. Apart from open spaces and the craft centre, they include a complex of monumental buildings, in which the central place was occupied by a four-staged ziggurat 12 m high. The whole complex had a front 55 m long. This was clearly a building of prestige, one might call it a bright symbol of a new era. These phenomena indicate that specific features of the early formation of an urban centre were clearly dependent on economic and social factors, sometimes, it is true, appearing in religious vestments.

In the northern regions of central Asia, the transfer to an economy based on food-production took place not only later than in southern Turkmenia but also under different ecological and ethnic conditions. Thus in Khoresm, in the second half of the second millennium B.C., the widespread Tazabagryab culture, whose economy was based on agriculture and animal husbandry, was essentially a local variation of the Andronovo culture, and was very much of north-western origin[30]. In spite of the presence at this time of irrigation agriculture[31], there was no sun-dried brick architecture in this area. The settlements consist of semi-subterranean dwellings with a central hearth which differed from the Kelteminar houses only by the presence of right-angles in the structure and the smaller dimensions of the houses (8 by 10 m; 8 by 12 m; 12 by 15 m). It would seem that these characteristics were associated both with the type of economy and with the natural environment, in that, in the conditions of the Akčadariya delta, settlements frequently had to be transferred from place to place. It is important to remember that the culture was part of the Andronovo tradition, in which semi-subterranean dwellings had developed because of ecological factors which subsequently became a distinctive cultural characteristic. In the first centuries of the first millennium B.C., this type of settlement was still in use. Semi-subterranean dwellings, approximately 75–100 m² in area, stretched out in a line along the banks of streams and channels. The large dimensions of the semi-subterranean dwellings suggest that they were habitations of extended family groups[32]. Only with the use of permanent irrigation systems in ancient Khoresm was this tradition finally changed, and houses built of sun-dried bricks.

Notes

1 Hole, F. and Flannery, K. V. (1967). The prehistory of south-western Iran: a preliminary report, *Proc. Prehist. Soc.*, **33**, pp. 162–63.
2 Okladnikov, A. P. (1949). *Tešik-Taš: paleolitičeskij Čelověk.* Moscow.
3 Suleimanov, R. K. (1968). Grot Obi-Rakhmat i opit matematičesko-statističeskovo izučeniya Obirakhmatskoi kulturi, *Avtoreferat Diss. Kandid. Nauk* (Ph.D. thesis). Taškent.
4 Hole, F. and Flannery, K. V. (1967). *op. cit.* pp. 163–64.
5 Lev, D. N. (1956). Drevnij paleolit v Aman-Kutane, *Trude UzGY*, **61**.
6 Okladnikov, A. P. (1956). Peščera Džebel, *Trude YuTAKE*, 7.
7 Markov, G. E. (1966). Grot Dam-Dam-Češme 2 v Vostočnom Prikaspij, *Sov. Arkh.*, **2**.
8 Lev, D. N. (1964). Poseleniye drevnekamennovo veka v Samarkande, *Trude Samark. GU*, **135**.
9 Ranov, V. A. (1964). Itogi razvedok pamyatnikov kamennovo veka na Vostočnom Pamire, *Mat. i Issled. po Arkh.*, **135**.
10 Korobkova, G. F. and Ranov, V. A. (1968). Neolit gornikh rayonov Srednei Azii, *Problemi arkheologij Srednei Azii.* Leningrad.
11 Tolstov, S. P. (1948). *Drevnij Khorezm.* Moscow. pp. 59–62; Vinogradov, A. V. (1968). *Neolitičeskiye pamyatniki Khorezma.* Moscow. pp. 31–35.
12 Vinogradov, A. V. (1968). *op. cit.* pp. 67–72.
13 Masson, V. M. (1964). *Srednyaya Asiya i Drevnij Vostok.* Moscow-Leningrad; Masson, V. M. (1961). The first farmers in Turkmenia, *Antiquity*, **139**, pp. 203–212; Mellaart, J. (1967). The earliest settlements in western Asia, *CAH*, pp. 51–58.
14 Rowe, J. H. (1963). *Urban Settlements in Ancient Peru.* Nawpa Pacha; Lanning, E. P. (1967). *Peru before the Incas.* New Jersey. pp. 32–37.
15 Masson, V. M. (1970). Poseleniye Džeitun, *Mat. i Issled. po Arkh.*, **180**.
16 Berdeyev, O. K. (1968). Noviye raskopki na poseleniyakh Pessedžik-depe i Čakmakli-depe, *Kara-kumskiye drevnosti*, **2**, pp. 12–13.
17 Berdeyev, O. K. (1966). Čagilli-depe—novij pamyatnik neolitičeskoij Džeitunskoij kulturi, *Materialnaya kultura narodov Srednei Azii i Kasakhstana.* Moscow.
18 Berdeyev, O. K. (1968). Čakmakli-depe—novij pamyatnik vremeni Anau IA, *Istoriya, arkheologiya i etnographiya Sredei Azii.* Moscow.
19 Sarianidi, V. I. (1962). Kultoviye zdaniya poselenij Anauskoi kulturi, *Sov. Arkh.*, **1**.
20 Khlopin, I. N. (1969). *Pamyatniki razvitovo eneolita yugo-vostočnoi Turkmenii.* Leningrad. pls. VIII, XVII.
21 Masson, V. M. (1966). K evolutsii oboronitelnikh sten osedlikh poselenii, *KSIA*, **108**.

22 Masson, V. M. (1960). Kara-depe u Artika, *Trude YuTAKE*, **10**.
23 Sarianidi, V. I. (1965). *Pamyatniki pozdnevo eneolita yugo-vostočnoi Turkmenii*. Moscow.
24 Eggan, F. (1950). *Social Organization of Western Pueblo*. Chicago. pp. 29–30; Haury, E.W. (1956). Speculations on prehistoric settlement patterns in the south-west, *in* Willey, G. R. (ed.). *Prehistoric Settlement Patterns in the New World*. New York. p. 6.
25 Childe, V. G. (1950). The urban revolution, *Town Planning Review*, **21**.
26 Adams, R. (1960). *City Invincible*. Chicago; Daniel, G. (1968). *The First Civilizations*. London. p. 25.
27 Masson, V. M. (1968). The urban revolution in south Turkmenia, *Antiquity*, **42**, p. 178.
28 Adams, R. McC. (1966). *The Evolution of Urban Society*. Chicago. p. 44.
29 Masson, V. M. (1967). Protogorodskaya civilizatsiya yuga Srednei Azii, *Sov. Arkh.*, **3**; Masson, V. M. (1970). *Raskopki na Altin-depe v 1969 g.* Aškhabad.
30 Tolstov, S. P. (1962). *Po drevnim deltam Oksa i Yaksarta*. Moscow. pp. 47–68.
31 Itina, M. A. (1968). Drevnekhorezmskiye zemledeltsi, *Istoriya, arkheologiya, i etnographiya Srednei Azii*. Moscow.
32 Itina, M. A. (1963). Poseleniye Yakke-Parsan 2, *Materiali Khorezmskoi ekspeditsii*, **6**.

JAMES MELLAART

Anatolian Neolithic settlement patterns

As the study of the Neolithic in Anatolia is still in its infancy future discoveries will undoubtedly modify, enlarge or even invalidate many of the conclusions reached in this short compendium of our present knowledge. This is inevitable—but should be borne in mind.

For the purpose of this paper I include the so-called Early Chalcolithic (*c.* 5600–5000 B.C.) with the Neolithic, as its economy is still purely Neolithic. The Anatolian Neolithic, as at present known, covers the sixth, seventh and probably the eighth millennia B.C., but little is known of the eighth millennium as its main site, Aşíklí Hüyük, has not yet been excavated. Three main phases can be recognized in the Anatolian Neolithic: an Aceramic phase before *c.* 6500, a phase with monochrome pottery between *c.* 6500 and *c.* 5500 and a phase often, but not always, characterized by red on cream painted pottery down to *c.* 5000 B.C.[1] To the Aceramic phase belong Aşíklí Hüyük, Aceramic Can Hasan, Suberde and Aceramic Hacílar. Of these the first is not yet excavated, and the second is in the course of excavation. No other Aceramic Neolithic sites are known to me. To the second phase belong Çatal Hüyük (East), Can Hasan 4–7, Reis Tümegi, and Keyren in the Konya Plain, and Erbaba, Çukurkent, Kovada, Alan Hüyük, Hacílar IX–VI, Kízílkaya, Ilícapinar, Mersin, Tarsus, etc., a total of more than fifteen sites, mainly known from archaeological surveys. To the third group belong Hacílar V–I, Çatal Hüyük (West), Can Hasan 3–2b, Mersin XXIV–XX, etc., as well as the monochrome cultures of Bucak, the Akhisar-Manisa plain, and possibly the Menteşe, Fikirtepe and Yarimburgaz sites in the Marmara basin, though an overlap into the Late Neolithic (from *c.* 5750 onwards) is not excluded.

There are some very serious gaps in our knowledge of Anatolian settlement patterns before the Neolithic period. Upper Palaeolithic and Protoneolithic (I prefer this term to Mesolithic) is only known in the Antalya region of the south coast, but is still almost absent on the Anatolian plateau and in the lowland regions of the west. Ceramic Neolithic remains are well attested in Cilicia, but in Pamphylia there are no mounds, and sherds come from the top layers in previously inhabited caves and rock shelters only. The chances that this material belonged to agricultural communities are very slim indeed, and the pottery may have been left by hunters or herdsmen who occasionally took refuge in such places. Pamphylia is still not a wheat or barley growing area, having been covered by forest until well into the last century.

In Cilicia, on the other hand, farming villages were established at least as early as *c.* 6250 B.C., but the earlier Aceramic phase is not yet attested, mainly for lack of exploration, I assume, for there is clear evidence that Central Anatolian obsidian was traded with Syria and Palestine and hence passed through this area which can hardly have been uninhabited.

Contrary to earlier theorizing, the richest evidence for the Neolithic in Anatolia comes from the plateau and all the sites so far found are mounds, i.e. permanent settlements which were inhabited over a considerable length of time, which may vary from a few centuries (Aceramic Hacílar) to a millennium or more (Çatal Hüyük, and probably Aşíklí). Settlements of hunters, fishermen, or nomads probably also existed, but few have been located (e.g. Mesolithic Baradiz on Lake Burdur), and none have been subjected to modern scientific excavation. Without excavation it is impossible to say whether a settlement like Ilícapinar, perched near the edge of a malodorous salt lake, was permanently occupied or served as a temporary settlement for salt traders. Yet perhaps the most striking feature of the pattern of Neolithic occupation of Anatolia is the absence of any settlements within the northern half of the country, apart from the Marmara region. Neither around Ankara, nor in the Halys basin, is there any evidence for Neolithic settlements, but directly south of the Halys between the salt lake and Kayseri they appear. The reason for this is hard to see, for the land on either side of the Halys is of the same type (survey by Todd[2]). Two possibilities suggest themselves; either the regions north of the Halys were covered by forest and were thus unattractive to farmers in search of land, or other tribes of hunters and fishermen inhabited these territories, and did not welcome the Neolithic southerners. Perhaps both factors together did stop Neolithic expansion northwards and deflected it to the north-west, i.e. towards the Sea of Marmara. At present we have no evidence for either suggestion.

Environmental factors played a decisive role in the choice of sites; the presence of a stream or lake and good arable land is common to all (or nearly all) the Neolithic sites found, and only Ilícapínar is an exception. A glance at the map shows that nearly all Neolithic sites are near important modern agricultural centres (Aşíklí near Aksaray, Can Hasan near Karaman, Çatal Hüyük near Çumra and Konya, Hacílar near Burdur, Mersin, Tarsus, etc.).

As we have no evidence for the existence of trade routes before the establishment of Neolithic settlements, the establishment of such routes was probably based on the pattern of settlement and not vice versa. The same communication system still exists today. A number of Neolithic settlements lie on natural routes: such as the Calycadnus (Gök Su) valley, leading from the Konya Plain to the Mediterranean coast; on the Aksaray-Kayseri,

Konya-Beyşehir, Isparta-Burdur roads, and the Burdur-Antalya road and the Maeander valley. The choice in this case looks like a combination of agricultural land (and water) and the advantage of a trade route. We have as yet no evidence for Neolithic mountain villages, built of wood on hilly slopes, but the presence of such settlements, which would have played their role in the exploration for raw materials (stone, metal ores, ochre, pigments, timber, fruit, nuts, etc.) may be surmised, unless such services were rendered by a more mobile population such as pastoralists. Transhumance is another factor that should be taken into account. Even today it is widespread, but involving only part of the settled population. As the corn ripens in the summer months and the marshes dry, the best pasturage for domestic animals is found in the uplands and mountains and there is some advantage in feeding the beasts while keeping them away from the cultivated fields. This presumably would take place after the spring festival and the replastering and other repairs necessary to the houses at the end of the spring rains. After the harvest the animals could be allowed to feed on the stubble upon their return from the mountains, and before the next round of sowing in the late autumn. If the Neolithic inhabitants practised transhumance one can easily account for their knowledge of natural resources and the beginning of trade with their neighbours in obsidian and doubtless other products. The evidence from Mureybit and Protoneolithic Jericho shows that as early as the second half of the ninth millennium B.C. such a trade was in operation. In southern Anatolia itself Central Anatolian obsidian reached the Antalya caves perhaps as early as the Late Upper Palaeolithic, but the exact dating has not been established. For the Late Protoneolithic of Beldibi obsidian has been reported, and during the Neolithic it was widespread. The sixth millennium sites in the Akhisar-Manisa plain also used obsidian, but this might be Melian rather than Central Anatolian, involving sea voyages in the Aegean, and at the same time Central Anatolian obsidian reached the sites of the Khirokitia culture in Cyprus, evidently from the Turkish south coast. North Syrian flint is common at Çatal Hüyük, but the typical Neolithic material is obsidian, with flint and chert playing a rather subsidiary role, except at Hacílar. The use of obsidian even in the Aceramic phase indicates early trading connections—as elsewhere in the Near East and the Aegean (Frankhthi cave, Argolid, late eighth millennium). Obsidian acts as a guide fossil in recognizing trade, most of which may have been in perishable products, such as domestic grain and animals, woven cloth, etc. This would explain the uniformity of early crops (mainly two-row barley, some emmer and lentils) *c.* 7000 B.C. from Aceramic Hacílar to Jericho, Beidha, Çayönü (no wheat) and Ali Kosh, all apparently grown by dry farming, a uniformity that is somewhat surprising. Only one of the Neolithic sites in Anatolia, Suberde, has not yet yielded definite evidence for the use of cereal food, but one should beware of calling Suberde a village of settled hunters until more is known. Settlement is nearly always associated with crop cultivation, even if the crops are still morphologically "wild", as at Mureybit (second half of the ninth millennium).

By Çatal Hüyük VI, *c.* 6000 B.C. and at Hacílar (VI–I), *c.* 5600–5000 B.C. irrigation agriculture had taken over from dry farming. When this started at Çatal Hüyük we do not yet know, as the lower layers have not yielded any carbonized grain. According to Angel[3], a side-effect of this was the appearance of falciparum malaria, carried by mosquitoes that bred in the stagnant flood pools of the river on which Çatal Hüyük was situated. The result was an increase in crops: instead of two-row barley as main crop, with some emmer wheat and some six-row naked barley and lentils (Aceramic Hacílar), the diet changed to emmer, einkorn, breadwheat, six-row naked barley, field pea, etc., grown in abundance and in separate plots. What we do not know is whether irrigation farming was general or confined to a number of sites; one suspects the latter, and the current excavations at Erbaba near Beyşehir, a region still dry farming, might elucidate this point. This early agriculture in Anatolia was of course facilitated by the fact that these southern regions were part of the natural belt of wild cereals: wild barley and einkorn wheat grow in the Taurus mountains, but the case is not so clear for emmer wheat and lentils, the home of which is generally sought further east, anti-Taurus and eastern Taurus for the former, the Zagros-Caucasus complex for the latter. If correct, emmer and lentils at Aceramic Hacílar *c.* 7000 B.C. might indicate eastern contacts. All this indicates the very reverse of a self-sufficient economy and the evidence for domestic animals points in the same direction. The earliest domestic animals in Anatolia are the dog (Aceramic Hacílar, Suberde, Çatal Hüyük) from *c.* 7000 B.C., then cattle at Çatal Hüyük from *c.* 6500 B.C. (which may explain the origin of the bull cult at this site). Dogs would have been very useful in herding. At Suberde, on the other hand, only the dog was domesticated and cattle, sheep, goat and pig were wild, like the red deer. Here then wild sheep were hunted *c.* 6800 B.C. or so, whereas further east, at Zawi Chemi, there is evidence for domestic sheep *c.* 9000 B.C. The goat was never very common in the Anatolian Neolithic settlements and the pig was not domestic till Can Hasan in the later (?) sixth millennium B.C. Hacílar unfortunately yielded too few bones for a statistical analysis. Whereas hunting was the main feature of Suberde, at other Neolithic Anatolian sites, hunting was definitely subordinate to domestication—which does not rule out the hunt as a subsidiary form of food supply, along with fishing and fowling and food collection (berries, nuts, fruits), e.g. Çatal Hüyük.

In general then, one obtains the impression that the Neolithic agricultural communities of Anatolia exploited their natural resources to the utmost and the evidence we have at hand does not support a theoretical development from village to town. Given the right circumstances and environment, it was possible for towns to grow as early as the eighth millennium B.C., as in the case of Aşíklí, or again in the seventh millennium at Çatal Hüyük. Yet it is clear that in the majority of cases one has to deal with large or small villages, which need not however be considered as self-sufficient. The hamlet of Suberde, for example, used imported Central Anatolian obsidian for its tool manufacture, rather than relying on local chert. On the other hand it may not yet have learnt to make proper pottery or metal trinkets which were already produced at its eastern neighbour, Çatal Hüyük. What strikes one about these Neolithic Anatolian sites is the diversity of economy and craftsmanship rather than a uniform pattern of development. This diversity is evidently the result of different environments, not yet studied. We are still woefully ignorant of what the environment was like, of climatic conditions, natural vegetation, fauna, hydrology, etc. Yet these early settlers knew what they were after; for example at Çatal Hüyük they had selected the best arable land, with backswamps inhabited by wild cattle, red deer, wild boar, and leopards and with onagers and gazelle (and possibly lion) not too far away to the arid north and bears within the range of their knowledge, yielding prey to their seasonal pursuits. It is already clear that around the larger centres there were smaller settlements, possibly owing allegiance and paying tribute to a central and almost certainly religious authority, as early as the seventh if not the eighth millennium B.C. Without such support the people of Çatal Hüyük, situated in an alluvial plain, would have to go without stone, metal, pigments, timber, etc., yet the very fact that these were available in abundance shows either their dependence on trade or tribute. It would be hard to find a better example of interaction in the seventh and early sixth millennium than Çatal Hüyük, and the same pattern probably applies to Can Hasan and Hacílar in the sixth millennium B.C.

Specialization in the seventh and sixth millennium is clearly recognizable; at Aşíklí there is obsidian working on a grand scale, at Suberde the emphasis is on hunting, at Aceramic Hacílar on building, at Aceramic Can Hasan on bone carving, at Çatal Hüyük on nearly everything, at Ilícapínar on the salt trade, at Hacílar on ceramic industries, at Çukurkent on fine stone carving (statuettes, pins, etc.), at Fikirtepe on fishing and shell-fish collecting and hunting, etc. These views are, of course, somewhat selective and open to re-interpretation, yet they emphasise the variety of Neolithic man's interests and predilections, and if nothing else, they tend to show that there was no such thing as a uniform Neolithic development, based on self-sufficient farming communities.

As it was, the processes of urbanization could operate in the eighth millennium B.C. (Jericho PPNA and Aşíklí) as well as in the seventh and sixth (Çatal Hüyük); neither of these were mere "villages" and the same applies to settlements like Beidha, Tell es-Sawwan, Can Hasan or Hacílar, for what they have produced does in no way fit the pattern of a "self-sufficient village". Sites with shrines, temples, workshops, residential quarters, etc., do not fit into "village life"; on the contrary, they represent something more, and closer to the classical concept of a town or city, the centre of authority, the nucleus of a state, however petty. Who could accept that the Neolithic sites like Çatal Hüyük, Can Hasan, Aşíklí, Çukurkent, Hacílar, Jericho, Beidha, Tell es-Sawwan, Chogha Mami, Ali Kosh, etc. were mere villages of peasants, rather than the local centres of culture, religion, and hence government? The difference between Çatal Hüyük and Protoliterate Uruk is one of degree, but basically the pattern remains the same, and one can have urbanization before the invention of writing, which is usually and mistakenly regarded as a prerequisite for conferring the "title" of civilization on a "culture". Urbanization as a development was not confined to late fourth millennium Lower Mesopotamia or early third millennium Egypt—it happened there, but it had precocious predecessors no less entitled to the rank of civilization. In archaeology as in human life, it is achievement which counts, not the arbitrary and often prejudiced bestowal of titles.

Notes

1 Mellaart, J. (1970). Neolithic Anatolia, *Cambridge Ancient History*, I, p. 316.
2 Todd, I. A. (1966). Asíklí Hüyük, a proto-Neolithic site in central Anatolia, *Anat. Studies*, **16**.
3 Angel, J. L. (1971). Early Neolithic skeletons from Çatal Hüyük, Demography and pathology, *Anat. Studies*, **21**.

JOHN MIDDLETON

Patterns of settlement in Zanzibar[1]

I

Zanzibar is an example of a situation in which the natural resources open to a society are in general ample but access to them is limited for topographical, demographic, and social reasons. A certain kind of social organization is consistent with this pattern, especially where areas suitable for actual settlement are few. Whether this kind of organization is a necessary concomitant is a problem that needs comparative research. In this paper I consider only the factors within the two islands of Zanzibar and Pemba. The field research on which it is based was carried out in 1958, before the Revolution of 1964; since then there have been several significant changes in population and economy, but I have few details. The paper thus refers to pre-revolutionary Zanzibar, although I use the "ethnographic present" tense.

The state of Zanzibar comprises the two islands of Zanzibar and Pemba. Zanzibar Island has an area of 640 square miles, with a population of 166,000 people, a third of whom live in the capital, Zanzibar City. The total density of population is thus 260 to the square mile; in the rural areas only it is 173 to the square mile. Pemba Island is smaller with an area of 380 square miles. The population is 134,000 living at a density of 350 to the square mile, and the townships of the island contain only a small proportion of the population. Pemba provides most of the clove crop on which the prosperity of the islands has depended for most of the past century.

The population of the islands is ethnically very mixed. The majority comprises the three (150,000) tribes of the Hadimu (mostly in Zanzibar), the Tumbatu (on Tumbatu Island off the north-west coast of Zanzibar Island), and the Pemba (on Pemba Island). Today these groups are known as Shirazi, claiming a Persian origin so as to distinguish themselves from the Arabs and others. Until the last century these groups had their own rulers, but were subjected by Arabs from Muscat who provided the ruling house of the Sultanate that was destroyed in the 1964 Revolution. There were in 1958 45,000 Arabs (including many recent and poor immigrants from southern Arabia), but most were deported or killed in 1964. The third main group comprises mainland Africans, including both descendants of former slaves and more recent immigrants: these number 85,000. There are also some 20,000 members of other ethnic groups, mainly various immigrants from the Indian sub-continent and the Comoro Islands. The total population in 1958 was 300,000.

There are a few points that may be made here about this total population. The first is that all the groups are ultimately immigrants, entering the islands at various times from the beginning of the Christian era to the twentieth century. All have thus had to make adjustments to the local situation in the sense of having to accept topographical and other limitations on their former notions of economy and settlement. A second point is that all of them maintain links both with one another within the islands and also with their parent communities elsewhere. These latter ties have been of least importance for the Shirazi and of most importance for the Arab, Indian, and European elements. This leads to the third point, that a principal distinction within the total society has been that between those groups ultimately protected by a strong external power (Muscat or Britain) and those not so dependent (Shirazi and Mainlanders). The former have controlled the external trade of the islands, the government from Zanzibar City, and have in general secured a main share in the wealth of the country; the latter have been peasant cultivators, fishermen, and labourers.

In this paper I shall deal mainly with the Shirazi, the majority of the population and the only group that is virtually entirely rural; all the others live also in the various urban centres of the islands and so present a very different kind of settlement pattern from that of the Shirazi.

II

There is a clear relationship between the pattern of settlement and the distribution of various types of soil in the two islands. Of course, this is always true to some extent in every society, but the link is particularly obvious and important in Zanzibar and Pemba. The islands are based on coral, and contain four main soil types:

(*a*) Over much of the coral, especially in Zanzibar Island and in eastern Pemba, are pockets of deep soils known as *kinongo*. These are found both along the coast and also inland. On them are built the Shirazi "towns".

(*b*) In the western half of Zanzibar Island and over most of Pemba are rich sandy soils. Before the mid-nineteenth century these supported dense forests; since that time the forests have been mostly cleared and these areas have supported the clove plantations on which the wealth of the islands has depended.

(*c*) In a few areas of Zanzibar Island, and more widely dispersed in Pemba, are stretches of heavier and richer soils that are used only for growing the staple food crop, rice.

(*d*) Finally, over 40% of Zanzibar Island and stretches of eastern Pemba consist of almost exposed coral rag, covered with shallow soil and scrub.

The settlement pattern is determined almost entirely by this distribution of soils. The Shirazi "towns" (*miji*) are large and permanent settlements set on the pockets of deep *kinongo* soil. They are densely populated and are finite in size—usually about a mile across—and there is no settlement of permanent houses beyond their edges, which are clearly marked by the appearance of the coral rag. Most are at least ten miles from neighbouring towns, although there are one or two clusters of nearby towns. Many of them have certainly been occupied for some thousand years, perhaps more. In the clove plantation areas (cloves do not grow in the town sites) there are no indigenous Shirazi towns, but more modern straggling villages and the scattered huts of the plantation squatters. The rice valleys and plains are too valuable to be built on. The coral rag areas contain scattered fields owned by the townsmen and growing grain crops.

In brief, the Shirazi occupy the traditional towns, which are geographically distinct from other settlements and are virtually closed to outsiders. The immigrant Arabs and mainland Africans occupy (or occupied until the revolution) the clove areas and the "Stone Towns" of the western coastline: Zanzibar City and its suburbs in Zanzibar Island, and Wete, Chake Chake, and Mkoani in Pemba.

III

I turn now to discuss the internal economic and social organization of the Shirazi towns in Zanzibar Island.

The town itself is a closely settled area of permanent houses made of coral blocks with palm leaf roofs, that are set in rows and blocks with alleys between them. Houses are often connected by grass or coconut leaf fencing and may consist of several interconnected rooms with a small patio, bath house, and kitchen. Many towns are well over a mile across and vary in population from 500 to 5000 people, the average being about 1500. The central buildings are the mosque (the oldest mosque still in use in Zanzibar dates from the ninth century), market place, coffee shop, and laundry place. Between the clusters of houses are gardens, for root-crops and tobacco, and coconut palms and fruit trees (except for cloves) are planted wherever there is a spare foot or so of space. The whole is a continual stretch of houses and trees, and always green and lush in appearance. The town is divided into several levels of local grouping, which are described below.

The members of these towns have certain resources open to them for exploitation. Most of these are outside the towns themselves, but all must be available to the town for it to be a viable social unit. In addition, since the islands themselves are not large and the total resources are limited by the fact that being islands there are no further adjoining areas to be conquered, there must be established rules to allocate the resources equitably between the towns themselves. To assist in this aim, as will be shown below, there is also a continual redistribution of the population itself: if resources cannot be moved, people can.

The most important and necessary resources comprise:

(*a*) The town itself, its mosque, other central buildings, graveyard, and houses.

(*b*) Drinking water. Water is generally not a scarce resource. Wells are dug in and near almost all towns, many of them having been in use for centuries.

(*c*) Coconut palms, a source of wood for building and for fuel, roof-coverings, coir fibre, copra, food, and many other lesser purposes. They are planted among the buildings of the towns, along the beaches, and wherever a patch of sandy soil is available.

(*d*) Gardens, set beneath the palms where the soil is deep enough, and used for root-crops and tobacco. Gardens are always set within the town boundaries.

(*e*) Fields, set in the areas of coral rag outside the town boundaries. Fields may be large and bounded for protection by stone walls, or may consist of a few square feet of earth laboriously carried to a place in the rag where there are already a few inches of soil. They are used mainly for maize, millet, and other grains.

(*f*) Rice plains and valleys. These are very unevenly distributed over Zanzibar Island, although more numerous and widespread in Pemba. The distribution of rice is today generally through market transaction, although many people try to rent portions of rice plains for a few seasons.

(*g*) Townlands in the coral rag areas. Besides containing the fields, townlands are used for grazing cattle, collecting wood for fuel and building, collecting coral for building and lime. Long ago they were used also for hunting. Townlands along the coast also supply mangrove poles for building and export.

(*h*) Fruit trees, especially the mango. Cloves are not grown in Shirazi towns but in the clove plantations only.

(*i*) Fishing areas. Most towns on or near the coast recognize boundaries within the sea itself that stretch from the beach to the reef about a mile offshore. The beaches are also used for coir making, the fibres being buried in the sand until the covering is rotten.

IV

A basic problem of production and distribution in this society is how to relate the differential territorial distribution of these widespread resources to the households in the towns that are the units of consumption. One way of achieving this end would be by the exercise of superior political authority by the central government. Today such authority is very weak at the local level and is not concerned with the allocation of these resources. The former indigenous rulers of the Hadimu, Tumbatu, and Pemba acted mainly as ritual figures and as representatives of their peoples to the outside world and the Arab Sultans at Zanzibar City. The indigenous rulers ceased to exist during the nineteenth century, and the Arab Sultan and his officials lived in the city with little direct contact with the Shirazi towns in the eastern parts of Zanzibar Island. The isolation of these towns has continued to the present day, as far as the exercise of everyday authority is concerned. Within a town there is a traditional council of four men, representing various local descent groupings, and a ritual figure concerned with protection of people and crops; but these are not concerned with allocation of resources. Also, feud and warfare among the Shirazi seem to have been minimal, despite the pressures of population; they did not provide sanctions for the allocation of land and resources.

Failing "external" sanctions for allocation, the system depends upon its own internal sanctions. The problem of resource allocation is resolved by the groupings of households and their members into different kinds and levels of local and kinship groupings, each of which is considered to hold rights of various kinds in one or more resources, and all of which are interlinked by cross-cutting ties of loyalty expressed in terms of propinquity, descent, and ethnicity.

Full details of local and kinship groupings need not be given here, but the principles are simple. The town is divided into villages, and then into wards or quarters; the town also has two moieties, usually with about the same number of villages in each. There are typically at least half a dozen villages in a town, each village comprising three or four wards. Each of these groups consists of people who reckon bilateral kin ties to each other, of increasingly narrow range. The largest of the kin groups is the *ukoo*. All members of a town are members of the same *ukoo*, but any one *ukoo* also has members living in other towns throughout the island (and often in Pemba and even the Tanzanian mainland as well). The *ukoo* includes several thousand people who reckon common descent, through both men and women, from a common founder who is said to have immigrated into the islands long long ago. Any person, of course, is a member of more than one *ukoo*, since descent is traced through both men and women. Every person, man or woman, has an inalienable right to be buried in his "own" town's graveyard (i.e. the town in which he has spent most of his life and, particularly, into which he was born); but otherwise he may claim rights in the lands of any other town with which he can claim *ukoo* membership. Villages and wards are occupied by smaller segments of the *ukoo*, known as *mlango* (door) and *tumbo* (womb).

Full details of the distribution of different kinds of rights in resources cannot be given in this brief paper. A cluster of rights is associated with each of the groups mentioned above. The largest is the town, to which is attached a stretch of land comprising the settlement itself, its gardens, and bushland in the coral rag; the members of the town hold rights in these as against other towns, there being traditionally no areas not claimed by one town or another. Within the town is the village, with its own defined residential area, its own gardens, and its own stretch of bushland within the townlands. Within the village is the ward, associated with a particular *tumbo* and its own residential area within the village area; the ward's gardens adjoin the residential area. A man who is a member of the town has rights in all these kinds of land. He may build and plant trees in part of his ward's area, whether or not the soil beneath them is claimed by someone else. He may plant crops in any spare piece of land in his ward's area, whether or not others have trees there. A man has certain rights also in regard to the townland as a whole; he may open fields anywhere, plant trees anywhere outside the settlement area proper, cultivate anywhere under standing trees belonging to others, dig a well anywhere, gather fuel, building wood, lime and coral, graze livestock, hunt, gather wild fruits, and gather fallen palm fronds, gather mangrove poles, and fish in the sea from shore to reef (although an individual may set up fishing fences that remain his property).

Under Muslim law, these various rights are heritable through both men and women. Marriage is with cousins (all except for mother's sister's child), who are in the opposite moiety. Thus these rights can quickly become dispersed throughout the town and beyond (since people may move to another town and carry their inheritance with them). In addition, most women marry more than once, marrying different cousins each time and bearing at least one child by each. A sibling may sell out his share of an inheritance to his other siblings as a group, a process that decreases the number of descent lines that hold inheritance rights in a particular resource; such a seller is typically a person who moves to another town to live, but he retains his right to burial in his original town graveyard.

In addition, a man has the right to be accepted as a kin-tenant in the village lands of any recognized kinsman, whether in the same town, another town in Zanzibar, or in Pemba. A tenant has the full rights of an ordinary inhabitant except that of burial. The amount of movement from one town to another by kin-tenants is very considerable, over half of the men moving in this way at least once in their lifetimes (but mostly returning to their original settlement when old). The consequences of this system must be seen as it works over time. The local and kin groups overlap and are open; they are not closed groups as are unilineal descent groups. A person is, or can claim to be, a member of more than one group, at any level. He may opt to live in the settlement associated with a distant kinsman but does not thereby forfeit membership in the others, and this membership is transmitted to his children unless he sells out to his siblings. Thus a person can claim the membership most beneficial to him and can change if he wishes to do so. The more usual reasons are to find adequate space for building and planting trees, to live near a rice plain or valley, to exploit fish or wood sources, or to marry. These moves thus continually create networks of descent links between various settlements. The distribution of population can change continually according to vicissitudes of birth and death rates, and the degrees of local soil fertility and overcrowding (both of which change from one year to another).

In brief, a given town is largely but not totally economically and socially self-sufficient although its constituent villages and wards are not so. Today towns are no longer self-sufficient, mainly as a consequence of the introduction of a money economy; a man needs money which he can obtain either by selling crops such as tobacco, copra, or coir, or by working as a temporary labourer on the (formerly) Arab-owned clove plantations. Also, a town is not self-sufficient in the staple crop, rice. But the island-wide network of towns is still an almost self-sufficient, self-balancing, and self-sustaining system in which provision is built-in for changes in wealth and population of the overcrowded towns.

V

I have not mentioned the settlements of the non-Shirazi elements of Zanzibar Island. With the exception of a few Arab shopkeepers and some Indian farmers, no non-Shirazi may settle in the Shirazi towns, and those few who do are allowed there with no firmly defined rights. In the plantation areas the land was given outright to the Arab followers of the Sultans during the past century; they cleared the forests and planted cloves, using slave labour. After the abolition of slavery the slaves mostly remained as squatters. Since then the number of squatters has increased, mainly by the immigration of Mainlanders. Shirazi leave their towns in the clove-picking seasons to work as casual labourers, but few remain as squatters. Squatters settle where they like, keep the ground clean beneath the trees by growing their own crops there, and help with picking when needed. This system of settlement is found only in these fertile plantation areas, and stands quite outside and apart from that of the Shirazi towns.

VI

A brief comparison of Zanzibar Island with Pemba Island is useful here. The situation along the eastern coast of Pemba—the coral rag area—is essentially similar. But in the remainder of the island—well over half of its surface—the various resources are found close together within any one town area, and the building of permanent settlements is possible almost anywhere. In place of the large towns there are many smaller and scattered settlements with room near them for trees, gardens, fields, and rice valleys; only fishing and mangrove forests are lacking to the inland settlements. With this system there would seem to be less need for widespread kinship ties; the main unit is the *tumbo*, a smaller grouping than the *ukoo*; descent groups are more closed than in Zanzibar Island; moieties are weak or non-existent; there is a lower incident of cousin marriage, and there is a strong skewing of inheritance to the patrilineal line only.

VII

The conclusion to this short paper is a simple proposition, based mainly on the one case of Zanzibar Island but perhaps meriting further refinement through comparative analysis which cannot be attempted here. In Zanzibar the environment has limited permanent settlement to a relatively few places that cannot themselves provide adequate resources for exploitation by the immediate inhabitants. These therefore establish social relationships and rules for their maintenance so that they can have easy and equitable access to the various resources of the total island. A necessary corollary to the wide dispersal of rights of various kinds is that the settlements and their divisions must be assured enough sense of cohesion, tradition, and perpetuity for the people fully to belong to them as active and responsible members. In addition, at the local level there is the virtual absence of the sanctions of overruling political authority or of warfare. This sense of cohesion is achieved by calling into operation different levels and kinds of groupings in different social situations and in response to different economic needs. The unit of analysis is not the individual Shirazi town, but the total system of the two islands, throughout which are recognized ties of kinship that permit continual individual mobility without the total or local structures being affected.

Note

1 Research in Zanzibar and Pemba was carried out in 1958, with the financial assistance of the Zanzibar Government and the Colonial Office, London. Further data may be found in my report entitled *Land Tenure in Zanzibar*, London, Colonial Office, 1961, which contains a bibliography of works on the topics covered here.

PEDER MORTENSEN

Seasonal camps and early villages in the Zagros

The early settlement and subsistence patterns of south-western Iran have been dealt with in several recent publications[1] by Frank Hole and Kent V. Flannery. Based particularly on excavations in the alluvial plain of Deh Luran, but with regard also to investigations in the inter-montane valleys of the Zagros range (southern Kurdistan and Luristan), they have been able to construct a model comprising four types of settlement characteristic of the early dry-farming era (*c.* 10,000 B.C.–*c.* 5500 B.C.) within this region[2]:

1 Permanent, mud-walled villages, based on an economy of farming, supplemented by hunting and collecting of wild local resources.
2 Semi-permanent seasonal camps or villages, occupied either by a group of persons during a planting-and-harvesting season, or by a herding unit sent out seasonally from a permanent village for the purpose of pasturing flocks.
3 Pastoral camps in caves, occupied seasonally by herding units from permanent villages.
4 Transitory camps made by herding units travelling from one valley to another.

The model illustrated in Fig. 1 is fundamentally a simplified reproduction of the system suggested by Hole and Flannery. But it diverges from that model in the attempt to establish a distinction between what might be called circulating subsistence patterns (stages I and II) and radiating patterns (stage III). In this connection it might perhaps be of interest to comment briefly upon the development from a semi-permanent seasonal camp (stage II) to a permanent village (stage III), as it is represented at Tepe Guran in Central Luristan[3].

Tepe Guran lies almost 950 m above sea level in the northern part of the Hulailan valley, not far from Jazman Rud, a small stream joining the Saimarreh river 7 km south of Guran. It was inhabited continually during a thousand years from about 6500 B.C. to about 5500 B.C. The first settlers, apparently, lived in wooden huts, the remains of which were recognized in our sections as very thin, dark-coloured layers. When these layers were carefully

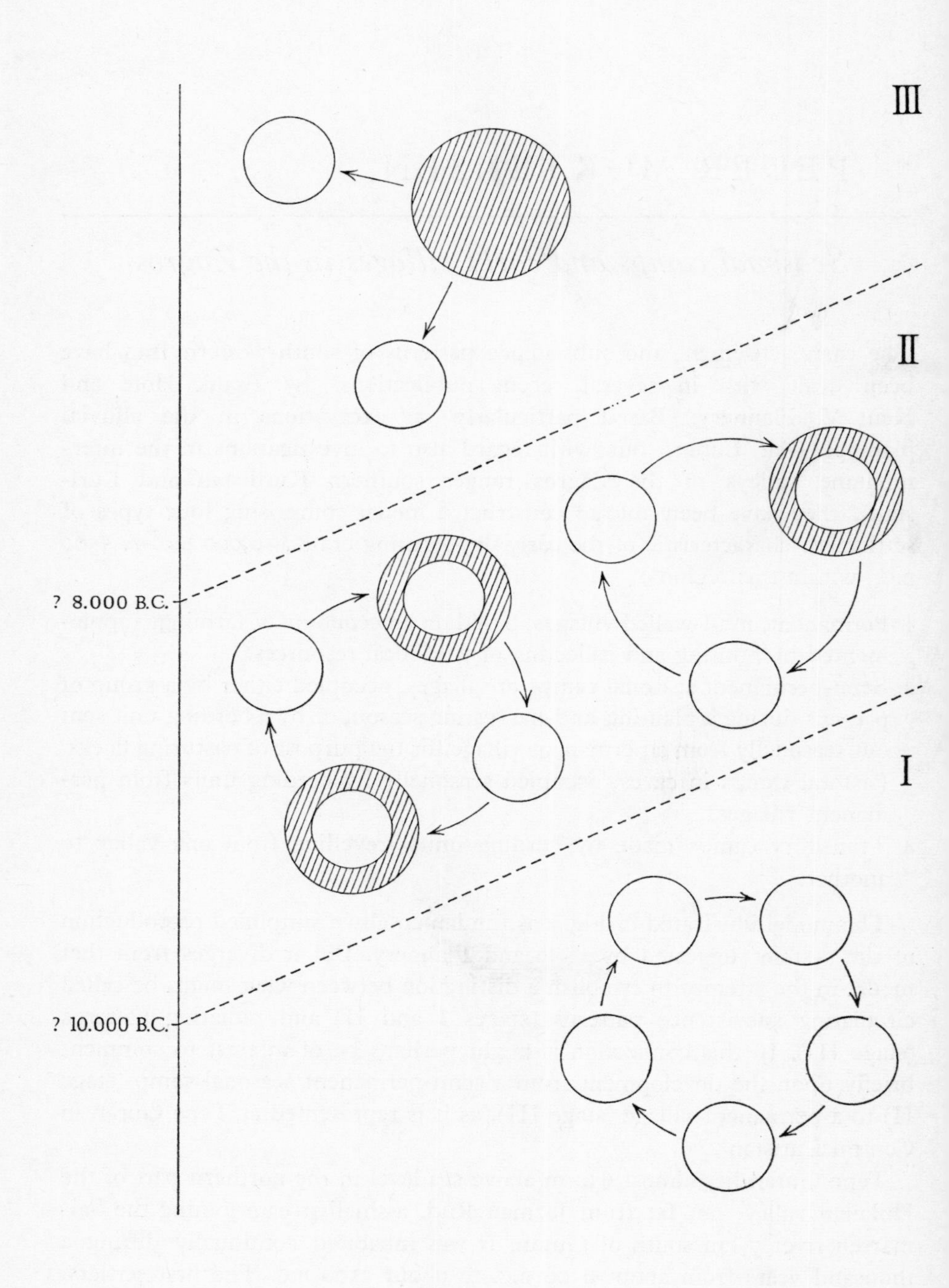
III
II
I
? 8.000 B.C.
? 10.000 B.C.

excavated horizontally, the contours of spaced huts with two or three small rooms, and with rectilinear or slightly curved walls appeared. There were traces of what must have been matting on some of the floors, and two open-air fire-places were found. In the later levels (from about 6200 B.C.) remains of mud-walled houses were found together with the wooden huts, the latter gradually disappearing. The upper part of the mound was characterized by mud-walled houses, built close together. The houses were divided into small rooms, the walls of which had recesses for low mud-benches or tables, and for openings to domed ovens. Walls and floors were faced with red or white plaster, and courtyards and floors were sometimes constructed in a kind of terrazzo-technique with small pieces of feldspar laid in clay coloured with red ochre. The fact that tools which can probably be related with agriculture (grinders, mortars and sickle-blades with glossy edges) were almost absent in the earliest layers of the mound gave rise to the idea that the people who first settled at the site were primarily herdsmen.

This hypothesis was later supported by Flannery's observations, during his study of the faunal remains from Tepe Guran[4]. It appeared that goat horn cores, sufficiently medially flattened to suggest the domestic form, were already present in the lower levels, and that goats in those levels represented anywhere from 80% to 100% of the ungulate remains. There was a relative scarcity of any hunted wild mammals except for gazelles and a few foxes and hares. On the other hand, there was a virtual concentration in the nine earliest levels of the remains of goose, crane, and heron—all migratory waterfowl visiting the middle course of the Saimarreh river during winter time.

With the reservation in mind that the sounding at Tepe Guran was rather small[5], the evidence seems to suggest that the early levels at Guran reflect a semi-permanent camp for a group of transhumant herders who stayed with

Fig 1 An explanatory model illustrating how the settlement patterns in the Zagros area may have developed from the Late Pleistocene Period to *c.* 5000 B.C.

Stage I represents the Zarzi Period, highly specialized hunter-gatherers moved from one seasonal camp to another, thus describing a circulating annual movement.
In Stage II the annual movement is still circular, but the number of base camps decreases, because the economy is supplemented by herding and/or farming so that one or two camps situated in favourable environments are occupied during a longer period (i.e. develop into semi-permanent seasonal camps).
Stage III shows a semi-permanent seasonal camp that has developed into a permanent village. At this point the circular annual pattern is replaced by a radiating movement represented, e.g. by butchering stations and pastoral camps for herdsmen sent out in the spring with flocks of goat and sheep to pastures in the higher valleys from permanent villages.

The present evidence clearly indicates that stages II and III overlap. In the future a similar overlapping will probably be determined between stages I and II in the 10th millennium B.C.

their goats during the winter in huts of wood, which was easily obtainable from the nearby hills. Occasionally, they hunted a few gazelles and some of the seasonally available waterfowl, but they did not range very far from home in search of wild game. It is likely that the herders who spent the winter in the Hulailan valley moved up to higher pastures in the summer.

About 6200 B.C. the situation changed. Mud-walled houses appeared, and there is abundant evidence of farming and collecting (two-row hulled barley and pistachio). At the same time an interesting change is taking place within the faunal remains. The number of gazelles relative to domestic goats increases steadily, and in the early sixth millennium there are as many gazelles as goats. Hunting of red deer, wild cattle, wild pig, fox, and wolf increased significantly, and *Helix salomonica*, a landsnail which was collected and eaten in enormous quantities by the early villagers in Kurdistan and Luristan, is represented by hundreds.

The change which can be observed in fauna and architecture seems almost certainly to reflect a transition from semi-sedentary to fully year-round occupation at Tepe Guran around 6200 B.C. When the village was firmly established, the herders and farmers exploited not only the valley floor, but they also hunted beyond the immediate vicinity of the village in the surrounding hills. In this context it is perhaps of interest to observe that Guran—at least in the seventh millennium B.C.—was the only permanent village in the Hulailan valley.

If one accepts the model outlined above (Fig. 1) it follows that the earliest villages in an area should be preceded by semi-permanent camps established by herders or farmers exploiting the ecological possibilities of the new environment (stage II). It has recently been suggested that the earliest habitations on the North Mesopotamian plain were seasonal camps for herders who were attracted by the Assyrian steppe because of its great potential as winter grazing land, and that the first permanent villages in the area (Hassuna and Matarrah)—like Tepe Guran—developed on the basis of such seasonal camps[6]. But unfortunately it is impossible at the moment to tell if the earliest habitation at Hassuna should be interpreted as a seasonal camp reflecting a circulating annual subsistence pattern (stage II), or whether, in fact, it was a pastoral camp connected with a permanent village of early Hassuna type (stage III); whichever, we still do not know where the earliest Hassuna people came from[7].

Notes

1 Hole, F. and Flannery, K. V. (1967). The prehistory of south-western Iran: a preliminary report, *Proc. Prehist. Soc.*, **33**, pp. 147–206; Hole, F. (1968). Evidence of social organization from western Iran 8000–4000 B.C., *in* Binford, S. R. and Binford, L. R. (eds.) *New Perspectives in Archaeology*. Chicago. pp. 245–65; Hole, F., Flannery, K. V. and Neely,

J. A. (1969). *Prehistory and Human Ecology of the Deh Luran Plain.* Ann Arbor. pp. 1–438.

2 Hole, F. and Flannery, K. V. (1967). *op. cit.* p. 166f.

3 Meldgaard, J., Mortensen, P. and Thrane, H. (1963). Excavations at Tepe Guran, Luristan, *Acta Archaeologica*, **34**, pp. 97–133; Mortensen, P. (1964). Additional remarks on the chronology of early village-farming communities in the Zagros area, *Sumer*, **20**, pp. 28–36; Mortensen, P. and Flannery, K. V. (1966). En af verdens aeldste landsbyer, *Nationalmuseets Arbejdsmark*, pp. 85–96.

4 The faunal information presented in this paper is based on two chapters by Kent V. Flannery on "Hunting and early animal domestication" and "The prehistoric environment of Tepe Guran" for the final publication of the excavations at Tepe Guran.

5 The sounding at Tepe Guran yielded roughly 15,000 fragments of animal bone and shell. But only about 20% of these were identifiable to genus or species.

6 Mortensen, P. (1970). Tell Shimshara. The Hassuna Period, *Hist. Filos. Skr. Dan. Vid. Selsk.*, **5**, pp. 1–148.

7 Mortensen, P. (1970). *op. cit.* p. 132.

JOAN OATES

Prehistoric settlement patterns in Mesopotamia

The archaeological record demonstrates that for thousands of years the two most powerful factors affecting settlement patterns in Mesopotamia were water and security, the former not only in the sense of adequate rainfall and wells but, in the south, in terms of the suitability of the terrain for the application of simple gravity-flow irrigation techniques. Before the mobility of resources, both plant and animal, that resulted from domestication, natural food supply was a further limiting factor. Permanent non-agricultural communities would in theory have been viable in the mountainous zones, which comprise the natural habitats of the potential domesticates, on the Assyrian steppe where winter rainfall and a profusion of wild game, including wild cattle, onager and gazelle, could have ensured an adequate diet, and in the marshes of southern Iraq where settled life based on an economy of hunting, fishing and gathering is still a practicable proposition. Of the first type of community we have evidence in the Zagros[1]. Of the second, evidence is strangely lacking but the existence of apparently non-agricultural villages along the Euphrates at Bouqras and Mureybit[2] makes it inconceivable that similar settlements should have been totally absent in the more hospitable upper Tigris valley. Equally, we lack evidence of early settlement in the south, but the pattern of shifting lagoon and marsh together with the heavy over-burden of silt seriously restricts the possibilities of prehistoric investigation there.

A fourth important factor affecting Mesopotamian settlement patterns lay in the trade routes many of which were in use at least as early as 7000 B.C. and probably far earlier. Trade in obsidian from Turkey, shells from the Gulf and semi-precious stones from the mountains to the east is attested at a very early date. In northern Iraq, where the river regimes have changed relatively little, there is evidence for virtually continuous settlement at important crossing points. The consequent growth of vast mounds like Nineveh has made extensive examination of their earliest levels impossible. To these bridgehead sites must be added many more large mounds at important road stations, such as Tell al Hawa, north-west of Mosul, Erbil, and Kirkuk, all on the Achaemenid Royal Road and all occupied, like Nineveh, from the period of the earliest farming villages. Reason suggests that these sites, even at this early period, were likely to have been more important than the scattered farming villages whose lesser position and lack of continued occupation make possible their present identification and investigation. This is an extremely important consideration and one which has far too often been overlooked. We cannot indulge in the economic and social analysis which is now so fashionable without a strong suspicion that our direct evidence relates only to peripheral material.

Although in the south the changing river regimes and the establishment of new routes along new canals resulted in more discontinuous occupation, an even more complicated situation obtains. Not only are many early prehistoric villages buried beneath vast mounds like Ur, Eridu and Nippur, but, owing to silting and wind deposition, it is almost impossible in most areas to find by surface survey small prehistoric mounds of the type that lie scattered over the northern plain. This problem is well illustrated by the modest village of Ras al ʻAmiya, not far from Babylon, occupied only briefly early in the fifth millennium B.C., the upper limits of which now lie at a depth of 1–2 m below the modern plain[3]. Space does not permit further elaboration but it is vital to appreciate the limitations imposed by the nature of the Mesopotamian evidence.

With these reservations we may examine some of the new evidence which has resulted from recent work in Iraq, in particular concerning the pattern of prehistoric settlement and the development of irrigation. The British School of Archaeology is carrying out surveys in two parts of Iraq, in areas chosen for their marginal character, both geographic and climatic. The first lies some sixty miles west of Mosul, on the north-eastern edge of the Jazirah, in the neighbourhood of Tell al Rimah[4], the second east of Baghdad, along the edge of the alluvial plain between the tip of Jebel Hamrin and Badra[5]. In the former area, although the survey is far from complete, we are in a position to give a rough estimate of the density and distribution of settlements of the Hassuna period, the earliest phase for which we have any record in the non-mountainous areas of Mesopotamia. Judging from the excavations at the type site and those more recently at Yarim Tepe[6] these were permanent villages with well-built *tauf* houses and an economy based on dry-farming and herding; their physical appearance was in many ways not unlike those of the modern countryside. In an area of approximately 350 km² twenty-three Hassuna-Samarra sites have been identified. This can be compared with the thirty-four somewhat later Susiana *a* sites mapped by Adams in a considerably larger area in Khuzistan[7]. Perhaps a dozen sites in the Rimah survey yielded Hassuna pottery without an admixture of Samarra, but on grass-covered mounds surface sherds are often scarce and one cannot say without excavation how many of these sites genuinely represent the earliest period of village occupation in the area. In general the sites are small, 2–4 m in height and a hectare or slightly more in area. Perhaps a half of their number, including the larger sites, lie on modern seasonal water-courses, but like the small modern villages, many must have relied on wells for drinking water, both for men and animals. At present neither the water-courses nor the wells could reliably support intensive settlement and there is serious crop failure roughly two years in five. There is no evidence for a major climatic change, but it is likely that tree cover on the hills overlooking the plain on the north and east would have had a beneficial effect not only on the reliability of rainfall but on the important factor of ground-water retention. Without pollen cores, however, this must remain speculative. This is an area where irrigation has never been practised on any considerable scale and settlement has retreated and advanced with the natural water supply, but only as security permitted. One would guess that the Hassuna villages may have had a *minimum* population of 50–100 persons, but in estimating population in Mesopotamia it is essential to consider the modern situation. Mud-brick houses have a maximum life span of about forty years and the more normal period of use is probably closer to twenty years. Modern practice tends to replace an ailing structure with one sited elsewhere. The result is that over a period of time the living village tends to move around the total village area but at no single time is the whole area inhabited. Excavation suggests that this was probably the ancient practice; certainly it is more economical of labour[8]. In addition, all the Hassuna sites yielded later prehistoric materials, thus making it impossible, without excavation, to establish even the limits of any Hassuna settlement.

The geographical distribution of Hassuna farming villages appears to coincide with the rolling Assyrian plains and the north-eastern fringe of the Jazirah, roughly within or at the edges of the present limit of rainfall adequate for dry-farming, at elevations roughly between 150 and 350 m, an area of rich winter grazing lands but oppressive summer heat. Up to now we have no evidence for the existence of pure Hassuna settlements in the mountainous homeland of the "Zagros group[9]", and the origins of Hassuna, which clearly represents a well-developed stage of farming village, apparently unrelated either to the early Zagros villages or to material so far found in Anatolia, must remain a mystery. One would expect a period of transhumance preceding permanent village life on the plain. Thus far only Hassuna itself has yielded any convincing evidence for semi-permanent encampments within this ecological zone[10], and these could, by modern analogy, represent seasonal Bedu camps contemporary with Hassuna or other as yet unknown settlements elsewhere in the area. Relevant to this is the fact that there is nothing in the archaeological evidence to necessitate the assumption that the people of the Ia campsite were the same as those who built the settled village, and there is some evidence to suggest that they were not[11]. Moreover, there is no reason to assume that because the techniques of domestication have been discovered, everyone of necessity adopted them. The prehistoric models that have thus far been offered seem to presuppose a unilinear development for which there is no real evidence, and we should certainly consider at least the possibility that already in the sixth millennium there may have been nomadic herdsmen contemporary with and probably in some sort of economic relationship with the settled villages, a pattern which is well-established by the time of the earliest written records. Relevant also to the interpretation of the site of Hassuna is the fact that it is climatically peripheral, that is, if our somewhat limited climatic evidence is correct, not in an area where one would expect to find remains of the earliest farming villages on the Assyrian plain; moreover, it is still a regular Bedu campsite. Of course, there may well have been, again by modern analogy, a transhumant element in the settled Hassuna economy; possible summer encampments may be seen at the site of Gird Ali Agha, on the Greater Zab in an area approaching the foothill zone, and cave sites like Baradost[12]. But none of these seasonal settlements need be regarded as antecedent to the Hassuna culture, and there is clearly much to be learned about the evolution of prehistoric settlement in northern Iraq.

Our second area of work, in particular around Mandali, has produced new evidence concerning the development of irrigation[13]. Excavation here and at Tell es-Sawwan[14] has confirmed that the Samarra assemblage must be considered culturally distinct from Hassuna although, on the basis of ceramic evidence, closely related to it. While Hassuna appears essentially to be an adaptation to the rain-fed northern plain, Samarra flourished further to the south, on the fringes of the alluvium. Binford has argued that pressures for the exploitation of new food sources and techniques would be felt most strongly around the margins of population-growth centres and not in the centres themselves[15]. The prehistoric Assyrian plain would seem to represent an 'optimal habitat' and the growth of population which the Rimah survey confirms may well have led to migration to the more marginal and climatically less favourable areas to the south. We cannot say how early such a movement might have taken place, however, and present evidence suggests that each of the distinct geographical areas of prehistoric Greater Mesopotamia—Assyria, middle Mesopotamia, Sumer, and Khuzistan—displays essentially a local though not isolated development. The earliest Samarra sites appear to have clustered along the Tigris, between Baiji (elevation 115 m) and Sawwan (65 m), well south of any conceivable limit of rainfall agriculture but where catch-crop cultivation and minimal flood-plain irrigation would have been feasible[16]. The botanical evidence from Sawwan[17] suggests that agriculture was probably conducted there on the basis of the seasonal floods of the river. Spill pools were exploited and run-off checked in favourable spots by primitive damming—generally the activities we may envisage as fore-runners of fully-fledged canal irrigation. Linseed was cultivated, a crop which could not have been grown there without artificial watering, but "the poor size of the cereal grains makes it improbable that regular canalization was instituted at the time"[18].

The distribution of sites north of Mandali takes a new turn. Here we have reasonably convincing archaeological evidence for irrigation channels (Fig. 1) and irrigated crops in the Samarra period. For the first time the line of settlement adopts an artificial pattern. No longer do sites necessarily follow the natural line of water courses but several are situated along a line parallel to the hills and at right-angles to the natural flow of water into the plain. There is clear evidence for at least two later canals running along this line (Pl. I) and the position of the Samarra villages together with the excavated evidence from one of the sites, Choga Mami, makes it virtually certain that at least elementary irrigation channels were in use here well before the end of the sixth millennium B.C. We may note that the area north of Mandali would have been particularly suitable for an early canal system since it lies in a triangle between two rivers, the Gangir and the Ab-i-Naft, of which the former provides a head of water and the latter, at a lower level, a natural drainage outlet. It is in fact the lower Tigris-Euphrates basin in miniature, with the problems of drainage and stagnation partially eliminated by the more rapid drop in land levels that result from its position at the edge of the alluvial plain[19].

Modern irrigation techniques nearly double the grain yield[20], and in the area around Mandali the increased yield *per capita* which must have resulted from even the simple irrigation techniques of the Samarran farmers can be seen in the density of the prehistoric population. Choga Mami is a large site by previous standards (5–6 ha). Within 4 km are three slightly smaller villages which were certainly occupied for at least part of the same period, and in the same area Samarra sherds can be picked up on one or two other small mounds. A similar population burst is apparent in the southern alluvium very shortly after this time in the vicinity of Ur and Eridu, where there occur a number of permanent settlements which of necessity depended on some form of irrigation[21]. By the Al 'Ubaid 3 phase (early in the fifth millennium) we have at Mandali probable evidence for a canal 4–6 m wide running from the Gangir past Choga Mami[22]. At the same time there would appear to have been a flurry of expansion, and sizeable settlements with identical assemblages occur over virtually the whole of the country. This development in northern Iraq ('Ubaid 3–4) suggests actual colonization, or certainly a movement of peoples, from the south. By the end of the fifth millennium (Al 'Ubaid 4) towns like Eridu, 'Uqair and Gawra are characterized by monumental architecture, clearly stratified society and other appurtenances of incipient "cities"; and here was set the pattern for future developments.

Any discussion of these developments is beyond the scope of this paper but a few of the immediate consequences of the new irrigation technology should be mentioned. One of the most obvious is the need for more efficient methods of digging and tillage. Not only does hoe or digging-stick agriculture severely limit the area of possible cultivation per individual but it is unsuited to irrigation agriculture with its expansion potential and the need not only to till the fields but to excavate irrigation ditches and run-off channels as well[23]. It is generally accepted that some form of plough was a prerequisite to successful expansion even for dry-farming[24], and it is perhaps no coincidence that

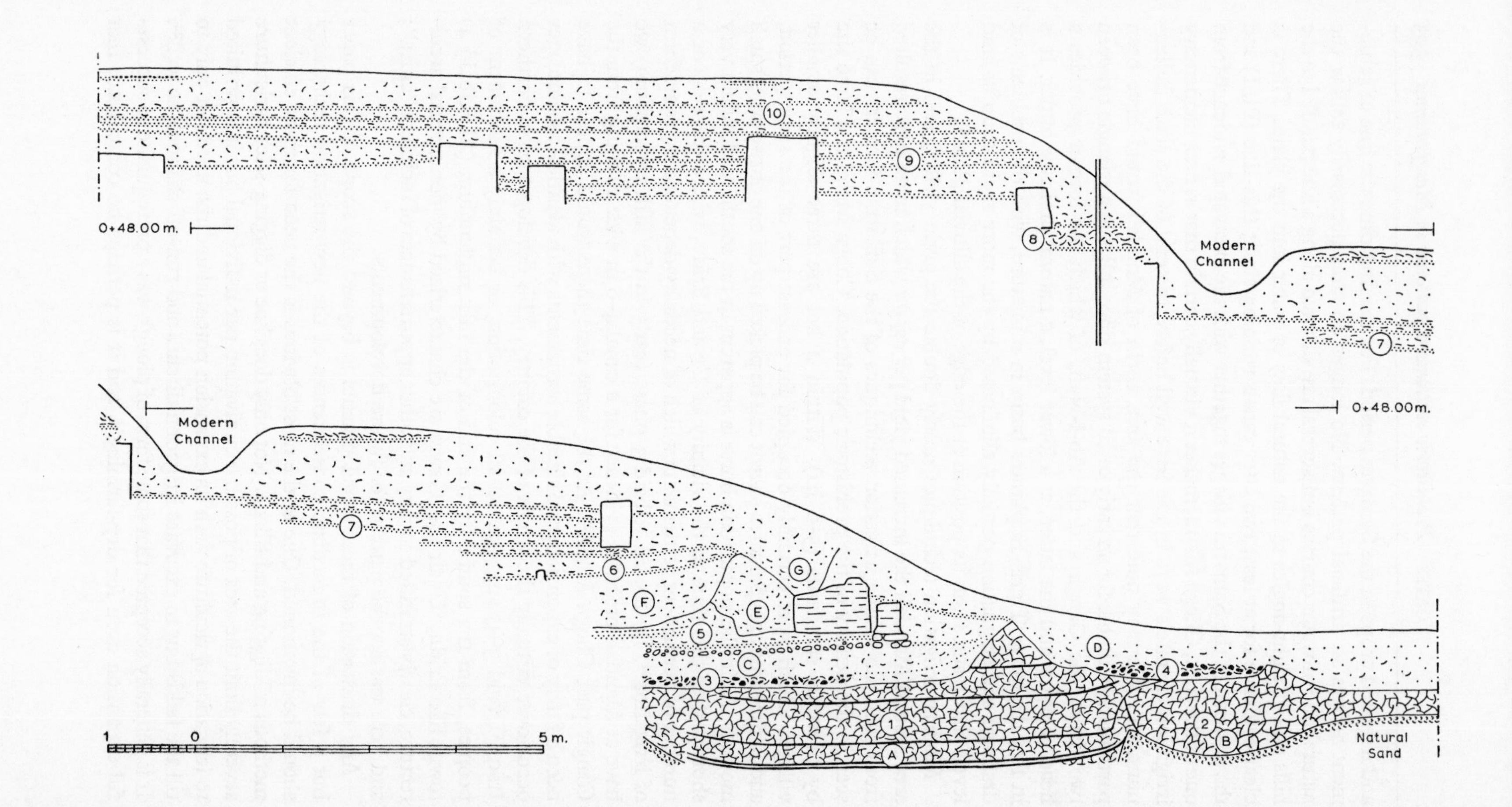

0+48.00 m.
Modern Channel
Modern Channel
0+48.00 m.
Natural Sand
1 0 5 m.

the first evidence in Mesopotamia for domesticated cattle, the first animal capable of pulling a plough, is found during the period of which we are speaking[25]. Concerning the earliest use of the plough, Reed has written that "lacking remains of ploughs, which were wooden, archaeologists might well look for changes in settlement-patterns and for possible population increases after the domestication of cattle"[26]. Samarran settlements in Mesopotamia provide precisely this picture. Moreover, shortly thereafter there is found extensive settlement in the south with an apparent high proportion of domesticated cattle[27].

Much has already been written concerning the social implications of irrigation[28] and it is agreed that the construction and maintenance of simple irrigation systems, utilizing flood water and small channels, require neither an elaborate administration nor labour resources larger than those available to small communities. Moreover, Adams has pointed out that although large-scale irrigation imposes technical and social demands of a different order, there is some evidence to suggest that "the introduction of great irrigation networks

---

Fig 1 Section illustrating irrigation ditches at Choga Mami:

1. The letters denote prehistoric water-channels while the numbers identify associated deposits or occupation levels.
2. A and B are the earliest identified channels, cut into virgin sand. They are both contemporary with Samarra occupation elsewhere on the site. The width of these channels is exaggerated on the drawing, since the trench cut diagonally through them; A was approximately two metres wide. Local conditions favoured aggradation rather than degradation in the water-courses at this time.
3. The thick sherd deposit in channel C indicates that the settled area was now close at hand.
4. Channel C was diverted to the line of channel D, probably to permit an extension of the settlement over its original course.
5. The earliest building in this trench overlies Channel C, and the associated levels (5) produced Samarra pottery.
6. The next two ditches, E and F, cut through occupation deposit 5 and the building associated with it. These ditches were above the contemporary plain level and thus must have been intended for irrigation.
7. Ditches E and F are overlaid by two occupation levels (6 and 7) both of which produced Samarra pottery.
8. Ditch G is contemporary with 6, and perhaps remained in use during the early part of 7.
9. A further extension of the settlement to the north pushed its boundary beyond the surviving limits of the mound; consequently no later irrigation ditches can be identified.

From D onwards the ditches are clearly artificial and we interpret the sequence as a series of occupation levels periodically cut by irrigation channels that followed the northern contour of the mound in precisely the same manner as does the modern channel shown on the section and visible in the foreground of Pls Ia and Ib. We cannot prove that the earlier channels were man-made but it seems not unlikely; although they appear to lie at ancient plain level they could have served to irrigate land to the north of the mound. (A fuller description can be found in Oates, J. (1969) Choga Mami 1967-8: a preliminary report, *Iraq, 31*, pp. 122-27.)

was more a 'consequence' than a 'cause' of the appearance of dynastic state organizations"[29]. In rightly dismissing "irrigational determinism", however, we must not forget that the process was clearly a spiral: the differentiation of society that was to culminate in the bureaucratic administrations of later Mesopotamia depended initially on food surpluses which irrigation served not only to increase but, for the first time, to make secure. This economic situation must have encouraged the political and social developments that in turn made possible more ambitious hydraulic schemes. Moreover, one must not forget that to exploit any considerable part of the potentially rich land between the southern reaches of the Tigris and the Euphrates demanded the resources of a large and well-organized community, implying an effective degree of social cohesion if not actually a central authority.

It is the beginning of this process that we can observe at Choga Mami and there are several apparently related factors that are worthy of mention if only as steps on the route society was about to follow. We have already referred to the concentration of population in and around Choga Mami, clearly indicating more efficient and reliable methods of securing food. Moreover, we believe we have found evidence suggesting the existence of rigidly observed property rights and of the possible existence of extended households in the sense of the Arabic *bayt*[30]. The recognition of private ownership is also suggested in the appearance at this time of the first stamp seals, and a more conscious specialization of labour in the use of potters' marks which can be observed at all Samarra sites. The extent of the ground stone industry at Tell es-Sawwan suggests a concentration of specialized craftsmen previously undreamt of. In addition, the peculiar nature of the graves and associated buildings at Sawwan strikingly suggests some form of religious centre. That all these apparent innovations should occur simultaneously, in what must have been a marginal habitat, is surely no coincidence.

It remains to point out that dependence on a common irrigation system has strong potentialities for antagonism and social disruption as well as for co-operation. This is abundantly clear in later Babylonian litigation, but at the same time the open terrain of the Mesopotamian plain has always acted against social isolation. Moreover, the consequences included, together with the obviously advantageous increases in food surpluses and opportunities for social and political development, the seeds of a damaging increase in the potential for environmental destruction. Large irrigation systems were to prove increasingly vulnerable to external disruption, while inadequate drainage, basically a product of the flatness of the alluvial plain, led to intolerable salt concentrations in the soil and resulted ultimately in large-scale abandonment of vast areas of the southern plain.

Notes

1 Braidwood, R. J. and Howe, B. (1960). Prehistoric investigations in Iraqi Kurdistan, *S.A.O.C.*, **31**; see also Mortensen, P., this volume, pp. 293-7.

2 Contenson, H. de (1966). Découvertes récentes dans le domaine du Néolithique en Syrie, *L'Anthrop.*, **70**; Contenson, H. de and Liere, W. J. van (1966). Premier sondage à Bouqras en 1965, rapport préliminaire, *Ann. archéol. arabes syriennes*, **16**; Loon, M. van (1968). The Oriental Institute Excavations at Mureybit, Syria: Preliminary report on the 1965 campaign, Parts I and II: Architecture and General Finds, *J. Near East. Stud.*, **27**; Zeist, W. van (1970). Part III: The Paleobotany, *J. Near East. Stud.*, **29**.

3 Stronach, D. (1961). The excavations at Ras al 'Amiya, *Iraq*, **23**, p. 98 and Fig. 5.

4 The survey has been conducted over the course of five years by the staff of the Tell al Rimah expedition under the direction of Professor David Oates, and especially by Mr. D. H. French, now Director of the British Institute of Archaeology in Ankara. Where possible at least two visits have been made to each recorded site. None the less many less prominent mounds have undoubtedly been overlooked, and these will add to the tally of purely prehistoric sites. A general description of the area can be found in Oates, D. (1968). *Studies in the Ancient History of Northern Iraq*. London. ch. 1; see also Oates, D. (1965). The excavations at Tell al Rimah, 1964, *Iraq*, **27** (2), pp. 62–7, 79–80.

5 Oates, J. (1966). Survey in the region of Mandali and Badra, *Sumer*, **22**; Oates, J. (1968). Prehistoric investigations near Mandali, Iraq, *Iraq*, **30** (1).

6 Lloyd, S. and Safar, F. (1945). Tell Hassuna, *J. Near East. Stud.*, **4**; Lloyd, S. (1938). Some ancient sites in the Sinjar district, *Iraq*, **5**, p. 138. The Russian excavations at the site of Yarim Tepe, north-west of Rimah, began in the spring of 1969 and continued in the spring of 1970, under the direction of Dr. Rauf Munchaev, Deputy Director of the Institute of Archaeology, Academy of Sciences, Moscow. We are indebted to him and all his staff, both for information and their very generous hospitality.

7 Adams, R. M. (1962), Agriculture and urban life in early south-western Iran, *Science*, **136**, p. 112 and Fig. 3. Of 138 sites surveyed in the region of Rimah, fifty-seven have yielded prehistoric materials. Unquestionably Samarra sherds occur at eleven of the twenty-three Hassuna-Samarra sites. All but four or five of the prehistoric sites produced painted pottery of the Al 'Ubaid period, and there are about thirty-five Halaf sites. Eight only have so far yielded the complete prehistoric sequence. Hassuna sherds have been found on over 50% of the Halaf sites.

8 Compare the results at Jarmo, Braidwood, R. J. and Reed, C. A. (1957). The achievement and early consequences of food-production: a consideration of the archaeological and natural-historical evidence, *Cold Spring Harbor Symposia on Quantitative Biology*, **22**, p. 26; also Braidwood, R. J. and Howe, B. (1960), *op. cit.* p. 159.

9 Mortensen, P. (1964). Additional remarks on the chronology of early village-farming communities, *Sumer*, **20**, p. 33. It has long been customary in identifying sites on survey to make little distinction between "Hassuna" and "Samarra" and indeed it is sometimes impossible to do so. Samarra material was clearly recognized by Lloyd and Safar as "extraneous" at Hassuna, but it is the recent work at Sawwan and Choga Mami that has emphasised the necessity of distinguishing the two ceramic styles. Thus

many sites originally described as "Hassuna" should properly be attributed to the Samarra phase. Hassuna pottery is found in the plain north of Jebel Sinjar and has been reported from near Ain Sifni, at the edge of the foothills north of Mosul (Lloyd, S. and Safar, F. (1945), *op. cit.* p. 260, note 1). There would also appear to be Hassuna material east of Jebel Hamrin, where the climate differs little from that of Hassuna itself (Oates, J. (1968), *op. cit.* p. 1, note 2), but it should also be noted that the sherds published from Nuzi (Starr, R. F. S. (1939). *Nuzi: report of the excavations at Yorgan Tepa near Kirkuk, Iraq.* Cambridge, Mass., pl. 46) like the majority of those from Matarrah and Shemshara are probably to be attributed to the Samarra assemblage. With reference to Shemshara, Mortensen remarks that no pottery earlier than Samarra has been found in the surrounding Rania plain: Mortensen, P. (1970). Tell Shimshara, the Hassuna Period, *Hist. Filos. Skr. Dan. Vid. Selsk.*, **5**, p. 120.

10 Lloyd, S. and Safar, F. (1945). *ibid.* Level Ia, comprising three successive camp-sites. The basal level at Matarrah is much more likely to represent a temporary builders' floor; we lack adequate information concerning the earliest levels at Nineveh and Telul eth-Thalathat in which, like Hassuna, only plain pottery has been found: Braidwood, R. J. *et al.* (1952). Matarrah. *J. Near East. Stud.*, **11**; Mallowan, M. E. L. (1933). The prehistoric sondage at Nineveh, 1931–32, *A.A.A.*, Liverpool, **20**, pl. 73; Egami, N., Sono, T., and Horiuchi, K. (1966). Brief report of the third season's excavations at Tell II of Telul eth-Thalathat and some observations, *Sumer*, **22** (Levels XV–XVI).

It should also be pointed out that the undistinguished character of the prehistoric chipped-stone tradition in Assyria makes difficult the survey identification of possible pre-Hassuna materials, if such exist. The site of M'lefaat however, situated not far east of Nineveh (elevation approximately 300 m), with chipped and ground stone industries resembling Karim Shahir and Zawi Chemi Shanidar, brings this earlier type of settlement for the first time within the Hassuna province. See Braidwood, R. J. and Howe, B. (1960), *op. cit.* pp. 27, 51–2.

11 There are distinctive differences both in the stone and ceramic materials; moreover, the two javelin points from the camp-site level suggest a Syro-Cilician connection that does not apply to materials from the settled village. See also note 23.

12 Braidwood, R. J. and Howe, B. (1960), *op. cit.* pp. 26, 37–8, 66; Safar, F. (1950). Pottery from the caves of Baradost, *Sumer*, **6**. It should now be noted that, since this article was written, excavated evidence for the long suspected pre-Hassuna occupation of the northern plain has been found at two different sites, one near Tell Afar and the other west of Hatra. Further information must of course await the publication of these new sites.

13 Oates, J. (1969). Choga Mami 1967–68: a preliminary report, *Iraq*, **31**, pp. 122–28, pl. 23.

14 El-Wailly, F. and Abu al-Soof, B. (1965). The excavations at Tell es-Sawwan, first preliminary report (1964), *Sumer*, **21**; Wahida, G. (1967). The excavations of the third season at Tell es-Sawwan, 1966, *Sumer*, **23**.

15 Binford, L. R. (1968). Post-Pleistocene adaptations, *in* Binford, L. R. and Binford, S. R. (eds.). *New Perspectives in Archaeology.* Chicago; see also the discussion in Flannery, K. V. (1969). Origins and ecological effects of early domestication in Iran and the Near East, *in* Ucko, P. J. and Dimbleby, G. W. (eds.). *The Domestication and Exploitation of Plants and Animals.* London.

16 A further group of Samarra sites which would repay investigation lies on low bluffs overlooking a lake about 15 km west of Hatra, again well

south of the isohyet representing the acceptable minimum of reliable rainfall. A pollen core from the lake bed would be a boon to all Mesopotamian archaeologists!

17 Helbaek, H. (1964). Early Hassunan Vegetable Food at Tell es-Sawwan near Samarra, *Sumer*, **20**.

18 Helbaek, H. (1964). *op. cit.* p. 47.

19 It is a situation comparable with that in the Deh Luran plain where Hole and Flannery have found evidence for irrigation at a slightly later period. Hole, F., Flannery, K. V. and Neely, J. A. (1969). Prehistory and Human Ecology of the Deh Luran Plain, *Mem. of Mus. of Anthrop.*, Univ. of Michigan, pp. 354f. Conditions advantageous to the development of irrigation techniques were also noted by Adams on the upper plains of Khuzistan. Adams, R. M. (1962). *op. cit.* p. 110.

20 Oates, D. (1968). *op. cit.* pp. 45–49; Adams, R. M. (1965). *Land Behind Baghdad*. Chicago, pp. 13–20; Adams, R. M. (1962). *ibid.*

21 Oates, J. (1960). Ur and Eridu, the prehistory, *Iraq*, **22**, p. 48.

22 Oates, J. (1969). *op. cit.* p. 128. We cannot as yet prove that this channel was artificial but it certainly continues the line of the canal to be seen in the background of Pl. Ia, along which lay sites yielding 'Ubaid and Uruk sherds.

23 It is perhaps relevant that the better-made Hassuna "hoes" appear to be confined to the "camp-site" level (Lloyd and Safar (1945). *op. cit.* p. 269); they have not been found at any of the Samarra sites nor in the levels so far excavated at Yarim Tepe I which have produced a mixture of Hassuna and Samarra pottery. (I am indebted to Dr. Munchaev for the latter information.)

24 *Inter alia*, Aschmann, H. (1962). Evaluations of dry land environments by societies at various levels of technical competence, *in* Woodbury, R. B. (ed.) Civilizations in desert lands, *Univ. of Utah Anthrop. Papers*, **62**, pp. 7–8.

Perhaps some other name should be applied to the ancient Mesopotamian implement usually called a "plough". In appearance it closely resembled a wooden "plough" still in common use in Iraq twenty years ago. This traced a shallow furrow which broke the caked surface of the soil, encouraging penetration of moisture in fallow seasons and the rooting of broadcast seed, but did not discourage weeds by turning the sod. In function it has largely been replaced by tractor-drawn cultivators or harrows. I agree with Allan, W., this volume, p. 224, that deep furrowing was only accomplished at a much later date with iron ploughs, but I very much doubt that such implements were in common use even with the more general introduction of iron in the first millennium B.C.

25 I am indebted to Dr. Sàndor Bökönyi for the identification of domesticated cattle bones in the Samarra levels at Choga Mami and for information concerning similar bones in comparable levels at Yarim Tepe.

26 Reed, C. A. (1969). The pattern of animal domestication in the prehistoric Near East, *in* Ucko, P. J. and Dimbleby, G. W. (eds.) *op. cit.* p. 375.

27 Flannery, K. V. and Cornwall, I. W. (1969). Fauna from Ras al Amiya, Iraq: a comparison with the Deh Luran sequence, *in* Hole, F., Flannery, K. V. and Neely, J. A., *op. cit.* Also, Flannery, K. V. and Wright, H. T. (1966). Faunal remains from the "Hut Sounding" at Eridu, Iraq, *Sumer*, **22**. In the Samarran levels at Choga Mami domesticated cattle bones constituted only 1% of the total, in the slightly later "Transitional" levels, 3%, in the still later Halaf well, 9%. A figure of only 1·2% for the 'Ubaid well is clearly out-of-line, but one suspects equally that the very high percentages, at least at Eridu, may to some extent reflect the size of the bones which would serve to make them more obviously and easily collected.

28 *Inter alia*, Wittfogel, K. A. (1957). *Oriental Despotism: a Comparative Study of Total Power*, New Haven, and more recently, Millon, R. (1962). Variations in social responses to the practice of irrigation agriculture, *in* Woodbury, R. B. (ed.) *op. cit.* pp. 56–88; and Adams, R. M. (1960). Early civilizations, subsistence and environment, *in* Kraeling, C. H. and Adams, R. M. (eds.) *City Invincible*, Chicago, pp. 269–95.
29 Adams, R. M. (1960). *op. cit.* p. 280.
30 Oates, J. (1969). *op. cit.* p. 143.

BARRIE REYNOLDS

Kwandu settlement: isolation, integration and mobility among a south-central African people

The Kwando River, a major tributary of the upper Zambezi, lies to the north of the Kalahari Desert (Fig. 1). Its middle reaches today mark both the border between Angola and Zambia and also the south-western boundary of the Barotse or Lozi Kingdom which is focused on the upper Zambezi flood plain, some eighty miles to the east. These reaches support a small population of intermingled groups of Bantu-speaking peoples together with a few Bushmen. Most of these groups belong to the Luyana complex that spreads throughout much of Barotseland and therefore speak related languages or dialects.

The Kwandu[1], who, as is common among Luyana, have adopted a locality name (AkaKwandu—people of the Kwando) based on the primary feature of their natural environment, form one of these groups. Like most of their neighbours they are a riverine people living in small scattered villages, extended families of between ten and one hundred adults and children. Traditionally, hunting and fishing dominated the economy but in this century dependence has moved steadily to cattle and gardens, initially hoe-cultivated but now increasingly worked with plough and oxen. Simple mat shelters have given place to roofed huts and the villagers are today abandoning their island homes deep in the reeds to settle along the banks of the river.

Economically speaking, the middle Kwando valley has little to offer a subsistence people that is not available, usually in greater abundance, on or closer to the upper Zambezi itself. The particular advantage of the Kwando lies in its isolation, both by a wide swathe of uninhabited and in parts waterless bush to the east and, within the valley itself, in the dense reed beds with their hidden islands and mounds linked by narrow twisting channels. For fugitives and small weak groups it was long a haven. When strong forces appeared in the valley, the response of the inhabitants has always been either to hide deep in the reeds or, if this were not feasible, to acknowledge the authority of the intruders only as long as they were present. The isolation of the Kwando ensured that such visits were, until recent years, infrequent and, in the intervals, the Kwandu continued little disturbed.

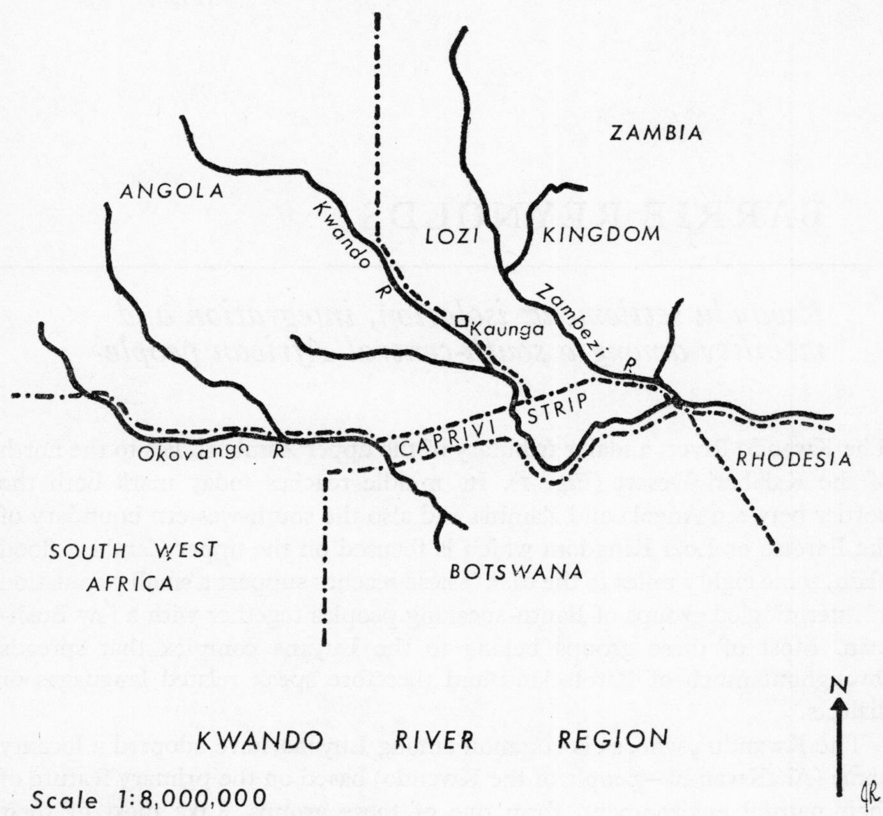

Fig 1 The Kwando River region.

The Kwandu, perhaps in common with many of their neighbours[2], are the product of various and varied immigrations, absorptions and intermarriages. They did not, as some of their myths of origin would have us believe, arrive in the valley as an already formed "tribe"; rather they have grown around a nuclear few Luyana immigrants who probably formed part of an early Luyana migration wave into the upper Zambezi region as a whole. These early immigrants drove out or, to a limited extent, absorbed the Bushmen already in the valley and established a ritual suzerainty over the land that is still recognized today. As other immigrants (often it would seem fugitives from political or minor upheavals in Barotseland) appeared, they were in their turn absorbed. Where a group was too large or unwilling to be so absorbed, they retained their own name (Shanjo, Fwe, Mbukushu, Simaa) and established separate villages.

The result today is a close intermingling of villages along the length of the valley, the inhabitants of which live amicably together and intermarry but still retain their separate traditions, language and identity. The extent of this intermingling may well be seen in Kaunga locality where of thirty-three villages the Kwandu provide fifteen; the remaining eighteen, which are interspersed with theirs, belong to eight other groups of both Luyana and non-Luyana stock. Between all except three of these (Lunda-Luvale villages) intermarriage and close friendships are common. The degree of intermarriage may also be seen from one Kwandu village in this same locality wherein, in 1962, of the eighteen womenfolk only five were Kwandu, the rest having been drawn from four other groups, some of them from distant parts of the valley.

It is perhaps this demonstrated ability to maintain, in spite of these influences, separate languages and identities that makes the Kwandu and their neighbours of such interest, though it should be noted that they are by no means alone in this in south-central Africa. The Kwandu do not function as a political or economic unit, being submerged in the common web of relationships that covers all groups, except perhaps Bushmen and Lunda-Luvale, in the middle valley. Their material culture and response to their environment are little different from those of their neighbours, cultural borrowings being common, though again a few are recognized as having come from a particular neighbour (for example, Mbukushu knives and baskets). Village structures are closely similar. Without local guidance, it would be difficult to distinguish ownership of old village sites or even of present-day gardens, except of course kitchen gardens around the villages themselves. For a child of a mixed marriage, membership of the Kwandu is again a matter of choice, depending on whether he wishes to stay in the village of his Kwandu parent or to attach himself to the family (Simaa, Kwandi, Fwe, etc.) of his other parent. Even where both parents are Kwandu, a man who moves far away from the river is deemed to be no longer Kwandu but AkaMushitu (people of the bush).

Earlier in this century the Kwandu were considered "very wild and at the same time, timid people—very rarely emerging from the swamps[3]. . . ." This timidity is understandable in view of the turmoil of nineteenth century Barotseland with its bloody history of invasion, insurrection, revolution and counter-revolution. Whilst the Kwandu were never directly involved, their valley was, for example, part of the no-man's land between the ousted Luyi to the north and the victorious Kololo to the south. Both exercised a casual authority over the middle Kwando. For the Kwandu, life must have been very much restricted to the islands whereon they established their small villages and cut tiny gardens. When danger threatened, the removal of their mat shelters and few belongings by canoe to other more inaccessible islands or mounds deeper in the reeds was a simple matter. Even in peaceful times shifts from island to island seem to have been very common.

Today, this restlessness is still apparent. At one point in the 1962 dry season sixteen of the thirty-three villages in Kaunga had moved or were planning to move. That a number, in the event, did not do so and that others moved abruptly, making the decision only a few days before their departure, underlines the restlessness that appears to be common among almost all people in the valley. Reasons for moves are various: one particular Kwandu village, the history of whose movements can be traced back to early in this century, began with the fission of the parent village on the west bank, the sons of the late headman moving away. Various moves in search of good village and garden sites, together with a brief move away from the river as followers of Lukama[4], and then, in recent decades, a gradual shift again out of the reedbeds to the east bank to avoid Portuguese control and, at the same time, to maintain contact with other shifting neighbours, summarizes a very complex movement history. Loneliness, tax dodging and escape from authority, freak floods, the search for better gardens and nowadays the desire to be closer to large ploughable sites in the bush and to the trade stores and migrant labour recruitment camps[5] all play their part as do also internal strains—fission, accusations of witchcraft and quarrelling—and even those created by too close a proximity to one's neighbours—conflict of economic interest and court disputes arising from cattle damage in gardens.

It is probable that with the shift from hunting and fishing to the cultivation of large bush gardens as the primary subsistence activity and with the concentration of villages east of the international boundary, movement may now be expected to diminish. Presumably, the change from mat shelters to roofed huts, with their greater investment of labour and materials, reduced to some extent the mobility of the Kwandu. By 1963 interest had begun in having Lunda-Luvale contract workers build more substantial rectangular huts for payment, either in money or cattle. It is reasonable to suppose that such very heavy investments form a strong argument against further sudden and frequent shifts.

Although relatively few in number and never of economic, political or military significance, the Kwandu are of interest in a consideration of settlement factors and patterns, especially in the way in which settlement in the valley has developed during the past two centuries and the response of the Kwandu to the particular natural and cultural stimuli that have helped to shape their present way of life; in the complexity of their inter-relationships with their neighbours yet maintenance of a separate language and identity; in their restlessness and the gradual changes in their subsistence activities and types of dwellings. These are all on-going processes and modern urban or white influences are merely a new group of factors in a long series. The Kwandu are adapting to them just as they have done to others in the past.

Notes

1 The work on which this paper is based was undertaken between 1961 and 1966 when the writer was on the staff of the Livingstone Museum, Zambia. The project was supported jointly by the Museum and by the Colonial Development and Welfare Fund.

2 On the rise of the Luyi (Lozi) see also White, C. M. N. (1962). The ethno-history of the upper Zambezi, *African Studies*, **21**.

3 Senanga District Notebook, 471.

4 The Lozi prince who ruled from Kaunga capital until 1925, when a clarification of the border dispute between the British and Portuguese placed the line definitely along the eastern side of the valley, thus separating the Lozi from the river; Lukama was then recalled.

5 Witwatersrand Native Labour Association recruiting for the mines of South Africa; this ceased in 1966.

BOHUMIL SOUDSKÝ and IVAN PAVLŮ

The Linear Pottery Culture settlement patterns of central Europe

Our knowledge of the settlement patterns of the Neolithic cultures of central Europe is based especially on evidence obtained from the Linear Pottery settlement of Bylany, 68 km east of Prague in East Bohemia. During fifteen seasons, an area of 6·5 ha has been excavated at this site. Only one fifth of the material obtained from these excavations has so far been analysed in detail. On the basis of this and the preliminary analysis of the rest of the material, in particular the pottery, it has been possible to date the various structures and features of the site to individual habitation phases. The evidence is particularly detailed for the Middle Neolithic (MN) period which in the west and middle of central Europe corresponds to Periods 2–4 of the Linear Pottery Culture (hereafter referred to as LnK), the Middle, Late and Latest LnK periods. In terms of the habitation phases at Bylany this corresponds to the period from "Violet I phase" to the "Dark Brown phase[1]".

The houses and their respective size

The settlements of the Linear Pottery Culture are characterized by the remains of large houses and contemporary features such as pits of various shapes, grain silos, ovens in pits, etc., which formed the domestic area of each house complex. The houses, judging from the evidence of the five rows of post-holes representing their frame, were rectangular in shape with a constant width of 5·5–7 m and varied in length from 7–45 m.

There are three main types of house-plan found in Linear Pottery settlements[2]:

(*a*) The most common and basic ground plan consisted of a tripartite house; the front section at the south-eastern end had the entrance, and possibly an upper storey for grain storage; the middle part was the main habitation area, and the back section at the north-western end was for storage and general domestic activities.

(*b*) A bipartite ground-plan which was derived from the basic tripartite plan. In this, the south-eastern part was merged with the central habitation area.

(*c*) The ground-plan consisted of only one section, the central habitation unit.

At each side of the houses were longitudinal construction pits used subsequently for domestic activities. Taking these into account, it is possible to calculate the external area of a house complex as well as the internal area indicated by the ground plan (Fig. 1):

(i) The external area of a house does not usually exceed 5 m width beyond each lateral wall of the house. No subterranean feature associated with a

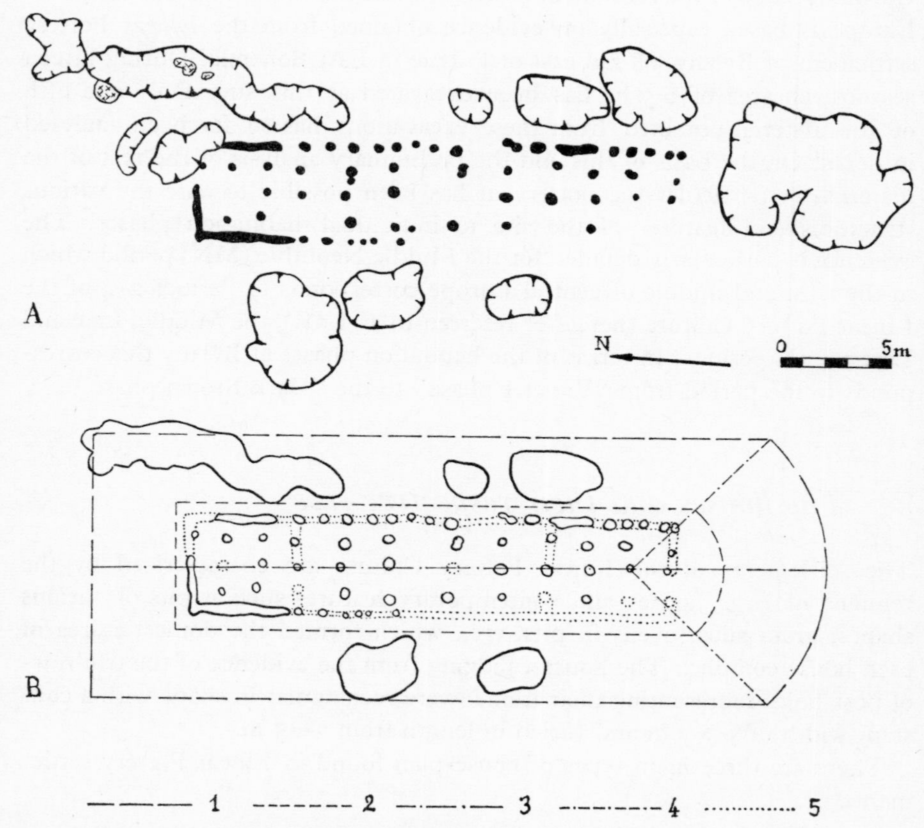

Fig 1 Bylany, house 88 (period II of Lnk, phase LR):

A. Situation: the house and pits in the vicinity; B. Reconstruction: tripartition of the internal area, pits belonging to the complex according to analysis; 1—limits of external area; 2—limits of the nearest external area; 3—longitudinal axis; 4—limits of the front area; 5—reconstruction of internal area of the house.

particular house (with the exception of sporadic drainage pits) was found within 1 m of the lateral walls of the houses.

(ii) The domestic activities carried out in front of a house took place in a semi-circular area whose centre was situated on the longitudinal axis of the house, and whose periphery cut the corners of the interior ground-plan of the houses. In this area there were no subterranean features. (Any pits which do occur in this area belong to other houses from other phases.)

(iii) The area at the back of the house was also never more than 5 m beyond the back wall of the house. It seems to have been used, however, only in the earliest phases (e.g. the pit with a pottery kiln behind the back wall of House 2223).

The houses of each phase at Bylany occupy an area of about 2½–3 ha. They are distributed at least 15 m from each other, and generally over 20 m. Occasionally they form a double semi-circle (Fig. 2), but generally they form staggered rows, which provided the maximum area of undisturbed space along the sides of each house, while at the same time preserving a relative coherence in the village. There were generally about eight houses in each habitation phase in the Middle period of the LnK (LGr. phase 8, DGr. phase 8, LR phase 7). In the Late period there were generally seven to nine. It has not been possible to ascertain the exact number of houses in the Latest period of the LnK at Bylany, although there were at least five in one phase, and at least six to seven in another.

The dominating feature of the settlement in each habitation phase is the clubhouse (communal house). This was identical in ground-plan to the houses of the basic tripartite type. Often this traditional method of constructing houses was retained in the clubhouse long after the feature of an upper storey granary had disappeared from dwelling houses (because grain was stored in subterranean silos). The converse also occurs, however, in that a long dwelling house (e.g. House 88 in the LR phase) may have no upper floor granary in a phase when this was a normal feature of dwelling houses. Generally the postholes of the clubhouses and the impressions of the posts within them were larger than those of normal dwelling houses. In addition, the houses were of abnormal length, generally 38–45 m long, although shorter clubhouses also probably existed. (It should be remembered, however, that (*a*) abnormally long dwelling houses also occur, e.g. at Postoloprty[3]; and (*b*) there is not always a correspondence between greater length and greater solidity or monumentality of construction, e.g. House 877 is 34 m long but not of solid construction, whereas House 679 is only 30 m long but of very solid construction.) The external area in front of the clubhouses seems to have been much larger than that of dwelling houses, and may have served as a meeting place. In two cases, and possibly in more, an enclosure was attached to one side of the clubhouse, which may have served as a communal cattle

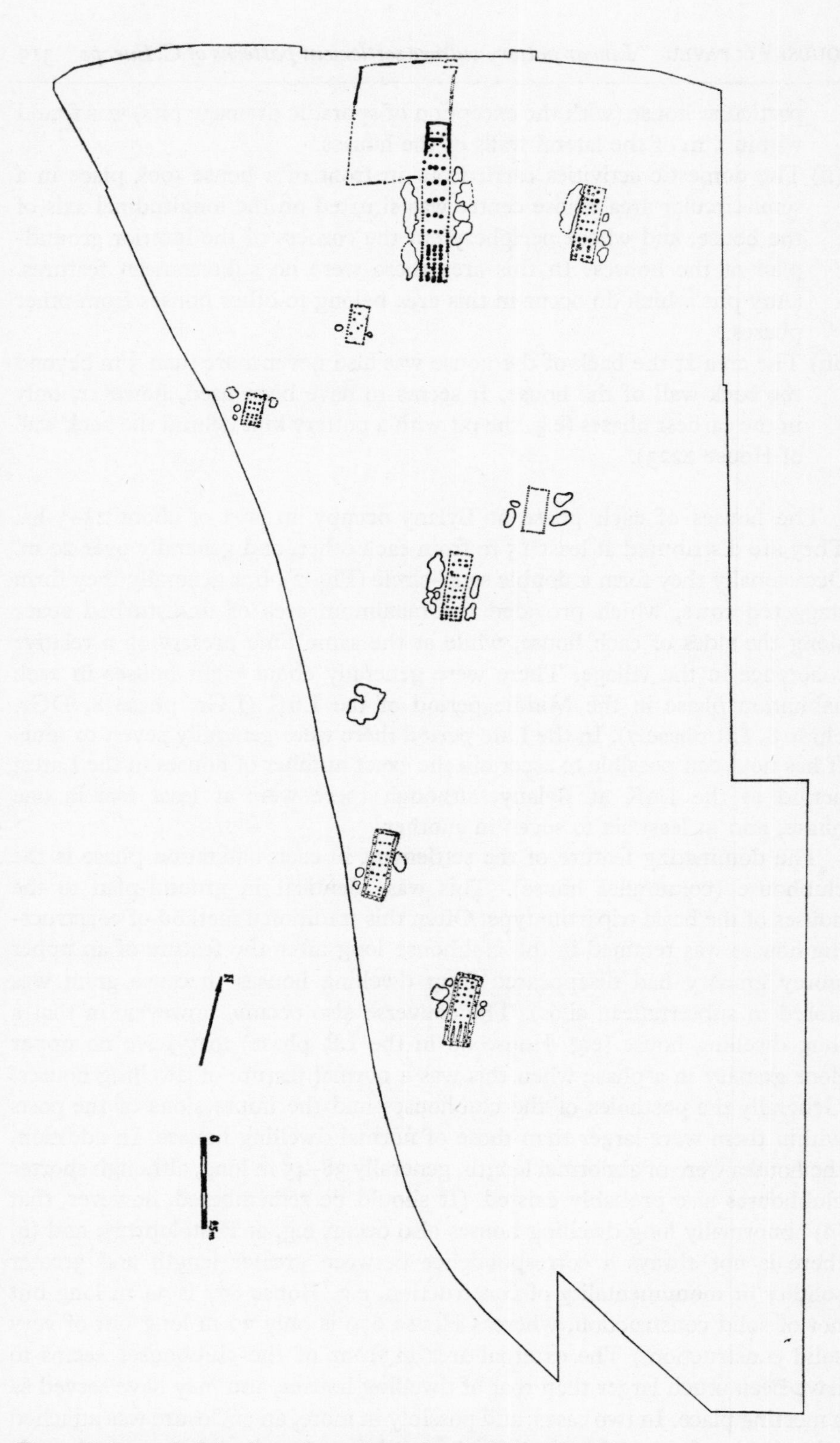

Fig 2 Bylany, habitation phase LGr (period II of Lnk).

kraal. One of these was 30×25 m, the other was 17×12 m. So far it is not possible to prove that these large clubhouses were located at the centre of the village, but the function of such a house is well-known from modern ethnographic societies as a place for meetings to discuss matters relating to the whole village (village councils, annual ceremonies, etc.), and also as a place in which activities which were beyond the scope of a large family household (e.g. manufacture of stone implements) were carried out. It is possible that the same functions applied also to the clubhouses of the Linear Pottery settlements.

The size of the houses and number of their inhabitants

The range in the length of the house in the three phases quoted in Table 1 is almost identical. The only possible interpretation is that the length of the houses depended on the number of families inhabiting them. The average minimum size for a one-family house in these three successive phases is 10 m, 11 m, 7–8 m. A comparison with houses of later phases supports this. For example, the minimum size in the YI phase (period III) is 10 m, in the YIII phase is 7–10 m, LBL phase is 6 m, and DBL phase 8–9 m.

The two-family houses should be double this minimum length, i.e. 14–16 m. There were probably other factors influencing the length of houses, such as the number of children or unmarried members of the household, the extent of supplies to be stored, etc. We see the same situation in the houses of the Tripolye culture settlement of Kolomiiščina in the Ukraine SSR, where the number of families is thought to correspond to the number of ovens. The length of the houses, however, does not entirely correspond to the number of ovens; there is a certain amount of overlap. For example, houses

(*a*) The distribution of houses in the LGr. phase (period II of LnK)

Clubhouse 41: 45 m long	House 211: 18 m long
House 680: 20 m long	House 683: 10 m long
House 999: 19 m long	House 427a: 9 m long
House 19: 19 m long	(?House Y: ?12 m long)

(The postholes of House Y were not preserved, but their presence could be reconstructed from the perfect system of construction pits.)

(*b*) The distribution of houses in the DGr. phase (period II of LnK)

Clubhouse 703: 45 m long	House 132: 19 m long
House 165: 32 m long	House 426: 16 m long
House 678: 28 m long	House 764: 16 m long
House 224: 20 m long	House 681: 11 m long

(*c*) The distribution of houses in the LR phase (period II of LnK)

House 702: 34 m long	House 277: 18 m long
House 88: 27 m long	House 85: 8 m long
House 174: 24 m long	House 739: 7 m long
House 581: 22 m long	

Table 1 House lengths

of the same length can contain two or three ovens. We have also attempted to calculate the number of families inhabiting a Linear Pottery house by the number of ovens originally situated in the house, on the basis of the distribution of oven fragments of baked clay in the postholes inside the houses. So far, this method has not proved too successful; for example, on the basis of its length, House 1111 in the YI phase would be interpreted as a four-family house (length 31 m), but on the basis of the number of reconstructed ovens, it should be a three-family house[4]. Thus our interpretation of houses of 6–11 m length as one-family houses, and those of 12–16 m length as two-family houses is very tentative. Similarly those houses of 18–24 m length would seem to be three-family houses, and those of more than 25 m would be houses for four or more families, unless they were clubhouses.

Phase	LGr.	DGr.	LR	YI[a]	YIII[a]	LBL[a]	DBL[a]
No. of houses							
One-family, 6–11 m long	2	1	2	2	3	3	3
Two-family, 12–16 m long	1	2	—	—	5	2	2
Three-family, 17–24 m long	4	2	3	3	—	2	1
Four-family, 25 m and longer	—	2	2	2	1	1	—
Total no. houses[b]	7	7	7	7	9	8	(6)
Total no. families	16	19	19	19	17	17	(10)
Clubhouse	1	1	?	?	1	1	1

(*a*) Y I phase:
House 1111: 31 m long
House 1246: 29 m long
House 16: 19 m long
House 982: 19 m long
House 1199: 17 m long
House 1116: 11 m long
House 1161: 10 m long

YIII phase: ?Clubhouse 679: 30 m long
House 877: 34 m long (? clubhouse)
House 993: 15 m long
House 245: 13 m long
House 149: 12 m long
House 580: 12 m long
House 740: 12 m long
House 225: (>7 m long)
House 263: 10 m long
? House 1240: 10 m long

LBL phase: ?Clubhouse 96: 23 m long
House 162: 28 m long
House 903: 23 m long
House 682: 18 m long
?House 983: (>8 m long)
House 959: 13 m long
House 79: 7 m long
House 190: 7 m long
House 250: 6 m long

DBL phase: Clubhouse 912: 40 m long
House 166: 23 m long
House 965: 14 m long
House 434: 12 m long
House 81: 10 m long
House 133: 8 m long
(House ?: 9 m long, no postholes, delimited by construction pits 1137, 1139)

(*b*) In the LGr. and DGr. phases, there is evidence of a clubhouse, but in each phase there is probably one dwelling house missing. In the LR phase, both the clubhouse and probably one or two dwelling houses are missing. In the YI phase we cannot identify a clubhouse, and several dwelling houses are missing. The YIII phase is probably the most complete, but it is unclear whether the clubhouse is House 679 or House 877. In the LBL phase it is difficult to identify the clubhouse, but it may be House 96, and there are several dwelling houses missing. The DBL phase is very incomplete; it has a clubhouse, but several dwelling houses are missing.

Table 2 Length of houses and corresponding number of families in various phases.

These calculations must, however, take into account the frequently tripartite and bipartite nature of the house plans, as described above. If only the central section of the house comprised the living quarters, the other domestic and storage sections of the house would not necessarily correspond in size to the number of families inhabiting the house. For example, a one-family house may not have all three sections, and occasionally the domestic sections of two-family houses may be longer than those of a three-family house.

From Table 2, it would seem that families in the LGr., DGr., LR and YI phases were distributed predominantly in three-family houses, less frequently in houses for more than three families, and very rarely in one- and two-family houses.

Phase	LGr.	DGr.	LR	YI	YIII	LBI	DBI
% houses							
One-family	28·6	14·3	28·6	28·6	33·3	37·5	50·1
Two-family	14·3	28·6	—	—	55·5	25·0	33·3
Three-family	57·2	28·6	42·9	42·9	—	25·0	16·6
Four-family	—	28·6	28·6	28·6	11·1	12·5	—

Table 3 Percentage of distribution of families to houses in various phases.

It is also interesting to note in Table 3 that there is a continuous and steady increase in the number of one-family houses. There is a corresponding decrease in the number of houses for three and more families. A departure from this "normal" distribution of families described above may be seen in the "anomalous" phases, when the extended families were abruptly reduced in size and occupied predominantly one- and two-family houses. These sudden reductions in the size of families must be indications of attempts to conserve a constant number of inhabitants in the settlement.

The economic and ecological factors affecting the size of the population of the settlement

We saw in the well-documented phases (cf. Table 2) that there was an average number of twenty-four families in each phase, which corresponds to about fifty adults and one hundred children, each nuclear family consisting of an average of two adults and four children[5].

The principal basis of the economy of the Bylany MN village was the cultivation of cereals, with rather less importance attached to legumes, and the supplementary basis was animal husbandry. The cereals comprise 80% emmer wheat (*Triticum dicoccum*), 15% einkorn wheat (*Triticum monococcum*), and 5% other sub-species of wheat (*T. compactum* or *vulgare*). Among the legumes there is evidence of peas and lentils. The exploited animals were almost exclusively domesticated, 80% of which were cattle, and the rest were sheep, goats, and pigs. Wild animal bones make up only 1%, which is hardly more than a random occurrence.

The size of the cattle herd can be calculated on the basis of the size of kraals or enclosures which in two cases have been excavated next to club-houses, as mentioned above. The smaller of these, from the DBL phase, is 17×12 m. In a kraal of this size, a maximum of fifty head of cattle could be stabled, on the basis of the modern standard of 4 m^2 ($2{\cdot}5 \times 1{\cdot}5$ m) per animal. In this herd, there would have been at least two bulls, and probably about sixteen cows and thirty-two young animals[6]. From this herd it would have been possible to obtain a maximum of 2260 kg of meat per year[7], i.e. 188 kg per month for 150 persons, or 1.25 kg per person per month. This would have provided sufficient proteins if supplemented by legumes. The calculation of a total of fifty head of cattle for the herd, however, would seem to be rather exaggerated. The actual number would probably have been between fifty and twenty-five. This would have provided each person with a ration of only about 1 kg per month, which is hardly the nutritional minimum. However, this small enclosure belongs to phase DBL, which is "anomalous", and probably represents a reduced population of the settlement. The second enclosure belongs to the LGr. phase, and is 30×25 m (750 m^2). This is one of the phases with the "normal" distribution of families to houses. The enclosure is over three times larger than that of the DBL phase and presumably represents the size of herd for the "normal" population. The size of the herd, calculated at $37{\cdot}5 \times 2{\cdot}5$, would be about ninety-four head of cattle, which would have assured a sufficient supply of proteins. The ratio between the size of the two enclosures indicates that the division or filiation of the village was associated with a division of the herd.

To calculate the ration of wheat, we must start from the number of inhabitants and from the fact that flour was probably their only source of carbohydrates. The poorest Russian peasant in the nineteenth century A.D. needed an annual 280 kg of rye and 210 kg of potatoes. This would be the nutrient equivalent of 270 kg of wheat (= 189 kg of flour). This could very likely be applied to the Neolithic population of central Europe. On the basis of detailed observations in Austria in the nineteenth century A.D., in the poor area of Steiermark, Hlubeck calculated that women consumed two-thirds of the men's ration, and children consumed half[8]. This calculation may be applied to evidence from the Bylany settlement: $(25 \times 270) + (25 \times 180) + (100 \times 135) =$ 24,750 kg = 247·5 quarts. From slash-and-burn intensive agriculture, one would expect relatively high yields. In the nineteenth century A.D. in Russia, a family produced, with slash-and-burn cultivation and without the use of a plough, 10–12 quarts of wheat from 1 ha of land. At Bylany, it is likely that the yield was about 10 quarts per ha. Of this about 2 quarts would have been reserved for the next sowing. Each family would have to cultivate and harvest 1.23 ha of land to obtain the necessary 9.9 quarts of wheat. This would have been quite possible for the inhabitants of Bylany[9]. Thus we may calculate that about 30 ha of land was needed for the cultivation area at Bylany, and 2–3 ha for the occupation area.

The 30 ha of land could be cultivated for only three to four years with a full yield. After a maximum of four years, another area of the same size would have to be prepared by slash-and-burning, that is if the village existed on the same site for more than four years. We know that the duration of a phase was approximately fourteen years. This has been deduced from the maximum number of attested re-linings of subterranean silos; it was definitely no less, and was probably not much more. During these fourteen years the yields would have been changed twice, giving a total of 60–63 ha. This is the exact area of the "a/b" site at Bylany, probably also the "d/e" site, and possibly the "c" site, assuming that the only soil cultivated by the LnK people was the typical chernozem soil on a loess base, without any loamification or illimerization. It should be added that there were numerous ovens for drying grains of wheat, particularly emmer wheat which has a tough hull. These ovens were not regularly distributed in the settlement, but occurred in clusters, presumably to serve the whole community rather than individual households.

The village as a micro-area

The alternation of two fields, one being cultivated and the other lying fallow, would only assure a high yield for a limited period, probably not longer than fifteen to eighteen years. It would then have been necessary to allow not only the soil, but also the forest, to regenerate. A new site would have to be found for the settlement, with the same extent of cultivable soil. At Bylany, we have discovered three to four settlements ("a/b", "d/e", "c" and "P" which possibly alternated with "c"). None of these sites was occupied at the same time, but in a cyclical sequence. The micro-area thus formed by all these sites, acting as the occupation area of one community was 1·2×0·8 km. The repeated abandonments of the sites are reflected in the record as hiatuses. Since the duration of one habitation phase was fourteen to fifteen years, the interruption in the occupation of each settlement must have lasted about double this period, i.e. thirty years. This would have enabled the oak trees to attain the thickness of 11 cm which appears as the normal thickness of posts in the house structures of the Linear Pottery settlements. This system of slash-and-burn agriculture has been referred to as cyclical. In each of the three sites in the cycle at Bylany, each family needed 2 × 1·23 ha. To calculate the total area of the micro-area of the Bylany village, this number would have to be multiplied by its twenty-four families. It is unlikely that the same number of families, or the same number of sites in the cycle, was constant in the MN villages of central Europe. The consistency would have been in the ratio between number of families and size of the cultivation area.

The MN village of Bylany compared with the EN and LN villages, and contemporary MN villages in other parts of central Europe

It would seem that the economy of Middle Neolithic settlements was fully food-producing, in contrast to the Early Neolithic settlements in which the main part of subsistence was provided by food-production sources, but a small part was still obtained by hunting and/or collecting. In the Late Neolithic, on the other hand, a slight superproduction was achieved, but generally this was not exploited by exchange of the surplus. If, however, we compare the Middle Neolithic situation, as seen in the evidence from the LnK settlement at Bylany, with that from the Early Neolithic evidence from the same site, there seems to be very little difference: the ratio of wild animal bones is more or less the same. There may be a difference in the number of families, but there is no definite evidence yet. Thus we must accept the economy of the late EN village at Bylany was already MN, and only the cultural superstructure remained Early Neolithic. This would indicate a certain amount of retardation, which is typical of secondary Neolithic cultures. Primary[10] Early Neolithic cultures, for example the early Körös culture, have a much larger percentage, about one-third, of wild animal bones.

The basis of the economy of the subsequent Late Neolithic Stroke-Ornamented Pottery culture (StK) is still obscure. There was probably an increase in stock-breeding, as may be seen for example in the Dutch settlements dating to the latest LnK/StK period (= Elsloo type), since the enclosures were much larger and more solid than in the MN village at Bylany. In the Middle StK village at Bylany, relatively wide deep ditches have been excavated (on the "d/e" site) surrounding what would have been an enormous enclosure. In addition there was an increase in the number of wild animal bones during this period. However, there need not have been any associated change in cereal cultivation, or in the distribution of houses.

At the end of the Late Neolithic, in the so-called Late Lengyel Horizon, it is possible that hoe-cultivation was introduced. The palisaded enclosure at the "a" site with its solid entrance, would have provided space for several hundred head of cattle. In addition, the three houses which have been found at this site were constructed several hundred metres distant from each other, possibly as a protection for the cultivated areas. We do not know whether this phenomenon is typical for the end of the Late Neolithic in general (i.e. for the rest of central Europe, including the Early Chalcolithic of Hungary and Slovakia) or for Bylany only. There was an associated change in the ground-plan of the houses in that, although they were still long and of large dimensions, they were built to a trapezoidal or an apsidal plan.

We are limited in any comparative study through the lack of research and large-scale excavations made on contemporary LnK settlements, e.g. Köln-

Lindenthal, Sittard, Müddersheim, Stein, and Štúrovo. However, it is interesting to note that, in the LnK settlements of West Germany, there is evidence that hunting was still important in the economy, as reflected in the high proportion of wild animal bones and the presence of hunting equipment such as arrowheads, which were absent in central Europe. This would indicate that although the cultural superstructure was clearly Middle Neolithic, the basis of the economy was not even Early Neolithic. Thus the LnK culture of West Germany may be termed a "tertiary cultural process"[11], when, because of unfavourable conditions, the economy lost some of the properties which in more advanced and more favourable regions are considered criteria for civilization.

Notes

1 In chronological order:

	Phase colours
Period II (middle)	(*a*) pre-optimal stage: Violet I, Violet II, Violet ?III, Orange, Grey I
	(*b*) optimal stage: Grey II (and III ?), Light Green (= LGr)
	(*c*) post-optimal stage: Dark Green (= DGr), Light Red (= LR), Dark Red (= DR)
Period III (late)	Yellow (= Y) I, II, III, Light Blue (=LBl), Dark Blue (DBl)
Period IV (latest)	Light Brown (= LBr), Middle Brown (= MBr), Dark Brown (= DBr)

This incorporates two changes to the previously published sequence:
(1) a new phase has been identified between R and YI phases. It is called LR on the "a" site and DR on the "b" site, and belongs definitely to the post-optimal stage of Period II.
(2) the LBr phase has now been definitely assigned to the Latest (Šárka type) period, whereas previously it was classified as final Late or transitional Late/Latest period.
The best-documented evidence of the settlement organization comes from the LGr and DGr phases of the Period II settlement, from the YI, YIII, LBl and DBl phases of the Period III settlement, and from the MBr and DBr phases of the Period IV settlement.

2 Modderman, P. J. R. (In press). Zur Typologie den linearbandkeramischen Gebäude, 1, Alba Regia Székesfehérvár.

3 Soudský, B. (1969). Etude de la maison néolithique, *Slov. Arch.*, 17 pp. 5–6.

4 Soudský, B. (1966). *Bylany*. Prague. pp. 33–36.

5 If there were three children of the same sex, this would give an average of six of both sexes to each family born to the preceding generation. To this number should be added dead children who would number rather more than one-third, i.e. about four. Thus each woman would have given birth to about ten children in her lifetime, which would seem acceptable for the average age of skeletons excavated in LnK cemeteries. Thus each family would have comprised two adults and four living children.

6 The ratio of adult cows to heifers and young bulls aged one to two years may be calculated on the basis of the fact that there have been two young

cattle to every adult cow, assuming that every cow had a calf per year. The herd would consequently have been composed of $(n+2)$ adults $+ 2n$ young animals:

$$\text{i.e. } 3n + 2 = 50, n = \frac{50 - 2}{3} = 16 \text{ adult cows.}$$

There would, therefore, have been about sixteen cows, two bulls, thirty-two young animals.

7 From the stock of fifty head of cattle, $\frac{n}{2}$ young animals were probably killed every year (assuming that $2n$ existed for two years, half of them being bulls which were not needed for breeding) and $\frac{n}{3}$ adult cows, assuming that they were killed after calving three times. It would seem from the evidence of the bones, that young bulls were killed when they were three years old, and could have rendered about 150 kg of edible meat. The adult cows would have provided about 200 kg of meat. Thus $\frac{n}{3}$ cows $+ \frac{n}{2}$ young bulls $= 2{,}260$ kg meat. For the calculation $37{\cdot}5 \times 2{\cdot}5$, i.e. average between 50 and 25 = 37·5, 3·88 (= 705 : 204) is reduced to 2·5 because the multiplication is surely not linear.

8 Hlubeck, F. X. (1846). Die Landwirtschaft des Herzogstumes Steiermerk als Festgabe für die Mittglieder der X. Versammlung deutscher Land- und Forstwirte, nach den Eingaben der Filialen der K. K. Steiermärkischen Landwirtschafts-Gesellschaft, *In Auftrage Sr. kaiserlichen Hochheit des Erzherzogs Johann Baptist*. Graz.

9 We would like to thank Dr M. Beranová, C.Sc., of the Institute of Archaeology, ČSAV, specializing into medieval agriculture; Ing Z. Tempír, C.Sc., Director of the Agricultural Museum, a specialist on prehistoric cereals, and another member of the Neolithic project of the Institute of Archaeology, ČSAV, Dr M. Zapotocká, C.Sc. Together with Doc Dr R. Pleiner, C.Sc., we held a special session at Bylany to discuss the problems of Neolithic agriculture, especially as applied to the evidence from Bylany. The calculations and conclusions quoted here are the result of this session.

The yields of about 5 quarts per ha of land were calculated from the account books of large eighteenth to nineteenth century A.D. estates (possibly compiled by managers who were defrauding) and reduced for earlier times. They were first published by Lom, F. (1960). Zur Entwicklung der Technologie der pflanzlichen Produktion im Feudalismus und im Kapitalismus (in Czech), *Sborník vysoké školy zēmědělské, A*, pp. 403–24; they were also cited in Kudrnáč, J. (1958). Die altslawische Getreidewirtschaft in den böhmischen Ländern (in Czech), *Pam. Arch.*, **49**, pp. 478–98. However, the calculations seem too low. Slash-and-burn agriculture is very productive, and intensive ploughless cultivation is more efficient than primitive plough agriculture. An area of 1·23 ha could be tilled by one family in 94·6 hours if working 1 km per hour with a digging stick, i.e. two weeks. The same area could be harvested in 311–414 hours on the basis of the calculation that 50 m^2 can be harvested with a flint sickle in 76–101 minutes. See Steensberg, A. (1957). Some recent Danish experiments in Neolithic agriculture, *Agric. Hist. Review*, **5**, pp. 66–73. For a family in which the man was also reaping, and the children were helping, this would have taken a maximum of 156–207 hours, i.e. two to three weeks; see also Neustupný, E. (1967). *K počátkům patriarchátu ve střední Evropě*. Prague.

10 Every culture, or sub-culture, is considered as primary to another, more peripheral, and as secondary to a less peripheral one.

11 Tertiary "minus" are cultures which lost civilization taxa, tertiary "plus" cultures obtaining them by acculturation.

W. WATSON

Neolithic settlement in East Asia

As regards primitive settled life the territory of China divides between the three main ecological zones of East Asia, which broadly correspond to the riverine systems of Amur, Huang-ho and Yangtze. It is only in the middle zone, the Huang-ho valley, that the majority of pre-metal settlements depended primarily on agriculture and only here do we find indications of the classical coincidence of food-production, pottery and stone-polishing in a revolutionary advance to Neolithic economy. Elsewhere, particularly in the north, fishermen's permanent settlements existed where no crops were planted or animals raised, so that acquaintance with agriculture cannot be inferred from the practice of other Neolithic arts. It is questionable, for example, whether agriculture was known in the earlier stages of the Neolithic succession of the Baikal region. Here the opening of the Neolithic period is fixed at the pre-ceramic Khinskii stage, known from sites on the 8–10 m terrace of the Angara river, and distinguished by the adoption of the bow and arrow. By the same token the appearance of bronze in the northern and southern zones cannot alone be taken to herald the dissolution of the characteristic Neolithic community. On the sites of the slab-grave culture in the north and in the late pre-Han culture of the southern Chinese seaboard bronze is too scarce to bear the inference of economic or social change in a significant degree.

In the Maritime Province of the Soviet Union comparatively advanced and settled communities of food gatherers were established on the flatter country of the left bank of the Amur and along the Ussuri, their livelihood based on large-scale fishing. Large fish, particularly *chevich*, *sima* and salmon teemed during the spawning season in the rivers flowing to the Pacific. The catch was exploited from fixed stations. It must be presumed that the fishermen also ventured out to sea (an undertaking only compatible with a settled life ashore) and were responsible for an eastward spread of fishing villages from the Amur to Sakhalin, Kamchatka and Japan. Even the bears of east Siberia took to a fish diet. In the archaeological record the evidence of the fishermen's activities lies in net-weights and stone clubs. In one of the earliest of these fishing settlements, excavated in 1934 on the island of Sucha in the Amur, the dwellings were constructed partly underground, wooden pillars rising from the floor of each pit to support a roof covered with earth. A village of slightly later date excavated on the same island consisted of hundreds of similar houses conglomerated into a honeycomb. Such sites are tentatively attributed to the third or second millennium B.C., though in view of the much earlier radiocarbon dating of some of the earliest sites of the Jōmon culture of Japan, the possibility of fishing settlements of greater antiquity on the Amur and in the Maritime province cannot be excluded.

Between the river Ussuri and the Amur mouth large shell-mounds accompany the settlements and seem to denote a decline in the standard of living as compared with such sites as those of Sucha island. Similar shell-mounds occur in Japan, apparently spreading through a region which earlier already shared cultural traditions. Why the mounds should appear at a date argued to be comparatively late in the history of the fishing village culture is a question not satisfactorily answered. Okladnikov suggests that they began to form in the late second and during the first millennium B.C. Some certainly date after 500 B.C. The stone working and pottery associated with these sites point to links along the Amur valley with Baikalian culture, just as the siting and nature of the settlements reflect a similar economic life. Their grooved pottery somewhat resembles the pottery of the Glazkovskii stage of the Baikal, while the shape of some handles and the remarkable use of red slip points to a connection with the eastern zone of the slab-grave culture of North China, and beyond to a technical influence from the Huang-ho Neolithic. There is a tazza reminiscent of the Chinese *tou*, pointing to an affinity with the Lungshan of eastern China, but no sign of the *li*-tripod which is held by some to be evidence of emigration into Manchuria from central China[1].

The southern part of the northern ecological zone was occupied during the first millennium B.C. by the slab-grave culture. The westernmost province of this culture is denoted by the distribution in eastern Transbaikalia (in the Selenga basin, along the Orkhon and Tola rivers, around Lake Konsogol) of the *mayaki* (beakon) tombs with their contents of Late Tagar affinity, and to these the tombs of the north Chinese province seem to stand in cadet relationship. An important distinction between the two provinces is the inclusion in the western one of a much stronger element of Scythic influence, as shown by knives and other bronze objects. But nomadic Scythic culture as such did not penetrate eastwards beyond the north-south line of the Ta Hsing-an range. To the east of these mountains the Scythic influence is perceptible only in a few inferior and atypical bronze knives of Scythic derivation, and in their bone and stone copies. Moreover this influence is attenuated progressively from west to east through the western, central and eastern sectors of the Chinese Slab-grave region. At some of the richer sites of this tradition all the signs are in favour of settled agriculture of long standing and of stockraising, in which hunting played a minor role, but elsewhere hunting and fishing may have predominated. The slab-tombs and their contents, allowance made for the eastward impoverishment, present a remarkable picture of cultural unity, over a vast region, which is somewhat at variance with the evidence for the character of economic life.

The slab-grave cemeteries of the western and central sectors, from Hopei to Chi-lin, are more or less uniformly sited on hill-tops or on steep slopes, and though the corresponding settlements have not always been identified, the clear example of the proximity of burial and dwelling places at two rewarding sites in the central sector suggest that this was the general practice. At Sao-ta-kou near Chi-lin in this sector the settlement is reported to have occupied an area of 52,000 m^2. At the village of T'u-Ch'eng-tzú in the same region were houses with sunken floors, having walls built of pisé in the Shang manner and the floors covered with a layer of white clay like the Lungshan houses of central China. Fireplaces inside the houses were built of stone slabs. Nearby were storage pits. The commonest domestic animals were dog and pig, but some cattle, sheep and a few horses were raised. Two kinds of millet have been identified in excavations. In addition to the normal polished stone-axes (*walzerbeil* cf. Yangshao) in the western and central sectors and oblong flat axes (cf. Lungshan in the eastern sector) there have been some isolated finds in the western sector of characteristic waisted flat stones, some 20–30 cm in length, which are interpreted as heavy hoes or ploughshares. Their presence in the west of the slab-grave region and not farther east perhaps indicates the more intense agriculture of the part most accessible to influence from central China, where however no such tool is found.

The signs of local variation are chiefly noted in the central sector. Mikami attaches importance to the observations that at T'u-ch'eng-tzú net weights were numerous and no stone sickles were found, while crescentic stone knives were abundant at Sao-ta-kou, net weights absent and arrow-heads fewer. He is inclined to account for this discrepancy by assuming that two populations inhabited the region. In that case we should have to assume also that they built the same type of tombs. More likely what is reflected is a local variation of economic activity, which would not disagree with the impression one gains from the tomb contents of a fluctuation of prosperity through a fairly uniform terrain, accompanied by some uniform features of material life. Stone knives do not however vouch by themselves for intensive agriculture, since their distribution, in the crescentic form known also to the Lungshan farmers, extends far to the north, e.g. Kamchatka, where the farming life can have offered little attraction in the middle centuries of the first millennium B.C. Villages placed on hilltops are perhaps a further indication of the limited nature of the slab-grave agriculture, particularly of stockraising. There is no suggestion, as there is in the case of the eastern Lungshan province of the Chinese central zone, that this siting was imposed by the impenetrability of the terrain[2].

The relative uniformity of this phase of settlement in the region of the slab-graves, albeit with an uneven reliance on food production, is in contrast to the variety and cultural heterogeneity of the earliest occupation of the Manchurian basin. Here, at a period prior to the rise of Slab-grave Culture (which takes in southern Manchuria but apparently not its northern part), a mosaic of cultures coexisted which were to varying degrees adapted to settled life and food-production. They retained features allying them to other cultures whose main distributions lie beyond the well-defined rectangle of the Manchurian lowland. The flat country along the Shara-Muren and Nonni rivers, with its pleasant seasonal rain and, for the north, more moderate winds, was fitted to attract the influence of the advanced Neolithic farming of the Huang-ho valley, but this influence did not extinguish local tradition. Unless one believes, with some Chinese scholars, that the presence of the *li*-tripod is itself evidence of specific migration from central China, there is no reason to postulate a massive transfer of central Chinese population to Manchuria during the Neolithic period.

The most striking evidence of early agriculture is found in the tract south of the Shara-Muren, in Jehol and Liaoning, particularly at the hillside site of Hung-shan-hou near Ch'in-feng on the Lao-ha river. Here remains of a black earth witness to a climatic phase at the end of the Pleistocene during which the region had forest cover. In the Hung-shan-hou period the climate was in all probability cool-temperate, with vegetation of the wooded-steppe kind, yielding grassland with a sparse growth of birch and spruce. The pottery of the habitation site betrays affinities to the potteries of both Yangshao and Lungshan (adding a difficulty to the current theory of the latter as wholly derivative and therefore wholly later than the former), but includes a form of straight-sided beaker with northern connections reaching to the lower Amur, Sakhalin and even Japan. The bones of sheep, pig and horse were excavated. The rectangular flat stone-axes of Lungshan lineage, the stone querns and rubbers, and, as found sporadically in the region, the stone 'ploughshares' are evidence of the primacy of agriculture in the economy. At Hung-shan-hou nothing could be deduced as to the forms of houses and village; but at the site of Tung-pa-chia in the same vicinity the stony foundations of round houses, with a preponderately large house at the centre, were surrounded by the remains of a rough stone wall enclosing a space measuring about 100 by 140 m. There was nothing in this to suggest southern architectural traditions.

On the north bank of the Shara-Muren are traces of habitation on the surface of the black earth, in the vicinity of the town of Lin-hsi, which reflect the adherence of the western part of Manchuria to the province of microlithic culture sited sporadically on the Gobi plateau. The pottery at Lin-hsi has net-markings, suggesting fishermen's work, which allies it to the Baikal region, and comb-marking which no less underlines a northern cultural affinity. While the finds indicate comparatively long settlement at the site, there are no signs of agriculture or even of polished stone. Between the Lin-hsi group of settlements and those of the Shara-Muren valley the eastern edge of the Mongolian plateau marked a sharp cultural boundary. Similarly in the upper Nonni valley the microlithic element is prominent in the stone equipment found on such a site as Ang-ang-hsi, where the carefully shaped arrow-heads and blades recall the stonework of the Baikal, as do no less the thick globular pots. The fourth division of Manchurian culture is typified by a site near I-lan on the Sungari, at Wei-k'en-ha-ta, where the stonework and pottery connect unambiguously with the shell-mound habitations of the Maritime Province in the Ussuri valley between Khabarovsk and Vladivostok.

There seems to be no clear reason of environment which would explain why the only considerable Neolithic development in Manchuria should have been confined south of the Shara-Muren river, and one is tempted to conclude that it was the occupation of the more northerly Manchurian lowlands by settled communities of food-gatherers with conservative tribal traditions which prevented the northward spread of farming Such conservatism would have been nourished by cultural affinities leading still farther north. But north Manchuria has been little explored by archaeologists[3].

Permanent villages were also a feature of the Jōmon Neolithic period of Japan, which from its beginnings in the eighth millennium B.C. to the arrival of the Yayoi rice-growers in the second century B.C. betrays no sign of food production. A complication of regional traditions is reflected in varying allegiance to sunken houses of round or square plan. On the whole the former belong to east central Honshū (Kantō) and the latter to west central Honshū (Kansai). The two forms never mix freely as they appear to have done on the Huang-ho. From a presumed family burial of five in a house at the Ubayama site, and from the six successive enlargements of a house at Kamifukuoka it has been calculated that the average houses had three to eight inhabitants and the settlements populations of 100 to 300. Excavations of settlements belonging to the middle three of the five divisions of the Jōmon period (i.e. embracing a period from the fourth to the second millennium B.C.) have demonstrated the persistence of a horseshoe plan. The houses are arranged along the edge of low plateaux and the area within the horseshoe is left free of buildings. At the Early Jōmon site of Minamihori in Hara-ken, two larger houses, holding, according to the formula, eighteen and twenty persons, were located towards the end of each branch of the horseshoe. At Togariishi in Nagano-ken, a site belonging to the Middle Jōmon, the line of houses along the south edge of a flattish promontory is bent in a wide curve, although the alignment on the north edge, 60–100 m away, is almost straight.

At these two settlements the shell heaps, of no great extent, were sited at some distance from the houses. In the Late and Final periods of the Jōmon vast accumulations of shell are the rule. They tend to be continuous along the contours of spurs and their position must reflect rather closely that of the dwellings. The hint of a duality in the village organization which is observed at Minamihori and Togariishi is repeated at the Late Jōmon settlement at Horinouchi in Chiba-ken. Two horseshoes measuring 200–250 m in length and opening respectively to north and south are located a short distance apart. The increase in the consumption of shell-fish, falling in second and early first millennia B.C., would appear to coincide approximately with the increase of shell mounds at the mouths of the Ussuri and the Amur in the Maritime Province. In Japan the change accompanies the shift from shallow-water to deep-sea shell-fish, and it is even argued that slight changes in the orientation of the horseshoe settlements are a consequence of the more serious sea-faring imposed by the altered demands of food-collecting. The new activity can only have increased the necessity of operating from fixed bases, and the Jōmon tradition of settled life was further confirmed[4].

The growth of Chinese Neolithic settlement in the central zone, the Huang-ho valley, was closely bound to the topography and the properties of the loess. In the primary distribution of this aerial deposit, in Kansu, along the Wei valley and the Huang-ho as far as the middle of Honan, the loess varies in depth from 300 to 50 ft, the greatest depths being in Kansu and northern Shensi. Through the peculiarity of its vertical fissure and its porosity the loess is apt to be deeply ravined, producing vertical cliffs separating many more or less horizontal areas. It is naturally very fertile and under rainfall has a capillary capacity whereby mineral elements in the subsoil are brought within reach of the roots of crops. Owing to its poor water retention it is unlikely that the loess was ever heavily forested, and movement over it even in early times must have been comparatively easy. Its porosity also controlled the siting of settlements, for small, easily approached and easily controlled streams are few. Agricultural use in the dry season demands much irrigation, yet the places on the banks where farmers might be tempted to settle are apt to be separated from the water by several metres of vertical cliff. These conditions obtain on the upper and middle Huang-ho and the tributaries in this region. To the east of the middle of Honan the riverine terrain consists in increasing measure of *secondary* loess, a fluviatile deposit of loess mixed with other soils and detritus. The fertility is no less and the accessibility to water greater, but the low or even raised banks of the river offer little protection against flooding and consequent meandering. The river's freaks must have caused no less travail to farmers of the Neolithic period than to those of more recent times.

The Yangshao red-pottery culture and the Lungshan black-pottery culture belong in the main respectively to the regions of primary and secondary loess. These cultures are defined primarily in terms of their ceramic traditions and the complexity of their origins is indicated by the very discrepant distribution of types of stone axes and knives, and methods of burial, none of these features coinciding geographically with the diagnostic pottery in the manner which might be expected in migratory settlement. The leading diagnostic potteries, the painted red and the polished black, are themselves by no means uniformly distributed within the Yangshao and Lungshan areas, and are variously associated with corded and smooth potteries representing cultural horizons of different import. Corded ware relates to an early and far-spread tradition, evidently pre-agricultural at first, which survives late in south-east and south China. The smooth grey pottery mainly distributed in central and east-central China, is present on many Lungshan sites, persists in modified forms in the Shang bronze age, and has given rise to the questionable concept of a "grey pottery people".

Within the area defined by their specific potteries the Yangshao and Lungshan cultures show divisions on which the problems of interpretation largely turn. Thirty years ago Wu Gin-ding[5] had shown that in central Honan a local sequence of Yangshao, Lungshan and Shang bronze age can be established stratigraphically. All the indications of diversity in the two traditions were then interpreted as evidence of two cultures independent in origin and progress, the Lungshan being generally later. The question of the Lungshan sequence, with its evident close connection to the Shang, was confused by Andersson's claim that the Kansu Ch'i-chia-p'ing culture of Lungshan affinity antedated the whole sequence of Yangshao stages he had identified. This has been rectified and the Ch'i-chia-p'ing shown to come later in the Kansu sequence of painted pottery, reinforcing the original view of the Lungshan as evolved in the settlement of the secondary loess on the lower course of the Huang-ho. While the point was not pressed by the earlier interpreters, it is clear that the geology of the Huang-ho mouth played an important part in limiting the extension as well as conditioning the growth of this cultural tradition. It is not certain that the lowest part of the fluviatile plain of the Huang-ho existed at all during the *floruit* of the Yangshao.

The interpretation of the Yangshao-Lungshan sequence currently adopted in China regards the whole as a single cultural affiliation, having its origin in a small area bounded by western Honan and southern Shansi. Sites with mixed contents which were previously accounted for in terms of "cultural contact" are now proposed as a transitional stage between the two main traditions ("Transitional Lungshan"), the new development having taken place in the same restricted region that saw the initial growth of the Yangshao. No site however has yet revealed a complete stratigraphical sequence from Yangshao through Transitional Lungshan to Lungshan, whereas in central Honan a fair number of sites show Lungshan overlying Yangshao without an identified Transitional level. The new interpretation facilitates a diffusionist account of Neolithic culture spreading in China from the heartland of the Han race, but tends to obscure the significance of local tradition, such as implied by the north-eastern connections of Lungshan, seen in its stonework and architecture, and makes somewhat unaccountable the fact that the most pronounced characteristics of the Lungshan in pottery and stone are confined to the east coastal region. Meanwhile with one exception it has not proved possible to make formal chronological subdivisions of either Neolithic tradition, so as to illustrate more fully the process of expansion and settlement, and no radiocarbon dating is available. We may assume that the antiquity of Neolithic tradition in central China is no less than in Japan, where radiocarbon dates in the late eighth millennium B.C. have been calculated for the beginnings of the Jōmon sequence. The exception concerns the western extension of the Yangshao into Kansu. The Yangshao culture of the Wei valley shows local variation but remains basically identified with the Yangshao of Honan. Its progress was arrested sharply on the north-south line of the eastern watershed of the T'ao valley in Kansu. In this valley and farther westwards into Kansu is located a province of painted-pottery Neolithic which by ceramic criteria appears closely related to the central Yangshao but still is clearly distinguished from it. On the borderline at least one site shows Kansu Yangshao indubitably stratified over central Yangshao, so that the relative lateness of the Kansu province is beyond doubt. In the latest stages of the Kansu painted-pottery culture bronze occurs in simple forms still insufficient to clarify a relationship to the Shang bronze age typologically or chronologically[6].

The Yangshao settlement of the primary loess is best studied in the Wei valley of south Shansi, where the line of the Ch'in-ling mountains marks the southern edge of the loess. South of the river the valley extends over distances from 20 to 50 km to the foot of the mountain range, and to the north expands up to 100 km, to end on a gentler upland slope. The Yangshao farmers usually chose sites on tributaries rather than near to the main stream, and for this reason sites were most numerous south of the river, where many short tributaries flow direct from the hills. Comparable positions to the north of the Wei necessitated still farther retreat from the main stream, and sites chosen there were fewer, lying in the western part of the Kuanchung passage (the right bank tributaries here being perhaps too torrential and high-banked) and far to the north on the upper waters and tributaries of the Ching river.

The sites are classified into three groups: (1) sites on low mounds, which are formed to some extent by the habitation refuse. These are mainly on the southern Wei tributaries, those nearest to the main stream being on the first terrace above the modern flood-plain; (2) sites to the north of the river, beyond its ancient flood plains and on the gentle slope of the much dissected Malan terrace of the loess. These positions correspond to those of the Kansu Yangshao sites west of the T'ao watershed. The sites are often at the edge of high loess cliffs, and their present surface is flat, lacking the mounds observed in the previous group; (3) sites in the Ching valley which sought out the headwaters of tributary streams in order to avoid the inconvenience of very high vertical banks in the loess[7].

The density of Neolithic settlement of the Wei valley was such that in places the reconnoitred sites are as numerous as the modern villages in a district, and it is evident that many modern villages occupy the same places as their Neolithic predecessors. The ancient settlements often occupied a position at the confluence of two streams, and often they were sited in pairs, facing each other across the stream at one or two kilometres' distance. Because of the lack of a relative dating of the settlements it is hazardous to generalize from the observations made at particular sites. There is little doubt however that the larger occupations on the middle Huang-ho fall towards the end of the Yangshao development. Some remains represent an agglomeration of villages. Often a settlement includes a distinct burial area, and some isolated cemeteries are thought to have served several neighbouring villages. The great depths of the deposits at many sites is evidence of prolonged and unbroken occupation. The signs of duration and of concentrated population raise questions of agricultural technique no less than of social order and cohesion. No trace of irrigation works has been found at excavated sites, yet it must be supposed that the large-scale cultivation of the staple millet (despite its drought-resisting properties) required regular organized effort to ensure the water supply from streams comparatively difficult of access

Only the village of Pan-p'o has been subjected to the large-scale excavation which the investigation of the problem warrants, one fifth of the habitation area of its 30,000 m^2 having been uncovered. Some thirty post-built huts of round and square plan were revealed and vouch for a total population of between 400 and 800, if we assume that the unexcavated part of dwelling area was similarly built on and that each house held two to four people. By the formula quoted above for the Jōmon villages of Japan the total would be nearer to 1000 inhabitants. The habitation area, with its houses, storage pits and quarters of domestic animals, was surrounded by a ditch 5 or 6 m wide. The signs of occupation extended beyond a further 20,000 m^2, forming an approximately oval area with its longer axis orientated north and south. In the northern part of this area was located the cemetery, and in the east were pottery kilns. The excavators do not hesitate, in the first sentence of their report, to designate the village a classical matriarchal gens commune, and point to the group burials in some graves, the separate burial ground of young children, and the absence of the elaborate tillage (irrigation?), the last indicating simple hoe cultivation and the specialization of women in food production. On the other hand evidence of hunting does not suggest that this—men's work—played a big part in the food quest. One unique square house with a floor space of about 100 m^2 is interpreted as a village hall.

The nature of Lungshan settlements is less well understood. The sites normally occupy the crest of low eminences on the plain, and there is no indication that the size or populousness of the villages approached those of the Yangshao area. The two most notable sites are in Shantung, each well furnished with the most diagnostic polished black ware and marking earlier and later stages of the east-coastal development. The late site of Ch'êng-tzŭ-yai (near to the eponymous Lungshan) is on a small earthen plateau raised some 15 m above the surrounding country and was closed by a wall

5 or 6 m wide at the base. There are indications in the pottery shapes and potters' marks, scapulimancy, etc. that the period of occupation lies close to the beginning of Shang. The typologically earlier site in Shantung, Jih-chao, shows closer analogies with the Lungshan culture of Honan as excavated at Miao-ti-kou and the neighbouring San-li-ch'iao, but no chronological criteria are available to tie the connection closer[8]. The lower strata at Miao-ti-kou (a site covering 24,000 m^2) contained traces of a Yangshao settlement with the classical painted pottery of the central tradition and the foundations of a square house, some 6 by 7 m, which closely resembled those of Pan-p'o. In the upper levels were Lungshan remains, including a house with round plan. The floor, sealed with white clay in the manner found farther east in the Lungshan sphere, was sunk about 1 m below the level of the ground outside. In contrast the Yangshao house at this site was lowered only some 20 cm below ground level. In the Yangshao tradition, as the excavations at Pan-p'o show, the sunken floor was still common, though it was not so deep in the more elaborate houses of the later levels.

The picture of Neolithic settlement which emerges from the considerations outlined above still fails to show satisfactorily how the eastern, Lungshan, province depended on the advanced farming of Yangshao. To treat the diffusion of advanced agriculture as the only issue is to gloss over all that separates the cultural traditions of the western and eastern regions of central China and to obscure complexities in the growth of food production and settled economy in the southern territories. Insofar as influence from a more advanced centre was the cause, food production can only have advanced from the middle Huang-ho to the south-west and south and, at a later stage, northwards. A wide zone extending south-eastwards from the middle Huang-ho, through Anhui to the lake region of Hupeh and Hunan and the mountainous Pacific seaboard of Fukien, is occupied from the earliest times by small settlements of hunters and fishers established for the most part on low mounds in flattish territory. To a varying degree many sites in Anhui and Chekiang show ceramic links with the Lungshan of the north, in the forms of vessels more often than the ware; but the most characteristic black burnished ware extends from Shantung only through Kiangsu to the Yangtze mouth, and is sparse. The difficulty in the way of deriving the southern agricultural economy from the north as a simple diffusion lies in the substitution of rice for millet as the staple, and all that this substitution portends in social and economic affinity to the southern sphere, which begins on the Yangtze and comprises all of lowland south China and much of lowland south-east Asia. Where palpable evidence for agriculture occurs in the south, a technical connection with the Yangshao is hard to discern. Thus in the lower level at Ch'ien-shan-yang in Chekiang, which had pottery showing resemblances to Lungshan, there was evidence of two sub-species of rice, pig and dog as the farming product, and of nuts, gourds and water plants as the yield of gathering or cultivation[9].

Up to the present such direct evidence for agriculture in the southern sphere has been too slight to warrant clear conclusions on the propagation of food-producing techniques. The importance of fishing and shell-gathering and of the hunt is everywhere demonstrated for archaeological levels which may reasonably be put in the Neolithic category. The total absence of chronological criteria in any part of the southern sphere before the general adoption of metal and the beginning of historical record still invalidates general reconstruction of the settlement pattern, but certain features already familiar must be accounted for in any scheme. The variously shaped stone knives associated with the northern Neolithic are not found south of the Yangtze. The only exception is the rectangular (and most widely spread) form which occurs in a limited area south of the Yangtze mouth. It is also found throughout the Yangshao realm. The crescentic type which relates most closely to the Lungshan distribution and extends far to the north, is hardly found south of Shangtung. The supine, east and north-east orientated burials and the flat trapezoid stone axes which are attributed to the Lungshan along a zone on the eastern seaboard, extend south of the Yangtze only in the region of the mouth. Conversely the segmented and shouldered stone axes of the southern sphere are on the Yangtze and south of it, and are quite unknown farther north.

South of the Huang-ho, apart from certain ceramic links in the northern part of this territory with the classical Lungshan, thus far only two archaeological horizons can be said to be generally valid. In the south-east the impressed pottery of ancient tradition (which shows no Lungshan affinities) changes rather suddenly from a soft, low-fired ware to a fabric verging on stoneware, the styles of decoration remaining little affected. This introduction of improved kilning seems to fall about the sixth-fifth century B.C. Of the two forms of southern stone axes the shouldered type is the later, and seems to have been first manufactured after the tenth century B.C. It has not yet proved possible to connect changes in patterns of settlement or food-production with these events, both of which would appear to have interesting economic implications. That no considerable village site has been discovered for excavation in the south may have less to do with the shifting nature of occupation than with the custom of building houses on piles. The recognition of habitation refuse in forms significant for settlement pattern calls for excavation on a scale not yet attempted in the southern sphere, where moreover the underground activities of insects and the chemistry of tropical soils in zones of heavy rainfall are peculiar challenges to the archaeologist.

The question of the mobility of the Neolithic communities within the cultural spheres outlined above has been variously answered. The suggestion of periodic occupation of the site accompanied by slash-and-burn cultivation was made briefly in the report of the Pan-p'o excavations, but not in the earlier report on the comparable sites of Miao-ti-kou and San-li-ch'iao. In fact the grounds for applying this concept to farming on the loess are quite dubious. The deposits of three and more metres' depth at the larger sites show no convincing signs of periodicity, and their contents, apart from the succession at places from Yangshao to Lungshan, are uniform. The self-renewing fertility of the loess would not be exhausted by primitive farming; the density of the farming population and the peculiar water requirements of the sites would discourage peripatetic tendencies, and the solid construction of the larger houses at Pan-p'o is itself a guarantee of permanent occupation. Communities of 500 to 1000 persons cannot have migrated so regularly as the slash-and-burn theory would suppose. Even at places in the northern zone the sites where agriculture predominated over hunting appear to be permanently settled. The possibility remains that the Neolithic peoples of the southern sphere led a more mobile existence, and in this region the sites where deposits suggest really prolonged occupation are comparatively few. But here too the change from a predominant reliance on hunting, etc. to reliance on rice cultivation can only have resulted in making village settlement permanent. In south China and Thailand, as elsewhere in the East, the most urgent problem facing the archaeologist is the dating of stages in the progress of rice cultivation and the construction of terrace irrigation. It is worth recalling Ssŭ-ma Ch'ien's description of the economy of the Yunnan kingdom of Tien as it was known to the Chinese in the second century B.C.[10] Crops were cultivated on the flats around lake Tien, but many tribesmen were nomadic, following their cattle in quest of pasture. The bronze models recovered from the necropolis of the Tien capital on Shih-chai-shan illustrate a village of houses raised on piles similar to those shown by pottery models of only slightly later date from south China. Such houses can never have been readily abandoned or transported. In the scenes cast on bronze vessels or represented by bronze figurines there is also the hint of an economic symbiosis of nomadic herdsmen and settled farmers which may hold a solution to the problem of agricultural organization in the uplands parts of the southern sphere. The difficulty of combining animal husbandry with rice cultivation is still fundamental in the agrarian economy of rice-eating East Asia.

Notes

1 cf. Okladnikov, A. P. (1956). Primorye v pervoe tysyacheletie do nashei ery, *Sovetskaya Arkheologia*, **26**, pp. 54f; Okladnikov, A. P. (1959). *Dalyokoye proshloye Primorya*. Vladivostock.

2 cf. Tsugio, Mikami (1961). *Mansen genshi funbo no kenkyū*. Tokyo; Watson, W. (in press). *Cultural Frontiers in Ancient East Asia.*

3 cf. Cheng, Te-k'un (1959). *Prehistoric China.* Cambridge. pp. 135f.

4 cf. Aso Masaru: Jūkyo to shūraku, in *Jōmon jidai* ed. Kamagi Yoshimasa, Tokyo (1965), pp. 232f.

5 Wu, G. D. (1938). *Prehistoric Pottery in China.* London.

6 cf. Chang, Hsüeh-chêng: Kan-su ku-tai wên-hua yi-ts'un, *K'ao-ku hsüeh-pao* 1 p. 60 no. 2 p. 11f.

7 Acad. of Sci. Institute of Archaeology, Peking (1963). *Hsi-an Pan-p'o* (The Neolithic village of Pan-p'o). Peking.
8 Ejusd (1959). *Miao-ti-kou yü San-li-ch'iao*. Peking.
9 Chekiang Cultural Properties Control Commission (1960). A report on the excavations of the first and second seasons of the site at Ch'ien-Shan-Yang in Wu-hsing-hsien, *K'ao-ku hsüeh-pao*, **2**, p. 84.
10 *Shih-chi* chüan **116**.

Part II: Factors influencing both non-urban and urban settlement

Section 1: Population, disease and demography

ROBERT BOYD

Urbanization, morbidity and mortality

It is sometimes supposed that an increase in population density will automatically lead to increased disease and a shortened life span. The qualitative and quantitative effects of urbanization on health are, in fact, extremely complex, depending on a large number of interrelated variables. Some of these variables will be considered in this extremely oversimplified account. An attempt will be made to decide what conclusions can be drawn about the health of a people from its pattern of urbanization.

Two scientific approaches have, over the last hundred years, produced a body of knowledge of the factors governing the amount of disease and resulting illness or death in a community. One approach is the experimental study of individual diseases in the laboratory or clinic. The other involves the collection of epidemiological data on mortality and morbidity. Although individual diseases may change in character and medical science may modify the effect of these factors, much the same principles apply to the spread of disease in London in 1970 as must have applied at other stages in man's history. I intend to consider three questions in particular. Is a rise in population density likely to lead to increased exposure to infection? Does increased exposure lead to increased death and disease? What sorts of data are available comparing rural and urban health?

First, it is necessary to consider one or two matters of definition. The death (mortality) rate is the number of deaths per year in a population of given size and composition. The impact of the death rate on the expectation of life in a society obviously depends not only on the total rate, but on the proportion of deaths occurring at different ages.

Many diseases clearly lead to temporary or permanent sickness short of death and the sum of sickness in a community can be referred to as the morbidity rate. In communities without modern medicine a large proportion of total mortality and morbidity is caused by infectious disease. In fact a "healthy" inhabitant of the tropics today is likely to be suffering from several chronic life-shortening infectious disorders simultaneously. These will be compounded with associated nutritional deficiencies. It is important to realise that infection does not always lead to disease. It is necessary to distinguish sharply between the incidence of infection and the incidence of death or disease.

Besides the "medical" morbidity considered here, urbanization may alter the morbidity and mortality from psychological stresses, accidents or chemical pollution. It is unlikely that these influences will be of quantitative importance except under unusual circumstances such as the very large-scale urbanization of recent times, or where the normally predominant infectious causes of death are suppressed by modern medical techniques.

Urbanization and infection

The spread of infection in a community depends on how easily one case leads to others[1]. If, on balance, one case leads to the infection of more than one other person (Fig. 1a) the infection will spread more and more widely.

Fig 1a *Fig 1b*

If Fig. 1b represents the state of affairs, the infection is likely to die out. The occurrence of situation A rather than situation B depends on the nature of the infection, the way it spreads and the opportunities offered to it. The balance between spread and limitation of spread is neither solely a feature of the disease nor of the population at risk but of their interaction. The main factors to be considered in predicting the result of this interaction are shown in Table 1.

Of the various social factors to be considered under heading C, increased population density will interact most clearly with the method of spread. As can be seen in Table 2, the method is very different for the different infections. The degree of crowding in houses, the disposal of sewage, the density of disease vectors and the amount of physical contact between individuals are respective crucial factors for the spread of each of these classes of infection.

A. Characteristics of a germ or parasite on which its spread depends	
1. Source and duration of excretion or infection	man, animals; cases briefly or permanently infective
2. Method of spread	droplets, food or water, vectors, soil, contact
3. Method of entry	inhaled, ingested, genitalia, skin
B. Resistance of exposed individual	
C. Social factors interacting with A and B	

Table 1 Factors in the spread of infection.

Let us first consider airborne infections. The spread of these depends rather precisely on the closeness and duration of proximity between infected and uninfected persons. When an infected individual sneezes millions of droplets containing bacteria are released into the air where they may be breathed in by someone else. Close proximity indoors is thus particularly relevant to the spread of such infections. It would be reasonable to deduce from evidence of crowded sleeping accommodation in a town or village that tuberculosis, measles or whooping cough would spread easily there, and there are modern reports of an association between crowded housing and a high incidence of these diseases[2].

The effect of urbanization on faecally transmitted infections will depend on the details of faeces disposal and especially its relation to the water supply. When water for a large number of people comes from a single source which may become contaminated with faeces, the risk of widespread outbreaks of this type of illness is high. If full details are available it is often possible to work out the chain of infection.

In the autumn of 1937 in Croydon there was an epidemic of 341 cases of typhoid with forty-three deaths. The water supply to Croydon at that time was from a number of chalk wells. All the cases occurring in the early weeks of the epidemic (that is before secondary cases had time to develop) had drunk water in the area supplied by one well. After exhaustive detective work it was eventually shown that one of eighteen workmen who had repaired that well shortly before the start of the epidemic was a carrier of typhoid and was passing the germ in his excreta. These had been the source of infection for 341 people[3].

1 Airborne (coughs and sneezes)	tuberculosis measles whooping cough
2 Faecal contamination, of food or water	typhoid diarrhoeal diseases intestinal parasites (worms and protozoa)
3 Vectors[4]	malaria (mosquito) filariasis (mosquito) plague (rat fleas)
4 Contact	syphilis, scabies, trachoma

Table 2 Predominant method of spread of some important infections.

Vector transmitted diseases are those which involve insects or animals in their spread. If urbanization leads to a better life for the vector it is likely to cause more human infection. As is well known, malaria spreads through the bite of a mosquito which breeds predominantly on water. The site and style of a town or village as well as the climate will all influence the amount of water available to the insect. The introduction of a plentiful supply of water without adequate drainage has led to the rapid increase of filariasis in some parts of India. This infection, the cause of elephantiasis, is transmitted by a mosquito much more commonly resistent to insecticides than the malarial mosquito. The effect of this particular social factor (e g. introduction of a water supply) can therefore be observed unclouded by the results of chemical control of mosquitoes[5].

It seems likely, on balance, that urbanization will lead to more spread of droplet and faecal infections. Any change in the incidence of vector and contact spread infections is less easy to predict. In all cases, the exact structure of the town or village both social and physical will be very important. The relation of individual diseases to various social factors is summarized in Table 3.

The relationship of domestic and other animals to the community will also be of importance in considering the many infections that can be acquired by animals as well as man.

Number of people sharing one enclosed sleeping place	droplet infections
Sexual partners and sleeping beds (shared or single)	syphilis, skin diseases
disposal of faeces, relation to water supply, preparation of food	diarrhoeal diseases and intestinal parasites
amount of stagnant and waste water	mosquito diseases

Table 3 Social factors of especial importance.

Infection and mortality[6]

Table 3 suggests a number of questions that can be asked in comparing life in a town with rural life or one type of town or village with another. It is possible to suggest that one situation will lead to more exposure to infection than another. However this does not necessarily mean more disease and a higher mortality rate. Although some diseases are commonly more severe than others, in almost every case there is a wide range of severity. A person infected with tuberculosis may die, he may become incapacitated by illness without dying or he may have no symptoms at all. The same is true for malaria, syphilis, cholera, typhoid, measles and a host of other infections. The severity of response to infection depends partly on the intensity of exposure to infectious agents. One million tuberculosis germs inhaled are more likely to cause grave disease than one thousand. However, various aspects of the exposed person which together make up his resistance to infection are as, or more, important. If urbanization is to make a large impact on mortality or morbidity rates it is more likely to do so by altering the resistance factors of the individuals in the town than by any changes in the amount of exposure to infectious diseases that may occur. These factors are summarized in Table 4.

1 Poor nutrition
2 Exposure at age of maximum susceptibility (varies with disease)
3 Pre-existing other infections
4 Absence of previous exposure to same infection
5 Other factors

Table 4 Factors reducing resistance to infection.

Measles is a well-documented example of resistance to the effects of infection being of much more importance than exposure in deciding the mortality rate[7]. Until the very recent introduction of immunization, virtually the whole world population was exposed to the virus and developed some degree of illness. Most children were infected before the age of ten years. Immediately before the introduction of immunization in this country the mortality rate was less than 0·1% of children infected, but in 1908 in Glasgow between 2% and 9% of children who caught the disease died. There has been little change in medical treatment of measles since 1908 and the decline in mortality was already well under way before the introduction of antibiotics.

The decreasing mortality cannot be attributed to a change in the virus either, for mortality figures are still very high in under-developed countries. Thus in Zambia "eight hundred children died in Lusaka Central Hospital last year. One third died of measles"[8].

It seems likely that improved nutrition is the most significant factor reducing the severity of response to measles infection and leading to the fall in measles mortality in this country since the turn of the century. It is likely that improved diet was also the most important single factor in the the decline of mortality from tuberculosis in nineteenth-century England[9].

Recently an attempt has been made to assess the impact of nutritional improvement on infections and other disease by direct experiment. Scrimshaw and his colleagues[10] studied three villages in the Guatemalan Highlands. The villages were kept under close health surveillance over a period of several years. In one village the children were given a dietary supplement, in one medical services were provided and the third "control" village was presented with various educational and recreational projects! The growth, health and mortality of village children was recorded. Compared with the "control" village, the "nutrition" village had a substantial reduction in child deaths from infection. However, it was striking that even given a large staff, a substantial budget, and a study performed at the time and not several thousand years later, it was very difficult to obtain clear-cut results. This example should be remembered before drawing firm conclusions from much more incomplete archaeological data. Nevertheless it is obviously of great importance to learn if urbanization leads to better or worse food supplies.

It has been seen that town life may lead to easier spread of infection and thus exposure to infections for the first time earlier in life. The effect of exposure at an earlier age on mortality is again complex. Measles is more severe at one year of age than later on, but is milder in young babies born, as they are, with some transient immunity to the disease[11]. The precise timing of exposure to infection is therefore important in setting the mortality rate for measles. If urbanization leads to exposure at three months of age the mortality rate might fall; if at one year the rate might rise! With tuberculosis the story is different and the mortality rate for children infected in pre-streptomycin days was more than twice as high in the first six months of life as a year later. This fits with data showing that crowded housing increases the death rate from this disease. Poliomyelitis infection on the other hand has a higher mortality rate in older individuals.

Other factors with a significant impact on resistance include the degree of infection with other diseases, past exposure to disease, sex and genetic make-up.

Urbanization and mortality

Quantitative data comparing urban and rural life are available in a number of forms. One approach is to study nineteenth-century mortality records. Figures are available for 1841 which show that the expectation of life in large conurbations was substantially lower than in rural areas (Table 5).

These figures are particularly valuable because they were collected at a time when medical influences were insignificant. Treatments were of little therapeutic value before 1900 and the sanitary reforms which led to good water and sewage systems did not get underway until well into the second half of the nineteenth century. Smallpox vaccination was the only effective medical tool available when registration of deaths in England and Wales began in 1837. These data indicate a very significant increase in mortality under conditions of active industrial revolution. Today the calculated expectation

London	35
Liverpool	25
Manchester	24
Surrey	44
England and Wales	40·2

After Glass, D. V. (1964). Some indicators of differences between urban and rural mortality in England and Wales and Scotland, *Population Studies*, 17, pp. 263–7.

Table 5 Expectation of life at birth (males) 1841 figures.

of life shows little difference between urban and rural England, the figure for London being 67.3 in 1950–2 compared with 66.4 for England and Wales as a whole.

Perhaps more surprisingly no increased urban mortality is found today in countries without widespread modern medical services. It is difficult to say if this is because death records are of uncertain reliability, because such medical facilities as there are are more effective in towns or because the English figures quoted dealt with large conurbations while the W.H.O. definition of urban includes much smaller towns with populations of about 5000 and over. There are twelve African Countries (excluding South Africa) for which W.H.O. provides statistics comparing rural and urban mortality. Of these twelve countries eleven have lower death rates recorded in urban areas and seven of nine with records have lower infant mortality rates in the town[12].

National statistics of uncertain validity can be filled out by more detailed comparisons made specifically to compare urban and rural health. One such study is that made by Mills in Sierra Leone. In 1959–61 he made a very thorough investigation of certain diseases in a sample of people from a town of 10,000 inhabitants. He compared them with the population of 1000 in four villages in the surrounding country and found better nutrition and less infection in the town[13].

Conclusions

In summary it seems likely that a change from rural to urban life will lead to more exposure to infection and perhaps to exposure earlier in life. As the severity of illness and the chance of death depends more on resistance to infection than on degree of exposure, this may have little effect on mortality. Altered resistance to infection through nutritional change is likely to override any direct effect of urbanization on the spread of infectious disease. Even this conclusion must be tentative, for in any medical analysis of this sort it is essential to remember the extreme complexity of the way in which one disease factor impinges on another making secure conclusions almost impossible.

Notes

1 Hobson, W. (ed.) (1969). *The Theory and Practice of Public Health.* 3rd ed. Oxford.
2 Mackintosh, J. M. (1952). *Housing and Family Life.* London.
3 Leff, S. (1957). *Recent Outbreaks of Infectious Disease.* London.

4 Vectors, especially insects, may transmit disease by inoculation or by contact. The infecting organism may or may not undergo an essential part of its life cycle in the body of the vector or of another host.
5 World Health Organization Technical Reports Series (1965). **297**, p. 50.
6 Scrimshaw, N. S., Taylor, C. E. and Gordon, J. E. (1968). *Interactions of Nutrition and Infection* (W.H.O. monog. series 57). Geneva.
7 Morley, D. (1970). Childhood infectious diseases, *in* Jeliffe, D. B. (ed.) *Diseases of Children in the Tropics and Sub-tropics*. London.
8 Savage, F. M. A. (1967). A year of measles, *Med. J. Zambia*, **1**.
9 McKeown, T. and Record, R. G. (1962). Reasons for the decline of mortality in England and Wales during the 19th Century, *Population Studies*, **15**.
10 Scrimshaw, N. S., Behar, M., Guzmán, M. A. and Gordon, J. E. (1969) Nutrition and infection field study in Guatemalan villages, 1959–64, *Arch. Environ. Hlth.*, **18**.
11 Morley, D. (1970). *ibid.*
12 U.N. Demographic Yearbook (1967). Tables 13, 18.
13 Mills, A. B. (1967). The effect of urbanization on health in a mining area of Sierra Leone, *Trans. Roy. Soc. Trop. Med. Hygiene*, **61**.

DON BROTHWELL

Community health as a factor in urban cultural evolution

"Disease has influenced the course of history, economic, social, and political".

Henry Sigerist. 1951. *A History of Medicine*

The main argument I wish to explore and develop here is that the people of earlier societies were caught in a demographic web, and that the common experiences of birth, death, and ill health had an influence on their behaviour and the way cultures evolved—perhaps especially the later urban ones. Medical history has long acknowledged its debt to archaeology for providing the matrix into which the epidemiological and other such data could be fitted. It is time now for archaeology to recognize, and draw into its historical reconstructions, the question of community health. Archaeology and medical history should have a mutual interest in ancient disease and demography. The difference in approach between the disciplines is that medical history is primarily concerned with the history of diseases and medicine *per se*, whereas the archaeologist is concerned with the relevance of health and disease in changing man, impeding settlement, influencing cultural procedures—economic, ritual, and so forth. Perhaps I should mention that palaeopathology, which has become fashionable again in recent years, is only a small part of this story.

In considering earlier societies, then, it is very relevant to ask such questions as: Did the community tolerate a heavy disease load? Which diseases may have prevailed and did any have a special influence? What was the life expectancy? How high was infantile mortality? Might there have been different disease patterns related to geographic background or social stratification? Could sanitation, habitation, diet, etc. have influenced community health? How did the community attempt to maintain and promote health or prevent disease (rational or ritual methods)? What of personal hygiene or public health practices? How did the group react to a disease (physical or ritual)? Did ill health of any sort initiate special "medical services"[7]? How large and concentrated were the populations, from the point of view of epidemic spread? Answers to these questions depend on a variety of information, including the actual skeletal evidence of age and disease, the ancient records, type of habitation, food debris, and special objects (such as surgical equipment) of the material culture. We can also make use, albeit guardedly, of recent "primitive" communities and by comparing broadly similar cultures and environments, fill in details on the possible health and medical limitations of early groups by reference to the modern ones. Comparison with modern advanced societies is not so meaningful. Kiev[1] concluded recently that illness has far greater social significance in "primitive" cultures than in our elaborate Western societies. In the former, he says, "illness often suggests that something is wrong in the community, the sickness of a single individual often arouses collective fear and community action".

What I want to achieve is not some sort of lengthy re-oriented medical history, but rather a brief review of some of the areas where community action is influenced by varying human fitness. Overlying most other factors, is of course the question of change in population number and concentration. I have discussed this to some extent previously[2], particularly in relation to the generally held view of a Neolithic population "explosion", and the beginnings of serious food mis-use and deficiency diseases. But the trends towards population increase in the early farming settlements, whatever the pace of increase, must have gradually brought other problems affecting the fitness of the community.

Beginning with congenital defects, the bringing together of larger numbers of people than was probably achieved in Palaeolithic economies could have decreased the chances of long-term close inbreeding and thus perhaps of the likelihood of increased frequencies of certain congenital defects. On the other hand, earlier urban societies were in a far better position to maintain such defectives in the community. Of course it is difficult to determine whether infanticide was practised, although it has been suggested from skeletal evidence, but certainly it is significant to see surviving in the late prehistoric and protohistoric groups, such defects as club foot, hydrocephaly and achondroplasia[3]—abnormalities which would have placed the individual at a severe disadvantage in a more primitive hunting and collecting society. Certainly, the occurrence of noticeable defects of this sort in earlier communities shows that they were tolerated. Certain evidence of congenital mental defect can so far only be demonstrated during the Anglo-Saxon period in Britain[4], but could have been associated with marked hydrocephaly as early as the European Neolithic. In such cases, there is no doubt that the attitudes of the group and their economy were sufficiently geared to support at least some with severe disability.

As regards textual evidence, congenital blindness is noted in early Mesopotamia, and indeed, second millennium texts from Chagar Bazar in northern Mesopotamia mention certain blind females who worked at reed mat making. Kinnier Wilson has also suggested that certain early Mesopotamian words indicate "deaf-mute" and congenital idiocy[5]. Again, the Bible and Talmud provide evidence that early Palestinians were also acquainted with a variety of congenital defects. Even haemophilia was noted, and it was advised that if two boys of one mother died following circumcision, the third should not be operated on[6].

One of the most serious threats to early urban communities was infectious disease. For the first time in the whole course of human cultural development, the agricultural revolution heralded the elaboration and increase of the size of societies to the extent that epidemic disease could cause devastation as never before. Although only hypothesis, it seems very probable that it was the enormity of the health threats in these later prehistoric/protohistoric communities which saw the differentiation of physicians, surgeons and other "medical" specialists from the ordinary sorcerer and witch classes.

The devastating and epidemic proportions of some of the early "plagues" might be exemplified. The great Plague of Athens (typhus or smallpox?) which came at the beginning of the Peloponnesian war, could well have helped the decline of Athenian greatness. A few decades later, in 396 B.C., it was the turn of the Carthaginians besieging Syracuse to be stricken by another plague—again perhaps smallpox[7]. Again, it was probably this disease which decimated an Abyssinian raiding party near Mecca some time between A.D. 569 and 571[8].

Other disease also changed the fortunes of war and, through the suffering and numerous dead, must have emphasized the continual need to strive against these common enemies. Dysentery must always have struck hard at susceptible groups (civilian migrants or troops) entering localities where it was maintained as a "grumbling endemic infection". Hare[9] recalls that the first Persian invasion of Europe under Darius was probably checked by this disease; that it probably influenced the retreat of Xerxes from the third invasion of Greece in 400 B.C.; and that it certainly influenced the activities of the Crusaders, especially during the Second Crusade. Shrewsbury[10] argues that the "plague" of the Philistines was also bacillary dysentery, and plausibly suggests that, in the long term, it was responsible for the restoration of the Ark of the Covenant to the broken and demoralized Israelites.

With respect to African cultural development, Darlington[11] has recently stated the view that any retardation which can be seen in African cultural history is to a large part to be blamed on the deterring and destructive effects of disease. Lambrecht[12] outlines convincingly how just one of the important African diseases—trypanosomiasis—could have deterred population development over large areas and over long periods of time.

Perhaps a counterpart of this in Europe and the Mediterranean world was plague. Its antiquity is uncertain, although rodent infestation (especially of *Rattus rattus*) of human habitations might have been early established in Old World settlements[13]. *Pasteurella pestis* (i.e. true plague), certainly appears to have been responsible for the Plague of Justinian in the sixth century A.D. The epidemic lasted fifty-two years and decimated the populations it struck, and according to the Byzantine historian Procopius, half the inhabitants of the Byzantine Empire had died by A.D. 565. It is no wonder that Hirst[14] and others believed that "this calamity played an important part in giving rise to the Dark Ages". A further case for the culturally stunting effect of a disease may be provided by the history of malaria in parts of the Mediterranean world. The disease certainly has a long history in the Old World[15]. Reference to fevers in both Greek and Roman literature have been studied in detail[16], and leave little doubt that malaria was the cause. It would thus seem reasonable to question, as some have done, whether this apparently common debilitating disease contributed to the decline of these early civilizations.

I have discussed at some length a number of historical examples where infection seems to have influenced the well-being or actions of a community. They were given merely as examples of the sort of situation which must have developed again and again, to a greater or lesser degree, in the expanding late prehistoric and protohistoric cultures[17]. More than one early writer refers to the problem of corpse disposal during an epidemic, and one can see that the feeding, diagnosis and maintenance of the sick would have posed considerable problems as well. It is therefore only a minor rational step for a community to set aside some of its resources to be concerned with the health of the group. This can take various forms of action, which could be cumulative to some extent. From Egypt, Mesopotamia and India we have a wealth of early medical observations to show the extent to which abnormality was being noted, usually with a view to its treatment[18]. In China also, by the second millenium B.C., "medicine men" were a part of the society, and by the sixth century B.C. specialized medical professions were well differentiated[19]. Also of note is the fact that by the third and fourth centuries B.C. there was some contact and interchange of medical ideas between Babylonia, India and China[20], a feature of earlier societies which need not have been restricted to those areas for which there is literary evidence.

In contrast to what is known about the development of medical treatment and the emergence of "physicians and surgeons", the evolution of special "temples" or hospitals for the sick has been poorly considered. Some degree of separation might well have been attempted early on in relatively densely populated areas. As recent "primitive" communities indicate, virulent disease is not only destructive of human resources and generally terrifying, but quickly leads to social chaos. The impact of smallpox among the Tiv "literally blows the society wide open"[21], and the community fragments in an attempt to avoid contamination. The alternative to this was the separation of the sick, an expedient which was certainly established in Mesopotamia some 3500 years ago. In particular the Mosaic law advises on the separation of "lepers"[22] from the community.

In the Mediterranean world, Greek medical and pseudo-medical practices preceded those of the Romans. The widespread and popular cult of Aesculapius had, at the most, psychotherapeutic value. By about 200 B.C., Greek medical practitioners were beginning to make their mark in Rome. Gask[23] suggests that hospitals probably began for military needs, a good example being that at Novaesium, near Dűsseldorf. However, it really depends upon how we define a hospital. If those expelled from the early cities of Babylonia and Assyria on account of disease gathered together in small settlements, then we have the beginnings of hospitalization (especially if the early physicians attended them there).

The prevention and treatment of wounds, and the development of surgery and surgical equipment, are fields which provide further evidence of the antiquity of man's concern for the maintenance of health. The seriousness of head injuries, for instance, and the need to evolve forms of head protection, must have been realized early in the history of organized warfare[24]. In the early Indian literature (Rigveda), reference is made to amputations, extraction of arrows, and even the fitting of artificial "limbs"[25]. A large range of surgical instruments was evolved in early Greece and Rome, clearly to satisfy a variety of different procedures[26]. Surgical instruments related to the problems of childbirth were in use in early Babylonia[27]. In the New World also, minor surgery and the rational treatment of wounds was well developed in at least some Amerindian cultures prior to European contact[28].

It seems possible that druglore, which must have undergone its basic development in Palaeolithic cultures, was evolved further and perhaps more rationalized with the emergence of the physician and other medical specialists. Sigerist[29] notes that the Swiss "lake-dwellers cultivated or gathered over two hundred different plants, among which are not a few that possess medicinal qualities". Incidentally, European folk medicine has embraced a very great variety of plants, so that one is tempted to question whether the sixty-three varieties in the stomach of the Danish Graubølle Man were more reasonably bound up with a population health ritual than the generally accepted one of land fertility.

Some plants were really of no medical value, but this was not the case with others. In Talmudic times, dates were used as a laxative and in the treatment of haemorrhoids[30]. In early Mesopotamia *Solanum* berries, which are well known to have antispasmodic properties, were used in certain cases of difficult childbirth[31]. From China, the *Book of Herbs* (perhaps a compilation dating back to 3000 B.C.), contained certain descriptions of many drugs, including Ch'ang Shang. This drug was claimed to be useful in treating fevers, and has in fact been shown to contain an active antimalarial alkaloid[32].

Some mention must be made regarding the development of public hygiene and sanitation. Whenever there is an increase in density of people living together, whether in a long house, village or city, there are likely to be increased health problems. These may arise from the ease with which infection can be transferred, from the elimination of waste (especially faeces and food waste), and from the attempt to maintain an ample and unpolluted water supply. The archaeologist could certainly provide far more information of this kind. Of course we know a little. The people of Mohenjo-daro and Harappa went to much trouble to plan baths, lavatories, drains and fresh water tanks, a fact which certainly reflects their concern for community sanitation and health. The canals constructed in Egypt, Assyria and Babylonia[33] were no doubt multi-purpose, including for the disposal of sewage. Ancient Hindu literature refers to "filth temples". The famous Roman city sewer, the *cloaca maxima*, was being built by about 600 B.C., similar constructions being undertaken in Carthage and Alexandria. From Mesopotamia through the Mediterranean world much attention was given to water supply. In particular, the Minoans seemed well aware of the value of clean piped water, baths and sewage disposal drains. Dry refuse appears to have been differentially buried[34]. Water contained by artificial man-made barriers may at times have become disease reservoirs, as well as perhaps encouraging malaria-carrying mosquitoes in some areas.

Finally, I have already mentioned the question of congenital mental defect in early societies, but of far more importance to the community as a whole is of course the question of mental disease in adults. In particular, mental abnormality affecting individuals in authority can have far reaching consequences for a culture. Examples of this during the present century are not difficult to find, and it would be wrong to consider that past cultures were free of them. However, it could be that it was not until the development of the larger stratified urban societies that mental "abnormality" became a more potentially important or dangerous phenomenon in society. It would seem reasonable to argue that in small hunting communities, however psychopathic an individual might be, the effects on the extended tribal group would be minimal.

Early records give some evidence to show the variety of conditions which were sufficiently marked to be noted. Some could have had far reaching consequences. Kinnier Wilson[35] notes that delusional misinterpretations, hallucinations and heightened religiosity occur in early Mesopotamia. The Babylonian textbook known as *Shurpu* also outlines a psychopathic state. Kinnier Wilson has also produced a good case for believing that the "War of the Sons of Light against the Sons of Darkness" in the Dead Sea Scrolls is concerned with a sect possibly dominated by a paranoic.

Moss[36] has also discussed mental disease in earlier societies, and gives as one example the sudden blindness of Epizelus, during the Battle of Marathon. This could well have been due to hysteria. A delusional state is seen in the physician Menecrates, who thought he was Zeus. Recognition of nervous tension, and the long-term dangers of this state, were mentioned by Amasis, a king of Egypt.

It seems possible, if Aldred and Sandison[37] are correct, that the Egyptian Pharaoh Akhenaten suffered from a chronic illness which might well have affected his pyschological make-up. Because of his considerable religious and artistic innovations, the complex question of the relationship between his health and mental outlook is not to be dismissed lightly. Aldred and Sandison in fact present a good case for believing that Akhenaten suffered from an endocrinopathy with hypogonadism and adiposity—possibly the result of a chromophobe adenoma.

The medical history of Herod the Great, King of Judea, has also been considered in detail, the various evidence being reviewed by Sandison[38]. In his later years, Herod became impatient, markedly paranoid, tending towards excesses, ruthlessly severe and vindictively savage. These are highly dangerous traits in a man with the social power which Herod enjoyed. If, as Sandison argues, they were to some extent the result of progressive hypertension, with developing cerebral arteriosclerosis, and because these are not uncommon conditions in more elderly individuals, it raises the question of how often in the past rulers have been affected by such conditions. These need not of course have been necessarily detrimental to the culture as a whole.

Mental disease is a relatively common, and to society an expensive, facet of modern civilization. Organic disease, social stresses and other factors, were probably no less common in early urban societies, and at times could clearly have led to mental states in individuals which in turn may have had marked cultural repercussions.

This can only be a brief outline of the interaction of early cultures with disease. What I have attempted to do is simply to point out that a consideration of community health has very real relevance to the proper study of early societies. I have also tried to point out that, whatever the early community might be, there are the common problems of *a.* the reaction of the population to congenital, infectious and other disease; *b.* their response to injury, particularly where large numbers may be involved together, as in group conflict; *c.* the extent to which health problems may be resolved either by rational or ritual measures; *d.* their exploitation of the environment in an attempt to maintain or restore health—whether in terms of druglore, surgical equipment, water supplies, waste disposal, protective clothing, or attempts to separate the sick in separate buildings.

Notes

1 Kiev, A. (ed.) (1969). *Magic, Faith and Healing: Studies in Primitive Psychiatry Today*. New York.

2 Brothwell, D. R. (1969). Dietary variation and the biology of earlier human populations, *in* Ucko, P. J. and Dimbleby, G. W., *The Domestication and Exploitation of Plants and Animals*. London.

3 Brothwell, D. R. and Powers, R. (1968). Congenital malformations of the skeleton in earlier man, *in* Brothwell, D. R. (ed.) *The Skeletal Biology of Earlier Human Populations*. London.

4 Brothwell, D. R. (1960). A possible case of mongolism in a Saxon population, *Ann. Hum. Genet.*, **24**, p. 141.

5 Kinnier Wilson, J. W. (1967). Organic diseases of Ancient Mesopotamia,

in Brothwell, D. R. and Sandison, A. T. (eds.) *Diseases in Antiquity*, Thames.

6 Sussman, M. (1967). Diseases in the Bible and the Talmud, *in* Brothwell, D. R. and Sandison, A. T. (eds.) *op. cit.*

7 Patrick, A. (1967). Disease in Antiquity: Ancient Greece and Rome, *in* Brothwell, D. R. and Sandison, A. J. (eds.) *op. cit.*

8 Hare, R. (1967). The antiquity of diseases caused by bacteria and viruses, a review of the problem from a bacteriologist's point of view, *in* Brothwell, D. R. and Sandison, A. T. (eds.) *op. cit.*

9 Hare, R. (1954). *Pomp and Pestilence. Infectious Disease, Its Origins and Conquest.* London.

10 Shrewsbury, J. F. D. (1964). *The Plague of the Philistines.* London.

11 Darlington, C. D. (1969). *The Evolution of Man and Society.* London.

12 Lambrecht, F. L. (1967). Trypanosomiasis in prehistoric and later human populations, a tentative reconstruction, *in* Brothwell, D. R. and Sandison, A. T. (eds.) *op. cit.*

13 I am not aware of any analyses of rodent skeletal remains from early settlements, with a view to assessing what disease carriers may have been about. This could be useful corroborative evidence.

14 Hirst, L. F. (1953). *The Conquest of Plague.* London.

15 Bruce-Chwatt, L. J. (1965). Palaeogenesis and palaeo-epidemiology of primate malaria, *Bull. Wld. Hlth. Org.*, **32**, p. 363.

16 Jones, W. H. S. (1907). *Malaria: A Neglected Factor in the History of Greece and Rome.* London.

17 Early communities of both the Old World and the Americas had to contend with a mosaic of diseases, but until post-Columbian times these mosaics may have been noticeably different in some respects.

18 See Note 5; Dawson, W. R. (1953). The Egyptian Medical Papyri, *in* Underwood, E. A. (ed.) *Science, Medicine and History.* London. Kutumbiah, P. (1962). *Ancient Indian Medicine.* Calcutta.

19 Needham, J. and Lu Gwei-Djen (1969). Chinese medicine, *in* Poynter, F. N. L. (ed.) *Medicine and Culture.* London.

20 Huard, P. and Wong, M. (1968). *Chinese Medicine.* London.

21 Alland, A. (1970). *Adaptation in Cultural Evolution. An Approach to Medical Anthropology.* New York.

22 The term 'leper' in this context may well indicate a variety of diseases affecting the external appearance of the individual, and may not even have included true leprosy.

23 Gask, G. (1950). *Essays in the History of Medicine*, London.

24 See, for example, Courville, C. B. (1949). Injuries to the skull and brain in Ancient Egypt *Bull. Los Angeles Neural. Soc.* **14,** 53.

25 See Kutumbiah, P. (1962). *ibid.*

26 Milne, J. S. (1907). *Surgical Instruments in Greek and Roman Times.* Oxford.

27 Kinnier Wilson, J. W. (1967). *op. cit.*

28 Sigerist, H. (1951). *A History of Medicine. I. Primitive and Archaic Medicine.* Oxford; Vogel, V. J. (1970). *American Indian Medicine.* Oklahoma.

29 Sigerist, H. (1951). *op. cit.*

30 Sussman, M. (1967). *op. cit.*

31 Kinnier Wilson, J. W. (1967). *ibid.*

32 Hare, R. (1954). *ibid.*

33 Krepp, F. C. (1867). *The Sewage System.* London.

34 Wylie, J. C. (1959). *The Wastes of Civilization.* London; also, Vallentine, H. R. (1967). *Water in the Service of Man.* London.

35 Kinnier Wilson, J. V. (1967). Mental diseases of Ancient Mesopotamia, *in* Brothwell, D. R., and Sandison, A. T. (ed.). *op. cit.*

36 Moss, G. C. (1967). Mental disorder in Antiquity, *in* Brothwell, D. R. and Sandison, A. T. (ed.). *op. cit.*
37 Aldred, C. and Sandison, A. T. (1962). The Pharaoh Akhenaten: a problem in Egyptology and pathology, *Bull. Hist. Med.*, **36**, p. 293.
38 Sandison, A. T. (1967). The last illness of Herod the Great, King of Judaea, *Med. Hist.*, **11**, p. 381.

ANTHONY FORGE

Normative factors in the settlement size of Neolithic cultivators (New Guinea)

In a paper published in 1953, Hogbin and Wedgwood provided a typology of local groupings in Melanesia that has, in part at least, been used ever since[1]. To discover what is an "Osculant Multicarpellary Parish" it is necessary to consult this paper. The present paper, however, will not be concerned with what anthropologists normally call social structure; indeed, one of the striking things about Melanesian material is how little variation in patterns of actual behaviour can be attributed either to whether the society is patrilineal, matrilineal or neither, or indeed to whether the settlement pattern is concentrated in a village or dispersed over a considerable area. These two distinguished and experienced experts in the area state "that Melanesian societies are minute in scale. The same culture and language are often shared by thousands, but the widest social unit possessing a coherent system for the maintenance of internal order consists of the seventy to three hundred persons resident within the boundaries of a clearly defined area seldom more than a few square miles in extent"[2]. The present paper is concerned with the size of the social units in Melanesia, accepting the criteria of internal order and a bounded territory. The usual explanations for the size of settlements are the difficulties of the terrain, the lack of easily storable crops, the exigencies of shifting cultivation, and so on; explanations that are basically ecological, and it is to such arguments that Hogbin and Wedgwood refer in passing. It is also implicit in many such ecological explanations that the ethnographer regards the ecology as a limiting factor, so that if it was not for ecological difficulties people would live in larger settlements. This assumption would appear to be justified in a few instances, e.g. the Mountain Arapesh who, Mead[3] makes clear, deplore their inability to maintain gatherings larger than the hamlet (largest size 85) as a group for more than a few days at a time. Such explicit sentiments are, however, rarely reported, it is indeed not unusual for ethnographers to make comments of the form—"they say they split because of quarrels and sorcery accusations but my analysis shows that they did it because of pressure on scarce resources". No one with any experience of Melanesian circumstances can doubt the great importance of ecological factors in every facet of society

Name of tribe or group A	Total population: Average size, or size of settlement studied B	Total population: Range of size of settlements C	Number of households: Average size, or size of settlement studied D	Number of households: Range of size of settlements E	Nucleated—N or Dispersed—D F	Source (see references, p. 366) G	Comment H
Abelam		300– 800			N	Forge, 1970	Average density 120–200/m² (see text)
Mountain Arapesh	85				N	Mead, 1935	
Asmat	400	50–1500			N	Gerbrands, 1967	
Banaro				9–48	N	Thurnwald, 1916	"Tribe" consists of four villages
Bena-Bena	232				N	Langness, 1964	Figure for the "second largest clan". Density up to 100/m²
Buka	136				N	Blackwood, 1935	Kurtatchi village only
Busama	600				N	Hogbin, 1951	Single village much larger than neighbours, split into two components under administration influence in 1945
Chimbu		75– 200			D	Brookfield and Brown, 1963	Density 320/m² for the area studied
Choiseul Island		300– 400	50		N	Scheffler, 1965	Estimate based on Goldie's house count of 1906
Daribi		100– 320			D	Wagner, 1967	"Community" composed of two to four clans Low density
Dobu		100– 600			N	Fortune, 1932	Local group "composed of four to twenty-five hamlets" of average size twenty-five
W. Elema	740				N	Williams, 1940	Low density
S. Fore	180	40– 525			D	Glasse, 1969	Two-thirds of settlements were in the range 100–200. Density 25–45/m²
Garia	166	100– 300			D	Hogbin and Lawrence, 1967	
Gnau	232	93– 500			N	Lewis, p.c.	Figures for the whole E.Au census district including non-Gnau speakers. Low density
Gururumba				25–30	N	Newman, 1965	
Huli	133	80– 186			D	Glasse, 1968	Density 30–70/ m²
Iatmul		200–1000			N	Bateson, 1936	Density less than 15/m²

Kapauku	120		15		N	Pospisil, 1958	Density $100/m^2$
Kuma		100– 300			D	Reay, 1959	Figures for Reay's "type A clan" having only one level of segmentation within it. Larger units with two or more levels of segmentation also occur, sizes up to 1700
Kyaka	200	40– 540			D	Bulmer, 1965	Density in the region of $80/m^2$
Lakalai	135				N	Chowning and Goodenough, 1965–6	
Mae-Enga	350	100–1100			D	Meggitt, 1965	Average density $120/m^2$ up to $250/m^2$ in places
Maenge		180– 200			N	Panoff, 1970	
Little Mala	250				N	Ivens, 1927	Estimate for the village of Sa'a in 1896
Manus	200				N	Mead, 1950	
Maring	350	100– 900			D	Rappaport, 1967	Density $97/m^2$ for total arable area
Mbowamb			56*		D	Strathern, 1966	Density $80/m^2$. *Married men of one clan with five segments
Orokaiva			71		N	Williams, 1930	Average number of households had fallen to twelve by 1930
Siane		200– 250			N	Salisbury, 1962	Density $80/m^2$
Siuai			13	10–25	N	Oliver, 1955	Figures for 1938. Pre-contact hamlets of one to nine households
Tangu	66			1–20	N	Burridge, 1969	Formerly in larger communities
Tchambuli	500				N	Mead, 1935	Village comprising three hamlets
Tifalmin		125– 150			N	Cranstone, 1968	
Tolai	200				N	Epstein, 1968	Salisbury 1970 gives a similar figure for Vunamami village in 1875, with a density of $150/m^2$
Waina-Sowanda		120– 250			N	Gell, p.c.	
Wogeo	60		10		N	Hogbin and Lawrence, 1967	Figures are for hamlets, the political units are larger, averaging 180 each

Table 1. Settlement size in New Guinea, the Bismark Archipelago and Solomon Islands (for notes see p. 369).

Note to Table 1 (pp. 364-5)

Table 1 (pp364-5) makes no claim to completeness. Over 50 ethnographies were consulted. About one third contained no information at all on size of settlement. Ethnographers who did give information varied greatly in the manner of presentation; some gave the total population of the settlement studied, some a range of sizes for the group concerned, others an average. Some gave numbers of households but not total population. As far as is possible figures represent the pre-contact or early contact settlement pattern. Information on the density of population is, in general, only available for recently studied Highlands populations. All information about density (Column H) is expressed as so many per square mile. These figures are not all strictly comparable some being based on the total area available others on the area actually cultivated, ignoring uncut virgin forest, grassland, etc. But the general comparative picture is clear. Especially in the cases of dispersed settlement there are problems in deciding which unit shall be isolated as the "settlement"; where there is doubt the criterion of normal daily face to face interaction and co-operation has been added to the criteria of bounded territory and internal order. I am grateful to Miss Rita T. Hayes for extracting much of the data in this table.

References to Table 1

Bateson, G. (1936). *Naven*. Cambridge.
Blackwood, B. (1935). *Both Sides of the Buka Passage*. Oxford.
Brookfield, H. and Brown, P. (1963). *Struggle for Land*. Melbourne.
Bulmer, R. (1965). The Kyaka of the Western Highlands, *in* Meggitt, M. J. and Lawrence, P. (eds.) *Gods, Ghosts and Men in Melanesia*. Melbourne.
Burridge, K. O. L. (1969). *Tangu Traditions*. Oxford.
Cranstone, B. A. L. (1968). War Shields of the Telefomin sub-district, *Man*, NS **3**.
Chowning, A. and Goodenough, W. H. (1965–66). Lakalai Political Organization, *Anthropological Forum*, **1**.
Epstein, S. (1968). *Capitalism, Primitive and Modern*. Canberra.
Forge, A. (1970). Learning to See in New Guinea, *in* Mayer, P. (ed.) *Socialization: The Approach from Social Anthropology* (ASA. 8). London.
Fortune, R. (1932). *The Sorcerers of Dobu*. London.
Gell, A. (1970). Personal communication.
Gerbrands, A. A (1967) *Wow-ipits: Eight Asmat Woodcarvers of New Guinea*. The Hague.
Glasse, R. M. (1968). *Huli of Papua*. The Hague and Paris.
Glasse, R. M. (1969). Marriage in South Fore, *in* Meggitt, M. J. and Glasse, R. M. (eds.) *Pigs, Pearlshells and Women*. Englewood Cliffs.
Hogbin, I. (1951). *Transformation Scene*. London.
Hogbin, I. and Lawrence, P. (1967). *Studies in New Guinea Land Tenure*. Sydney.
Ivens, X. (1927). *Melanesians of the South-East Solomons*. Oxford.
Langness, L. L. (1964). Some Problems in the Conceptualization of Highlands Social Structures, *Amer. Anthrop.*, **66** (4, Part 2).
Lewis, G. A. (1970). Personal communication.
Mead, M. (1935). *Sex and Temperament in Three Primiiive Societies*. London.
Mead, M. (1950) *Male and Female*. London.
Meggitt, M. J. (1965). *The Lineage System of the Mae-Enga of New Guinea*. Edinburgh and London.
Newman, P. L. (1965). *Knowing the Gurumumba*. New York.
Oliver, D. (1955). *A Solomon Island Society*. Cambridge, Mass.
Panoff, M. (1970). Food and Faeces: a Melanesian Rite, *Man*, NS. **5** (2).
Pospisil, L. (1958). *The Kapauku Papuans and Their Law*, University Publications in Anthropology, No. 54. Yale.
Rappaport, R. A. (1967). *Pigs for the Ancestors*. New Haven and London.
Reay, M. (1959). *The Kuma*. Melbourne.
Salisbury, R. (1967). *From Stone to Steel*. Melbourne.
Salisbury, R. (1970). *Vunamamii*. Berkeley and Los Angeles.
Scheffler, H. (1965). *Choiseul Island Social Structure*. Berkeley and Los Angeles.
Strathern, A. (1966). Despots and Directors in the New Guinea Highlands, *Man*, NS. **1**.
Thurnwald, R. (1916). Banaro Society, *Memoirs of the Amer. Anthrop. Assoc.*, **3**.
Wagner, X. (1967). *The Curse of Souw*. Chicago and London.
Williams, F. E. (1930). *Orokaiva Society*. Oxford.
Williams, F. E. (1940). *The Drama of Orokolo*. Oxford.

and indeed there is every reason to suppose that ecology provides limiting factors throughout New Guinea; the important question is how narrow are those limits? Another factor almost invariably cited as a determinant of settlement size is warfare, the need for defence giving great advantage to large concentrated settlement units. Here too is a factor obviously of the greatest importance. It has in general been assumed by all the writers who have considered the size of settlements in Melanesia that these two factors—ecology and warfare—plus any special specific local factors are sufficient to explain the range of figures they have reported. I shall suggest in this paper that there is evidence that other factors may be involved, and present some alternatives as to what these factors might be.

Size of settlement

It is first pertinent to ask to what extent Hogbin and Wedgwood's assertion is true, especially as it was made at a time when there were very few reports available about the settlements of the New Guinea Highlands. Furthermore the amount of anthropological work done in Lowland and Island Melanesia must have about doubled the amount of data available since Hogbin and Wedgwood's paper.

The most complete demographic study undertaken on a large scale refers to west Irian and was undertaken by the Dutch in 1961. Although it restricted itself to Papuans, it covered all settlements, thus including the six major towns with populations of over 2000. Nevertheless 52·8% of all settlements had populations falling within the range 75–300, while 50·1% of the total population lived in such settlements[4]. The average size of all settlements is given as 159. Similar figures are not available for the other half of the Island but it is possible to deduce from the figures published in a survey of indigenous agriculture in the Territory of Papua and New Guinea that in the same year 1961 the average size of settlement was 169·7[5]. Such figures of course tell us very little, but they are not incompatible with the 70–300 range we are considering.

Table 1 sets out in highly compressed form information on size of settlement drawn from some standard ethnographies, the list makes no claim to be exhaustive but covers more or less adequately a range of societies from New Guinea and the Solomons with varying ecologies, subsistence crops, and so on. It is of course impossible to apply any statistical operations to such a list, but certain trends are worth comment.

Although there are several highland groups whose settlement ranges go greatly above the 300 point, in only two cases (Mae-Enga and Maring) does the average reach 350, whereas there are several lowland, river, lakeside and coastal settlements considerably larger (e.g. Asmat, W. Elema, and Iatmul). These water-based settlements are considered below. There is some suggestion that high density goes with larger settlements (Mae-Enga, Kuma, Abelam) in both highlands and lowlands. But there are plenty of exceptions (Tolai, Kapauku, and especially Chimbu with the highest density). War in pursuit of territorial expansion as well as for the more usual human reasons is not unexpectedly found in areas of high population density, but it certainly does not automatically produce easily defended nucleated settlements (e.g. Mae-Enga and Chimbu) nor necessarily even large settlements (Chimbu).

Water-based villages

The Dutch survey of West New Guinea includes, of all the highland areas, only the immediate area of the Wissel Lake in its figures, that is the Kapauku and related groups; in the two sub-divisions that make up this area at least 90% of the settlements are shown as of 300 people or less. Almost a third of the enumerated population of west Irian is shown as living in riverside and lakeside settlements (see Table 2); these areas and particularly the great flat river systems of the south-west such as the Asmat show a substantial increase in the size of settlements compared with the other two classes of "coastal" and "other interior settlements".

	River and lakeside	*Coastal*	*Other interior*	*Total*
Less than 300	54·2	51·6	80·3	60·4
300 to 1000	41·2	21·5	18·7	27·1
Over 1000	4·6	26·9	1·0	12·5

Table 2. Population by location of settlements. Percentage of each class living in settlements according to numbers of inhabitants[a].

(*a*) Source: Groenewegen, K. and Kaa, D. J. van de (1964). *Nieuw-Guineas als gebied voor demografische onderzoekingen*; vol. 1 of *Resultaten van het demografisch onderzoek Westelijk Nieuw-Guinea*. The Hague. Table x.

There are only twenty-six settlements in the over 1000 class containing 12.5% of the total population and as all or almost all of these are the product of Colonial rule they will be ignored in the subsequent discussion. The tendencies revealed by the table are reflected in the figures for the Asmat area itself: 26·9% of the population living in settlements of less than 300 persons, while 62·1% fall in the range of 300-1000. These riverain villages, and the Iatmul and other groups of the Lower and Middle Sepik river system should be included, all have certain distinctive features and must be considered separately.

Rivers were the only effective means of communication over more than a few miles in the whole of New Guinea; the sea by comparison was an unreliable and a comparatively speaking little exploited means of communication[6]. The rivers, flowing through huge flood plains, provide ample quantities of fish and Crustacea, while their swamps contain vast quantities of naturally regenerating sago palm. Probably something in the region of 90% of the diet is thus obtained by methods which strictly considered are hunting and gathering. However coconuts are often planted and need special care if they are to survive flooding, pigs are frequently kept, trees whose leaves or fruit are used for food or ritual are planted at village sites, and some Iatmul at least, plant a few yams and other tubers during the season when the floods have gone, so that these very large groups can be saved for the horticultural category. The rivers were extensively used for warfare, and raiding parties in a fleet of canoes could enjoy the benefits of surprise and mobility without any sacrifice in numbers. Under these circumstances only large villages could survive, and smaller settlements, for instance of many of the peoples of the Karawari tributaries to the Sepik, lived back from their rivers on hilltops, cut off from, or at least severely handicapped in, exploiting ample supplies of the river simply because they could not support the raiding, often over long distances, by the large riverain villages to the north. The easy subsistence of the river systems then is only available to those who organize themselves into large villages, and it in itself gives men the leisure to indulge in the frequent fighting and raiding that produces the pressure to live in large villages. Whether all these large villages support Hogbin and Wedgwood's criterion of "a coherent system of the maintenance of internal order" is perhaps questionable. Certainly all for which we have adequate ethnographic information had complex internal structures of groups and sub-groups usually of a binary form; while Bateson reports from the Iatmul intra-village killings that were treated more as a matter of credit, that is comparable to killing enemies, than as murder, that is reprobated killings within the basic unit. The large Iatmul villages are all composed of two or more clearly separated residential units, in which most ritual and mundane interaction was concentrated. The basic residential unit is therefore a half or a third of a village, and there is evidence of similar components in other large riverain villages.

Tuber-dependent villages

The vast majority of the population of New Guinea and island Melanesia was traditionally dependent on the cultivation of some form of tuber crops: sweet potato in the highlands, yam and taro in the lowlands; most too relied exclusively on domestic pigs and hunting for supplies of animal protein. In terms of most of the institutions with which anthropologists are concerned these populations vary widely, some are patrilineal, some matrilineal, some have age grades, others do not, and so on and so forth. In general there is a very great diversity of custom and language throughout the islands, but despite this there are some sociological constancies, the most striking is perhaps that all these settlements are egalitarian, that is they are virtually, or totally, devoid of ascribed status or formal hierarchy, there are no offices and birth entails, no advantage or disadvantage. Another universal feature is that all the adult men of a village are capable of forming a defensive unit and that at times of crisis this obligation overrides all others. There is no necessary similar obligation to join together to form an offensive unit, although presence at set battles (occasions closer to a contemporary football match than war as conceived in our present civilization) which had elements of both offence and defence, are more or less mandatory. As a corollary of the egalitarian nature of the societies, effective leadership is exercised by men of high prestige, and hence influence, who are usually called big-men.

Had Hogbin and Wedgwood written their paper fifteen years later, when data on the central highlands and other high density groups was available, it is possible that they might have extended the numbers within their range to 400 or even more; one suspects that they would have been happy to do so since absolute size is in no way essential to the classification of local groups that they present. I want to discuss qualitative rather than quantitative factors for the rest of this paper and attach no special importance to the figure 300. But before we leave the Hogbin and Wedgwood range altogether, it is perhaps worth noting that of the twenty-two administrative sub-divisions of Dutch New Guinea, which formed the basis of the 1961 survey, only three have average settlement sizes that fall outside the 70–300 range, all being above this figure. Of these one is the Asmat, already discussed, another the capital, then called Hollandia, and its immediate environs, and the third is Japen-Waropen, an area of long and intense contact.

Few of the settlement sizes in Table 1 are under 150, indeed most populations which have been studied are those with settlements which only rarely fall below 200. Now this of course reflects the source of information—anthropologists' field reports—and the belief that when the basic residential unit is 100 or below the anthropologist is in for a rough time; such groups are in low density areas, people wander about and nothing very much happens, the village often seems or actually is deserted for long periods, and the anthropologist spends a lot of his time walking around trying to catch up with something or somebody to study. Such cultures tend to be "poor" cultures often occupying inferior land, and with rudimentary subsistence techniques; they are rarely involved in large-scale ritual or exchange, and as a correlate there is no social demand for surplus production of basic foodstuffs, and this means of course that they are more seriously affected by bad seasons than cultures which aim to over-produce. Anthropologists tend to avoid such "poor" groups and concentrate their efforts on larger scale societies often in areas of medium or high population density. "Rich" cultures with larger settlements tend to have elaborate ritual and exchange institutions, there is a consistent social demand for surplus production of basic crops and indeed intense competitive pressure in all forms of production and exchange of food, luxuries and wealth. For anthropologists there is more to study, struggles for prestige can be carried on in many different fields, and so on. Behind the anthropologists' preference lurks, I suggest, a real difference in society. A settlement size of about 150 suggests, at a generous estimate, thirty-five adult men, and it seems to me that somewhere around this figure is a critical point in the type of social structure which is appropriate. I should emphasize here that I am talking about the basic unit, that is men in more or less daily contact in relationships that involve co-operation and probably rivalry. Much below thirty-five men basic relationships of kinship and affinity, together with reciprocal assistance, are quite capable of ordering social relationships. Not of course that elaborate ritual classification with a descent ideology is in any way precluded, it is simply that such classification is not necessary and often does not occur. It is here worth comparing the Australian aborigines of the central desert, who spend most of the year in small groups but maintain a most elaborate totemic system, with Neolithic tuber cultivators who may actually reside in groups of as many as 70–80 (e.g. Mountain Arapesh) but maintain no such system. The point is that although the central desert aborigines may be smaller in group size and much lower in density than the horticulturalists, their system requires, and their ecology enables, them to come together at certain times in very much larger groups to perform ritual, and further that all those who come together have an essential part in the ritual. In comparison small groups in New Guinea may have visitors when they put on ritual but the visitors are essentially spectators.

With the populations in daily interaction with more than thirty-five adult men it seems that some form of segmentation always occurs. By this I mean that the group is divided up into sub-groups of a higher order than households or families; these segments will be part of a symbolic order and be constituent and distinguished groups in rituals, those involving prestige and exchange as well as those involving the supernatural. The ideology of the symbol system differentiating these constituent groups, frequently in New Guinea clans and sub-clans, either patrilineal or matrilineal, will usually involve wider extension than the basic settlement. Yet that settlement will be clearly demarcated both by such things as boundaries and by symbolic foci such as men's or ceremonial houses, dancing grounds, etc. There seems to me some evidence in the sparse and unsystematic data that I have been able to assemble that such basic residential groups in "rich" cultures have populations in the range 150–350, that is with 35–80 adult men. Further where the basic political (war-making) units are much larger than this and in some cases when they are within the higher end of the range, the political unit is composed of two or more basic units clearly demarcated residentially and in land-holdings as well, holding some form of ritual in common and differentiated by at least one symbolic parameter from each other. I should perhaps add that I made no distinction between, for instance, alleged ancestry, lineage names and totems, all are equally symbols in this context.

I believe that there are normative factors, whose possible nature will be discussed below, that tend to produce among egalitarian Neolithic communities of New Guinea and Melanesia a basic residential and face to face interaction unit of 250±100 in total population; ecological circumstances may vary the physical distance between such units and warfare may cause them to collect together for mutual defence, but whether distant or contiguous that remains the optimum size of unit to which all others tend.

The Maprik sub-district of the Sepik district illustrates the process very well. The Abelam[7], numbering in all some 30,000, live in basic residential units almost all falling in the range 150–250, which for analytical purposes I call the ritual group. As the name implies it was this group that performed all the major ritual, each ritual group having an internal structure which formed the basis of the organization of ritual. Each ritual group was composed of a series of hamlets, usually running along the top of a ridge, and this was the stockaded group in the days of fighting. The ritual group, even though it had its own stockade, was not however the war-making unit—it was apparently too small. All the war-making units were composed of at least two ritual groups. It so happens that among the Abelam and their immediate neighbours, there is a wide variety of both population density and warfare practices, and although the methods of cultivation are basically the same, the variation in soil quality and pressure on land combine to produce considerable variety in the composition of the subsistence diet throughout the year. These factors vary from village to village but to illustrate the points at issue I shall take three contrasted parts of the Abelam area. In the furthest east: population density less than 100 persons per square mile, moderate soils, yams (the preferred food throughout the area) provided basic subsistence for eight months of the year, supplemented by taro and sago for the rest, villages mainly of two ritual groups each, with temporary alliances and fairly heavy fighting but no occupation of the territory of the enemies; northern Abelam: larger villages, mainly three or four ritual groups each, population density 100–150 per square mile, better soils, yams eaten throughout most of the year with some supplementation, fighting less heavy and no occupation of territory, alliances stable but villages tended to fight on their own; central Wosera (south-west Abelam): very high density, up to 400 per square mile, fighting between large confederations of six to eight ritual groups each, territorial expansion actively pursued, gradually at the expense of neighbouring confederations, more effectively by the rest of the confederation turning on one of its members and taking over all their land, soils poor and deteriorating under cultivation pressure, yams barely adequate for six months of the year, supplementation not always adequate to prevent hunger. Throughout all this variation in ecology and war, the size of the ritual group remains in all examples in the range of 150–250. Further, where the pressure of war is mitigated (as in the Mamblep (northern Abelam), a valley system protected by mountains on three sides, which has about 2000 inhabitants forming a single confederation against two very large villages to the south), the ritual group emerges as a separate unit not contiguous to other like units for protection. However land holdings and old residence sites show that until the stable confederation was formed, these now spatially separate ritual groups were grouped together in twos and threes like the rest of the Abelam. Lest it should be thought that this constancy is peculiar to the Abelam the same range of figures is to be found among the neighbouring southern and plains Arapesh, although their language is totally unrelated and the institutions on which they found their settlements different. Indeed a southern Arapesh village, Ilahita, is one of the largest traditional villages in lowland New Guinea, 1200 plus, apparently having been formed under the pressure of Abelam expansion. Although the housing appears almost continuous, this complex is made up of at least six ritual groups each of which maintains its identity. It is only when the Arapesh reach the cold and poor soils of the higher parts of the coastal range that their settlement size falls to eighty and less.

The danger with normative factors is that they tend to present a picture of static societies. Before leaving the Abelam, therefore, I should mention the way in which new groups are formed. Each of these ritual groups has a complex internal structure, but when their population rises much above 300 and a split becomes inevitable, they do not split along any of the existing lines of internal cleavage, rather each constituent part splits so that from one entity with a complex internal structure arises two entities with internal structures identical with each other and the entity from which they came. The process is like cell division. Since the evidence suggests that the Abelam population has been growing for some time this is a fairly frequent process, and despite the varied ways in which they combine to form villages all Abelam ritual groups have the same basic internal structure.

What could the factors be?

It would be possible to go on discussing examples almost indefinitely, but I suspect that if I have not carried conviction so far, I am unlikely to do so by boring the reader. As already stated, it seems to me that there is a limited range of size of the basic face to face and residential group among these Neolithic, egalitarian societies that is not totally explained by ecology and warfare, but rather by a normative factor basic to the very nature of New Guinea society, or perhaps even to all Neolithic cultivators without hierarchical institutions. After all, it is generally assumed that societies with these specifications, although not of course necessarily tuber cultivators, were the originators of immobility and at the same time the first of the hominids to live at a high density; if there are, or could be, general factors which affect the size of permanent settlements they should be of wide interest. There is a need at this stage to amplify each of the terms of our specification. Neolithic refers to the basic assemblage of tools, but it is worth pointing out that with the same basic tool array there can be vast differences in the efficiency and productivity of horticultural techniques, nor does the apparent excellence of the tools correlate with the excellence of productive technology. For instance, small groups on the Upper Keram river have ample supplies of excellent stone, and make beautiful square-sided adzes of large size, yet their gardens are poor and small and their subsistence precarious; in contrast the Abelam, and related groups 100 miles or so to the north, have no local stone and make unsightly crude adzes of anything that comes to hand and can take an edge, but have successfully cleared hundreds of square miles of tropical rain-forest and consistently produce a surplus. In fact all the high density groups of Melanesia have refined gardening techniques often giving quite startling results per acre without modifying their basic tool array. In other words productivity and subsistence is determined more by social than purely technological factors.

Egalitarian societies, in the true sense of absence of any elements of ascription in status, seem to be confined to hunters and gatherers and Neolithic cultivators. Despite the well documented existence in Africa of acephalous political systems, none of them so far as I know is devoid of some ascription of status, even if only of institutionalized mediators, e.g. the Leopard-skin chief of the Nuer. It is possible that there is a technological component here; in societies such as those of Australia and New Guinea, the technology is open and available to all, there are differences in skill, but not in understanding or comprehension of what is involved. The technology is based on the exploitation of the obvious properties of naturally occurring things, there is an absence of the recondite specialized knowledge associated with for instance metal smelting and working[8]. The only source of inequality and difference in such societies is the universal difference between the sexes. Egalitarian societies can only be maintained at the cost of continuous vigilance by their members. This usually takes the form of highly aggressive competition which takes a wide variety of institutionalized forms in Melanesia but is present in all the societies under consideration (that is those with basic residential units of 100 plus). The same principles that maintain a more or less egalitarian balance within the basic unit are used to maintain the same balance between units. It seems not unlikely that for such systems to work the numbers of competing adult males must fall within a certain range. If there are too few, less than thirty over a range of age of say twenty to sixty, personality factors would, I suggest, tend to dominate, there would be an insufficiency of challengers to men of strong personality. Over seventy-five to eighty adult men the numbers of players in the game becomes too many and each player having deficient information the game becomes disorganized and unbalanced and the players split into two games of manageable size. This hypothesis is basically sociological and states that when the players in the game (of prestige) are all defined as potentially equal the game becomes static, i.e. unplayable, if the numbers are too low, and impossible if the numbers are too high.

It is of course possible to say basically the same thing in bio-genetic terms. Such a hypothesis would run: *Homo sapiens* can only handle a certain maximum number of intense face to face relationships, successfully distinguishing between each. When the number of relationships he is involved in rises above this figure he can only continue by using classification of relationships to cut down the total number of different relationships he has to act in and carry information about around in his head. This leads us to the next point which applies to either and both hypotheses. Man is undoubtedly an inherently classifying animal, this is of course one of the prime characteristics of all language, which remains a distinctive human feature. I have already said that these Melanesian groups with basic units of 150 plus, use internal segmentation (that is classification) at a higher than household level, as an essential part of their structure, and that this structure is often the basis of ritual. However, within these basic units the classification does not help much in handling relationships, which tend to remain on an individual face to face plane. Where the classification is useful is in dealing with relationships across the boundaries of the basic unit, so that, for instance, an individual Abelam carries and uses full information about all the adult males of his ritual group; with some of the members of the other ritual group of his village he will have similar individual relationships, but the rest he will interact on the basis of their clan or sub-clan membership. With the neighbouring village he will have an even smaller proportion of individual relationships, and he will have other relationships on the basis of clan and sub-clan, some other relationships only on the basis of ritual group membership; with more distant villages only the village will be considered as the basis of the relationship. Further, our individual knows that within his ritual group (basic unit) men frequently change their clan or sub-clan thus producing additional complexity. He has similar but less complete information about members of other ritual groups of his village, but as distance, social and physical, grows, he assumes that such changes do not occur so that members of other villages can be treated wholly as operating the ideal system, and the additional complexities, due to the flexibility with which the system is everywhere treated, are simply ignored. Distance, in short, lends perfection to social structures.

The use of the classification of human beings and relationships to simplify and render manageable human social life is then to be seen in New Guinea societies, but the two most important dimensions of classification in human society, after such natural ones as sex, are not used in these communities. I refer to classification by occupation and classification by ascription of status in some hierarchical scheme; usually of course these two dimensions are linked. My hypotheses would imply that either or both of these dimensions which remove the egalitarian specification, and are therefore classifications which operate within the basic unit as well as outside, are necessary if the size of the basic unit is to consistently be maintained at above 350–400 total population, or eighty to ninety adult males.

Notes

1 Hogbin, H. I. and Wedgwood, C. H. (1953). Local grouping in Melanesia, *Oceania*, **23**; Hogbin, H. I. and Wedgwood, C. H. (1953). Local grouping in Melanesia, *Oceania*, **24**.

2 Hogbin, H. I. and Wedgwood, C. H. (1953). *op. cit.* p. 2.

3 Mead, M. (1938). The Mountain Arapesh, 1. An importing culture, *Anthrop. Paper Amer. Mus. Nat. Hist.*, **26**.

4 Groenewegen, K. and Kaa, D. J. van de (1964). *Nieuw-Guinea als gebied voor demografische onderzoekingen*; vol. 1 of *Resultaten van het demografisch onderzoek Westeliik Nieuw-Guinea*. The Hague.

5 Walters, C. L. (1963). *Survey of Indigenous Agriculture and Ancillary Surveys 1961–1962*. Konedobu. However the basis of the sample used in this study is not entirely clear.

6 The southern coast of Papua is to some extent an exception to this generalization, not only did the Motuan traders whose spectacular *lakatoi* voyages are well reported, tend to live in large villages, but their customers, notably in the Gulf of Papua, lived in very large concentrations, e.g. western Elema with several villages of 1000+. Maritime trade however does not in itself produce large settlements, the participants in the *kula* ring for instance are almost entirely within Hogbin and Wedgwood's range.

7 I am most grateful to the Horniman Scholarship Fund of the Royal Anthropological Institute and the Bollingen Foundations, New York, for financing my two field trips to the Abelam and the Sepik.

8 The only transformation of natural materials (in the sense that man changes their natural properties) that takes place in New Guinea is in the making of pottery. Throughout the area pottery-making tends to be concentrated in certain villages and pottery to be extensively traded. The pots are very low-fired and there seems little mystique involved. The transformation of nature is minor and indeed familiar even in non-pottery villages through the widespread use of clay and earth for fireplaces.

ROBERT LAYTON

Settlement and community

Gemeinschaft and Gesellschaft

When investigating the nature of social relations within settlements it is useful to study these relations under two headings. First, one may attempt to discover how far social interaction tends to have any constant features. Second, one should try to show how relations in a particular settlement are channelled and restricted by conventions more or less unique to that particular social setting. The extent to which the members of a settlement may be said to participate in a common community, and the conditions under which they are able to take concerted action as a group, will certainly vary from one instance to another. It can, indeed, be argued that in exceptional circumstances a settlement need be nothing more than an aggregate of people resident in a particular locality.

Tönnies attempted in the 1880's to distinguish between "natural" forms of interaction and deliberately constructed social conventions. He contrasted village life with the activities of a commercial town. "Neighbourhood", he wrote, "describes the general character of living together in the rural village. The proximity of dwellings, the communal fields, even the mere contiguity of holdings necessitate many contacts of human beings and cause inurement to and intimate knowledge of one another"[1]. From such intimate knowledge a sense of communal sentiment was supposed to emerge which Tönnies described as "Understanding". When there were positions of superordination in the community they were modelled on the paternal authority of the father, and each individual was linked to others by ties of mutual service and dependence.

"Those who love and understand each other remain and dwell together and organize their common life"[2], their activities revolving around the possession and enjoyment of common property. To Tönnies, such a community (called by him a *Gemeinschaft*) seemed the natural or original form of human society. Tönnies contrasted this perhaps rather idealized picture of community life with what he named the *Gesellschaft* (association or society) exemplified by the processes of life in an urban environment. In the *Gesellschaft* no actions are undertaken by one individual on behalf of those others who participate with him in the same society; each individual exists in a state of competition with all others. Transactions of goods and labour are carried out on a rational, commercial basis, and each member of the *Gesellschaft* is interested in others "only in so far as and as long as they can further his own interest"[3]. If there is consensus in the *Gesellschaft* it is based not on implicit understanding, but on a formally specified contract.

What Tönnies saw in the rural village was, then, a particular form of interaction; a form which he believed to be a "natural" aspect of human society. That there may be a degree of truth in his conclusion is suggested by a number of more recent studies, studies which appear to show that ties of mutual service and dependence, based on a common understanding of certain values, are to be found in quite diverse contexts.

More recent studies

In his study of the function of marriage exchanges in small-scale societies, Lévi-Strauss took up an argument earlier proposed by Mauss, maintaining that in general "exchange in primitive societies consists not so much in economic transactions as in reciprocal gifts . . . this primitive form of exchange . . . is an event which has a significance that is at once social and religious, magic and economic, utilitarian and sentimental, jural and moral"[4].

Lévi-Strauss goes so far as to maintain that the existence of extensive social relationships in "primitive" societies depends on the presence of certain universal structures of the human mind; among which he would include "the notion of reciprocity regarded as the most immediate form of integrating the opposition between the self and others" (i.e. of establishing social relationships) and the belief "that the agreed transfer of a valuable from one individual to another makes these individuals into partners, and adds a new quality to the valuable transferred"[5]. According to Lévi-Strauss the generation of ties through the reciprocal exchange of valuables is thus "a universal mode of culture, although not everywhere equally developed"[6].

One of the peoples discussed by Lévi-Strauss are the aboriginal inhabitants of north-east Arnhem Land, in Northern Australia. These nomadic hunters and gatherers live in small, scattered bands. During the rainy months of the year the local group consists generally only of a few adult brothers with their wives and children. When the dry season arrives and natural conditions improve, the small bands come together to form rather larger groups, of thirty to forty people[7]. Despite their dispersal, these people possess elaborate patterns of reciprocal exchange. "The native sometimes exchanges things he possesses for the things he needs and cannot get from his own territory; but the larger basis of trade in north-eastern Arnhem Land is one of social reciprocity which establishes a social bond between the traders and enlarges the social periphery of each. A trade between two people is considerably ritualized, and may be an exchange of the same objects, such as carved spears, spear-throwers, or baskets"[8].

Material evidence of trade among Palaeolithic or Mesolithic populations presumably demonstrates that notions of reciprocity existed among prehistoric peoples. It thus appears that the presence of settled villages is not a necessary precondition for patterns of social exchange to develop. It seems rather that if, for other reasons (economic, political or geographical), settled villages are established, then fresh opportunities emerge for the formation of such relationships.

Passing from simple to complex societies, Tönnies' characterization of the *Gemeinschaft* is echoed in Blau's recent study of interpersonal relationships in Western culture[9].

Blau's analysis is based largely on observation of the personal relationships established (outside the formal ranking system) in offices and institutions in the United States, but just as the *gemeinschaft* was supposed to be a natural state, so the relationships described by Blau are supposed by him to develop whenever individuals engage in the exchange of goods and services on an informal, face-to-face basis. Such "social exchange", Blau writes, is distinguished from purely economic exchange by the fact that it is not based on a precise evaluation of the worth of any of the commodities that are involved. Obligations are definitely created, but they are of an unspecified kind: "while there is a general expectation of some future return, its exact nature is definitely not stipulated in advance. . . . Since there is no way to assure an appropriate return for a favour, social exchange requires trusting others to discharge their obligations"[10]. If two persons who engage in such an exchange are both able to fulfil the obligations they incur the relationship that emerges is one of equality and friendship. If one partner is unable to reciprocate but nevertheless prepared to enter into the relationship then it becomes one of super- and subordination.

The possibilities of engaging in face-to-face, interpersonal relationships with other members of one's society are limited, and Blau argues that where large numbers of people are involved, there must be "mediating structures" to channel and direct social behaviour. While some forms of social organization thus emerge directly from the processes of exchange, other forms are established explicitly, in order to achieve particular ends. "In these formal organizations special mechanisms exist to effect the co-ordination of tasks . . . in pursuit of given objectives"[11]. Such formal structures may be compared with the contracts of Tönnies' *Gesellschaft*.

Foster has suggested that personal knowledge of all other adults will be difficult in a settlement of more than 1500 people[12]. One useful definition of the village is as a settlement small enough to house a single community. Nevertheless, it is possible for communities to emerge within certain parts of a larger settlement.

Bott, in a study of the East End of London, concluded that interpersonal ties were best able to proliferate within a residential area where the inhabitants had uniform interests, where physical and social mobility were at a minimum, and where people were most dependent on their friends and relatives as intermediaries in obtaining employment[13]. These requirements were found to be best developed in working-class neighbourhoods: the structure of their occupations made the development of a homogeneous residential area of doctors, lawyers or accountants unlikely, and their opportunities for employment seemed to depend more on the possession of particular skills and training than on personal contacts[14].

The relevance of these concepts to life in actual settlements

Interpersonal ties are unlikely, alone, to give any great cohesion to a settlement. As Tönnies himself acknowledged, those who frequently come into contact are as likely to quarrel as to co-operate: "Not only does constant proximity and frequency of contact mean mutual furtherance and affirmation, but inhibition and negation also become real possibilities or probabilities"[15]. Nor do interpersonal ties alone generate corporate groups; they lead to the formation of *networks* in which each individual is linked perhaps to several others, but there is no possibility of common action being taken by all those so related, nor of one individual legitimately being able to represent the interests of the others. If interpersonal ties generate any cohesion, it is a solidarity that depends on a multiplicity of cross-cutting ties and temporary allegiances.

Like many such dichotomies, however, the concepts of *Gemeinschaft* and *Gesellschaft* appear in fact to relate not to distinct types of settlement but to forms of organization that co-exist in the majority of actual instances. Not only may a community develop within one sector of an urban centre, but even small and isolated settlements appear to codify all procedures to a greater or lesser degree. The notion of reciprocity may be universal, but what exactly is held to constitute a reciprocal exchange will differ from one culture to another. Local standards appear to colour even the most implicit of values. Almost invariably, moreover, the rights and duties accompanying certain roles will be precisely specified; particularly those that allow the administration of resources and the delegation of authority to representative officials. If the settlement is part of a wider State, then this administrative system is likely also to provide for offices intermediary between local settlement and central government, the latter taking over some of the activities that, in an autonomous unit, would have been performed locally.

If a settlement acts as a corporate unit it is, then, likely to be through the medium of its formal administrative system, and one would want to know in what areas does this operate: *are* important resources exploited on a corporate basis? To what extent is authority delegated by the body of the community to its representatives? What are the channels for reaching a consensus of opinion within the group? This is not simply because interpersonal ties alone are insufficient to generate a bounded group. Both Blau and Tönnies, as mentioned above, maintained that they were describing patterns of relationship that emerge wherever individuals are able to engage in face-to-face interaction: because they are kin or because they live or work together. Such relationships *cannot* therefore alone be called on to explain any of the diversity that may be found in settlement patterns.

The relevance of these concepts to social change

It is, of course, probable that any one settlement will adopt a variety of administrative systems in the course of its history, in response perhaps to new means of exploiting resources, or (if it participates in a wider State) as a result of deliberate policy by the central power. It might, in consequence, be possible to demonstrate that the settlement has at different times had a greater or a lesser ability to engage in corporate action, or to channel the activities of individual members within formally-constituted limits. It also appears that an increase in size, in diversity of occupation or in the geographical mobility of the population is likely to result in a reduction in the *Gemeinschaft*-like aspects of informal interaction between members of the settlement.

Notes

1 Tönnies, F. (1957). *Community and Society, Gemeinschaft und Gesellschaft*. Michigan. p. 43.
2 Tönnies, F. (1957). *op. cit.* p. 48.
3 Tönnies, F. (1957). *op. cit.* p. 57.
4 Lévi-Strauss, C. (1969). *The Elementary Structures of Kinship*. London. p. 52. French edn. (1949).
5 Lévi-Strauss, C. (1969). *op. cit.* p. 84.
6 Lévi-Strauss, C. (1969). *op. cit.* p. 53.
7 Warner, W. L. (1964). *A Black Civilization*. New York. pp. 127–8. 1st edn. (1937); see also Lee, R. B., this volume, p. 181.
8 Warner, W. L. (1964). *op. cit.* p. 450.
9 Blau, P. (1964). *Exchange and Power in Social Life*. New York.
10 Blau, P. (1964). *op. cit.* pp. 93–4.
11 Blau, P. (1964). *op. cit.* p. 199.
12 Foster, G. (1960). Interpersonal relations in peasant society, *Human Organization*, **19**, p. 178. I myself would guess that 500 would be a more accurate figure; see also Forge, A., this volume, pp. 374-5.
13 Bott, E. (1957). *Family and Social Network*. London. pp. 103, 104, 112.
14 Bott, E. (1957). *op. cit.* p. 113.
15 Tönnies, F. (1957). *op. cit.* p. 44.

COLIN RENFREW

Patterns of population growth in the prehistoric Aegean

The estimation of settlement density and of population density on the basis of survey data from prehistoric sites is one of the most perilous exercises in prehistoric archaeology. Indeed there is perhaps only a single statement to be made about survey data whose truth cannot be denied: that the picture obtained is an incomplete one. The nature and extent of the lacunae are matters for speculation. Many of the techniques of locational analysis devised by geographers are not appropriate (*a*) when the data are known to be incomplete, and (*b*) when no reliable estimate of the proportion of unrecorded sites has been made.

Many studies of prehistoric settlement pattern, in consequence, very naturally study the ecological and social environment of single or idealized sites or site-types. With rare exceptions[1] the problem of comparing settlement density, or population growth in absolute terms is avoided.

In the treatment which follows an attempt is made to overcome some of these difficulties by comparing the patterns of growth over two millennia in adjacent or nearby areas of the Aegean. Basic underlying factors may then be outlined which could account for the different growth patterns observed.

Using these apparent regularities, together with two basic assumptions, hypothetical figures may be obtained for the prehistoric population of the various Aegean regions in question. These figures reinforce the view that prehistoric Aegean civilization was fundamentally a civilization without cities. Minoan-Mycenaean civilization was notable not for the scale of its monuments or the size of its centres, but for its internal organization and the products of that organization, both reflected in the Minoan-Mycenaean palace.

The limitations of survey data

Even when a systematic survey for prehistoric sites has been undertaken within a specified region there are several cautionary factors to be contemplated:

(i) Members of the original population of sites may have been destroyed or obscured, whether by geological agency (erosion, aggradation, etc.) or human activity (e.g. deep ploughing).
(ii) Only part of the *surviving* site population is represented in the survey. Moreover there is no easy way of establishing the relationship between total and sample populations.
(iii) Known sites are generally most numerous in areas where archaeological activity has been most intense.
(iv) Entire categories of site may sometimes escape recognition while other categories, easily recognizable (e.g. tells) are disproportionately well represented in the survey.
(v) The direction of archaeological research may positively favour the recording of sites of a particular category.
(vi) The dating of prehistoric sites on the basis of surface finds implies that those periods lacking in characteristic pottery, flint or other artefact types are likely to be under-represented in the survey.

Systematic search within a given region can diminish the effect of some of these factors by ensuring that small and inconspicuous sites are not overlooked. In general, however, and this is certainly true for the Aegean, most surveys for prehistoric sites in a given region set out by including all known sites, however they have been located. A search operation is then carried out for further sites. Often the lands chosen for survey are those which have proved most productive in the past, in an effort to maximize the total number of sites discovered. More rarely, sites of a different kind are sought, in different terrain, to maximize the range of settlement types located, but in general all sites already known, through aerial photography or chance discovery or whatever means, are included. The "sample" of sites yielded by the survey is simply the total of sites located in the area in question. Of course a more systematic procedure is to select an area and survey it intensively, in this way eliminating the accidents of chance discovery. But complete survey of the terrain is so time-consuming (especially in hilly country, where air photographs are more difficult to interpret archaeologically) as to be scarcely practical for areas greater than a few hundred square kilometres.

These limitations make the comparison of settlement density figures between regions especially dangerous. That the recorded settlement density in late Bronze Age Crete, for instance, is nearly eight times that in Central Macedonia, may simply reflect the very much greater archaeological activity in Crete since the early years of this century. Evident differences in settlement pattern—most known Macedonian prehistoric sites are tell mounds, while only one tell is known from Crete—make this a difficult situation to control.

A convenient solution, and it is only a partial one, is to defer the comparison of observed settlement densities in different regions, comparing instead the observed settlement densities for different periods within the *same* region.

This will at least avoid some of the ill-effects of differing survey intensities, although not those arising from the preferential inclusion of certain specific classes of site (for example tell mounds) in a single region if these were in fact a preferred settlement type in one period rather than another. Instead of absolute figures for settlement density, we compare in the first place the differing growth patterns for the various regions. Period-by-period ratios for each region are compared, rather than absolute figures. In the case of the Aegean the results are so striking as to encourage a further examination of the observed absolute density figures themselves.

Patterns of growth in the Aegean

Site surveys in recent years in various areas of the Aegean make possible a comparison of the growth in settlement numbers in each from the Neolithic to the Late Bronze Age. From north Greece, figures are available from central Macedonia[2]; from the southern Aegean from Crete[3], the Cycladic Islands[4], Euboia[5] (off the coast of Attica) and Laconia[6], a coastal district of the eastern Peloponnese. Data are available also from Messenia[7], in the western Peloponnese, which faces westwards to the Ionian Sea. In each case most indentifications classed as doubtful have been accepted and included with the exception of those sites in Euboia which were identified as Neolithic simply on the basis of a scatter of obsidian. Generally assignment to period is on the basis of pottery found on the surface. It should be noted, of course, that sites may not have been occupied throughout the duration of the periods in which they are represented by surface finds. The relative length of the periods is clearly a factor here.

The figures given in Table 1 express observed settlement densities in terms of the total land area (and see Fig. 1).

Certain striking regularities are at once apparent, the data from southern Greece clearly yielding two different growth patterns: A, with steady increase in settlement numbers from Neolithic to Late Bronze Age, and B, with a

	Number of sites per thousand km²				
Region	*Neo.*	*E.B.A.*	*M.B.A.*	*L.B.A.*	*Total*
Crete	5·1	13·6	23·2	34·7	46·2
Cyclades	4·0	20·5(50·9)	7·2	12·9	33·0(63·8)
Messenia	5·9	11·9	36·6	69·0	73·6
Laconia	2·8	11·1	7·1	15·9	16·7
Euboia	4·5	15·2	9·5	13·8	23·0
C. Macedonia	2·8	3·7	2·3	4·8	6·6

Table 1 Period-by-period comparison of observed settlement densities in selected regions of Greece. (The figure for the Early Bronze Age Cyclades excludes locations known exclusively through cemeteries: the total, including such cemeteries, is given in brackets).

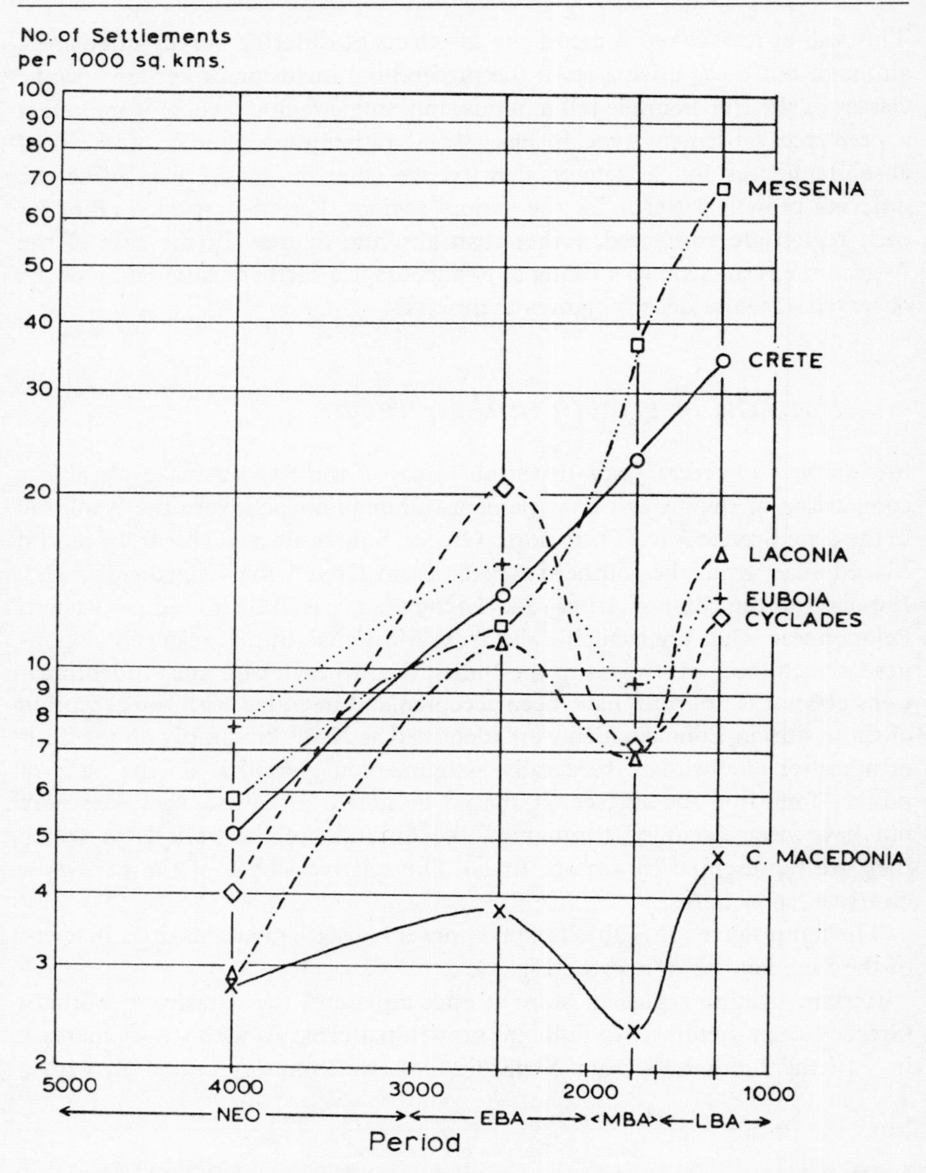

Fig 1 The growth in number of settlements recorded in different areas of the prehistoric Aegean, on the basis of recent site surveys (See Table 1).

marked increase from Neolithic to Early Bronze Age followed by a decrease in settlement density in the Middle Bronze Age and a striking recovery in the Late Bronze Age.

These differences, and the anomalous behaviour of central Macedonia (the only region of north Greece represented in the surveys) are brought out by the growth factors given in Table 2.

The two regions whose growth may be described as Pattern A, namely Crete and Messenia, have a growth factor (i.e. a ratio of increase of observed settlement numbers) of between two and three from Neolithic to Early Bronze Age, of at least six from Neolithic to Late Bronze Age, and consequently of at least 2.5 from Early to Late Bronze Age.

The three regions whose growth may be said to conform to Pattern B, namely Laconia, Euboia and the Cyclades show a growth factor of at least three from Neolithic to Early Bronze Age, yet of less than 1·5 from Early to Late Bronze Age. In addition they show a strong reduction in settlement numbers in the Middle Bronze Age.

Region	*Crete*	*Cyclades*	*Messenia*	*Laconia*	*Euboia*	*C. Macedonia*
Growth factor Neo. to E.B.A.	2·6	5·1(12·7)	2·0	3·9	3·4	1·3
Growth factor Neo. to L.B.A.	6·8	3·2	11·6	5·6	3·1	1·7

Table 2 Growth factors (i.e. ratio in number of settlements recognized) from Neolithic to Early Bronze Age and from Neolithic to Late Bronze Age, for selected regions in Greece.

Central Macedonia stands apart from the rest by virtue of its very low growth factor from Neolithic to Early Bronze Age. As in the regions conforming to Pattern B, there is a decrease in settlement numbers in the Middle Bronze Age.

Pattern A, a straight line on a logarithmic scale, approximates to an exponential increase in the number of settlements (see Fig. 2). This would seem to imply an uninhibited growth of settlement numbers, without severe restraint imposed by very limited food supplies or other constraining factors. Such patterns of settlement and population growth are familiar in many parts of the world at different times. Viewed in these terms the growth of settlement may be seen as a natural one, without striking discontinuity.

Pattern B indicates a severe decline in the number of settlements recorded for the Middle Bronze Age (see Fig. 2). This may in part be due to the shorter duration of the period. And we shall see below that while the number of settlements may have been smaller, their individual size may have increased. No decline in human population is necessarily implied, although the growth rate certainly diminished. The striking disparity in the growth patterns at this time suggests the operation of factors in the regions conforming to Pattern B which did not apply to those of Pattern A.

Perhaps the most striking feature of these figures is the high growth factor for all the regions of southern Greece over the time span Neolithic to Early Bronze Age. In the one region of northern Greece under consideration, central Macedonia, the factor is 1·3. If the figures from the remainder of Macedonia and from Thrace were comparable to those from central Macedonia, and

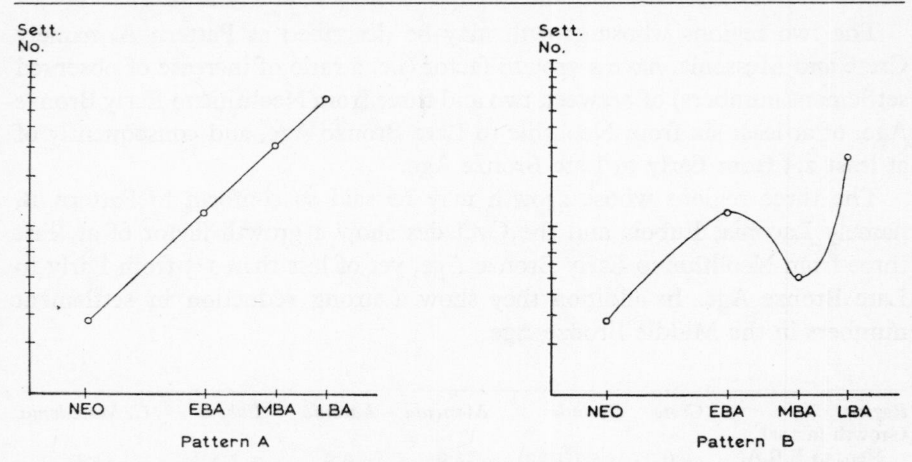

Fig 2 Patterns of growth: two different growth patterns for settlement numbers in the prehistoric Aegean. Pattern A (left): exponential growth (Messenia and Crete). Pattern B (right): retarded exponential growth (Laconia, Euboia, Cyclades).

this remains to be demonstrated, factors must have come into operation during the Neolithic/Early Bronze Age serving to differentiate between settlement densities of northern and southern Greece.

To recapitulate, certain conclusions may be drawn from this analysis (see Fig. 2):

1 Two very different growth patterns are obtained for settlements in the prehistoric southern Aegean, patterns A and B.
2 Pattern A presents a picture of exponential growth in settlement numbers.
3 Pattern B shows an increase severely inhibited in the Early Bronze Age/Middle Bronze Age.
4 A distinction is drawn between north Greece, where growth is slow throughout, and south Greece, where growth is rapid from the Early Bronze Age onwards.
5 The growth inhibition of Pattern B, in Early Bronze Age/Middle Bronze Age times is so marked in the Cyclades and Euboia that the number of settlements recorded in Late Bronze Age times is less than that for the Early Bronze Age.

These conclusions, and the figures upon which they are based, are, of course, subject to many of the sources of error discussed in the preceding section. They should avoid, however, the grosser inconsistencies arising from the different level of archaeological activity in the various regions in question. The low *absolute* figures for prehistoric settlement density in central Macedonia may simply reflect the absence there, until recently, of systematic survey work. The very slow growth in settlement numbers observed cannot be explained in this way.

Factors influencing growth

The explanation of events or trends in the prehistoric past, and this doubtless applies to explanations for observed settlement densities or settlement patterns, remains an art rather than a science. Event X is often said to be "explained" by event Y if they are approximately contemporary (strictly Y should precede X, but this is not always easy to show in the prehistoric past) and if Y may be regarded as independent of X. Causal links are so difficult to demonstrate that the dependence of X upon Y is often simply assumed.

Preferably, of course, the explanation should be in the form of a generalization: events of type X are generally preceded by (and hence "caused" by) events of type Y. This gives the explanation a predictive content theoretically open to testing. In most cases, however, no satisfactory test case is immediately available, so that the choice of explanation is left to subjective criteria. Often the most satisfactory explanations are those adducing widespread and independently documented changes outside the field of human activity (e.g. worldwide climatic changes). In the absence of such a cause, an explanation in terms of factors endogenous to the culture system is generally preferable to an appeal to outside "influences", unless these can fully and convincingly be documented.

The explanation here proposed for the marked growth in settlement numbers from Neolithic to Early Bronze Age in various regions of the southern Aegean conforms to these criteria. Evidence from the site of Sitagroi in north Greece shows that the wild grape was being utilized there from about 4500 B.C. Morphological developments in the preserved remains indicate that by the early third millennium B.C. the grape had been domesticated. Finds of olive stones from Early Minoan contexts in Crete, supplemented by traces of olive oil, together with two oil lamps, from an Early Cycladic grave at Spedhos in Naxos, show that the same is true of the olive. It is here proposed that the marked increase in settlement numbers at this time in the southern Aegean was made possible by the development of Mediterranean polyculture, that is to say the intensive cultivation of Mediterranean arboreal crops (olive, grape, fig) in addition to the typical Neolithic agricultural products, cereals and legumes (see Fig. 3).

The significance of olive oil in the diet of southern Greece is documented by recent figures for Crete[8] where it contributes 29% of the total calorie intake (as against 33% for cereals), with a production of 1·7 million calories per acre, compared to 0·99 million for barley and 0·95 million for pulse crops.

In terms of this explanation, a marked increase in settlement numbers is to be expected in regions of true Mediterranean climate, where Mediterranean polyculture can most successfully be practised. The restricted increase observed in central Macedonia, which lies outside this region, is in conformity with this explanation.

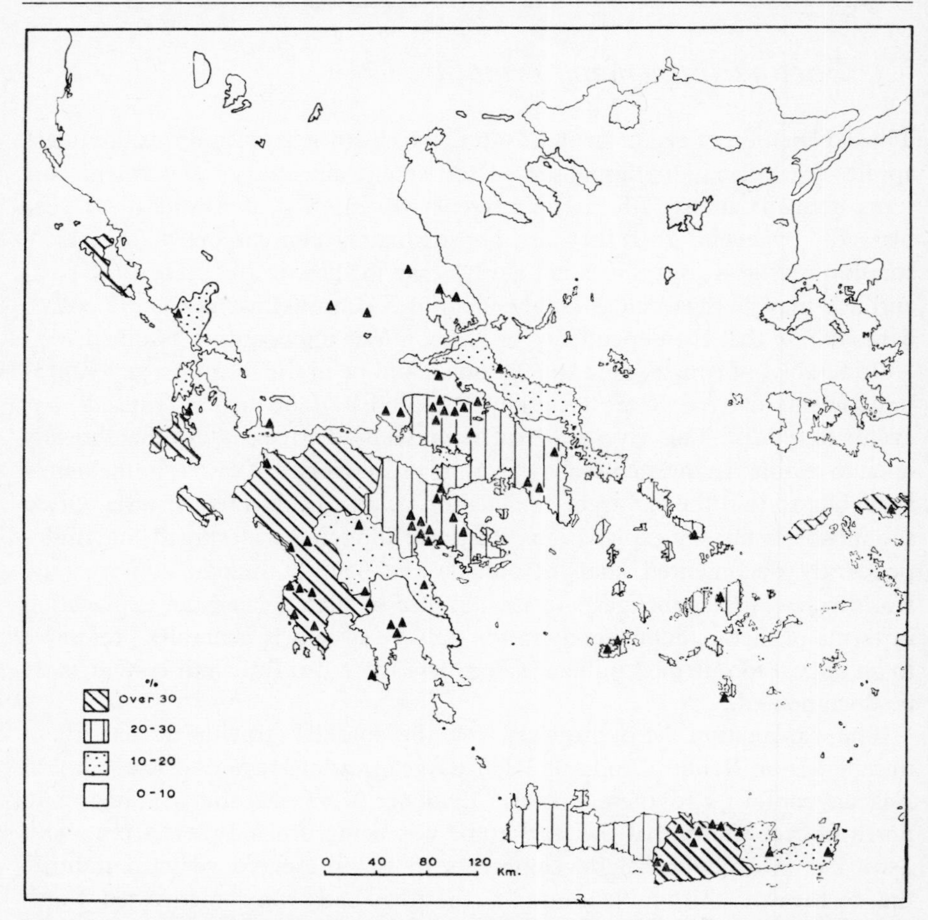

Fig 3 The correlation between the distribution of major Minoan-Mycenaean sites in Greece and the practice of Mediterranean polyculture, as represented by modern viticulture. Percentages indicate the proportion of land in each region given over to cultivation of vines in A.D. 1937.

A second explanation is required to account for the differences between growth pattern A and pattern B. The sustained growth in Crete and Messenia during the Middle and Late Bronze Ages is in part explained by the productivity of these regions in cereals, olives and vines. But this does not account for the marked reduction in the number of settlements of the Middle Bronze Age in areas of growth pattern B.

The explanation here must be in social terms. It is indeed during the Early Bronze Age that the first good evidence for warfare is seen in the Aegean. Copper and bronze daggers are found in considerable numbers in most regions of the southern Aegean as well as in the Troad. Fortifications are now seen in the Cyclades (Chalandriani in Syros, Panormos, Spedhos and possibly Kastraki in Naxos), in Euboia (Manika), in mainland Greece

(Lerna, Askitario, Aegina), and in the Troad and offshore islands (Troy, Thermi, Poliochni, Emborio).

It is here suggested that the principal danger was from maritime attack—and both incised representations and actual models of ships, dating from exactly this time[9], indicate the efficacy of sea movement. Eloquent testimonies exist, both from the Classical period and the Middle Ages[10], to the depredations wrought by pirates upon the inhabitants of the Cyclades and other maritime regions, with consequent depopulation. It is suggested that similar conditions came into operation in the Aegean during the later part of the Early Bronze Age. Coastal settlements were now highly vulnerable to attack by sea-raiders or pirates, and only well defended sites, or those situated some kilometres inland, could survive.

The suggestion is supported not only by the fortifications which appeared at this time, but by a concomitant change in settlement type. In many areas the favourite location is no longer a low promontory with easy access to the sea, like Pyrgos in Paros or Manika in Euboia. Instead defensible hilltop sites were preferred. In some cases, like Rizokastelia in Naxos, and like many medieval villages in the Cyclades (e.g. Seriphos, or Plaka in Melos), these were remote from the sea and out of reach of pirates. In others, like Chalandriani in Syros or Spedhos in Naxos, they immediately overlooked the sea, but from steep cliffs. In the Middle Bronze Age Cyclades, in contrast to the Early Bronze Age, there was generally a single important settlement in each island. Often, as at Phylakopi in Melos or Akroterion Ourion in Tenos, it was fortified by thick walls.

Crete and Messenia, the two regions surveyed whose development conforms to pattern A, will also have faced these problems. Indeed there is evidence for a choice of hilltop settlement sites in Messenia, such as Malthi, in the Middle Bronze Age. Yet of the several regions surveyed, only Crete and Messenia have extensive cultivable lands remoted from the sea (further than about 10 km) and out of reach of pirates. This advantage may have been decisive, especially when olives formed a major part of the agricultural produce. If the olive trees are destroyed, at least a decade is needed before effective production can be re-established, and the destruction of crops is much more serious to Mediterranean polyculture than to simple cereal farming.

A further causal factor in these regions was the developing social order which reached its acme in the Cretan palaces and the Messenian palace at Pylos. Of course the increasing population and the developing social order were mutually favourable causal factors: there was positive feedback, and neither can be regarded as the independent variable (see Fig. 4).

This mutual dependence prevents our regarding the social structure as the "cause" of the population and settlement growth, in the sense discussed at the beginning of this section. Yet the complete absence of fortifications in prehistoric Crete is one of the most notable characteristics of Minoan culture.

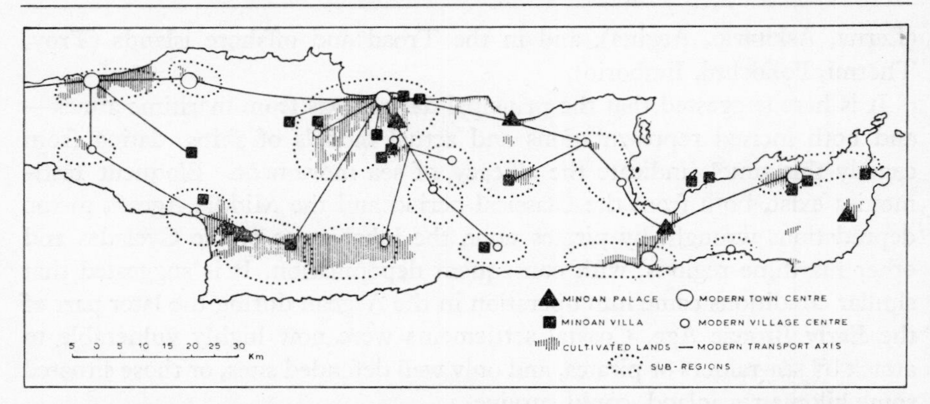

Fig 4 Settlement hierarchy in prehistoric and modern Crete. The late Minoan distribution of palaces and villas compares strikingly with that of modern towns and village centres. (The numerous smaller prehistoric sites and modern small villages are omitted.) Both relate logically to sub-regions and to cultivable land.

In prehistoric archaeology, where discussion in systems terms is becoming common, the *explicans* may be as dependent on the *explicandum* as vice versa. In Crete, as probably in Messenia, the social system and the population density developed together: neither could have advanced without the other.

Towards an estimate of population

Two essential steps have to be taken before the observed settlement figures for each period, as determined by site survey, can be used as the basis for an estimate of population. First some way has to be found of estimating the original settlement density in prehistoric times from the density observed through recent survey of prehistoric sites. Secondly an estimate must be made of the mean population of each settlement, on the basis of observations of settlement area, spacing of dwellings and so forth.

A notable feature of the site surveys (cf. table 1) was the very high settlement density obtained in the Cyclades for the Early Bronze Age period if all known Early Cycladic cemeteries were included, as well as sites known from settlements or both cemetery and settlement. The total of fifty-one Early Cycladic settlements known is raised to 127 if all the cemeteries are included. In the foregoing discussion the former figure was preferred, giving a settlement density more in keeping with the figures obtained from other regions, although actually still in excess of them.

This high figure for the Cycladic cemeteries is not difficult to explain. It arises partly from their construction, and partly from the great interest in Cycladic antiquities, which for fully a century now have attracted archaeologists. This has ensured that most finds made during this century have been reported in the archaeological literature. The graves themselves are generally simple cists, built of flat slabs of marble or schist, which admirably preserve the grave goods. They are easily discovered and observed in the course of ploughing.

Undoubtedly, every cemetery originally had an accompanying settlement, and clearly a better estimate of settlement numbers, in absolute terms, is obtained by including sites known from the cemetery only. This leads one to suggest that the survey figure for Cycladic settlements (fifty-one) be multiplied by a figure of three, yielding a total slightly in excess of the total number (127) of sites, including cemeteries without settlements, which are known.

In order to reach some sort of absolute figure, however approximate, it is proposed to generalize this result, applying it to other regions of Early Bronze Age Greece. The observed settlement densities will thus be multiplied by three to give the estimated original settlement density. Since Late Bronze Age pottery is very characteristic in southern Greece, and, of course, near the surface on deeply stratified sites, a factor of only two is proposed for the Late Bronze Age, and a factor of 2·5 for the Middle Bronze Age. For the Neolithic period, where sites are not always easy to recognize, a factor of four is proposed.

It is necessary, also, to guess how many of the settlements observed in each period were actually occupied at the same time. For the Neolithic, Early and Late Bronze Age, which are each 500 years or more in duration, a figure of 75% is suggested. For the Middle Bronze Age, which was shorter, the assumption is made that all the settlements observed were simultaneously occupied.

The first problem has thus been resolved, largely by guesswork, following up the hint offered by the Cycladic cemeteries. To estimate the original settlement density, we shall therefore multiply the observed settlement density for Neolithic sites by a factor of 3·0; for Early Bronze Age sites by 2·3; for Middle Bronze Age sites by 2·5; and for Late Bronze Age sites by 1·5.

The second problem, to estimate the mean population of each settlement, has to be approached by the estimation of settlement area. A study of Aegean settlement sites[11] suggests a typical Neolithic village size in the range of 4000 to 8000 m^2, reaching 10,000 m^2 at times, a size comparable with those of very early farming villages in the Near East such as Jarmo or Ali Kosh. The area of proto-urban "towns" in the Aegean ranges between 4500 m^2 at Thermi to 15,000 m^2 at Poliochni, and a figure of 10,000 m^2 has been chosen. This may be compared with 70,000 m^2 for pre-Dynastic Uqair, and areas of the same order for Chalcolithic Can Hasan and Neolithic Çatal Hüyük (although these last are based on the area of the tell mound).

The area of a typical Late Bronze Age Aegean major settlement may be set somewhere between 10,000 and 40,000 m^2 (1 to 4 hectares; 2½ to 16 acres). The settlement at Pylos, for instance, covered an area of 12,600 m^2, and 38,500 m^2 were enclosed within the citadel at Mycenae. These may be compared with the area of 600,000 m^2 for Early Dynastic Ur, and 4·5 million m^2 for Uruk.

As a working basis for the area of settlement, 5, 10, 15 and 20 thousand square metres will be used for the Neolithic, Early, Middle and Late Bronze Age.

In converting these areas to population figures we may refer to Frankfort's estimate for Mesopotamian urban sites[12], where he suggested a population density of 400 persons per hectare. Even in Late Bronze Age times in the Aegean, the population was probably not so closely packed together as in a Sumerian town. So 300 persons per hectare will be taken for the Bronze Age periods, when many Aegean settlements were of urban or proto-urban nature. For the Neolithic, when houses were sometimes more widely spaced, a figure of 200 persons per hectare is suggested.

This yields a suggested mean population of 100 persons for Neolithic settlements, 300 for the Early Bronze Age, 450 for the Middle Bronze Age, and 600 for the Late Bronze Age. These figures do, however, rest on the assumption that the mean Early Bronze Age settlement occupied twice the mean area of Neolithic settlements, and half the mean area of Late Bronze Age ones. It may be that the size estimates discussed above have emphasized the larger settlements of the bronze age at the expense of the smaller ones. A more modest estimate would be 100, 200, 275 and 360 persons respectively as the average settlement population at the times in question. In such a case the population densities given below would be too large: on the line of reasoning here adopted they are more likely to be too high than too low.

On the basis of these figures, and the observed settlement densities quoted in Table 1, population density estimates may be obtained, as quoted in Table 3 (and see Fig. 5).

The populations of Crete and Messenia (pattern A) increase almost exponentially. Those of Laconia, Euboia and the Cyclades (pattern B) increase markedly from Neolithic to Early Bronze Age, and thereafter only slightly. Except in the case of the Cyclades, however, the increase in the size of individual settlements offsets the decrease in settlement numbers in the Middle Bronze Age. Central Macedonia, outside the region of Mediterranean polyculture, experiences a much more gradual population growth.

Region	*Neo.*	*E.B.A.*	*M.B.A.*	*L.B.A.*
Crete	1·53	9·18	26·1	31·3
Cyclades	1·20	13·8	8·1	11·6
Messenia	1·77	8·04	41·2	63·3
Laconia	0·84	7·50	8·0	14·3
Euboia	1·35	10·3	10·7	12·4
Central Macedonia	0·84	2·5	2·6	4·3

Table 3 Estimated approximate population densities for various regions of prehistoric Greece, expressed in number of persons per square kilometre.

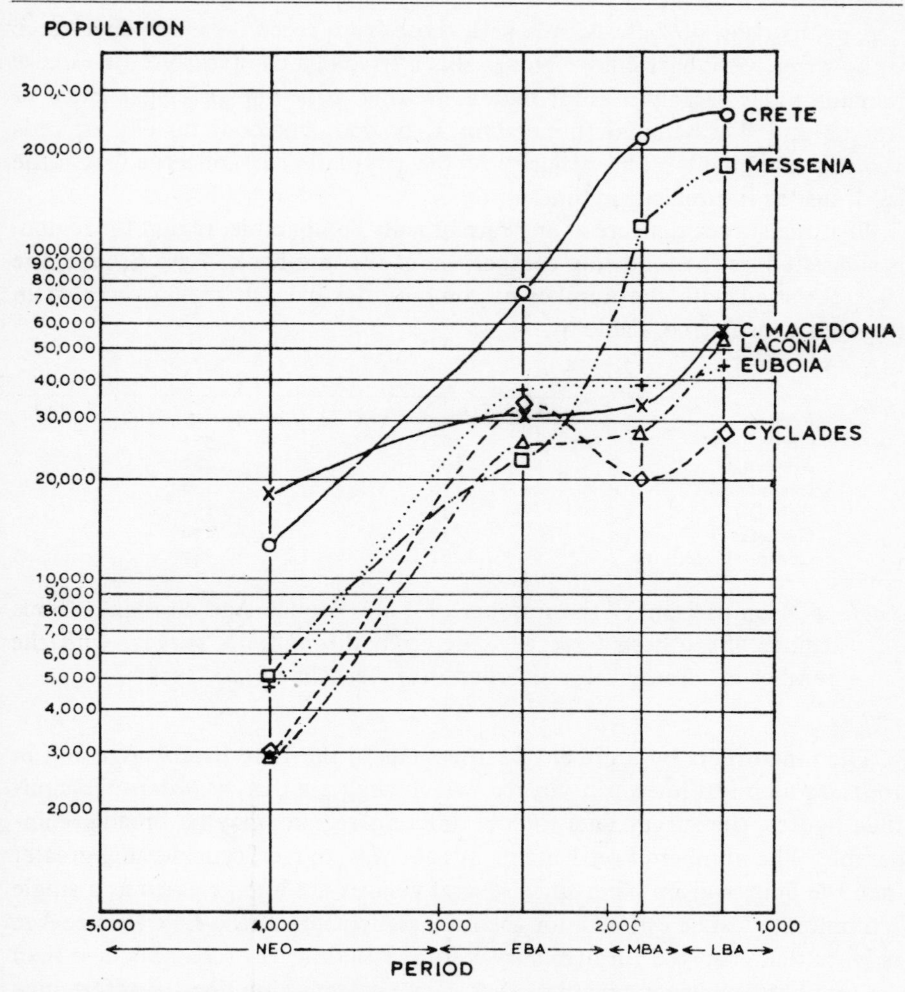

Fig 5 Estimated growth of population in the prehistoric Aegean (cf. Tables 3 and 5).

Prehistoric and recent population growth in the Aegean

In attempting to evaluate these results, several factors must be borne in mind. In the first place, for the Bronze Age periods they may err on the side of generosity since the population figures for average settlements taken for those periods may in fact refer to the larger settlements in the size range. Secondly, no allowance has been made for the varying intensity of survey: the consistently high figures for Messenia may in part reflect the recent intensive archaeological work there, just as the low ones for Macedonia mirror the relative lack of archaeological interest in that area until recent years.

A comparison of these figures with those from recent censuses in Greece is, however, very heartening. Nor is this a frivolous comparison, for Greece remains today largely a rural nation. Setting aside the principal cities of Athens and Thessaloniki (population 1,500,000; 200,000) no city exceeds 100,000 inhabitants. The majority of the population is concerned with the land, just as in prehistoric times.

That the survey data are at any rate broadly comparable, region for region, is suggested by the following comparison of the number of Late Bronze Age sites observed with the number of communities in each region revealed in the 1961 census (see Table 4).

Region	*L.B.A. sites observed*	*Communities in* 1961
Crete	284	581
Messenia	198	282
Laconia	56	166
Euboia	49	159
Cyclades	32	119
Central Macedonia	61	347

Table 4 Comparison of the number of Late Bronze Age sites in various regions of southern Greece, as detected through site survey, with the number of communities recorded in the census of A.D. 1916.

The rank orders by number of settlements in the Late Bronze Age and in 1961 are identical for south Greece, which suggests that, as order-of-magnitude figures, the survey data for the different regions may be broadly comparable. The number of settlements in 1961 was, in fact, considerably greater than the figures given here, since several villages are often classed as a single "community". The estimate for the total settlements in the Late Bronze Age as twice that observed through survey must still, in every region, be less than the number of villages existing today. The confrontation does therefore give ground for hope that the survey results for the different regions may be very approximately comparable. Again one is led to suspect that Messenia has been rather more efficiently surveyed than the other regions. Comparison in Macedonia is made difficult by the presence there of the major urban centre Thessaloniki.

The prehistory and recent history of Greece suggests a parallelism in its development, even before reference is made to the population figures. Culturally there are obvious comparisons between the newly awakened Greece, just over a century ago, and Greece in the Early Bronze Age. This idea first presented itself through a consideration of the Cyclades which have always flourished at a time when Greek prosperity was just developing, and have then taken second place when the Aegean was unified under a central power. The prosperity of the Cycladic Islands in the Early Bronze Age, or in the sixth century B.C., was seen again the middle of the last century, when

Syros was, for a while, the principal port of Greece. Athens was not yet dominant—Nauplion was even the capital of Greece for a brief period—and local centres were of great importance. With the unification of Greece, and full security on land, regions such as the Cyclades declined in importance, just as they did in the Minoan-Mycenaean empire of the Late Bronze Age and in the Athenian empire of the fifth century B.C.

It is tempting, then, to compare modern Greece with the Late Bronze Age, and Greece shortly after the liberation of A.D. 1821 with the Early Bronze Age. The modern census figures, and the estimated prehistoric population figures (cf. Table 3) are given in Table 5 (and see Fig. 6).

Region	*Population* *1838*	*Population* *Est. EBA*	*Ratio* *EBA/1838*	*Population* *1961*	*Population* *Est. LBA*	*Ratio* *LBA/1961*
Crete	c. 130,000	75,000	0·58	526,000	256,000	0·49
Messenia	70,000	23,000	0·33	252,000	178,000	0·70
Euboia	43,000	36,000	0·84	193,000	44,000	0·23
Laconia	67,000	26,300	0·39	136,000	50,000	0·37
Cyclades	97,000	34,000	0·35	141,000	29,000	0·21

Table 5 Comparison of the estimated populations for various regions of southern Greece in the Early and Late Bronze Age with census figures for 1838 and 1961 respectively. The ratios of Early Bronze Age to 1838 population, and of Late Bronze Age to 1961 population are also given.

The general similarity in growth over the two time ranges is very striking: the population total for the regions in question had increased by a factor of 3·1 over the 123 years between A.D. 1838 and 1961. The estimated population total had increased by a factor of 2·8 over the two millennia between Early and Late Bronze Age.

Even the growth pattern shows similarities. In the 130 years following the liberation of Greece[13], Crete has shown steady, indeed exponential, population growth over this period, just as in prehistoric times. This is again growth pattern A. The Cyclades, on the other hand, reached an early maximum in the 1860s and then declined slightly, only surpassing this figure in the present century. This at first sight resembles growth pattern B, although the inhibitory factor here is not lack of military security but depopulation through emigration to Athens or overseas. It would be wrong to exaggerate the significance of this general similarity: one factor highly relevant in the nineteenth century A.D. and probably much less so in prehistoric times is emigration, which has undoubtedly contributed to the relative depopulation of the Cyclades. Yet at a superficial level at least, the general analogy with the prehistoric pattern of growth is striking, and of course emigration is itself a response to factors evidently more keenly felt in the Cyclades than in other regions of Greece.

The comparison between prehistoric and modern population growth

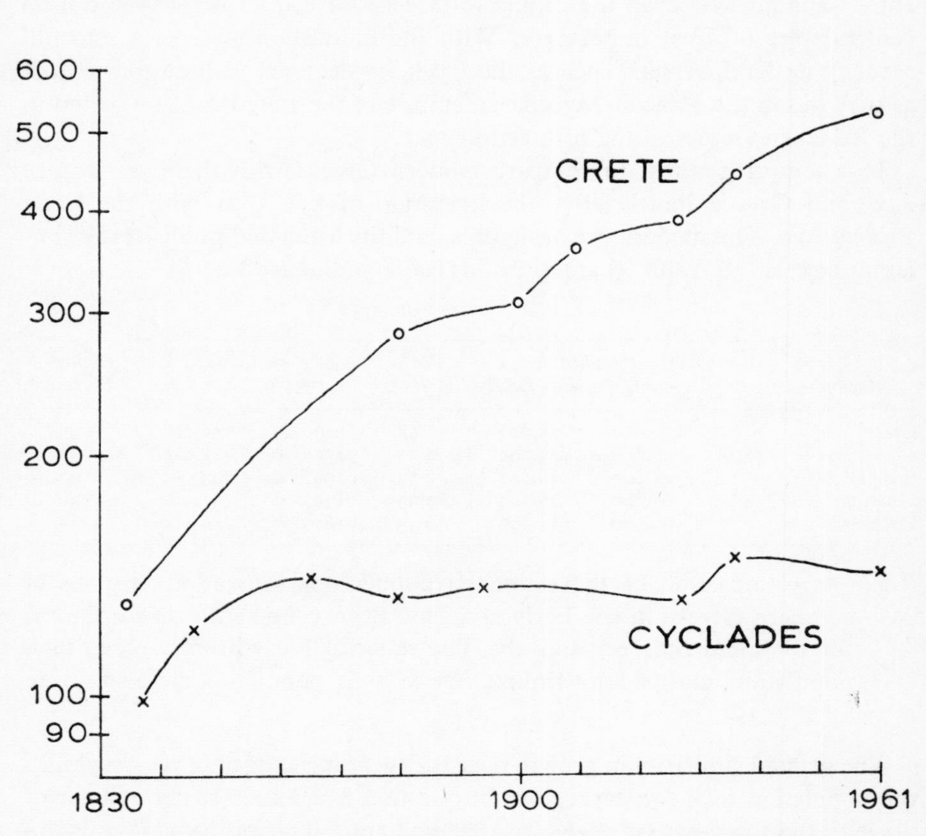

Fig 6 The growth of population in Crete and the Cyclades compared over the period A.D. 1830 to A.D. 1961. The general analogy with the prehistoric growth patterns is of note.

patterns is an interesting one. Moreover, in non-industrial regions, the area-by-area comparison of site survey data and modern settlement figures offers the hope of pinpointing significant differences, and hence of revealing either inconsistencies of survey or real underlying factors which have affected settlement growth.

The estimation of absolute population figures is a much more difficult problem. How are they to be evaluated? For a century now scholars have tried to estimate the population of Greece in classical times, without notable agreement. The problem is worse for the prehistoric period. The arguments outlined above lead to an estimate of about 200,000 for the population of Crete in Middle Minoan times (*c.* 2000 to *c.* 1550 B.C.), and, as already suggested, it appears that this may be too generous an estimate. Yet Hutchinson[14], in the best available review of prehistoric settlement in Crete, preferred a figure of about 500,000. This would imply a population in Late Minoan times greater than that today. This is a highly unsatisfactory situation (although a healthy reminder of the hypothetical nature of our figures), since a much better understanding of economic and social factors in prehistoric times would flow from a plausible estimate of the absolute population figures.

Notes

1 e.g. Hole, F. and Flannery, K. V. (1967). The prehistory of south-western Iran: a preliminary report, *Proc. Prehist. Soc.*, **23**, p. 199; Adams, R. McC. (1965). *Land behind Baghdad*. London and Chicago. p. 115.
2 French, D. H. (1967). *Index of Prehistoric Sites in Central Macedonia* (duplicated and circulated privately); Heurtley, W. A. (1939). *Prehistoric Macedonia*. Cambridge.
3 Hemingway, E. M. M. (1969). *Prehistoric Settlement Patterns in the Aegean in the Neolithic and Early Bronze Age* (unpublished undergraduate dissertation, Univ. of Sheffield); Pendlebury, J. D. (1939). *The Archaeology of Crete*. London; Hood, M. S. F. (1965). Minoan sites in the far west of Crete, *Ann. Brit. Sch. Archaeol. Athens*, **60**; Hood, M. S. F. and Warren, P. M. (1966). Ancient sites in the province of Aghios Vasilios, *Ann. Brit. Sch. Archaeol. Athens*, **61**; Hood, M. S. F., Warren, P. M. and Cadogan, G. (1964). Travels in Crete 1962, *Ann. Brit. Sch. Archaeol. Athens*, **59**.
4 Renfrew, C. (In press). *The Emergence of Civilisation: the Cyclades and the Aegean in the Third Millennium B.C.* London.
5 Sackett, L. H., Hankey, V., Howell, R. J., Jacobsen, T. W. and Popham, M. (1967). Prehistoric Euboia: contributions towards a survey, *Ann. Brit. Sch. Archaeol. Athens*, **61**.
6 Waterhouse, H. and Hope-Simpson, R. (1961). Prehistoric Laconia, Part II, *Ann. Brit. Sch. Archaeol. Athens*, **56**.
7 McDonald, W. A. and Hope-Simpson, R. (1969). Further explorations in the south-western Peloponnese, *Amer. J. Archaeol.*, **73**.
8 Allbaugh, L. G. (1953). *Crete, a Case Study of an Underdeveloped Area*. Princeton. p. 126.
9 Tsountas, C. (1899). Kykladika, *Archaiologike Ephemeris*, pp. 86 and 90; Renfrew, C. (1967). Cycladic metallurgy and the Aegean Early Bronze Age, *Amer. J. Archaeol.*, **71**, pl. 1 and 3.
10 Thucydides, *Peloponnesian War*, I vii–viii; Miller, W. (1908). *The Latins in the Levant*. Cambridge and New York. pp. 580, 590 and 598.
11 C. Renfrew (In press). *ibid.*
12 Frankfort, H. (1950). Town planning in ancient Mesopotamia, *Town Planning Review*, **21**, p. 103.
13 Figures from publications of Ethniki Statistiki Yperesseia, Athens. Figure for Crete in 1834 quoted by Pendlebury, J. D. (1939). *op cit.* p. 303.
14 Hutchinson, R. W. (1950). Prehistoric town planning in Crete, *Town Planning Review*, **21**, p. 206.

VERNON REYNOLDS

Ethology of urban life

The most important first step in the ethology of any species is to observe the behaviour of a wild group surviving in its natural habitat; after this has been done, it is useful to study representatives of the species in other settings. In the wild situation, the observer watches the occurrence of behaviour patterns in the context in which they have survival value. Much ethological theory is based on comparisons between the behaviour of animals in the wild and differences observed in captive or experimental conditions.

The crucial feature of a wild population is that it is living in circumstances to which it has become finely adapted as a result of natural selection occurring over a very long time-span, perhaps hundreds of thousands or, more likely, millions of years. During this time there has been, in every generation, selection for those physical and behavioural characteristics that have survival value for the individual or the group, and the behaviour we actually see is the outcome of this selection.

Ninety-nine per cent of man's evolutionary history has been spent living off local resources, hunting animals and gathering vegetable foods. He has evolved a body, a set of behaviour patterns and a society purpose-built to live in this particular way. A few human populations living as hunter-gathers still exist today. A number of them have recently been re-studied with a new emphasis on ecology and behaviour[1].

The similarities in the social systems of present-day hunters and gatherers, particularly in *de facto* social groups, in spite of vast differences in cultural ideals and avowed norms, is striking. Groups as geographically disparate as the Eskimo, American Indians, Pygmies, Hadza, Bushmen, Indian Birhari and Australian Aborigines, show many common patterns of organization[2]. These common social patterns of hunter-gatherers provide a baseline from which to measure the effects of pressures brought to bear on our species in recent times by becoming dependent on economic systems other than the one to which it is adapted[3].

The history of man's economic development from hunting and gathering, through a period in which there was seasonal settlement and food storage, to domestication of animals and agriculture, towards urbanization and industrialization has had many corollaries in terms of human social behaviour. What are these? To answer this question we need first to know what the basic evolved patterns of human society are, second the way recent economic changes have altered man's relation to his habitat and third how, or if, behaviour and social organization have managed to adjust to the new circumstances.

Nomadism

Nomadism is an essential adaptive element of a subsistence economy based on hunting and gathering[4].

Personal antagonisms can be solved by going off to join another band, as happens for example among the Hadza[5] and the Bambuti Pygmies[6]. The frequent striking of camp and moving off elsewhere renews hope and motivation, and even if there is a well-worn round of camping areas as described for the Netsilik Eskimos[7], the routine punctuates the year with changing conditions and different companions. Settled living tends to eliminate all these social advantages. However, it is interesting that as the nuclear family becomes economically more independent and mobile once again within modern industrial society, some of the "psychology" of nomadism has returned; nuclear families strike camp and move on in the hope of better luck at the next hunting ground.

Fission and flux among bands in the community

Correlated with nomadism and the independence and mobility of the individual and the nuclear family among hunter-gatherers is the flexibility of band and community structure[8]. By a system of visits by families to the parents of each partner, or to a brother or sister in another band or another regional community, or by sending a child to live for a while among a relative's family, the whole nexus of relationships is converted into a social security system, insurance against localized disasters.

The most difficult thing about localized resources, be they domestic stock, agricultural produce or coastal fishing grounds, is that large groups of people have to live together and organize and co-operate whether they like each other or not, because they are all dependent on the same resources, and interdependent on each other. The built-in safety valves of the flexible hunter-gatherer community cannot operate. There is considerable evidence that it is dependence on, or exploitation of, localized resources that underlies the significant step from the nomadic flexible non-territorial groups of the generalized hunter-gatherer adaptation, to the settled, organized and territorial groups found elsewhere. Flannery[9], drawing conclusions from the archaeological data of Mesoamerica and the Near East, suggests that sedentary life began with the establishment of ownership of limited areas of high resource potential, and this applies to many other forms of economy as well as agriculture. For example Glover[10] cites evidence that bands on Little Andaman which possess a coastal range build permanent huts, whereas the inland groups tend to be more nomadic. He also points to ownership of resources as an important factor. MacNeish[11] makes the important point that changes in social organization took place gradually over thousands of years as revealed in the archaeological record of the valley sites he has studied in Mexico. He suggests that many factors, such as technological refinements, changes in species hunted and increased population, led to a gradual process of regulating the seasonal cycle and defining tribal territories, and with the start of planting seeds and fruits, seasonal bases became established. This kind of half-way economy can be called a 'hunter-planter' economy. Further examples of a half-way stage between nomadism and settlement can be seen in the Plains Indians of the last century. These were war-like and closed in tribal boundaries, though hunters. However they followed the buffalo herds and were based on fixed winter and summer camps in which the old people and some women and children were left behind[12].

When groups have to stay together because of a vested interest in land or other fixed resource, this leads to tensions which, since they cannot be resolved by fission of the group, have to be resolved in other ways. Studies of other primate species in conditions of enforced permanent togetherness have shown that they resolve tension by the formation of a more or less rigid dominance hierarchy.

Wild chimpanzees, for example, live in flexible open communities with very impermanent and changing band structures. There are differences of status between adult males but there is no role of leadership, no structure of authority. In a thirty-acre site containing thirty chimpanzees, the dominant male stopped fights, inspected all the females each morning, protected young or subordinate chimpanzees in quarrels, and determined whether a strange chimpanzee introduced into the consortium should be accepted or rejected. And when the dominant male was removed one day, the next dominant immediately took over the complete role[13]. In a cage in the London Zoo, with even less social space for manoeuvre, one despotic male chimpanzee maintained a rule of terror[14].

In the case of Man, enforced permanent togetherness has not merely resulted in the formation of behaviour-based power structures of the animal kind, but of institutionalized authority structures as well. Ever since early man began to become aware of, and to use, symbols to categorize his pre-existing social structure, he has formulated rules of social conduct and re-formulated them.

Size of groups

Birdsell[15] has shown that considering the local group of Pleistocene generalized hunters as an equilibrium system, and taking into account ecological and demographic pressures, a figure of around twenty-five individuals would be a viable number for an average local band, within a regional nexus of about 500 individuals. These figures are modal to the actual figures found in studies of modern hunter-gatherers. It is likely in the long evolution of man that his psychology and motivational system also became adapted to working most efficiently within groups of this size; it is possible that when a group rises beyond the size at which the individual can have a personal relationship with every other individual (which must be at some point above a few hundred persons) the members lose their sense of involvement in the group and it is likely to split into sub-groups.

It is difficult to point to more than observational or experiential evidence for the loss of involvement in large human groups, but free-ranging primate groups have been studied in the process of splitting into two groups[16] and population expansion of rhesus monkeys on Cayo Santiago has led to fission and the formation of new groups[17]. In this context, it is interesting that in the book of Genesis, Abraham and Lot decide to split up not because pasture is insufficient but because the herdsmen are quarrelling. Several anthropological reports state that the size of the residential group cannot easily be correlated with ecological factors or economic viability. Most of the examples quoted, however, are drawn from societies without a centralized political power hierarchy, where small scale agricultural and stock-keeping communities are self-supporting, and can regulate themselves according to face-to-face relations. For example Smith[18] writes: "When Amazonian villages reach a certain size (usually about 200–250 persons) there is a tendency to fission due to internal frictions". Harris[19] writes: "Why tropical swidden populations should normally stabilize at levels below the ecologically permitted maximum is obscure, but the explanation probably relates to the difficulty of generating appropriate social controls to integrate larger populations". Forge[20], discussing the same phenomenon in New Guinea, suggests one vital factor in determining social structure is the number of adult males who interact daily in co-operation and rivalry. He suggests that in small settlements of around 150 persons, with no more than forty adult males, normal bonds of interaction such as kinship, affinity and personal friendship, are enough to order the relationships within the community and give it structure. He finds that in settlements where the population (and therefore the number of adult males) is much in excess of these figures, some form of segmentation into separate ritual groups with their own focus of activities always occurs. "When their population (i.e. of the ritual groups) rises much above 300 . . . a split becomes inevitable". Forge goes on to hypothesize very tentatively that man can handle only a limited number of face-to-face relationships, and that it is these constant personal interactions that maintain an egalitarian society. As group size increases beyond the face-to-face interacting group, he suggests that unity can only be maintained by classification of relationships and ascription of status in a hierarchical system, which supports the point made in this paper. This point can be related to the idea that where there is a face-to-face group with a shared aim, whether it is the livelihood of the village, a herd of sheep, or based on political motives, this generates a group dynamism, or collective energy, which disappears with the imposition of a centralized hierarchy.

Thus to relate the ethological and anthropological data to present-day urban contexts, most individuals do in fact deploy their energies in small face-to-face local groups. But where these groups have a political aim a paradoxical situation arises: in order to affect the political and social organization of the urban or national group, either the face-to-face group has to expand to a size where it has to impose a hierarchical structure on itself, or it has to join with an existing hierarchical system. On the other hand, if the small dynamic group remains locally based, and remains outside the larger social and political organization, it becomes one of the main sources of contemporary urban violence, as it seeks to establish its political identity.

A solution which arises out of these behavioural analyses would be to give direct representation at the urban level to a delegate of each self-made common-interest group registered as having 300–500 active members; at the regional level to a delegate from a federation of similar groups, and at the national level, Parliament would be made up of the national delegates of each and every federation of community-based groups whether concerned with political aims, sport, work, art, culture, flat earth or pigeon fanciers. The only criterion would be proof of an actively participating local group; Parliament would be huge, but non-party, and individual delegates could be changed at any time by direction from the grass roots membership. Such a system would make sense in the context of human ecology. Whether it would work as a form of national organization for the country is another matter, but it has never been tried.

Violence

Apart from the violence caused in urban society by politically dynamic face-to-face groups, adolescent and young adult males are responsible for most urban violence. This fact alone should make us realize that the cause of the problem lies in the total *social organization* rather than specifically in poor education, overcrowded living conditions, poverty, parental behaviour, communist infiltration, etc.

In primates, male adolescence and young adulthood is a time for self-assertion and competition with peers; then the young male attempts to stand up to some of the older males in the group, eventually winning a position in the male dominance hierarchy. This process has been observed closely for over ten years now in the groups of rhesus monkeys on Cayo Santiago. Sade[21] found that nearly half the orphan male monkeys left their natal troop at adolescence, whereas only one in ten with mothers still living did so. In macaques, a mother supports her son in his encounters with other males, and a high ranking mother helps to ensure a high rank for her son in the adult male dominance hierarchy. The fact that so many orphan males choose to leave the group (they are not driven out) rather than spend their life on the periphery, indicates the importance of status for the male monkey. In the captive situation, of course, the young peripheral male cannot leave the group; correlated with this is the fact that captive macaques are more aggressive than free-ranging groups, their leaders are more despotic, and there are sometimes outcast omega animals.

Among generalized hunter-gatherers, the problem of status is not a big one. Life-stages are clearly demarcated and personal relationships broadly defined by family, sex and age. As personal property is minimal, and a man has no use for a huge collection of spears or nets, acquisition of large amounts of property is pointless and is not an attribute of status. An adult male who can hunt and beget children has full stature.

Agricultural and pastoral societies have structured adolescence very clearly, by age set systems, early betrothal or marriage, prolonged initiation, warfare, etc., and have thus brought under control adolescent male energies. Eventual achievement of full and respected adult status with a say in the affairs of the community is certain in such societies. But in large urban societies adolescence is largely unstructured except for the minority who continue to higher education, and there are no clearly demarcated stepping stones ahead to status and power in the community. Despite this situation the majority of male adolescents are unable to get *away* from the situation, because they are economically dependent on their family. The situation where sexually mature male adolescents live with their parents is not normal for primate species in general, nor for man; we need to recognize this as many tribes do by setting aside adolescent houses where the young can live apart. Why not self-determining adolescent residential campuses of up to 500, which could be integrated into the system of political representation outlined in the previous section?

Some writers blame overcrowding as an important source of stress leading to violent and destructive behaviour by adolescents. However the salient factors are the *captivity* and political frustration of the urban-dwelling adolescent—the sense of not being able to get away from antagonistic family situations or hopeless job conditions; and no opportunity for achieving status or functional significance in the community. But it is not just the adolescents living at home who are captive; whole classes of people have become captive to the urban complex. Rouse[22], discussing the criteria of advanced cultures, makes the useful distinction between professional co-peoples and sustaining co-peoples. This vital distinction still exists and it is the sustaining co-peoples who are captive to the city, economically dependent upon the work of industry, and generally living close to their work. The professional co-peoples on the other hand, are able to enjoy the nomadic life previously discussed. They visit the city for competitive games with other males, then retire to family security in the suburbs, or at least to a high-rise apartment with a view. It is the captive workers who are most subject to breakdown of social patterns, analagous to captive primates. Grove[23] asks the question whether, if we succeed in planning rationally, the age-old attractions of the city will disappear to our loss? The answer must be to retain at all costs the city as the centre of commerce, sophisticated entertainment, factories, services, operations, the organizational nucleus of the communities within a region. But people must leave the city after work and return to well-planned residential and social communities outside. The age-old attractions of the city have always been greater for the visitor who is economically independent of its machine, and who can leave at will.

Territory and war

Contrary to the arguments of Ardrey[24], man cannot be shown to be territorial by the definition applied in the study of animal behaviour, that is, exclusive use of, and defence of, a particular area by an individual or group. Ardrey saw humans as possessing both the pair territory typical of many birds, and the group territory as found in some species of American monkey and other species. However, the human nuclear family household, far from being an exclusive and defended area, is in most societies only a small unit within a kinship nexus, and everywhere its members interact in the community as a whole. In discussing group territories, Ardrey likens the face-to-face endogamous group of monkeys or lemurs with the complexities of the human nation-state. But hunter-gatherers are not territorial in this way. Bands share a sense of belonging to an identifiable region but hunting parties are allowed into other regions if pursuing prey, and visiting parties frequently cross from region to region, which are linked one to another by kinship ties.

The history of territorial-like violence between human groups since domestication, agriculture and urbanization can be seen as a response to the ecological exigencies of economic systems to which they are poorly adapted, as has been shown by Russell[25].

But there seems also to be another important factor. Just as dependence on land or stock caused groups to remain permanent, with the result of increasing institutionalization of systems of status and authority, so different nations now, in the face of their common dependence on world resources, are in the throes of institutionalizing a world system of authority. And in the light of what we know of human group behaviour there can be no possible end to the international status struggles until a world institution of authority backed up by power has emerged. At one time a nation could live quietly on its own but now like settled man, nations have become captive to their shared resources and the same features of institutionalization and reorganization have to be developed to keep viable the whole interacting world system. Although to some extent territory in the modern world gives *status*, other forms of property, such as shares in oil wells, are equally powerful; the struggle basically is about control of world resources. It is resolved in the only possible way, by a status-ranking or dominance hierarchy of nations.

Notes

1 Lee, R. B. and DeVore, I. (eds.) (1968). *Man the Hunter*. Chicago.
2 And see Glover, I. C., this volume, pp. 157–64.
3 Lee, R. B. and DeVore, I. (1968). Problems in the study of hunters and gatherers, *in* Lee, R. B. and DeVore, I. (eds.) *op. cit.* pp. 3–12.
4 Lee, R. B. and DeVore, I. (1968). *op. cit.*
5 Woodburn, J., this volume, p. 199.
6 Turnbull, C. (1968). *The Bambuti Pygmies*. London.
7 Balikci, A. (1968). The Netsilik Eskimos: adaptive processes, *in* Lee, R. B. and DeVore, I. (eds.) *op. cit.*
8 Hiatt, L. R. (1968). Ownership and use of land among Australian Aborigines, *in* Lee, R. B. and DeVore, I. (eds.) *op. cit.* pp. 99–102; Lee, R. B. and DeVore, I. (1968). *op. cit.* pp. 3–12; Rose, F. G. G. (1968). Australian marriage, land-owning groups and institutions, *in* Lee, R. B. and DeVore, I. (eds.) *op. cit.* pp. 200–208; Turnbull, C. (1968). The importance of flux in two hunting societies, *in* Lee, R. B. and DeVore, I. (eds.) *op. cit.* pp. 132–7; Woodburn, J. (1968). Stability and flexibility in Hadza residential groupings, *in* Lee, R. B. and DeVore, I. (eds.) *op. cit.* pp. 103–10.
9 Flannery, K. V., this volume, p. 28.
10 Glover, I. C., this volume. *ibid.*
11 MacNeish, R. S., this volume, pp. 67–93.
12 Point, Father N. (1968). *Wilderness Kingdom: Indian Life in the Rocky Mountains* 1840–1847. New York.
13 Reynolds, V. (In preparation). *Final Scientific Report to USAF.*
14 Personal observation.
15 Birdsell, J. B. (1968). Some predictions for the Pleistocene based on equilibrium systems among recent hunter-gatherers, *in* Lee, R. B. and DeVore, I. (eds.) *op. cit.* pp. 229–40.
16 DeVore, I. and Hall, K. R. L. (1965). Baboon Ecology, *in* DeVore, I. (eds.) *Primate Behaviour*. New York. pp. 20–52.
17 Koford, C. B. (1966). Population changes in rhesus monkeys: Cayo Santiago, 1960–4, *Tulane Stud. in Zoology*, **13** (1), pp. 1–7.
18 Smith, P. E. L., this volume, p. 414.
19 Harris, D. R., this volume, p. 256.
20 Forge, A., this vol., pp. 371-73; and see Layton, R., this vol., pp. 377-81.
21 Sade, D. S. (1968). Inhibition of mother-son mating among free-ranging rhesus monkeys, *Science and Psycho-analysis*, **12**.
22 Rouse, I., this volume, p. 100.
23 Grove, D., this volume, p. 563.
24 Ardrey, R. (1967). *The Territorial Imperative*. London.
25 Russell, W. M. S. (1967). *Man, Nature and History*. London.

PHILIP E. L. SMITH

Land-use, settlement patterns and subsistence agriculture: a demographic perspective

With each new assessment of the problems of settlement patterns in archaeology there has been increased awareness of the complexity of the subject and of the many variables that must be taken into account. These assessments, particularly in the two volumes edited by Willey and Chang[1], have served to reveal the variety of approaches or perspectives that can be adopted in studying prehistoric settlement patterns. I do not intend to present any summary of these various approaches in this paper but to concentrate on one perspective that has not, I think, been sufficiently emphasized in the past. In reviewing recently a considerable number of the publications in this field I was struck by the relatively low degree of interest paid to the demographic aspect and in particular to the role that changes in population size, density and pressure can play in influencing the settlement patterns of ancient societies. While many archaeologists do often pay a certain lip-service to demographic factors it is true, this rarely goes beyond a superficial level. Most typically they are concerned with estimating the vital statistics of the group under study, including such data as total population size or age and sex distributions.

To some extent this relative neglect of demographic factors in the explanatory field is characteristic of most branches of anthropology where, as Fried has suggested, there has been a reluctance to accept demographic phenomena as causal factors in cultural development[2]. In recent years, however, a number of writers, particularly those with an ecological bent, have pointed out the necessity of considering population density, pressure and size in trying to understand cultural processes of various sorts[3]. Hence it may be an opportune moment to see how this approach can be profitably used in evaluating the processes determining or influencing settlement patterns, especially among simple agriculturalists.

On the whole there has been little attention paid to the details of how the existence (or absence) of population pressures influences the way in which agricultural land is used and how this in turn affects the localities and forms of the settlements. It is true that many, perhaps most, writers no longer stop at considering population growth or increasing density simply in terms of expansion or migration. Increasingly there are references in the publications to the association of growing social, political or religious complexity with changes in population size and/or density at given points in local sequences. An increase in population is nearly always seen axiomatically as a consequence of enlarged food resources or of new means of extraction; that is, as a variable dependent mainly on agricultural output—which is, of course, partly true as scholars recognized long before Malthus. But often no clear distinction is made between population size or growth on the one hand, and population density on the other, while the role of the latter as a causal factor rather than as simply a result of something else is generally underplayed. Normally it is granted that increasing population size means that communities will often be larger and more numerous with a wider base for more complex political, economic and other superstructures. But there is another viewpoint involving a reciprocal relationship (or, in the idiom of the day, a feedback situation) that has usually been neglected: the influence of changes in population pressures on the available resources as they bring about in some cases developments in the technology and exploitative efficiency of the groups concerned, with consequent repercussions on social and cultural phenomena. That is to say, the demographic fact of changing population pressure can itself be regarded in certain circumstances as an independent variable rather than simply a dependent one as archaeologists have traditionally tended to consider it[4]. (It is important, however, to remember that it is *changes* in population pressure, and not simply population increase or population density *per se*, that have to be examined.)

What are the real determinants of settlement patterning among "traditional subsistence" (palaeotechnic) cultivators? Although one must hold sight of the many direct environmental factors (soil, water, terrain, etc.) as well as factors such as religion or the relationship to larger political units if they exist locally, nevertheless the primary determinants seem to reside in the mode of subsistence and the way this mode is exercised. Among traditional agriculturalists this involves above all the quantity of cultivable land and the degree of ease of working it. Here we touch on the problem of land-use in such societies.

There has been a tendency among archaeologists (and ethnologists as well) to think of the cultivation systems in use, and particularly of the degree of intensification of cultivation, as largely determined by the variations in soil and climatic characteristics and by the technological capacities of the cultivators. The various levels of intensification have usually been considered relatively static adaptations to local environmental conditions combined with the available techniques, and the agricultural systems themselves as rather stable and fixed in the absence of external cultural influences or environmental changes. No one will of course deny that such factors as soils and climate *are* important to traditional agriculturalists and that the cultivation practices of a group *can* be affected by outside peoples and events. But what is too frequently overlooked is that each agricultural system often reveals component subsystems with much diversity, and that the pattern of land-use is normally highly flexible and adaptable in the face of stresses. The different types of agricultural land-use are in fact not primarily adaptations to local geographical conditions but, within certain limits, reflections of decisions by the cultivators regarding food-production and labour input; and, to return to the theme of this paper, they are resilient in the face of demographic pressures caused by high man-to-land ratios.

In an essay written primarily for agricultural economists but which has great potential value for archaeologists, Boserup has been concerned to demonstrate that under pre-industrial agricultural conditions the supply of labour available is the final determinant of the production and level of development attained[5]. Among traditional cultivators fallowing is normally an integral part of the pattern of land-use. Boserup emphasizes that there are disadvantages to shortening these recovery intervals between crops in favour of more intensive forms of land-use, for although the absolute output of food may very well increase the output per man-hour of work is more likely to decline. Thus the cultivators, who should not be supposed to be ignorant of the ratio between their labour input and productivity, will not willingly intensify their agriculture (thereby sacrificing part or all of their traditional periods of leisure) although they may be aware of the possibility of producing more and even of creating an economic "surplus". One of the forces that may compel them to intensify is, of course, domination by another tax-imposing group. But under ordinary circumstances it is only when a certain critical level of population density has been reached, the carrying capacity of available land is approached and they face a decline in living standards if the traditional form of land-use is maintained, that cultivators find it profitable to change in the direction of more intensive cultivation. Naturally there are other ways of avoiding this decision, including emigration, predation against neighbours, development of other resources such as fishing or trading, as well as methods deliberately stabilizing the population by spacing births, e.g. through contraception, abortion, infanticide, intercourse taboos, prolonged lactation, or by such means as delayed marriage. Nevertheless it is not always considered possible or desirable to adopt these alternatives, whereas intensifying the agriculture by shortening the fallow periods is a fairly simple and obvious solution that can be introduced gradually without abnormal disruption and can be reversed if circumstances change in the group's favour.

Boserup has suggested five main types of pre-industrial land-use along a spectrum of shorter fallow periods and increasing intensification. Beginning at the simplest level with two varieties of Long Fallow (Forest Fallow and Bush Fallow, often called shifting or swidden agriculture), the latter merges with Short (or Grass) Fallow, is often followed by Annual Cropping and finally, in exceptional circumstances, by Multicropping. To some extent, it might be mentioned in passing, these types of land-use may be associated with certain tools for soil preparation, since digging sticks are often characteristic of Long Fallow, hoes of Short Fallow and ploughs of Annual Cropping and Multicropping. This can be shown rather schematically as follows[6].

LONG FALLOW

Forest Fallow	Bush Fallow	Short (or Grass) Fallow	Annual Cropping	Multi-cropping
1	2	3	4	5

——————— population pressure increase ———————>

<——————— population pressure decrease ———————

—— digging stick— – –

– – – ——— hoe ——— – – –

– – ——— Plough ——

Production should increase as cultivators move along the spectrum from type 1 to type 5, but each type requires more work per unit of land and worker than did the previous one. The reasons for this are various but involve such things as the greater prevalence of weeds and tough grasses, heavier sod cover, the need for artificial fertilizer to counteract lessened soil fertility even when burning is practised, and the need to care for draught animals when the traction plough is used. Thus it is estimated that the shift from extensive agriculture (types 1 and 2) with long fallow periods, to multicropping without chemical fertilizers, can involve an increase from 1000 hours of work annually per family to something like five times that figure[7]. Conversely, a move "backward" along the scale, from more to less intensive cultivation, involves less labour input; and there are plenty of ethnographic examples of groups happily reverting to longer fallow periods once the compelling forces of population pressure on land or of outside domination were removed or lessened[8].

In contrast to most who have treated this question, Boserup sees the land-use types outlined here as flexible and fluctuating in response to factors that can be defined in terms other than simply soil quality, technological level or vaguely defined cultural preferences. From the archaeological viewpoint the value of her approach is not merely that it permits us to regard soil-preparation implements in a more useful light, as skilfully selected devices for particular circumstances and needs. More important, her model enables us to examine more closely the causal principles involved in the change from one type of land use to another. It offers a new dimension to the ecology of traditional cultivators, as a number of geographers and ethnologists have recently noted[9]. If the task of the geographer and agronomist is to elucidate the structure and function of the agricultural systems and subsystems[10], the least the archaeologist can do is to keep the diversity of land use practices in mind in evaluating his own data, particularly those data that bear on settlement and community patterning.

I have already suggested that settlement patterning is governed primarily by the mode of subsistence and associated land-use among the kinds of societies being examined here. Community patterning, on the other hand, seems to be determined to a large degree by the social organization of the group, especially the kinship grouping and the residence and descent rules[11]. But there is a growing body of evidence to indicate that these determinants of settlement and community patterns are in turn strongly influenced at times by changes in the demographic circumstances[12]. It seems reasonable to suggest, therefore, that archaeologists interested in these problems should go considerably beyond attempting to identify the subsistence modes and social organization of past societies—praiseworthy as these attempts are—and should be more aware of the ways through which demographic factors may cause variability in subsistence and social organization. Indeed, this is one way to investigate the provocative question that Willey raised recently: to what extent does settlement patterning itself influence or determine rather than merely reflect social organization?[13] It may be useful to examine Willey's query by going back a step further. Since we may assume that there tends to be *some* coincidence between types of land-use and settlement-community patterns (though not a direct or close one in all cases), we can ask: what factors, in what order of importance, influence the cultivation systems and land-use among a given group? How do changes in agricultural land-use patterns tend to affect social and political organization? And what changes in turn might be expected to come about in settlement and community patterns as a result of the social and political modifications? If one accepts the proposition that the sources of social and cultural changes are to be sought primarily in economic, technological, environmental and demographic features[14], then these are legitimate and perhaps even profitable questions to ask.

In this paper there is no opportunity to discuss the first question, that is, the entire range of factors that influence cultivation systems and land-use. These are multiple and complex, as already indicated. At the risk of appearing to overstress one aspect of a complicated problem, the aims of this paper are restricted to one narrow focus: to examine the settlement patterning among pre-industrial cultivators, particularly early and simple ones, as seen through the twin lenses of land-use and changes in population pressure. The paper tries to draw attention to the probability that an important dimension has often been neglected in the past by archaeologists who have been attempting to understand the processes of change in agricultural societies and in the evolutionary developments in various parts of the Old and New Worlds[15].

Population pressure as a variable: discussions and illustrations

In this part of the paper I offer some illustrations, with a number of examples culled from ethnographic sources, of how increasing or declining population pressure in its influence on patterns of agricultural land-use may act as a variable in the determination of human settlement patterns.

First of all, the matter of the permanence of settlements under various types of land-use. Archaeologists have long had a tendency to think of extensive agriculture as being associated with shifting settlements and intensive agriculture with stable ones[16]. In fact, this correlation is not a very precise one. Shifting (long-fallow) cultivators can at times have more or less permanent villages of respectable size. Carneiro describes an Amazonian swidden-farming village as being occupied for ninety years and including up to several thousand persons, although villages of this size are admittedly rare[17]. The Ashanti occupied villages under shifting cultivation for generations and even centuries[18]. A distinction should clearly be made, in discussing shifting cultivators, between those where only the fields or gardens are moved, and those where villages move as well. One should also remember that there may be intermediate positions where villages do move but at a slower rate than do the fields. "Shifting cultivation often suggests shifting settlement, but even when most of the characteristics of shifting cultivation are present the settlement site may be permanent"[19]. The factors involved in the degree of sedentariness achieved by a group are many and involve such things as soil fertility, distance from the fields or gardens, warfare, the depletion of natural food resources in the vicinity and the degree of ease with which new houses can be built. And of course at times the reasons for shifting a settlement may have nothing to do with subsistence; thus Carneiro's Kuikuru village in Amazonia was abandoned and rebuilt several times in the same locality for supernatural reasons[20]. But in many cases the degree of mobility or of sedentariness seems to be a function of the population density in relation to the amount of cultivable land available and the kind of agricultural cycle practised. When Amazonian villages reach a certain size (usually about 200–250 persons) there is a tendency towards fission due to internal frictions, while village movement can also result in order to reduce walking time to fields when the nearby ones are in long fallow. If land is not easily available (here "available" refers not only to land naturally present but also to whether those using it are safe from attack by hostile neighbours), the entire village may be abandoned, particularly if the houses are easily rebuilt; or part of it may bud off and migrate some distance away. In such cases the settlements will be dispersed over the landscape with no very large concentrations anywhere and, at least in the cases of the Amazonian villages mentioned above, no social or political organization more complex than those based on lineages. On the other hand, where the ratio of cultivated land to fallow land is high and population pressure exerts itself on the available land there is a tendency not only for the fallow periods to be shortened in response but also for villages and compounds to become almost or completely permanent sites. Increased stability of settlements obviously is of great significance for archaeologists interested in settlement and community patterns, including, as Whiting and Ayres suggest[21], the possible effects on house forms and character which may tend to be larger, more solidly built and often rectangular in shape[22]. It is difficult, of course, to separate completely sedentariness and size of settlements. As Carneiro argues, sedentariness involves not only how long a settlement remains in one place but also how large it is[23]. Thus a large village might be considered more sedentary than a smaller one occupied for the same length of time.

A dispersed habitation pattern is not always associated with extensive agriculture nor is nucleated settlement always a concomitant of more intensive agriculture. Although in principle dispersed settlement may be advantageous for shifting cultivators who find distance from their land holdings inconvenient (walking to and from fields must be considered part of the labour input), other factors also enter into account and the result usually is a compromise. Completely dispersed settlements of households and nuclear families are probably very uncommon except in modern states with commercial agriculture[24]. As already mentioned, settlements of long-fallow cultivators tend to be small, though the total population of the region may be surprisingly large. Long-fallow cultivation does not so much limit the size of the total population (within the limits of the environment's carrying capacity, of course) as it limits the size of the local units. The role of fission due to internal friction was indicated earlier; but with increasing population pressure on cultivable land the factor of fission may be overruled by the threat of hostile or competing neighbours. Warfare encourages a build-up of the local unit for defensive or aggressive purposes and simultaneously discourages expansion into nearby areas still permitting long-fallow cultivation. In such cases there may be a trend towards larger and more nucleated settlements with shorter fallow periods in the cultivation system. It is not a universal rule that more intensive cultivation calls for larger and nucleated settlement, however. It may lead to dispersed homesteads if land tenure rules are also changed, as Netting illustrates among the Kofyar of Nigeria under conditions of high population density and competition for land[25]. Residence in close proximity to the cultivated land may not only be desirable because intensively cultivated plots require more care, but also because residential occupation is a way of asserting individual control over the land. So in certain cases agricultural intensification associated with high total population density may stimulate dispersal rather than nucleation of residence, while less intensive cultivation may be represented by intermediate population density with village settlement[26].

One other point is that increased population pressure usually leads to a greater variability of settlement patterns than exist in areas of low density and pressure. The so-called hydraulic farmers, especially, may reside in a great variety of settlement types[27]. Increasing pressure beyond a certain point may lead not only to intensification of cultivation and enlargement of sites in the traditionally occupied sectors of the region but also to expansion into formerly marginal sectors of nearby mountains, swamps or deserts, often to introduce intensive methods into these areas eventually, e.g. by terracing. (It may also lead to the introduction or expansion of long-fallow types of cultivation in such marginal areas, but this is not as paradoxical as it might seem, for the land, regardless of the length of the fallow periods, is being used more intensively than when it was vacant.) That is, a new range of land use is present although it is not always easy to recognize this archaeologically if the cultivators continue to reside in their traditional settlements unless such constructions as terraces or canals are preserved or unless temporary or seasonal houses are found[28].

Land tenure, which seems to be one of the main factors influencing the degree of nucleation or dispersal of settlements, is also one of the aspects of agricultural systems most susceptible to changes in population pressures. This can take a number of forms but one of the most common is the splitting up of the large blocks of cultivable land formerly controlled by the community into smaller holdings owned or controlled by families or individuals following the reduction of the fallow periods[29]. Rules of land tenure as well as of recruitment norms of descent groups may become more rigidly interpreted[30]. With the gradual shortening of the fallow periods and the reduction of the non-rotated part of the territory the attachment of individual families to particular plots becomes more and more important; and less land will be available for redistribution by the chief or other authority[31]. One obvious consequence of the trend to permanent land rights is the emergence of institutions or individuals with privileged positions and greater resources, which might be translated in archaeological terms by differences in architectural forms and building locations within the community. (Needless to say, several different systems of land tenure can coexist within the same grouping, with short-fallow lands being held under permanent cultivation rights by particular families while more extensively used land is more freely obtained by any family with customary general rights to do so.)

Carneiro has discussed the effects of population pressure combined with environmental and social circumscription and resource concentration on the political and social organization of long-fallow cultivators[32]. In cases where villages are not environmentally circumscribed they tend to be autonomous with little supra-village integration, to be smaller in size and more dispersed, less permanent, with little or no tendency towards more complex political and social organization. Chagnon has described the situation in Amazonia where villages in regions of high population density are socially circumscribed and find it difficult or impossible to move because of pressure by neighbouring groups[33]. In such cases the villages in the centre of the territory will cluster together closer than those on the peripheries and will thus impinge more on each other, resulting in further conflict. Villages tend to be larger as a defence against expected attacks, war leaders become more powerful, there is greater emphasis on offensive and defensive alliances and a stronger trend towards a higher form of political organization. There is also a greater commitment to sedentariness, to cultivation rather than to hunting-collecting and to specialization in the production of certain items. Here we have an instance of the interplay of demographic factors, land-use and concomitant settlement pattern influencing the social and political organization along the lines of the query by Willey already referred to.

Several writers have pointed out how fixed cultivation tends to lead to closer identification of people with a particular locality and to a weakening of their ties with a kinship group whose membership is growing larger and more dispersed[34]. This effect on social organization of movements caused by population pressure on cultivable land is also illustrated by Netting among the Kofyar[35]. Groups of formerly intensive cultivators on the plateau changed to shifting cultivation with longer fallows when the pressure of population density forced a budding-off onto the less densely occupied plains. The former pattern of dispersed homestead settlement typical of the plateau hardly changed with the move, but in contrast there were significant changes in social organization because of the new conditions of land-use. On the much larger bush farms of the plains a larger resident labour force, such as the extended family, offered a much higher production potential, so the plains households became larger with more polygynous marriages and greater frequency of extended families, unlike the situation on the plateau where the pattern of small co-resident households was the rule. It is possible that shifts such as these in community patterning resulting from changes in land-use might be archaeologically significant if the processes of the change are kept in mind[36].

Some cases of archaeological relevance

The inferences to be drawn from Boserup's model are of use to archaeologists both synchronically and diachronically. Although the model was not devised initially with an evolutionary aim in mind, its value in terms of change through time is implicit in her argument. A number of other writers, however, have been more explicit in pointing out the implications of a succession of agricultural land-use types for explanation in evolutionary terms[37]. Elsewhere, in addition, the writer and Young have attempted to apply a modified form of Boserup's model to the sequence in Greater Mesopotamia from the period of earliest food production to full urban times[38], but only the settlement aspects of this are relevant here. We may now consider a few of the archaeological cases where it is perhaps possible to see changes in settlement patterns in terms of changes in population densities and pressures reacting on patterns of agricultural land-use (or, in a few instances, on patterns of non-agricultural land-use). These cases are offered as being suggestive but by no means conclusive; they are based on a rapid survey of the available publications and on data which I am in most cases not competent to evaluate in detail and are often probably susceptible to several interpretations. Nevertheless they may serve to illustrate, if not to prove, the point I am trying to make in this paper.

Judging by the published data, it is possible that the changes observed by Ritchie in prehistoric New York from the Early Owasco to the Middle Owasco period might be profitably reviewed in the light of Boserup's model[39]. Apparently a relatively large increase in population took place during the Owasco period (an increase which perhaps had its beginning in somewhat earlier times), along with more intensive use of agriculture and a progressive development of larger and more numerous villages. Possibly as a result of the rapidly multiplying population and increasing pressure on available agricultural land, there is a trend from unfortified villages in Early Owasco to fortified ones in defensible positions in Middle Owasco. It is interesting to note, however, that this trend in settlement patterns due to demographic and ecological factors seems to have been reversed by a subsequent *political* innovation, the formation of the League of the Iroquois. The League, in spite of continuing high population density, appears to have provided greater security for its members who were enabled to move their villages from the defensible localities of Owasco times to lowland open sites.

Again, it is possible to speculate that somewhat similar forces were operating in the Middle Mississippian situation in the Midwest (*c.* A.D. 600–1000) where, judging from Fowler's recent opinion, there seems to have been a major demographic shift with evidence of increased population, larger settlements, greater sedentariness and fortified towns accompanying a move to far more intensive cultivation utilizing the hoe rather than, as before, the digging stick[40]. Although Fowler apparently sees the direction of the process as resulting from the hoe cultivation which provided sufficient food resources to permit the stability and more complex developments of Middle Mississippian, it can also be argued that it was the pressure of population on the limited available land in an environmentally circumscribed zone that induced more intensive cultivation with a different implement and brought in its train the changes in settlement pattern.

The evolution of prehispanic Peru also offers a long succession, from pre-agricultural times to the development of large urban and territorial units, in which we can see various shifts in settlement patterns coming about with changes in subsistence and political, social and religious organization which were probably intimately linked with demographic pressures on land resources[41]. If one accepts Lanning's argument that a climatic change led to desiccation of the *lomas* or fog-meadows on the coast and thus to a forced sedentarization of the hunting-gathering groups who had exploited them, then the initial pressure of population was on non-agricultural resources. The alleged desiccation of the *lomas* had the effect of concentrating the population within a smaller area and thus increasing the competition for the remaining resources. Subsequent exploitation of the rich maritime resources nearby in turn permitted a larger population with more extensive permanent settlements and even ceremonial centres. While some food production was adopted early by these fishing groups, agriculture was not significant in the coastal region until the growing population pressure on the best fishing localities led to the "excess" population expanding into the coastal valleys and developing new subsistence patterns based now on more intensive cultivation. But in time in the valleys also growing pressures on the naturally available agricultural land led to the introduction of terracing and intensive irrigation, and eventually political integration developed as defeated competitors were brought under control. In terms of settlement it is possible to see these political and economic developments as ranging from the small coastal villages based largely on marine resources to the communities concentrated in the river valleys relying on plant cultivation, and culminating in the elaborate spectrum of settlement and community patterns characteristic of the state level of integration.

In Mesoamerica, too, the recent analysis by Sanders and Price[42] suggests that the events, particularly in the Valley of Mexico, might profitably be viewed in the light of increasing population pressure exerting itself so as to produce more intensive forms of land-use and cultivation, especially in environmentally circumscribed zones. Sanders and Price have, more explicitly than most writers, applied the concept of population pressure and competition for cultivable land to explain the evolution of urban and non-urban civilization in Mesoamerica; thus it is not necessary to repeat here their arguments which in many respects are similar to those proposed earlier by Boserup.

Another instance from North America where changes in areal settlement patterns follow trends of population increase and improvements in agricultural production is suggested by Sears' data from the south-eastern United States[43]. Here one suspects that the development from what he calls the Priest State, with a single major ceremonial-centre type of settlement, to the Militaristic State with multiple centres, large populations and sometimes truly urban ceremonial and population centres, is similarly susceptible to further analysis in the light of the hypothesis outlined above.

Greater Mesopotamia between the tenth and fourth millennia B.C. appears to offer a rather good illustration of how changes in population pressures lead to changes in agricultural land use and in settlement patterning. Before about 7000 B.C. settlement by food-producers was apparently almost completely restricted to the highland zone (the High and Inner Zagros and the Piedmont) where dry farming was possible. Villages and hamlets with perhaps seasonal camps seem to have been the rule, under conditions of Long or Bush Fallow probably. After 7000 B.C. as the population increased and agriculture became more intensive (probably with Short Fallow developing in many areas), the areas of settlement spread not only within the highland zone but also into the fringes of the Mesopotamian alluvium and the Zagros foothills to the south. Eventually, by about 5500 B.C., the southern Mesopotamian alluvium, the north Mesopotamian Plain and the Iranian Plateau were also occupied. As colonization filled up the more favourable areas of cultivable land, methods such as irrigation, perhaps even shorter fallows in places and (probably) the plough helped to relieve the growing population pressure on land. Just how the events of this period were reflected in the settlements of the highland zone (where the population probably stabilized on a Short Fallow level and perhaps even declined in the late fourth millennium) is not yet well understood, but at least in Lower Mesopotamia (a naturally circumscribed zone, it should be remembered) there is an obvious trend towards nucleation of towns by Uruk times. In terms of settlement this may represent a change from dispersed villages and small towns to a pattern of small sites clustering around much larger ones, with a gradual decline of the small settlements in some areas in Early Dynastic times. Indeed there seems to have been an abandonment of entire districts of Lower Mesopotamia as overcultivation and subsequent salination under the spur of population pressures made some agricultural land unprofitable. Already rules of land tenure may have been modified in the direction of concentration in institutional and individual hands, while the internal structuring of the cities (insofar as we know them) seem to reflect a high degree of social stratification. The competition between towns for cultivable lands undoubtedly was partly responsible for the development of walled and fortified centres at this time in Lower Mesopotamia. In northern Mesopotamia, on the other hand, pressures on cultivable land never reached the level of the south (in part, perhaps, because emigration westward was easier), and the pattern of dispersed villages and small towns seems to have persisted[44].

The degree to which we can see developments in settlement patterns in Europe, Africa and the Far East as related to changes in demographic pressures remains an almost virgin topic for investigation. Over much of these areas the study of settlement patterns is still in an embryonic state, and little can be suggested at the present time. One may suspect that Boserup's general model of agricultural development is applicable to parts of prehistoric Europe, nevertheless. Thus the change-over from Cortaillod and Michelsberg hoe-using (short-fallow?) cultivation to the plough-using types of land-use of Corded Ware and Urnfield groups seems to have coincided with the growth of larger population concentrations[45]. If Boserup's general model is pertinent to explain the stages of agricultural succession in Europe, then we might expect the changes in settlement patterns to be related in some degree to the changes in population growth and land-use. As for China, it is tempting to see the earliest Neolithic, Yangshao, as long-fallow cultivators under conditions of relatively stable population growth, and the succeeding Lungshan expansion as the result of population build-up as groups simultaneously became somewhat more sedentary (in certain areas) and, in other areas, moved into lightly-populated and environmentally uncircumscribed regions where there was ample opportunity to spread out without recourse to drastically shorter fallow periods[46].

A final instance may be cited from New Guinea. Here Waddell[47] has recently proposed that the evolution of agricultural settlement patterns in the highlands follows a broad agricultural sequence in which changes in population density played an important role very much as Boserup outlined. He postulates a settlement trend from nucleated settlements through a relatively unstable stage to a stable, dispersed pattern with residences located progressively closer to the open fields. This trend in settlement runs in tandem with a trend in agricultural land-use from simple shifting cultivation with mixed gardens at the beginning, to grassland expansion following the growth of population and, about 2000 B.C., the development of intensive cultivation in open fields for certain staple foods while mixed gardens continued for subsidiary crops. In other words, one sees the progressive substitution of labour for land as a factor in food production under the spur of increasing population density and competition for cultivable land.

Conclusions

An agricultural community, particularly one of the types discussed here, does not necessarily have a single economic-subsistence pose. It may have several. It is necessary to consider the marked diversity within such agricultural systems and sub-systems and to elucidate the effects of these on the settlements and communities responsible for them. A better understanding of the processes by which land-use types are altered and transformed is an essential step to reaching an understanding of the changes and functions of the settlement patterns associated with the groups concerned. Up to recent years archaeologists have rarely gone beyond the level of conjectures based on such broad macro-types as shifting or semi-sedentary or fixed cultivation. It is surely time now to study finer divisions of land use along the lines Boserup has suggested, with full allowance of course for the variability created by different environmental settings and the different food staples produced. But this in turn awaits, among other things, a re-study or evaluation of the role of demographic factors as they interact with land use among cultivators. The strategies of the cultivators undoubtedly involves game theory in selecting the proportions of land allotted to each kind of use and it is in this light that we should try to view the decisions regarding settlement patterning, for the placing of sites also involves game theory to invoke the best payoff under given circumstances[48].

I have made no attempt here to suggest a newer typology of settlements and communities appropriate for each of the land-use types mentioned. It is perhaps too soon to try this, and in this paper I have been more concerned to emphasize the point that the approach outlined here, based as it is on a kind of cultural uniformitarianism, provides a fertile field for the production and examination of hypotheses by archaeologists—hypotheses that might otherwise be obscured by imprecise understanding of the workings of simple agricultural systems. It provides us with material to construct better models which can then be evaluated by the means archaeology is now beginning to develop.

The study of settlement patterns in the light of demographic forces and of land use seems to me to avoid the dilemma inherent in either of the extreme positions taken by some discussants of settlement patterns in recent years: the one position that seems to stress natural environmental factors beyond credible limits, and the other that overemphasizes the inferences on social, religious and political organization. Demography and land use, as integral elements of human ecology, promise to provide a link or articulation between these two extremes. There is always risk involved, of course, in selecting one variable out of many and thus oversimplifying what must have been a complex series of processes. Nevertheless I suggest that many puzzling or anomalous data in the archaeology of simple cultivators might be more usefully examined in the perspective outlined here. At least the application of a model based on this hypothesis permits us to introduce a greater degree of determinism into the situation, and this presumably is part of what archaeological explanation is all about.

Notes

1 Willey, G. R. (ed.) (1956). Prehistoric settlement patterns in the New World, *Viking Fund Publ. in Anthrop.*, **31**; Chang, K. C. (ed.) (1968). *Settlement Archaeology*. Palo Alto.

2 Fried, M. H. (1967). *The Evolution of Political Society. An Essay in Political Anthropology*. New York. p. 197.

3 Bartholomew, G. A., Jr. and Birdsell, J. B. (1953). Ecology and the protohominids, *Amer. Anthrop.*, **55**, pp. 481–98; Carneiro, R. L. (1961). Slash-and-burn cultivation among the Kuikuru and its implications for cultural developments in the Amazon Basin, *in* Wilbert, J. (ed.) The evolution of horticultural systems in native South America, *Antropológica*, supp. no. 2, Caracas, pp. 47–67; Dumond, D. E. (1961). Swidden agriculture and the rise of Maya civilization, *S.W.J. Anthrop.*, **17**, pp. 301–16; Dumond, D. E. (1965). Population growth and cultural change, *S.W.J. Anthrop.*, **21**, pp. 302–24; Lanning, E. P. (1967). *Peru before the Incas*. Englewood Cliffs; Binford, L. R. (1968). Post-Pleistocene adaptations, *in* Binford, S. R. and Binford, L. R. (eds.) *New Perspectives in Archaeology*. Chicago, pp. 313–41; Sanders, W. T. and Price, B. (1968).

Mesoamerica. The Evolution of a Civilization. New York; Flannery, K.V. (1969). Origin and ecological effects of early domestication in Iran and the Near East, *in* Ucko, P. J. and Dimbleby, G. W. (eds.) *The Domestication and Exploitation of Plants and Animals.* Chicago. pp. 73–100; Harner, M. J. (1970). Population pressure and the social evolution of agriculturalists, *S.W.J. Anthrop.*, **26**, pp. 67–86.

4 Variables are designated as "dependent" or "independent" only as a convenience to the analyst, of course, and we are usually dealing with functional inter-relationships rather than with unilateral causation. But, while it is undoubtedly true that population growth and subsistence expansion are interdependent and mutually reinforcing variables, with neither really independent, it is nevertheless useful to remember that the degree of dependence or independence can vary. In other words, some independent variables can be more independent than others. In this paper I take the position that population pressure on cultivable land is the critical variable, "the engine which sets in motion adaptive changes in a set of related technological and social variables among subsistence cultivators", Netting, R. M. (1969). Eco-systems in process: a comparative study of change in two West African societies, *in* Damas, D. (ed.) *Contributions to Anthrop: Ecological Essays*, Nat. Mus. of Canada Bull., **230**, p. 109.

5 Boserup, E. (1965). *The Conditions of Agricultural Growth. The Economics of Agrarian Change under Population Pressure.* London.

6 This is by no means a clear-cut association, for there are some exceptions and much overlapping. Nevertheless the sequence of association of tools with cropping types seems broadly valid, and in conjunction with other techniques promises to permit archaeologists a greater precision in determining prehistoric land-use than is usually the case. In this paper I have modified somewhat the correlation of tools with land-use types suggested by Boserup.

7 Boserup, E. (1965). The interrelation between population trends and agricultural methods, *United Nations World Population Conference*, Belgrade, WPC/WP/318.

8 Gourou, P. (1956). The quality of land-use of tropical agriculture, *in* Thomas, W. L., Jr. (ed.) *Man's Role in Changing the Face of the Earth.* Chicago. p. 345; Netting, R. M. (1969). *op. cit.* p. 106.

9 Clarke, W. C. (1966). From extensive to intensive shifting cultivation: a succession from New Guinea, *Ethnology*, **5**, pp. 347–59; Netting, R. M. (1969). *op. cit.*; Waddell, E. (1969). Agricultural evolution in Melanesia, with special reference to the New Guinea Highlands, *Paper delivered to McGill Ecology Seminar, 1969*; Denevan, W. M. (1970). Aboriginal drained-field cultivation in the Americas, *Science*, **169**, pp. 647–54.

10 Waddell, E. (1969). *op. cit.*

11 The terms settlement and community patterns are used here in the sense of Chang, K. C. (1962). A typology of settlement and community patterns in some circumpolar societies, *Arctic Anthrop.*, **1**, pp. 28–41.

12 Oberg, K. (1965). The marginal peasant in rural Brazil, *Amer. Anthrop*,. **67**, pp. 1417–27; Carneiro, R. L. (1961). *op. cit.*; Watson, J. B. (1965). From hunting to horticulture in the New Guinea Highlands, *Ethnology*, **4**, pp. 295–309; Fried, M. H. (1967). *op. cit.*; Netting, R. M. (1969). *op. cit.*; Harner, M. J. (1970). *op. cit.*

13 Willey, G. R. (1968). Settlement archaeology: an appraisal, *in* Chang, K. C. (ed.) *op. cit.* pp. 221–2.

14 Harris, M. (1968). *The Rise of Anthropological Theory. A History of Theories of Culture.* New York.

15 There is no reason to suppose that population pressure on food resources was not an important feature in determining settlement patterns among

less complex groups than those examined in this paper. The problem of demographic pressures bearing on hunting and gathering societies, or marginally food producing ones, is a particularly important one. See, for instance, the possibility that increasing population densities during Early Woodland times in the Lower Illinois Valley were in part instrumental in inducing the shift to more intensive food-gathering (with changes in the subsistence-settlement system from small, scattered and impermanent villages on the bottomlands to larger and more numerous sites of several types of exploiting a larger number of micro-environments) that Streuver documents for the Middle Woodland, Streuver, S. (1968). Woodland subsistence-settlement systems in the Lower Illinois Valley, *in* Binford, S. R. and Binford, L. R. (eds.) *op. cit.* pp. 285–312.

16 e.g., Childe, V. G. (1953). Old World pre-history: Neolithic, *in* Kroeber, A. L. (ed.) *Anthropology Today*. Chicago, p. 198.

17 Carneiro, R. L. (1961). *op. cit.*

18 Nye, P. H. and Greenland, D. J. (1960). The soil under shifting cultivation, *Tech. comm.* **51**, *Commonwealth Bureau of Soils*, Harpenden, pp. 5–6.

19 Brookfield, H. C. and Brown, P. (1963). *Struggle for Land. Agriculture and Group Territories among the Chimbu of the New Guinea Highlands.* Melbourne. p. 164; see also Conklin, H. C. (1961). The study of shifting cultivation, *Curr. Anthrop.*, **2**, pp. 27–61; Dumond, D. E. (1961). *op. cit.* pp. 305–6.

20 Carneiro, R. L. (1961). *op. cit.* p. 48.

21 Whiting, J. W. M. and Ayres, B. (1968). Inferences from the shape of dwellings, *in* Chang, K. C. (ed.) *op. cit.* pp. 117–33.

22 The primary importance of sedentariness from the viewpoint of this paper lies in the fact that it removes some of the restrictions on the size of the local groupings, and permits the development of new forms of social and political organization. The awareness of the advantages of larger numbers and the commitment to the existing social structure might be powerful incentives to the groups concerned to maintain as high a degree of sedentariness as conditions allowed. Agriculture is of course one means of stabilizing settlement and it is very likely that the selective advantages inherent in sedentariness were among one of the chief, if not the most important, of the spurs to the development of food production in many parts of the world.

23 Carneiro, R. L. (1968). The transition from hunting to horticulture in the Amazon Basin, *Proc. VIII Int. Cong. Anthrop. and Ethnol. Sci.*, Tokyo-Kyoto, **3**, pp. 244–8.

24 Sanders, W. T. and Price, B. (1968). *op. cit.* p. 194.

25 Netting, R. M. (1969). *op. cit.* p. 108; Sahlins, M. D. (1957). Land-use and the extended family in Moala, Fiji, *Amer. Anthrop.*, **59**, pp. 449–62.

26 For a discussion of this point see Waddell, E. (1969). *op. cit.* As archaeologists we should not leap to the conclusion that an increase in the number of extensive-cultivation sites in a given area necessarily reflects a decrease in population—the reverse may be true. The key to the apparent paradox is that we must focus attention on population *density* in relation to available land, and not merely on population growth *per se.* As an obvious corollary, it is dangerous to rely simply on the number of sites from one period to another to determine precisely increasing or decreasing population density in a region. We must also consider the sizes of the sites and the degree of permanence involved.

27 Sanders, W. T. and Price, B. (1968). *op. cit.* p. 195.

28 See Morgan, W. B. (1953). The Lower Shire Valley of Nyasaland: a changing system of African agriculture, *Geogr. J.*, **121**, p. 465; also Flannery, K. V. and Schoenwetter, J. (1970). Climate and man in Formative Oaxaca, *Archaeology*, **23**, p. 151.

29 Morgan, W. B. (1953). *op. cit.*; Morgan, W. B. (1955). Farming practice, settlement pattern and population density in south-eastern Nigeria, *Geogr. J.*, **123**, pp. 320–33; Carneiro, R. L. (1958). Agriculture and the beginning of civilization, *Ethnographisch-archäologische Forschungen*, **4** (1–2), p. 26; see also Demangeon, A. (1927). La géographie de l'habitat rural, *Ann. de Géogr.*, **36**, pp. 1–23, 97–114, for a somewhat schematic but useful discussion based largely on medieval European data.
30 Kelly, R. C. (1968–9). Demographic pressure and descent group structure in the New Guinea Highlands, *Oceania*, **39** (1), p. 59.
31 Boserup, E. (1965a). *op. cit.* p. 81.
32 Carneiro, R. L. (1961). *op. cit.*; Carneiro, R. L. (1970). A theory of the origin of the state, *Science*, **169**, pp. 733–8.
33 Chagnon, N. A. (1968). The culture-ecology of shifting (pioneering) cultivation among the Yanomamö Indians, *Proc. VIII Int. Cong. Anthrop. and Ethnol. Sci.*, Tokyo-Kyoto, **3**, pp. 249–55.
34 e.g. Carneiro, R. L. (1958). *op. cit.* p. 26.
35 Netting, R. M. (1969). *op. cit.* pp. 108–10.
36 Harner, M. J. (1970). *op. cit.* has recently offered an evolutionary model of descent through various phases of population density and competition for agricultural land, from cognatic or bilateral to lineages to non-localized clans or sibs, and finally to stratified society with hereditary class distinctions.
37 Carneiro, R. L. (1958). *op. cit.*; Watters, R. F. (1960). The nature of shifting cultivation, a review of recent research, *Pacific Viewpoint*, **1**, pp. 59–99.
38 Smith, P. E. L. and Young, T. C., Jr. (In press). The evolution of early agriculture and culture in Greater Mesopotamia. A trial model, *in* Spooner, B. J. (ed.) *Population Growth: Anthropological Implications*. Philadelphia; see also Young, T. C., this volume, pp. 827–42.
39 Ritchie, W. A. (1965). *The Archaeology of New York State*. New York. p. 280; also Whallon, R. (1968). Investigations of late prehistoric social organization in New York State, *in* Binford, S. R. and Binford, L. R. (eds.) *op. cit.* pp. 223–44.
40 Fowler, M. L. (1969). Middle Mississippian agricultural fields, *Amer. Antiq.*, **34** (1), pp. 365–75.
41 Lanning, E. P. (1967). *op. cit.*
42 Sanders, W. T. and Price, B. (1968). *op. cit.*
43 Sears, W. (1968). The state and settlement patterns in the New World, *in* Chang, K. C. (ed.) *op. cit.* pp. 134–53.
44 This is a highly simplified version of the evolution of settlement in Greater Mesopotamia. The reader is referred to Smith, P. E. L. and Young, T. C., Jr. (In press) *in* Spooner, B. J. (ed.) *op. cit.* for the archaeological details, and to Young, T. C., this volume, p. 832, especially for the events in Lower Mesopotamia after Ubaid times.
45 Higham, C. F. W. (1968). Trends in prehistoric European caprobovine husbandry, *Man* (n.s.), **3**, pp. 64–75; Higham, C. F. W. (1969). Towards an economic pre-history of Europe, *Curr. Anthrop.*, **10** (2–3), 139–50.
46 Chang, K. C. (1968). *The Archaeology of Ancient China* (2nd ed.). New Haven, pp. 133, 147.
47 Waddell, E. (1969). *op. cit.*
48 Barth, F. (1959). Segmentary opposition and the theory of games: a study of Pathan organization, *J. Roy. Anthrop. Inst.*, **89** (1), pp. 5–21; Gould, P. R. (1963). Man against his environment: a game theoretic framework, *Ann. Assoc. Amer. Geographers*, **53**, pp. 290–7 (reprinted in Vayda, A. P. (ed.) (1969). *Environment and Cultural Behaviour. Ecological Studies in Cultural Anthropology*. New York. pp. 234–51).

Section 2: Territoriality and the demarcation of land

R. D. MARTIN

Concepts of human territoriality

The ethological viewpoint

The points raised in this paper are concerned with a simple hypothesis: that human beings are related to animals within the general framework of organic evolution, that in all animals there is some species-specific hereditary contribution to the determination of overt behaviour, and that therefore predetermined behavioural components (such as behaviour patterns and motivational tendencies) of animal origin may be expected to contribute to the structuring of human activities. Among other aspects of behaviour, patterns of settlement might (according to this hypothesis) be reasonably expected to involve predetermined tendencies or actual activities typical of the human species as a whole.

This kind of hypothesis is primarily proposed by *ethologists*—students of animal behaviour. The ethologist's point of view has often been criticized, and almost as often it has been misunderstood. In order to avoid at least some of the misunderstanding from the outset, two main points must be made: firstly, no trained ethologist[1] has ever proposed that *all* human behaviour is predetermined through genetic programming ("innate determination", "instinctive control", etc.). In most cases, all that has been said is that certain contributory elements of human behaviour have probably been laid down during a long period of evolution and that such elements make some contribution to the overall repertoire of human activities. In the latter stages of human evolution, there has evidently been pronounced increase in the scope of *intelligent control* and *modification* of behaviour, and this means that in modern man any species-specific behaviour programmed by evolutionary processes may easily be enhanced, overlain, modified or suppressed by individual participation of human reasoning. Secondly, it must be emphasized that any statements about genetically programmed components in human behaviour are necessarily hypothetical at this stage of our knowledge. At best, our evidence is usually circumstantial; at worst, it is no more than anecdotal. It is only in the popular works of some ethological authors[2] that one is given the impression that arguments about fixed elements of human behaviour are based on established fact. Although ethologists working with the behaviour of various animal species are generally convinced that there is some genetically determined foundation of behaviour, nobody has yet been able to show, in detailed terms of central nervous processes, what "innate behaviour patterns" and "specific motivation" really represent. Conclusions about their existence are generally derived through inductive reasoning, rather than through scientifically conducted exhaustive tests, and therefore explanatory systems of the type provided by Lorenz[3] can only rank as working hypotheses. Many alternative provisional hypotheses have been proposed to explain the bases of human behaviour—ranging from theories of exclusive conditioning to dogmas of absolute intellectual mastery—and we are faced with a highly controversial situation in which opinion is far ahead of established scientific fact. As working hypotheses, however, ethological systems of explanation have a great deal to offer. The main lesson which should be drawn from the recent spate of popular books about the possible ethological background of human behaviour is not that the claims of a particular author are right or wrong, but that detailed scientific investigation of this background should provide much useful information. This information could eventually play an important part in a general, pan-disciplinary theory of human behaviour.

Most ethological research involves observation of intact animals, and conclusions about species-specific operation of the central nervous system are derived largely by *implication*. Some attempts have been made to study the integrated operation of the nervous system itself (e.g. by von Holst[4]); but so far this has only led to general confirmation—rather than detailed verification—of ethological analyses of behavioural organization. The main strength of the explanatory systems proposed by ethologists is their ability to provide logical connections between various observations, and the fact that the validity of such systems can be tested to varying extents by appropriately conducted experiments. Because of the difficulty involved in demonstrating integrated processes within the central nervous system, however, we are still at a stage where a large number of competitive, apparently equally logical, hypotheses can be presented to explain particular behavioural activities. This is inevitable in a situation where one must choose between studies of the behaviour of unrestricted intact animals and direct study of the operation of the nervous system in animals whose activities must to some extent be confined, and perhaps disrupted, in order to permit investigation. For some time to come, ethological theory will inevitably be largely dependent upon inductive reasoning.

A main point of controversy in ethological studies has been the use of the word "innate", and many highly sophisticated experiments have been conducted in order to define exactly what this term implies. Such experiments have greatly enriched ethological theory; but they have also added enormous complexity to what was originally a quite simple concept. In order to sidestep this complexity, one can adopt a more pragmatic approach and talk in terms of *species-specific behaviour*. After all, if one is simply concerned with an animal's behaviour under natural conditions, the main concern is whether every healthy member of the species behaves in the same way, rather than the exact proportions of nature and nurture which produce such stereotyped behaviour. As a general rule, one can assume that species-specific characteristics in behaviour have *some* genetic basis (e.g. in the form of genetic determination of the structure of the nervous system and its auxiliary organs or genetically determined factors canalizing the development of behaviour), though one must be careful to exclude the possibility of traditional transmission of behavioural traits from one generation to another. For practical purposes, this kind of approach is sufficient for a general examination of the ethological background of human behaviour, since our knowledge in this field is still extremely restricted. This does of course mean that a number of important questions (e.g. whether man can successfully change his behaviour by mere force of will) are left untouched; but we may nevertheless be able to arrive at some general guidelines to typical behavioural features of the human species.

In actual fact, this approach is not very far from that used by many other branches of study concerned with human behaviour. Through observation of what some human beings do in certain situations, the attempt is made to produce a general theory describing, and perhaps to some extent explaining, human behaviour. And all such approaches encounter the same major obstacle —the fact that human cultural influences are so pronounced that the potential similarities between surviving human populations in different areas are vastly outweighed by culturally-determined differences. If the possibility of some intangible extraneous influence on human behaviour is left aside, the similarities which *do* exist must reflect something inherent in "human nature", and this is precisely what the adherents of various disciplines are attempting to identify. Within this framework, prehistoric archaeological material plays a special part, in that it provides "fossil" evidence of human activities in the past. However, the interpretation of such activities often involves extrapolation from what is known about recent human activity (through introspection or comparison), and thus a circular argument can emerge when archaeological evidence is used to support theories about the factors controlling surviving human societies. For this reason, it is essential to maintain a clear distinction between *description* and *interpretation* in considering archaeological material as evidence for past human activities, and this is particularly necessary where "territorial demarcation" is involved.

The biological approach to human behaviour depends, like other approaches, upon comparative study of human beings in isolation and in groups—and the reliability of any conclusions is proportional to the range of human beings covered. (Hence the major drawback of Morris's popular writings[5]; they refer exclusively to western European and American society.) One of the main sources of information about typical, identifiable features of human behaviour is thus to be obtained from broad, cross-cultural comparisons (*cross-cultural behaviour*)[6]. Since the influence of cultural factors exerts such a strong pressure towards diversity, it is highly likely that any similarities between different human societies reflect inherent behavioural tendencies (ranging from pre-determined patterns of behaviour to predisposition towards stereotyped response to a given set of environmental factors), rather than chance similarities.

Further information can be obtained in cases where, for some reason, the usual cultural influences and/or normal adult development of rational capacities have been excluded. Studies of newborn babies and young infants can be presumed to show up some inherent aspects of human behaviour, since the objects of study are too young to have learnt very much and can generally be assumed to be largely dependent upon automatic, pre-programmed responses (*infant behaviour*). However, such studies tell us little about most adult patterns of behaviour, since some patterns are specific to early life, whilst others are specific to the adult. Nevertheless, some information can be obtained about behavioural features which appear to have a continuous development (e.g. development of speech responses to visual and vocal signals from conspecifics; utilization and defence of specific living areas). Another category in which there is exclusion of the intelligent control usually evident in adult human beings is provided by pathological cases where various essential sense-organs (eyes, ears, etc.) are non-functional, or where the brain is in some way abnormal (*derangement behaviour*). In such cases, the subjects may "revert" to more automatic, pre-programmed types of behaviour. Normal human beings may temporarily exhibit the same phenomenon in cases of extreme physical and mental stress, for example, under extended battle conditions (*stress behaviour*).

In principle, all of the above categories of evidence are accessible to every discipline involving comparative study of man. But the biological approach benefits from the additional fact that all human beings belong to a single animal species—*Homo sapiens*. Although there are a number of pronounced morphological characters distinguishing human races (e.g. skin pigmentation, bodily dimensions and skeletal characters), these races can potentially interbreed and produce fertile offspring, thus fulfilling the biological criterion of the species. This fact is very important, since every animal species has a specific complement of inherited characters—affecting both morphology and behaviour—and since this complement is essentially typical for all members of the species. In association with this, there are two outstanding characteristics of the evolved complement of characters:

1 Those characters must have been selected on the basis of their *survival value* under natural conditions (as has been pointed out by Reynolds[7]).
2 The selection processes concerned must have operated under a given set of environmental conditions (the *natural environment*) and the value of the resultant characters is therefore relevant to that particular environment.

Apart from providing insights into the possible evolution of early human behaviour, these two points also open the way to comparative evolutionary considerations drawing upon studies of man's closest living relatives within the Order Primates (lemurs, lorises, tarsiers, monkeys, apes and man). In particular, investigation of the *apes* (gibbon, orang-utan, gorilla and chimpanzee) can provide strong indications about behavioural tendencies and patterns likely to have been present in the early evolution of the Family Hominidae. Since the main feature of human evolution, within the restricted framework of diversification of the ape/man (Hominoid stock), has been replacement of programmed behaviour by insight-controlled activities, such indications can only be used as a basis for a maximizing hypothesis about present-day behavioural control of human activity (*hominoid behaviour*) Studies of the possible pathways of behavioural evolution in the Primates can provide two types of information:

1 If a particular behavioural character proves to be typical within a group of closely related species (e.g. the apes), there is a reasonable probability that this character is typical of other annectant species (e.g. man).
2 Given a particular set of environmental conditions and a certain set of inherited features, it is likely that a characteristic behavioural development will occur. Thus, a comparison of Primates living under similar ecological conditions to those encountered by man during the main phase of his evolution (e.g. taking baboons in savannah habitats) may provide insights into the early canalization of human behaviour. This point, together with the earlier reference to the importance of the *natural environment* in evolution, serves to emphasize the need to study *human ecology* as an integral part of the study of man. (It also highlights the difficulty of interpreting the inherent behavioural characters of human beings now living in artificially-created environments.) This is actually very close to the concept of ecologically-determined *convergence* or *parallelism*, relating to cases where similar environmental conditions or requirements seem to have led to similar evolutionary developments. A number of striking cases have been demonstrated by Dubost for mammals[8], and even within the Order Primates there are several instances (e.g. parallel behavioural developments in the squirrel monkey and the talapoin).

Application of the evolutionary viewpoint

It has become increasingly common for those already committed to the ethological point of view to assume that one can simply extrapolate from the behaviour of living monkey and ape species—as observed in the field—to that of human beings. The clearest example of this approach is given by Morris[9] in "The Human Zoo", where he draws heavily upon the behaviour of rhesus monkey groups in attempting to explain the fundamental organization of human social behaviour. This attitude has, of course, been criticized by numerous authors; but such criticism is frequently based on a refusal to accept any degree of evolutionary determination of human behaviour, and thus derives from individual belief in the all-pervading intelligence of man rather than from scientific exclusion of the possibility of pre-programming of human behaviour. This kind of criticism is therefore suspect. But serious criticism can be advanced from a *methodological* point of view. Derivation of the behaviour of one living species from the behaviour of another living species can only be justified if it is assumed that the latter species has retained its behaviour in unchanged form from the common ancestor of the two. This concept of the "living fossil" has bedevilled every aspect of evolutionary comparison, and it has repeatedly emerged that a particular living animal supposedly representing a "primitive archetype" is in fact extremely specialized for its particular way of life. If one considers a given group of related animal species from an evolutionary point of view, it is far more useful to consider each living species as a specialized form adapted to a particular way of life rather than to assume, in *a priori* fashion, that a particular species represents the unchanged common ancestral type. In most cases, one can safely assume that the common ancestral type died out some time ago and that each of its living descendant species is specialized in its own particular way. It is therefore far more likely that rhesus monkey behaviour and human behaviour both represent specialized offshoots from a common ancestor, and that the behaviour of the common ancestor was not identical with that of the living rhesus monkey. When looked at in this light, it is obvious that many authors have been using the evolutionary concept without actually considering evolutionary processes—that is, the *time factor* has been ignored, and comparison of living species has been used as an unjustifiable short-cut.

This question is best considered in terms of a diagram (Fig. 1). For any given evolutionary comparison, one can think in terms of an "ancestral stock" giving rise to all of the living forms under comparison. This ancestral stock may have given rise to a living form which has remained primitive in certain (but not all) respects—the so-called "primitive survivor". It could also have given rise to a further stock ("later stock"), which itself produced a somewhat more advanced form ("less primitive survivor") and an extremely advanced form ("advanced survivor"). The short-cut which has been made by many

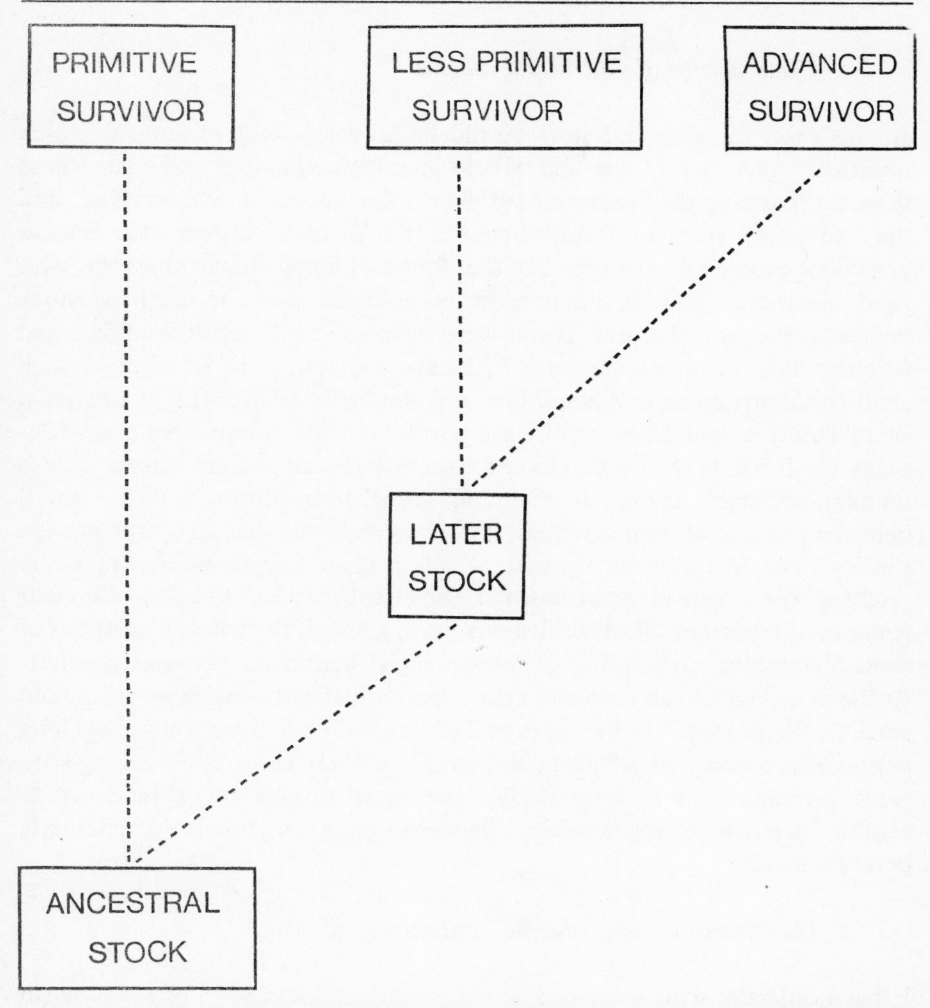

Fig 1 Diagram illustrating the simplistic assumption that descendants from ancestral stocks may survive in unchanged form, thus permitting the investigator to make direct evolutionary derivations from existing forms.

authors is the simple equation of the "primitive survivor" with the ancestral stock and the "less primitive survivor" with the later stock, followed by explanation of evolution of the three living forms in terms of derivation of one living species from another. Obviously, the only reliable approach is attempted reconstruction of the characteristics of the ancestral and later stocks themselves. The concept of the "living fossil" is extremely suspect, and it involves —among other things—the implication that the key ecological factors have remained unaltered for vast periods of geological time.

This diagram can be applied to two particular cases involved in considerations of the evolution of human territorial behaviour:

1 The evolution of Old World simians

In this case, the ancestral stock would be an early stock of simians which eventually gave rise to the Old World monkeys, the apes and man (these three representing the "primitive survivor", the "less primitive survivor" and the "advanced survivor", respectively). If it is to be argued that human behaviour evolved from a type like that found in living rhesus monkeys, with rigid hierarchies and so on, it must be assumed that the ancestral stock possessed the same features. If this were the case, it is difficult to understand why the "less primitive survivors" (the apes) should show no signs of such rigid social structuring. The gibbon and the orang-utan form, at the most, small family assemblages, whilst the gorilla and the chimpanzee form relatively small bands with much looser structure than that found in the rhesus monkey society[10]. If one is really to argue from rhesus monkey social behaviour to that of human beings, it must be assumed that all four living ape species have lost various features which human beings have supposedly retained. If one considers, in addition, the fact that rhesus monkey behaviour is not even typical of Old World monkeys in general, the whole hypothesis of rigid hierarchical structuring of human social behaviour becomes obsolete. At the most, one can assume that man's derivation from a common, Hominoid stock (="later stock" of Fig. 1) provided him with a heritage entailing a long period of protection of offspring and strong mother-infant bonds, along with some predisposition to form fairly loose social groups and a tendency to exhibit dominance relationships between group members (particularly between males).

2 The "hunting and gathering" ancestry of mankind

It has been said, with some conviction, that man spent 99% of his time (in terms of evolution of the species) as a hunter and gatherer, and that other forms of human activity represent relatively recent developments. In terms of complexity, it indeed seems likely that early groups of *Homo sapiens* would have lived on a hunting and gathering basis, prior to the establishment of agriculturally based settlements. However, it is not completely justifiable to conclude from this that *modern* hunting and gathering societies are primitive in every respect (e.g. in frequently exhibiting nomadic habits). It has, for instance, been argued that the apparent absence of marked territorial behaviour in Eskimos and Kalahari Bushmen shows that early *Homo sapiens* was non-territorial, since "hunting and gathering"—in a very general sense—was the ancestral way of life[11]. To take the diagram once again, one can take early *Homo sapiens* as the "ancestral stock", giving rise to the few "primitive survivors" represented by still viable hunting and gathering communities. The "less primitive survivor" could be represented by a simple, agriculturally based settlement, and the "advanced survivor" would accordingly be an urban community. Once again, it seems highly likely that surviving hunting and gathering communities would represent specialized adaptations to particular conditions, rather than pure "primitive social groups". It is not justifiable to assume, as Morris does, that studies of such communities are irrelevant in studying the evolution of human behaviour—they are in fact highly relevant. On the other hand, however, one cannot simply assume that modern hunters and gatherers are "living fossils". The lack of territoriality in some groups may represent a specialized adaptation to particular environmental conditions. Such communities may even be retrogressive derivatives from more complex human societies[12].

Territoriality in mammals

There has been an enormous expansion of research into the behaviour of mammals under natural conditions in recent years, particularly regarding various species of Primate and Ungulates. However, such research has not yet reached the point where a general synthesis is possible, so that one could speak of typical characteristics of territorial systems of mammals in the wild state. Nevertheless, some broad categories of behaviour relating to territoriality can be defined, and these provide an indication of *what to look for* as possible mechanisms determining similar behaviour in human beings.

1 *Territory, home-range and social space*

With every new species of mammal studied, the difficulty of defining these concepts is multiplied, particularly since far-ranging or even nomadic species necessitate some concept of "mobile territoriality". Perhaps the major point to note is that in each case it is obvious that the system of spatial use, demarcation and defence in each mammal species represents a pattern of adaptation to the particular ecological requirements of that species. It is also important to underline that closely related species may differ quite extensively with respect to spatial utilization systems, so that it is obviously a relatively simple matter (in evolutionary terms) for the details of such systems to be varied. Accordingly, one should only expect the broad outlines of spatial utilization systems to be laid down in evolution, and relatively superficial features (such as the development of a rank order) should be treated with caution when broad comparisons are made. Most mammals exhibit a tendency to form a rank order when confined in a relatively small space (e.g. in a zoo), and a species which is typically "territorial" under natural conditions will form some kind of dominance hierarchy in close confinement. Thus, one can postulate a general tendency to form a rank order in mammals, and formation of a social hierarchy of some kind can be viewed as an outcome of localized social grouping rather than as a major contributory factor of social life. In some social species, this ranking tendency can, of course, be developed so that it operates in a controlled manner as an integral aspect of social life (e.g. in the rhesus monkey). This tendency would adequately explain the tendency for human groups to exhibit different degrees of status differentiation under different conditions.

Whatever the spatial system adopted by an animal species, there is usually some kind of *social space* surrounding each individual—an area within which approaches by other members of the species are not usually tolerated, usually evoking attack or flight. (Mating represents an exception and requires special aversion-reducing mechanisms.) But this area is primarily associated with the body of the individual concerned, rather than with requirements from the external environment. As far as the latter is concerned, two concepts have generally been employed to describe mammalian spatial utilization systems. The concept of the *home range* usually implies the area which the animal normally covers in its everyday activities, without implying the exclusion of conspecifics, and it primarily concerns the natural resources exploited by the animal (food, water, sleeping facilities, etc.). The concept of *territory*, on the other hand, usually implies some kind of *defence* of a given area, with partial or total exclusion of conspecifics. In most cases, only certain categories of conspecifics (e.g. adult males, as opposed to adult females and juvenile animals) are excluded by territorial defence. Such defence of a particular area is probably correlated in every case with exclusive use of certain resources (e.g. a guaranteed food-supply for the territorial animal and/or the offspring) and may be concerned with more effective protection (e.g. a preferred area may be one which is better protected against predation). There is also the fact that a territory may be used by a male for advertisement to receptive females; this may be a primary or secondary attribute of territoriality. Territoriality in mammals is in most cases related in some way with reproduction, and it is typically based on exclusive use of an area for this purpose and/or for providing adequate quantities of particular natural resources. If such resources are abundant, and if the requirements of reproduction (including some kind of sexual selection) are met in other ways, clear-cut territoriality becomes superfluous.

The main lesson to be drawn from presently available studies of home-range and territoriality in mammals is that one cannot provide a complete *a priori* pattern for all species, apart from saying that a given mammal will tend to restrict its activities to a given area if it is a sedentary rather than a migratory species (particularly since familiarity with a given area provides subsidiary advantages such as established escape routes and efficient movement patterns), and that it is likely that at least some categories within the species (e.g. adult males) will defend a small part of this area against the intrusion of at least one category of conspecifics. Since early mammals (and early Primates) were probably arboreal, nest-living and primarily insectivorous, it is probable that a sedentary habit associated with some form of territoriality (ensuring an adequate diet, a protected nest-site and the appropriate level of sexual selection) was evolved early in mammalian evolution. Some mammals have secondarily lost most features associated with such a way of life, but it is likely that any species which adopts a sedentary habit where resources are restricted, or where a particular mechanism for sexual selection is required, will exhibit roughly the same kind of territorial system. Thus, it is of paramount importance to form some idea as to whether early human populations were nomadic or sedentary, and whether the necessary resources were in any way limited, if any evidence regarding human territoriality is to be deduced from "Hominoid Behaviour". A nomadic early human population would have been unlikely to exhibit clear-cut territorial behaviour, whilst sedentary populations might well have done. Of course, it is also possible that when the presumed nomadic hunting and gathering stage gave way to human settlements, there was a re-emergence of behavioural features laid down in the early stages of Primate evolution. But such features would be extremely general in nature and would not involve anything as specific as a defined rank order and specific patterns of territorial defence.

It is important, too, to consider the *seasonal* aspect and the possibility of *local variations* in social structure, since failure to consider these two aspects can lead to a totally false idea of the "typical" behaviour of a species. Because of its specific functions (e.g. individual nourishment, mating, rearing of the young), territorial behaviour is likely to change markedly over the course of the year where these functions are seasonal in nature, and there may similarly be seasonal variations in the size of home-ranges. It is not generally realized that the most frequently cited case of "pure territoriality" in animals is rigidly tied to certain seasonal factors: the male stickleback sets up a territory at a certain time of the year (primarily in association with reproduction) and develops specific signals at that time in order to "inform" other members of the species that he has become territorial. At other times of the year the stickleback moves around in anonymous groups without any indications of territoriality other than the defence of "social space". In view of this, it is unlikely that "pure territoriality" is particularly common in the animal kingdom; it appears sporadically as a special adaptation to special requirements and does not exclude non-territorial behaviour in other contexts. One must therefore also consider the possibility that early human populations exhibited some seasonal or other temporal variation in their spatial utilization system[13].

Thus, simply considering *individual* aspects of human space utilization, the following points must be raised in any consideration of human territoriality:

(*a*) Were the earliest populations of *Homo sapiens* sedentary or nomadic?
(*b*) Were there any seasonal factors affecting diet, cover and/or reproduction?
(*c*) What functions might have necessitated *territories* rather than fixed or labile home ranges lacking territorial defence?

(*d*) How were exogamy and sexual selection determined in early human populations?

2 *Social behaviour*

One must also consider the size and structure of social groupings in order to analyse human territorial behaviour, since social mammals inevitably exhibit a far more complex pattern of spatial and behavioural inter-relationships. In view of what is known about human, ape and monkey behaviour in general, it can be safely assumed that early *Homo sapiens* lived in social groups of some kind, and that individuals in such groups exhibited strong bonding patterns (e.g. social grooming) and extensive parental care. Beyond this, there is a number of important questions to ask about the actual structure of these social groups. In discussing the evolution of social groups, the following evolutionary sequence is often suggested as a basic pattern for the emergence of complex mammal societies:

1 All adults solitary; male and female meet only for mating. (This presumes some kind of spacing mechanism, probably with both males and females exhibiting territorial defence against all conspecifics outside the breeding periods.) Infants remain with the mother until sexual maturity, or perhaps earlier, and then move out to establish their own home-ranges or territories.
2 Infants remain with the mother *after* the attainment of sexual maturity, producing groups based on a family system. Associated with this is the prior or coincidental development of adult male/adult female bonds, which may be specific pair bonds, such that there may be several breeding adults in a social group.

(NB. It has also been postulated that a specific pair bond stage may represent an intermediate between the solitary adult system and the social group system, giving the following sequence: solitary adults—family groups (1 adult male, 1 adult female and immature offspring)—social groups (several sexually mature adults with their offspring). The development of social groups could accordingly be interpreted as based upon the retention of infants beyond sexual maturity, through extension of parent-infant bonding patterns).

However, this concept of the evolution of social groups is to some extent based upon the (western) concept that a healthy male-female bond is an essential feature of social groups. Indeed, it has been suggested that the fundamental structure of Primate societies is basically composed of sexual bonding. Although theories exclusively based upon sexual bonding have generally been eliminated[14], there has been an obvious tendency to attempt to explain the evolution of social behaviour in Primates on the basis of *one* primary factor. Many authors who rejected a unitary sexual-bond theory replaced this with a unitary dominance relationship theory, assuming that the cohesion of all Primate social groups is dependent upon dominance and submission among the members. But it has now emerged, with the appearance of many new field studies of Primate societies, that each particular species exhibits an equilibrium between a number of different factors, and that sexual bonding and dominance effects may or may not count among these factors. What is more important is that some Primate societies—particularly those of Old World monkeys and apes—have recently been discovered to exhibit a wide range of *flexibility*, such that social structure may vary with local conditions. This means that it is misleading to think in terms of a rigid blueprint of social behaviour for each Primate species; at the most, there is a programmed set of guidelines leading to a particular social structure under a particular set of conditions. This also means, of course, that any programming of territorial behaviour operating between social groups of a Primate species is likely to be flexible as well. In fact, groups of Old World monkeys and apes show little sign of clear-cut territorial behaviour (in terms of defence of particular areas against conspecifics) under natural conditions, and it may be that there is no more than a programmed *propensity* to exhibit territorial behaviour when the pressures upon a social group (e.g. shortage of natural resources) require it. Most Old World monkeys and apes seem to be dependent upon resources which are relatively abundant, such that a free-ranging system with tolerated overlap of social groups is quite adequate. It is probably only in a few cases where (for example) a particular preferred food is in short supply that a basically territorial system involving small groups (e.g. in the gibbon and orang-utan) appears. It is therefore extremely important to know what the probable diet of early *Homo sapiens* was, and whether the dietary components were likely to have been abundant or sparsely distributed, in order to guess whether or not there was pronounced territoriality[15]. Even if, as it has been suggested, early hunting and gathering human communities were essentially non-territorial, it is possible that the switch to localized settlements with some kind of agricultural basis necessitated the emergence of some kind of territorial pattern, and that programmed propensities for such behaviour once again became overt. Accordingly, it is possible that any territorial features exhibited by early human settlements (and their successors), although they would not incorporate actual behaviour patterns inherited from a Hominoid ancestor, might incorporate certain *propensities* for territorial activity derived from man's early Primate ancestry.

As far as the origin of Primate group structure is concerned, recent evidence suggests that the popular evolutionary sequence outlined above (solitary animals—family groups—social groups) may well be without foundation, because of its inherent assumption of the early appearance of specific male-female bonds. A more likely evolutionary sequence would seem to be:

1 All adults solitary and territorial; male and female meet for mating; mother retains offspring for some time between birth and the attainment of sexual maturity.

2 Female groups are established, at least on the basis of regular non-hostile encounters, and solitary males each have an association with one or more of these female groups. Mothers retain offspring for some time between birth and the attainment of sexual maturity within a loose social framework; at sexual maturity males leave the group and females are usually associated with the exclusive female adult society.

3 Bonds between males and females become stronger, such that different kinds of social system are produced:

  (i) single male stays with female group and young conspecifics of same sex; other adult males are excluded (harem system)
  (ii) several adult males remain with a female group and young conspecifics of both sexes (multi-male social group); specific pair bonds may be formed between adult males and females
  (iii) the female groups may disintegrate in favour of exclusive single male/female bonds (pair system). The rarity of this type of system among Primates indicates that it is more likely to be a secondary rather than a primary phenomenon.

This sequence of social evolution provides a built-in mechanism for ensuring genetic variety (through exogamy), since in stage 2 maturing males move out of the basically female group and have to set themselves up elsewhere in association with a different female group. It also allows for enhanced sexual selection of the males, since only those which establish an association (e.g. a territorial relationship) with one or more of the female groups can mate, and there would be a distinction between established males and peripheral males. This is found in many Primate species, either on a clear-cut territorial basis or in the form of peripheral location of many adult males in social groups. Such a system allows easy replacement of any adult male which disappears (through predation, etc.) from an established position associated with adult females, and it also permits migration of younger adult males between social systems during the period between severance from the mother and establishment of a status with breeding potential. It has been repeatedly observed that males of intermediate ages are more likely to pass between Primate social groups than are females.

It is rather important to establish which of the two proposed evolutionary sequences led to human social behaviour as we know it, since the expected patterns of behaviour are different in the two cases. One must bear in mind the possibility that early *Homo sapiens* social systems were based on central harems each led by a single breeding male (or a small nucleus of males), with solitary or group-living "bachelor" males living on the peripheries of such harem groups. Such a structure would offer certain advantages for division of labour within early hunting and gathering communities, with bachelor males going through an early non-breeding initiation period in association with the accomplishment of the more dangerous tasks (e.g. inter-group combat). It is automatically assumed by some modern authors[16] that human social systems are inevitably based on pair-bonding between adult males and females, but this (even if it does eventually emerge that such pair-bonding really *is* typical of all contemporary human societies) could well be a relatively recent addition, evolved as human societies grew from small hunting and gathering bands to larger settlements. A harem system with peripheral males would solve the problem of exogamy far more easily than a system based on groups with established male-female bonds and general retention of infants within a social group.

Another possibility which must be considered, once one has a society incorporating several adult individuals of the same sex, is the formation of a dominance hierarchy. In general, it seems that in most mammal species there are separate hierarchies for males and females. One would expect such hierarchies to develop where there is some kind of competition for one commodity or another, such that the established order reduces conflicts to a minimum within groups. If the harem system represents a transitional stage between solitary life and full societies in the evolution of any animal species, one would expect female hierarchies to be more sophisticated, since they would have been undergoing evolution for some time before evolution of male hierarchies began. The fact that human societies generally exhibit a situation where some kind of hierarchy occurs between males could be taken to indicate that full human societies containing several adult males, as well as females, have been a going concern for some time. However, it can also be argued that female hierarchies are generally better equilibriated and less prone to disruption, and this may indicate that the earliest stages in the evolution of human groups involved harem systems rather than small groups containing several adult males along with several adult females and their offspring.

This discussion underlines the importance of seeing the evolution of human societies as a continuing process, with evolution of specific social mechanisms and behaviour patterns occurring at particular levels. Just for the sake of illustration, one might suggest the following pattern of evolution as a working hypothesis:

1 Harem groups of females each led by a single, dominant male, with bachelor males living in groups both adjacent to and loosely associated with harem groups.
2 Offspring are retained in the harem groups for ever-increasing lengths of time, until there is a system where the young stay on in the group after reaching sexual maturity. The effect of the previous stage is still evident in the general predominance of one male over the whole system, the more balanced relationships within the female grouping and exaggerated rivalry between young males and older ones. There would also be a tendency for young ("bachelor") males to wander off, perhaps remain solitary for a while, and then join other groups. These larger social groups would be generally

sedentary, protecting the core of their group home-range against intrusion of other groups if resources were scarce.

3 As population sizes and complexity of behaviour increase, such full social groups tend to live in increasing proximity, with the consequent development of some kind of acceptance behaviour between adjacent groups. This would lead to three levels of distinction among conspecifics: "member of home group", "member of adjacent group", "member of foreign group". With the development of speech and adornment patterns, such stratified social relationships become more and more complex, incorporating an increasing proportion of "cultural" factors.

Thus, in the analysis of the contribution of evolved elements to human territorial behaviour as it now appears, there is a double problem. First, one is faced with the difficulty of extricating evolved elements from the complex picture incorporating vast cultural, environmental and intellectual factors. Second, one must distinguish the various levels at which certain patterns of behaviour may have evolved, and the levels at which they may have become suppressed and modified.

Territorial marking

It can be regarded as fairly well established that early *Homo sapiens* must have been living in social groups of some kind, and that these groups probably defended some kind of territory against intruding conspecifics (though this may have been a tendency correlated with abundance and localization of resources, rather than a homogeneous ubiquitous feature). Some light may be thrown upon the persistence of automatic patterns and motivational elements of territorial behaviour in human beings by considering the evidence for territorial demarcation in human beings at the present time.

In general, mammals can demarcate their territories with scent, vocal signals and visual signals (including mere presence). Mammals produce a number of essentially excretory products (urine, faeces, sweat, etc.) which can easily be converted to vectors of specific odours. Marking with urine and/or faeces is extremely common in mammals, and a vast variety of body areas have been modified to produce "scent glands" derived from original sweat glands, etc. In many species, vocal signals are used as a supplement, with calls given at certain times of the day at specific points of the territory. However, such vocal demarcation of territory is relatively uncommon. Several Primate species employ this means of demarcation, but it seems likely that this was unimportant in the evolution of human territorial demarcation. One of the apes—the gibbon—*does* employ vocal territorial demarcation of an obvious kind, but this is immediately obvious from the highly specialized calls and the special vocal sac, and it may well be an adaptation to forest conditions where visibility is poor. In the evolution of human territorial demarcation, as with the Primates in general, there seems to have been a primary shift of emphasis from olfactory/auditory to visual signals. But this does not mean that olfactory signals have disappeared altogether; they are simply likely to be less important.

The switch to visual signalling in human societies also opened the door to a vast range of cultural and environmental modification, such that it is nowadays extremely difficult to identify any common underlying patterns or tendencies. Wickler and Eibl-Eibesfeldt have made out a fairly good case for the evolution of *genital display* as a territorial factor in human evolution[17], and a fair amount of research has been carried out on conspicuous morphological features, such as the male beard[18]. However, much remains to be done to identify the basic common features of human signalling behaviour related to territoriality as such. One field of research which could be extremely fertile is the analysis of artefacts where the signals have been transferred from the human body to an object which is used in a territorial context. In addition, much can be obtained from studies of "Infant Behaviour", "Derangement Behaviour" and "Stress Behaviour" in drawing conclusions about territorial demarcation and signalling. Studies of these three kinds of behaviour also indicate that olfactory signalling (e.g. through urine deposits) still plays some part in human territorial behaviour.

Studies of children and psychiatric patients have also provided some information about the way in which *individuals* demarcate "territories" in interaction with their fellows. In particular, it has been suggested that the presence of recognizable discontinuities in the environment (e.g. an outcrop in the wall of a children's nursery) are necessary for the establishment of "territories"[19]. Without such discontinuities, little in the way of defence of a particular area can be observed. Some studies can be very illuminating at this individual level, and can lead to insights into conditions necessary for optimal living conditions in modern societies; but such territorial behaviour should not be confused with the territorial behaviour exhibited between human *groups*, which may have an essentially distinct basis.

Territoriality and stress

A final word should be said about the possibility of social stress[20] emerging in groups living at densities above the average for which the behaviour of the species was evolved. It is difficult to consider this problem with respect to human societies, because of the gradual evolution of societies of ever-increasing size; but it can reasonably be assumed that in recent times human populations have evolved culturally at rates far in excess of the rate of evolution necessary for behavioural adaptation to keep pace. In addition, developments in agricultural techniques may have removed pre-existing physical barriers to population concentration. Experiments with animals show that certain pathological effects appear at population densities above that for which each species has been adapted, and it is quite conceivable that similar effects may eventually emerge in human societies, such as:

1 general individual weight loss
2 digestive disorders and kidney ailments
3 circulatory disfunction (e.g. high blood pressure)
4 disruption of parental behaviour
5 disruption of social behaviour
6 decline in fertility
7 premature death

For this reason alone, it is extremely important to attempt to determine the sort of group size and structure and the kind of territorial system (or range of territorial systems) for which *Homo sapiens* was adapted during his early stages as a hunting and gathering species.

Notes

1 Tinbergen, N. (1968). On war and peace in animals and man, *Science*. **160**; Lorenz, K. (1966). *On Aggression*. New York; Wickler, W. (1967), Socio-sexual signals and their intra-specific imitation among primates, *in* Morris, D. (ed.) *Primate Ethology*, London; Morris, D. (1967). *The Naked Ape*, London.
2 Ardrey, R. (1967). *The Territorial Imperative*. London; Morris, D. (1967). *ibid*.
3 Lorenz, K. (1970). *Studies in Animal and Human Behaviour*. I (trans. Martin, R. D.). London.
4 Holst, E. von and Saint Paul, U. von (1963). On the functional organization of drives, *Animal Behav*., **11**.
5 Morris, D. (1967). *ibid*.; Morris, D. (1969). *The Human Zoo*. London.
6 Eibl-Eibesfeldt, I. (1967). Neue Wege der Humanethologie, *Homo*, **1**.
7 Reynolds, V., this volume, p. 401.
8 Dubost, G. (1968). Les mammifères souterrains. *Rev. Ecol. Biol. Sol*, **1**; Dubost, G. (1968). Les niches écologiques des forêts tropicales sud-americaines et africaines, sources de convergences remarquables entre rongeurs et artiodactyles, *La Terre et La Vie*, **115**.
9 Morris, D. (1969). *ibid*.
10 This same point was emphasized by Vernon Reynolds in discussion.
11 This suggestion was supported by J. Woodburn in discussion. Lee, R., this volume, p. 177 has rejected the "rigid territorial model" of Radcliffe-Brown, proposed for Australian Aboriginals. Forge, A., in discussion, pointed out that Australian hunter-gatherers are "not entirely blasé about territory".
12 Lewis, I., introduced this possibility in discussion, in asking whether existing hunter-gatherers might represent "drop-outs" from pre-existing agricultural societies.
13 This point is important in view of R. Lee's discussion, this volume, p. 181, of the fact that !Kung Bushmen and Eskimo hunter-gatherers exhibit seasonally correlated changes in social group size—small groups for economical hunting and large groups for "public functions" with social advantages.

14 Gartlan, J. S. (1968). Structure and function in primate society, *Folia primat.*, **8**.
15 For a discussion of the relationship between ecology and social structure in Primates, see: Crook, J. H. and Gartlan, J. S. (1966). Evolution of primate societies, *Nature*, **210**.
16 e.g. Morris, D. (1969). *ibid.*
17 Eibl-Eibesfeldt, I. and Wickler, W. (1968). Die ethologische Deutung einiger Wächterfiguren auf Bali. *Z.f. Tierpsychol*, **25**.
18 For the measurement technique used for measuring human response to proximity and to the presence of a beard, see: McBride, G., King, M. G. and James, J. W. (1965). Social proximity effects on galvanic skin responses in adult humans. *J. Psychol.*, **61**. McBride (personal communication) states that a bearded face evokes a higher skin response than a clean-shaven face.
19 Hutt, C. and Hutt, S. J. (1965). Effects of environmental complexity on stereotyped behaviour of children, *Anim. Behav.*, **13**.
22 Martin, R. D. (1970). The workings of social stress, *New Society*, **16**.

M. J. ROWLANDS

Defence: a factor in the organization of settlements

The description and classification of settlement patterns can be achieved only through the study of the influences that determine them[1]. Defence is one of a range of interacting factors that may influence settlement form and has been isolated for study in this paper.

Defence is defined here as any action that constitutes resistance against attack. It is limited therefore to the influence that cultural responses to aggressive behaviour by human agencies have on settlement patterns. Responses to protection against other factors, for example climate, disease or wild animals, have not been specifically included in this study.

The data on which this study is based are drawn principally from small-scale societies, and no attempt has been made to include comparisons from societies with complex political structures and specialized military organizations. The paper has been divided into four sections (1) the aims of defence, (2) factors that influence the manner of defence, (3) cultural responses to defence, (4) influence of warfare on settlement patterns.

1 The aims of defence

A tendency exists to view defence as a need requiring the erection of physical barriers against aggression. In many cases, however, defence against attack need not necessarily result in fortification and the presence of physical barriers in or around settlements need not necessarily be due to defence. It is necessary therefore to establish the aims of defence in order to understand the multi-functional roles that might be involved in responses to aggression.

economic: The protection of the minimum requirements to maintain life and which have to be defended to prevent social disintegration. This may include the protection of gardens and fields, particularly during the planting and growing seasons, the protection of stored food, the protection of animals, the protection of tools, working areas, sources of raw materials and communications for trade. An obligation for defence of areas outside the settlement may therefore be involved, for example fields, grazing areas, water supplies and hunting areas.

territorial: Emotional and historical ties may bind people to a particular region. Such a region may be regarded as a refuge (e.g. Stanner's "heartland" concept[2]) and form a minimal unit to which people may be reduced by aggression before active defence is adopted. An offensive response against infringement of territory is often characteristic of complex political systems possessing the capacity for its defence.

political: The preservation of political autonomy based on the belief that conquest requires political disintegration, in addition to territorial occupation, to be successful. The degree of specialization in political institutions found may govern the efficiency with which resources can be organized for defence. The growth of alliances and confederacies also assumes the surrender of political autonomy by their constituent groups in anticipation of mutual benefits for defence.

ritual: The identification of a people with particular sacred areas, shrines, localities figuring prominently in myth cycles and places where ancestors are buried may contribute to the idea that some places are vital to social well-being and must be defended.

The reasons that influence a people to defend a particular place or territory are, therefore, likely to be complex. It is essential to realize that people may defend a locality for a number of different reasons and that the relative importance of these different aims will influence the manner and intensity with which this is carried out.

In addition, physical barriers may be erected for other reasons besides defence. The moats around villages in southern Japan, for example, were erected principally for protection but also held fish which were a valuable source of food; they were also used for irrigation during the dry season and for drainage during times of flooding[3]. In New Guinea, Dani compounds are surrounded by a ditch and wooden fence yet these compounds are not attacked in fighting and it seems that the ditch only serves to drain and irrigate the gardens and the fence serves only to contain pigs[4]. In many societies the provision of walls around dwellings and courtyards serves to retain the privacy of their occupants. Such barriers can also be used as a means of regulating social relations and preventing the entrance of undesirable visitors into certain sections of a dwelling or settlement.

The symbolic significance of walls and gates around cities has been shown by Wheatley, particularly in his description of the religious significance of the multiple walls surrounding the sanctuaries of "temple cities" in southern India[5]. In Bali, the walls around a compound are said to be able to exclude "evil spirits" with the entrance being the only means by which such spirits can get in. Since these spirits are said to be able to turn only with difficulty, the entrance never leads directly to the open courtyard and is tortuously arranged to prevent their admission[6]. There is therefore ample evidence to demonstrate that physical barriers can serve a number of different functions besides that of defence, nor need they necessarily have anything to do with defence at all.

2 *Factors that influence the manner of defence*

It is suggested that the aims of defence will predict a variety of cultural responses that would theoretically satisfy a need for security in a range of given situations. The extent to which a particular response could be realised, however, would depend on a set of limiting factors which would influence or even play a selective role in the implementation of a potential solution.

environment: Differences in topography, climate, vegetation and soils influence the siting and arrangement of settlements. Environment may provide natural advantages for defence and can influence the choice of site for a settlement. The Yanomamo (Venezuela) site their villages so that geographical barriers such as swamps, rivers and rugged hills will fall between them and an enemy village[7]. The Diola (Sierra Leone) site their habitations on the edge of the forest to be near their fields and yet have easy access to the forests as a refuge[8]. In some cases, the environment forms such a good natural defence that little else is required for the defence of a settlement. Labouret describes a general village plan in "Equatorial Africa" where impenetrable vegetation limits access to certain routes which can be blocked with barricades or with guard houses in emergencies[9].

The degree of concern with which people will defend everyday habitations may be influenced by the availability of natural defences. Firth emphasises that the concentration of "Pa Maori" (hillforts) in the northern part of North Island, New Zealand, should not be understood as an indication of more intensive warfare but a clue to the frequency of naturally defended sites, suitability of soils for entrenching and a higher population density[10]. The lower frequency of hillforts in the centre of North Island, for example, is due to the tendency for inhabitants to seek refuge in dense forests and steep mountain ranges where ambushes could be used as a mode of defence against attack[11].

The "Pa Maori" also illustrates the point that the area enclosed by fortification need not necessarily be related to the size of the group occupying it or their economic circumstances. The "Pa", for example, was constructed to increase the defensive advantage of a natural position and its fortification followed natural contour lines, enclosing vulnerable points and, if possible, a water supply. It would often enclose therefore a much larger area than needed by its occupants and its size would not be a reliable guide to the number of people using it or to the activities being carried on within it.

Environmental factors, however, work both for and against the needs of defence. Settlements sited for defence need not be in the most convenient position for social and economic activities and with increased security there is a likelihood of finding the movement or dispersal of settlements to sites better adapted to everyday living. In some areas there may be no natural advantages for defence that can be exploited. Siting will therefore play a less important role in the defence of settlements and the absence of natural refuges may tend towards the elaboration of artificial defences around settlements. Jackson, for example, describes a mountain fortress at Natewa in Fiji which had been built as a refuge by people living in an undefended village on the nearby coast. He also visited Tokotoko, an elaborately fortified town on flat delta land to the south where he complained of the presence of innumerable ramparts and ditches extending for four miles, from which he could not find his way out[12].

technology and raw materials: A simple technology associated with mobility of population and the use of temporary or portable habitations encourages the use of natural defences and the dispersal or concentration of people in periods of insecurity. On the other hand, stability of population, the construction of permanent habitations associated with a more complex technology encourages an investment in the construction of more complex defences.

Lack of technical skill can be overcome by other means, in particular the use of large groups of labour. The Maori, for example, used digging sticks, wooden shovels and baskets for building a 'Pa' and, considering that these forts were not kept in permanent repair, must have relied on a very large labour force for their construction (e.g. Pa Ke-rewa-ki-runga is said to have been occupied by about a thousand warriors in addition to non-combatants).

Structures associated with defence, either separate or integrated with a settlement, tend to constitute an extension of existing technology in small-scale societies. The technical skills and materials used for the construction of habitations thus have a tendency to be adapted for the needs of defence. In some cases, the defence of settlements might require the use of different materials than normally used for building habitations; the Muntifiq of the Euphrates delta, for example, live in reed huts that are vulnerable to fire and construct small mud forts beside each house for protection against attack[13]. The raw materials used may limit or encourage variation in fortification; technical skill in the use of stone and brick, for example, allowing the construction of complex structures[14]. The materials used may modify the form of structures of similar function; for example in Sierra Leone wood was used for stockades around villages in the south, and stone or pisé used in the north[15].

Finally, aggression can create an environment in which new forms of defence associated with different technical skills and materials may be adopted, particularly in the face of an enemy using new military tactics and weapons[16]. Changes in fortification may therefore be more rapid than found for other structures in a settlement and some care may be required in interpreting the implications of such changes.

subsistence economy: Although warfare is no less prevalent among hunting and gathering peoples than elsewhere, the high degree of mobility, impermanence of settlement sites, simple technical skills used in the construction of habitations and a tendency towards a small size in the co-operating group, tends to limit the variety to responses possible for defence. The literature abounds with references of temporary camps set in or near natural refuges, the concealment of camps and fires and the movement of camps away from sources of danger with mobility itself tending to be a form of defence.

Mobility is also an advantage in defence for pastoralists, particularly since animals can be moved away from danger if sufficient warning is given of an impending raid. The Nandi (Kenya), for example, used to move their cattle into forested and hilly regions when raided by the Masai[17]. Pastoralism tends however to favour the habitation of grassland environments where groups of people would be particularly vulnerable to attack. Also the protection of fixed and often limited resources, such as grazing areas and water supplies, particularly when in competition against other pastoralists, tends to favour the concentration rather than the dispersal of population for common defence[18]. The temporary settlements of the pastoral northern Somali, for example, are normally dispersed in balance to the amount of pasture and water available, but during fighting, lineages become more closely localised and settlements as large as a hundred 'hamlets' may develop[19]. The importance of animals and their protection can also be seen in the arrangements of the compounds of the Sotho and Tswana (Bantu) where the 'kraal' forms a nucleus around which the homestead is built and is a focus for homestead life[20].

The rate of movement of groups practising swidden cultivation can affect the degree to which inhabitants may be willing to defend their settlements. Both the Wangullam (New Guinea) and the Yanomamo (Venezuela), for example, have more than one territory where they can move and establish new gardens if warfare becomes too intense[21]. The ease and opportunity for movement will therefore not only affect the permanence of a settlement but also the amount of time and energy that people would be willing to invest in defending it against attack. This is illustrated by the Tanala (Madagascar) who formerly practised dry rice cultivation and lived in villages that were abandoned after one or two years[22]. These villages were fortified with a ditch and stockade, without involving a vast amount of labour. Organized warfare was rife, since an enemy had a fair chance of penetrating these fortifications. When the Tanala adopted wet rice cultivation, permanent villages were established and the inhabitants proceeded to invest them with impregnable fortifications, so successfully in fact, that warfare degenerated into petty raids on stragglers and people working in the fields.

The erection of complex fortifications around settlements tends therefore to be linked with people living, either permanently or over a long period of time, in agglomerated settlements or with people combining to erect and defend a permanent refuge. Such circumstances may occur with intensive hunting/fishing or mixed pastoral and agricultural economies but are most frequently realised by societies practising sedentary agriculture. The protection of fields, stored food, tools and simply the greater complexity of economic and social life that permanent settlement seems to produce, acts as an incentive and provides the capacity for the construction of large and elaborate defence structures.

socio-political organization: The structure of settlements may reflect the size of the group that will co-operate with each other against aggression and that will contribute labour and resources for the erection of defence works. The size of this group need not be the same as that which occupies the settlement, nor the same as that which co-operates in everyday activities; for example, individual units or sections within a settlement may be responsible for their own defence or a number of separate settlements may combine into an alliance for defence. Each nuclear family in a Nimboran (New Guinea) village, for example, is responsible for its own defence and each house is separately defended against attack[23]. Each longhouse of the Iban (Borneo) contained an ambilineage and was responsible for its own defence[24]. Best says that a Maori "Pa" could be occupied by anything from a full tribe, made up of a number of clans, to a clan section, which in turn would be composed of extended families that would normally be dispersed in small undefended hamlets near their fields and would only combine to occupy the "Pa" in emergencies[25]. Lewis' description of the agnatic lineages of the northern Somali illustrates the ability of related descent groups, dispersed due to environmental and economic factors, to combine into larger units for defence[26].

Autonomous settlements may enter into loose alliances with each other for their mutual security, involving agreements on siting of settlements, warning systems, restrictions on inter-village hostilities and the building of common defence works. A loose alliance may obviate the need to defend the settlements of all its members and it may be thought simpler to combine labour and resources in the defence of the settlement of some of its members or to build a common refuge. Among the Hopi, one of the villages situated on top of a mesa was considered a guard village and was located to protect the rest[27]. The defence of more than one settlement can be even more effectively pursued under the control of a single political authority. In this case, patterns of defence may range from examples such as the chiefdoms of the Chagga (Kenya) where the chief's compound formed a rallying point and refuge for his followers, to confederacies such as in Fiji where the chief town was undefended and relied on the network of fortified towns of its allies for protection. As has been seen, increased political control tends to be more efficient in administering the resources necessary to maintain a specialized organization for military offence and defence, which can act as a sort of "protective umbrella" for the society within its political domain.

incidence of warfare: A tacit understanding, for their mutual benefit, between hostile groups as to when and when not to indulge in warfare seems quite widespread in small-scale societies. Such an understanding implies a recognition of relatively safe periods (e.g. during the planting and growing season), in contrast to relatively dangerous periods (e.g. after the harvest), with a resulting effect on people's attitudes and activities at different times of the year or economic cycle. Iban men, for example, would go off on trade journeys during the planting and growing seasons and only return for the harvest and the raiding period afterwards[28]. Such an understanding sets indirectly a limit on the scale of destruction possible from warfare. For example, an implicit safeguard on growing crops may explain why little care may often be taken to protect fields but, on the other hand, the concern that may be taken over the protection of granaries.

Greater insecurity and a concern for defence will be found therefore when raids are carried out by outsiders practising a different economic livelihood or who may not be limited in warfare by the same economic regime. Adam, for example, contrasts the presence of villages in southern Morocco, fortified with ramparts and towers against nomadic desert raiders, to undefended villages further north where restrictions on inter-village hostility prohibited attacks on settlements[29].

Restrictions or tacit understandings on the intensity of hostilities between culturally related groups can therefore lessen the risk of destruction and loss of life and affect the degree to which people feel they must defend themselves. Sweet describes certain "rules of etiquette" that governed raids between tribal chieftains of the north Arabian Bedouin, which prevented the destruction of complete encampments and which protected women, children and the property of craftsmen and guests. No such niceties, however, were observed by the Bedouin in raids on oasis dwellings, settled villages and caravans[30].

influence of weapons and military tactics: One of the functions of defences around settlements is to prevent the possibility of surprise attack. Chagnon has described a reliance on surprise attack, the small size of groups involved and the short duration of active hostilities as three defining elements of warfare in small-scale societies[31]. The need for some sort of barrier around a settlement may be determined therefore by the possibility of surprise attack and yet need not be too durable, since any attack would be unlikely to last long.

People could use a number of different methods to prevent a surprise attack. The Dani for example constructed artificial ponds in their gardens so that the ducks settling there would rise in the air when disturbed by intruders. The Maori used to place skilled warriors on watch in their "Pa" who could distinguish the screech of alarm of different parrots as intruders passed by. In heavily wooded areas, where approaches to settlements are restricted to certain routes, barricades, pit-falls and traps could be set to impede an enemy and give warning of their approach.

Individual variations in the form of fortifications may only be understood in relation to the tactics and weapons used in offence and defence. In particular, a reliance on projectiles by attackers can have a significant effect on the structure of defences. Multiple lines of defences may be built to put attackers out of range of habitations and non-combatants. The erection of flanking towers and bastions are related to the use of projectiles whereas the construction of a stockade, behind which to hide, may not be necessary if throwing weapons are not used.

Finally, changes in tactics and weapons can produce changes in defence structures. The Maori adapted their fortifications to the introduction of guns and artillery by moving the "Pa" on to low, flat ground and building trench complexes and bastions to suit a new type of warfare. The appearance of mercenary armies and the development of siegecraft produced radical changes in the defence of some Greek cities in the fourth century B.C.[32]. Such swings in advantage between offensive and defensive tactics and technology illustrate Andrzejewski's contention that the predominance of attack over defence tends to diminish the number of independent political units within a given area, but when tactics and techniques favour defence, the size of political units in a given area tends to be smaller and they are more numerous[33].

tradition: The acceptance of a traditional mode of defence implies that a set of constraints and incentives are still operating to make that response viable, although the factors involved may not be rationally analysed by the members of each succeeding generation. There would thus be a strong inclination for an existing mode of defence to be retained and modified or even adapted to new needs, particularly since a number of limiting factors would presumably still be operating.

The possibility exists of defensive patterns being retained in the siting and structure of settlements after the need for them no longer exists. A village form in Slovakia, for example, is made up of houses grouped in a circular or polygonal pattern which is said to follow the same pattern as the camps of former clans, laid out in a defensive pattern[34]. Settlements that originally developed because of some defensive advantage may survive in their original form through their capacity to adapt to new functions. In other cases, however, settlements, formerly used as strongholds in periods of insecurity, may now be much reduced in size and their population dispersed in smaller and more numerous settlements.

3 Cultural responses to defence

It can be seen that various factors interact with one another in contributing to the form of defence that a particular group will adopt in the face of aggression. The need for defence may involve living in inconvenient and inhospitable places, it may conflict with the requirements of other social and economic activities and may necessitate entering into unwelcome alliances and honouring burdensome obligations. The erection of fortifications may limit mobility and individual freedom, cause extra work and necessitate the supply of extra food and materials. The form of defence adopted would therefore have to be carefully considered in the light of other factors and constantly adjusted depending on changes in circumstances.

siting: A settlement can be most easily protected by siting it in or near a naturally defended position, for example on high ground, amongst rocks, in dense vegetation or on river bends. In lowland forest areas, such as those of New Guinea and South America, they are usually near rivers, which are the highways, but are often set back a little and concealed by a belt of vegetation for fear of raiders. For similar reasons coastal villages in Melanesia were often sited to be invisible from the sea. Only very powerful and confident people built openly on the beach. In the mountainous parts of New Guinea, villages are often sited on grassy spurs, with steep open slopes on three sides and a narrow ridge path as the approach. With the advent of settled conditions the need for concealment or defence vanishes. Bush villages tend to move down to the coast. The New Guinea mountaineers leave their ridges and settle either nearer to their gardens or, more often, nearer to a water supply[35]. Artificial sites for settlements may be constructed in particularly favourable locations, for example in shallow lakes, swamps and lagoons.

The spacing of settlements may reflect a need for security. It may be desired to keep a certain distance from the settlements of enemies and be as near as possible to the settlements of potential or actual friends and allies. On a broader scale, settlements can be spaced to give maximum support to each other and for mutual territorial defence. Siddle, for example, describes how the Yalunko of Sierra Leone constructed a series of evenly spaced fortress towns capable of withstanding the well-organized military pressure of the Fulani[36].

Relocation of settlements and/or the concentration of population in larger settlements is often found as a consequence of aggression or fear of attack. The dispersal or concentration of dwelling units for protection is found to be most fluid in societies where there already exists a high degree of population mobility. But such solutions may have far-reaching consequences due to the fact that the balance between settlement and the exploitation of its natural environment may be upset. A "war town" in Sierra Leone, for example, contained a larger population than could be supported by its immediate environment and had to establish satellite villages to provide it with food but these settlements were manned by slaves because of the danger involved. Consequently the need to acquire slaves acted as an incentive for warfare and added to the general insecurity[37].

defence of dwelling units: Dwellings may be built on stilts, piles, in trees or on platforms and raised terraces as protection against wild animals and intruders. Villages in the Upper Volta (West Africa) formerly contained subterranean houses for protection although it is not clear if these were used permanently or not[38]. The internal plan of houses may facilitate protection against attack. The use of concealed or double entrances, blind walls and a confusion of passageways can help to impede an intruder.

A group whose members normally co-operate with each other and support each other against aggression may choose to inhabit a single defended dwelling unit. Cranstone[39] has pointed out the defensive functions of the longhouse in Borneo and the tendency for these communities to break up into separate domestic units with increased conditions of security. It is possible to trace historically the building of courtyard houses in Japan, Ceylon and part of West Africa to periods of intense hostility[40]. Finally, ring-homesteads, a characteristic feature of the western Sudan, are formed from the round huts of a family or extended family which are placed in a circle, with the entrance to each hut pointing inwards and the gaps between filled with thornbush or stone walling[41].

defence of settlements: The degree to which settlements may be defended seems to depend on (*a*) their permanence, (*b*) the availability of alternative refuges, (*c*) the intensity of warfare, (*d*) the value attached to staying in and defending a settlement by its inhabitants.

A number of different methods, besides fortification, can be found for the protection of villages against general insecurity and particularly against surprise attack. As has been seen, the presence of traps, pit-falls and barricades along the approaches to a village may be thought sufficient to impede attackers and give a warning to defenders. The Mossi (Upper Volta), for example, would abandon the village and flee into the bush if an enemy managed to penetrate this defensive screen[42].

A defensive function may be found in the arrangement of dwellings in a settlement. Schachtzabel described the organization of villages in Bechuanaland as an irregular layout, formed deliberately to create confusion to anyone not living there[43]. A similar sort of defensive maze, found in villages in Sierra Leone, has been described as "disordered nucleation" by Siddle and interpreted as deliberately formed to impede the progress of intruders[44]. The location of houses on the periphery of a settlement so as to form a defensive wall was described by both Plato and Aristotle as the traditional manner of defending settlements in Greece. Plato advocated a return to this principle for the defence of the ideal city, instead of building walls[45]. The use of houses to form a defensive wall is still found today in many parts of the Mediterranean coastland, for example in Algeria, where only recently have the blank external walls of houses on the perimeter of villages begun to sprout doorways and windows, as a response to greater security[46].

A defensive wall or stockade around settlements does allow dwellings to be arranged more conveniently inside and prevents some houses being put at greater risk than others. But its erection does require organization, labour and extra materials, and communal agreement over its erection, defence and upkeep. As has been seen, the complexity of such fortifications may vary depending on topography, materials used and technical skill available and the military tactics and weapons being used in warfare. It should be stressed that any interpretation of fortification based on formal criteria without explanation, would be unlikely to provide adequately for the variety of modifying factors likely to be involved.

defended refuges: Settlements located in particularly vulnerable positions or social groups that combine for mutual defence, may construct refuges for defence in emergencies. An alliance of settlements may also choose to build a separate refuge since this could be abandoned and the alliance broken up when there was no longer any danger. However the ability to retire to a refuge assumes the presence of a warning system that will give people time to make the necessary preparations for defence.

In some cases, such refuges may be quite specialized and only intended to protect valuables or certain categories of non-combatants in warfare. In both Algeria and Morocco, for example, communal granaries or "guelaa" are built by villages on high cliffs where they store their grain after the harvest[47]. The grain is kept there whilst the villagers take their flocks to the summer grazing areas but these redoubts were also used in the past as refuges against attack and often little care would be taken over defending the village. The security of non-combatants is often a serious problem and during hostilities they may be sent to a specially prepared refuge which could also be used as a rallying point in defeat. Finally the construction of such refuges may be an indication of the intensity of hostilities. The Murut (Borneo), for example, would defend their longhouses during intra-tribal warfare but would combine to build a separate refuge in wars against other tribes or against the British[48].

defence of territory: The territory defended may be a small area enclosing a single settlement or may be larger and include a number of settlements which contribute to its defence either voluntarily or under a common political control.

The principle of a system of defence based on the protection of territory rather than of individual settlements and habitations tends to allow greater freedom of movement and provide greater security for social activities. The Dani (New Guinea), for example, are protected against raids by a system of watchtowers which allow the movement of compounds and people between compounds in search of new gardens and pig-rooting areas[49]. In Tahiti, a collection of gardens, half a mile square, was surrounded by high stone walls and dwellings were scattered inside, situated on each family's individual plot[50]. The clan territory of the Moufou (northern Cameroun) was defended by seven successive stone walls, behind which their ring-homesteads were loosely scattered with the chief's homestead guarding the entrance of the first wall[51].

The maintenance of armies and the erection of specialized fortifications tends to relieve individual settlements of the responsibility of their own defence. This enlargement of the "protective umbrella" provides the necessary security for wider and more complex social and economic activities to be developed. Garlan, however, has argued that the erection of walls around some of the Greek city states of the fourth century B.C. was encouraged by the economic self-sufficiency of such cities[52]. The importance of trade and production and the rise of mercantile and artisan classes in Athens, for example, encouraged an awareness of the city forming a focal point for defence and no longer required the defence of its surrounding territory in order to ensure supplies of food.

warning systems: The need for defending settlements can be avoided by circumstances allowing the development of a warning system of impending attacks which would rob aggressors of the element of surprise and allow counter-attacks to be organized or a retreat to be implemented.

The success of such a system seems to depend on (a) topography allowing rapid communications, (b) the length of time given by the warning, (c) relations of a group with neighbouring settlements who may or may not pass on warnings of an approaching attack. Best for example says that hillforts were not built by the Maori in the Wellington district of North Island, New Zealand because the people were closely related to adjoining clans who warned them so successfully of any impending attack that an enemy could never hope to take them by surprise[53]. Fathauer emphasizes the importance in Mohave warfare of a system of spies that watched all trails and waterholes for intruders. The system gave such efficient warnings of attack that counter-attacks and ambushes could be organized to meet any aggressors and allowed non-combatants to live securely in dispersed hamlets near the fields[54]. In both these cases the element of surprise in attack would be lost and the defenders would have sufficient time to meet the attackers without having to organize their settlements for defence.

4 *Influence of warfare on settlement patterns*

The intensity of warfare can quite clearly have a considerable impact on the socio-political structure of societies and it is intended here to discuss some of the effects that this impact might have on settlement patterns.

As has already been mentioned, warfare may encourage the concentration of people into larger and less numerous settlements in a given area. Among the Yanomamo (Venezuela), for example, a greater intensity of warfare between villages is found in the centre of the tribal area than between villages at the periphery. Villages at the centre are larger, nucleated, palisaded and enmeshed in alliances with other villages, whilst villages at the periphery are more widely spaced and isolated, are smaller and are not nucleated and inter-village alliances are less common[55]. Similarly Blanchard has compared the distribution of dispersed hamlets in the northern part of the French Alps with the pattern of larger and more concentrated villages to the south and has explained the latter as due historically to the threat of attack from Arab colonies on the coast during the Middle Ages[56]. In some cases warfare may impede the fission of settlements as they reach a critical size and are no longer in balance with the economic resources available. The maximum size of a settlement practising bush fallow cultivation in Sierra Leone, for example, is about sixty huts (about five hundred inhabitants) farming an area within one and a quarter miles of the village. In periods of social upheaval, normal processes of growth and sub-division were impeded and settlements of up to twice this size developed which intensified soil deterioration and necessitated the formation of "slave villages", for the provision of food, as satellites to the main town[57]. Such a situation might suggest that there could be some difficulty in deciding whether environmental deterioration has been a factor contributing to or a result of warfare and a concentration of population for defence.

As Trigger has noted, there may also be a limit on the size of settlement that can be effectively defended by the labour and resources available[58]. Warfare may therefore encourage the fission of large settlements and possibly their relocation in more easily defended positions. It may also be a factor controlling the undisciplined spread of settlement growth and be responsible for setting physical limits to settlement size. The erosion of political control through warfare may rob settlements of a wider defence system and villages and towns that had tended to develop into a sprawling scatter of buildings, could suddenly find themselves responsible for their own defence and have to hastily erect or repair defences and abandon areas vulnerable to attack.

Warfare may cause the temporary evacuation of settlements, particularly when raiding is short lived and when no intention exists on behalf of the aggressors to establish permanent rights over land or other resources. Evacuation of a settlement is often favoured when alternative refuges exist where animals, food and other valuables can be hidden in anticipation of attack. However, the scale of warfare can threaten the socio-political unity of a group and it may be forced to move and establish a new settlement elsewhere. Service, for example, has described the mass migrations and the disruption of settled agriculture in central and southern Africa in the last century, due to the penetration of European and Arab slave traders instigating an unprecedented series of tribal wars[59]. Even more dramatic examples occur of people changing their subsistence economy and social structure in the face of overwhelming military aggression. The Chippewa of the Georgian Bay region, Ontario, for example, responded to the military aggression of the Iroquois by adapting to a hunting, gathering and fishing economy and dissolving their territorial groups to live in discrete family units.

Warfare can impose limitations on social activities which may be reflected in the structure of settlements. The Diola, for example, are organized into small political units comprised of the village or quarters of large villages which were in competition with each other for land, water and grazing rights. Such a high degree of settlement autonomy tended to contribute to a diversity of house form from one village to another and to the other cultural and linguistic differences present[60]. The erection of fortifications is, in fact, the antithesis of communication and tends to impose limitations on social activities and alter the arrangement of dwellings that might be found in undefended settlements. The Maori tended to take greater care over the arrangement of dwellings inside a "Pa" than was usual in an undefended village and also had to overcome problems like sanitation which were never normally necessary. Warfare also disrupts socio-economic activities such as trade; in cases where people import material, general insecurity may require the adoption of new materials and technical skills for the erection of habitations and other structures.

Conclusion

The objective of the paper is to establish a procedure for explaining the adoption of particular solutions to defence against aggressive attack. It is argued that the aims of defence will predict a certain range of possible cultural responses but the extent to which these will be realized will depend on another set of limiting factors that influence the manner of defence. For theoretical purposes, these two sets of causal and modifying factors have been isolated as analytical categories, so that changes, either physical or in the value/importance attached to them, might be seen to produce consequent changes in the cultural response. These two sets of "constraints and incentives" could conceptually be understood as two separate stages in a process of decision making and implementation, although in any empirical situation the factors involved would interact continuously within the cultural experience of the participants.

The manner in which people evaluate these two sets of "constraints and incentives" will thus depend on a third set of factors; the cultural context within which they are operating. A traditional way of life, composed obviously of a wider range of elements than has been outlined in this paper, creates particular orientations which will influence the way in which people evaluate choices of responses for defence or maintain or modify a traditional response as a viable solution to their needs.

Notes

1 Demangeon, A. (1963). The origins and causes of settlement types, *in* Wagner, P. L. and Mikesell, M. W. (eds.) *Readings in Cultural Geography*. New York. pp. 506–516.

2 Stanner, W. E. H. (1965). Aboriginal territorial organization: estate, range, domain and regime, *Oceania*, **36**, pp. 1–26, cf. Martin's discussion of the home range/territory concept, Martin, R. D., this volume, p. 436; also Flannery's observations on the investment of territory with facilities, Flannery, K. V., this volume, p. 28.

3 Hall, R. B. (1931). Some rural settlement forms in Japan, *Geogr. Rev.*, **21**, pp. 93–123.

4 Heider, K. G. (1967). Archaeological assumptions and ethnography, *S.W.J. Anthrop.*, **23**, pp. 52–64.

5 Wheatley, P. (1969). *City as Symbol*. London.
6 Tan, R. Y. D. (1967). The domestic architecture of Bali, *Bijdragen tot der Taal-, Land- en Volkenkinde*, **123**, pp. 442–75.
7 Chagnon, N. A. (1967). Yanomamo social organization and warfare, *in* Fried M., Harris, M. and Murphy, R. (eds.) *War, the Anthropology of Armed Conflict and Aggression*. New York.
8 Thomas, L. V. (1964). Pour une systématique de l'habitat Diola, *Bull. Inst. Franc. Afrique Noire*, **26**(B), pp. 78–118.
9 Labouret, H. (1931). L'habitation indigène en Afrique occidentale et équatoriale, *in* Bernard, A. (ed.) *L'habitation indigène dans les possessions françaises*. Paris.
10 Firth, R. (1927). Maori hillforts, *Antiquity*, **1**, pp. 66–78.
11 Best, E. (1927). The Pa Maori, *Dominion Mus. Bull.*, **6**, pp. 37–8.
12 Erskine, J. E. (1853). *The Islands of the Western Pacific*. Appendix: narrative of John K. Jackson, pp. 411–27 (cf. esp. pp. 429–30 and 459).
13 Thesiger, W. (1964). *The Marsh Arabs*. London. pp. 213–4.
14 Trigger, B. G. (1968). The determinants of settlement patterns, *in* Chang, K. C. (ed.) *Settlement Archaeology*. Palo Alto. cf. p. 57.
15 Siddle, D. J. (1968). War towns in Sierra Leone. A study in social change, *Africa*, **38**, pp. 47–56.
16 Andrzejewski, S. (1954). *Military Organization and Society*. London. pp. 75–6.
17 Huntingford, G. W. B. (1953). *The Nandi of Kenya*. London.
18 Trigger, B. G. (1968). *op. cit.* p. 69.
19 Lewis, I. M. (1961). *A Pastoral Democracy*. London. p. 63.
20 Walton, J. (1956). *African Village*. Pretoria. Part IV: Bantu houses and kraals.
21 Ploeg, A. (1969). *Government in Wangullam*. The Hague. pp. 13–14.
22 Linton, R. (1936). *The Study of Man*. New York. pp. 348–52.
23 Kouwenhoven, W. J. H. (1956). *Nimboran*. The Hague. pp. 13–14.
24 Leach, E. R. (1948). Some features of social structure among Sarawak pagans, *Man, JRAI*, **48**, note 103.
25 Best, E. (1927). *op. cit.* p. 24.
26 Lewis, I. M. (1961). *op. cit.*
27 Farmer, M. F. (1957). A suggested typology of defensive systems of the South West, *S.W.J. Anthrop.*, **13**, pp. 249–66.
28 Morgan, S. (1968). Iban aggressive expansion; some background factors, *Sarawak Museum J.*, **16**, pp. 141–85.
29 Adam, A. (1951). La maison et le village dans quelques tribus de l'anti-Atlas, *Collect. Hésperis*, **13**, pp. 39–441.
30 Sweet, L. A. (1965). Camel raiding of the North Arabian Bedouin, *Amer. Anthrop.*, **67**, pp. 1132–50.
31 Chagnon, N. A. (1967). *op. cit.* pp. 112–13.
32 Garlan, Y. (1968). Fortifications et histoire grecque, *in* Vernant, J.-P. (ed.) *Problèmes de la guerre en Grèce ancienne*. Paris.
33 Andrzejewski, S. (1954). *op. cit.* pp. 75–6.
34 Puskar, I. and Thurzo, I. (1967). Peasant architecture of Slovakia, *Architect. Rev.*, pp. 151–3.
35 I am indebted to Mr. B. A. L. Cranstone, personal communication, for this information about New Guinea.
36 Siddle, D. J. (1968). *op. cit.* p. 48.
37 Siddle, D. J. (1968). *op. cit.* p. 47.
38 Moal, G. (1960). Les habitations semi-souterraine en Afrique de l'ouest, *J. Soc. Africanistes*, **30**, pp. 193–203.
39 Cranstone, B. A. L., personal communication.
40 Hall, R. B. (1931). *op. cit.*; Chalvadurai-Proctor, R. (1927). Some rules and precepts among Tamils for the construction of houses, towns and cities

during the Medieval Age, *J. Ceylon Branch Roy. Asiatic Soc.*, **30**, pp. 337–57, Thomas, L. V. (1964). *op. cit.* p. 87.
41 Beguin, J.-P. and Kalt, M. (1952). *L'habitat au Cameroun.* Paris.
42 Skinner, E. P. (1964). *The Mossi of the Upper Volta.* Stanford. p. 105.
43 Schachtzabel, A. (1912). Die Siedlungsverhaltnisse der Bantu-Neger, *Suppl. to Int. Archiv. f. Ethnographie*, **20**, p. 22.
44 Siddle, D. J. (1968). *op. cit.* p. 48, fig. 1.
45 Plato. *The Laws, Book VI*, 778–80; Aristotle. *Politics*, 1330–1.
46 Masqueray, E. (1886). *Formation des cités chez les populations sédentaires de l'Algérie.* Paris. pp. 86–7.
47 Bernard, A. (1931). L'habitation indigène en Afrique du nord, *in* Bernard, A. (ed.) *L'habitation indigène dans les possessions françaises.* Paris. pp. 16–17.
48 Rutter, O. (1929). *The Pagans of North Borneo.* London. Chap. 10.
49 Gardner, R. and Heider, K. G. (1968). *Gardens of War.* New York. pp. 141–2.
50 Best, E. (1927). *op. cit.* p. 311.
51 Beguin, J.-P. and Kalt, M. (1952). *op. cit.* pp. 13–30.
52 Garlan, Y. (1968). *op. cit.* pp. 256–7.
53 Best, E. (1927). *op. cit.* pp. 3–4.
54 Fathauer, G. H. (1954). Mohave warfare, *S.W.J. Anthrop.*, **10**, pp. 97–118.
55 Chagnon, N. A. (1967). *op. cit.* pp. 113–14.
56 Blanchard, R. (1929). *Les Alpes françaises.* Paris. pp. 74–75.
57 Siddle, D. J. (1968). *op. cit.* pp. 50–1; Alternatively, Allen has suggested that in northern Ghana, fear of attack whilst moving to new fields may have encouraged the development of manuring practices in order to maintain soil fertility and permanent occupation. Allan, W. (1965). *The African Husbandman.* New York. p. 246.
58 Trigger, B. G. (1968). *op. cit.* pp. 65–6.
59 Service, E. R. (1967). War and our contemporary ancestors, *in* Fried, M., Harris, M. and Murphy, R. (eds.) *op. cit.*
60 Thomas, L. V. (1964). *op. cit.* p. 99.

RUTH TRINGHAM

Territorial demarcation of prehistoric settlements

The first furrow with the plough. A field. Space comes into being.
The trees fall. The woods are opened up. Space comes into being.
Manmade space.

You dig a hole.
Pile up some rocks.
Put up a pole.
Architecture comes into being.

Hans Hollen, *Space in space in space*

Territory has been defined as "a defended area"[1] ("variations in readiness to fight, depending on the place and on various local factors inhibiting the fighting urge"[2]) and as an "area of space, whether water, earth, or air which an animal or group of animals defends as an exclusive preserve"[3]. Thus, animals, including human beings, which defend a particular space are called "territorial"[4]. Territoriality, therefore, may be regarded as a spacing mechanism in that it separates certain members of a species from their conspecifics[5]. The degree to which this separation is carried out, i.e. at which point along the continuum from barely effective spatial separation or demarcation to physical defence of territory, depends firstly on the degree of attachment to the particular space (and its contents), and secondly on the degree to which the excluded conspecifics threaten to enter the particular space.

Studies of animal behaviour have shown that the readiness to defend a territory is the greatest in the centre of that territory, that is, where the animal feels the most secure[6]. It is very clear, however, that this concept cannot be simply applied to human territoriality, and that the defence and demarcation of territory is very much more complex than a mere increased security-increased aggression correlation[7]. Similarly, the concept that "the desire for domination by force and if necessary killing is . . . as deeply seated a human emotion and urge as any other", or that an "evil intra-specific selection must have set in from the Early Stone Age", or that Man is "innately aggressive" would seem to have no support in either archaeological or ethnographic evidence; it is rather the simplistic ethnocentric concept of sophisticated guilt-ridden Western investigators[8]. By this concept, they would justify the physical defence and demarcation of space by humans as normal behavioural phenomena. It is clear from archaeological and ethnographic evidence, however, that the defence and physical demarcation of territory is by no means universal, and is dependent on a large number of interrelated factors.

The archaeological term "settlement" has been defined as the archaeological equivalent of the sociological or social anthropological term "community"[9]. Whether or not we are in agreement with this definition, in its broadest terms an archaeological "settlement" may correspond to the area (comprising in archaeological evidence various sites or locations) inhabited *and* exploited by a particular social unit. The size and composition of a "settlement", therefore, varies greatly according to the subsistence economy and seasonal mobility of the social unit. In addition there are (or must have been) varying degrees of territoriality attached to different parts of the settlement area; the hunting area ("home-range"), pasture area, cultivated fields, habitation locale would be variously demarcated and defended according to how much value was attached to the space and its contents, and to what degree these were threatened by excluded conspecifics. The degree of territoriality would by no means be constant or permanent, but irregular and temporary; scarcity of land and food would, for example, increase the threat of human predation which could result in increased territoriality and corresponding physical defence of the space or its contents. In view of the very specialized evidence produced by archaeological investigations, this paper is restricted to a discussion of the methods of demarcating the habitation area, the "home base" or "core" of prehistoric settlements[10]. It is hoped to provide a range of interpretative models from evidence of the demarcation of ethnographic settlements for features which are blindly assumed by most archaeologists to be defensive fortifications. It is possible that the ditches, walls, banks, etc. excavated around prehistoric settlements were defensive, but unless there is literary evidence or evidence, for example in the form of piles of decapitated bodies, as at Maiden Castle, we cannot assume that a barrier was designed to prevent human attack. It is surely safer to consider a range of possible interpretations of the features, and to examine as many factors as possible which may have restricted their form and content.

Within a habitation area, space-consciousness or territoriality can operate at the level of the individual[11], and sub-residential unit (i.e. individual household). At all levels the human beings identify themselves as part of a social hierarchy within a particular space or territory, which may or may not be demarcated and defended, to a greater degree at certain levels than at others.

The method by which a habitation area can be demarcated varies considerably from physical barriers such as visible structures (e.g. walls, fences, banks) and modifications of the earth (ditches and moats), to invisible, non-physical barriers. These latter are generally known only to the inhabitants of the house or settlement concerned, and include imaginary lines between natural features such as trees and rivers, and also between houses or artificial mounds[12]. The former are much more likely to be a clear demarcation of the territory of one residential unit from that of other residential units, so that the demarcated area is obvious not only to those inside, but also to excluded conspecifics. Physical demarcation barriers may have secondary and tertiary functions, such as protection from wild animals, control of domestic animals, defence against human predators, or even exhibitionism. As has been shown by Rowlands, however, defence of territory need not be by means of the construction of a physical barrier[13] and, as will be shown below, such physical barriers were not always constructed for the defence of territory from human attack.

In modern ethnographic studies of small-scale societies, it has frequently been difficult enough to identify the means by which a habitation area was demarcated, especially if, as among the Tiv, it consists of an imaginary line beyond which a member of the group feels insecure and on the defensive. The reconstruction of non-physical demarcation lines of prehistoric habitation areas, such as the use of natural features, is, at the most, speculative. Such reconstruction can normally only be attempted when the method of demarcation is very obvious, such as by a wall, bank or ditch. Even substantial physical barriers, if made of perishable materials such as fences constructed of thorn brush, like those round Masai kraals[14], or of small branches, wattling or stalks which surround Shilluk compounds[15], would not normally survive in the archaeological record.

It can be seen from present-day small-scale societies that a number of interdependent factors restrict the range of possible methods of demarcating a habitation area. These include the natural environment and available raw materials, the level of technological skill with which these can be exploited, the subsistence economy of the settlement and its relative degree of mobility, as well as socio-cultural factors. The first three factors can be applied to the archaeological evidence, whereas the effect of the fourth factor is much more difficult to reconstruct.

I *Environment*

It is very unlikely that convenience in demarcating territory played a significant part in determining the choice of the location of a habitation site. As a result of the choice of site, however, various limits and potentialities in the means of demarcating the habitation area may be observed. The inhabitants of sites with distinctive natural features, such as a clearing in a forest[16], an upland area with outcrops of rocks and knolls or sharply defined river valleys, are unlikely to employ an artificial encircling demarcation barrier. The same applies to settlements located on a site whose boundaries are naturally defined, such as a coastal or riverine bay, a narrow river bank or coastal plain[17], an alluvial mound in a river flood plain[18], a small island[19], a cave[20], or a promontory. On a wide grassland plain or plateau, however, such as the American prairies and Russian steppes, it is difficult to find in the wide expanse of uniform topography any natural means of demarcation, so that some kind of artificial means of expressing the identity and territory of the settlement is necessary. Solutions which have been found to this problem include encircling the settlement with a fence[21], or arranging the habitation structures in concentric circles or closed squares so that they themselves form the demarcation barrier[22].

It is possible that these observations may throw some light on the earliest "fortified" settlements of temperate Europe, dated to *c.* 3800 B.C., when agricultural settlements had already been established in Europe for at least 1500 years. In the Ukraine and Moldavia in this period, settlements were located on higher terraces, the tops of hills and the edges of plateaux, possibly as a result of the greater exploitation of sheep who thrive best in drier grassland areas. Two distinctive types of settlement have been excavated, with a very similar material culture[23]. The first are in the east, in the Ukraine, on the edge of the grassland steppe plateaux; the houses were arranged in concentric circles with an open space in the middle[24]. No artificial demarcation barrier has been excavated, nor were there any natural means of demarcation available. It seems likely, therefore, that the arrangement of the houses themselves demarcated the area of the habitation site, as among the Plains Indians. Further west, in Moldavia, where the loess plateau is more frequently dissected by river valleys, the settlements were located on promontories[25]; the promontory acted as a well-defined natural "territory", apart from the neck, which in each case was dissected by a shallow ditch. It has been suggested that, from their location and the presence of a ditch, these settlements should be interpreted as defensive sites. Although the ditch may have had a secondary function of protecting the settlement and its contents from wild animals or human aggressors, it is just as likely that the shallow ditch served to demarcate that side of the habitation site which was not demarcated by natural features.

From these observations we may conclude that the minimum requirements for the demarcation of a habitation territory may be met by natural boundaries or, where these are not available, by artificial barriers. If the minimal means of demarcation appears to have been supplemented by further artificial demarcation barriers, it is likely that these were measures taken against threats and aggression by animal or human agencies, that they had a defensive or protective function in addition to passive demarcation of the habitation area. Clear examples would be an island surrounded by a fence[26], a hilltop or promontory settlement surrounded by a ditch, bank or wall[27], a settlement comprising concentric circles of houses which is further surrounded by a strong physical barrier[28], or a mound surrounded by a physical barrier.

2 *Technology and raw materials*

These factors are closely associated with those affecting the methods of house construction[29]. A lack of suitable environmental conditions, technological skill and raw material to make clay bricks will limit the complexity in shape, height and thickness of any demarcation barrier, as for example in temperate Europe where this material was not used until the La Tène period[30]. An alternative to clay bricks may be provided by stone, so long as the raw material and skill in stone-working techniques exist. Barriers of stone, however, take much longer to construct and require greater manpower to quarry and manufacture. Apart from exceptional settlements in Iberia and Greece[31], dated *c.* 2300 B.C. by C14, where complex encircling bastioned walls of stone were built, factors of time, skill and manpower seem to have prevented the exploitation of stone in the construction of demarcation barriers in Europe until the Iron Age and the Roman Conquest, when there was a need for strong indestructible walls to defend against large-scale human attacks with complex war-machinery[32].

In temperate Europe, the most common materials used in house construction were wattle-and-daub and clay mixed with chaff on a wooden framework. The use of physical barriers constructed in these materials, in particular as defensive barriers, is limited by the height of tree trunks, by the relatively flimsy structure of fences, and by the ease with which they can be set alight. Without any additional support, a fence of upright wooden posts may serve to control the movements of domestic animals, to protect them from wild animals, and to defend the inhabitants from sporadic raiding or surprise attack. Fences also serve as very obvious demarcation lines, but are not effective barriers against a concerted human attack. One of the methods used to strengthen wooden fences in prehistoric temperate Europe was to construct a bank of earth backing on to the fence[33]; in later prehistoric sites, there is evidence that a wooden fence was incorporated into a higher earth bank, acting as its framework[34].

The excavation of ditches and moats surrounding habitation sites depends on their intended function, whether to protect from animal or human attack, or whether only to demarcate the settlement. The width and depth also depend on the soils or rocks into which the the ditches are dug, and the tools used in their excavation.

3 *Economy*

The choice of physical or non-physical means to demarcate a habitation site and the form, construction and dimensions of the barriers depends to a certain extent on the degree of mobility or stability of the inhabitants.

Frequently, however, no correlation can be made between the stability of settlement and the solidity and size either of structures or of territorial demarcation barriers. Even so, it would seem that in general the temporary habitation areas of mobile hunters and gatherers either were not demarcated or were demarcated by no more than imaginary lines drawn between natural features[35]. This lack of emphasised or clear demarcation of habitation territory among hunters and gatherers need not only be a result of relatively greater mobility and temporariness of settlement. It may also be due to the relatively low concentration of commodities such as stored food or domestic animals in their habitation areas. Their main valued commodity is the "home-range" which is exploited for wild plants and animals, and which may or may not be demarcated. Even among more sedentary hunters and gatherers, for example those whose economy is based predominantly on fishing, the habitation area itself is rarely demarcated by physical barriers, since the valued commodities (in this case the fishing grounds) are generally situated at a distance from the habitation areas[36].

With societies whose subsistence economy is based predominantly on the exploitation of domesticated plants and animals, the situation may be much more complex. Clear physical demarcation of the habitation areas of many pastoral settlements occurs among small-scale societies[37]. As explained above, however, this may be explained by the fact that many pastoralists inhabit flat grassland areas. Territoriality *per se* is not a characteristic of pastoral societies, whose values are attached more to their domestic animals than grazing land. The domestic animals, however, act as an encumberance in that they must be kept in close proximity to the habitation area (within the habitation area at certain times), their wanderings must be controlled, and they must be protected from predatory wild animals. In addition, domestic animals provide a most accessible and tempting prey for human raiders, since they are self-transporting and self-productive. Thus most physical demarcation barriers of pastoral settlements have the secondary function of the protection and defence of the contents. The demarcation-defence barrier may surround only the herd, leaving the habitations themselves undemarcated[38]; or the protected area for the animals may be incorporated in the human habitation area and whole unit is surrounded[39].

Many prehistoric sites at which surrounding ditches have been excavated have been interpreted as "cattle kraals" (as an alternative to "fortified settlement" or "ritual enclosure")[40], but it is difficult to think of any modern analogies for this practice. At other sites, areas enclosed by wooden fences (from the evidence of a bedding trench or a line of post-holes) have more realistically been interpreted as for the control of domestic animals[41]. Without greater evidence of the relative importance of domestic plant and domestic animal resources in prehistoric settlements, it is difficult to calculate how much value would have been attached to a domestic herd of animals; for example, at Bylany in Czechoslovakia there is still little evidence to show whether the small fenced enclosure attached to the Early Neolithic house was to control domestic animals, or to enclose a plot of cultivated land to protect it from wild or domestic animals[42].

It is perhaps easier to understand the increased "territoriality" of agriculturalists, if it is noted that not only are they more sedentary, but also their major resource area (i.e. cultivation area) is more densely concentrated around the habitation area; in addition, facilities for the storage of agricultural products are provided within the habitation area. Without effective means of land-transport, it is difficult to remove the spoils of raids on agricultural settlements; thus, in general, unless there are other highly valued commodities (slaves, etc.) or effective means of removing them with or without land-transport machinery, agricultural settlements need not feel threatened and need not be defended. The factors mentioned by Flannery, however, could lead to an increased feeling of territoriality in the "core" of a settlement, and demarcation of agricultural settlements by physical barriers does occur[43]. Demarcation-defence barriers of the cultivated fields themselves also occurs in some small-scale societies, and, in certain fortunate cases, can be recognized in the prehistoric evidence[44].

Evidence of the physical demarcation of habitation areas has been excavated in the early (but not the earliest) agricultural settlements of south-east Europe[45]. The dwellings of the early agricultural Near Eastern "tell" settlements comprise agglomerated rooms arranged in a "courtyard" plan so that the territory of the settlement may have been demarcated by the outside walls of the structures, as in the modern Yoruba compounds[46]. It has been suggested for the settlement of Çatal Hüyük in eastern Turkey that the "settlement presented a solid blank wall and further defences were evidently deemed unnecessary. Defenders . . . were evidently a match for any marauding band that would dare attack the city and no traces of massacre have been found"[47]. Is it not equally possible that Çatal Hüyük was never threatened and that the blank outer wall served to demarcate the core area representing the territory of the settlement?

A similar situation may possibly have existed with the enigmatic wall and tower excavated in the pre-Pottery Neolithic village at Jericho[48]. This has been assumed to be a fortification for the defence of the settlement of Jericho. And yet there is no evidence that the habitation area of Jericho contained any commodity of value which could have been removed by raiders in bulk with the existing level of land-transport technology to justify the construction of such a wall and bastion. Nor is there any evidence that Jericho was attacked and raided during the existence of the wall. It has also been suggested that the "fortifications" at Jericho were constructed to control the trade of resources from the Dead Sea to other parts of the Near East. It is possible, if this were the case, that the wall and tower at Jericho were not so much for defence against human attack, although this may have been a supplementary function, but for the demarcation of a market area. If so they may have controlled the entrance and exit of those who wished to participate in the market, as indeed was the function of the surrounding walls of many Yoruba settlements[49].

4 *Socio-cultural factors*

As a general principle, the decision to demarcate a habitation area with a physical barrier is entirely the result of social or cultural choice. In no case is the choice to demarcate the area in which a social group dwells determined by the environment or subsistence or technology, apart from the need for shelter from the weather and protection from wild animals. Thus in some cases the environment or economy might necessitate the construction of protective barriers for human individuals or small groups (what might be termed the "sleeping unit"), or their domestic animals or stored food, but never for the whole residential unit (unless this comprises one sleeping unit). The very fact that "territoriality" involves threats from excluded conspecifics or high evaluation of a space or its contents by the inhabitants, both of which involve social forces, means that any attempt to express territoriality of a habitation area by the construction of a physical barrier is the result of cultural choice. Once the choice to construct a physical demarcation barrier has been made, a further choice exists as to the means by which the barrier is constructed, a choice which is selected from the range of possible methods determined by the interdependent factors of environment, technology and economy described above.

The choice of whether or not to enclose a habitation area with a physical barrier, whose significance would be clear to the inhabitants and the excluded conspecifics, depends on how much the space and its contents are valued by the inhabitants and/or threatened by excluded conspecifics; it therefore depends on the amount of political and social cohesion which exists among the inhabitants, and the social and political relationships between the inhabitants and neighbouring settlements.

It is interesting to note that the demarcation of settlements was almost completely absent in the Arctic and sub-Arctic parts of the world[50]. The reasons for this lack are not environmental or technological; as is shown by their architecture, the raw materials and techniques to build enclosing structures existed; some of the inhabitants of these regions lived in settlements which contained raidable commodities such as domestic animals and stored food and the land-transport techniques would have facilitated their removal[51]. Apart from protecting these from wild animals, no attempts were made to enclose them in demarcation-defence barriers. Nor was this because raiding of stores was unknown, although raids on domestic animals were extremely rare among the Lapps and Samoyeds (ownership of reindeer was strictly legalized and expressed by the use of ownership marks on the animals themselves). Raiding of stores outside and inside the habitation area, however, was quite common, especially in the periods of near-starvation. The raiding was not reciprocal, however, nor even punishable, and no precautions were taken against it. A description of such a raid among the Tungus Samoyeds may shed some light on the attitude of sub-Arctic hunters to their property: "My grandfather . . . Ngoibo . . . made his living by plundering raids. However, he differed from the others insofar that, when raiding, he would not kill anyone. One lean year . . . he found a Yakut cache. However, despite hunger, he did not raid it, but moved into an earthen hut next to it. Close to the store house there lay many bows . . . and these had sinew bowstrings. To keep from starving, Ngoibo's family cut off the bowstrings and ate them. Meanwhile the owner of the storehouse arrived . . . He gave Ngoibo the neck of a wild reindeer carcass. When Ngoibo went off to cook the meat, the Yakut started to load a large quantity of fish from storehouse on to his sledge . . . Ngoibo went to the Yakut and began to strike him. The Yakut fell unconscious and Ngoibo carried all the fish into the earthen hut. After a while, the Yakut came to, but being frightened did not say anything and drove home . . . Ngoibo ate the fish . . ."[52].

The situation in the Arctic and sub-Arctic is in direct contrast to the compulsive raiding of the Plains Indians and some of the north-west coast Indians, and the Iroquois and Huron[53]. According to the participants of these raids, their purpose was not economic gain so much as social prestige and (in the case of the north-west coast Indians) political domination. It would seem, however, that economic gain in the form of captured horses and women and children, may have been at the root of some of the social pressures. Certain of the habitation areas of the Iroquois and north-west-coast Indians, where the capture of human beings was one of the aims, were defended by wooden fences, but by no means all of them. Among the Plains Indians, however, apart from the circular arrangement of the tents, there was no demarcation of the settlement and there were no protective barriers for the human or horse population.

On the other hand there are examples of settlements which are demarcated by large solid barriers, which by archaeologists would generally be assumed to be defensive, but which, on ethnographic and historical evidence, had nothing to do with any threat from human attackers. In certain cases, admittedly, the original purpose of the barrier was the defence of the habitation area, after which the barrier has been allowed to remain and may even have been enlarged or repaired[54]. The choice to demarcate a settlement when there is no direct threat to the inhabited area or its contents may be the result of the attempt of one settlement to establish an illusion of impregnability and invincibility in the early stages of the formation of a political hierarchy of settlements (for example, in the establishment of the early urban centres). In addition, such physical barriers, which can be used for defence if necessary, serve as a clear demarcation of the administrative/political/religious/market centre of a settlement hierarchy, the urban centre. In some cases, the actual habitation area of the organizers of the centres is further demarcated by clear barriers, as with the early medieval castles. This demarcated nucleus, or "citadel", can act not only as the hub of power, but also as the main defended refuge, particularly of urban centres whose demarcation walls are so long that it is difficult to defend them effectively. Demarcation of political centres is characteristic of settlements in which the hierarchy is still unstable and insecure; in cases when it is not only still necessary to be able to defend the settlement, but also to demonstrate potential power.

Notes

1 Morris, D. (1967). *The Naked Ape*. London. p. 30.
2 Lorenz, K. (1966). *On Aggression*. London. p. 28.
3 Ardrey, R. (1967). *The Territorial Imperative*. London. p. 13.
4 Morris, D. (1967). *op. cit.* 30, 148.
5 Martin, R. D., this volume, pp. 435–38.
6 Lorenz, K. (1966). op. cit. p. 28.
7 Rowlands, M., this volume, pp. 447-62.
8 This is the general theme of the cited works of Morris, D., Ardrey, R. and Lorenz, D., and has been taken up by some archaeologists, e.g. Piggott S. (1965). *Ancient Europe*. Edinburgh. p. 15.
9 Chang, K. C. (1968). Toward a science of prehistoric society, *in* Chang, K. C. (ed.) *Settlement Archaeology*. New Haven. p. 3; Rouse, I., this volume, p. 96; Trigger, B. (1968). The determinants of settlement patterns, *in* Chang, K. C. (ed.) *op. cit.* p. 60; Trigger, B. (1967). Settlement archaeology—its goals and promise, *Amer. Antiq.*, **32** (2), p. 151.
10 "Home base" generally refers to hunters and gatherers: Isaac, G.Ll., this volume, p. 175; "core area" is a concept described by Flannery, K., this volume, p. 28.
11 e.g. Martin, R. D., this volume, *ibid.* Douglas, M., this volume, p. 514; Woodburn, J., this volume, p. 198.
12 Although such barriers apply more to the resource area of a settlement, such as hunting territory, grazing territory, and agricultural land (e.g. Hopi settlements: Forde, C. D. (1963). *Habitat, Economy and Society*. London. p. 226), it can apply also to the habitation area and adjacent pastoral or cultivation area, as with the Tiv settlements: Bohannan, P. (1968). *Tiv Economy*. London, or the winter settlements of the pastoral Kazakhs: Forde, C. D. (1963). *op. cit.* p. 333.
13 Rowlands, M., this volume, *ibid.*
14 Forde, C. D. (1963). *op. cit.* Figs. 93, 94.
15 Seligman, C. G., and Seligman, B. (1932). *Pagan Tribes of the Nilotic Sudan*. London. Pl. VI.
16 e.g. the Boro settlements, S. America: Forde, C. D. (1963). *op. cit.* pp. 133–4; the Mbuti Pygmy settlements: Turnbull, C. M. (1965). *Wayward Servants*. New York. Plan 7; Fraser, D. (1968). *Village Planning in the Primitive World*. London. Figs. 13–14. This may have been the case with some of the early agricultural settlements in temperate Europe which may have been situated in clearings in light, deciduous forest, e.g. the settlements of the Linear Pottery culture, such as Bylany (Soudský, B. and Pavlů, I., this volume, pp. 317-28).
17 e.g. Haida villages: Forde, C. D. (1963). *op. cit.* p. 73; Papuan coastal villages, New Guinea: Fraser, D. (1968). *op. cit.* Figs. 26–27; also see for example the prehistoric fishing settlement on a riverine bay of the Danube

at Lepenski Vir: Srejović, D. (1969). *Lepenski Vir*. Belgrade; Tringham, R. (1971). *Hunters, Fishers and Farmers of Eastern Europe*. London. Pl. I.

18 e.g. the marshland settlements of certain Dinka: Seligman, C. G. and Seligman, B. (1932), *op. cit.* p. 136.

19 e.g. the prehistoric settlement of Cascioărele on an island in the R. Danube: Dumitrescu, V. (1965). Cascioărele: a Late Neolithic settlement on the lower Danube, *Archaeology*, **18**, pp. 34–40.

20 e.g. the Cliff Palace in the Mesa Verde National Park: Fraser, D. (1968). *op. cit.* Figs. 5–6.

21 e.g. the large number of examples of "kraals" on the African grassland plateau; some have a separate fence around the houses and central area, e.g. the Masai compound: Forde, C. D. (1963). *op. cit.* Figs. 93, 94; others have a fence in the gaps between the houses, e.g. Massa and Mousgoum villages in Cameroun: Fraser, D. (1968). *op.cit.* Fig. 53; and Shilluk villages: Seligman, C G. and Seligman, B. (1932). *op. cit.* Pl. V.

22 e.g. the camp circles of the Plains Indians: Grinnell, G. (1923). *The Cheyenne Indians*. New Haven. p. 90.

23 The middle phase of the Tripolye culture: Gimbutas, M. (1956). *The Pre-history of Eastern Europe*. Cambridge, Mass.

24 e.g. Kolomiiščina, and Vladimirovka: Passek, T. S. (1949). Periodizatisya Tripolskikh Poselenii, *Materiali i Issledovaniya po Arkheologii SSSR*, **10**, Figs. 36, 71; Piggott, S. (1965). *op. cit.* Fig. 23; Tringham, R. (1971). Fig. 32c, p. 178.

25 e.g. Truşeşti, Habaşeşti, Tirpeşti, Radovanu. For a description in English see, Piggott, S. (1965). *op. cit.* Fig. 22; Tringham, R. (1971). *op. cit.* Fig. 32: a, b, p. 162. Also see Florescu, A. (1969). Befestigungsanlagen der spätneolithischen Siedlungen im Donau-Karpatenraum, *Študijné Zvesti*, **17**, pp. 111–24.

26 The site of Onondaga, an Iroquois fortified village, was almost an island and was surrounded by a high fence (Fraser, D. (1968). *op. cit.* Fig. 78).

27 The large numbers of "hillforts" dating to the European Iron Age are examples of these, but it is questionable to what extent the fortifications acted as physical barriers to halt human attack, and to what extent they acted as an expression of prestige and of military and political power. Fortified hill-top villages also occurred in New Zealand (see Firth R. (1927). Maori Hillforts, *Antiquity*, **1**, pp. 66–78; and Rowlands, M., this volume, *op. cit.*).

28 For example many of the South African and East African "kraals", e.g. Ambo kraal, South West Africa (Fraser, D. (1968). *op. cit.* Fig. 49a); Masai kraal: Forde, C. D. (1963). *op. cit.* Fig. 93–4.

29 Cranstone, B. A. L., this volume, pp. 487–503; Hodges, H. W. M., this volume, pp. 523-30; Rapoport, A. (1969). *House Form and Culture*. Englewood Cliffs.

30 Piggott, S. (1965). *op. cit.* Pl. XXXVI.

31 Piggott, S. (1965). *op. cit.* pp. 76–7;; Blance, B. (1961). Early Bronze Age colonists in Iberia, *Antiquity*, **35**; Renfrew, A. C. (1967). Colonialismus and Megalithismus, *Antiquity*, **41**, p. 278.

32 For a general discussion of the construction of the fortifications of European Iron Age hillforts, see Piggott, S. (1965). *op. cit.* pp. 199–207; 216–20; Filip, J. (1960). *Celtic Civilization and its Heritage*. Prague. pp. 120–33.

33 e.g. the eneolithic settlement of Homolka, Czechoslovakia, a hill-top settlement which was ringed by two such stockades with ditches: Ehrich, R. and Pleslova, E. (1968). Homolka, an eneolithic site in Bohemia, *Bull. Amer. School Prehist. Research*. **24**, pp. 29–45.

34 e.g. Early Iron Age hillforts (Late Urnfield culture), see Piggott, S. (1965). *op. cit.* Fig. 115; and the *murus Gallicus* type of fortification seen in La Tène oppida: Piggott, S. (1965). *op. cit.* pp. 216–17.

35 e.g. see Woodburn, J., this volume, pp. 193–206; Hiatt, L. R. (1968). Ownership and Use of Land among the Australian Aborigines, *in* Lee, R. and De Vore, I. (eds.) *Man the Hunter*. Chicago. The Plains Indians are clear exceptions to this pattern of "non-territoriality", but the large numbers of horses and the value and prestige attached to them by these communities involves similar factors to those pertaining to pastoral societies, in that here was a raidable commodity, on the one hand, and one which could be used in the transport of other stolen deadweight commodities. It is likely that before the introduction of the horse, raiding was carried on on a much smaller scale, if at all (Forde, C. D. (1963). *op. cit.* pp. 66–7). An interesting study of pre-European contact warfare and raiding between the Iroquois and Huron with an economy which was changing from one based predominantly on hunting to one based predominantly on agriculture may be found in Trigger, B. (1967). *op. cit.* pp. 153–8.

36 e.g. N. W. Coast Indians: Forde, C. D. (1963). *op. cit.* pp. 92–3.

37 e.g. "kraals" of many East and South African settlements, e.g. the Shilluk, the Masai, and Swazi, Zulu and Ambo (Fraser, D. (1968). *op. cit.* Figs. 47–9).

38 e.g. Kazakhs, whose animal corrals may be some distance from the main habitation area: Hudson, A. (1964). Kazak Social Structure, *Yale Univ. Publics. in Anthrop.*, **20**, p. 29. Zulu kraals: Fraser, D. (1968). *op. cit.* Fig. 48; and, during the breeding season, in certain areas domesticated reindeer calves are kept in a corral, Forde, C. D. (1963) *op. cit.* p. 354.

39 e.g. Masai: Forde, C. D. (1963). *op. cit.* Figs. 93–94, where the different species of domesticated animals are kept separate; Shilluk, Dinka, and Nuer in which the cattle are kept not only within the habitation area, but are housed at night in special byres: Seligman, C. G. and Seligman, B. (1932). *op. cit.* pp. 38–9; 137; 209.

40 Piggott, S. (1965). *op. cit.* pp. 59; and for the so-called "causewayed-camps" of the Early Neolithic in southern England, see Piggott, S. (1954). *The Neolithic Cultures of the British Isles*. London; Smith, I. (1968). *Windmill Hill*. London.

41 e.g. the concentric rings of a palisade at Anlo, Holland, which contained no dwelling structures (Piggott, S. (1965). *op. cit.* Fig. 42) and that within the Early Iron Age settlement at Goldberg (Piggott, S. (1965). *op. cit.* Fig. 112), and the "palisaded settlements" referred to in Ritchie, A., this volume, p. 541.

42 Soudsky, B. (1966). *Bylany*. Prague. Fig. 16, 18; Soudský, B. and Pavlů, I., this volume, p. 319.

43 Flannery, K., this volume, *op. cit.* p. 28; settlements surrounded by physical barriers include temporary palisades round villages among the Boloki: Weeks, J. (1909). Notes on the Bangala, *J. Roy. Anthrop. Inst.*, **39**, 109–10; and the Yoruba villages. Forde, C. D. (1963). *op. cit.* pp. 152–3; these often very elaborate physical barriers were constructed as defences against highly organized large-scale raids, but they also acted to demarcate the centre of administrative and market control of a settlement.

44 e.g. the demarcation of yam-gardens by high fences in Melanesia including New Guinea, (Forde, C. D. (1963). *op. cit.* p. 188; Malinowski, B. (1935). *Coral Gardens and their Magic*. London) which had the important secondary function of protecting the gardens from raiding pigs. In prehistoric evidence, the demarcation of cultivation areas occurs for example as linear ditches which it is thought demarcated areas of arable land from grazing areas in later prehistoric Britain, in which case the ditches would also presumably have had the secondary function of protecting the cultivated plants from the predations of domestic animals (Ritchie, A., this volume, pp. 541-4); also the so-called "Celtic field-systems", which are shallow ditches demarcating much smaller arable areas (Piggott, S. (1965). *op. cit.* pp. 250–1).

45 e.g. evidence of wooden "palisades" around the Late Neolithic settlements of Azmak, Vinitsa and Russe in Bulgaria, dated by C14 to *c.* 3700 B.C.: Georgiev, G. and Angelov, N. (1957), Raskopki na selištnata mogila do Ruse prez 1950–3 godina, *Izvestia na Arkheologičeski Institut*, **21**, Figs. 15–16; Georgiev, G. (1969). Die Aneolithische Kultur in Südbulgarien im Lichte der Ausgrabungen vom Tell Azmak bei Stara Zagora, *Študijne Zvesti*, **17**, p. 142.

46 Forde, C. D. (1963). *op. cit.* p. 152; also see Rowlands, M., this volume p. 448, and reconstructions of *pueblos*: Fraser, D. (1968). *op. cit.* Fig. 7.

47 Mellaart, J. (1965). *Earliest Civilizations of the Near East.* London. p. 81, Fig. 68.

48 Kenyon, K. (1960). *Archaeology in the Holy Land.* London; Mellaart, J. (1965). *op. cit.* pp. 33–6.

49 Forde, C. D. (1963). *op. cit.* p. 153.

50 Driver, H. and Massey, W. (1957). Comparative Studies of North American Indians, *Trans. Amer. Philosophical Soc.*, **47:2**, pp. 310–13.

51 e.g. Lapps: Vorren, O. and Manker, E. (1962). *Lapp life and Customs.* London. pp. 25–27; Samoyeds: Popov, A. A. (1966). *The Nganasan.* Indiana. pp. 77–8.

52 Popov, A. A. (1966). *op. cit.* p. 107–8.

53 Forde, C. D. (1963). *op. cit.* pp. 66–7; 93–4; Trigger, B. (1967). *op. cit.* pp. 154–5.

54 e.g. Yoruba town wall: Forde, C. D. (1963). *op. cit.* p. 153; e.g. the European Iron Age hillfort of Heuneburg, southern Germany in which the east Mediterranean method of fortification was imported, but the bastioned wall would not have been very effective as a defence: Piggott, S. (1965). *op. cit.* pp. 204–5.

W. A. WEST

The effect of private and public law on the use and deployment of land

Introduction

The emergence of a legal system from a regimen of caprice in ancient societies was characteristically associated with the development of religion and the priest caste. This identification of law with religion may account for the rigidity and formalism of early law; the commonly held idea that one can find a simple system of "natural law" by going back far enough in history is almost certainly erroneous. Indeed, we are probably evolving towards it rather than away from it. In relation to the transfer of property for example, the ceremony was elaborate and, whatever the wishes of the parties, the transaction could be nullified by the slightest departure from its solemnities. In the words of Sir Henry Maine "Ancient law uniformly refuses to dispense with a single gesture, however grotesque; with a single syllable, however its meaning may have been forgotten; with a single witness, however superfluous may be his testimony".

The customs which became the common law were probably handed down through a limited class of citizens but in due course—probably linked with the discovery of writing—codes have been compiled in most civilizations.

Although English law has never undergone a code stage (except in isolated parts of the law and even that came only in the nineteenth century), the development of legal systems in most countries follows a broadly similar pattern. First of all customs evolve which win common acceptance; in due course these are formally recognized by the courts, which themselves have been evolving into recognizable institutions administrating a general system of justice. Early legal procedure is usually highly formalistic and as a consequence this customary or common law tends in time to establish itself into an over-rigid pattern.

The rigidity was mitigated by a fairly well recognized system of legal fictions (for example, in Roman law, the non-traversable averment that a plaintiff was a Roman citizen even if he were not) but, as the scope for this was clearly limited, further development to meet changing conditions came from an entirely open and avowed intervention in various areas of the law by a system claiming a higher degree of sanctity (by reason of the nature of its principles). The Praetorian Edicts fulfilled this function in Roman law. In England the Sovereign's instrument was the Lord Chancellor and the system which was thus engrafted on to common law became known as Equity. In turn, this too crystallizes and the further development of law becomes the function of the legislative body.

At the present time the areas of law which are most important in influencing our environment are those grouped under the heading of Administrative Law. Such statutes as the Town and Country Planning Acts, the Housing Acts, the Highways Acts and the Public Health Acts dominate the use and development of land. But this type of influence is a relatively recent factor; it belongs almost entirely to the last hundred years and the greater part of it has appeared within the last fifty years.

The common law period

Most of the law of contract and the law of tort belong to this period and the ancient tort of nuisance (usually the use of land in such a way as to diminish one's neighbour's enjoyment of his land) must have had a considerable effect in segregating the various types of land use, so that, for example, disagreeable uses such as tanneries or glue or tallow making could not be started too close to existing residences. Wealthy people could thus, to a greater or less extent, protect their homes from incongruous neighbouring uses, but the courts were fairly robust in deciding the permissible extent of an individual's sensitivities and of course the majority of people had little opportunity for access to the civil courts.

Other torts may have had a marginal effect upon land use and development, but the one which probably could have substantially affected the situation developed much too late for its effect to be felt to any extent. This was the rule in Rylands v. Fletcher which imposes a strict liability upon any landowner who brings on to his land anything likely to do damage if it escapes. In the case which established the rule in 1866, the danger was a large volume of water in a reservoir, but the rule has been applied to escapes of gas, chemicals and many other things (including, perhaps anomalously, vibrations, electricity, and even gypsies).

Real property

An owner of land owns an estate or interest in that land and annexed to this are a miscellany of right and privileges. He may dispose of some out of this bundle of rights voluntarily, one of the most common cases being the relinquishment of his right to possession by the granting of a lease, or he may be divested of some of his rights compulsorily, as where planning restrictions prohibit some form of development. He may even have his whole interest taken away from him by the exercise of compulsory purchase power.

The development of land has usually been dependent upon the facility with which it could be transferred, whether on death or *inter vivos*. The path to free alienability has been a long one and in early law the picture is one of elaborate ceremonies, restrictions ranging up to outright prohibitions, mandatory rules of descent (such as primogeniture) and complicated forms of landholding. All of these have tended to sterilize land use.

Servitudes

The property rights known as easements and profits have since Roman times been recognized as an extension of a landowner's rights. Typical easements are rights of way over another's land, the right to take water from another's land, the right to receive support to land or buildings and the right to receive light over another's land. Riparian owners could also acquire rights to obstruct and to pollute a stream in addition to the natural right to take a reasonable amount of water from it. (Pollution and abstraction are now the subject of statutory control).

The protection of ancient lights is much more strict in England than in most other countries—in America rights to light virtually do not exist. The code in England must have restricted a considerable amount of development particularly in relation to high building. In certain cases, however, the courts have been prepared to compel an overshadowed owner to accept monetary compensation rather than to require the removal of the offending building.

Rights of common are another form of right *in alieno solo*. Many are survivals of the manorial system under which the wasteland of the manor was open to the commoners to graze their cattle. Other common rights have come into existence by grant or by prescription. A considerable number of these common rights have dropped into obsolescence and the common lands have remained more or less as public open spaces. Many commons were enclosed, usually under statutory powers, during the great enclosure period from the middle of the eighteenth century to the middle of the nineteenth century.

Certain customary rights enabled areas of land to be kept open for public enjoyment, the most familiar being the village greens.

Private rights of way were a necessary incident of the strip agricultural system and they have also undoubtedly played a considerable part in urban development. More important in this context is the public right of way.

Whenever highways were established for military or administrative reasons, they inevitably created a pattern of development. Some of our most famous roads were built by the Romans; the origin of many others is unknown. In more recent times, the acquisition of highway rights has been by express dedication or by implication from long use—twenty years being the accepted period. Highway rights have had a decisive effect on the location of new development—examples can be found in the many roadside villages. (It was not until 1935 that Britain first controlled ribbon development.) Early developers could use existing highway facilities and they hardly ever made any provision for the making-up of new highways. They carried out their development and merely dedicated the spaces as highways, leaving it to the highway authority to take them over and make them up. A statute in 1835 finally put a stop to this, and in due course the various private street works codes have enabled highway authorities to recover the cost of making up streets from the frontagers of new estates[1].

Methods of land ownership

Forms of land ownership assume a considerable role in the evolution of development in this country. The feudal order began to crumble after the prohibition of subinfeudation by the statute Quia Emptores in 1290 but a new peril to the flexibility of the property system came from another method by which landowners could achieve their dynastic aspirations—the strict settlement. Between the thirteenth and fifteenth centuries it was the ordinary course of events for landowners to entail their land. This led to a considerable amount of sterilization of realty and because of inter-marriage real property became concentrated into fewer and fewer hands. In the fifteenth century the lawyers managed to produce a legal device which enabled entails to be barred and thus the owner for the time being could dispose of the estate. The landowners riposte came in due course and by means of resettlement procedure the land could again be tied up, despite the evolution of the Perpetuity Rule (which otherwise limited the inalienability of land to an average of eighty years or so). The industrialization of the country in the eighteenth and nineteenth centuries made the problem of land availability an acute one as the settlement and resettlement prevented the owner for the time being from disposing of more than his life interest; changed economic conditions made it urgent that land should be available for economic exploitation, and of course for recharging with capital. The practice arose for private Acts of Parliament to be promoted to curtail individual settlements but the expense of this was prodigious. A Public General Act was clearly the only solution, and the first of these was passed in 1856. An Act with much wider powers was passed in 1882 and in 1925 the final freeing of the land from the Strict Settlement was effected by the Settled Land Act of that year. This enabled tenants for life to dispose of the whole fee simple, the interests of subsequent beneficiaries being protected by having their interests attached to the proceeds of sale.

Covenants

Covenants restricting a landowner's rights to do as he likes with his land have played an important part in the pattern of land development. The law has traditionally divided these into two kinds: firstly the leasehold covenant under which a lessor, usually the freeholder, could impose upon his lessee covenants restricting the use of land, for example prohibiting commercial buildings or imposing positive duties upon the lessee such as to build a house according to a particular specification. Leases can be for any period, (they are commonly three years, seven years, twenty-one years, ninety-nine years and nine hundred and ninety-nine years) and such covenants would bind not only the original lessee but all his successors in title for the benefit of the reversioner for the time being. Although leaseholds have existed since the late twelfth century, this type of covenant only became important when building development accelerated under the impetus of industrialization. Covenants undoubtedly served a very useful purpose in preventing the worst sort of slum development and indeed some of our finest urban developments come from this source—parts of Westminster, Bloomsbury, Bath and many other places were thus given and were able to retain their character. Restrictive covenants between adjoining *freeholders* made a much later appearance owing to a lacuna in our legal doctrines and it was not until as late as 1848 that the courts finally recognized that a freeholder could be bound by a covenant entered into by his predecessor in title. Even then the ambit of such covenants was much more limited than for leasehold covenants; in particular, only *negative* covenants could be enforced against the freehold. The test as to whether a covenant was negative was one of substance rather than of form, and typical examples of such restrictive covenants are—not to use the premises for commercial purposes, not to build more than a fixed number of houses on the land, not to build any house other than one according to specification. A further product of this kind of covenant was the Building Scheme under which purchasers from a common vendor could enforce restrictive covenants not only against the vendor, but against one another. Many existing residential estates still rely on this form of covenant to preserve their character.

In due course, obsolete restrictive covenants became a serious impediment to development. There were great difficulties in releasing land from such covenants without statutory intervention and various different statutory powers were given which enabled restrictive covenants to be removed, the most important of these being Section 84 of the Law of Property Act, 1925.

English Law has never recognized the right to a view as being an incorporeal hereditament, and indeed there has never been any formal restriction on the owner's right to develop any of his land as he pleases until the emergence of the current body of administrative law. Limited ownership, rights *in alieno solo* and restrictive covenants were more or less incidental restrictions on development.

Mortgages

The development of the mortgage system (the loan of money on the security of a landholding) was an important influence at a time when money was in some ways scarcer than land. The present form of mortgage had evolved by the middle of the fifteenth century but the protection of the landowner's right to redeem even after his contractual repayment date had passed did not come until the beginning of the seventeenth century. The mortgage has thus for over three centuries played a crucial part in the process of recharging land with money. Building societies began to play an important part in the nineteenth century and local authorities have been empowered to grant mortgages during the whole of the twentieth century. Nearly nine million houses are owner-occupied and, at the present rate, owner-occupied houses will exceed half the total of all houses within a year or two; nearly all of these are acquired by means of a mortgage.

Corporations

The doctrine of corporate status has had a profound effect in shaping our man-made environment. Under this doctrine a group of persons could, by following a recognized legal process, create a new body with a legal existence separate from that of the individual members. The procedure for achieving this was originally a prerogative of the Sovereign, who could confer corporate status by means of a Royal Charter, and thus create a new legal entity with powers, rights and duties of its own. All of our Boroughs came into existence in this way, although the original charters of many of them have been lost. Once the citizens of a particular town could get a Royal Charter, they were able to buy lands for the town and to carry out many acts of local government long before the present system of local authorities was set up in the nineteenth century. The only limiting factors on the powers of these chartered corporations were the terms of their Charters. Their corporate status enabled them to influence the development of their towns in important ways such as in controlling markets, conferring of trade privileges, providing a court system, and the Private Acts of Parliament promoted by many of them in the early nineteenth century became the precedents for nearly all of our Public Health law and much of our other municipal law.

Although some early trading companies owed their incorporation to Royal Charters, the real flowering of the trading corporation came in the nineteenth century with the *statutory* company, by which groups of individuals could partake in a venture without incurring more than a fixed limit of liability if it went wrong.

Town and country planning

Hippodamus in the fifth century B.C. is thought of as the first named town planner and the system of planning using the criss-cross grid still bears his name. Some form of town planning has undoubtedly been practised in all civilizations and many of the great cities of the past—Memphis, Babylon, Harappa and Mohenjo-daro are examples—were clearly laid out on the basis of preconceived plans. Many of the Greek and Roman settlements (but little of Athens or Rome themselves) were laid out on a rectangular framework with high standards of public health.

Medieval towns were mainly haphazard and irregular developments but the bastide or new town also made its appearance at this time. These conformed to a plan essentially defensive in concept. Some of these have remained small settlements—such as Aigues Mortes—whilst others grew to become great modern cities—such as Strasbourg and Hull. Military considerations dominated planning at this stage—Caernarvon in Britain, Krak des Chevaliers in Syria and Carcassonne in France are random examples.

From the sixteenth century onwards was a period of expanding trade and consequent urban expansion which took the form of town improvement more than of new town construction, at any rate in Europe. One may mention one great lost opportunity—Wren's plan for rebuilding London after the Great Fire—and one great fulfilled opportunity—Haussman's Paris, two centuries later. St. Petersburg (now Leningrad) in Russia and Washington in America were built at the beginning and end of the eighteenth century respectively.

Law and planning—early planning must have been the result of sovereign command (Pericles, Edward I, Louis XIV, Peter the Great, Napoleon III are some of the great known names) but we have to wait until modern times for any systematic planning code to be enacted by the legislative function of government.

Today Town and Country Planning restrictions are familiar in most advanced countries and this country now has the most comprehensive system of planning control in the world. Germany was in the forefront of municipal administration for town and country planning; the first Act in this country did not come until 1909, and even this was largely ineffective, partly because of the limitations in its scope but more important because the 1909–10 Finance Act[2], and shortly afterwards the First World War, caused a major slump in estate development. There were a number of Acts in the inter-war period dealing with planning, the most important being in 1932, which extended planning effectively for the first time to urban areas. It was not until

1947 that there was created the comprehensive planning control which we have today. This Act has been considerably amended (and it has been consolidated), but the basis of control has largely survived. The main effect of the planning control provisions is that permission has to be sought to develop any property or to change its use. Certain minor developments are permitted by a General Development Order, but otherwise individual application has to be made for each proposed development. In Green Belt areas development is virtually forbidden. This of course has had the effect of driving up the prices of developed property in Green Belt areas.

Elaborate financial provisions were enacted which in effect imposed a tax on development. The 1947 provisions relating to development charge were repealed after six years, but another statute in 1967 made a fresh attempt to cream off some of the profits of betterment. These provisions are in the course of repeal at the present time.

Development plans are prepared by the local planning authorities for each area showing the manner in which land is to be used and the stages by which development is to be carried out. Applications for planning permission will normally be considered in the light of what the plans provide.

Development plans are now (since 1968) to be prepared at two levels:

1 a Structure Plan which sets out the local authorities' policy on the broad lines for future development in the area (including improvement of the physical environment and control of traffic). "Action areas" may be defined indicating that particular parts of the area are to be given comprehensive treatment by positive action. The Structure Plan requires confirmation by the Central Government;
2 a Local Plan, which is more or less entirely under the control of the local authority. Broadly speaking, this fills in the details of the Structure Plan.

Very wide powers of compulsory acquisition for planning purposes are given by the various Acts.

One of the problems associated with Town and Country Planning control is that planners have to plan today for tomorrow on yesterday's data. For this reason, a great number of pitfalls have trapped the planners over the years. Up to a generation or so ago, planning was based upon the assumption of a diminishing population. Hurried revisions were made, and planning was then based upon a greatly increased population. Today we see a scaling down again. The dominance of the motor-car in our lives was almost entirely overlooked by the planners until far too late. Massive measures were taken to prevent a population drift to the south-east of England, but the south-east drift turned out to be a myth. Measures were also taken to induce industry to move to development areas at a cost which is said to be out of all proportion to the result. It is claimed that the Industrial Development Certificate system merely inhibits development in some areas without shifting it to others.

New towns

Since the World War, as the result of the recommendations of the committee set up under Lord Reith, new towns have been built by statutory development corporations created for the purpose. They were intended to be self-contained balanced communities but they have tended to consist exclusively of lower middle-class tenants living under somewhat restricted social and cultural conditions. There are now about thirty new towns in various stages of development in Great Britain, the public investment cost being about one thousand million pounds.

Town development

In 1952 an Act was passed which provided machinery for expanding existing towns mainly by exporting people from overpopulous areas. The machinery has on the whole been difficult to operate largely because of the complex consultation procedure with the various public authorities.

Public health legislation

The first general Public Health Act was passed in 1848 after a severe cholera epidemic, but until 1875 most of the legislation on the subject of Public Health was local legislation promoted by particular towns. In 1875, a Public General Act of the greatest importance was passed dealing with Public Health. Although it proved to be somewhat too rigid in operation, it undoubtedly had a profound effect on all subsequent development. It imposed by-law requirements on the construction of all new building and on the making-up of streets. It required the provision of sewers and of water supplies. It made however, no provision as to the *siting* of developments and of streets.

Housing

Housing legislation dates back to 1851, but the first really effective Housing Act did not reach the Statute Book until 1890. The Housing Acts of the nineteenth century dealt mainly with the condition of existing houses and the control of lodging houses, and it was not until 1919 that the present system of subsidized council housing appeared. Since then, local authority housing has increased to such an extent that now over a quarter of all our households live in council houses. The other main housing role of the local authorities is the clearance of slums.

Rent control of private lettings has also existed in most countries for over half a century. An unfortunate by-product, which is also true of council housing, is that the labour force tends to lose some of its mobility when occupation of low-rented housing is at stake. One of the consequences of the Public Health and Housing legislation together with the Rent Acts has been the virtual withdrawal of private enterprise from the provision of rented housing accommodation, and the main role which the private sector now fulfils in housing is in providing houses for owner-occupiers.

Unless there is a revolutionary change in financial policies the future housing stock is likely to be almost entirely made up from council houses and owner-occupied houses in the proportion of 1 : 2, with privately rented housing more or less extinguished.

Private law affecting use and development of land is now completely overshadowed by public law. The local authorities' role in positive development is increasing year by year, not only in the realm of council housing estates but in such "commercial" activities as town centre developments. Their "negative" role in controlling all new buildings and other development (not merely on public health grounds but on planning grounds) has been a major feature of the environmental scene ever since the end of the Second World War. The picture in the future seems to be one of continuing increase in this dominance of administrative law.

Notes

1 By then of course the original developer was out of the picture, and it was not until 1951 that statute compelled the developer to make provision for street works costs before he started building.

2 An example of the influence of fiscal policies on development.

Section 3: Techniques, planning and cultural change

B. A. L. CRANSTONE

Environment and choice in dwelling and settlement: an ethnographical survey

Few sorts of artefacts illustrate more clearly than do dwellings the positive influence of environmental factors on material culture. The environment influences dwellings in two main ways: it imposes a need for shelter from cold, rain or sun; and by its effect on the economic basis of life it creates a need for a particular type of dwelling (e.g. portable or permanent) and makes other sorts inappropriate. It provides raw materials, and when climatic conditions are rigorous or certain materials lacking, it excludes some sorts of dwelling. Whether, and in what way, the opportunities it offers are exploited depend on a variety of non-material factors.

Environment, ways of life and techniques[1]

Hunters and gatherers

Peoples who live by hunting and gathering usually have to range over an extensive territory during an annual cycle, exploiting animal or vegetable resources as they become available. Since such peoples are frequently on the move any permanent dwelling is out of the question, and since they usually have no animal transport they cannot carry portable dwellings with them. So they make their shelters afresh from locally gathered materials at every camping site. In the modern world most such peoples inhabit forests or somewhat arid areas in the tropical or sub-tropical zones: the Andaman Islanders, the Malayan aborigines, the Bambuti of the Congo are examples of the former; the Australian aborigines, the Bushmen of South West Africa and the Hadza of Tanzania of the latter. The forest peoples require shelter from rain, but not usually from cold. Their hunting shelters are made from saplings or branches, covered with leaves; camps occupied for longer periods may have more commodious huts built on similar principles, as in the Andamans[2]. Many of the large forest leaves make excellent roofing material. The Australian aborigines, in the desert or semi-desert areas, and the Bushmen require shelter from the sun, and (especially the aborigines) also from the cold, which can be quite severe at night. Their shelters are often no more than windbreaks, propped at an angle; but sometimes they have the form of roughly dome-shaped structures of branches, the butts stuck into the ground in a circle and the ends tied together or interlaced where they meet, and covered with grass or whatever herbage is available. These provide shade, some warmth and shelter from light rain. The sites are not occupied for long at a time, though they may be re-occupied year after year and generation after generation.

This picture of the hunting-and-gathering life is not, however, universally valid. A few peoples with this sort of economy are able to settle more permanently. The North-West Coast Indians of North America provide an example. Living on the apparently inexhaustible stocks of salmon, supplemented by sea-fishing, hunting and collecting wild fruits and roots, they built great timber houses which accommodated more than a hundred people. Among the Kwakiutl the sides were 40-60 ft long[3]. But like the other North-West Coast tribes the Kwakiutl were not entirely sedentary. The winter was spent in the principal villages, but at other seasons they migrated to fishing camps, often taking with them (used as rafts) the wall planks of their houses, with which they erected temporary dwellings[4].

The Eskimo of Alaska partly excavate their winter houses, build up the walls with stones or sods, and roof them with whale ribs or driftwood supporting two layers of seal-skins with a layer of moss between them. The Greenland Eskimo use driftwood, which is there more plentiful. Only the central Eskimo make the elaborate snow-houses lined with skins, with storerooms and interconnecting tunnels, which are regarded as typical of this people. Like the excavated house, the snow-house is approached by a passage which is below floor level, so that cold air drains away, and for the same reason the sleeping platform is above floor level. Heated by blubber lamps, the snow-house is relatively warm and the inhabitants are sometimes half naked. Living in one has been described as like having your head in the tropics and your feet in the Arctic. Other Eskimo make small snow-houses as one-night shelters when travelling.

The snow-house is unique in a number of ways. It is a dome built of snow blocks, laid not in horizontal courses but spirally, and is self-supporting during construction. After a little use the warmth inside, melting the inner surface which freezes again, converts it to ice, and it is said that a polar bear can walk over it without breaking through. Clear ice windows are often inserted. It is the only sort of house the material of which disappears rapidly and without trace; and it has no hearth to provide evidence, though of course there may be occupation débris. If it is built on the sea ice even the débris vanishes in the spring[5].

Many Siberian peoples use semi-subterranean houses, especially during the winter. The depth of excavation depends on the depth of the permafrost, and on drainage. Some use these houses during the winter, and pile houses during the summer (Pl. 6)[6].

Nearly all Eskimo migrate seasonally, exploiting different resources, and many use skin tents from spring to autumn. The Caribou Eskimo of the Canadian tundra use tents the whole year round and many never visit the coast. When the Eskimo migrate they can take their dwellings with them because they possess, in addition to their skin-covered boats, a very efficient form of land transport—the dog sledge.

The plains Indians of North America also carried their tents, which many of them used throughout the year. Their movements were dependent on those of the bison. Their *tipi* was made from four poles tied together near the top with thongs. These formed the basic framework, against which additional poles were laid. Over this was fitted a shaped cover of bison skins (Pl. 7a). On the open plains material for the framework was not available, so the poles had to be transported. Again dogs provided the means. Two poles were crossed over a dog's shoulders and held by a harness. The ends trailed on the ground. Other dogs carried small bundles. When the Indians acquired the horse this *travois* was transferred to the new animal[7].

On the eastern borders of the plains bark sheets were substituted for the skin covering of the *tipi*, the pole framework being retained (Pl. 7b).

Pastoral nomads

Pastoral nomads also lead a migratory life in search of pasture for their herds. The more specialized of them live in deserts or steppes, where materials for making shelters afresh at each new camp are lacking, so they must transport their dwellings. Since they possess numbers of animals this is not difficult.

Many reindeer herders of northern Asia use tents of the *tipi* type[8]. The Lapps have a form in which the poles are forked at the end so that they lock together easily. These are used by young men tending the herds. The Lapp family tent is more elaborate in construction. The wooden framework is illustrated in Fig. 1. Against this further poles are leaned. In recent times the tent is covered with horse-blankets in winter, canvas in summer; formerly reindeer skins and birch bark were used[9].

The Kazak and other horse and sheep herders of the south Siberian steppes have a different sort of tent, the *yurt*. The framework is of willow, obtained in the river valleys; the covering is of felt, made from fleeces (Fig. 2)[10].

Both Lapps and Kazak have more substantial winter houses. Among the Lapps the elderly and some women and children stayed at the church villages,

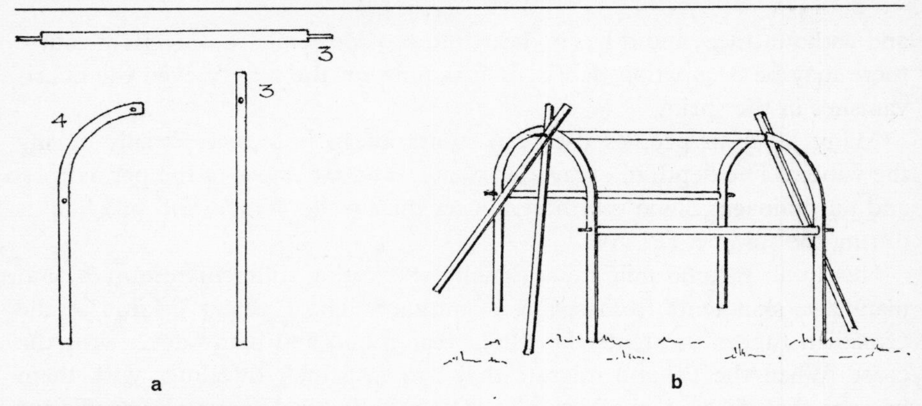

Fig 1 The Lapp tent.

a. The component parts of the basic framework; the figures denote the number of each required. *b.* The framework assembled, with the entrance to the left. Further poles are leaned against this frame, all round, to support the cover.

in huts of sods supported by a birch frame[11]. The Kazak winter settlements were in the river valleys. Typically their houses were partly excavated, with walls of sods or stones, roofed with poles covered with brushwood and finally with turves. Sheets of felt lined the walls internally. A group of such houses was surrounded by a turf or reed fence against which shelters were built for the more delicate livestock[12].

The Bedouin of Arabia use their tents throughout the year. Basically these consist of a strip of goat-hair cloth (obtained from village Arabs) supported

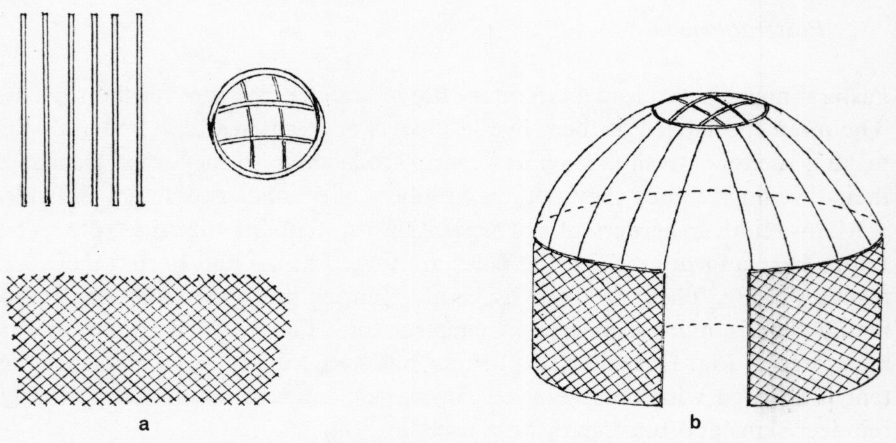

Fig 2 The *yurt* of the Siberian horse- and sheep-herding nomads.

a. The component parts of the framework: a lattice of willow rods, fastened with thongs where they cross, which forms the wall; a number of rods which are fastened to the top of the lattice, and to a wood ring which forms the smoke-hole; and the ring. *b.* The framework assembled. A three-piece door frame is usually added, and the tent is covered with sheets of felt.

by pairs of poles with guy-ropes. They may, however, be quite elaborate, with sub-divisions for women, guests and storage[13]. Similar tents are used widely in the Middle East and North Africa.

Agricultural peoples

Plough agriculture is often linked with permanent settlement, but long fallow (swidden) cultivators usually move their dwelling sites as the soil within a convenient distance is exhausted: sometimes every two or three years. Some, more powerful or warlike, move steadily forward. The Iban (Sea Dyak) of Borneo are an example. Others move within a tribal territory. Each Zemi Naga community, living in the rugged Barail Range of Manipur, had a number of village sites round which it moved in a cycle covering a long period of years[14]. Such peoples are often in a state of endemic warfare and for safety it is necessary to return from the fields to the settlement each night. If they are too far away the time taken by daily travelling to and fro becomes uneconomic. Besides, such travelling is very dangerous. An estimate of the population of such an area based on the number of village sites could therefore be greatly exaggerated (see below). It should, however, be remembered that the empires of West Africa and of the Maya were based on forms of long-fallow cultivation, and that up to modern times long-fallow cultivation has been practised by most of the cultivating peoples of America, of Oceania, of Africa south of the Sahara, and of many of the mountainous parts of Asia.

Plough cultivation is accompanied by the use of draught animals and often by mixed farming, and manure is therefore available. The generally higher level of agricultural technique makes it possible to preserve the fertility of the soil and so really permanent settlement becomes possible. Investment in such improvements as irrigation or drainage works or terraces may further tie the people to the land.

So great is the variety of materials and methods used by agricultural peoples in building their houses that the only way in which the subject can be surveyed in a short space is by discussing in turn materials and the main techniques associated with them.

Stone. However important it may be archaeologically, the use of stone for building dwellings—as distinct from ritual or defensive structures—is relatively uncommon in an ethnographic context. Its use by some Eskimo and by the Kazak and other Siberian peoples has already been mentioned. Stone houses of rectangular plan, the shorter walls often raised to form gables, and roofed with timber, have a wide distribution in northern Europe, the Mediterranean area, the Near and Middle East and the slopes of the Himalayas. In this type the walls support the lower ends of the rafters and the ridge pole is often supported by the peaks of the gable walls. In fact the walls as a whole support the roof, which is a separate structure.

The other main method of building in stone is corbelling, in which the walls and roof are one structure. Ethnographically it extends down the eastern side of Africa, though there are gaps in its recorded occurrences in this area. African corbelled huts are circular or oval, and owing to the limitations of the method are small in size. They are found mainly in the more arid areas[15]. The method does not necessarily dictate an oval or circular form; in the Mediterranean area rectangular corbelled buildings have survived to the present day[16].

Mud. The material is carefully selected and prepared, though not so carefully as for pottery-making.

There are two main methods of using mud as a building material, apart from its uses as a plaster. The first is the making of mud-bricks dried in the sun. The techniques of building with them resemble those of building in stone or fired brick.

The second method is building in plastic mud; the wet material is built up in courses, each course being allowed to dry before the next is added. It is a modelling technique, often done entirely with the hands. The Kofyar, a northern Nigerian plateau people, build two-storied domed houses in mud[17]. A course of large stones is laid first as a foundation. Often the lower courses are of mud and stones alternately, the mud being mixed with straw, but the arches which support the upper floor, and the dome, are of mud alone. Storage bins for grain, floors, sleeping benches and other fittings are modelled during building. Finally, to protect the material from rain, the dome is covered with thatch. This is a remarkable instance of virtuosity in a material sometimes rather despised. A point of possible archaeological interest is that when a new house is to be built an old house is often demolished, the material pounded into powder, mixed with water and re-used[18]. The Kofyar are familiar with stone-building techniques and build substantial goat-pens and defensive walls of stone.

As a building material mud is most suitable for arid areas: the margins of the Sahara and the Arabian Desert, for instance. However it will stand quite heavy rain if it can dry out in hot sun between showers and if the water does not penetrate the interior of the wall. So mud buildings are usually well thatched, and the eaves project outwards and downwards so as to throw the rain clear. The walls may be built on a footing of stones to ensure that they do not stand in moisture.

Reeds. Reeds (including dry-land cane-grasses) are used extensively as a supplementary building material, for making screens, walls and fences and for thatching. Only locally, where other materials are lacking, are they used for making the principal structure. The best instance of their use is by the marsh Arabs (Madan) of the marshes at the junction of the Tigris and Euphrates (who are not primarily agricultural). The reeds here grow to great heights. They are bound in bundles, nine feet or more in circumference at the bottom, and are then used in two ways: they are erected vertically, as if they were timber posts, or they are curved over in a series of arches to make long guest-halls of barrel-vaulted form[19].

Wood. Wooden buildings are much more varied than those in other materials. Wood is usually a preferred material where it is available. It is easily procured, easily worked and adaptable, and apart from shaping with wood-working tools needs little special preparation or treatment.

The type of hut sometimes called "beehive", especially typical of parts of southern Africa, is made in two ways. In both the plan is usually circular, and the framework is made from saplings or light flexible poles. In the first method the poles are driven into the ground in a circle and the ends are bent down to the centre where they are overlapped and tied. In plan they are radii of the circle. In the second method the poles are bent so that both ends are inserted in the ground (or two are tied where they overlap, which has the same effect). The hoops so formed are parallel to each other, the highest being the centre one. A second series is often inserted at right-angles to the first, and they are lashed where they cross. In this method only the centre poles of each series cross at the centre of the circle. In each case the hut is hemispherical in shape and is covered with grass thatch (Fig. 3). Both methods are also recorded in Siberia[20].

Another common type in the lower-rainfall parts of Africa is made by inserting poles upright in the ground, usually closely spaced in a circle. On to the cylinder so formed a conical roof is placed. Sometimes the roof is made separately on the ground; sometimes it is made *in situ*. In either case there is usually no centre post, though when the roof is made on the hut a temporary centre post may support it during construction, to be removed when it is finished[21]. Melanesian round huts, on the other hand, usually have centre posts, which sometimes project through the roof to form decorative finials.

The Bemba of Zambia (who are long-fallow cultivators) make houses of this type and plaster them with mud. A shallow circular trench is dug, and the light posts are driven in about an inch apart. The women collect and puddle clay. Both sexes then throw this against the walls, externally and

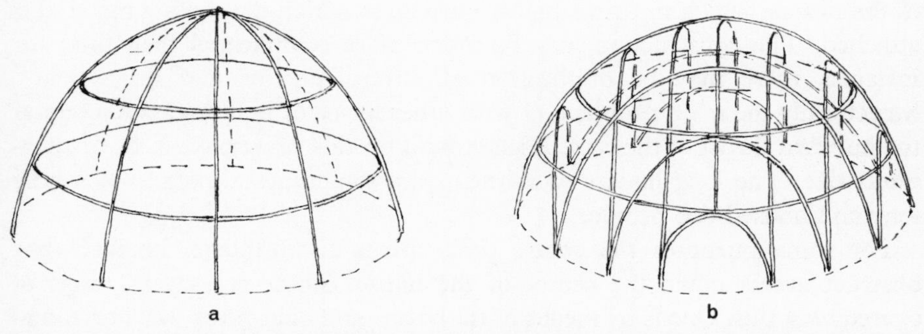

Fig 3a and b Two methods of making the framework of a "beehive" hut.

internally, so that it adheres and fills all the chinks. After a period of drying it is smoothed with wood scrapers. The thatch of the roof projects all round the eaves to form a working space and to protect the walls. The Bemba, living in savanna country, have difficulty in finding sufficient straight timbers of adequate length, for the savanna is frequently burnt and what trees survive are stunted and gnarled. In fact villages are sometimes sited because suitable timber is available nearby. A house lasts about five years, by which time the soil has lost fertility and it is time to move[22].

The American log cabin derives from a form of dwelling which is very old and widely-spread in Europe and Asia. Requiring plentiful and easily trimmed straight logs, it is closely associated with the pine belts. The logs are laid horizontally, one above another, and where the walls meet and the logs cross the ends are cut away so that they fit snugly. Chinks are plugged with mud or moss. The whole structure forms a solid box, the walls supporting the roof. This is still an important type in northern Asia and the Himalayan pine belt. The method of construction does not necessarily dictate a rectangular form. The logs can intersect at more than 90°; in northern Asia multangular forms are common[23]. There is no structural factor which makes the insertion of posts in the ground essential: the box can rest on the ground surface. There need therefore be no post-holes or foundation trenches for the archaeological record.

Ethnographically the method is unusual in other parts of the world. The Massim, at the east end of New Guinea, build yam stores in this way, but they do not cut away where the logs cross, so that there is a space the width of a log between each. This is done deliberately: it ventilates the store; and it allows the yams to be seen, a good stock of yams being important to the prestige of the chief[24].

The largest group of wood houses is that in which the roof is supported at a number of points by posts, and the walls are a skin which can be added last of all. The main interest here is in the different ways of supporting the roof. The simplest is by means of centre posts, two or more, which support a ridge pole. Corner posts, and often further posts along the walls, bear wall plates. The ridge pole and the wall plates support the upper and lower ends of the rafters, which in turn support purlins, to which the roofing material is attached. The construction may be much more complicated than this; for instance, there may be duplication of rafters, purlins and ridge pole[25]. Structurally the main problem, as with other forms of building, is to counter the outward thrust of the roof. Houses built on this principle can be of enormous size. The long-houses of Borneo were sometimes a quarter of a mile long and housed 600 people.

For some purposes the centre posts are a disadvantage, because they obstruct access down the centre of the house. There are several ways of overcoming this. One is by means of tie-beams and king-posts, the horizontal tie-beam bracing the walls against the outward thrust of the roof, the king-post set vertically on it and supporting the ridge pole. This method is used, for example, in the Solomon Islands canoe houses, where it is important to be able to run the canoes into the centre of the house because of their high bow and stern pieces, and in some Polynesian assembly houses (Pl. 8).

Another way of dispensing with centre posts is to set a row of posts on each side of the centre line, a method used in the communal houses of western Papua[26] and in the great men's houses of the Purari Delta[27], which may be 80 ft high at the front gable. Each row bears horizontal poles. Rafters are supported by these poles and the wall plates. At the apex where the unsupported ends of the rafters meet there is a light ridge-pole; but this is merely a means of finishing the junction neatly (Fig. 4). The purpose is to give unobstructed access along the centre line to the family or lineage areas to each side.

There are several other ways of building wood houses. Cruck houses and the medieval and later carpentered frame houses, filled in with wattle-and-daub, brick or other materials, are familiar examples.

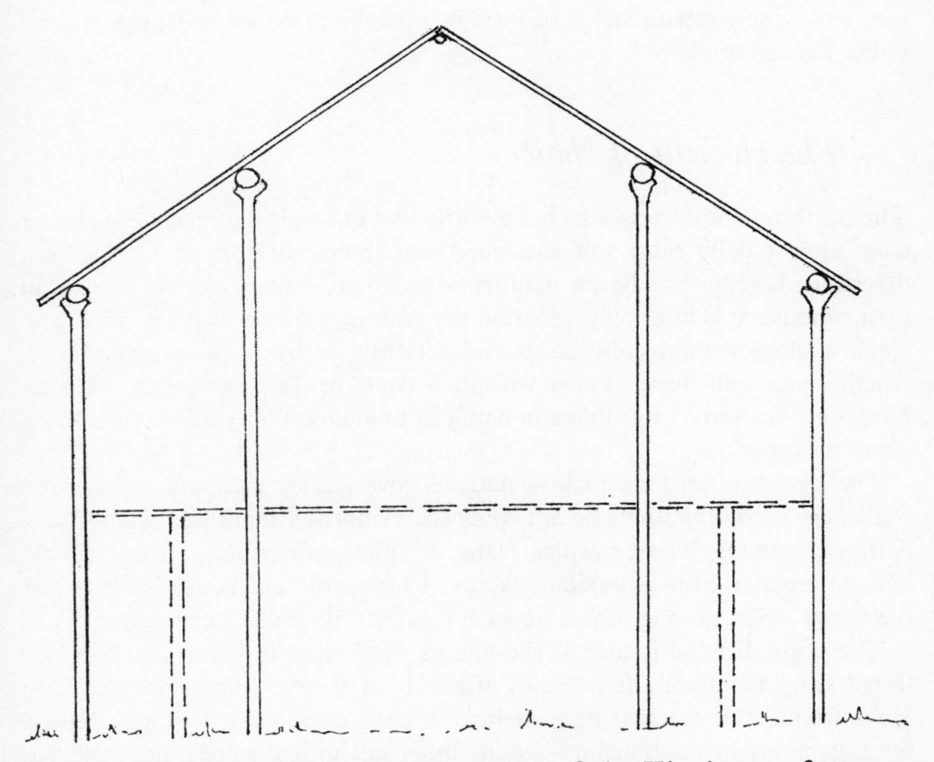

Fig 4 The communal house or men's house of the Kiwai area of western Papua (transverse section).

The wall-posts and the two inner rows of posts support the rafters, which support a light ridge-pole. The floor is separately borne on piles. Both communal houses and men's houses are made in this way.

Techniques

Roofs are most commonly covered with grass or leaf thatch. Palm leaves are often doubled over a length of cane or their own split mid-rib. This sort of thatch is often made up into sheets on the ground and then lifted to the roof and lashed to the purlins. For thatching a large wood or bone needle is frequently used to pass the lashing material between workers outside and inside. Roofs can also be covered with planks, wood shingles, half bamboos, sheets of bark, or stone or pottery tiles. Walls too can be thatched; the division between roof and wall is not always clear. Walls can be made from planks, split with wedges or adzed; covered with matting of interlaced bamboo strips, flattened reeds or leaves; or be of wood in various forms, plastered with mud. Nails are rarely used in ethnographic contexts; if they are available at all they are too precious. Wood pegging, as used in Britain in timbered houses, is also uncommon: drilling the holes presents difficulties. The same applies to carpentered joints. The usual method of fastening is by lashing, using creepers, vines, rattan, strips of bark or manufactured vegetable-fibre cord, either twisted or plaited.

The element of choice

The environment decrees that people who live in tropical forests need shelter from almost daily rain, and use wood and leaves to make it. It does not determine whether the house shall be rectangular, square, circular or oval in plan, whether it is built on piles or on the ground, whether it is occupied by a single biological family, by an extended family or by a whole community, whether men will share a house with their wives or sleep separately, whether houses are scattered or grouped in hamlets, to mention only some of the more common variations.

The lay-out of settlements is sometimes governed by the nature of the site, but where such limitations do not apply social motives are often paramount[28]. Although among many peoples clans or other groups have their clearly defined areas and the alignment, aspect or elaboration of houses may reflect the social order, among others no such clear-cut determinants operate.

The unpredictable nature of the factors sometimes involved can be illustrated from the Bambuti pygmies of the Ituri Forest in the Congo. They move frequently, erecting new shelters at each camping site. There are no set alignments or relationships for the huts, but to make one's hut with the entrance directly facing the entrance of another implies a very close friendship with its occupants. Turnbull describes an episode. A arrived first and his wife built his hut. B arrived and his was constructed with the entrance directly facing A's. A disapproved of the fact that B was sleeping with his younger wife (A's sister) too soon after the birth of a baby, and altered his entrance so that it faced C's. C, not wishing to take sides, altered his entrance so that it faced diagonally across A's but into D's. D then altered his—and so on[29].

Pile houses. Building on piles has a number of advantages. Over water, it gives very easy access by boat, and if the water is flowing or tidal it facilitates the hygienic disposal of refuse. If the houses can be built even a short distance out from the shore, mosquitoes will not reach them. On land, pile building enables rough or sloping ground or ground liable to flooding to be used. It is the only means by which areas like the Purari Delta in New Guinea (which is rich in sago) can be inhabited. Some protection is provided against snakes and dangerous animals, and the space beneath can be used for storage or for keeping livestock. It allows circulation of air, and it has defensive advantages. It is common in Asia and Melanesia. Frequently the floor is supported separately, not by the wall posts (Fig. 4). Sometimes the floor is built first. In the building of a *ravi* in the Purari Delta it provides a platform on which the men work when erecting the enormous front posts (Fig. 5).

Houses built in trees occur locally, but widely. Their purpose is usually refuge from wild animals or enemies; they are not permanent or regular dwellings in settled times.

Communal houses. Among many peoples one house accommodates several families or a whole community. The houses of the north-west coast tribes of North America have been mentioned above; other North American Indians

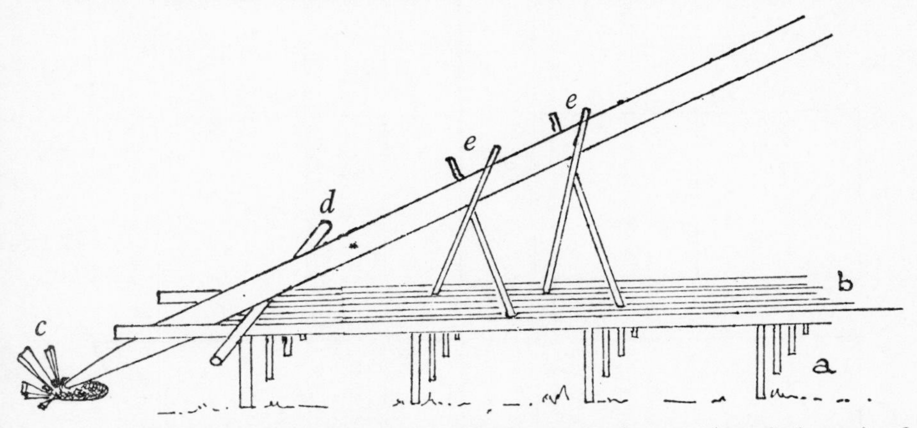

Fig 5 The erection of one of the main front posts of a *ravi* (men's house) of the Purari Delta, Papua.

The piles supporting the floor (*a*) are erected and temporary flooring (*b*) is laid to form a working platform. The hole to take the post is dug and some short posts (*c*) driven in on its rim to guide the end of the main post. As many men as can find space weigh the end of the post down. Others raise it with the pole (*d*) and pairs of crossed poles (*e*) which are moved forward as the post rises.

After Williams, F. E. *The Natives of the Purari Delta*. Port Moresby, sketch VIII, 1924.

had them, constructed in various ways, and so do many tribes of the western Amazon basin. The Boro house, in the latter region, holds up to about 200 people. It is square in plan with sides up to 70 ft long, and is about 30 ft high. The ridge pole is supported by centre posts, and the roof, thatched with palm leaves, comes down to within about 3 ft of the ground[30]. The Kiwai communal house accommodates mainly women and children; the men have a separate but similar house.

In south-east Asia a totally different sort of communal house is found. The Borneo long-house is really a form of terrace development. The house is built on piles, and consists of a large number of family dwellings side by side. Each family has a living room, which occupies the rear half of the roofed area. Above is a loft for storage, where girls sometimes sleep. In front, still under the roof, but open for the whole length of the house, is first an area used for family chores such as pounding rice, then a communal passage, then an area where ritual is performed, men sit and male visitors sleep. The floor is often continued beyond the roof to form an uncovered platform, used for drying rice and domestic work. The headman and important people have their apartments near the centre; the poor, widows and later recruits to the community live near the ends (Fig. 6).

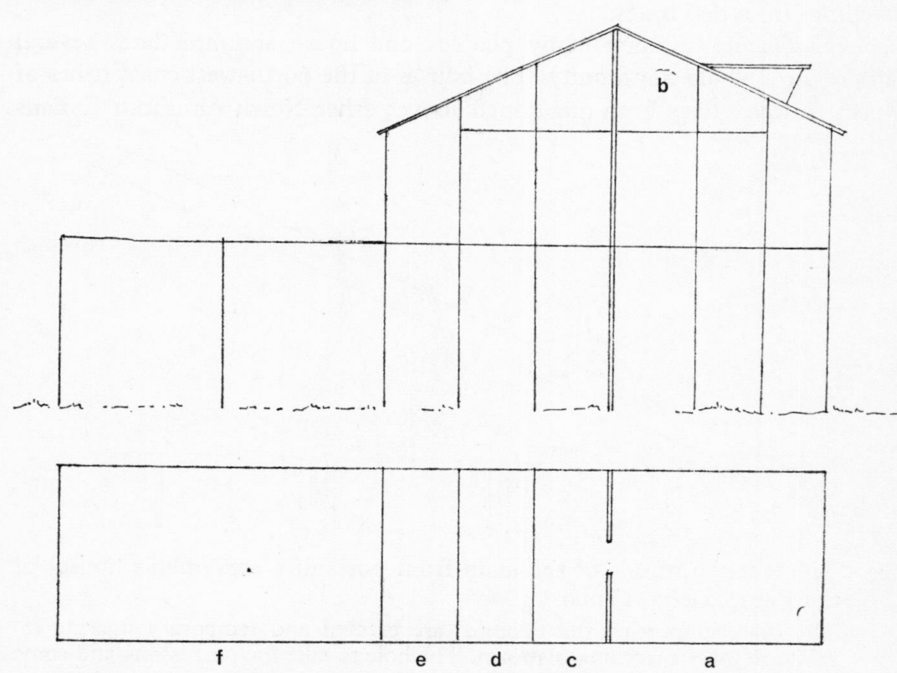

Fig 6 The Borneo long-house: transverse section and plan of a family apartment.

a. The family room; *b.* loft; *c.* family working area; *d.* communal thoroughfare; *e.* space where men conduct affairs and male guests sleep; *f.* open platform.

The main timbers, which are of durable wood and very substantial, are erected by a group of men working together, but each family is responsible for finishing its own section and keeping it in repair. If a family suffers misfortune, such as the loss of a man, their house may be unfinished when those on each side are completed; so there is a gap in the long-house, the communal passage being continued by a plank laid across.

These houses are usually built beside rivers, which are the main means of communication. The long-house community, living by long-fallow cultivation, is forced to move at intervals, and when it does so the main timbers are sometimes floated to the new site[31].

The traditional house of the Karen of the Pegu Hills, in Burma, was basically similar. It too was raised on piles and consisted of rows of family apartments, but here the apartments opened from both sides on to a central corridor. An interesting feature of the Karen house was that it was entirely of bamboo. The main structure was of bamboos 20 ft long or more, and up to 6 inches in diameter. Bamboos formed floor beams (the floor was not a separate structure), joists, wall-plates, rafters and purlins. The floor was of bamboo, split and flattened, except the floor of the central corridor which, because of its continual use, was of entire bamboos. Walls were of flattened bamboos set vertically, with small bamboos interlaced horizontally. The roof was covered with half bamboos placed overlapping and alternately concave and convex. The house was surrounded by a bamboo stockade set with bamboo spikes. In the rooms were bamboo water vessels, bamboo baskets, bamboo shelves and fittings (Fig. 7)[32].

Men's houses. Men's houses are a common feature of tribal societies in south and south-east Asia and Melanesia. Their social significance and function vary. In parts of New Guinea, especially in the Highlands, there is a feeling that too much association with women is inimical to the important male qualities, such as aggressiveness, cunning, force of personality, warlike prowess —qualities essential to success in societies lacking hereditary chieftainship, where power is in the hands of self-made "big-men". In such societies men sleep not in their family houses but in separate men's houses to which women are not admitted. In these, initiation rites are performed, policy matters discussed, ceremonies prepared, and ritual objects kept. The only males who sleep in the family houses are those so young or so old that they do not count as men in a social sense.

In the Purari Delta the *ravi* houses the cult objects and is the centre of ceremonial life; it is both a sacred building and a club house. Bachelors sleep there. If a man wishes to achieve status he must spend much time in the *ravi* and be prominent in discussion. The graded societies round which society is organized in the New Hebrides also have their houses, and in the south-east Solomons the canoe-house fulfilled a similar function. In southern Asia[33] it is usually the young bachelors who live in these houses, though older men may frequent them by day. Apart from its great importance in ritual life,

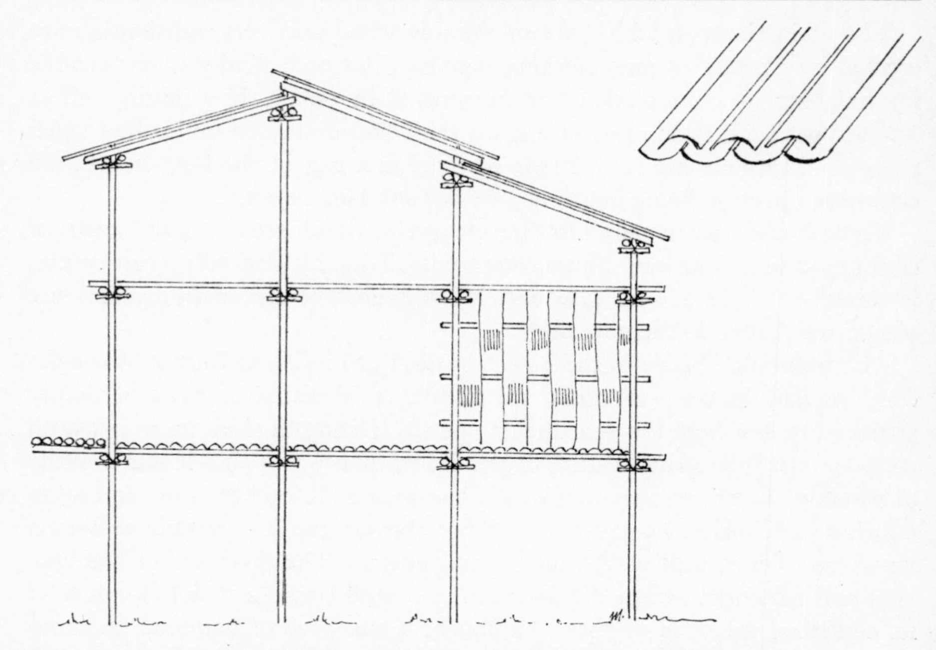

Fig 7 The traditional house of the Karen of the Pegu Hills, Burma.
Transverse section of a family apartment and the central passage; another apartment opens on to the passage from the left. All the structural parts shown are of bamboo, either entire, split, or flattened.
After Marshall, H. I., *The Karen People of Burma*. 1922, Ohio, p. 65.

the institution has very practical advantages. It provides a place where visiting males can be entertained, at the charge of the community, where they will be under watch if there is suspicion of treachery, and where they will not be in too close contact with the local girls. Above all, it ensures that the fighting men are together, with their weapons handy, in case of surprise attack.

The interpretation of house forms

There are many instances of houses similar in size and plan but serving different purposes. The Tifalmin, in the mountains at the sources of the Sepik and Fly Rivers in New Guinea, have men's houses, family houses and an ancestor-cult house, identical in size and construction. All are 12–14 ft square. The floor is raised about 1–$1\frac{1}{2}$ ft above ground level, and in the centre is a clay hearth. The sides are of light poles or split timbers, driven a few inches into the ground, and the roof is supported by these wall poles; there are no centre posts. Since men must not sleep in their family houses, every hamlet must have at least one men's house. Hamlets vary in size from about ten houses to two. Two is the minimum number, since even if there is only one family there must be a separate men's house. Each hamlet also has at least one women's house to which women withdraw periodically and at child-birth; these are usually low huts with an earth floor. Garden houses are also built. Sometimes these are crude sheds to provide shelter and shade during the day, but in the more distant gardens they are proper houses which may be inhabited for weeks at a time.

Instances of houses of different form used by the same community at the same time for the same purpose, where European influence has demonstrably not been a factor, are more difficult to find. The Kukukuku of New Guinea provide an example. Their house is typically circular and has two, concentric, walls. The entrance to the inner living space is at the opposite side of the circle to the house entrance, so that an intruder has to crawl half-way round the house under the inspection of the inmates before gaining access to the interior: a wise precaution for this exceptionally ferocious and blood-thirsty people. Blackwood illustrates a Kukukuku village with both circular and rectangular houses; it seems that in this case the village was so situated as not to be vulnerable to raiding, and some people felt they could dispense with the inner passage[34]. Among the Waka of the Western Highlands District (New Guinea) men's houses can be either rectangular or oval. Nearby are the women's houses, enclosed by a fence. Sometimes there is another men's house within the fence near the women's house, for use on cold or wet nights when the man does not wish to go to the main men's house[35]. The Laiap Enga of the same district have two sorts of men's houses: one resembles those of the Mae Enga, which are rectangular with an apsidal end, the other is circular. They also have some women's houses built in the style of the Hagen-Mendi tribes, but most are in the Mae Enga style. One Mae Enga clan had nineteen men's houses and forty-five women's houses[36].

House styles often change very rapidly under Western influence. When different styles co-exist, and Western influence is unlikely, it seems usually to be in a border area, such as that of the Laiap Enga—an area of intensive cultural contact. Coexistence of two forms is probably temporary; one is in the process of superseding the other.

In some contexts it would be legitimate and reasonably accurate to estimate a population by estimating the average size of a family and multiplying by the number of houses. In other contexts this could be very misleading. Take the case of the Tifalmin. One hamlet had seven family houses and one men's house. Another had six family houses, three men's houses and an ancestor-cult house. The different sorts of houses have identical plans and therefore would be indistinguishable archaeologically, so an estimate based on the number of houses would considerably exceed the true figure. In the case of the two-house hamlets—family house and men's house—the error would be 100%. A further point is that hamlet sites are changed every few years: until the Administration introduced them there were no latrines, and the people say that after a short time pigs and babies began to do badly, giving this rather than exhaustion of land as a reason for moving. The move was sometimes a matter of only a few hundred yards. Some hamlets were sited where erosion is very rapid and so disappear quite quickly[37].

Population estimates based on the number of houses of the Waka, where men have a choice of two men's houses, could be equally misleading. So could estimates of Kiwai population, where the men's house and the communal house are similar. In the villages of northern Bougainville, in the Solomons, men's houses and family houses are said by Blackwood to be made in exactly the same way, though those with local knowledge realize that the family houses are carefully aligned whereas the men's houses stand apart from the alignment[38].

Notes

1 Several of the peoples mentioned here are discussed, and their dwellings described and illustrated, in Forde, C. D. (1963), *Habitat, Economy and Society*, London, where extensive reference lists will be found. Since this (a reprint of a work first published in 1934) is readily available I refer to it when possible, adding more recent references if necessary.

In selecting the references listed in succeeding footnotes I have given preference to works in English and to those likely to be accessible.

2 Radcliffe-Brown, A. R. (1933). *The Andaman Islanders*. Cambridge. pp. 30–6, 409–15, pls. VI–VIII.

3 Forde, C. D. (1963). *op. cit.* pp. 72–6.

4 The Hadza of Tanzania are not usually compelled to move by food shortages. See Woodburn, J. C., this volume, p. 202.

5 For Eskimo dwellings see Forde, C. D. (1963). *op. cit.* ch. VIII; and Birket-Smith, K. (1959). *The Eskimos*. London. pp. 120–6.

6 Many types of Siberian dwellings are described and illustrated in Levina, M. G. and Potapova, L. P. (1961). *Istoriko-etnograficheskу Atlas Sibiri*. Moscow and Leningrad. The numerous resemblances between Siberian and North American dwellings are noteworthy.

7 Forde, C. D. (1963). *op. cit.* pp. 60–2.

8 Levina, M. G. and Potapova, L. P. (1961). *op. cit.* pp. 161–223 passim; Forde, C. D. (1963). *op. cit.* p. 104.

9 Collinder, B. (1949). *The Lapps*. Princeton. Ch. VI; Birket-Smith, K. (1960). *Primitive Man and His Ways*. London. pp. 126–30.

10 Forde, C. D. (1963). *op. cit.* ch. XVI; Levina, M. G. and Potapova, L. P. (1961). *op. cit.* p. 215.

11 One type of Lapp wood-and-turf house is supported by a frame almost exactly similar to that of the tent mentioned above. See Birket-Smith, K. (1960). *op. cit.* pp. 127, 129.

12 Forde, C. D. (1963). *op. cit.* p. 334.

13 Forde, C. D. (1963). *op. cit.* pp. 317–8; Musil, A. (1928). *The Manners and Customs of the Rwala Bedouins*. New York. Ch. IV.

14 Bower, U. G. (1950). *Naga Path*. London. pp. 149–53.

15 Walton, J. (1951). Corbelled stone huts in southern Africa, *Man*, **51**, No. 82; Walton, J. (1956). *African Village*. Pretoria.

16 Walton, J. (1969). Megalithic building survivals, *in* Jenkins, G. (ed.). *Studies in Folk Life: Essays in Honour of Iorwerth C. Peate*. London.

17 Netting, R. (1968). Kofyar building in mud and stone, *Expedition: The Bull. Univ. Mus. Univ. Pennsylvania*, **10**.

18 As by the Hausa of northern Nigeria. See Schwerdtfeger, F. W., this volume, p. 553.
19 Thesiger, W. (1964). *The Marsh Arabs*. London. Especially pp. 215–7 and pls. 93–104.
20 Walton, J. (1956). *ibid.*; Levina, M. G. and Potapova, L. P. (1961). *op. cit.* pp. 198, 221.
21 Walton, J. (1956). *ibid.*
22 Richards, A. I. (1950). Huts and hut-building among the Bemba, *Man*, **50**, Nos. 134, 162.
23 Many examples are illustrated in Levina, M. G. and Potapova, L. P. (1961). *op. cit.* pp. 174–82, 216–9.
24 Seligmann, C. G. (1910). *The Melanesians of British New Guinea*. Cambridge. pls. LXXVI, LXXVII.
25 See, e.g. Roth, G. K. (1954). Housebuilding in Fiji, *J. Roy. Anthrop. Inst.*, **84**; Hiroa, Te Rangi (P. H. Buck) (1930). *Samoan Material Culture*. (B. P. Bishop Mus. Bull. 75), Honolulu.
26 Landtman, G. (1927). *The Kiwai Papuans of British New Guinea*. London. Ch. 2.
27 Williams, F. E. (1924). *The Natives of the Purari Delta*. (Territory of Papua, Anthropology Report No. 5), Port Moresby. pp. 28–31.
28 For a general survey with further references see Fraser, D. (1968). *Village Planning in the Primitive World*. New York.
29 Turnbull, C. M. (1965). *Wayward Servants*. London. pp. 102–5. For a similar example from the Hadza, see Woodburn, J. C., this volume, *ibid.*
30 Forde, C. D. (1963). *op. cit.* pp. 133–5.
31 The Borneo long-house varies in detail from tribe to tribe. See, e.g. Hose, C. and McDougall, W. (1912). *The Pagan Tribes of Borneo*. I. London. pp. 50–5; Roth, H. L. (1896). *The Natives of Sarawak and British North Borneo*. II. London. ch. XVI.
32 Marshall, H. I. (1922). *The Karen People of Burma*. Ohio. ch. VIII.
33 See, e.g. Mills, J. P. (1926). *The Ao Nagas*. London (indexed under *Morung*).
34 Blackwood, B. (1950). *The Technology of a Modern Stone Age People in New Guinea*. (Occasional Papers in Technology, 3. Pitt Rivers Museum) Oxford. (frontispiece); Blackwood, B. (1939). Life on the Upper Watut, New Guinea. *Geogr. J.*, **94**, p. 18. Also personal communication.
35 Meggitt, M. J. (1956). The valleys of the Upper Wage and Lai Rivers, Western Highlands, New Guinea, *Oceania*, **27**.
36 Meggitt, M. J. (1957). House building among the Mae Enga, Western Highlands, Territory of New Guinea, *Oceania*, **27**.
37 I have discussed the Tifalmin population evidence in more detail in Cranstone, B. A. L. (In press). The Tifalmin: a Neolithic people in New Guinea, *World Archaeol.*
38 Blackwood, B. (1935). *Both Sides of Buka Passage*. Oxford. pp. 382–8.

18 [illegible] Northern Rhodesia [illegible]
19 [illegible] (1964) [illegible]
20 [illegible] (1964) [illegible]
21 [illegible] (1962) ibid.
22 [illegible] (1960) Huts and [illegible] *Nature* [illegible]
23 [illegible]
24 [illegible]
25 [illegible]
26 [illegible]
27 [illegible]
28 For a general survey, with further references [illegible] (1953) [illegible]
29 Turnbull, C. M. [illegible] similar example from [illegible]
30 Forde, C. D. [illegible]
31 The [illegible]
32 [illegible]
33 Marshall [illegible]
34 Blackwood, B. [illegible] *The Technology of a Modern Stone Age People in New Guinea*. Occasional Papers on Technology, [illegible] Pitt Rivers Museum, Oxford. [illegible]
35 Meggitt [?], [illegible]
36 [illegible] Highlands [illegible] New Guinea [illegible]
37 I have discussed the [illegible]
38 Blackwood, B. [illegible] Oxford [illegible]

GÜNHAN DANIŞMAN

The architectural development of settlements in Anatolia

Introduction

As a result of ever increasing evidence about Anatolian prehistory and its earlier archaeological periods, it is becoming possible to indicate the development of an indigenous architecture in Anatolia from the earliest traceable building activity onwards[1]. It is also becoming increasingly clear that this architecture had from time to time an influential effect on the development of the architecture of other centres in the ancient Near East.

The beginnings

The excavated evidence of the first building activities in Anatolia indicates that the beginnings of architecture took place earlier than the domestication and exploitation of plants and animals. At a period which corresponds to the emergence of agriculture in this part of the Old World the Anatolian building activity illustrates a high degree of proficiency. All excavated buildings from this period have rectangular plans and there is no doubt that the rectangle is an "advanced" form in architecture[2]. Furthermore, the walls were constructed of sun-dried rectangular mud-bricks which were usually reinforced with timber. The use of these building materials definitely points to an advanced stage in the development of earth-based buildings[3]. Flat timber roofs covered with packed clay for which there is good evidence would probably have served as additional living space on the buildings, an architectural sophistication that is still used in present-day rural Anatolia. The raised hearths with curbs and kilns with separate fire boxes that were provided within residential buildings also indicate a developed system of cooking and heating, as well as artificial lighting. Thus, it seems necessary, at least in Anatolia, to search for some other explanation than the emergence of agriculture in order to understand the reasons behind man's impetus to create permanent settlements.

Structural and non-structural detailing in Anatolian architecture

Meticulous attention to detailing has characterized Anatolian architecture at all periods. This fact becomes clearer when the methods of construction of the various structural and non-structural elements of buildings are observed individually.

Foundations

Stone foundations are unknown in the earliest excavated buildings in Anatolia. Instead, several courses of mud-brick sunk below the floor-level were used for foundations at certain sites, e.g. at Çatal Hüyük[4]. As the need for more substantial foundations arose, especially from the fifth millennium B.C. onwards, the use of stone in the construction of the foundations became the general practice (Fig. 1a). As the system of construction became more complicated, especially at the end of the fourth millennium B.C. and the beginning of the third, such techniques as laying tree-trunks longitudinally beneath the walls and holding them in position with stones, and the use of undressed stone foundations reinforced at intervals with timbers were developed, e.g. at the site of Beycesultan in south-western Anatolia[5]. At this same site a further step towards greater stability was taken by laying tree-trunks transversely to the direction of the walls, projecting up to 60 cm beyond the wall faces (Fig. 1d). The normal practice would have been to fill the spaces in these foundation trenches with rubble up to the pavement level; instead, however, they were roofed over with small joists and brushwood, thus creating small tunnels around the bases of the walls, which communicated with each other beneath the lintels of the doors. It has been suggested that the purpose of these passages was for heating or ventilation[6].

Walls

From the beginning, the walls of all buildings were constructed of sun-dried rectangular mud-bricks that were hand-shaped or moulded. The wall faces were covered with mud-plaster, and were usually whitewashed. From very early periods, the mud-brick walls were reinforced by the incorporation of timber beams and posts within their structure. This practice was improved as the methods of building developed. For example, at the site of Beycesultan, the earlier practice of strengthening the walls by inserting rows of runner-beams at regular intervals held in position by cross ties (Fig. 1c) was later elaborated by means of vertical posts, which extended from the foundations to the roof. At the contemporary site of Kültepe in central Anatolia[7], similar means of reinforcing walls with timber insertions have been observed, but

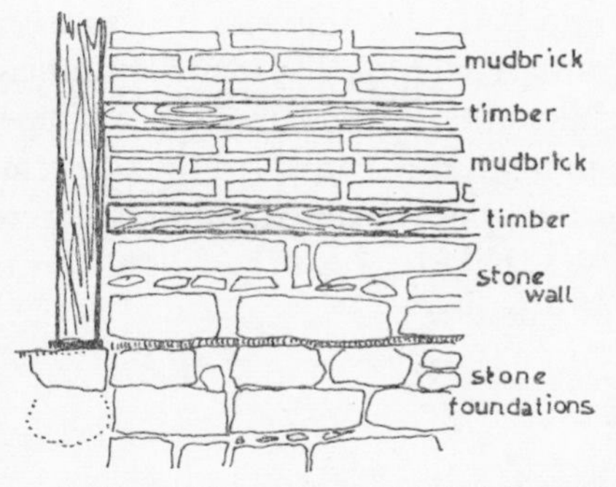

a. WALL DETAIL FROM MEGARON II A AT TROY SHEWING TIMBER REINFORCINGS WITHIN THE MASONRY (after NAUMANN, R., 1955, *Architektur Kleinasiens - von ihren Anfängen bis zum ende der Hethitishen Zeit*, Tübingen)

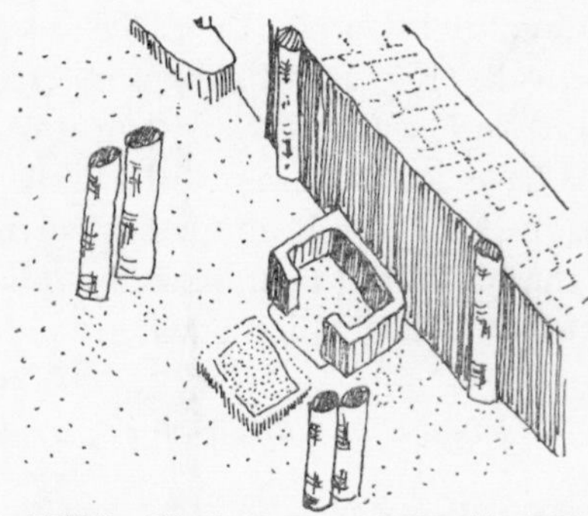

b. DETAIL OF HACILAR LEVEL VI ARCHITECTURE SHEWING THE INCORPORATION OF TIMBER COLUMNS (reconstruction after MELLAART, J., 1961, Excavations at Hacılar, 1960, fourth preliminary report, *Anatolian Studies*, 11, Fig. 3, p. 44)

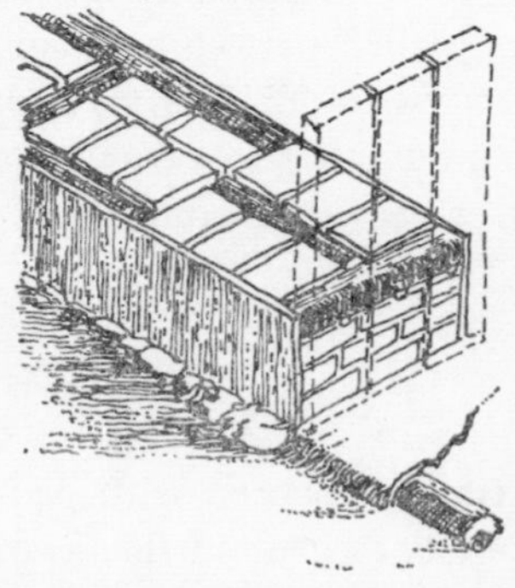

c. DETAIL OF PORCH WALL IN MEGARON A, LEVEL IX AT BEYCESULTAN (after LLOYD, S. & MELLAART, J., 1962, *Beycesultan, I*, London, Fig. 23, p. 60)

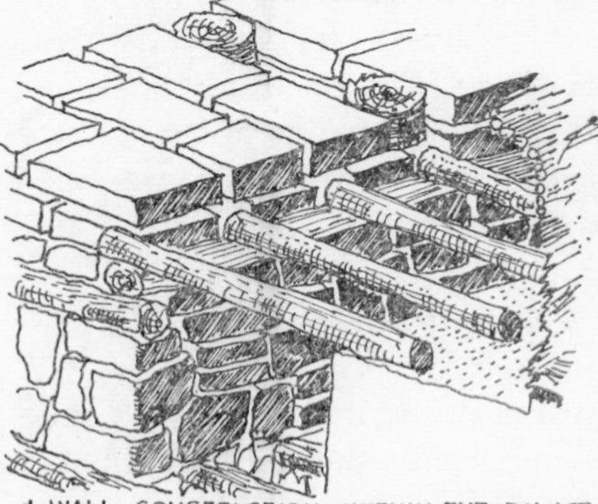

d. WALL CONSTRUCTION WITHIN THE PALACE AT BEYCESULTAN (after LLOYD, S. & MELLAART, J., 1965, *Beycesultan, II*, Fig. A. 10, p. 20)

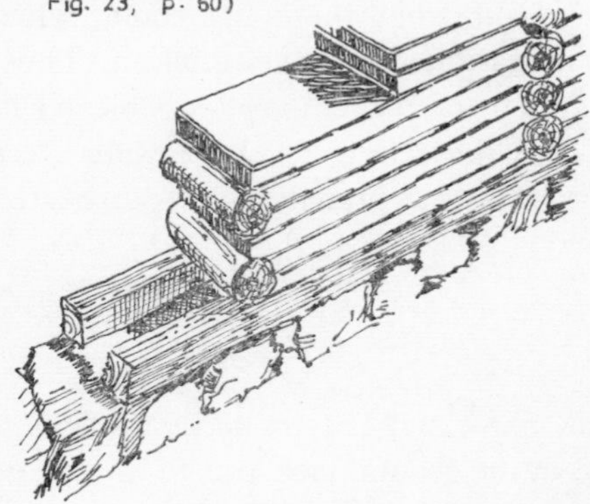

e. WALL CONSTRUCTION AT KÜLTEPE (after ÖZGÜÇ, T., 1959, *Kültepe - Kaniş*, Ankara, Fig. 21, p. 22)

f. USE OF AUXILIARY TIMBER POSTS AT KÜLTEPE (after ÖZGÜÇ, T., 1959, *Kültepe-Kaniş*, Ankara, Fig. 20, p. 21)

Fig. 1 Structural detailing in Anatolia architecture

in this case by piling logs one on top of the other from the foundations to the ceiling, giving a similar effect to vertical posts used within the structures of the walls (Fig. 1e). Thus, most excavators have identified systems of timber-framework within the structures of buildings from Çatal Hüyük to Kültepe, and have accepted mud-brick walls as infilling material in this frame system. Claims have also been made that the elasticity thus obtained in the structure of the buildings, in spite of the inherent fire hazards, was to resist the effects of earthquakes which occur frequently in Anatolia[8].

Columns and upper-storey structures

At Kültepe examples have been recorded of posts, not this time embedded within the thickness of the wall, standing on small stone bases against the wall surfaces (Fig. 1f). The use of free-standing timber columns interpreted as supporting ceilings and upper storeys has been well attested in other sites such as contemporary Beycesultan, and in earlier periods at Can Hasan[9] and Hacilar[10] (Fig. 1b). Additional evidence of fallen débris from upper floors and carbonized remains of timber structures have led to various claims for the existence of lightly built upper floors over heavier ground floors at most early Anatolian sites.

The construction of roofs

In most early Anatolian sites flat roofs of timber covered with packed clay were supported either on stout mud-brick walls of ground floors, or on top of main beams in one direction and smaller beams in the other, carried by timber uprights (Figs. 2b and 2c). However, lack of direct evidence regarding the means by which rain-water was led off from these flat roofs, when cellular construction was employed, has necessitated some speculation. Thus, at the site of Çatal Hüyük a peculiar system of corbelling was achieved in timber:—above each horizontal beam near the ceiling level the walls were brought forward up to 23 cm. This has been interpreted as a sophisticated means of rain-water disposal (Fig. 2a).

Doors and windows

In the earliest periods, especially at sites like Çatal Hüyük and Can Hasan, there was no provision for door-openings at the ground level due to the agglutinative planning. It has generally been suggested that entrance openings were left at roof level and that descent into the living area was made via wooden stairs. However, during the next phase of building activity, i.e. during the Chalcolithic and Early Bronze Ages, doors were provided for the buildings. Timber reveal-linings, thresholds and sills, as well as parts of pivoted doors have been unearthed at various sites (Figs. 2d and 2e). The excavated

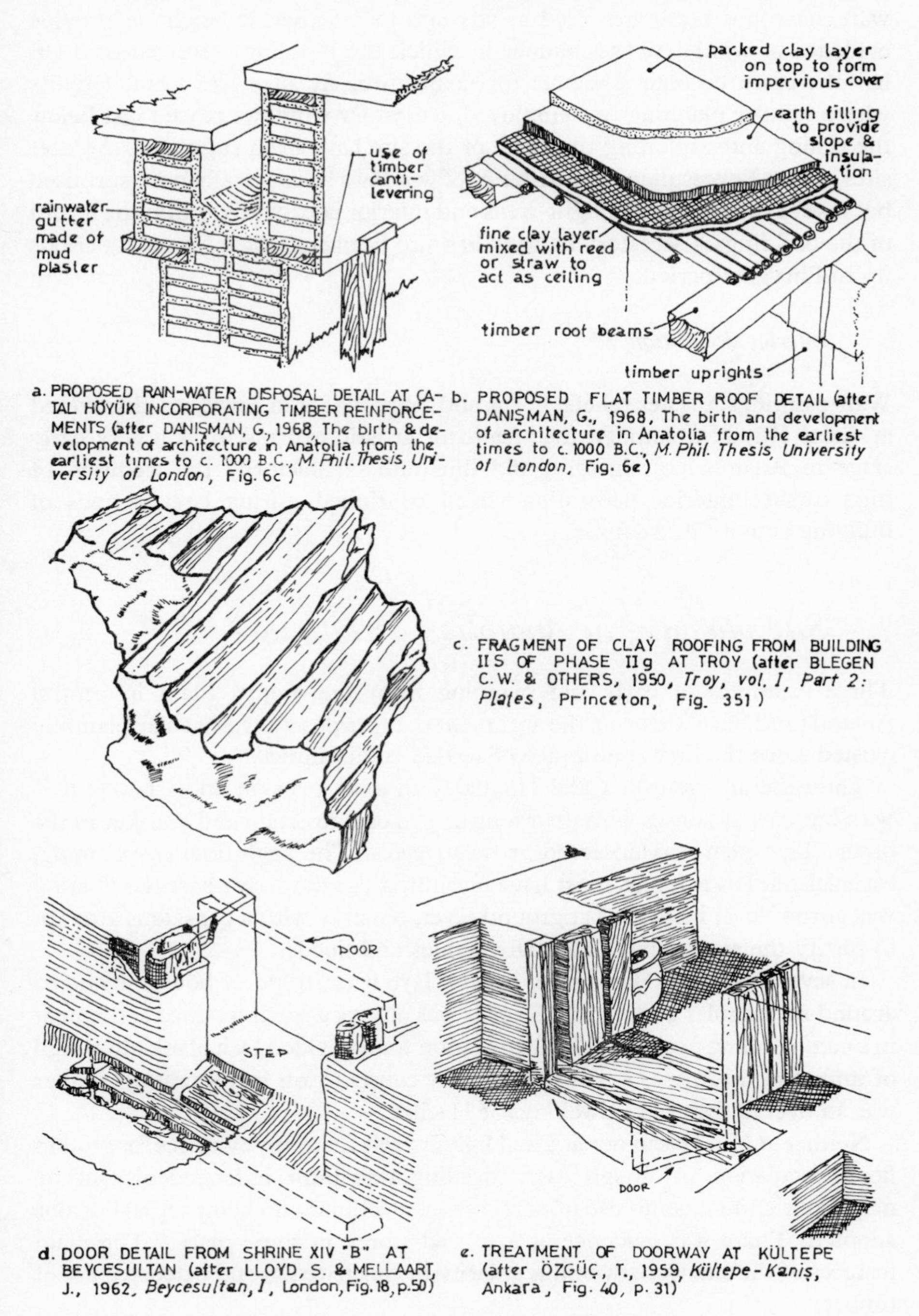

a. PROPOSED RAIN-WATER DISPOSAL DETAIL AT ÇATAL HÖYÜK INCORPORATING TIMBER REINFORCEMENTS (after DANIŞMAN, G., 1968, The birth & development of architecture in Anatolia from the earliest times to c. 1000 B.C., *M. Phil. Thesis, University of London*, Fig. 6c)

b. PROPOSED FLAT TIMBER ROOF DETAIL (after DANIŞMAN, G., 1968, The birth and development of architecture in Anatolia from the earliest times to c. 1000 B.C., *M. Phil. Thesis, University of London*, Fig. 6e)

c. FRAGMENT OF CLAY ROOFING FROM BUILDING IIS OF PHASE IIg AT TROY (after BLEGEN C.W. & OTHERS, 1950, *Troy, vol. I, Part 2: Plates*, Princeton, Fig. 351)

d. DOOR DETAIL FROM SHRINE XIV "B" AT BEYCESULTAN (after LLOYD, S. & MELLAART, J., 1962, *Beycesultan, I*, London, Fig. 18, p. 50)

e. TREATMENT OF DOORWAY AT KÜLTEPE (after ÖZGÜÇ, T., 1959, *Kültepe-Kaniş*, Ankara, Fig. 40, p. 31)

Fig 2 Nonstructural detailing in Anatolian architecture

walls have not been preserved at any site to a sufficient height to provide evidence of windows; the manner in which the buildings received daylight has therefore become a matter for conjecture. At sites like Çatal Hüyük, where cellular planning was employed, clerestory windows placed just below the ceiling and exploiting the slope of the site have been suggested. At later sites, e.g. at Beycesultan, the existence of windows has naturally been surmised because of the presence of light-wells and interior courtyards within the layout of the buildings. Clerestory lighting has also been suggested for most periods up to Phrygian period.

Interior decoration

Wall-paintings, plaster reliefs, horns and bucrania set in rows, etc. discovered at Çatal Hüyük exemplify the wealth of artistic expression at a very early stage in Anatolian prehistory, and illustrate satisfactorily the tradition of high quality interior decoration which continued during later periods of building activity in Anatolia.

Site planning in Anatolian architecture

Three examples of early site planning from the Konya Plain in central Anatolia suffice to illustrate the argument that purposeful vernacular planning existed since the first appearance of settled communities.

The general layout at Çatal Hüyük is an axial plan on an east-west axis with one row of houses with their long axis in one direction and one row in the other. This plan was observed at most levels. The agglutinative planning, especially at lower levels, must have meant that, in the early phases circulation was at roof level instead of at ground level, possibly with a conscious attempt to fortify the site against potential enemies and floods.

In several levels at Hacilar the general layout consisted of houses arranged around rectangular courtyards. Hacilar was always a small fortified settlement in a commanding position. The fortification was provided by a blank outer wall of immense thickness. Accurate and neat construction incorporating timber was an overall feature of dwellings at Hacilar.

Neither at Can Hasan nor at Çatal Hüyük, was there any evidence for ground floor circulation. Although each dwelling was built independently of its neighbour and made no use of party walls, a cellular and compact layout was adopted. There was evidence of a second storey in some places. The main features of the site were overall regularity in construction and extensive use of timber.

Conclusions

The review of this evidence from Anatolia suggests that from its first conception, architecture in Anatolia, whether in the layout of complete settlements or in the development of structural and non-structural details (for example, timber reinforcement of walls and foundations), reflects planned growth. The settlements were built on an artificial grid, or were based on preconceived frameworks or function-oriented plans. This, however, did not preclude the organic growth of the settlements, but rather facilitated their growth and development.

Notes

1 This theme has been treated summarily by Lloyd, S. (1963). Bronze Age architecture of Anatolia, *Proc. Brit. Acad.*, **49**, pp. 153–76. For a more comprehensive discussion on this subject refer to Danişman, G. (1968). The birth and development of architecture in Anatolia from the earliest times to 1000 B.C., *M. Phil. Thesis, Univ. of London.*

2 Lloyd, S. (1963). *op. cit.* p. 154.

3 Danişman, G. (1968). *op. cit.* pp. 49–56, in which the origins of building activities in Anatolia and the construction of mud-brick buildings are discussed at length.

4 Mellaart, J. (1967). *Çatal Hüyük, a Neolithic Town in Anatolia.* London. p.55.

5 Lloyd, S. and Mellaart, J. (1965). *Beycesultan*, II. London. pp. 62–63

6 e.g. Mellaart, J. (1967). *ibid.*

7 Özgüç, T. (1959). *Kültepe-Kanis.* Ankara.

8 Lloyd, S. and Mellaart, J. (1965). *op. cit.* p. 66.

9 French, D. (1961–9). Preliminary excavation reports, *Anatolian Studies*, **11** to **18**.

10 Mellaart, J. (1970). *Hacilar*. Edinburgh. Also Mellaart, J. (1957–60). Reports of excavations from Hacilar, *Anatolian Studies*, 7 to **10**.

MARY DOUGLAS

Symbolic orders in the use of domestic space

It is not necessary to warn archaeologists against naiveté in the sociological dimension. Nearly 100 years have passed since Morgan wrote *Houses and House-life of the American Aborigines*[1]. To have been the inventor of a vanished culture of High-Bank Pueblos in Ohio, whose spacious courtyards and strategically well-designed houses now turn out to be burial mounds, can stand for any archaeologist's nightmare (Figs. 1 and 2). But alas for our wish for a sounder basis of sociological interpretation—the lesson remains cautionary and negative.

The anthropologist's approach to symbolic principles in the ordering of domestic architecture derives squarely from the volumes of *L'année sociologique*, particularly *De quelques formes primitives de classification* (1901–2) by Durkheim and Mauss[2] and the *Essai sur les variations saisonnières des Sociétés Eskimos* by Mauss and Beuchat[3]. The latter especially is an explicit attack on

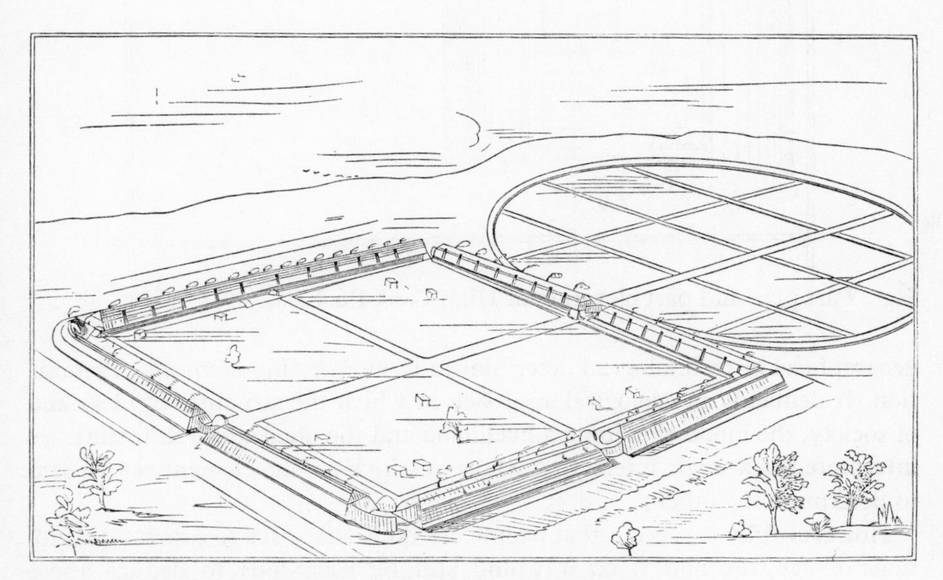

Fig 1 Morgan's reconstruction of High Bank Pueblos.

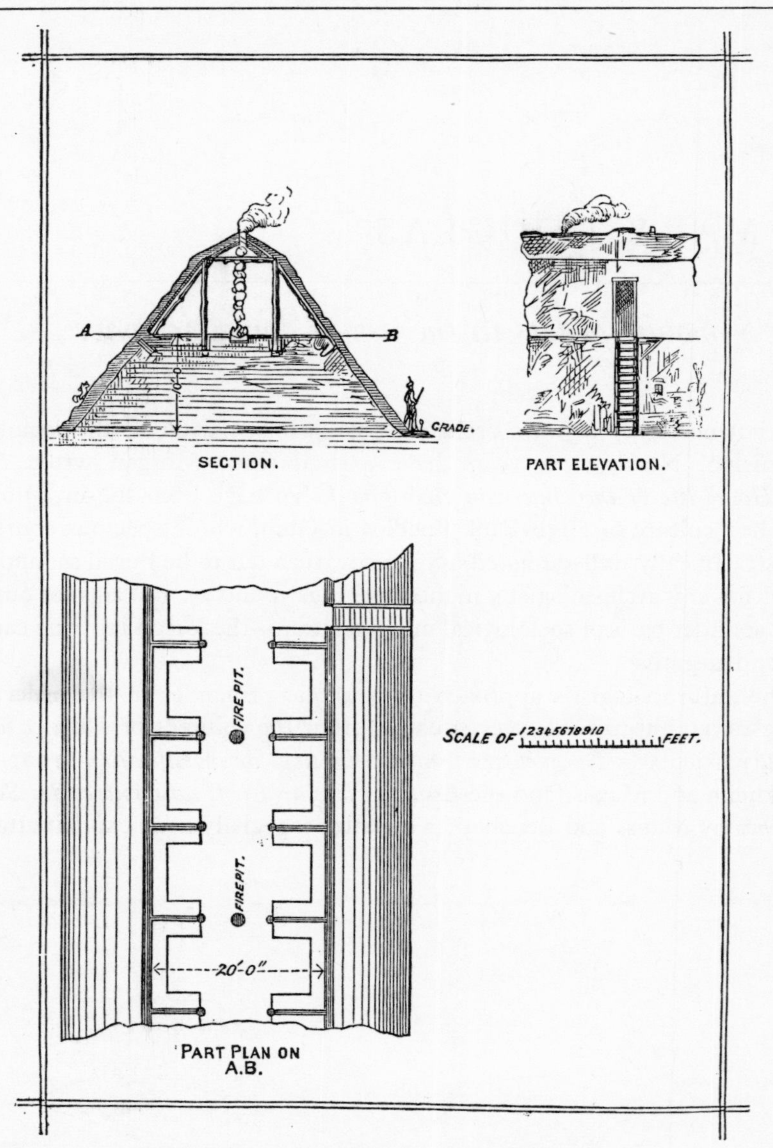

Fig 2 Part plan and part elevation of High Bank Pueblos.

geographical or technological determinism in interpreting domestic organization. It demands an ecological approach in which the structure of ideas and of society, the mode of gaining a livelihood and the domestic architecture are interpreted as a single interacting whole in which no one element can be said to determine the others.

However fully we accept that people use domestic space to express distinctions of age, sex and rank, it would still be hazardous to deduce these symbolic orders from the material remains alone. No material remains would

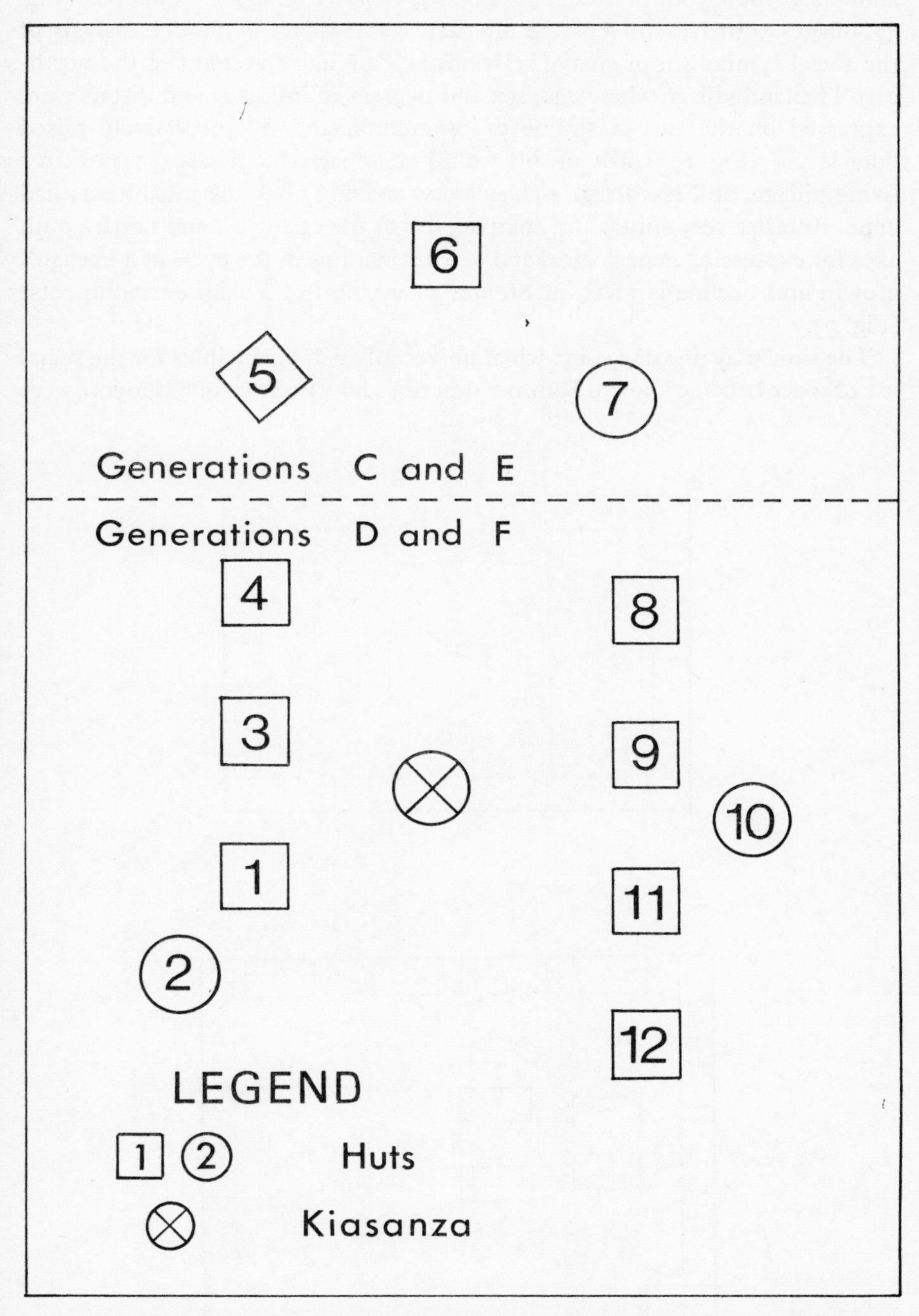

Fig 3 Central African (Kaonde) village plan showing separation of lineages and generations.

show how the layout of Kaonde[4] (Fig. 3), Lele[5] (Fig. 4), or Ndembu[6] (Fig. 5) villages contrives to separate alternate generations. A model analysis of the social symbolism of spatial relations is Tambiah's account of the north-east Thailand village where age, sex and degrees of intimacy and distance are expressed on the east–west line of orientation and by successively raised floor levels[7] (Fig. 6). Little of this would be apparent to a casual visitor to a living village, still less to an archaeologist arriving after the inhabitants had gone. Another very subtle and complex use of the east–west and north–south axes for expressing genealogical and age distinctions in the cycle of a lineage's growth and decline is given in Stenning's account of Fulani encampments[8] (Fig. 7).

The time may one day come when universally valid principles for the social use of space become known. Sommer describes how twenty-four students were

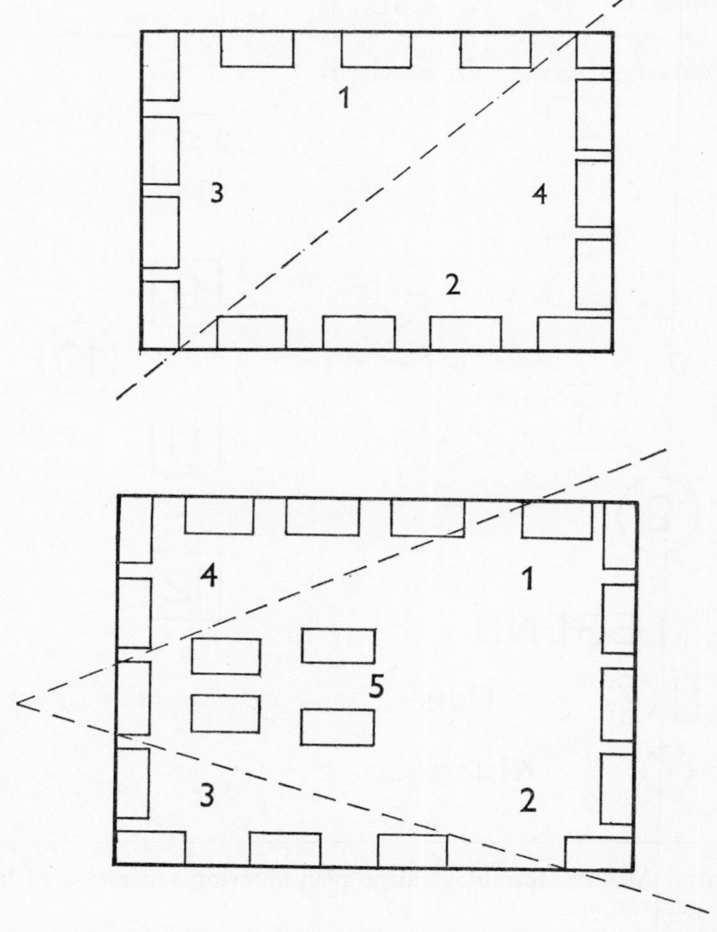

Fig 4 Central African (Lele) village plan showing alternation of generations.

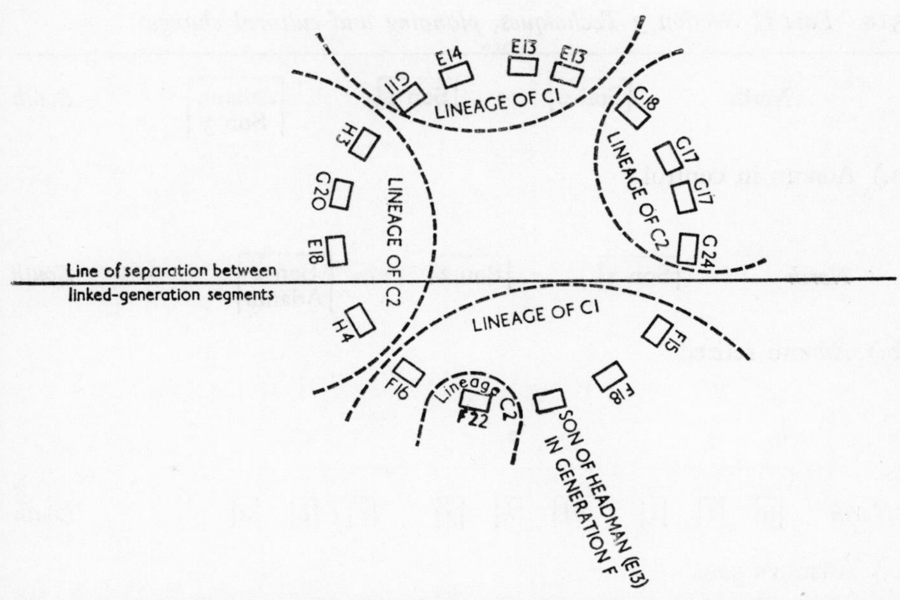

Fig 5 Central African (Ndembu) village plan showing separation of generations.

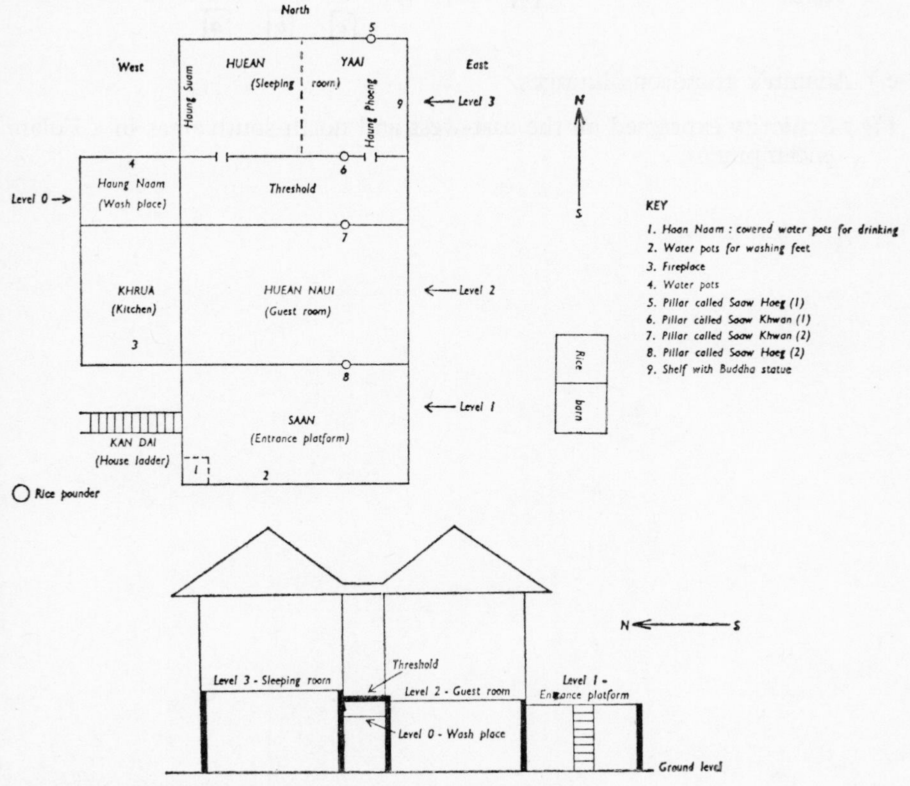

Fig 6 Elevation and plan of N.E. Thailand village.

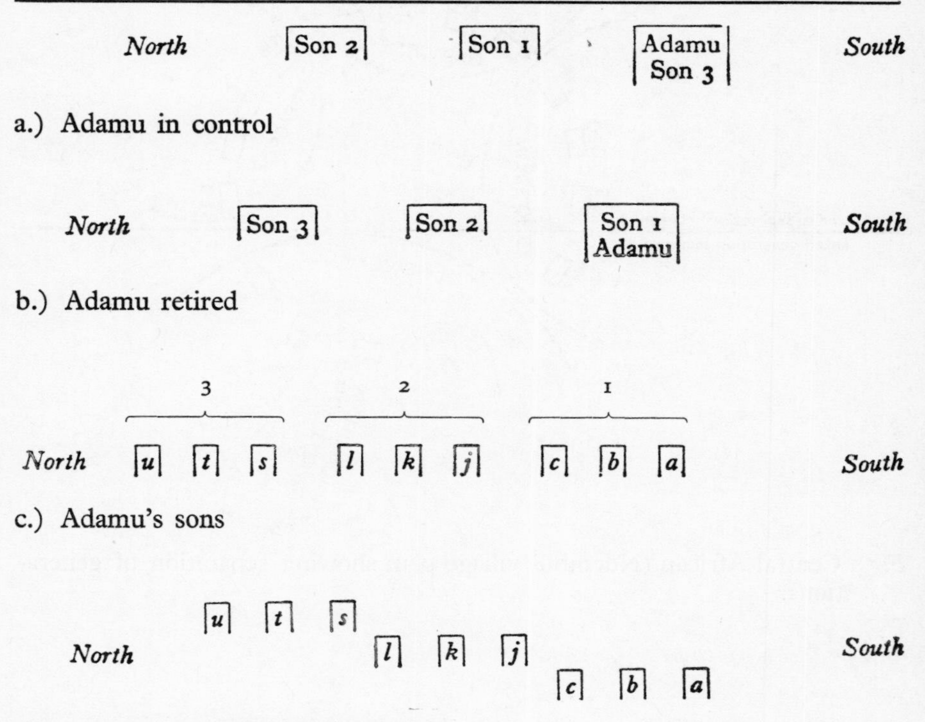

Fig 7 Seniority expressed on the east-west and north-south areas in a Fulani encampment.

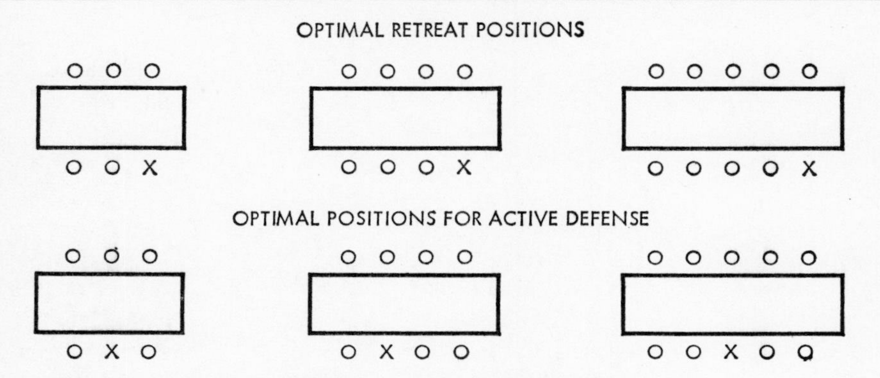

Fig 8 Students' choice of 'optimal retreat' positions.

asked to choose positions on a long table best to secure "freedom from the distractions of other people". The "optimal retreat position" turned out to be the far corner[9] (Fig. 8). This recalls the ranking of families in the Nootkan Indians' longhouses on the coast of Vancouver. According to Drucker[10] the two highest ranking families take the two corners furthest from the door and the lesser families fill in the interstices.

It is not safe to assume that relative size is an index of wealth. Gellner reports a Bedouin group which combined relatively greater wealth with relatively lower status, and which adopted an unnecessarily modest style of building to disguise its affluence[11]. Even when able to question informants, we do not always get straightforward replies. Lévi-Strauss has recalled the two different designs of village layout given to Radin by Winnebago Indians[12] (Fig. 9) and argues that a consistently different pattern is seen by members of different phratries. For a last example, take Hill's study of Ghanaian cocoa farms[13] (Fig. 10). Aerial surveys and ground investigation revealed two patterns of farm-holding, one in long strips, one in a mosaic of squares (Fig 11).

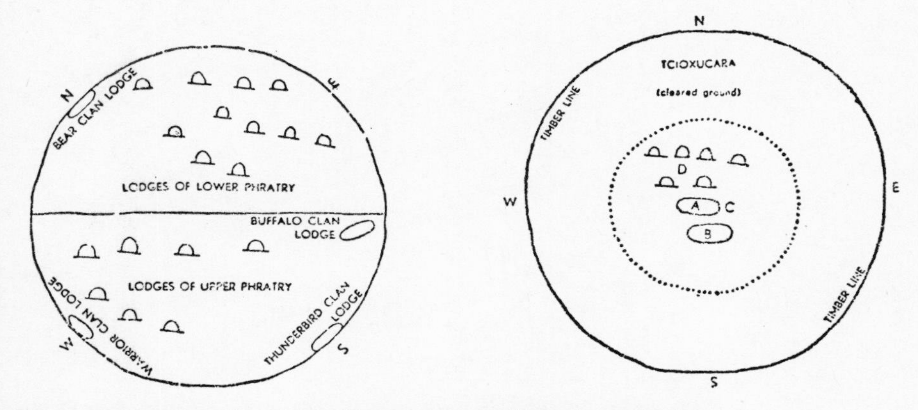

Fig 9 Alternative views of village layout, Winnebago Indians.

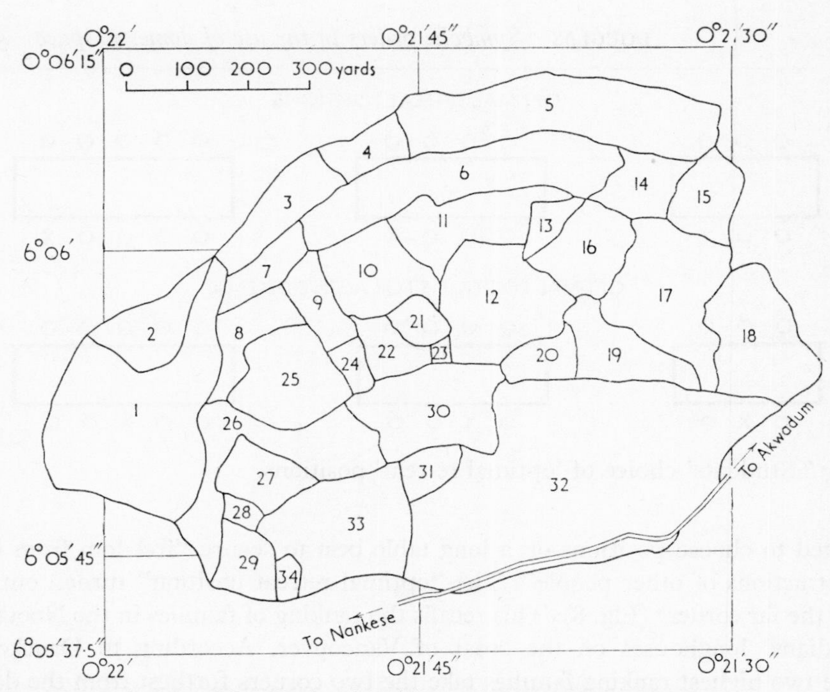

Fig 10 Boundaries of family land holdings (Akwapim, Ghana).

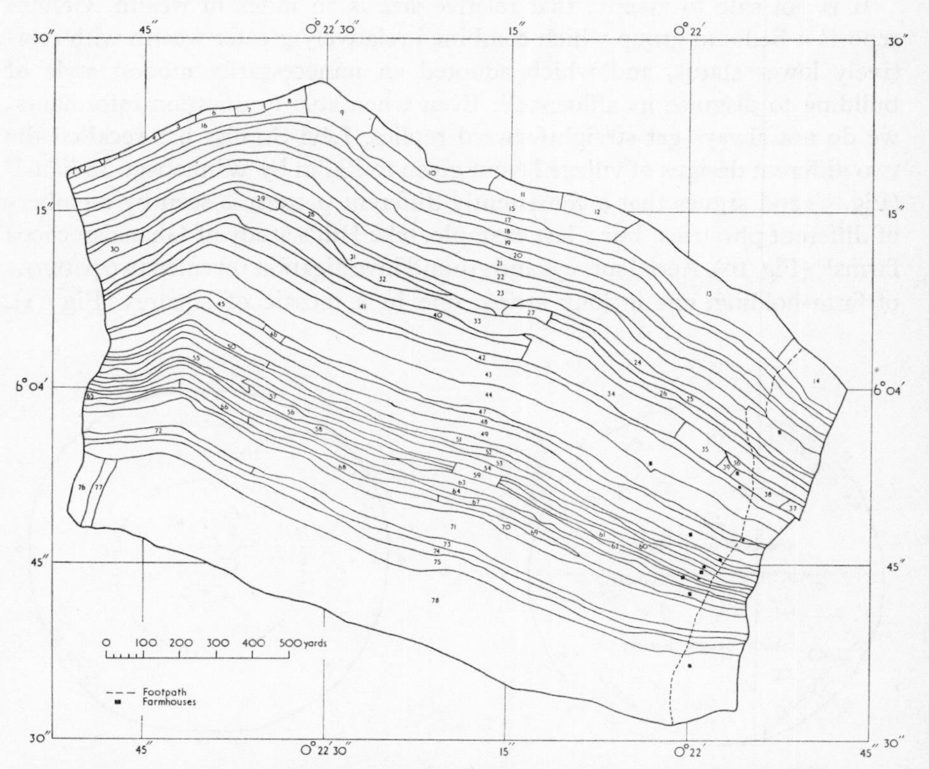

Fig 11 Boundaries of company finance strips of land (Akwapim, Ghana).

Nothing in the system of cultivation, quality of soil, type of crops, nothing economic or physical explained it, only a difference in financing arrangements for the original purchase. As Dumont has powerfully argued for the case of India, the organization of thought and of social relations is imprinted on the landscape[14]. But, if only the physical aspect is susceptible of study, how to interpret this pattern would seem to pose an insoluble problem.

Notes

1 Morgan, L. H. (1881). Houses and house-life of the American Aborigines, *in* Bohannan, P. (ed.) *Contributions to North American Ethnology*, **4**. Republished 1965, *Classics in Anthropology*. Chicago.
2 Durkheim, E. and Mauss, M. (1903). De quelques formes primitives de classification, Paris, *L'année sociologique* 1901–2, **6**. *Primitive Classification*, trans. 1963 by Needham, R. London.
3 Mauss, M. and Beuchat, M. H. (1906). Essai sur les variations saisonnières des sociétés Eskimos, *L'année sociologique*, (1904–5), **9**.
4 Watson, W. (1954). The Kaonde village, *Human Problems in British Central Africa*, **15**.
5 Douglas, M. (1963). *The Lele of the Kasai*. London.
6 Turner, V. W. (1955). The spatial separation of generations in Ndembu village structure, *Africa*, **25**.
7 Tambiah, S. J. T. (1969). Animals are good to think and good to prohibit, *Ethnology*, **8**.
8 Stenning, D. (1959). *Savannah Nomads, a Study of the Wodaabe Pastoral Fulani*. London.
9 Sommer, R. (1969). *Personal Space, the Behavioural Basis of Design*. New Jersey.
10 Drucker, P. (1951). The Northern and Central Nootkan Tribes, *Smithsonian Institution, Bureau of American Ethnology, Bull.*, **144**, p. 71.
11 Gellner, E. (1969). *Saints of the Atlas*. London.
12 Lévi-Strauss, C. (1956). Les Organisations Dualistes Existent-ils ?, *Bijdragen tot de taal - land - en Volkenkunde*, **12**, Trans. in *Structural Anthropology*, 1963.
13 Hill, P. (1963). *The Migrant Cocoa-farmers of Southern Ghana, a Study in Rural Capitalism*. Cambridge.
14 Dumont, L. (1966). *Homo hierarchicus, essai sur le système des castes*. Paris.

H. W. M. HODGES

Domestic building materials and ancient settlements

This paper is concerned solely with domestic buildings. Its aim is to demonstrate the complex interplay between the characteristics of raw materials, technological skills and social factors. As I hope to demonstrate, the social uses to which buildings are put seem to have much more influence on the style of building adopted than do the building materials used.

The building materials available to early man can be conveniently classified as belonging to one of four groups: (1) light vegetable materials—reeds, rushes, wickerwork, bamboo, etc.; (2) timber, which must include some of the heavier bamboos; (3) stone; (4) clay, pisé and unfired mud-brick. Combinations of these materials can, of course, be used together. Most of these materials were available for use in prehistoric times, except in areas of desert, bare rock, true swamp or tundra, none of which were likely to attract early settled communities.

For the purpose of this paper we can treat each domestic structure as having three essential components:

(*a*) *Floors*. As a general rule floors were made either of compacted mud or clay or stone paved at a level approximating to that of the ground outside the building. When one examines modern ethnographic parallels, one finds that raised floors are used mainly for a limited number of reasons: (1) for ventilation in tropical zones; (2) to avoid the damage that can be done by termites; (3) to minimize the effects of flooding; and (4) to raise the building above the expected winter level of snow. Generally one finds such buildings set on wooden or stone piles, although solid stone plinths are also known[1]. Raised floors are, again generally, made of wood.

(*b*) *Walls*. The merits and disadvantages of the different building materials are given in Table 1. In this instance load-bearing refers particularly to the ability of a wall to withstand the lateral thrust of a pitched roof, while stability is of importance in areas where tectonic movements may be expected.

(*c*) *Roofs*. The properties of the materials are given in Table 2. Here load-bearing refers to the ability to withstand deep falls of snow or the effect of becoming saturated during heavy rainfall. The actual pitch given to a roof will, of course, be dictated by the nature of snow and rainfall on the one hand and the nature of prevailing winds on the other. Generally, the more violent the wind, the lower will be the pitch of the roof.

It is clear from these data that there is no one material which has an overwhelming advantage over all others in a wide geographical context; and viewed even on a narrow regional basis there must have remained a very real choice of building materials, the advantages and disadvantages of the one pretty well balancing those of the others.

	Survival	*Insulation*	*Working*	*Load-bearing*	*Stability*
Light vegetation	poor, prone to biological attack	moderate	easily cut, carried and manipulated	poor	excellent
Timber	moderate, prone to biological attack	moderate	requires heavy tools	moderate	good
Stone	excellent	poor, allows rising damp	difficult unless rock already fractured; laborious	excellent	poor
Mud	moderate, prone to weathering	good, but allows rising damp	easily shaped and manipulated	moderate	poor

Table 1 Materials used for wall building.

A further important set of considerations concerns the functions of the structures. Dwellings and food stores or silos need to be not only weatherproof but also well insulated; byres can be less effective in either of these respects. Inflammability in the building of kitchens and many workshops may be important, the latter also requiring adequate fenestration. The type of work to be carried out within a structure may dictate particular dimensions and these in turn may demand the use of particular materials. As I have already said, social factors may outweigh such physical considerations; thus, in the Trobriand Islands yam stores are built on piles with spaces left between the timber elements to allow ventilation. Pride, however, demands that the spaces should also be large enough to allow the size of the yams to be seen, the largest invariably being stacked against the walls of the store[2].

	Survival	*Insulation*	*Working*	*Load-bearing*	*Stability*
Pitched roofs—leaves or thatch	moderate	excellent	easy	moderate	moderate
Pitched roofs—shingle, slates	good	poor	easy	good	poor
Flat roofs—mud or bitumen	poor	good	easy	poor	good

Table 2 Materials used for roofing.

The ubiquitous mud-brick

Familiarity with early settlements of the Middle East leads one to the almost inescapable conclusion that, save in most unusual circumstances, mud and mud-brick were the only building materials employed for floors and walls, and that in all probability flat roofs were used from the very beginning of the Neolithic period. Indeed, the archaeological record makes it look almost as though the idea of mud-brick manufacture descended miraculously from heaven, and this alone demands some explanation. In effect one must ask whether it is reasonable to suppose that the development of settlements and the invention of mud-brick went hand-in-fist, or whether mud structures had not been in use long before the development of settled communities.

The latter point appears clear from the archaeological evidence. In the earliest settlements known the technique of building in mud-brick was already a sophisticated one. Bricks might be hand-modelled[3] or moulded[4] but in any area they conformed to a uniform pattern: they were made of a material or mixture of materials that did not crack unduly on drying; and applied mud surfaces were frequently burnished. All these factors suggest an existing familiarity with the use of mud structures. One should, therefore, examine other structures that could have been made of mud or mud-brick and which might conceivably have been used by communities living in a peripatetic manner. If one looks at the ethnographic material, one finds that there are indeed such structures. Not only are mud walls the common form of field boundary throughout areas such as the Punjab, where domestic mud-brick buildings are in use, but similar walls are also built by the nomads of Mongolia to create yards in which they pitch their yurts during winter[5], a conglomeration of these yards incidentally providing a plan of orderly "streets". In both Persia[6] and the Punjab, moreover, it is not uncommon to find stacks of foodstuffs, particularly fodder, encased in a layer of mud as a protection from the elements. Cylindrical clamps of this sort with conical roofs look deceptively like silos which are also found in the same regions. Whether structures such as these were in use before the development of settlements is a matter of pure conjecture, unlikely to be resolved save by pure chance, although a satisfactory method of grain storage must have been a prerequisite of settled agriculture.

It is not difficult to see why mud-brick became so popular. Cheap and easy to produce, it allowed house-building with a minimum of skilled labour. The material insulates excellently and is very easily repaired, despite the fact that it weathers badly. The poor load-bearing quality of mud-brick is perhaps its greatest practical disadvantage, so that to add a pitched roof to a mud-brick building requires a fairly substantial number of tie-beams to overcome lateral thrust.

Attempts to reconstruct early Neolithic dwellings, with few exceptions[7], envisage a flat roof. The supposed precursors of such houses, tent-like or yurt-like dwellings, have pitched or domed roofs; there is in fact no *a priori* reason why the walls discovered during excavations should have supported a roof at all. Thus, in the modern villages of the Yami of Botel-Tobago wooden single floor dwellings and wooden two-floor workshops are constructed within a network of substantial stone-walls the upper surfaces of which serve as a pavement. The heavy timber posts incorporated into the mud-brick walls of the surviving buildings at Çatal Hüyük[8] might thus well have supported either a pitched roof, or even conceivably a second timber-built storey. At Can Hasan there is, anyhow, sufficient evidence to suppose that some buildings were of two storeys[9].

Concrete evidence of flat roofs only appears in the form of pictures and models at a period when the development towards urbanization had gone a long way[10] and one is tempted to see the flat roof as a feature dictated by urbanization itself. With increasing cost of land within the city only the wealthy could afford gardens or courtyards, yet a clean level area was vital to the economy, for only in such spaces could many domestic chores such as beating bedding or drying fruit be carried out. Indeed, in many parts of the Middle East today the flat roofs perform not only such functions as these but, because they are all at much the same level, they provide a means of moving about the town or village clean shod, which the streets do not.

The flat roof has, of course, the great disadvantage that it needs regular repair, and even then is prone to leak and may require patching. Inconvenient as it may be, a leaking roof is no disaster in a dwelling where the occupants will only too soon become aware of the defect and put it to rights. For the foodstore or silo, however, it would be a disaster, and it is particularly relevant to this argument that the earliest known representations of silos from Mesopotamia show them to have had domed roofs[11], although exactly what method of construction was employed escapes us. The fact remains, then, that the use of flat roofs on dwellings must have been dictated by other criteria than ability to shed water efficiently.

In parenthesis it should be noted that the flat roof today in a large part of the Middle East is regarded as a mark of poverty, and that the pitched tiled roof has become virtually a status symbol[12]. In some instances house owners have even been prepared to suffer the inconvenience of having no clear space in which to work in order to demonstrate their "superiority" by giving their dwellings pitched roofs. Nor has the flat roof invariably been accepted throughout the Middle East. In Persia, for example, the town house with a domed or barrel-vaulted mud-brick roof has a long history[13].

As we have already seen in the construction of a foodstore, matters of social habit may cause the design to be modified, but certain fundamentals have to be adhered to: the silo must be weatherproof, pest-proof, and for grain storage, at least, well insulated. These are features shared with the ideal dwelling, the major difference lying in the absence of windows and large doorways. Indeed in China the distinction made by archaeologists between tomb models representing silos and dwellings is based on the absence or presence of windows[14]. Economically speaking, therefore, the silo was as expensive to construct as a dwelling, and while in a purely agricultural community one might expect to find each individual household maintaining its own silo, in the city one would imagine the communal silo to have been a more economic way of storing food. Cities in which each household maintained its own silo would suggest some unusual economic situation such as ownership by individuals of outlying farmlands.

One final advantage of mud-brick construction needs to be mentioned namely the speed with which, if needs be, buildings can be constructed. As the inhabitants of Ankara are well aware, a few men can put up a modest two-room dwelling overnight. The advantage of such "instant building" insofar as urban life is concerned, lies in the fact that commercial activities need to be little interrupted by any programme of rebuilding, and even if a large part of a city were to become devastated the period of reconstruction could be brief.

Building in wood

There are, of course, many ways of building with timber, but essentially one can recognize four main types:

(*a*) Stave building in which whole or cleft timbers are buried for a part of their length vertically in the ground.
(*b*) Log-cabin building in which timbers are laid horizontally, the ends often being notched to receive timbers lying at right-angles to them.
(*c*) Lashed frame building in which a framework is made by lashing timber together, the lacunae normally being filled with other materials.
(*d*) Jointed frame building in which a timber framework is constructed largely using mortise and tenon joints.

Of these, it will be noticed all but the stave buildings are normally set either on piles, footings or a solid plinth; and given the usual ravages of time it is these only which survive. Reconstruction from such meagre remains is extremely hazardous, but generally it is seen in terms of frame building of some kind, or even as a building in mud-brick. Log-cabin construction is normally associated with the northern European countries in the popular mind, overlooking the fact that such buildings, especially as foodstores, are still commonly in use from the area of the Black Sea[15], the Hindu Kush and other forested zones of central Asia. While direct evidence of this style of building can be traced only as far back as the Phrygian period in Anatolia[16], the simplicity of construction and its great stability suggest that it may well have been in use in far more remote periods. Indeed, the woodworkers capable of shaping the trees from, for example, Çatal Hüyük, should have found log-cabin building well within their technological grasp. Equally, one has no evidence to suggest when and where jointed frame buildings began to be made. Their recent distribution suggests that they are more commonly found in areas of deciduous rather than pine forest. Again, insofar as technical skill is concerned, a carpenter capable of making the earliest known ploughs and wagons would certainly have been able to construct a simple timber frame building[17].

These arguments may well seem to be special pleading. They are, however, put forward because even amongst some of the technologically most undeveloped societies one finds far more complex buildings than one might expect, very often with more than a single floor; and a very common arrangement is one in which the ground floor is given over to cooking, workshop and storage, while the upper floor, or floors, are used as living or sleeping quarters. Particularly common in areas of the Middle East and Central Asia today is a style of building in which the ground floor is constructed in stone or mud-brick with an upper floor of timber of either log-cabin or frame construction[18]. How far back into antiquity such a type of building goes is a matter of guesswork, but possibly to a far more remote period than we would like to suppose. It is well worth asking oneself whether the mud-brick structures discovered during excavation represent the whole building or merely the ground floor of a more complex one. The difference between the two interpretations will, of course, make a vast difference when one attempts to answer sociological, demographic or logistical questions relating to ancient societies.

A similar problem confronts us when we attempt to reconstruct buildings which we know to have been entirely of timber. Post-holes, for example, are almost invariably interpreted as evidence of vertical supports for a roof ignoring the possibility that they may have been piles that supported an upper floor. Thus published reconstructions of the dwellings of early Danubian villages[19] represent, as it were, a lowest common denominator of what possibly stood on the footings.

Round or rectangular?

It is obvious that certain types of construction dictate a rectangular building plan, particularly log-cabin construction and, to a lesser degree, frame building. All other materials could, in theory, have been used with equal ease to create dwellings of either circular or rectangular plan. It is also fairly self-evident that houses of rectangular plan fit better into the overall street pattern of a large settlement than do circular ones, despite the fact that in Africa, for example, large towns have been built of dwellings of round plan[20]. The limitations of the circular building, however, lie rather in the restricted span of unsupported roof that this kind of construction will allow. Thus both the conical thatched roof and the simplest domed roof will exert considerable lateral thrust which is difficult to eliminate by the use of tie beams which must cross at the centre of the building.

It is presumably for this reason that circular buildings with flat roofs are an ethnographic rarity. Furthermore, attempts to create larger buildings by the simple expedient of conjoining one or more circular buildings present a roofing problem in that between any two conical or domed roofs there must inevitably be a valley, a potential source of weakness which, even in a modern building, tends to admit moisture.

Admittedly there is today a close association between the use of a light timber framework daubed with mud and houses of circular plan, especially in tropical and sub-tropical Africa; and the argument could be advanced that, using this type of building material, the house of circular plan is simpler to construct than any other, and no doubt the builders themselves would insist that this were so. In fact it is just as easy, with light flexible timbers, to build a house of rectangular plan, using as a basic structure a series of inverted U-shaped supports, similar to the tilt frame of a covered wagon, a style of construction known from many parts of the world[21].

One is tempted, therefore, on purely technological grounds, to see the house of circular plan as a totally unsatisfactory structure for urban development. And yet the house of circular plan had a long history in Europe, remaining a fairly common form not only throughout the prehistoric periods, but also well into the early Christian period in many areas[22], while in Africa its history is even longer. In neither case can its long life be attributed to the use of a single method or material of construction, and one can only infer that its survival resulted from a matter of social habit. Thus, in most areas in which circular dwellings remain in use ethnographers have recorded a strict allocation of space for various purposes within the house[23].

In general, therefore, we must conclude that the building materials used and the manner of their assembly appear only to have had a marginal effect upon the shape and style of building in antiquity. There were, of course, exceptions where locality dictated the style of construction absolutely, typified today in the buildings of the Ma'adan; but, as today, these cases represented a small minority, and should be seen as exceptional.

Notes

1 Hassam-al-Mulk, S. and Staley, J. (1968). Houses in Chitral: traditional design and function, *Folklore*, **79**.
2 Commentary to the film *The Trobriand Islands* made by Dr. H. Powell.
3 Ghirshman, R. (1938). *Fouilles de Sialk*. Paris. pl. 58; Kenyon, K. (1957). *Digging up Jericho*. London. pl. 33.
4 Mellaart, J. (1967). *Çatal Hüyük*. London. pp. 55–6.
5 Bourlière, F. (1965). *The Land and Wild Life of Eurasia*. London. p. 97.

6 Michigan State University (1968). *International Program in Agricultural Engineering*. London. p. 3.
7 Lloyd, S. and Safar, F. (1945). Tell Hassuna, *J. Near East Stud.*, **4**, Fig. 36.
8 Leach, E. R. (1937). The Yami of Koto-sho, *Geogr. Mag.*, **5**, pp. 417–34.
9 Personal communication from Mr. David French.
10 Evans, A. J. (1921). *The Palace of Minos*. I. London. Fig. 226; Davies, N. G. (1929). *The Town House in Ancient Egypt*. London. pp. 246–7.
11 Amiet, P. (1961). *La Glyptique Mesopotamienne Archaïque*. Paris. pl. 16, nos. 267–9.
12 Kuran, A. (ed.) (1965). *Yassihöyük: a Village Study*. Ankara. pp. 107–13.
13 Wulff, H. E. (1966). *The Traditional Crafts of Persia*. Teheran. p. 105.
14 Shimada, S. and Hamada, K. (1933). *Nan-Shan-Li*. Tokyo. pls. 20–2.
15 Koşay, H. Z. (1951). *Alaca Höyük*. London. pl. 2, no. 7; Hassam-al-Mulk, S. and Staley, J. (1968). *op. cit.*
16 Kuran, A. (ed.) (1965). *op. cit.*
17 Heinrich, E. (1957). *Bauwerke in der altsumerischen Bildkunst*. Berlin. p. 59.
18 Gurney, O. (1952). *The Hittites*. London. pl. 30.
19 Buttler, W. and Haberey, W. (1937). Das Bandkeramische Dorf Köln-Lindenthal, *Germania*, **21**, pp. 213–17; also Bersu, G. (1936). Rössener Wohnhäuser vom Goldberg, OA. Neresheim, Württemberg, *Germania*, **20**, pl. 48.
20 Denfield, J. (1947). The Practical Pagan, *Geogr. Mag.*, **20**, pls. 1–8.
21 Wulff, H. E. (1966). *op. cit.* p. 103, Fig. 151.
22 Waterman, D. M. (1956). The excavation of a house and souterrain at White Fort, Co. Down, *Ulster J. of Archaeol.*, **19**, pp. 73f.
23 Phillips, E. D. (1969). *The Mongols*. London. p. 30.

J. G. HURST

The changing medieval village in England

Until recently it was generally supposed that medieval village plans were unchanged and demonstrated a steady growth from an original Anglo-Saxon or later settlement. Geographers studied village plans and classified them, assuming that the basic layout was original[1]. Historians took the earliest large scale maps which survive from the sixteenth century and demonstrated the continuity in plan and in minute details of property boundaries[2].

During the past twenty years the position has been completely transformed by the work of both historians and archaeologists. In the first place Postan demonstrated that there was no steady development in the medieval period but there were periods of expansion when new settlements were formed and contraction when villages shrank or were deserted[3]. Beresford took up the evidence for desertion and showed that the economic changes of the fourteenth and fifteenth centuries led to the desertion of nearly 2000 villages in England[4].

During the 1950's and 1960's, following the formation of the Deserted Medieval Village Research Group[5], archaeologists took an increasing interest in medieval settlements. Previously what little work there had been on medieval archaeology was mainly concentrated on castles, monasteries and manor houses. There was little interest in how the medieval peasant lived and almost nothing was known about the type of house he lived in. Now, after nearly a generation of active work, it is possible to say much more about the medieval village but many problems still remain to be solved. The reports on these excavations are widely scattered in local and national journals but a general synthesis of the archaeological results of the last twenty years' work, together with the latest historical evidence, is now available[6].

Excavations on deserted village sites all over England have demonstrated that there were three basic medieval peasant house types (Fig. 1), though there were many variations and sub-types: (1) The small cot which was a single- or two-roomed house, usually about 5 m by 3·5 m to 10 m by 4 m. This was the hut of the cottar or bordar who had no land of his own. (2) The long-house in which the medieval villein lived. This would have at one end a living-part, often divided into two rooms, with a byre at the other end, usually separated from the living-part by a cross-passage, but always with

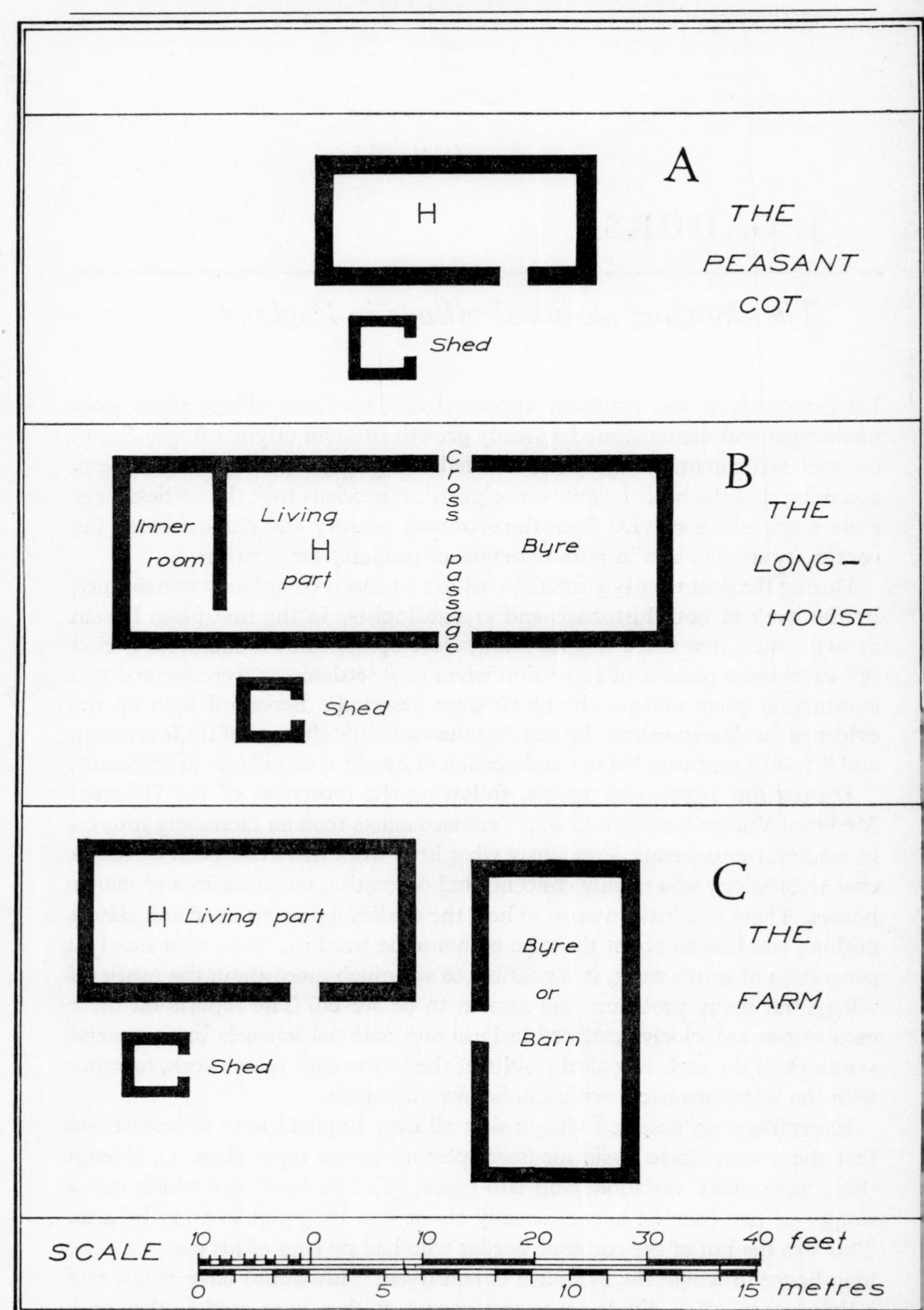

Fig 1 The three main types of medieval peasant house.

access between the two without going outside. These could vary in size from small (10 m by 4 m) to a more normal size of 15 m long, but there were many examples which were as long as 25 or 30 m. (3) The third main type was the farm in which the byre or barn was separated from the main living-house and placed in a distinct building, usually at right angles to the other, forming the basis of a rectangular courtyard.

Excavations so far do not suggest any major regional variations in plan. The size of the various houses was dependent on the prosperity of the peasant building the house. Likewise, although there were more farms in the late medieval period than earlier, the change from long-house to farm was more a matter of prosperity than date or region. In many areas long-houses survived on into the sixteenth century and later. All Anglo-Saxon and early medieval peasant houses were constructed of either timber or turf. There was a general trend in the later twelfth and thirteenth centuries for stone houses to be built in regions where stone was available, while in other areas the timber houses were built on stone foundations if possible.

Whether they were built of timber or stone they were poorly built since excavations show a constant rebuilding about every generation. It is not yet clear why this should be but it may be that peasant houses were intentionally built in a flimsy do-it-yourself manner in the knowledge that the son would wish to rebuild his house when taking over from his father. Another possibility is that, with the continued clearance of woodland, there would be less and less good quality timber available. Air photographs show that in many parishes medieval field systems covered the whole area. These were not necessarily all in use at the same time but even if an in-field out-field system was in operation, or cultivated areas were allowed to go to waste, there might only be scrub growing, especially with the constant grazing of stock.

More surprising than the constant rebuilding of the peasant houses has been the discovery that they were often rebuilt not on the same foundations but in a new position. In many cases this was not just a few feet to one side but on a quite different alignment, often at right angles to the earlier building. In other villages house alignments changed gradually in a clockwise or anti-clockwise direction. Excavations on a larger scale than single crofts have shown that this constant change did not apply to the peasant houses only. More fundamental changes have been found and property boundaries have been drastically replanned. This goes far beyond any subdivision or amalgamation of plots. There is now no doubt that many villages were constantly changing their plan and this is perhaps the most important result of recent research into the medieval village.

For example the excavation of a sample toft (or backgarden) at Wharram Percy (Yorkshire E.R.) demonstrates the complexity of medieval village sites. It was not possible to simply excavate House 10 as the whole area was found to be covered by a palimpsest of superimposed structures on differing alignments in succeeding periods (Fig. 2). If an interpretation of this is made

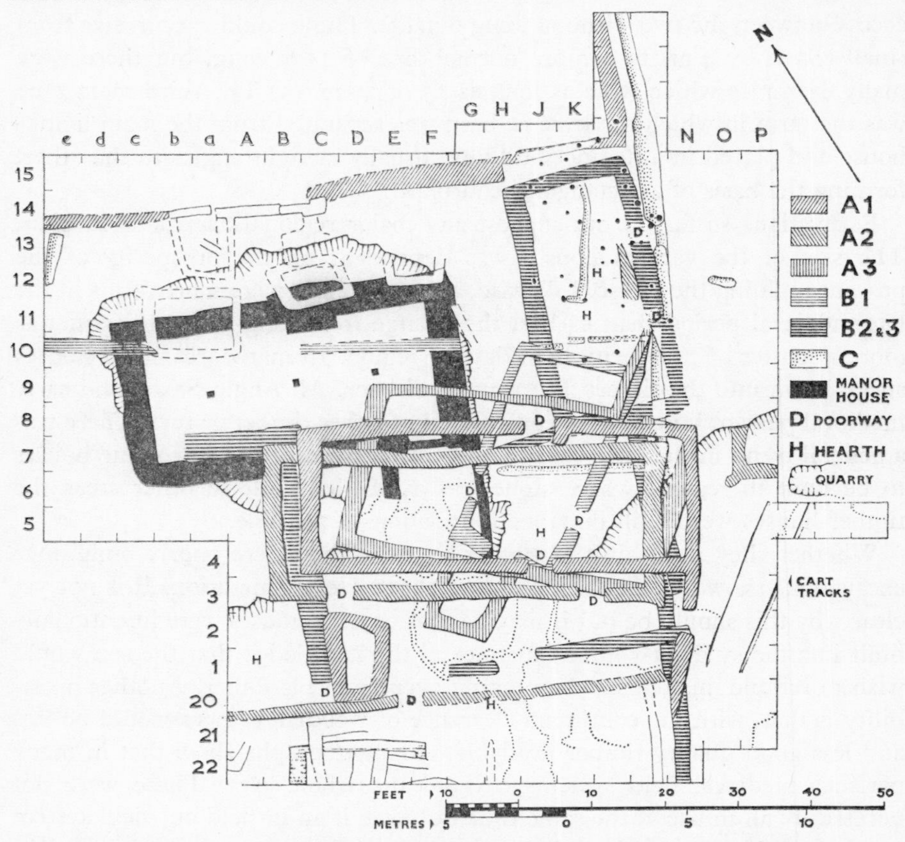

Fig 2 Wharram Percy, Yorkshire East Riding. Excavation plan of the Area 10 complex. See *Fig 3* for interpretation.

it is seen that the use of this plot of land changed fundamentally during medieval times (Fig. 3). In the Anglo-Saxon period this land was still outside the village, presumably cultivated as part of the open fields (Fig. 4). In the twelfth century the Percy family extended the village to build a manor house on the site. In the thirteenth century, when the village was extended further, the manor site was moved further north and this plot was occupied by two peasant houses set parallel with the street in two separate tofts. This demonstrates the pressure on land at the time. In the fourteenth century these two tofts were combined and a single house built across the two on the same general alignment. In the fifteenth century there was a final major replanning with the house turned round at right angles to the street.

On a wider basis, Fig. 4 shows how Wharram Percy developed from a small Anglo-Saxon nucleus in the valley by the church, through a larger twelfth century phase with an haphazard growth of houses, to a major planned extension in the thirteenth century at the time of the greatest expansion of the village. Fig. 5, on the other hand, appears to demonstrate the same sequence

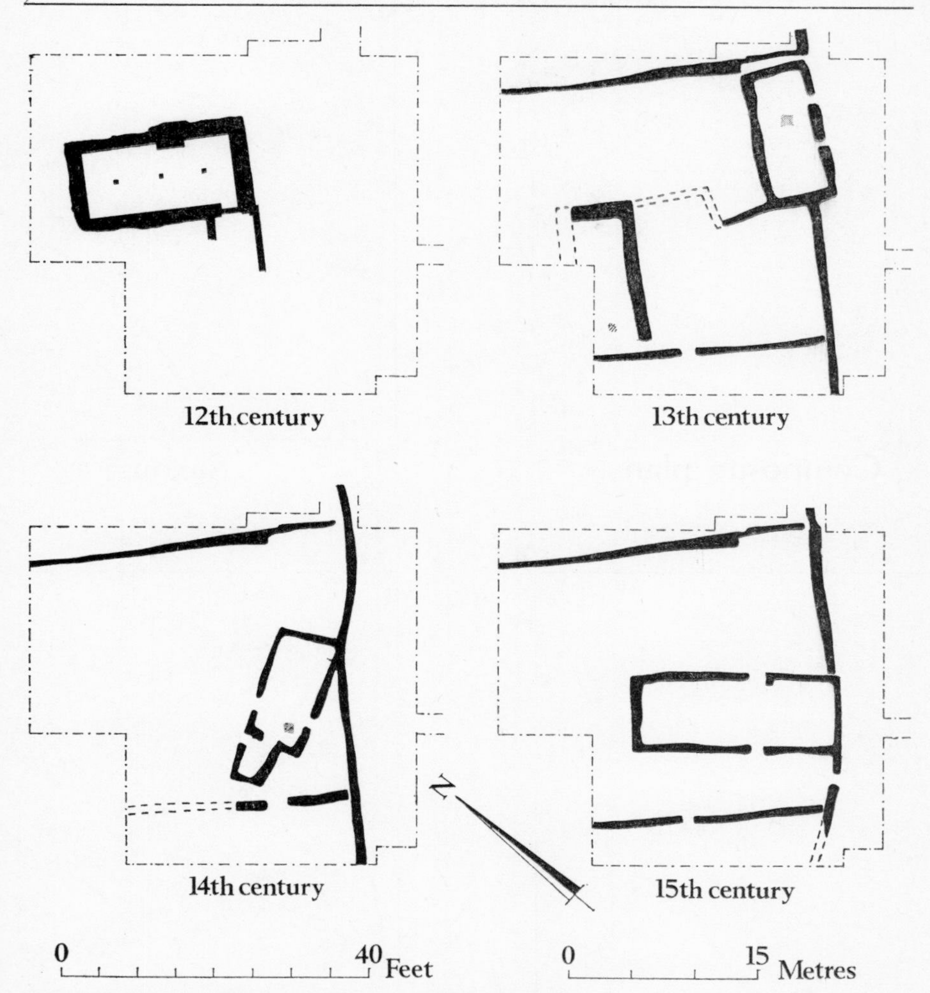

Fig 3 Wharram Percy, Yorkshire East Riding.
Interpretation plan of Area 10 showing the twelfth-century manor house, followed by two peasant houses in separate tofts in the thirteenth century. In the fourteenth century there was a single house in the same area and in the fifteenth century the house was rebuilt at right angles to the earlier one.

at Wawne (Yorkshire E.R.) with an haphazard growth of houses round a rectangular green area with a planned linear expansion to the south. Excavation, however, demonstrated that these two phases succeeded each other (Fig. 6). In the fourteenth century the northern area was completely abandoned and replaced by the linear street village to the south. This shows the need for care in interpreting surviving remains since at Wharram Percy and Wawne the same superficial resemblance was shown to be caused by quite different factors and reinforces the dangers of interpreting surviving remains or plans without a thorough investigation.

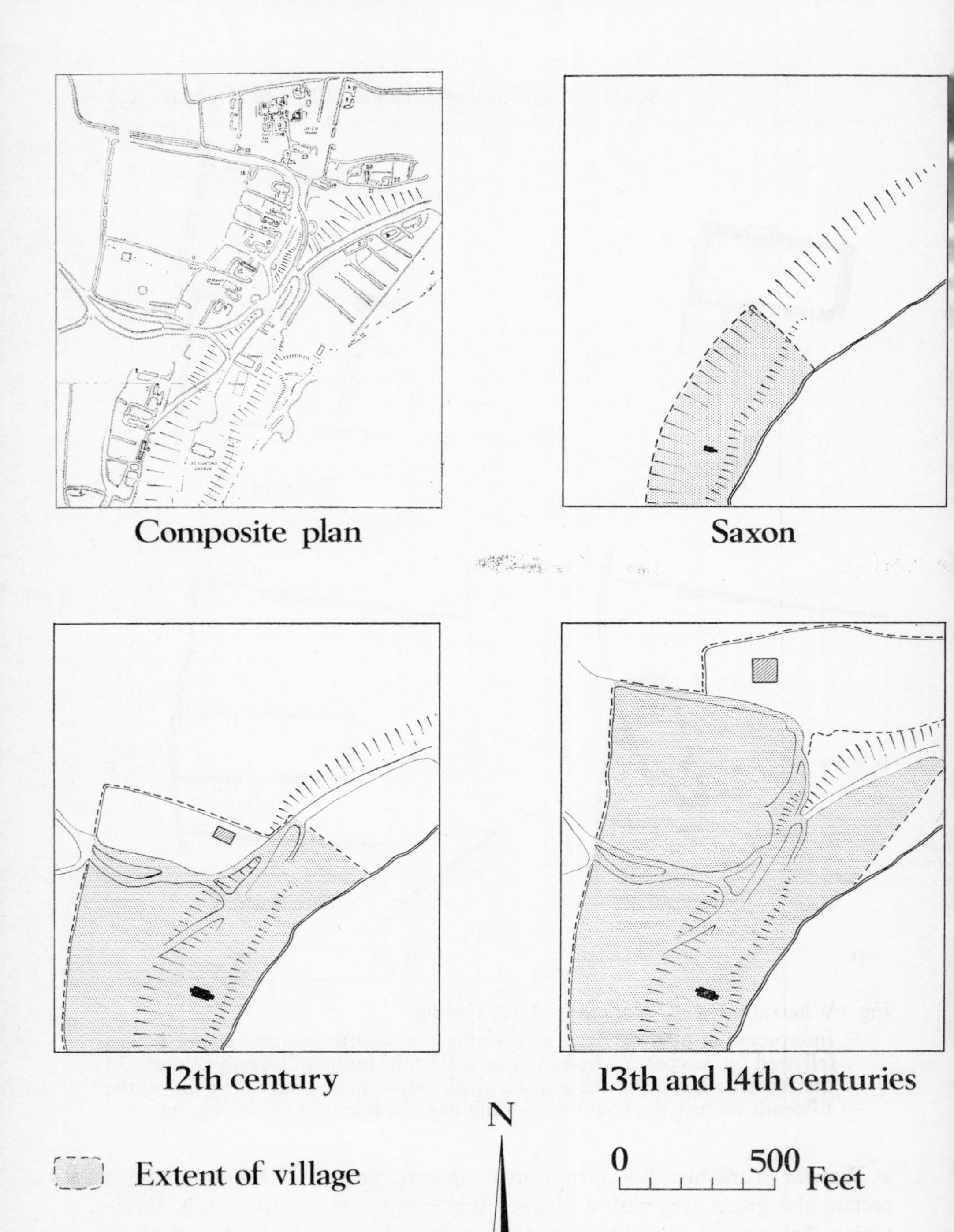

Fig 4 Wharram Percy, Yorkshire East Riding.
Series of plans showing the present earthworks and how excavation has demonstrated a steady growth from a small Saxon settlement to the large fourteenth-century village.

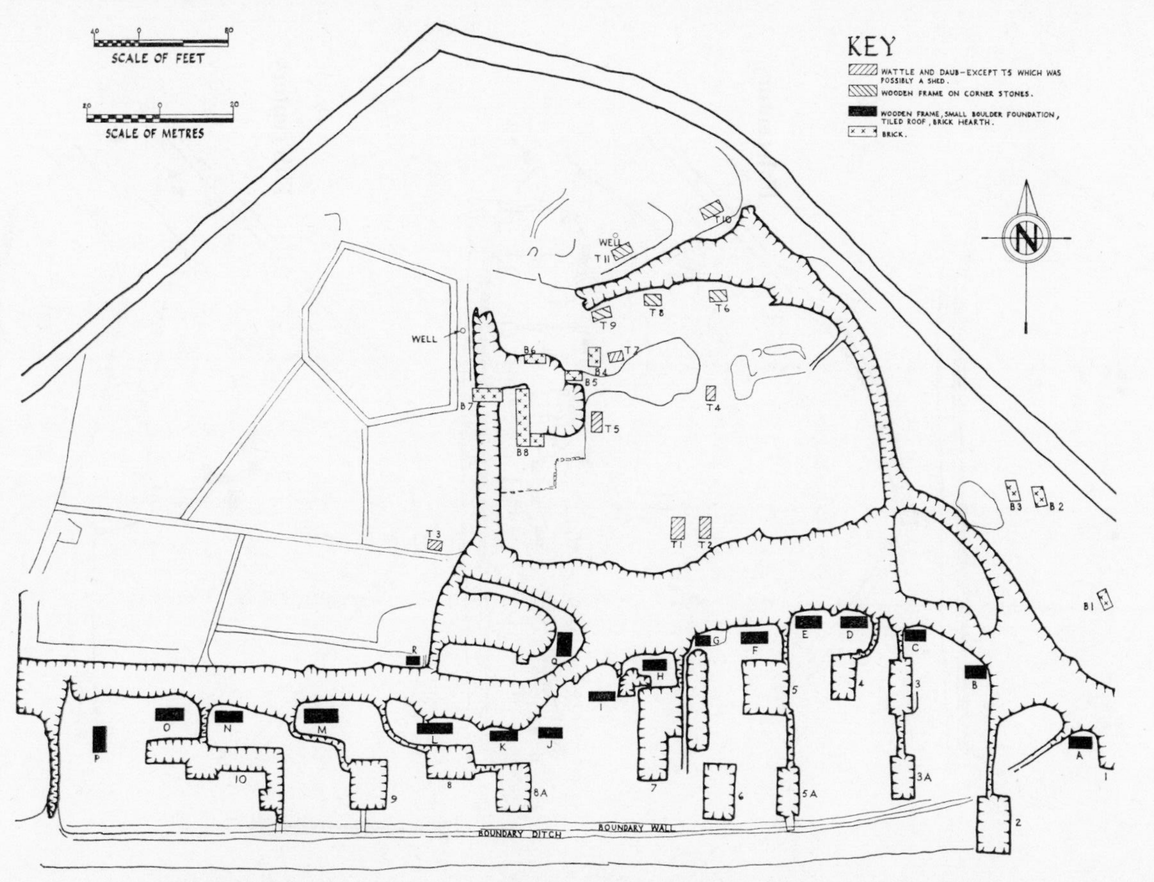

Fig 5 Wawne, Yorkshire East Riding. Plan of earthworks and excavated buildings. See *Fig 6* for interpretation.

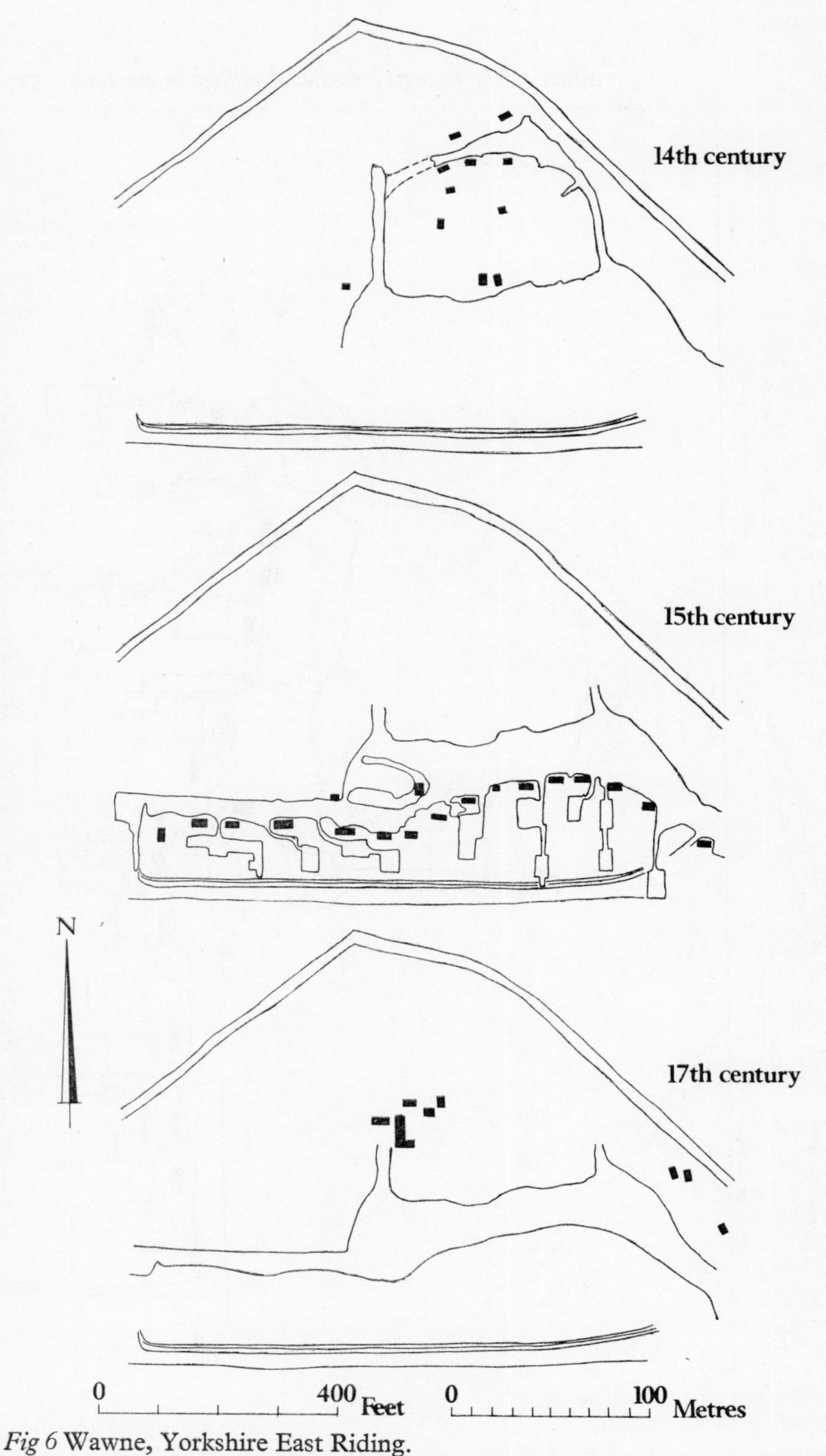

Fig 6 Wawne, Yorkshire East Riding.

Interpretation plan showing that, unlike Wharram Percy, the earthworks do not show a steady development but the fourteenth-century village was completely rebuilt on a new site in the fifteenth century and the earlier site abandoned. By the seventeenth century the village had shrunk to two farm complexes.

It is now possible to take another look at deserted village plans and see different types of layout suggesting how the site might have changed or developed over the centuries. Several green villages have been shown to be quite recent in origin and it is unlikely that many of them go back earlier than the thirteenth century. There is little doubt that seigneurs were replanning their villages on a large scale. This was taking place in the thirteenth and fourteenth centuries in the country at the same time as the new towns were laid out with their regular plans[7]. All this would be much easier in the early medieval period when most of the peasants would be tenants and could therefore be moved at will. Later with the growth of freeholders, and the breakdown of manorial control, it would be more difficult to make fundamental changes. It is likely that few villages have changed significantly in plan since the sixteenth century, except for the special case of emparkings. This was the period when large scale maps were first made so it was quite natural for historians to be misled into thinking that, having traced the plan back so far, this was the original layout of the village. On reflection it is unlikely that a present village plan will be the same as at the time of its Anglo-Saxon foundation. Besides all other factors it must have been much smaller when originally laid out and it is most unlikely that at this time any thought would be given to future expansion.

Basic changes in village plan did not only affect the peasant houses and crofts. There are many cases of manor houses being built over earlier peasant houses showing that seigneurs cleared away parts of the village so they could have the best site, in the same way that town houses were removed to make way for castles. Church sites are more permanent but many new churches, founded in the twelfth century, were built on the sites of earlier houses. Both these activities would lead to large-scale replanning of the rest of the village. These changes must have taken place throughout Anglo-Saxon and medieval times.

Recent work has concentrated on deserted medieval villages since these are the only sites where medieval remains can be studied uncontaminated and undisturbed by later activities. It must be remembered as well that, even before the fourteenth century, there was not a constant pattern of expansion. The discovery of nearly 100 Anglo-Saxon settlement sites, in what is at present open country, shows that there were desertions before the Norman conquest too. It has long been a puzzle that so few medieval villages have Anglo-Saxon remains underneath. This cannot be entirely because of the slight nature of the remains or the fragile pottery which might not survive. There does seem to be growing evidence in many areas for some fundamental changes in settlement patterns in the later Saxon period. This would account not only for new settlements, which could be further colonization, but for these early desertions which imply a more fundamental change in settlement patterns than is apparent from the historical record.

Recent work on rural settlements since the Roman period has, therefore, dramatically demonstrated that settlement patterns were just as fluid then as they were in most of the prehistoric period in many parts of the world. The idea that the medieval village was planned at the time of its original settlement in Anglo-Saxon or early medieval times, and has since simply grown, has been shown to be false. This has important implications for anyone attempting to study Saxon or medieval settlement patterns from present maps or surviving earthworks; the more so if this information is then used to elucidate less certain evidence for earlier periods or other parts of the world. Only archaeological excavation can demonstrate the full story. Medieval sites, in view of their complexity, are very expensive to dig and take a long time. This means that it will be several generations, with the present limited resources which are available for research, before many of the outstanding problems can be solved. But at least in the last twenty years some very important discoveries have been made which will form the basis for future work on rural settlement.

Notes

1 Thorpe, H. (1961). The green village as a distinctive form of settlement on on the North European Plain, *Bull. Soc. Belge d'Etudes Géogr.*, **30**, pp. 93–134; Thorpe, H. (1949). The green villages of County Durham, *Trans. Inst. Brit. Geogr.*, **15**, pp. 155–80.
2 Beresford, M. W. and St. Joseph, J. K. S. (1958). *Medieval England: an Aerial Survey*. Cambridge.
3 Postan, M. M. (ed.) (1966). *Cambridge Economic History of Europe*, I. (2nd ed.) Cambridge.
4 Beresford, M. W. (1954). *The Lost Villages of England*. London.
5 Deserted Medieval Village Research Group. (1953–69). Annual Reports, **1–17**.
6 Beresford, M. W. and Hurst, J. G. (eds.) (1971). *Deserted Medieval Villages*. London. See Chapter 2, pp. 76–144 for full references and further supporting evidence for most of the statements given.
7 Beresford, M. W. (1967). *New Towns of the Middle Ages*. London.

ANNA RITCHIE

Inferences from settlements in Britain in the first millennium B.C.

A cultural pattern based on structural evidence such as the method of enclosing a site, house-plans and features connected with agriculture can be used more reliably to demonstrate social and economic development than artefacts of potentially swift stylistic change such as pottery or bronzes. The imperfect survival of domestic structures during earlier periods in Britain allows the use of structural evidence in this way only in the first millennium B.C. and, in southern England, towards the end of the second millennium; during this period, the development of structural tradition among non-defensive settlements illustrates a continuity and a unity which provide convincing arguments against the "invasion hypothesis"[1]. The ideas offered here are based upon a detailed survey of the form of non-defensive settlements throughout Britain, with the exceptions of the area north of the Forth-Clyde line and much of Devon and Cornwall, where the nature of settlement is such that it cannot usefully be compared with contemporary tradition elsewhere[2]. Regional cultural divisions indicated by artefacts, particularly fine pottery, have been ignored deliberately in order to discover independently whether any such divisions existed on a more fundamental social level than those of trading activities and changing stylistic preferences.

Non-defensive sites may be divided into three major types: those enclosed by timber palisades, those enclosed by earthen banks and ditches, and open sites. This primary classification based upon construction method (factual information) may be sub-divided secondarily according to the social unit represented by each site within each major class (interpretative information). The estimation of social unit is necessarily somewhat crude, for the archaeological record for many sites is incomplete, but its crudity may at least avoid the false conclusions which would inevitably accompany a more elaborate classification. Quite simply, a site consisting of one to three houses is taken to represent the homestead of a single family unit, while a site consisting of more than three houses represents a settlement occupied by more than one family unit. The settlement class includes, therefore, everything from small hamlets to villages, but detailed interpretation of the social unit represented is more safely carried out, where possible, in relation to the individual site. The number of houses is derived from excavation evidence or surface traces; it cannot be estimated from the size of the enclosure, for this can be very misleading[3].

In order to explore the problem of continuity between the later Bronze Age and the pre-Roman Iron Age, it is necessary to deal separately with the domestic sites belonging to each of these two periods, taking a dividing date of *c.* 650 B.C. After that approximate date, the distribution of palisaded sites shows a concentration in northern England and southern Scotland, with sufficient examples further south to permit recognition of the type as common throughout the area. A more widespread distribution is likely in view of the fact that more intensive fieldwork has been carried out in the north than in the south, while the peculiar quality of the Cheviot turf of southern Scotland is particularly susceptible to the survival of visible surface traces. Earthwork enclosed sites are also distributed throughout Britain but are concentrated in the south. Unenclosed sites occur south of a line from northern Wales to Norfolk, but this distribution is governed closely by accidental discovery in the course of modern building and agricultural activities. All three major types of site occur in a wide variety of topographical contexts which can have only local implications, but all show a marked preference for light soils, thus reflecting the limitations of contemporary agricultural implements.

As a type, open sites are not closely datable and seem to have been in use throughout the middle and later first millennium B.C. The chronology of sites enclosed by timber palisades or earthworks reveals, however, a pattern of development in the prevalent social unit. Analysis of the dating of homesteads and settlements shows that homesteads appear from the fourth century B.C. onwards, whereas the larger social units represented by settlements tend to be confined to the preceding period from the seventh to the fourth centuries. This development may also be demonstrated by examples of successive reduction in the size of the area enclosed on sites occupied over long periods[4], and it may be suggested that the role of the larger settlements was taken over by hill-forts. Large settlements then reappear in the first century B.C. as a specialized form of earthwork enclosed site (complex ditch settlements[5]), associated with the use of massive linear earthworks and with two minor types of curvilinear enclosure (banjo and spectacles enclosures[6]) which may sometimes be interpreted as cattle-pens. This complex represents a structural innovation, primarily on chalk plateau areas of southern England, which is associated with the Belgic cultural horizon in the first centuries B.C. and A.D.

The dating of the major structural forms and the sequences of building demonstrable on individual sites allow a general relative chronology of types to be devised. In those cases where different structural traditions are associated, palisades precede earthworks and are often preceded themselves by open sites[7]. This pattern holds good throughout the period for which there is detailed information about the form of prehistoric settlement in Britain;

this period begins in the last third of the second millennium B.C. with the cultural complex in southern England known as Deverel-Rimbury, in which the same three major types of site were in use and the method of enclosure initially adopted was that of timber palisades[8]. The decision to enclose, at any period, reflects a change in social and economic conditions; more permanent settlement connected with the development of agriculture and an increase in stock-rearing which necessitates protection against marauding animals and human raiders.

There is little in the way of structural tradition on British Early Iron Age sites which cannot be matched on earlier domestic sites of the later second and early first millennia B.C. Settlement archaeology has been largely neglected in discussions of the question of continuity or innovation as explanations for the Early Iron Age, primarily because it has been accepted that there is no site continuity between the Bronze Age and the Iron Age in Britain[9]. It can be argued, however, that some sites belonging to the Deverel-Rimbury complex continue in occupation well into the first millennium B.C.[10], while associations of pottery of Deverel-Rimbury tradition with pottery of Early Iron Age tradition are well known[11]. Sites such as West Harling in Norfolk[12] and Amberley Mount in Sussex[13] are clearly transitional between the two cultural horizons in terms of pottery, and West Harling retains an archaic structural tradition in the houses set within individual earthwork enclosures. There is also a small number of domestic sites which cannot be linked with Deverel-Rimbury tradition but which belong to the transitional eighth and seventh centuries B.C.[14].

The gap in the record for settlements in the early first millennium B.C. is thus more apparent than real, and it is possible to trace both site continuity, and continuity of structural as well as potting tradition between the Bronze Age and the Iron Age in Britain. The only real innovations of the mid-first millennium B.C. in terms of settlement sites are an increase in the size of the social unit represented and a change from a predominantly sub-rectangular to a predominantly curvilinear plan among enclosed sites. Neither innovation need involve population change or intrusive influence, for they reflect social and economic development of a purely insular nature.

All the basic features of the "Little Woodbury type" of economy current in the second half of the first millennium B.C.[15] were already present in the early first and late second millennia: storage pits, timber granaries and storehouses, drying racks, working hollows and field-systems. Economy provides a useful source of evidence for continuity which is often neglected.

Throughout the first millennium the economy of southern Britain was mixed, but it is possible to discern tendencies towards the predominance of either agriculture or pastoralism. The appearance of linear earthworks forming large grazing paddocks and often cutting across abandoned field-systems[16], and the use of stock enclosures[17] show the increased importance of pastoralism during the later phase of the Deverel-Rimbury cultural complex, in contrast to the earlier predominance of agriculture. By the mid-first millennium B.C., large systems of linear earthworks were no longer being constructed, although in some cases existing earthworks continued in use. This, together with the appearance of very much larger storage pits[18], indicates a return to an essentially agricultural economy. This form of economy flourished in southern Britain until the end of the millennium, by which time the construction of specialized types of cattle or sheep enclosure indicates a swing back towards the importance of stock-rearing and a renewed appreciation of the value of manure.

As well as acting as inter-farm boundaries, it is clear that linear earthworks were also constructed as territorial boundaries[19]. Their proportions indicate that no intention of defence was involved. One of the few innovations of the mid-first millennium was the construction of a new type of linear earthwork, the cross-dyke; this type of earthwork provided a stronger barrier than the old "ranch boundaries" and fulfilled a definite localized function. Normally comprising a ditch with either one or two banks, the cross-dyke spans ridges and spurs of land, controlling traffic, dividing up the ridge-top areas[20] and often acting as outworks for hill-forts[21]. They are part of the earthen fort-building horizon in general, and occur in topographically suitable areas of southern England and on the Cheviot hills of southern Scotland and northern England[22].

Since the form of domestic sites in the middle and later first millennium B.C. is essentially insular in origin, the predominance of the circular timber house in Britain at this period, in contrast to certain continental European areas, may be seen not as a problem but as a logical continuance of building tradition. Advanced types of circular house represent a refinement of existing tradition, while the relatively few examples of rectilinear timber structures are best interpreted as barns or work-sheds rather than as dwelling-houses, for they are a natural development from the small rectangular or square structures currently interpreted as over-head storehouses for meat, skins, grain or similar commodities.

Notes

1 Clark, J. G. D. (1966). The invasion hypothesis in British archaeology, *Antiquity*, **40**, p. 185f.

2 Ritchie, A. (1969). *Settlements and Economy in Britain during the First Millennium B.C.* Unpublished Ph.D. thesis, University of Edinburgh.

3 e.g. Wainwright, G. J. (1968). The excavation of a Durotrigian farmstead near Tollard Royal in Cranborne Chase, southern England, *Proc. Prehist. Soc.*, **34**.

4 e.g. Castlehill, Peebles: Royal Comm. on Ancient and Hist. Monuments of Scotland. (1967). *Peeblesshire*. Edinburgh. No. 195.

5 e.g. Hanging Langford Camp, Wilts: Crawford, O. G. S. and Keiller, A. (1928). *Wessex from the Air*. Oxford. pl. XVII.

6 e.g. banjo at Blagden Copse, Hants.: Stead, I. M. (1968). Excavations in Blagden Copse . . . 1961, *Proc. Hants. Field Club*, **23**; spectacles enclosures at South Tarrant Hinton Down, Dorset: Sumner, H. (1913). *The Ancient Earthworks of Cranborne Chase*. London. p. 41f., pl. XX.
7 e.g. Jobey, G. (1962). An Iron Age homestead at West Brandon, Durham, *Arch. Aeliana*, **40**; Alcock, L. (1960). Castell Odo: an embanked settlement on Mynydd Ystrum, near Aberdaron, Caerns, *Arch. Cambrensis*, **109**.
8 e.g. New Barn Down: Curwen, E. C. (1934). A Late Bronze Age farm and a Neolithic pit-dwelling, *Sussex Arch. Coll.*, **75**.
9 Hawkes, C. F. C. (1966). British prehistory: the invasion hypothesis, *Antiquity*, **40**, p. 298; Burgess, C. B.(1968). The Later Bronze Age in the British Isles and north-western France, *Arch. J.*, **125**, p. 30.
10 The pottery from New Barn Down, for example, includes forms which should be dated to the seventh or sixth centuries B.C. and which indicate a late phase of occupation at the site.
11 Cunliffe, B. and Phillipson, D. W. (1968). Excavations at Eldon's Seat, Encombe, Dorset, *Proc. Prehist. Soc.*, **34**, p. 230f.
12 Clark, J. G. D. and Fell, C. I. (1953). The Early Iron Age site at Micklemoor Hill, West Harling, *Proc. Prehist. Soc.*, **19**.
13 Ratcliffe-Densham, H. B. A. and M. M. (1966). Amberley Mount: its agricultural story from the Late Bronze Age, *Sussex Arch. Coll.*, **104**.
14 The most important of these is Minnis Bay, a two-period site which was initially open and subsequently enclosed by a timber-laced bank: Worsfold, F. H. (1943). A report on the Late Bronze Age site excavated at Minnis Bay, Birchington, Kent, *Proc. Prehist. Soc.*, **9**.
15 Piggott, S. (1958). Native economies and the Roman occupation of North Britain, *in* Richmond, I. A. (ed.) *Roman and Native in North Britain*. Edinburgh. p. 3f.
16 There is little evidence surviving to support the idea that these earthworks acted as divisions separating arable from pasture.
17 e.g. Harrow Hill, Sussex: Holleyman, G. A. (1937). Harrow Hill Excavations, 1936, *Sussex Arch. Coll.*, **78**.
18 Grain storage need not have been the sole function of these pits but it was evidently a major one.
19 e.g. "Old Ditch", Wilts.: (1957). *Victoria County History of Wiltshire*. I. London. p. 123, no. 58.
20 Fowler, P. J. (1964). Cross-dykes on the Ebble-Nadder ridge, *Wilts. Archaeol. Nat. Hist. Mag.*, **59**.
21 e.g. Whitesheet Hill, Wilts., where there is a univallate cross-ridge dyke either side of the hill-fort; Harehope Rings, Peebles, where an unfinished dyke guards the approach to the fort.
22 The cross-dykes of Yorkshire and Glamorgan belong to a later phase of dyke construction in the first centuries B.C. and A.D.

FRIEDRICH W. SCHWERDTFEGER

Urban settlement patterns in northern Nigeria (Hausaland)

The object of this paper is to discuss urban settlement patterns in northern Nigeria and the factors which have contributed to their development.

Nigeria is one of the most densely populated countries in tropical Africa, and according to the 1952 census[1] nearly 6,000,000 people live in urban centres of 5000 or more inhabitants[2]. The level of urbanization varies considerably from region to region. The figures for 1952 show the ratio of urbanization as 49% in the western region (Yorubaland), 14% in the eastern region and 9% in the northern region. Hausaland, with which this paper is concerned, occupies a considerable part of northern Nigeria, and has an urbanization ratio of approximately 12%, for which Kano (127,205 inhabitants in 1952) and its closely settled zone is largely responsible.

Most northern Nigerian towns have two different types of settlement. The first is a traditional, pre-industrial urban area and the second an adjoining new and rapid growing industrial belt. Each type has a distinct socio-economic and cultural background, and the existence of this typical duality with its varying rates of economic development and distinctive patterns of life is a potential source for local conflict.

In order to comprehend the present urban settlement pattern in northern Nigeria it is essential to understand its historical development. The Moslem Arabs who conquered the Mediterranean coast of North Africa in the eighth century A.D. established a wide network of trading routes which led southward from the coast to the Sudan. The trade which flowed along these routes probably determined the siting of most of the major trading centres on both sides of the Sahara and established their power and wealth. In the Sudan one can distinguish two zones of city development, one situated on the edge of the desert including such famous towns as Timbukto, Gao and Agades which functioned as trading depots; the other, as Bovill mentioned[3], is a chain of cities further to the south roughly on the twelfth parallel and includes Bamako, Djenné and Wagadugu. These latter handled the inter-regional trade with the forest area. Although the wealth of all these towns depended largely on the trade and safety of the different caravan routes, arts and crafts flourished

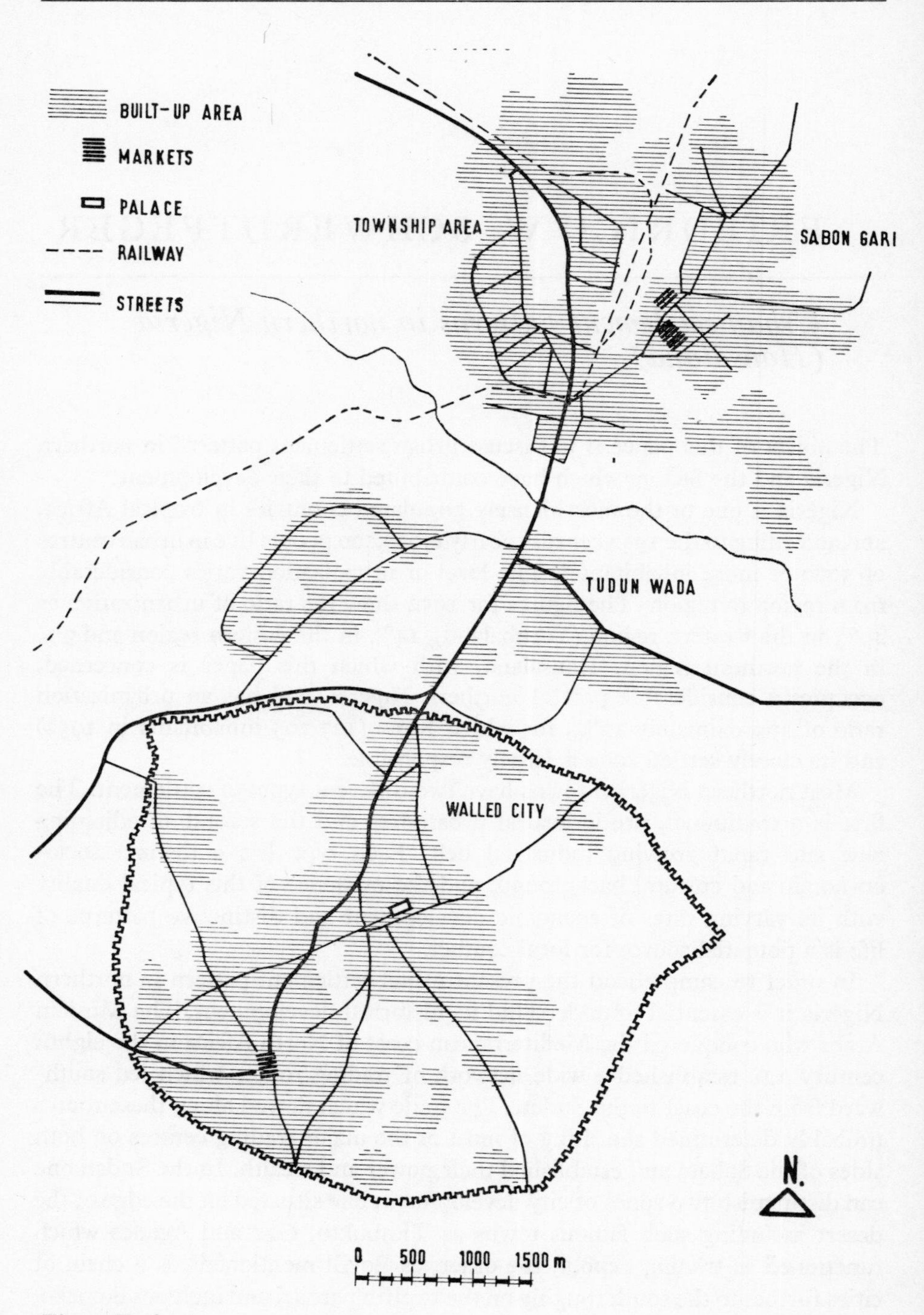
BUILT-UP AREA
MARKETS
PALACE
RAILWAY
STREETS
TOWNSHIP AREA
SABON GARI
TUDUN WADA
WALLED CITY
N.
0 500 1000 1500 m

Fig 1 Zaria.

and towns like Katsina and Kano were well known for their excellent products of leather and cloth. Thus by 1901–3, when British forces under Sir Frederick D. Lugard advanced into northern Nigeria, such towns as Sokoto, Katsina, Kano and Zaria had long achieved a high degree of socio-economic complexity and were clearly distinct from the surrounding rural communities[4].

The British colonial administration undoubtedly had a great impact on the existing towns and their subsequent development. To show the extent of this impact I will trace the spread and development of Zaria, a traditional Hausa town situated 110 miles south-west of Kano. Zaria can be divided into three areas: (1) the European township area; (2) the Sabon Gari or new town inhabited by immigrants from southern Nigeria, and; (3) the walled Hausa city including the settlement of Tudun Wada (see Fig. 1). I will analyse these three areas briefly before concentrating on the settlement type inside the walled Hausa city and the lesser known changes that have taken place behind the high mud walls surrounding its residential units.

In 1904 a Government station was set up approximately two miles to the north of the walled city and when the railroad arrived from Lagos seven years later it formed the nucleus for a new town. By 1917 the government deemed it necessary to pass a Township Ordinance to regulate such developments[5]. Three categories of townships were created by the Ordinance: a first-class township governed by a town council, second and third-class townships administered by officers appointed by the Governor. Zaria was placed in the second category.

In the next thirty-five years Zaria grew rapidly but unevenly. By 1952 the Sabon Gari had reached a population of 10,717 or 20% of the total at Zaria, and with an annual increase of 3 to 4%, was by far the most rapid growing community in the area. This substantial increase was mainly due to the high level of immigration from southern Nigeria. In 1952 the township area had a population of 2281, of whom the majority were non-African. The walled Hausa city, after a decrease of population in the early days of British rule, had grown very slowly reaching in 1952 a population of 32,559 or 60·3% of the total at Zaria. It has since grown at a rate of approximately 1·4% per year, a slow increase which is due to the still relatively high infant mortality and the emigration of some young people to new urban centres such as Kaduna and Jos which offer better employment opportunities.

The population of the Sabon Gari consists mainly of Yorubas and Ibos from the south who settled there to work either as skilled or semi-skilled labourers in jobs offered at local factories and the railway yard, or to trade by importing palm oil and cola nuts from southern Nigeria. The streets of the Sabon Gari are lined with one-storey houses built on small rectangular plots similar in style and construction to those of larger Yoruba towns such as Ibadan in the south (see Fig. 2). House walls are constructed from mud or concrete blocks, roofed with corrugated iron and windows have wooden shutters. The population density in these houses is often substantial and may

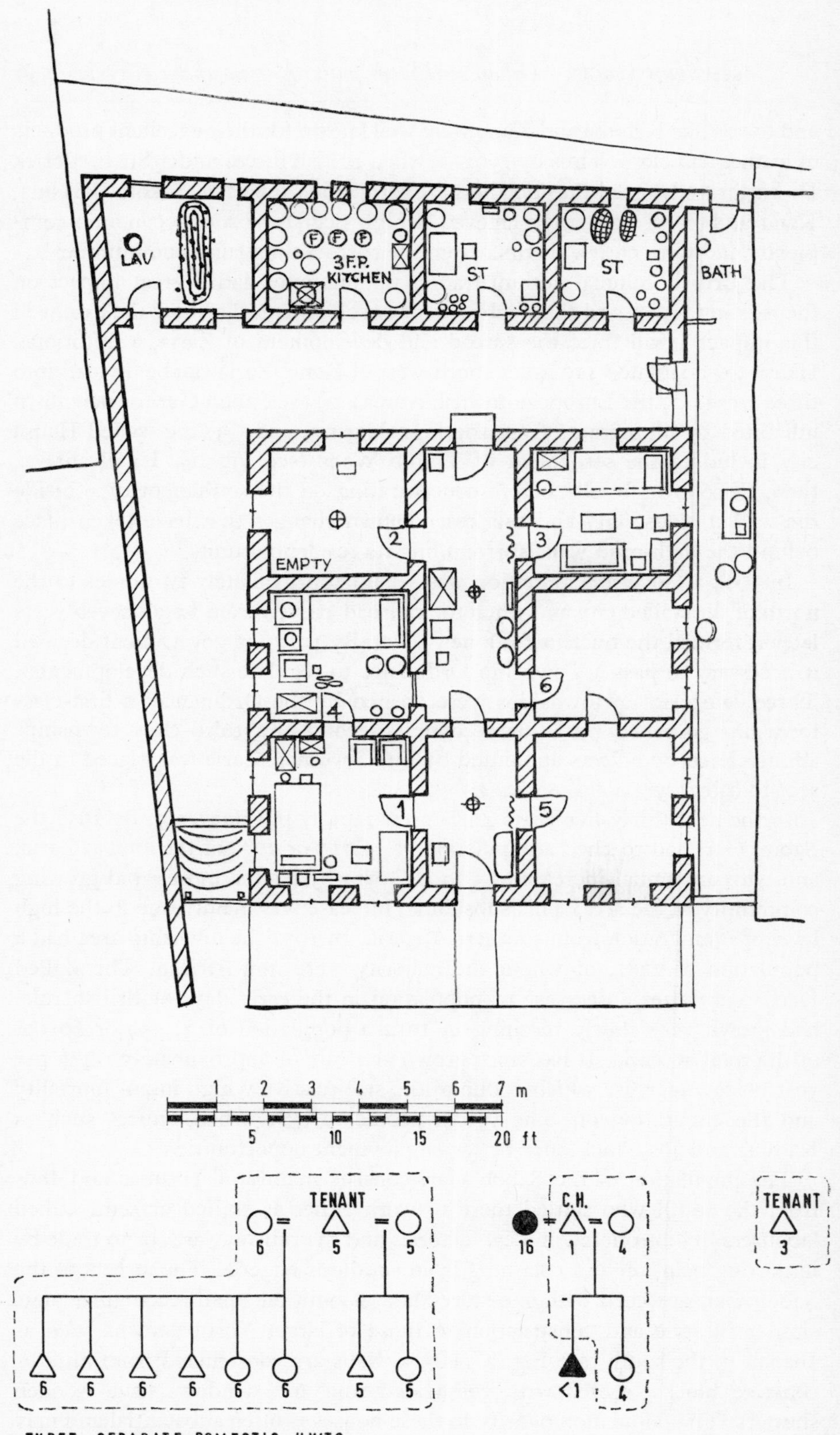

Fig. 2 House No. 34.

reach four to six persons in an average room of approximately 10 m^2. Shops or wooden stalls fronting most houses on the main streets attract to this area many customers and considerable trade from the walled city market.

The European township area is situated to the west of the Sabon Gari. Its residential quarters, which are grouped around a club-house with a racing-track, polo field and a golf course, are well served by a network of wide streets shaded by trees. Detached houses built in the familiar colonial style are equipped with all necessary amenities and surrounded by well-maintained gardens.

Before proceeding to the walled city I should mention the settlement of Tudun Wada. Tudun Wada is situated just outside the walled city to the north and was founded in the early twentieth century. Hausa traders, from other parts of the north, and such northern minority groups as Kanuri and Tiv who were unable to settle in the walled city because of restrictions imposed on immigrants by the Emir's administration, settled here to seek a living in the walled city market and the Sabon Gari. The houses of Tudun Wada are generally smaller than those found in the walled city but are of similar design.

The people of the walled Hausa city are mainly engaged in retail trade, agriculture and traditional crafts. The majority of its compounds, connected by narrow winding footpaths, cluster around the most important institutions common to all Hausa capitals: the palace and the central mosque which together form the politico-religious centre, and the market, the socio-economic centre. All compounds found in the walled city are surrounded by high mud walls. Their only entrance is a *zaure* or entrance hut that leads into the fore-court (*kofar gida*), from which one has to pass through a *shigifa* (second entrance hut) to reach the centre of the compound, the *cikin gida*, where the rooms of the household head, his wives and other close relatives are located (see Fig. 3). Nearly all houses are built from mud and the majority have

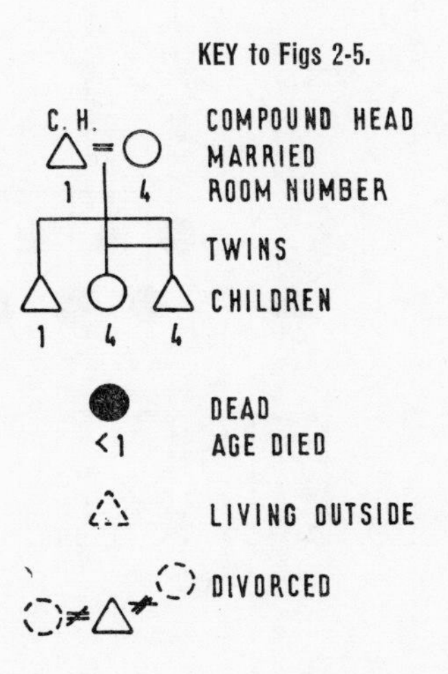

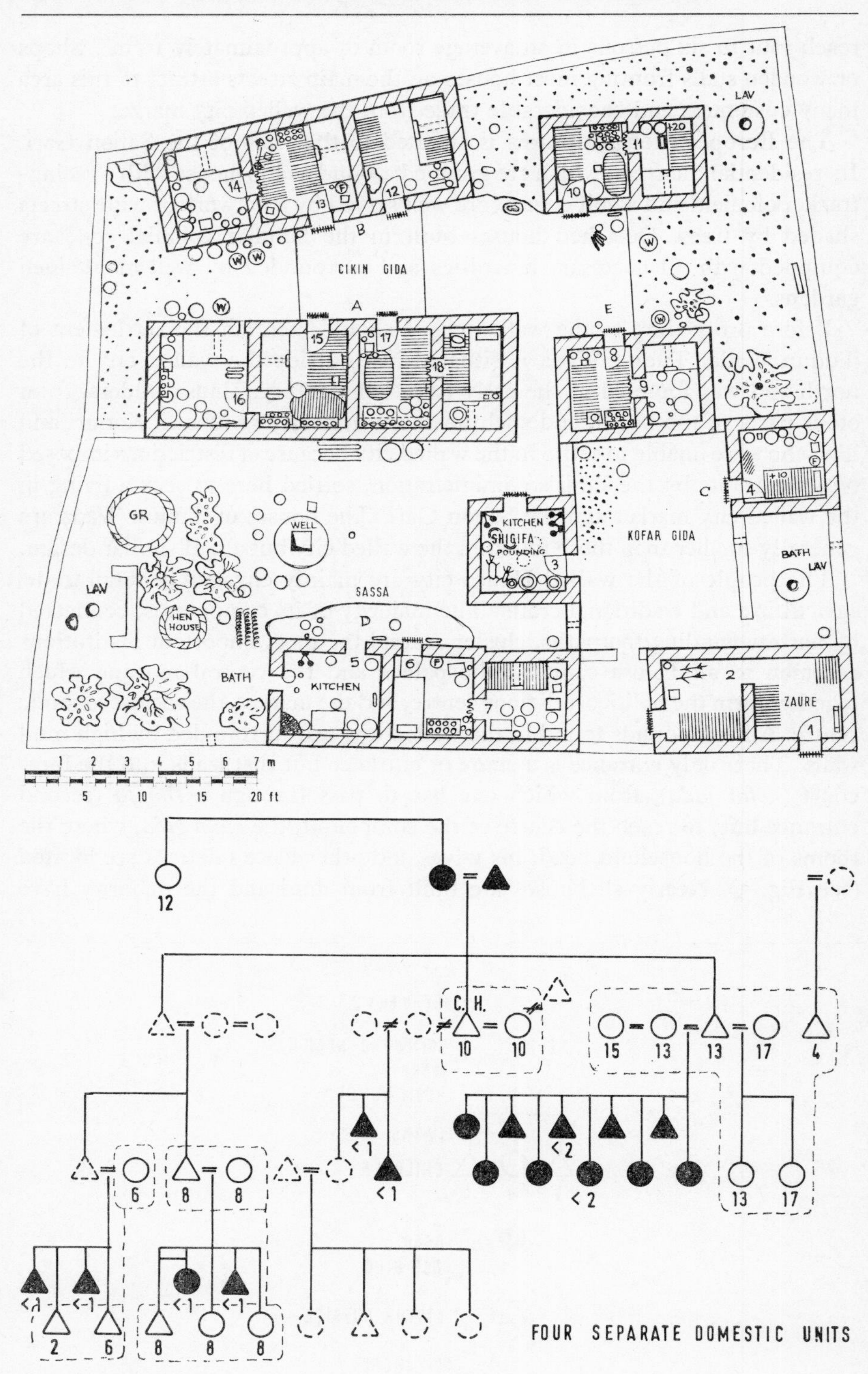

Fig 3 House No. 74.

vaulted or flat mud roofs. Most compounds have their own well and several pit latrines, but some also have piped water, and in a few cases electricity. Such Hausa and Fulani women who live under the arrangement of *purdah*, are not allowed to leave the compound; male strangers are not permitted beyond the forecourt. Compounds average approximately 5–600 m² in size and twelve to fourteen persons in population, the inhabitants being close family and kin.

For centuries, huts in Hausa compounds have been constructed on a common basic design and left to decay and collapse after the death of their occupants. But in the last twenty years there have been many changes. Perhaps the most obvious is the virtual disappearance of the traditional round thatched hut, which has now been replaced by a larger square building with big windows and a flat or vaulted mud roof. The introduction of Western domestic furniture, such as beds, tables, chairs and shelves, clearly reflects the influence of the Europeans. Developments in local education and economy have also contributed in two ways to these changes. Firstly they provide higher earnings some of which are spent on building improvements that use cement instead of the traditional mud plaster and corrugated iron sheets as roofing instead of thatch. Secondly higher incomes encourage the division of larger families into smaller units, who then construct their own compounds, often with permanent building materials such as concrete blocks or reinforced concrete. This development reduces the adaptability of the traditional compound built of mud. The present situation is characterized by the fact that mud huts can be constructed relatively quickly and cheaply on demand and disintegrate as rapidly if not occupied for any length of time. The mud from the disintegrated huts provides raw material for new houses.

The most common household composition among the immigrant population of the Sabon Gari is the nuclear family consisting of parents and their children. A co-residential unit may comprise several unrelated nuclear families of the same ethnic origin. Each family constitutes a separate household unit, each operating its own domestic economy and occupying one or two rooms for which they pay a weekly rent.

In the walled Hausa city the majority of co-residential units are extended family groups based on agnatic kinship. These family groups can be roughly divided into three categories: the nuclear family—with or without additional kin, such as the compound head's widowed mother or other divorced or single relatives—and composite units of two types: the compound head's family, his married sons and their descendants and collateral agnates and their dependants. The two latter types are usually divided into a number of households, each forming a separate unit of domestic economy and occupying a section of the *cikin gida* known as *sassa*. If the inhabitants of a compound depend on agriculture as a means of livelihood, they may form a single work-unit or *gandu* which consists of all the adult males and their families. Individual households within a *gandu* farm and cook together, their common

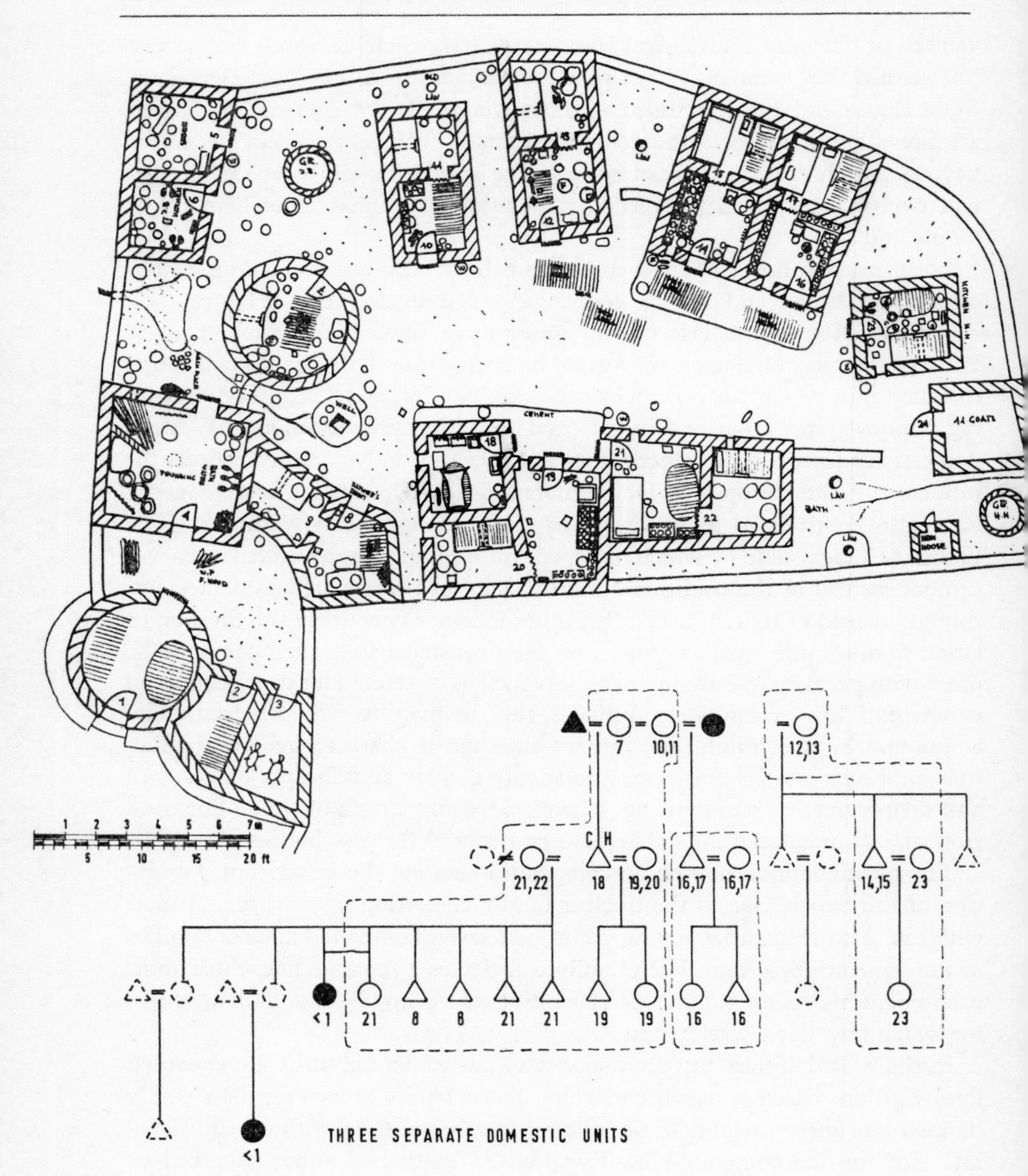

Fig 4 House No. 71.

meals being prepared in turn by each of the members' wives. In the dry season each individual household head may pursue his own occupation such as weaving, embroidery, trading or teaching the Koran, and at this time each household operates a separate domestic economy[6]. Adoption (*tallafi*) is widely practised but limited to children from the wider kinship group, women usually adopting girls and men boys.

The compound shown above is situated near the walled city market and has belonged for several generations to a butcher's family (see Fig. 4). This compound was chosen to demonstrate the close relationship between the changes in family size and composition and the changes of the compound layout resulting from the need for new huts to house the growing family. As mentioned earlier huts are abandoned and allowed to collapse after the owner's death, so providing space and raw material (mud) for new development.

According to the present compound head his grandfather rebuilt the compound around 1900 shortly before the British occupation. He lived with his wives in three double round huts (*adada*) and had also a round store—indicated by dotted line in the plan—the exact positions of which are no longer known (see Fig. 5). There was also some uncertainty about the number

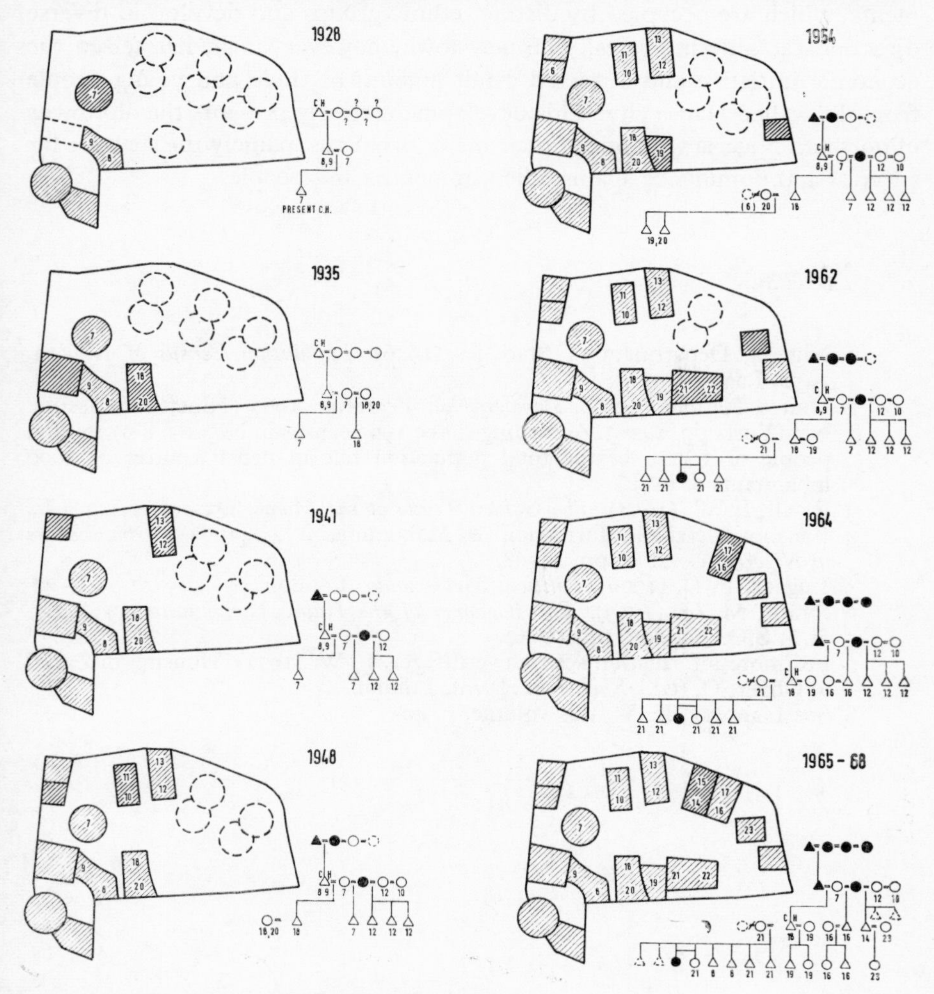

Fig 5 Changes in compound organization.

of wives that he married and the number of children by these marriages. Nevertheless from the data collected it was possible to produce a series of plans showing the development of the compound from 1928 to 1968. The kinship patterns were compiled independently over the same timespan and later compared with the structural development of the compound, any inconsistency found was discussed with the compound head and corrected when necessary[7].

It is worth noting that due to family expansion, substantial structural changes inside the compound took place within a single generation. These changes include not only the addition of new buildings and the abandonment of others but also show a growing tendency to build rectangular instead of round huts[8].

To conclude, contemporary Zaria contains three different types of settlements, which are occupied by distinct ethnic groups and develop in diverse directions at differing rates. The new town, however small, has seized the economic initiative and attracted a fair amount of trade and young people from the walled Hausa city. This development clearly presents the old towns of northern Nigeria with one of their major problems, namely the struggle for survival and dominance against alien influences and peoples.

Notes

1 Nigeria, Department of Statistics (1956). *Population Census of Nigeria, 1952.* Lagos.
2 United Nations (1968). *Demographic Yearbook,* 1967 (Nineteenth Issue). New York. pp. 142–3. According to the 1963 census in Nigeria, 8·97 million people or 16·1% of the total population live in urban centres of 5000 inhabitants or over.
3 Bovill, E. W. (1958). *The Golden Trade of the Moors.* Oxford. pp. 241–2.
4 For more detailed information see Mabogunje, A. L. (1968). *Urbanization in Nigeria.* London. pp. 44–68.
5 Lugard, F. D. (1906). *Political Memoranda.* London. pp. 405–22.
6 Smith, M. G. (1955). *The Economy of the Hausa Communities of Zaria.* H.M.S.O. London. pp. 19–40.
7 For more information see Schwerdtfeger, F. W. (1971). Housing in Zaria, *in* Oliver, O. (ed.) *Shelter in Africa.* London.
8 See Flannery, K. V., this volume, p. 40.

Part III: Urban Settlement

Section 1: Development and characteristics of urbanism

DAVID GROVE

The function and future of urban centres

This paper attempts no more than to set down the suggestions of a practising planner on the definition and classification of towns and some personal reflections on the future of urban form.

Words equivalent to the English "town" (or "city") and "village" have similar connotations in most languages. This must reflect an almost universal perception that most permanent nucleated settlements belong to one or the other of two contrasted classes. Everybody knows the difference between them—yet it is difficult to establish an objective criterion.

I have written[1] about this problem as it presents itself in a relatively undeveloped tropical country but I am convinced that its nature is essentially the same everywhere.

Density—or some other measurable physical attribute—will not serve as a criterion for there are many countries (England included) where towns and villages have much the same sort of layout and land-use.

Size (area or population) clearly enters into the popular notions of town and village—but no satisfactory threshold has been established, even for a single country. The commonly accepted minima of 5000 or 10,000 people for defining a town may be adequate for broad statistical analysis but hardly reveal the specific character of urban settlements. Whatever threshold one selects, there will be "towns" with a smaller population and "villages" with a larger one. In County Durham, for instance, there are lots of pit "villages" with more people than some of the country "towns"; the colloquial use of these words recognizes some quality more fundamental than size.

Another possible criterion is the proportion of the working population engaged in those primary economic activities (farming, forestry, fishing, mining) with which villages are traditionally associated. But in this respect too it is hard to know where to draw the line, and the criterion breaks down completely in face of the modern phenomenon of commuter villages.

However, with the last suggestion, we are getting warmer. It recognizes that towns and villages should be distinguished by *function*. The historical origin of towns lies surely in the need to concentrate in one place functions related to a wider area than a village, such as markets, administration or defence. Thus we may define towns as settlements offering a given variety or level of certain characteristic services. The precise threshold may vary with the culture and level of development of a country or region.

Clearly, towns differ in the variety and level of services they provide. Broadly speaking, the greater the variety and the higher the level of services in a town, the larger is its area of influence. Hence arises the well-known concept of the *urban hierarchy*. In this context the village may be regarded as the lowest rank in the hierarchy and can thus be said to differ only in degree from a town. However, it can also be said to differ in kind because a village provides services only for its own inhabitants (and those of scattered home-steads in its "parish"); this produces a significant difference in patterns of movement and activity.

It is true that not all towns originated as service centres and most of those that did have acquired other functions. Mining, manufacturing and resort towns grew up to perform operations of national or international rather than local or regional significance. Yet all but a few such towns have subsequently become service centres for a surrounding area.

In Britain it is usual to distinguish five or six ranks (or orders) in the urban hierarchy. London, with its unparalleled variety of services, stands alone. Then come the second order centres, like Liverpool, characterized, *inter alia*, by such features as a daily morning newspaper and a selection of department stores. In the third order are towns, like Chester, usually with a Marks and Spencer's and a general hospital and often with a daily evening newspaper. The fourth order includes most of the long-established market towns, like Mold (Flintshire) which generally have a Woolworth's as well as a distinctive range of professional services. In the fifth order are places, often small industrial towns like Buckley (Flintshire), with a noticeably smaller variety of shops and other services. The towns named in this paragraph are representative of the hierarchy in an area for which my firm has recently completed a planning study[2].

The ranking of towns should ideally be based on a complete inventory of services, but this is difficult to obtain and laborious to analyse. Many short cuts have been used, including single indicators such as bus services, telephone connections, newspaper circulations and sales of durable goods. Any system will reveal many deviations from an ideal pattern. No two towns have precisely the same complement of services. Areas of influence are different for different services; they frequently overlap; and they may be different for different socio-economic groups. A new town like Crawley, for instance, will attract working class custom from a very wide area, some of whose middle class residents will prefer the higher quality shops of Horsham and East Grinstead.

But these deviations do not invalidate the concepts of hierarchy as an underlying tendency in the distribution of settlements; they can generally be explained by specific historical, geographical, social, economic or political influences. The broad orders may be subdivided[3] to take account of variations (e.g. 3A, 3B, and 3C centres); but it seems to me that it is more helpful to think in terms of a "norm" for each level of the hierarchy, with deviations from it in individual places.

Strictly speaking, we are dealing with a hierarchy of *service centres*, not of towns, a centre being a group of activities serving roughly the same area and located together to enjoy the external economies arising from physical proximity. Larger places contain several centres of different levels. London has suburban centres from the third order down, Liverpool from the fourth order down, and so on. But much in the character of a place stems from the status of its principal, and usually original, centre.

At the beginning of this paper I used the words "town" and "city" as equivalents. It would be possible to adopt such terms as metropolis, city, town, township for different ranks in the urban hierarchy. But it seems more meaningful to describe the first to fifth orders successively as "national", "provincial" (or "regional"), "regional" (or "sub-regional"), "district" and "local" centres.

Of course urban settlements may be classified in several other ways according to their physical and socio-economic characteristics[4]. Such classifications may assist planners in coming to grips with the precise character of each individual place. But the concept of a hierarchy has proved to be a powerful general-purpose tool—helping regional planners to devise the most satisfactory pattern of settlements and routes, and helping urban planners to foresee the appropriate variety and level of services in individual centres.

This approach to the definition and classification of towns depends upon the practicability either of drawing a line around an individual settlement or of identifying an individual centre. If some present trends continue and if some contemporary planning doctrines prevail, a day may come when this will no longer be possible.

Different cultures and periods have produced many different types of town. Yet almost all historical towns have had one common feature: they have been compact settlements sharply distinguished, physically and socially, from the surrounding countryside. Towns have grown both by extending their area *and* by increasing their density, especially in the central areas where the greater part of their economic and social activities has always been concentrated. Such compactness and high density are a necessary condition of urban life when the vast majority of people are able to move about only on foot.

Perhaps the historians at this seminar will confirm my impression that horse-drawn (or man-drawn) vehicles were always confined to a small minority. If so, the external economies that led a variety of functions to locate in the same place could only be realized when they were in close physical proximity.

The first mechanical means of transport—the railway—made little difference to this necessity for compact high density towns: most people still had to be within walking distance of a station.

The need for compactness has meant that the status of a town in the hierarchy of centres has usually been reflected in its physical form. The higher the rank of a centre, the greater the variety of buildings and spaces within a small compass, and the more complex the relationships between them. Equally so, the greater is the variety of people who rub shoulders there. And these features (to return to the starting point of this paper) are surely the most widely perceived criterion of a town. Perhaps they are what most people *mean* by a town.

While compactness and high density were determined by technical factors, they are often considered to have had important social and psychological advantages. They gave the town the sort of "life" that it is hard to envisage being generated by any other form. But modern technology is rapidly changing the parameters of urban planning. Will it dictate a new form—or does it present us with a choice?

In the coming era of almost universal car ownership (plus electricity, radio, telephone, television and other "piped" services) compact high density centres will no longer be a necessary condition for realizing external economies. Some planners would go so far as to assert that these traditional urban virtues will soon be transformed into vices. On the one hand, with modern means of communication, the required face-to-face contacts can be quickly contrived from widely spaced locations. On the other hand, the most economical manner of providing for universal private transport is to disperse activities widely over a regular grid of roads; this minimizes the need for expensive multi-level intersections and car parks, and makes the maximum use of road space. It provides the highest possible degree of accessibility between *any* two points in an area. It can also provide a spacious setting for homes and other urban uses, allowing plenty of room for expansion and for green landscaping.

In 1940, in one of the few "popular" books ever written about town planning, Thomas Sharp[5] quoted a long passage written by H. G. Wells in 1900[6]. Wells envisaged that "Neither-Town-nor-Country" would be the most appropriate form of human settlement for the motor age. At about the same time Frank Lloyd Wright was designing "Broadacre City". These concepts were anathema to Sharp and a whole generation of planners; but in recent years we have come much closer to them both in theory and practice.

Seen in this broad context, the first round of British post-war new towns, for all their virtues, are an unhappy compromise—too dispersed to generate concentrations of activity on the traditional pattern; not dispersed enough to allow a free flow of motor traffic.

The nearest we have come in England to Broadacre City is the recently published plan for Milton Keynes. This proposes a loose framework within which most urban activities can be freely located; the residential density is to average eight dwellings to the acre—about half that in the Mark I new towns. It is said that the road network proposed for Milton Keynes will encourage the provision of fast and convenient bus services between all parts of the town as well as permitting unrestricted private movement. But it remains to be seen whether such a dispersed low density community will generate enough passengers to make public transport a paying proposition or, alternatively, whether the majority will be prepared to subsidize services for a minority.

At present it is intended that Milton Keynes should present a definite edge to open countryside like any traditional town. But other major urban expansions are taking place not far away; already suggestions have been made to link them together in a vast loosely structured conurbation. "Neither-Town-nor-Country" is coming closer.

It should not be thought that this kind of development would mean the end of the countryside we know and love. Lionel March[7] has shown that the whole of the expected population increase up to the end of the century and beyond could be housed in the London-Liverpool "backbone", at a very low density, and still occupy only a fraction of the surface. The rest of rural Britain could remain much as it is now. It would mean, though, that the age-old excitement of passing sharply between town and country would be a rare experience for future generations.

The plan for Milton Keynes shows a city centre as well as several sub-centres—though the planners will admit to doubts whether the large area they have reserved will be fully taken up by central activities. There is no technical reason why dispersed cities should not have centres. They could be partly covered pedestrian precincts with adequate (and not necessarily ugly) car parks around their edges or under a deck. They would certainly be easier to get to and pleasanter to use than most existing city centres. Londoners will soon be able to see such a centre under construction at Brent Cross.

But while they might be *in* the town, would they truly be *of* it? In an area designed to encourage disperal, such centres would hardly attract the volume and variety of activities found in traditional centres. With low residential densities few people would be living close to the centre, or would go there on foot, or pass through it to go elsewhere. Milton Keynes centre may not even have the city's railway station. Most people would have to make a special journey to the centre; they might not do so very frequently as they would have no reason to prefer it to many competing attractions. There would be little going on there in the evening or on Sundays.

Only the compact higher density plans for Hook, Cumbernauld and

Runcorn have tried to overcome these difficulties. Hook was never built, Cumbernauld is still grappling with the special problems of its site, Runcorn is too new to draw conclusions.

It may be that a Milton Keynes is the best we can do in this century to try and provide for the growing and conflicting needs of a wealthier, more mobile and more aspiring population. Given our present resources and their distribution, our present ideologies and their reflection in what people think they want, perhaps it will be the most generally acceptable pattern of urban life. But will it suit people of the twenty-first century, when there should be much greater resources to devote to the urban environment, when there may have been a reaction to home-based and car-based living, when (let us hope) people may not just have more leisure time but live more leisurely lives?

It may be that generations brought up in these conditions will find new satisfactions to replace the old thrill of exchanging town for country, the old excitement and surprise of a centre that is truly the culmination of urban living. But suppose they don't. Suppose it turns out that the traditional city offered individual and social fulfilment of a kind that no other urban (or suburban) form could replace. Is there any alternative to Broadacre City? Can we restore the qualities of the medieval or Renaissance town on a higher level that recognizes and exploits the benefits of modern technology?

I have suggested elsewhere[8] that "if nobody had to work for more than three days a week or six months a year, and if every family could afford two homes, entirely different physical forms would be possible and necessary . . . Wealthy individuals have nearly always had a town house and a country house; perhaps in a wealthy society everybody will". Is it accidental that some of the finest urban creations—Bath, Brighton, Bloomsbury, for instance—were designed for people with two homes and the leisure to enjoy them? When such advantages cease to be the privilege of a minority perhaps we can once again build compact cities where people can both live and work, can move around on foot or by new forms of public transport, can meet together, purposefully or casually, in a truly urban environment.

A two home/long weekend culture may not appeal to everybody, even in the twenty-first century. Perhaps there are other social bases on which urbanness could be rebuilt. What seems certain is the need to re-examine our planning philosophy.

New forms would require *collective* decisions of far-reaching significance. Urban planners today tend to be obsessed by the doctrine of providing for *individual* freedom of choice. The current doctrine is to plan only the essential framework of roads and engineering services and to do it in such a way that it can support a variety of urban forms resulting from a mass of individual decisions.

To allow the maximum freedom for individuals, families and organizations to make their own choices is an estimable aim—provided they are made within a framework which ensures that the total result is not something nobody would have wanted. This, I take it, is what planning is about. I wonder whether we are striking the delicate balance at the right point.

Certainly the great historical cities have been moulded by social decisions, though usually, it is true, decisions of a dominant minority who imposed their will on society. The challenge of our time (and not only in urban planning) is surely how to democratize social decision-making, how to obtain a consensus on those matters that should be decided collectively, while still leaving adequate scope for individual choice and variety. Unless we meet the challenge, our settlements (and much else besides) will follow in the wake of technology rather than using technology to achieve the ends we set ourselves.

Meanwhile, though we may be right to build Milton Keynes and other cities like it, we should beware of accepting them as a long-term solution, and we should do everything possible to conserve the physical fabric and economic viability of our traditional centres.

Notes

1 Grove, D. J. and Huszar, L. (1964). *The Towns of Ghana*. Accra.
2 Shankland, Cox & Associates (1970). *Deeside*. London.
3 Smith, R. D. P. (1968). The changing urban hierarchy, *Regional Studies*, **2**, pp. 1–19.
4 e.g. Moser, C. A. and Scott, W. (1961). *British Towns: a Statistical Study of their Social and Economic Differences*. London.
5 Sharp, T. (1940). *Town Planning*. London.
6 Wells, H. G. (1900). *Anticipations*. London.
7 March, L. (1969). The spatial organization of hyperurban societies, *Proceedings of Town and Country Planning Summer School*.
8 Grove, D. J. (1968). Physical planning and social change, *in* Young, M. (ed.) *Forecasting and the Social Sciences*. London.

M. G. SMITH

Complexity, size and urbanization

If we are to explore the dynamic relations of the terms complexity, size and urbanization, we must first distinguish their alternative meanings. Since size is clearly the simplest of these conceptions, we may conveniently discuss this first; and since the urbanization to which "complexity" and "size" refer in this essay relates to phenomena that characterize a given class of human settlement, it is to these latter that the concept of size attaches. What then does size denote in relation to such settlements? Their territorial extent, number and density of occupied buildings, number and density of population, or some unspecified combination of some or all of these variables? It is evident that criteria of size may be geographic, demographic or mixed. However, if I interpret correctly the continuing debate about the urban or other status of such pre-colonial Yoruba settlements as Ibadan and Oshogbo[1], sociologists and social anthropologists do not agree that simple territorial extent, number and density of buildings, population size and density, or any constant function of these variables provide unambiguous criteria or indices of urbanization. There remains the possibility that some probably variable combination of these measures is normally associated with urbanization, either as a requisite, a correlate or a consequence.

Dichotomies are beloved of typologists; comparative and theoretical discussions of urbanization abound in them. University courses and learned journals bearing such titles as Urban or Rural Sociology illustrate the assumption that these two classes of human settlement are clearly distinct and house generically contrasting systems of social organization. Despite their fertility, the validity of this dichotomy and its underlying assumptions remains doubtful. Even if we ignore the variety of rural settlements, for which the ecological axioms that underlie this dichotomy imply a corresponding diversity of social systems, it is not always easy to distinguish village from town or town from city. The "rurban" communities of North America mentioned below present other intriguing problems.

These difficulties arise because the urban units to which the concepts of urbanization and urbanism attach are simultaneously conceived as types of settlement and types of society. Sociologists have widely adopted the geographers, empirical distinction between urban and rural settlements, and have employed this as the basis for two contrasting models of forms of social life. Empirical differences in the demographic size, density and divisions of labour that commonly distinguish urban and rural aggregates in Western societies are interpreted as evidence that the geographic, demographic and sociological features that identify and distinguish each of these models bear necessary relations with one another. Inadequate attention to urban aggregates in prehistoric periods and exotic lands has reinforced the axiomatic equations that underlie this antinomy. But since census classifications of settlements as rural or urban according to size and density of their population are patently unsatisfactory, and since parallel classifications that employ criteria of territorial extent or number of buildings would be equally indiscriminate, sociologists committed to the dichotomy that assumes these necessary relations seek their contrasting intrinsic features in social and cultural spheres. The greater complexity of urban social systems is then invoked without adequate attempts to demonstrate universal co-variations of complexity, population size, density or extent. Thus the material minima of urbanization remain obscure.

In part this obscurity reflects the obscurity of the notion of urbanization itself. The term has at least three distinct sets of meanings, which should be distinguished. Like stratification, organization and several other sociological terms in common use, urbanization simultaneously refers to certain types of social process on the one hand, and to certain conditions or states of affairs, presumably the products of such processes, on the other. It may also refer to individuals, social groups and categories whose "urbanization" need not involve their residence in a town[2]; or to settlements that experience the processes or exhibit the characteristics of urbanization, however those are defined; but as the "urbanization" of individuals, social groups, or rural populations presupposes an urban process or centre from which such influences radiate, I shall direct my remarks to the urbanization of settlements. This confronts us once more with the multiple criteria by which urbanization as a process or a state may be defined, namely those geographic, demographic or sociological indices cited above. For example, if urban units are identified geographically by a certain minimum extent and number or density of occupied buildings, urbanization as a process denotes the development of settlements to and beyond this minimum, while urbanization as a state denotes the characteristics of settlements having these attributes. If we define urban units demographically by certain population minima of size and/or density, then as a process the term denotes demographic expansion to or beyond these minima, and as a state or condition the characteristics of units with these attributes. In like fashion, if we define urban units by sociological criteria, urbanization as a process refers to the sequence by which local populations develop these attributes, while the state of urbanization denotes the condition in which such criteria are crystallized.

If sociologists and anthropologists are correct in rejecting material criteria of a demographic or geographic kind as adequate indices of urbanization, they none the less disagree about the particular social and cultural criteria that distinguish the concept as process or as state, and also about their fit with the empirical data. To summarize and exhibit the greater organizational complexity of urban communities, social scientists elaborate polar models that contrast the mobile, secular, differentiated or heterogeneous *gesellschaft* qualities of urban society with the immobile, isolated, sacred, weakly differentiated *gemeinschaft* qualities of non-urban units[3]. Durkheim's powerful contrast between social systems characterized by mechanical and by organic solidarity, and Max Weber's contraposition of traditionalistic and bureaucratized (legal-rational) societies are also relevant[4]. But so too is the distinction that Sjoberg draws between pre-industrial cities in which the contexts, forms and degrees of differentiation, mobility and secularization vary widely, and industrial cities in which, until *apartheid* at Johannesburg demonstrated otherwise, it was assumed that these qualities were necessarily more developed. If we admit the critical relevance and variable socio-cultural organizations of pre-industrial cities for the study of urbanization, we can provisionally reformulate our problem to ask what degrees of differentiation, mobility, secularization and contractual freedom distinguish urban from rural units, and how do these minima relate to such factors as demographic size, density, geographical extent, or the division of labour. This reformulation neither assumes the validity of the conventional rural-urban dichotomy, nor the association of demographic, geographic and socio-cultural variables that underlies it.

Differentiation, mobility, secularization and contractuality are commonly linked with one another as correlates of an advancing division of labour; and division of labour itself is often employed as an exact representation of the levels of social complexity. We must therefore consider whether complexity in social organization provides an appropriate index of urbanization as a process or a state, and try to indicate the conditions that promote or generate it. But to what does such complexity initially refer? Culture, social relations, economy, technology or society? Citing Raymond Firth's exceptionally fine and detailed reports on the Polynesians of Tikopia, Max Gluckman remarks that "The organization which was required to hold together a thousand people on a South Sea island was almost as complicated as that which rules a city like London . . . even though these island societies . . . did not have a cultural apparatus as complex as ours'[6]. Thus Gluckman contrasts complicatedness of social organization, conceived as "networks of social relationships"[7], with complexity of culture, conceived as "the content of those relations . . . the . . . accumulated resources, immaterial as well as material, which the people inherit, employ, transmute, add to and transmit"[8].

We may combine Gluckman's casual distinction between complexity and complication with another that he draws between simplex, single-stranded or functionally specific personal relations, and those which are multiplex, many-stranded or functionally diffuse[9]. Multiplex relations are normally ascriptive, perduring and valued as ends in themselves, while single-purpose relations are commonly optional, transient and perceived as means to other ends. Thus while multiplex relations have greater content and complication than simple ones, social systems with high densities of multiplex relations must normally be simpler than those with equivalent densities of simplex ones. This is demonstrated by the fact that we need to take into account an exponentially increasing number of variables, situations and relations in order to understand or analyse any specific single-stranded relation as social systems increase in the number and diversity of these constituents. Thus complexity and complicatedness may illustrate a continuum in which the number, diversity and proportion of simplex interpersonal relations declines as their multiplexity and complicatedness increases, together with the organizational simplicity of the social system. This ratio is obviously linked with the variable number of individuals with whom the average or typical male adult member of the social system interacts in institutionalized roles and situations. The fewer the number of differentiated roles and role-situations, the greater the ratio of multiplex perduring relations, the higher the complicatedness of the social organization and the lower the complexity of the social system.

Some writers discuss these issues as aspects of social scale[10]. But since scale has unavoidable connotations of size, the term complexity is preferable, and not only because it more directly describes the system properties under discussion, but also because it avoids any positive implications about the relationships between complexity and size of social units. Nonetheless, if social systems vary in complexity as a function of the number and variety of single-interest role relationships and situations they embrace, and if these functionally specialized relations increase in number and variety with the size and density of the interacting population, social complexity should correlate with the size and concentration of social aggregates, and thus with urbanization. We may thus reformulate the general problem of the relations between complexity, urbanization and size to ask what conditions underlie or are associated with sufficient increases in the number and variety of single-stranded relations to exemplify urbanization.

Sociologists have offered several answers to this general question. These answers include the relative decline of ascriptive roles and increase of contractual ones; relative increases of the material and moral densities of social units, increases of secularization, mobility and so on. Since differentiation, heterogeneity, individuation and the division of labour are merely alternative aspects or descriptions of social complexity, they may be employed to measure or illustrate it but not as explanations. While the degrees of differentiation, heterogeneity, secularization, mobility and contractuality that characterize pre-industrial cities are normally less than those to be found in contemporary industrial ones, they are considerably greater than those that characterize the rural populations of the societies to which these pre-industrial cities belong. None the less, it seems unlikely that the role-structure of many pre-industrial cities exceeds those of "rurban" communities in contemporary industrial societies in the number and variety of their differentiated relations. If so, we have to deal not with a dichotomy but a continuum of societal complexities scaled in terms of role components and situations, as some sociologists have argued[11]. This simplifies our problem in several ways and suggests that the relative complexity of social systems is neither an invariable nor a simple function of density and size; for if the "rurban" sectors of contemporary industrial societies are more complex than pre-industrial cities in role structure and organization, neither urbanization nor demographic and geographic factors are directly requisite for such complexity, however commonly associated with it.

To indicate the structural requisites of increasing role differentiation and systemic complexity, we need only contrast the bases of multiplex and simplex relations current in contemporary complex societies. There, most multiplex relations are familial, kin-based, and restricted to the domestic or private domain, while most simplex relations are occupational and economic components of the public domain[12]. Such considerations led Durkheim to identify the developing division of labour as the source and condition of social complexity; but perhaps the reverse is nearer the truth, as Maine's epigram—"the movement of progressive societies has hitherto been a movement from *status* to *contract*"[13]—suggests.

Maine exemplified social systems in which status and role were prescribed by corporate kinship groups by the patrilineal organization of early Rome. He went on to show how these corporations lost their legal sovereignty as their members were emancipated by the development of legally valid procedures and criteria for the differentiation of personal status and roles. The displacement of kinship by territoriality as the basis for political community illustrates one aspect of those general processes of societal transformation and development which accompanied the evolution of the city-states of classical antiquity. It facilitated immigration, and thus the material growth of these city-states, while converting their lineage consociations into stratified orders whose central law regulated individual rights and preserved individual opportunities for the creation of further rights by contractual relations or other procedures of role differentiation. Maine's argument and insights may help us to clarify the tortuous relations of complexity, urbanization and size if they are supplemented by the observations of Ibn Khaldûn and Vico.

Ibn Khaldûn and Vico independently stress the prominence and consequences of social stratification within archaic cities.[14] Vico shows how the population of ancient Rome expanded by attracting immigrants who were differentiated categorically by culture and social conditions as plebs from patricians, and individually as clients from patrons. Ibn Khaldûn attributes the creation of cities to the power and wealth of ruling dynasties; but he dwells at length on the consequences of conquest for those invading groups that impose themselves as ruling strata on cities subjugated by force of arms. In either case the initially rigid stratification of patricians and plebs, of rulers and ruled, erodes as linguistic and cultural assimilation develops within a context of increasing competition for support, prestige and advantage in either stratum. In consequence, as Ibn Khaldûn remarks, "genealogies become confused," thus dissolving the initially exclusive ascriptive division of the community into two structurally distinct blocs, which were either endogamous or aligned by hypergamy. Dissolution of these exclusive strata proceeds by the gradual extension of symmetrical connubium throughout the population, by the gradual if incomplete extension of legal and political rights and obligations to all as citizens, and by their ultimate incorporation in a common cult. Thus, whether the original concentration of population arose through synoecism or voluntary immigration, granted the juxtaposition and interaction of two markedly unequal and closed social orders within a city, their continuous co-operation in economy, defence, and the enforcement of order generates novel interests, activities, situations and social relations which are inconsistent with the original rigid separation of these strata in connubium and kinship.

At Rome the kinship groups whose transformation and dissolution proceeded simultaneously with the growth of centralized city-state were originally exclusive corporations whose members were ascriptively regulated in their internal and external activities by particularistic genealogical relations. Their kin-bound role structures were no less proscriptive than complicated, perduring and multiplex. They restricted individual capacities for independent action because they subsumed and regulated all the essential requirements, interests and rights of individuals as lineage members in such matters as residence, inheritance, succession, property, personal security, socialisation, cult, association, representation and dispute settlement; and since such structures were replicated among all units of identical base, they effectively excluded the differentiation of personal relations by voluntary individual action oriented to other situations and interests. None the less, internal pressures and the exigencies of their continuing association as an exclusive stratum together modified these ascriptive and all-encompassing patrician lineages by eroding the bases of the particularistic and multiplex relations that underlay their mutual exclusions, while generating alternative or supplementary bases and forms of social relation.

As new scope for voluntary and conditional individual association outside the traditional contexts of kinship and marriage developed, so did the number and variety of differentiated roles, situations and procedures for interpersonal association, and with this the complexity of the social systems of which they were increasingly common and prominent components. The ultimate effect of these developments was to render individual identifications with, or obligations to, the previously prescriptive corporations based on kinship and differential status conditional on such factors as individual volition, experience, opportunities and the development of public law. As individuals gradually recognized that the traditional corporations, whether based on kinship, jural status, locality, cult or occupation, were clearly conditional, they also realized their new capacities to create other types of corporation and association to subserve interests of a specifically economic, social, educational, religious or political kind by allocating resources and designing role structures for these particular ends. Thus the prescriptive corporations of uniform base and type that initially encapsulated individuals from cradle to grave in perduring nets of mutually exclusive and multiplex relations were initially supplemented and then supplanted by instrumental, voluntary associations of dyadic or multilateral kinds which individuals were variously free to join, modify, develop or dissolve according to their particular interests. Clientage was a particularly important mode of contractual association. As such processes generated new roles and situations, the range, variety and specificity of interpersonal relations proliferated, together with the diversity in the normative and technical components of these relations, and the complexity, lability, and integration of the system that embraced them. In pre-industrial populations, urban milieux, defined as unusually large and dense concentrations of people, provided the typical matrix for these developments, which normally proceed at rates that reflect the relative strength of the internal and external forces at work to promote and obstruct them.

Notes

1 Bascom, W. B. (1955). Urbanization among the Yoruba, *Amer. J. Sociol.*, **60**; Bascom, W. B. (1957). Urbanization as a traditional African pattern, *Sociol. Rev.*, 7; Bascom, W. B. (1962). Some aspects of Yoruba urbanism, *Amer. Anthrop.*, **54**, Schwab, W. B. (1965). Oshogbo—an urban community?, *in* Kuper, H. (ed.) *Urbanization and Migration in West Africa*. Berkeley and Los Angeles. pp. 85–109.

2 Miner, H. (1965) Urban influences on the rural Hausa, *in* Kuper, H. (ed.) *op. cit.* pp. 110–20.

3 Tönnies, F. (1955). *Community and Association*. London; Becker, H. and Barnes, H. E. (1938). *Social Thought from Lore to Science*. New York. ch. 1; Redfield, R. (1947). The Folk Society, *Amer. J. Sociol.*, **52**.

4 Durkheim, E. (1947). *The Division of Labour in Society*. Glencoe, Ill.; Weber, M. (1947). *The Theory of Social and Economic Organization*. Edinburgh. ch. 1.

5 Sjoberg, G. (1960). *The Pre-industrial City*. New York.

6 Gluckman, M. (1954). Political institutions, *in* Evans-Pritchard, E. E. (ed.) *The Institutions of Primitive Society*. London. pp. 66–7.

7 Radcliffe-Brown, A. R. (1952). *Structure and Function in Primitive Society*. London. pp. 190–3.

8 Firth, R. (1951). *Elements of Social Organization*. London. p. 27.

9 Gluckman, M. (1955). *The Judicial Process among the Barotse of Northern Rhodesia*. London. p. 155 and passim; Gluckman, M. (1962). *Essays on the Ritual of Social Relations*. Manchester. pp. 26f; Bailey, F. G. (1964). Two

Villages in Orissa (India), *in* Gluckman, M. (ed.) *Closed Systems and Open Minds: the Limits of Naiveté in Social Anthropology*. London. pp. 73f; Parsons, T. and Shils, E. (1951). Values, motives and systems of action, *in* Parsons, T. and Shils, E. (eds.) *Towards a General Theory of Action*. New York, pp. 83f.

10 Wilson, G. and Wilson, M. (1945). *The Analysis of Social Change*. London. pp. 26f; Benedict, B. (1966). Sociological characteristics of small territories and their implications for economic development, *in* Banton, M. (ed.) *The Social Anthropology of Complex Societies*. London. pp. 23–35.

11 Miner, H. (1952). The folk-urban continuum, *Amer. Sociol. Rev.*, **17**.

12 For the distinction between private and public domains, see Smith, M. G. (1969). Institutional and political conditions of pluralism, *in* Kuper, L. and Smith, M. G. (eds.) *Pluralism in Africa*. Berkeley and Los Angeles. pp. 38–9.

13 Maine, Sir H. S. (1905). *Ancient Law*. London. p. 141.

14 Ibn Khaldûn (1958). *The Muqaddimah: An Introduction to History*. (Trans. from the Arabic by Franz Rosenthal). New York. Vol. I, pp. 258–300, 311–19, 330–80. Vol. II, pp. 111–56, 235–43, 270–308; Bergin, T. E. and Fisch, M. H. (1961). *The New Science of Giambattista Vico*. New York. pp. 1–18, 135–207.

BRUCE TRIGGER

Determinants of urban growth in pre-industrial societies

Introduction

Anthropological studies of the origins of urban life have been bedevilled by an undue preoccupation with unilineal evolution and evolutionary typology. Anthropologists who have been attracted to these concepts have attempted to discover a single process that would explain the development of all complex societies. Confronted with historical and ethnographic evidence of much greater variation in the economic, social and political organization of non-industrial civilizations than can be accounted for by a single line of development, they have advanced a series of doubtful arguments to sustain the importance of their formulations.

Some have claimed that "pristine" cases of social evolution (i.e. those uninfluenced by more advanced societies elsewhere) are fundamentally different from all others and that understanding them is vital if later developments are to be explained[1]. Grave doubts exist about which civilizations are "pristine" and about how "pristine" a civilization must be to count as "an independent recurrence of cause and effect"[2]. In any case, the resulting sample is too small and too poorly documented to support the ambitious fabric of speculation that has been built upon it. Other evolutionists have deliberately limited their search for regularities to civilizations that have developed in broadly similar environments, such as river valleys located in arid or semi-arid regions[3]. These societies as well are generally assumed to have enjoyed a priority in the development of civilization. Yet another example of efforts to use facile ecological explanations to restrict the theoretical significance of observed structural variation is the recent suggestion that greater centralization took place in the Inca Empire than in the Aztec one because of the smaller size and wider spacing of central Andean population clusters[4]. Fortunately, no modern anthropologist has carried unilineal ideas as far as has Gideon Sjoberg[5], who assumes so close an association between writing and urbanism that he feels free to postulate, without additional supporting evidence, that cities are present in all literate cultures and to explain away urban phenomena wherever they occur in the absence of writing.

The shortcomings of a unilineal approach have not been significantly overcome by token efforts to attribute the origins of civilizations to a limited range of primary causal factors, such as irrigation for some and trade for others[6]. What seems to be required is a more piecemeal and institutional approach to complex societies. It is clear, for example, that while the state is a necessary concomitant of urban life, many states have existed without cities[7]. This suggests that for some purposes the development of the city and of complex forms of political organization may profitably be discussed separately. Such investigations will help to determine both the complete range of factors which promote the growth of various institutions and the variety of ways in which these evolving institutions interact with one another.

Not long ago it seemed possible that the discussion of whether or not civilizations can exist without cities might lead anthropologists in this direction[8]. Unfortunately, no intensive studies were made to determine why some civilizations should develop cities and others should not, and recent discussions have sufficiently blurred the distinction between cities and ceremonial centres that the latter concept now appears to have little independent meaning[9]. Moreover, archaeological discoveries have demonstrated the urban status of many key sites formerly believed to be only sparsely or periodically inhabited ritual centres[10].

Possibly the weakest aspect of current theorizing about the origin of complex societies is the inadequacy of the archaeological evidence which is used to support and nourish it. Virtually nothing is known about the early phases of some civilizations and for most, if not all, our knowledge remains fragmentary, equivocal and will undoubtedly be subject to major revisions for years to come[11]. Because of this, I propose initially to examine those factors which promote the growth of urbanism in pre-industrial societies from a structural rather than from a genetic or historical point of view. In the following discussion, no distinction will be drawn between city states and cities which are either founded or grow up spontaneously within pre-existing political structures. In this way it is possible to make use of information about a large number of contemporary or historically well-documented cities. In the final section an effort will be made to set these factors into a tentative structural-developmental framework.

Theoretical orientation

In spite of clear-cut and often highly technical definitions of cities in specific cultures, neither anthropologists nor geographers have been able to agree on a generally acceptable cross-cultural definition of urbanism[12]. Arbitrary definitions based on population size or population density have not won general acceptance, nor can it be agreed that only communities that have a sizable majority of their inhabitants engaged in varied non-agricultural pursuits can be classified as towns or cities[13]. The majority of inhabitants of even the principal Yoruba communities in West Africa worked as farmers, as did many of the inhabitants of the largest communities in ancient Sumer and prehispanic Mexico[14]. On the other hand, in at least some complex societies, many small communities, usually classified as villages, derive most of their income from services they provide for a rural hinterland and few, if any, of the inhabitants of such communities need be agriculturalists. I am thinking in particular of the British village replete with its pub, general store and church. Such communities are functionally distinct from some predominantly agricultural peasant communities and from self-sufficient Neolithic settlements, although all these types of communities are called, depending on their size, villages or towns[15]. It is clear that the latter terms have no precise functional or evolutionary significance. Moreover, size alone cannot serve as an index of specialization, at least for small and medium-sized communities.

Armed with this understanding, it is possible if not to define the city at least to suggest a crude working model that has cross-cultural significance. It is generally agreed that whatever else a city may be it is a unit of settlement which performs specialized functions in relationship to a broader hinterland [16]. These functions distinguish urban-type communities from self-sufficient Neolithic villages or hunting bands, whose closed economies produce only for internal consumption. Moreover, while numerous inhabitants of a city may engage in food production, it is agreed that the specialized functions of a city are not agricultural in nature[17]. Food production is often a highly specialized activity and it may be hierarchically organized as, for example, on Roman *latifundia* or with plantation agriculture generally. The specialized relationship between agriculture and the land tends, however, to concentrate similar rather than different specialities in any one area at one time, while surplus produce is marketed at specific points rather than being distributed to a surrounding hinterland. Moreover, in the absence of opposing tendencies, agricultural activities encourage the dispersal rather than the concentration of producers in order to locate them in maximum proximity to the soil they cultivate. These tendencies are the contrary of those involved in the evolution of urban groupings[18].

On the other hand, the city appears to differ from the small English village mentioned above mainly in the number and complexity of the functions it performs and in the size of the hinterland it influences. Service villages, towns, and cities are at best arbitrary divisions of a continuum. As Vining[19] has put it "Like pond, pool and lake the terms hamlet, village and town are convenient modes of expression but they do not refer to structurally distinct entities". Definitions based on arbitrary quantitative measures are possible[20], but for a dicussion of factors that promote the growth of urbanism such distinctions are of little importance.

Finally, to round out our model we may note that cities or villages differ from military camps or monasteries, which also perform specialized functions within a larger social setting, by having a community structure composed of all ages and both sexes. Such institutions also differ specifically from cities in the limited range of their functions.

In so far as specialized non-agricultural functions are among the generally accepted attributes of urbanism, any discussion of the development of such communities can profitably be set within a framework of more general theories of settlement location as these have been developed by human geographers[21]. Among the premises underlying these theories which are of special importance for understanding urban growth are the following:

1. *There is a tendency for human activities to be hierarchical in character and for this to be reflected in spatial organization*

As cultures become more complex, activities which take place within and concern only a single community are supplemented by ones which influence a wider area. These may concern productive, administrative, social, religious or military matters. As a rule, the wider the area that a function performed in a single location influences, the less frequently is the function performed in the landscape. Hence, with increasing complexity, a hierarchy of locations may develop with respect to any one kind of activity, the higher or more specialized functions being performed from a smaller number of centres.

2. *With increasing complexity there is a tendency for activities and social institutions to be more clearly defined and for their personnel to be more highly specialized*

In complex societies there is an intensive division of labour with individuals specializing in particular types of production, distribution, administration, religious, military and service activities. The organization of these activities, as noted above, tends to be hierarchical in varying degrees. While some hierarchies, such as those of a craft guild, may be co-extensive with no more than a single community, others, such as a government bureaucracy, may extend over a much broader society. Within these hierarchies a larger number of people perform lower level functions than perform higher level ones.

3. *Human activities tend to be focal in character in order to take advantage of scale economies*

In order to increase efficiency, activities susceptible to varying degrees of interrelationship tend to be concentrated at a single point. This tendency is governed by rules of accessibility and movement-minimization as they relate to the activities concerned[22]. In accordance with these rules, locations which serve one kind of function frequently tend to serve another. In combination with the hierarchical premise outlined above, such tendencies give rise to a hierarchy of locations varying in terms of accessibility and the size of the area they serve and influence. The scale and density of such a hierarchy of centres will vary according to geographical conditions affecting accessibility and societal factors such as overall population density.

4. *The size of communities tends to vary with the number of functions they perform*

Various studies which have correlated community size either directly with the number of functions performed or with related measures such as "trade area" or "total population served" have confirmed this observation[23]. This principle, in combination with the hierarchical premise, explains why larger communities are less frequent in a society than are smaller ones. Larger centres perform most types of specialized functions that are performed in smaller centres, but in addition they discharge more specialized functions which are carried out at a more limited number of places. Various studies have investigated regularities in the rank-size of communities in different parts of the world, but mathematical variations between cultures indicate that a considerable range of factors influences rank-size[24].

Determinants

In this section I wish to examine the range of factors which have been noted as promoting increases in the size of urban populations. These factors include, among others, ones which have led to the formation of cities. They do not include, however, those which influence either the specific location of cities or their internal layout. While many or even all of these factors may influence the size of any single community, my aim is to discuss each factor separately, as far as this is possible. Hence, qualifications which are ignored in one section may be found covered in another.

Among the factors which are, or have been alleged to be, correlated with an increase in the population of pre-industrial urban centres are the following:

1. *An increase in food supply.* It is axiomatic that any community that is to survive must be able to provide a reliable source of food for its members; but how it does so depends upon a variety of economic and political factors. Among some urban groups a considerable amount of food is produced by the inhabitants of the city working on either a full-time or part-time basis in the surrounding countryside. We have already observed that in pre-colonial times nearly all of the Yoruba engaged in agriculture to some degree and Sanders[25] has estimated that a high percentage of the inhabitants of the

Aztec capital did the same. During certain periods an "appreciable proportion" of the inhabitants of most Islamic Middle Eastern cities have been farmers cultivating adjacent fields[26]. This pattern has continued to the present in some cities such as Damascus[27] and Kerman, Iran (where, in 1956, 13% of the workers were engaged in agriculture[28]). An alternative is for city dwellers to purchase food from farmers in return for manufactured products and professional services. This kind of reciprocal relationship between city and countryside has played an important role in theorising about the development of European cities[29]. Still others have obtained supplies of food in an exploitive manner. The government officials, soldiers, landlords and skilled artisans who inhabited the traditional Chinese *ch'eng* or administrative towns produced few goods for sale in the surrounding countryside but were fed with food collected from there as rent or taxes[30].

In pre-industrial societies transportation requires the expenditure of considerable time and human energy; hence even where extensive river systems make transportation relatively easy[31] there is a tendency for bulk items, such as food, to be produced as near as possible to where they are consumed[32]. As demand grows, significant increases in productivity can be achieved through more intensive forms of cultivation, involving processes such as irrigation, fertilization and systematic crop rotation[33]. The efficiency of these more intensive methods of agriculture can be measured as a function of the additional labour that is required compared with the costs of collecting food over a larger hinterland. The ancient trading cities of Arabia, many of which were located astride highly profitable trade routes in the desert, are an extreme example of such developments[34]. There the profits to be derived from trade made a heavy investment in vast catchment areas and systems for flash-flood irrigation worthwhile. These systems, which were expensive not only to build but also to maintain, were abandoned with the collapse of the trade routes which supported them. Other factors may reinforce a trend towards more intensive agriculture. Around Hausa cities in northern Nigeria one finds unusually dense farm populations growing food for an urban market[35]. In this instance intensive agriculture is stimulated not only by demand but also by the additional protection that a farmer receives as a result of living near a city.

Pre-industrial technologies by their very nature impose narrow limitations on the degree to which agricultural production may be intensified and goods may be transported. In the absence of additional complicating factors, these two factors suggest an optimal size to which individual cities may grow and still be able to feed themselves. Under certain circumstances, however, particularly with the growth of empires, the cost of transportation may be rendered less restrictive as far as the feeding of dominant cities is concerned. In the third millennium B.C., the Mesopotamian city of Lagash claimed to have exacted about 10,800 metric tons of grain from the neighbouring city state of Umma as the indemnity for an unsuccessful rebellion[36] and the Aztec capital of Tenochtitlán is estimated to have received 52,600 tons of foodstuffs annually as tribute, in addition to vast amounts of clothing and luxury items[37]. All of these goods had to be transported to the latter city by human carriers. The imperial *annona* of Rome imported vast amounts of grain each year, much of it brought across the Mediterranean from Egypt and North Africa. This grain was paid for out of taxes and tribute that flowed into the imperial treasury and was distributed free to all resident citizens of Rome who wished to apply for it, possibly more than 100,000 people[38]. Foodstuffs from China appear to have supported the Mongol capital of Karakorum, which was founded in the pastoralist heart of a vast empire[39]. Needless to say, the collapse of political power eliminates the economic base which supports such populations.

2. *Increasing population and/or rural unemployment.* All cities depend on an external population as suppliers of food and raw materials and as consumers of their goods and services. A sparse hinterland population increases the cost of transport and communication and this factor is probably important in explaining the general failure of urban-type communities to develop among food collectors, pastoralists or extensive agriculturalists[40]. It would appear that, within the limits of effective agricutural production, the denser the overall population is, the more urban centres it can support.

On the other hand, there does not appear to be any simple mathematical relationship between the population density of a region and the extent of urbanization. The Ibo, for example, did not develop cities although their overall population was about the same as that of the Yoruba and their population density about double[41]. As will become apparent below, urbanization is influenced by a variety of additional factors. Moreover, under special circumstances it is not impossible for cities to develop in areas with low population densities. Trade cities, such as Petra and Palmyra, grew up in the desert and Karakorum, which was supported by an imperial tributary system, was founded in a predominately pastoral region. It is significant, however, that cities of both these types ultimately depended upon regions that were densely populated.

Cities also act as an outlet for surplus rural population. This may come about, as it does in many underdeveloped countries, from the failure of more intensive agriculture to absorb a growing rural population[42]. Overpopulation may also come about as a result of new techniques of agriculture displacing a traditional farm population. The development of slave-worked *latifundia* in Roman Italy, as the peasants were conscripted into the army, or the displacement of peasants during the period of the enclosures in Tudor and Stuart England provide examples of such processes[43]. It has been noted for many regions that an influx of rural population has been essential to maintain the population of towns and cities, where high death rates were brought about by poor sanitary conditions and were raised sharply every few years by outbreaks of plague, cholera and other epidemics[44]. It is curious that anthropologists have paid relatively little attention to this factor in building models of rural-urban relations. Sometimes, however, the rural population entering a city exceeds the capacity of the urban economy to provide employment for it. Such immigrants as may remain form a class of urban unemployed; thus constituting a separate category contributing to urban growth. Many tens of thousands of the inhabitants of imperial Rome belonged to this class[45].

Finally, while urban colonies such as those the Greeks established around the Black Sea and in the western Mediterranean may have been founded mainly for trade, the pressure of population in encouraging this process is not to be ignored. The urban nature of these colonies reflects the deeply established tradition of urban life in the Greek homeland[46].

3. *Craft specialization.* Handicraft production is an extremely important urban activity and the population of many non-industrial cities has supported itself almost entirely with earnings from it. It has been observed that the nature of manufacturing in urban centres is inherently different from that which occurs elsewhere. My aim is to examine briefly the nature of this difference.

Cities occur in complex, socially stratified societies. In even the most puritanical of such societies, the powerful and the wealthy seek to possess clothing, jewellery, utensils, furniture and houses that are more luxurious than those of ordinary people. These serve both for comfort and as status symbols. The poorer classes may also desire such goods and obtain some of them on a limited scale unless prevented from doing so by sumptuary laws or grinding poverty. The result is a distinction between the everyday tools and implements of the lower classes, which usually are manufactured by craftsmen who produce a wide range of goods, and élite items which are produced by specialists who are more highly skilled because they concentrate on a narrower range of work. For example, a Berber housewife may tan skins for her own use but the production of Morocco leather in Fez requires twenty consecutive operations each performed by a different, highly trained work group[47]. Likewise, a master carver may design and add the finishing touches to a piece of work, leaving to apprentices the tedious but less demanding job of chiselling out the design[48]. This kind of a division of labour results not only in a higher level of technical perfection but also in a saving in production costs, since the work can be divided among more and less skilled craftsmen. To co-ordinate production of this sort an effective hierarchical organization is necessary. This may vary in scale from a master with his journeymen and apprentices to a guild organization which regulates the production of a particular line of goods throughout an entire city or region, setting standards, prices and possibly even production quotas.

In light of the premises outlined above, two rules can be deduced as governing production of this sort: (1) the more individuals are involved in a production unit, the more they will produce and the larger market they will require and (2) the more specialized and expensive a good is, the fewer people are able to buy it, hence the fewer are the centres in which it can be produced economically. This is true whether the goods are consumed within the community where they are produced (as they may be, for example, where a royal court creates a large internal market for luxury goods) or are exported elsewhere.

Many forms of specialization do not of themselves necessarily induce the development of larger communities. In the following paragraphs I wish to digress briefly to discuss alternative situations in which craft specialization may be carried on. What is pointed out here also applies to the specialized activities associated with administration, religion, defence and education, which are discussed below. Limitations of space prevent a similar detailed examination of each of these topics under its respective heading.

Even the most primitive craft production is specialized, if only along age and sex lines. However, specialization beyond what is possible in the closed economy of a Neolithic village can be achieved in other ways than by centralizing it in towns and cities. In various parts of the world, neighbouring villages participate in trading networks which provide a wider market for the agricultural produce and manufactured goods of all concerned[49]. Such an arrangement allows the craftsmen in any one village, who may also be part-time farmers, to concentrate on the production of a single line of goods such as pottery, leather objects or metalware, thus permitting specialization resembling that found in larger centres. In terms of the specialization involved, very little would appear to separate a well-developed system of this sort from the Yoruba situation, where although craft production occurs in an urban setting it is carried on largely by part-time specialists. Yet another way to increase markets if demand is low is for itinerant craftsmen to wander from group to group offering their services[50].

Nevertheless both methods suffer from certain inherent limitations: (*a*) the necessity for craftsmen to transport either their produce in bulk to market or their tools to wherever their services are required and (*b*) their failure to bring potentially related crafts into proximity with one another, thus limiting the degree to which interdependence and further specialization can develop. By its very nature, itinerant production is associated with relatively simple technologies which require little permanent equipment. The specialized and bulky tools needed for crafts such as glassmaking would almost certainly preclude this pattern of production[51].

Even in urbanized societies goods are not infrequently produced in non-urban settings. This may be for economic or institutional reasons or both. In Roman Britain, tile factories and leather and metal working establishments were attached to certain military camps in order to supply the needs of the army[52]. In part, this was done to avoid the transportation costs involved in moving goods from the civil zone of southern England into the military zones in the north and west. Yet it may also reflect the desire of a major institution, the Roman army, to economize and at the same time to be more independent by supplying its own necessities. A similar desire for self-sufficiency can be noted (this time in an urban setting) in the Sumerian temple communities[53], or on Roman estates which often had their own mills, iron foundries and potteries. It has been pointed out that while the manufactured products of these estates were occasionally exported, manufacturing was generally carried out to supply internal requirements. Thus there was little impetus for improving techniques or increasing productivity[54].

On the other hand, the dispersal of some only moderately specialized industries, such as the potteries producing Castor Ware, which were spread out over 20 square miles in the Nene Valley in Roman Britain, may have been carried out for solid economic reasons; in this case to achieve savings by locating close to sources of bulky raw materials such as clay and wood used for firing kilns[55]. The nearby town of Durobrivae served as a marketing centre for this industry. Likewise, the production of carpets in many parts of the Near East, while organized by city merchants, is carried on as a cottage industry in the villages[56].

Yet in spite of these partial exceptions, it would appear that in most industries the degree of specialization which is possible and desired in complex societies is most frequently achieved only by the concentration of different kinds of manufacturing in a common centre. The proximity of related trades encourages greater interdependence and hence more specialization. For example, in such situations the production of tools and the working up of raw materials for various crafts can be left to other specialists, whereas if each craft is geographically separate workmen normally prepare their own tools and materials[57]. Such concentration reflects the general rule that non-agricultural activities tend to be focal in character to take advantage of scale economies.

Another result of scale economies is that, whatever else major cities manufacture, each tends to specialize in the production of a limited range of items that have a national or international reputation. In the ancient Mediterranean sphere, cities were famous "for the presence within their walls of one or two main technologies such as the making of pottery or glassware, or the manufacture of jewellery . . ."[58]. Similar observations have also been made concerning Islamic, Chinese and South Asian cities and also for cities in the New World[59]. Low-cost handicrafts for which there is a popular market tend to be produced in a large number of cities and because they do not repay transportation costs they normally do not circulate beyond the local hinterlands of these cities. However, as we have already noted, the more specialized goods become, the fewer are the centres in which any one item can be produced economically. This gives rise to a hierarchy of cities which produce for local, local and regional, and local, regional and international markets respectively; with the size of the city normally varying according to its position in the hierarchy (size being roughly determined by the number of functions a city performs). In medieval Fez, for example, over 20,000 persons were employed in weaving, but much of the cloth produced was either used locally or sold in the surrounding countryside. On the other hand, a limited amount of very expensive cloth and most of the production of the city's famous tanning and leather goods industries were exported all over the Arab world[60]. By itself, this hierarchical tendency does not appear to generate a pyramidal structure with one exceptionally large city in each major region producing the greatest amount of specialized goods. Instead, as has been observed in the Middle East and the Mediterranean[61], it tends to produce a network of major cities each of which produces one, or at most a few, major specialities which are then exchanged throughout the system. At a higher level of specialization and on a larger overall scale, such cities are analogous to the villages participating in a reciprocal trading pattern discussed above.

4. *Marketing and Trade.* More specialized production increases the need for mechanisms to distribute what is being produced. Where craftsmen are attached to the royal court or are employed by institutions such as the church or army, these organizations normally supervise the distribution of goods. In this manner, such organizations become major subsystems within the economy and their activities give rise to corps of specialists concerned with storage, transportation and keeping accounts.

Bilateral exchange takes place at the most basic level by a producer selling his goods either in his own shop or in the market place. Increasing complexity tends, however, to produce specialists who are concerned exclusively with the buying and selling of produce. These may be either market women travelling between country and town, retail merchants who sell goods within the community or wholesale import-export merchants who carry on long-distance trade. Auxiliary to these merchants are men who transport goods and craftsmen who construct and maintain the means of transportation. In the thirteenth century, the population of Cairo was reported to have included the crews of 36,000 Nile boats and 30,000 renters of mules and donkeys[62].

Even in urban societies trade, like production, goes on in non-urban settings. Many rural Celtic temples in Roman Britain were the site of periodic fairs[63]. These temples were often located near tribal borders and in pre-Roman times they had probably served as neutral meeting places between rival groups. In the Roman period, the rural trade at such temples was supplementary to that in the cantonal (tribal) capitals[64]. Long-distance trade was centred in the towns rather than at these rural sites because the former were also centres of production, because of the scale economies that were gained through centralization and because of the greater security and protection that urban centres offered.

Because of the high cost of transportation in non-industrial societies, long-distance trade is restricted mainly to goods and materials that are of great value or can be produced only in limited areas. Otherwise local copies would drive the more expensive imported items off the market[65]. In such societies, leading import-export merchants are often among the wealthiest members of an urban community. Some may be resident foreigners and occasionally such merchants may live in special communities on the outskirts of native cities, as Assyrian traders did at Kanesh and other places in Anatolia early in the second millennium B.C.[66]. Even in Mesopotamian cities, traders appear to have lived in a special "harbour" district in a manner which suggests that the society as a whole perceived a significant cleavage between the activities of the merchants and those of the city proper. Oppenheim[67] suggests that the failure to discover references to overland trade in the texts from Sippar may be because once merchants became rich they invested their money in land before moving from the unexcavated "harbour" area into the city. In spite of many privileges, Aztec long-distance traders had a similar, somewhat culturally anomalous relationship to their society[68]. In medieval Europe, on the other hand, rich merchants often played a key role in the political life of towns and cities.

Numerous studies indicate the importance of long-distance trade for the development of urbanism in many parts of the world. The cities of the Levant appear to have flourished as trading centres, first between Mesopotamia and Egypt and later between the Mediterranean and regions farther east[69]. In addition to being collecting centres for the incense trade, the cities that developed in the first millennium B.C. in south Arabia were entrepôts in vast networks of trade that embraced India, East Africa and the Mediterranean[70]. Long-distance trade also may have played an important role in the development of civilization in the Aegean area, although in pre-Classical times it appears to have supported urban populations on only a modest scale[71]. Henri Pirenne[72] has brilliantly documented the importance of trade for the revival and development of urban life in medieval Europe. The effectiveness of trading cities often has depended upon their independence of rival foreign powers; hence we have examples of powerful states recognizing the freedom of vulnerable coastal trading cities, such as Sidon and Tyre, or "desert ports", like Petra and Palmyra[73]. In so far as a city is dependent upon long-distance trade, its prosperity and population can be adversely affected if political or economic changes lead to a decline in trade or if alternative trade routes are discovered. Where trade constitutes the main resource of a city such developments may lead to its total collapse.

5. *Landlords*. In pre-industrial cities live numerous families and individuals whose ancestors made large fortunes and whose income is derived largely or wholly from investments in land and other forms of real estate[74]. In China, many landlords lived in the walled *ch'engs* or administrative towns. They were supported by the income from their lands but preferred to reside in the cities because life there was more interesting and because these well-garrisoned

towns offered protection to their lives and household property against peasant uprisings[75].

6. *Administration.* Urban centres require administration to co-ordinate the various groups who inhabit them. Within a city one finds government officials, legal officers, police and sometimes public utility workers whose basic concern is with maintaining order and keeping the public sector operating. Large communities are frequently divided into districts along kinship, ethnic or occupational lines. In these cases the city government may be organized on a two or more tiered system, with perhaps only the officials on the highest level being full-time specialists[76].

Cities also frequently serve as centres of administration for larger political units. Occasionally, where the political structure is weak and decentralized, as it was in medieval Europe, rural castles rather than cities served as administrative centres for the surrounding countryside. Even in feudal societies, however, the ruling élite frequently prefer to live in cities[77].

In city-states, such as those in ancient Mesopotamia, Greece or prehispanic Mexico, the city administration may also be the government of the entire state. In such cases, the added role of defending the state may provide employment for considerable numbers of professional soldiers and artisans who specialize in making military equipment and these activities may require considerable hierarchical organization[78]. An even greater need for administrators, as well as for soldiers and artisans, is found in city-states that have acquired political hegemony over their neighbours or in cities that are founded as administrative centres within larger pre-existing political units. In these cities, and particularly in those that contain a royal court, there is also a thirst for ostentation which provides employment for large numbers of retainers and artisans. The palace staff at Susa in Early Dynastic III times numbered about 950 men and women; after Mesopotamia was unified by Sargon I in the succeeding period, 5400 men are reported to have eaten daily in the royal palace[79].

The increased population of capital cities is supported by tribute which may be collected in the form of food, raw materials and manufactured goods. The control that rulers exercise over the distribution of this tribute constitutes an additional source of power for them and helps produce a more stratified society than might otherwise be possible or tolerated. For example, one of the functions which enhanced the power of the Aztec emperor was his ability to distribute food and clothing to the inhabitants of Tenochtitlán during one entire month each year[80]. It would appear that anthropologists have generally underestimated the role of conquest as a means by which a particular city or ruling group can increase its access to wealth. Oppenheim has argued that "real prosperity came to a Mesopotamian city only when it had in its midst the palace of a victorious king"[81]. While an ordinary city was poor and a prey to invading armies, the sanctuary and palace of a dominant city were sumptuously adorned and traders, craftsmen and retainers flocked thither to serve its ruler. Not infrequently, the rulers of conquered regions, along with their retainers, were settled for safe-keeping in these cities as well. As a result of their role as centres of administration, political control and court life, cities that were the capitals of successful empires or tributary systems often came to exceed in size any other cities within their sphere of influence[82].

Some capitals, such as Rome and Tenochtitlán, began as city-states but as they acquired control over an increasing number of other city-states and regions they expanded greatly in size. Other cities are founded, or designated by decree, to be the centres of administration and court life for existing states. Babylon grew to be the largest city in Mesopotamia after it was selected to be the capital by the second ruler of the Hammurabi dynasty[83]. When the city of Fez, which had hitherto been a centre of craft production, was designated as the capital of the Marinide empire in the thirteenth century, its rulers left the old city largely intact but founded a smaller new district about 750 yards to the east of it. This new city, later called Fez Jedid (New Fez), contained the palace of the sovereign, the residences of principal court dignitaries and two quarters occupied by foreign troops loyal to the king, as well as eventually the Jewish ghetto, which was under royal protection[84]. While the Old City was governed by functionaries appointed by the king, who acted in consultation with leaders nominated by the notables of the community, Fez Jedid was under direct royal rule. When the king was away fighting or touring the provinces, the official city was largely abandoned.

While cities may grow to exceptional size as the actual centre or as symbols of political power, the withdrawal of such support can very quickly undermine their ability to support the population they have acquired. Assur was abandoned with the fall of the Assyrian Empire[85] and following the disintegration of Roman power in the West in the fifth century A.D. Rome declined, to become a small town. Even more dramatic fates may befall capitals in states which remain large and prosperous. The ancient Egyptian city of Akhetaton was founded, overtly for religious reasons, as a new court centre. Within a few years its houses, palaces and temples sprawled along eight miles of river front and the city functioned as the administrative centre of the Egyptian empire. Yet, following the rejection of its founder's religious innovations, the court left the city and only a few decades after the city had begun the site was once again uninhabited[86]. Likewise, the vast city of Samarra was founded in A.D. 836 by the Caliph Mu'tasim on a site removed from the major trade routes of the Near East but also remote from the political turmoil of the previous capital of Baghdad. Despite the compulsory settlement of thousands of merchants and artisans in the new city, it endured for less than fifty years before another political shift led to its abandonment[87].

In political units that are larger than city states there is a tendency for secondary administrative centres to arise. In keeping with a hierarchical political organization, these cities usually contain a smaller number of soldiers and administrators than does a capital, hence they are usually smaller.

Like capitals, they may be new foundations built for an express purpose or their administrative functions may have been grafted on to an existing community.

Depending on the general nature of the political organization, secondary administrative centres serve a variety of different functions[88]. The Chinese *ch'eng* was inhabited by the emperor's representative and government officials, with a garrison to support them. Often there was arable land within the city walls so that food might be grown in time of siege. The purpose of such a city was to uphold royal authority in the region and to see that taxes were collected. Fei[89] has described them as instruments of power in the hand of the ruling classes, a symbol of power and a necessity to keep it. On the other hand, the largest towns in Roman Britain, apart from London, which was the commercial and administrative centre for the whole province, were mainly cantonal (tribal) capitals. These towns, which had been founded by the Roman conquerors as centres of regional self-government and which were the meeting places of the cantonal councils, were provided with civic centres and market places, baths, temples, official inns and some place of amusement. Apart from their minor political functions these communities had obviously been founded as centres of Roman life for the acculturating, and hopefully ever more pro-Roman, tribal aristocracy of Britain. In spite of the development of considerable handicraft industries in some of these towns, none ever acquired a particularly large population nor were the cantons ever viewed as city-states by the inhabitants of these communities. Instead the towns continued to be viewed in a totally un-Roman way as regional centres[90]. Secondary administrative centres also frequently offer inns and other amenities for travelling officials and for the central courier service, and these provide employment for numerous servants who are also urban dwellers[91].

7. *Defence*. The defence of cities is sometimes a source of specialized employment for some of their inhabitants, although the men so employed may not necessarily be distinct from the soldiers who are part of the administrative-military complex discussed above. Throughout most of the Old World, cities were surrounded by walls which protected them against attacks and political coercion or made them a strong point for dominating the surrounding countryside[92]. The absence of walls in this area can generally be correlated with a strong central government which was capable of defending and policing its entire territory in a rigorous manner[93]. In the Near East, the different wards of a city were often separated by walls pierced only by narrow gates that could be barred at night[94]; thus supporting the observation that walls are constructed not only as defences against an external enemy but also to protect a city against internal disorder and brigandage. Such concerns with internal security reflect the general inefficiency of law enforcement in many pre-industrial societies.

The security against external threats that large communities can offer men and property accounts in part for the tendency of farmers to settle in or near urban centres, in so far as this is possible. Wars in the early nineteenth century stimulated the growth of the large Yoruba cities at least partly in this manner[95]. Insecurity in the Islamic world explains the presence of a farming population in cities such as Damascus[96] and similar factors have been seen at work in the development of urbanism in Mesopotamia[97]. Even more urgently, however, wherever there is danger palaces, large temples, valued industries and the houses of the rich seek the protection of large centres. While the protection of such units is possible in a dispersed setting, as can be seen in the case of Egyptian temple complexes with their fortified temenos walls, considerable economies can be effected by concentrating as much valuable property as possible in a single fortified location. Where there is danger, the natural course seems to be for the specialized institutions of society to huddle together behind some sort of fortification.

Cities also grow through the forced settlement of dangerous elements within them. We have already noted the forced removal of hostages, or entire defeated élites, to the cities of their conquerors. The Bedouin population of the Near East is difficult to control and represents a threat to authority; hence powerful sedentary governments have frequently attempted to make them settle down, either as farmers or in cities as government-employed mercenaries[98].

Large states often seek to protect their frontiers with complex systems of fortifications. The ancient Egyptian cataract forts, the Great Wall of China and the Roman *limites* provide examples of such fortifications[99]. As much as 5% of the total population of Roman Britain may have been employed as soldiers along the northern and western frontiers of the province[100]. The nature of these latter defences required that most soldiers be dispersed in a large number of small camps. While such an arrangement does not give rise to large urban concentrations, the British defence complex did support several towns which functioned as supply centres and gave rise to four colonies of veterans. Small settlements of merchants and craftsmen also grew up outside many of the forts and some of these achieved the status of self-governing towns (*vici*)[101]. Many famous Arab cities, such as Basra, Kufa, Fustat and Kairouan began as military camps which the Arab armies founded on the edge of the desert and from which they dominated conquered territory[102].

8. *Religion.* In complex societies, official religious organizations, like those of a governmental or military nature, tend to be hierarchized, with the upper echelons administering a heavy investment in property and trained personnel. While religious concepts may dictate that certain temples and other religious buildings be located in rural settings, the major institutions are usually situated in urban centres. The largest and most important temples are often in the centre of the city, thus expressing in spatial terms the theological rationalization of the unity of the city or state[103]. In turn, the various divisions of the city may have their own subsidiary cult centres, each located in its respective

quarter. Altogether, such temples support a considerable number of priests and clerks and provide work for numerous urban craftsmen who might otherwise not find employment.

In ethnically heterogeneous empires or areas where an international religion predominates, the correlation between the political and religious hierarchies may not be so close as it is where national or ethnic religions prevail. In order to assure greater independence, the religious organization may prefer to have its headquarters elsewhere than in an important secular centre. In such cases, the presence of religious administrative and cult functions may contribute greatly to the prosperity and population of a smaller urban community. Medieval London and Canterbury provide an example of such a contrast between a secular and a religious centre.

Religious pilgrimages may also be a factor contributing to urban growth. The inhabitants of Mecca derive much of their income from the annual Moslem pilgrimage, mainly from the fees they charge for the food and accommodation they provide for pilgrims. Mecca was both a local cult centre and a trading community in pre-Islamic times, but it is doubtful that today without the pilgrimage it would be anything more than a small village, if it existed. Other cities in the Near East and Europe derive a supplementary income from religious tourism which occurs either periodically or throughout the year[104].

9. *Secular tourism.* Secular travel and relaxation also provide a livelihood for urban dwellers. Inns, either privately or officially managed, are a feature of many towns and cities. These provide accommodation for travelling merchants and government officials and also for the individual travelling for pleasure. On occasion in the ancient world even whole towns served as tourist resorts. Roman examples include the aristocratic beach resort of Baiae in Italy and British towns such as Bath, or Corbridge and Carlisle, which in addition to being military storehouses functioned as holiday towns for soldiers stationed along Hadrian's Wall[105].

10. *Education.* Another source of population for urban centres is provided by educational institutions. Formal instruction appears to constitute a part-time or full-time profession in all ancient civilizations, but some towns, such as Athens in the Hellenistic and Roman periods, attracted teachers and students in large numbers[106]. In the Middle Ages, Islamic theological schools housed many thousands of students, while their teachers played an important role in the political and intellectual life of the cities in which they were located. For the most part these schools were located in the political capitals, which have always been the main cultural centres in the Islamic world. In Europe, major educational institutions also developed in smaller communities, such as Cambridge, Salamanca, Coimbra and Uppsala, where their contribution to the support of a local population was proportionately much greater[107].

11. *Retainers.* It is also in the nature of urban life to give rise to considerable numbers of people whose sole occupation is to provide personal services for other people. Some of these services are of a utilitarian nature, such as the water supplied by 12,000 camel-driving haulers in medieval Cairo[108]. Other less utilitarian ones reflect the tendency of the wealthy and powerful to indulge in luxury and conspicuous consumption. As a result of this, cooks, household servants, personal attendants and professional entertainers swell the population. The numbers of such retainers tend to be correlated with the general prosperity of the city and hence they occur with the greatest frequency in court centres and cities whose economy is swollen by conquest. Huge numbers were incorporated into the population of imperial Rome[109].

Integration

Krader[110] has recently reminded us that the state has had not one origin but many; the same is certainly true of the city.

Among the conditions which have been seen as necessary for the development of urbanism is the achievement of a population density which is sufficiently great that it must be dependent upon intensive agriculture. In a recent survey, Murdock[111] found that 56% of societies practising intensive agriculture have cities larger than 50,000 whereas only 18·5% live entirely in villages of 200 persons or less. Only a tiny number of groups classified as dependent on horticulture or extensive agriculture have towns of 5000 or over. This suggests that intensive agriculture is a necessary but not a sufficient condition for the development of urbanism. In at least some regions, more intensive agriculture appears to be a result rather than a cause of increasing urbanism[112].

The state is also a necessary but not a sufficient condition for, or the concomitant of, the development of urbanism. The functioning of a city requires the development of a political apparatus at least as complex as that of a city state. On the other hand, states controlling a vast area have developed in Africa and elsewhere without the development of cities. The variety of ways in which states evolve provides numerous institutional milieux in which cities may develop.

Cities have been defined as population centres which perform a variety of functions in relationship to hinterlands of varying size. As the existence of cityless states indicates, however, even in some complex societies the forces which tend to concentrate functions at a single point may not be strong enough to give rise to cities. We have already noted the variety of ways in which specialized production can be carried on at a village level or in country estates. Trading can take place at periodic markets or through redistributive networks embedded in the social fabric, administration be centred in a feudal castle and defence provided by a local hill-fort. Collective religious observances may be held in rural shrines where the worshippers gather periodically. In theory, each of these functions could be carried out at a separate centre or group of centres, some of which would require no full-time personnel and none of which would have to be larger than a village.

Nevertheless, such an arrangement has inherent limitations. Even when transportation is relatively easy, such decentralization limits the efficiency with which individual functions can be interrelated and this in turn imposes restrictions upon the specialization which any one function can achieve. Therefore, with increasing cultural complexity one would expect different functions to converge on a common centre, even though the pattern of convergence may differ considerably from one society to another. By concentrating a number of specialized functions in one place movement is minimized, greater interdependence is possible and the more valuable property and personnel of a society can be protected more easily. In complex societies the tendency can be noted for a number of different functions to be performed even in very small centres (i.e. villages). Given sufficient complexity, this focal tendency appears likely at some point to give rise to the large multi-functional units we call cities. At what point this will happen depends not only upon the overall complexity of the society but also upon factors such as the ease or difficulty of transportation or the degree of security that can be offered on a regional as opposed to a local basis. There is little doubt, for example, that the security offered by a central government in Egypt was less conducive to the clustering of population than were the competing city states of ancient Mesopotamia[113].

It has been noted that in a general way the size of urban populations tends to vary in relationship to the number of functions they perform. However, the number of different types of functions that any one city acquires, and the order in which individual cities do so, vary widely. For example, the co-ordination of specialized production units and competition over scarce local resources are seen as two of the principal factors leading to the development of city states[114]. There has, however, been considerable speculation about the types of community that serve as the nuclei for such developments. Agricultural villages with an expanding division of labour, periodic markets, cult centres and forts, possibly inhabited by a local headman and his retainers, have been suggested as possibilities and perhaps, under differing circumstances, each of these types has served as the starting point for a city[115]. Warfare is clearly a potent force promoting the consolidation of such centres, often driving even farmers to seek protection within a city's wall. The urban centres of city states which embark on successful military careers grow in size as the army and bureaucracy expand. Such growth is amplified as the requirements of the army and a royal court lead to further increases in the numbers of artisans and retainers who inhabit the city.

Trade has also stimulated the development of cities in many parts of the world, sometimes even in infertile regions where great effort is required to feed a large population. Many Near Eastern trading communities were dominated by a powerful leader who was responsible for the defence of the city and its trade routes. In western Europe, where the countryside was controlled by feudal lords, cities tended to specialize as centres of trade and craft production, with auxiliary religious and educational functions. It is worth noting, however, that at least some of these towns developed out of communities which had survived through the Dark Ages as religious administration centres.

Some states which control large areas arise prior to urban development. Such growth may result from efforts to control trade routes or to extract exotic goods as tribute from neighbouring groups[116]. The first cities which develop in such states tend to be garrison towns and centres of administration and court life, around which traders, religious specialists, craftsmen and retainers may congregate as appendages of royal power. The demands of the royal court and of the army may be a strong inducement to economic specialization in such societies.

Within a large, well-established state additional cities are founded as centres for administration, defence, production, trade, religion, education and tourism. The tendency noted above, for centres serving one function to acquire others, frequently results in cities founded for very different reasons growing functionally more similar as they increase in size. In spite of this, differences in patterns of city growth and in the overall social milieu in which this growth takes place may continue to be clearly marked in their social structure. For example, craft production played an important part in the development of many cities in medieval Europe and the Near East and the guild structure of both is roughly similar. On the other hand, because Moslem cities were also chosen as the principal residences of powerful dynastic rulers they failed to develop traditions of self-government, unlike European cities which grew up outside the general feudal political order.

Finally, it is clear that some factors promote the growth of cities more powerfully than do others. Retainers are, in effect, a measure of the affluence of a non-industrial city's upper-class. Likewise, a city's ability to support the unemployed within it is related to its general economic well-being. Finally, while the relative security of a city is a very important factor in attracting settlement, this security ultimately reflects the administration of a state in relationship to its broader political environment. On the other hand, tourism, education, landlord settlement and religious activities constitute the most important functions of certain towns and cities and are very important auxiliary functions in others. Yet, on the whole, these functions can sustain only relatively small centralized populations. Craft production and trade support cities of various sizes but, as we have seen, the normal distribution of functions tends to produce a number of more or less equal-sized cities at the top end of the scale; these being of medium range for non-industrial cities as a whole. The largest cities develop as centres of administration and court life and, in particular, there is a tendency as empires develop for a single capital city to tower in size above all others. These giant capitals evolve as a result of the tendencies towards conspicuous consumption, economic centralization, bureaucratization and economic planning that are characteristic of large political units[117]. They also reflect the ability of the dominant power to command the taxes and tribute necessary to support such agglomerations. In mature pre-industrial urban societies one frequently encounters an apical pattern of administrative functions superimposed over a more truncated hierarchy of centres of craft production and trade[118].

Notes

1 Fried, M. H. (1966). On the evolution of social stratification and the state, *in* Diamond, S. (ed.). *Culture in History: Essays in Honor of Paul Radin*. New York. p. 729; Adams, R. M. (1966). *The Evolution of Urban Society*. Chicago. p. 21. For a critique, see Trigger, B. G. (in press). Archaeology and ecology, *World Archaeology*, 2, (3).

2 Steward, J. H. (1955). *Theory of Culture Change*. Urbana. p. 182.

3 Steward, J. H. (1955). *op. cit.* pp. 178–209.

4 Sanders, W. T. and Marino, J. (1970). *New World Prehistory: Archaeology of the American Indian*. Englewood Cliffs. p. 105.

5 Sjoberg, G. (1960). *The Preindustrial City*. Glencoe. pp. 32–4, 38.

6 Steward, J. H. *et al.* (1960). *Irrigation Civilizations: A Comparative Study*. Washington, Pan American Union, Social Science Monograph 1.

7 Krader, L. (1968). *Formation of the State*. Englewood Cliffs; Trigger, B. G. (1968). *Beyond History: The Methods of Prehistory*. New York. pp. 52–3.

8 Wilson, J. A. (1960). Egypt through the New Kingdom: civilization without cities, *in* Kraeling, C. H. and Adams, R. M., *City Invincible*. Chicago. pp. 124–64; Coe, M. D. (1961). Social typology and tropical forest civilizations, *Comparative Studies in Society and History*, **4**, pp. 65–85.

9 Sanders, W. T. and Marino, J. (1970), *op. cit.* pp. 7–8.

10 Millon, R. F. (1967). Teotihuacan, *Sci. Amer.*, **216** (6), pp. 38–48; Haviland, W. A. (1969). A new population estimate for Tikal, *Amer. Antiq.*, 34, pp. 429–33; Parsons, J. R. (1968). An estimate of size and population for Middle Horizon Tiahuanaco, Bolivia, *Amer. Antiq.*, **33**, pp. 243–5.

11 Some authorities are even more pessimistic. Concerning Mesopotamia see, Oppenheim, A. L. (1969). Mesopotamia—land of many cities, *in* Lapidus, I. R. (ed.). *Middle Eastern Cities*. Berkeley and Los Angeles. p. 4.

12 Mabogunje, A. L. (1962). *Yoruba Towns*. Ibadan. p. 3.

13 Many medieval European cities had only a few thousand inhabitants while settlements classified as villages in Eastern Europe had populations that were equal or larger. In parts of China and Indonesia rural population densities have long exceeded 1000 per square mile, the figure that many sociologists have accepted as being indicative of urban settlement. Cressey, G. (1955). *Land of the* 500 *Million*. New York. p. 15; Huntington, E. (1956). *Principles of Human Geography*. New York. p. 428; cf. Burgess, E. W. (1926). *The Urban Community*. Chicago. pp. 118–19. Fei, H. T. (1953). *China's Gentry*. Chicago. p. 95, points out that the population density of many Chinese cities is no greater that that of rural areas beyond the walls.

14 Bascom, W. (1955). Urbanization among the Yoruba, *Amer. J. Sociol.*, **60**, pp. 448–9; Bascom, W. (1969). *The Yoruba of Southwestern Nigeria*. New York. pp. 18–28; for a more recent discussion of Yoruba urbanism, see Wheatley, P. (1970). The significance of traditional Yoruba urbanism,

Comparative Studies in Society and History, **12**, pp. 393–423; Frankfort, H. (1951). *The Birth of Civilization in the Near East*. New York. pp. 61–2; Sanders, W. T. (1956). The Central Mexican symbiotic region, *in* Willey, G. R. (ed.). *Prehistoric Settlement Patterns in the New World*, Viking Fund Publications in Anthropology, **23**, p. 122.

15 In southern Italy, Sicily and Sardinia, for example, peasant villages or "agro-towns" not infrequently reached a size of 10,000 persons or more. Chisholm, M. (1962). *Rural Settlement and Land Use*. London. pp. 60–1.

16 Mabogunje, A. L. (1962). *op. cit.* p. 3–4.

17 Burgess, E. W. (1926). *op. cit.* p. 118. For a critical case, see Mabogunje, A. L. (1962). *op. cit.* pp. 3–4.

18 Netting, R. M. (1969). Ecosystems in process: a comparative study of change in two West African societies, *in* Damas, D. (ed.). Contributions to Anthropology: Ecological Essays, *National Museums of Canada Bull.* 230, pp. 102–12. I am not, however, at this point trying to draw a precise distinction between rural and urban, a point about which many scholars have failed to agree.

19 Vining, R. (1955). A description of certain spatial aspects of an economic system, *Economic Development and Cultural Change*, **3**, p. 169.

20 Garner, B. J. (1967). Models of urban geography and settlement location, *in* Chorley, R. J. and Haggett, P. (eds.). *Models in Geography*. London. pp. 324–6.

21 Garner, B. J. (1967). *op. cit.* pp. 304–5.

22 Garner, B. J. (1967). *op. cit.* pp. 304–5 (premises 2 and 3).

23 Garner, B. J. (1967). *op. cit.* pp. 322–5.

24 Garner, B. J. (1967). *op. cit.* pp. 326–9. In Europe and the U.S. the size of cities follows the rule $S_R = \frac{A}{R} n$, where A is the size of the largest city, R the rank of a given city and S_R is the size of the city of that rank. Issawi, C. (1969). Economic change and urbanization in the Middle East, *in* Lapidus, I. R. (1969). *op. cit.* p. 115.

25 Sanders, W. T. (1956). *op. cit.* p. 122.

26 Issawi, C. (1969). *op. cit.* pp. 105–6. In 1877, 57% of Cairo's economically active population was listed as engaged in farming, although this figure may be inflated by too-generous boundaries. By 1910, less than 10% were full-time farmers. Abu-Lughod, J. (1969). Varieties of urban experience: contrast, coexistence and coalescence in Cairo, *in* Lapidus, I. R. (1969). *op. cit.* p. 164.

27 Issawi, C. (1969). *op. cit.* p. 106.

28 Gulick, J. (1969). Village and city: cultural continuities in twentieth century Middle Eastern cultures, *in* Lapidus, I. R. (1969). *op. cit.* p. 124.

29 Weber, M. (1958). *The City*. New York. pp. 65–7.

30 Fei, H. T. (1953). *op. cit.* pp. 91–9. For the Near East, see Issawi, C. (1969). *op. cit.* p. 105.

31 In his "isolated state" model, von Thünen estimated the cost of river transport at only one-tenth that of land transport. Henshall, J. D. (1967). Models of agricultural activity, *in* Chorley, R. J. and Haggett, P. (1967), *op. cit.* p. 444.

32 Heichelheim, F. M. (1958). *An Ancient Economic History*. Leiden, I. pp. 116–38; 222–49.

33 Netting, R. M. (1969). *op. cit.*

34 Bowen, R. L. (1958). Ancient trade routes in South Arabia, *in* Bowen, R. L. and Albright, F. P. (eds.). *Archaeological Discoveries in South Arabia*. Baltimore. pp. 35–42.

35 Netting, R. M. (1969). *op. cit.* pp. 109–10.

36 Adams, R. M. (1966). *op. cit.* p. 149.

37 Adams, R. M. (1966). *op. cit.* p. 165.

38 Carcopino, J. (1946). *Daily Life in Ancient Rome*. London. pp. 16–21.
39 Krader, L. (1968). *op. cit.* p. 102.
40 Murdock, G. P. (1969). Correlations of exploitative and settlement patterns, *in* Damas, D. (ed.). *op. cit.* pp. 129–46.
41 Bascom, W. (1955). *op. cit.* p. 452.
42 Netting, R. M. (1969). *op. cit.* pp. 110–11; Issawi, C. (1969). *op. cit.* p. 106.
43 Heichelheim, F. M. (1956). Effects of classical antiquity on the land, *in* Thomas, W. L. (ed.). *Man's Role in Changing the Face of the Earth.* Chicago. p. 170.
44 Issawi, C. (1969). *op. cit.* p. 106; Forbes, T. R. (1970). Life and death in Shakespeare's London, *Amer. Sci.*, **58**, pp. 511–20.
45 Carcopino, J. (1946). *op. cit.* pp. 173–4.
46 For a discussion of the geo-ecology of Greek colonial expansion see McEvedy, C. (1967). *The Penguin Atlas of Ancient History.* Harmondsworth. pp. 10–11.
47 Coon, C. S. (1962). *Caravan: The Story of the Middle East.* New York. p. 242.
48 Coon, C. S. (1962). *op. cit.* p. 238.
49 Coon, C. S. (1962). *op. cit.* pp. 178–9; Benet, F. (1957). Explosive markets: the Berber Highlands, *in* Polanyi, K. *et al.* (eds.). *Trade and Market in the Early Empires.* Glencoe. pp. 188–213; Malinowski, B. (1922). *Argonauts of the Western Pacific.* London.
50 Hodges, H. (1970). *Technology in the Ancient World.* London. pp. 212–13; Coon, C. S. (1962). *op. cit.* p. 305.
51 Hodges, H. (1970). *op. cit.* p. 213.
52 Frere, S. (1967). *Britannia: A History of Roman Britain.* London. pp. 226–8.
53 Frankfort, H. (1956). *op. cit.* pp. 64–73.
54 Hodges, H. (1970). *op. cit.* p. 197; Heichelheim, F. M. (1956). *op. cit.* p. 174.
55 Gilbert, E. W. (1951). The human geography of Roman Britain, *in* Darby, H. C. (ed.). *An Historical Geography of Britain Before A.D. 1800.* Cambridge. p. 76; Phillips, C. W. (1956). *Map of Roman Britain.* Chessington, Ordnance Survey. Fig. 3.
56 English, P. W. (1966). *City and Village in Iran.* Madison. p. 67.
57 Le Tourneau, R. (1961). *Fez in the Age of the Marinides.* Norman. pp. 90–1.
58 Hodges, H. (1970). *op. cit.* p. 212.
59 Coon, C. S. (1962). *op. cit.* pp. 229–30; Fei, H. T. (1953). *op. cit.* pp. 91–9; Crane, R. I. (1955). Urbanism in India, *Amer. J. Sociol.*, **60**, pp. 463–70; Sanders, W. T. and Price, B. J. (1968). *Mesoamerica: The Evolution of a Civilization.* New York. pp. 30–33.
60 Le Tourneau, R. (1961). *op. cit.* pp. 88–93.
61 Coon, C. S. (1962). *op. cit.* pp. 229–31; Hodges, H. (1970). *op. cit.* p. 212; for a discussion of factors distorting this pattern, see Issawi, C. (1969). *op. cit.* p. 231.
62 Coon, C. S. (1962). *op. cit.* p. 231.
63 Collingwood, R. G. and Richmond, I. (1969). *The Archaeology of Roman Britain.* London. pp. 158–9; Rivet, A. L. F. (1964). *Town and Country in Roman Britain.* London. p. 134.
64 This arrangement corresponds with G. W. Skinner's model of periodic central places. See Garner, B. J. (1967). *op. cit.* pp. 320–2.
65 Frere, S. (1967). *op. cit.* pp. 289–93.
66 Özgüç, T. (1963). An Assyrian trading post, *Sci. Amer.* **208**, (2), pp. 96–106; Lloyd, S. (1961). The early settlement of Anatolia, *in* Piggott, S. (ed.). *The Dawn of Civilization.* London. pp. 170–4; 188.
67 Oppenheim, A. L. (1969). *op. cit.* pp. 6, 11.
68 Soustelle, J. (1962). *Daily Life Among the Aztecs.* New York. pp. 59–65.

69 Culican, W. (1961). The sea peoples of the Levant, *in* Piggott, S. (ed.). *op. cit.* pp. 151–60.
70 Heichelheim, F. M. (1958). *op. cit.* pp. 236–8.
71 Renfrew, C. (1969). Trade and culture process in European pre-history, *Curr. Anthrop.*, **10**, pp. 131–69.
72 Pirenne, H. (1925). *Medieval Cities*. Princeton.
73 Revere, R. B. (1957). "No Man's Coast": ports of trade in the eastern Mediterranean, *in* Polanyi, K. *et al.* (eds.). *op. cit.* pp. 38–63; see also Chapman, A. M. (1957). Port of trade enclaves in Aztec and Maya Civilizations, *in* Polanyi, K. *et al.* (eds.). *op. cit.* pp. 114–53.
74 Coon, C. S. (1962). *op. cit.* p. 245.
75 Fei, H. T. (1953). *op. cit.* pp. 91–9.
76 For an example, see Le Tourneau, R. (1961). *op. cit.* pp. 36–43.
77 Issawi, C. (1969). *op. cit.* p. 105.
78 Adams, R. M. (1966). *op. cit.* pp. 139–40.
79 Adams, R. M. (1966). *op. cit.* p. 143.
80 Soustelle, J. (1962). *op. cit.* p. 85.
81 Oppenheim, A. L. (1964). *Ancient Mesopotamia*. Chicago. p. 117.
82 See also Bray, W., this volume, pp. 909–26.
83 Lambert, W. G. (1964). The reign of Nebuchadnezzar I, *in* McCulloch, W. S. (ed.). *The Seed of Wisdom*. Toronto, pp. 3–11.
84 Le Tourneau, R. (1961). *op. cit.* pp. 15–19.
85 See also Oates, D., this volume, pp. 799–804.
86 Pendlebury, J. D. S. (1937). *Tell el-Amarna*. London.
87 Adams, R. M. (1965). *Land Behind Baghdad*. Chicago. p. 90.
88 In a recent study, C. Morris and D. E. Thompson have concluded that Huanaco Viejo, an Inca community in the North Central Highlands of Peru, was a regional centre created by the Inca rulers to serve a series of functions vital to the perpetuation and expansion of the state. Their statement that "the image of an artificial, essentially imposed, population centre... contrasts markedly with some of our notions of 'urban' and 'city'" provides evidence of the narrow basis on which many archaeologists' concepts of these entities are founded. Morris, C. and Thompson, D. E. (1970). Huanaco Viejo: an Inca administrative centre, *Amer. Antiq.*, **35**, pp. 344–362.
89 Fei, H. T. (1953). *op. cit.* p. 95.
90 Rivet, A. L. F. (1964). *op. cit.* pp. 72–98; Frere, S. (1967). *op. cit.* pp. 239–63; Collingwood, R. G. and Richmond, I. (1969). *op. cit.* pp. 95–132; Richmond, I. A. (1963). *Roman Britain*. Harmondsworth. pp. 66–108.
91 Collingwood, R. G. and Richmond, I. (1969). *op. cit.* pp. 122–3.
92 It is curious that while the city states of prehispanic Mexico resembled those of ancient Mesopotamia in many ways, they appear only rarely to have been walled (although some were fortified in terms of their location). To my knowledge, the reasons for this difference have not been discussed.
93 Few urban settlements in Roman Britain were fortified before the second century A.D. For reasons for fortifying, see Frere, S. (1967). *op. cit.* pp. 248–57.
94 Le Tourneau, R. (1961). *op. cit.* p. 26; Issawi, C. (1969). *op. cit.* pp. 107–8.
95 Mabogunje, A. L. (1962). *op. cit.* pp. 8–10.
96 Issawi, C. (1969). *op. cit.* pp. 105–6.
97 Falkenstein, A. (1954). La cité-temple sumérienne, *Cahiers d'histoire mondiale*, I, p. 810.
98 Oppenheim, A. L. (1964). *op. cit.* pp. 57, 118.
99 For Egypt, see Kees, H. (1961). *Ancient Egypt: A Cultural Topography*. Chicago. p. 317.
100 Frere, S. (1967). *op. cit.* p. 309.

101 Rivet, A. L. F. (1964). *op. cit.* p. 66.
102 Issawi, C. (1969). *op. cit.* p. 107.
103 Cf. Soustelle, J. (1962). *op. cit.* p. 8; Oppenheim, A. L. (1964). *op. cit.* p. 130. The classic study of the relationship between the religious and civil institutions of a city remains Fustel de Coulanges, N. D. (1864). *La Cité Antique.* Paris. Paul Wheatley notes that it is no surprise that in China, with its worldly Great Tradition, the centrally situated temple of South Asian cities was replaced by the seat of secular authority. Wheatley, P. (1969). *City as Symbol.* London. pp. 12–13.
104 Issawi, C. (1969). *op. cit.* p. 106.
105 Frere, S. (1967). *op. cit.* p. 240.
106 Graindor, P. (1934). *Athènes sous Hadrian.* Cairo.
107 Issawi, C. (1969). *op. cit.* p. 118.
108 Coon, C. S. (1962). *op. cit.* p. 231.
109 Carcopino, J. (1946). *op. cit.* pp. 68–72.
110 Krader, L. (1968). *op. cit.* p. 106.
111 Murdock, G. P. (1969). *op. cit.* p. 146.
112 Murdock, G. P. (1969). *op. cit.* p. 147; Netting, R. M. (1969). *op. cit.* pp. 109–10.
113 Frankfort, H. (1956). *op. cit.* pp. 97–8.
114 Sanders, W. T. and Price, B. J. (1968). *op. cit.* pp. 170–210.
115 Trigger, B. G. (1968). The determinants of settlement patterns, *in* Chang, K. C. (ed.). *Settlement Archaeology.* Palo Alto. pp. 72–3.
116 Trigger, B. G. (1969). The personality of the Sudan, *in* McCall, D. F. *et al.* (eds.). *East African History.* New York. pp. 88–98.
117 Issawi, C. (1969). *op. cit.* p. 117.
118 I wish to thank my colleagues D. Aronson, P. C. Salzman and G. M. Woloch for helping to clarify certain references used in this paper. The author's participation in this seminar was made possible by a travel grant from the Canada Council.

PAUL WHEATLEY

The concept of urbanism

"Urbanism" is one of the most protean of terms. In one or other of its inflections[1] it is customarily used to denote sets of qualities possessed by certain of the larger and more compact clusters of settlement features that at any particular moment in time represent centroids of continuous population movements[2]. It is also often held that these larger nodes in the settlement pattern are theatres for the acting out of a distinctive manner of life characterized as "urban". It is known that nodes of this order of dominance first appeared in the settlement hierarchy some five thousand years ago in the course of the transformation of relatively egalitarian, ascriptive, kin-structured groups into socially stratified, politically organized, territorially based societies[3], since when they have progressively extended the scope and autonomy of their institutional spheres so that today they mould the actions and aspirations of vastly the larger proportion of mankind.

During the five millennia of their existence these clusters of institutions have taken a great variety of forms which differ widely both among themselves and from most present-day urban settlements. Indeed, as far as morphology is concerned ancient and modern cities share only traits of so general a character that they are virtually useless for classificatory or analytical purposes. There is, in fact, no *a priori* reason to suppose that all the multifarious groupings of population both past and present that are and have been conventionally designated as "urban" should necessarily be subsumable within a single logically coherent field[4], but if structural regularities are ultimately elucidated, then it is practically certain that they will be manifested in shared functions and in trends in systemic change rather than in form.

In any case, it is not particularly profitable for a social scientist to attempt to discuss the nature, the essential quality, of urbanism. That is a metaphysical question more amenable to philosophical enquiry than to the empirical methods of the social sciences[5]. In this paper I shall therefore adopt a point of view that Karl Popper would categorize as "nominalist", and attempt to answer the questions, "When we use (or have used) the term 'urban', what do we mean (or have we meant)?" and "How do those aggregations of functionally interrelated institutions that we call (or have called) a city (or a town or an urban form) behave? What is their function?" The question "What is urbanism?" is one that I shall leave to Popper's "methodological essentialists"[6].

Attempts to classify urban forms and functions have inevitably, and I believe properly, reflected the interests of the classifiers. In classifying in the interests, as it were, of different interests, by emphasizing the aspects of urban life that they individually have considered significant, scholars have sought to introduce order into the immense (though often still poorly documented) variety of urban forms that have emerged during the past five thousand years. There have been, I think, five conceptually distinct, though in practice substantially overlapping, approaches to the investigation of these forms, namely (1) reliance on ideal-type constructs, (2) formulation of ecological theories, (3) delineation of trait complexes (4) conceptualization of the city as a centre of dominance, and (5) an expediential approach usually based on the size of the urban population. In what follows I propose to discuss some of the more important and representative examples of these approaches.

Ideal-type constructs: folk-urban and rural-urban dichotomies

Ideal-type constructs have been invoked most frequently in attempts to discriminate urban society in its totality from pre- or non-urban (folk[7]) society. The germ of this conceptualization is very ancient. In the early literature of the western Semites, for example, moral valuations were attached to very clear perceptions of the differing life-styles of the urbanized Canaanites and certain nomadic Hebrew tribes[8]. For Lucretius and the Epicureans much the same distinction was subsumed by the paired terms *concordia* and *justitia*, for Mencius by court (*ch'ao t'ing*) and village (*hsiang tang*)[9]. During the nineteenth century several pioneers in the development of the social sciences gave formal expression to these distinctions. Sir Henry Maine, for instance, distinguished between kin-based social organizations in which position was ascribed through "Status", and territorially-based societies in which position derived from "Contract"[10]. Morgan symbolized this distinction between an intimate and traditional mode of village association and a formal and contractual type of urban association under the terms *Societas* and *Civitas*[11], Tönnies as *Gemeinschaft* and *Gesellschaft*[12], and Weber as "Traditional" and "Rational"[13]. Other scholars have from time to time proposed similar dichotomous constructs[14], but the most influential, so far as urban theory is concerned, has probably been Durkheim's distinction between societies of mechanical and organic solidarity[15], between societies integrated through a sharing of sentiments and those integrated through a complementarity of functions. These polar types were based predominantly on societal characteristics. Couched in formal morphological terms, they took little account of the roles assumed by exchange in different societies. Marcel Mauss, Durkheim's pupil and subsequently his collaborator, extended the theory to explain how exchange functioned as the cement binding functionally differentiated societal parts into organic wholes[16], but it was Robert Redfield and Milton Singer who, in 1954, formulated a typology of urban forms that took cognizance of the changing place occupied by modes of economic integration in society as a whole. At one pole of their classification were those cities in which economic institutions were subordinated to the religious and moral norms of society, and which were mostly administrative and political foci diffusing traditional culture. At the other end of the spectrum were those cities in which the market system was autonomous and self-regulating, cities of the entrepreneur where the values of society were structured about expediential norms that manifested themselves in a consensus appropriate to the technical, rather than the moral, order. In the first type of city change was mediated by literati according to the mores of a classical tradition, and was consequently felt to be an inevitable outgrowth of the past, whereas in the second type of city change was generated by conflict and dissent, cultures disintegrated, and an intelligentsia propagated heterodoxies that not infrequently assumed the complexion of heresy. Rootlessness and anomie were prevalent in such cities and, as new cultural integrations were forged, there developed marked and often painful discontinuities between past and future. On the basis of the manner in which these two types of city mediated change Redfield and Singer classified them as cities of, respectively, orthogenetic and heterogenetic transformation.

The folk-urban ideal-type dichotomy was relatively easily devised for it opposed two markedly distinct levels of socio-cultural integration. The usefulness of such constructs was sometimes questioned, but there was usually a fair degree of consensus as to the qualities by which they were categorized. Adequately to discriminate component entities *within* an urbanized society, to formulate a rural-urban ideal-type dichotomy, however, was a rather more subtle exercise. It is true that the two forms of settlement are still readily apparent in patterns of land utilization, but their relevance to the ordering of society, particularly the highly complex industrial societies of the modern world, has been—and indeed still is—a matter for debate. It is in some ways anomalous that the most widely cited attempt to devise such a construct is associated with the name of Robert Redfield, for the urban component of Redfield's model was defined only implicitly. His main interests, deriving from his researches in Middle America, were with what he called "folk" and "peasant" societies, by which he meant respectively pre- or non-urban groups and the rural component in an urbanized society. The former he described as:

> small, isolated, non-literate, and homogeneous, with a strong sense of group solidarity. The ways of living are conventionalized into that coherent

> system which we call "a culture". Behaviour is traditional, spontaneous, uncritical, and personal; there is no legislation or habit of experiment and reflection for intellectual ends. Kinship, its relationships and institutions, are the type categories of experience and the familial group is the unit of action. The sacred prevails over the secular; the economy is one of status rather than of the market[18].

Urban life was conceived as being virtually the opposite of this polar construct, and Redfield did not pursue the matter significantly farther except to define the peasant as an urbanized countryman.

> There were no peasants before the first cities. And those surviving primitive peoples who do not live in terms of the city are not peasants . . . he [the peasant] is long used to the existence of the city, and its ways are, in altered form, part of his ways. The peasant is a rural native whose long established order of life takes important account of the city.
> . . . Peasantry then, whether Mexican or Chinese or Polish, is that style of life which prevailed outside of the cities and yet within their influence during the long period between the urban revolution and the industrial revolution[19].

In his later work Redfield went further and envisaged city-dweller and peasant as elements in a continuum[20]. Clearly the assemblages of traits by which he characterized these complementary components of urbanized society were high-level abstractions designed as heuristic aids in the discrimination of broad stages or types in a spectrum of societal evolution. Subsequently, however, these constructs sometimes came to be regarded, despite Redfield's protestations, as generalizations derived from empirical investigations, as the results of research rather than as tools with which to undertake it. For instance, we find Queen and Carpenter writing that "there is a continuous gradation in the United States from rural to urban rather than a simple rural-urban dichotomy and . . . as human communities are arrayed along this rural-urban continuum, consistent variations occur in patterns of behaviour"[21].

Criticisms of the concepts of a rural-urban dichotomy and a rural-urban continuum have been based on four main grounds: (i) that both parts of the dichotomies "represent confounded variables and, in fact, complex systems of variables which have yet to be unscrambled"; (ii) that they fail to recognize the co-existence of both urban and folk or peasant-like characteristics in numerous empirical situations[22]; (iii) that they are incapable of accommodating the presence of certain types of communities which are commonly found to exist in both societies. In particular they fail to take sufficient account of the wide range of levels and types of societies that exist in association with urban forms[23]; (iv) that the hypotheses are structurally unable to incorporate the discontinuities that are observed to exist in both rural and urban contexts. In fact, the so-called "theories of contrast"[24] are of so general a nature that they can be of little analytical use in the study of processes of change, while they are often too inaccurate to be reliable guides in the study of societies in equilibrium[25]. However, although their heuristic value as research tools has never been proven[26], although the relationship between the various characteristics of populations ordered by settlement size is seldom linear, and although the formulations probably tend to confuse rather than illuminate issues in the investigation of short-term change, nevertheless, in principle the constructs may still have some utility in discriminating gross categories of social development. What is necessary to render them viable is, as Hauser has specified, the devising of "well-designed empirical researches in which deviations from the constructs are noted in greater detail and with greater precision than are now available. Such research would better illuminate the nature of diverse social orders and, in the process, perhaps lead to the construction of an ideal typology more useful than that which is now available as prolegomena [*sic*] to empirically based generalizations"[27].

The relationship between urbanism and urbanization on the one hand and economic development on the other has evoked a good deal of discussion, particularly among students of the "modernization" process[28]. The popular belief that development of cities is a necessary condition of economic growth would seem to be supported by the fact that almost invariably there is a higher density of high-order central places (defined in the next section) in countries with higher incomes than in countries with lower incomes, and that, generally speaking, the population of higher-order central places constitutes a larger proportion of the total population in countries with higher incomes than in countries with lower incomes. Moreover, the high degree of specialization of labour and management involved in self-sustaining growth is often held to be attainable only in a relatively highly urbanized society[29], and it is similarly claimed that the universalist, achievement-oriented values that are a prerequisite for rationalization of production, and hence for industrialization, flourish best in an urban milieu[30]. However, Hoselitz has pointed out that, even if the preceding propositions are generally true, there are numerous examples of cities which impede economic growth, and he has proposed the ideal-type constructs of the generative and the parasitic city. A city is generative 'if its impact on economic growth is favourable, i.e., if its formation and continued existence and growth is [sic] one of the factors accountable for the economic development of the region or country in which it is located". A city is considered parasitic if it exerts an opposite impact[31]. The model has recently been tested in two African contexts by Barber (Rhodesia and Zambia)[32] and Mabogunje (Nigeria)[33], with the latter author paying particular attention to the time element inherent in the construct[34].

Ecological theories of urban development

The ecological approach has come closer than any other to providing a systematic theory of urbanism. It took its inception towards the end of the nineteenth century in the works of sociologists, social workers and philosophers who opposed both the romantic notion of the "unnaturalness" of the city and certain theories holding that it functioned literally as an organism or a person[35], but the bench mark from which it really developed was Park's paper on "The city: suggestions for the investigation of human behaviour in the urban environment" in 1916[36]. In endeavouring to specify a set of basic patterns and postulates according to which people and institutions were distributed within the city, Park and his students at the University of Chicago drew upon the principles of plant and animal ecology.

> Hence, the human ecologist defined the city primarily as a natural environment. Within it he expected to study the effects of ecological forces and processes. Within it, he assumed, people and their social institutions would be interdependent. The city, in other words, was seen as an ecological unit in which patterns and processes could be discerned by the same techniques and from the same perspective as those in nature[37].

Perhaps the best known formulations of members of the school of urban ecology were their descriptive models of the internal structure of cities, some of which are mentioned on page 607 below. Although they underwent continual modification, these models never managed entirely to accommodate the trenchant criticisms that were levelled at them, notably by Firey, who pointed out that the ecologists' investigation of land utilization patterns as the consequences of subcultural and unplanned processes was not only untrue to reality but also distorted the effects of culture and generally non-economic motivation. In his famous study of Boston, for example, Firey demonstrated that urban land was developed largely in accordance with cultural demands set by community sentiments and symbols[38]. In other words he castigated the ecologists for over-simplifying their problems by the exclusion of complex cultural variables.

In the face of these criticisms, urban ecologists tended to reduce the scope of their theoretical formulations, and concentrated their studies on such limited empirical investigations as the incidence of various forms of maladjustment, migration, mobility and so forth. Ultimately Park himself came to deny that he was attempting to devise a theory at all[39]. Subsequently a school of what has been called Neo-ecology arose, principally under the stimulus of Amos Hawley's book *Human Ecology*, published in 1950. For Hawley the task of the ecologist was to describe the characteristics of the population aggregate, to analyze both the community structure through which it functioned and the tendencies within the structure inducing differentiation and specialization, and to discern the effects of internal and external changes on the organization of the human aggregate[40]. Duncan, taking up Firey's criticism of the manner in which earlier ecological theories had excluded considerations of culture, has argued that "the functional and analytical approach of human ecology involves a concern not with culture as an undifferentiated totality but with aspects of culture as they play into the process of adaptation"[41]. Together with Schnore, Duncan has postulated an ecological complex that owes a good deal to Durkheim's *De la Division du Travail Social*[42]. The functionally interrelated basic components of the complex are environment, population, technology, and social organization, with the last conceived as "an adaptation to the unavoidable circumstance that individuals are interdependent and that the collectivity of individuals must cope with concrete environmental conditions"[43]. On this view the study of urbanism (or of urbanization) involves specification of "the precise technological, demographic, and environmental conditions under which various urban forms of organizations may be expected to appear and—once established—to develop at given rates"[44]. In so far as urban studies are concerned, the major advantage of the "ecological complex" is that it permits the deployment and analysis of large quantities of numerical data. Sjoberg has contended that this approach is overly materialisti and is deficient in that it rejects consideration of value systems[45]. Moreover, the four components of the ecological complex require more rigorous definition than they have so far received, particularly the one labelled "social organization", which Duncan and Schnore appear to all intents and purposes to equate with division of labour. Recently Wheatley has attempted to take account of these criticisms in applying the theory to the problem of urban genesis[46]. Closely related to the Duncan and Schnore model, though employing a neo-deductive methodology and stressing the notion of "sustenance", is the approach employed by Gibbs and Martin[47]. At present this theory is descriptive rather than explanatory, and tends to confuse "materialism" with "ecology". However, it does incorporate both certain types of knowledge and beliefs in its definition of technology[48].

It is probably true to say that in whatever form an ecological theory has been presented it has been directed primarily towards the study of social organization as a response to pressures of environment, and as such has often disseminated a whiff of biological determinism. Nevertheless, the urban ecologists have made signal contributions to our understanding of the city. Reissman has summarized these contributions in the following words:

> The ecological period in the history of urban sociology was as valuable as it was necessary. Its value derived from the quantity of information gained about the city. It was necessary because, as in the development of any science, the more apparent clues have to be investigated and evaluated before more complex abstractions are possible. A theory of the city, if it is at all possible, can be achieved only by recognition of the

complexity of the urban environment and creation of concepts that are abstract and general enough to deal with it[49].

Trait-complex approaches

One of the more productive essays in this genre—because it is one which has provoked a great deal of discussion—has been Louis Wirth's minimal definition of a city as "a relatively large, dense, and permanent settlement of socially heterogeneous individuals"[50]. Wirth further attempted to associate a way of life with these postulated external criteria of size, density, permanence and heterogeneity, and in so doing went some way towards converting a simple aggregate of features into an ideal-type construct:

> The contacts of the city may indeed be face-to-face but they are nevertheless impersonal, superficial, transitory and segmental . . . The superficiality, the anonymity, and the transitory character of urban social relations make intelligible, also, the sophistication and the rationality generally ascribed to city-dwellers . . . Whereas the individual gains, on the one hand, a certain degree of emancipation or freedom from the personal and emotional controls of intimate groups, he loses, on the other hand, the spontaneous self-expression, the morale, and the sense of participation that comes with living in an integrated society. This constitutes essentially the state of anomie, or the social void, to which Durkheim alludes[51] . . .

There is reason to believe that when Wirth used the phrase "socially heterogeneous" he was concerned more with class differentiation than with ethnic diversity, but he was not explicit on this point. In any case, both social and ethnic heterogeneity are at best relative criteria and consequently difficult to apply cross-culturally. This led Bascom, who was attempting to evaluate the representative Yoruba city in terms of Wirth's model, to reject the criterion of heterogeneity in favour of the notion of formalized government, that is mechanisms of control on a secondary, supra-kinship level[52]. More recently Krapf-Askari, also attempting to accommodate the Yoruba city to the Wirthian paradigm, similarly rejected social heterogeneity as a necessary criterion of urbanism:

> In the face of so many factors potentially separable from the settlement pattern as such, it might in the last resort be simpler to revert unambiguously to size, density, and permanence as criteria; i.e. to retain the loose, common-sense meaning of the words "town", "city", and "urban" to refer to all large, permanent, closely nucleated settlements. "City" and "town" might have quantitative referents, if these could be agreed upon; if not, they could continue to be used as quasi-synonymous. We could then

rely on one or more sets of descriptive categories to indicate the variables that may affect urban social life[53].

The fundamental difficulty with this approach arises, however, not from ambivalences in the concept of heterogeneity but from the assumption that size and density of population necessarily induce variations in behavioural patterns, in short that size generates differentiation in communities. In fact, there is little empirical evidence for the hypothesis that increases in size and density of a population promote anonymity and, together with a more intense division of labour, create social heterogeneity. Duncan's quantitative analysis of data from the 1950 census of the United States failed to reveal any close correlation between magnitude of population and such characteristics as relative size of income and age groups, mobility of population, extent of formal schooling, size of family, and proportions of non-white, foreign-born, white-collar or women workers[54]. Reiss has gone farther in claiming that, "empirically, at least, "urban" can be independent of size and density. If this is true, then large size and high density of settlement are not always conditions for an urban way of life in any given community"[55]. The truth of the matter surely is that Wirth's formulation, like Redfield's, in so far as it related to a way of life was not an inductive generalization from research findings but an emotionally charged ideal-type construct descriptive of the urban aggregations of Western Europe and North America at the end of the nineteenth, and in the earlier decades of the twentieth, century[56].

Wirth's view of the city was based essentially on moral considerations. "On the basis of the three variables, number, density of settlement, and degree of heterogeneity, of the urban population", he wrote in 1938, "it appears possible to explain the characteristics of urban life and to account for the differences between cities of various sizes and types"[57]. The city was for him a crucible for the transformation of personality, producing such characteristics as reserve, blasé attitudes, indifference, sophistication, cosmopolitanism, rationality, relativistic perspectives, tolerance, competitiveness, self-aggrandizement, exploitative attitudes, frustration and irritation, nervous tension, instability, insecurity, tolerance of eccentricity and novelty, approval of efficiency and innovation, and a high degree of personal disorganization. From the same three population variables also stemmed such qualities of social organization as the importance and interdependence of specialists rather than of individuals; impersonal, transitory, superficial, segmental, and utilitarian social contacts; deterioration in the integration of social organization; a pecuniary nexus; an emphasis on time; predatoriness; formal controls; anonymity; a flexible caste structure counterbalanced by an intensified and ramified differentiation by income and social status; heightened mobility; involuntary segregation of racial, linguistic, income, and class groups; a high incidence of tenancy; diversification of, and rapid turnover in, group membership; importance of symbols and stereotypes; standardization of products and processes; gearing of facilities and institutions to the average user; subordination of individuality; weakening of kinship bonds; decline in the social significance of the family; loss of traditional bases of solidarity; disappearance of the neighbourhood; and replacement of territorially based social units by interest groups[58]. With this formulation Wirth grappled with what he called "the central problem of the sociologist of the city", namely "to discover the forms of social action and organization that typically emerge in relatively permanent, compact settlements of large numbers of heterogeneous individuals"[59].

It will have been noted that numerous of the characteristics cited above are cultural in nature and not specific to urban life. Invention, for example, is not restricted to urban milieux. Nor is literacy, and sacred ties may well be stronger in some cities than in many villages and hamlets. Diversity of language and religion may also be greater in certain rural areas than in some urban communities, and complex technology may be employed in some farming areas while remaining relatively undeveloped in some cities. The ethnic heterogeneity of large American cities may well have been determined more by their settlement history and immigration policies than by their size. More important, perhaps, is the lack in the thesis of an adequate personality theory, which results in a deterministic attempt to deduce individual psychological traits from the urban environment. Lewis (1965, p. 497) has put this criticism succinctly enough:

> The city is not the proper unit of comparison or discussion for the study of social life because the variables of number, density and heterogeneity as used by Wirth are not the crucial determinants of social life or of personality. There are many intervening variables. Social life is not a mass phenomenon. It occurs for the most part in small groups, within the family, within households, within neighborhoods, within the church, formal and informal groups, and so on.
>
> Any generalizations about the nature of social life in the city must be based on careful studies of these smaller universes rather than on *a priori* statements about the city as a whole. Similarly, generalizations about urban personality must be based on careful personality studies.

It was probably the failure of this attempt to link analysis of a social system and conjecture about individual personality that led Wirth to attribute so many contradictory features to urban life. Presumably he regarded both élites and intellectuals on the one hand and slum dwellers on the other as sharing similar personalities, and he certainly viewed the city as a focus of learning, innovation, progress, and higher standards of living at the same time as it was a place of slums, poverty, crime and, above all, disorganization. It was at once the culmination of man's achievement and symbol of his greatest degradation, an antinomy that was subsequently to prove intractable in analytical studies and impossible to apply cross-culturally.

In assembling their clusters of diagnostic criteria some proponents of the trait-complex approach have, like Wirth, emphasized the role of institutions, while others have paid more attention to their material expression. Notable among these latter was Max Weber who, nearly half a century ago, wrote:

> Denn dazu gehörte, da*β* es sich um Siedelungen mindestens relativ stark gewerblich-händlerischen Charakters handelte, auf welche folgende Merkmale zutrafen: 1. die Befestigung, —2. der Markt, —3. eigenes Gericht und mindestens teilweise eigenes Recht, —4. Verbandscharakter und damit verbunden, —5. mindestens teilweise Autonomie und Autokephalie, also auch Verwaltung durch Behörden, an deren Bestellung die Bürger als solche irgendwie beteiligt waren[60].

Weber realized that the inclusion of autonomy and autocephaly among these criteria disqualified from full urban status a very high proportion of those settlements of the traditional world which have customarily been designated as cities or towns. Complete autonomy has been comparatively rare in the history of urbanism and, city-states apart[61], has usually fallen to a city accidentally and temporarily during periods of political disruption. Outside Europe truly privileged cities, enclaves of urban law islanded in territories subject to the common law of the state, have seemingly occurred only in fourteenth-century Japan, in the Nile delta during the interludes between the Kingdoms, and from time to time during, and in various parts of, the Hellenistic empires[62]. Weber was quite prepared to accept the far-reaching implications arising from his adoption of these criteria of urbanism: namely, those settlements which failed to incorporate the salient features of the European city failed in a greater or lesser measure to qualify for urban status. "Eine Stadtgemeinde im vollen Sinn des Wortes," he wrote in 1921, "hat als Massenerscheinung vielmehr nur der Okzident gekannt. Daneben ein Teil des vorderasiatischen Orients (Syrien und Phönizien, vielleicht Mesopotamien) und dieser nur zeitweisse und sonst in Ansätzen"[63]. Moreover, Weber did not explain the specific type of market that he had in mind, but presumably it was the autonomous, price-fixing market of Western classical economists, which we now know was by no means universal in the traditional world[64]. Nor did he distinguish systematically between market and trade, and he would doubtless have been surprised to learn of the volume of exchange which was conducted in some ancient cities without benefit of a formal marketing system.

It is probably true to say that archaeologists, by reason of the tools and techniques that they employ, have tended to mould their discussions of cities on the principles elaborated by Lewis Henry Morgan, that is in terms of an easily recognizable set of traits supposedly indicative of a stage of social development. In other words they have selected as allegedly diagnostic a cluster of criteria from the constellation of features exhibited by early settlements which they have already designated as urban. This was indeed the method by which the late Gordon Childe arrived at the ten indices whose concurrent appearance in the archaeological record he held to signify the advent of urban forms into the world, namely: the concentration of a relatively large number of people in a restricted area; craft specialization; the appropriation by a central authority of an economic surplus; monumental public architecture; developed social stratification; the use of writing; the emergence of exact and predictive sciences; naturalistic art; foreign trade; and group membership based on residence rather than kinship[65]. As these criteria were explicitly delineatory rather than explanatory, it is not surprising that Childe was unable to establish functional relationships between them. In fact, although he was ostensibly investigating a *process*—that which he termed the Urban Revolution—he succeeded only in demarcating a *stage* of development. It will be observed, too, that some of his criteria, particularly naturalistic ("representational" would probably have been a more apposite term) art, appear to have been of minimal functional significance in the process of urban evolution[66]. Moreover, not all of the criteria occurred in all early cities (e.g., writing was lacking both in the Inca cities, foci of the only true empire in the pre-colonial Americas and among the dense aggregations of population in the Yoruba territories[67]), so that it is to be presumed that Childe himself regarded them less as indices of urbanism than as components in a constructed, or ideal-type, city. Yet other of the ten criteria are not specific to cities. Monumental architecture, for instance, is to be found in both pre-urban and non-urban contexts. It is also evident that from an operational point of view, Professor Childe's criteria of urban status comprise one primary and several secondary variables, and he leaves us in no doubt that he regarded the progress of technology, resulting in the augmentation of food surpluses, as the dependent variable in the process of urban generation. Finally, the cluster of features that Professor Childe invoked to discriminate early cities bears no great resemblance in form or function to those of present-day urban settlements, and there is so far no incontrovertible support for the teleologically inspired implication that they were the first in a series of functionally interdependent institutional aggregations which, through succeeding centuries, approximated progressively more closely to those of the contemporary city.

Another archaeologist, John Rowe, has recently based a study of cities in prehispanic Peru on a somewhat unusual complex of allegedly urban traits.

> For the purpose of the present argument, an urban settlement is an area of human habitation in which many dwellings are grouped closely together. The dwellings must be close enough together to leave insufficient space between them for subsistence farming, although space for gardens may be present. In the case of a site where the foundations of the dwellings have not been excavated, an extensive area of thick and continuous habitation refuse provides a basis for supposing that the settlement was an urban one.

> The intent of this definition is to exclude clusters of dwellings so small that they could be interpreted as belonging to the members of a single extended family. Twenty dwellings is perhaps the minimum number which would provide this exclusion[68].

Rowe goes on to classify urban forms as *pueblos* and *cities*, using the former "to designate an urban settlement in which all the residents are engaged in hunting, fishing, farming or herding at least part of the time", and city "to designate one which includes residents engaged in other activities (manufacturing, trade, services, administration, defence, etc.)"[69]. According to this interpretation the earliest known urban form in Peru was the pre-ceramic habitation site at Haldas, but the definition is sufficiently idiosyncratic not to recommend itself for cross-cultural studies. In the Middle East at least one archaeologist has ascribed urban status to a settlement solely by virtue of its being enclosed within a wall. The early settlement at Tell es-Sultan, Kenyon has written, "is quite clearly on the scale, not of a village but of a town. Its claim to a true civic status is established by the discovery . . . that it possessed a massive defensive wall"[70]. Needless to say, reliance on a single morphological feature of this character would, if pushed to its logical extreme, make most of the world urban for most of its history.

In the nineteen-forties the trait-complex approach was applied to contemporary urban forms by Smailes, who investigated clusters of key services which he regarded as indicative of various grades of urban development. Full urban status ("a fully-fledged town" in his terminology) was denoted by the combination of (i) a shopping centre with a full range of specialized retail services (reflected in the presence of at least three banks and a Woolworth's store), and (ii) a set of institutions providing social, educational, and health services for the surrounding district (reflected in the presence of a cinema, a weekly newspaper, a secondary grammar school, and a hospital). Other clusters of features were stipulated for "sub-towns", "urban villages", "cities", and "major cities"[71]. Similar investigations subsequently undertaken in Germany, the United States, and India modified the several sets of criteria to accord with local conditions. Somewhat earlier Dickinson had devised a similar but simpler classification of East Anglian towns on much the same principle[72].

Cities as centres of dominance

Recently Horace Miner has considered the role of the city as a power phenomenon, a centre of dominance[73]:

> As social units become differentiated in function, a hierarchy of power relations arises from the greater inherent influence of some functions, and particularly from the emergence of units to co-ordinate the differentiated parts. When the United States is regarded as such a system of

> interdependence, research results clearly support the contention that "the units which mediate and control these dependency relationships tend to be localized in the large urban centres" . . .
>
> When we examine the logic which draws political and economic power together in the city, we find that, to a degree, they tend to have a single locus because they are different aspects of the same thing. But even when separate systems of control are discernible, one kind of power is used to secure another, and various sorts of control become concentrated in the same individual through his positions in the different systems. On the other hand, this very fact complicates the maintenance of the system boundaries . . .[74]

As explicated by Miner this conceptualization of urban function would seem to be descriptive rather than explanatory. In many ways closely similar is John Friedmann's notion of the city as a generator of effective space. "The hierarchy of urban places", this author writes, "represents the ultimate means for organizing a geographic area into its component social, political-administrative and economic spaces . . . A mere area becomes effective 'space' . . . solely through the agency of urban institutions which extend their influences outward, binding the surrounding regions to the central city and introducing to them urban ways of thought and action"[75]. One sector of the spectrum of relationships focused in the city in this way, the provision of tertiary goods and services, has served as a basis for one of the best developed bodies of theory in urban studies, namely the Theory of Central Place.

The pattern of settlement in a region typically consists of three component elements: a linear pattern arising from break-of-bulk and allied services provided by transport foci disposed in relation to communication routes; a cluster pattern made up of centres concerned with specialized activities that may range from mining and manufacturing to the provision of recreation and religious facilities; and a uniform pattern consisting of settlements each of which performs comprehensive services for a surrounding area, and which is consequently customarily referred to as "a central place"[76]. To date, no model has been constructed capable of accommodating the generation and location of all three components of a settlement pattern, but a great deal of research has been devoted to elucidation of the arrangement and distribution of central places. Although the idea of a more or less regular distribution of such central places had been advanced in embryonic form by Léon Lalanne as early as 1863[77] and by C. J. Galpin in 1915[78], the foundations of a central-place theory were laid by Walter Christaller in 1933[79]. On the basis of the fundamental spatial concept of the range of a good, and using purely economic arguments, Christaller was able to define an optimal spatial arrangement for a hierarchy of settlements developed on the assumption of an isotropic surface[80]. In the model derived under these conditions settlements are regularly spaced to form a triangular lattice, and are centrally located within hexagonally shaped market areas, this shape of area requiring the least average distance of movement by consumers to the central settlement[81]. The hierarchical dimension in this pattern arises from the assumption that higher-order central places supply all the goods and services offered by lower-order central places together with a number of higher-order goods and services distinctive to themselves. Higher-order central places therefore provide wider ranges of goods and services, are loci for larger clusters of establishments, house bigger populations, serve more extensive market areas and market-area populations, and engage in greater volumes of business than do lower-order central places. It follows that this hierarchical arrangement is expressed spatially in higher-order central places being more widely spaced than lower-order central places, and in lower-order central places "nesting" within the market areas of higher-order central places according to a definite rule. Because of the competitive basis to this solution, in which it is assumed that all areas are able to be served from a minimum of central places, Christaller described the system as organized according to a "marketing principle". The progressive increase in numbers of market areas at successively higher levels of the hierarchy by a rule of threes led subsequent workers to refer to the hierarchy developed on this principle as a K-3 network[82]. Christaller himself postulated two alternative hierarchies, arranged according to (i) a "transport principle" which permits the hierarchy to maximize the number of central places located on major transport routes and which results in a nesting of complementary regions according to a rule of fours (K-4); and (ii) an "administrative principle" requiring political-social separation of complementary regions, which is achieved when each central place controls six lower-order places. Nesting is accordingly by rule of sevens (K-7). In each case the whole hierarchical system is integrated with, and is given coherence by, an interconnected set of regional and national metropoleis.

During the last thirty years Christaller's exposition of Central Place Theory has undergone considerable modification and elaboration. August Lösch, in particular, has greatly strengthened its theoretical foundations[83]. Whereas Christaller had assumed that, once generated by regional development, K values would remain constant, Lösch incorporated in his theoretical statement all possible hexagonal solutions, with the fixed-K assumptions of the marketing, traffic and administrative principles represented only as special limiting cases. By rotating the nest of possible hexagonal patterns all centred on one point—which he designated the metropolis, the highest order of central place in his system—Lösch was able to effect the greatest degree of coincidence in the locations of cities, the minimization of the aggregate distance between all settlements, and the maximization of the quantity of goods that could be supplied locally. The pattern that emerges from this manipulation of the hexagonal market areas comprises twelve sectors of alternating many and few central places, each of which exhibits considerable variation in incidence of urban forms with distance from the metropolis[84]. This hierarchy is less rigid than that developed by Christaller, and consists of a more or less continuous sequence of central places in contrast to the distinct tiers of settlements postulated in the earlier model[85]. It follows, therefore, that cities of the same size do not inevitably exercise the same functions, and that the functions of larger central places do not necessarily subsume all those of smaller central places. It would appear that Christaller's formulation is likely to throw more light on the distribution of retail and service business, while Löschian analysis is probably more relevant to an understanding of the spatial distribution of market-oriented manufacturing[86]. It is also possible that Christaller's model is better adapted to study of systems of cities founded in areas of sparse settlement, whereas that of Lösch provides a more satisfactory framework for analysis of settlement patterns in regions of dense, long established settlement[87].

It will have been noticed that the formulations of both Christaller and Lösch were developed on the basis of economic theory without reference to the behaviour of retailers and consumers. In an attempt to develop an operationally more expedient formulation, Curry sought to shift the emphasis of investigation from its preoccupation with an economic rationale to a description of general patterns of behaviour. To this end he substituted random variables descriptive of shopping habits, elements of retailing policy, and suchlike for Christaller's economic postulates[88]. Using a characteristic spectral-density function to describe the gross periodic activities of consumers, he then adduced the ergodic hypothesis to relate temporal behaviour to spatial form. In other words, he treated the regularities of consumer behaviour through time as possessing the same statistical properties as does the spatial pattern of retail locations.

Christaller seems to have been more than a little inclined to view his formulation as a general theory of urban location:

> Für das Entstehen, die Entwicklung und das Vergehen von Städten ist es ganz eindeutig ausschlaggebend, ob die Bewohner der Stadt hier ihre Erwerbsmöglichkeiten finden und ob ein Bedürfnis nach den Dingen, die die Stadt als solche zu bieten hat, besteht: somit sind aber wirtschaftliche Tatsachen entscheidend für das Vorhandensein von Städten—für die Existenz der ländlichen Siedlungen,deren Häuser ja alle gleichzeitig Produktionsstätten sind, sind wirtschaftliche Gründe ja ohne weiteres maßgebend—; daher ist die Siedlungsgeographie ein Teil der Wirtschaftsgeographie, es ist, wie überhaupt in der Wirtschaftsgeographie, die ökonomische Theorie heranzuziehen, wenn das Städtewesen erklärt werden soll. Wenn es nun in der ökonomischen Theorie Gesetze gibt, so muß es auch Gesetze in der Siedlungsgeographie geben, und zwar ökonomische Gesetze von besonderer Ausprägung, die man speziell als wirtschaftsgeographische Gesetze ansprechen könnte[89].

It would appear from this statement that Christaller regarded the economy as an autonomous agent in the generation of the settlement hierarchy, but it is likely that in pre-modern societies more often than not it was only a subsystem of the more inclusive category of culture[90]. Indeed Parsons and Smelser adopt this point of view even in relation to contemporary society[91]. In fact, it is now accepted that Central Place Theory is concerned only with the location of tertiary economic activity, but even in this restricted context it provides a useful theoretical basis for a broad sector of the field of urban studies.

A great deal of recent work in central-place studies has been concerned with attempts to test the several theoretical formulations against data derived empirically from contemporary situations. Faced with the difficulty of matching in the so-called real world the isotropic surface postulated by Christaller, experimental investigators have tended either to view the inevitable discrepancies—such as irregularities in the distribution of population and purchasing power, diversified terrain, unequal incidence of resources and transport facilities—simply as operational reasons for the incongruence of actual spatial patterns with those predicted by theory, or to incorporate negative evidence in the formulation as morphological variants. More than a decade ago Berry and Garrison somewhat simplified the problem when, using only the concepts of range and threshold, they were able to restate the theory in a manner that no longer required the assumptions of a uniform distribution of population and a hexagonal shape for market areas, thus permitting its application to intra-urban and highway ribbon development[92]. Subsequently, various probabilistic interpretations have been introduced into the theory. Thomas, for example, examined Christaller's concept of "centres of the same population-size class" in a stochastic context. By using a normal curve of error to determine whether cities were of *significantly* similar populations and at *significantly* similar distances apart, he was able to establish the relative stability of their population-size and distance relationships[93]. More recently Dacey has conceived Central Place Theory in terms of an equilibrium solution for a spatial arrangement of settlements subject to the influence of external forces, but with inconclusive results[94]. In fact, the error-term incorporated in the data from Dacey's study area in Iowa (presumably induced by non-economic forces) is so large that it is unlikely ever to be susceptible to treatment by the classical theory of errors.

Recently Harvey has drawn attention to an even more fundamental problem in the search for empirical evidence with which to test Central Place Theory. This formulation, he points out, is derived from a theory of demand which Clarkson has demonstrated to be inherently unverifiable by reference to empirical evidence[95]; therefore it cannot itself yield empirically testable hypotheses. "The only alternative", he writes, "is to regard settlement location as an economic process to which non-economic processes contribute a noise' or 'error-term' element"[96]. Perhaps in this conclusion he is being unduly pessimistic. Possibly that error-term can eventually be incorporated in an expanded Theory of Central Place. Possibly it may ultimately prove feasible to devise a model of the urban hierarchy on broadly cultural and socio-economic principles rather than on narrowly economic ones, to construct a theory of urban size and location in which, say, the administrative principle is not, as in Christaller's model, an alternative to the economic principle but in which both principles complement one another. Certainly settlement hierarchies based on other than an economic rationale have been observed in various parts of the world. The imposed hierarchy of administrative central places which existed side by side with, and reinforced, the indigenously generated market-based hierarchy in China has been the subject of frequent comment[97]. In quite a different context von Grünebaum has sketched the outline of a sacral hierarchy of cities in the Muslim world. Interestingly enough, he emphasizes that cities performing higher-order sacral functions invariably perform lower-order sacral functions in addition[98].

The principle of dominance has also been invoked in discussions of the internal structure of cities. In the modern city, it is alleged, the multitude of individual decisions which combine to create the pattern of land utilization are regulated in the main by economic processes operating in society[99]. Competition for the use of available locations results in the occupation of each site by the firm able to derive the greatest utility from it and, therefore, able to pay the highest rent. The ideal-type pattern generated by this principle is one in which rents are maximized throughout the system and all activities optimally located. High land values in central zones are associated with ease of intra-urban accessibility, savings in transport costs being set against higher rent payments for central locations. At different times different models have been proposed to depict the spatial patterns of land utilization thus induced, among which the best known are (i) the concentric model of Burgess, which assumes that accessibility and land values decline with equal regularity in all directions from a common central point[100]; (ii) the sector model of Hoyt, which takes account of differences in accessibility and, therefore, in land values along routeways radiating outwards from the city centre[101]; and (iii) the multiple nuclei model of Harris and Ullman, which develops zones of land-use around discrete centres arising from peculiarities of site and evolution[102]. These models are not mutually exclusive, and elements of all three can be discerned in many urban forms, especially in the conurbations created by the fusion of several cities. In fact Marble has proposed a model that takes account of just such fusions as radially patterned growth around individual nuclei intersected by axial growth extending outwards from the city centre[103]. The resulting pattern is one in which discrete social, economic and technical zones are integrated into a star-shaped urban form.

Partial, as opposed to comprehensive, theories of urban internal structure such as have just been discussed have been concerned mainly with the distribution of population densities and the deployment of commercial enterprises. Somewhat surprisingly patterns of land utilization appear to exert relatively little control over population densities which, generally speaking, decline as a negative exponential function of distance from the city centre[104]. There is reason to think that older and larger industrial cities tend to exhibit higher overall population densities[105], and that larger cities with low transport costs, dilapidated central sectors, and dispersed employment centres tend to be more compact[106]. Winsborough has shown that population densities in the central sectors are a function of overall population density within the city, which can itself be correlated fairly directly—and not altogether unexpectedly—with the total population, its status as a manufacturing centre, and the proportionate age of its dwellings[107].

It is implicit in the concentric, sector, and multiple-nuclei theories that all substantial retail outlets are subsumed within the Central Business District in the heart of the city. The pioneer investigations of urban commercial structure were directed towards elucidation of its morphological aspects[108], but more recently Berry has devised a descriptive model which recognizes three basic conformations, namely shopping centres providing convenience and shopping goods primarily for local purchasers, ribbon developments catering mainly for demands generated by traffic flows along major transport routes, and specialized centres providing for specific consumer needs[109]. The arrangement of shopping centres has usually been examined within a framework of Central Place Theory, but evidence is accumulating to suggest that more than one hierarchy is present in any one city. Berry, Tennant, Garner and Simmons, for example, basing their analysis on the technique of factor analysis, have shown in 1963 that there is likely to be a relatively higher duplication of functional types in lower-income enclaves where demand is both quantitatively less and qualitatively poorer.

Investigations of urban internal structure have been confined almost entirely to the so-called Western world, Japan[110], and the metropoleis of the developing countries[111], to the virtual exclusion of the traditional city—where, incidentally, the current emphasis on autonomous economic processes may well prove to be analytically less profitable. In many traditional urban forms, for example, the pre-eminence of the central sector over the periphery derived less from economic and technological considerations than from a principle that may conveniently be termed proximity to the focus of power in both its secular and sacred, material and ritual, aspects[112]. In such cities rent-distance relationships appear to have been structured on principles that generated a zoning of land-uses somewhat different from those commonly encountered in the modern city. Not only were location patterns based primarily on localized vertical organizations and product groupings (represented typically by producer-retailers in traditional-style bazaars[113]), but the central precinct was also often reserved for ritual purposes, a situation which tended to create central population-density craters that persisted in many cases until the onset of modernization[114].

The expediential approach

This has been the preferred approach of both demographers and administrators, and has usually been based on a classification of population size. The main difficulties arise from the fact that absolute size is not necessarily directly related to structure, form or function but varies widely in its significance from culture to culture and from time to time, with the result that the classification is often arbitrary. In Denmark, Sweden, and Finland, for example, a settlement of a mere 250 people is accorded urban status, whereas in Canada the relevant figure is 1,000, in Venezuela 2,500, in Ghana 5,000, and in Greece, Spain and Switzerland 10,000. The Indian Census is more ambitious, combining size with population density and employment characteristics in its definition of urban form. Before a settlement can be so classified it must house more than 5,000 inhabitants, clustered so as to produce a density exceeding 1,000 to the square mile. Additionally, more than 75% of adult males must be engaged in work other than agriculture. However, the size factor appears to be regarded as normative rather than essential for, of the 2,700 urban settlements recorded in the 1961 *Census of India*, no less than 268 have populations fewer than 5,000 (p. 261). Furthermore, as many as 4,198 settlements which would seem to qualify as "urban" by reason of population size are in fact classified as villages (p. 243).

In an attempt to obviate the deficiencies of fixed population densities, which are liable to obscure both significant regional and cultural variations in urban incidence, Grytzell has employed a sliding-scale of population density to demarcate the city areas of New York, London, Paris, Stockholm and Copenhagen[115]. He argues that the "urbanness" of a settlement resides in its population density *relative* to surrounding territories. Administratively inspired classifications have usually been of a simpler nature than Grytzell's, but an exception is provided by the United States Bureau of the Census which publishes urban data for four different levels of organization: (i) *the urban place*, which is a settlement of 2,500 or more inhabitants; (ii) *the incorporated city*, with 2,500 or more inhabitants, and which also has a distinct political identity; (iii) *the urbanized area*, which comprises a city of 50,000 or more inhabitants together with its urban-fringe zone; (iv) *the standard metropolitan statistical area*. Cast in terms of density, function, and level of integration, this is the most complex definition of urban status employed in any country. An SMSA is based on at least one city of 50,000 or more inhabitants, to which are added adjacent counties that are "metropolitan" in character. By this is meant that they contain at least either 10,000 non-agricultural workers or 10 per cent of the non-agricultural workers of the SMSA, *or* have at least 50% of their population residing in civil divisions both contiguous to the central city and containing population densities of 150 or more per square mile, *and* that non-agricultural workers constitute at least two-thirds of their total employed population. In addition there are mandatory criteria of integration between outlying counties and that one in which the central city is located: namely, 15% of the labour force in an outlying county must work in the county containing the central city, and the number of telephone calls per month to the central-city county must be at least four times the number of subscribers in the outlying county[116]. But even these apparently comprehensive criteria cannot be applied in New England where an exception has to be made to meet the exigencies of an historically entrenched "township" system; and it is only too evident that a definition as distinctively "American" as this would require considerable modification before it could be used for cross-cultural purposes. Nevertheless, the International Urban Research Group of the University of California at Berkeley has attempted a generalized definition of the term "metropolitan area" applicable to the whole world, and their somewhat involved conclusions run to no less than twelve pages of their report[117].

Generally speaking, these expediential definitions based on numerical size (and sometimes grandly referred to as demographic classifications) are more useful in the study of *urbanization* than in a discussion of *urbanism*. Of course, such operational classifications need not be formulated in terms of numbers of population. For example, various types of legal definitions are in current use in different parts of the world. In Turkey, Czechoslovakia, the United Arab Republic, and the Dominican Republic officially the sole criterion of urbanism is the performance of a specific administrative function; while in Algeria, Japan, Tunisia, and the United Kingdom urban status is dependent upon the existence of a particular mode of local government. In both cases urbanism is decreed rather than defined, and its ecological characteristics subordinated to the needs of administrative expediency[118].

* * *

It will have become apparent in the foregoing pages that the five approaches discussed there are not conceptually distinct and by no means mutually exclusive. A simple trait complex, for instance, borders on an ideal-type whenever its author associates it with a mode of behaviour, an elaborate census definition designed as an operational expedient can come very close to qualifying as a trait complex, and an ecological approach is always likely to expound its discoveries in terms of dominance. However, although the *strategies* are complementary, or at least mutually supportive, they are jointly directed towards four seemingly contradictory—or at best competing —*conceptions* of urbanism in terms of (i) an interactional model which emphasizes the growth and structure of specialized networks of social, economic, and political relationships focused in cities; (ii) a normative model in which urbanism is viewed as a way of life. Sometimes this life-style is viewed as an abstract entity divorced from the features customarily associated with dense aggregations of people. For example, we have come to speak of the urbanization of rural areas bringing increasing urbanism to the countryside while the density of the farm population declines[119]; (iii) an economic model, concerned primarily with productive activities in a spatial context; and (iv) a demographic model, which treats urban forms essentially as aggregations of population in restricted areas[120].

These conceptualizations of urbanism may or may not be useful in comparative studies, but there is no doubt that the attributes of which each is compounded have varied in form and function from culture to culture. Certainly the differentiation and development of urban political, social, economic, and religious institutions did not proceed symmetrically in each of the major cultural realms, nor was the relationship between those institutions and their material expression in the landscape anything like as direct as the several indices of urbanism discussed above might be held to imply. The cores of functionally related institutions which are compositely designated by terms customarily rendered into English as "town" or "city"—*alum*, *asty*, *balig*, *burgh*, *ch'eng*, *città*, *ciudad*, *gorod*, *happira*, *ilu*, *madīna*, *mu'ang*, *polis*, *pura*, *qaṣaba*, *shi*, *Stadt*, *t'inh*, *urbs*, *uru*, *ville*, and the like—*may* ultimately prove to exhibit certain functional regularities irrespective of their cultural settings, but the particular *forms* of service they provide are integral to specific cultural contexts. In the representative city of ancient Mesopotamia during the Early Dynastic period, for instance, power was distributed unequally between three major institutions: the community of citizens constituted as an assembly, a temple organization, and a palace organization[121], but I suspect that to the inhabitants of such a city it was the temple of the city's tutelary deity which conferred urban status upon the settlement. In the city which provided the Hellenistic Greek with the good life that was inseparable from law and justice[122], the criteria of urbanism were the presence of a *prytaneion*, a *gymnasion*, and a theatre[123]. For the traditional Muslim the city was necessary as the only locale in which the life prescribed by the Book of God and the Tradition of the Prophet could be lived out to the full, and the features predicated of his city were a Friday mosque (*jāmi'*), with its adjunct of the public bath (*ḥammām*), and a permanent market (*sūq*)[124]. In traditional China the representative city was first and foremost the seat of a representative of the imperial government, which secondarily attracted to itself a class of literati and a suite of service industries[125]. In the Maya, as in the Khmer realm perhaps the sole criterion of urban form was a temple; in Mauryan India probably both a temple and a palace were required. Among the traditional Yoruba of western Nigeria the city, the *ilu*, was not conceptually distinguishable from its agricultural hinterland, but both were perceived as a unitary whole radiating outwards from the seat of a crowned ruler. In these circumstances the sole criterion of urban status was the existence of an *afin*, the palace of such a ruler (*ọba*)[126]. For the urbanist of today, with his enhanced perception of the range of past human experience and his self-conscious awareness of the diversity of institutional forms, the city is everything—and a great deal more—that has been discussed above. Faced with this diversity of phenomena that have at one time or another been designated as "urban", it is, in the present state of knowledge, impossible to do more than characterize the concept of urbanism as compounded of a series of sets of ideal-type social, political, economic and other institutions which have combined in different ways in different cultures and at different times. It is not unlikely that the only feature which such congeries of institutions will ultimately prove to have in common is the fact of their aggregation.

Of the five approaches to the study of urbanism discussed above, only the expediential does not attempt in one way or another to deal with the city in contexts extending beyond its morphological limits. This surely implies a lack, or at least a relative feebleness, of specifically urban theory, of an independent nexus of concepts and generalizations applicable to the city as an analytically discrete entity. In fact a high proportion of urban studies have been directed towards the investigation of a totality of events within a city, impliedly assuming that it was the fact of "urbanness" which unavoidably determined all activity within the urban enclave. As Reiss has phrased it, ". . . much of the research simply considers the urban community to be a context within which a particular kind of theoretical problem is studied, but the context itself is not often made the object of investigation"[127]. I think that possibly it is this lack of attention to the urban *context* which has done more than anything else to inhibit the development of cross-cultural and comparative urban studies, and the ultimate justification of any of the constructs discussed above is its utility for precisely this purpose.

Notes

1 The term *urbanization*, mentioned only incidentally in this paper, has connotations as diffuse as those associated with the word "urban". For some sociologists, notable among whom are Kingsley Davis and his collaborators, it refers to the ratio obtaining between urban dwellers and total population, and can be expressed algebraically as

$$u = \frac{Pi}{Pt}$$

where u = degree of urbanization
Pi = number of urban dwellers
Pt = the total population.

[Cf., *int. al.*, Davis, K. and Golden, H. H. (1954). Urbanization and the development of pre-industrial areas, *Economic Development and Cultural Change*, **3**, pp. 6–24]. For other sociologists, and for most anthropologists and geographers, by contrast, urbanization denotes the rate of change in the ratio of city dwellers to total population, which in practice means a change in the number and size of cities. From this standpoint the term may be said to subsume three analytically separable conceptions, each appropriate to a particular genre of questions and each characterized by its own peculiar difficulties of definition and measurement: (i) the behavioural conception, which views urbanization as an adjustment of personal conduct, as a transformation of attitudes and values experienced by individuals over time. (ii) The structural conception which focuses attention on the patterned activities of whole populations, generally in situations where subgroups from mainly agricultural communities are being absorbed into larger, more complex, non-agricultural communities. It follows that this conception of urbanization tends to concentrate on the differential ordering of occupations or industries within a specified territorial space. (iii) The demographic conception which, while also focusing attention primarily on aggregate groupings within defined spaces, largely ignores both individual behaviour and occupational structure, and concerns itself almost exclusively with the process of population concentration. Consequently this conception recognizes only two significant variables: population and space. On this theme see Lampard, E. E. (1965). Historical aspects of urbanization, *in* Hauser, P. M. and Schnore, L. F. *The Study of Urbanization*. New York, London and Sydney. pp. 519–20. See also Lampard, E. E. (1963). Urbanization and social change: on broadening the scope and relevance of urban history, *in* Handlin, O. and Burchard, J. (eds.) *The Historian and the City*. Cambridge, Mass. pp. 225–47; Tisdale, H. (1942). The process of urbanization, *Social Forces*, **20**, pp. 311–316; Gibbs, J. P. (1966). Measures of urbanization, *Social Forces*, **45**, pp. 170–7. For the suggestion that indices of the degree of urbanization of different sectors of the population within a city might be based on the available "range of alternatives for individuals in most aspects of living" see Lewis, O. (1965). Further observations on the folk-urban continuum and urbanization with special reference to Mexico City, *in* Hauser, P. M. and Schnore, L. F. *op. cit.* p. 499.

The concept of *over-urbanization* has received some attention in recent years, e.g. Davis, K. and Golden, H. H. (1954). *op. cit.* esp. pp. 16–20; Sovani, N. V. (1964). The analysis of "over-urbanization", *Economic Development and Cultural Change*, **12**. pp. 113–22; Kamerschen, D. R. (1969). Further analysis of over-urbanization, *Economic Development and Cultural Change*, **17**, pp. 235–53; Hauser, P. M. (ed.) (1957). *Urbanization in Asia and the Far East*. Calcutta. pp. 8–9; Ness, G. D. (1962). Asian over-urbanization and the industrial distribution of the labour force, *Proceedings of the International Association of Historians of Asia: Second Biennial Conference* 1962. Taipei. pp. 657–75. Basically this concept is concerned with the relationship between the proportion of urban dwellers in a country's population and the apportionment of that country's total labour force between agricultural and non-agricultural occupations. In other words it represents a fusion of the demographic conception (viewing urbanization primarily in terms of a spatial index) with the structural conception (emphasizing mainly the occupational patterns of a total population). Difficulties in the application of this notion of over-urbanization arise mainly in the process of deciding what should be regarded as the normative relationship between these two indices, and to this problem there have been two approaches. The first (represented by the work of Davis, K. and Golden, H. H. (1954). *op. cit.* pp. 16–20) is based on cross-sectional analysis of such data as censuses provide for the contemporary world. The second relies on historical analogy and seeks to derive its norm from the proportion of the labour force engaged in other than agricultural occupations in the presently industrially advanced countries at a time when their levels of urbanization were generally comparable to those of the underdeveloped world of today. This approach is represented by the relevant sections of Hauser, P. M. (ed.) (1957). *op. cit.*

2 For the settlement pattern as representing the centre of gravity of continuous population movements see Hägerstrand, T. (1957). Migration and area. Survey of a sample of Swedish migration fields and hypothetical considerations on their genesis, *Lund Studies in Geography, Series B, Human Geography*, **13**, pp. 27–8.

3 The most succinct account of this transformation is that provided by Adams, R. McC. (1966). *The Evolution of Urban Society. Early Mesopotamia and Prehispanic Mexico*. Chicago.

4 It is Reissman's explicit view that they are not. "The ancient city," he writes, "the medieval city, and the industrial city are . . . quite different social phenomena", and, "Although cities in different periods resemble each other in some ways (e.g. size), the social bases of urban organization have differed': Reissman, L. (1964). *The Urban Process. Cities in Industrial Societies*. New York. pp. 16 and 19.

5 On the general problem of elucidating the nature of urbanism see specifically McTaggart, W. D. (1965). The reality of "urbanism", *Pacific Viewpoint*, **2**, pp. 220–1, and more generally Jones, W. T. (1965). *The Sciences and the Humanities. Conflict and Reconciliation*. Berkeley and Los Angeles. Ch. 2.

6 Cf. Popper, K. (1957). *The Poverty of Historicism*. London. pp. 26–34.

7 For a definition of this term see p. 603.

8 Cf. Jeremia, W. R. (1958). *Handbuch zum Alten Testament*, **12**. Tübingen. pp. 207 ff; Ben-gavriel, M. Y. (1962–3). Das nomadische Ideal in der Bibel, *Stimmen der Zeit*, **88/171**, pp. 253–63; Hertzler, J. O. (1936). *Social Thought of the Ancient Civilizations*. New York. pp. 298ff. A similar recognition of distinct "integrated patterns of dominant attitudes", or styles of life, is evident in Hesiod's description of the countrymen of Boeotia in the sixth century B.C. [See Francis, E. K. L. (1945). The personality type of the peasant according to Hesiod's *Works and Days*, *Rural Sociology*, **10**, pp. 275–95. First cited in this connection by Robert Redfield: cf. Note 19] and in Plato's several discussions of the virtues of urban life in *Critias*, *The Laws* and *The Republic*.

9 Cp. "In the court [that is, urbanized society] it is nobility that holds the first place, in the village it is age" (II, ii, 3, 6).

10 Maine, Sir Henry (1861: here cited from 1963 reprint). *Ancient Law. Its Connection with the Early History of Society and its Relation to Modern Ideas*. Boston. p. 165.

11 Morgan, L. H. (1877). *Ancient Society, or, Researches in the Lines of Human Progress from Savagery through Barbarism to Civilization*. New York.

12 Tönnies, F. (Eighth revised edition, 1935). *Gemeinschaft und Gesellschaft*. Leipzig.

13 Weber, M. (1947). *The Theory of Social and Economic Organization*. Transl. by Henderson, A. M. and Parsons, T. New York. pp. 136–8, 329ff and 341ff.

14 Notably Herbert Spencer (1892). *Principles of Sociology*. 1. New York ["Military" as opposed to "Industrial"]; Becker, H. (1950). Sacred and secular societies, *Social Forces*, **28**, pp. 361–76; Becker, H. and Barnes, H. E. (1952). *Social Thought from Lore to Science*. Washington. ch. 1 ["Sacred" as opposed to "Secular"]. Also Becker, H. (1957). Current sacred-secular theory and its development, *in* Becker, H. and Boskoff, A. (eds.) *Modern Sociological Theory in Continuity and Change*. New York. The most comprehensive treatment of these paired terms is provided by Becker, H. (1950). *Through Values to Social Interpretation*. Durham North Carolina.

15 Durkheim, E. (1893). *De la Division du Travail Social: Etude sur l'Organisation des Sociétés Supérieures*. Paris. Especially chs. 2 and 3. For an application of this ideal-type construct to urban forms among the Maya and the

Khmer see Coe, M. D. (1961). Social typology and the tropical forest civilizations, *Comparative Studies in Society and History*, **4**, pp. 65–85.

16 Mauss, M. (1925). *Essai sur le Don, Forme archaïque de l'Echange*. Paris.

17 Redfield, R. and Singer, M. B. (1954). The cultural role of cities, *Economic Development and Cultural Change*, **3**, pp. 53–73.

18 Redfield, R. (1947). The folk society, *Amer. J. Sociol.*, **52**, pp. 293–308.

19 Redfield, R. (1953). *The Primitive World*. Cornell. pp. 31 and 53. This characterization probably owed something to an earlier definition of peasant society by Kroeber, A. L. (1948). *Anthropology*. New York. p. 284. Note especially: "Peasants are definitely rural—yet live in relation to market towns; they form a class segment of a larger population which usually contains urban centres, sometimes metropolitan capitals. They constitute part-societies with part-cultures." Of course, by insisting on viewing urban society as simply the antithesis of folk society, and thereby denying real independence to typically urban features, Redfield went a good way towards subverting the utility of his construct.

20 In fact in these later works Redfield adopted a neo-evolutionary point of view and tended to identify "folk" with "pre-urban" society; e.g. Redfield, R. (1953). *The Primitive World and its Transformations*. Ithaca; N.Y. Other works in which Redfield invokes or elaborates on the folk-urban or rural-urban dichotomies and continua include: Redfield, R. (1941). *The Folk Culture of Yucatan*. Chicago; Redfield, R. (1953). The natural history of the folk society, *Social Forces*, **31**, pp. 224–8; Redfield, R. (1954). *The Little Community. The Gottesman Lectures, V*. Uppsala; Redfield, R. (1956). *Peasant Society and Culture*. Chicago.

21 Queen, S. A. and Carpenter, D. B. (1953). *The American City*. New York. p. 38.

22 Cf., for example, Tax, S. (1941). World view and social relations in Guatemala, *Amer. Anthrop.*, **43**, pp. 27–42; Caplow, T. (1949). The social ecology of Guatemala city, *Social Forces*, **28**; Lewis, O. (1952). Urbanization without breakdown: a case study, *The Scientific Monthly*, **75**; Lewis, O. (1951). *Life in a Mexican Village: Tepoztlan Restudied*. Urbana; Lewis, O (1955). Peasant culture in India and Mexico, *in* Marriott, McK. Village India, *Amer. Anthrop.*, **57**, *Memoir* 83, esp. p. 165; Bopegamage, A. (1957). *Delhi: a Study in Urban Sociology*. Bombay. pp. 93–103; Desai, I. P. (1964). *Some Aspects of Family in Mahuva*. Bombay.

23 On this point see particularly Foster, G. M. (1953). What is folk culture? *Amer. Anthrop.*, **55**, pp. 159–73.

24 Reissman's phrase [Reissman, L. (1964). *The Urban Process*. Glencoe, Ill. p. 123]. Hofstee used the less complimentary phrase "vulgar Tönniesism" in referring to the polar typologies discussed here [Hofstee, E. W. (1960). Rural social organization, *Sociologia Ruralis*, **1**, pp. 105–17].

25 All, and more, of these criticisms of the folk-urban and rural-urban dichotomies and continua are to be found in, *int. al.*, the following works: Spaulding, I. A. (1951). Serendipity and the rural-urban continuum, *Rural Sociology*, **16**, pp. 29–36; Miner, H. (1952). The folk-urban continuum, *Amer. Sociol. Rev.*, **17**, pp. 529–37; Mintz, S. W. (1953). The folk-urban continuum and the rural proletarian community, *Amer. J. Sociol.*, **59**, pp. 136–43; Mintz, S. W. (1954). On Redfield and Foster, *Amer. Anthrop.*, **56**, pp.87–92; Duncan, O. D. (1957). Community size and the rural-urban continuum, *in* Hatt, P. K. and Reiss, Jr., A. J. (Second edition) *Cities and Society. The Revised Reader in Urban Sociology*. Glencoe, Ill. pp. 35–45; Duncan, O.D. and Reiss, Jr., A. J. (1956). *Social Characteristics of Urban and Rural Communities, 1950*. New York and London. pp. 37–40; Reiss, A. J. (1959). Rural-urban and status differences in interpersonal contacts, *Amer. J. Sociol.*, **65**, pp. 182–95; Stewart, C. T. (1958). The urban-rural dichotomy: concepts and uses, *Amer. J. Sociol.*, **64**, pp.

152–8; Dewey, R. (1960). The rural-urban continuum: real but relatively unimportant, *Amer. J. Sociol.*, **66**, pp. 60–6; Benet, F. (1963). Sociology uncertain: the ideology of the rural-urban continuum, *Comparative Studies in Society and History*, **6**, pp. 1–23; Reissman, L. (1964). *The Urban Process. Cities in Industrial Societies*. New York. pp. 122–38; McGee, T. G. (1964). The rural-urban continuum debate, the pre-industrial city and rural-urban migration, *Pacific Viewpoint*, **5**, pp. 159–81; Lewis, O. (1965). Further observations on the folk-urban continuum and urbanization with special reference to Mexico City, *in* Hauser, P. M. and Schnore, L. F. *op cit.* pp. 491–503; Hauser, P. M. (1965). Observations on the urban-folk and urban-rural dichotomies as forms of Western ethnocentrism, *in* Hauser, P. M. and Schnore, L. F. *op. cit.* pp. 503–17; Pahl, R. E. (1966). The rural-urban continuum, *Sociologia Ruralis*, **6**, pp. 299–329; Ommen, T. K. (1967). The rural-urban continuum re-examined in the Indian context, *Sociologia Ruralis*, **7**, pp. 30–48.

26 Lewis, O. (1965). *op. cit.* p. 491.

27 Hauser, P. M. (1965). *op. cit.* p. 514.

28 Defined by Daniel Lerner as "the social process of which development is the economic component": Lerner, D. (1967). Comparative analysis of processes of modernization, *in* Miner, H. *The City in Modern Africa*.New York, Washington and London. p. 21.

29 See, for example, Lampard, E. E. (1955). The history of cities in the economically advanced areas, *Economic Development and Cultural Change*, **3**, pp. 81–136.

30 See, for example, Kolb, W. L. (1955). The social structure and functions of cities, *Economic Development and Cultural Change*, **3**, pp. 30–46.

31 Hoselitz, B. (1955). Generative and parasitic cities, *Economic Development and Cultural Change*, **3**, pp. 278–94.

32 Barber, W. J. (1967). Urbanization and economic growth: the cases of two white settler territories, *in* Miner, H. *op. cit.* pp. 91–125.

33 Mabogunje, A. L. (1968). *Urbanization in Nigeria*. London. pp. 21–6 and 315–19.

34 Space forbids that I enter into discussion of the various philosophies of organicism that have been applied to the city, and of which the most recent is Lewis Mumford's metaphorical conception of the city as a neurotic person [Mumford, L. (1961). *The City in History. Its Origins, its Transformations, and its Prospects*. New York].

35 Cf. Coker, F. W. (1910). *Organismic Theories of the State: Nineteenth Century Interpretations of the State as Organism or as Person*. Columbia University Studies in History, Economics, and Public Law, 38. New York.

36 Park, R. E. (1916). The City: suggestions for the investigation of human behavior in the urban environment, *Amer. J. Sociol.*, **20**, pp. 577–612.

37 Reissman, L. (1964). *The Urban Process. Cities in Industrial Societies*. New York. p. 99. Chapter 5 of this work affords one of the most succinct and perspicacious evaluations of the contributions of the school of urban ecologists.

38 Firey, W. (1946). *Land Use in Central Boston*. Cambridge, Mass.

39 Park, R. E. (1939). Review of Alihan's *Social Ecology*, *Ann. Amer. Acad. Political and Social Science*, **202**, pp. 264–5.

40 Summary by Reissman, L. (1964). *op. cit.* pp. 115–16.

41 Duncan, O. D. (1959). Human ecology and population studies, *in* Hauser, P. M. and Duncan, O. D. *The Study of Population*. Chicago. p. 682.

42 Duncan, O. D. and Schnore, L. F. (1959). Cultural, behavioral, and ecological perspectives in the study of social organization, *Amer. J. Sociol.*, **65**, pp. 132–46; Schnore, L. F. (1958). Social morphology and human ecology, *Amer. J. Sociol.*, **63**, pp. 620–34. For a critique of the Theory of the

Ecological Complex, see Willhelm, S. M. (1964). The concept of the "ecological complex": a critique, *Amer. J. Econ. and Sociol.*, **23**, pp. 241–8.

43 Duncan, O. D. (1959). p. 683.

44 Duncan, O. D. and Schnore, L. F. (1959). *op. cit.* p. 138.

45 Sjoberg, G. (1965). Theory and research in urban sociology, *in* Hauser, P. M. and Schnore, L. F. *op. cit.* p. 166.

46 Wheatley, P. (1967). Proleptic observations on the origins of urbanism, *in* Steel, R W. and Lawton, R. (eds.). *Liverpool Essays in Geography. A Jubilee Collection.* London. pp. 315–45; Wheatley, P. (1971). *The Pivot of the Four Quarters. A Preliminary Enquiry into the Origins and Character of the Ancient Chinese City.* Edinburgh and Chicago.

47 Gibbs, J. P. and Martin, W. T. (1959). Toward a theoretical system of human ecology, *Pacific Sociol. Rev.*, **2**, pp. 29–36; Gibbs, J. P. and Martin, W. T. (1962). Urbanization, technology and the division of labour: international patterns, *Amer. Sociol. Rev.*, **27**, pp. 667–77. In this latter paper, the authors propound the following propositions, which are sufficiently indicative of the scope and direction of their work:

IA The degree of urbanization in a society varies directly with the division of labour;

IB The division of labour in a society varies directly with the dispersion of objects of consumption . . .

IIA The degree of urbanization in a society varies directly with technological development;

IIB Technological development in a society varies directly with the dispersion of objects of consumption.

48 Gibbs, J. P. and Martin, W. T. (1962). *op. cit.* p. 672.

49 Reissman, L. (1964). *op. cit.* p. 120.

50 Wirth, L. (1938). Urbanism as a way of life, *Amer. J. Sociol.*, **44**, p. 8. A similar deductive argument had been advanced by Simmel at least as early as 1908, except that he had used social rather than ecological characteristics as the basis from which to derive psychological features: Simmel, G. The metropolis and mental life. Reprinted in Wolff, K. H. (1950). *The Sociology of Georg Simmel.* Glencoe, Ill. Cp. also Davis who followed Wirth's model more closely: Davis, K. (1948). *Human Society.* New York.

51 Wirth, L. (1938). *op. cit.* p. 13.

52 Bascom, W. (1955). Urbanization among the Yoruba, *Amer. J. Sociol.*, **60**, pp. 446–54.

53 Krapf-Askari, E. (1969). *Yoruba Towns and Cities. An Enquiry into the Nature of Urban Social Phenomena.* Oxford. p. 157.

54 Duncan, O. D. (1957). *op. cit.*

55 Reiss, Jr., A. J. (1955). An analysis of urban phenomena, *in* Fisher, R. M. (ed.) *The Metropolis in Modern Life.* New York. p. 43. It is true, though, that Dewey, R. (1960). *op. cit.* pp. 63–4, like Bascom, W. (1955). *op. cit.* and Krapf-Askari, E. (1969). *op. cit.*, has taken the opposite view: ". . . it is probably futile to argue for the abandonment of the terms "'rural' and "urban" as indicators of size and density of population. There is little, if any, evidence that the established usage of these words is in process of change: their abandonment would require the invention of a new pair of terms to denote the large, densely associated communities, distinguished from the small and sparsely settled groups of the world." Yet the fact remains that the use of these terms in analytic investigations is subverted by an almost complete absence of quantitative empirical confirmation.

It may be remarked parenthetically that the term "urban" has sometimes been applied to settlements such as those of the Classic Maya, of Classical Cambodia, of China prior to the Spring-and-Autumn period, Athens prior to 431 B.C., and Etruscan Rome, which were not characterized by high densities of population at their core. Such settlements have been termed "extended boundary towns" by Miles, S. W. (1958). An urban type: extended boundary towns, *S. W. J. Anthrop.*, **14**, pp. 339–51, and "synchorite urban settlements" by Rowe, J. H. (1963). Urban settlements in ancient Peru, *Ñawpa Pacha*, **1**, p. 3. For the possibility that such settlements represent a functional and developmental stage of urban evolution, see Wheatley, P. (1971). *op. cit.* Ch. 3.

56 As Krapf-Askari, E. (1969). *op. cit.* p. 22 points out, the Wirthian model looks back beyond Durkheim to Rousseau; "it is redolent with suppressed nostalgia for a subjacent 'integrated society' in which contacts were personal, deep, permanent, and involved the whole person. Even *anomie* implies that a *nomos* has, somewhere along the way, been lost." In America this point of view found powerful proponents in, among others, Jefferson, Emerson and Dewey: cf. White, M. and L. (1962). *The Intellectual versus the City*. Cambridge, Mass.

57 Wirth, L. (1938). *op. cit.* p. 18.

58 These characteristics implicit in Wirth's formulation are here listed as summarized by Dewey, R. (1960). *op. cit.* pp. 61–2.

59 Wirth, L. (1938). *op. cit.* p. 9.

60 Weber, M. (1921: cited in 1925 reprint). *Grundriss der Sozialökonomik, III. Abteilung, Wirtschaft und Gesellschaft*, 2. Halbband. Tübingen. p. 523. For an "Americanized" adaptation of Weber's definition see Munro, W. B. (1926). *The Government of American Cities*. New York. p. 13.

61 The city-state appears to represent a developmental phase of socio-political organization in which, it is true, the city exercises the sovereign powers of a state government. But this type of city constitutes a special case, coming as it does at that extremity of the spectrum of political dependence at which the territory of the city is coincident with that of the state. The German *Freistädte* and some Italian cities towards the end of the Middle Ages (when the authority of the German Emperor had become merely nominal in Italy) have sometimes been classed as quasi-sovereign, but such clearly do not represent a developmental phase, and should more properly be classed among those cities to which some degree of autonomy has fallen accidentally.

62 The three instances of wholly privileged cities mentioned in the text are discussed in Gonthier, A. (1954). Les villes japonaises. Histoire des institutions administratives et judiciaires, *Recueils de la Société Jean Bodin* (Bruxelles), **6**, pp. 241–8; Gonthier, A. (1957). Le droit privé urbain au Japon, *Recueils de la Société Jean Bodin*, **8**, pp. 111–14; Pirenne, J. (1956). Les institutions urbaines dans l'ancienne Egypte et dans le pays de Sumer, *Recueils de la Société Jean Bodin*, **6**, pp. 27–48; Pirenne, J. (1957). Le droit privé dans l'ancienne Egypte, *Recueils de la Société Jean Bodin*, **8**, pp. 25–44; Préaux, C. (1956). Les villes hellénistiques, principalement en Orient. Leurs institutions administratives et judiciaires, *Recueils de la Société Jean Bodin*, **6**, pp. 69–134. Concerning the varying status of the Hellenistic city see also Jones, A. H. M. (1940). *The Greek City from Alexander to Justinian*. Oxford.

63 Weber, M. (1921). *Wirtschaft und Gesellschaft*, p. 523. For a critical summary of Weber's conception of the non-European city see Eisenstadt, S. N. (1955). Social problems of urban organization and planning in underdeveloped countries, *Atti del Congresso Internazionale di Studio sul Problema delle Aree arretrate*, **2**, pp. 887–900.

64 Cf. Polanyi, K., Arensberg, C. M. and Pearson, H. W. (eds.), (1957). *Trade and Market in the Early Empires*. Glencoe, Ill. For a comprehensive, though highly critical, survey of the discussions provoked by this book see Cook, S. (1966). The obsolete "anti-market" mentality: a critique of the

substantive approach to economic anthropology, *Amer. Anthrop.*, **68**, pp. 323–45.

65 Childe, V. G. (1950). The urban revolution, *Town Planning Review*, **21**, pp. 9–16. The process of urban genesis as Childe envisaged it in Mesopotamia is treated in more detail in Childe, V. G. (1952). *New Light on the Most Ancient East*. London. Chap. 7.

66 In its anthropomorphic manifestations representational art may have been a result either of the crystallization of social stratification or of the elaboration of a religious tradition. The former would appear to have been the case in early Mesopotamia where gods—or, at least, recognizably anthropomorphized deities—were portrayed at a somewhat later date than human figures. In the mural art of Teotihuacán, by contrast, gods appear to have been portrayed at least as early as, and probably prior to, the representation of men. In Shang China animal forms seem, on present evidence, to have preceded the portrayal of men, and there are no recognizable depictions of deities. Generally speaking, rather than an index of urbanism, representational art would seem, like urbanism itself, to be merely one specific phenomenon within the inclusive category of civilization.

67 In Mesoamerica only limited forms of writing developed so that these societies are perhaps categorized most appropriately as oligo-literate.

68 Rowe, J. H. (1963). Urban settlements in ancient Peru, *Ñawpa Pacha*, **1** p. 3.

69 Rowe, J. H. (1963). *Ibid.*

70 Kenyon, K. M. (1957). *Digging up Jericho: the Results of the Jericho Excavations 1952-1956*. New York. pp. 65–6.

71 Smailes, A. E. (1944). The urban hierarchy in England and Wales, *Geogr.*, **29**, pp. 41–51.

72 Dickinson, R. E. (1932). The distribution and functions of the smaller urban settlements in East Anglia, *Geogr.*, 7, pp. 19–31.

Exigencies of space prohibit discussion of the numerous classifications of cities which might appropriately be included in this section, but the following references would provide an introduction to what has become a very extensive literature: Harris, C. D. (1943). A functional classification of cities in the United States, *Geogr. Rev.*, **33**, pp. 86–99; Steigenga, W. (1955). A comparative analysis and a classification of Netherlands towns, *Tijdschrift voor Economische en Sociale Geografie*, **46**, pp. 106–12; Pownall, L. L. (1953). The functions of New Zealand towns, *Ann. Assoc. Amer. Geogr.*, **43**, pp. 332–50; Nelson, H. J. (1955). A service classification of American cities, *Econ. Geogr.*, **31**, pp. 189–210; Smith, R. H. (1965). Method and purpose in functional town classification, *Ann. Assoc. Amer. Geogr.*, **55**, pp. 539–48; Moser, C. A. and Scott, W. (1961). *British Towns*. London; Alexander, J. W. (1954). The basic-nonbasic concept of urban economic functions, *Econ. Geogr.*, **30**, pp. 246–61; Alexandersson, G. (1956). *The Industrial Structure of American Cities*. Lincoln, Neb.; Andrews, R. B. (1953–56). Mechanics of the urban economic base, *Land Economics*, **29**, pp. 161–7, 263–8, 343–9; **30**, 52–60, 164–73, 260–9, 309–19; **31**, 47–53, 69–84; (Reprinted *in* Pfouts, R. W. (ed.) (1960). *The Techniques of Urban Economic Analysis*. New York. pp. 6–17, 40–183,; Ullman, E. L. and Dacey, M. F. (1960). The minimum requirements approach to the urban economic base, *Lund Studies in Geography, Series B, Human Geography*, **24**, pp. 121–43; Isard, W. and Kavesh, R. (1954). Economic structural interrelations of metropolitan regions, *Amer. J. Sociol.*, **60**, pp. 152–62.

73 Miner, H., (ed.) (1967). *The City in Modern Africa*. New York. pp. 5–10. For the concept of dominance Miner refers to Amos H. Hawley's discussion of the term in (1950). *Human Ecology. A Theory of Community Structure*. New York. p. 221. Cf. esp.: "Dominance attaches to the unit that controls the conditions necessary to the functioning of other units. Ordinarily that

means controlling the flow of sustenance into the community. Any alteration of the sustenance flow requires immediate readjustment on the part of all other units. Such influence may be exercised directly or indirectly through control over the allocation of space to different activities, the determination of who shall be employed, the regulation of credit, the censoring of news and information reaching the community, and in many other ways".

The notion of the city as a centre of dominance had been implicit in the writings of numerous earlier authors ranging in time at least from Ibn Khaldûn in the fourteenth century A.D. and Giovanni Botero in the sixteenth century, to Donald Bogue (1949). *The Structure of the Metropolitan Community: a Study of Dominance and Subdominance*. Ann Arbor. Miner, however, as far as I have been able to ascertain, was the first to give the concept formal expression.

74 Miner, H. (ed.) (1967). *op. cit.* pp. 6–8.

75 Friedmann, J. (1961). Cities in social transformation, *Comparative Studies in Society and History*, **4**, p. 92. The author has further elaborated the concept of "effective space" in Friedmann, J. (1961). L'influence de l'intégration du système social sur le développement économique, *Diogène*, **33**, pp. 80–104, and applied it to problems of development in preprints of two papers so far unpublished: Friedmann, J. (1969). A general theory of polarized development. School of Architecture and Urban Planning University of California at Los Angeles [mimeo.] Friedmann, J. *et al.* (1970). Urbanization and national development: a comparative analysis. School of Architecture and Urban Planning, UCLA [mimeo].

76 Harris, C. D. and Ullmann, E. L. (1945). The nature of cities, *Ann. Amer. Acad. Political and Social Science*, **242**, pp. 7–17. The term "central place" is apt because, to provide anything like comprehensive services with maximum efficiency, a city must be located close to the point of minimum aggregate travel within its tributary area. The term was introduced into American geography by R. S. Platt (personal communication from Norton S. Ginsburg).

77 Lalanne, L. (1863). Essai d'une théorie des réseaux de chemin de fer, fondée sur l'observation des faits et sur les lois primordiales qui président au groupement des populations, *Comptes Rendus Hebdomadaires des Séances de l'Académie des Sciences*, **42**, pp. 206–10.

78 Galpin, C. J. (1915). *The Social Anatomy of an Agricultural Community*. Research Bulletin 34, Agricultural Experiment Station of the University of Wisconsin. Madison.

79 Christaller, W. (1933). *Die zentralen Orte in Süddeutschland: Eine ökonomisch-geographische Untersuchung über die Gesetzmässigkeit der Verbreitung und Entwicklung der Siedlungen mit städtischen Funktionen*. Jena.

80 I.e., assuming a uniform distribution of population and purchasing power, uniform terrain and resource distribution, and equally developed transport facilities in all directions.

81 See Hilbert, D. and Cohn-Vosson, S. (1957). *Geometry and the Imagination*. New York. p. 35.

82 K in hierarchies based on the marketing principle = the bifurcation ratio k for central places + one; that is, it refers to the number of settlements at a given level in the hierarchy served by a central place at the next higher level.

A statement of the simplest model of a hierarchy of cities in mathematical form has been provided by Beckmann, M. J. (1958). City hierarchies and the distribution of city size, *Econ. Development and Cultural Change*, **6**, pp. 243–8; and Prost has compared the model with results obtained empirically by Christaller in Germany: Prost, M-A (1965). *La Hiérarchie des Villes*. Paris. pp. 70–4. As early as 1931 W. J. Reilly advanced a "Law of Retail Gravitation" which included a formula providing a ready approximation of the extent of market areas: Reilly, W. J. (1931). *The Law of Retail Gravitation*. New York.

83 Lösch, A. (1944). *Die raumliche Ordnung der Wirtschaft*. Jena. Cf. also Valavanis, S. (1955). Lösch on location, *Amer. Econ. Rev.*, **45**, pp. 637–44; Nourse, H. O. (1968). *Regional Economics*. New York. Thompson, A. C. (1970). Some comments on Lösch, *Geogr. Analysis*, **2**, pp. 397–400, has criticized Lösch's basic assumption that establishments normally exist at their thresholds, and proposed an alternative solution, which he claims is more consonant with empirical findings. According to the Löschian analysis, in a state of saturation establishments are of minimum possible scale with their demand curves tangential to their long-run average-cost curves. In Thompson's solution, however, the demand curve of each establishment is tangential to its short-run average-cost curve, but above its long-run average-cost curve, so that establishments generally exceed the minimum possible scale.

84 Bogue has discerned similar sectors in a study of population densities around the 67 largest metropolitan centres in the U.S.A. The hinterland of each metropolis was found to comprise three *types* of sector: *route* sectors containing a major highway, *subdominant* sectors each of which contained one city of 25,000 or more inhabitants, and *local* sectors which constituted the rest. Urban densities proved to be highest in the subdominant sectors, lower than might have been expected in the route sectors, and well below both in the local sectors. See Bogue, D. J. (1949). *op. cit.*

85 During the past thirty years there has been considerable debate as to whether the settlements of a region are disposed in an orderly manner among size categories as is implicit in Christaller's hypothesis or, alternatively, are ranged along a continuum. The latter distribution, which can be made conformable with a variety of explanatory hypotheses, has been generalized by Zipf as the Rank-Size Rule

$$r \cdot P^q = K$$

where q (which approximates to unity) and K are constants for a given group of cities, and P is the population of a particular city of rank r in the descending array of settlements. Simply stated this means the size of the *n*th city in the hierarchy approximates one-*n*th the size of the largest city. The distribution underlying this rule can be regarded as lognormal and obeying the law of proportionate effect. Although the rule is essentially an empirical finding rather than a theoretical proposition, Beckmann has been able to show that with the addition of a random variable the discrete size categories of Christaller's hierarchy merge into a virtual rank-size distribution. There is still some doubt as to the precise significance of the several forms of city-size distribution, but Simon's suggestion that rank-size regularities are generated by some stochastic process has gained considerable favour. In testing the hypothesis for the distribution of city sizes in the state of Washington, U.S.A., Berry and Garrison found a reasonably close agreement between observed and predicted frequency distributions. Berry has subsequently argued that rank-size (lognormal) distributions are typical of areally extensive countries with long traditions of urbanism in economically and politically complex contexts. Primate distributions, in which a stratum of small urban forms is dominated by one or more very large cities, by contrast, are allegedly characteristic of small countries recently urbanized and with relatively simple political and economic institutions. However, just under a quarter of the thirty-eight countries studied yielded intermediate distributions. and the classification of several in the two main groups can be justified only by invoking the operation of an array of cultural and historical factors. The presence of Korea and El Salvador in the first group and of Thailand

Uruguay and Denmark in the second are especially intractable features of the classification.

For general discussions of the points touched on above see: Auerbach, F. (1913). Das Gesetz der Bevölkerungskonzentration, *Petermanns Mitteilungen*, **59**, 74–6; Zipf, G. K. (1949). *Human Behaviour and the Principle of Least Effort*. Cambridge, Mass.; Allen, G. R. (1954). The "Courbes des Populations": a further analysis, *Bull. Oxford Univ. Inst. Statistics*, **16**, pp. 179–89; Vining, R. (1955). A description of certain spatial aspects of an economic system, *Econ. Development and Cultural Change*, **3**, pp. 147–95; Stewart, C. T. (1958). The size and spacing of cities, *Geogr. Rev.*, **48**, pp. 222–45; Simon, H. A. (1955). On a class of skew distribution functions, *Biometrica*, **42**, pp. 425–440; Berry, B. J. L. and Garrison, W. (1958) Alternate [*sic*] explanations of urban rank-size relationships, *Ann. Assoc. Amer. Geogr.*, **48**, pp. 83–91; Berry, B. J. L. (1961). City size distributions and economic development, *Econ. Development and Cultural Change*, **9**, pp. 573–88; Beckmann, M. J. (1958). *op. cit.* Cf. also Berry, B. J. L. (1964). Cities as systems within systems of cities, *Papers and Proceedings of the Regional Science Assoc.*, **13**, pp. 147–63.

86 This is the view of Von Böventer, E. (1962). Towards a unified theory of spatial economic structure, *Papers of the Regional Science Association*, **10**, pp. 163–87. However, Tinbergen's model of market-oriented location, developed on an isotropic plain and minimizing total costs of production and transport, generates a tiered hierarchy not dissimilar from that of Christaller. See Tinbergen, J. (1961). The spatial dispersion of production: an hypothesis, *Schweizerische Zeitschrift für Volkwirtschaft und Statistik*, **97**, pp. 412–19; Tinbergen, J. (1964). Sur un modèle de la dispersion géographique de l'activité économique, *Rev. d'Economie Politique*, **74**, pp. 30–44.

87 Recent papers dealing with the geometry of Central-Place Theory and mixed hierarchies include: Dacey, M. F. (1965). The geometry of Central-Place Theory, *Geografiska Annaler*, **47B**, pp. 111–24; Hudson, J. C. (1967). An algebraic relation between the Lösch and Christaller central place networks, *Professional Geographer*, **19**, pp. 133–5; Woldenberg, M. J. (1967). The identification of mixed hexagonal central place hierarchies with examples from Finland, Germany, Ghana and Nigeria, *Harvard Papers in Theoretical Geography*. Paper No. 5 in the series *Geography and the Properties of Surfaces*. For general comments on recent theoretical departures see Berry, B. J. L. (1967). *Geography of Market Centers and Retail Distribution*. Englewood Cliffs, N. J.

88 Curry, L. (1962). The geography of service centres within towns: the elements of an operational approach, *Lund Studies in Geography, Series B, Human Geography*, **24**, pp. 31–53.

89 Christaller, W. (1933). *op. cit.* p. 3.

90 Cf. Polanyi, K., Arensberg, C. M. and Pearson, H. W. (eds.) (1957). *Trade and Market in the Early Empires*. Glencoe, Ill.; Cook, S. (1966). The obsolete "anti-market" mentality: a critique of the substantive approach to economic anthropology, *Amer. Anthrop.*, **68**, pp. 323–45.

91 Parsons, T. and Smelser, N. J. (1956). *Economy and Society. A Study in the Integration of Economic and Social Theory*. London.

92 Berry, B. J. L. and Garrison, W. L. (1958). Recent developments of Central-Place Theory, *Papers and Proceedings of the Regional Science Association*, **4**, pp. 107–20. The operational use of the restated theory is demonstrated by Berry, B. J. L. (1959). Ribbon developments in the urban business pattern, *Ann. Assoc. Amer. Geogr.*, **49**, pp. 145–55.

93 Thomas, E. N. (1961). Towards an expanded central place model, *Geogr. Rev.*, **51**, pp. 400–11; Thomas, E. N. (1962). The stability of distance-population-size relationships for Iowa towns from 1900–50, *Lund Studies*

in Geography, Series B, Human Geography, **24**, pp. 13–30; Thomas, E. N. (1962). The spatial behavior of a dispersed non-farm population, *Papers and Proceedings of the Regional Science Assoc.*, **8**, pp. 107–33; Thomas, E. N. (1964). The comparative spatial behaviour of two dispersed populations, *J. Regional Science*, **4**.

94 Dacey, M. F. (1966). A probability model for central place location, *Ann. Assoc. Amer. Geogr.*, **56**, pp. 550–568. Cp. especially p. 568: "(1) a displaced central place model describes the Iowa urban pattern though the Iowa pattern is in a strong state of disequilibrium, or (2) the utility of this stochastic interpretation of central-place theory is limited to description of the urban pattern but has no explanatory implications to the locational process underlying the urban pattern. The present analysis does not permit the objective evaluation of the relative merits of these two contradictory conclusions".

95 Clarkson, G. P. E. (1963). *The Theory of Consumer Demand: a Critical Appraisal*. Englewood Cliffs, N.J.

96 Harvey, D. (1969). *Explanation in Geography*. London. pp. 138–9; Dacey, M. F. (1966). *op. cit.* p. 550, had already pointed out that the theory as explicated by Christaller and Lösch was algebraic, and hence deterministic rather than probabilistic; that because it did not allow for deviations from precisely stated relations and locations, there was no chance of accumulating evidence verifying the theory.

97 Cp. *int. al.*, Skinner, G. W. (1964). Marketing and social structure in rural China. Part I, *J. Asian Studies*, **24**, p. 31. On succeeding pages Skinner discusses the social dimensions of Chinese marketing systems.

98 Von Grünebaum, G. E. (1962). The sacred character of Islamic cities *Mélanges Ṭāhā Ḥusain: offerts par ses Amis et ses Disciples à l'occasion de son 70ième anniversaire*. Cairo. pp. 25–37, especially p. 27.

99 This principle was enunciated and elaborated in its classic form by Ratcliff R. U. (1949). *Urban Land Economics*. New York. The operation of the urban land-market has been partially elucidated by Alonso, W. (1960). A theory of the urban land-market, *Papers and Proceedings of the Regional Science Association*, **6**, pp. 149–57, and Wingo, L. (1961). *Transportation and Urban Land*. Washington, D.C. Cf. also Haggett, P. (1968). The spatial structure of city regions, *Quarterly Bull., Research and Intelligence Unit, Greater London Council*, **4**, pp. 28–34.

100 Burgess, E. W. (1024). The growth of the city: an introduction to a research project, *Publications of the Amer. Sociol. Soc.*, **18**, pp. 85–97. Reprinted in Park, R. E., Burgess, E. W. and McKenzie, R. D. (1925). *The City*. Chicago. pp. 47–62. See also Quinn, J. A. (1940). The Burgess zonal hypothesis and its critics, *Amer. Sociol. Rev.*, **5**, pp. 210–18; Alihan, M. A. (1938). *Social Ecology: a Critical Analysis*. New York. pp. 224–225; Davie, M. R. (1937). The pattern of urban growth, *in* Murdock, G. P. (ed.). *Studies in the Science of Society*. New Haven. pp. 133–161; Schnore, L. F. (1965). On the spacial structure of cities in the two Americas, *in* Hauser, P. M. and Schnore, L. F. *op. cit.* pp. 349–56. For partial explanations of the concentric arrangement in terms of the substitution of rents for transport costs, see Berry, B. J. L. (1959). Section *in* Garrison, W., Berry, B. J. L., Marble, D. F., Nystuen, J. D., and Morrill, R. L. *Studies of Highway Development and Geographic Change*. Seattle; Isard, W. (1956). *Location and Space-economy. A General Theory Relating to Industrial Location, Market Areas, Land Use, Trade, and Urban Structure*. Cambridge, Mass. pp. 200–6. Alonso, W. (1960). *op. cit.* and Wingo, L. (1961). *op. cit.* have partially elucidated the rationale in the concentric model of the poor living on land of high value close to the city centre and the rich living on cheaper land at the periphery. If the theories of the Chicago School of urban ecologists be thought of as representing the confluence of streams of thought deriving

from nineteenth-century Social Darwinism and Classical Economics, it is not surprising that their authors should have attempted to explain human action in terms of impersonal competition in either its economic (Burgess) or social (Park) manifestations.

101 Hoyt, H. (1939). *The Structure and Growth of Residential Neighborhoods in American Cities*. Washington, D.C.

102 Harris, C. D. and Ullman, E. L. (1945). The nature of cities, *Ann. Amer. Acad. Political and Social Science*, **242**, pp. 7–17.

103 Marble, D. F. (1959). Section *in* Garrison, W., Berry, B. J. L., Marble, D. F., Nystuen, J. D., and Morrill, R. L. *op. cit.* The spread of housing during the past twenty-five years in an area to the south of Cambridge has been adduced as evidence from England tending to support Marble's interpretation of American data. Cf. Haggett, P. (1965). *Locational Analysis in Human Geography*. London. pp. 180–1. See also King, L. J. (1967). Discriminatory analysis of urban growth patterns in Ontario and Quebec, 1951–61, *Ann. Assoc. Amer. Geogr.*, **57**, pp. 566–78.

104 This relationship is usually generalized as

$$P_d = P_o e^{-gd}$$

where P_d is the population density at distance d from the centre of the city; $-g$ is the slope of the density decline curve, and P_o is the density of the central area as extrapolated from the slope for outer areas. See Clark, C. (1951). Urban population densities, *J. Royal Statistical Soc., Series A*, **114**, pp. 490–6; Clark C. (1958). Urban population densities, *Bull. Institut International de Statistique*, **36**, pp. 60–68; Tanner, J. C. (1961). Factors affecting the amount of travel, *Road Research Technical Paper No. 51*. Department of Scientific and Industrial Research, London; Sherratt, G. G. (1960). A model for general urban growth, *Management Sciences, Models and Techniques. Proceedings of the Sixth International Meeting of the Institute of Management Sciences*, **2**. New York. pp. 147–59. In this paper Sherratt suggests that urban population densities decline exponentially as the square of distance such that

$$D_d = D_o e^{-cd^2}$$

where D is the population density at distance d from the centre of the city, D_o is the density at the centre of the city, and $-c$ is a measure of the rate of change of the logarithm of density with distance squared; Berry, B. J. L., Simmons, J. W., and Tennant, R. J. (1963). Urban population densities: structure and change, *Geogr. Rev.*, **53**, pp. 389–405; Casetti, E. (1967). Urban population density patterns: an alternate [*sic*] explanation, *Canadian Geogr.*, **11**, pp. 96–100. Recently B. Newling has used a quadratic exponential model to derive a density-profile classification of stages of urban growth: Newling, B. E. (1969). The spatial variation of urban population densities, *Geogr. Rev.*, **59**, pp. 242–52.

105 Berry, B. J. L., Simmons, J. W. and Tennant, R. J. (1963). *op. cit.* pp. 389–405; Berry, B. J. L., Tennant, R. J., Garner, B. J. and Simmons, J. W. (1963). *Commercial Structure and Commercial Blight*. University of Chicago, Department of Geography, Research Paper 85.

106 Muth, R. F. (1962). The spatial structure of the housing market, *Papers and Proceedings of the Regional Science Association*, **7**, pp. 207–20.

107 Winsborough, H. H. (1962). City growth and city structure, *Regional Science*, **4**, pp. 35–49.

108 e.g. Proudfoot, M. J. (1937). City retail structure, *Econ. Geogr.*, **13**, pp. 425–8; Mayer, H. (1942). Patterns and recent trends of Chicago's outlying business centres, *J. Land and Public Utility Economics*, **18**, pp. 4–16; Kelley, E. J. (1956). *Shopping Centres*. Saugatuck, Conn.

109 Berry, B. J. L. (1959). Ribbon developments in the urban business pattern, *Ann. Assoc. Amer. Geogr.*, **49**, pp. 145–55. Perhaps it should be noted that this idea had already been propounded in less formalized terms by Louis

Wirth and C. C. Colby, both of the University of Chicago. A discussion of the functional structure of each conformation is incorporated in Berry, B. J. L., Tennant, R. J., Garner, B. J., and Simmons, J. W. (1963). *op. cit.* Cp. also Garner, B. J. (1966). The internal structure of shopping centres, *Northwestern University Studies in Geography*, **12**.

110 Numerous references to studies of the internal structure of Japanese cities are to be found in a report commissioned by the *Nihon Chirigakkai* (Association of Japanese Geographers) and entitled *Toshika Kenkyū Shiryō* (*Research Materials on Urbanization*), 1959. There is also a comprehensive and up-to-date review of Japanese urban geography by Ginsburg, N. S. (1965), Urban geography and "Non-Western" areas, *in* Hauser, P. M. and Schnore, L. F., *op. cit.* pp. 319–27.

111 Representative of such works in three different culture realms are the chapters on Ibadan and Lagos in Mabogunje, A. L. (1968). *Urbanization in Nigeria.* London; McGee, T. G. (1967). *The South-east Asian City.* London; Singh, R. L. (1955). *Banaras: a Study in Urban Geography.* Banaras; Berry, B. J. L. and Rees, P. H. (1969). The factorial ecology of Calcutta, *Amer. J. Sociol.*, **74**, pp. 445–91; Prakasa Rao, V. L. S. (1962). Macro-urban analysis: geographers' contribution. Presidential Address to the Council of Geographers (India), *Annual Proceedings.* Cuttack [mimeo].; Kar, N. S. (1962). Urban hierarchy and central functions around Calcutta, in Lower West Bengal, India, and their significance, *Lund Studies in Geography, Series B, Human Geography*, **24**; Liang, Chi-sen, (1966). Urban land-use in Hong Kong and Kowloon, *Chung Chi J.*, **6**, pp. 1–24, **8**, 107–32; Liang, Chi-sen (in progress). *The Factorial Ecology of Hong Kong.*

112 See Wheatley, P. (1969). *City as Symbol.* London: Wheatley, P. (1970). The significance of traditional Yoruba urbanism, *Comparative Studies in Society and History*, **12**, pp. 393–423, esp. p. 421.

113 There is a preliminary discussion of the bazaar economy of the traditional world in Sjoberg, G. (1960). *The Preindustrial City.* Glencoe, Ill. pp. 204–9, and detailed analyses from two different points of view in Geertz, C. (1963). *Peddlers and Princes. Social Change and Economic Modernization in two Indonesian Towns.* Chicago and London. pp. 30–47, and Darwent, D. F. (1967). Towards a general theory of urban development in the Middle East, *Aspects of Central Place Theory and the City in Developing Countries.* Institute of British Geographers Study Group in Urban Geography. Durham Conference. No pagination.

114 In this respect this type of traditional city is the exact opposite of the Western-style city, where it is alleged central density craters develop only during hypothesized evolutionary stages of Late Maturity and Old Age: See Newling, B. (1969). *op. cit.* For a conspectus of current views on the nature of the traditional city, see Sjoberg, G. (1960). *op. cit.* with comments by Wheatley, P. (1963). What the greatness of a city is said to be, *Pacific Viewpoint*, **4**, pp. 163–88.

115 Grytzell, K. G. (1963). The demarcation of comparable city areas by means of population density, *Lund Studies in Geography, Series B, Human Geography*, **25**,

116 Office of Statistical Standards (1958). *Criteria for Defining Standard Metropolitan Areas.* Washington, D.C.

117 International Urban Research (1959). *The World's Metropolitan Areas.* Berkeley and Los Angeles. pp. 20–31.

118 For the several ways in which this may occur, see International Urban Research (1959). *Ibid.* Cf. also Goudswaard, G. (1954). Quelques notes sur les concepts de "ville" et "agglomération", *Proc. World Population Conference*, **4**, New York. pp. 685–93.

119 My attention was first drawn to the implications of this turn of phrase by Miner, H. (1967). *op. cit.* p. 4.

120 For a similar discussion of ways of conceptualizing "urbanization" (as opposed to the urbanism with which we are concerned here) see Tilley, C. (1967). The state of urbanization, *Comparative Studies in Society and History*, **10**, p. 102.
121 Oppenheim, A. L. (1964). *Ancient Mesopotamia. Portrait of a Dead Civilization.* Chicago. Chap. 2.
122 Cf. Aristotle. *Politics*, I. i. 8, 11, 12, 1252b–1253a.
123 Cf. Pausanias the Periegete (d. A.D. 176). *Description of Greece*, X, iv. 1. Transl. W. H. S. Jones (1918–35). Cambridge, Mass. For comprehensive evaluations of the Greek mode of urbanism, see Martin, R. (1956). *L'urbanisme dans la Grèce antique*. Paris; Jones, A. H. M. (1940). *The Greek city from Alexander to Justinian.* Oxford.
124 Von Grünebaum, G. E. (1955). Die islamische Stadt, *Saeculum*, **6**, pp. 138–53; Marçais, G. (1945). La conception des villes dans l'Islam, *Rev. d'Alger*, **10**, pp. 517–33; Lapidus, I. M. (ed.) (1969). *Middle Eastern cities.* Berkeley and Los Angeles; Hourani, A. H. and Stern, S. M. (eds.) (1970). *The Islamic city. A colloquium.* Oxford and Pennsylvania.
125 Balazs, E. (1954). Les villes chinoises. Histoire des institutions administratives et judiciaires, *Recueils de la Soc. Jean Bodin*, **6**, pp. 225–39; Eberhard, W. (1956). Data on the structure of the Chinese city in the pre-industrial period, *Econ. Development and Cultural Change*, **4**, pp. 253–68; Gernet, J. (1970). Note sur les villes chinoises au moment de l'apogée islamique, *in* Hourani, A. H. and Stern, S. M. (eds.). *op. cit.* pp. 77–85; Tuan, Y. F. (1968). A preface to Chinese cities, *in* Beckinsale, R. P. and Houston, J. M. *Urbanization and its problems. Essays in honour of E. W. Gilbert.* Oxford. pp. 218–253.
126 See Krapf-Askari, E. (1969). *op. cit.* pp. 25–26; Wheatley, P. (1970). *op. cit.*
127 Reiss, Jr., A. J. (1957). "Introduction" to Hutt, P. K. and Reiss, Jr., A. J. *Cities and Society. The Revised Reader in Urban Sociology.* Glencoe, Ill. p. 10.

Section 2: Regional and local evidence for urban settlement

A: The Nile Valley

AHMED M. ALI

Meroitic settlement of the Butana (central Sudan)

This paper deals with the area of the Butana, which lies between the Atbara and the Blue Nile, east of the main Nile, i.e. the traditional "Island of Meroe" as it was known to the Classical world (Fig. 1). The period covered is the Meroitic period—seventh century B.C. to fourth century A.D. During the eighth and seventh centuries B.C. the founders of the Meroitic Kingdom followed the Nile down and established their rule over Kush and Egypt. A century later they lost Egypt, but they were able to consolidate and extend their power southward to include most of the central Sudan, including the Butana region[1].

The Nile played a profound role in the rise of the urban centres along its banks, and in the shaping of the culture which developed in its valley. Similarly, the region outside and in this case the Butana region, seems to have had no less profound influence in the shaping of the settlement pattern during the Meroitic period.

In Lower Nubia a system of hill-forts (Qasr Ibrim, Jebel Adda) and fortified settlements (Karanog, Faras, etc.) was kept for obvious military as well as commercial purposes. Along the Nile Valley in Upper Nubia and southward there arose great urban centres which were also established religious centres. Such centres are Gebel Barkal, Kawa, Argo, and Meroe which were primarily permanent settlements of the Nile agriculturalists. Each town contained temples (designed according to the ancient Egyptian plan of temples) a domestic quarter and an extensive cemetery. The history of most of these urban centres goes back to the second millennium B.C., during the New Kingdom in ancient Egypt 1500 to 1000 B.C.

The history of the Meroitic occupation of the Butana, as revealed by the

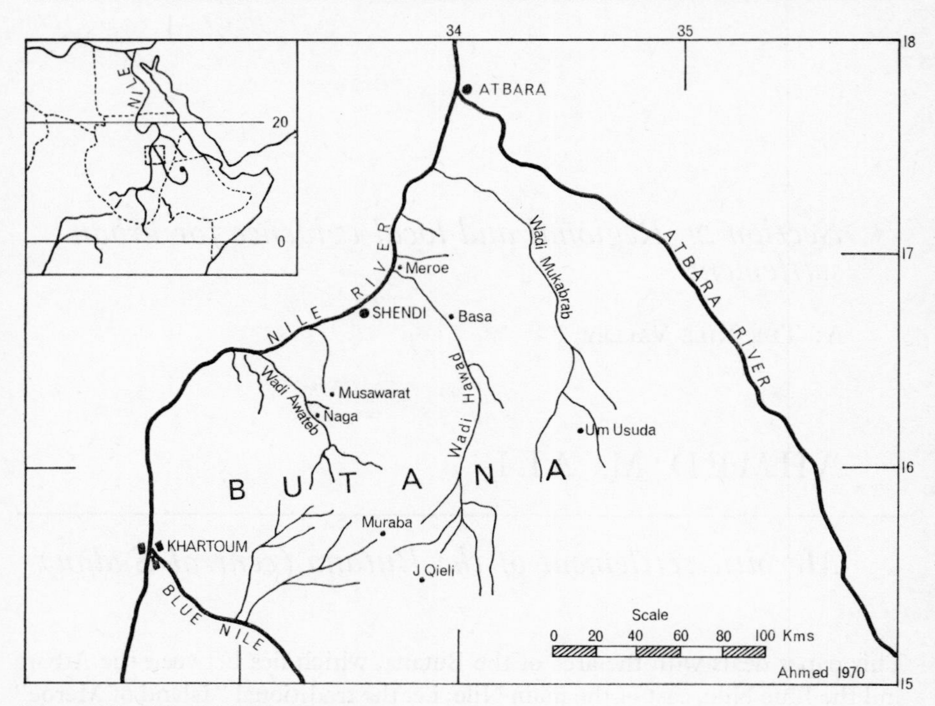

Fig 1 The Meroitic Butana.

recent excavation of Musawwarat-es-Sufra, goes back to the beginning of the Meroitic culture. Musawwarat became an important religious centre with at least one elaborate, but single-chambered, temple early in the sixth century B.C. The god closely associated with it is Apedemak, who was a traditional titulary god of Naga. In the succeeding century and a half, Musawwarat became a religious centre not only of the Butana but of all the kingdom. Naga became a permanent administrative centre with temples, domestic buildings and an attached large cemetery with at least two *hafirs*. It was only during the last two centuries B.C. that other sites were established at such sites as Basa, Um-Usuda, Geheid, Muraba. However, it is possible that many sites were under the influence of the Butana occupants before the temple/*hafir* was constructed.

In the following brief survey, an attempt will be made to distinguish and analyse the characteristic features of the Meroitic occupation of the Butana.

Geographically, the area can be divided into three sub-areas: the first is the Nile Valley, where the river directly influenced the rise of such centres as the capital city—Meroe, the town of Wad-ban-naga, and the provincial centres of Nubia. These exhibit characteristics typical of the Nile Valley settlement pattern; the second sub-area is the plain, comprising an eroded sandstone plateau and gravel ridges dissected by numerous valleys and flood channels; the third, and perhaps the most important, sub-area is the sandstone plateau, where the hills and raised plateau at the end of the plain helped in the formation of large valleys, which consist of four main wadis, formed by the convergence of various tributaries at the catchment area (the plateau), whence they cut through the plain to drain into the Nile. These are Wadi Mukabrab, Wadi Hawad, Wadi El-banat and Wadi Awatib. The Meroitic settlements considered here are all contained in the latter two sub-areas—the plain and the wadis[2] (Fig. 2).

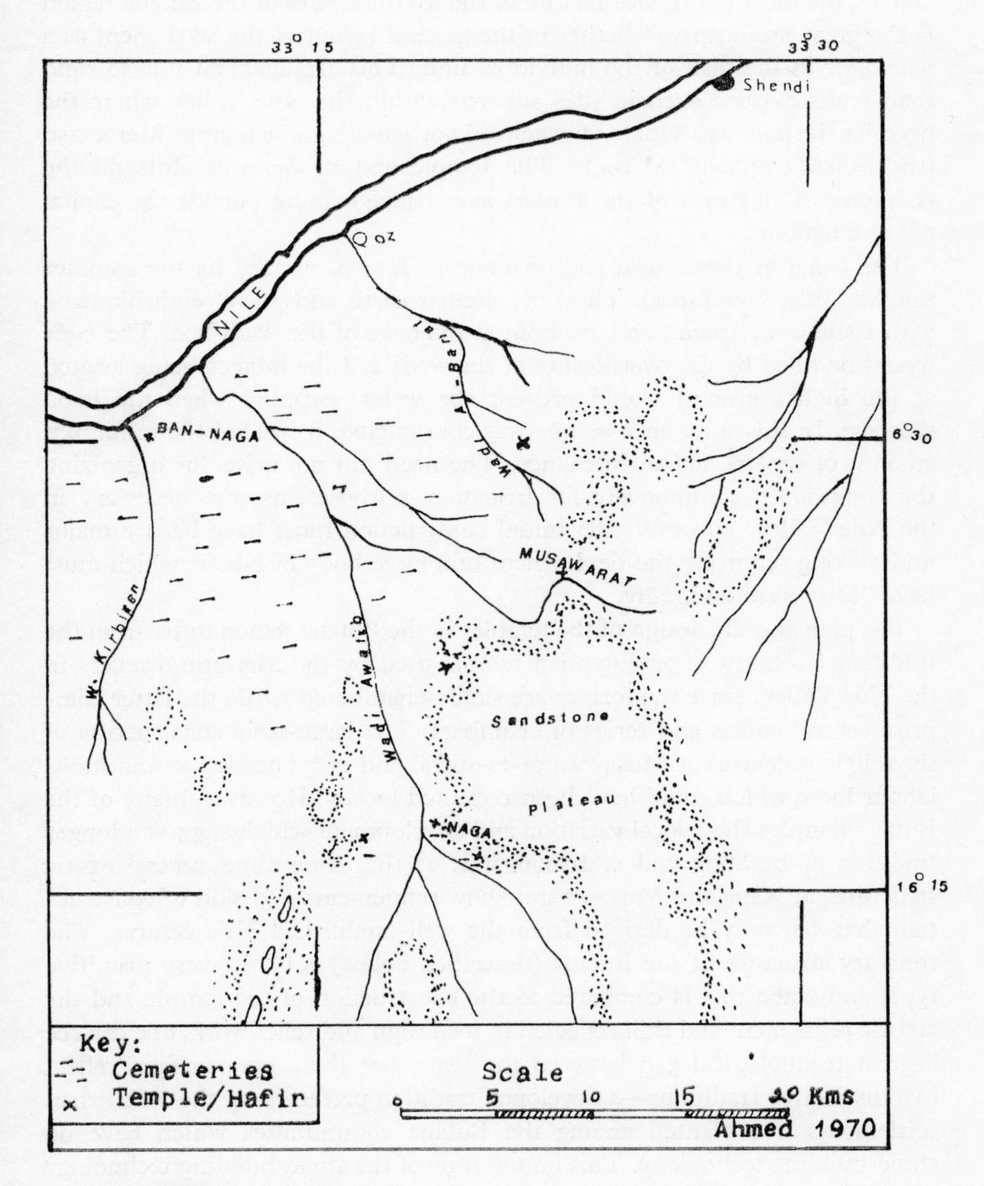

Fig 2 Wadi al-Awatib.

The settlement unit under consideration is very simple and characteristic of the Butana region. It includes a single-chambered temple and a *hafir*—an earth embankment for storing a large quantity of water (Fig. 3). However, the two most important sites in the region, Naga and Musawwarat-es-sufra, each has two *hafirs* and as many as eight temples. The rest of the sites conform to the typical unit of one temple and one *hafir*. No matter how many temples are present, the temple retains its simplicity and character of a single chamber. In fact, the most impressive feature of the Meroitic sites of the Butana region is the great uniformity, whether in the general layout of the settlement as a whole, or in the plan of the individual unit. This arrangement was so rigid that it was applied even in sites situated within the Nile Valley where the need for the *hafir* as a source of water did not arise (e.g. the temple Meroe 250 (the Sun Temple of Meroe)). The temple and its *hafir* exhibits all the characteristic features of the Butana sites, though lying outside the capital of the kingdom[3].

The *hafirs* of these sites, and of modern Butana, are fed by the summer rainfall (July/September). They are circular earth and gravel embankments with a stone revetment, and are built at the edge of the wadi bed. The *hafir* would be filled by the overflowing of the wadi and the intakes being located at the higher ground would prevent the water escaping when the flood recedes[2]. In this way, once a *hafir* was constructed, it needed the minimum amount of repairs and maintenance. The need did not arise for organizing the community continuously for irrigation purposes, as was necessary in the Nile Valley. However, the initial construction must have been a major undertaking requiring the deployment of a large body of labour which must have been recruited locally.

The plan and the design of the temples in the Butana region differ from the traditional concept of an Egyptian temple used by the Meroitic dwellers in the Nile Valley, since the former are single-chambered while the latter comprise several courts and series of chambers. The large-scale construction of the religious centres of Musawwarat-es-sufra[5] and Naga needed an enormous labour force which could have been recruited locally. However, many of the Butana temples show local variation and development which suggest a longer tradition of building and craftsmanship; at the same time several exotic structures at Naga and Musawwarat show refinements and skill of construction that can only be derived from the well-established Nile centres. The funerary structure of the Butana (described below) is of the very primitive type, and when this is compared to the construction of the temple and the artistic refinement and skill reflected in its design and relief work, one can see a great technological gap between the two; that these two, in fact, reflect two distinctive traditions—a developed tradition perfected by the Nile urban settlements transplanted among the Butana communities which have no stone-building technology. This importation of the stone-building technology from the Meroitic Nile Valley, and the absence of other comparable structures outside these religious centres, show that the local inhabitants had no tradition of building permanent structures.

A characteristic feature of the Meroitic sites near the River Nile is the thick stratified deposit built up as a result of the continuous occupation of the sites, e.g. at Meroe (10–15 m deep) and Kawa[6]. The absence of this thick stratified occupational débris is a marked feature of all the Butana sites. The same thing was reported at Basa, Um-Usuda, Muraba, etc. and was noticed recently during the excavation of Musawwarat-es-Sufra[7].

However, in all the sites of the Butana there is a marked absence or scarcity of domestic buildings or any other evidence of permanent occupation for a large population which would be comparable with the religious structures and their importance. The one exception to this is Naga. Naga, however, is exceptional in that it is the only site of the Butana where an Egyptian type of temple can be found[8]. The walls of the structures at Naga are connected to temples dated to the first century A.D. onwards, but the site might have been begun earlier, like any other site in the region, and then developed later into a political and administrative centre modelled after the Nile Valley examples[9].

However, Naga shared with the other Meroitic sites of the Butana a complete lack of any fortifications. Warfare and defence do not seem to have had any effect in the shaping of the Butana settlement as a whole or on the individual site or structure. The sites are scattered over a large region and far removed from the Nile[10].

The fact that the Butana sites, apart from Naga, were not permanent settlements, becomes clearer when the distribution of the cemeteries in this area is considered. The higher ridges on the plain between Wadi Awatib and Wadi Kirbikan are occupied by scattered cemeteries consisting of one to 100 graves. Each grave is indicated by a circular mound of stone rubble or a tumulus of gravel and earth. Similar cemeteries can still be seen on the ridges bordering Wadi Hawad, Wadi Mukabrab and their branches[11]. However, strikingly enough, the pyramidal form of superstructure characteristic of Meroitic cemeteries at the capital and in Nubia does not occur in these wadi cemeteries. The presence of these burial grounds, removed from the temple, points to the fact that the majority of the population did not develop any attachment to a particular place and burial took place near where death occurred. This is indicated by the small size of each cemetery and their scattered distribution. In fact, this distribution seems to be connected more with the grazing ground than the permanent source of water, i.e. the *hafir*.

However, in spite of the nomadic character of the settlement there are several factors which have a stabilizing effect. The presence of the only water source—in this case the *hafir*—in these selected sites transformed them into centres attracting, and sometimes limiting, the movement of the population to their environs, at least during the better part of the year. The construction of other *hafirs* eased watering problems and made it possible to expand to others sites since the *hafir* can cope with a limited number of men and their

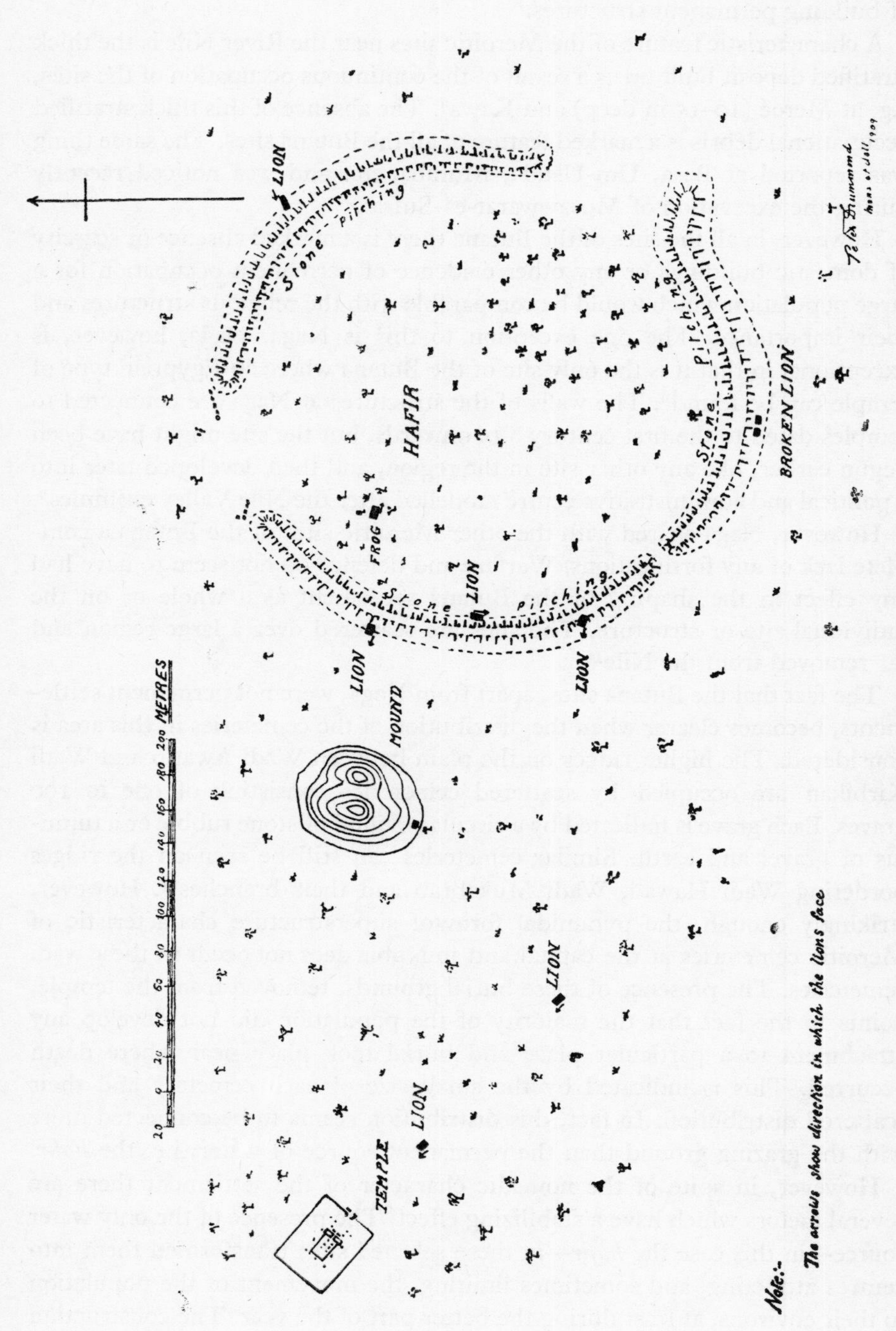

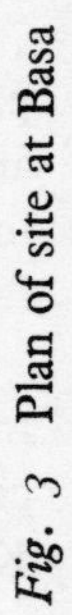
Fig. 3 Plan of site at Basa

beasts. Hence, expansion does not mean the expansion of a single unit of settlement but in fact means duplication of these units at other sites.

Religious importance was perhaps the strongest element in the stabilization of such a settlement pattern. The temples were the focal point of their respective surroundings, exerting great influence and power. No doubt, it is through such religious institution that the central government was able to effect its maximum power and control on the populace. On the walls of the temples, the Meroitic Royal Family is represented in the reliefs as the only worshippers of the gods of the Butana, and its lesser members formed the chief officiants. After all, the temple—and most probably the *hafir* also—were built in the name of the reigning monarch. In this way, the central government through these politico-religious institutions retained its maximum effectiveness of control at these places, where maximum gatherings regularly occurred. This is presumed to have continued as long as the central government remained strong and the structure of the Meroitic kingship and rule continued.

Conclusion

The sites in the Butana region were permanent centres for semi-nomadic pastoralists. There were no changes in the settlement pattern and even the individual structures—the temple and the *hafir*—remained basically unchanged, showing a continuous occupation by the same cultural community. They formed an integrated part of the Meroitic kingdom whose other centres lay within the Nile Valley. The Butana sites depended upon the Nile centres for political leadership, economic and trade outlet, and more important, for the supply of building technology. At the same time they in their turn were important religious centres of the kingdom and provided the government with political and economic power.

Notes

1 Arkell, A. J. (1961). *History of the Sudan from the Earliest Times to 1821.* 2nd ed. London. pp. 110–37; Shinnie, P. L. (1965). *Meroe, a Civilization of the Sudan.* London. pp. 29–98.
2 Barbour, K. M. (1964). *The Republic of the Sudan.* London. p. 215.
3 Garstang, J. (1911). *Meroe, the City of the Ethiopians.* Oxford. pl. II. A special study is being undertaken by the present writer on this temple as part of his thesis at the University of Cambridge.
4 Crowfoot, J. W. (1911). *The Island of Meroe.* London. pp. 9–29; Hintze, F. (1959). Preliminary report of the Butana Expedition, *Kush*, 7, p. 171.
5 Hintze, F. (1962). Musawwarat-es-Sufra, Preliminary report on the excavations of the Institute of Egyptology, Humboldt University, Berlin, *Kush*, **10**, p. 170; Hinzte, F. (1963). Musawwarat-es-Sufra, *Kush*, **11**, p. 217;

Hintze, F. (1968). Musawwarat-es-Sufra . . . 1963 bis 1966, *Wissenchafliche Zeitschrift der Humboldt Universität zu Berlin, Ges. Sprachw.*, **17**, p. 667.

6 Garstang, J. (1916). Fifth interim report on the excavation at Meroe in Ethiopia, *Liverpool Ann. of Archaeol. and Anthrop.*, **7**; Macadam, F. L. and Kirwan, L. P. (1955). *The Temples of Kawa*, II. Oxford, pp. 207–37.

7 Hintze, F. Personal communication, January 1970.

8 Budge, W. E. A. (1907). *The Egyptian Sudan, II.* London. p. 139.

9 Ali, A. M. (1963). *Naga in the Meroitic Times.* Unpubl. thesis in the Dept. of History, Univ. of Khartoum.

10 Crowfoot, J. W. (1920). Old sites in the Butana, *Sudan Notes and Records*, **2**, p. 85; Addison, F. and Whitehead, G. O. (1926). Meroitic remains, *Sudan Notes and Records*, **9**, p. 51.

11 Crowfoot, J. W. (1911). *op. cit.*; Hintze, F. (1959). *op. cit.*; Adam, A/ Rahman, and Smith, H. F. C. (1950). Four ancient sites in the Island of Meroe, *Sudan Notes and Records*, **31**, p. 303.

D. M. DIXON

The disposal of certain personal, household and town waste in ancient Egypt[1]

No prehistoric settlement site in Egypt has produced any evidence for the systematic disposal of waste; the waste was simply left to accumulate on the spot. The general pattern revealed by excavation shows layer on layer of village rubbish, of varying thickness, extending over the settlement area. These deposits included not only organic matter such as ash, charcoal and dung, but also such items as sherds, broken implements, stone grinders and rubbers, ear-studs, shells and lumps of malachite. Grain-pits have also been found, filled with miscellaneous rubbish[2]. As a result of re-use of sites, village waste and grave goods were inextricably mixed.

By the Archaic period, however, it would appear that improvement in living standards had produced among the upper classes a marked change in Egyptian attitudes to waste—at least in the matter of their disposal, particularly those of a personal nature.

During the first three dynasties, graves grew from their simple beginnings in the pre-Dynastic period into monumental structures. The embryonic single-roomed structure evolved into multi-roomed structures with separate units for sleeping, cooking, etc. By the latter half of the Second Dynasty the common type of deep rock-cut tomb substructure had evolved, which was large and had numerous rooms[3]. This subterranean "house" followed a stereotyped pattern which probably reflected the general arrangement of domestic dwellings of the time. It is in these large tombs of the Archaic period that the earliest evidence has been found for the existence of sanitary installations and waste disposal methods. While the ordinary Egyptian doubtless continued when out-of-doors to perform his toilet in a river or canal, the fastidious noble performed his in the comfort and privacy of his house. Particularly instructive is the substructure of the large Second Dynasty Mastaba-tomb 2302 at Saqqara[4] where there was a bathroom with anti-splash plastered walls. In addition many of these early tombs possessed lavatories. The ancient Egyptians did not defecate in a crouching position, but seated. In essentials the ancient Egyptian latrine consisted of two low, parallel walls or sides, between which was placed a pottery vessel half-filled with sand, which was removed and emptied when necessary, the contents being exposed to the hot sun which rapidly desiccated them. The latrine in tomb 2302 at Saqqara was situated at the extreme east end of the chamber facing south. From later periods portable latrines of wood or pottery[5] are known.

These elaborate sanitary arrangements seem to have been used only by the upper classes. The houses of workers at Twelfth Dynasty Kahun and Eighteenth Dynasty el Amarna contain no provisions for the evacuation and disposal of human bodily wastes.

Household and town wastes can be taken to include a great variety of items: rags from old linen and clothing, sandals; items of personal adornment such as amulets and beads; toilet objects such as combs and mirrors, tweezers, pins and razors; toys; pieces of wood and reed furniture; faience and glass vessels; broken and worn utensils such as pottery vessels and lamps, stone vessels, old baskets, sieves, mats; domestic tools and implements such as knives, scrapers, grinders; old rope and cordage; and materials resulting from the processes and activities of everyday domestic life: soot, charcoal, ashes, sawdust, etc. One should also mention remains of food and drink, such as grain, rind, pips, sediment, animal and fish bones. Such things were dis-disposed of in a number of ways. No doubt much was thrown into the river, and, as in ancient Greece and Rome, some domestic, as well as personal, wastes were probably tossed into the streets. A very large proportion of town and domestic refuse, however, found its way, as at the present time, on to outdoor rubbish-heaps; and such mounds were doubtless as characteristic a feature of Egyptian towns and villages in ancient times as at the present day.

Where possible, deserted buildings in the town would be used as rubbish depositories in order to save the effort of conveying it to the heaps, particularly if these lay some distance away. Refuse was also tipped into pits and sometimes burnt there. All these methods of disposal were frequently practised at the same time.

At Kahun in the Twelfth Dynasty a town was specially constructed to house the workers engaged on the construction of the nearby pyramid of Sesostris II at Lahun[6]. Evidence found by Petrie indicated that the town was occupied for a relatively short period, and was then abandoned[7]. While it was in use, domestic and town rubbish was either dumped in heaps beyond the northern wall of the town[8] or in abandoned buildings within the town itself. For example, a "guard's house" south of the "Acropolis" was used as a dumping ground for pottery by those who dwelt in the rest of the town[9]. The rubbish mounds north of the town contained a wide variety of remains, among them pottery, including fragments of Aegean ware found "in and under" the heaps and therefore contemporary with the period of Sesostris II[10].

At el Amarna "on the desert, about three furlongs from the palace", and on the nearest open ground to it, was "a wide stretch of waste heaps" containing refuse from the palace, "though probably mixed with waste from other large houses in the neighbourhood. The extent of the heaps was about 600 ft by

400 ft, and the depth varied from 4 ft to a mere sprinkling, probably averaging more than 1 ft". "Scattered throughout the whole area (were) dozens of objects with the names of the royal family, and hundreds of pieces of imported Aegean pottery". The heaps also contained hundreds of fragments of glass vessels. According to the excavator, much more broken glass "must have been thrown elsewhere, probably into the river with the kitchen waste, while a few bits went into the dust-hole, to be brought out here with the solid stuff and potsherds"[11]. Pits created by digging for sand to mix with Nile mud in constructing the houses in the city were utilized for dumping miscellaneous rubbish, apparently not domestic refuse, but that of building workers. As el Amarna expanded, rubbish heaps were levelled. Pits too were filled up and new estates and houses laid out over them. It is interesting to note that the few fragments of el Amarna letters recovered by Petrie had been tossed into two rubbish pits which had been filled in prior to the erection of a mud-brick structure[12]. The north suburb of the city was still being developed northwards when the town was abandoned, and every stage of building was found there, including levelled dumps and filled-in pits. Some of the latter had even been "disinfected" by burning[13].

At Deir el-Medina on the west bank at Thebes a village was constructed early in the Eighteenth Dynasty to accommodate the men engaged in making and decorating the royal tombs in the Valley of the Kings[14]. A remarkable point about this village is that, although it was occupied for 400 years, its floor level was never raised and, when houses were rebuilt, they were erected on the same foundations. This suggests that some degree of control was exercised over building and planning—and waste disposal. The first village, built against Gurnet Murai, was surrounded by a stout brick wall. The long, narrow houses opened on to a single street which followed the line of the old path through the valley. Rubbish was dumped in heaps outside the walls and, as the village expanded, the dumps were levelled and built over.

Notes

1 See Dixon, D. M. (In preparation). *The Wastes of Ancient Egypt*, for more details and references. Excluded from the present article, but treated in Dixon, D. M. (In preparation). *ibid.*, is the disposal of waste water, whether from personal ablutions and domestic use or from floods and rainfall.

2 cf., for example, Brunton, G. and Caton-Thompson, G. (1928). *The Badarian Civilization*. London. pp. 43–8, 69–116; Brunton, G. (1937). *Mostagedda and the Tasian Culture*. London. pp. 7–25, 75–82.

3 cf. Emery, W. B. (1961). *Archaic Egypt*. Harmondsworth. pp. 158–9.

4 Quibell, J. E. (1923). *Excavations at Saqqara (1912-1914). Archaic Mastabas*. Cairo. pls. 30 and 31, 3; p. 29.

5 e.g. from houses at Deir el-Medineh (Eighteenth Dynasty): Bruyère, B. (1939). *Rapport sur les fouilles de Deir el Medineh* (1934–5). Cairo; from the tomb of Kh'a: Schiaparelli, E. (1927). *La tomba intatta dell'architetto Cha*. Turin. p. 117.

6 Petrie, W. M. F. (1890). *Kahun, Gurob, and Hawara*. London. pp. 21 ff.
7 Petrie, W. M. F. (1890). *op. cit.* pp. 31–2.
8 Petrie, W. M. F. (1891). *Illahun, Kahun and Gurob*. London. p. 9.
9 Petrie, W. M. F. (1891). *op. cit.* p. 6.
10 Petrie, W. M. F. (1891). *op. cit.* pp. 9–10. pl. 1.
11 Petrie, W. M. F. (1894). *Tell el Amarna*. London. pp. 15–6.
12 Petrie, W. M. F. (1894). *op. cit.* p. 23.
13 Frankfort, H. and Pendlebury, J. D. S. (1933). *The City of Akhenaten*. II. London. p. 3.
14 Bruyère, B. (1939). *Rapport sur les fóuilles de Deir el Medineh*, 1934–5. Pt. III, Le village, les décharges publiques . . . Cairo.

BARRY J. KEMP

Fortified towns in Nubia

From the point of view of settlement pattern the garrison or colony established in a foreign country is of interest in revealing the extent to which familiar schemes are adapted to changed circumstances. At one extreme will be the complete transplantation of a type of community plan irrespective of new and different surroundings, and at the other the carefully devised response to a correct assessment of the new situation in all its aspects.

In the particular case of the ancient Egyptian settlements in the conquered territories of Nubia this sort of approach is hampered by the lack of information on settlement patterns in Egypt itself, to the extent that the most pressing question becomes: how much of Egyptian town design and fortification can we see reflected in these colonial establishments?

The question is easier to answer for the New Kingdom, roughly the sixteenth to twelfth centuries B.C., when the Egyptians effected a fairly rapid takeover of the Nubian Nile valley as far south as the Fourth Cataract[1]. The design of the new towns established by the Egyptians in this area seems to have been fairly uniform (Fig. 1). The main part was square or at least rectangular enclosed within a thick wall of mud-brick. Square brick towers were added to the exterior of the wall at intervals along the sides, at the corners, and on either side of stone-lined gateways. These gateways received no additional protection, nor was there apparently a ditch or outworks to hinder mining or scaling of the walls.

Inside the enclosure the town was laid out around a system of narrow streets whose adherence to a strictly rectangular grid varies somewhat. But at Amara West in particular and to some extent at Sesebi where erosion has left little but the lowest level, it can be seen that even this began rapidly to break down by piecemeal rebuilding on rising floor levels into an "organic" layout. In this way the essentially bureaucratic nature of town planning in ancient Egypt, for reasons not of urban aesthetics but of speed and economy in the initial construction on a new site, is underlined. The interior of the enclosure housed in general three types of building. Most prominent was a stone temple of characteristic Egyptian design which, as at Soleb, could rival some of the most splendid in Egypt itself. Associated with it were blocks of long and

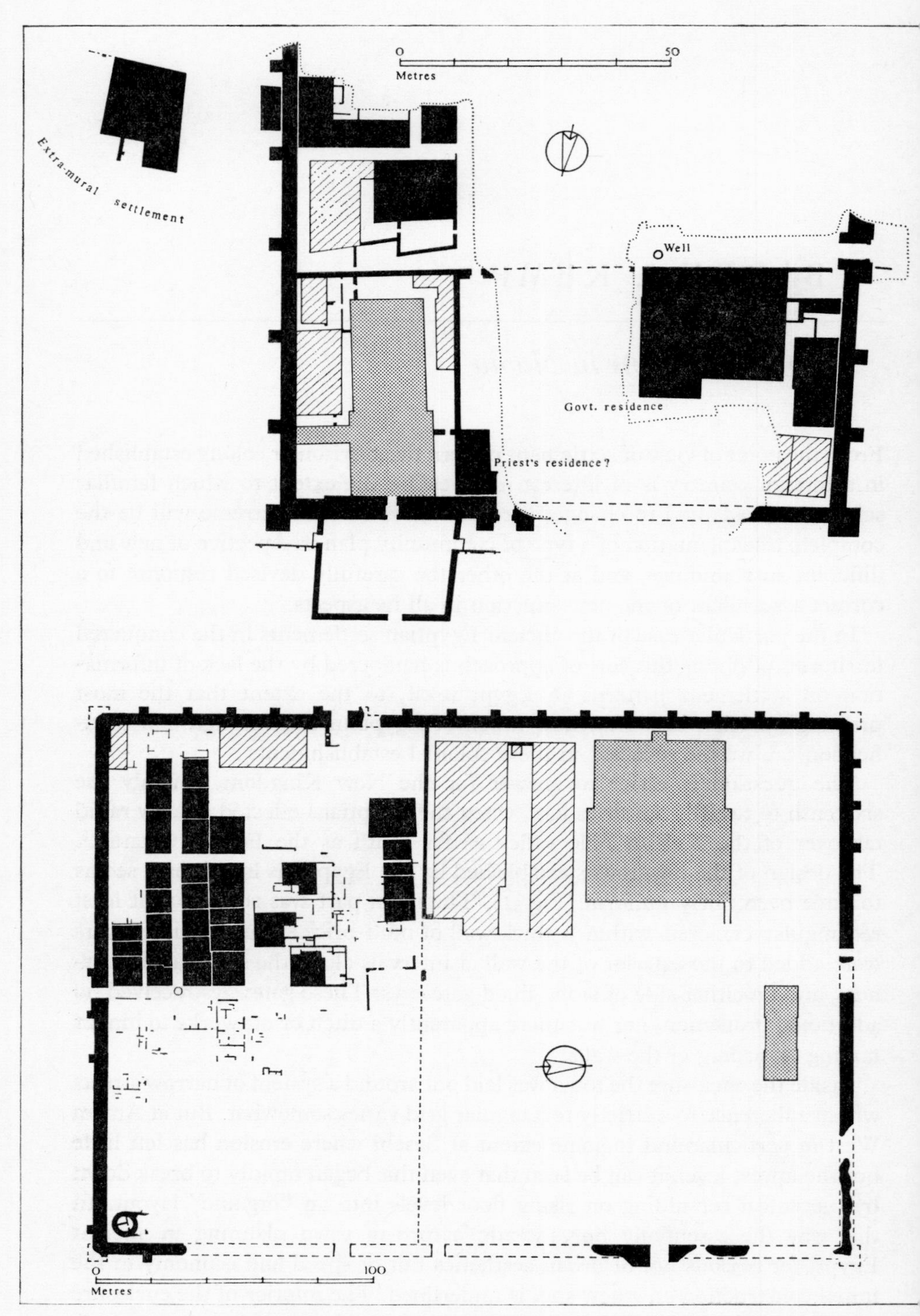

Fig 1 Two new Kingdom towns in Nubia, Amara West (above), and Sesebi (below). Shaded according to use: residential by black, storage by hatching, religious by dots.

narrow storerooms, probably largely for farm produce and in some cases perhaps too for raw materials gained from local exploitation or from trade beyond the borders. The remainder of the area was taken up with mud-brick domestic and administrative buildings, including the civil government residence. At Amara West excavation outside the town walls also brought to light parts of a thinly developed extra-mural settlement of small brick house units, one of which was built up against the main town wall.

The uniformity of this general design throughout the Nubian territories is emphasized by the conversion of an old citadel at Buhen, dating from an earlier period of Egyptian expansion, to conform to the appearance of these new towns.

The fact that the fortifications around these towns were of a rather perfunctory nature has been pointed out before by others. What has not been made clear is that they appear to be unmodified copies of a type of temple enclosure wall in Egypt itself. This is evident from the plans of excavated examples[2], and from contemporary representations[3]. The most explicit of these is a limestone libation tank of the nineteenth dynasty found at Memphis which is modelled to show a wall with towers along the sides and at the corners, the whole capped with battlements (Plate 2)[4]. The sides are inscribed with prayers to the god Ptah of Memphis, one of them reading: "Praise to thee at the great rampart; it is the place where prayer is heard". To emphasize this a human ear has been carved at the top of each tower. There can be little doubt that this castellated structure represents the main enclosure wall of the temple of Ptah at Memphis, into the interior of which the general public was not admitted. Interestingly the word translated above as "rampart" was used also by Ramses III in his description of the enclosure walls of his mortuary temple at Medīnet Habu, Thebes, which is one of the excavated examples of towered walls[5].

The castellated appearance of these temple walls must be assumed to be largely symbolic. In those cases where the front pylon of the temple interrupted the wall and bore giant scenes of the king symbolically vanquishing Egypt's foes in the presence of one or more of the gods the towers and battlements on either side must have appeared a logical continuation[6]. The intention was presumably to convey the impression that the temple was the impregnable core of Egypt's defence against the disorder and hostility of the foreign lands since it was the gods who dwelt inside who bore the ultimate responsibility for history, the temple providing them with a microcosm of the universe for a suitable home[7].

When these Egyptian temple enclosures are compared with the Nubian towns one important difference does emerge, however. This is the inclusion in the latter of a large area for residential buildings, whereas in Egypt the enclosure wall surrounds little more than temple and storerooms. Little excavation has been conducted outside temple enclosure walls in Egypt but, as I have discussed in my other paper—Temple and town in ancient Egypt[8]—

one must probably envisage extra-mural settlements spreading around and engaging in economic activities both with the temple and with independent or semi-independent centres within themselves. Thus, in Nubia the Egyptians attempted to turn what must have seemed to them an unbearably backward land into an extension of their own country, building temple-centred towns of the type which probably now formed the backbone of urbanism in Egypt. But, because of the dangers from raids, real or imagined, the very real defensive possibilities provided by what in Egypt were primarily symbolic fortifications were utilized to give protection to the houses which in Egypt spread around outside the walls.

In the preceding Middle Kingdom, roughly the twentieth to eighteenth centuries B.C. there is much less that can be said with confidence. The political and military situation was very different. The Egyptians controlled the Nile valley only as far upstream as the Semna Gorge in the Second Cataract region, whilst to the south were a number of native kingdoms which were eventually to coalesce and, as the Kingdom of Kush, to become a significant threat to the Theban seventeenth dynasty during the Second Intermediate Period which separated the Middle from the New Kingdom.

The communities established by the Egyptians were of two main varieties, corresponding to the great differences in the topography of the Nile valley between the First and Second Cataracts. In that part to the north of Wadi Halfa, where there are long stretches of flat or gently shelving banks, extensive towns were built. The central part was a rectangular citadel surrounded by a massive mud-brick enclosure wall with square towers along the sides and at the corners and with a ditch and, in some cases, an embanked counterscarp and glacis. Special attention was paid to the fortification of the gateways, and a stone tunnel was constructed from inside down to the river to ensure a safe water supply. Most striking of all was the special line of defence to protect the main enclosure wall from mining, battering or scaling[9], a loopholed parapet being built along the inner lip of the ditch, at intervals interrupted by rounded bastions. Inside the citadel the ground was densely built up, often with multi-storey units, following a strict grid plan which was bounded by a pomoerium road beneath the main wall, presumably to make successful tunnelling operations more readily visible. Conspicuous by their absence are the large stone temples of the New Kingdom type, the brick chambers which can be allotted to this function being scarcely distinguishable from other buildings. In some cases at least the citadel was surrounded by an extensive town area with fortifications of its own.

This type of fortification is suited only to flattish ground. In the Second Cataract area the topography is, by contrast, of a very rugged nature, and in the generation of forts which later extended Egyptian control through this area we find serious modifications of the plains type of fortification. The shape of each fort was tailored to a polygonal plan which matched the irregularities of a hilltop, with solid spur walls to cover narrow ridges. The natural fall of the ground then served to protect the main towered wall, rendering further outworks unnecessary. Internally the layout was modified only as little as was necessary to fit a rectangular grid inside an irregular polygonal figure, and the stone passage down to the water was retained even where a most precipitous descent was necessary.

The striking business-like appearance of these Nubian forts obviously rules out an adaptation of an essentially symbolic form of architecture such as seems to have been the case in the New Kingdom. But the question still remains, as to how far they represent something specially devised to meet a specific military threat in Nubia which the plains type of fortification would suggest was of a sophisticated nature, probably involving siege engines. The answer may lie in the fact that the abandonment of the loopholed parapet protecting the base of the main wall in the later generation of Second Cataract forts was not entirely a matter of changed topography. The fort at Semna, on the actual frontier itself, is actually built facing on to relatively flat ground. But whilst the walls and towers are as massive as ever they face, especially on the north and west sides, only a broad flat ditch and rough stone glacis, ignoring whatever sort of attack the loopholed parapet was designed to thwart[10]. This suggests that the plains type of fortification may represent the use, during an early phase of conquest, of a fortifications system devised for the defence of towns in Egypt, perhaps during the civil wars of the First Intermediate Period. But the truth of this can be tested only by excavations in Egypt, bearing in mind the probability that it would have been politic for the kings of the Middle Kingdom, once their authority was sufficiently established, to see that provincial fortifications were demolished.

Notes

1 For the general historical background to this paper see Säve-Söderbergh, T. (1941). *Ägypten und Nubien.* Lund; and Arkell, A. J. (1961). *A History of the Sudan.* (2nd ed.) London. A useful summary of fortified sites in Nubia is given by Lawrence, A. W. (1965). Ancient Egyptian fortifications, *J. Egypt. Archaeol.*, **51**, pp. 69–94, with a site bibliography on pp. 93–4. Further evidence from excavations has since been published in volumes of the journal *Kush*, and a major addition has been Dunham, D. (1967). *Uronarti Shalfak Mirgissa.* Boston.

2 In chronological order: the smaller Aten temple at el-'Amarna, see Pendlebury, J. D. S. *et al.* (1951). *The City of Akhenaten*, Part III. London. p. 92, pl. XVI; the Temple of Seti I at Abydos, see Caulfeild, A. St. G. (1902). *The Temple of the Kings at Abydos.* London. p. 12, pl. XXIII; Ghazouli, E. B. (1964). The palace and magazines attached to the Temple of Sety I at Abydos and the facade of this temple, *Ann. Serv. Antiq. Egypt.*, **58**, p. 111, Fig. 3, 156; the nineteenth dynasty temple at Hermopolis, see Roeder, G. *et al.* (1937). Bericht über die Ausgrabungen der deutschen Hermopolis-Expedition 1935, *Mitt. dt. Inst. äg. Alt. Kairo*, 7, pp. 18–19, plans I and II; Roeder, G. (1952). Zwei hieroglyphische Inschriften aus Hermopolis (Ober-Ägypten), *Ann. Serv. Antiq. Egypt.*, **52**, p. 362, Fig. 2;

the mortuary temple of Ramses III at Medīnet Habu, Thebes, see Hölscher, U. (1941). *The Mortuary Temple of Ramses III*, part I. Chicago. pp. 60–1. The first and last examples appear in Figs. 3 and 4 in Kemp, B. J., this volume, pp. 665 and 669.

3 For example, the scene of the High Priest Herihor offering a model of the temple of Khensu at Karnak, see Lepsius, C. R. (1897–1913). *Denkmaeler aus Aegypten und Aethiopien*, part III. Berlin. pl. 244; Hölscher, U. (1910). *Das hohe Tor von Medinet Habu*. Leipzig. p. 60. Fig, 56.

4 Jacquet, J. (1958). Un bassin de libation du nouvel empire dédié à Ptah; Wall-Gordon, H. (1958). A New Kingdom Libation Basin dedicated to Ptah, *Mitt. dt. arch. Inst. Abt. Kairo*, **16**, pp. 161–75, pls. XII–XIII; Anthes, R. (1965). *Mit Rahineh 1956*. Philadelphia. pp. 72–5, pls. 24–25.

5 The word is *tsmt*. For a discussion of its meaning see Gardiner, Sir A. H. (1947). *Ancient Egyptian Onomastica*, II. Oxford. pp. 213* (A 445), 324*. Strictly it seems to have referred to only an aspect of wall fortifications, probably the battlements, but a wall possessing this aspect could also receive the term.

6 Strikingly visible in the reconstructions of Medīnet Habu, see Hölscher, U. (1934). *General Plans and Views*. Chicago. pls. 22, 23, 30; Hölscher, U. (1951). *The Mortuary Temple of Ramses III*, part II. Chicago. pl. 6.

7 See, for example, Brunner, H. (1970). Die Sonnenbahn in ägyptischen Tempeln, *in* Kuschke, A. and Kutsch, E. (eds.) *Archäologie und altes Testament*. Tübingen. pp. 33–4 and the references in note 39; also section III of Smith, H. S., this volume, p. 713.

8 This volume, p. 657.

9 On the state of siege warfare at this period see Yadin, Y. (1963). *The Art of Warfare in Biblical Lands*. London. pp. 65–71. An important eleventh dynasty tomb scene at Thebes, showing a form of wheeled siege tower in use is published in Arnold, D. and Settgast, J. (1965). Erster Vorbericht über die vom Deutschen Archäologischen Institut Kairo im Asasif unternommenen Arbeiten, *Mitt. dt. arch. Inst. Abt. Kairo*, **20**, Fig. 2 opposite p. 50, and comments on pp. 50–1.

10 Borchardt, L. (1923). *Altägyptische Festungen an der zweiten Nilschnelle*. Leipzig. pls. 1, 19, 20; Dunham, D. and Janssen, J. M. A. (1960). *Semna Kumma*. Boston. pls. 1, 5A, plans III, IX–XIV.

BARRY J. KEMP

Temple and town in ancient Egypt

The relationship between these two entities, temple and town, appears to have been fundamental in the fabric of ancient Egyptian society, at least in those periods where adequate documentation is preserved. Whilst archaeology obviously provides the basic evidence for the physical association between the two, the full extent of the economic and psychological forces which bound temples and urban communities together is apparent only from ancient written sources. I will begin this paper, therefore, with an outline reconstruction of a temple economy as it is known from one particularly well-documented period, the New Kingdom, roughly the sixteenth to twelfth centuries B.C. The diagram (Fig. 1) which appears below to summarize this reconstruction is a considerable over-simplification, but it should at least serve to focus attention

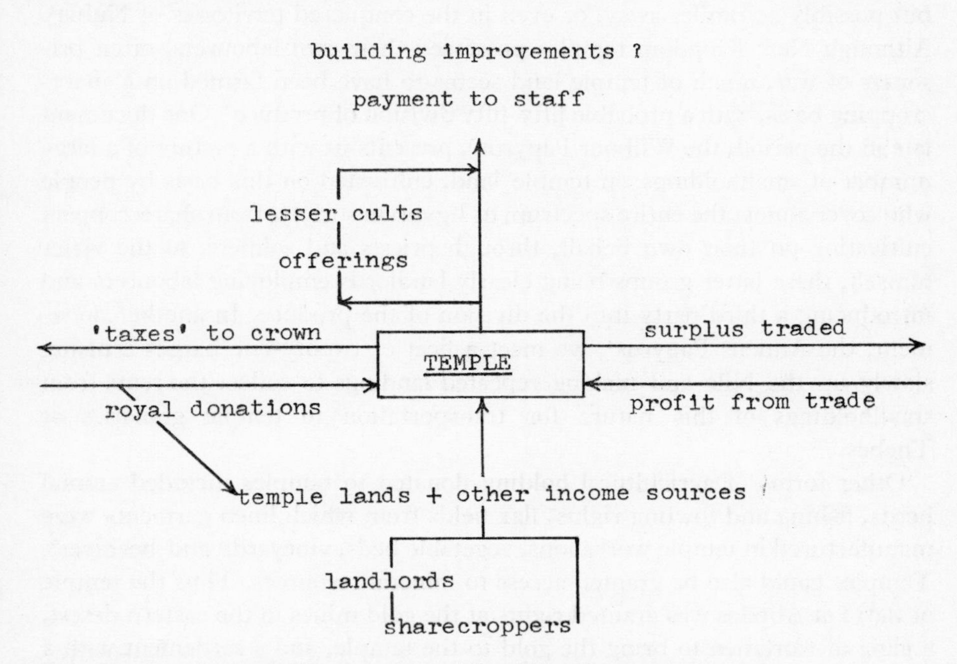

Fig 1 Economic flow diagram for a New Kingdom temple.

on the likely flow of economic activity which the documents suggest a temple might become responsible for in this period of Egyptian society; though even here one might object to treating this period as a unity since much of this documentation comes only from its latter years when it seems likely that temples, especially those at Thebes, were actually enlarging their degree of influence.

Whilst Egyptian theology of the New Kingdom was inclined towards universalism and pantheism it remained a basic tenet that a god's particular home was in a statue which represented him in characteristic appearance, usually an anthropomorphic one, and bore his name, and through which he was accessible by mortals. The temple itself was conceived, with some literalness, as a house for the statue, and the gods who dwelt inside it as in need of regular food offerings. These offerings derived from productive sources owned by the temple. But this was not the only function of temple property. For it also bestowed on the gods a status which corresponded to power and importance on a strictly material scale. In a sense, the gods were given the status of landed nobility, which suited some of the Egyptians' concrete conceptions of divinity, and we find the material enrichment of divine property a major theme in texts dealing with the duties of kingship.

The riches bestowed on the gods seem to have been drawn from the full diversity of Egyptian financial resources, both durable forms of wealth, and permanent sources of revenue. Foremost amongst the latter was, of course, cultivable grain land, not necessarily in the vicinity of the temple itself, but possibly 200 miles away, or even in the conquered territories of Nubia[1]. Although New Kingdom temples possessed their own labourers, often prisoners of war, much of temple land seems to have been farmed on a sharecropping basis, with a probable fifty-fifty division of produce[2]. One document late in the period, the Wilbour Papyrus[3], presents us with a picture of a large number of smallholdings on temple land, cultivated on this basis by people who cover almost the entire spectrum of Egyptian society, from sharecroppers cultivating on their own behalf, through priests and soldiers, to the vizier himself, these latter groups being clearly landlords employing labourers and introducing a third party into the division of the produce. In another document, the Amiens Papyrus[4], we meet a fleet of twenty-one barges cruising slowly up the Nile and making repeated landings to collect the rents from smallholdings of this nature for transportation to temple granaries at Thebes.

Other forms of agricultural holding donated to temples included animal herds, fishing and fowling rights, flax fields from which linen garments were manufactured in temple workshops, vegetable beds, vineyards and beehives[5]. Temples could also be granted access to mineral resources. Thus the temple of Seti I at Abydos was granted rights at the gold mines in the eastern desert, a gang of workmen to bring the gold to the temple, and a settlement with a well at the mines themselves[6]. The temple of Amen at Thebes seems to have had a similar arrangement for gold mining in this area, and another for acquiring galena, used for eye pigment and as a medicament, also probably from the same region[7]. Direct gifts of precious stones and metals also appear as a regular expression of royal piety.

All of these various types of wealth, from beehives to boats, were designated by a common word for "offerings", and what was actually presented to the god during the offering ceremonies must have been regarded merely as tokens.

Temple income seems to have been disposed of in three ways. Although it is hard to find much explicit testimony, temples do seem to have paid a tax on farmed grain to the crown[8], perhaps a tenth, and were often made responsible as well for administering a class of crown land, called *khato*-land[9]. How far other forms of temple wealth were taxed is unknown, but from the crown's point of view it must have been grain which counted most since this must have been the principal commodity of payment to its own servants, building workers, the army, and so on. But as temples received, as pious gifts, a share of the crown's income of precious materials, agricultural taxation on the temples must have been, in effect, a reciprocal exchange of wealth.

But here, in envisaging an exchange between two separate bodies, one encounters a considerable difficulty. For at times the temples emerge clearly as just one branch of government administration. At the beginning as at the end of the New Kingdom we find that the person ultimately responsible for the management of temple economic affairs is the vizier, at least in Upper Egypt, and that other officials may bear titles which show that their careers encompassed both spheres of administration[10].

This absence of demarcation becomes very prominent in considering an actual example, the payment of the necropolis workers at Thebes whose principal task was the preparation of the royal tomb in the Valley of Kings. Near the end of the Twentieth Dynasty, apparently a time of economic difficulty at Thebes, we find corn from tax assessments on various temples, and from *khato*-land administered by them, taken to western Thebes for the necropolis workmen and stored in granaries under the charge of the mayor of western Thebes[11]. This seems logical enough. But, somewhat earlier, we also find the same official blamed for not having paid these workmen from the "offerings" of the century-old mortuary temple of Ramses II at Thebes, and the demonstration by these men outside others of these temples suggests that they too regarded them as possible sources of income[12], something largely confirmed by a few surviving pay records[13]. At other times these men were paid from the temple of Maat at Karnak, across the river[14]. The ultimate responsibility for payment is clearly the vizier's, and one has the impression that the various granaries at Thebes, no matter to which institution they ostensibly belonged, were administered as an integrated unit under a central accounting system, with crown employees on the payroll. The vizier had overall charge, and under him people like the mayors of Thebes and the High

Priest of Amen. The independence of the temples as owners of wealth was probably very much a matter of theological nuance.

A second form of expenditure was in the form of overheads for running the temple establishment, principally the payment of staff in kind. Here the actual offerings themselves made a contribution, being first presented to the god, then taken before the statues of lesser cults, including the king and his ancestors and honoured private individuals, and then finally divided amongst the priests themselves. Since these were only tokens of the total temple revenue the term "offerings", as noted above, was applied to all commodities paid out, as in the case of the wages of the Theban necropolis workmen. How far temples paid for their own aggrandisement and fabric maintenance is unknown. The building and enlargement of temples was another traditional duty of kings, but since temple income was initially bestowed by the king the use of this to pay for improvements would not have been inconsistent with the claim for royal responsibility, and in any case, his authority was presumably necessary for any major alteration in the disposal of temple income. Furthermore, as the Wilbour Papyrus shows, one means by which a temple could temporarily increase its income was by renting lands from another temple.

It seems clear from ancient written testimony as well as from general considerations that temple economies functioned at a profit, producing a surplus of income. Some must have been stored to cover future expenditure, but a portion must have remained over to be traded at a profit, particularly perishables, food, linen, and so on. Whether raw materials from mines and quarries, for which crown and temples seem to have had monopoly rights of exploitation, were also traded is nowhere definitely attested but must be counted as likely. For efficient exploitation of markets temples possessed their own merchant ships to reach wider markets, not only in Egypt but also abroad. The temple of Seti I at Abydos, for example, was given a sea-going ship for foreign trade equipped with "traders" by Ramses II[15]. These "traders" seem to have been a regular feature of temple staffs and presumably had the practical responsibility for these merchandizing activities[16]. A papyrus in the Turin collection[17] has preserved a fragment of a set of accounts of one such ship belonging to the temple of Amen at Thebes and administered by the High Priest. It had apparently set out from Thebes with a cargo of linen garments, presumably manufactured by the "weavers of the estate of Amen". Two months later we find it moored at Memphis and then its suburbs, "trading" linen garments for various commodities, principally sesame oil, but also things like corn and papyrus rolls. We do not know what subsequently happened to the sesame oil, but we may guess that it was either retained as a means of future payment to temple workers, or else retraded again at a profit for other commodities with the ultimate purpose of converting it into a more durable form of wealth, either copper vessels and suchlike, or the silver "pieces" which seem to have been functioning as a sort of proto-money[18]. In modern terms this was a way of increasing one's liquidity.

It can be seen from what has so far been said that temples were potent sources of economic activity. They were employers of labour, partly permanent administrative and productive personnel, partly building workers engaged on enlargements, all receiving regular wages in kind from which further goods and services were to be obtained by barter. The temples themselves were active traders, and with their income sometimes comprising raw materials as well, were also patrons of manufacture, though whether or not they also sold materials for private manufacturing is uncertain. That the temple should have provided a powerful motive for settlement, both of its employees and of those who included the temple or its personnel in their own economic cycle, seems a reasonable enough conclusion. Though where temple assets lay some distance away a dispersal of settlement units related to temple activities must have occurred.

Turning now to the archaeological evidence two different situations must be distinguished: (a) where the temple was a new foundation on a hitherto non-urban site and where it was to remain the principal or even sole reason for a settlement's existence; and (b) where the temple was one part of a community which had other independent bases for its economic life.

The most obvious category in situation (a) is the royal mortuary temple. If we turn for the moment to periods earlier than the New Kingdom the scanty evidence available suggests that, architecturally at least, the royal mortuary temples attached to pyramids were the most lavishly endowed temples in the land. Elsewhere, on the sites of provincial temples, available evidence points in many cases to modest buildings of mud-brick, with decorated stonework at only limited points, such as door frames, and more rarely colonnaded halls. Documents from as early as the late Fifth Dynasty[19] (*c.* 2350 B.C.) suggest that the economic cycle sketched above was already in existence for these mortuary temples, and several mud-brick settlements have been excavated which must have housed the personnel involved in the running of them[20]. The largest and best known is Kahun dating to the Twelfth Dynasty, initially from about 1895 B.C.[21]. It forms a large-scale example of grid-iron planning, of the sort which occurs in contemporary Nubian forts. The question of urban planning in ancient Egypt will be taken up again later, but it should be pointed out here that the function of planning seems to have been not to create an ideal townscape but simply to produce a settlement of a given density as rapidly and economically as possible. Given the need to build a settlement paid for from a single source in one stage it is hard to see what alternative to a grid-iron layout could have existed. That it may not have been, in the New Kingdom at least, regarded as an ideal will become apparent from subsequent discussion.

The relationship at Kahun between the town and pyramid of King Senwosret II and the large lower mortuary temple now destroyed to its very foundations can be seen from Fig. 2. The town is the largest known of its type. But since some documents from here deal with work being carried out on a construction of King Amenemhat III, possibly part of his pyramid complex, its functions seem to have extended beyond the upkeep of the local mortuary cult. The town was quite separate from the large lower mortuary temple which probably incorporated within itself its principal magazines and granaries. About half of the town appears to have been lost under cultivation, and the preserved area seems to be almost exclusively occupied by houses. Of this a good half is taken up with small often back-to-back houses of the ordinary inhabitants. The remainder consists largely of nine much larger house units on either side of a single street whose internal layouts have obviously been given a great deal of thought. One of these lay on a projecting spur of higher ground, called the Acropolis by the excavator, and here we may probably recognize the civil government residence which appears in the legal papyri, where there was located the "office of the vizier" for his periodic visits and other "offices" where legal proceedings were held. In an open square opposite was a building of which only part has survived but which, in its plan, resembles a building very tentatively identified as a temple in the contemporary Nubian fort at Uronarti[22]. This recalls the fact that the mortuary temple of Senwosret II was not a place of popular worship and that documents from Kahun show the town to have possessed at least one temple of its own, dedicated to Sepdu, Lord of the East, with its own part-time priesthood.

Another document from Kahun indicates that the mortuary temple may not have been the only centre of economic life. This is a fragment of an account of land—composed of a number of separate plots—in the hands of a man in the priesthood apparently of the local Sepdu cult which seems to include a plot given to him from land belonging to the mortuary temple itself[23]. The document in its damaged state is hardly conclusive but in the light of what we know of New Kingdom arrangements we may perhaps legitimately wonder if a part of the mortuary temple lands was leased out on a sharecropping basis. That the occupants of the large house units may have been not merely administrators but also lessee landlords might explain why a standard part of these units is a granary block of a design exclusive to this period[24]. The likelihood of this should be judged in relation to the discussion below on the economy of New Kingdom el-ʿAmarna. But if true, then we must regard some of the small houses as being for the employees of these landlords, paid for from their own granaries, rather than for people directly employed by the mortuary temple itself. Since one of the house lists, of a lector priest, contains the names of a fair number of "serfs", at least twenty-two, most of them females (for weaving ?), this should not be thought too unlikely[25].

Nevertheless, the subsequent history of Kahun illustrates the ultimate total dependence of the town on the mortuary temple. Although endowed in perpetuity the mortuary temples of the Twelfth Dynasty proved to be no

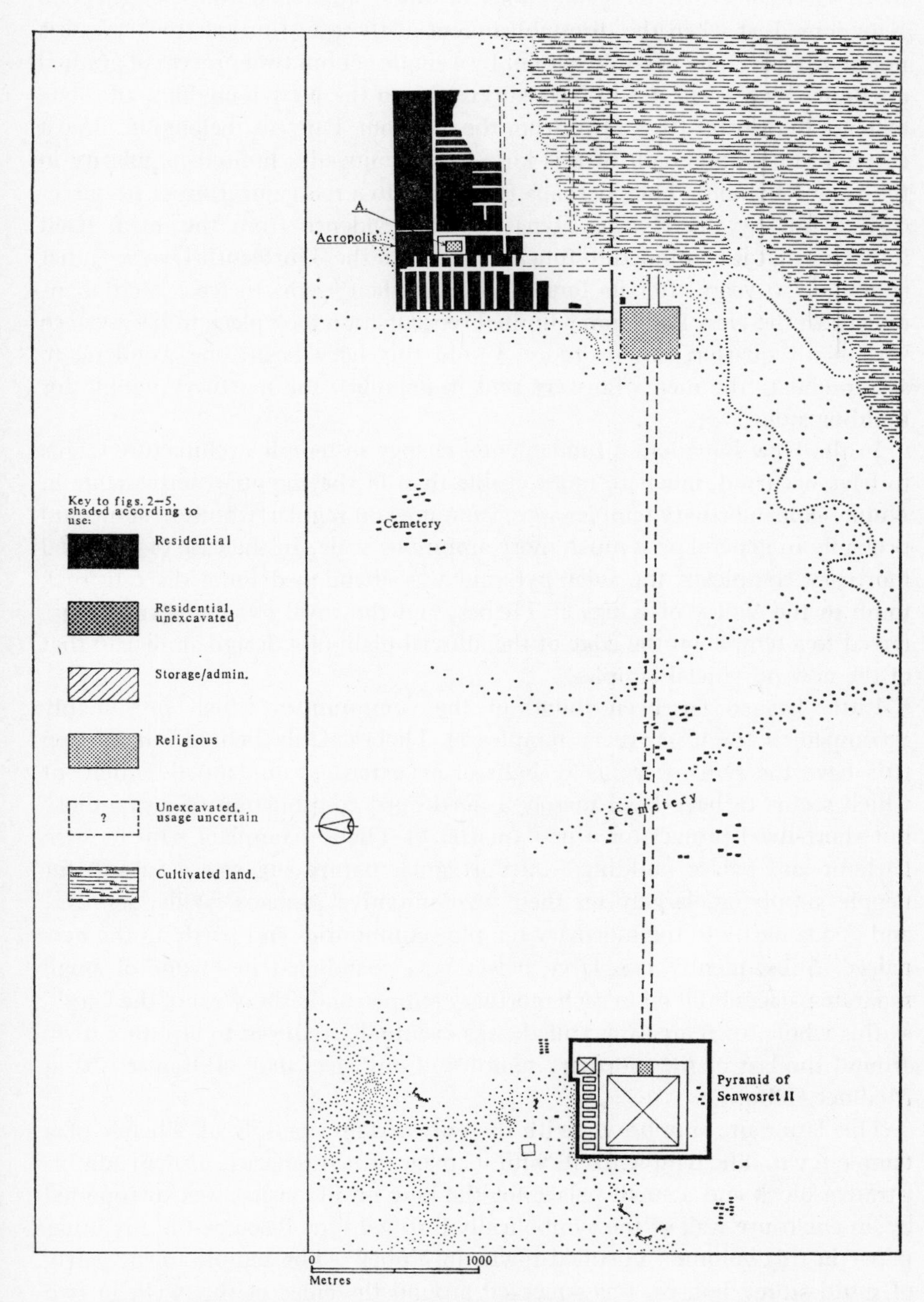

Fig 2 Block plan of Kahun town, Middle Kingdom. A: Temple of "Sepdu Lord of the East"?

more sacrosanct to later generations of rulers than had those of the Old Kingdom. Just when the disestablishment of these institutions took place is not quite clear, nor even if it was not by a single act but by a process of gradual attrition. They seem not to have survived into the New Kingdom, an estate of King Senwosret III, known from the Wilbour Papyrus, belonging almost certainly to a statue cult of this king which enjoyed a limited popularity in the New Kingdom. They perhaps fell victim to a reorganization of resources carried out by the Hyksos Dynasty. The evidence from the town itself indicates that it continued to function well into the Thirteenth Dynasty, that is about 150 years after its foundation, but then seems to have been abandoned. In the New Kingdom a limited reoccupation took place in the western workmen's quarter, but no more. Could this have been, one wonders, to accommodate the men who were sent to demolish the mortuary temple for building stone?

In the New Kingdom a fundamental change in temple architecture seems to have occurred, nowhere more visible than in the Egyptian settlements in Nubia. Non-mortuary temples were from now on regularly built of stone and probably in general on a much more ambitious scale. In the case of the royal mortuary complexes the solar pyramid was abandoned for a discrete rock tomb in the Valley of Kings at Thebes, and the royal mortuary cult transferred to a temple on the edge of the alluvial plain of a design similar to that of the new provincial temples.

Little has so far been found of the communities which presumably accompanied these mortuary temples at Thebes. Only behind Amenhetep III's have the remains come to light of an extensive unplanned settlement which seems to have been merely a northward continuation of a grandiose but short-lived project for a new quarter of Thebes complete with its own harbour and palace buildings[26]. Its irregular nature suggests a suburb for people supplying, largely on their own initiative perhaps, skills, services, and goods partly to the mortuary temple communities and partly to the new palace. Subsequently this large palace was abandoned in favour of small separate palaces built on to each mortuary temple, and "the West of the City", as this whole town area was called, was eventually reduced in size to a town around the last of the mortuary temples of any size, that of Ramses III at Medīnet Habu.

This latter site supplies us with the only certain example at Thebes of a temple town. The temple itself, with its magazines, granaries, a small administrative block and a small palace for the king on his visits, was surrounded by an enclosure wall of the symbolically fortified type discussed in my other paper in this volume "Fortified towns in Nubia". The temple town, partly of multi-storey houses, was squeezed around the sides of this wall, in two continuous rows separated by a street with a few larger houses towards the front. Presumably as a reflection of the troubled times in which it was built this town was itself surrounded by an even more massive wall which effectively

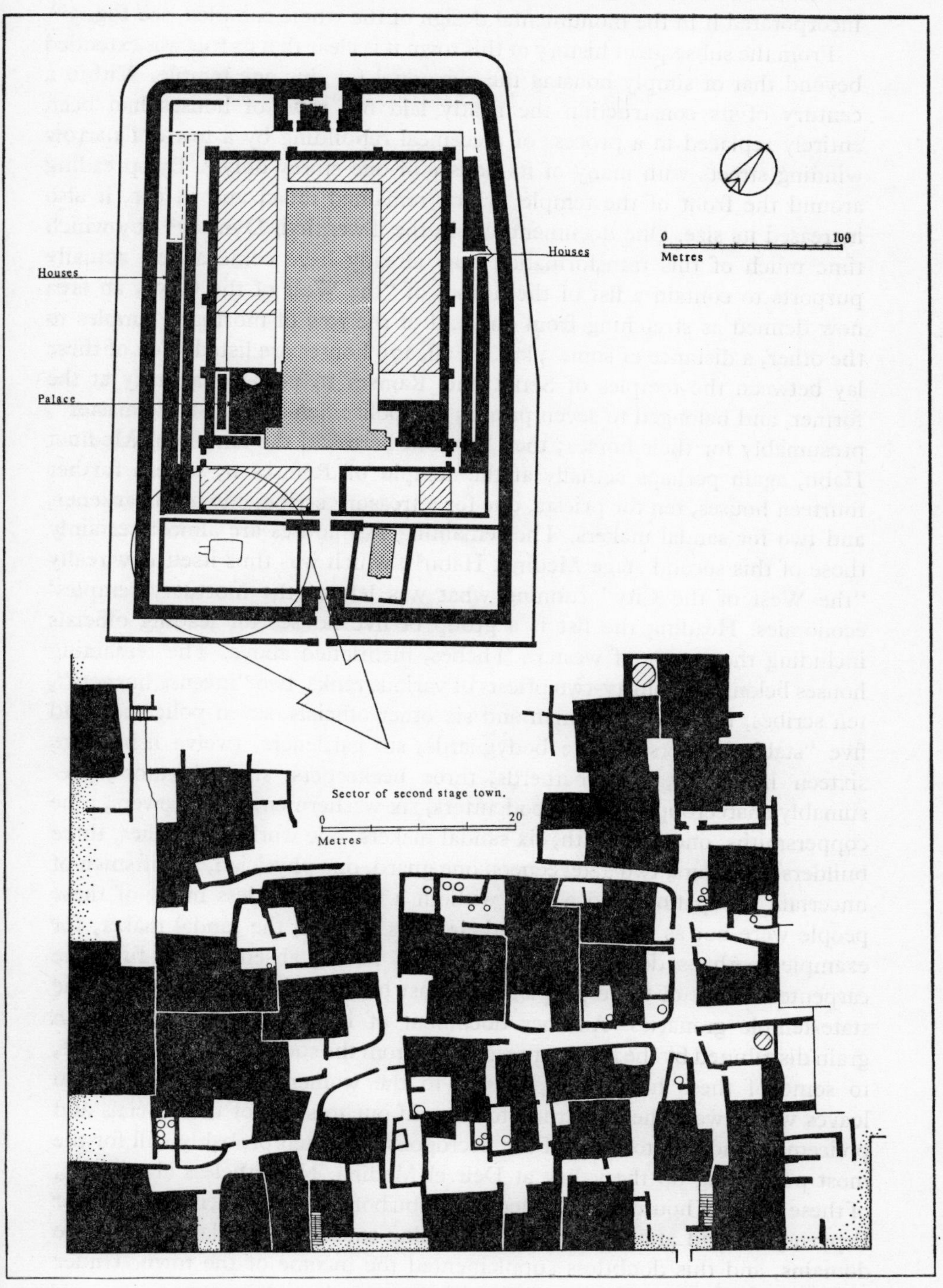

Fig. 3 The mortuary temple of King Ramses III at Medinet Habu, western Thebes

incorporated it in the monumental design of the whole complex, see Fig. 3[27].

From the subsequent history of this town it is clear that its role was extended beyond that of simply housing the personnel for this one temple. Within a century of its construction the neatly laid out rows of houses had been entirely replaced in a process of piecemeal rebuilding by a town of narrow winding streets with many of its houses of one storey only[28]. By spreading around the front of the temple, which remained intact and in use, it also increased its size. One document late in the Twentieth Dynasty[29], by which time much of this transformation may already have come about, actually purports to contain a list of the houses of "the West of the City", an area now defined as stretching from one end of the row of mortuary temples to the other, a distance of some 3 km. In all, 179 houses are listed. Ten of these lay between the temples of Seti I and Ramses II, perhaps actually at the former, and belonged to seven priests, two policemen and a "stable master", presumably for their horses; then between here and the temple of Medīnet Habu, again perhaps actually at the temple of Ramses II, were a further fourteen houses, ten for priests, one for a treasury scribe, one for a gardener, and two for sandal makers. The remaining 155 houses are almost certainly those of this second stage Medīnet Habu[30], which was thus itself now really "the West of the City" running what was left of the mortuary temples' economies. Heading the list is a group of five houses for leading officials including the mayor of western Thebes, mentioned above. The remaining houses belonged to thirty-two priests of various ranks, two "incense burners", ten scribes, two warehousemen and six other officials, seven policemen and five "stable masters", three bodyguards, six gardeners, twelve fishermen, sixteen herdsmen, two goatherds, three beekeepers, six cultivators, presumably sharecroppers, two woodcutters, six washermen, four brewers, nine coppersmiths, one goldsmith, six sandal makers, one worker in rushes, three builders or potters, two gatekeepers, one guard, one physician, a craftsman of uncertain occupation, and a man without a title. Doubtless many of these people were not as specialized as their titles suggest, the sandal maker, for example, perhaps doing odd-job carpentry in the absence of a full-time carpenter. Many of these people, too, must have received a salary from the state/temple granaries. Another document of the same date[31] deals with grain distributed by the mayor, presumably from the sources referred to above, to some of these households, mostly to the womenfolk, for baking into loaves which were then returned to be paid out to some of the officials and policemen, and also to some of the necropolis workmen probably still for the most part living up the valley at Deir el-Medina. Nevertheless about 70% of these heads of households at Medīnet Habu bore titles which, in the Wilbour Papyrus, we find attached to people holding small plots of land on temple domains, and this doubtless supplemented the income of the town. Under the name of Jeme it lasted until the eighth or ninth century A.D.

In further consideration of situation (a) (temple towns on "new" ground),

I would suggest that many of the New Kingdom settlements in Nubia be seen in this light. The appearance of these walled towns with their stone temples as the dominant feature closely resembles the symbolically fortified temple enclosures of the New Kingdom with the difference that a large part of the actual town was placed inside, presumably for defensive purposes. The most prominent non-temple building was the civil government residence, the "viceroy of Kush" and his "deputies" here replacing the vizier as senior minister, whilst the priests seem to have had, in the earlier phases at least, a secondary role. But from what has been said already on the relationship between temple and state this is no evidence for the status of the actual temple. Written documents from Nubia are very few, and archaeology has done little other than demonstrate that by the reign of Ramses II temples appear to have become the dominant physical feature of all Egyptian town sites in Nubia of any importance. Consequently a dominant economic role for them must remain a matter of conjecture, though the presence of large magazine blocks at Sesebi adds weight to the conjecture.

But the advantages of making a temple economy the basic unit for the administration and exploitation of Nubia would appear to have been considerable, since it provided a ready-made self-sufficient unit integrated with the fabric of the Egyptian state and capable, through the institution of *khato*-land, of administering crown lands as well. Indeed, one tract of Nubia was granted to a temple in Egypt, that of Seti I at Abydos[32]. This is not to deny that some, perhaps a great many, people living in and around these Nubian temple towns were not temple employees, for the leasing of temple lands seems to have been a very widespread practice. There are even a few explicit references to land donated to or being farmed by private individuals[33].

Nor, in considering the general economic role of these temple towns in Nubia, can one quite exclude aspects which to us belong to the realms of religion. For if, to an educated Egyptian, the fertility of his country based on the annual inundation of the Nile was to be ascribed to the beneficence of well-contented gods, then the poverty of Nubia without, for the most part, an inundation might well have appeared to be an attribute of godlessness. The apparently non-ecologically based distribution of temple towns into the impoverished area between the Batn el-Hagar and the Third Cataract looks suspiciously like the result of an over-assessment of agricultural potential based on a false understanding of the processes of nature. Their survival must have depended on subsidies from land owned by them in more fertile areas. Here one must emphasize that in trying to elucidate the motives behind a piece of ancient planning one must certainly envisage a far from simple decision-making process in which doctrinaire considerations advanced by people in powerfully entrenched positions must have vied with expert testimony from local investigations.

For situation (b), where the temple was but part of a much larger community, the archaeological evidence is very scanty since little excavation has been conducted outside temple enclosure walls. "The West of the City" during the reign of Amenhetep III might perhaps be counted. Otherwise there is only one site where there is enough to enable any sort of analysis to be undertaken, el-'Amarna, the new capital built by King Akhenaten from about 1365 B.C. and abandoned some fifteen years later. Although there were obviously special factors at work here I believe that they can be largely isolated and that much can be learnt about basic urban organization in New Kingdom Egypt. It is, moreover, highly unlikely that there will ever be another opportunity of seeing a contemporaneous area of an Egyptian city on anything approaching this scale. One may regret that the promising pre-war excavations have not been continued[34].

The city extended for some 7 km along one river bank, and was unfortified. The smaller temple in the centre was given symbolic fortifications and at the far northern end what was probably a major palace was surrounded by a towered wall of similar design[35], but this was all. Akhenaten's boundary inscriptions seem to indicate that his new sun temples here were endowed with a massive grant of land in the opposite bank[36]. The principal ones appear to have been two in the central part of the city (Fig. 4), and they were given storage and administrative buildings to handle their income, though for reasons deriving from a change in cult practices these buildings were placed outside the enclosure wall instead of being inside, so emphasizing, incidentally, that the function of these massive enclosure walls was not primarily to protect temple wealth. Since this was also the capital the central area contained one or two palaces, government offices including one dealing with foreign diplomatic correspondence, and police and military barracks. Since many of these buildings had a rectangular plan of their own a certain regularity of overall layout was perhaps inevitable though the degree of real planning seems to have been small. Palaces and temples were built to a common street frontage and intervening buildings took their alignment from them, but further back alignment seems to have been dictated most by the incoming streets from the South Suburb which arrived at an angle. Residence in this central area—presumably for workmen, craftsmen and some lesser officials—was provided by a few houses interspersed amongst the official buildings and by the Central Village, a block of back-to-back houses, perhaps originally totalling about 150 (Fig. 4). This is one of the only two fully planned residential developments at el-'Amarna, the other being a necropolis workmen's village out in the desert to the east.

To the north and to the south of this central area stretched the main residential areas, the North and South Suburbs. Beyond them lay further palaces and temples, including the far northern palace with its fortified wall which was perhaps, at least in the latter years at el-Amarna, the main royal residence. Behind it were at least two very large private estates, perhaps for princesses[37]. The South Suburb ran directly into the central area and its streets influenced the alignment of some of the central buildings. Whether

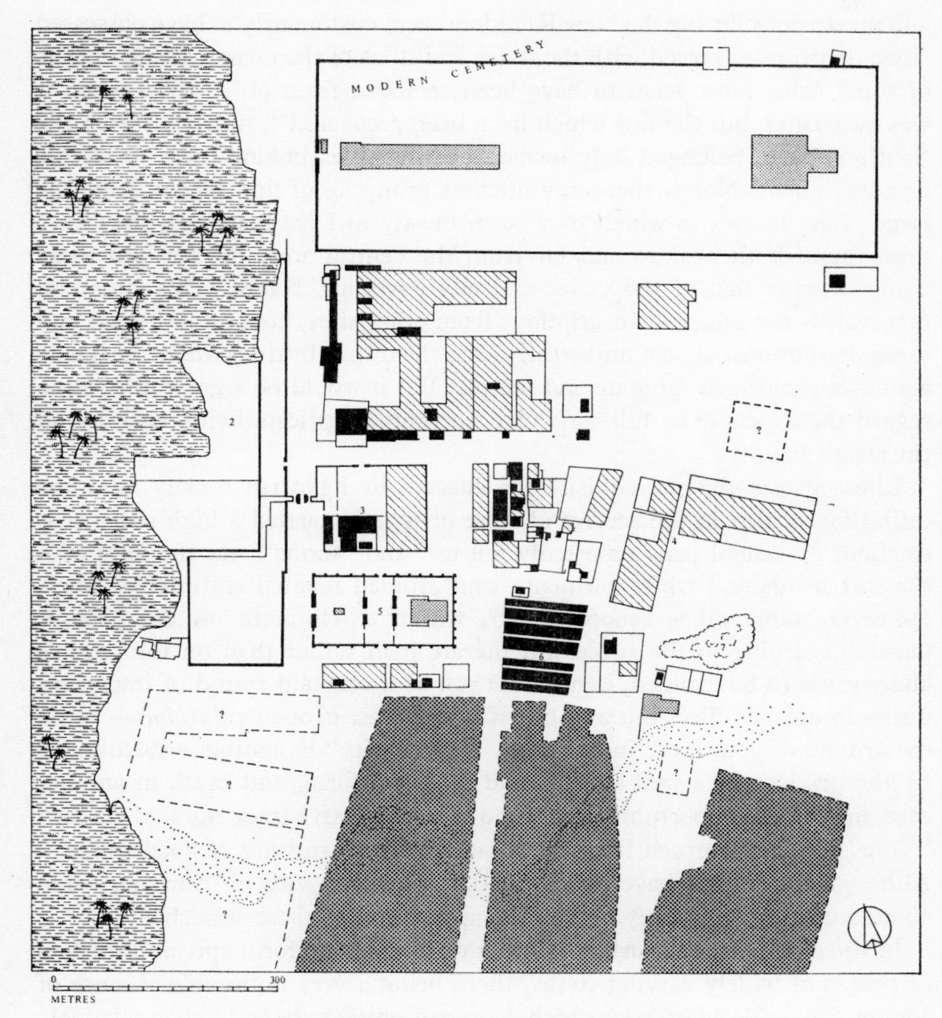

Fig. 4 The central area of Amarna

1. Larger Aten sun temple; 2. "The Great Palace", possibly a temple and its dependencies; 3. King's residence in the central area; 4. Police and military barracks; 5. Smaller Aten sun temple; 6. Central village; 7. Rubbish heaps.

the North Suburb did likewise has been obscured by a modern cemetery which prevented excavation north of the larger temple enclosure. The considerable extent of these two suburbs in relation to the area covered by the surviving administrative buildings prompts the important question: was the whole of el-'Amarna's population sustained by the economic activity of court and temples which, for reasons inherent in Akhenaten's policies, were probably even more completely integrated than normally? Was el-'Amarna a "company town"?

At first sight the answer might be thought to be "yes". Egyptian houses of any pretensions during the New Kingdom seem customarily to have possessed stone doorframes carved with the name and titles of the owner. Being objects of some value most seem to have been removed from el-ʿAmarna when it was evacuated, but the few which have been recovered[38], nearly all from the South Suburb, belonged only to men bearing titles linking them to temple or government. Nor is there any obvious grouping of find places. Like the large estate houses to which they seem mostly to have belonged, they have been found both near to and far from the central area, the furthest away being actually that of the vizier and city governor, Nakht. This situation, however, is not unusual. Inscriptions from other sites, house doorframes and tomb decorations, almost universally class the owner by titles which designate activities performed for state and temple. But it would be a great mistake to regard these people as full-time civil servants or priests living entirely on the state's bounty.

Literacy was the prerequisite for success in Egyptian society, and the utilization of this in the service of state or temple carried a high social distinction. As school texts repeatedly tell us[39] the "scribe" was the man with the soft hands and white garments who avoided manual and menial work. However, some other school texts[40], which dwell more on the positive benefits accruing to the successful literate man rather than on the terrible alternatives to be avoided, cast his lot not as a constant round of important duties in official offices but as a life of bucolic ease in one's villa (*bḫn*)—"upon the ground of your city" in one case, "in your city" in another—surrounded by the produce of a well-stocked and managed farm, and even, in another case, finding the opportunity to engage in a speculative trade mission abroad: "Your ship has returned from Syria laden with all manner of good things". Although these texts have no connection whatever with el-ʿAmarna, being about a century later, they fit well the appearance of these suburbs.

In the el-ʿAmarna suburbs we find a remarkably uniform spread of houses of people of widely varying status, there being a very noticeable absence of zoning. House designs show a high degree of uniformity and only a relatively modest progression of size to reflect the owner's wealth. The largest houses remain fairly constantly in the 250 to 500 m^2 class, with only a very few larger exceptions, as in the case of the vizier Nakht (800 m^2), and the far northern houses U.25.7 and 11—for princesses?—(about 580 and 960 m^2). In contrast to this is the much wider variation in the size of the compounds surrounding these houses and the ways in which they were utilized. Compounds are also found surrounding a minority of houses in a smaller class of 75 to 150 m^2. When the contents of these compounds are examined too their variation begins to look like a reflection of highly personalized household economies based on agriculture.

This variation and some idea of its consequences can be most clearly traced in the distribution of grain silos. These had a circular plan with an average diameter of about 2·5 m. Their original appearance is preserved in numerous contemporary representations. Their capacity must have been about 9500 litres, equivalent in ancient terms to about 125 *khar* of emmer. At the Theban necropolis workmen's village of Deir el-Medina a skilled artisan's annual wage for his family included a maximum of 48 *khar* of emmer, whilst the foreman received 66. For guards and porters the annual rate was 24 and 12. The payments of barley for beer seem to have been about a third[41]. If, as I believe, these silos were primarily for the storage of an annual harvest then two or three of them, a not uncommon number in these el-'Amarna compounds, would appear just to suffice a family with a purchasing power somewhat larger than a Deir el-Medina foreman, probably supplemented by payments for official duties, employing a few servants whose houses must be recognized partly within the compounds of the larger houses, but much more in the multitude of small dwellings which cluster around the larger ones. Others who would, in effect, have been supported by these silos would have been independent suppliers of goods and services both to the large houses and to those of their servants. As a measure of comparison one can quote the second stage Medīnet Habu town where only a very small minority of houses possessed silos, and then only a single one of about 2·0 to 2·5 m in diameter.

Two or three silos, however, was a common rather than a constant number. House T.35.9 in the North Suburb (Fig. 5), whilst in the "large" range, possessed a small compound, one half containing a little chapel and perhaps a garden, the other no less than eleven silos and a devious entrance system. The excavators suggested that this was the house of a "corn merchant", by which one can understand either a retailer who has purchased from elsewhere or, more likely in my opinion, a landowner specializing in corn growing and bringing a large part of his harvest to the capital for trading. A similar layout can be observed in house N.49.18, a corner site in the South Suburb occupied by a "large" house in a small compound possessing six silos, of which two were later replaced by rectangular magazines, which probably held more. Here we are fortunate in knowing who the owner was, a man of high military rank: "First charioteer of His Majesty, officer of horses of the entire stable Ra-nefer"—a man who may even have spent a period of office as governor in one of the conquered Palestinian provinces"[42].

Even Ra-nefer and the owner of house T.35.9, however, may have been some way from the top of the incomes scale. For in some houses these circular silos, very convenient for storing limited amounts of grain, have been replaced, sometimes physically as in the cases of T.35.16, N.47.1, L.48.2, and probably partially by Ra-nefer himself, by the long rectangular magazines familiar from temple sites. Since these were suitable for storing other commodities as well—contemporary pictures of temple magazines show them full of meat, fish, wine, minerals, precious metals, and so on—the corn income of their owners cannot reasonably be estimated. But house V.36.7, 12, 13, for example, must have had a maximum capacity for some 2800 *khar*. And, as the stockyards and cattle sheds in many of these compounds show, corn was only one item of agricultural income.

I would suggest that Ra-nefer and his neighbours at el-'Amarna, whilst holding public offices for which doubtless they received a stipend, possessed as their basic source of income agricultural lands, and that a portion of the harvest in excess of the needs of their households was transported annually to the capital to be traded at a profit, this being left, perhaps, in the hands of a "trader" such as we find attached not only to temples, as already mentioned, but also to private households[43]. Much was doubtless traded to the occupiers of the numerous small houses in their neighbourhoods.

This touches upon the difficult topic of the extent and nature of private landholding in New Kingdom Egypt. Hereditary possession of land is definitely attested, though on what precise legal basis is uncertain. The Inscription of Mes[44] introduces us to a parcel of land granted by King Ahmose at the beginning of the New Kingdom to an "officer of ships". About three centuries later, in the reign of Ramses II, we find the same land still in the hands of the same family, descendants of the original "officer of ships", now squabbling over its division into lots, and appealing to a duplicate set of government land records, kept by the treasury and by the department of the granary, wherein was recorded the history of ownership. Papyrus Valençay I of the reign of Ramses IX is even more explicit in drawing a distinction between *khato*-lands of Pharoah and privately owned land whose taxes were paid independently to the treasury[45]. And there is, of course, the evidence for the very widespread leasing of temple lands, with hints that this might be an hereditary practice. Together the available documents suggest a highly complex pattern of land-holding where one is led to envisage a "farm" as being not a single discrete area of agricultural land, but a whole series of scattered plots held in several ways, under private ownership or rented on a sharecropping basis from a temple or from some other landowner, the farm centre being often actually in a town, even the capital itself. O'Connor has analysed the written evidence from the Wilbour Papyrus for "villas" or farm centres (*bḫn*) as one class of settlement in the countryside, but even so it shows the greatest number located in the area which also contained the important provincial town and nome capital of Ninsu (Herakleopolis)[46].

A landholder such as this would find himself operating an economic cycle more or less identical to that described for a temple. Part of his produce would go in rents and taxes, part on payments to his servants and managers and on feeding and maintaining his own family. But for any luxuries and to provide some liquidity as a safeguard against an emergency the landholder automatically put himself in the position of a trader employing part of his farm produce as a medium of exchange for services and goods not otherwise at his command. For someone possessing a shrewd eye for business the situation logically led to one where, for example, a surplus of grain or meat one year could be used to purchase a cargo of, say, imported Syrian beer or

Cypriote base-ring juglets and then retailing them at a profit[47]. Private enterprise trading must have been inherent in the economic structure of ancient Egypt.

The picture of the el-'Amarna suburbs as a series of self-contained and economically largely self-sustaining neighbourhoods is amplified by the distribution of industry in the city as a whole. Take the case of glass and glaze manufacturing. Petrie's excavations located a large factory area in the centre. But on the block of houses across the street from Ra-nefer's house was a smallish house of a normal plan, M.50.15, which had, in its courtyard, a small glaze factory with a kiln and the remains of slag. For whom was it worked? We cannot tell for sure. More illuminating are the sculptors' workshops. Again, as might be expected, one was found in the central area, opposite to the larger temple. But the excavation of the South Suburb has so far brought to light probably at least four others, situated well within the residential neighbourhoods. Part of the work was carried on in separate courtyards, but some seems to have been done in houses and compounds of otherwise normal type. Again, the question arises, for whom were they working? For the state and temples? Or could these institutions have been just one patron of relatively independent craftsmen who were also accepting private commissions for sculpture for tombs and for the chapels which adorned many a compound garden? Amongst the débris from the workshops were plaster casts of the faces and statues of private individuals, and, in one case, 0.49.14, an unfinished statuette of the goddess Isis and an alabaster pedestal for a statuette of Ptah, which can scarcely have been made for a state following a path of religious exclusiveness at this time. But whatever the nature of the patronage it begins to look as though a fairly considerable number of people in the smaller houses of the suburbs were living adjacent to their source of employment which did not lie in the central area.

The layout of these suburbs seems to have accentuated this pattern of separate neighbourhood units which must have been almost villages. As far as the excavated area allows us to judge these neighbourhood units were divided, on a north-south axis, by continuous wide spaces, some 25 or 30 m across, which served also as thoroughfares to the city centre, linked by narrower cross streets at intervals. Within this loose framework each neighbourhood was allowed to find its own plan. A series of compounds, seldom quite rectangular, was laid out for the larger houses, and the smaller houses filled up the intervening spaces in an irregular manner. On the outer edge of the city, where most excavation in the suburbs has been concentrated, the density of housing was still fairly low when the city was abandoned, with many open spaces remaining. But elsewhere, and especially in the central part of the North Suburb (Fig. 5), we find an "organic" town plan developing of the sort familiar from medieval European and Arab towns: a network of narrow twisting streets, the juxtaposition of houses of people from all positions on the social scale, the claustrophilia which is the essence of

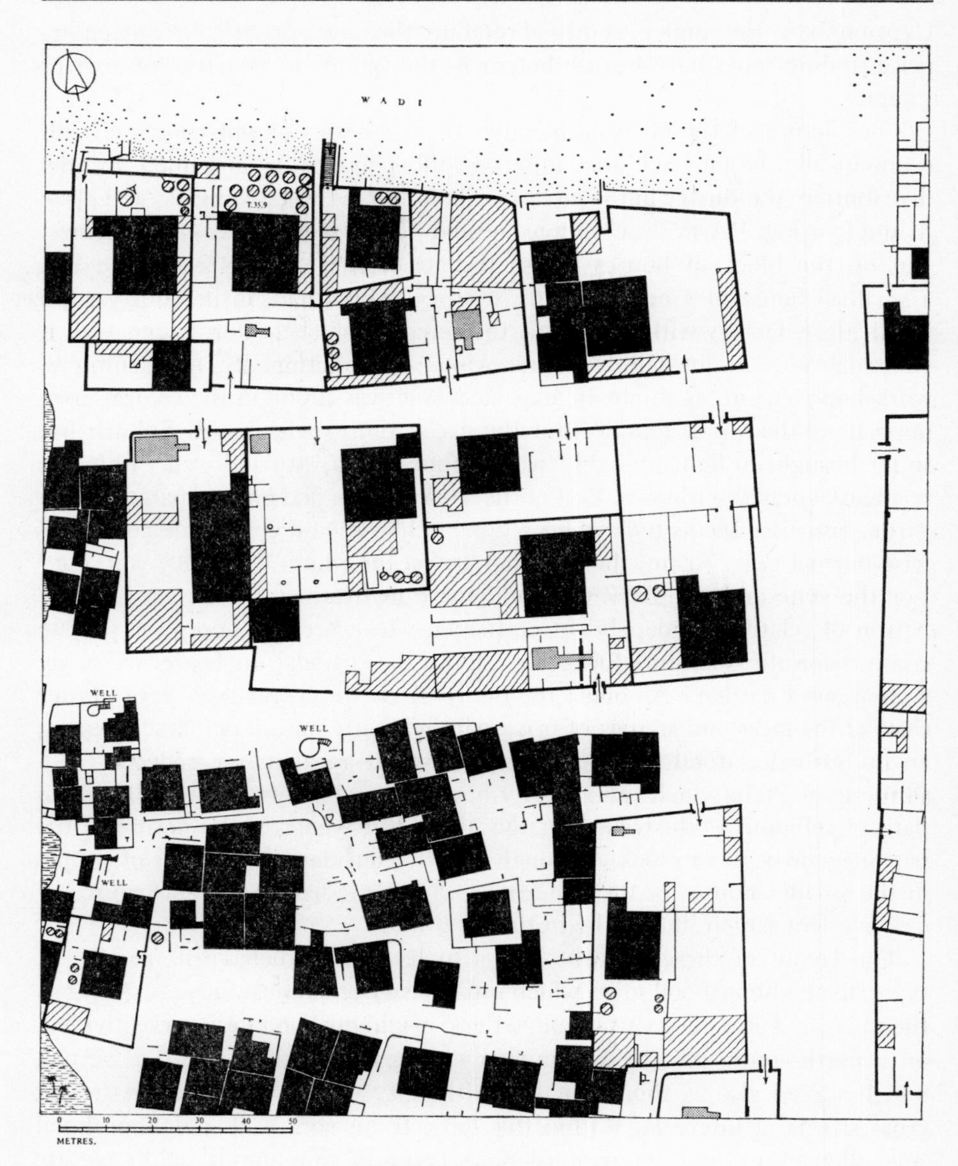

Fig 5 Part of the North Suburb at el-'Amarna.

successful community living. Some of the apparent open spaces probably represent nothing more than areas where the house remains have entirely disappeared, but some, particularly those around the public wells, must have been left as open spaces for marketing, and it would doubtless have been here that much of the farm produce from the neighbouring large compounds would have been disposed of.

The excavators termed this particular area a "slum", but whilst some of the houses were small enough to justify this description, most of them, as far as can be seen from the badly destroyed wall remains, were of medium size, a few even possessing small compounds with outbuildings. Here, I believe, we have some insight into the New Kingdom Egyptians' own sense of urban aesthetics and the value which they placed on strict rectangular residential planning. To be effective in providing an acceptable urban aesthetic, simple geometrical layouts require some variation in the architecture itself to avoid inhuman monotony. This is something which Egyptian domestic architecture patently lacked, always presenting blank walls to the street with the interest exclusively on the interior. But with a winding and closely packed street layout an endlessly variable juxtaposition of flat planes and complex chiaroscuro emerges, giving to each part an identity of its own. The combination of these organic neighbourhood units with a system of wide thoroughfares leading to the city centre represents a probably very effective and acceptable type of urban layout[48].

It remains to consider how far the nature of el-'Amarna differed from other towns and cities in Egypt with a more normal history of development. From what has been said earlier the temple would appear to have been a centre of economic life in a town, as well as the dominant architectural feature in a society with little tradition of non-religious public buildings. The el-'Amarna temples were unusual in not possessing processional routes down to the river and in having their subsidiary buildings outside the enclosure wall. In normal cases these buildings would have been placed inside, thus removing one source of rectilinear layout from the town plan. The presence of the temples, government offices, indeed the court itself, at el-'Amarna was obviously the reason for the presence of the owners of the larger houses, all probably holding some position in the government, temples or army. They and the inhabitants of the Central Village and probably many others living in the suburbs must have been involved in or sustained by the economic cycle of these central institutions, principally the temples which had initially received an enormous grant of land on the opposite side of the river. However, as Smith and O'Connor point out[49], this situation, on a smaller scale, must have been repeated in the provincial towns of Egypt because they were normally the local administrative and cult centres, where the chief townspeople must have been also the local government and temple officers. But, the el-'Amarna suburbs also show clearly a large number of small economic centres formed by the fact that the larger houses were also farm centres—maybe even renting some of the new temple lands on the opposite bank—which were apparently involved in manufacturing industries as well, though the precise structure escapes us. This gave to el-'Amarna the appearance of a series of villages, which is something emphasized by the overall layout adopted. It has often been pointed out that since el-'Amarna was for the most part built upon an open stretch of desert a degree of spaciousness could be allowed which was not normally available in towns built on cultivable land. Supporting this are the contemporary representations in a few Theban tombs of multi-storey houses which are presumably the town houses of Thebes[50], contrasting with the bungalows at el-'Amarna. But to assume that the other essential feature of an el-'Amarna large house, its compound for farm produce, was also abnormal in an urban context is to assume that el-'Amarna had a radically different economic structure from other Egyptian towns and cities. I find this an implausible assumption. Instead, I would suggest that if one takes the central part of the North Suburb (Fig. 5) where the sizes of the compounds have been reduced to an absolute minimum, replaces some of the bungalows with multi-storey houses, and probably narrows the wide north-south thoroughfare, one would have a close approximation to a typical urban layout in New Kingdom Egypt. Indeed, in the tomb of Anena at Thebes we have a depiction of just this combination: a multi-storey house in a farm compound[51]. The edges of towns presumably resembled el-'Amarna even more closely.

Thus towns and cities of New Kingdom Egypt would appear to have had as their principal characteristic a two-tier economic structure based on farm centres, the temple—an adjunct of the state—being a major one, with below it others represented by private landholders. The extent to which each developed must clearly have varied according to the history of the individual site. This pattern was necessary because the nature of economic transactions required farm produce to be close at hand all the time. It would be an important matter to discover how ancient a pattern this was in view of one recent theory[52] that the domestication of plants and animals was primarily an innovation of cities rather than their prerequisite.

Notes

1 For the latter see the Nauri Decree of Seti I, published by Griffith, F. Ll., (1927). The Abydos Decree of Seti I at Nauri, *J. Egypt. Archaeol.*, **13**. pp. 193–208; Edgerton, W. F. (1947). The Nauri Decree of Seti I, *J. Near East. Stud.*, **6**, pp. 219–30; Gardiner, Sir A. H. (1952). Some reflections on the Nauri Decree, *J. Egypt. Archaeol.*, **38**, pp. 24–33.

2 On this and many other aspects of the New Kingdom economy see the excellent article by Baer, K. (1962). The low price of land in ancient Egypt, *J. Amer. Res. Centre*, **1**, pp. 25–45. Revealing for a slightly later period are the documents published by Hughes, G. R. (1952). *Saite Demotic Land Leases*. Chicago. For a case where a lessee enjoyed the produce of livestock (geese in this case) owned by a temple see the documents published by Malinine, M. (1968). Un contrat démotique de société (Pap. Loeb n°n° 47 et 46), *in* Helck, W. (ed.) *Fest. für Siegfried Schott*. Wiesbaden. pp. 87–93; Malinine, M. (1968). Une livraison d'oies au domaine d'Amon, *J. Egypt. Archaeol.*, **54**, pp. 188–92.

3 Gardiner, Sir A. H. and Faulkner, R. O. (1941–52). *The Wilbour Papyrus, I–IV*. The Brooklyn Museum. See especially vol. II, pp. 75–84; also O'Connor, D., this volume, especially table 5.
4 Gardiner, Sir A. H. (1941). Ramesside texts relating to the taxation and transport of corn, *J. Egypt. Archaeol.*, **27**, pp. 37–56.
5 Considerable information on temple possessions will be found summarized in the five volumes of Helck, W. (1960–64). *Materialien zur Wirtschaftsgeschichte des Neuen Reiches*. Mainz, being *Abh. Akad. Wiss. u. Lit. Mainz, Geistes- u. Soz. Kl.* (1960), **10, 11** (1963), **2, 3** (1964), **4**.
6 Schott, S. (1961). *Kanais. Der Tempel Sethos I. im Wadi Mia*. Göttingen, being *Nach. Akad. Wiss. in Gött. Phil.-hist. Kl.* (1961), **6**, pp. 143–59.
7 Helck, W. (1967). Eine Briefsammlung aus der Verwaltung des Amuntempels, *J. Amer. Res. Centre*, **6**, pp. 135–51.
8 Baer, K. (1962). *op. cit.* pp. 31–3; Caminos, R. A. (1954). *Late-Egyptian Miscellanies*. London. pp. 18–20.
9 Gardiner, Sir A. H. and Faulkner, R. O. (1941–52). *op. cit.* vol. II, pp. 161–90, 209–10; Gardiner, Sir A. H. (1941). *op. cit.* pp. 23–4.
10 See the considerable biographical material collected by Helck, W. (1958). *Zur Verwaltung des mittleren und neuen Reichs*. Leiden-Köln. pp. 285–533; also Davies, N. de G. (1943). *The Tomb of Rekh-mi-Rēʿ at Thebes*. New York.
11 Gardiner, Sir A. H. (1941). *op. cit.* pp. 22–37; also Černý, J. (1965). *Egypt from the death of Ramesses III to the end of the Twenty-First Dynasty*. Cambridge. pp. 29–30.
12 Edgerton, W. F. (1951). The strikes in Ramses III's twenty-ninth year, *J. Near East. Stud.*, **10**, p. 144.
13 Some quoted in Helck, W. (1960–64). *op. cit.* vol. III, pp. 267–8 and vol. IV, p. 410; also Gardiner, Sir A. H. (1948). *Ramesside Administrative Documents*. London. p. 64, line 12 to p. 65, line 4.
14 Peet, T. E. (1930). *The Great Tomb-robberies of the Twentieth Egyptian Dynasty*. Oxford. p. 12, note 1. One source of this statement is presumably Botti, G. and Peet, T. E. (1928). *Il Giornale della Necropoli di Tebe*. Turin. Tav. 13 and 14; another, O. Turin 108, see Helck, W. (1960–64). *op. cit.* vol. IV, pp. 405–6, also p. 410.
15 Breasted, J. H. (1906–7). *Ancient Records of Egypt, I–V*. Chicago. vol. III, p. 113, para. 274.
16 Janssen, J. J. (1961). *Two Ancient Egyptian Ship's Logs*. Leiden. pp. 101–2.
17 Janssen, J. J. (1961). *op. cit.* pp. 53–95.
18 For a transaction of this nature see Peet, T. E. (1934). The unit of value *šʿty* in Papyrus Bulaq 11, *Mém. Inst. Fr. Archéol. Orient. Cairo*, **66**, pp. 185–99.
19 Posener-Kriéger, P. and Cenival, J. L. de (1968). *Hieratic Papyri in the British Museum*. Fifth Series, The Abu Sir Papyri. London; Posener, P. (1963). Les archives du temple funéraire de Neferkareʿ à Abusir, *Chron. Egypt.*, **38**, pp. 257–8.
20 The most important are: at the tomb of Queen Khent-kawes, Giza, Old Kingdom: Hassan, S. (1943). *Excavations at Gîza*. Cairo. vol. IV, pp. 35–50. At the valley temple of Menkaura's pyramid at Gîza: Reisner, G. (1931). *Mycerinus*. Cambridge, Mass. pp. 49–54; Hassan, S. (1943). *op. cit.* contains additional material. At the valley temple of Sneferu's pyramid at Dashur, Old Kingdom: Fakhry, A. (1959). *The Monuments of Sneferu at Dahshur*, I. Cairo. pp. 114–17, Fig. 62; the pottery is dealt with in vol. II, part II, pp. 105–40. At the valley temple of Senwosret II's pyramid at el-Lahun, see note 21. At the pyramid of Amenemhat III at Dashur: Morgan, J. de (1903). *Fouilles à Dahchour en 1894–95*. Vienna. pp. 100–2, pl. I; Morgan, J. de (1897). *Carte de la nécropole memphite*. Cairo. pl. II.

At the cenotaphs at Abydos South: Ayrton, E. R., Currelly, C. T. and Weigall, A. E. P. (1904). *Abydos*, part III. London. pp. 37–8, pl. LIII. Early Eighteenth Dynasty, and probably earlier.

21 Petrie, Sir W. M. F. (1890). *Kahun, Gurob, and Hawara*. London. pp. 21–32, pl. XV; Petrie, Sir W. M. F. (1891). *Illahun, Kahun and Gurob*. London. pp. 5–15, pl. XIV; Petrie, Sir W. M. F. (1904). *Methods and Aims in Archaeology*. London. Fig. 23; Petrie, Sir W. M. F., *et al.* (1923). *Lahun II*. London. pp. 39–40, pls. II, XXXIII; Ricke, H. (1932). *Der Grundriss des Amarna-Wohnhauses*. Leipzig. pp. 51–5.

22 Dunham, D. (1967). *Uronarti Shalfak Mirgissa*. Boston. pp. 5, 8–10, 13, map III, rooms 7–12, 30–4.

23 Griffith, F. Ll. (1898). *Hieratic Papyri from Kahun and Gurob*. London. pp. 52–4, pl. XXI, lines 1–14.

24 These are the blocks of interconnecting square rooms with thick walls, which also seem to have been the principal storage units in the Nubian forts of this period, where corn was not only used to feed the garrison but also distributed, possibly in the course of trading, to groups of visiting Nubians, see Smither, P. C. (1945). The Semnah Despatches, *J. Egypt. Archaeol.*, **31**, pp. 3–10.

25 Griffith, F. Ll. (1898). *op. cit.* pp. 25–9, pls. X–XI. On the occupations of members of large households of this period see Hayes, W. C. (1955). *A Papyrus of the Late Middle Kingdom in the Brooklyn Museum*. The Brooklyn Museum. pp. 103–8, where weaving is the commonest female occupation.

26 Robichon, C. and Varille, A. (1936). *Le temple du scribe royal Amenhotep fils de Hapou*. Cairo. pls. V, VIII, IX, pp. 33–4; Hölscher, U. (1934). *General Plans and Views*. Chicago; Hölscher, U. (1939). *The Temples of the Eighteenth Dynasty*. Chicago. pp. 68–73; for the palace of Amenhetep III see the bibliography in Porter, B. and Moss, R. L. B. (1964). *Topographical Bibliography* (2nd ed.), I, part 2. Oxford. pp. 778–81; also Smith, W. S. (1958). *The Art and Architecture of Ancient Egypt*. Harmondsworth. pp. 160–72.

27 Hölscher, U. (1934). *op. cit.*; Hölscher, U. (1951). *The Mortuary Temple of Ramses III*, part II. Chicago. Also Černý, J. (1940). "The Temple", . . . as an abbreviated name for the Temple of Medînet Habu, *J. Egypt. Archaeol.*, **26**, pp. 127–30; Eric Uphill's paper, "The Concept of the Egyptian palace as a 'Ruling Machine' ", this volume pp. 721–34.

28 Hölscher, U. (1934). *op. cit.*; Hölscher, U. (1954). *Post-Ramessid Remains*. Chicago. pp. 1–13. *A terminus ante quem* for this phase in one part of the town is provided by the house of the scribe Butehamen at the very end of the Twentieth Dynasty, see Wente, E. F. (1967). *Late Ramesside Letters*. Chicago. pp. 5–17. I would regard this second stage town as the result of a slow, piecemeal, and essentially peaceful alteration to the town as laid out which may well have begun fairly soon after it was first completed. The excavator's interpretation that it was only rebuilt after a single violent destruction at the end of the Twentieth Dynasty rests on the mistaken assumption that town planning was a way of regulating urban life rather than a bureaucratic short-cut and that, conversely, the disintegration of a plan represents a period of decline or the aftermath of a riotous destruction. Proponents of this view may be identified as the "squatter school". Smith, H. S., this volume, p. 705, illustrates one reason for constant house alteration—to adjust to constantly changing ownership of often quite small parts of a house which might represent a person's major financial asset.

29 Peet, T. E. (1930). *op. cit.* pp. 83–5, 93–102. On the dating of this text see Wente, E. F. (1967). *op. cit.* p. 2, note 4.

30 That this list refers to Medīnet Habu is confirmed at one point by the listing of the house of the necropolis scribe Djehutymes, the father of Butehamen, the inscribed door frame of whose house was actually recovered from a later Coptic house in Medīnet Habu, see Hölscher, U. (1954). *op. cit.* p. 5, note 25; Porter, B. and Moss, R. L. B. (1964). *op. cit.* p. 777; Černý, J. (1936). Une famille de scribes de la nécropole royale de Thèbes, *Chron. Egypt.*, **11**, p. 249. The composition of this list also makes it clear that the necropolis workmen were still a separate group living elsewhere, presumably at Deir el-Medina. That some of their officers should have chosen to live separately in Medīnet Habu near the officials of "the West of the City" is hardly surprising. Another document mentions a set of chambers for the vizier, See Černý, J. (1940). *op. cit.* p. 129. Could they have been inside the inner enclosure wall?

31 Peet, T. E. (1930). *op. cit.* pp. 55–6, 64–71.

32 See the references in note 1.

33 Crum, W. E. (1893). Stelae from Wady Halfa, *Proc. Soc. Bib. Archaeol.*, **16**, pp. 18–19; Helck, W. (1957). *Urkunden der* 18. *Dynastie*, **19**. Berlin. p. 1637, lines 11–14; Breasted, J. H. (1906–7). *op. cit.* vol. IV, pp. 231–5; Helck, W. (1960–4). *op. cit.* vol. II, pp. 1077–9.

34 For bibliographies and general surveys see Porter, B. and Moss, R. L. B. (1934). *Topographical Bibliography*. vol. IV. Oxford. pp. 192–236; Pendlebury, J. D. S. (1935). *Tell el-Amarna*. London; Fairman, H. W. (1949). Town planning in pharaonic Egypt, *Town Planning Review*, **20** (1); Badawy, A. (1968). *A History of Egyptian Architecture: the Empire*. Berkeley and Los Angeles; Uphill, E. P. (1970). The Per Aten at Amarna, *J. Near East. Stud.*, **29**, pp. 151–66.

35 Largely unpublished. For preliminary reports see Pendlebury, J. D. S. (1931). Preliminary Report of Excavations at Tell el-'Amarnah 1930–1, *J. Egypt. Archaeol.*, **17**, pp. 242–3, pl. LXXVIII; Pendlebury, J. D. S. (1932). Preliminary Report of the Excavation at Tell el-'Amarnah, 1931–2, *J. Egypt. Archaeol.*, **18**, pp. 143–5, pls XII and XIII. An aerial photograph showing its relationship to the large houses (for princesses?) outside to the east is published in Bille-de Mot, E. (1966). *The Age of Akhenaten*. London. pl. 26. Unpublished plans exist on lantern slides in the possession of the Egypt Exploration Society, duplicates being in the Faculty of Oriental Studies, University of Cambridge. The towered wall possessed no ditch, and seems to have been merely the outer skin of a cavity wall of the type used at the contemporary temple site at Soleb in Nubia, see Giorgini, M. S. (1962). Soleb. Campagna 1960–1, *Kush*, **10**, pp. 154–5, 159.

36 Davies, N. de G. (1908). *The Rock Tombs of el Amarna*, vol. V. London. pp. 19–34.

37 For "households" or "estates" of princesses at el-'Amarna see Peet, T. E. (1930). Two letters from Akhetaten, *Ann. Archaeol. Anthrop.* (Liverpool), **17**, pp. 82–97; Redford, D. B. (1967). *History and Chronology of the Eighteenth Dynasty of Egypt*. Toronto. p. 103, note 69; Helck, W. (1960–4). *op. cit.* vol. II, p. 994.

38 Porter, B. and Moss, R. L. B. (1934). *op. cit.* pp. 199–207.

39 Caminos, R. A. (1954). *op. cit. passim*; Gardiner, Sir A. H. (1935). *Hieratic Papyri in the British Museum*, third series, Chester Beatty Gift. London. p. 41.

40 Caminos, R. A. (1954). *op. cit*, pp. 137–43, 164–8, 400–10, 412–19.

41 Černý, J. (1954). Prices and wages in Egypt in the Ramesside Period, *Cah. Hist. Mond.*, **1** (4), pp. 903–21; Helck, W. (1960–64). *op. cit.* vol. IV, pp. 390–408.

42 Schulman, A. R. (1964). *Military Rank, Title, and Organization in the Egyptian New Kingdom*. Berlin, pp. 68, 145 (375), 162 (480); Albright,

W. F. (1946). Cuneiform material for Egyptian prosopography, *J. Near East. Stud.*, **5**, pp. 19–20, (*49).

43 Janssen, J. J. (1961). *op. cit.* pp. 101–3.

44 Gardiner, Sir A. H. (1905). *The Inscription of Mes.* Leipzig.

45 Gardiner, Sir A. H. (1951). A protest against unjustified tax-demands, *Rev. Egypt.*, **6**, pp. 115–24; Gardiner, Sir A. H. and Faulkner, R. O. (1941–52). *op. cit.* vol. II, pp. 205–6; Baer, K. (1962). *op. cit.* p. 26, note 10; note also "fields of your own acquisition", Caminos, R. A. (1954). *op. cit.* pp. 401, 403, note on 9, 2.

46 O'Connor, D., this volume p. 691.

47 For foreign trade conducted by private individuals see Merrillees, R. S. (1968). *The Cypriote Bronze Age Pottery Found in Egypt.* Lund. pp. 173–74.

48 The "neighbourhood unit" has become a concept of modern town planning. It is interesting to see it applied in the Arab world as a deliberate attempt to continue the local traditions of organic town growth and as a reaction against the indiscriminate use of grid-iron planning. See, for example, Shiber, S. G. (1969). *Recent Arab City Growth.* Kuwait. especially p. 500, with illustrations of examples at Kuwait on pp. 307–19; Fathi, H. (1969). *A Tale of Two Villages.* Cairo. This documents the brilliant but ill-fated New Gourna project, an attempt to infuse into a new housing scheme in Egypt the essential characteristics of a traditional village.

49 Smith, H. S., this volume, p. 713; O'Connor, D. this volume, p. 687.

50 Davies, N. de G. (1929). The town house in ancient Egypt, *Metropolitan Mus. Stud.*, **1** (2), pp. 233–55; Desroches, C. (1938). Un modèle de maison citadine du nouvel empire, *Rev. Egypt.*, **3**, pp. 17–25.

51 Wreszinski, W. (1923). *Atlas zur altaegyptischen Kulturgeschichte.* Leipzig. vol. I, pl. 60; Davies, N. de G. (1963). *Scenes from some Theban Tombs.* Oxford. p. 20, pl. XXIII.

52 Jacobs, J. (1970). *The Economy of Cities.* London.

DAVID O'CONNOR

The geography of settlement in ancient Egypt

The Aethiopians lead for the most part a nomadic and resourceless life, on account of the barrenness of the country and the unreasonableness of its climate and of its remoteness from us, whereas with the Aegyptians the contrary is the case in all these respects; for from the outset they have led a civic[1] and cultivated life and have been settled in well-known regions, so that their organizations are a matter of comment[2].

When Strabo wrote these words in approximately A.D. 19 his impression of the "civic and cultivated life" of Egypt was based, like that of other Greek and Roman writers, primarily on the undoubtedly urbanized Persian, Ptolemaic and Roman periods. Nevertheless such writers had some knowledge of the pre-Persian period and appear to project their impression of an Egyptian society in which urbanism in some form was important far back into time. They knew for example, that Memphis, the national capital of the Old Kingdom (*c.* 2686–2181 B.C.) and administrative centre of northern Egypt for two millennia, had traditionally been founded by Egypt's first historic king, Menes (Early Dynastic Period; *c.* 3100 B.C.)[3]. Modern research suggests that this tradition is correct[4] while the actual remains of Early Dynastic towns have been found at Nekhen (Kom el-Ahmar), Enkhab (El Kab) and Ebodju (Abydos)[5] (Fig. 1). Prehistoric settlements, some of apparently large size, are also known, but denudation and the lack of recent excavation limit our understanding of their nature[6].

After Early Dynastic times (*c.* 3100–2890 B.C.) textual sources often use terms translated with some justification by Egyptologists as "town", "city", "village", "fortress", etc. and reveal a complex society in which certain settlements clearly had special functions which are found on a much reduced scale or not at all in the hundreds of other settlements known to have existed. The archaeological evidence is very incomplete but decisive. The physical reality of Egyptian towns has been revealed by the rare excavation of settlements on the desert adjoining the cultivation. Although most are laid out in a rigorously systematic way untypical of most Egyptian settlements, Kemp has shown[7] that Amarna (Fig. 1) illustrates the more typical

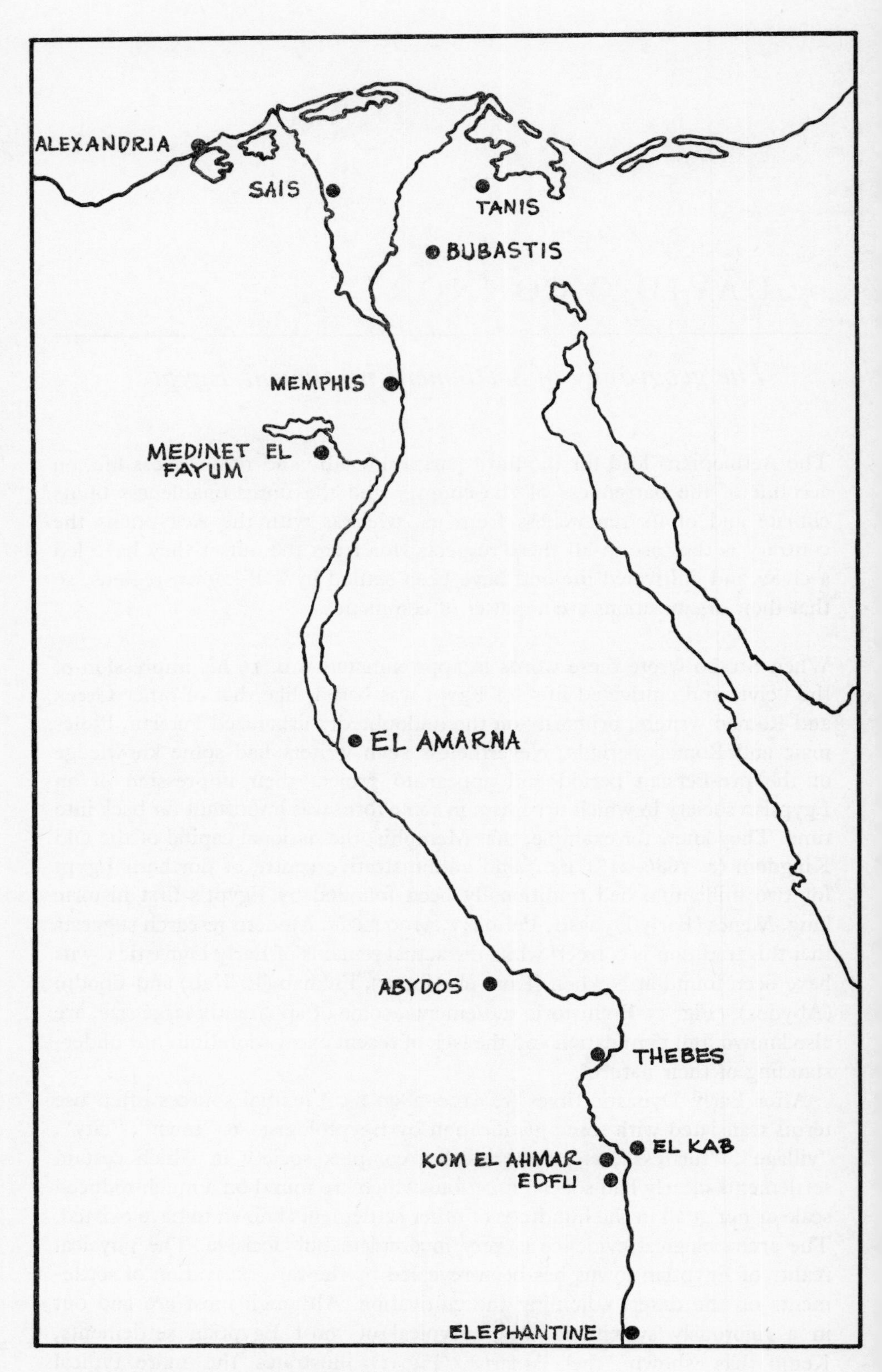

Fig 1 Egypt; places mentioned in the text.

accretion of a variety of residential types around already established administrative and religious centres. In one region the cemeteries of the local communities enable one to detect a substantial concentration of population in the vicinity of a known provincial capital at certain periods and other regions, if properly surveyed, would surely reveal similar patterns[8]. In addition, the tombs of provincial administrators from the later Old Kingdom on are not infrequently found concentrated near the known locations of important towns[9]. Finally, the sites of many and probably most settlements which textual sources show to have been important for administrative, economic or religious reasons were once marked by substantial town-mounds, now often largely destroyed (e.g. Bubastis, Medīnet el-Fayum and Ebodju [Abydos]) but sometimes still surviving in part (e.g. Tanis, Memphis) (Fig. 1)[10]. Such mounds have never been excavated on a large scale but in the exposed sides of the mound at Edjbo (Edfu; Fig. 1) it is possible to trace the transformation of an Old Kingdom town into a fortified town of the First Intermediate Period (*c.* 2181–2040 B.C.) and observe continuing occupation of the site into Christian times[11].

Therefore the definition once given of Egypt as "civilization without cities"[12] can only be accepted if "city" is understood in a most narrow and specialized sense; a more broadly defined type (or types) of urbanism was certainly characteristic of historic Egypt.

The location of the most important towns of the Middle (*c.* 2040–1674 B.C.) and New (*c.* 1567–1087 B.C.) Kingdoms can be fixed through textual and topographic evidence, at least as far as Upper Egypt is concerned[13]; the settlements of the broad Delta, cut by numerous waterways and exposed to a variety of external influences and pressures, would probably have formed distribution patterns different from those of the narrow southern valley but the poor evidence concerning the Delta prevents one from developing the comparison further. Even in Upper Egypt the archaeological evidence is inadequate for the plotting of the distribution of the numerous smaller and less important settlements. However, the pattern of settlement within one area of Upper Egypt at a particular moment in time is revealed by an analysis of the Wilbour Papyrus and it can be suggested that this pattern may have a wider geographical and chronological significance.

The most important towns of Egypt were of course the national capitals or administrative centres, of which only the aberrantly located Akhetaten (Tell el-ʿAmarna) can be studied in detail[14]. For nearly two millennia the chief centres were Memphis and Thebes, dominating respectively the northern and southern parts of Egypt; their histories and extant remains suggest that at times both Memphis and Thebes were probably large enough in area and population to be termed "cities", at least when compared with the other chief towns, and indeed the Egyptians often referred to Thebes simply as "the (southern) town" (an implied contrast with the major northern town, Memphis), further identification being thought unnecessary[15]. Later, in

	Nome Capital (Upper Egypt)	Modern Name	Karnak List Senwosret I (*c.* 1971–1928 B.C.)	Ramesseum Onomasticon (*c.* 1786–1633 B.C.)	Taxation List Rekhmire (*c.* 1504–1450 B.C.)
		Gez.Biggeh			Senmet
1	Yebu	Gez.Aswan	O	X	X(mayor)
(105.5)		Kom Ombo			Ombi (mayor)
2	Edjbo (Behdet)	Edfu	X	X	X (mayor)
(21)					
3	Nekhen	Kom el-Ahmar	O	O	X (mayor)
	Enkhab	El Kab	X	X	X ? (mayor ?)
		Komir			Pi-meru
		Esna			Inyet
					3 ?names lost
(78.6)		Gebelein			Pi-Hathor
		?			I-m-iotru
		Armant			Iwny
4	Weset	Luxor (Thebes)	X	O	End Southern Series
					Begin Northern Series
(45.5)		Kus			Gasy (mayor)
5	Gebtyu	Kift	X	X	X (mayor ?)
(22)					
6	Inu (En-t-entore)	Dendera	X	X	X (mayor ?)
(55)					
7	He-sekhem	Hu	O	X	X (mayor)
		?			He-waret Amenemhet (mayor)
(50)		?			Wah-sut-Kharkaure (mayor ?)
8	Ebodjdu	Abydos	X	X	X (mayor)
(53.5)		?			This (mayor)
9	Khant-Min (Ipu)	Akhmim	X	X	X (mayor ?)
(58.5)				ENDS	
10	Tjebu (Dju-ka)	Qau el-Kebir	O		X (mayor)
(44)		?			Pi-mut-neb-megeb (mayor)
11	Shashotep	Shutb	(Mikr)		X (mayor)
{ (?)					
(5.5) { 12	Pr 'Anty	?	O		O
13	Siyawti	Asyut	O		X (mayor)
(67.5)					ENDS
14	Kos	El Kusiyah	X		
(42)					
15	Khnum (Unu)	Ashmunein	?		
(37.5)		?			
16	Hebnu	Kom el-Ahmar	X		
(62.7)					
17	Hardai	Sheik Fadl	(Hnw)		
(8.8)					

Table 1 Town lists

Abydos List Ramses II (*c.* 1304–1237 B.C.)	Medīnet Habu List Ramses III (*c.* 1148–1166 B.C.)	Onomasticon Amenope (*c.* 1100 B.C.)	Known status of certain New Kingdom Towns (excluding Rekhmire list)	Edfu List Ptolemy IV (222–207 B.C.)
X	X	X	mayor	X
X		X		X
X	X	X	mayor	X
O	O	X	mayor	O
			mayor	
	End Southern Series			
X		O	mayor	X
X		X		X
X		X		X
	Begin Northern Series			
X	X	X	ww-district	X
X	X	X		X
X	X	X	kḥt-district	X
ENDS				
	O	X	mayor	X
	O	X		X
	(Dju-fyet)	X		X
	X	X		X
	O	X		X
	X	X		X
	LOST			
			Nefrusy:mayor	
		X		X
		X	mayor	(H-nesu)

	Nome Capital (Upper Egypt)	Modern Name	Karnak List Senwosret I (*c.* 1971–1928 B.C.)	Ramesseum Onomasticon (*c.* 1786–1633 B.C.)	Taxation List Rekhmire (*c.* 1504–1450 B.C.)
18	H-nesu	Kom el-Ahmar	X		
(34)		Sawaris			
19	Spermeru	?	(Wnsy)		
(30.8)		?			
20	Ninsu	Ihnasyah	O		
		Med. Ghurob			
		Nr. Med. Ghurob			
		Med. Fayum			
(39.6)					
22	Tpehu	Atfih	O		
(9.9)		Meydum			
21	Shena'-khen or Smen-Hor	Kafr Ammar	O		
(38.5)					
L.E. 1	Mennefer	Memphis			

Table 1 Town Lists (continued)

response to changed historical circumstances certain Delta towns achieved also the status of national administrative centres, namely Tanis, Bubastis, Sais and finally, under the Ptolemies, Alexandria (Fig. 1).

Next in importance were the capitals of the nomes or provinces into which Egypt was divided from the Old Kingdom onwards. The evidence for this importance is derived from a series of town lists (the relevant data from which are summarized in Table 1)[16] which, since they are not entirely unambiguous, need to be supplemented by other sources. The lists are all, in one way or another, products of the central administration and therefore of a high degree of reliability *if* the purpose of each can be properly understood. All were inscribed on temple walls with the exception of the two Onomastica, which include a town-list in a catalogue "of things arranged under their kinds". The Senwosret I and the Ptolemy IV (Edfu) lists give only the nome-capitals; most of these capitals occur also in the late Middle Kingdom and New Kingdom lists, but with a number of other towns which are not differentiated from them and which themselves often differ, at least in name, from one list to the other[17]. Two questions then must be considered:

1 Are these lists of nome-capitals, like some nome-lists[18], archaizing and no longer a reliable guide to the contemporary situation?
2 Are the other towns found in some lists identical in function and importance with the nome-capitals?

There seems no doubt, on the basis of contemporary documents, that the Ptolemaic nome-capitals did dominate the administrative, economic and

Abydos List Ramses II (*c.* 1304–1237 B.C.)	Medīnet Habu List Ramses III (*c.* 1148–1166 B.C.)	Onomasticon Amenope (*c.* 1100 B.C.)	Known Status of certain New Kingdom towns (excluding Rekhmire list)	Edfu List Ptolemy IV (222–207 B.C.)
		O		(H-boinu)
		X	mayor	X
			Keep of Onayna: mayor	
		X	mayor	X
			Miwer: mayor	
			She: mayor	
			Shede: mayor	
		X	mayor	X
			Mertum:mayor	
		O		X

religious activities of the nomes[19] and it seems likely that the, in many cases identical, capitals (Table 1) of the Senwosret I list had a similarly dominant position. The interest of early Dynasty XII kings such as Senwosret I in the reorganization of the administration of Egypt is well documented[20] and included in his list is factual data on the length of each nome and the height of the Nile inundation in various parts of Egypt which suggests that the towns given are not simply religious centres[21].

In the New Kingdom all the nome-capitals included in the incomplete Rekhmire list[22] (Table 1) are identified as centres for the collection of the annual taxes on cereals, animals and other products while it is striking that most of the known mayors (*h3ty-r*; a key figure in the provincial administration of the New Kingdom[23]) are associated with nome-capitals; others without known mayors have "rural Districts" (*ww* or kṛḥt) which are thought to be characteristic of nome capitals[24]. (Table 1: "Known status" column; most of the towns in the Rekhmire list also have mayors). Other towns, it is true, also collect taxes for the state and were ruled by mayors; but in our one detailed source on New Kingdom provincial administration, the Wilbour Papyrus, it is evident that the nome-capitals (in this case Ninsu and Hardai) are administratively and, in terms of the amounts of land owned or controlled by their temples and officials (Table 3 and Fig. 3a), economically much more important than any of the other towns of the nomes, even though some of these also have mayors[25]. In fact, some nearby nome-capitals appear to have become insignificant (18; H-nesu) or not yet developed (19; Spermeru, 21;

Shena'–khen)[26] and it must be realized that administrative changes could affect the importance of a nome-capital[27].

Nevertheless, it is clear that from the Middle Kingdom onwards (and probably in fact even earlier[28]) the settlement pattern of Upper Egypt was dominated by a network of major towns the functions of which were to exploit the agricultural and human resources of the country and to control its chief means of communication, the Nile[29], in the interests of a highly centralized government[30]. The distribution of these towns illustrates the essentially utilitarian role of most of them. The physical character of Upper Egypt is remarkably uniform[31], the chief variation being in the area and quality of cultivable land available; as Wilson pointed out this in ancient as in modern times probably affected both the absolute size and the density of population from area to area[32] (Fig. 2a). The distances by river between each nome-capital are in this context revealing (cf. kilometres in brackets, Table 1 and Fig. 2b, c). In the south, with its at times dense but numerically small population

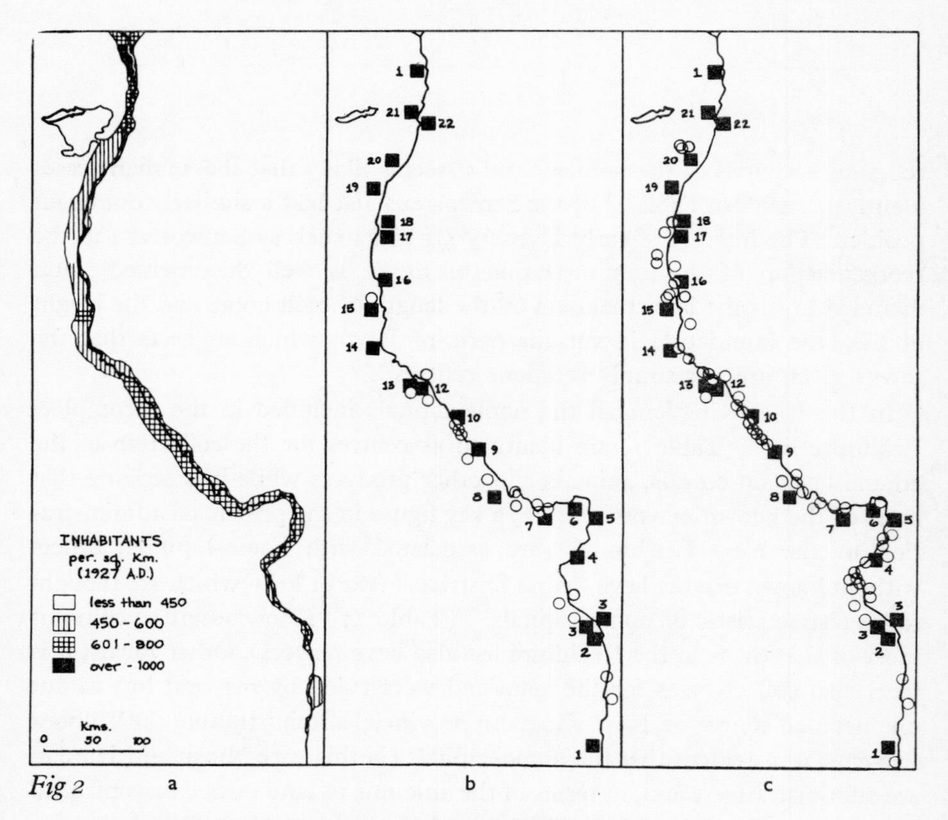

Fig 2

a. Distribution of population in Upper Egypt. A.D. 1927
b. ■ Upper Egyptian nome capitals.
○ Other towns in the Medinet Habu list.
c. ■ Upper Egyptian nome capitals.
○ Other towns in Onomasticon Amenope.

the capitals are far apart, the very ancient religious centres of Enkhab and Nekheb being an exception. In the more heavily populated area north of Thebes the nome-capitals are closer and regularly spaced; Gebtyu is anomalous because it controlled the mouth of the Wadi Hammamat, a source of stone and a major link with the Red Sea and the gold mines of the Eastern Desert, and the concentration of capitals near Siyawti may reflect the strategic and economic importance of this area[33]. North of Hardai the nome-capitals tend to be closer together but as we have seen above not all of them were necessarily active at the same time.

In times of stable, centralized government therefore the Middle and New Kingdom nome-capitals were the dominant towns of their regions and probably, like the Ptolemaic capitals "artificially regulated all economic enterprise into specific channels . . . [leaving] little chance for other settlements, possibly with better location, to succeed them[34]." In more disturbed periods the pattern may have been significantly different. It is striking (Table 2 and Fig. 2b, c) that while the number of important towns in Upper Egyptian nomes 1–6 remains stable throughout the New Kingdom, there is a marked increase in those of nomes 7–15 at the end of Dynasty XX[35].

	Ramesseum Onomasticon	Abydos list	Medīnet Habu list	Onomasticon Amenope
Nomes 1–6	16	15	7 (Nomes 4–6 not documented)	18
Nomes 7–15	5 (Nomes 10–15 not documented)	7 (11 ?)	19	27
Nomes 16–22	Lists no longer comparable			

Table 2 Number of important Towns in different nomes

At this point a period of political disintegration, accompanied by disturbances which included a virtual civil war in southern and northern Egypt[36], was beginning and it is possible that a larger number of towns came to enjoy a form of administrative independence as centralized control weakened and/or the population was beginning to concentrate into larger units for defensive purposes. The latter process can be traced archaeologically in one region of Upper Egypt at a slightly later period[37]. Such developments would be more likely to occur in the fertile and heavily populated area north of nome 6 rather than in the less densely populated area south of it (Fig. 2a), where there would be few towns capable of developing a new importance and the existing centres would presumably prove sufficient as refuges. The end result of this process can be seen in the inscription of Piankhy (*c.* 751–730 B.C.) describing his conquest of Egypt; parts at least of Upper Egypt and the Delta at this time were divided up amongst a number of petty rulers, each based on a heavily fortified town which in some cases was an old nome-capital and in some not[38]. Under Dynasty XXVI (*c.* 664–525 B.C.) however the importance of the nome-capitals was reasserted and they continued to dominate the settlement pattern until the administrative reforms of Diocletian in A.D. 307–310.

An analysis of the Wilbour Papyrus suggests what the effect of this administrative and economic dominance was on the general pattern of settlement. This document is a record of the measurement and assessment for tax purposes of hundreds of plots in Middle Egypt, carried out in year 4 of Ramses V (1156 B.C.)[39]. The papyrus falls into two parts. Text A concerns land owned or more rarely simply administered by a number of institutions in four separate but adjoining zones; the greater part of the land is owned by important temples outside of the region, but often administered by local officials, and most of the remainder belongs to local temples. Text B concerns royal land administered by a number of usually local officials and priests and covers a region which largely but not entirely coincides with that of A. This unique document gives, for each plot, its position (fixed by reference to some 400 named settlements), size and calculated yield, the name of the institution, priest or official who owned or administered it and (mainly in Text A) the name and occupation of its holder who was directly or indirectly responsible for its cultivation. These data are not only an important source of economic and social history but also enable one to plot the pattern of settlement and occupation types throughout the region. Certain limitations however must be noted. The opening of Text A has been lost so that the statistics used below cannot be completely accurate. Further, Texts A and B each deal with an extraordinarily small percentage (certainly less than 10%) of the probable amount of cultivable land available in the region in which the survey was carried out. Fairman has convincingly suggested that this is because A deals with *sharaki* and B with *gezirah* land[40]. *Sharaki* land, in the days of basin (instead of the modern perennial) irrigation, was restricted to the banks of the river and perhaps some other large water ways; *gezirah* land is, by definition, the river banks. We have therefore only a sample of the total settlements and population, but one which reveals patterns characteristic of the whole region.

Although Gardiner was able to establish the approximate limits of each zone and to fix on the ground the locations of the major local land-owning or administering temples and towns, as well as the approximate positions of the minor ones, a detailed topography of the *c.* 416 settlements used as survey points proved impossible. It is only possible to assign each settlement named in Text A, as well as many of those in Text B, to one of four zones. Helck's ingenious attempt to reconstruct the topography of the Wilbour Papyrus stretches the evidence too far, especially as it ignores Fairman's valuable suggestion that the settlements concerned were distributed along the river and perhaps also along other waterways[41].

It is, however, possible from an analysis of the settlement pattern of

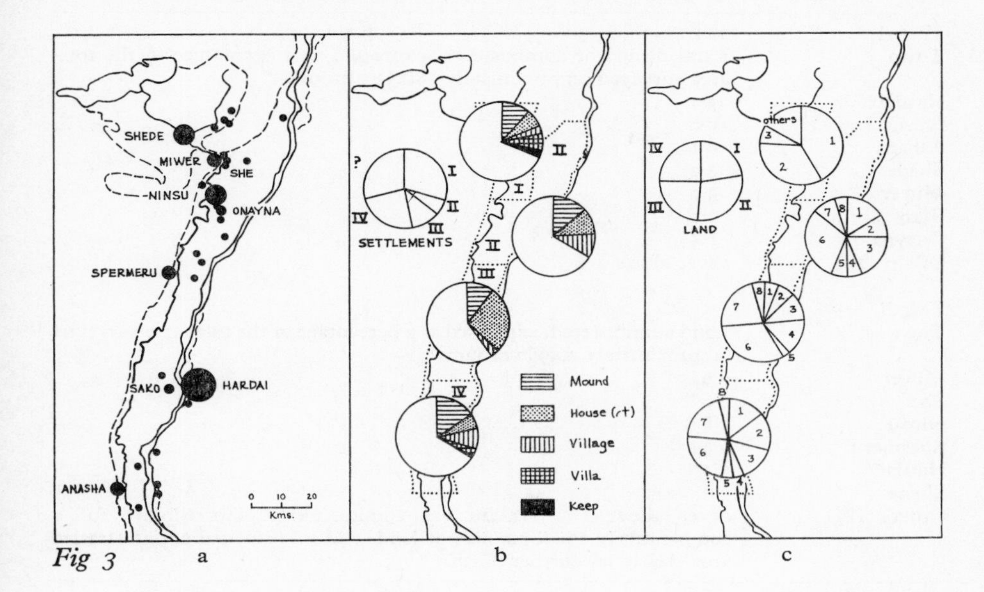

Fig 3 a b c

a. Wilbour A towns graded according to the amount of land they own or control.
b. Distribution of settlements and of certain settlement-name types.
c. Amount of land surveyed in each zone of Wilbour A. Distribution of types of occupation: 1. Cultivator. 2. Priest. 3. Lady. 4. Herdsman. 5. Scribe. 6. Stable-master. 7. Soldier. 8. Sherden (foreign mercenary).

Wilbour to establish the relative importance of the major towns of the region. Hardai, Ninsu and Tpehu are clearly referred to as major administrative centres in the papyrus and Gardiner's analysis of this and other evidence suggests that at this time Hardai was nome capital of Upper Egyptian nome 17 and Ninsu and Tpehu of 20 and 22 respectively. Nomes 18, 19 and 21 appear to have had no administrative importance and some may not have existed as separate entities. Wilbour also shows that Shede, Miwer, She, Onayna, and Spermeru were important since each, like Hardai, Ninsu and Tpehu, had a mayor (*ḥ3ty*-r), a key figure in the provincial administration of the New Kingdom[42].

Further information about the relative importance of the regions' larger towns can be gained by calculating how much *sharaki* and *gezirah* land each town (as represented by its temples, priests and officials) owned or administered (see Table 3).

In general therefore Texts A and B show that the administratively and economically important towns are concentrated mainly in zones I and IV; Text A further, since it reflects the actual land owned by the towns (apart from Hardai and Ninsu, which are often named in an administrative as well as an owning capacity), gives us a clearer idea of their relative importance (Fig. 3a).

Text A

Town	Land owned or administered, expressed as a percentage of the total area surveyed (approximately 28,255·7 arouras)[a]
Hardai	31·7
Ninsu	9·6
Ashana	6·1
Shede	4·2
Miwer	3·2
Sako	2·1
Onayna	1·5
Others (29)	never above 1%

Text B

Town	Land administered, expressed as a percentage of the total area surveyed (approximately 20,987 arouras)
Miwer	6·9
She	6·6
Ninsu	5·8
Spermeru	6·4
Hardai	3·1
Shede	3·0
Others (18)	never above 1% (Tpehu has administrative responsibility for a comparatively small amount of land, and its principal administrative area clearly lay further north)

[a] The aroura = 2735 m²

Table 3. Ownership and administration of land

In studying the general distribution of the, approximately, 416 smaller settlements named as survey points and never as land-owners or administrators the following data is significant:

Distribution of settlements in the zones (Fig. 3b)

Zone I	123	(29·6%)
Zone II	22	(5·3%)
Zone III	46	(11·1%)
Zone IV	102	(24·5%)
?	123	(29·5%)

Amount of land surveyed in each zone (Text A; the same calculation cannot be made for Text B) (Fig. 3c)

Zone	Arouras	Percentage of total
I	6232 (+)	22·05
II	8368·06	29·67
III	6612·08	23·40
IV	7025·52	24·86
Total	28,255·66	

These figures suggest that the densest settlement was in zones I and IV. However, it must be noted that 29·5% of the settlements cannot be assigned to a zone, but it may be safely presumed that these settlements were roughly divided amongst all four zones. Approximately 50% of the settlements can be assigned to zones on the basis of Text A alone, and since the four zones in Text A all contain roughly the same amount of surveyed land, the distribution pattern of the 50% seems certain. However, while many of the Text B settlements can be assigned to one of the zones, it is less easy to control the distribution of the surveyed land of B, and it is possible that there *may* here be a source of distortion. Finally, although the settlements are probably confined to a narrow strip of land running along the river and perhaps other waterways, they are probably a fair indication of the settlement density within the unsurveyed hinterland; the high ground along the river was a favourite place of settlement, in order to mitigate the effect of the inundation[43].

The smaller settlements of the Wilbour Papyrus bear a variety of names, amongst which are a number compounded with the words for "(ancient) mound" (*i3t*), "house" (*ꜥt*), "village" (*wḥyt*), "villa" (*bḫn*; "villa" is preferable to Gardiner's "castle" as a description of these large estates) and "keep" or "fortress" (*p3 sg*(3))[44]. It is of course true that some of these names, while related to original and earlier functions or status of settlements, may not reflect their contemporary functions or status, but analysis does yield suggestive results (Table 4; Fig. 3b).

Taking into account the relative numbers of settlements in each zone, ancient "mounds" (i.e. settlements built on earlier sites) and "villages" seem evenly distributed, while the greater importance of zones I and IV is perhaps indicated by the comparatively high numbers of "villas" (estates of noblemen) and "keeps" (fortresses) that can be with certainty assigned to them. Most interesting, however, is the distribution of the name "house (rarely: houses) of so-and-so"; the word employed (r*t*) is not an old one and Gardiner believes that it was applied literally to existing individual houses, presumably standing alone[45]. "Houses" are a low percentage of the total settlements in zones I and IV, a higher one in II and markedly higher in III, suggesting, as did the general settlement pattern, a less dense settlement pattern in zones II and III.

Finally, it is important to examine the occupations of the majority of land "holders" (sometimes called "agents" or "cultivating agents")[46] who were responsible for the cultivation of the land and were in most cases inhabitants of the region. Since nearly all of them are found in Text A, we can be sure of having directly comparable samples from each zone (each of which represents roughly identical amounts of surveyed land). The following occupations are by far the commonest: "cultivator", "priest", "lady" (citizeness), "herdsman", "scribe", "stable-master", and "Sherden"[47]. Excluding other, much smaller categories of holders, these occupations represent 140 individuals in zone I, and 426, 629 and 376 in zones II, III and IV respectively[48]. The distribution of occupation types is as follows (see Table 5 and Fig. 3c):

Even allowing for the fact that the record for part of zone I is lost, a fairly clear-cut distribution pattern emerges. "Cultivators" and "priests" dominate in zone I and occur in significantly large numbers together with "ladies" in zones II and IV; all three however are markedly low in III. "Stable-masters"

Type	No.	I No.		I, II III	I–IV	II No.		II ?	II–III IV	III No.		III–IV IV	IV No.	
Mound	51	14	11·4*	2	1	3	13·6*	2	1	5	10·9*	1	16	15·7*
House(s)	37	8	6·5		1	2	9·1	2		12	26·1		6	5·9
Village	29	4	3·3	3	2	2	9·1	7		1	2·2		8	7·8
Villa	17	8	6·5	1				1	2				2	2
Keep	7	3	2·4					1	1				1	1

* Percentage of total settlements in I, II, III and IV respectively.

Table 4. Nomes and functions.

	I	% of total individuals in I	II	% of total individuals in II	III	% of total individuals in III	IV	% of total individuals in IV	Text B zone uncertain
"Cultivator"	58	41·4	65	15·3	32	5·1	50	13·3	
"priest"	48	34·3	37	8·7	46	7·3	56	14·9	
"lady"	10	7·1	83	19·5	66	10·5	61	16·2	
"herdsman"	7	5	24	5·6	94	4·9	15	4	1
"scribe"	3	2·1	27	6·3	16	2·5	20	5·3	4
"stable-master"	6	4·3	131	30·8	213	33·9	91	24·2	1
"soldier"	2	1·4	41	9·6	132	21	74	19·7	
"Sherden"	6	4·3	18	4·2	30	4·8	9	2·4	24

Table 5. Occupations.

are predominant in II, III and IV, "soldiers" occur frequently only in III and IV and "herdsmen" are common only in III.

Gardiner had already pointed out that three of the settlement names in Wilbour suggested the presence of military settlements in the general region[49] (one, "Village of the Soldiers" is in zone II and another, "Village of the Army" in III) and the predominance of soldiers in III and IV clearly indicates a major military settlement in this area. This then is an unexpectedly concrete illustration of the securely attested but poorly documented existence of New Kingdom settlements for veterans and their families[50].

It is also significant that while sedentary occupations ("cultivators", "priests", "ladies" and (settled) "soldiers") taken together predominate in zones I, II and IV, "stable-masters" and "herdsmen", both concerned primarily with the pasturing of animals, are predominant in zone III, with "stable-masters" also well represented in zones II and IV. This certainly suggests more open land, and hence a lesser settlement density in zone III and parts of zones II and IV.

The evidence then reveals two dense concentrations of, predominantly, agriculturalists around the nome-capitals of Ninsu and Hardai; between these concentrations lies a region with a smaller, more scattered population concerned mainly with the pasturing of animals. There are no known changes in soil, vegetation, etc., that could explain this distribution and it may well have been the result of deliberate policy. The close control of the national government over agriculture and grazing is well documented; its aim was not only to ensure an adequate food supply for the local population but also to produce a surplus, collected annually in the form of taxes, which was at the disposal of the state. Clearly it would be more convenient to have the bulkier produce such as cereals produced near main collection centres such as the nome-capitals while the more easily moved animals were pastured some distance away. A concentration of population around administrative centres would also facilitate the control of human resources, either for the purposes of forced labour, or in the case of military settlements, of raising troops rapidly. The preferences of individuals are also likely to have led to concentrations around towns for as Smith and Kemp have emphasized[51] private commercial exchanges, mainly of produce and manufactured goods, played an important if subordinate role in the Egyptian economy. Town dwellers owned country estates (*bḫn*) which presumably they would prefer to have close to the town, while the inhabitants of the countryside must have been well aware of the markets open to them in the towns. Towns had further attractions for individuals; ready access to the courts and officials who settled disputes and registered wills, agreements and other documents and to the chief temple of the nome. Temples were not only themselves economically significant but, as Professor Smith brings out[52], provided a focus for important desires and activities which were not, in our terms, "utilitarian".

Given the uniformity of physical and cultural conditions in the New

Kingdom the pattern of settlement revealed by the Wilbour Papyrus, and the explanations suggested for it, may well be typical for most of Upper Egypt.

Notes

1 $\pi^{o}\gamma$itik$_{\hat{\omega}}$s.
2 Jones, H. L. (trans.) (1949). *The Geography of Strabo*. Cambridge, Mass. 17.1.3.
3 Edwards, I. E. S. (1964). The Early Dynastic Period in Egypt, *Cambridge Ancient History*, I (rev. ed.) Cambridge. p. 10.
4 Edwards, I. E. S. (1964). *op. cit.* pp. 10–11.
5 El Kab: Capart, J. (1937). Les fouilles d'el Kab, *Chron. Egypt.* **24**, pp. 133–4. *Fouilles de El Kab. Documents. Livraison III* (1954). Bruxelles; Hieraconpolis: Weeks, K. (1969). Excavations at Hieraconpolis, 1969, *American Research Centre in Egypt, Inc. Newsletter*, **70**, pp. 2–3; Abydos: Petrie, F. (1902). *Abydos I*. London. pp. 9–26.
6 Butzer, K. (1960). Archaeology and geology in ancient Egypt: Geomorphological analysis permits reconstruction of the geography of prehistoric settlement, *Science*, **132**, pp. 1617–24.
7 Kemp, B. J., this volume, especially note 19, p. 677.
8 O'Connor, D. (in press). A regional population in Egypt: prehistoric times to *c.* 500 B.C., *in* Spooner, B. (ed.) *Population Growth: anthropological implications*.
9 See, for example, Porter, B. and Moss, R. (1934–7). *Topographical Bibliography of Ancient Egyptian Hieroglyphic Texts, Reliefs and Paintings*, Oxford. IV. pp. 134–8, 148–60 (U[pper] E[gyptian] nome 16); 177–82 (U.E. 15); 247–54 (U.E. 14); 242–4 (U.E. 12); 261–4 (U.E. 13); V. pp. 7–14 (U.E. 10); 19 (U.E. 9); 177 (U.E. 3); 200–1 (U.E. 2); 231 ff (U.E. 1); Fischer, H. (1968). *Dendera in the Third Millennium B.C.* New York, p. 187. (U.E. 6).
10 These mounds are sometimes illustrated in excavation reports (e.g. on Edfu; see note 11) and I examined a number personally during a survey trip in 1965–6.
11 Bruyère, B. *et al.* (1937). *Fouilles Franco-Polonaises rapports I. Tell Edfou 1937*. Cairo. p. 2–23, 59–99; Michalowski, K. *et al.* (1938). *Fouilles Franco-Polonaises rapports II. Tell Edfou* 1938. Cairo. pp. 1–29.
12 Wilson, J. (1960). Egypt through the New Kingdom, civilization without cities, *in* Kraeling, C. and Adams, R. Mc. (eds.). *City Invincible*. Chicago, p. 124.
13 Gardiner, A. (1947). *Ancient Egyptian Onomastica*. Oxford; Montet, P. (1957). *Geographie de l'Egypte ancienne*, I. Paris; (1961). *Geographie de l'Egypte ancienne* II. Paris.
14 Kemp, B. J., this volume, *ibid.*
15 Sethe, K. (1907). Die Namen von Ober-und Unterägypten und die Bezeichnungen für Nord und Süd, *Z. fur Ägypt. Sprache*, **44**, pp. 5–6.
16 Gardiner, A. (1947). *op. cit.* I, pp. 40–62; Montet, P. (1957). *op. cit.* pp. 11–22. Most of the data used for Table I are found in Gardiner, A. (1947). *op. cit.* III, pls. XXIV–XXVI. For the Senwosret I list see Lacau, P. and Chevrier, H. (1956). *Une chapelle de Sesostris Ier a Karnak*. Cairo, pp. 220–30. The data on "known status of certain New Kingdom towns" will be found in Helck, W. (1958). *Zur Verwaltung des mittleren und neuen Reichs* (*Probleme der Ägyptologie*, 3). Leiden-Köln. p. 224. For the Ptolemaic nome-capitals see also, Butzer, K. (1960). Remarks on the geography

of settlement in the Nile valley during Hellenistic times, *Bull. de la Soc. de Geogr. d'Egypte*, **23**, p. 12.

17 The reason for including these other towns is only clear in the Rekhmire list (Table I), where they are identified as tax collecting centres; the towns given in the other lists are therefore not included in Table I.

18 Y(oyotte), J. (1959). Nome, *in* Posener, G. *Dictionary of Egyptian Civilization*. New York. p. 191.

19 Butzer, K. (1960). *op. cit.* pp. 5–17.

20 Hayes, W. (1961). The Middle Kingdom in Egypt, *Cambridge Ancient History*, I. (rev. ed.) Cambridge. p. 35.

21 The tombs of Middle Kingdom provincial governors (*h3ty-ꜥ*) have been found near several of the nome-capitals named in the Senwosret list. See note 9.

22 Helck suggests that the Rekhmire list is actually an archaizing copy of a late Middle Kingdom original; Helck, W. (1958). *op. cit.* p. 214. However, while the Rekhmire material in general is derived in part from earlier sources, it is also regarded as a valid source of information on the New Kingdom; Hayes, W. (1962). Egypt: internal affairs from Tuthmosis I to the death of Amenophis III, part I, *Cambridge Ancient History II* (rev. ed.). Cambridge. p. 45. I know of no reason why the tax collecting centres cited in Rekhmire should not have been functioning as such in his lifetime.

23 On the role of the mayors; Helck, W. (1958). *op. cit.* pp. 223–6; Hayes, W. (1962). *op. cit.* p. 47.

24 Helck, W. (1958). *op. cit.* p. 226.

25 Gardiner, A. (1948). *The Wilbour Papyrus*, II. pp. 18, 39–41, 49–55, 162.

26 Gardiner, A. (1948). *op. cit.* II, p. 54.

27 Gardiner stresses the likelihood of such developments; Gardiner, A. (1947). *op. cit.* II, p. 57*.

28 On the Old Kingdom (*c.* 2686–2181 B.C.) provincial administration and the development of the nomes see Helck, W. (1958). *op. cit.* pp. 194–203.

29 The comparative unimportance of land transport in Egypt is shown by the fact that wheeled vehicles were rarely if ever used to transport produce or goods and by the many references to officials carrying out their duties by means of boats.

30 On the centralization of authority in the New Kingdom especially see Hayes, W. (1962). *op. cit.* parts 1 and 2; Edgerton, W. (1947). The government and the governed in the Egyptian empire, *J. Near East. Stud.*, **6**, pp. 152–60.

31 Butzer has argued persuasively that there was a significant rise in the level of the annual Nile inundation during the New Kingdom; Butzer, K. (1959). Studien zum vor-und frühgeschichtlichen Landschaftswandel der Sahara III. Die Naturlandschaft Ägyptens während der Vorgeschichte und der Dynastischen Zeit, being *Abh. Akad. Wiss. u. Lit. Mainz, Math.-naturwiss.* Kl. jahr. 1959. nr. 2. Weisbaden. pp. 113–4. The effects of this however would have been uniformly experienced throughout Upper Egypt.

32 Wilson, J. (1955). Buto and Hierakonpolis in the geography of Egypt, *J. Near East. Stud.*, **44**, pp. 209–31. Fig. 2a is based on Simons, P. (1968). Die Entwicklung des anbaus und der Verbreitung der Nutzpflanzen in der ägyptischen Nilstromodse von 1800 bis zur Gegenwart, *Kölner Geograpische Arbeiten* Heft 20. Weisbaden. Karte 8. The ancient population would not of course have been so dense, but the general pattern of density distribution may be assumed to have been similar.

33 Kees, H. (1961). *Ancient Egypt. A Cultural Topography*. London. pp. 99–100, 105.

34 Butzer, K. (1960). *op. cit.* p. 11.

35 It is assumed that the inclusion of these towns in the New Kingdom lists cited in Table I means that they were considered important; perhaps, like the towns of the Rekhmire lists and those with mayors in the Wilbour Papyrus, they were administrative centres of less importance than the nome-capitals themselves.
36 Černý, J. (1965). Egypt from the death of Ramses III to the end of the twenty-first dynasty, *Cambridge Ancient History*, III (rev. ed.). Cambridge. p. 23 f.
37 O'Connor, D. (in press). *ibid.*
38 Gardiner, A. (1961). *Egypt of the Pharaohs.* Oxford. pp. 335–40; Helck, W. (1958). *op. cit.* pp. 228–30.
39 Gardiner, A. (1948). *op. cit.* II, III; Faulkner, O. (1952). *The Wilbour Papyrus*, IV, Indices. Oxford.
40 Fairman, H. (1953). Review of Gardiner, A. (1948). *op. cit.* and Faulkner, O. (1952). *op. cit. in J. Egypt. Archaeol.*, **39**, pp. 119, 122.
41 Gardiner, A. (1948). *op. cit.* II, pp. 36–55, 173–8; Helck, O. (1961). Materialen zur Wirtschaftsgeschichte des Neuen Reiches II, being *Abh. Akad. Wiss. u. Lit. Mainz, Geistes-Soz. kl.* **11**, pp. 298–324; Fairman, H. (1953). *op. cit.* p. 119.
42 Gardiner, A. (1948). *op. cit.* II, pp. 18, 39–41, 49–55, 162.
43 Butzer, K. (1965). Physical conditions in eastern Europe, western Asia and Egypt before the period of agricultural and urban settlement, *Cambridge Ancient History*, I, (rev. ed.) Cambridge. p. 34.
44 Gardiner, A. (1948). *op. cit.* II, pp. 32–5.
45 Gardiner, A. (1948). *op. cit.* II, p. 34.
46 Gardiner, A. (1948). *op. cit.* II, pp. 75–79.
47 Gardiner, A. (1948). *op. cit.* II, pp. 79–84.
48 It is true that some of the commoner personal names may have referred to the same person, but this seems to have happened comparatively rarely.
49 Gardiner, A. (1948). *op. cit.* II, p. 33.
50 Schulman, A. (1964). Military rank, title and organization in the Egyptian New Kingdom, *Münchner Ägyptol. Studien*, **6**, pp. 54–55.
51 Kemp, B. J., this volume; *ibid.*; Smith, H. S., this volume, p. 717.
52 Smith, H. S., this volume, *ibid.*

J. D. RAY

The House of Osorapis

The first Greek travellers found in Egypt a land of wonders, and two things most attracted their attention: the worship of animals and the elaborate care of the dead[1]. The House of Osorapis was not a house but a temple, and Osorapis (a contraction of the two names Osiris and the sacred bull Apis) was not a man, but a dead animal. The temple in question stood in the desert to the west of Memphis, over the subterranean galleries where the mummified bulls were buried. The Greeks, who identified Osiris-Apis with the god Sarapis[2], called the sacred place τὸ πρὸς Μέμφει μέγα Σαραπιεῖον, "The Great Serapeum which is near Memphis[3]." There were other Serapea, notably in Alexandria and at sites on the Greek mainland, but in the Ptolemaic period the Memphite temple remained the Serapeum *par excellence*. At the beginning of the nineteenth century native diggers discovered somewhere among its ruins a large number of papyri, which are now divided among various collections. The Greek papyri were collated and published by Wilcken; their demotic counterparts are scattered through various publications, or not published at all.

The name "House of Osorapis" was extended to cover the entire northern stretch of the Memphite necropolis from the Serapeum proper across to the Anubieion, another temple area on the edge of the cultivation. This is a distance of well over a kilometre[4]. Between the two ran the great dromos lined with sphinxes, which acted both as a processional way and as a central road for the adjoining houses. On the west, the town extended to a desert wadi, which provided another approach to the dead Apis (see Fig. 1).

That the settlement was named after the temple is significant. It was devoted to the burial, maintenance, and worship of the sacred bulls in their underground sepulchre[5]. Above ground stood the temple, similar to the mortuary temples of dead kings, in which offerings were made to the god, and a daily ritual enacted around his image. In addition to a regular priesthood, many other persons were attached to the cult, as acolytes, mourners, watchmen, and the like. In the Hellenistic period we hear also of κάτοχοι, people under a restraint, to whom the god in some way had indicated that he had need of their service[6]. At an unknown date, but certainly earlier than the

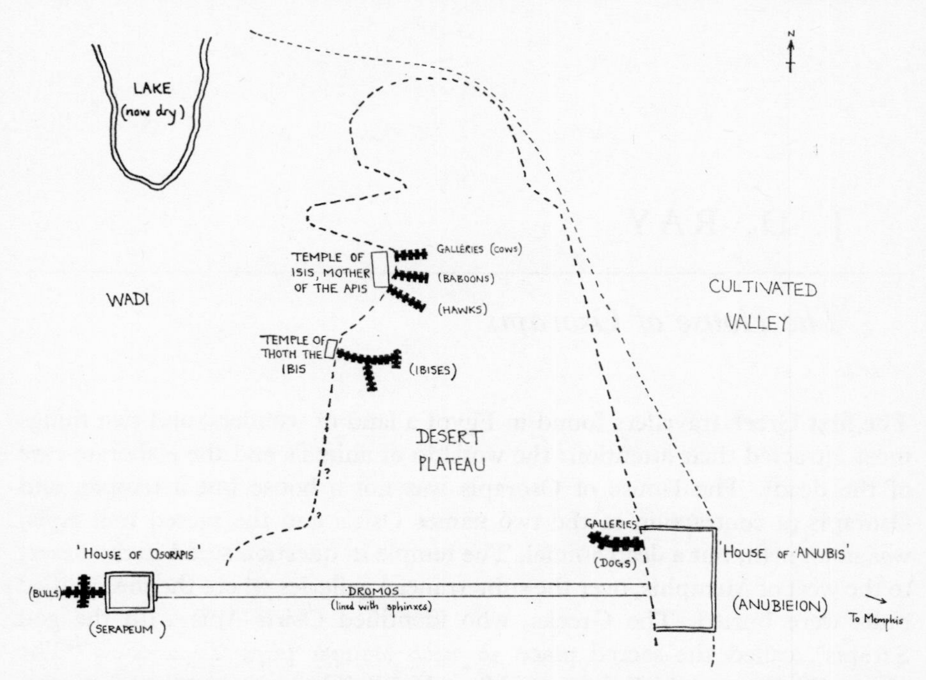

Fig 1 This sketch is approximate only, and the galleries in particular have been simplified. No attempt has been made to show features whose location is doubtful, such as the "great street" and the sacred rams. The length of the dromos is 600 m approx.

Persian period[7], the Mother of the Apis received her own cult and hypogeum to the north of her offspring. This has recently been unearthed by the Egypt Exploration Society[8].

Many other cults followed. A list of these is given by Guilmot[9], where they are located within the temenos of Osorapis itself. However, the cult-centre of Thoth the Ibis can now be located well to the north of the main temple, and the same may well be true of the "Compound (*sbtt*) of Isis" described in two demotic papyri now in New York[10]. Both of these are said to be "in the house of Osorapis", but this may simply mean that they were within the confines of the town. It is tempting to identify this "compound of Isis" with the brick walls at the edge of the wadi, recently excavated by Emery, which have been found to adjoin the galleries of the Mothers of the Apis[11]. The other compound of the papyri, dedicated to Horus[12], would then be the enclosure to the south of this which faces the hawk- and ibis-catacombs.

Another animal which we should place in this area is the mysterious *Gm*. This too had a resting-place in the Serapeum[13] and on the basis of a hieroglyphic sign was thought to have been a kind of ibis. Reich, on philological grounds, identified it as the statue of a bull, taking one of its final signs as the determinative for "Silver"; however, it may just as easily be the sign denoting an animal. Apart from the divine termination, it seems to be the same as a word in Pap. Berlin 15831, recently published by Zauzich and there translated "calf"[14]. Whatever links existed between the *Gm*-calf and the "Children of the Apis", whose cult was also established in the region, remain unknown.

Enough has been said to show the remarkable nature of this settlement. Around the cemeteries of the Apis and its associated cult-animals grew up communities of priests, and accompanying these were the quarters of the men employed to maintain these cults; stone-masons to cut the galleries and fashion the sarcophagi[15], sculptors, scribes, porters, watchmen, makers of temple equipment, feeders of ibises, people to embalm dead feeders of ibises, and relatives. Only the embalming of the Apis itself was carried out elsewhere, by the temple in the valley where the bull had spent its life.

We have no way of tracing the history of these growing settlements, and we do not know to what extent the central government exercised planning, or allowed buildings to spread haphazardly between the religious complexes. In this second case, the "House of Osorapis" would have resembled el Amarna in the eighteenth dynasty; but perhaps a closer parallel can be seen in the workmen's villages at Kahun and Deir el-Medina[16]. Certainly the dromos seems to have been used as a main street and business-area[17], and the demotic house-documents, beside describing a *great street* which intersected this at right-angles, are normally careful to list the neighbouring houses. This gives an impression of some regularity, even if this must have tailed off into narrow lanes and alleyways. The "House of Osorapis" therefore probably resembled the contemporary town at Jeme in western Thebes, or indeed most Egyptian towns of the day, and owed little or nothing to Greek ideas of town-planning. Unfortunately, hardly anything remains of the houses themselves, which were made of perishable mud-brick.

The papyri allow several glimpses of the central authority; Ptolemaic kings, on their visits to the Serapeum, seem to have had some kind of residence, with a window in which they received petitions[20]. The Anubieion (a similar complex at the end of the dromos which Mariette was even led to term the "Greek Serapeum") also had its prison[21]. But like most Egyptian temples under the Ptolemies, the Serapeum was left to organize its own industries and estates; even an "Orchard of Osorapis" bloomed in this desert place[22].

The very existence of such a city, devoted to its animal-cults, might seem a little vicarious, if Osiris-Apis and the "gods who rest with him" had not rapidly developed into one of the leading oracles of Egypt and the whole Greek world. Pilgrims, increasing with the fame of the site, came to put their questions to the god, spend their money, and depart; hence the guest-houses (καταλύματα) of the Serapeum, an institution paralleled in Greece at Olympia, Epidaurus, and elsewhere. There were shops to cater for the needs of pilgrims, and the papyri acquaint us with a large number of highly specialized occupations[23]. We find butchers, bakers, inn-keepers, launderers, porters, and water-carriers[24]. The existence of so many traders suggests that pilgrims had to provide their own means of support while seeking the god's advice. As in Greece, consideration may have been given to more needy cases.

Thus the Serapeum possessed a "floating population", which during the great festivals must have increased markedly. The funeral of the Apis, the installation of his successor, and the many calendar feasts, must all have brought new visitors. Any attempt to estimate the size of this population, however, must remain conjecture.

Most of the industry naturally concentrated on the cults themselves. A visitor coming from Memphis might pass the lake, journey along the *wadi* and approach the temple of the Mothers of the Apis. Perhaps he might find lodgings nearby, while petitioning the goddess. Continuing towards the Serapeum itself, he would reach the temple of the Ibis, dedicate one of the birds, which, elaborately bandaged and mummified, lined the galleries in their millions, and purchase a bronze statue to be offered in his name as an *ex-voto*. Arriving at the Sanctuary of Osiris-Apis, he might choose to sleep there in the hope of seeing the god in his dreams[25]. Professional priests were also in the habit of consulting the god in this way, and many acted as the interpreters of other people's dream-experiences. The best known of these, the κάτοχος Ptolemaeus, recorded dreams in both Greek and Egyptian. The experiences of both Ptolemaeus and his younger brother Apollonius can be read in Wilcken, and much information is contained in Reich's inimitable article on the quarrel between Apollonius and the sons of Djeḥebes[26]. Among such a heterogeneous population, often living in cramped and overcrowded conditions, such quarrels must have been frequent. Friction cannot have been alleviated by the presence of a large number of "unofficial" persons—private dream-interpreters[27], conjurers, musicians, writers of petitions, exorcists, astrologers, ecstatics. The status of these people remains a little uncertain; it seems probable that a licence, written or verbal, would have to be obtained from the temple, but that the State, as often in ancient Egypt, would not intervene provided that the expected quota of revenue was forthcoming. The Serapeum anyway was a second home for magicians; had not Nectanebo himself, the arch-magician and the last native Pharaoh, flown over the sea by night to become the father of the conquering Alexander[28]? He built the temple of the Apis, also that of his mother, and his silver statues were still to be seen there[29]. In such a place, the traditions of Egypt and Greece must have found inspiration in each other. Apollonius copies out Euripides while suffering the onslaughts of the donkey-boys; Hermes Trismegistos gives his oracles alongside Thoth the Ibis[30].

Many questions remain unanswered. One would like to know, for instance, how such a settlement fitted into the Ptolemaic monopoly system. Temple land (γῆ ἱερατική) formed a class apart, and its industries were largely exempt from the usual assessments[31]. Yet this was a settlement whose very fame attracted a surplus of wealth. Another handicap is our ignorance of the internal organization of the town. It seems to be true that embalming and its associated industries furnish the clearest examples of "private enterprise" in ancient Egypt, and we may expect the Serapeum to shed more light upon this; perhaps publication of more of the demotic material will help. One interesting feature can be seen from the many graffiti left by priests and workmen in the various galleries belonging to the animal cults. Servants of the Ibis tend to have names comprising that of the god Thoth, those of the Mothers often bear the name of the divine cow. Fathers, sons and grandsons appear constantly, and even the mothers of these men bear the same gods in their own names. Such functions were family concerns, and a detailed study might show much about inheritance, exclusiveness, specialization, and family relationships in Hellenistic Egypt. Yet these graffiti have never been published.

The Serapeum is a unique example of urbanization, whose roots lie amid the deepest religious ideas of the Egyptians. That these ideas were considered sufficient foundation for a city is significant in itself, a remarkable affirmation of the continuity of living things; *mortua aetas nos peperit morituros.*

Notes

1 Herodotus, *Histories*, **II**, 38 (Apis), 65–76 (sacred animals), 85 (the dead).
2 For the complex question of the origins and nature of Sarapis, cf. Sethe, K. (1913). *Sarapis*. Göttingen. Greek papyrological evidence in Wilcken, U. (1922). *Urkunden der Ptolemäerzeit*. Berlin-Leipzig.
3 Reich, N. J. (1933). New Documents from the Serapeum of Memphis, *Mizraim*, **I**.
4 For the topography cf. Guilmot, M. (1962). Le Sarapieion de Memphis. Etude Topographique, *Chron. Egypt.*, **37**; Mariette, A. and Maspéro, G. (1882). *Le Sérapéum de Memphis*. Paris.
5 Otto, E. (1938). *Beiträge zur Geschichte der Stierkulte in Ägypten*. Leipzig.
6 Sethe, K. (1913). *ibid.*; Wilcken, U. (1922) *op. cit.* **I**, p. 62 f.
7 The earliest date yet known is 513 B.C. on an unpublished ostracon from the site.
8 Forthcoming preliminary report in *J. Egypt. Archaeol.*
9 Guilmot, M. (1962). *op. cit.* p. 366.
10 Reich, N. J. (1933). *op. cit.* p. 63.
11 Emery, W. B. (1969). Preliminary report on the excavations at north Saqqara, 1968, *J. Egypt Archaeol.*, **55**: and Emery, W. B. (1970). Preliminary report on the excavations at north Saqqara, 1969, *J. Egypt. Archaeol.*, **56**.
12 Pap, B. M. 10236; Reich, N. J. (1933). *op. cit.*, p. 65.
13 Reich, N. J. (1933). *op. cit.*, p. 83; Guilmot, M. (1962). *op. cit.*, p. 366.
14 Zauzich, K. Th. (1969). Ein Kaufvertrag aus der Zeit des Nektanebos, *M.D.A.I.K.*, **25**, esp. note p. 226.
15 cf. the "Serapeum Stelae"; Vercoutter, J. (1962). *Textes Biographiques*. Paris; Bresciani, E. (1967). Nuovi testi demotici dal Serapeo di Menfi, *Studi in Honore di G. Botti*. Rome.
16 cf. Fairman, H. W. (1949). Town planning in Pharaonic Egypt, *Town Planning Review*, **20**.
17 Reich, N. J. (1933). *op. cit.*, p. 38.

18 Guilmot, M. (1962)., *op. cit.*, p. 374.
19 el-Amir, M. (1959). *A Family Archive from Thebes*. Cairo.
20 cf. esp. Pap, Gr. Vatican 2303—Wilcken, U. (1922). *op. cit.*, no. 15.
21 Wilcken, U. (1922). *op. cit.*, I, pp. 13–15.
22 Spiegelberg, W. (1909). *Demotische Papyri . . . du Cinquantenaire*. Brussels. Pap. no. 3, p. 10.
23 Guilmot, M. (1962). *op. cit.*, p. 372.
24 Water, as at Deir el-Medina, had to be brought some distance across the desert. Perhaps the lake, now drained, but marked on old maps at the mouth of the wadi, is the λίμνη of the papyri. See Mariette, A. and Maspéro G. (1882). *op. cit.*, maps at end of text volume.
25 For the subject of incubation, cf. Wilcken, U. (1922). *op. cit.*, pp. 31f.
26 Reich, N. J. (1933). *op. cit.*, p. 147.
27 e.g. the stela of a Cretan dream-interpreter published in Roeder, G. (1960). *Kulte und Orakel*. Zürich-Stuttgart. pl. 14.
28 Pseudo-Callisthenes, I, 8.
29 Meulenaere, H. de (1960). Les monuments du Culte des rois Nectanébo, *Chron. Egypt.*, **35**, p. 92.
30 Skeat, T. C. and Turner, E. (1968). An oracle of Hermes Trismegistos at Saqqara, *J. Egypt. Archaeol.*, **54**.
31 Bevan, E. (1927). *A History of Egypt under the Ptolemaic Dynasty*. London. Chapter 5.

H. S. SMITH

Society and settlement in ancient Egypt[1]

Introduction

Excavated Dynastic settlements in the Nile valley provide an uneven and inadequate basis for assessing the nature and variety of Egyptian urban organization. The majority of the population lived then as now on the alluvial flood-plain, but for practical reasons excavators have mainly examined sites on the desert edges outside the alluvium. Work on sites within the alluvium has tended to be concentrated on major building complexes, especially temples; these were made of stone, and have therefore survived the ravages of annual inundation, digging for fertiliser (*sebbakh*), and continuous occupation better than the civil buildings of mud-brick. At present there is no published plan of a Dynastic settlement within the alluvium complete enough to be of real use for analytical purposes. Valuable evidence from desert edge settlements, notably the capital city at el Amarna and the fortified towns in Nubia, is utilized in various ways by my colleagues in this Seminar. The majority of these settlements were however laid out by the Pharaonic government as officialfoundations on virgin sites. They do not therefore necessarily represent by the aspect of the normal settlement in the valley which grew up through time gradual accretion and alteration. This paper turns, therefore, to documentary sources that may illuminate life in Egyptian towns and villages.

I

Legal documents of the fourth and third centuries B.C. written on papyrus in demotic Egyptian allow some insight into the history of certain town properties in Thebes. In 324 B.C. the carpenter of the temple estate of Amun Djufachi drew up legal instruments having the effect of a will, dividing his house between his family. To his eldest son Kludj and Kludj's son Pasemteu he gave its southern portion, and to each a half share of the portico and its upper storey, of the stairway and roof terrace, of the harem, and of the southern end of the courtyard. The northern end of the courtyard was given to

Kludj's daughter Mut. To his younger sons Phib and Petechons he gave joint shares in the northern portion of the house, together with its booth, probably a summerhouse. They were instructed to close the doorway leading from their portion of the house into the portico given to Kludj and Pasemteu, and to open a new doorway northward on to the street of Pharaoh. In 315 B.C. Petechons included his half of the northern share of the property in his marriage settlement upon his wife Taesi, from whom it passed by inheritance to their son Puertiu before 301 B.C. In 297 B.C., however, Puertiu returned it to his mother Taesi, who being now widowed and in financial straits quoted it in 295 B.C. as security for a loan to a man named Pleehe, who was not a member of the family. In 293 B.C. Taesi, unable to repay the debt, ceded the property to Pleehe. In 290 B.C. Pleehe made a legal agreement with a woman called Tahib, who had acquired Mut's share of the original property, authorizing her to build against his western wall provided that she inserted no timbers in it in addition to those already present, and that she left a light-well of specified width opposite his windows to protect his "ancient lights". In 284 B.C. Pleehe made over his property to his wife Teihor, but two months later, with her permission, quoted it as security to a man named Oseruer. This time, however, there was no foreclosure, for five years later in 279 B.C. Teihor sold it with her father's consent, her husband Pleehe having presumably died in the meantime, to a woman named Teianti, who continued to pay the government taxes on the property down to 274 B.C., though she leased it to her sister Teiba in 275 B.C. Meanwhile Phib's share of the northern portion of the original property had been sold in 301 B.C. by his heirs, two daughters, to a stranger called Harsiese, who despite sundry claims against him by others retained possession until 279 B.C. The southern portion of the property apparently remained in Kludj's legal possession until 288 B.C., but was actually occupied by a man named Petenefhotep, who is not stated to have been related to the family but may perhaps have been Kludj's son-in-law or grandson-in-law; by 284 B.C. he was the legal owner. Thus over a period of forty-five years a dwelling-house, probably not of unusual size or character, passed from occupation by a single nuclear family to divided occupation by four separate but related family groups, and finally to occupation as four separate units by unrelated families[2].

A neighbouring house across the street of Pharaoh was occupied in 324 B.C. by a woman named Tetenefhotep, daughter of Djeho. In 319 B.C. she gave a third of her house, measuring 150 cub^2 in area, in equal portions to two sons of her second marriage, Petenefhotep and Thotane. The whole house thus occupied 450 cub^2, approximately 125 m^2, a modest area. In 314 B.C. she gave the remaining two-thirds of the house to Tamin, the wife of her son Djeho II by her first marriage, on condition that she undertook the expenses of Tetenefhotep's mummification and burial. This was a normal condition in bequests, and two-thirds was the normal share of an eldest son; why Tetenefhotep left it to Djeho's wife when he was himself still alive we do not know, but we may suspect that it was to protect the rights of Djeho's eldest son Oseruer against the claims of his children by a second wife, Teiba. Oseruer duly inherited in 282 B.C., and his son Djeho III had inherited in turn by 241 B.C. In that year Djeho III also purchased Petenefhotep's one-sixth share of the house from Petenefhotep's son Amenhotep. Meanwhile half of the one-sixth share of Thotane, the son of Tetenefhotep's second marriage, had passed back to the major line and was in the possession of three daughters of Djeho II. Thus in 241 B.C. five-sixths of the original house was owned by the direct heir, Djeho III, one-twelfth by his aunts, and one-twelfth presumably by descendants of Tetenefhotep's second marriage. These minute patrimonies remind one irresistibly of the joint ownership of palm-trees in modern Nubia, where shares of a sixteenth or even a thirty-second of the yield were not uncommon.

In 307 B.C. Paret, Tetenefhotep's husband, acquired another house across the street of Pharaoh next door but one to the house of Djufachi, discussed earlier. He did not, however, leave it to his children by Tetenefhotep, perhaps because he considered them already provided for, but to his son Thotew by another wife, on condition that Thotew paid for his funeral. In 287 B.C. Thotew found himself compelled to sell the house to cover his father's burial expenses, and had to obtain a formal renunciation of claim from his sister to enable him to do so. He sold it to his stepmother Tetenefhotep's grandson Oseruer, from whom it passed to his widow, who bequeathed half of it to her eldest son Djeho III and half presumably to her other children[3].

Much similar evidence of this period could be quoted for towns and villages in various parts of Egypt. Earlier evidence is meagre; but government household censuses of the eighteenth century B.C. from Kahun show the large size and mixed character of many of the households there[4]. In the twenty-first century B.C. Hekanakhte, a mortuary priest of the Vizier Ipi, housed under his roof in the village of Nebeseyet, south of Thebes, at the cost of much bickering, his mother, his concubine, five men who may have been his sons, one being married with a family, three women who may have been his daughters, a female dependent relative, a confidential agent and his family, and at least two maidservants[5]. These conditions are reminiscent of those prevailing in the mud-brick domestic properties in Egyptian villages and provincial towns today. Houses are constantly being added to, amalgamated, partly demolished, partly reconstructed, divided and partitioned, according to family considerations and pressing domestic and financial needs. A network of streets, alleys and courts results; with blocks of contiguous houses, yards, workrooms, shops, byres, lean-tos, roof shelters, gardens and dovecotes, built round and against and between one another out of a supremely cheap and easy building material—mud-brick. The whole has a casual appearance, but is regulated by a legally defined and jealously guarded pattern of ownership and mutual obligation.

Most Dynastic towns and villages in the alluvium may well have contained congeries of property of this character, either forming whole quarters or interspersed among public buildings and larger self-contained properties. The Theban houses quoted above were situated in a quarter known as "The House of the Cow", north of the great temple of Amon at Karnak and west of the temple of Montu; the quarter was named after the temple of the sacred cow, mother of the divine bull of Thebes. Many of its known residents were necropolis workers or held minor secular posts in the Amon temple. At Memphis also the necropolis workers tended to congregate in a single quarter, Anubieion, round the temple of the embalmer god Anubis. Necropolis workers and embalmers may have been a special case, since tabus attached to some parts of their activities, and there is evidence from a document of the second century B.C. from Asyut that they were organized in tight professional groups, who had legal agreements with each other concerning their precise sphere of professional activity and mutual obligations. Secular temple personnel probably clustered round their temples for convenience's sake, while the priests or at least their higher echelons, having to observe purity regulations, were provided with houses within the outer girdle walls of temple complexes. Carian and Ionian mercenaries had special quarters of their own at Memphis from the seventh century B.C., while the Jewish enclave at Elephantine in the fifth to fourth centuries is famous for preserved Aramaic papyri. In Greek cities of the Ptolemaic period, there were native quarters, and it is probable that in native cities the Greeks tended to congregate. But in general there does not appear to be much evidence of zoning of settlements along craft and professional lines; as Kemp's analysis of neighbourhoods at el Amarna emphasizes[6], there is reason to believe rather that the houses of people in subordinate positions tended to cluster round the sources from which they obtained their income and supplies.

Occasionally fragmentary insights into the topography of one of these living quarters can be gained because the boundaries of houses and plots are given in the documents; an example is the village of Jeme which had grown up in and about the disused mortuary temple of Ramses III at Medīnet Habu in western Thebes by the Ptolemaic period. One house "by the market of Jeme" had the street of Pharaoh on its south and the stone wall of the temple on its north side; the market therefore must have lain between the south wall of the temple and its inner girdle wall. Another house "in the south-east district of Jeme by the Great Wall of Jeme" had the "Great Wall" on its west and the "Sacred Way" on its east; it must have lain outside the great outer fortified temple enclosure near the temple quay. A house in the south-western district of Jeme within the wall had on its south the street of Pharaoh, on its east and west vacant plots, on its north the pottery. A chapel of Isis was in the northern district of Jeme with the stream of Jeme on its south, the sacred lake on its north, the dovecote on its east, and the way of Amun of Jeme on its west[7]. These disjointed details give an interesting picture of the casual way in which this residential settlement grew up around the ancient temple and on the ruins of the temple town; the rigid rectilinear axes and carefully articulated planning of the original was disregarded but standing features like the temple and the great girdle walls were used for abutment, the sacred lake utilized no doubt as village wash-place and duck-pond, ruins and open areas for essential village industries like the pottery.

Such evidence suggests a contrast in the Dynastic towns of the Nile valley between those portions laid out by the government, namely the temple, administrative and military quarters and their service areas, and the domestic quarters in which the lower middle and lower classes lived. The fortified garrison towns of the twentieth and nineteenth centuries B.C.[8], exemplify most strikingly the planning principles adopted when the government built for official purposes on a virgin site; the rectilinear fortified enclosure, exactly measured and symmetrically balanced on all sides with schematically placed gateways; the parallel axial paved main streets equipped with central gutters, the narrower cross streets at regular intervals; the symmetrical building blocks planned as units for particular functions, commandant's palace, administrative buildings, barracks, stores, officials' houses, temple quarter; the standardized planning of the sub-units, and their conformity with the axial orientation of the whole, determined by the great traffic artery, the Nile. The capital city of el Amarna vividly illustrates this contrast, as Kemp's analysis shows[9]; the great and small temples, the "Great Palace", the royal residence and the royal road were laid out in positions of maximum advantage in relation to each other and to the Nile; the departments of state were then fitted in according to the axial pattern in rectilinear blocks around and behind the royal residence. But high officials then evidently had the privilege of choosing desirable sites for their villas with no great regard for the axes of the official quarters; round them cluster the lesser houses of their subordinates, their scribes, their craftsmen, their servants, packed into the vacant areas as building space, economy and whim suggested, without observing axes, street plan or neighbours' convenience of access; that is, in all probability according to normal custom in the domestic quarters of agricultural towns and villages. To be in contact with one's master, one's source of supply, one's family and working groups were probably the main preoccupations; formal considerations of town planning were absent.

Where, however, the state itself provided for the artisan class, an interesting combination of dirigiste and individualist tendencies can be observed. The village of Deir el-Medina in the Theban necropolis was founded for the gang of workmen who constructed the royal tombs in the Valley of the Kings. It was built under state authority in the reign of Tuthmosis I, whose stamp appears on the bricks of the enclosure walls, and continued in occupation from the sixteenth to the eleventh century B.C.[10]. The gang or crew, usually of sixty men, was divided into a "right" and a "left" side, presumably according to the side of the tomb which they excavated; their walled village, approximately rectangular but accommodated to the conformation of the narrow uneven desert valley in which it was situated, was also divided by a longitudinal axial street into two sides which housed the personnel of these administrative divisions and their families. Each was headed by a foreman and his deputy, who had houses of greater size at the southern end of the village, as did the scribe who was responsible for recording the work. There were about seventy workmen's houses, built in blocks with party walls; a few houses, probably later additions, were built outside the village wall on the north. The workmen worked at the tomb for nine consecutive days, lodging overnight in temporary huts above the Valley of the Kings; on the tenth, twentieth, and thirtieth days of the month, they rested in their village at Deir el-Medina, where they also celebrated the great religious festivals of Thebes and national holidays such as royal accessions and their anniversaries, New Year's day and the epagomenal days, the birthdays of the gods. The work was strictly supervised by the foreman and the scribe, who noted absentees with reasons for absence, any irregularities in the work or its conduct, the exact weight of the metal tools issued by the government and the numbers of lamp-wicks for underground work; they measured progress periodically and sent frequent reports to the office of the vizier, who was ultimately responsible to the Pharaoh for the work. The vizier or a royal butler made regular inspections.

Here we see at work a system of direct employment, maintenance and supervision of a specialist labour force by the state, with living quarters provided by the state and erected under its authority. Yet within the individual houses all the marks of private enterprise are to be seen. Front rooms were used as shops, workshops or beerhalls; cellars and silos were improvised, stairways moved, doors blocked and opened, windows pierced, partitions added and removed, houses amalgamated or radically altered. Disputes among the villagers were settled by a village tribunal, composed of a foreman or scribe and senior villagers; this tribunal convicted and sentenced, subject only to the vizier's confirmation in capital cases and to his prerogative of pardon. Village sanctuaries stood on the slopes outside the wall, dedicated to deified kings and queens and local tutelary gods and goddesses. Here prayers and petitions were made and oracle questions asked at great feasts, often for the purpose of resolving disputes or obtaining or reversing legal decisions. The workmen themselves served as priests in these shrines. When on several occasions in the twelfth and eleventh centuries the government authorities were late in issuing their monthly rations, the workmen felt themselves strong enough to go on strike, march in procession to one of the main temples and demand an immediate grain issue; generally after parley with the priests they were given a partial issue and a promise that their complaints would be referred to the vizier. Černý concludes that "the king's workmen thus enjoyed a remarkable degree of self-government in both civil and religious matters, an interesting feature in a monarchy governed by a highly developed officialdom and an economically strong priestly class"[11]. As evidence is lacking, he does not discuss whether this was due to the fact that progress in the royal tomb and its security was dependent upon the royal workmen's co-operation, or whether it reflects general conditions in New Kingdom Egypt. Certainly there is much evidence for the exercise of legal and priestly office locally, but regularly by people occupying a position in the official hierarchy; one doubts whether the fellahin of an agricultural village would have exercised such functions. Whether local dignitaries owed their local eminence to their state offices, or their official titles to their local importance clearly varied throughout Egypt's long history according to the relationship between state and local administrations, and no doubt also to individual factors. In other settlements built by the state authority, for example the nineteenth century town at Kahun or the fourteenth century workers' village at el Amarna, there are the same signs of individual enterprise in altering the arrangements of houses as at Deir el-Medina, and documents from Kahun suggest that the inhabitants had some say in conducting their affairs. Even in the military garrison towns of the twentieth and nineteenth centuries B.C., where the rule of military discipline is attested by the frequent whitewashing of the ditch at Buhen, houses were adjusted to individual requirements. But even the most authoritarian regimes of history have failed to restrict private enterprise completely.

II

In theory, the land of Egypt belonged to Pharaoh as heir of the gods. If the temples owned land it was through initial endowment by a Pharaoh, and there is some evidence to show that such endowments were renewed by each successive Pharaoh. If individual families owned land and built up estates and derived wealth, power, and some independence through this, they originally acquired their title to it by grants from Pharaoh in recognition of service to the state or its gods. Though the succession by a son to his father's office was regular practice, such appointments required formal confirmation by Pharaoh. If an individual or family was disgraced his property returned to the state. The produce of all land, including temple land, was subject to state taxation unless exempted by royal decree, and all labour likewise to corvée. Large cultivable areas in addition to all deserts and waste lands were state property, and the working of certain natural resources, for instance fine quality stones and metals, appear to have been state and temple monopolies in all times of centralized government. That all streets in towns were regularly referred to as "the street of Pharaoh" in demotic documents emphasizes this principle of royal ownership. Indeed, public building works were regarded as commissioned directly by the Pharaoh and built to his orders. Temple building inscriptions are always in his name, and couched in the first person, though the name of the official who executed the work at his command may also be mentioned. In theory, the King always "stretched the cord" for a temple (i.e. founded it) and dedicated the completed building, and no doubt often did so in practice. The texts of the dedicatory rock stelae proclaiming the foundation and bounds of the new capital city of el Amarna are couched in Akhenaten's own name, and there can be no doubt that the New Kingdom temple towns in Nubia were ordained by royal decree.

Each royal project was carried out through a high official entitled "the overseer of all the works of the King", or in the case of a great state temple "the overseer of all the works of Amon-re', King of the gods". Such master architects had under their direction chief masons, sculptors, painters, draughtsmen, scribes, plasterers, carpenters, quarrymen, stone-handlers, boat captains, each at the head of a body of skilled craftsmen working in the service of the state and bearing official titles. In addition they had a corvéed labour force of requisite size. All these were supported out of the resources of the state or temple treasuries, granaries, clothing and other stores. Attached to all royal quarrying and mining expeditions, and doubtless also to building projects, was a highly ranked treasury official with his own scribes responsible for logistics and commissariat. The overseer of works was directly responsible through the vizier to Pharaoh for the satisfactory completion of the work. Architects' scale working drawings have survived from as far back as the nineteenth century B.C., and we know that plans were often changed during construction, doubtless owing to the displeasure of the King or his envoy during an official inspection.

As part of his duty to gods and people, each successive Pharaoh was anxious to extend, elaborate, and beautify the monuments of the divine benefactors to Egypt, and in so doing had no compunction in demolishing or altering the work of his predecessors. The great temple of Amon-re' at Karnak for instance, like a medieval cathedral, must have constantly rung to the mason's chisel and mallet. Evidence concerning palaces is sparser, but suggests that Pharaohs, like Assyrian kings, tended when practicable each to build a new palace. Striking examples in the New Kingdom, apart from el Amarna, are the palace of Amenophis III at Malkata in western Thebes, which seems to have been utilized little after his reign and mainly for jubilees during it, and the successive Delta residences of the Ramessides. But even in the Old Kingdom during the third millennium B.C., the chief royal residence may have changed from reign to reign with the site of the royal pyramid; the names of pyramid cities appear later, significantly, as names for quarters of Memphis or for the whole city. In Near Eastern monarchies it was always desirable for a new king to remove from the influence of his predecessor's harem and palace officials. In the New Kingdom the King always had major residences at Memphis and Thebes, and in the Ramesside period in the Delta as well. There were also small palaces attached to major temples in the provinces, where the King and his immediate family could lodge during festivals; for in Egypt, as elsewhere in the Near East, the court was where the King was and accompanied him even on major military campaigns. In a country with the Nile as a spinal traffic artery, the court and high officials of state travelled commodiously in rich caparisoned barges of the type of the modern *dahabiya*.

Other types of government and official building may have had more stable histories. But *nomarchs* and great dignitaries with provincial estates probably vied with each other in building fine residences and official quarters in towns, as they did in building fine tombs. During the inundation season there was no work on the land, a labour force was at hand, and transport on the flood waters was simple. The tendency of bureaucracies to grow by their own momentum is well exemplified in successive periods in Egypt. In all probability the capitals of nomes, ports and frontier towns had official quarters of some size, generally situated by the Nile or main canal, providing a focus for the residential quarters. Numbers of "new towns" were founded by the government in times of prosperity, not only in outposts like Nubia but in Egypt also. For it was constantly necessary for the state, in order to be able to support and reward its officials, civil and military, with grants of land, and to feed an increasing population, to bring unproductive areas under cultivation by marsh clearance and irrigation works; new foundations on marginal land were the means of achieving this. Thus while the old agricultural villages may have been little affected by government building, royal cities, the centres of great cults, the capitals of nomes, ports and frontier towns, settlements associated with royal mortuary cults, and a proportion of marginal agricultural settlements must to a greater or lesser degree have exhibited the influence of official planning.

But social, economic and political factors are not the only ones to affect the nature of human settlements. Men's ideas of what should be have a powerful influence on what is. We must consider, however superficially, Egyptian ideas of the nature of their world.

III

In Egyptian thought life was a microcosm reflecting a macrocosmic process. The cycle of human birth, maturity and death mimicked that observed in the powers of nature; the rising, fullness and setting of sun and moon; the waxing, culmination and waning of the Nile flood; the seasons of the year; the burgeoning, flowering and death of vegetation; the birth, life and death of the animal kingdom. In these examples, rebirth followed death in a pre-ordained cycle which had come into being at the creation of the universe; this was the basis of the human hope of afterlife. The physical world itself was considered the replica of a pre-existing pattern; the geography of the land of Egypt corresponded to that of the eternal realms. Units of space and time were not only chronicled by the sun, moon and stars, but were microscopic repetitions of the aeons since creation and the distances of the heavens. Kingship had been handed down in divine succession from the creator through dynasties of gods; the rule of righteousness (*M3ʿt*) by which the King ruled as the rule of order (*M3ʿt*) by which the Lord of All guided the universe.

The temple, as "the mansion of the god", was the replica on earth of the god's eternal dwelling, while the mortuary temples of kings were referred to as "Mansions of Millions of Years". This is why all temples were built of durable stone, whereas civil and domestic buildings, including royal palaces, were built of brick. Egyptian myths of creation are manifold, and present many images of the process at different stages, which are variously combined according to the requirements of particular cult centres: but a pattern is revealed. Before the beginning there was Nun, the waters of chaos, existing as a limitless, dark and timeless void. The waters receded, and an island emerged, just as each year the first islands of mud emerged from the receding waters of the inundation. This island (*iw*), often called "that which has risen" (*ḫʿy*), "the upraised land" (*t3-tnn*), "the mound" (*i3t*) or "the hill" (*k3y*), was the place of creation. Light split the darkness, and out of chaos came forth primeval beings, variously described, who were present at the coming into being on the island of the demiurge; this is portrayed in many ways. The demiurge then by creative thought and word or physical act brought into being the panoply of the universe and created the first generation of gods. The sanctuary of each major Egyptian temple was thought of as being founded upon this primeval island or mound at the hub of the universe on which creation had taken place; the temple was the abode of the creator.

This is most graphically shown in the Building Texts of the Ptolemaic Temple of Edfu[12]. The Falcon-god Horus of Edfu, the ruler of the sky, says: "They found my house as the work of antiquity, and my sanctuary as the work of the Primeval Ones". The nucleus of the temple was "made like unto that which was made in its plan of the Beginning"; its adyton was the god's "genuine great seat of the first occasion". The temple was built at the dictates of the Ancestors "according to that which was written in this book which descended from the sky to the north of Memphis", elsewhere perhaps referred to as "the book of planning the Temple". In the representations of the foundation ceremony of the temple the god Thoth is shown arriving at the sacred temple site and saying: "I come here in my true form unto the foundation ground of the Great Seat of Harakhte. I cause its long dimension to be good, its breadth to be exact, all its measurements to be according to the norm, all its sanctuaries to be in place where they should be, and its halls to resemble the sky". This statement is significant, for the sky was the realm of the gods; the temple domain is referred to as *niwt* "city", and there is a persistent pun in religious texts upon *niwt* "city" and *nwt* "sky". A regular word for the portico or front hall of a temple is *h3yt*, which also means "sky". Egyptian temple halls often show on their roofs (also called *h3yt*) a representation of the night sky and its constellations. The arrangement of open papyrus-flower and papyrus-bud columns in these halls has been interpreted, doubtless correctly, as the representation of the papyrus thicket in the Delta where in myth Isis protected the child Horus from his enemy Seth while she suckled him. The gradual rise in the floor level of temples from entrance to sanctuary by a series of ramps or stairways represents the primeval island or mound itself.

According to the Edfu myth, two beings emerged from the waters of chaos, landed on the reedy margin of the primeval island, and planted a slip of reed in the water. The falcon Horus, flying out of space, perched on the reed, which became the god's first home. Gradually the "Great Primeval Mound" emerged from the waters around it, creating a new domain (*niwt*). On this was set up a reed-hut for the falcon within a reed fence to protect the god from the cthonic enemy-snake. When creation was completed, a temple was laid out by builder gods in the presence of Reʿ, the sun-god, according to a book named "The Specification of the Sacred Mounds of the Primeval Age". The sanctuary represented the Falcon's perch, the temple his hut, the girdle wall the reed enclosure. The dimensions given for this "Great Seat of the First Occasion" are exactly those of the standing Ptolemaic temple of Horus of Edfu. This temple was thus the perfect replica of the god's eternal mansion[13].

Similar myths could be quoted from other temples. Even subsidiary elements of the temple complex were thought of as planned by the god himself, as is clear from Queen Hatshepsut's text engraved in the fifteenth century B.C. on the obelisks she had erected before Karnak temple: "I made these with a willing mind for my father Amon, for I entered in to meet him on the First Occasion and learnt his effective power; I have not forgotten the plan of what he ordained. My Majesty knows well that he is divine; indeed it is under his command that I have acted, it is he who has led me. It is not the case that I myself planned the work as a thing not of his making; but it was he himself who gave the design. . . . I have not neglected the Domain (*niwt* "city") of the Lord of All, but have given it (all) my attention; for I know that Karnak is the horizon upon earth, the August Hill of primeval time, the Sacred Eye of the Lord of All, his favoured abode that lifts on high his beauty, that encloses the gods who are in his train"[14]. Thirteen centuries earlier than the Edfu texts we meet here the claim that the temple is the god's original abode, couched in similar language.

This cast of thought, familiar from oriental civilizations ancient and modern[15], explains the sanctity of temple sites and the uniformity of temple plans, at least from the New Kingdom to the Roman period. Indeed, the earliest representations of temples preserved, dating from the end of the fourth and early third millennium, show reed hut-shrines, sometimes covered by an animal skin, and reed enclosures of the sort described by the Edfu texts; and these illustrations probably do not represent contemporary buildings but more primitive, prehistoric shrines. But it is also significant that the word *niwt* "domain" (of the god) is the normal Egyptian word for city from the earliest times, and the word *ḥwt-nṯr* "mansion of the god" is the same as that used in *ḥwt-nsw* "mansion of the king". The hieroglyph for *ḥwt* shows a rectangular fortified enclosure with towered gateways, within one corner of which is shown a rectangular residence. The type is known from vast brick enclosures at Hierakonpolis, and from the dummy palaces of brick erected by the kings of the first and second dynasties in the necropolis at Abydos to be their eternal residences[16]; in the Step Pyramid complex at Saqqara, the type is translated into stone, with elaborate internal arrangements, some undoubtedly based on the interior arrangements of royal palace compounds, others perhaps on cosmogonical concepts, for it is possible that the Step Pyramid itself represents the hill of creation. This royal palace type undoubtedly forms the model for the typical fortified temple enclosure defining the god's domain. And, despite the modifications caused by the processional and ritual functions of temples, the pylon, courtyard, portico, hypostyle halls and sanctuary of the temple can be shown to derive from the towered gateway, court, loggia, audience halls and private sleeping quarters of the royal palace. The temple's sacred lake and grove represent the palace pool and garden, the temple granaries, butcheries, storehouses and offices —the domestic quarters of the palace compound. Egyptian royal mansions of this sort were not simply houses for the King and his family; they comprised state audience halls, scribal offices, harems, nurseries, accommodation for official guests, treasuries, granaries, byres, cooking and workshop courts, and gardens and summer houses for recreation. The way in which the houses and estates of royal kindred, high officials, courtiers and clients were grouped in early times round the royal mansion can probably be deduced with confidence from the famous pyramid complex of Cheops; the Queens' pyramids and the great *mastaba*-tombs of the royal princes are ranged rectilinearly round the King's pyramid, while beyond, a strict grid of parallel streets contains the *mastabas* of officials, mainly in descending order of rank as they become more distant from the pyramid. The same principles of basic rectilinear planning and the clustering of lesser round greater tombs can be observed in the archaic necropolis of north Saqqara, which was in use from the end of the fourth millennium until late in the third. The relation of the developed Dynastic temple town like Medīnet Habu (twelfth century B.C.) within its fortified enclosure to the royal palace complex is clear.

The origin of rectilinear planning is more difficult to decide, as it is already exhibited by the earliest royal and religious buildings for which we possess evidence. Technical factors played an important part. Details of the decoration of the dummy buildings in the Step Pyramid complex of Zoser prove conclusively that the prototypes of Egyptian palace architecture were reed buildings of forms remarkably similar to those constructed by the marsh arabs of southern Iraq to this day. These depend basically on a framework of arched bundles of reeds, and therefore necessarily conform to a rectangular plan. When mud-brick was adopted with wooden framing elements, the plan was retained as mud-bricks lend themselves most easily to rectilinear building.

The form of major individual buildings may well have influenced the street plan. However, the hieroglyphic ideogram of the word *niwt* "city, town" shows a circular enclosure, with axial streets crossing at right angles in the centre. Though an oval temple platform of Late pre-Dynastic or Early Dynastic date was discovered at Hierakonpolis, there is as yet no positive evidence for circular town enclosures in early Egypt. A recent suggestion that the sign represents the four quarters of the universe, defined by the four pillars of heaven at the cardinal points, within the circle of the celestial boundary, may perhaps be right[17]. If so, Egyptian ideas of the form of the universe might have influenced their concept of a settlement at a very early period of their history.

However, in Dynastic times this symbolic tradition affected only temple complexes, not the civil areas of towns and cities. But the temple held a central place in the social, cultural and administrative life of towns. Private persons came to the temples on the days when the god appeared to address petitions, to obtain succour and healing, to ask oracle questions, to dream and have their dreams interpreted, and to have foreknowledge of the future, as Ray shows[18]. Scribal schools and scriptoria were usually housed within the temple enclosure, and the temple was a major employer of artists and craftsmen. Civil and criminal courts were held at temple gateways and at the angles of temple enclosures, not only because priests often functioned as dignitaries of the courts, but because oaths by the god and recourse to the divine oracle were integral parts of legal procedure. The scribes who drew up legal documents naturally took up their station near the temple gate, as did the letter writers, the writers of pleas and of petitions, the casters of accounts. For reasons of convenience, governmental administrative residences and offices, police barracks, military barracks, treasuries and storehouses must often have been sited near the main temple, as is the case at el Amarna. The economic function of the temples as great landowners and producers and the part their treasuries and stores played in the local economy must have attracted commercial activity to the temple areas. The great stone temple quays, at which the divine barks drew up at festivals, must have formed the focus for commercial quays, and a deal of daily trafficking. At el Amarna, I believe that the alignment for the temple was first decided (probably by astronomical means), and that this dictated the alignment of the royal street and of the royal residences and government buildings. In a city like Thebes, where straight stone paved processional ways lined by avenues of sphinxes led from Karnak to Luxor, from Karnak to the river and across the river to the temples of western Thebes, and each lesser temple complex boasted its own sacred ways, these main arteries can hardly have failed to influence the whole aspect and orientation of the central quarters of the city. Even in valley towns that had grown up through history in their own quiet fashion, the effects of official ideology and practice must not be discounted.

Much more of interest for the social, economic and ideological background of settlement in Egypt can be derived from ancient texts. It should be possible to study urban and village life in Egypt as an integral part of the country's whole social history. Until a range of representative settlements in the Nile valley have been completely excavated to their full depth and over their total area, and meticulously recorded and analysed, this opportunity will be denied us. Time is short.

Notes

1 This paper attempts to present to those unfamiliar with Egyptological literature certain social factors relevant to Dynastic settlements. There is no original work of mine in the paper. It is so generalized that I have given references only where I have borrowed illustrative material directly from other scholars' work. For the inadequacy or error of the views expressed I am responsible.

Terminology

In this paper I have been wilfully imprecise in my use of the English terms city, town and village. The Egyptians appear to have been similarly imprecise; the common words *níwt* and *dmí* cover a wide range of settlements. We do not possess adequate evidence to assess the relative populations and the religious, economic and administrative functions of various types of settlement. "City" might reasonably be applied to state capitals with royal residences, and to certain national religious centres; "town" to the capitals of nomes and perhaps of lesser administrative districts, and to ports and frontier posts; "village" to the majority of agricultural settlements in the alluvium without regard to probable population. Such a subjective categorization tends to emphasize the futility of attempting to apply the terminology of one culture to a different cultural situation.

2 Excerpted from Glanville, S. R. K. (1939). *Catalogue of the Demotic Papyri in the British Museum, vol. 1: A Theban Archive of the Reign of Ptolemy Soter*. London.

3 Excerpted from Mustafa el-Amir (1959). *A Family Archive from Thebes*. Cairo.

4 Griffith, F. Ll. (1898). *Hieratic Papyri from Kahun and Gurob*. London.

5 James, T. G. H. (1962). *The Hekanakhte Papers and Other Early Middle Kingdom Documents*. New York.

6 Kemp, B. J., this volume, pp. 670-76.

7 Mustafa el-Amir (1959). *op. cit.*; see Uphill, E. P., this volume, pp. 723-33, for an account of Medinet Habu in Ramesside times.

8 Kemp, B. J., this volume, *ibid.*

9 Kemp, B. J., this volume, *ibid.*

10 Excerpted from Černý, J. (1965). *Egypt from the Death of Ramses III to the End of the Twenty-first Dynasty*. Cambridge Ancient History, rev. ed. vol. II, ch. 35. Černý's brief summary of his life's work on this subject should be read as a whole. See also Bruyère, B. (1924–48). *Rapport sur les Fouilles de Deir el-Medineh*, 1922–40. L'Institut d'Archéologie Orientale, Le Caire.

11 Černý, J. (1965). *ibid.*; Bruyère, B. (1924–48). *ibid.*

12 Excerpted from Reymond, E. A. E. (1969). *The Mythological Origin of the Egyptian Temple*. Manchester, which gives summaries and partial translations of the texts concerned. My précis travesties the complexity both of the texts and of Dr. Reymond's analysis.

13 Reymond, E. A. E. (1969). *ibid.*

14 Translated from Sethe, K. (1906). *Urkunden der 18 Dynastie*, II. Leipzig. pp. 363–4.
15 Wheatley, P. (1969). *City as Symbol*. London.
16 Kemp, B. J. (1966). Abydos and the Royal Tombs of the First Dynasty, *J. Egypt. Archaeol.*, **52**. For this paragraph generally see Lauer, J. P. (1962). *Historie Monumentale des Pyramides d'Egypte*, 1. Paris; Smith, W. S. (1958). *The Art and Architecture of Ancient Egypt*. Penguin; Hölscher, U. (1941). *The Excavation of Medinet Habu*, 3. Chicago Oriental Inst. publ. 54.
17 Leclant, J. (1969). Espace et temps, ordre et chaos dans L'Egypte Pharaonique, *Revue de Synthèse*, III, Nos. 55–6, p. 223.
18 Ray, J. D., this volume, p. 702.

ERIC UPHILL

The concept of the Egyptian palace as a "ruling machine"

To understand the design and role of an ancient Egyptian palace it is necessary to completely disabuse one's mind of any preconceived ideas founded upon western buildings. To the occidental the word "palace"—derived from the Palatine and the residences of the Roman emperors—conjures up a picture of what is essentially a large house, a dwelling which, however enormous, is yet specifically one building and usually contrived under one roof. Typical examples are the Vatican, Versailles or Hampton Court. Such was not the the case in the orient where a royal residence constituted rather a collection of different buildings often very loosely strung together and generally contained within an outer enclosure wall, which last sometimes gave the various blocks of hall, pavilions and courts the only unity they possessed.

Architecturally a much closer equivalent would have been the medieval or Tudor palace, like the Richmond of Henry VII, where many acres of buildings were grouped to form what was in effect a complete community in itself. Similarly it must be remembered that all government buildings, offices, soldiers' barracks, great store rooms and arsenals were grouped around, if not actually inside, the palace quarter. It took four square miles to house the emperor and his entourage in his "Imperial City" in Tang China[1] while the Caliphs of Baghdad and other semi-divine rulers in the Near East were not far behind[2]. The actual palace proper of course only occupied a part of this great area, but was still vast by Western standards.

If such rulers as these required a governmental city to house them, and bedecked it with every conceivable kind of ornament and architectural trick in order to produce a suitable feeling of awe in their subjects and those visitors favoured enough to have audience of them, how much more so did the Pharaoh need and make use of such a system to enhance his rule.

No rulers before or since have been so shrouded with divinity or so hedged about with the hieratic trappings of state. Pharaoh, at least in official matters, not only summed up all the power and attributes of the state, but also acted as the supreme intermediary between heaven and earth, having direct relationship with all the known gods and goddesses as their "son", and being also in his own right the only acknowledged god on earth.

Hence his very physical presence was treated as that of a god, his throne being literally set up in a shrine like those of the principal deities in the great temples. He was so encompassed about with divinity that it is almost impossible today to understand what this entailed. Every object he touched was deemed sacred, everywhere he sat must be placed at a higher level upon a dais, thus commanding the place where others stood. His diet and daily habits were carefully regulated as were his clothes and royal regalia, of which last he had more than is conceivable even by the standards of other monarchs. His every action was recorded in detail and minute observation, being carved and painted with infinite labour and care upon the walls of temples and palaces, while his likeness was copied and made known to millions of subjects and neighbouring peoples by such a quantity of statuary and figures, named articles and commemorative objects, as to make the work of modern publicity agents look puerile.

His very appearances in public were in fact not just recorded events but veritable epiphanies. Yet while much is known and has been published about the palaces and homes of other monarchs of the ancient Near East, those of Pharaoh remain neglected, a strange thing indeed considering that this very title is derived from the word *palace*, i.e. great house, itself. We know of the splendours of the Assyrian kings with their vast residences at Nimrud, Ashur and above all Nineveh, guarded by huge stone bulls up to 20 ft high. It is still easy to see the kind of impression that must have been made on visitors as they were swept up ramps and through gates guarded by such creatures, into an endless succession of courts and halls, all proclaiming from their sculptured walls the might and majesty of an Ashurnazirpal or Tiglath Pileser[3]. Similarly Nebuchadnezzar at Babylon and Darius at Susa or Persepolis had equally grand palaces which are also well documented[4].

But what of Ramses II or Amenhotep III? It seems evident that if King Khufu for example could put up six or seven million tons of masonry for his tomb he must surely have lived in a comparable house. Unfortunately most of the evidence for these palaces lies buried, but what has been excavated shows us that although mainly constructed of mud-brick and timber like the homes of their subjects, the residences of the Egyptian kings were at least of comparable grandeur to those of other contemporary rulers and in some periods infinitely more splendid.

Zoser's pyramid enclosure has long been recognized to be a stone copy of a third dynasty palace enclosure, probably the one which once stood just below in the valley at Memphis, and it covers about 37½ acres[5]. The excavated portions of the immense palace buildings of Amenhotep III at Thebes, today called Malkata, cover 80 acres and this is probably only a fragment of the whole. His son Akhenaten is known to have had five palaces and villas at el Amarna alone. But it is certain that there also others in every major city throughout Egypt, with villas and country estates undoubtedly scattered all over the countryside in specific areas, an incredible diversity of royal lodgings.

Faced with this wealth of material one example must suffice to demonstrate the manner in which the palace was used as a device to emphasize the king's power and enhance his rule by creating a feeling of awe in the minds of subjects through a series of architectural illusions. Much of this is purely theatrical in nature and shows how a feeling of subservience was created psychologically by remarkable stage management.

Medīnet Habu is both temple and town, the funerary foundation of Ramses III, *c.* 1195 to 1163 B.C., and the best surviving example of such building complexes. It was built in *two distinct* phases, the first belonging to the earlier part of the reign contains the temple itself, a slightly smaller edition of the Ramesseum but still nearly 500 ft long. Around this heart were grouped vast magazines, stores, offices and other appendages, all served by a system of well-paved streets and enclosed within a massive brick wall, fortified and once nearly 50 ft high, the whole occupying about 6 acres. Without and around this temenos was later built a much greater enclosure protected by a huge girdle wall, the area that this in turn contained being filled with houses for temple officials, administrative buildings, a garden and other royal establishments. What is of special interest is the fact that this great complex contained not one, as usually stated, but no less than three palaces for the king's use when he visited the temple.

In ancient times the visitor would arrive by a canal connected with the Nile (Pl. 3), or else by a road bordering it, which ended in a square basin or dock excavated with much labour from the rock before the main eastern gate. Quite sizable boats could use this harbour as it was 150 ft or more across[6]. With the water standing at low level the façade above presented a colossal sight, for those seeking an audience with Pharaoh must first climb forty stairs leading up and around the sides of the 17 ft high stone quay platform. Next came a walk up further steps and sloping dromos to a low massive stone wall that guarded the lower rampart around the entire enclosure[7]. All the front of this last was carved with religious scenes brilliantly coloured and showing the king with the gods, as were the squat and massive towers flanking the entrance, each 20 ft high to the top of their battlements.

Beyond lay one of the two mighty gate towers (Pl. 4a), this east one being balanced by a similar but more massive one at the west end of the outer enclosure[8]. It is a major construction in itself as can be seen even today, a solid cubical mass of mud-brick work partly cased in stone, measuring about 75 ft^2 at the base and nearly 70 ft high in front. To either side of it stretched the stupendous "Girdle Wall', 60 ft high and no less than 34 ft thick at the base. After passing through the outer gate and crossing the broad rampart here paved with stone, the visitor found that there was no gate immediately in front between the towers at either side of the High Gate as might be expected, but that the road passed straight on into a small courtyard where today stand two seated statues of the lion-headed goddess Sekhmet, serving as guardians. On looking upwards he would see that the figures carved high on the walls above appeared to be coming to life. This was because at either side, i.e. north and south, three rows of four captive figures seemed to be lying on the floors of rooms inside the tower at anything from 30 to 40 ft above[9], with their heads protruding out of windows or apertures[10] (Pl. 4b).

These figures represented Negroes, Asiatics and other foreign peoples that the Egyptians ruled or claimed to have control over. Above them so that his feet appeared to be resting on them were statues of the king, portrayed in the age-old ritual act of slaying an enemy chieftain whom he held by the hair. These statues are missing today but must have been flat rather like deep reliefs although almost 10 ins deep, and measuring about 10 ft high with their crowns so that they stood out from afar. All these dramatic representations were brilliantly coloured, and traces of the colours still remain on some of the captive heads and the necks show the pattern of their robes in detail. They inevitably lead the eye to the rear of the gateway and up to a point immediately above the single-leafed door.

The tradition of gateways as places of audience and positions from which rulers could hear and receive petitions from their more humble subjects is extremely ancient in the orient. The very name of the Turkish Sultans' government "Sublime Porte" really reflects this concept, and the domed gate of Felicity in the palace at Constantinople was the place where the Sultans received visitors at their coronations while enthroned beneath it[11]. Above the 18 ft high gate at Medīnet Habu there are significantly two large and wide windows, both of which could have been used as "Windows of Appearance" as well as serving to light the rooms on the first and second floors and affording the occupants wonderful views of the Nile valley to the east. As the king used these rooms when in residence he must have often looked down on people standing or perhaps specially gathered in the court below, and would himself have been visible far out across the fields in the valley below. I would therefore suggest that this double window system is analagous to the audience-hall and gate of petitions mentioned above.

Even more astonishing than the size of this gatehouse was the unbelievable richness of its original decoration, for besides being carved with reliefs and painted in bright colours, it was also treated as so much of the temple itself, becoming a field for decoration which can only be likened to the jeweller's art. It was encrusted with polychrome ornament that must literally be compared with examples of New Kingdom jewellery now preserved in museums. Some of the scenes were enhanced with faience inlay and ornamental metal strips with gold or gilding, particularly on the high spots such as window- and door-frames. The whole gate structure thus formed what was really one gigantic piece of jewellery 70 ft high, a frame for the king's windows inside and in vivid contrast to the white wall at either side.

The reason for this fairy-tale ornamentation equalling anything in the

Arabian Nights was that within it was a small palace termed "High Pavilion" by the excavators[12]. This set of apartments designed purely for pleasure was really an aerial palace and could only be approached by a winding ramp on the south side of the gate tower. Upon ascending this, the king would arrive at a door built at approximately 24 ft above ground level and would enter a building that must have appeared to be floating in the air, although really very solidly built beneath the light ornamentation and firmly tied within the mass of the gate. A chain of five or six rooms with a passage leading to a winding staircase filled the first floor at the rear of the tower. These chambers were partly constructed in the stonework and partly within the great brick mass of the gate. All of them were vaulted except for the central one which had a flat ceiling with wooden beams and which measured about 13½ by 14½ ft. The others were longer and narrower but their numerous windows gave a feeling of lightness and space to the whole structure.

The second floor covered the whole area of the building and comprised twelve further rooms, while the uppermost floor contained only two turret chambers, a sort of eyrie for seeing afar. There were thus about twenty rooms in all or nearly as many as in the main palace within the enclosure, and all were richly decorated. In contrast to the warlike and religious scenes outside, all the reliefs remaining here are of a peaceful nature showing the private life of the king. They are very naturalistic and charming and show him seated on a throne or state chair (one of which is a folding chair of a movable type much favoured by the Egyptians), with the girls of his harem[13]. These companions present him with fruit, flowers and gifts in baskets. He caresses one girl under the chin and plays a game resembling draughts with another. Today these girls appear nude but it is possible that the artist may have originally painted diaphanous robes over them, they certainly wear rich headdresses and diadems as well as sandals. One little girl shown in one of the side chambers appears to be a princess or daughter rather than a wife or concubine as she only stands as high as the king's waist when he was seated.

In the rooms immediately above the gate passage and in the doorways the king is shown walking with sceptre and royal regalia and accompanied by a fan bearer, suggesting a more official purpose. In the side doorways and window-openings the wall surfaces, against which the door or window leaves opened, are usually painted with baskets of flowers and fruit such as must once have stood here. It is the ceilings of the window niches and embrasures which have best preserved their coloured decoration and which are consequently the most informative about the original fittings, for they were mostly painted with tapestry patterns. Rich woven fabrics with designs made up from circles, flower and rosette bands, papyrus loops and other intricate elements, recapture the carpets and hangings that once adorned the furnished apartments, and which help to give them the light and almost tent-like feeling already remarked on. Window soffits and door lintels show vultures with outstretched protective wings, and friezes of hanging lotus blossoms festoon the upper parts of the walls.

The western gate somewhat larger in area had a similar palace pavilion inside it, equally richly wrought with decoration and containing polychrome glazed tiles representing prisoners similar to those of the main palace here.

Leaving the area of the gate house the visitor would next be aesthetically overwhelmed as for the first time he became fully conscious of the scale of the architectural layout encompassing him on all sides (Fig. 1). Around about him stretched the great wall seemingly engulfing nearly the whole horizon by shutting off everything outside and dwarfing all the buildings within the $16\frac{3}{4}$ acre enclosure.

This ring of wall was in turn dominated by the magnificent view ahead to the west, where the western hills of Thebes formed an incomparable backdrop. The vista ahead constituted the architectural climax of the whole for there stood the façade of the great temple, the only building not overshadowed by the outer wall.

The more sacred character and privacy of the inner enclosure was further stressed by a system of houses and storerooms for the priests and inhabitants of this great foundation (Fig. 2). Two rings of dwellings completely encircled the inner temple enclosure on three sides, north, west and south, and were separated from it by a completely encompassing road system, stone-paved and running all round this enclosure.

The siting of these dwellings is curious at first glance but is really quite logical. Inside the great girdle wall was a broad raised way, a kind of military road such as is found in Roman forts where it acted as a quick unobstructed route for soldiers to reach any part of the walls, hence it is here termed "Pomoerium" by the excavators. The embankment of this massive construction was made of gravel filling and measured 13 ft wide. As it was really a continuation of the much wider outer rampart it lay at an elevation of anything up to $7\frac{1}{2}$ ft (5 cubits) above the general ground level inside the enclosure, and therefore needed a massive retaining wall which itself measured about 5 cub thick at the base. The houses could thus be contrived to stand directly against this important feature, but significantly without having direct access to the pomoerium. Two cross-passages ran between the alley that lay between the two rows of houses on the south side of the enclosure, and between the inner ring road and the pomoerium on the north. These passages seem to have been the only means of reaching the fortified area from this township, and the ingenious contrivance of the system not only put all the parts within access by the use of the very minimum amount of passage way, but also distributed the rows of houses and their approaches in different sectors. Consequently it will be seen that the floors of the outer rows of houses were standing at a level almost equal to the whole height of the ground floor rooms below the road on the pomoerium, at least at the east end of the site, for the roads sloped up as they ran westwards following the natural rising ground level.

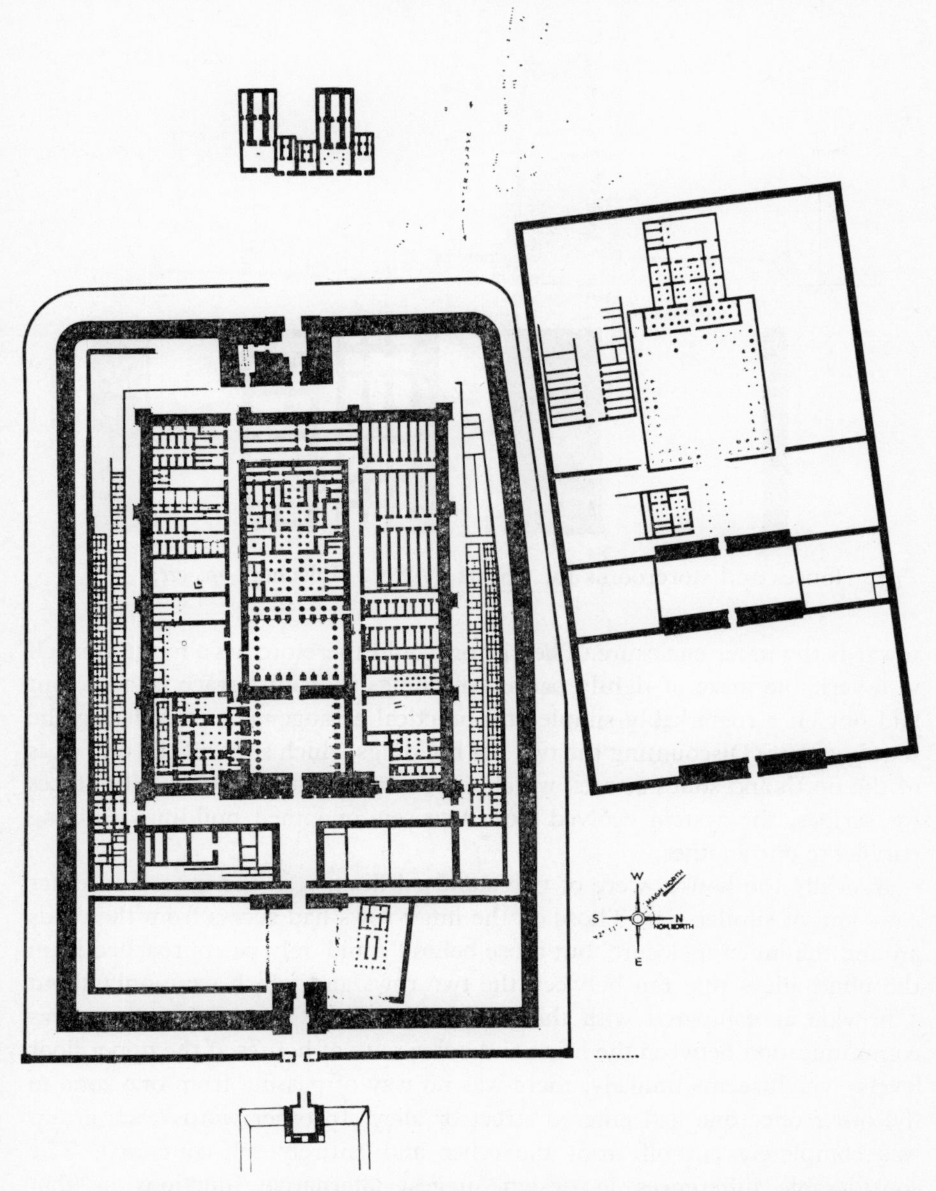

Fig 1 General plan. (After Hölscher, U. *Medinet Habu*, Vol. IV, Part II. University of Chicago Press, 1941.)

Therefore by merely continuing the retaining wall of the pomoerium upwards, at least a first if not other upper floors could easily be added.

To fit the maximum number of houses into the space available the architect had to make each one long and narrow, and they were accordingly laid out in extended straight lines with the entrances in the longer sides facing inwards

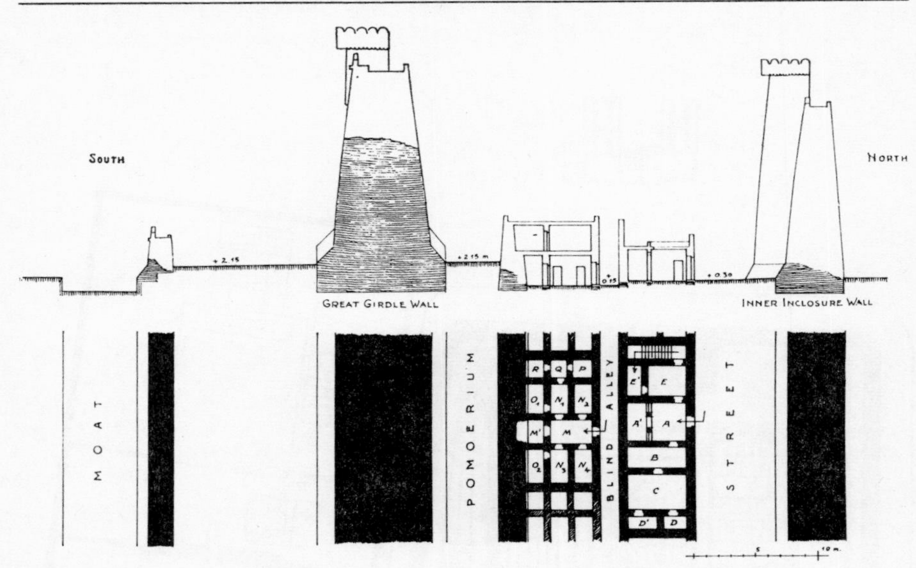

Fig 2 Houses and storerooms for priests. (After. Hölscher, *op. cit.*)

towards the inner enclosure. The visitor would therefore have found himself in a veritable maze of tightly packed dwellings built in terrace fashion, but laid out on a remarkably simple and practical arrangement considering the requirements. Discounting the two big buildings which shut off the east ends of the north and south streets, which were administrative blocks with offices for scribes, the system evolved from two sets of joined buildings running parallel to one another.

Basically the houses were of two standard designs in the inner and outer rows and of similar size. Those on the inner rows had access from the roads around the inner enclosure, but those behind could only be approached from the blind alleys that ran between the two rows and which were only about 5 ft wide as compared with the 20 ft wide roads. Now unless there was communication between the inner and outer rows of houses at the upper floor levels, which seems unlikely, there was no way of passing from one zone to the other once one had entered street or alley. In other words each group was completely cut off from the other and entirely self-contained. The considerable differences in design suggest alternative functions or that different classes of people lived in the two zones. The approach route from the fortified area to the outer homes possibly indicates their use by the guards or temple patrols, while those within looking towards the temple would seem to be connected with priests or temple servants. Such arrangements would fit not only a mortuary temple such as this but also any royal residence or palace sited within a temple enclosure throughout Egypt. The houses themselves were substantial and well-built, the inner ones measuring about 53 by 21 ft and outer 35 by 21 ft.

To reach the breath-taking vision which the temple façade must have presented, the visitor would have had to traverse a 260 ft long raised paved highway, passing through two further gates and crossing a transverse passage on the way. On the left lay a large walled enclosure containing a sunk garden with trees and a rectangular pool or well system about 65 to 70 ft long and at least 30 ft deep. On the right to the north lay the old Eighteenth Dynasty temple of Amun within its own enclosure. Passing through the next brick pylon and the third gate beyond, he arrived at a small square cunningly made by the architect to afford only partial glimpses of the whole temple façade, but also completely dominated at the western end by the temple gateway flanked by the two pylons. To the left lay an enclosure perhaps containing the royal stables and chariot house, to the right another enclosure with a pleasant grove of trees planted out in three rows in deep pits filled with Nile soil and with low mud-brick protective walls built around them. Here was also a building that Hölscher has suggested was a stable and yard for the cattle offered up in the temple services.

No more effective way could have been contrived for showing the power and might of the king. Imagine emerging on to a restricted space, 50 or 60 ft wide and about 86 ft long, hemmed in on three sides by walls about 20 ft high, with the full sweep of the 80 ft high temple pylon in front, a dazzling white mass with brilliantly coloured relief and inscriptions all over the surface. The north tower had a gigantic portrait of the king slaying his enemies before the sun god Rē'-Harakhti, the southern balanced it with a similar scene before Amūn. Red flagstaffs with gaily coloured pennants rose far above the roofs of the towers. A wide but not very deep esplanade ran across the whole of the front of the inner enclosure separating this more sacred area from the outer environs.

Passing through the great gate of the temple the visitor entered the first court which really served the palace as much as the shrine itself, and was therefore almost an atrium. The scenes on the walls were thus of a more secular kind than those to be found further inside the temple. This court measured about 130 ft broad and 112 ft deep. To the left was the south colonnade stretching across the front of the palace which occupied a whole court beyond it on the south behind the stone wall of the temple and comprising eight massive open bell topped papyrus columns encrusted with coloured decoration and bearing the king's names. Opposite was a truly startling reminder of the man whose presence he was now near to approaching, seven huge pillars with gigantic 30 ft tall "Osiride" figures of the king fronting each of them. Even today in their decay they are impressive and one can only imagine what an effect they would have made when new and coloured so brightly as to be almost garish. Although constructed of sandstone courses, they were painted to look as lifelike as possible and as one piece, the deep red brown of the flesh contrasting with the white clothing and many hues of the jewels. All were alike and all wore the feathered Atef crown, being not really mummiform but shown with crossed arms holding crook and flail. Tiny figures of prince and princess beside the legs were on closer examination really life-size and still give the scale.

To the west a ramp led up to higher levels and to the second court which was the beginning of the temple proper. Here one might catch a glimpse of two colossal limestone figures 34 ft high, such portraits when painted being the largest realizations of the royal features possible, for if seated they were proportionately more immense.

It was, however, the palace façade which would arrest the attention most, for here was situated the "Window of Appearances" (Pl. 5) at which the king manifested himself to the more priviledged of his courtiers and visiting peoples who would wait humbly in the court, most probably gathering under the colonnades. The immensely thick stone wall was pierced here by a window and three doors. Under the lower frame of the window more captive heads were arranged, six under the opening and seven at each side, twenty in all, representing all the peoples and nations of the known world, Negroes, Libyans, Semites and Aegeans. These were again coloured to be as realistic as possible, and were arranged so that the king's feet would actually rest upon a ledge over them, an allusion to an age-old African custom in which the chieftain steps over the recumbent bodies of his foes or the conquered, as a visible symbol of their submission to his might. It should be pointed out that the window was really a doorway with either a rail or a cushioned balcony for the king to lean over so that the illusion in the former case was complete. Doubtless the window recess which was approached by a staircase on the other side of the wall, was curtained off or closed with shutters until the proper moment.

And what a blaze of glory surrounded it! Here was concentrated all the splendour of Pharaonic Egypt. The whole façade and especially the window and great doorframes were decorated with inlays of brilliant tiles, glass, faience and polychrome stone set in position with a jeweller's art. The stone being of a somewhat rough quality was unsuitable for producing the finest surfaces for carving and was therefore first coated with a fine layer of gesso as a base. This itself could be worked into the very finest and most delicate relief in contrast to the rougher and deeper carving of the stone generally, and was then painted in a variety of colours, red, dark lapis, lazuli blue, light turquoise blue, yellow, green and so on.

The doorframes had a wonderful golden yellow background used to imitate real gold, which was also doubtless used freely as overlay. Larger figures were carved in relief in stone and had the exposed parts of the body, i.e. face, hands and feet, made up from coloured inlays such as red faience. Blue crowns and red sun disks were also made up in this fashion, and robes and clothing inlaid with small pieces of glass and faience, as were the inscriptions. This was the overlay while large hieroglyphs and ornamental borders were incised in front and inlaid with pieces of glass and other rich materials like alabaster bedded in gypsum. Large pieces were fixed with copper nails as well. Some were grouped like mosaics, others were set in the manner of the goldsmith's art, by the use of wooden cloisons to hold the variegated pieces in strips.

Far from neglecting the bases of the doorways the Egyptians covered each jamb to a height of about 2 ft with sheet metal probably worked to show a panelled design. These panels were then inlaid with a wonderful series of ceramic tiles showing foreign prisoners, each wearing a robe inlaid with innumerable tiny pieces of coloured glass and other materials. The metal strips and plates which protected the door frames from damage seem to have been gilded thus helping to contribute to the effect that caused the window to be referred to in texts as "golden"[14].

And now at last the visitor was to enter the palace itself and see the king within his actual dwelling. Two palaces were built here during the reign of Ramses III, this, the second, although small by residence standards of the time, yet covered most of a rectangular area of about 25,000 ft^2 or a little over half an acre. Like most oriental palaces and private houses of any pretensions it could be divided into two main parts, the public outer rooms and the private and intimate quarters at the rear[15]. It was entered by either of two very small vestibules. A stairway ascended from the eastern one suggesting the possibility of there once having been a second storey over parts of the palace building, otherwise it must have led to the roof and perhaps to some light pavilions for sitting aloft. Hölscher pointed out the fact that as each of these vestibules had an arched roof known to have been 24 or 25 ft high, their interiors would present a somewhat disproportional effect if there was no first floor, but no actual archaeological evidence was found for flooring between two levels. It may be remarked that if such did exist the effect of the contrast between these small chambers and the large reception halls beyond would have been much more striking and truly dramatic.

Between the vestibules running from east to west was a spacious hall with two columns about 14½ ft wide and 34 ft long. From this hall a double stairway led up to the raised platform that formed the inner approach to the Balcony of Royal Appearances.

A magnificently decorated doorway led to the Throne Hall beyond on the south side, the principal and central hall of the palace, whose roof and that of the first hall were raised higher than those of the surrounding apartments. Its ceiling was vaulted and was upheld by six palm (or open papyrus) columns, and it measured 34 ft wide by about 46 ft long and no less than 34 ft high, the columns being of stone and over 24 ft tall. The decoration of this main hall was of standard type, the walls were plastered, then brilliantly coloured and decorated with painted scenes and the main features inlaid with faience, gold and other ornamentation. The bases of the column shafts had small panelled patterns depicting Negro and Libyan captives. The stone window grilles, although rather crudely carved at times, were worked into fabulous designs showing the king's names in cartouches, "lucky" hieroglyphic signs and protective bird deities with outstretched wings, the whole being painted in red, yellow and blue. One measuring about 4 ft^2 was found by the excavators. At the rear of this hall against the further wall stood the throne resting upon a magnificent dais, an alabaster base measuring about 6 ft across and rising about 2 ft above the floor so that it had to be approached from the front by a short ramp. This base was decorated around the sides with little prisoner figures set in panels and probably once had a little pavilion or baldachino erected over it, if contemporary wall scenes are any guide. Behind the throne there also probably stood a double false door similar to those still to be found in temple shrines such as those in the temple of Seti I at Abydos, a true reminder as we have seen of the king's function as a divine ruler.

Our visitor would not have been allowed to approach the divine presence in the direct way just described. After waiting his turn in the outer hall or antechamber he would have been summoned by means of numerous court officials and attendants to the throne hall which was divided down the middle by screens set in front of the second pair of columns and resting in holes let into the floor for their posts. This light barrier must have added much to the mystique of approaching the dread presence of the "good god". The central opening of this barrier was presumably either curtained or else had folding gates, so that while the upper parts of the hall were visible the throne was not, and at a given signal these would be thrown open and the visitor could advance and throw himself prostrating on his face, or as the Egyptian texts say, he "smelt the earth" with his nose[16].

As a rule visitors would not be allowed to enter further into the palace nor to know whence the six doors that opened off the hall led, for beyond this point was the private palace open only to the royal family, servants in immediate attendance and perhaps a few very important ministers on occasion. Three doors opened to the right or west, two to the left and one behind the throne wall balanced as always by a niche on the other side for symmetry. The design was a remarkably systematic and functional one, the two major staircases flanked the throne hall at either side, while the rear door led to a vestibule with bathroom opening off it and an inner hall the private living room also with a throne. This hall was of different design having two columns and an alabaster throne dais (still in position today) against the rear wall, a rectangular platform 6½ by 5 ft, and a little over 2 ft high. A cavetto cornice ornamented the top and stairs ascend at the front and sides, the latter being probably only for the sovereign's use. Opening off this room was the bedroom with a stone-paved niche for the king's bed again approached by three steps, and almost certainly with a windchute above it projecting high above the roof to collect air and cool the bedroom below. The bed space was rather short but quite normal in design. Most of the other rooms need not concern us here; there were three Harem houses at the rear of the main palace building, each a self-contained apartment with several rooms including a small bathroom. Another vestibule on the west side led to another hall with a third throne dais, the wall between these rooms seems to have once had another window of appearances looking out on the court at the north. A garden lay west of the palace. An eastern vestibule was most imposing and had two columns with simplified palm leaf capitals, their grooves filled with gypsum perhaps inlaid with faience and/or glass[17]. The position of this room, or better from a room or the roof above it, would seem to have been the best place to view the wonderful hunting scene carved on the rear or southwest face of the southern pylon of the temple.

In all the main building of the palace contained over twenty rooms on the ground floor.

This is but a glimpse of one of the palaces of Ramses III as it functioned in the days of its original glory. Similar palaces and residences are known to have existed in every other major city in Egypt, some much larger, while the residence city of Per Ramses must have surpassed them all. Add to these the many lesser dwellings of the king and it becomes apparent that what has been described here was not a hundredth part of the whole.

Notes

1 For this layout and city plan see Willetts, W. (1958). *Chinese Art*. II. London. Ch. 8, pp. 658–60, Fig. 92.
2 Creswell, K. A. C. (1958). *A Short Account of Early Muslim Architecture*. London. pp. 161–82, and esp. pp. 259–66.
3 Mallowan, M. E. L. (1966). *Nimrud and its Remains*. 3 vols. London; Gadd, C. J. (1936). *The Stones of Assyria: the Surviving Remains of Assyrian Sculpture*. London; Loud, G. (1936–8). *Khorsabad*. 2 vols. Chicago.
4 Schmidt, E. F. (1953). *Persepolis*, I. Chicago.
5 Windsor Castle is about 13 acres. Among ancient palaces, Persepolis terrace about 33 acres, Nebuchadnezzar's palace in Babylon about 12½ acres, Khorsabad platform nearly 25 acres, N.W. Palace of Nimrud 6 acres, and Knossos only 3–6 acres.
6 Hölscher, U. (1951). *The Excavation of Medinet Habu*. IV: The Mortuary Temple of Ramses III. Pt. II. Chicago. pp. 11–13, Figs. 10–13, pls. 1, 2, 13B.
7 Hölscher, U. (1951). *op. cit.* pls. 6, 12.
8 Hölscher, U. (1951). *op. cit.* Figs. 1, 14, pls. 7–11, 40.
9 Hölscher, U. (1951). *op. cit.* noted the idea of using perspective with these figures placed at different levels to make the gate look deeper and larger.
10 Hölscher, U. (1951). *op. cit.* pls. 3, 7, 8, 16B, 21B, 25H.
11 See Hoag, J. D. (1963). *Western Islamic Architecture*. New York. pp. 45–6. The Abbaside Caliphs also had domed audience-halls over their palace gates as shown by the description of the Round City of Baghdad in Creswell, K. A. C. (1958). *op. cit.* p. 165, Fig. 32. Also Hoag, J. D. (1963). *op. cit.* p. 16, note 9. This feature is also found in Fatimid Egypt where the palace in Cairo had a "sakifa or porch where the Caliph listened every evening, while the oppressed and wronged came and cried aloud"—Lane-Poole, S. (1898). *Saladin*. London. p. 114. Of a comparable nature was the porch for judging of Solomon mentioned in the Bible.
12 Hölscher, U. (1951). *op. cit.* pp. 6–8, pls. 16A, 18B, 20, 21, 22, 23, 24.

13 A contemporary papyrus preserves interesting details of the acquisition of concubines for this king at Thebes.
14 Hölscher, U. (1951). *op. cit.* pp. 38–42, Figs. 51–4, pl. 5.
15 Hölscher, U. (1951). *op. cit.* pt. 1, pp. 7–8, 37–59.
16 As stated in many Egyptian texts and depicted in many reliefs and scenes. For this type of division see also the arrangements at Alalakh in the palace of Yarim-Lim, Woolley, Sir L. (1953). *A Forgotten Kingdom*. London. pp. 73–4, Fig. 12 and in some of the smaller palaces in Crete, Evans, Sir A. (1928). *The Palace of Minos*. II, pt. II. London. pp. 391–4: "House of the Chancel Screen", pp. 396–413, esp. p. 412: "Royal Villa".
17 A similar treatment to that described in Petrie, W. M. F. (1894). *Tell el-Amarna*. London. p. 10, pl. VI.

B: Western Asia and the Aegean

ROBERT McC. ADAMS

Patterns of urbanization in early southern Mesopotamia

In any period or locality, the phenomenon of urbanization is only to be understood within a broad spectrum of closely interrelated social and cultural processes. Truly urban agglomerations depend upon the institution of the state as a political form, and the emergence of the latter is but an aspect in turn of the formation of stratified class societies. Also linked to the emergence of cities are increased levels of specialization in labour; innovations like writing that facilitate administration and assure the pre-eminence of a few of the many strands of unrecorded tradition; and symbolic affirmations of the newly achieved status of kings and cults in the form of monumental art and architecture. The sequence and relative importance of these and similar features naturally varies from case to case, but V. Gordon Childe's description of their common outcome and cumulative effect as an "Urban Revolution" continues to be an illuminating one[1].

The study of urbanization thus must be carried on within a context of numerous, interrelated developmental processes. However, the essence of any analytical procedure is not to dwell descriptively on the uniquely occurring combinations of properties characterizing individual cases but to disentangle the web into its components. Surely the geographic aspects of the process of urbanization form one such component or avenue of approach, on which many new data have recently become available and on which the comparative perspectives generated by the presence of a large group of participants is particularly useful. Hence the geographic aspects of the initial onset of urbanism in southern Mesopotamia are the central emphasis in this paper.

Essential to a geographical perspective on urbanism is a regional rather than a site-limited framework. Cities are never independent of their hinterlands, even if developmental forces in the two run at times in the same and at times in an opposite direction. Moreover, regions are seldom if ever uniform in their resources, and the disequilibrium of regional components can be a powerful stimulus or deterrent to urban growth.

In these circumstances, it is most unfortunate that the traditional sources on the growth of early Mesopotamian civilization are urban-centred and consequently ill-adapted to formulating a more balanced regional view. Excavations at sites of the fourth and third millennia B.C. are virtually limited to major towns and cities. Within this highly selective group, the prevailing emphasis has been not on the living arrangements of the mass of the population but on public buildings and their associated archives and works of art. Non-artefactual materials, particularly relevant to questions of the subsistence economy, have been almost wholly neglected. Rich as they are for other purposes, the written sources obviously suffer from most of the same biases; in addition, the early cuneiform texts are subject to many limitations and ambiguities of their own.

Archaeological reconnaissance constitutes the only significant exception to the presently discouraging state of the evidence for regional patterns of urbanization in southern Mesopotamia. There has never been a comprehensive programme, but within the last decade and a half surveys of an at least preliminary character gradually have formed a patchwork extending virtually the entire length of the alluvial plain. While it is not within the scope of this paper to summarize the results and limitations of every phase of past reconnaissance in detail[2] the cumulative effect has been to provide a rough but serviceable indication of changes in regional configurations of settlement that reflect—and that must have contributed to—the initial establishment of cities.

It is important to outline certain points of view and technical constraints within which most surveys of the Tigris-Euphrates flood-plain heretofore have either elected or been forced to operate. To begin with, although varying considerably in intensity and systematization, surveys to date have tended to emphasize not the exhaustive study of small districts but the preliminary blocking out of spatial and temporal settlement patterns affecting fairly large areas. The availability of only relatively small-scale aerial photographs (1 : 35,000) and a map grid of variable and generally inferior quality has been one consideration leading to this approach; political uncertainties that militate against long-range project designs have been another. In any case, the outcome is that individual surveyed sites have had to be dated on the basis of rapid, casual, and essentially subjective assessments of the presence or absence of a limited number of ceramic "index fossils" rather than through more time-consuming, quantitative studies of exhaustive collections made within randomly selected areas. Perhaps as a result the chronological intervals used for dating are broader and less reliable than they might have been otherwise, although even this is by no means obvious in view of the absence of quantitatively studied ceramic sequences from excavations in this time range against which surface collections might be calibrated.

A second set of constraints is more geographically selective. As a result of rapid advances in technique, some of the earlier surveys discriminate less adequately than more recent ones between the successive time intervals into which the Urban Revolution can now be divided. This reduces the utility of data drawn from the two territorially most extensive surveys—the only ones, as it happens, in the northern part of the alluvium; one covered most of the region of ancient Akkad between the Tigris and Euphrates, while the other included the lower flood-plain of the Diyālā River[3]. In both cases, moreover, a considerable part of the reconnaissance had to be conducted in areas which are at present cultivated, where later experience has made it clear that only much more intensive survey techniques than were then employed would permit the detection of many small, low sites. In short, improvements in technique and localized differences in preservation limit the comparability of surveys in different areas.

A related difficulty arises from topographic conditions first observed around the ancient city of Uruk, near the southern end of the alluvium. A detailed survey in that area furnishes the largest body of data yet available on changes in settlement patterns accompanying the development of urbanism. Close scrutiny of geomorphic processes there makes clear, however, that the land surface surveyed by the archaeologist is a complex product of variable, interacting agencies rather than the outcome of uniform, general alluviation as has generally been assumed. Both wind erosion and wind deposition have played a very significant part, with the latter in particular probably serving far more effectively than alluviation to obliterate surface traces of smaller settlements. As a consequence, there have been unknown—but probably considerable—losses from the configurations of settlement that can be plotted by surface reconnaissance for successive phases of urbanization[4]. Similar findings have been made still more recently further to the north, around ancient Nippur, in the first stage of what has been planned as a progressively more intensive and systematic reconnaissance.

A further limitation concerns the interpretation of multi-period sites. "Index fossils" brought to the surface through many kinds of disturbances have been shown to indicate with considerable accuracy the span of periods during which an occupation of some kind occurred. The area and character of the terminal phase of occupation can be determined from the surface in such cases, just as in the case of single-period sites. Statistical sampling procedures, or even carefully followed non-quantitative inspection, may in addition disclose areas abandoned in a previous phase and not re-occupied. But only by making a number of assumptions that are difficult to test, because of the virtual absence of excavations in this time range directed to smaller sites, can we deal with the question of site area during a succession of earlier periods of occupation. Yet reliable detailed information on long-term changes in occupational areas is precisely what is needed if we are to deal effectively with a theme like urbanization.

One further reservation is to be noted, of an entirely different character from the preceding ones. The Tigris-Euphrates alluvium is a distinct physiographic unit, and the history of its human occupation provides numerous instances of considerable cultural and political autonomy. However, an understanding of the historic development of civilization there can only be partial and distorted if it ignores interactions on a still wider geographic scale. Evidence is accumulating for a number of close relationships between the alluvium and adjacent areas. To the south, for example, it is now clear that there were numerous small settlements in the late fifth and early fourth millennia B.C., both along the northern part of what is now the Saudi Arabian coastline and further inland. Permanent settlements seemingly disappeared in this now-desert area, not long after the Ubaid period, giving rise to the possibility of a considerable movement of population northward into the alluvium just as the process of urbanization was getting under way. Similarly, details are currently emerging on highly developed caravan and coastal trading networks of a slightly later period, specializing in commodities like raw and processed steatite and extending not only across the Iranian plateau but along much of the length of the Persian Gulf. What role this trade may have had in generative processes leading to Mesopotamian urbanization is admittedly not yet known. But to dismiss long-distance trade on *a priori* grounds, as derivative from the growth of urban civilization rather than having perhaps helped to bring the latter into existence, seems clearly premature. In short, these examples suggest that in terms of process the Mesopotamian alluvium should never be described as a fully self-contained, bounded ecosystem.

While the combined effect of all these difficulties may be considerable and cannot be ignored, it is equally true that surveys to date have shed light on a number of aspects of the urbanization process and the natural conditions under which it occurred. We turn next to a summary of findings which appear to have been at least tentatively established. It will then be possible to consider in a concluding section how, through the design of more comprehensive future programmes involving both reconnaissance and excavation, present hypotheses can be supplemented and tested.

Perhaps the most important point to be noted at the outset is that there is considerable variability within the alluvium as to the timing and extent of urbanization trends, as well as to the relationship between the growth of particular centres and their smaller, more numerous outliers. Conditions were not homogeneous even within relatively small districts, so that in at least some cases closely neighbouring city-states followed markedly divergent paths of development. In the classic heartland of southern Sumer, for example, what is known of the histories of Uruk, Ur, and Umma, all within a day's foot journey of one another, seems presently to defy the application of a single paradigm of urban growth.

Uruk was a substantial ceremonial centre, probably approaching urban proportions (although its size has not yet been clarified by excavations or detailed topographic reconnaissance) by the mid-fourth millennium B.C. and perhaps earlier. Around it lay a very large number of small towns and villages, unimodally distributed in size rather than forming a differentiated, tiered hierarchy suggestive of an economically or administratively differentiated network centred on Uruk. At around the beginning of the third millennium the city seems to have expanded rapidly in size by a full order of magnitude, perhaps housing as many as 40,000 to 50,000 inhabitants on the 400 ha within its newly constructed defensive wall. The evidence of simultaneous, widespread abandonment of small settlements in its hinterlands leaves little doubt that this extraordinary growth was essentially an implosive process, transferring rural population into a new, urban setting in response to some combination of internal tendencies toward the consolidation of political leadership and external military threats.

Initial agricultural settlement in the district around Ur may have been earlier and denser than around Uruk. In spite of this, Ur never attained more than an eighth of at least the physical size of Uruk, and seems to have crossed the threshold of urbanism several centuries later. Its rural sustaining area was correspondingly smaller and more poorly developed, and a forced transfer of population to the city from its hinterlands either did not occur or was apparently much less significant than in the case of Uruk[5]. We may need to consider as an alternative the ebb and flow of population between coeval city-states as their respective political fortunes advanced or declined, for at the time of Ur's ascendancy Uruk seems to have been less extensively occupied than it had been previously.

Umma is much less well understood than either of the previous cases, both because it has never been scientifically excavated and because much of its environs is shrouded in dunes. Surface reconnaissance suggests that its period of urban growth coincided more nearly with that of Ur than with the earlier growth of Uruk, and that it is intermediate in size between the two. In other respects, however, the pattern of its growth appears to be an entirely distinctive one. The immediate district was almost unoccupied for most of the fourth millennium, then suddenly and intensively settled with dense clusters of large towns that in some cases were almost adjacent to one another. Such town-clusters had no counterparts around Uruk or Ur, and probably were totally absorbed into the population of Umma when the latter joined the growing ranks of city-states before the mid-third millennium B.C.

To complete this brief array of local contracts, we can turn to conditions further north. Around Nippur, near the centre of the alluvium, early settlements of the Ubaid period are much less well attested than in the south. The subsequent pattern converges on that already described for the area around Uruk, although the abandonment of outlying villages and small towns in favour of nucleated, urban sites may even have begun slightly earlier. In the

Jemdet Nasr period toward the end of the fourth millennium, a time of maximal rural dispersion around Uruk, the Nippur countryside had already been virtually depopulated. There is, however, some evidence suggesting that a formerly important channel of the Euphrates in the vicinity fell permanently into disuse at about this time. Hence Nippur and its neighbouring urban centres did not necessarily emerge as an immediate consequence of rural abandonment. Pending clarification through further excavations of Nippur's size and composition in the late fourth millennium, it is also possible that there was an intervening stage in which a few main towns like Nippur struggled on while most of the rural population simply abandoned the district.

A still further contrast is provided by patterns of settlement at the northern end of the Mesopotamian plain, on the alluvial fan of the Diyālā River. True cities did not appear in that district during the span of time under discussion here, although towns embodying many of the same functions on a smaller scale were present already by the late fourth millennium. Stable groupings of towns and smaller dependencies constituted an undoubtedly viable alternative to the emphasis on urbanism further south, for the distribution and increasing number of sites suggests that the density and frontiers of settlement in this area continued to expand vigorously until at least the late Early Dynastic period.

In spite of the sharply divergent paths of development followed by different cities and regions, what does seem to be uniform are the basic features of the alluvial landscape. Delineation of the riverine features of that landscape depends primarily on the assumption—amply documented in the fully preserved systems of settlement recorded for later periods—that Mesopotamian communities have always gravitated to positions on levee-summits adjoining permanent, dependable water-courses. From this it follows in turn that the approximate courses of at least the larger, more significant rivers and canals can be traced by the adjoining lines of contemporaneous sites identified through surface reconnaissance. In addition, as noted earlier, it has recently become apparent that in certain areas, watercourses dating from the fourth and fifth millennia can be observed and mapped directly, as a result of their recent re-exposure through wind erosion.

The common characteristics of the late prehistoric river system, in all parts of the Mesopotamian plain, are those of an essentially natural alluvial regime rather than one extensively modified by large-scale irrigation. Channels as a rule were multiple, small, unstable, meandering and anastomosing. Settlement and irrigation were confined to relatively narrow bands along the backslopes of natural levees, and to favourably situated tracts along the margins of the numerous permanent or seasonal swamps. Not all areas could have been reached at any one time by the prevailingly natural drainage pattern, so that in places semi-arid steppe was to be found that was suitable only for grazing. The wide and fairly uniform distribution of late fourth millennium sites implies, however, that unwatered steppe occurred in scattered small pockets rather than in great undifferentiated zones.

A second uniform feature of the setting and process of urbanization involves the configuration of settlement existing before its onset, in the late Ubaid period at roughly the beginning of the fourth millennium. All known sites of the time can be classed as small towns, villages or hamlets, not exceeding 10 ha or so in area. Present evidence suggests, to be sure, that the earliest widespread agricultural occupation occurred in the southern part of the alluvium, but even this apparent regional disparity in development may yet prove to be at least partly an artefact of the generally greater depth of alluviation near the upper end of the plain. More striking is the characteristically dispersed distribution of Ubaid sites in all areas, scattered at fairly uniform intervals rather than arranged in clustered enclaves as later became the prevailing pattern. It is also noteworthy that, with the possible exception of a handful of centres like Uruk that may have emerged as major ceremonial nuclei not long after the Ubaid period, Ubaid settlements seem to have been slightly larger on the average than their immediate successors. The meaning of this early tendency toward nucleation is still quite uncertain. Perhaps it is somehow associated with the initial colonization of the plain itself, in which sedentary communities successfully pioneered a new subsistence regime involving the balanced pursuit of fishing, herding, and irrigation agriculture.

A further regularity appears following the Ubaid period. The Ur enclave constitutes an apparent exception and elsewhere the time of onset varies, but in general there was a major, rapid increase in the number of small settlements. In the absence of a well-controlled time sequence, and of excavation concerned with subsistence and demographic questions, we can only posit alternative explanations for this phenomenon without being able to narrow the choices among them. A "natural" multiplication of population, the budding-off of daughter-communities as a successful subsistence adaptation spread, certainly is one. Another explanation involves the conversion to sedentary life during the Uruk period of what were formerly semi-sedentary fishing, shell-fish collecting, hunting and herding folk, whose earlier presence somehow failed to leave superimposed mounds of débris for the archaeologist to detect (although remains of shell-fish, common at many small Uruk sites, should have led to the identification of earlier sites also). But still a third, and perhaps the most likely explanation if the increase occurred within a few generations, involves a substantial process of immigration into the alluvium. Obviously this possibility touches upon what has been traditionally formulated as the "Sumerian Problem", but the available data, from excavations and surface reconnaissance alike, do not permit us to advance further at present towards its solution.

An additional feature of this increase in the number of settlements and the density of rural population is a pronounced shift toward highly clustered site distributions. Distribution maps make it apparent that the tendency toward clustering occurs fairly generally throughout the Tigris-Euphrates alluvium, although only in the Uruk-Umma region has this subjective impression yet been tested quantitatively. Nearest Neighbour coefficients in the order of 0·3 to 0·4 occurring there in the late Uruk and Jemdet Nasr strongly confirm such a tendency. The usefulness of the statistic is somewhat vitiated, however, by two problems alluded to earlier: uncertainties in delimiting the boundaries of the settled area to be considered, partly due to possible losses in areas with overburdens of wind-laid silts; and uncertainties in calculating even very roughly the area of multi-period sites during their earlier phases of occupation. Possibly more useful is an alternative statistical approach developed in botanical ecology[6] in which the actual distribution of coeval sites of all sizes within a geometric checkerboard of quadrats covering the survey area is compared with a random Poisson distribution. By either approach, the fact of a contagious distribution of small, rural settlements is made immediately clear.

Its explanation is another question, and probably not one to which there is a unitary, simple answer. Some clusters, consisting only of a pair of sites of approximately equal size, may represent no more than a single community's seasonal alternation between opposite banks of a watercourse. Others, more dispersed and with more numerous components, recall complex herding and farming adjustments along the seasonally expanding and contracting margins of modern Iraqi marshlands. Still others seem to form substantial agricultural enclaves, with numerous small villages and sometimes a single, larger one which may have housed the functional equivalent of a nineteenth century A.D. shaykh in his mud-walled fortress. But if the individual patterns and explanations vary, they are alike in suggesting an intensely rural adaptation in which the recognized social units were prevailingly small and highly localized.

This is the landscape of settlement upon which urbanization was then rapidly imposed. Significant exceptions have been noted, but during roughly the first half of the third millennium wide areas of the alluvium were heavily depopulated as the overwhelming majority of their inhabitants were persuaded or compelled to take up residence in politically organized city-states. In the general vicinity of Uruk, where this process can be illustrated best with the data at present available, the number of villages and towns falls precipitately from 146 recorded for the Jemdet Nasr period, to 76 during the succeeding Early Dynastic I period, to 24 in later Early Dynastic times. Meanwhile the number of cities—here defined, in purely spatial terms, as exceeding 50 ha in size—climbs from two, to four, and finally to eight. What is involved in urbanization is clearly the movement of population into newly founded or greatly enlarged urban settings, often at the cost of abandoning large, formerly fertile districts. There is nothing whatever to suggest any growth in the aggregate population of the alluvium during or after this transfer was consummated.

Massive structural changes naturally were going on simultaneously in the newly emergent cities. Little discussion of them is needed here, since the urban origin of virtually all of the textual and archaeological data has meant that the formation or great expansion of the urban centres is, if not without major themes of controversy, at least relatively much better understood. Perhaps most obvious is an increasing emphasis on militarism, reflected in the construction of fortifications; in the mustering and equipping of large bodies in militia; in the emphasis on martial equipment in the so-called "Royal Tombs"; and in a host of myths, epics and historical inscriptions recounting the internecine struggles of the city-states themselves. Responsibility for the conduct of military affairs was lodged not in the older, economically and ritually oriented, institutions of the temples but in the newly emergent dynastic state. And closely linked to the latter was the increasing stratification of the society at large. This involved, *inter alia*, the decay of kin-based social units; increasing concentrations of private wealth in land; and the employment of modest numbers of war captives and other slaves not only for household service and agricultural subsistence, but also for commodity production.

Turning once again to the countryside from which the urban population was recruited, we may ask in more detail what the effects of urbanization were. Most immediately apparent is the virtual absence of outlying villages or towns within ranges of 5 to 15 km depending upon the area (and presumably population size) of the individual city. This, one must assume, was the zone within which cultivation was carried on by the urban population. As is documented in abundant detail in textual sources, the Mesopotamian prototype of urbanization did not provide the conditions for the overwhelming bulk of the new city-dwellers to shift from primary subsistence activities into the crafts or service occupations.

This emphasizes once again the essentially artificial character of early Mesopotamian cities, at least from a socio-economic viewpoint. They were amalgams brought together to increase the economic well-being and offensive and defensive strength of a very small, politically conscious superstratum. In the view of the mass of the population, there is ample evidence (at least from somewhat later times) that the advantages and disadvantages of urban life were more precariously balanced. On the one hand, cities provided relatively greater security than was available outside the walls, not only for one's person and household goods, but also for institutionally maintained stockpiles of foodstuffs sufficient to meet prolonged military crises or natural disasters. On the other hand, they exposed the individual to heavy demands for taxes and military and corvée service. The net effect of these opposing considerations was that the greater part of the population was tied to the city only under varying degrees of duress; was only marginally affected by many of the most characteristically urban institutions; and can be described only with some risk of hyperbole as having been significantly urbanized in outlook.

Several further effects of cities can be traced on both their immediate and more remote hinterlands. Defensive considerations and the drawback of long commuting time would have placed a premium on cultivating areas as close as possible to the city walls. This may have led, although as yet we can show nothing of how and to what degree it did so, to new forms of capital investment in land such as improved irrigation works to enlarge and sustain agricultural output. At the same time, there is some textual evidence for new forms of temporary settlement that would have been needed beyond convenient daily commuting range particularly during the harvest season. Whether these quite ephemeral groupings around threshing floors and similar installations can ever be detected through techniques of surface reconnaissance remains to be established.

At still greater distances from the cities, land in many cases apparently passed out of cultivation altogether. The high *edin* is perhaps the best example of great steppe marches that were formed in this way along the frontiers between periodically contending city-states[7]. Such zones had the secondary advantage of serving as buffers, augmenting a defence in depth against overland military encroachments. Much increased in size over earlier scattered pockets of steppe-land, they also provided a base of operations for non-sedentary, perhaps ethnically distinct, herdsmen who were increasingly less dependent on their close symbiotic ties to cultivators.

At least in the classic heartland of southern Sumer, these developments led to a concentration of by far the greater part of the region's entire population within city-states of varying size. A disproportionately large number of small communities apparently were the first to be abandoned. As a result, there emerged fairly rapidly a more differentiated, tiered hierarchy of non-urban sites. From a large number of small settlements with a unimodal size distribution centring on a mean of 2 ha or so in the late Uruk period, by the late Early Dynastic period the countryside between Uruk and Umma saw a much smaller number of settlements distributed more or less evenly between small hamlets and towns of widely varying size. Moreover, it can be shown that the larger towns generally lie at greater distances from the cities; in other words, cities had an inhibiting effect on the growth of towns in their immediate vicinity. Quite possibly the size of the towns correlates with progressively higher steps in a four- or five-stage administrative hierarchy by which rural areas were controlled from the cities[8]. On the other hand, this interpretation must be qualified by the fact that much of the former rural population had been brought directly into the cities. In addition, the existence of deep rivalries between city-states would have seriously jeopardized durable administrative ties with fairly remote towns. Hence the explanation for the latter—at any rate, in the Early Dynastic period—seems more likely to lie in the maintenance of a considerable degree of autonomy by medium and large-size towns rather than in their intermediate position in an urban-centred administrative network.

This point touches upon an important theoretical and methodological issue, the relevance of the classical Central Place Theory to early Mesopotamian settlement patterns. Perhaps the ultimate test will be a rigorously empirical one, involving the degree of fit between formal, urban-centred hexagons and the known distribution of outlying smaller sites. But there are a number of intervening considerations that induce scepticism as to the explanatory utility of models available at present.

One argument involves the availability of a substantial time perspective in the Mesopotamian data, as contrasted with the essentially synchronic perspective of the formal Central Place models of Lösch and Christaller. The latter are hardly germane to the hyper-developed urbanism of the late Early Dynastic period, when the overwhelming preponderance of cities is illustrated in the Uruk-Umma area by the fact that cities constituted one-quarter of the total *number* of all known settlements, from the smallest hamlet onwards. Nor are the constructs of Central Place Theory very helpful for the Uruk period, perhaps three-quarters of a millennium earlier; at that time very small settlements were numerous, but a graded hierarchy of intermediate-sized towns was lacking. Having questioned the applicability of Central Place models to both the lower and the upper end of our time continuum, however, it must then be asked what even a fairly precise fit of formal hexagons with the intervening Jemdet Nasr and Early Dynastic I period settlement patterns would really signify. If the whole configuration of settlement was demonstrably in transition between two states, to both of which the formal models are not applicable, then an intervening period of congruence does not necessarily confirm any of the alternative steady-state constructs that Central Place Theory specifies.

The foregoing argument applies only to those portions of the alluvium in which defensive and other considerations led the great bulk of the population to congregate in cities. In Akkad and the Diyālā region, occupying the northern part of the plain, we have seen that stable configurations were established involving hamlets, villages, towns of varying size, and a few, widely scattered cities. In districts like these, Central Place Theory may well be more pertinent. There is, however, a further set of negative considerations that appears to apply fairly uniformly to all parts of the alluvium.

Central Place Theory posits an explicit set of underlying assumptions, including a plain surface offering uniform inducements for settlement and opportunities for communication proportional only to distance. In a number of highly significant respects, the Mesopotamian plain fails to meet these conditions. One steep internal gradient is based upon proximity to the permanent watercourses upon which agricultural subsistence has always depended. Another follows an upstream/downstream axis along any particular channel, with more adequate, less vulnerable water supplies for irrigation being concentrated near the former[9]. In addition, textual sources make clear that overland transport was minimal until long after the periods we are concerned with here. Hence relatively fixed and by no means uniformly distributed watercourses suitable for barge transport were the indispensable avenues of commerce and communication. This strongly suggests that it is navigable channel distance, rather than abstract spatial co-ordinates, that should govern in any Central Place models applied to early Mesopotamian conditions.

Similar problems arise from the setting aside of large steppe areas as border marches, suitable only for non-sedentary herdsmen. The effect of such differentiated patterns of land-use and intensity of settlement is to require at least partial modification of the usual parameters of a Central Place landscape. The goal of detecting regularities in settlement through the application of a body of formal theory, and in this way of testing the relative importance of factors in urbanization, of course remains as valid as ever. But it is equally important that the assumptions on which formal theory rests be congruent with the natural conditions for settlement that Mesopotamia offered.

While the regularities discussed above indicate that some progress has been made during recent years in understanding certain processes associated with urbanization, it is also clear that few or no generalizations have been adequately demonstrated. In important respects, all of the proposed regularities remain undetected because methods have not yet been employed that will elicit and test them. Hence a brief listing is appropriate of currently outstanding problems that can be studied with techniques either already available or in prospect.

An improved approach to problems of changing population density is of a high order of importance. There is growing interest in the hypothesis that population growth has been a major, independent variable conducive to socio-economic development, and in particular that it has played a part in evolutionary change leading to the rise of primitive states[10]. Yet we have been constrained to work heretofore with the relatively crude assumption of a direct, invariant relationship between the area of an archaeological site and the number of its inhabitants. Such an assumption takes no account of the possibility of differences in density related both to functional specialization and to increasing size, although scattered contemporary data strongly suggest that such differences exist.

In this respect, one urgent requirement is improved Middle Eastern ethnographic data with which archaeological findings might be compared. The only substantial body of relevant information now available, at least with the requisite degree of detail, involves a small area in south-western Iran that may not be entirely applicable to conditions in the Tigris-Euphrates alluvium[11].

Another requirement, crucial in its own right if there is to be any significant advance in knowledge on the problem of urbanization as a whole, is a programme of excavations on the Mesopotamian plain directed to a representative range of fourth millennium sites. Large-scale as well as carefully sampled small-scale exposures in smaller sites are particularly needed, but, in addition, attention must be given to non-ceremonial precincts of major centres. Building on the results of such a programme, for example, the question of regional population density can be much less speculatively approached than at present through controlled, systematic surface collections. Replacing qualitative *index fossil* methods of the past with a rigorously quantitative, sampling approach, available surface collecting techniques now also allow us to deal with related problems like subsistence specialization and seasonal variation within and between clusters of sites.

Refinements in excavation and survey techniques that will permit a closer specification of time intervals are both necessary and practical. As adumbrated earlier, choices between major processual alternatives often depend on the length of the time span with which a transformation like the emergence of cities occurs. Sequences of stratigraphically secure, radiocarbon-dated samples may provide the absolute chronological framework for this. But such information can be effectively tied to the results of surface reconnaissance only if pottery and other artefacts in the associated levels are fully—i.e. quantitatively and not selectively—reported.

As indicated earlier, multi-period sites offer a number of complex problems of interpretation that at present force us to depend on many untested assumptions. Hence another objective of an excavation programme ought to be a systematic scrutiny of a representative sample of such sites. Any estimate of population, for example, depends upon whether it is more accurate to regard the numerous multi-period sites as continuously occupied or periodically re-occupied. Similarly it would be useful to know whether there was some fairly stable relationship between the occupied area of earlier periods and that of the terminal period. If no such stable relationship is found to exist, then it becomes increasingly important to develop hypotheses relating earlier size to potentially observable features of surface collections, and to test these hypotheses with subsequent excavations. Promising beginnings along these lines have already been made, and as usual appear to be heavily dependent on the adoption of a quantitative approach[12].

Needed in addition are re-studies of areas like those reported earlier, in which preliminary surveys have already been completed. A programme of examining small, randomly selected sample areas should indicate the extent to which the rapidity and relative superficiality of previous, large-scale studies have led to the omission of easily recognized sites. Equally important would be a detailed assessment of the extent to which recognizable, built-up centres of occupation were supplemented by scattered traces of rural settlement in various periods. For this purpose, only an extremely intensive gridding of small areas on foot will suffice, and it is worth noting that the adoption of such procedures in other regions has led to an increase in the estimated number of settlements by a full order of magnitude[13]. Particularly important for such studies will be areas in which preliminary observation has indicated that ancient land surfaces currently are being re-exposed by wind erosion.

Much of the foregoing discussion has been concerned with the need for a shift from an approach tied to the distinctive (usually monumental) features of individual sites to one concerned with regional themes of development. The study of settlement patterns is obviously central to such a shift. Settlement patterns are, however, more limited than regional patterns. They deal with nucleated sites at relatively limited, fixed locations, not with the more extensive patterns of activity by which varying types of settlements were sustained and articulated with one another. Hence, if we are to understand the processes of urbanization, it is essential that research programmes be undertaken allowing us to come to grips with, for example, varying patterns of land-use and the characteristics of irrigation regimes.

To be sure, techniques for doing so are considerably less developed than those discussed previously. With regard to land-use, there are serious, unsolved problems in equating the soil characteristics of a specific stratum in a presumed former field with the use to which that soil was put in a particular time horizon, independently of a complex, shifting succession of later uses and natural events. Nevertheless, it is a pleasure to note that systematic attempts are finally under way to do just this[14].

With regard to irrigation, prospects are equally promising. It has been shown that at least some characteristics of ancient Mesopotamian watercourse régimes can be deduced from studies of the associated molluscan fauna[15]. Calculation of volume of flow, based on channel slope and cross-sectional area, is a routine problem in civil engineering. Hence it is not at all visionary to suppose that before long we will be able to deal quantitatively not only with settlement patterns and their associated populations, but also with the irrigation supplies they deployed and the areas and crops they cultivated.

In short, our future task is to integrate reconnaissance with excavation in a systematic approach to urbanization as a regional process. To do so, it will not be enough merely to apply present methods and approaches to previously unexcavated sites and unsurveyed areas. Methodology cannot remain static unless the field is to remain static, but instead it must change progressively as problems change. Clearly, what has been reported here are only the first steps in what should become a multi-stage research design. Now it is time to move on to the later stages.

Notes

1 Childe, V. G. (1950). The urban revolution, *Town Planning Rev.*, **21**, pp. 3–17; Adams, R. McC. (1966). *The Evolution of Urban Society: Early Mesopotamia and Prehistoric Mexico*. Chicago.

2 Adams, R. McC. (1970). The study of ancient Mesopotamian settlement patterns and the problem of urban origins, *Sumer*. **25**, pp. 111–24.

3 Adams, R. McC. (in press). Settlement and irrigation patterns in ancient Akkad, *in* Gibson, McG. The city and area of Kish, *Field Research Projects*. Coconut Grove; Adams, R. McC. (1965). *Land behind Baghdad: a History of Settlement on the Diyala Plains*. Chicago.

4 Adams, R. McC. and Nissen, H. J. (in press). *The Uruk Countryside*. Chicago.
5 Wright, H. T. (1969). The administration of rural production in an early Mesopotamian town, *Anthrop. Papers, Univ. of Michigan*, **38**.
6 Greig-Smith, P. (1964). *Quantitative Plant Ecology*. New York.
7 Jacobsen, T. (1958). La géographie et les voies de communication du pays de Sumer, *Rev. d'Assyriologie*, **52**, pp. 127–9; Sauren, H. (1966). *Topographie der Provinz Umma nach den Urkunden der Zeit der III. Dynastie von Ur. Teil I: Kanäle und Bewässerungsanlagen*. Heidelberg.
8 Wright, H. T. (1969). *ibid.*
9 Fernea, R. A. (1970). *Shaykh and Effendi: Changing Patterns of Authority among the El Shabana of Southern Iraq*. Cambridge.
10 Boserup, E. (1965). *The Conditions of Agriculture Growth*. Chicago; Spooner, B. (ed.). (in press). *Population Resources and Technology*. Philadelphia.
11 Gremliza, F. G. L. (1962). *Ecology of Endemic Diseases in the Dez Irrigation Pilot Area*. A report to the Khuzestan Water and Power Authority and Plan Organization, Government of Iran, New York.
12 Redman, C. L. and Watson, P. J. (1970). Systematic, intensive surface collection, *Amer. Antiq.*, **35**, pp. 279–91.
13 Plog, F. T. (1969). An approach to the study of prehistoric change, *Ph.D. diss., Dept. of Anthropology, Univ. of Chicago.*
14 Wright, H. T. Personal communication.
15 Harris, S. A. and Adams, R. McC. (1957). A note on canal and marsh stratigraphy near Zubediyah, *Sumer*, **13**, pp. 157–63.

KEITH BRANIGAN

Minoan settlements in east Crete

Although three of the four major Minoan palaces yet discovered lie in the central part of the island of Crete, and in spite of the recent excavations conducted on parts of the towns surrounding these three palaces, it remains true that our only clear impressions of Minoan towns as a whole must be gained in east Crete. The British and American excavations there during the early years of the present century revealed large parts of three Late Bronze Age towns—Pseira, Palaikastro, and Gournia—as well as more fragmentary remains of the Early and Middle Bronze Ages on the same sites and at Vasiliki. Unlike palatial architecture, however, these settlements have been little studied as examples of urbanization in Minoan Crete, and even Richard Hutchinson's pioneer study of Minoan "town-planning" devoted far more space to a discussion of the architecture of individual buildings than to more general topics such as population, roads, drainage, siting and water supply[1]. It is with these and similar matters that this paper is largely concerned. We may begin, however, by discussing earlier developments in urbanization in eastern Crete.

Of the Neolithic settlements we know little, and there were probably very few of them to judge by the scarcity of even Late Neolithic pottery in the east of the island[2]. The only settlement excavated, at Magasa, was seemingly an isolated farmhouse and nowhere in the east are there neolithic cemeteries to suggest the existence of larger communities[3]. The same is true of the sub-Neolithic and Early Minoan I periods in eastern Crete, and here there is certainly a contrast for in central Crete there is sufficient information from Knossos, Phaistos and Ellenes to recognize the emergence of village communities, to which the cemeteries of Mesara type tholoi also bear witness[4]. It is not until Early Minoan II (*c.* 2600 to 2200 B.C.) that communities of this sort begin to appear in eastern Crete on Mochlos and at Priniatiko Pyrgos. Contemporary with them, however, are settlements of a type not found (to date) in central Crete—the "mansions" at Vasiliki, Fournou Korifi (Myrtos) and possibly Palaikastro[5]. Of the latter we have only tantalizing fragments, though sufficient to recognize an exceptionally large building, and there are good reasons for thinking that much remains to be found at Vasiliki too.

No one, however, disputes that Vasiliki is a "mansion"—the property of a single, wealthy man. This is not the case with Fournou Korifi, and here I find myself in a minority (perhaps of one) in identifying this complex as a "mansion". Certainly the excavator of the site, Dr. Warren, believes it to be a precursor of the Late Bronze Age towns like Gournia and Palaikastro[6]. There is insufficient space here in which to argue the point at length, but the single building complex of more than eighty rooms, the absence of intervening roadways or alleys, the appearance of specific workshops and magazines, and the situation of the most finely-built rooms facing the sun and the Libyan Sea are to my mind suggestive of a single economic and architectural unit belonging to a single man or family.

Whatever the status of these settlements at Vasiliki and Fournou Korifi they enable us to make some comments on early developments towards large communities in east Crete. Both settlements were situated on hill-tops, with resulting problems of fresh water supply, resolved at Vasiliki by the digging of a well, and at Fournou Korifi by the use of many large pithoi as rainwater butts. In neither case was any evidence found of a system of built drains, and similarly extensive excavation of the hillside at Vasiliki and exploration of the surrounding area at Fournou Korifi failed to reveal any suggestion of a cemetery in the proximity of the settlements. Finally, it is perhaps important to note that neither of these sites is close to a good anchorage, and that both seem to have prospered on the produce of industry and the land[7]. We may compare this situation with that of the two contemporary villages in eastern Crete, Mochlos and Priniatiko Pyrgos, both of which were exceptionally good harbour sites but possessed little cultivable land. I believe the differences in the economic basis of these Early Bronze Age settlements may do much to explain the development of two contrasting types of settlement.

Unfortunately both Vasiliki and Fournou Korifi were partially destroyed by fire during Early Minoan II, the latter site then being completely and finally abandoned and the former going through a period of more squalid occupation before the establishment of a small village or town there in Middle Minoan I (*c.* 2000 B.C.). Of this Middle Minoan settlement, and those now appearing at Palaikastro, Pseira, and Gournia, we again know very little, largely as a result of extensive rebuilding, demolition, and adaptation at the beginning of the Late Bronze Age. All three of the new settlements however are situated on low hills or peninsulas like the Early Minoan settlements. For the first time in eastern Crete we can speak of stepped and/or surfaced roads (Vasiliki, Palaikastro, Gournia) which presumably carry with them the implication of a certain degree of central or communal planning and organization. As yet, however, there seem to be no drainage systems. Apart from the appearance of laid and built roadways, the main development would seem to be in the size of the communities.

The "mansions" had been small settlements—Fournou Korifi occupied about 1300 m^2, Vasiliki (as it is known) rather less, although in its entirety it might have been twice as large—and Early Bronze Age buildings at Palaikastro, Priniatiko Pyrgos and on Mochlos seem to have been thinly distributed over smaller areas than those occupied by the later towns on the same sites. Unfortunately the remains are so fragmentary that even an approximate figure cannot be given for the area of occupation, but at Palaikastro and Vasiliki certainly, Middle Minoan buildings extended over greater areas than had been previously occupied, and the same may be said of the settlement on Pseira. Following Seager's statement that the Middle Minoan town on Pseira occupied about half of the area covered by its successor, we can give an approximate figure for the area of Middle Minoan Pseira—approximately 7000 m^2. The expansion of the harbour towns during the period Early Minoan III to Middle Minoan I is to be expected in eastern Crete, since this is the period when commercial contacts with the Levant are made and strengthened[8]. Pseira, Gournia, and Pachyammos are indeed not merely expanded during this period but to all intents and purposes are founded then, since earlier occupation on these three sites had been sparse indeed.

The towns reached their greatest extent however in the Late Bronze Age, during the period Middle Minoan III to Late Minoan I (*c.* 1700 to 1450 B.C.), and it is on the towns of this period that we may concentrate now. The first point to be made, though by now it should be self-evident, is that none of the east Cretan Late Bronze Age towns are new foundations; all were founded in or before Early Minoan III to Middle Minoan I. Thus, their siting was predetermined and was in no way related to any new factors operative in the seventeenth century B.C. All three of the excavated towns (Pseira, Palaikastro, and Gournia) stood on low hills (Pseira on a steep-sided peninsula) close to a good harbour or anchorage. That all three were in fact harbour-towns we cannot doubt, and their function will have to be borne in mind in any discussion of their nature, since the two are inevitably related to one another.

To a large extent, the town plans of the Late Bronze Age were as predetermined as were their situations, for at Gournia and Pseira it was possible at several points to demonstrate that the roads antedated the Late Bronze Age buildings[9]. That is to say, the road systems of these towns were established in the Middle Bronze Age and remained basically unchanged during the succeeding period. There is no suggestion of town-planning in the pattern of roads and alleys but a single, simple principle may perhaps be recognized—that the main road or street should be so situated that it could serve as much of the settlement as possible. At Palaikastro it seems to run along the long axis of the town, and may originally even have formed a complete circuit, at Gournia it forms a ring-road with houses both within and without the ring (Fig. 1), and at Pseira it can be traced running through the long axis of the settlement (a ring-road here would have been difficult to construct and wasteful of valuable space on the peninsula). From these major roads smaller ones run at (approximately) right-angles, and it is normally

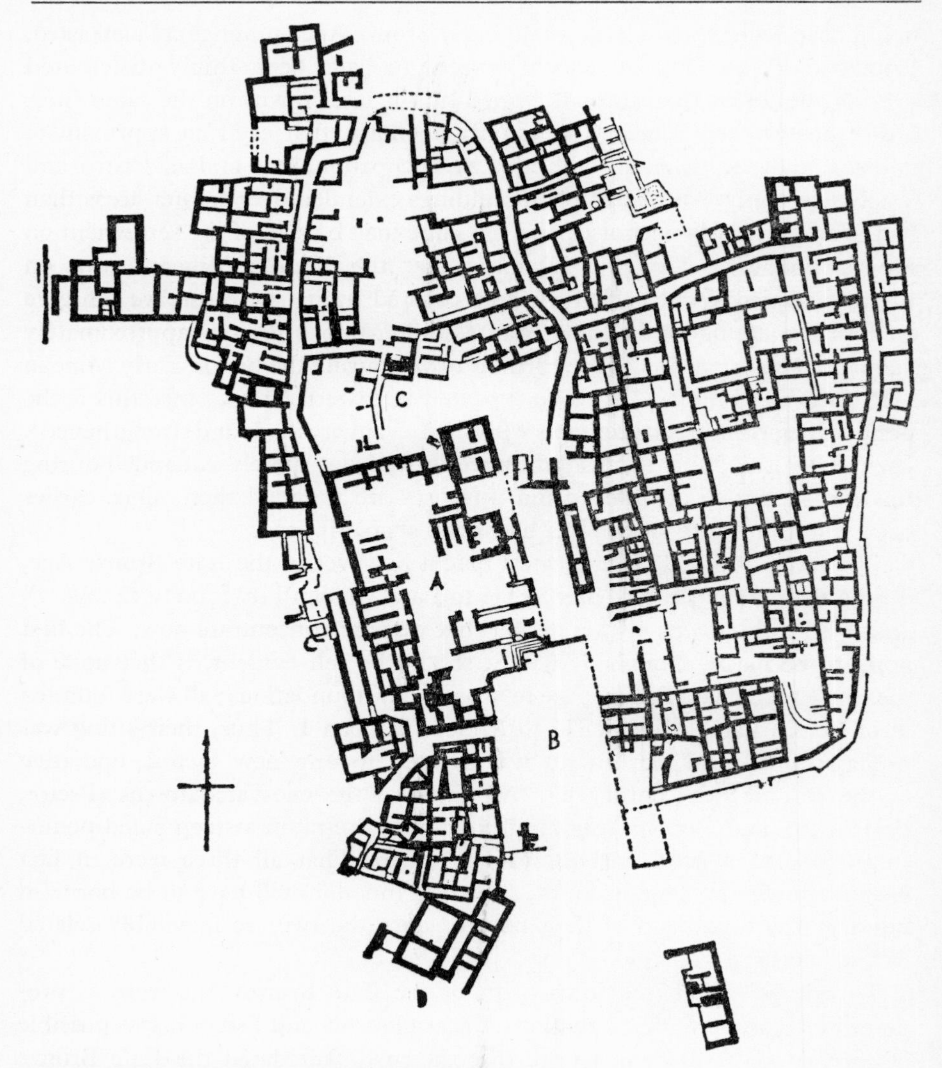

Fig 1 The Late Bronze Age town of Gournia.

these which negotiate the steeper slopes of the hillsides. This means that many of the minor roads or streets are stepped at intervals or occasionally along most of their length; at Gournia and Palaikastro they are also surfaced with stones and earth, but at Pseira most were simply cleared rock surfaces.

These roads, particularly the high streets, certainly imply a degree of communal organization and control in Minoan town development, and this is emphasized at Palaikastro where between blocks D and M the "high street" runs straight and true, broad and well-surfaced, and flanked by impressive façades. Public works extended to the drainage system too, since the streets of both Gournia and Palaikastro were equipped with stone-built water channels, fed from the houses by clay pipes[10]. That public works of this sort were directed by a central authority is perhaps more than a probability, since at Gournia where we have a small "palace", one of the palatial storerooms contained a supply of new pipes[11].

On the other hand, the problems of water supply which the situations of the towns presented in no case seems to have been tackled as a communal problem. At Gournia houses seem to have relied on cisterns, small stone tanks, and pithoi, and the same seems to be true of Palaikastro[12]. Seager noted cisterns at Pseira too, but here there was a well of uncertain (but pre-Roman) date although this was situated below the town, close to the shore.

The east Cretan towns of the Late Bronze Age are, in modern terms, small —Pseira approximately 15,000 m², Gournia approximately 25,000 m², and Palaikastro perhaps as much as 55,000 m² [13]—but they were of course much larger than anything previously known in eastern Crete and they were quite densely populated. Few areas where it was possible to build were left open, and apart from the "high streets" access to houses was mainly by means of narrow alleys. On the basis of the excavated areas and the topography of the sites, it is possible to estimate that Pseira and Gournia each contained something like a hundred houses during the Late Bronze Age. Gournia seems to have occupied more space partly because it contained some quite extensive houses, and partly because it possessed a "palace" with a large courtyard attached to it. The number of houses at Palaikastro is much more difficult to estimate, since a much smaller proportion of the town appears to have been excavated. The excavated area includes perhaps fifty houses, but the original number would have been many more than this, particularly as there is reason for thinking that the houses in the most intensively excavated area are considerably larger than those in the southern sector, which has been examined on a more limited scale.

What these figures mean in terms of population cannot be objectively calculated, and the population figures which I suggest for the east Cretan towns are little more than guesses. I have however studied each building individually (in so far as the excavation reports allow this to be done) and tried to assess the area given over to living quarters as opposed to that occupied by storerooms, kitchens, workshops, and vestibules. I have also taken into account the proportion of large houses to small ones, and at Gournia the additional factor of the "palace" and its inhabitants. With these various factors in mind I would suggest the following population figures for the east Cretan towns of the Late Bronze Age—Pseira approximately 400, Gournia approximately 700, Palaikastro, as excavated, approximately 500, but perhaps three to four times as many in the entire town[14].

At Palaikastro and Gournia, and to some extent at Pseira, the essential character of the town was determined by the street system, which divided the town into a number of insulae. At Gournia there were six, at Palaikastro many more (more than a dozen were discovered in the excavated area). For the most part these insulae are irregular in shape and in size, and in several cases their shape has clearly been determined by the existing course of the Middle Minoan streets which flank them. On the other hand the many odd projecting corners which break up the outlines of these insulae are of course characteristic of Minoan architecture as a whole, and perhaps serve to emphasize the paradox inherent in Minoan architecture—that each house is built as a functional entity within itself, yet that houses are built in blocks or groups and not as free-standing structures. This conflict in attitudes to domestic architecture does much, I think, to explain the "untidy" appearance of Minoan settlements (and indeed, Minoan palaces). Whether the Minoans preferred to build their houses in great blocks or insulae because they were gregarious by nature is uncertain, but certainly the adoption of this method of building allowed for a much greater density of population than would have been possible with other types of development. In block C at Gournia for example, we find not only fifteen houses strung around the perimeter of the block, but five or six built at its very centre, and reached by the narrowest of alleys or passages. The alternative use of the insulae method, with longer but narrower houses reaching back to the centre of the block was never adopted, and the absence of this type of building from Minoan Crete suggests that the Minoans simply did not like it, but preferred the house which more closely approximated to a square.

At Palaikastro it is perhaps possible to identify the most favoured area within the town, since each of the insulae bordering the "high street" contains one large house which fronts on to the "high street" and presents an imposing façade. At some stage in the development of the town it is clear that men of wealth gained possession of the land on either side of the main street at Palaikastro. Furthermore the uniform appearance of a single large house and several smaller ones in each of the relevant insulae might even suggest that each of the insulae was owned by a single man, and that the other householders were what we might loosely describe as "leaseholders". This can only be suggested for the insulae bordering the "high street", since the other insulae have only been sampled by excavation. That the best properties lay alongside and close to the main street however is certain, and excavations in the "suburbs" at Palaikastro—blocks Κ, Λ, Σ, Υ—revealed houses which were much smaller than those found at the town centre.

The same might be said of Pseira, where large houses were fewer in number and smaller in size than at Palaikastro, but again seemed to be concentrated in the centre of the peninsula—the original focus of the town. At Gournia on the other hand most of the larger houses were found in the insulae which lay outside the ring-road—houses Hf, Ab, Ec, for example—while the centre of the town was dominated by the "palace". All of the biggest houses, however, again fronted on to the main street so that the basic principle would seem to hold good and the wealthiest men built their homes alongside the main road. At Gournia it is possible, perhaps, to go a little further and recognize the less fashionable quarter of the town in insula C. We remarked earlier on the way in which houses are crammed into every available space in this large block, and it is noteworthy perhaps that none of these houses is notable for either its size or its quality of construction.

One is tempted to recognize the larger and better houses in these towns as the homes of successful merchants, such as one might reasonably expect in settlements whose prosperity was seemingly related to the proximity of a good harbour or anchorage. The smaller houses one might ascribe both to the men who farmed the adjacent land and fed the settlement, and to those who were in various ways employed by the merchants. It is surprisingly difficult however to recognize some of the other characteristic buildings of a small harbour town—shops, hostels, and workshops—or even to identify any part of the towns given over largely to these establishments. The presence of magazines, groups of pithoi, oil separating installations, or wine presses is insufficient evidence on which to identify specialist buildings, since many of the dwelling houses appear to have possessed some or all of these things. It is only where we find evidence of more specialized occupations than agriculture that we can be reasonably certain we are dealing with shops and workshops.

At Palaikastro Dawkins identified one shop fronting on to one of the side streets (Block Γ, 37–8)[15]. This little two-roomed building was open to the street, while the back room contained four large pithoi (one holding peas) and a stone sink with drain. The building next to it may also have been a shop, since it had a very wide doorway and only one large room apart from two small storerooms to the rear containing masses of domestic pottery. Unfortunately there were no excavations further south along the same side of the street, but on the opposite side of this street is another small building with a 2 m wide doorway and one large room with two small storerooms along one side. On one side of the large room was a massive stone platform on which stood two querns, each situated above the open-mouth of a pithos[16]. This building too seems likely to have been a shop, and street Γ–E begins to look as if it may have contained several shops or workshops. Uncertainty about the position of doorways in rooms 26–34 of block Ξ further south along the same road prevents us from identifying this block of rooms as either a long narrow house (which seems unlikely) or as a group of three small shops. Elsewhere in Palaikastro the evidence for shops and workshops is scarce but block X is perhaps of some interest. This was partly comprised of a large, well-built house, and partly of a maze of small irregularly-shaped rooms or buildings. Among the latter we find one containing a mass of crushed murex shells, and next to it a room with a stone tank (dye-works?), and another room facing on to the "high street" containing four carpenter's tools[17]. In block E a cloth-works might be suggested in house 5 (rooms 36–43) which produced two large hoards of loomweights[18]. Tantalisingly isolated finds include a potter's wheel (block N, room 9) and a stone mould for metalwork (block Σ, room 5). From these scant indications of workshops and shops at Palaikastro it is difficult to draw any conclusions, except that such buildings might perhaps have tended to group together (but in several locations) and that it was rare for shops to be situated on the "high street", which seems to have been preserved (at least in the excavated area) for the façades of large houses.

At Gournia the identified workshops seem to concentrate at the north end of the settlement. Alongside the west road are two metalworkers' shops, one in Fh (producing several moulds), and another in Ea (with moulds for thirteen tools)[19]. A third metalworkers' shop may have been situated in block C, room 24, which produced a crucible. Fronting on to the east road in block F was a sizable house with two entrances, one into a paved courtyard, the other into a basement workshop from which were recovered a group of eight carpenter's tools[20]. The only industrial establishment at Gournia yet identified outside of the northern quarter is the potter's workshop with eight wheels, found in the southern part of the town. Of the small two- and three-roomed buildings like the shops at Palaikastro there is no trace at Gournia except perhaps alongside the east ascent, where a group of such buildings might be recognized (A, rooms 34–41). Unfortunately excavations here were incomplete so that the relationship between the various rooms is uncertain.

Considering the way in which the houses at Gournia, Palaikastro and Pseira are crowded together, it is surprising perhaps that these towns reveal little trace of communal activity. At Palaikastro and Pseira there is no trace whatever of either public squares or temples. Only at Gournia can these features be identified. Here, the "palace" courtyard is uniquely open to access from at least two public streets, and was rightly labelled as the "public court" by the excavator. To the rear (north) of the palace was situated the only public shrine in the town, but this was a small building with only a narrow alleyway before it, so that the ceremonies performed there could never have been truly communal ones. The east Cretan towns are also notable for their lack of defence/enclosure walls and the excavations at Gournia and Palaikastro have not yet demonstrated to what extent houses were built beyond the main concentration of settlement, although trial excavations across the east valley at Gournia revealed traces of buildings there[21]. At Pseira houses certainly spread beyond the peninsula and on to the adjoining hillside. The possibility of suburban developments might perhaps explain why cemeteries were commonly situated at some distance from the town—the cemeteries for Pseira, Palaikastro, and Gournia were all about half a mile from the towns themselves—but it seems likely that this tradition goes back to the Early Bronze Age (if not earlier) in view of the absence of burials near Vasiliki and Fournou Korifi. In this respect there would seem to be a clear contrast between the east Cretan settlements and those of southern and central Crete[22], and this is not the only difference between eastern and central Cretan towns. A comparative study of these, however, must be postponed for another occasion.

Notes

1 Hutchinson, R. W. (1950). Prehistoric town planning in Crete, *Town Planning Review*, **21**.
2 For general distribution of late Neolithic sites in Crete see Weinberg, S. (1965). *The Stone Age in the Aegean*. Cambridge. map 3.
3 Dawkins, R. M. (1905). Excavations at Palaikastro, IV, *Ann. Brit. Sch. Archaeol. Athens*, **11**, pp. 260–68.
4 For discussion of Early Minoan I settlements in central Crete see Branigan, K. (1970). *The Foundations of Palatial Crete*. London. pp. 39–42, and for the tholoi and their relationship to the settlements see Branigan, K. (1970a) *The Tombs of Mesara*. London. chapter 7.
5 Discussed, with full references, in Branigan, K. (1970). *op. cit.* pp. 43–8.
6 A view expressed in private correspondence, but implicit in his remarks in Warren, P. (1969). Minoan village on Crete, *Ill. London News*, Feb. 8, 1969, p. 27.
7 The situation at Fournou Korifi is fully discussed in Warren, P. (1969). *op. cit.* and Warren, P. (1968). A textile town—4,500 years ago? *Ill. London News*, Feb. 17, 1968.
8 Branigan, K. (1967). Further light on prehistoric relations between Crete and Byblos, *Amer. J. Archaeol.*, **71**.
9 Boyd-Hawes, H. (1908). *Gournia*. Philadelphia. p. 21.
10 Boyd-Hawes, H. (1908). *op. cit.* p. 28; Dawkins, R. M. (1905). *op. cit.* p. 290.
11 Boyd-Hawes, H. (1908). *op. cit.* p. 28, from room G.24.
12 Boyd-Hawes, H. (1908). *op. cit.* p. 26; Bosanquet, R. (1902). Excavations at Palaikastro, I, *Ann. Brit. Sch. Archaeol. Athens*, **8**, pp. 310 ff.
13 The situation at Palaikastro is uncertain since excavations in blocks N, Σ and Γ suggest that a great deal of the town may yet remain to be excavated.
14 Cf. Renfrew's estimates, arrived at independently and in an entirely different way, for population density of Aegean LBA towns. On the basis of his figures, Pseira would have 450 people, Gournia 750, and Palaikastro 1650.
15 Dawkins, R. M. (1904). Excavations at Palaikastro, II, *Ann. Brit. Sch. Archaeol. Athens*, **10**, pp. 290–92.
16 Dawkins, R. M. (1903). *op. cit.* p. 295.
17 Dawkins, R. M. (1905). *op. cit.* p. 276.
18 Dawkins, R. M. (1904). *op. cit.* p. 207.
19 Boyd-Hawes, H. (1908). *op. cit.* p. 26.
20 Boyd-Hawes, H. (1908). *op. cit.* p. 23.
21 Boyd-Hawes, H. (1908). *op. cit.* p. 20.
22 Cf. Branigan, K. (1970a). *op. cit.* p. 132.

H. E. W. CRAWFORD

Stimuli towards urbanization in south Mesopotamia

The aim of this note is to draw attention to one particular factor which may have contributed to the urbanization of southern Mesopotamia. This is the paucity of all but the most basic raw materials, i.e. reeds, mud and a little soft-wood in the area. It is arguable that this was the decisive factor in pushing Mesopotamia towards what is conventionally called civilization. This lack and the resulting economic imbalance, to be a spur, must be coupled with a knowledge of the uses and whereabouts of the lacking raw materials; contacts must also be shown to exist between south Mesopotamia and the sources of supply before the onset of urbanization so that the postulated knowledge can be proved. We must in fact show that there was a recognized lacuna in the economy and the knowledge to fill it.

These conditions are fulfilled whether the earliest inhabitants of south Mesopotamia are regarded as having come from the north or the east in the era of incipient urbanization. From an early period the obsidian trade provided links between Mesopotamia and the area west of the Caspian Sea[1]; finds of carnelian and turquoise in the graves at Tell es-Sawwan also indicate contacts with the area east of the Caspian as early as the Hassuna period[2]. Evidence from Tell es-Sawwan also indicates links with the Deh Luran area where there is sufficient evidence to show that copper and stone were used as early as the Ali Kosh phase, and most probably wood too[3].

It can then be assumed that these raw materials were known, and indeed used, in very limited amounts in Mesopotamia before the period of urbaniza-tion. Once an economic surplus existed an obvious use for it would have been to exchange it against the much needed raw materials. The supply thus obtained would surely have created a greater demand as the properties of the new materials were more widely appreciated. It is suggested that to supply this demand people began to group together in order to facilitate the produc-tion of a surplus and its barter for foreign products. The gradual expansion of such a trade would have increased the need for specialists, both on the production and selling sides; specialists must have non-specialists to provide for them in order to survive and so the process of accretion continues, leading to the formation of central-place and distributive centres[4]. Adams suggests that this process had already begun in the Ubaid periods[5].

Some small measure of support can be adduced for this theory; many of the earliest towns are in areas where trade communications are easy, for example, the cluster on the Euphrates of Ur, Uruk and 'Uquair. It is in this period of urbanization that south Mesopotamia, where the shortage of materials is most acute, began to forge ahead of the north. A similar pattern may also be seen in the emergence of the Indus civilization, but here diffusion cannot be discounted[6]. Egypt's economic self-sufficiency on the other hand may be a factor in explaining her apparently late development towards urbanization. If one considers the fashionable comparison with the South American civilizations, here again each area seems to suffer from some deficiency and according to Adams the market had a central role in each town[7].

It is obviously facile to reduce the stimuli for the complex process of urbanization to one factor, or even two, but the absence of raw materials coupled with a high agricultural potential sets Mesopotamia apart from other early agricultural areas in the Near East and this perhaps had a greater significance than has hitherto been realized. It could perhaps be cited as an example of Toynbee's classic Challenge and Response formula.

Notes

1 Renfrew, C., Dixon, J. E. and Cann, J. R. (1966). Obsidian and early cultural contacts in the Near East, *Proc. Prehist. Soc.*, **32**.
2 El-Wailly (1965). Excavations at T. es Sawwan, *Sumer*, **21**.
3 For example, annular deformation of the skull. Flannery, K. V., Hole, F. and Neely, J. A. (1969). *Prehistory and Human Ecology of the Deh Luran Plain*. Ann Arbor.
4 Morrill, R. (1965). *Migration and the Spread and Growth of Urban Settlement*. Lund.
5 Adams, R. McC., this volume, p. 741.
6 Allchin, B. and Allchin, F. (1968). *The Birth of Indus Civilization*. London.
7 Adams, R. McC. (1966). *Evolution of Urban Settlement*. London.

S. C. HUMPHREYS

Town and country in ancient Greece

This paper is concerned with Greek ideas and institutions rather than with archaeological evidence for settlement patterns. Archaeological research of this type is still backward in Greece[1]; and in so far as archaeologists have worked on settlement patterns, the two subjects which have recently attracted most attention—town planning and land tenure—are precisely those in which ideas and institutions imposed a geometrical pattern on town and landscape.

Halbwachs distinguished three sub-sections in his treatment of social morphology[2]: the religious, the political and the economic. In examining Greek notions of the relation between town and country it will be useful to add two more: military organization, as a special subdivision of political structure with its own characteristics, and the distinction between rural and urban culture, in the sense of education and style of living. It is obvious that societies can vary in their polarization or disregard of the potential opposition between town and country in these different spheres. The city may be regarded as a holy place[3], an enclave of "burghers" with privileged political status, a centre of commerce and industry, a fortress, the home of education and sophisticated manners; it may equally be associated with sin, revolution, unemployment, vulnerability in war, illiteracy or bad drains.

To the ancient Greeks the city (*polis*) was above all a political conception[4]. In normal usage, *polis* meant a city-state, both territory (*chōra*) and the conurbation at its centre (sometimes also called *asty*) where one would find the basic Greek political institutions—magistrates, council and citizen assembly—and the public buildings in which they were housed. As a type of political organization and settlement the *polis* was contrasted with the *ethnos*, a cultural unit either ruled by a king or acephalous, with perhaps some form of loose or intermittent federal organization. Such societies were settled in villages, *kata kōmas*. When Thucydides calls classical Sparta and other Greek states in earlier times "*poleis* not synoecised but settled *kata kōmas*"[5], there is an element of paradox in the description; these communities had the political institutions of the city but lacked the architectural characteristics—walls, monumental public buildings and density of settlement—which were normally associated with them[6].

The term "synoecism" which Thucydides uses here (literally, "settling together") carried implications both of state-formation and of urbanization. The Athenian state came into being, according to tradition, through the creation in Athens of a central government for the twelve towns of Attica which had previously been autonomous, turning to the king of Athens only in moments of danger. The growth of Athens as a city was for Thucydides (II.15) largely the result of this change in political structure. The former towns, however, continued to exist as settlements and local centres of the subdivisions of the citizen body. Most Greek cities had such smaller settlements in their territory, and the distinction between the *polis* and *ethnos* forms of political organization should not be exaggerated[7]. Tribes, phratries and local centres were vital parts of the *polis* with their own cults and autonomy in managing their own affairs; on the other hand, the existence of cities influenced the institutions of those parts of Greece which were still settled in villages. The same segmentary structure can be seen in both types of organization (even if the relation between the tribes of the *polis* and the sub-groups of the *ethnos* remains unclear), despite the greater centralization of government in the *polis*. In consequence, the notion of political autonomy and the use of the term *polis* were apt to fluctuate according to context, and there was no sharp dividing line between city and countryside (*chōra*) either in political status or in religion. The Greek city, unlike the Roman, was not surrounded by a sacred boundary[8]. Some of its gods had their sanctuaries in the city, others in the countryside or villages.

There was indeed a tendency in Greek political thought towards eliminating altogether the distinction between town-dwellers and country-dwellers. Plato, in the *Laws* (745b), prescribed that the territory of his ideal city should be divided into lots of which each household should have two, one near the city and the other near the borders. And Cleisthenes of Athens, in 507 B.C., had already created a new system of tribes and subdivisions designed to eliminate regional factionalism, in which each tribe contained areas from the "city", "coast" and "inland" regions of Attica[9].

But outside Utopia, the attempt to override geographical factors could hardly be entirely successful. Division of the territory into equal parcels of land to be distributed by lot, which seems to have begun with the occupation of conquered land (neighbouring or colonial), and which became part of the general Greek conception of the process of founding or reforming a state, often did not extend over the whole area controlled by the city[10]. The lots of the Spartans occupied an area of the central plain of Lacedaemon, while the rest of the territory was settled by "*perioikoi*" whose communities were autonomous in local affairs but subject to Sparta in foreign policy. This was probably also the position of some native settlements in territory controlled by Greek colonies[11]. The difference between *perioikoi* ("subjects") and neighbouring allies was often ill-defined. In some colonies there were disputes between the original settlers' descendants and late-comers settled in outlying parts of the territory, who might be in an inferior position both economically and politically. Here the spatial definition of an original *polis* which, one might almost say, extended the idea of town planning to the countryside[12], led to a later differentiation between this area (called *politike chōra*, polis-country, by the Spartans) and a remoter countryside beyond its borders[13].

Although the Greek acropolis has often been characterized as a *Fluchtburg*, medieval comparisons, in this as in other respects, are dangerous. Not every Greek city was built on or round a hill. Sparta long remained unwalled, claiming that the spears and shields of her hoplites were sufficient defence. Plato thought city walls only made the citizens cowardly[14]. Although Greek myths told of famous attacks on cities (Troy, Thebes), the tradition of the early *polis* wars (seventh and sixth centuries B.C.) was to stress the conquest of land and the defeat of enemy armies in pitched battle. The members of the Delphic amphictyony swore an oath not to destroy each other's cities or interrupt water-supplies (Aeschines II.115). Siege warfare and elaborate city fortifications developed only gradually from the fifth century B.C. onwards. Pericles' policy in the Peloponnesian war of concentrating the population of Attica inside the walls linking the city to Piraeus and a supply of food from overseas, and abandoning the countryside to Spartan invasion, was new and unpopular. Until the fourth century B.C., when the use of mercenaries began to spread, the Greek army was composed of citizens and to a large extent of peasants; they fought to defend their land.

The development of Athens in the fifth century B.C. as a large urban settlement depending on imported corn, and the striking contrast between Athens and Sparta in economic, political and military organization, introduced new elements into the Greek conception of the relation between city and country. Right-wing Athenians, like Xenophon and Plato, associated oligarchy, subsistence farming and country life with the courage, physical fitness and obedience required of the soldier; urbanization and seapower led to democracy, mercenary self-interest and weediness. In this contrast between different types of economic activity and ethos, the most extreme form of the urban type was associated with the growing harbour towns, which in a number of Greek states (Athens, Megara, Colophon, several Cretan cities) were situated at some distance from the city itself. An attempt was thus made to detach the developing economic conception of the city from the idea of the city as a political centre, leading to debates in political philosophy on the need to separate the commercial market from the political *agora*[15], the correct distance to be maintained between the ideal city and the sea[16], and the desirability of granting citizenship to traders and craftsmen.

Here again, however, as in the distinction between *polis* and *ethnos*, we have a contrast between different Greek states and not between town and country within the state. The Athenians distinguished the city type, the *asteios*—sophisticated, witty, cunning—from the rustic, *agroikos*[17]; but there is little sign of a grouping of interests, of a conscious solidarity, corresponding to the division of economic activity and manners. Many of the city craftsmen and traders were resident non-citizens, or semi-independent slaves; the struggle of the poorer citizens to emphasize the status distinction separating them from these categories inhibited the development of a breach between town and country[15].

The situation became quite different later in the Hellenistic kingdoms, where Alexander's conquest led to the creation of a plural society in which the Greek cities differed in ethnic origin, political status and culture from the surrounding countryside. The Romans too granted privileged status to cities and towns. We find the distinction in political status between town and country again in the medieval "Stadtluft macht frei"[19]. Perhaps consideration of the differences between the Greek city-state and the later pattern may provide a starting-point for a comparative discussion of the conditions which make for identification or opposition of town and country. Other papers in this volume raise the question whether the urban/rural distinction should be treated as a dichotomy or as a continuum[20]; taking what might perhaps be called a structuralist approach, I have here treated it as a potential contrast which in any given society may be minimized (continuum) or emphasized (dichotomy). In comparing the treatment of the contrast in different societies, it would be necessary to consider (*a*) the extent of polarization both in "folk" concepts of town and country and in the measurable clustering or even gradation of settlement sizes and densities; (*b*) the hardness of boundaries between town and country, both physical (presence or absence of town walls, suburban fringe, etc.) and institutional (discontinuities in legal jurisdiction, political rights, allegiance to religious centres), and the frequency of traffic across the boundaries; (*c*) the consonance or dissonance of the different types of density (political, economic, religious, military, cultural).

Classical Athens and Hellenistic Alexandria—cities comparable in size, density and level of economic specialization—may serve to represent the ideal types of urban-rural continuum and urban-rural dichotomy. The Athenian countryman had a close and direct relationship with the city; he voted in its assembly, bought and sold in its markets, took part in its religious festivals, sued in its courts, had the same political rights and obligations—including that of military service—as the urban population. As we have seen, the Athenians did their best to ignore the contrast of town and country. Hellenistic Egypt, on the other hand, is a typical example of a plural society in which the boundary between the Greek city and the Egyptian countryside was underscored by structural, social and cultural divisions[21]. Nevertheless, a marked urban/rural dichotomy is neither essential nor peculiar to plural societies. As shown above, it is found also in medieval Europe. What the classical Greek material suggests is that the nature of the country-dweller's relations with the city and its institutions—direct or mediated through middlemen—is of crucial relevance. The middleman—whether feudal lord, patron or economic entrepreneur—not only derives his position from the existence of a dichotomy, but also helps to preserve it.

Notes

1 See Humphreys, S. C. (1967). Archaeology and the economic and social history of classical Greece, *Parola del Passato*, **22**. The available archaeological material is collected in Philippson, A. (1950–9). *Die griechischen Landschaften*. I–IV. Frankfurt a. M. and in Pauly-Wissowa. *Real-encyclopädie d. class. Altertumswiss;* see names of ancient cities and regions. See also Robert, L. (1965). *Villes d'Asie Mineure*. Paris; Robert, L. and J. *La Carie*, II. Paris. Important discussions (1971) *in La Citta e il suo Territorio, VII Convegno di Studi sulla Magna Grecia*. Naples.

2 Halbwachs, M. (1938). *Morphologie sociale*. Paris. English translation (1960). *Population and Society*. Glencoe.

3 Wheatley, P. (1969). *City as Symbol*. London; Tritsch, F. (1929). Die Stadtbildungen des Altertums und die griechische Polis, *Klio*, **22**.

4 For reasons of space I can only give here a simple outline which does not do justice to the complexity and variety of the data. The bibliography touching on the subject is immense; see (apart from the studies cited below) the recent collection of basic articles *in* Gschnitzer, F. (ed.) (1969). *Zur griechischen Staatskunde*. Darmstadt; and Kirsten, E. (1956). *Die griechische Polis als historisch-geographisches Problem des Mittelmeerraumes*. Bonn.

5 Thucydides, I.10, cf. I.5.1.

6 Bölte, F. (1929). Sparta (Topographie), *in* Pauly-Wissowa, *RE* III A 6, s.v. The circumference was about six miles.

7 Gschnitzer, F. (1955). Stammes- und Ortsgemeinden im alten Griechenland, *in* Gschnitzer, F. (1969). *op. cit.* contains valuable observations, but starts from the hardly justifiable assumption that the etymology of state-names indicates the historical process of state-formation. The analysis of Kachin society by Leach, E. (1964). *Political Systems of Highland Burma*. London, might well be applied to the relation between *polis* and *ethnos* organization in Greece.

8 The prohibition of burial within the city imposes itself only gradually; see Young, R. S. (1951). Sepulturae intra Urbem, *Hesperia*, **20**; Vallet, G. (1971), *in La Cittá e il suo Territorio*. For the Romans, the concept of *urbs* might indeed be defined by the performance of the correct foundation rites (Varro, *De Lingua Latina*, V.143), though this was perhaps a juridical as much as a religious way of thinking. The earliest Greek description of city-foundation begins with a wall: Homer, *Odyssey*, VI.9 ff.

9 Leveque, P. and Vidal-Naquet, P. (1964). *Clisthène l'Athénien*. Paris.

10 Asheri, D. (1966). *Distribuzioni di terre nell' antica Grecia*. Turin. Areas marked out in "lots" have recently been explored in the Crimea (Chersonesos, near Sebastopol) and in the territory of Metapontum in southern Italy. See Wasowicz, A. (1966). À l'époque grecque: le peuplement des côtes de la mer noire et de la Gaule méridionale, *Annales* (*ESC*), **21**; Uggeri, G. (1969). *Kleroi* arcaici e bonifica classica nella *chora* di Metaponto, *Par. del Pass.*, **24**; Pecirka, J. (1970). Country Estates of the Polis of Chersonesos in the Crimea, *in Ricerche storiche ed economiche in memoria di Corrado Barbagallo I*. Naples; Dufkova, M. and Pecirka, J. (1970). Excavations of farms and farmhouses in the chora of Chersonesos in the Crimea, *Eirene*, **8**.

11 Cf. esp. *La Cittá e il suo Territorio*.

12 For the beginnings of Greek town planning (eighth to seventh centuries

B.C.) see now Drerup, H. (1969). Griechische Baukunst in geometrischer Zeit, *in* Matz, F. and Buchholz, H.-G. (eds.) *Archaeologia Homerica*. Göttingen, II.O; Vallet, G. and Villard, F. (1969). Mégara Hyblaea IX. Les problèmes de l'agora et de la cité archaïque, *Mélanges Ec. Fr. Rome*, **81**.

13 For the opposition between mountainous frontier territory as "bush", and cultivated land see Vidal-Naquet, P. (1968). The Black Hunter and the origins of the Athenian ephebeia, *Proc. Camb. Philol. Soc.*, **14** (French text *Annales* (*ESC*), **23**); Vidal-Naquet, P. (1970). Grèce. Une civilisation de la parole politique, *in Encyclopaedia Universalis*, 7, Paris.

14 *Laws* 778d–779a, with a reference also to the fortification of frontiers. On Greek warfare, fortifications, etc. see Vernant, J.-P., (ed.) (1968). *Problèmes de la guerre en Grèce ancienne*. Paris. On early city walls Drerup, H. (1969). *op. cit.* Fortifications may have played a greater part in Mycenaean warfare.

15 Aristotle, *Politics*, 1331a 30 ff. Note also that the beginning of structural differentiation of law from politics made it possible to treat litigiousness as an urban vice (Aristophanes; Polybius, IV.73. 6–10), and often to associate it closely with commerce.

16 Momigliano, A. (1944). Seapower in Greek Thought, *in* Momigliano, A. (1960). *Secondo Contributo*. Rome.

17 Lammermann, K. (1935). *Von der attischen Urbanität und ihrer Auswirkung in der Sprache*. Diss. Göttingen; Ribbeck, O. (1888). Agroikos, eine ethologische Studie, *Abh. Kgl. Sächs. Ges. d. Wiss.* (ph.-h. Cl), **10,** pp. 1–68.

18 See further Humphreys, S. C. (1970). Economy and society in classical Athens, *Annali Scuola Norm. Pisa*, **39.** The urban/rural boundary did not coincide with the opposition between rich and poor even when upper-class Athenians attempted to separate hoplites from sailors. The political philosophers' idea of the peasant proprietors as a "middle class" (Mossé, C. (1962). *La fin de la démocratie athénienne*. Paris. pp. 247–53) was based on the traditional identification of citizenship and the right to own land. Aristophanes' frequent play on the urban/rustic contrast belongs mainly to the years when the rural population of Attica was evacuated to the city, and is cultural rather than political; on the false picture of Aristophanes as champion of a conservative peasantry see Gomme, A. W. (1938). Aristophanes and Politics, in Gomme, A. W. (1962). *More Essays in Greek History and Literature*. Oxford.

19 See Nicholas, D. M. (1969). Medieval urban origins in northern continental Europe: state of research and some tentative conclusions, *Stud. Mediev. Renaissance Hist.*, **6**.

20 e.g. Ward-Perkins, J. B., this volume, pp. 867-82.

21 For the *chōra* of Alexandria see Jones, A. H. M. (1940). *The Greek City from Alexander to Justinian*, Oxford. p. 20.

GREGORY A. JOHNSON

A test of the utility of Central Place Theory in archaeology

Much analysis of archaeologically known settlement patterns has dealt with the effects of environmental factors on site location. Of these factors, resource localization has perhaps been most emphasized. Differential distributions of floral and faunal food resources, and of organic and inorganic raw materials, have been associated with differential site locations. This man-land approach has been highly productive, especially when dealing with cultural systems of relatively low complexity. I would suggest however that when dealing with systems of greater complexity, man-man relationships take on increasing importance in the determination of the spatial distribution of activity loci and thus of settlements. In emphasizing basically economic man-man relationships, Central-Place Theory furnishes a potentially useful analytical model for dealing with these systems. In the present paper the utility of certain aspects of Central-Place Theory in the analysis and interpretation of archaeological settlement pattern data is tested. The Early Dynastic I (E.D.I) settlement pattern from the Diyālā Plains, Iraq (*c.* 2800 B.C.) is used as a test case. The application of a central-place model to these data will be considered useful if it leads to the formulation of new, potentially important, and archaeologically testable hypotheses relative to the operation of processes resulting in the observed settlement pattern.

In his initial formulation of Central-Place Theory, Christaller proposed ways in which a settlement system associated with a modern market economy could be spatially organized to perform certain types of work most efficiently[1]. This work involved the production and distribution of goods and services. Briefly, in a settlement system organized according to Christaller's "Marketing Principle" central places or towns of the same functional size are equidistant from one another and when most efficiently located have a hexagonal distribution. Each town serves a surrounding, hexagonally shaped complementary region. Smaller central places and associated complementary regions may be hierarchically nested within this system to form an intricate settlement lattice.

Within Christaller's model, the form of this lattice may be altered under conditions in which efficiency of transportation or administration are the primary factors in the determination of settlement locations. Christaller emphasized that the distribution of actual settlements should not be expected to conform exactly to any portion of his model. Deviations could result from the operation of a combination of his ordering principles or from the influence of variables not included in his model.

Generalizing from this model, it may be suggested that settlement distributions similar to those Christaller proposed should appear (1) to the extent that least effort considerations influence the spatial organization of the production and distribution of goods and services; and (2) that the operation of other variables does not obscure the influence of these considerations. In that effort minimization may appear to a greater or lesser extent in the context of market, redistributional or mixed economies the application of the model need not be restricted to the modern market situation for which it was designed.

The following assumptions are implicit above. (*a*) That there is a tendency for agglomeration of human activities due to gains in efficiency obtainable by concentration of related activities at the same spatial locus. (*b*) That locational decisions are made in general to minimize energy expended in movement. (*c*) That all locations are accessible, but some are more accessible than others. (*d*) That structural hierarchies within settlement systems may, but do not necessarily, appear as a function of the interrelation of activity agglomeration, movement minimization and differential accessibility[2].

The model upon which the proposed settlement lattice for E.D.I sites in the Diyālā is based may be considered a special case of the combination of Christaller's marketing and transport principles[3]. The known E.D.I settlement pattern is presented in Fig. 1[4]. Table 1 lists the thirty-nine sites in the sample with their respective estimated areal sizes. In this analysis areal size is considered to be directly proportional to functional and population size[5].

Site Number	Site Size (ha)	Site Number	Site Size (ha)	Site Number	Site Size (ha)
421	25·00	355	3·68	362	·91
244	19·60	359	3·14	259	·84
515	13·50	458	3·00	381	·76
397	9·62	633	2·40	77	·74
109	9·45	56	2·16	364	·74
264–5, 7	8·15	396	2·06	637	·73
366	7·07	517	2·01	341	·72
590	7·00	450	2·00	634	·67
113	6·25	372	1·47	262	·50
16	4·91	31	1·26	357	·20
568	4·91	102	1·01	270	·20
520	4·20	160	·95	344	·17
462	3·68	176	·95	350	·13

Table 1 Early Dynastic I sites on the Diyālā Plains.

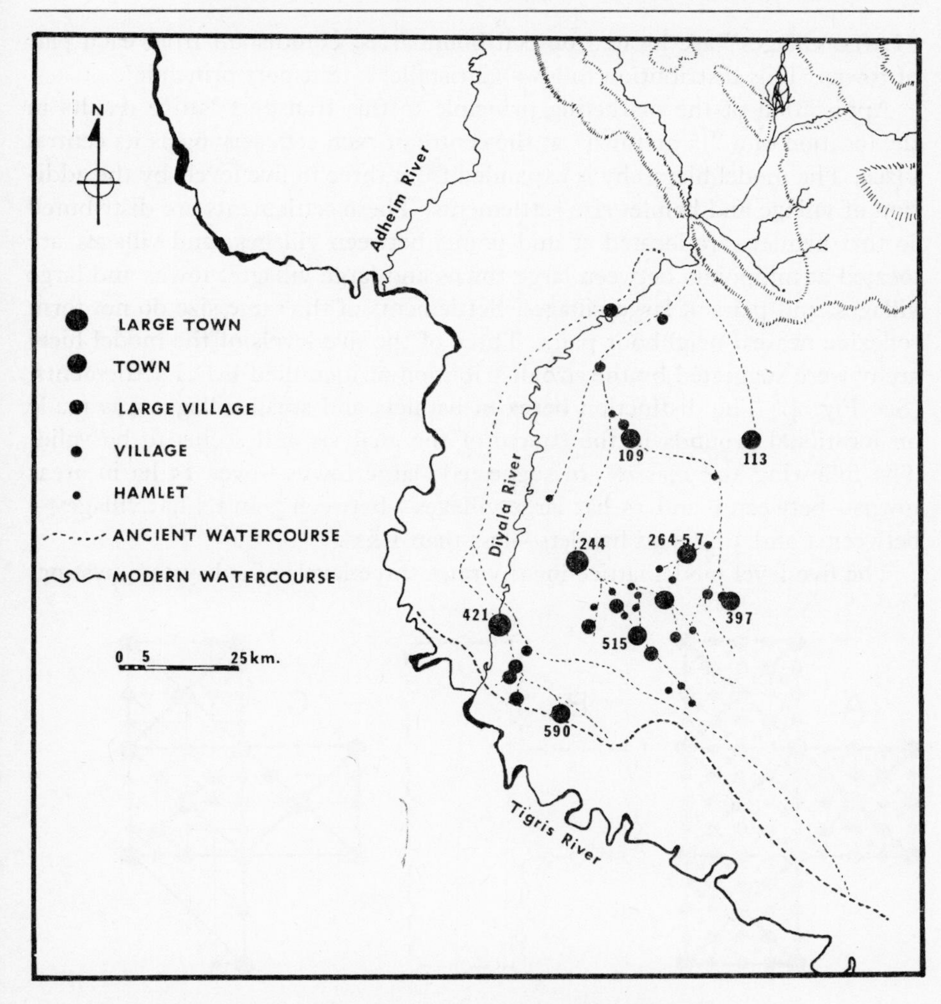

Fig 1 Early Dynastic I settlement pattern. Diyālā Plains, Iraq.
(After Adams, R. McC. (1965). *Land Behind Baghdad: A History of Settlement on the Diyālā Plains*. Chicago and London.)

The model used in this paper and the proposed E.D.I settlement lattice are presented in Figs. 2 and 3.

Most central place studies have concentrated on hexagonal distributions, although rhomboidal patterns have been noted[6]. The roughly parallel paths of major watercourses in the Diyālā were apparently responsible for the rhomboidal distribution which seems to occur there. Thus for present purposes, Christaller's distributions have been adapted to a rhomboidal model.

This model consists of four lattice cells. A cell is defined as the complementary region of a central place of highest order in the system. The corners of each cell are defined by a distribution of four "town" size settlements.

"Large villages" are located on cell boundaries, equidistant from each pair of towns. This distribution follows Christaller's transport principle[7].

Application of the marketing principle to this transport lattice results in the location of a "large town" at the centre of each cell, serving as its central place. The model hierarchy is expanded from three to five levels by the addition of village and hamlet size settlements. These settlements are distributed so that hamlets are located at mid-points between villages; and villages are located at midpoints between large towns and large villages, towns and large villages, and pairs of large villages. Settlements of the same size do not form reflexive nearest neighbour pairs. Three of the five levels of the model hierarchy were suggested by the size distribution of identified E.D.I settlements. (See Fig. 4.) The distinction between hamlets and small villages was made on locational grounds in the course of the analysis and seems to be valid. The following size classes are suggested: large towns—over 15 ha in area; towns—between 6 and 15 ha; large villages—between 3 and 5 ha; villages—between 1 and 3 ha; and hamlets—less than 1 ha[8].

The five level model lattice incorporates three levels of subsystem nesting.

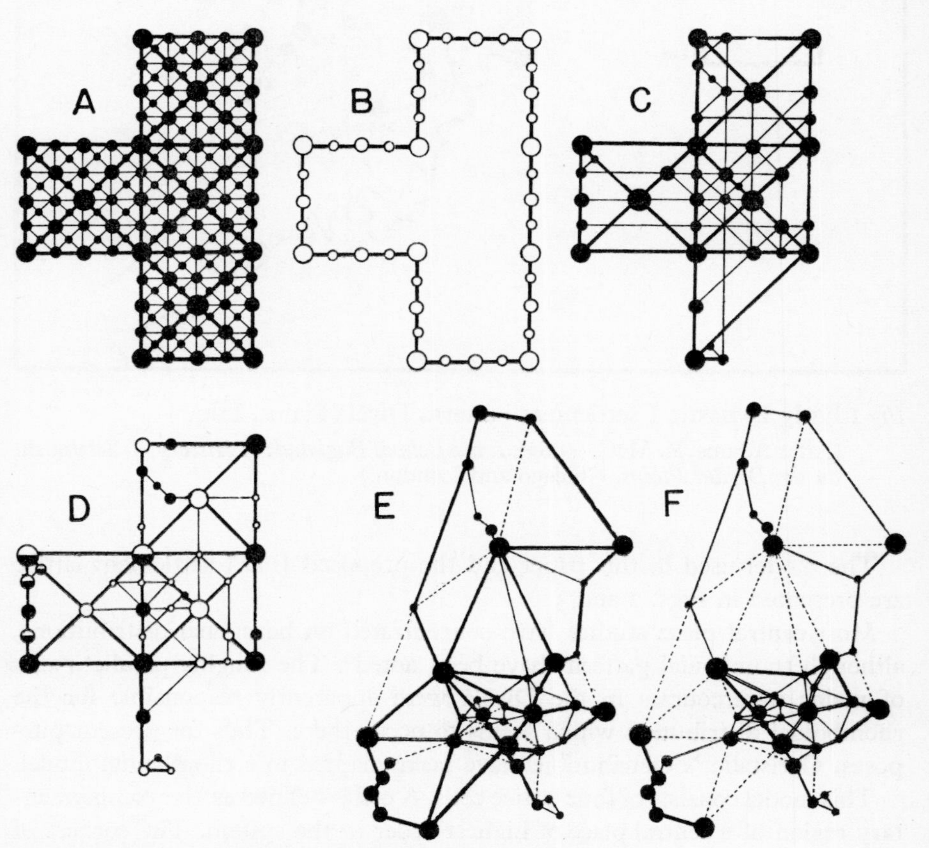

Fig 2 Derivation of the proposed settlement lattice.

These are: a town level, in which a large town serves as a central place for surrounding towns; a large village level, in which a town serves as a central place for surrounding large villages; and a village-hamlet level, in which a large village serves as a central place for surrounding villages and hamlets.

Derivation of the proposed lattice is illustrated in Fig. 2 A-F. Fig. 2A presents the model lattice. Fig. 2B indicates sites which due to their truncated complementary regions might be expected to be smaller than predicted by the model. Fig. 2C illustrates the locations of observed sites relative to site locations predicted by the model. Fig. 2D indicates deviation of observed from predicted site size in terms of the five level hierarchy. Filled circles represent an accurate size class prediction. A half-filled circle indicates that a site is larger than predicted. "Most reasonable" connectivity lines are retained. Fig. 2E introduces deviations in relative site to site distance from that predicted by the model[9]. Additional connectivity lines (dashed) are added on a subjective basis. Fig. 2F represents the proposed lattice. Fig. 3 presents an enlarged version of this proposed lattice with site reference numbers indicated.

Having derived the proposed settlement lattice through a less than rigorous procedure, an initial quantitative evaluation of similarity or dissimilarity between the model and the proposed lattice is required. The model was primarily constructed on the basis of theoretical transport considerations. Relative success of the model in predicting relationships within the proposed lattice would provide evidence for the operation of similar processes in the Diyālā. The following procedures test the fit between the observed distribution of site sizes, and the distribution predicted by the model.

(i) Each site was assigned a category rank from 1 to 5 on the basis of its position in the proposed functional settlement hierarchy. Large towns were assigned rank 1; towns, rank 2; large villages, rank 3; small villages, rank 4; and hamlets, rank 5. (ii) Each site was assigned a second category rank based on its hierarchical position predicted by the model. (iii) A Kolmogorov-Smirnov non-parametric one-sample test was performed to compare the observed and predicted cumulative site size category rank distributions. (iv) A parametric product-moment correlation coefficient (Pearson's r) was calculated for the observed and predicted site size-category rank distributions. (v) The analysis was repeated for each of the three major cells of the proposed lattice. This allows evaluation of possible differences in fit for major lattice components.

Sample sizes for various aspects of this analysis were determined by the number of settlements in each cell of the proposed lattice. These cells are defined from the model as follows. The Tutub cell, named after the ancient town of Tutub (site 421), containing the following sites; 421, 244, 515, 590, 568, 462, 355, 359, 458, 450, 362, and 357. The Eshnunna cell, named after the ancient town of Eshnunna (site 244), containing the following sites: 244, 515, 397, 264–5, 7, 366, 355, 396, 517, 372, 362, 259, 381, 364, 341, 262, 344, and 350. The Northern cell, contains the following sites: 244,

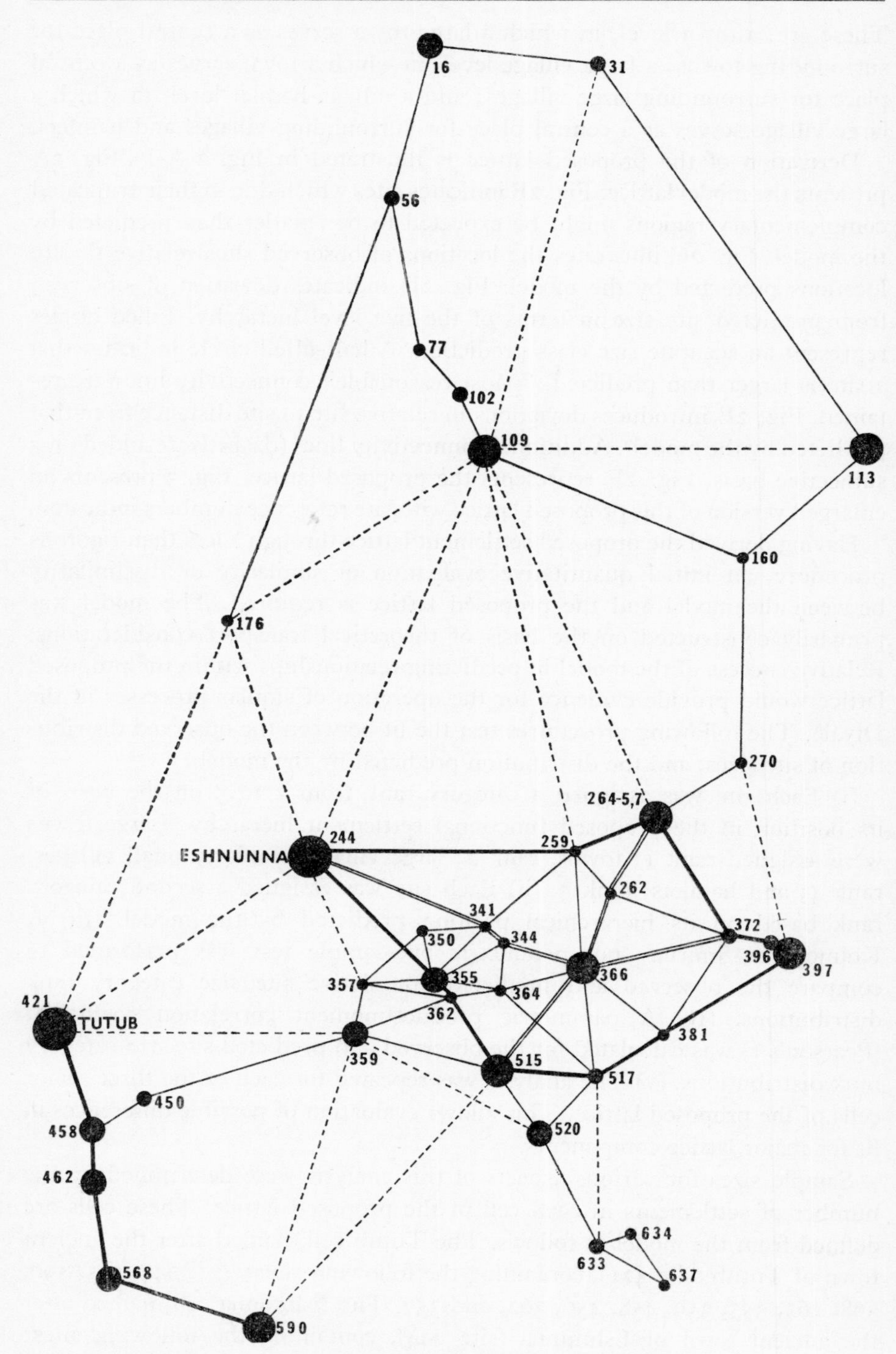

Fig 3 Diyālā Plains, Iraq. Early Dynastic I. Settlement lattice.

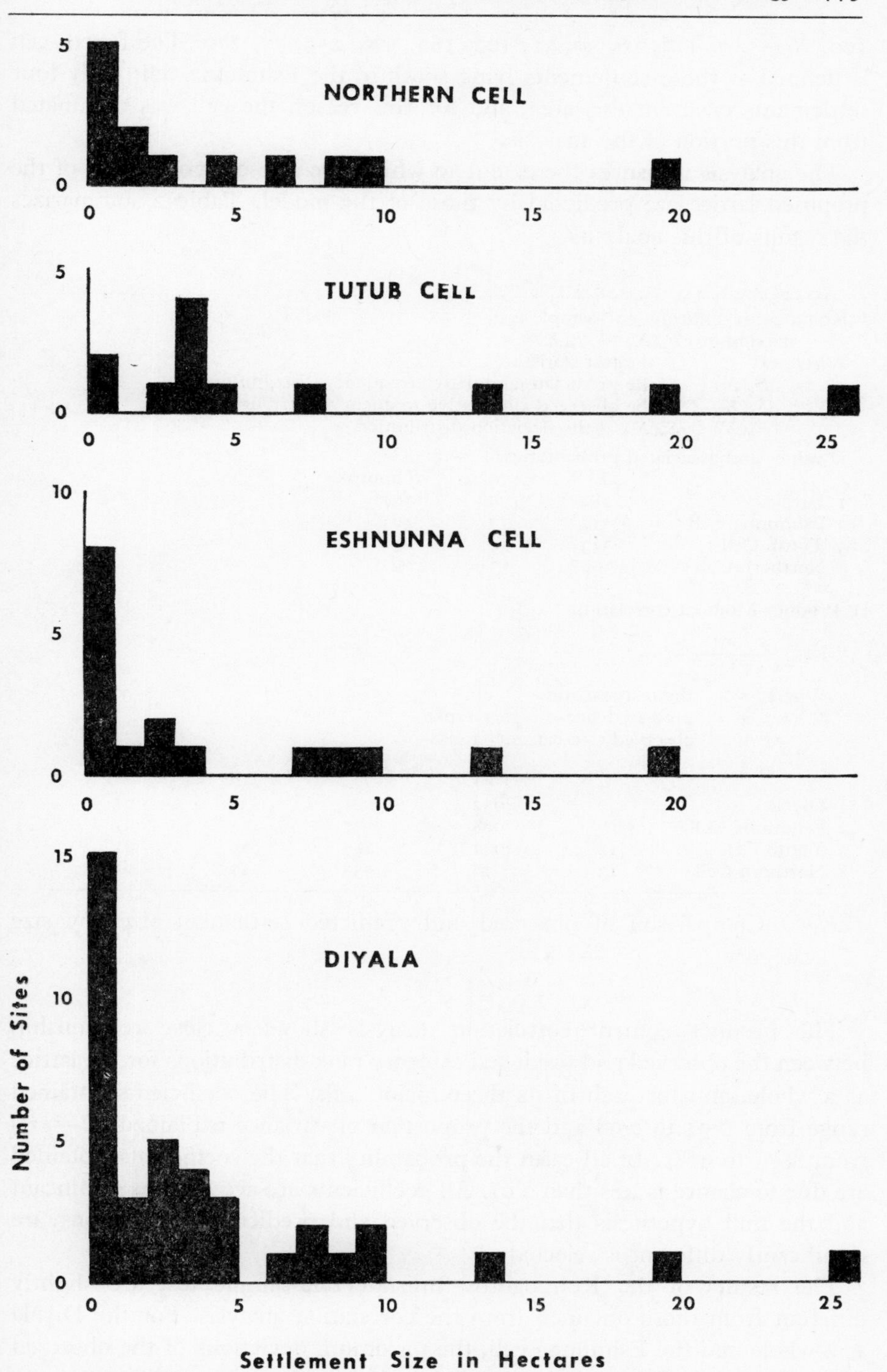

Fig 4 Histograms of settlement size.

109, 264–5, 7, 113, 16, 59, 31, 102, 160, 176, 259, 77, 270. The fourth cell is defined as those settlements lying south of the Eshnunna cell. Only four settlements occur in this area, and for this reason the cell was eliminated from this portion of the analysis.

The analysis measures the extent to which site size-category ranks of the proposed lattice are predicted by those of the model. Table 2 summarizes the results of this analysis.

I Kolmogorov-Smirnov one-sample test.

$D = \text{maximum} \mid F_0(X) - S_n(X) \mid$

Where: D = the test statistic
$F_0(X)$ = the predicted cumulative frequency distribution
$S_n(X)$ = the observed cumulative frequency distribution
$F_0(X) - S_n(X)$ = the deviation distribution

D values and associated probabilities

	D	*N*	*P* approx.
1 Diyālā	·282	39	·01
2 Eshnunna Cell	·412	17	·01
3 Tutub Cell	·333	12	·10
4 Northern Cell	·308	13	·15

II Product-Moment correlation

$$r = \frac{xy}{\sqrt{(x^2)\,(y^2)}}$$

Where: r = the test statistic
x = predicted size-category ranks
y = observed size-category ranks

	n	*r*	*R*	*df*	*P*
1 Diyālā	39	·952	·901	37	·01
2 Eshnunna Cell	17	·968	·937	15	·01
3 Tutub Cell	12	·941	·885	10	·01
4 Northern Cell	13	·976	·953	11	·01

Table 2 Comparison of observed and predicted settlement sizes by size category[10].

The product-moment correlation analysis shows a close relationship between the observed and predicted category rank distributions for the lattice as a whole, and for each of its three major cells. The coefficients obtained range from 0·94 to 0·98 and the proportion of variance explained ($R = r^2$) from 88% to 96%. In all cases the probability that the coefficients obtained are due to chance is less than 0·01. All coefficients are accepted as significant and the null hypothesis that the observed and predicted distributions are significantly different is rejected.

The results of the Kolmogorov-Smirnov one-sample test are slightly different from those obtained from the correlation analysis. For the Diyālā as a whole and the Eshnunna cell, the maximum deviations of the observed from the predicted cumulative frequency distributions are significant at the ·01 level. For the Tutub and Northern cells, probability values of approximately ·10 and ·15 respectively are not sufficient for the null hypothesis of no significant difference between the observed and predicted distributions to be rejected.

The Eshnunna cell is the most complex area of the Diyālā. Increasing complexity may be associated with increased deviation in the distribution of size values from the predicted distribution. This reflects upon the power of the present model to predict size distributions in complex situations. Deviations for the Diyālā as a whole appear to be largely a function of deviations in the Eshnunna cell.

Despite these deviations, a generally good fit is indicated between the size distributions of the model and the proposed lattice. This suggests that processes analogous to those postulated by Christaller's transport principle may have been operative in the Diyālā during the E.D.I period.

Moving to a more detailed consideration of individual lattice cells, certain statistical regularities may be noted in the distribution of small settlements within the Eshnunna cell. These regularities seem to support the contention that the lattice reflects the operation of transport considerations on settlement location.

Fig. 5 indicates one component of the positional deviation of nine small villages and hamlets from their predicted positions on the paths of shortest straight-line distances connecting the five towns of the Eshnunna cell. This deviation is indicated by a line originating at the actual position of each village or hamlet and terminating at its perpendicular intersection with the shortest distance path between two towns.

These deviations were measured in millimetres from the map and correlated with the areal sizes of the nine associated settlements. With all nine site sizes and positional deviations included, a correlation coefficient of ·41 was obtained. This value does not reach even the .05 significance level. Inspection of the scatter plot of the data reveals that one site (396) has a considerable effect in lowering the coefficient obtained. 396, a "small village" of 2·06 ha is located very close to a town size settlement (397). The proximity of 396 to this town suggests that they are in some special relationship, or perhaps are the same settlement. If 396 is deleted from the analysis a coefficient of ·81 between site size and positional deviation is obtained. This is significantly above the .05 level. Connectivity lines between towns show deviation from their expected straight line paths in direct proportion to the sizes of the villages or hamlets located on these lines. If these connectivity lines are interpreted as rough transport routes, it is evident that greater effort is made, in terms of distance travelled, when these routes pass through larger villages and hamlets than through smaller settlements. Note that in general deviations from expected site location are greater on the periphery of the cell than in the interior.

The analysis thus far has considered only villages and hamlets located on connectivity lines between towns. If all such settlements are considered, the

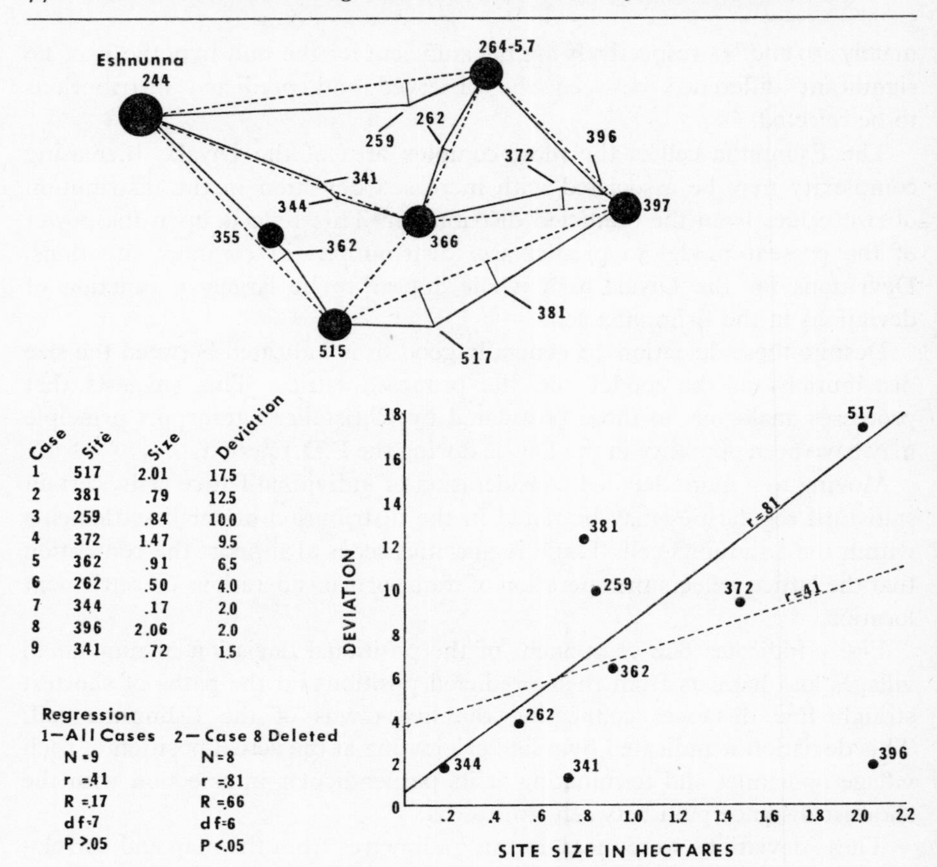

Fig 5 Eshnunna enclave. Relationship of village size to deviation from expected location.

sample size is increased to twelve. This sample was divided into two groups based on settlement location. Group 1 consists of sites located on connectivity lines linking the four peripheral towns of the system. Group 2 consists of sites located off these lines. Stratified in this way settlements of Group 1 are consistently larger than those of Group 2. (See Fig. 6.)

If site size is taken as a measure of settlement importance, the relative importance of connectivity lines may be associated with the relative size of settlements located on these lines. This suggests that the primary connectivity lines of the Eshnunna cell are those between its four peripheral towns. Lines passing through the central town (366) are less important.

If the interaction, economic or otherwise, among the five towns of the area were uniform, the connectivity lines among these towns would be expected to be of roughly equal importance as reflected by the importance (size) of the lesser settlements located along these lines. As this is not the case, it is suggested that interaction was not uniform throughout this system.

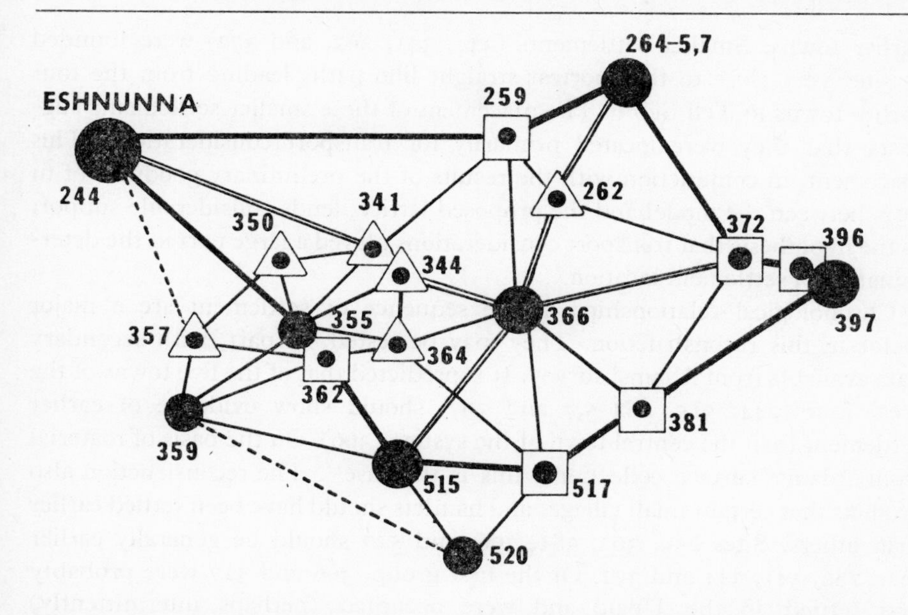

Group 1			Group 2		
Rank	Site	Size	Rank	Site	Size
1	396	2.06	7	364	.74
2	517	2.01	8	341	.72
3	372	1.47	9	262	.50
4	362	.91	10	357	.20
5	259	.84	11	344	.17
6	381	.79	12	350	.13

Fig 6 Eshnunna enclave. Definition of primary connectivity lines.

The central town of the system (366) does not appear to function fully as the focus of interaction for the cell. This situation might have come about as follows. During the Early Dynastic I period the lattice proposed for the Eshnunna cell was in the process of development. Initial major settlements in the area appeared along two major watercourses or branches thereof. To the east, one watercourse connected Eshnunna (244) and Tell Asmar (515). To the west, the second major watercourse connected Tulūl Dhiba^{c}i (264–5,7) and Abū Yiwālik (397). Smaller settlements along these watercourses probably appeared at approximately the same time as the larger settlements. Increasing integration of the activities of the four towns led to the settlement of new sites favourably located for the mediation of this interaction. A town, Tell Seb^{c}e (366) appeared at a position roughly equidistant from the four earlier towns. Smaller settlements (344, 341, 262, and 372) were founded at sites very close to the shortest straight line paths leading from the four earlier towns to Tell Seb^ce. The placement of these smaller settlements suggests that they were located primarily for transport considerations. This placement, in conjunction with the results of the preliminary goodness of fit tests between the model and the proposed lattice, lends considerable support to the hypothesis that transport considerations played a large part in the determination of settlement location.

Chronological relationships in the sequence of settlement are a major factor in this reconstruction. They may be tested, in part, with secondary data available from Adams' survey. It is predicted that of the five towns of the area, four (244, 515, 264–5,7 and 397) should show evidence of earlier settlement than the central town of the system (366). On the basis of material from Adams' surface collections, this is the case[11]. The reconstruction also predicts that certain small villages and hamlets should have been settled earlier than others. Sites 259, 362, 381, 396, and 517 should be generally earlier than 262, 341, 344 and 372. Of the first group, 396 and 517 were probably first settled in the Ubaid and were occupied (perhaps intermittently) through E.D.I. Of the four sites predicted to have later initial occupations, three were settled during E.D.I, and one shows evidence of an Ubaid occupation. The prediction of relative chronology of villages and hamlets is thus partially borne out by secondary data.

Locational relationships among the towns of the Eshnunna cell have been discussed as an isolated system. It appears however to be a subsystem, the upper level of a central-place hierarchy in the area. Two additional nested central-place subsystems may be proposed. (See Fig. 7.)

A large village subsystem includes three large villages (359, 355, and 520); three small villages and hamlets (362, 364, and 517); and one town, Tell Agrab (515). A small village (517) is located at a position predicted for a large village. At 2·01 ha however, it is one of the larger settlements of its size class. Tell Agrab is in a position and of a size to indicate a central-place function for this subsystem. Large villages 359 and 520 are located outside the proposed Eshnunna cell but are apparently included within its central-place hierarchy.

A small village subsystem includes six small villages or hamlets (350, 341, 344, 362, 364 and 357); and one large village, Tulūl Shilbiyat (355). Site 355 is indicated by its position and size to have functioned as a central-place for this subsystem.

In summary, the Eshnunna area seems to have had a highly complex locational structure, determined to a significant extent by "transport" and "marketing" considerations and including hierarchically nested central-place subsystems.

Tutub (421) and the sites which lie to the south in a generally linear fashion have usually been described as a settlement enclave[12]. The apparent isolation of these settlements results from the virtual absence of sites in the area between

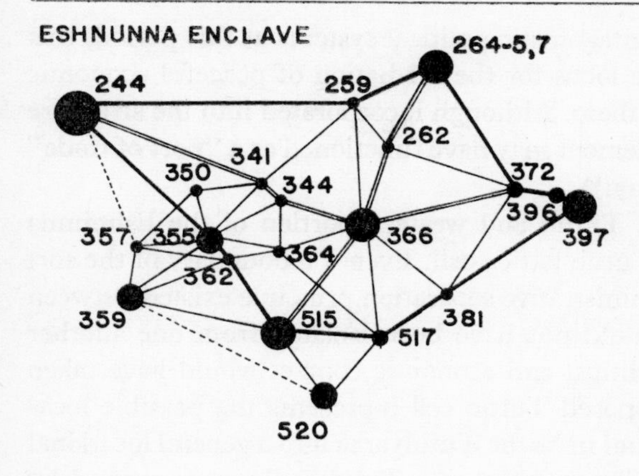

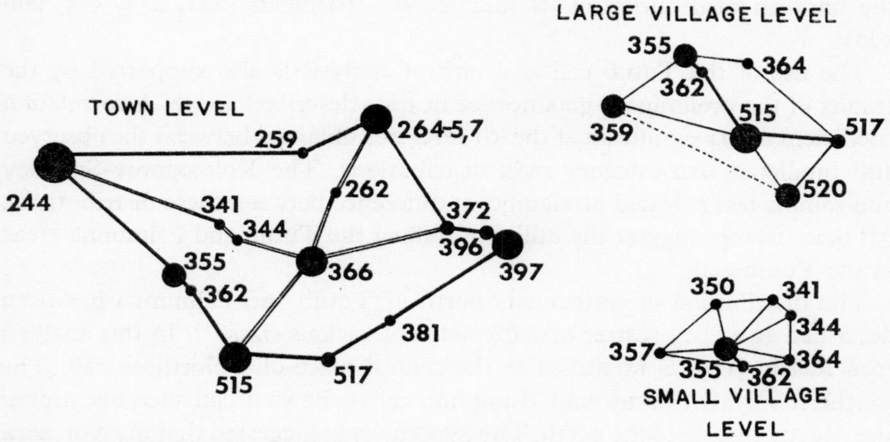

Fig 7 Subsystem nesting. Eshnunna enclave.

this "enclave" and the Eshnunna area. This is perhaps the most intensively surveyed area of the Diyālā. It has been suggested that silting and levee formation from the late Narawān canal which passes through this area may have obscured evidence of local settlement. Adams feels however that deposition from this source had little effect in masking evidence of settlement in this area[12]. Much of the area is simply devoid of Early Dynastic I settlement.

This has led to the hypothesis that some sort of political boundary lay between Tutub and Eshnunna and their associated settlements. The area would then be interpreted as a buffer zone between two antagonistic states. At present there is little evidence bearing on this question. Preliminary examination of published artefactual material from Tutub and Eshnunna gives no indication of stylistic differences which might be indicative of minimal or reduced contact between these areas.

The large village (359) located in the interior of the proposed Tutub cell is particularly interesting. If Tutub and Eshnunna were the administrative centres of rival and perhaps antagonistic political systems at this period, this village might have served as a locus for the mediation of peaceful economic and political contact between them. Although incorporated into the structure of the Eshnunna area, the settlement may have functioned as a "port of trade" of the sort suggested by Polanyi[14].

In the present analysis the Tutub and western portion of the Eshnunna areas are linked forming the Tutub lattice cell. Even if a boundary of the sort suggested by Christaller's administrative separation principle existed between these areas, they probably would not have been isolated from one another in a systemic sense. Some political and economic contact would have taken place between them. The proposed Tutub cell represents the possible locational aspects of this contact, and links the Tutub area into a general locational system of the entire Diyālā. The proposal of a Tutub cell was suggested by the uniform distribution of its four major settlements (241, 244, 515, and 590).

The use of the Tutub cell as a unit of analysis is also supported by the results of the preliminary goodness of fit tests described above. A correlation coefficient of ·94 significant at the ·01 level was obtained between the observed and predicted size category rank distributions. The Kolmogorov-Smirnov one-sample test revealed no significant difference between these distributions. All these factors suggest the utility of linking the Tutub and Eshnunna areas in the Tutub cell.

The distribution of settlements north of Tutub and Eshnunna has been described as a ". . . scatter of settlements of various sizes"[15]. In this analysis Abū Rasain (109) is identified as the central place of a Northern cell. The southern Diyālā (Tutub and Eshnunna cells) shows a considerably higher site density than does the north. The hypothesis is suggested that the Northern cell represents occupation of the area by a population expanding from the south. Two possible reasons for such settlement are suggested. One, to open new agricultural areas. The apparent availability of considerable unused agricultural land in the south suggests that this was not a major factor in the settlement of the northern area. Two, the northern area was possibly settled, in part, to establish centres to mediate the importation of raw materials from source areas in the Zagros to the area of primary population concentration in the southern plain.

The principal raw material resources of the Mesopotamian alluvium were soil, water and reeds. Most other materials had to be imported from other areas. Wright includes wood, bitumen, flint, copper, alabaster, and similar stones for vessels, and basalt and similar stones for grinding slabs among the principal bulk imports into the alluvium[16]. Lapis lazuli, carnelian, agate, and obsidian for lapidary purposes; gold, silver, and lead among other materials appear in smaller quantities[17].

As the Zagros was a primary source area for many of these materials, the existence of specialized centres for their importation into the Diyālā area would be expected. Adams reports that after prolonged search only Early Dynastic materials were found on Abū Rasain (109). Evidence of a defensive wall and "... greater quantities of worked flint of all kinds than have ever been encountered by the author on an alluvial site" are also reported[18]. Jacobsen, as reported by Adams, visited Abū Salabikh (113) the other town size settlement of the northern area. Again only Early Dynastic materials were found, with "many large unworked stone blocks ..."

The presence of these concentrations of raw materials associated with large E.D.I settlements supports the hypothesis that sites of the Northern cell served as mediation points in the importation of these raw materials. Available evidence also supports the hypothesis that population expanded into the northern area from the south. Adams notes that, "Ubaid farmers entered from, or at least had their primary affiliations with, areas farther south in the Tigris-Euphrates flood-plain"[20]. This suggests that the southern Diyālā was occupied first, with subsequent northern expansion.,

The fourth proposed lattice cell remains to be discussed. It contains only four sites, located south of the Eshnunna cell. These include a small village (633); two hamlets (634 and 637); and a large village (520). 520 has been associated with the large village subsystem of the Eshnunna area. These four sites may represent a budding off phenomenon: an initial stage of population expansion from the Eshnunna cell. Possibly they represent an entry point into the Diyālā for materials from further south. Additional data are required to test such hypotheses.

Central-Place Theory postulates regular spatial patterns of the differential distribution of activities related to the production and distribution of goods and services. In theory activity loci are so distributed that energy expended in these activities is minimized. In practice it appears that energy output minimization need not be assumed as necessary for the appearance of spatial distributions predicted by the theory.

In the present paper, the distribution of Early Dynastic I settlements in the Diyālā, and by implication activities, has been shown to approximate distributions predicted by various aspects of Central-Place Theory. The presence of such distributions need not imply that they are in whole or in part the result of the operation of processes postulated by the theory. A different set of ordering principles may be reflected in the observed distributions.

For example Adams has pointed out that on the basis of textual evidence overland transport was minimal in the Diyālā in the Early Dynastic period[21]. This would suggest that either the interpretation of the present analysis is basically in error, or that overland transport was more significant and/or intense than present evidence would indicate. The present analysis should in no way be construed as providing even a partial explanation of settlement locations in the Diyālā. It does, however, provide a series of hypotheses relative to the operation of processes which may have contributed to the determination of settlement locations. Further data on differential activity distributions are required to test these hypotheses. These data must reflect activities and their organization both within and among the settlements of the area.

For example, the arrangement of settlements in the Eshnunna enclave suggests that nested central-place subsystems existed there in the E.D.I period. Material imported from outside the system would be expected to show a differential distribution among its three proposed levels. Raw materials may have been imported in large blocks to settlements of the highest level in the hierarchy, then subdivided for distribution to lower level settlements. Such subdivision might be expected to occur at each hierarchical level. In this case, the size and/or quantity of waste material from subdivision and artefact fabrication found on sites of different hierarchical position would be expected to be directly associated with these differential positions.

Though tentative and incomplete, the present analysis does indicate that Central-Place Theory and locational analysis in general may be a valuable aid in archaeological hypothesis formation. This type of analysis does not offer the archaeologist concerned with settlement systems a panacea for all his problems, but may make some of them less formidable.

Notes

1 Christaller, W. (trans. by Baskin, C. W.) (1966). *Central Places of Southern Germany*. Englewood Cliffs.

2 These assumptions are adapted from Garner, B. J. (1967). Models of urban geography and settlement location, *in* Chorley, R. J. and Haggett, P. (eds.) *Models in Geography*. London. pp. 304–5.

3 As the proposed model is based on a combination of transport and marketing considerations, it cannot be assigned a K-value.

4 Adams, R. McC. (1965). *Land Behind Baghdad: A History of Settlement on the Diyālā Plains*. Chicago and London. Additional data for this study were obtained from Dr. Adams' original survey notes which he very kindly made available to me. It should be pointed out that these data are known or suspected to be biased in several ways. At present little of this bias can be controlled. The reader is referred to Adams, R. McC. (1965). *op. cit.* p. 120 ff. for a discussion of these problems. Due to these biases, the results of the present analysis should be considered with caution.

5 This is most probably an erroneous assumption. It is however necessary due to the present lack of more detailed data.

6 Berry, B. J. L. (1967). *Geography of Market Centres and Retail Distribution*. Englewood Cliffs. p. 40.

7 Berry, B. J. L. (1967). *op. cit.* p. 79.

8 The sample size of E.D.I settlements used in this study is not sufficient to establish statistical modalities in their size distribution.

9 Due to the lack of appropriate data, distance transformations were not attempted in this analysis.

10 See Siegal, S. (1956). *Nonparametric Statistics for the Behavioural Sciences*, New York, Toronto, London, pp. 47–52, for a discussion of the Kolmogorov-Smirnov statistic. The use of Pearson's r is not strictly appropriate in this case, and should be considered here as a goodness of fit test.

11 Adams, R. McC. (1965). *op. cit.* p. 135 ff.

12 Wright, H. T. (1969). The administration of rural production in an early Mesopotamian town. *Anthropological Papers, Museum of Anthropology, University of Michigan No. 38*. Ann Arbor. p. 30.
13 Adams, R. McC. (1970) personal communication.
14 Dalton, G. (ed.) (1968). *Primitive, Archaic and Modern Economies: Essays of Karl Polanyi*. Garden City. p. 23 ff.
15 Wright, H. T. (1969). *op. cit.* p. 30.
16 Wright, H. T. (1969). *op. cit.* p. 31.
17 Delougaz, P., Hill, H. D. and Lloyd, S. (1967) Private houses and graves in the Diyālā region. *The University of Chicago Oriental Institute Publications, Vol. LXXXVIII*. p. 25 ff.
18 Adams, R. McC. (1965). *op. cit.* p. 139.
19 Adams, R. McC. (1965). *op. cit.* p. 140.
20 Adams, R. McC. (1965). *op. cit.* p. 36.
21 Adams, R. McC., this volume, p. 745.

ANTHONY McNICOLL

The development of urban defences in Hellenistic Asia Minor

In this paper I shall examine the principal strategic and tactical developments in the urban defences of Hellenistic Asia Minor, with brief reference to some of the political, economic and military conditions governing them[1].

Rostovtzeff has described the Greek conflict between the rulers or leagues on the one hand and the cities on the other as union versus particularism[2]. During the struggles of the Successors of Alexander there was an intensification of one aspect of this conflict, namely urban unification, which was economically advantageous for all and politically expedient for the rulers. In Asia Minor the first man to enforce this kind of urbanization was the Hecatomnid Mausolus of Caria (377/6–353/2 B.C.), "the greatest genuine forerunner of the Hellenistic Age"[3]. Mausolus created a large, economically forceful community, gathering at Helicarnassus the native Lelegian inhabitants of the peninsula to populate his new city by the process known as *synoecism*.

The contrast between Mausolus' new city and the Lelegians' towns is marked[4]. The towns are small, standing on isolated hill-tops. Their fortifications are characterized by fairly small towers, curving curtains and rather crude styles of masonry. Helicarnassus' fortifications on the other hand are of more regular masonry, with larger (though not very large) towers.

Most important, the Helicarnassus circuit is sited to control forward ground, i.e. to dominate all ground beyond the walls. Because the city itself was on low ground by the harbour, the search for height and clear fields of fire led the military architects to build a wall-line enclosing some 350 ha, a much greater area than was actually inhabited. This disparity between enclosed and occupied areas is a common feature of the "great wall" sites.

"Great wall" circuits controlling forward ground are in fact considerably older than the fourth century, e.g. the Bronze Age fortifications of Boğazköy. However, during the fourth century in addition to the ram and sap, artillery began to assume importance in siege-warfare as a means of breaching walls. The development of the torsion catapult in the later half of the century[5] made the construction of walls controlling forward ground almost vital.

Although neither synoecism nor long walls were beyond the capabilities of a democracy, e.g. Rhodes, in the Hellenistic period, both frequently reflect a social organization led or driven by an autocrat with a standing army. Such organizations flourished in the time of the Successors, thus facilitating establishment or reconstitution of large cities and the construction of modern defences of enormous length, with or without synoecism. Allegiances to the rulers were rarely strong, and by concentrating the populations of several towns into larger units economic viability could be increased and garrison forces concentrated.

Among the Successors, Lysimachus, Seleucus and Antigonus were the principal builders of "great circuits"; the Ptolemies, perhaps because of their military system, seem to have built few if any of these long enceintes. Perhaps too the comparative flatness of a site such as Alexandria made anything but the shortest line of defence pointless (cf. the desert wall of Dura Europus). The coastal regions of Asia Minor are, by contrast, generally hilly, and "great circuits" of the late fourth and early third centuries are found at Heracleia-on-Latmus, Magnesia-on-Maeander, Cnidus, Myndus, Colophon, Smyrna, Ephesus and Alexandria Troas.

Control of forward ground was expensive and therefore difficult for those cities which had to finance the building of their own fortifications. By the end of the fourth century the torsion ballista was widely used in attack, and wherever it could be afforded in defence. If it was reckoned impossible to dominate the ground beyond the wall by building the defences on or in front of a crest, an alternative, more passive method of defence was used. This entailed building enormously thick walls. For example, at Erythrae (*c.* 330 B.C.) the south wall, dominated in places from higher ground, is 4·80 m wide[6]; the east wall of Cnidus (*c.* 320 B.C. ?) is about the same width[7]. With the thick Erythraean wall went small towers, about 5 m^2, in size comparable to Colophon's half-round towers (*c.* 310 B.C.)[8]. As both these cities' defences appear to have been built voluntarily by the community rather than by decree, it seems that large torsion ballistas (which would require large towers) commonly existed in professional standing armies, but were not available to all citizen forces.

The "great circuit" type culminates in western Asia Minor in the walls of Ephesus (*c.* 290 B.C.). Here Lysimachus enforced the synoecism of Colophon, Lebedus and Old Ephesus to inhabit his new city Arsinoeia (the name did not outlive him) in a site brilliantly selected for trade but militarily necessitating a defence line more than 9 km long, estimated by Miltner to contain at least 200,000 m^3 of stone, not including the towers[9]. There are more than forty of these on the 385 m high Bülbüldaği (Mt. Coressus), which dominates the city from the south; their average size is about 10 m^2. Just above ground level in the flanks of a number of the towers large windows survive, showing that heavy artillery was extensively used. Height of wall and towers does not appear to have been any more important at Ephesus and its contemporary,

Heracleia-on-Latmus, than it was at earlier sites such as Alinda (360–340 B.C.).

An interesting feature of the Ephesian defences is the large number of posterns on Coressus. The defence was planned to allow harassment of the attackers by infantry sorties as well as by artillery bombardment. Such "aggressive defence" points to a professional force. We may compare it with Dura Europus (*c.* 300 B.C.), the Dema in Attica (*c.* 335 B.C.), the land-wall at Iasus, the work of Philip V (*c.* 200 B.C.), and the Miletus cross-wall (*c.* 100 B.C.). Notice, however, that the Iasic and Dema fortifications are not urban[10].

As the third century advanced, long walls became increasingly impractical. Through exhaustion of finances or man-power, or both, enormous fortifications could not be professionally manned. This is apparent from the walls of Heracleia-on-Latmus (*c.* 300 B.C.)[11]. Here the earlier circuit was abandoned and a *diateichisma* or cross-wall built, cutting 2 km off the total length (? mid-third century). The same expedient was used at Assos to shorten the defences[12], and perhaps also at Theangela[13].

By the end of the third century the concept of control of high points enclosed by long walls was largely superseded by the idea of building the shortest possible fortifications, which are, of course, a series of straight lines bearing a direct relationship to the size of the city. A number of these "straight line" fortifications exist in Pamphilia and Pisidia[14]—at Perge (*c.* 200 B.C.), Side (*c.* 180 B.C.), Sillyum and the site called by Paribeni[15], Pednelissus (both late second century). Side's and Perge's fortifications are characterized by tall curtains and towers three or four storeys high. The top floors of the towers are designed to contain heavy artillery to outrange the attackers' weapons, while batteries of lighter engines, small scorpions and three-span arrow-firers, could be established along the curtains.

On the west coast of Asia Minor the "straight line" defence was also used, for example at Teos (*c.* 200 B.C.) and at Ceramus (? mid-second century). At Teos possibly width rather than height was preferred as a means of defence; the fortifications are reminiscent of Erythrae's, having small towers, and curtains 4·20–4·50 m wide.

With the shortening of the circuits we find in a number of fortifications possibly experimental features designed to further impede the enemy in front of or on the actual wall without greatly increasing costs. The Miletus cross-wall of *c.* 100 B.C. could be classed as a "straight line" for very active defence, although it is also a *diateichisma* in the tradition of the later wall at Heracleia, renouncing the high ground of Kalabaktepe in favour of a shorter line[16]. It is probably a new development, as, in quite a different way, is the passive "straight line" at Oenoanda with its round and pentagonal towers (*c.* 150 B.C.)[17]. Perhaps also new is the idea employed in the fortifications of Cadyanda[18] (second century) which probably had wooden wall-walks supported on buttresses behind a curtain only 1·20 m wide.

Finally it should be pointed out that the "straight line" never totally replaced the "great circuit". Where there was sufficient power and money, long lines continued to be built, e.g. by Eumenes II at Pergamum (early second century) and by Amyntas at Isaura Vetus (late first century).

The outline presented here will, I hope, need only minor modifications. Our state of knowledge concerning the vast mass of potential evidence is likely to remain unsatisfactory until excavators expend more effort on obtaining accurate stratigraphy relating to fortifications of Classical and Hellenistic sites.

Notes

1 Some of the dates proposed for the fortifications mentioned in this article are generally accepted. I have reached others in the course of study for a D.Phil. degree at Oxford.

2 Rostovtzeff, M. (1941). *Social and Economic History of the Hellenistic World*. Oxford. p. 36.

3 Judeich, W. (1892). *Kleinasiatische Studien*. Marburg. p. 236.

4 The most recent study of Helicarnassus and the Lelegian settlements is by Bean, G. E. and Cook, J. M. (1955). The Helicarnassus Peninsula, *BSA*, 50.

5 Some authorities have held that torsion artillery was invented in the early fourth century. However, the time of Philip II seems more probable for its inception. See Marsden, E. W. (1969). *Greek and Roman Artillery: Historical Development*. Oxford. p. 60.

6 See Weber, G. (1901). Erythrai, *AM*, XXVI. The inscriptions relating to the walls are re-examined by Maier, F. G. (1959). *Griechische Mauerbauinschriften I*. Heidelberg.

7 Rostovtzeff, M. (1941). *op. cit.* p. 1266 quotes F. Kirschen, who reckoned the east wall of Cnidus to be 4·50 m wide.

8 For a general description of Colophon's walls, see Schuchhardt, C. (1886). Kolophon, Notion und Klaros, *AM*, XI. For their date see Maier, F. G. (1959). *op. cit.* pp. 223–4.

9 Miltner, F. (1958). *Ephesos: Stadt des Artemis und des Johannes*. Vienna p. 14. On the date of the fortifications see Maier, F. G. (1959). *op. cit.* p. 238.

10 On the Iasus land-wall, see Bean, G. E. and Cook, J. M. (1959). The Carian Coast III, *B.S.A.*, **52**, pp. 100–5, and an important note by Winter, F. E. (1963). The chronology of the Euryalus Fort, *Amer. J. Arch.*, **67**, pp. 374–5, note 38. Winter points out the Hellenistic features of the Iasus land-wall, but offers no precise date for the complex. I ascribe it to the period of Philip V's occupation on architectural and historical grounds. On the Dema, see Jones, J. E., Sackett, L. H. and Eliot, C. W. J. (1957). Το Δέμα: a survey of the Aigaleos-Parnes Wall, *B.S.A.*, **52**, pp. 152–89, and McCredie, J. R. (1966). Fortified military camps in Attica, *Hesperia Suppl.*, **11**, pp. 63–6.

11 See Krischen, F. (1922). Die Befestigungen von Herakleia am Latmos, *in* Wiegand, Th. (ed.). *Milet III*, 2. Berlin.

12 See Koldewey, R., Clarke, J. T. and Bacon, F. H. (1902–21), *in* Bacon, F. H. (ed.). *Investigations at Assos . . .* 1881–2–3. Cambridge, Mass.

13 Theangela is described and discussed by Bean, G. E. and Cook, J. M. (1955). *op. cit.* and Bean, G. E. and Cook, J. M. (1957). *op. cit.*

14 For the cities of Pamphylia and Pisidia see Niemann, G. and Petersen, E.

(1890–92), *in* Lanckoronski, K.g. (ed.). *Städte Pamphyliens und Pisidiens* I, II. Wien.

15 Paribeni, R. (1916–20). Di una città della Pisidia: forse Pednelissos, *Annuario*, III.

16 Described by Gerkan, A.v. (1935). Die Stadtmauern, *in* Wiegand, Th. (ed.). *Milet II, 3*. Berlin. Redated by Kleiner, G. (1968). *Die Ruinen von Milet*. Berlin. p. 17.

17 On the ruins of Oenoanda see Petersen, E. and Luschan, F.v. (1889). *Reisen im Südwestlichen Kleinasien* II. Vienna.

18 For Cadyanda see Benndorf, O. and Niemann, G. (1884). *Reisen im Südwestlichen Kleinasien* I. Vienna.

HANS JÖRG NISSEN

The city wall of Uruk

Introduction

In ancient times the city wall of Uruk was considered to be the work of Gilgamesh, the great hero and early king of Uruk. Unfortunately the only information we have about the height or general aspect of the wall is the statement in the Gilgamesh Epic that the wall enclosed an area of 3 sar[1], that is, twice as much as the 5·5 km² actually encompassed by the wall. Excavations of portions of the wall give us an idea of the structure: 9·5 km long with towers, rectangular at first but later round, spaced every 9 m [2]. There were at least two main gates, in the north and the south; other gates presumably existed but have not been found. Parts of the wall are still preserved up to a height of 7 m, but the actual upper edge has not been found anywhere. The remains of another wall have been found 10 m in front of this one; though much less well preserved and without towers, it was probably part of the same structure.

Parts of the uncovered portions of the wall were built in the Old Babylonian period in the nineteenth century B.C., but all the above-mentioned features belong to the earliest wall, which by its building material can be dated to the Early Dynastic period (*c.* 3000 to 2300 B.C.)[3]. Unfortunately more accurate dating within these 700 years is impossible on the basis of the associated artefacts as the main group, the pottery, was not given proper attention. Since it is not possible to get an exact date from internal evidence, it may be possible to date the wall on the basis of what we know about the early history of Uruk and its hinterland, and by the recognition of a time at which a city wall would have been necessary.

Early history of Uruk[4]

It has been suggested that Uruk grew out of two settlements, Kullaba and Eanna. The earlier one, Kullaba[5], has recently revealed structural remains dating to the Ubaid period (*c.* 4500 to 3500 B.C.)[6]. Later buildings on top of these consist of a sanctuary whose ground plan and construction on a platform show a close relationship to the early temples at Eridu (*c.* 5500 to 4500 B.C.). A deep sounding sunk down to virgin soil in the main area of the Eanna district failed to produce anything earlier than material from the latest phase of the Ubaid (*c.* 4000 to 3500 B.C.)[7]. The Eanna temples also show a different plan and are built directly on the ground. Until the end of the Early Uruk period (*c.* 3300 B.C.) Eanna may not even have had a substantial central sanctuary which could rival that of Kullaba, because only from Archaic level VI onwards do we have remains of monumental architecture[8]. After this time, however, Eanna apparently took the lead and developed fast with a sanctuary larger than Kullaba ever had[9]. At some time in the same period a huge terrace of unknown purpose was constructed in the open space between the two sanctuaries, actually touching both of them[10]. This terrace came so near the platform of the temple of Kullaba that simply by its height it deprived this temple of its uniqueness, and thus may be taken as an argument for the decline of the Kullaba sanctuary as the Eanna one expanded. This development seems to have come to an end during the last phase of the Late Uruk period when the terrace between Kullaba and Eanna was enlarged both in height and in extent so that the remains of the last temple on top of the Kullaba platform were enclosed and buried completely[11].

Probably connected with this development there were also radical changes in Eanna during the very last phase of level IV, when all buildings in Eanna were destroyed[12]. This apparently was planned destruction because bricks of torn-down walls were used to fill in the spaces between the remaining stumps in order to level the ground for the construction of a sanctuary of a completely different plan. For the first time Eanna was centred around a temple on a platform[13]. From this time on, that is from the beginning of level III or Jamdet Nasr period (*c.* 3100 to 3000 B.C.) there seems to have been only one central sanctuary in Uruk, which now covered a smaller area and was more centralized, with only one religious structure surrounded by many administrative or storage rooms[14]. We do not know much more about this period nor about the following Early Dynastic period, as these levels in the vicinity of the main temple suffered much from the building and digging activities of later times. However, we know slightly more about the first part of the Early Dynastic period through various test excavations and sporadic surface surveys; we can be almost sure that the entire area enclosed by the city wall[15] and even beyond[16] was occupied, in total exceeding the area occupied during earlier periods.

Generally speaking, it seems that opposite an older settlement with a sanctuary a new settlement was founded, which, after a period of time, developed its own sanctuary. This one seems to have been differently organized and to have grown faster in importance than the old one. At one point efforts were apparently made to connect both sanctuaries, ending, however, in a merger of the two. The result was a single sanctuary on the site of the younger of the two previous ones, but differently organized. The entire development seems to have arisen from a struggle for hegemony between the two sanctuaries, by then the two central economic powers, ending in a complete take-over by the presumably economically stronger Eanna.

Early history of the hinterland of Uruk[17]

The developments in the hinterland of Uruk are mainly of interest here because, through the interactions between Uruk and its hinterland, they may reflect certain changes in Uruk itself, and thus may contribute indirectly to our understanding of its early development.

In the Early Uruk period (*c.* 3500 to 3300 B.C.) Uruk itself (or rather some parts of it) was already a flourishing economic centre at a time when the occupational level of the surrounding land was still rather low. A large body of evidence from the hinterland is, however, available from the Late Uruk (*c.* 3300 to 3100 B.C.), the Jamdet Nasr (*c.* 3100 to 3000 B.C.) and the Early Dynastic I (*c.* 3000 to 2800 B.C.) periods, because apart from Uruk there were 100 settlements of the Late Uruk period, ninety of the Jamdet Nasr period and seventy of the Early Dynastic I period in the area to the north, north-east and east of Uruk itself. At first glance a map of all settlements of the period would seem to show that the Late Uruk settlements were evenly distributed throughout the area. However, on closer examination we find regional differences. In the northern part of the area, in a belt from approximately 15 to 30 km from Uruk, there were clusters of four to eight settlements each, whereas in the belt immediately adjacent to Uruk (approximately 12 km or less from the city) we find none of these clusters. Though they obviously existed there is no positive evidence so far of the interactions between Uruk and its hinterland.

Among the settlements of the following Jamdet Nasr period, the differences between the areas of the hinterland become even clearer since the situation in the inner belt remains the same, whereas in the northern area only one or two settlements of the above-mentioned clusters continued to be inhabited; they had now become much larger than before. This was obviously the result of a centralization process which seems to have been confined to the outer belt. As there seems to be a critical distance of 12 to 15 km from the city which was almost free of settlements and marked a kind of no-man's land between the two belts, it is clear that any explanation of the different development in the two belts has to focus on the relations between Uruk and its hinterland. As we have seen above, there were also indications in Uruk of a trend towards centralization, and it is suggested here that this was not an underlying trend but that both developments were interdependent. Although other factors were certainly involved as well (the rise of certain central functions, concentration along the fewer waterways, etc.) it is suggested that an overall reorganization with concentration of power in Uruk—of which the reorganization of the sanctuary was the only part visible to us—triggered off or speeded up the process of centralization in the hinterland.

If we look briefly at the later developments during the Early Dynastic I period we see that this trend towards centralization continued in the northern part, with the additional feature that the small emerging nuclear centres moved from an almost even distribution to a concentration along the two main channels of the Euphrates, which probably only became established as such during that time. Uruk as the main centre was naturally opposed to this trend and attempted to stop this reaction, but it could fight it effectively only in areas which were directly under its control, i.e. in the inner belt. The pressure and hostilities at first resulted only in a speeding up of the centralization of the hinterland. However, these newly formed centres could in no way be a real threat to Uruk since they were very much smaller. Yet we may assume that the concentration of economic power in these towns could not leave Uruk unaffected. Owing to their favourable location and their distance from Uruk, however, the trend continued until some of these centres expanded so much in size and power—and therefore in influence—that they could rival Uruk. At this stage it is very likely that the situation changed and hostilities were reciprocated.

Conclusion

From Christaller's Central Place Theory we would expect the new centres to be at such a distance from Uruk that there would be no conflict of interests. Yet if we attribute to the new centres, especially Umma, areas of interest comparable with what we saw for Uruk from the Late Uruk to the Early Dynastic I periods (i.e. a sphere of direct influence up to 12 km from the centre, and a sphere of indirect influence from 15 to 30 km), then we see that the old and new centres would be contiguous, and that their spheres of indirect influence would even partially overlap. Leaving aside the nature of this influence—though I think it could be nothing other than economic domination—the emergence of new full-sized urban centres would result in conflicts of a form not known before. In fact, the records of the subsequent Early Dynastic II and III periods are full of local fights between neighbouring cities, such as the one known from contemporary sources between the old centre of Lagash and the new centre of Umma over a stretch of borderland. It is probable, therefore, that the first city wall of Uruk was a response to feared or actual hostilities from the new emerging centres. This would make the earliest wall of Uruk a product of the changes in the distribution of political power at the end of the Early Dynastic I period.

Notes

1 Pritchard, J. B. (1955). *Ancient Near Eastern Texts*. Princeton. p. 97, lines 308ff.

2 Heinrich, E. (1932). Die Untersuchung an der Stadtmauer, *in* Noeldeke, A. 4. *Vorl Ber. Uruk/Warka*. p. 24; Haller, A. (1936). Die Stadtmauer, *in* Noeldeke, A. 7. *Vorl Ber. Uruk/Warka*. pp. 41–5; Haller, A. (1938). Die Stadtmauer, *in* Noeldeke, A. 8, *Vorl. Ber. Uruk/Warka*. pp. 5–7.

3 In Babylonia the so-called planoconvex bricks come into use during the first phase of Early Dynastic I and go out of general use during the last phase of Early Dynastic III. See Delougaz, P. (1933). Planoconvex Bricks and the Methods of their Employment, *Stud. Ancient Or. Civ.* 7, *Chicago*.

4 The preliminary reports rarely give more than an actual description. More generalizing publications are: Heinrich, E. (1949). Die Stellung der Uruktempel in der Baugeschichte, *Z. f. Assyriol. N.F.* **15**, pp. 15ff; Lenzen, H. J. (1941). *Die Entwicklung der Zikurrat*. Leipzig; Lenzen, H. J. (1949). Die Tempel der Schicht Archaisch IV in Uruk, *Z. f. Assyriol. N.F.* **15**; Lenzen, H. J. (1951). Zur Datierung der Anuzikurrat, *Mitt. d. Deutschen Morgenl. Gesell.*, **83**; Lenzen, H. J. (1955). Mespotamische Tepelanlagen . . . , *Z. f. Assyriol. N.F.* **17**.

5 Wilcke, C. (1969). *Das Lugalbandaepos*. Wiesbaden. p. 29.

6 Personal communication from the present director of the excavations in Uruk, Prof. J. Schmidt.

7 Jordan, J. (1932). Die praehistorischen Perioden der Tiefgrabung in Eanna, *in* Jordan, J. 3. *Vorl. Ber. Uruk/Warka*. Berlin. pp. 29–31 with the plates 10–13; Haller, A. (1932). Die Keramik der Archaischen Schichten von Uruk, *in* Noeldeke, A. 4. *Vorl. Ber. Uruk/Warka*, Berlin. pp. 31–5.

8 In Archaic level VI were found layers of clay cones, comprising the elements of large mosaics which, according to their use in the succeeding periods, were confined to temples or other representative buildings (3. *Vorl. Ber. Uruk/Warka* p. 30).

9 The actual enclosure wall of the holy district of Eanna of Archaic level V has not so far been found. Yet the existence in this period of temples in the outer parts of what later in level IV was the holy district, points to the level V holy district being already as large as the later one.

10 Lenzen, H. J. (1951). *ibid.*

11 Heinrich, E. (1937). Die Grabung im Planquadrat K XVII, *in* Noeldeke, A. 8. *Vorl. Ber. Uruk/Warka*. Berlin. pp. 27–9.

12 Lenzen, H. J. (1968). Uruk IVa, *in* Lenzen, H. J. 24. *Vorl Ber. Uruk/Warka*. Berlin. pp. 13–18.

13 Lenzen, H. J. (1941). *op. cit.* p. 11.

14 Lenzen, H. J. (1964). Uruk III, *in* Lenzen, H. J. 20. *Vorl. Ber. Uruk/Warka*. Berlin. pp. 11–18.

15 See my forthcoming reports, Nissen, H. J. (in press), Grabungen in den Quadraten K/L XII, *Baghd. Mitt.* **5**; Nissen, H. J. (in press), Kurzgrabung im Quadrat I XIII, *in* Schmidt, J. 26. *Vorl. Ber. Uruk/Warka*, Berlin. Early Dynastic I surface material has been found in quantities in the north-western and north-eastern parts of the city area and in the southern-most part, to a lesser degree in the low-lying areas of the eastern part.

16 Much of the pottery from the so-called "Archaic Settlement" in the plain to the north-east of Uruk (see Strommenger, E. (1963). Archaische Siedlung, *in* Lenzen, H. J. 19. *Vorl. Ber. Uruk/Warka*, Berlin. pp. 45–55) must be assigned to the Early Dynastic I period, contrary to the statements by the excavator. In fact, Early Dynastic I pottery spreads out from the

city wall towards the north-east, covering an area of several hectares which includes also the site of the Archaic Settlement.

17 The following remarks are based on information collected during an archaeological surface survey of the area around Uruk in 1967, conducted by Dr. Robert McC. Adams and myself. The results will be published in detail in Adams, R. McC. and Nissen, H. J. (in press). *The Uruk Countryside*. Chicago.

DAVID OATES

The development of Assyrian towns and cities

In this paper I offer some observations on factors which have influenced the rise and the fortunes of towns and cities in Assyria. Too few sites have been excavated, and even on these the evidence is insufficient to permit any exhaustive analysis of such factors. All we can do at present is to isolate a few obviously relevant physical and psychological considerations. To this end I have used brief case-histories of three sites, Assur, Nineveh and Tell al Rimah. I have attempted no definition of "town" or "city" but there is no doubt that, by any definition, all these sites justified one description or the other at some periods in their history[1].

Assyria, in the sense in which the term is used here, is a rolling plain bounded on the south and south-west by the central steppe of Iraq (the "Jazira") and on the north and east by the mountains of Kurdistan. It is intersected by intermittent hill-chains, outliers of the Kurdish mountains, and by the valleys of the Tigris and its eastern tributaries. Westwards the plain extends towards the Khabur valley and Syria, and this fertile corridor was both a major trade route and, intermittently, one of the main arteries of Assyrian imperialism. There have, so far as we know, been only marginal changes of climate in the last eight thousand years. Agriculture has been the basis of the economy, dependent on rainfall which varies from an average annual aggregate of 1000 mm in the high mountains to the minimum acceptable for agriculture, 200 mm, on the edge of the Jazira. I refer to the land inside the 200 mm isohyet as the rainfall zone. Irrigation is only possible in the flood-plain of the Tigris and its tributaries and has never been of great economic importance.

The history and, we may suspect, the prehistory of Assyria were strongly influenced by its geographical position. Its heartland supported a prosperous agricultural population, hemmed in on the north-east by mountain tribes and on the south-west by nomad herdsmen in the steppe, the modern Bedouin. The farmers and their successive governments have had to control, submit to or compromise with these unruly neighbours, whose interests are often hostile to their own. In the most fertile and thickly populated nucleus, an area of perhaps 80 km radius around Mosul, the effect of hostile pressure has normally been infiltration rather than invasion and the resulting changes in society have been gradual, in economy negligible. Outside the nucleus, particularly on the borders of the Jazira where farmers were exposed to the uncertainties of rainfall as well as Bedu incursions, the vicissitudes of settlement have been more dramatic. The two variables, rainfall and security, have in the past produced widely different population densities in the region. The 1947 census, the last to be taken before the mechanisation of agriculture radically changed the traditional pattern of village settlement, showed an average of two to three persons per km^2 on the border of the Jazira, and up to twenty-nine in the districts immediately east of the Tigris near Mosul. Such figures have no direct bearing on ancient situations, but they demonstrate the important point that the region is not uniform, and that the interaction of climatic and political factors is likely to have affected settlement patterns in ancient as in modern times.

Assur lies 100 km south of Mosul, on a high spur overlooking the Tigris from the west, and very close to the modern limit of reliable rainfall. Rain-fed crops in the vicinity are liable to failure and are supplemented by irrigation in the flood-plain of the river, which here forms a bowl some 5 km in diameter, but the irrigable area is small. The income of modern village families is supplemented by outside work, either in the Kirkuk oilfields or in archaeological excavations all over Iraq. In fact, the agricultural resources of the immediate area would not support more than scattered villages. But two other factors have played a material role in the history of the site. Firstly, the most direct road from southern to northern Mesopotamia, in the days before motor transport, followed the Tigris northwards until it entered the rainfall zone. Thereafter supplies of water and forage were more readily available, and a number of alternative routes branched north and north-west across the plain, leading eventually to Anatolia. Assur was the obvious road-station at the focus of these routes, and of equally obvious military importance to any power that wished to control them. Secondly, it lies open on the west to the Jazira and its inhabitants. When the government of the settled lands was strong, the Bedouin were controlled. When government was weak, Assur was one of the first points of Bedu penetration into the rainfall zone, first to levy tolls on passing caravans, then to provide a settled residence for the tribal leaders. At the end of the last century the only building on the *tell* was a Turkish police post; the present administrative headquarters of the district lies 3 km to the north, and close by is a mansion built by the first paramount sheikh of the Shammar to settle his tribe

Excavations on the site of Assur have yielded no evidence of prehistoric occupation, but exploration of the early levels was inhibited by the massive remains of the Assyrian city, so the evidence is inconclusive. There was an Early Dynastic settlement, with provincial Sumerian sculpture in the temple of Ishtar. There are no documents to define the function of this settlement, but the foreign character of its art suggests that it was a commercial and perhaps military outpost of the southern civilization. It continued into Akkadian times, when the hegemony of the south was for the first time assumed by a Semitic dynasty (*c.* 3270 B.C.) and was apparently destroyed during the turmoil of the Gutian invasions that overthrew established authority in Mesopotamia. It was probably at this time that the Semitic-speaking people known to us as the Assyrians, a tribe or confederacy of desert origin, first established themselves on the site. When discipline was restored under the third dynasty of Ur (*c.* 2113 to 2006 B.C.), the first well-documented Mesopotamian empire, Assur was ruled by an Ur governor. Southern dominance ended with the fall of Ur. The Assyrian dynasty achieved independence, and within a hundred years had established Assur as a major trading power, with merchant communities under government control as far away as Kültepe in Cappadocia.

From this time another important factor influenced the history of the city. The Assyrian kings derived their formal authority from their position as priests and viceroys of the god Assur. The site itself had been called Assur at least as early as the Akkadian period, but the worship of a god Assur is first attested in theophoric personal names, in documents of the time of the third dynasty of Ur. Present evidence suggests that the god was a personification of the city, and he certainly shared his first known temple with Enlil, chief god of the Sumerian pantheon. As the fount of royal authority he soon achieved independence, and his city remained paramount in religious importance until the fall of the Assyrian kingdom in 612 B.C. For more than a thousand years, until the early ninth century, it was also the administrative capital, although its situation on the southern border of Assyria was not ideal for this purpose. Its fortifications and temples were maintained even after the royal residence had been transferred, first to Nimrud, then to Khorsabad and finally to Nineveh, all within the agricultural heart of Assyria.

After the collapse of the Assyrian Empire the site declined dramatically in importance. It was the first of the great Assyrian cities to fall to the Medes, two years before Nineveh. There is no trace of occupation under the Achaemenid Empire, when even its function as a road station seems to have been usurped by a site a little farther to the north, nor under the relatively short and militarily insecure Seleucid domination of Mesopotamia. Not until Parthian times, when discipline in the Jazira was re-established under the Arab client kings of nearby Hatra, do we again find a flourishing town on the site. When Hatra fell to the Sassanians *c.* A.D. 250, Assur once more disappeared.

Even this brief historical outline suffices to define the motives that led to the establishment of a town, later a capital city, on a site whose agricultural resources held no promise of wealth. The military and commercial attractions of a strongpoint at the focus of important trade routes appealed both to the civilized powers of the south and to the Assyrians in search of a foothold in the settled lands. Its aggrandisement was partly inspired and largely maintained by its special religious significance. When this significance disappeared, the site was once again subject to the vagaries of external, secular authority which permitted it only one short-lived revival.

Nineveh was the last and greatest of the Assyrian capitals but its history, although again based on fragmentary evidence, presents a striking contrast to that of Assur. Its geographical situation is markedly different. It lies at the centre of the fertile and populous Assyrian plain, and still commands the only bridge over the Tigris in northern Iraq. Mosul, on the opposite bank of the river, inherited the commercial advantages of its position and, until the coming of modern frontiers, traded with Aleppo, Erzurum, Tabriz and Baghdad. By nature, Assur was an exposed road-station while Nineveh was a great emporium, with the crops and the people to ensure its economic stability and military defence.

The exploration of its early history, as at Assur, has been hampered by the presence of later monumental buildings, but a sounding to virgin soil in a very small area demonstrated that it had been continuously occupied from at least the sixth millennium B.C. If the early levels could be explored more extensively, it would not be surprising to find a considerable settlement of a much earlier date. The prehistoric sequence is characteristically North Mesopotamian with, as on other important northern sites, evidence of close contact with Sumer as early as 3000 B.C. In the absence of documents we cannot infer that this contact was political until the time of the Akkadian dynasty, when Manishtusu of Agade restored the temple of Ishtar of Nineveh. The restoration of temples was, *inter alia*, an act symbolizing the assumption of royal functions in the city concerned, and emphasizes its political prestige. A further restoration of the same temple was carried out by Shamshi-Adad I, founder of the first Assyrian empire *c.* 1800 B.C., whose building inscription reflects the importance he attached to the control of Nineveh. Although the archaeological evidence is still meagre, there can be no doubt that it was the principal town of this wealthy region, which may now have come under the control of Assur for the first time.

In the middle of the second millennium B.C. all Assyria came under the hegemony of the kingdom of Mitanni, but the continuing prestige of Ishtar of Nineveh, and by implication the importance of her town, is attested by the action of one of the Mitannian rulers who sent the statue of the goddess on a diplomatic visit to the Egyptian court. With the rise of the Late Assyrian Empire in the first millennium, Nineveh was intermittently a royal residence even before it became the centre of administration *c.* 700 B.C. Its extant monuments belong to this phase, and embody a vision of imperial grandeur that must have been produced and supported on resources far greater than those of metropolitan Assyria, in fact on the tribute of an empire that now embraced the whole of the Fertile Crescent and Egypt. Its history after it fell to the Medes in 612 B.C. makes an interesting comparison with that of Assur, for occupation was continuous and apparently prosperous until it was progressively superseded by Mosul in the Sassanian and Early Islamic periods. It had a city constitution on the Hellenistic model in the Parthian period, presumably of earlier origin, and the latest description of the temple of Ishtar occurs in the Life of Apollonius of Tyana, written *c.* A.D. 200.

In Nineveh, in fact, we have an example of a site whose importance rested on the twin advantages of a good commercial position and a prosperous agricultural hinterland. Its religious prestige reflected its natural importance and long prosperity, but did not, as in the case of Assur, inflate its status beyond the limits of its natural resources. When such inflation did occur, as a reflection of the ambition of the Late Assyrian kings, the subsequent deflation was not fatal.

Tell al Rimah was a less important site than either Assur or Nineveh, and merits briefer consideration, but it illustrates certain artificial factors in the creation of a town more precisely. It is situated some 80 km west of Mosul, close to the border of the rainfall zone. It was originally a prehistoric settlement, dating back at least to the fourth millennium B.C., and had grown into a large village by the end of the second millennium. There are many such sites in the surrounding countryside. About 1800 B.C. the site of the existing settlement, by then a sizable mound, was used as the foundation for a monumental temple complex which is one of the most sophisticated products of early Mesopotamian architecture. At about the same time the site was surrounded by a defensive wall enclosing an area some 600 m in diameter, and at least part of this area was built upon; the only building that has been explored was an administrative headquarters. This startling change in the character of the site reflects expenditure far beyond the means of the earlier village and we must infer that it was paid for by some outside patron; archaeological and documentary evidence points to Shamshi-Adad I of Assyria. Shamshi-Adad was a usurper on the Assyrian throne, and his rebuilding of the temples of Assur at Assur and of Ishtar at Nineveh has an obvious political motive as a bid for the loyalty of these two important centres. In creating a new town and a grandiose religious foundation at Tell al Rimah—not necessarily an isolated gesture—he was probably reinforcing his position in an area vital to the communication and the security of Assur.

Shamshi-Adad's authority ended with his death, and his empire, which had included Mari on the middle Euphrates and the Khabur valley as well as Assyria, disintegrated. Tell al Rimah became the capital of a petty principality, and its gradual reversion to the status of a country town is marked by the progressive decay of the great temple. In common with half a dozen other neighbouring sites it seems to have maintained this position until *c.* 1200 B.C. Then it was suddenly abandoned, and not reoccupied until the ninth century, when the Late Assyrian kings carried out a deliberate policy of resettlement over a large area of their western province. The resettlement, obviously dependent on Assyrian authority, came to an end with the fall of the empire. The reasons for the abrupt break in occupation after 1200 B.C.

are obscure and cannot be discussed here; a marginal climatic variation cannot be ruled out, and another factor must have been the widespread political turmoil in the Near East at this time, marked among other events by the appearance of Aramaean tribes in the Syrian desert. Tell al Rimah is of especial interest, if my interpretation of the evidence approaches the truth, as an example of an unpredictable element in urban development, the impact of an act of policy by one man in an unconventional situation.

To sum up, these case-histories present us with a series of factors in the history of individual town sites, of which some are obvious in the sense that they arise from the facts of geography and climate and might therefore be deduced from an intelligent examination of maps, climatic data and the distribution of ancient sites. Certain past studies which still have wide acceptance have been vitiated by failure to take account of some or all of this material, but we may hope that modern investigation avoids such pitfalls. What is perhaps less obvious, and therefore more important, is the incidence of non-material factors that could not be deduced by the application of commonsense. The effect of the ambition of Late Assyrian kings in creating great cities to express their concept of royalty has enough modern analogies to suggest its relevance to the ancient situation even if the record of their activities had not survived. But it is unlikely that we should have been able, without documentary evidence, to identify the special religious position of Assur as an explanation of its long prosperity on a relatively ill-favoured site, or to guess at the reason for the dramatic change in the fortunes of Tell al Rimah that elevated it to a fortified town and a major religious centre. We must remember that when only a minute proportion of sites has been excavated, we cannot begin to define the "typical" town or city, and hence that idiosyncracies of this sort in the record of a known site affect our overall view to an unpredictable and indeed unacceptable degree.

Note

1 A fuller discussion of geographical and climatic conditions in relation to settlement in Assyria, and of the history of Assur and Nineveh, may be found in Oates, D. (1968). *Studies in the Ancient History of Northern Iraq*. British Academy, London. pp. 1–37 and bibliography, pp. 163–5. The excavations at Tell al Rimah have been published in annual preliminary reports in *Iraq*, **27–32**, 1965–70; the historical implications of each season's work are summarized at the end of each report.

PETER J. PARR

Settlement patterns and urban planning in the ancient Levant: the nature of the evidence

Settlement patterns and urban planning are subjects which have hitherto received scant attention from archaeologists involved with ancient Syria and Palestine. There is good reason for this, since—as we hope to indicate—the data for a meaningful study of settlement patterns and urban planning are still most inadequate, and scholars, realizing the inadequacies of the evidence, have proceeded with caution. In recent years, however, interest in these subjects has developed and a review of the available evidence is now both possible and desirable[1]. Such a review cannot be attempted here; instead the writer intends to take what may be criticized as a negative, indeed a destructive, attitude, and restrict himself to a number of admonitory remarks, pointing out some of the shortcomings of the available evidence and stressing the dangers inherent in its study. He makes no apology for this, believing as he does that the most important, as it is the most neglected, part of the archaeologist's task is the critical assessment of his data. Archaeology can be said (not entirely facetiously) to be like history, too often "a hard core of interpretation surrounded by a soft pulp of disputable facts"[2], and it is time we all paid more attention to the pulp.

Settlement patterns

Even in a country such as Palestine, one of the most intensively excavated areas of the Near East, a study of settlement patterns depends almost entirely upon evidence derived from surface surveys, and this dependence is all the greater in regions such as Anatolia or Mesopotamia. It is in these latter areas that the most significant recent developments in surface exploration have occurred, but the credit for introducing this technique of what Adams calls "topographic archaeology"[3] must go to Nelson Glueck, who commenced his surface reconnaissances of eastern Palestine in 1932, and in later years extended them to include the southern part of Israel. Practically everything we know about the ancient cultures of Transjordan—an area at least as important archaeologically as Palestine itself—derives from Glueck's work, since only eight of the hundreds of sites discovered by him have been even partially investigated by excavation. Glueck himself interpreted the primary data from his explorations in no more than a very general fashion, and based upon it broad historical deductions which, for the most part, appear correct[4]. The uniqueness and extent of his material, however, makes it a constant temptation to the scholar wishing to study settlement patterns and urban development in greater detail, while, at the same time, there has been a recent renewal of interest in surface explorations in the area, particularly by Israeli archaeologists working since 1967 in Sinai and the Golan Heights, which stems directly from the impetus of Glueck's earlier work. It is because of this that the following "cautionary tales" are deemed worthy of record. For it is obvious that all of the topics subsumed under the heading "settlement patterns"—nature, continuity, and permanence of settlement; territoriality; defence, etc.—can only be fruitfully discussed if the primary identifications of the type and date of individual sites recorded in the field are reliable; and it is this problem of the reliability of surface exploration which is the present issue.

Four examples from Glueck's work in Transjordan will make the writer's position clear:

1 *Dhiban*, the ancient Dibon, capital of Moab in the first millennium B.C. During his first visit to the site Glueck failed to find Iron Age sherds[5], though its identification with Iron Age Dibon was certain from epigraphic discoveries. At a later date he did pick up Iron Age material, but nothing earlier, though he surmised that an Early Bronze Age settlement must have existed there also[6]. In fact excavations have since proved that this was an important Early Bronze Age town[7]. Glueck's intuition has been proved correct, though his data were defective.

2 *Khirbet Iskander*. Soundings conducted at the site by the writer have shown that, as in the case of Dhiban, the evidence from Glueck's surface investigation was only partially valid[8]. In the first place, the late third millennium pottery (his "E.B.IV–M.B.I" pottery), which is all that he found there, belongs, in fact, to only the final phase of occupation at the site, and there is a significant basal deposit of late fourth millennium material which only excavation revealed. Secondly—and perhaps of greater significance in this context—the excavations showed that, on the one hand, the strong defensive wall which Glueck identified as enclosing the late third millennium settlement does not exist, being nothing more than a continuous line of house walls, while, on the other hand, the earlier fourth millennium site, of which he saw no trace, is indeed provided with a massive enclosure wall[9].

3 *Bab edh-Dhrá*. Another Early Bronze Age site, discovered by Albright in 1924, who stated that it was not a town since there was no deposit of

débris at the site, although he recognized the presence of a walled acropolis[10]. His interpretation of the remains as those of an open-air camping site used by religious pilgrims was accepted by Glueck[11], as was his date of "around 2000 B.C.", based on the surface pottery. Recent excavations by Lapp, however, have shown that the site is an important walled settlement, spanning the greater part of the third millennium[12].

4 *Tawilan.* The excavation of this site by Crystal Bennett in 1968–9 has shown that the thirteenth to twelfth century date given by Glueck to the distinctive painted pottery found on the surface there—and at many other sites—is too early by some 400 years[13]. Glueck's error seems to have stemmed from his ignorance of east Palestinian ceramics—quite excusable in view of the complete lack of stratified material from the area in his day—and from his consequent reliance upon what were, as it now appears, false analogies with west Palestinian material.

It should be stressed that the above examples are in no way intended to be an attack upon Glueck's scholarship or integrity; his stature amongst Palestinian archaeologists is great and is irreducible. Their purpose has been to illustrate the serious weaknesses in the method of topographic archaeology, weaknesses which stem alike from human frailty—the fact that visual evidence can easily be misinterpreted—and natural chance—the fact that all periods of occupation of a site need not necessarily be represented amongst the surviving surface sherds. Nor is it intended to deny to surface exploration a useful, indeed vital, function in archaeological research. However, the fact that, of the eight sites dug in east Palestine, at least four have been shown to be of very different character from what was suspected from the results of the surface investigation alone, must surely warn us that such results have to be used with great caution. The dangers are, as we have seen, particularly great where multi-period sites are concerned, and single-period sites, with little depth of deposit, should be easier to interpret from surface reconnaissance. But even here there are pit-falls, not least of which is the assumption, too readily made, that sites characterized by identical surface pottery were in use contemporaneously. It is too often forgotten that the ceramic periods distinguished by Near Eastern archaeologists normally each cover a number of centuries, and that a precise determination of the date of a particular settlement is rarely, if ever, possible from ceramic evidence alone. There can thus be a complex history of shifting occupation, of abandonment and resettlement, amongst sites identified, from the surface, as being of the same "period". Such considerations should be taken into account when assessing the meaning of, for example, the dozens of Chalcolithic sites recently found by Rothenberg in Sinai[14].

Town planning

The primary evidence for the history of ancient town planning is, of course, the plans of towns, and these can usually only be known from excavation. In this respect Palestine itself is seemingly well provided with evidence, since the number of excavated sites is greater per square kilometre than anywhere else in the Near East. (It should be noted, however, that this does not apply to the rest of the Levant where, in the much larger area of Lebanon and western Syria, only a handful of sites have been dug. Moreover, it must be acknowledged that Palestinian archaeologists have on the whole paid more attention to the remains of domestic and military architecture—central to our subject of town planning—than have their Mesopotamian colleagues, who have been more often seduced by temples and palaces.) But even from Palestine our evidence is so defective as to make any useful discussion of the types and evolution of town plans virtually impossible. There are, perhaps, three main points to be made here:

1 First, it has to be recognized that even in the relatively favourable conditions pertaining in Palestine the sampling of ancient town plans by excavation has not been large enough to permit valid generalization, and each new excavation seems to reveal new types and new situations. Thus, in the Early Bronze Age, the newly excavated plan of Tell Arad[15] is quite unlike contemporary plans at Tell Far'ah, Megiddo, or Jericho, for example, while the development of the whole concept of town planning and civic consciousness also seems to have had a different history in each of these sites[16]. In other words, there was no standardization, and no generalizations are therefore possible, at least not in the present state of knowledge.

2 Second, there is the problem, related to the first, of the proportion of the town plan of any particular site which has been revealed by excavation. Even in the case of sites such as Megiddo, Far'ah or Tell Beit Mirsim, where excavation has been carried out over a relatively wide area, this area is still usually small in relation to the total area of the settlement[17]. Any attempt to reconstruct a total town plan must therefore rely very largely upon the hypothetical extension to the unexcavated portions of the site of conclusions drawn from the partial excavations.

3 Third, even where an extensive town plan has been uncovered, as at Beth Shemesh or Megiddo in the second and third millennia levels, or to look further afield, Byblos in the third millennium, there is the problem of how reliable the published plans are. Undoubtedly the publications provide enough evidence to suggest, for example, a radial layout at Beth Shemesh and Tell Beit Mirsim throughout much of their history, in contrast to, say, the more orthogonal arrangement at Megiddo. But a closer look at the published plans shows that they are, by and large, meaningless and un-

workable as they stand, and the sort of re-analysis which is required to make sense of them has been shown by the work of Kenyon and Müller on Megiddo[18], and by yet unpublished work on the Byblos plans[19]. Once again, our conclusion must be that the evidence for reconstructing the actual town plans of ancient Levantine cities is woefully inadequate.

In order not to conclude this paper on a totally pessimistic note, perhaps the author might add, finally, a few words on what appears to him to be the one feature of ancient Palestinian town planning which might well be the subject of further research, even in the present state of knowledge; namely, the orthogonal layout at Megiddo, already mentioned. The most striking appearance of this layout is in Stratum II, but the same arrangement of *insulae* is already in evidence in III, and the two strata are considered very largely as one long period of uninterrupted occupation by the excavators[20]. The initial date of this period is in dispute; American and Israeli scholars would place it at *c.* 915 B.C., after the destruction of the previous city by the Aramaeans of Damascus[21], while Kenyon argues for a date in the second half of the eighth century, perhaps after the invasion of northern Palestine by Tiglath-pileser III in 732[22]. In any case the Assyrian origin of this orthogonal planning is generally accepted, the conception of the grid being considered a product of the Assyrian military mind. Yet there seems to be little if any evidence for such a grid plan in Assyrian cities of this or an earlier period, and Oppenheim[23] can cite only the Urartian acropolis of Zernaki[24] as an example. This is dated (by surface sherds!) to the eighth or seventh century B.C., and is therefore no older than Megiddo III. It might be worth investigating whether the orthogonal plan of the Palestinian city did not owe something to the earlier conception of monumental "royal quarters", of strict rectangularity, at Samaria and Megiddo itself, dating from the tenth century, and whether these concepts of town layout were not, like so many features of the associated architecture, of Phoenician and Canaanite origin.

Notes

1 Ruth Amiran and Paul Lapp have both written recently on the subject of Early Bronze Age urbanization, *in* Sanders, J. A. (ed.) (1970). *Near Eastern Archaeology in the 20th Century*. New York. pp. 83–100 and 101–31; while Lampl, P. (1969). *Cities and Planning in the Ancient Near East*. London, gives a brief survey of Levantine towns, useful mainly for its illustrations. One of my research students, Mr. Svend Helms, is working on the problem of urban development—both economic and architectural—in Early Bronze Age Palestine.
2 Carr, E. H. (1961). *What is History?* London. p. 18.
3 Adams, R. McC. (1965). *Land behind Baghdad*. Chicago. p. 119f.
4 Glueck's contribution on the Negeb, *in* Kraeling, C. H. and Adams, R. McC (eds.) (1960). *City Invincible*. Chicago. pp. 46–51, is such a broad, and somewhat unsatisfying, treatment of the subject.

5 Glueck, N. (1934). Explorations in eastern Palestine I, *Ann. Amer. Sch. Oriental Research*, **14**, p. 51.
6 Glueck, N. (1939). Explorations in eastern Palestine III, *Ann. Amer. Sch. Oriental Research*, **18–19**, p. 115.
7 Morton, W. H. (1955). Report of the Director of the School in Jerusalem, *Bull. Amer. Sch. Oriental Research*, **140**, p. 6.
8 Glueck, N. (1939). *op. cit.* pp. 127–9; For the writer's excavations see Parr, P. J. (1960). Excavations at Khirbet Iskander, *Ann. Dept. Antiq. Jordan*, **4–5**, pp. 128–33.
9 So little of this "enclosure wall" was exposed in the exiguous soundings that its meaning and purpose remains unclear.
10 Albright, W. F. (1924). The archaeological results of an expedition to Moab and the Dead Sea, *Bull. Amer. Sch. Oriental Research*, **14**, pp. 5f.
11 Glueck, N. (1935). Explorations in eastern Palestine II, *Ann. Amer. Sch. Oriental Research*, **15**, p. 5.
12 Lapp, P. W. (1966). Bab edh-Dhra[c], *Rév. Biblique*, **73**, pp. 556–61. See also Lapp's study of the Early Bronze Age, *in* Sanders, J. A. (ed.) (1970). *op. cit.* pp. 101–31. Unfortunately much of the data from these important excavations remains unpublished and deductions which cannot be checked against primary data must remain hypotheses.
13 These excavations are as yet unpublished. I am indebted to Mrs. Bennett for discussing the results with me and permitting me to mention them here.
14 Rothenberg, B. (1970). An archaeological survey of south Sinai, *Palestine Explor. Quart.*, p. 4f.
15 See, most conveniently, the discussion by the excavator, Ruth Amiran, *in* Sanders, J. A. (ed.) (1970). *op. cit.* pp. 84–100.
16 The present writer cannot agree with Mrs. Amiran's optimistic statement (Sanders, J. A. (ed.) (1970). *op. cit.* p. 96) that "we may safely assume . . . that the information we have gleaned from Arad about city structure should be considered not as isolated data, but as characteristic of cities of EBII all over Canaan". To take up one specific point made by her (on p. 84) it might well be that the town of Arad was in advance of its contemporaries as regards urban development precisely because it was in a marginal region, and thus was exposed to greater challenges both from natural and human agencies, to which it reacted with an untypical development. In the present state of knowledge both hypotheses are equally possible, and this is the main point to be made.
17 Some of these plans can most conveniently be referred to in Lampl, P. (1969). *op. cit.*
18 Kenyon, K. M. (1958). Some notes on the Early and Middle Bronze Age Strata of Megiddo, *Eretz-Israel*, **5**, pp. 51–60; Kenyon, K. M. (1969). The Middle and Late Bronze Age strata at Megiddo, *Levant*, **1**, pp. 25–60; Müller, U. (1970). Kritische Bemerkungen zu den Straten XIII bis IX in Megiddo, *Z. Deutschen Palästina—Vereins*, **86**, pp. 50–86.
19 See the forthcoming doctoral thesis on third millennium Byblos by Miss Muntaha Saghieh.
20 Laman, R. S. and Shipton, G. M. (1939). *Megiddo I.* Chicago. pp. 62–87, with relevant plans.
21 Aharani, Y. and Amiran, R. (1958). A new scheme for the sub-division of the Iron Age in Palestine, *Israel Explor. J.*, **8**, p. 177.
22 Kenyon, K. M. (1964). Megiddo, Hazor, Samaria and Chronology, *Bull. Inst. Archaeol.*, **4**, table on p. 148.
23 Oppenheim, L. (1964). *Ancient Mesopotamia.* Chicago. p. 137 f.
24 On this see, most recently, Nylander, C. (1965–6). Remarks on the Urartian Acropolis at Zernaki Tepe, *Orientalia Suecana*, **14–15**, pp. 141–54.

J. N. POSTGATE

The role of the temple in the Mesopotamian secular community

The study of the temple in Mesopotamia is particularly affected by the usual drawbacks which are common to all Mesopotamian written sources—a plethora of relevant texts, and a dearth of secondary studies. This has excluded any original work on my part, and I have therefore tried to discuss the temple's function in only the most general terms, giving specific instances in principle only where secondary material is available to be consulted. It is hardly necessary to add that, like all general statements, what follows does not apply automatically to all periods or places, but it must be stressed that the paper is not based on primary sources (except occasionally), and that it will therefore be subject to correction in points of detail. Further, it must be recalled that the bulk of texts relevant to the temple economy has still not been made the subject of adequate secondary study, and that when such studies are made some of the suggestions made here may require revision.

The sources

Cuneiform records in Mesopotamia are first attested *c.* 3000 B.C., and continue well into the Parthian period. Throughout this time, except perhaps at the very end, there are to be found documents relating to the temples and their administration. By far the greatest number of these comes from Babylonia (comprising the earlier Akkad in the northern alluvial plain round modern Baghdad, and, to the south, Sumer), while some additional archives are known from Assyria, in the lands now in the north of Iraq. Outside this central area cuneiform writing was only in use at certain periods, and for limited, generally non-utilitarian, purposes. In these peripheral areas—such as Syria, Palestine, and Anatolia—the cuneiform records do give us sporadic information about temples, but, dealing as they do with situations foreign to Mesopotamia, they are not taken into consideration here. In general, statements made below without further qualification refer to Babylonia, and datings are given only where a specific text or groups of texts is concerned.

The temple was one of the major components of Mesopotamian society, and texts which give us information about them may derive from a multitude of different situations. Thus practically any so-called "literary" text will contain some reference to the exercise of religion, and texts classed as religious naturally provide much evidence. Some texts describe the rituals to be performed by priests in the temples, while others more specifically prescribe the offerings to be made to the several deities on given days. On the whole, however, information derived from this sort of text is rather limited. The most direct evidence for the temple economy comes from the administrative archives of the temples themselves[1]. The Mesopotamian scribe was a master of exact accounting, and at every period the larger temples kept detailed records of the amounts brought in (either as offerings or for other reasons), and of the issues to gods or private persons. Daily quantities were recorded, and at the end of the year a complete balanced account could be prepared, usually on the basis of interim monthly totals. In addition to the simple recording, the temple scribes might prepare estimates and assessments for offerings, and they were also responsible for documenting any business activities which the temple might undertake.

One other source which deserves special mention is the royal inscriptions. The king was the religious as well as the military and legal head of the state, and the royal duties included the care of the estates of the gods (by whom, of course, he had been installed). In origin the so-called historical inscriptions were records of the king's building activities, and in particular of the fulfilment of his pious duties by the erection and renovation of temples. The kings also relate how they made rich individual gifts to the gods, or substantial grants of lands or people. Occasionally also they give details of offerings which they have arranged should be regularly made to the temples.

General considerations

In historical times it appears that every settlement of any size must have boasted its own temple, and what archaeological evidence there is supports the assumption that the same was true in prehistoric times. In the most primitive stage, the temple must have been no more than a building or even room where the village community was accustomed to worship and make offerings, and we may guess that the office of priest, and other temple functions, were the part-time activities of senior members of the secular community. Clearly at such a (hypothetical) stage, the temple was no more than an organ of the community, and any wealth that it may have acquired would in effect be the joint property of the community.

I know of no good reason to suppose that the temple in Mesopotamia could have preceded the settlement itself, or that temples ever grew independently of a normal community. Babylonia (and to a lesser extent Assyria) is singularly poor in natural features which could appear to have numinous qualities, and the sort of shrine which might be accommodated on a mountain top, or at such a site as Delphi or Yazılıkaya, seems not to have existed[2]. The temple always owed its existence to the community, and not vice versa. Naturally therefore the size of the temple depended on the size of the settlement it served, and although we have one or two references to village temples[3], since archaeology has been concerned with the important cities, the archives recovered mostly belong to much bigger institutions, which had become an entity largely independent of the secular community.

The area associated with each temple naturally varied in accordance with the importance of the settlement in which it was situated. This was true not only on a minor scale in terms of villages and towns—where the villages, although themselves having temples, also recognize in some way the temple of the town on which they depend—but also in political terms. So when Sumer was composed of city states (in the Early Dynastic period), the main temple of the main city of each state would function as the central temple for the whole area, while each separate state acknowledged the Ekur at Nippur as the central shrine for the whole of Sumer[4]. Thus a temple in a city-state capital would serve, in different ways, not only the city in which it lay, but the surrounding country, and the whole area of the state including other cities. This feature makes it very difficult to assess the relative importance of the temple in the urban economy, since in the case of such large cities about which we are informed, the resources of the temples were drawn from beyond the immediate region of the city itself[5]. To take an extreme example, the large complex of temples in neo-Assyrian Assur was not dependent on local support, but was built and maintained as a national monument, from the resources of the whole Assyrian empire. Only in the very simplest of circumstances can we hope to assess the quantitative function of the temple, and it is there that the evidence is missing. Efforts have been made, however, to make such an assessment in the case of Early Dynastic Lagash, and I shall return to these attempts later.

Relations between the temple and the settlement

At all periods the Mesopotamian temple was an integral part of the community which it served. Although the relationship between the secular population and the temple varied in the course of time, in origin at least the temple was administered by the community for the community. In economic terms, the temple could not survive without the co-operation of the community, and in return, the temple was able (and probably expected) to provide certain services to the settlement or group of settlements which it served. As would be anticipated, the temple was responsible for the care of the gods, and for the financing and organization of festivals, but it could also undertake privately commissioned ceremonies, such as burials and perhaps extispicies[6]. The temple also played a part in the administration of the law; it organized the taking of oaths and the river ordeal, and, at some dates, judges were drawn from the ranks of the priesthood[7]. Together with the palace, the temple was always concerned with the education of scribes and the maintenance of scribal traditions, without which the business life of the community could not have persisted. In social terms the temple also constituted an outlet for the employment and support of the indigent, and of those for whose support the family structure of the time made insufficient provision[8].

The temples also played a very considerable role in the community's economic life. Whether because of religious ideology, or an awareness of its creation by the community in the first place, the temple acted to the members of the community as a sort of "wealthy neighbour". This is most clearly stated in a passage from the Laws of Hammurapi (*c.* 1800 B.C.), which specifies that if a man has been ransomed from abroad by a merchant and brought back to his village or town, if his family is unable to refund the merchant the local temple should pay the amount, and, if the temple is unable to, the palace (i.e. secular authorities) should do so[9]. The temples also made loans of corn to small farmers or peasants to tide them over until harvest time, both in Babylonia and Assyria[10]. In this way, they functioned as "emergency granaries" for the communities, and, if justly run, prevented profiteering at the expense of the peasant.

In the large cities, where the people's means of livelihood were more varied than simple agriculture, the temple's range of activities broadened accordingly. On a larger scale, the temple still acted as a stabilizing force in the economy, and it was able to supply capital for business transactions as well as for the population's basic needs. An archive from Ur shows that in the Ur III period (*c.* 2150 to 2000 B.C.) the Nanše temple itself financed sea trading with Makkan, employing the seamen and traders as its own subordinates[11]; in this way they may have enabled the community to find an outlet for its goods and thus, by using the available capital, made a whole new range of ventures possible. Later, in the Old Babylonian period, most of the Persian Gulf trade was in private hands, although payments made by the traders to the temples suggest that they still had some hand in it[12]. At the same date we find over-land merchants borrowing their capital from the temple[13], although our texts seem to show that they more often turned to the palace or to wealthy private houses for support of this kind. In any case, whether the temples were facilitating trade, or simply making corn loans to avert hardship, we may suggest that the community, by establishing the temple, has in effect created a sort of "joint company" from whose capital the community which supports it is able to profit.

The temple's sources of wealth

Obviously if the temple was to act as the community's "wealthy neighbour", it needed a secure source of wealth. In the first instance this can only have been ceded to the temple by the will of the community, whether in the shape of offerings, or as outright gifts, but although such sources of income were not denied to the temples in historical times, the great temple institutions of which we know relied much more on their own permanent possessions for the maintenance of their economic strength.

Gifts and offerings: Offerings (i.e. more or less obligatory payments to the temples) and gifts (unsolicited) certainly went a long way towards supplying the temple's everyday sustenance. It is to be presumed that food offerings to the gods (which are the most frequent type) were at least partly reused for the feeding of the temple personnel, although this is hard to prove[14]. In addition to this the chief priest(s) might themselves receive payments personally, whether as a regular payment or in return for a specific service. Again, we can only guess, although quite plausibly, that the obligations to supply offerings to the temple were shared between the whole community; this is one of the features of daily life which was taken so much for granted that it finds no mention in our sources. The kings do occasionally tell of arrangements they made for the provision of offerings to a deity[15], and these and other passages tend to show that the regular offerings were the common burden of all ordinary citizens. In the larger temples it is not, of course, likely that the offerings went directly from offerer to deity: all offerings will have been received and checked in at the temple storehouse, and the necessary quantities issued from there as required to the priests in charge of the various shrines where the actual offering ceremony took place[16].

Free-will gifts were of course much rarer, but could be more valuable. The ordinary person will have risen usually no higher than the small and cheap *ex voto*[17], but kings and wealthy citizens might dedicate very valuable items, which would be deposited in the temple or its storerooms. In this way the temple might accumulate a vast treasure, which like the chryselephantine statue of Athene constituted both a symbol of the city's wealth, and an emergency reserve. The sacking of the temples of Sumer (e.g. at the end of the Ur III dynasty, *c.* 2000 B.C.) therefore not only shocked the religious confidence of the land, but also struck at its prestige and destroyed its reserves of wealth[18].

Occasionally private individuals, and more often the kings, made grants to the temples of people or of lands[19]. The royal grants of land are quite well attested, while evidence for the king's presentation of slaves (or virtual slaves) to the temples is less frequent. Perhaps this happened most commonly when a large number of captives had been taken in war, and it has been suggested by Gelb that this is the explanation for a series of texts of the

Ur III period which give long lists of women and children, without any adult males[20]. Private persons might also present slaves to the deity, but, for a variety of reasons, they might also dedicate members of their family or even themselves[21]. Finally, of course, the temple often served as a home for orphans and foundlings, being the obvious recipient both in practical terms, and as the organ of the gods who, at least theoretically, protected the orphan and widow.

Temple estates

The mention of lands and people leads us on to the temple's own estates, which (as said above) constituted the real basis of its wealth and power. Even where the temple belonged to a small village, we must imagine that it held land, since in a village community especially the ownership of land is a sign of dignity, and as such could hardly be denied to the god (and his priest). It was the produce from the temple lands which (combined with payments of naturalia as offerings) enabled the temple to act as the community's emergency granary, too. Originally the temple will only have held land within the area of the community it actually served, but with larger temples at least this must have altered in the course of time, with different gifts and even ordinary purchases leading to the temple estates' being dispersed over different parts of the country. At Khafajeh (in the Diyālā region of northern Babylonia, *c.* 1800 B.C.), we find that one of the city's temples not only had granaries in the city itself, but also owned other depots elsewhere, in villages or towns which must presumably have had their own temples as well[22]. This is part of the reason for the number of borrowers from the temple who lived away from Khafajeh itself.

We cannot hope to illustrate how the temples did in fact acquire their land-holdings. While grants, as we have seen, might be contributory factors, the temples in Sumer already owned large estates at an early date, and the development towards this position must lie in the pre-literate period. The pictographic tablets from Warka, which are the earliest written archive known, came from a temple precinct, and although they cannot be read, they demonstrably deal with agricultural affairs, so that the most we can say is that temples presumably had considerable land-holdings as early as 3000 B.C.[23] However the lands were acquired, it is clear that the temple's character as a permanent institution favoured it with regard to the retention of its holdings. Private estates were always liable to break up in the course of time (since the laws of inheritance prescribed the partition of property between brothers, in equal or unequal parts), whereas the temples were subject to no such restrictions, and there was perhaps at least an unwritten provision that priests could not lightly sell off temple lands. Of course the temples, like private houses, suffered during times of war and famine, but even so their additional resources must have made them more resilient, and (except in cases of downright barbarian invasion) they were to some extent protected by their sanctity. Indeed it is safe to say that the temples in Babylonia enjoyed a far longer continuous tradition than any secular institution, and so were largely responsible (together with the scribal class) for the continuity of culture.

The use to which the temple lands were put does not require special description, since it followed the usual practices of the country. Naturally corn growing was the major activity, with other vegetables and especially date-groves in addition[24]. Alongside their agricultural activities, they also maintained herds of animals, partly of course (in the case of oxen and donkeys) to assist in the work of the fields, but also as a stock-breeding concern in its own right. Flocks of temple sheep and goats are well attested, with smaller herds of cattle, both usually pastured in the uncultivated land and therefore placing little extra burden on the resources[25]; at some periods the temples also kept doves and flocks of domestic birds (duck and geese)[26], while in the south in the third millennium B.C. the salt and freshwater fisheries were much exploited. The larger temples in the cities also ran workshops, again best attested in the third millennium, particularly for precious metal and stone-working, but also for the processing of the temple's own raw products such as leather and wool[27]. Finally, as we have seen, the temples might concern themselves with trade; it is generally felt that after about 2000 B.C. most of foreign trade was either in private hands or conducted on an "inter-governmental" level, but there is some slight evidence that the temples also took some part in it[28]. Before that date private trade was certainly less developed, and it seems likely that both palace and temple were responsible for promoting the interchange of goods both within and outside the borders of Sumer.

The temple's labour force

Every temple, except for the very smallest, must have had some full-time employees, and when the temple's activities included business and agriculture, the number of persons involved must have been considerable, quite apart from the "religious" personnel, who also presumably varied in number according to the temple's material wealth. Most of those permanently in the temple's service were probably hereditary servants of the god—whether of priestly families, or slaves (or the children of slaves) in the temple's possession[29]. Others could join the temple's "family" as children, whether as orphans or foundlings, or because they were dedicated to the god by their parents; occasionally their stay in the temple might be only temporary[30]. The temple required as full-time employees not only "chief priests" and lower religious officials, but also scribes, bird-keepers, courtyard-sweepers, etc.; the employment for which children would be designated was certainly largely determined by the status of the family concerned, and the terms under which they came to the temple. Other services required by the temple might be rendered rather by private persons on a sort of "contract" basis (e.g. shepherds).

When we turn to work on the temple lands the situation is rather more complicated. In theory at least the land was the property of the god, and while it was obviously the duty (as well as the privilege) of the priest to see that it was adequately cultivated, the community in general was also responsible for seeing that its gods were properly served. Up to a certain point, therefore, the temple may have been able to exact from each member of the community a certain amount of labour on the land, but in practice different arrangements must have obtained in historical times. A full discussion of the possibilities is impossible here, particularly as there is still very much work to be done on the subject, but it may be useful to list here briefly the three main ways in which the temple might secure its labour force:

1 by owning, permanently, men (women and children) who are given rations and are effectively slaves of the temple, whether so named or not. In Early Dynastic Lagash such slaves were apparently not used for agricultural work, although they might work in orchards; they do not seem any at time to have accounted for the majority of the temple's agricultural workers[31];
2 by letting plots of temple land to private individuals, who in return cultivate it and pay to the temple either a fixed sum, or a proportion of their produce. Often such people were in some way dependent on, or in the employ of, the temple, but this was not essential. In neo-Babylonian times at least, the larger temples owned whole villages as well as simple land, and although the legal relationship of the inhabitants to the temple is uncertain, their land-tenure was organized on this sort of basis[32];
3 by employing casual labour as and when the need arose. Workmen are required for the preparation of the ground, for the sowing, and for the harvest, but not all year round; in harvest time especially the demand for labour is known to have been high, and it was customary to hire harvesters. The temple may have had traditional claims on the community for some work, as we have said, but another way of securing harvesters was to make work in the fields one of the conditions of a loan, often in place of interest[33]. In any case, the seasonal nature of agricultural work explains why the slaves, available all year round, were not used for it, and Gelb has distinguished between permanent and non-permanent personnel (receiving monthly rations for only four months of the year) who were employed by the household of Bau at Girsu[34].

The old Sumerian temple

So far I have written with the tacit assumption that the temple was always separate from the main body of the secular community. This may sound self-evident, but it is opposed to the earlier view of the Sumerian city-state, which maintained that it was the temples which owned most of the land, which commanded the services of the majority of the population, and which therefore presumably ran the whole life of the community. Thus the chief authority on the period was able to state that ". . . the entire economic life of the time was a temple economy . . . (the temple personnel) certainly constituted the major part of the population of the town-area of Lagash. The power and possessions of the temple will have been still greater in earlier times"[35]. The implications of this doctrine, which was generally accepted until recently[36], are obviously very far reaching for the process of urbanization in Mesopotamia. Since Early Dynastic Lagash, on which Deimel's conclusions were based, is indeed the only case where even a vague attempt can be made to assess the quantitative position of the temple within the state, and in view of the early period to which the archive belongs, it is perhaps worthwhile to repeat the evidence here very briefly.

The archive concerned is a large one, and covers a relatively short period of time; it consists of some 1,500 texts, written *c.* 2400 B.C., which form part of the accounts of the temple of the goddess Bau in the south Sumerian town of Girsu (modern Tello), in the city-state of Lagash[37]. Using the information contained in this archive, and allowing for similar situations in the other known temples of the city, Deimel calculated that the temples owned from 200 to 300 km^2 of land, and this, he maintained, would cover the entire area of the (admittedly reduced) state of Lagash[38]. This excessively low estimate has been effectively rebutted by Diakonoff, and I need only say here that his estimates are rather nearer 3,000 km^2 for the entire territory of the state, and, at the absolute lowest, 1,000 km^2[39]; a major factor in this revised figure is the realization that the state must have included not only the area round Tello itself (which Deimel thought incorporated both Lagash and Girsu), but the sites of El Hiba (ancient Lagash) and Zurghul (ancient Nina or Sirara)[40]. In view of this, we should obviously also raise the estimate for the area of land controlled by the temples, but even so no stretch of the imagination can allow that "approximately the whole of the land was in the possession of the temples"[41].

However, it remains true that the temple was probably a more important sector of the state in Early Dynastic times than at any later date, and even if it controlled no more than a tenth of the land and population, its economic and social influence must have been considerable. One reason why earlier scholars were able to suppose that the temple constituted the entire state, is the great variety of "secular" undertakings revealed by the Tello archive.

Besides the normal crops of corn (wheat, barley, and emmer being attested), vegetables and date-palms were also cultivated; in stock-breeding too, the situation is much as would be expected—there is no evidence for birds, but sheep and goats, oxen, pigs, and donkeys were kept, and extensive fishing was also undertaken. Craftsmen of various kinds are mentioned in the texts, but they are not particularly numerous or frequent, and there is little reason to suppose that the temples (or at least the Bau temple) ran large workshops of their own. The occasional reference to trade both within and without Sumer seems to show that it was undertaken by the temple in co-operation with the palace[42].

All these activities of the temple required a labour force, whether skilled or unskilled, and of course it is essential, in order to gauge the social significance of the temple, to establish not only the area of land in temple hands, but also who worked on those lands, and on what terms. It appears that the relatively few slaves in temple ownership were not used for work in the fields, as I mentioned above. It was therefore necessary for the temple to arrange for the land to be cultivated by those not directly under its authority; the position seems to have been that some land was allotted as a privilege to the (non-slave) temple personnel, while the majority of land was distributed between free "tenant-farmers", who probably owned none of their own. There is no sure indication in the sources whether the temple was able to exact work on its land from the free members of the community as a duty. Thus most of the work on temple lands was done by people who were dependent on the temple for their livelihood, but only in this indirect manner. Evidence, both from Lagash (the city of Girsu), and other Early Dynastic sites, shows that there were simultaneously large areas of land in private hands, usually owned jointly by a "clan" or extended family. The large areas concerned suggest that these "clans" must, like the temples, have employed outside or "client" labour to cultivate their lands to the full. Presumably if private families could own large estates, the ruler (*ensi*) and his family must also have had their own property independently of the temples, and this still further reduces the quantitative importance of the temple estates.

The temples and the state

I have just suggested that the *ensi* of Lagash had his own secular estates. At this date, however, the separation of political power from the religious duties and privileges of the ruler was still rudimentary, and by virtue of his office as *ensi*, he was also head of the temples. According to doctrine, it seems, the *ensi*'s function in the city-state was to act as steward for the city god, who, in the case of Girsu, was Ningirsu; similarly his wife was steward for Ningirsu's wife, Bau[43]. Hence she would act as the "owner" of the Bau temple at Girsu, and many of the documents from there testify that she was no idle figure-head; the same will have been true of the Ningirsu temple. It is at present uncertain whether these temple-states were effectively treated as the private property of the *ensi* and his family, and, if so, what arrangements obtained for the other temples which did not come under their immediate control.

In course of time, the secular duties of the ruler (especially with regard to foreign relations) seem to have loosened the bond between him and the temples, and to have allowed the temples to gain a certain degree of independence. Although the kings probably retained at least nominal authority (and at times much more) over the temples, the growth of the average political unit precluded any tight control, and on at least one occasion we know that the temples were strong enough to challenge the power of the king[44]. However, such tensions do not frequently appear in the sources, and they seem to have been resolved in two main directions: *either* (1) the king assumed full responsibility for the administration of the temples, scrutinizing their accounts and sometimes appointing his own administrative officers to form part of the temple hierarchy[45]; *or* (2) the temples themselves were able to assume some degree of political power; this will of course have happened mainly when central authority was weak, and should not really be confused with the situation in Early Dynastic Umma, where a priest is found leading the army[46]. In general the temples did not rely on military force of any kind.

The royal control of the temples was not entirely a selfish one. To the last the kings were acutely aware of their responsibilities towards the gods[47], and the appointment of administrators to the temples suggests not so much that they were after the temples' wealth, as that they were concerned to see just and efficient organization for the benefit of the whole community; Babylonian priests can have been no more incorruptible than any others. Further, the prosperity achieved under a strong monarchy can only have benefited the temples; although the general trend in Sumer/Babylonia was towards stripping the temples of their secular powers, some of the larger temples remained extremely wealthy and, as we have seen, were even better able to withstand periods of unrest than the royal houses themselves[48].

Assyria

The majority of references and statements in this paper refer to Babylonian temples. In Assyria the documentation starts only about 1450 B.C., and is only at all abundant in the neo-Assyrian period (900 to 600 B.C.). In general there seems to be no past history of extensive temple land-holdings—either in Assur itself, or in the lands further north—and what evidence we have suggests that the large temples which might be compared with those of contemporary Babylonia were very largely dependent on the generosity of the kings to maintain them. It is quite certain that the influence of the temples and the priests on politics was much less in Assyria than further south[49].

Notes

1 Although detailed references are given below where relevant, it may be helpful to list here the major archives: Protoliterate period: tablets from Warka; Early Dynastic III: archive from Bau temple at Tello; Old Babylonian period: Harris, R. (1955). The archive of the Sin temple at Khafajah (Tutub), *J. Cuneiform Stud.*, **9**, pp. 31–45 (the provenance in Mound D is as yet unpublished); neo-Babylonian period: archives from the Eanna temple at Warka and the Ebabbara temple at Sippar (see San Nicolò, M. (1941). *Beiträge zu einer Prosopographie neubabylonischer Beamten der Zivil- und Tempelverwaltung*, (Sitzungsberichte der Bayerischen Akad. der Wissenschaften, phil.-hist. Abteilung, **2**, ii), pp. 8f; Assyria (first millennium): Nabû temple at Nimrud (Parker, B. (1957). The Nimrud Tablets, 1956, *Iraq*, **19**); Balawat (Parker, B. (1963). Economic tablets from the temple of Mamu at Balawat, *Iraq* **25**). Smaller groups do of course come from other periods, but it may be worthwhile to point out that publications such as Torczyner, H. (Tur-Sinai) (1913). *Altbabylonische Tempelrechnungen*. Vienna and Reisner, G. A. (1901). *Tempelurkunden aus Telloh*. Berlin, deal mainly with texts which cannot be shown to have belonged to a temple archive.

2 The deification of rivers and mountains is attested, but there is no evidence known to me that they were worshipped in shrines of their own. Note, in the general context, that when the cults of the major gods were borrowed from one city to another, the new temple was equally the god's abode, and the original site of his worship does not seem to have commanded any special devotion on account of its priority. The special veneration of e.g. Ishtar of Nineveh in the second millennium may well be due to extra-Mesopotamian influences.

3 Assyria: Elumu, village near Carchemish with sixteen families (Woolley, C. L. (1921). *Carchemish II*, London. pp. 135f.); for Babylonia see note 9.

4 This has therefore been taken as grounds for postulating that Nippur was once the political centre of an alliance encompassing the whole of Sumer (see Jacobsen, Th. (1957). Early political development in Mesopotamia, *Z. f. Assyriologie*, **52**, pp. 104 f.).

5 Note that for the purposes of argument I have ignored the fact that large cities housed many separate temples.

6 Burials: Cones B and C of Urukagina VI. 4–14 (see Kramer, S. N. (1963). *The Sumerians*, Chicago. pp. 317–18). Extispicy: see the comment of Oppenheim, A. L. (1964). *Ancient Mesopotamia*. Chicago. pp. 107–8.

7 Harris, R. (1961). On the process of secularization under Hammurapi, *J. Cuneiform Stud.* **15**, pp. 117–20; for oaths in the temple see Falkenstein, A. (1956). *Die neusumerischen Gerichtsurkunden*, I. Munich. pp. 64 f. (Ur III); Driver, G. R. and Miles, J. C. (1935). *The Assyrian Laws*. Oxford. pp. 416–17 (Sect. 47; Middle Assyrian). For priests acting as judges: Edzard, D. O. (1968). *Sumerische Rechtsurkunden des III. Jahrtausends*. Munich. No. 78a (p. 133), Nos. 84–5 (pp. 138–41) (all Agade period); Falkenstein, A. (1956). *op. cit.* p. 31 (Ur III); Lautner, J. G. (1922). *Die richterliche Entscheidung und die Streitbeendigung im altbabylonischen Prozessrechte*, Leipzig. pp. 73 f. (Old Babylonian period). For the transaction of business in the temple see Oppenheim, A. L. (1954). The

sea-faring merchants of Ur, *J. Amer. Oriental Soc.*, **74**, p. 12; Walther, A. (1917). *Das altbabylonische Gerichtswesen* (Leipziger semitistische Studien **6** (4–6), pp. 210–12.

8 See in general Oppenheim, A. L. (1964). *op. cit.* p. 107.

9 Driver, G. R. and Miles, J. C. (1955). *The Babylonian Laws*, II, Oxford. pp. 22–3 (Sect. 32); the passage indicates that villages would be expected to have one temple, even if it might be very poor.

10 For Babylonia the best evidence comes from Khafajah (=Tutub), see Harris, R. (1955). *op. cit.* pp. 31–45, esp. p. 39; the same author refers to another archive with temple loans, *op. cit.* p. 35 (note 28), from Ischali or Tell Asmar; loans of wool, silver or corn from the Shamash temple at Sippar: see Kohler, J. *et al.* (1909–23). *Hammurabi's Gesetz*, III-VI, Nos. 217, 914, 915, 917, 1127–9; loans of silver with indefinite repayment date: Nos. 164, 189, 1501. Assyria: corn loans from the Nabû temple at Nimrud (Parker, B. (1957). *Iraq*, **19**, pp. 125–38; note low or no interest, date of loan usually in spring when stocks run low, and small amounts); Balawat: silver and copper loans from temple officials, usually low or no interest (except No. 113) (Parker, B. (1963). *Iraq*, **25**, pp. 86f, Nos. 101–3, 113, 120, 127–8).

11 Oppenheim, A. L. (1954). *op. cit.* pp. 13–14.

12 Oppenheim, A. L. (1954). *op. cit.* pp. 13–14; for the payment of dues (tithes) by the traders to the temples see the comments of Leemans, W. F. (1960). *Foreign Trade in the Old Babylonian period*. Leiden. pp. 31–2, giving a slight correction to Oppenheim's original position.

13 e.g. Kohler, J. *et al.* (1911). *Hammurabi's Gesetz*, V, No. 1130. In the Old Assyrian trade with Cappadocia the word *ikribu* refers to temple property which may be used as capital by merchants.

14 For the neo-Assyrian period this is indicated by the terms *rēhāti* and *tayyār(t)u*, "remnants" and "return(ed offerings)"; for a special situation with reference to the king in Babylonia, see Oppenheim, A. L. (1964). *op. cit.* p. 189 and Millard, A. R. (1970). Review *Orientalia*, **39**, p. 448. The assignment of offerings to priests is attested by the text Messerschmidt, L. and Ungnad, A. (1907). *Vorderasiatische Schriftdenkmäler*, I. Berlin. No. 35, and by King, L. W. (1912). *Babylonian Boundary Stones*. London. No. 36 col. iv. 47– v. 38 (a reference I owe to Miss B. Parker).

15 Neo-Assyrian period: Postgate, J. N. (1969). *Neo-Assyrian royal grants and decrees*. Rome. Nos. 27–38, 42–54; neo-Babylonian period: see Brinkman, J. A. (1968). *A Political History of Post-Kassite Babylonia*, (Analecta Orientalia, **43**) Rome. p. 290 (note 1180).

16 This at least was the position at Ur in the Ningal temple (Old Babylonian period), see Figulla, H. H. (1953). Accounts concerning allocation of provisions from offerings in the Ningal temple at Ur, *Iraq*, **15**, pp. 88–122, 171–192.

17 For "collecting-boxes" in the temples see Oppenheim, A. L. (1964). *op. cit.* p. 106 (neo-Babylonian period).

18 See the description given in Sumerian literary texts of this event, e.g. in the "Lamentation over the destruction of Ur" *in* Pritchard, J. B. (ed.) (1955). *Ancient Near Eastern Texts*. Princeton. pp. 455–63, esp. lines 275–81; more explicit is the sack of Lagash described in Thureau-Dangin, F. (1907). *Die Sumerischen und Akkadischen Konigsinschriften*. Leipzig. pp. 57f. (k).

19 Private land grants: Harris, R. (1955). *op. cit.* pp. 101–2 (No. 98) (with slaves, old Babylonian); Kohler, J. *et al.* (1909). *Hammurabi's Gesetz*, III, No. 453 (foundation of temple with a plot of building land; Old Babylonian); Gurney, O. R. and Hulin, P. (1964). *The Sultantepe Tablets*. II. London. Nos. 406–7 (grant to Ishtar temple in Huzirina, land and people; neo-Assyrian). Royal grants of land: *Assyria*: see in general Postgate,

J. N. (1969). *op. cit.*; *Babylonia:* see Brinkman, J. A. (1968). *op. cit.* p. 290 (note 1180).

20 i.e. the male prisoners were despatched, as being too difficult to control (oral communication at the XVIII *Recontre Assyriologique* in Munich, June–July 1970).

21 For slaves see already note 19; dedication of free persons is for two basic reasons: (1) to save the votary's life and remove a burden from the family (see above, note 8, also Harris, R. (1955). *op. cit.* pp. 42–3); (2) for reasons of prestige; in this case the dedicated person was generally a daughter (e.g. *Iraq*, **16** (1954), p. 40, ND 2316). A special case was the dedication of a daughter of the reigning monarch to the god Sin at Ur, as was the institution of the "cloister" of *nadītu* women of the Old Babylonian period, which may have arisen from some peculiar social or economic conditions of the day (see in general, Harris, R. (1963). The organization and administration of the cloister in ancient Babylonia, *J. Economic and Soc. Hist. of Orient*, **6**.

22 Harris, R. (1955). *op. cit.* p. 39 also p. 41.

23 For the texts see Falkenstein, A. (1936). *Archaische Texte aus Uruk.* Berlin–Leipzig.

24 On date-cultivation see recently Cocquerillat, D. (1968). *Palmeraies et cultures de l'Eanna d'Uruk* (559–20). Berlin.

25 e.g. San Nicolò, M. (1948). Materialien zur Viehwirtschaft in den neubabylonischen Tempeln, *Orientalia*, **17**, pp. 273–93 esp. p. 275 (general summary), with (1949). *Orientalia*, **18**, pp. 295–306, and (1950). *Orientalia*, **20**. pp. 134–9 (neo-Babylonian); for a temple flock in Assyria see the text Friedrich, J. *et al.* (1940). *Die Inschriften vom Tell Halaf* (Archiv f. Orientforschung, Beiheft 6), No. 106.

26 San Nicolò, M. (1950). *op. cit.* pp. 129–34; for the species see Landsberger, B. (1966). *Die Welt des Orients*, **3**, pp. 252f.

27 Text giving terms of employment of jewellers, carpenters, and stone carvers in Eanna (Uruk), *in* Weisberg, D. B. (1967). *Guild Structure and Political Allegiance in Early Achaemenid Mesopotamia.* Yale. pp. 5–9.

28 Temples had merchants (if not traders) in their employ, at least in neo-Assyrian times: e.g. Parker, B., (1963). *op. cit.* p. 90, a merchant of Ishtar of Arbela.

29 There is sporadic evidence at most dates that the post of chief priest (*SANGA*, *šangû*) might be inherited: see Schneider, N. (1947). Der sangu als Verwaltungsbehörde und Opfergabenspender im Reiche der dritten Dynastie von Ur, *J. Cuneiform Stud.*, **1**, p. 131 (Ur III); Harris, R. (1963). *op. cit.* p. 156 (old Babylonian); King, L. W. (1912). *Babylonian Boundary Stones.* London. No. 24 (pp. 96f) (Middle Babylonian); on the other hand the commoner non-hereditary procedure (at least in large temples) is given by Schneider, N. (1947). *ibid.*, and has obvious practical advantages.

30 See note 21.

31 Into this category would fall the *a-ru-a*; "dedicated" people, who are found working for the temples in the Ur III period.

32 Schwenzner, W. (1924–5). Zum neubabylonischen Latifundienwesen, *Archiv f. Keilschriftforschung*, **2**, a text recording the assignment of an area of about 80 km^2 land to two persons, together with cultivators and agricultural equipment; the cession is made by Nabonidus, but the lands actually belong to the Eanna temple at Uruk, and it is to the temple that they are to make repayment (of about 7,500,000 litres of corn and 3,000,000 litres of dates). The two men concerned are rather tax contractors on the Roman pattern than temple employees (see note 48).

33 See in general, Lautner, J. G. (1936). *Altbabylonische Personenmiete und Erntearbeiterverträge.* Leiden.

34 Gelb, I. J. (1965). The ancient Mesopotamian ration system, *J. Near East Stud.*, **24**, pp. 241–2.
35 Deimel, A. (1931). *Šumerische Tempelwirtschaft* (Analecta Orientalia, **2**) Rome. pp. 78.
36 See, for example, Falkenstein, A. (1954). La cité-temple sumérienne, *in Cahiers d'histoire mondiale*, I, 4, pp. 784–814; more recently, Roux, G. (1966). *Ancient Iraq*. Harmondsworth. p. 123.
37 The basic studies of the archive remain those of Deimel, A. (1931). *op. cit.* pp. 71–113; a very useful and detailed break-down of the economic activities of the temple written by an economist of the time is Schneider, A. (1920). *Die Anfänge der Kulturwirtschaft, die sumerische Tempelstadt*. Essen—although written as a complement to Deimel's work, the book retains today much basic information used here.
38 Deimel, A. (1931). *op. cit.* p. 79.
39 Diakonoff, I. M. (ed.) (1969). *Ancient Mesopotamia*. Moscow. pp. 173–4 (translation of a paper covering the same ground as the same author's fundamental book, *Obshchestvenniy i gosudarstvenniy stroy drevnego Dvuryechya. Shumer* (Society and state in ancient Mesopotamia. Sumer)). Diakonoff's position is also stated succinctly in Kramer, S. N. (1963). *The Sumerians*. Chicago. pp. 75–7.
40 See now Falkenstein, A. (1966). *Die Inschriften Gudeas von Lagaš*, I, (Analecta Orientalia **30**). Rome. pp. 17f.
41 Deimel, A. (1931). *op. cit.* p. 79.
42 See Lambert, M. (1953). Textes commerciaux de Lagash, *Revue d'assyriologie et d'archéologie orientale*, **47**, pp. 57f. for the texts.
43 See, on the theory of this, Jacobsen, Th. (1949), *in* Frankfort, H. *et al. Before Philosophy*, Harmondsworth, pp. 202 f.
44 The challenge to Năbonidus by the priests of Marduk (see *Cambridge Ancient History*, Vol. III, pp. 218f.); it is noticeable that this tension concerned primarily religious matters (at least ostensibly). See also Oppenheim, A. L. (1964). *op. cit.* pp. 108–9.
45 Schneider, N. (1947). *op. cit.* p. 130 (Ur III; appointment of priest and auditing of accounts by *ensi*); for neo-Babylonian times see the works by San Nicolò and Schwenzner quoted in notes 1 and 32 respectively.
46 Priest at Umma: Thureau-Dangin, F. (1937). Une tablette en or provenant d'Umma, *Rev. d'assyriologie et d'archéologie orientale*, **34**, pp. 181; Il, roi d'Umma (1941), **38**, p. 90. Priests in politics: see Brinkman, J. A. (1968). *op. cit.* p. 300 (with notes 1970–2).
47 cf. the royal grants to temples referred to in note 19.
48 A feature of the Babylonian (but not Assyrian) temple which I have not been able to discuss is the system of prebends—whereby posts or parts of posts in the temple hierarchy could be bought, bringing with them automatically the right to collect temple dues. The system has already begun in the Old Babylonian period, and reaches its height in the late first millennium B.C.; the position in note 32 is related to this system and is regular for neo-Babylonian temples (similar text for Babylon quoted *ibid.* by Schwenzner). On the general practice of prebends see Oppenheim, A. L. (1964). *op. cit.* pp. 190, 282–3.
49 For many detailed points to do with the temple in the Ur III and Old Babylonian periods, see Kraus, F. R. (1954). Le rôle des temples depuis la troisième dynastie d'Ur jusqu'à la première dynastie de Babylone, *in Cahiers d'histoire mondiale* [=Journal of World History], **1**, (3), pp. 518–45.

T. CUYLER YOUNG, JR.

Population densities and early Mesopotamian urbanism

Introduction

This paper examines briefly some of the possible relationships between population growth (or decline), shifts in population densities, and the development of early Mesopotamian urbanism. The data are drawn almost exclusively from a study of variations in settlement patterns.

Our concern is with the third of the three levels on which Trigger has suggested settlement patterns may be examined profitably: "the manner in which communities are distributed over the landscape"[1]. Several scholars have shown such an approach to be stimulating and useful for the study of ancient Mesopotamia and, as will become clear, any further discussion along these lines must follow their lead and rely heavily on their data[2]. Yet although population estimates and the documentation of fluctuations in population size and density have always been an important concern of past discussions, they have sometimes been neglected as a possible means of explaining, at least partially, why the economic, social and political developments which characterize early Mesopotamian urbanism occurred[3]. Most scholars concerned with settlement archaeology around the world have tended to consider population variations (when they dealt with them at all) almost exclusively as a result of changes in subsistence systems within a particular environment[4]. Here we shall take the opposite approach, admitting that we do so rather single-mindedly. This paper is certainly too brief and the data are too vague and incomplete to resolve this "chicken and egg" issue. Nevertheless at the least we hope to demonstrate the value of pondering further whether population fluctuations might play a major causative role in the development of early urbanism in Greater Mesopotamia.

The basic assumption here is that human population tends over the long run to increase naturally, and that these increases lead to population pressures. Thus it becomes important to define both the chronological and geographical limits of the argument.

Chronologically the discussion covers the Ubaid, Uruk, Jemdet Nasr and

Early Dynastic periods of Mesopotamian history. The Ubaid begins sometime in the second half of the sixth millennium B.C. and the Early Dynastic period ends *c.* 2400 B.C.—altogether some 3000 years of human history. Geographically we are dealing with the region often called Greater Mesopotamia[5], including the upland plains of north Mesopotamia, the southern Mesopotamian alluvium (with Khuzistan) and the highlands of the Zagros mountains[6]. Here is something of a natural geographical (and cultural) area. It is bounded on the east by the central deserts of Iran, on the south by the Persian Gulf and the central Arabian desert, on the west by the upper Arabian and Syrian deserts, and on the north by the knotted mountains of eastern Anatolia. Within this region communications are comparatively easy and the several zones have operated throughout history as economically if not always culturally interconnected. The Mesopotamian plains and alluvium provide the region with its richest and most productive agricultural land, the highlands of the Zagros supply natural resources such as bitumen, stone, metal and wood and excellent grazing for sheep, goats, cattle and horses. This natural symbiotic relationship between the zones is best reflected in the periods of greatly increased prosperity which occurred whenever it was possible to combine the highlands with the lowlands politically[7]. The outstanding example of this is, of course, the Sassanian period when the three zones of Greater Mesopotamia were most closely linked politically and when the economic development of the region reached its height.

In short, Greater Mesopotamia, considered as a whole, is a relatively circumscribed geographic unit within which to study the possible effects of increasing population and of the pressures arising therefrom[8].

The theoretical framework

The classic explanation of the relationship between technological change, cultural history and population growth, albeit in simplified form, is that technological developments lead to altered subsistence systems. These create economic surplus, make possible a more "civilized" or at least a culturally more complex way of life, and eventually encourage population growth[9]. A number of scholars have questioned this explanatory model and particularly the role of population in the equation[10]. Of these studies, the most forceful and convincing is Boserup's detailed argument that population growth is the autonomous or independent variable in determining agricultural development and productivity and thus in bringing about technological, economic and social change[11].

Her central observation is that the important issue for the primitive agriculturalist is the frequency with which the land is cropped. She defines five main types of land-use: (1) Forest-fallow cultivation; (2) Bush-fallow cultivation; (3) Short-fallow cultivation; (4) Annual cropping; and (5) Multicropping. As one moves up the scale from Forest-fallow cultivation to Multi-cropping, the period during which the land lies fallow decreases. Under shifting cultivation (Types 1 and 2) land may remain fallow for as long as twenty years, under Multi-cropping two or more successive crops are obtained each year. The length of time the land lies fallow determines the degree of intensification of the agricultural subsistence system. To move up the scale of intensification requires a steady and considerable increase of labour applied to the land, and though this produces more food, it does so increasingly out of proportion to the labour input—that is, with shorter fallow periods it may take three times as much labour to produce only twice as much food. Not only is the additional input of labour necessary for increased production, but under primitive agricultural conditions such a supply of labour is what determines the level of agricultural intensification reached. Under these circumstances man will not choose to shift to shorter fallow periods unless population pressure demands it, since it does not pay in terms of the input/output ratio; but, since such intensification when it is demanded will require a considerable increase in labour, the enlarged population also supplies what is absolutely necessary to achieve intensification[12]. Finally, and of prime importance for the archaeologist, is the fact that agricultural technology in terms of tool types, irrigation methods, and so forth can also be seen as developing under population pressures as the agricultural system intensifies since certain technologies tend to be associated with particular fallow systems.

In sum, Boserup's crucial point is that economic intensification—and we argue here that this point applies to more than mere agricultural intensification and changes in fallow systems—is a course of action which man will pursue only when forced to, that population pressure has often supplied the force, and that it is an increased labour supply available because of population growth which ultimately permits intensification.

Clearly a given population may opt for means other than agricultural or economic intensification to control natural population increase. It may choose to limit artificially the actual increase itself through delayed marriage, contraception, abortion, or infanticide; it may force marginal members of the group to migrate or may in some cases migrate *en masse*; or it may make war on a neighbour in order to gain more land for continued extensive exploitation (a solution often associated with migration). If, however, for one reason or another none of these options is open or attractive, a population can fall back on seeking a solution to increasing population pressure through economic intensification. It should also be noted that sometimes comparatively rapid artificial increases in population densities can occur in certain areas as a result of human actions whose causes can often be traced ultimately to a natural uncontrolled long-term natural population growth. In such situations the available solutions to the population problem will tend to become limited to intensification. A case in point from the Uruk area of southern Mesopotamia is described below.

Thus we argue that increased population densities, caused in the long run by an independent growth in population but sometimes caused by artificial short-term increases in population within a given area, are potential independent variables which we should look for in attempting to explain the technological and cultural revolution which accompanies the development of urbanism in prehistoric Mesopotamia.

The argument

In the seventh millennium B.C. we find the population of Greater Mesopotamia confined primarily to the highland zone with a few sites scattered in areas considered comparatively marginal for dry farming. Population densities are certainly higher in the better watered upland valleys than along the fringes of the lowlands[13]: some fifteen or more sites are known in the highlands, whereas only three sites are found in marginal Deh Luran and a few others in the area of the alluvial fans and eastern borderlands of the southern Mesopotamian alluvium[14].

Over the course of the next two and a half millennia population increased enormously in Greater Mesopotamia and, apparently for the first time, became distributed over the whole of the area. Turning first to the highland zone in the earlier part of this time range (the Hassuna and Halaf periods of the north Mesopotamian sequence), we find some forty-six sites distributed from as far north as the Lake Urmia basin to the borders of Khuzistan in the south, and east to the central deserts of the Iranian plateau[15]. Clearly, the highland zone was experiencing a marked increase in population. In response to this growing pressure considerable numbers of people perhaps emigrated from the highland zone to areas well beyond those previously occupied. Spread across northern Mesopotamia in areas available for dry farming we find scores of sites of Hassuna-Halaf date[16]. More important for the future, however, were contemporary developments in the south. Hole and Flannery suggest a doubling of the population in Deh Luran by the Sabz phase[17], and Adams has reported some thirty-four sites in this period (roughly Susiana a) in the northern half of the Khuzistan plain[18]. Down to the mid-fifth millennium the population in this section of southern Mesopotamia continues to grow. Twelve sites are reported for Deh Luran in the Khazineh phase and some 102 sites in northern Khuzistan in the Susiana b period[19]. Finally, for the first time we find occupations reported from southern Mesopotamia proper in the areas of Ur and Warka and further north in the Diyālā basin[20]. A simple site count suggests population in Greater Mesopotamia increased some thirteen-fold between *c.* 6000 and 4500 B.C.[21]. More important, that population spread out and now occupied at least parts of all three zones within the region, indicating that migration was an important mechanism in this period for relieving the pressures of population growth.

We have argued elsewhere that in the highland zone and in north Mesopotamia these same population pressures were causing shortened fallow periods and more intensive forms of agriculture. In the north this intensification may have led to the introduction of the plough at this time[22]. In southern Mesopotamia, on the other hand, the principal technological innovation was clearly the "discovery" of irrigation. The technology of early irrigation in Mesopotamia, the extent to which it represents an intensification of agriculture by considerably increasing crops yields, and the effects that its practice has on local settlement patterns have been treated in detail elsewhere[23]. Here we wish to stress Boserup's argument that the introduction of irrigation as a means of agricultural intensification tends to occur when fallow periods have already been considerably shortened under population pressure. She then points out that even when the irrigation system is very simple, based on gravity flow and requiring little maintenance (as was the case in southern Mesopotamia), "total labour input per crop hectare of a given crop may be twice as high as for dry farming"[24]. Thus not only does irrigation tend to come into use when population growth has already shortened fallow periods, but also as a method of intensification it requires the availability of increased amounts of labour. In Greater Mesopotamia the development of irrigation not only made it possible to handle larger populations through agricultural intensification, but, of equal importance for this discussion, *at first* also helped maintain existing population densities by making possible more widespread colonization.

On the whole, southern Mesopotamia seems to have had a relatively stable or at least a slowly growing population during the late Ubaid. In Khuzistan Adams reports some 100 sites in the Susiana c period, and 116 sites for Susiana d times[25]. For the Mehmeh phase in Deh Luran the population is estimated as slightly over twice the density of the Sabz phase and was somewhat higher yet by the end of the Ubaid[26]. In southern Mesopotamia proper, population densities were apparently not high[27]. Some fourteen sites are reported for the Ur-Eridu area[28], a "handful" for the Warka region, and only nine in the Diyālā basin[29]. The slightly greater number of sites in the far south (the Ur-Eridu area), considered in the light of the much more extensive occupation at this time in Khuzistan, may reflect a pattern of migration from south to north[30]. Given these facts one suspects that there was little intensification of agriculture in southern Mesopotamia in the Ubaid beyond what resulted from the introduction of irrigation.

Elsewhere in Greater Mesopotamia the population continued to increase during the Ubaid. There are some sixty-two known sites in this time range in the highland zone, up fourteen from Hassuna-Halaf times. Also one notes certain sites in the highland zone with cultural ties to the Ubaid of northern Mesopotamia so close as to suggest some movement of population between the two areas[31]. In northern Mesopotamia the population appears to have grown considerably; we have a great many more sites than in the Hassuna-Halaf periods[32]. Both the accelerated pace and the magnitude of this increase are underscored if we recall that the Hassuna-Halaf periods cover approximately twice the time span of the northern Ubaid. Some of this increasing population may have been absorbed through emigration into parts of the highland zone or perhaps westward into Syria and the Levant where Ubaid and Ubaid-related materials are found[33]. It is also possible, though not demonstrable, that some population was moving into northern Mesopotamia from the south, given the comparatively close correlations in material culture between these two areas at this time[34]. Such a pattern might, of course, tend to mask population growth in the latter zone.

Over the next millennium and a half the long-term trend in north Mesopotamia was probably toward continued population increase[35], but there is a possibility that in the highland zone a decline in population may have begun in Uruk times. In the latter area we have only thirty-six sites so dated as opposed to sixty-two for the Ubaid period. Only thirty-three sites are known in the highlands in the Early Dynastic time range. It is tempting to suggest that the marked increase in population in southern Mesopotamia in these periods, discussed below, might be attributed in part to immigration from the highlands.

Turning to the south, we find there is now sufficient evidence that for the first time we can trace in the record what common sense tells us is true of the whole region in all periods—developmental trends are not uniform even within a given zone. Yet since time and space preclude any detailed discussion of each sub-area within the south, we shall concentrate primarily on two contrasting areas, the Diyālā basin and the Warka region.

In the early Uruk period the area around Warka shows an increase in the number of small rural settlements and Warka itself approaches monumental size. The population is clearly growing. This trend apparently continues through the middle Uruk period, but in the late Uruk (*c.* 3300 to 3100 B.C.) there is a swift rise in population by a full order of magnitude. (This is indeed a remarkable increase considering that we are now speaking of a mere two centuries.) Presumably as a result of this massive growth in population in a comparatively short period of time, the local settlement pattern shifts from one of dispersed villages and small towns, which characterized early Uruk times, to a cluster pattern of small sites.

In the following Jemdet Nasr and Early Dynastic periods marked changes take place in this settlement pattern. Warka reaches a maximum size following rapid growth in the Early Dynastic I period and other sizable towns appear in the region[36]. More significantly, from Jemdet Nasr times onward a "balling" or concentration of population into the larger urban centres appears to take place at the expense of the countryside. Down to the end of Early Dynastic times this sort of urban development results in a steady decline in the number of occupied sites in the area, with abandonment of whole districts previously occupied and under cultivation.

A different pattern emerges in the Diyālā basin. Here there is a considerable increase in population beginning in Uruk times, but perhaps accelerating in the Jemdet Nasr period. Some forty-three sites are known from these periods in contrast to nine sites assigned to the Ubaid period[37]. By late Jemdet Nasr times some towns with public buildings appear. This trend of increasing population continues throughout the Early Dynastic period, by the end of which we find a total of ninety-six sites with ten large towns, nineteen small towns and sixty-seven villages in the area[38]. In the Uruk and Jemdet Nasr periods sites are distributed over the landscape in an enclave or cluster pattern, and the expansion of occupation into the western part of the region, where only three Ubaid sites are located, is evidence for continued colonization of the immediate region[39]. Despite continued population growth in the Early Dynastic period this cluster settlement pattern is maintained. Certainly there is no evidence for any "balling" of the population in greatly enlarged urban centres at the expense of the countryside as was the case in the Warka region. Instead, the population growth is spread over the landscape in roughly the same proportions as earlier, and there is some evidence for the further colonization of land previously uncultivated[40]. No single site attains the massive size of Warka[41].

Thus two contrasting historical patterns emerge. In the Warka region a massive increase in population accompanies the development of a cluster settlement pattern and continued population growth is associated with a shift in that pattern to the concentration of population in ever larger centres at the expense of the countryside. In the Diyālā basin a similar (but perhaps not so large and chronologically slightly later) increase in population also accompanies a cluster settlement pattern. Population continues to rise in the next period, but a "balling" of that population does not follow.

Why the difference? Given the theoretical framework of this discussion, one is inclined to answer, because in the Diyālā basin, despite population growth, population pressures were never so intense as to demand anything more than comparatively extensive forms of agriculture and social organization since the area was never fully occupied in these periods. In the Warka region, on the other hand, the option of colonization or migration as a means of maintaining existing population densities and related economic and social systems in the face of population growth either became more difficult to take up or ran out entirely. Thus Carneiro's conditions for an area of *circumscribed agricultural land* obtained[42]. This situation led to an increase in the level of both intra- and inter-community competition for the available land. And at this juncture trends toward shorter periods of fallow and more intensive agriculture, with their concomitant technological and social developments, became an increasingly attractive alternative for relieving the pressure and controlling the conflict.

Let us examine this proposition in greater detail. In some instances fallow periods may have become sufficiently shortened to have approached a form of annual cropping adapted to the particular conditions of an arid region[43]. Along with this intensification would come the introduction of the plough, which, in fact, does appear in the archaeological record for southern Mesopotamia in the Uruk period[44]. More important for the development of urbanism was the slow breakdown of simple social organization (perhaps an essentially egalitarian grouping based on kinship) and its replacement by a more complex form of ranked society in the face of the need to organize the larger blocks of labour now available and necessary because of increasing population densities and more intensive methods of cultivation[45].

Two examples of how this need perhaps arose may illustrate the point. Irrigation represents a more specific investment in a particular parcel of land than is necessary under dry farming conditions. However simple methods of irrigation were at first in southern Mesopotamia, the hunger of a growing population would have led gradually to improvements of that system in order to get more water to the land and to more land. One need not envisage massive state-supported irrigation works[46] in order to argue that under population pressure the southern Mesopotamian farmer would do all he could to develop the irrigation system which nature gave him. He would thus intensify his use of the land under cultivation. Thereby his sense of investment in the land would rise, his willingness to move would decline, and his determination to defend that investment would increase. Closely related to improvements in irrigation techniques would be a threat of salinization, well documented as one of the recurring headaches of the Mesopotamian farmer[47]. Increased salinization would tend to decrease crop yields on some land, and, in other cases, would actually decrease the amount of land under cultivation. Any restriction in the amount of good land available for use would, of course, have enhanced a man's attachment to and sense of investment in what high-yield land he controlled. Such conditions would lead to internal conflict between neighbouring farmers, conflict which would probably demand more complex methods of litigation and record-keeping[48]. Furthermore, the almost inevitable uneven resolution of such conflicts might gradually lead to an economic and hence a social regrouping of the growing population into larger land holders, smaller land holders and landless labourers[49].

Improved irrigation techniques, probably coupled with other technological developments under conditions of more intensified cultivation, would have made possible the use of agriculturally more marginal areas previously unoccupied. One is tempted to wonder whether the history of early settlement in the area south of the city of Umma is not a case in point. Here we observe a rapid and intensive occupation only in Jemdet Nasr and Early Dynastic I times—rather later than in the neighbouring Warka area. From the beginning the settlement pattern is one of densely clustered sites, many of which are substantial towns, and there is a canal some 15 km long where no previous water course had been. Adams has remarked that there is a "more consciously planned aspect to this pattern"[50]. Certainly any effort to apply the new technologies to a virgin area along these lines would require closer labour organization within more complicated social and political structures than would the spread of more extensive systems of cultivation and less developed methods of irrigation by simple emigration.

Such developments were probably at first mainly gradual and almost imperceptible in the Warka area as the population increased in the Early and Middle Uruk Periods. Indeed there may have been a balance in these periods between seeking solutions to population pressure through agricultural intensification and through further environmental "in filling" with ever more marginal land brought under cultivation. Changes in existing political and social systems may not have been so radical or so sudden as to be "revolutionary", and we might suspect that the trend of events was not so fixed as to be irreversible. Yet the rapid natural increases in population in the Late Uruk period, perhaps augmented by some immigration from other areas within Greater Mesopotamia, would have caused comparatively sudden and more irritating pressures on the system. Perhaps equally important, we suspect there is evidence that the local environment as exploited under the existing technology may have begun to contract at the same time. In his effort to maintain extensive farming methods, man had brought under cultivation ever more marginal land. This land was poor to begin with and in the end failed to produce consistently high yields. Also, the water resources in these areas would have been poorer, so that eventually man perhaps over extended his still relatively simple irrigation system. Finally, these marginal areas may also have been those most susceptible to salinization. In short, a complex of factors perhaps combined to produce a collapse inward of the environment available for agricultural exploitation, with those areas colonized last being the first to experience declining production and eventually enforced abandonment.

These trends toward contraction, involving man's relationship to the environment, were encouraged by political developments under sudden and acute population pressures. The spread into marginal lands brought different social and political groups into increased contact and hence into increased conflict. The decline in the amount of land available for cultivation—without any commensurate decline in population, let alone in the face of a rapidly increasing population—would of course have aggravated this threat of conflict. In response to such conflict, man increasingly sought protection in walled towns and cities. At the same time, the abandonment of outlying marginal areas tended to lessen the chances of conflict by creating buffer zones between warring groups[51]. Such a development would appear reasonable and would cost little since the abandoned areas were of declining economic importance and perhaps ready to be deserted. All of these factors can be seen as arising from the sudden massive increases in population in late Uruk times[52]; in them we might seek the reasons behind the "balling" of population and the creation of the more concentrated settlement pattern in the Warka area in Jemdet Nasr and Early Dynastic I times described above[53].

How much more rapid ("revolutionary" ?), more forced, and perhaps more uniform must all the trends towards economic intensification and cultural and political complexity have become under the pressures which obtained in the Warka area as the population gradually contracted into ever smaller units with ever higher densities in Jemdet Nasr and Early Dynastic times. Whatever the causes of the "balling" may have been, since much more population was now concentrated on far less land, intensification of agriculture became obligatory[54]. But the issue was no longer just food. The leaders of the "ranked" community who had been coming to power during the earlier elaboration of social and political structures needed to consume things other than bread.

Thus further agricultural intensification, an even larger and more concentrated labour force, much of it perhaps drawn from the landless poor, the need to maintain and expand levels of "luxury", the demands for protection in an atmosphere of increasing conflict, and the more compact blocks of people living together—with all of the disagreement and internal conflict such a situation creates—were cumulatively issues which, we suspect, made inevitable a commitment to urban patterns of life and the development of a truly stratified society in the Warka area in the early third millennium. All were factors which can be related directly to the considerable increases in population densities which followed the concentration of population after the Early Uruk period, and that "balling" in turn, we have argued, was the direct result of earlier gradual, but eventually more rapid, natural increases in population within an increasingly circumscribed environment.

To return all too briefly to the counter example, the Diyālā basin, the history of settlement patterns in this area indicates that until Parthian times at the earliest, and perhaps not until the Sassanian period, the local environment was never pushed to its limits by population growth. Local colonization or migration was still a possible "out" for such pressures as there were in the area as late as the Early Dynastic period. Population pressures in relation to the environment never reached levels which induced the "balling" of people into larger cities at the expense of rural occupation[55]: the event marking the threshold of true urbanism in the Warka region. The reasons why this was so are important to the historian of ancient Mesopotamia. The nature of the landscape itself, the position of the Diyālā basin within Greater Mesopotamia as a whole and particularly its relationship to nomadic groups entering the Tigris and Euphrates valleys from the west or to mountain peoples in the immediately neighbouring highland zone, and the dynastic interests of controlling groups centred elsewhere are all involved. To continue such a discussion, however, would carry us beyond our immediate task.

Conclusions

All of the above is undoubtedly too schematic and leaves too much unsaid. So in conclusion it might be valuable to select from things unsaid three important issues for brief mention.

First, we have not dealt with trade and the role it may have played in the development of early urbanism in Greater Mesopotamia[56]. Even when occupation was apparently confined to the highland zone fairly extensive patterns of trade developed to link areas with variable natural resources for the benefit of all. Nevertheless, given a simple economy, individual mountain valleys in the Zagros are comparatively self-sufficient. This is not so of southern Mesopotamia (or even, to a lesser extent, of northern Mesopotamia). Earth, water, reeds, date-palm logs and excellent agricultural land almost completes a list of the natural resources available in the southern alluvium. We discussed above the natural symbiotic relationship between the several zones of Greater Mesopotamia. The need to maintain, indeed to expand, the advantages of that relationship grew with the colonization of the south. And those needs intensified as new technologies developed (for example, the increased importance of metallurgy in the Early Dynastic period) and new demands for luxuries accompanied urban growth (for example, stone for monumental buildings). The less simple the economy and the social structure, the more the south demanded from the rest of Greater Mesopotamia. There can be no doubt that the trade (and commerce) which made this interchange of goods possible required more organization as the demands for goods arose; this need may be as closely related to the development of more stratified societies in Mesopotamia as the need to organize larger blocks of labour working on the land[57].

Secondly, we have not discussed the archaeological details of the possible relationships between different peoples in different areas. We have stated (with some qualification) that people emigrated from the highlands into north Mesopotamia and Khuzistan and that they eventually spilled over into southern Mesopotamia without any effort to trace in the archaeological record possible connections between the material cultures of these areas. For example, it was suggested above that the colonization of southern Mesopotamia proceeded from south to north broadly speaking. If so, then we might expect that areas further south experienced the effects of population pressure before areas further north. This may be the case. In Khuzistan the rural population, as we saw, reached its height in the late Ubaid period, while the Ubaid population elsewhere in the south was low and thinly spread. In the Uruk and Jemdet Nasr periods the average size of settlements in Khuzistan increased, but the number of sites dropped rapidly (almost suggesting depopulation—emigration to elsewhere in southern Mesopotamia?). Thus a cluster pattern of settlement, followed by a "balling" at the expense of the countryside, appeared first in Khuzistan[58]. That all these developments occur rather later in the Warka region and only in part still later and still further north in the Diyālā basin may support the suggestion of a general south-to-north movement of the population. Any attempt to *prove* this suggestion, however, would require a detailed study of the relationship between the material culture of Khuzistan and the Warka region in the time range in question. If past efforts along these lines are any measure, we may not hope for conclusive results in such an attempt. A healthy scepticism, such as Mortensen displays when discussing the relationships between the highlands and north Mesopotamia in an earlier period, should probably prevail[59]. Perhaps new data, collected with new questions and new possibilities in mind, are what is needed.

Thirdly, and of greatest importance, we have not discussed the problem of feedback—the reciprocal relationship between population growth and economic productivity. As more labour becomes available through natural population growth, agricultural systems are intensified. This intensification eventually demands a still larger labour supply, and the birth rate, to choose one possibility, increases accordingly[60]. One doubts if there ever has been a fairly complex society, and particularly a society that is increasing in complexity, that has managed to bring these reinforcing issues into perfect balance (though here pessimism may come from a view of modern Western man and his failings)[61]. Nevertheless, as all men are created equal but some are created more equal than others, so all factors in the equation are independent variables but some are more independent than others. We argue that, in the end, independent (i.e. non-imposed) agricultural intensification, and its closely related developments in technology, society and politics, cannot take place without population growth, but that population growth can and does occur without intensification. Thus population growth remains the more likely independent variable.

As population grew in Greater Mesopotamia prior to *c.* 7000 B.C., when settlement appears to have been confined primarily to the highland zone where dry farming was possible, it pressured existing agricultural systems. This pressure was relieved in two ways: locally, where possible, and unevenly, the fallow period was shortened; and some people emigrated. As more people were pushed into ever more marginal areas—marginal from a dry farmer's point of view—pressure eventually led to the "discovery" of irrigation. Once irrigation was understood, migration again became an excellent way of relieving continued population pressures since large areas of southern Mesopotamia became available for colonization. Eventually, though there were very important local variations in the pattern, the inhabitants of Greater Mesopotamia began to reach the limits of colonization as a solution to the population problem given existing subsistence systems. At this point a trend toward intensification on all levels appeared, including improvements in the irrigation system and a shortened fallow period. In time and in certain circumstances the amount of land available for cultivation actually began to decline. Conflict became more common. Both factors accelerated the need for intensification. When and where the pressures were greatest urbanism appeared on the scene as a means of organizing and controlling the increased population, the intensified economic structure which supported that population, and the labour force which made that intensification possible. In time there followed royal kings and empires, "Fear'd by their breed and famous by their birth, Renowned for their deeds . . ." and all out of "this teeming womb".

Notes

1 Trigger, B. G. (1968). The determinants of settlement patterns, *in* Chang, K. C. (ed.) *Settlement Archaeology*. Palo Alto. p. 55.

2 The most important of these studies are by Adams, e.g. Adams, R. McC. (1965). *Land Behind Baghdad*. Chicago and London; Hole, F. and Flannery, K. V. (1967). The pre-history of south-western Iran: a preliminary report, *Proc. Prehist. Soc.*, **33** (9), pp. 147–206; and Wright, H. T. (1969). *The Administration of Rural Production in an Early Mesopotamian Town*. Ann Arbor.

3 This is somewhat less so for recent studies of earlier developments in Greater Mesopotamia in the Neolithic: e.g. Flannery, K. V. (1969). Origins and ecological effects of early domestication in Iran and the Near East, *in* Ucko, P. J. and Dimbleby, G. W. (eds.) *The Domestication and Exploitation of Plants and Animals*. London. pp. 73–100. See also, Harner, M. J. (1970). Population pressure and the social evolution of agriculturalists, *S.W. J. Anthrop.*, **26**, pp. 67–86.

4 e.g. Trigger, B. G. (1969). *op. cit.* p. 61.

5 Flannery, K. V. (1965). The ecology of early food production in Mesopotamia, *Science*, **147**.

6 Each of these zones could, of course, be further sub-divided and often has been in discussions of this kind. See Flannery, K. V. (1965). *ibid.*; Smith, P. E. L. and Young, T. C., Jr. (In press). The evolution of early agriculture and culture in Greater Mesopotamia. A trial model, *in* Spooner, B. (ed.) *Population, Resources and Technology*. Philadelphia.

7 Viewed from the perspective of Iranian history, Baghdad (or the urban equivalent in whatever period) can almost be called the natural centre of the Iranian world.

8 In a discussion such as this the importance of defining a circumscribed area is stressed in Carneiro, R. L. (In press). A theory of the state, *Science*.

9 e.g. Childe, V. G. (1951). *Social Evolution*. Cleveland and New York; Childe, V. G. (1936). *Man Makes Himself*. London. A good example of this explanation applied to the development of early urbanism is Davis, K. (1959). The origin and growth of urbanization in the world, *in* Mayer, H. M. and Köhn, C. F. (eds.) *Readings in Urban Geography*. Chicago. pp. 59–60. Childe himself apparently began to change his mind somewhat on the possible role of population in the model: Childe, V. G. (1950). The urban revolution, *Town Planning Review*, **21**, p. 3.

10 One of the most important of these recent discussions is Dumond, D. E. (1965). Population growth and cultural change, *S.W. J. Anthrop.*, 21. For a more detailed discussion of recent literature, see Smith, P. E. L. and Young, T. C., Jr. (In press). *ibid.*

11 Boserup, E. (1965). *The Conditions of Agricultural Growth*. Chicago. See also, Smith, P. E. L., this volume, pp. 411-12.

12 It should be stressed that: (1) the reverse process can also obtain—that is, a decrease in population pressure can lead to a drop down the scale of agricultural intensification to longer fallow periods; and (2) the transition from one system to another is probably a slow process and at any given time the whole range of cultivation systems might be found coexisting in a single area.
13 Flannery, K. V. (1969). *op. cit.* p. 89.
14 Hole, F. and Flannery, K. V. (1967). *ibid.*; Oates, J. (1968). Prehistoric investigations near Mandali, Iraq, *Iraq*, **30** (1), pp. 1–20. For details on the sites in the uplands and a discussion of the level of intensification which may have obtained in this period, see Smith, P. E. L. and Young, T. C., Jr. (In press). *ibid.*
15 It must be stressed that site statistics given for the highland zone in any period are highly inaccurate. Survey work in these zones has been random and has never aimed at illuminating the issues discussed here. The figures used are based partly on scattered publications, partly on personal communications from others, and primarily on personal field work in the area.
16 A yet more cautionary note must be sounded about north Mesopotamian site statistics. If those for the highland zone are "highly inaccurate" (see note), those published for north Mesopotamia are hopelessly out of line with reality. Statements regarding possible population densities in north Mesopotamia are based on published data and on information and cautionary tales kindly supplied by Professor David Oates and Dr. Joan Oates. I am pleased to have shared with them some of their intimate knowledge of this important area, but hasten to add that they cannot be held accountable either for such data as I supply here or how I use it.
17 Hole, F. and Flannery, K. V. (1967). *op. cit.* p. 188.
18 Adams, R. M. (1962). Agriculture and urban life in early south-western Iran, *Science*, **136**, p. 112.
19 Hole, F. and Flannery, K. V. (1967). *op. cit.* p. 189; Adams, R. M. (1962). *op. cit.* p. 113.
20 Ur-Eridu region: Wright, H. T. (1969). *ibid.*; Diyālā: Adams, R. McC. (1965). *ibid.* All data presented here on the Warka region have been kindly supplied by Robert McC. Adams. I am greatly indebted to Professor Adams for his permission to use these materials, drawn partly from personal discussions but primarily from the manuscript of the concluding chapter of a book, soon to go to press, which he circulated to members of the University of Pennsylvania conference on Population, Resources and Technology held in March, 1970. Though the data are Adams', any illusions therefrom are my own.
21 One need hardly stress that simple site counts are a poor way of arriving at population estimates.
22 Smith, P. E. L. and Young, T. C., Jr. (In press). *ibid.*
23 Adams, R. McC. (1965). *op. cit.* pp. 7-8 and Flannery, K. V. (1969). *op. cit.* pp. 89–94.
24 Boserup, E. (1965). *op. cit.* p. 39.
25 Adams, R. M. (1962). *op. cit.* p. 110 and Hole, F. and Flannery, K. V. (1969). *op. cit.* p. 191.
26 Hole, F. and Flannery, K. V. (1969). *op. cit.* pp. 191 and 196.
27 One is particularly suspicious of site counts in this area and of this period because most sites were undoubtedly small and because at least in northern and central southern Mesopotamia alluviation has been considerable over the years. The accidental discovery of a completely buried site such as Ras al'Amiya underscores the point and forces one to make statements on settlement patterns in the south for the Ubaid most hesitantly. See Stronach, D. (1961). Excavations at Ras al'Amiya, *Iraq*, **23** (2), pp. 95–137.

28 Wright, H. T. (1969). *op. cit.* Fig. 2.
29 Personal communication from R. McC. Adams correcting figures for the Diyālā basin as published in Adams, R. McC. (1965). *op. cit.* p. 34.
30 See also Buringh, P. (1957). Living conditions in the lower Mesopotamian plain in ancient times, *Sumer*, **13** (1–2), pp. 30–46.
31 For example, Pisdeli Tepe. Dyson, R. H., Jr. and Young, T. C., Jr. (1960). The Solduz valley, Iran: Pisdeli Tepe, *Antiquity*, **34** (133), pp. 19–28.
32 For such data as is published, see Lloyd, S. (1938). Some ancient sites in the Sinjar district, *Iraq*, **5**, pp. 123–142; Abu al-Soof, B. (1968). Distribution of Uruk, Jamdet Nasr and Ninevite V pottery as revealed by field survey work in Iraq, *Iraq*, **30** (1), pp. 74–86.
33 A recent summary of Ubaid materials in Syria and the Levant is found in Mallowan, M. E. L. (1967). The development of cities from al-'Ubaid to the end of Uruk 5, *Cambridge Ancient History*, Fasc. 58. Part 2.
34 For example, one could almost argue that the architect who built the Northern Temple at Tepe Gawra in level XIII came from Warka.
35 The data for north Mesopotamia are at their worst in the Uruk and Early Dynastic periods.
36 Adams (unpublished ms., see note 20) has estimated the population of Warka at about 10,000 in the Jemdet Nasr period and at about 40,000 in Early Dynastic I times.
37 Adams, R. McC. (1965). *op. cit.* p. 37; See also note 29.
38 Adams, R. McC. (1965). *op. cit.* p. 39.
39 Adams, R. McC. (1965). *op. cit.* p. 37.
40 See especially sites in the neighbourhood of modern Balad Ruz: Adams, R. McC. (1965). *op. cit.* Fig. 2.
41 All ten of the large towns in the Diyālā with occupations dating to Early Dynastic times together contained less population than Warka in the Early Dynastic I Period. Adams, R. McC. (1965). *op. cit.* Table 10, p. 39.
42 Carneiro, R. L. (In press). *ibid.*
43 Annual cropping as practised in temperate climates is not possible in most of the semi-arid Near East. The local adaptation of the system involves allowing a field to lie fallow every other year, or two out of three years, but ploughing that field even so. Often, in fact, it is ploughed several times while fallow. This keeps down the moisture-absorbing weeds and breaks the soil so that precipitation can get into the ground, for the eventual moisture of the soil is the crucial question when crops are planted. See Smith, P. E. L. and Young, T. C., Jr. (In press). *op. cit.* note 16.
44 The plough appears in the pictographic script of the Uruk period. The southerners may have borrowed it from the north, where we have suggested it perhaps appeared somewhat earlier.
45 cf. Dumond, D. E. (In press). Population growth and political centralization, *in* Spooner, B. *op. cit.*
46 For example along the lines of Wittfogel's "hydrolic hypothesis": Wittfogel, K. (1957). *Oriental Despotism*. New Haven.
47 The earliest well-documented period of declining crop yields resulting from salinization falls between 2400 and 1700 B.C.: Jacobsen, T. and Adams, R. M. (1958). Salt and silt in ancient Mesopotamian agriculture, *Science*, **128**, pp. 1251–58. That similar problems with salt, perhaps on a smaller scale and subject to variable local conditions now difficult to reconstruct, obtained in earlier periods seems clear from the evidence of a complete shift to more salt-resistant barley as *the* cereal crop in Mesopotamia by the beginning of Early Dynastic times: Helbaek, H. (1960). Ecological effects of irrigation in ancient Mesopotamia, *Iraq*, **22**, pp. 194–5.
48 Sheer numbers of people involved in the economy might, of course, necessitate the development of written records. Our first documents from

Mesopotamia date to the Uruk period, and two of the earliest genres of texts which appear are economic and legal documents.

49 For much the same argument relating restriction of land to the development of social stratification, see Flannery, K. V. (1969). *op. cit.* pp. 92–95 and, in another context, Flannery, K. V., Kirkby, A. V., Kirkby, M. J. and Williams, A. W., Jr. (1967). Farming systems and political growth in ancient Oaxaca, *Science*, **158**, pp. 445-54. See also Carneiro, R. L. (1967). On the relationship between size of population and complexity of social organization, *S.W. J. Anthrop.*, **23**, pp. 234–43; Dumond, D. E. (In press). *ibid.* It is tempting to speculate whether the "temple economy", which was *one* way in which the more complex economies of early urban Mesopotamia were organized, developed in part because it provided a neutral, impersonal means of resolving disputes over productive land.

50 Adams, R. McC. (unpublished ms., see note 20).

51 Trigger, B. G. (1968). *op. cit.* p. 69.

52 Of course the trend might also be reinforced somewhat by the growing self-interests of the leaders of the ranked society which had already developed in response to the needs for more complex social and political organizations. It should be noted that if the most marginal agricultural people were those earliest pulled into the cities, they would also perforce be the poorest people in the overall economy. Once concentrated in the city these people would probably be landless and would create something of an "urban proletariat". Thus a concentrated and malleable labour force would be available to any economic élite for its needs.

53 Such a "balling" of population in towns and cities is also characteristic of the development of urbanism in late medieval Europe, where it is associated by some scholars with a prior increase of population in the countryside and the development of more intensive systems of cultivation. White, L., Jr. (1962). *Medieval Technology and Social Change*. Oxford. pp. 67–9.

54 Another case (one amongst many) of "balling" in the face of potential conflict which led to agricultural intensification is the example from West Africa discussed in Dumond, D. E. (1961). Swidden agriculture and the rise of Maya civilization, *S.W. J. Anthrop.*, **17** (4), p. 303.

55 ". . . it is still very difficult to see in these figures [for the Early Dynastic period] any evidence that "population pressure" in the sense of land or water shortage in relation to existing subsistence needs, was an important or widespread factor on the historical and social scene. Adams, R. McC. (1965). *op. cit.*, p. 42.

56 Field work by H. T. Wright currently in progress is designed to throw considerable light on the role of trade in early Mesopotamian urbanism. For a possible relationship between the development of trade and population densities, see Sanders, W. T. and Price, B. J. (1968). *Mesoamerica: the Evolution of a Civilization.* New York. pp. 189–91.

57 It is assumed that the southerners paid for these natural resources primarily with agricultural products. Such an exchange with the north in the early Ubaid is perhaps documented by the appearance in the north of six-row barley, a type characteristic of the south. Helbaek, H. (1960). *op. cit.* p. 190.

58 Adams, R. M. (1962). *op. cit.* and note 20.

59 Mortensen, P. (1964). Additional remarks on the chronology of early village-farming communities in the Zagros area, *Sumer*, **20** (1–2), pp. 28–36.

60 We build super highways because there are so many cars, but we also buy more cars because we now have super highways.

61 One suspects that feedback played its greatest role in the Warka area in the late Uruk period, thus in part explaining the remarkable population growth in that area and period and setting the stage for the "balling" which characterized the following centuries.

C: Western Europe

JOHN ALEXANDER

The beginnings of urban life in Europe

Evidence from several disciplines suggests that a reconsideration of the beginnings of urban life in Europe is desirable. Geographers and sociologists have in recent years analysed more fully the nature of the pre-industrial (revolution) town and have suggested new and more flexible definitions of urbanization and more detailed analyses of the origins of towns[1].

Archaeologists have uncovered evidence of urban life in many parts of the world which must force them to reconsider their traditional definitions of it in Europe. It is now known that prehistoric populations in Europe were more numerous and more sophisticated than had previously been realized. Historians of the Roman and Early Medieval periods seem agreed that many pre- or non-Roman traits, especially in architectural, social and artistic fields, survived in and after the Empire.

The problem of definition

Most archaeological definitions have, since Childe's studies, been firmly linked with a complex of cultural achievements in the "Urban Revolution"[2]. This concept based mainly on western Asiatic evidence, was once widely accepted by sociologists and geographers; Sjoberg as recently as 1960 could say "the city and civilization are inseparable"[3]. African and American archaeologists have been perforce less rigid in their definitions and there has been increasing discussion of "urbanism" as a phenomenon separate from "civilization".

Current geographical definitions rely much less on specific sizes, densities of population or particular cultural achievements, and more on the attitude of mind of the communities. Beaujeu-Garnier and Chabot, for example, can end a wide-ranging discussion with the statement "a town is when people feel themselves to be in one", and Jones says that "a town is what is implied by the *local* people when they call a locality a town"[4]. There has also been increased realization that in all continents men have, for a long time, had an affinity for life in large communities and that, therefore, regional varieties of urbanism are to be expected.

Among the changing geographical definitions two points remain fixed: that urban life only seems possible when well-developed agrarian societies exist, and where there is permanence in the settlements.

In Europe it is now known that agriculture associated with long-occupied settlements is to be found in some regions from the sixth, and in all regions by the end of the second millennium B.C. Theoretically, the development of urban ways of life could have taken place at any time after these dates. It is now proposed to take a number of the currently accepted geographical and sociological definitions of urban communities and apply them to the evidence from prehistoric Europe.

Size and density definitions

Size and density are now seen to be dangerous yardsticks when used in isolation. Official modern demographic definitions of the size of urban communities vary between 200 and 10,000 persons[5], and in Europe, Scandinavia accepts a community of 200 as having urban status, whilst undoubted towns of 400 exist in France (see Appendix 1). Acceptable urban densities also vary greatly in modern definitions; whereas 3000 to 4000 per square mile would seem reasonable to many archaeologists, 1000 per square mile is widely in use. In some areas, where distance between houses is used to define "urban", 70 m and even 200 m have been adopted[6].

If these criteria are applied to prehistoric Europe, many settlements, in spite of incomplete excavation, fall within their range. Perhaps in the past too much attention has been paid by archaeologists to definitions of this kind and evidence of other kinds should be looked for.

Definitions of functions

Three main categories of functions have been distinguished by urban geographers: political-administrative, commercial-industrial and cultural-religious.

The *political-administrative* role of towns has been much stressed especially as a cause of urban growth[7]. Large settlements can develop from the seat of a chief or the headquarters of a tribe or community, and once established grow at a greater rate than other settlements. Literary evidence from before the

Roman Conquest shows that considerable tribal organizations and confederations, presided over by powerful chiefs, existed from the Lower Danube Basin to the Atlantic coasts[8]. A number of them are known to have supported large settlements, which acted as tribal capitals[9], and it may be assumed that this function was present and active in Europe from at least the first millennium B.C.

The *commercial* role of towns has also been considered at length and Central-Place theories underline the importance of markets based on local agricultural surpluses[10]. The discussions of distances between markets and the analyses of the needs of farmers' families, and of where they are satisfied, seem most relevant to prehistoric studies[11]. Local marketing centres will always be difficult to recognize archaeologically, for even today they leave little material evidence but there are hints however that they existed as early as the third millennium B.C.[12]. In view of the general size of settled populations at the time of the Roman expansion, it seems likely that local markets were, in the first millennium B.C. and probably long before that, a factor in settlement growth.

Long-distance commerce, often between more and less elaborately organized societies, has been shown to play a large part in town development, whether as permanent staging or trans-shipment posts, or as "temporary towns" as some of the great fairs and markets might be called[13]. There is no doubt that commodities were moving over considerable distances in Europe in the third millennium B.C. and that in the second and first millennia B.C. trade routes linked Mediterranean centres with northern Europe[14]. Along these routes, at sites indicated by geographical definition, are large settlements containing exotic commodities. The widespread development of local coinages in the centuries preceding the Roman Conquest must have been a by-product of this commerce. In later times in Europe regional fairs accelerated the development of permanent towns[15]. It would be interesting to know if fairs played a similar part in earlier prehistoric times.

The *industrial* factor—production on a scale too large for personal needs—has been shown to be of less importance in the development of towns than was at one time supposed[16]. The extractive industries in particular have no inevitable association with large settlements, and manufacturing at a craft level can, but need not, be important. It has been pointed out that the solitary craftsman or the small workshop is normal in the pre-industrial (revolution) town, that they can well be part-time specialists and that agriculture can be accepted as an occupation for townspeople. The importance of the specialist craftsman as a factor in urbanization may have been overstressed.

In prehistoric Europe, links between some considerable third millennium B.C. settlements and mining activities have been suggested, whilst from the first millennium B.C. there are a number of highly suggestive, but no definite, associations[17]. Manufacturing industries are recognizable over a wide area in the second millennium B.C. and, when concentrated in specific localities within first millennium B.C. settlements, they must have been a factor in growth[18].

Cultural traits, other than those used by sociologists and social anthropologists concerned with the structure of societies, seem little used in recent definitions. Since it has become obvious that no one set of cultural traits can be linked with urban living all over the world, the older definitions must be recognized as regional ones. Childe's and Braidwood's criteria, based on the western Asiatic Bronze Age, and linking "urban" and "civilized" with literacy and large-scale public works, may be seen as a regional definition and there seems no reason to accept it rigidly in Europe or elsewhere. A number of distinctive cultural characteristics are present in Europe before the Roman Conquest, for example, the scale and nature of the fortifications to settlements; the many indigenous metal weapons and ornament types and "new directions given to artistic expression", which was one of Childe's criteria[19].

Sociological definitions centre round the class structure of urban settlements[20]. A heterogeneous class structure is considered essential and without literary evidence this is difficult to recognize in archaeological contexts. It is known from Greek and Roman writers that, at the time of the Roman Conquest, a class structure of some complexity existed in Europe, but these accounts do not define the social, or the contract, status of, for example, craftsmen[21]. Archaeology, and in particular the quality of surviving objects, attests that a wide range of specialists existed and later traditions from areas never conquered by Rome have been used to suggest that they had a secure social position. The presence of concentrations of industrial débris in some large pre-Roman settlements[22] suggests that they were a factor in the development of the settlements which, therefore, are likely to have contained a very wide social spectrum. The development of urbanism in Europe might perhaps be considered in the following way.

Europe before the Roman Empire

Considerable agricultural settlements existed in south-eastern Europe from the sixth millennium B.C., and by the late third to second millennium B.C. some were truly urban and associated with a bronze-using (Minoan-Mycenaean) civilization. Elsewhere, only in northern Italy and Iberia might third to second millennial settlements be considered to possess urban characteristics.

By the early first millennium B.C. fully urban settlements were in existence in the eastern and central Mediterranean, and were associated in Greece and Italy with iron-using civilization. Similar, if less sophisticated, developments were taking place by the middle of the same millennium in the Balkans, Iberia and southern France. In all these regions, by the fourth to third centuries B.C., substantial indigenous settlements developed into recognizable towns without the establishment of Greek, Roman or Punic colonies[23].

North and west of the Alps in Austria, southern Germany and Switzerland, also in the second half of the first millennium B.C., a version of the same process may be observed. Many of the larger settlements fall within the definitions of urban discussed above. These settlements, many of them destroyed by the Roman Conquest, cannot be considered copies of the Mediterranean (polis) towns[24] and seem to represent an indigenous and separate urban tradition. The main characteristics which seem to distinguish this from Mediterranean urbanism as well as from many Asian and African developments might be summarized as follows:

(i) Growth was due to trade and commerce, both local and long distance, but not apparently related to immigrant trading colonies from more developed societies.
(ii) Strong links with tribal systems but not perhaps with particular potentates. There seems to have been no tradition of arbitrarily founded towns.
(iii) Communal enterprise shows particularly in the elaborate defences of the settlements. Both the architecture and the methods of construction seem indigenous.
(iv) A particularly ingenious and inventive metallurgical tradition whose craftsmen came to be concentrated in the settlements. Glassworking, especially enamelling, later developed similarly. The crafts seem based on strong indigenous traditions.

Europe under the Roman Empire

Those parts of Europe beyond the Mediterranean shores which came within the Empire were transformed either by decree or by imitation so that the urban settlements resembled those of the Hellenistic world[25].

Europe after the Roman Empire

In studies of urban development in Europe, it has been common to begin with the towns of the Medieval period[26]. Either directly or by tacit assumption the towns of the Roman Empire, modelled on the Mediterranean polis, have been accepted as the origin of the Medieval ones, even in those regions never occupied by Rome. This assumption has been made despite the nature and organization of many northern and western European towns, which from their beginnings were little like those of the Roman Empire, even when on the same sites.

In many ways the Medieval towns of central and north-western Europe with their emphasis on the commercial and political functions discussed above seem to resemble their prehistoric predecessors more than their Imperial ones. In a number of specialized fields, e.g. architecture, religion, art history, social organization, the reappearance or, in the west, re-introduction of prehistoric traits has been increasingly recognized by medieval historians[27]. It may be of value to consider medieval urban development as derived in some part at least from an indigenous urban tradition which was developing in prehistoric times and which reappeared in the west after the Roman interlude.

The only region for which this has been partially accepted is in eastern and north-eastern Europe. Here the indigenous if later development of towns perhaps shows something of what might have been happening earlier in western Europe.

Appendix A modern French example of a "fully urban" small settlement Menèrbes (Vaucluse)

Resident population approximately 400 (doubled in summer)

Density: contiguous houses with small gardens or courtyards (approximately twenty per acre)

Contains:

- two fortified palaces (one outside walls)
- town walls
- two churches
- town hall
- post office
- primary school
- two agricultural co-operatives for neighbourhood (replaces market)
- public drinking fountains, washing place (lavoir) and lavatories
- public statues to its own poet, political martyrs and ideals (La Republique)
- public arena (outside walls)
- two solicitors (notaires)
- two bakers
- two butchers
- three grocers
- one draper/haberdasher
- one hardware shop
- one garage
- one carpenter's workshop (three men)
- two builders
- one electrical repair shop
- one bank
- two insurance agencies (part-time)

one café plus stationery, tobacco, newspaper, confectionery
one hotel
mains water, electricity and drainage to all houses
weekly rubbish collection
annual fête (three days) with public processions, displays, etc.

Notes

1 Beaujeu-Garnier, J. and Chabot, G. (1967). *Urban Geography*. London; Reissman, L. (1964). *The Urban Process*. London; Jones, E. (1966). *Towns and Cities*. Oxford; Smailes, A. (1960). *The Geography of Towns*. London; Sjoberg, G. (1960). *The Pre-Industrial City*. Illinois. Wheatley, P., this volume, pp. 601-37.
2 Childe, V. G. (1951). *Man Makes Himself*. London; Childe, V. G. (1950). The Urban Revolution, *Town Planning Review*, **21**; and defended in Childe, V. G. (1957). *Antiquity*, **30**, p. 36.
3 Sjoberg, G. (1960). *op. cit.* p. 5.
4 Beaujeu-Garnier, J. and Chabot, G. (1967). *op. cit.* p. 30; Jones, E. (1966). *op. cit.* p. 5.
5 *United Nations Demographic Year Book.* (1960). New York.
6 Beaujeu-Garnier, J. and Chabot, G. (1967). *op. cit.* ch. 2.
7 Jefferson, M. (1939). The law of the primate city, *Geogr. Rev.*, **29**.
8 Piggott, S. (1965). *Ancient Europe*. Edinburgh. ch. 6.
9 Piggott, S. (1965). *op. cit.* pp. 217–8. Good examples are *Colchester* (see Hawkes, C. F. C. and Hull, M. R. (1947). *Camulodunum*. Oxford); *Hradiště na Zavist* (see Prošek, F. (1948). *in Pamatky Archaeologické*, **39**); *Manching* (see Kramer, W. (1962). *in Germania*, **40**, p. 293); *Entremont* (see Benoit, F. (1957). *Entremont*. Marseilles); *Mont Beauvray* (see Bulliot, R. and Dèchelette, J. (1927). *Mont Beauvray*. Paris); *Ullestrett* (see Arribas, A. (1966). *The Iberians*. London. p. 102).
10 Christaller, W. (1933). *Die Zentralen Orte Suddesutschlands*. Jena. His table (quoted by Jones) is of interest to archaeologists.

Settlement	*Av. pop.*	*Km apart*	*Size of tributary area* (km^2)	*Pop. of trib. area*
market hamlet	800	7	45	2,700
township centre	1,500	12	135	8,100
country seat	3,500	21	400	24,000

11 Jones, E. (1966). *op. cit.* p. 88. The present needs of a farmer's family graded by frequency and distance:

Food, sweets, tobacco	village/small town	3–8 miles distant
*Repair of farm equipment		
Household goods, working-clothes	small town	10 miles distant
*Market for crops or stock		
Children's clothes, better clothes	larger town	10–30 miles distant
Display clothes, better furniture	provincial centre	30–100 miles distant

*possible additions to Jones' list
12 Buttler, W. (1938). *Der donaulandische und der westische Kulturkries der jungeren Steinzeit*. Berlin. p. 30. The results from Köln-Lindenthal were expecially interesting. Peacock, D. (1968). A petrological study of certain

Iron Age pottery from western England, *Proc. Prehist. Soc.*, **34**, p. 44. These pots should perhaps be considered in terms of what they may have contained, not as a trade in pottery. See Clark, J. G. D. (1952). *Prehistoric Europe*. Cambridge. p. 25.

13 Clark, J. G. D. (1952). *op. cit.* ch. 9.

14 Illustrated by sites like Donja Dolina (on the Middle Danube) Manching (on the Upper Danube) and Mont Lassois (near the Seine-Rhone watershed).

15 Whilst the importance of these is well known from the ninth century A.D. onwards, there has been little discussion of a possible earlier existence, e.g. Pirenne, H. (1925). *Medieval Cities*. New York.

16 Beaujeu-Garnier, J. and Chabot, G. (1967). *op. cit.* pp. 148–67.

17 e.g. Los Millares and other settlements in Almeria: Savory, J. (1968). *Spain and Portugal*. London. pp. 154-5; Hallstatt: Morton, F. (1955). *Hallstatt und die Hallstattzeit*. Hallstatt.

18 e.g. *Manching:* Kramer, W. (1962). *op. cit.*; *Mont Beauvray*: Bulliot, R. and Dèchelette, J. (1927). *op. cit.*

19 For example, wooden ramparts of "box" (timber-frame) construction were widespread in Europe in Late Bronze Age and Iron Age times. Compare *Castione* (Säflund, G. (1939). *Le Terremare*. Uppsala. pls. 85 and 95) with *Biskupin* (Kostrzewski, J. (1936). *Biskupin Pozham*; summarized in (1938). *Antiquity*, **12**, p. 33). For later Iron Age examples see Piggott, S. (1965). *op. cit.* p. 216; Jacobstahl, P. (1940). *Early Celtic Art*. Oxford.

20 Sjoberg, G. (1960). *op. cit.* ch. 5.

21 Filip, J. (1960). *Celtic Civilization and its Heritage*. Prague. pp. 97 and 113–18.

22 e.g. *Manching*: Kramer, W. (1962). *op. cit.*; *Mont Beauvray*: Bulliot, R. and Dèchelette, J. (1927). *op. cit.*; *Hradište na Stradonice:* Filip, J. (1960). *op. cit.* p. 124.

23 e.g. *San Jullia de Ramis:* Arribas, A. (1966). *op. cit.* pp. 97 and 101; *Podradje Kod Benkovac*: Wilkes, J. (1969). *Dalmatia*. London; *Tarquinia*: Hencken, H. (1963). Tarquinia, Villanovans and Etruscans, *Bull. Amer. Prehist. Research*, **23**.

24 e.g. *Hradiště na Stradonice:* Filip, J. (1960). *op. cit.*; *Manching*: Kramer, W. (1962(. *op. cit.*; *Colchester*: Hawkes, C. F. C. and Hull, M. R. (1947). *op. cit.* The size and nature of Colchester are particularly interesting.

25 Hassell, M. W. C. This volume, pp. 857-81.

26 Beaujeu-Garnier, J. and Chabot, G. (1967). *op. cit.* p. 49; Mumford, L. (1961). *The City in History*. London. ch. 1; Weber, M. (1960). *The City*. London; Toynbee, A. (1961). *A Study of History*, vol. 12. Oxford. pp. 389 and 521.

27 Mundy, J. and Reisenberg, P. (1958). *The Medieval Town*. New York; Pirenne, H. (1925). *op. cit.*; Jones, J. (1968). *A History of the Vikings*. Oxford. p. 166.

KEITH BRANIGAN

Verulamium and the Chiltern villas

This brief communication has four objectives: to describe the pattern of settlement, to outline its historical development, to suggest the factors which determined the pattern, and to stress the relationship between the rural settlements and the nearest major town.

The Chilterns form a broad chalk shelf, bounded on the north by the scarp (and the Vale of Aylesbury beyond), on the south and west by the Thames, and, for our purposes, on the east by the Colne—an area of about 600 square miles. If we plot on to the map of this area all of the known villa sites and all of the find-spots of Roman material, then a pattern very clearly emerges, with sites spaced relatively evenly along the valleys of the dip slope, usually about two Roman miles apart (Fig. 1). Other villa sites are found strung along the foot of the scarp and the north bank of the Thames. With few exceptions, all of these sites are found within 400 yards of running water and on small areas of level ground a little above the valley floor. This pattern is that which was established by the mid-second century A.D. and persisted until at least the end of the fourth century and probably somewhat later[1].

It is now clear, however, that the pattern of settlement began to evolve in the half century preceding the Roman invasion of A.D. 43. More than a dozen of the villa sites plotted on our first map have produced late Belgic pottery (Fig. 2), and excavations at Latimer, Park Street and Lockleys have produced evidence to demonstrate that this material represents a permanent occupation of the sites[2]. On the other hand, on none of these sites can continuity of occupation from native farm to early villa be demonstrated, and at Latimer the abandonment of the one and the building of the other are separated by a period of at least thirty years. Latimer, however, need not be typical of all the Chiltern villas, since it was not built until the mid-second century. The earliest villas, replacing the Belgic farms, appear in the last thirty years of the first century A.D.[3], and are found notably in the eastern part of the region, along the foot of the scarp, and along the Thames (Fig.2). West of Akeman Street the villas are built much later, seemingly in the period *c.* A.D. 150 to 180[4]. With the building of these latter villas, the pattern is completed but for an occasional addition like Harpsden (mid-third century)[5].

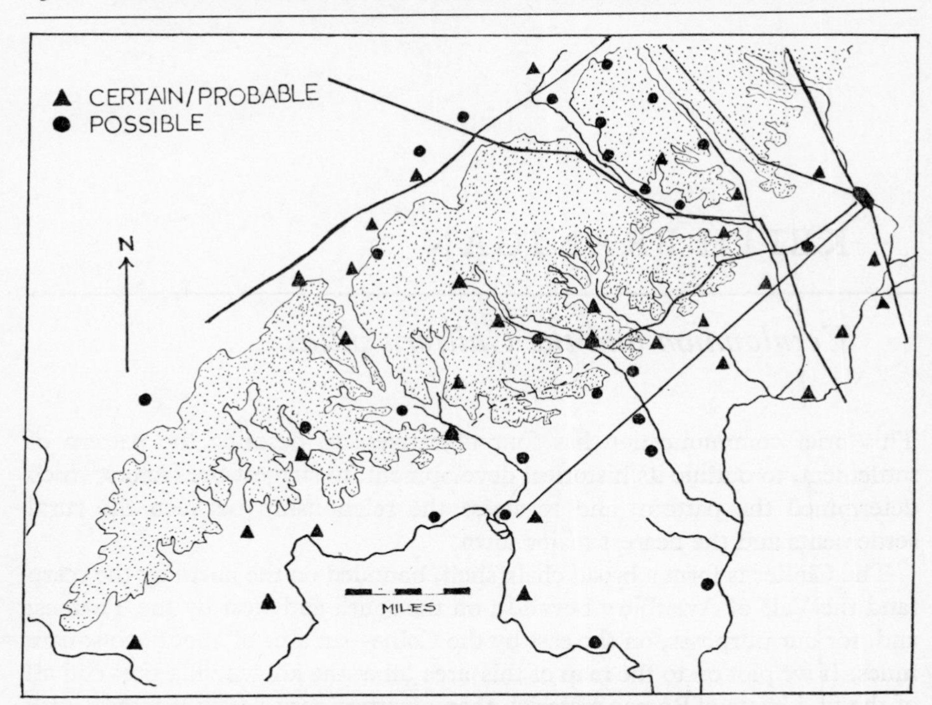

Fig 1 Villas in the Chilterns (Third and fourth centuries A.D.).

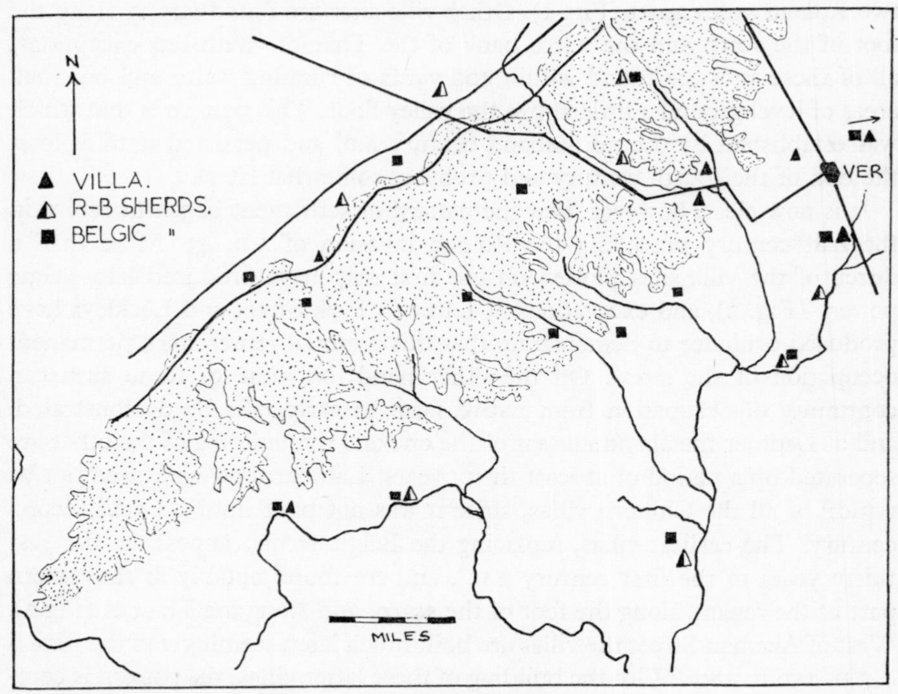

Fig 2 The Chilterns in the mid-first century A.D.

The overall pattern of settlement has clearly been determined by the topography of the region, the narrow river valleys allowing no deviation from a linear pattern based on the north-west to south-east flow of the rivers. In the first century A.D. there can be little doubt that the extent to which the pattern was developed is closely related to existing lines of communication between the villa estates and their economic and social focus at Verulamium. If we look again at the distribution of first century villa sites we can see that they all lie close to an early road, a trackway or a major river (Fig. 2). In the heart of the Chilterns, west of Akeman Street, farms like Latimer were still native in character, and there may have been some larger communities like the village on Lodge Hill and the largely unexplored settlement in the hill-fort at Cholesbury (both of which produce pottery similar to that found associated with the native farm at Latimer)[6]. The factor or factors which led to the rather sudden establishment of further villas, in the western Chilterns, in the mid-second century, cannot be identified with certainty. The building of a road running through the dip-slope valleys from Verulamium to Silchester is the sort of development one might expect, and the Viatores have indeed traced the line of two such roads[7], but with insufficient evidence either to confirm their existence or to date them.

Certainly the cantonal capital at Verulamium must be seen as the social and economic focus of the villa estates in the Chilterns, and the development of the pattern of settlement, and the nature of that settlement, were closely related to the development and nature of the urban centre. This can be demonstrated most clearly by reference to four key periods in the history of Verulamium.

1 *c.* A.D. 70 to 100: The town was rebuilt following the Boudiccan revolt and shows an earlier commitment to romanization and a less insular approach to it than the other towns of Roman Britain. The forum, with temples attached in the Gallic manner, the elaborated Romano-Celtic temple and temenos area, and the masonry market hall, all belonging to this period, were followed later by the theatre (juxtaposed to the temple in Gallic manner) and by the triumphal arches. This early commitment to the Roman way of life is reflected in the contemporary establishment of so many villas in those areas where rapid communication with the cantonal capital was possible at this time. Furthermore some at least of these early villas already possessed stone-built bath suites[8].

2 *c.* A.D. 150 to 180: A fire *c.* A.D. 155 resulted in large-scale rebuilding, notable for the appearance of extensive private houses equipped with mosaic floors and hypocausts[9]. There is a parallel development in the villas, both the new villas (e.g. High Wycombe) and the existing ones (e.g. Park Street, Boxmoor) being given these same refinements[10]. Elsewhere in the province, mosaics and hypocausts are abnormal in second-century villa contexts.

3 *c.* A.D. 270 to 300: The effect of the economic and political crises of the late third century can be recognized at Verulamium by a general lack of maintenance of the public and private buildings[11]. In the villas we find several reflections of the same situation, most clearly at Latimer, where the villa was abandoned for a short period, and at Park Street and Saunderton where there was certainly decay and probably abandonment[12]. Other villas suffered less severely (e.g. Gadebridge Park) but in no case can we point to repairs, rebuildings, or improvements during this period.

4 *c.* A.D. 400 to 450: This is perhaps the most significant period of all in terms of the relationship between Verulamium and the Chiltern villas. If the romanized life and institutions of Verulamium had been brought to an abrupt end following the end of the Roman occupation, then the Chiltern villas too would have seen an abrupt abandonment. But Professor Frere's excavations at Verulamium, particularly in insula XXVII, have demonstrated that civic authority and the maintenance of public works continued at least until the middle of the fifth century[13]. That is, the social and economic focus of the Chiltern villas was maintained. This, I am convinced, explains why the villas themselves were maintained as long as possible, and why their final abandonment was followed by further occupation of the same site in newly erected buildings, albeit of greatly reduced scale and complexity. The slow decline of the villa buildings themselves has been demonstrated at Totternhoe and Latimer (and possibly Saunderton)[14] and on both of these sites post-villa occupation in the vicinity of the courtyard gateway has been found. At Latimer four superimposed building phases of the post-villa period suggest that occupation of the site continued as long as that in insula XXVII at Verulamium[15]. Even at this late stage in the history of the villas and the town, their relationship would seem to be maintained.

Notes

1 Branigan, K. (1968). Romano-British rural settlement in the western Chilterns, *Arch. J.*, **124**, pp. 134–8.

2 To the sites listed in Branigan, K. (1968). *op. cit.* p. 134, add Saunderton, Cuttle Brook, Great Missenden, and Sarrat. Permanent Belgic occupation at Lockleys has been disputed recently, see Webster, G. (1969). The future of villa studies, *in* Rivet, A. L. F. (ed.) *The Roman Villa in Britain.* London. pp. 243–46.

3 Branigan, K. (1968). *op. cit.* p. 135, which gives full references to the relevant sites.

4 This date is slightly later than that given in Branigan, K. (1968). *op. cit.* p. 137, since the evidence at Latimer has now been both increased and re-evaluated.

5 Rivers-Moore, C. N. (1951). Further excavations in the Roman house at Harpsden Wood, Henley-on-Thames, *Oxoniensia*, **16**.

6 This pottery is discussed fully in Saunders, C. (forthcoming) The pre-villa

phases, *in* Branigan, K. *Latimer, The Excavations of 1864, 1910–12 and 1964–70*. Bristol.

7 The Viatores (1964). *Roman Roads in the South-East Midlands*. London. pp. 136–49, 210–14.

8 This is proved at Gadebridge and Park Street, is suggested by the evidence from Boxmoor, and seems likely at Saunderton. For Saunderton see Branigan, K. (1969). The Romano-British villa at Saunderton reconsidered, *Records of Bucks*, **18**, p. 269.

9 Frere, S. S. (1967). *Britannia*. London. pp. 246–47, pls. 21, 26.

10 Branigan, K. (1968). *op. cit.* p. 145. The evidence for the mosaics and their dating is discussed in Branigan, K. (forthcoming). Pavements and poverty in the Chiltern villas, *Britannia*.

11 The situation in Verulamium in the late third century A.D. is still uncertain. On the basis of his own excavations, Frere has largely disclaimed the gloomy picture painted by Richmond. cf. Frere, S. S. (1967). *op. cit.* p. 254, and Richmond, I. A. R. (1954). *Roman Britain*. p. 99. Until Frere's excavations have been fully published however the situation cannot be accurately assessed.

12 Branigan, K. (1968). *op. cit.* p. 147.

13 Frere, S. S. (1960). Excavations at Verulamium, 1959. Fifth Interim Report, *Antiq. J.*, **40**, pl. 8.

14 Matthews, C. L. (1963). *Ancient Dunstable*. Dunstable. p. 64; Branigan, K. (1970). Latimer Villa, *Curr. Archaeol.*, **20**, pp. 241–42, Branigan, K. (1969). *op. cit.* pp. 269–71.

15 Branigan, K. (1970). *op. cit.* pp. 243–44.

M. W. C. HASSALL

Roman urbanization in western Europe

During the reign of Augustus (29 B.C. to A.D. 14) Rome first began seriously to settle and exploit western Europe beyond its Mediterranean littoral. There followed a period of rapid expansion which was only checked during the reign of the emperor Hadrian (A.D. 117–138) who consolidated the conquests of his predecessor Trajan. This brief period not only saw the acquisition of vast areas of new territory, but the creation of towns and cities among peoples who, for the most part, were only urbanized in a rudimentary sense. This paper attempts to assess first of all the impact of this process on pre-existing urban communities and secondly the role played by the different types of major urban centres in the western Empire.

1 Roman impact on pre-existing urban centres and the policy of urbanization

The Roman armies came into contact with many communities in western Europe which can fairly be claimed to have had an urban character—thus Numantia near the headwaters of the Douro, which fell to Rome in 133 B.C., was a strongly fortified town with city walls, paved streets and stone-built houses with cellars[1]. The hilltop strongholds that Caesar and his successors found in Gaul, and later Britain, were described by them as towns—*oppida*—and excavations at sites such as Alesia[2] or Hod Hill[3] show that the term in a broad sense is justified. The fate of such towns after the conquest would depend on circumstances: Numantia remained in occupation, though a shadow of its former self[4]; the Gallic Gergovia was gradually abandoned in favour of a new foundation in the valley below; at Hod Hill a ten-acre Roman fort was planted uncompromisingly in one corner of the native hill town and the huts of the original population abandoned. However, there was a world of difference between these urban centres before and after the conquest. Sometimes, as stated above, the original site of the town would be abandoned for a more convenient though less defensible site nearby, as Maiden Castle[5] was abandoned for Dorchester—Durnovaria. The initiative for such a change did not necessarily come from the Roman side. In A.D. 78 the magistrates and councillors of the hilltop town of Sabora in south-west Spain successfully petitioned Vespasian to be allowed to rebuild their town "under his name" on the plain[6]. Apart from any change in site the whole character of such towns would alter. The raw new towns such as Roman Verulamium[7] or Augusta Raurica (Augst)[8] with their piped-water supplies, street grids delimiting the new building plots, timber-framed houses with garishly painted interiors, covered sidewalks, classical town halls and temples, are far closer to their equivalents in the American west than to any Celtic predecessor. Such towns differed from their American equivalents, however, in that the "natives" were deliberately encouraged to take part in them. Even in the most Romanized of all, the *coloniae* (see below), they had a place although here they were definitely regarded as second-class citizens and the Salassi[9] settled at Aosta would have been little better off than the Trinovantes at Colchester[10]. As the latter case shows, this could lead to trouble, for the Trinovantes were some of Boudicca's most enthusiastic supporters during the great revolt of 60-61. However, for one such failure there were countless successes. The aim behind this policy of urbanization in Britain is clearly stated by Tacitus. It was to "persuade a people previously scattered, uncivilized and, therefore, prone to fight, to grow used to peace through material well being"[11]. In Gaul epigraphy provides a striking instance of a descendant of one of Caesar's bitterest enemies, Lucterius, playing an active role in the new municipal life[15].

2 *Types of urban centre in the western Roman provinces*

(*i*) *Cantonal capitals*

It was a basic feature of Roman imperial administration to allow the provincials to continue to manage their own affairs. In the eastern Empire, this meant that the Hellenistic city-states retained control over the territory attributed to them. In the west each tribe was reckoned as a *civitas*, i.e. a semi-autonomous political unit. The term in Latin always has an urban connotation (from it is derived the English word city), and in the new order of things pre-existing tribal centres were encouraged or created. The surrounding countryside was not "attributed" (i.e. subordinated) to the city as in the eastern Empire, but each tribesman, whether or not he might reside in a chief city, was a citizen (*civis*) of the tribe. These tribes had largely been ruled by kings. Under the Empire they received magistrates and councils modelled on the urban constitution with which Rome was familiar in Italy. They naturally exercised their functions in the urban centres. In some cases inscriptions show that these tribal functionaries retained their old titles—thus vergobrets are known in Gaul in the same way that sufets are known in

North Africa. In one case only is this municipal machinery recorded as having been imposed by force upon a tribe—the Frisii[14]—but one can suspect that this was often the case. If there was no suitable pre-existing tribal centre one was created: the names of such new towns often reveal the imperial interest taken in their creation, e.g. Asturica Augusta (Astorga), Juliobriga (Rimosa), Augustonemeton and many others. It is worth noting that Vespasian gave permission to rebuild Sabora "under his name". Sometimes a district previously under military control would receive *civitas* status on the withdrawal of the army and a fort site, often a road centre, might be handed over by the government to the new *civitas*. Thus, in Germany the newly created *civitas Ulpia Nicretum* takes over the fort site of Lopodunum (Ladenburg), the *civitas Taunensium* occupies Nida (Heddernheim)[15], while in Britain important military sites like Viroconium (Wroxeter) and Corinium (Cirencester) become the cantonal capitals of the *civitas Cornoviorum* and *Dobunnorum* respectively[16].

(ii) Coloniae—"passive"

Augustus claimed to have settled vast numbers of citizens in the provinces, mostly the landless veterans of the civil wars[17]. Such communities could be established on new sites but were often attached to pre-existing communities[18]. In western Europe such colonies are not found beyond the Mediterranen littoral of Gaul and Spain and these lie outside the scope of this paper. They are here called "passive" to distinguish them from the para-military settlements found in the rest of western Europe which are described below.

(iii) Coloniae—"active"

Other settlements of veterans had a semi-military nature—for example, before their reduction by Augustus, strategically placed colonies had been established both north and south of the Alps at Turin, Aosta (succeeding Eporedia) and Aquileia, and at Lyons, Nyon and Augst[19]. Their function was, like Narbonne (Narbo Martius), Rome's earliest colony outside Italy, to act as a "watch-tower and bulwark of the Roman people and a barrier of defence over against the barbarians"[20]. The colonies of Olisipo (Lisbon), Scallabis and the significantly named Pax Julia in the newly formed Augustan province of Lusitania had a similar function. Such colonies could be established on the site of legionary fortresses when they took over their function releasing the legion concerned for another theatre—as Tacitus rather obliquely says happened in the case of Camulodunum (Colchester)[21]. The western legionary base of Gloucester eventually itself became a colony when the legion again moved to Caerleon. Similarly Lincoln was made a colony when the legion in garrison moved to York, and York *could* have been made a colony by Severus when it seemed likely that York legion would take up permanent quarters in the newly built stone fortress of Carpow on the Tay. Similar sequences could be adduced from other provinces.

(*iv*) *Titular colonies and municipia*

Certain native communities were sometimes given the title of *colonia* or *municipium*. The exact status of so-called "titular colonies" is in dispute and anyway may have varied. They were probably similar to *municipia* whose citizens had some, but not all, of the privileges of the Roman citizens. Both titles were eagerly sought and deputations to the emperor to secure the honour, as that sent from Volubilis in Mauretania to the emperor Claudius, will have been frequent[22]. In Spain something like 350 towns were given the title by Vespasian (A.D. 70 to 79)[23], testimony both to the degree of Romanization that these cities had attained, and the liberal policy of this emperor.

(*v*) *Provincial capitals*

The capitals of particular provinces invariably had higher status than that of the ordinary tribal centres. They might be colonies—either "active" like Colchester, Merida, or Lyons, or titular like Trier, the capital of Belgica. They might be entirely new cities like Emerita or based on tribal centres like Trier. Occasionally like Camulodunum, they had enjoyed something more than local pre-eminence before they became provincial capitals with the coming of Rome[24].

3 Conclusions

The differences in status between the various types of city described above, though real at first, gradually counted for less and less as time went on, but they are important if one wishes to understand the character of Roman urbanization in western Europe. Rome was *not* motivated by altruism in her creation of an urban society. The new towns were founded with very definite objectives—as administrative centres or as quasi-military strong-points. In the background was the feeling, certainly correct, that the newly created urbanized society led by comparatively small groups of influential citizens—the magistrates and councillors—whose interests were identified with those of Rome, would be more amenable to the Roman Peace. Yet, whatever motives she may have had in fostering the cities, they were an essential characteristic of Roman civilization and one of the greatest gifts of Rome to western Europe.

Notes

1 Schulten, A. (1933). *Geschichte von Numantia.* Munich.
2 Brogan, O. (1953). *Roman Gaul.* London pp. 104–5.
3 Richmond, I. A. (1968). *Hod Hill.* London. vol. 2.
4 Brogan, O. (1953). *op. cit.* p. 106.
5 Wheeler, R. E. M. (1943). *Maiden Castle, Dorset.* London.
6 I.L.S. 6092—Lewis, N. and Reinhold, M. (1955). *Roman Civilization.* New York and London. vol. 2, p. 341.
7 Frere, S. S. (1964). Verulamium—then and now, *Bull. No.* 4 *of Inst. of Archaeology.*
8 Lauer-Belart, R. (1966). *Führer durch Augusta Raurica.* 4th ed. Basel.
9 I.L.S. 6753. This inscription shows that some of the Salassi *did* take part in the settlement at Aosta, despite the statement of Strabo, *Geography,* 4.6.7, that the Romans sold 44,000 Salassi into slavery and that the tribe as such was wiped out.
10 Tacitus, *Annals,* 14.21, for the high-handed treatment of the local natives by the new settlers.
11 Tacitus, *Agricola,* 21.
12 I.L.S. 7041. Compare Caesar, *Gallic War,* 8.32.
13 Caesar, *Gallic War,* 1.16, for annual dual magistrates elected by the Aedui. Bronze coins of the Lixovii near the mouth of the Seine give the names of vergobrets. I.L.S. 7040 (Saintes) for a vergobret after the conquest. Several North African inscriptions attest suffets. Of particular interest is AE 1916.42 (Volubilis). The man honoured in this inscription was apparently Sufes. When the town became a *municipium* as a result of a deputation undertaken by him to Claudius, he became one of the first duoviri.
14 Tacitus, *Annals,* 11.19 (A.D. 47).
15 Stein, E. (1932). *Römische Beamte und Truppenkörper in Deutschland.* Berlin. pp. 15–16.
16 Webster, G. (1966). Fort and town in early Roman Britain, *in* Wacher, J. S. (ed.) *The Civitas Capitals of Roman Britain.* Leicester.
17 *Res Gestae* 3. Augustus claimed to have allotted lands or granted money to 300,000 veterans who were settled in colonies or sent back to their own municipalities.
18 At Valentia the veterans and the original inhabitants kept their identities, cf. C.I.L. 2.3739, where both "veterani" and "veteres" are mentioned. There were two town councils representing both groups, C.I.L. 2.3745.
19 Turin and Aosta were Augustan Foundations. Aquileia had been founded in 181 B.C. Eporedia in 100 B.C. Lyon, Nyon and Augst were planned by Caesar and founded after his death by L. Munatius Plancus, I.L.S. 886.
20 Cicero, *Pro Fonteio,* 5.13.
21 Tacitus, *Annals,* 12, 32. Colchester was founded as a "bulwark against revolt and to instill in the allied tribes the idea of constitutional government".
22 See note 13.
23 Sutherland, C. H. V. (1939). *The Romans in Spain.* London. p. 186.
24 Camulodunum was the capital of Cunobelinus who had established a hegemony over the tribes in south-east Britain. Compare Iol-Caesarea, Capital of Juba II of Mauretania and Sarmizegethusa, chief town of the Dacian king, Decebalus, both of which became provincial capitals.

DAVID STURDY

Correlation of evidence of medieval urban communities

The history of medieval European towns can be divided into three broad phases: (*a*) emergence in the period A.D. 700 to 1000, (*b*) development and expansion in A.D. 1000 to 1300, and (*c*) general decline in A.D. 1300 to 1500.

Earlier stages of settlement, such as form the greater part of the present series of studies, must be reached primarily through material remains and only occasionally through written evidence. For more recent periods the written word normally forms the primary evidence. Research is confused by the acute language problem of historians talking history, archaeologists talking archaeology and, on occasion, architects talking architecture, with marked disregard of each other. And yet the problems generally interlock and the solutions depend on mutual collaboration, and an understanding of all the local and broader contexts in some depth.

These problems can be considered in the context of the final century of the phase of "emergence" in England. Here, and over much of northern Europe, we can identify in A.D. 900 places that are little more than fortresses, or, rarely, sea-ports; by the year A.D. 1000 we can recognize enough industrial commercial and administrative factors to form an urban entity. The century could be claimed, in very simplified terms, as marking the true beginning of the process which has led to our modern urban civilization.

The number of places where a significant number of these factors can be traced in the tenth century A.D. is between fifty and sixty and as many again display a few indications of urban status. Over the countryside as a whole, among an approximate total of 10,000 rural settlements there was only one urban nucleus of any degree to every eighty or 100 settlements. This is in marked contrast to the ratio in A.D. 1300 when a countryside with a further 700 market towns had a centre with some sort of urban characteristics to every twelve settlements. These figures give only a rough indication of degree and must conceal many regional contrasts which only detailed analysis will emphasize.

The study of this first critical period of medieval urbanization depends on a static body of documentary evidence, much of it casual annalistic references; advances from additional evidence will accrue largely through material evidence from excavations or recovered by chance in the massive building-works by which modern redevelopment is overwhelming our ancient towns[1].

For later medieval periods the abundance of title-deeds and rent-rolls in certain towns is making it possible to relate the known tenants to excavated structures. This elaborate research will enable us to correct and amplify social and economic assumptions in a way similar to recent work drawn from detailed family studies in the sixteenth to nineteenth centuries[2].

For the early middle ages such "personalized" archaeology is not usually possible, but the conclusions made from material remains alone, in their proper context, are no less significant. The range of concepts and conclusions that can be made from unsupported archaeological data is considerable. At Thetford[3], for example, from the plan of an excavated area of some 10,000 m^2, we can suggest that towns of the tenth century were not built up along the street front but had large enclosed properties with the buildings dotted about. By the twelfth century, from standing buildings at Lincoln or documents at Oxford, we know that tight-built urban frontages were already established. How and why did this striking change take place?

At Oxford, from rescue work at Woolworths, Littlewoods and Marks and Spencer we can say that the tenth century forerunners of the main shopping street of medieval and modern times was much wider and on a slightly different alignment. By the late twelfth century the present line was established, on clear structural and documentary evidence. A massive act of re-planning must be explained.

The kinds of questions which we ask about medieval towns and can discuss on a basis of combined material and documentary evidence concern the siting, size and relative wealth, layout and structure of complete streets and properties, their complex activities and contacts.

The basic plotting of evidence has been published for six out of 100 or more towns[4]. But we must remember that our understanding of the town is not complete without setting it in its regional context. The rural landscape for which the town was a focus can again be plotted from combined evidence such as the Domesday Book or surviving field-systems. Against this background we can say that the southern capitals of Winchester and London both seem to lie in hinterlands of thin population and low agricultural prosperity. What were the factors of siting and communication that enabled them to maintain their position?

Or we can comment that Norwich and Thetford both appear to have developed out of an agglomeration of villages, a pattern familiar in central Europe. Or does this reflect the great population density of East Anglia?

Another aspect of the regional and national setting is the system of main roads which had developed by the twelfth century. In some cases the town depends on the road; in others the road was caused by the towns. The key to our knowledge of national communications lies in detailed research into the town.

It is clear that we must have more research units comprehending the total scientific analysis of the town in all aspects, as so well exemplified at Winchester.

Notes

1 Biddle, M. (1968). Archaeology and the history of British towns, *Antiquity*, **42**, p. 114.
2 Laslott, P. (1965). *The World We Have Lost*. London.
3 Davison, B. K. (1967). The Late Saxon town of Thetford, *Medieval Archaeology*, **11**, p. 189f.
4 Biddle, M. (1968). *op. cit.*; Addyman, P. V. and Hill, D. H. (1968). Saxon Southampton: a review of the evidence, I. *Proc. Hants. Field Club*, **25**, p. 61f.

J. B. WARD-PERKINS

Central authority and patterns of rural settlement

One of the many aspects of any given pattern of rural settlement about which the archaeological record alone can never provide more than an informed guess is the extent to which it reflects the presence or absence of a strong central authority. This is a problem which has bulked large in the interpretation of the results of the field survey that the British School at Rome has been conducting in south Etruria since 1955, and the paper that follows is an attempt to illustrate something of the broad conclusions which have been taking shape in the course of that survey with regard to this particular problem.

A word of introduction about the survey itself[1]. The area principally concerned lies immediately to the north of Rome, comprising a geographical unit some 400 square miles in extent and corresponding closely to the territories of the ancient Veii, Capena, and Falerii. To the south and east it is bounded by the lower Tiber valley, to the north-west and west by the belt of volcanic hills and lakes which today include Lake Bracciano and Monte Cimino, and which in Early Classical times was still covered by the natural barrier of the Ciminian Forest. Except along the eastern fringes, around Capena, this is all country of very recent geological formation[2], and for the purposes of a topographical survey it offers many advantages. It is a terrain that is physically retentive of the superficial traces of human activity; until very recent times it had suffered very little from modern development; and over the greater part of the area it seems that the earliest human settlement barely antedates the earliest written record. As a result, the surviving physical remains of antiquity are not only numerous and articulate in their own right, but they can also be studied within the framework of an independent documentary tradition. Except for the first few centuries this tradition is in fact substantially continuous over the whole of the three thousand odd years covered by the survey.

The late prehistory and early proto-history of central Italy contain much that is still highly controversial, but with much of this controversy our area is fortunately only marginally concerned. Although the Tiber valley itself was a long-established route, with many prehistoric settlements along it and

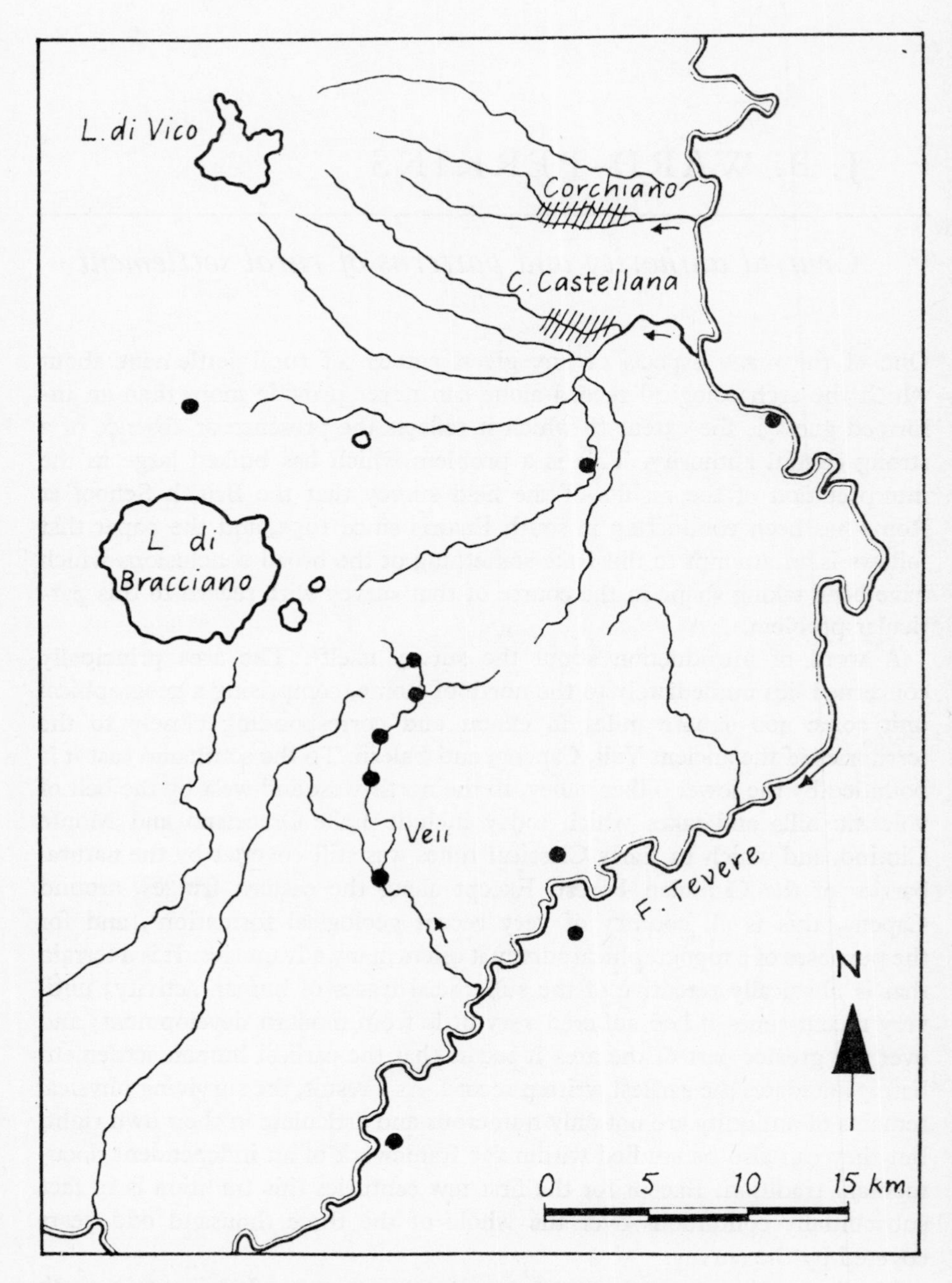

Fig 1 Late Bronze Age settlement in south-eastern Etruria.

many links northwards and eastwards towards central Italy, there seems to have been very little penetration into the forest until quite late in the second millennium B.C. There is a well-defined belt of Neolithic and Bronze Age settlement linking the middle Tiber valley, around Orvieto and Bagnoregio, with the coast around Vulci and Tarquinia, by way of the rivers Fiora and Marta and Lake Bolsena. But south of this belt most of the way to Rome, there was a very large tract of primeval forest-land[3], which has yielded little or no trace of permanent settlement before the Early Iron Age, except on the fringes, towards the coast, and, within our area, towards the Tiber valley. Here there is a thin but consistent pattern of Late Bronze Age penetration up the west-bank tributaries of the Tiber (Fig. 1), a pattern which is almost certainly to be interpreted as the work of small groups of pioneers moving up into what at the time was still almost virgin territory[4].

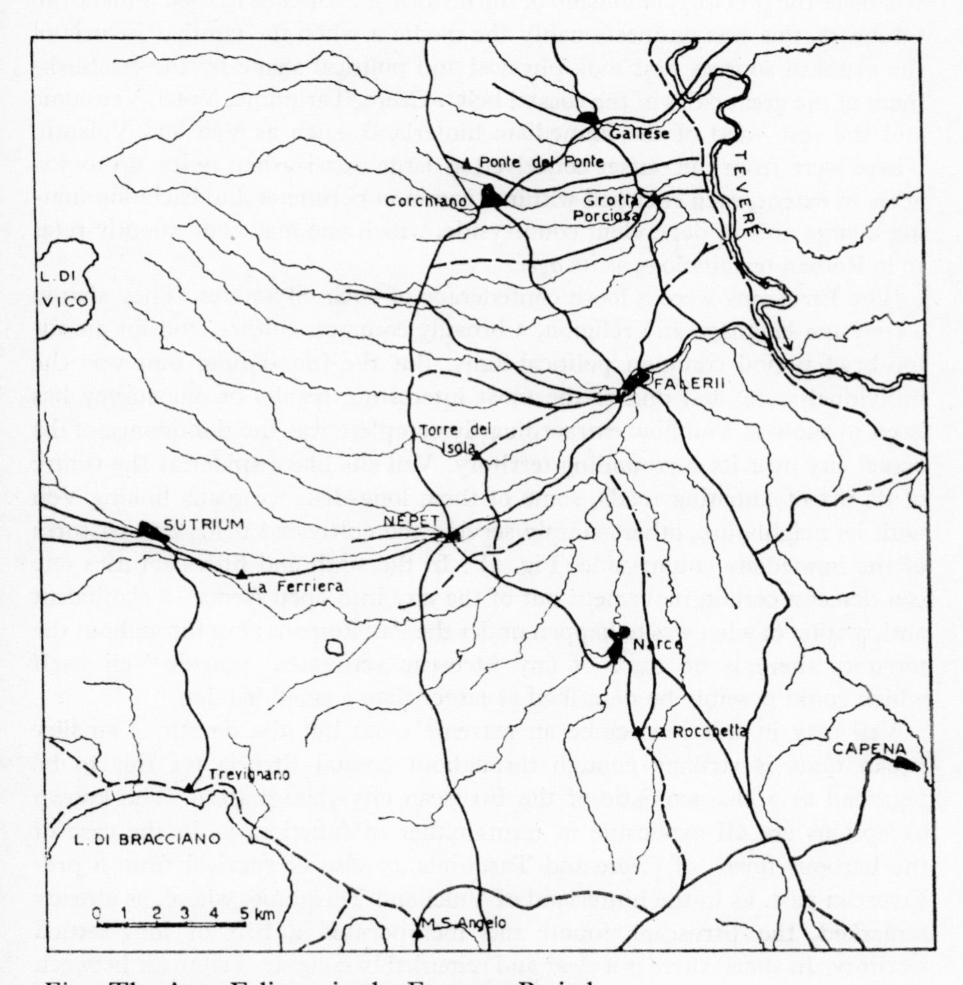

Fig 2 The Ager Faliscus in the Etruscan Period.

Where, as around Lake Bolsena and in the Ager Faliscus, there were enough of these pioneers, and where they had time to put down roots, they have left their mark on the subsequent settlement patterns. Throughout their independent history the Faliscans, the historical descendants of these Bronze Age settlers, remained a rather loose confederation of small towns and villages (*pagi*) (Fig. 2). The capital, Falerii, was the political and religious centre, but it was never more than *primus inter pares*. In the Ager Veientanus the story was very different. Here the Bronze Age population was totally submerged by the next wave of immigrants to the area, the Villanovans, who crossed the forest from the west to found Veii, and who throughout Etruria established the pattern of settlement which was to dominate the area for the next half-millennium.

With the Villanovans we are on the threshold of written history; and whatever the precise relationship of the historical Etruscans to their Villanovan forebears, this was unquestionably the moment when the familiar Etruria of the classical sources first took physical and political shape by the establishment of the great cities of the coastal belt—Caere, Tarquinia, Vulci, Vetulonia and the rest—and of the immediate hinterland, such as Veii and Volsinii. These were from the outset conceived as large quasi-urban units, up to 500 acres in extent, each enclosed within a common perimeter and each dominating a large area of dependent countryside, which one may conveniently refer to in Roman terminology as its *ager*.

The Etruscans were a loose confederacy of such city-states. They shared a common language and religion, a broadly common culture, and for an all-too-brief period common political aims. But the functioning unit was the individual state, and one of the most interesting results of our survey has been to show at Veii how extraordinarily complete was the dominance of the actual city over its surrounding territory. Veii sits like a spider at the centre of its net of radiating roads, some of them long-distance roads linking Veii with its neighbours, others simply serving the cultivated fields and pastures of the immediate countryside (Fig. 3)[5]. In the sixth and fifth centuries one can detect a certain movement out of the city into open farms[6], a significant anticipation of what was to happen under the Pax Romana; but throughout the territory there is no trace of any Etruscan settlement outside Veii itself which could possibly be described as larger than a small hamlet.

Veii may in this respect be an extreme case; but the dearth of smaller urban units is striking enough throughout coastal Etruria for this to be regarded as a characteristic of the Etruscan city-state pattern. The known exceptions are all explicable in terms either of function, as in the case of the harbour towns of Caere and Tarquinia, or else of survival from a pre-Etruscan past, as in the hinterland of Vulci and Tarquinia, where, as already remarked, the Etruscans found and incorporated a belt of long-settled territory. In short, there is a clear and remarkably consistent contrast between the prehistoric village pattern, as we find it still functioning in the historical

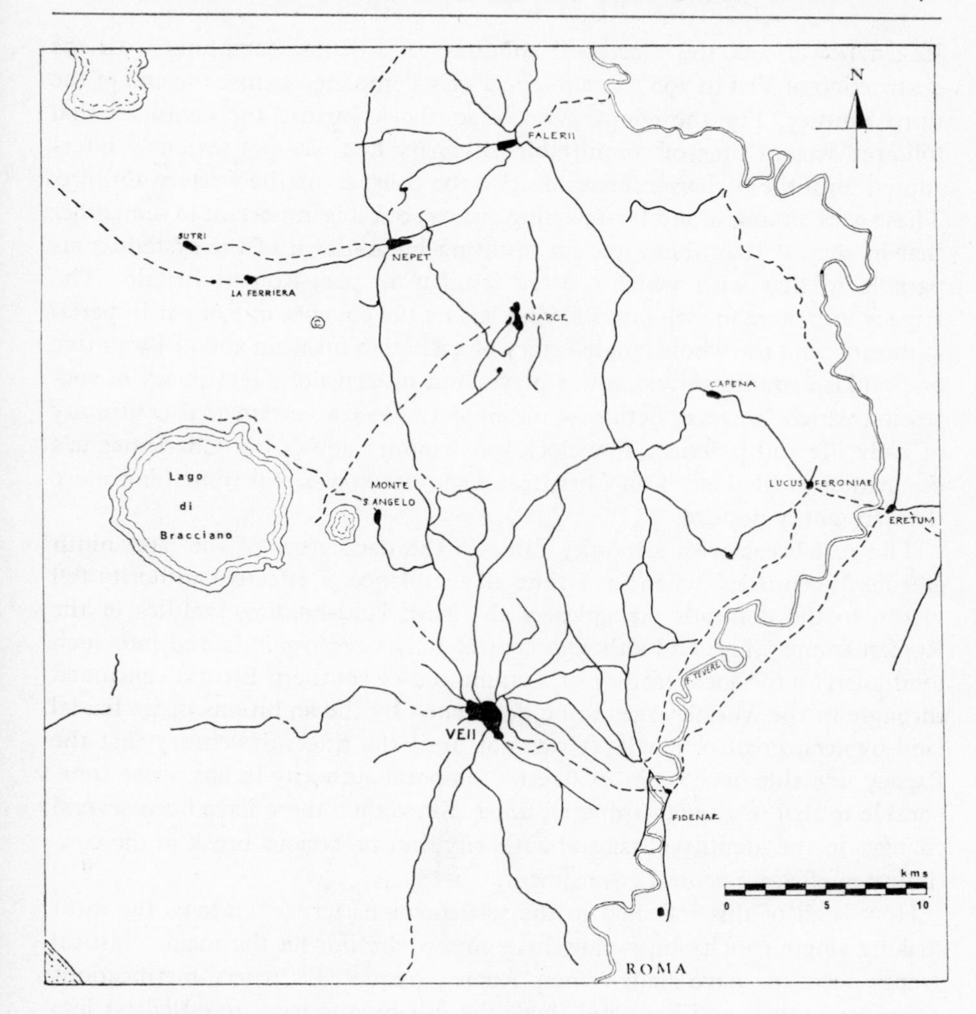

Fig 3 South-eastern Etruria in the Etruscan period.

Ager Faliscus, and the city-state pattern which the Villanovans imprinted on most of the rest of southern Etruria. Since physically and geographically the territories in question are virtually indistinguishable, the contrast must be due to extraneous factors; and in the light of what follows it does seem very likely indeed that it reflects the differences between the rather loose political association of the Faliscan settlers and the far more authoritarian organization of the fundamentally aristocratic Etruscan communities[7].

So far we are dealing predominantly with inference from the archaeological data. With the Etruscans of the seventh century we come within the scope of a documentary history which, for all its gaps and imperfections, was henceforth to be continuous; and it will be convenient at this point to call to mind a few salient historical facts. The first was the emergence of Rome as a major

Italian power and the piecemeal conquest of Etruria, beginning with the destruction of Veii in 396 B.C. and effectively completed before the end of the third century. For the countryside of southern Etruria the centuries that followed were a time of comfortable obscurity that was not seriously interrupted until the barbarian invasions and the collapse of the western Empire. These were dramatic and far-reaching events; but it is important to remember that in central Italy there was no resulting break-down of the system comparable to that with which we are familiar in post-Roman Britain. The Papacy was there to step into the void left by the collapse of Roman Imperial authority; and the whole tangled story of barbarian invasion and of Byzantine or Frankish counter-invasion was played out in terms of a legitimacy of succession which, however fictitious, did in fact reflect a certain real continuity of daily life and practice. The clock was winding down; but Charlemagne's coronation in St. Peter's on Christmas Day of 800 was still something more than an empty gesture.

The real break with antiquity came in the dark years of the later ninth and tenth centuries, when for a time all semblance of effective authority fell victim to the domestic struggles of the great land-holding families of the Roman countryside; and although central Italy never again lapsed into such total anarchy the local history of Latium and of southern Etruria continued throughout the Middle Ages to be dominated by the ambitions of its feudal land-owners, great or small. It was not until the fifteenth century that the Papacy was able once again to assert a temporal authority in any sense comparable to that of Roman Imperial times. Since then there have been several changes in the identity of central authority, but no serious break in the continuity of effective central government.

How is all of this reflected in the settlement patterns? Perhaps the most striking single conclusion is how little immediate impact the major political events seem to have had. When Rome conquered Etruria fortifications were dismantled, and here and there the inhabitants were resettled on less defensible sites, as happened at Falerii after its final capture in 241 B.C. But in the countryside of Veii, Rome's bitter rival, the archaeological record makes it quite clear that everyday life went on very much as before. Some of the farms were abandoned and others no doubt received new owners. But there is a broad continuity of settlement which accords well with Livy's statement[8] that many of the inhabitants were quietly absorbed into the Roman polity. Veii itself was sacked and may conceivably for a short while have been depopulated; it certainly lost all political power. But as the hub of the only effective (and highly efficient) road-system it inevitably retained (or very soon regained) its position as the economic centre of the Ager Veientanus[9]. For nearly two centuries after the violent substitution of Roman for Etruscan authority, one can detect hardly any significant change in the pattern of settlement.

The changes, when they came, were due principally to two factors[10].

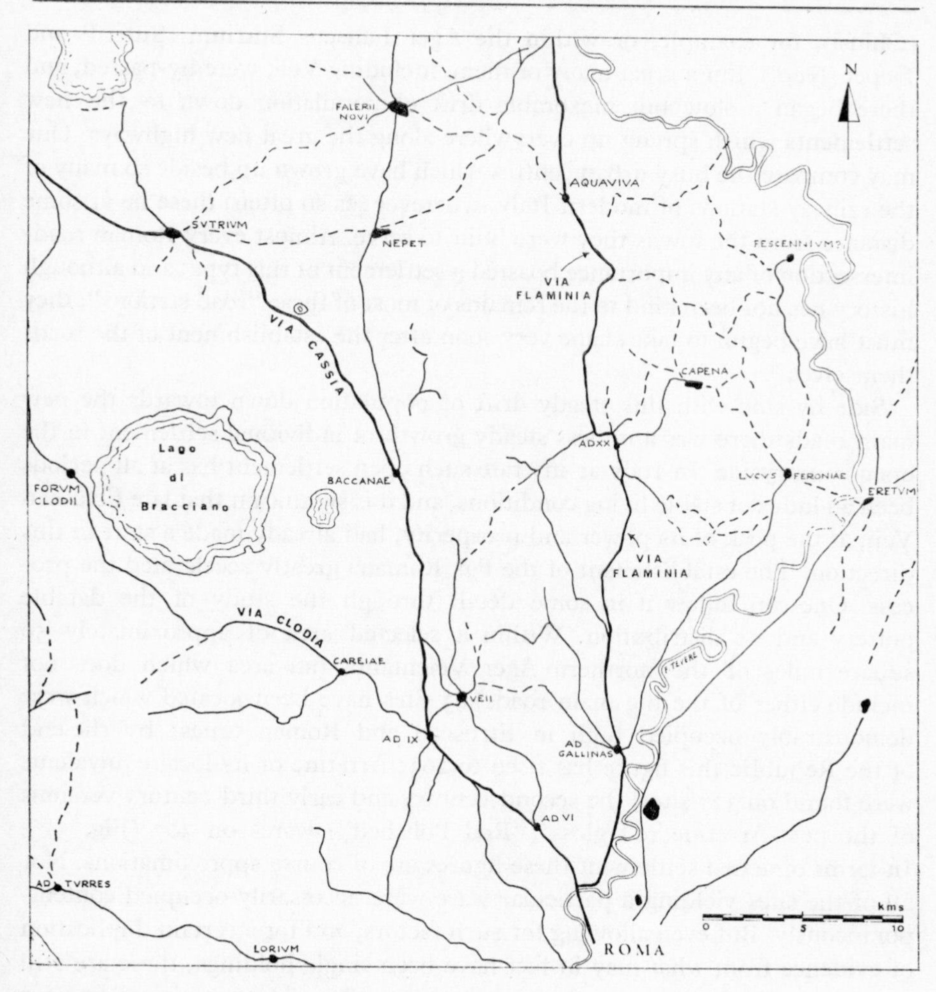

Fig 4 South-eastern Etruria in Roman times.

One was the building through the area of two of the great military roads to the north, the Via Flaminia in 220 B.C. and the Via Cassia some fifty or sixty years later (Fig. 4). Unlike the Etruscan road-system, which had grown up to serve local requirements, these were long-distance highways, designed to carry troops and equipment to districts far beyond southern Etruria. They were planned and built without any regard for local requirements, and for that very reason they were, unintentionally but inevitably, powerful instruments of social change. They did, of course, incorporate such stretches of the existing road-system as served their purposes (a good example is the old ridgeway track from Rome to Falerii, which the Via Flaminia followed, and still follows today, almost all the way to Civita Castellana), and a few of the old Etruscan towns which lay on the new roads were the gainers thereby: Clusium

(Chiusi), for example, or within the Ager Faliscus Sutrium (Sutri)[11] and Nepet (Nepi). But a great many of them, including Veii, were by-passed, and there began a slow but inexorable drift of population down to the new settlements which sprang up everywhere along the great new highways. One may compare the busy urban centres which have grown up beside so many of the railway stations of modern Italy, wherever (as so often) these lie at some distance from the towns they were built to serve. Almost every Roman road-intersection of any importance boasted a settlement of this type; and although history has not been kind to the remains of most of these "road stations", they must have begun to take shape very soon after the establishment of the roads themselves[12].

Side by side with this steady drift of population down towards the new main roads there was a no less steady growth of individual settlement in the open countryside. In Italy at any rate such open settlement has at all periods been an index of stable living conditions, and it is significant that late Etruscan Veii, at the peak of its power and prosperity, had already made a start in this direction. The establishment of the Pax Romana greatly accelerated the process. One can follow it in some detail through the study of the datable pottery and its distribution. Within a selected area of approximately 50 square miles of the northern Ager Veientanus (an area which does not include either of the big main roads) 63 sites have been located which were demonstrably occupied both in Etruscan and Roman times; by the end of the Republic this figure has risen to 263; Arretine or its local equivalents were found on 327 sites, the second century and early third century versions of the post-Arretine red gloss ("Red Polished") wares on 307 (Fig. 5)[13]. In terms of actual settlement these figures are of course approximations. Not all of the sites yielding a particular ware were necessarily occupied contemporaneously. But even allowing for such factors, and for a certain duplication of evidence from what may in fact have been single holdings, these are still formidable figures, figures which even today, after an intensive programme of post-war land redistribution, it would be hard to match. They do much to explain how Veii itself, which at the end of the Republic was still a substantial enough community to be created a *municipium*, vanished from history after the early fourth century[14].

This is inevitably an over-simplified picture of a complex process of change. There were many forces at work, but the factor common to nearly all of them was the security and the opportunity afforded by the establishment of the Pax Romana, imposed by a strong central authority. Sometimes the intervention of this authority was direct and explicit. The recent land-distribution schemes referred to above, for example, were adopted in conscious imitation of late Republican Roman practice, a practice which we know from Cicero[15] was under consideration in our area. In both cases it represents a deliberate political attempt to reverse the natural tendency for small-holdings to disappear in favour of larger, economically more viable estates. Quite

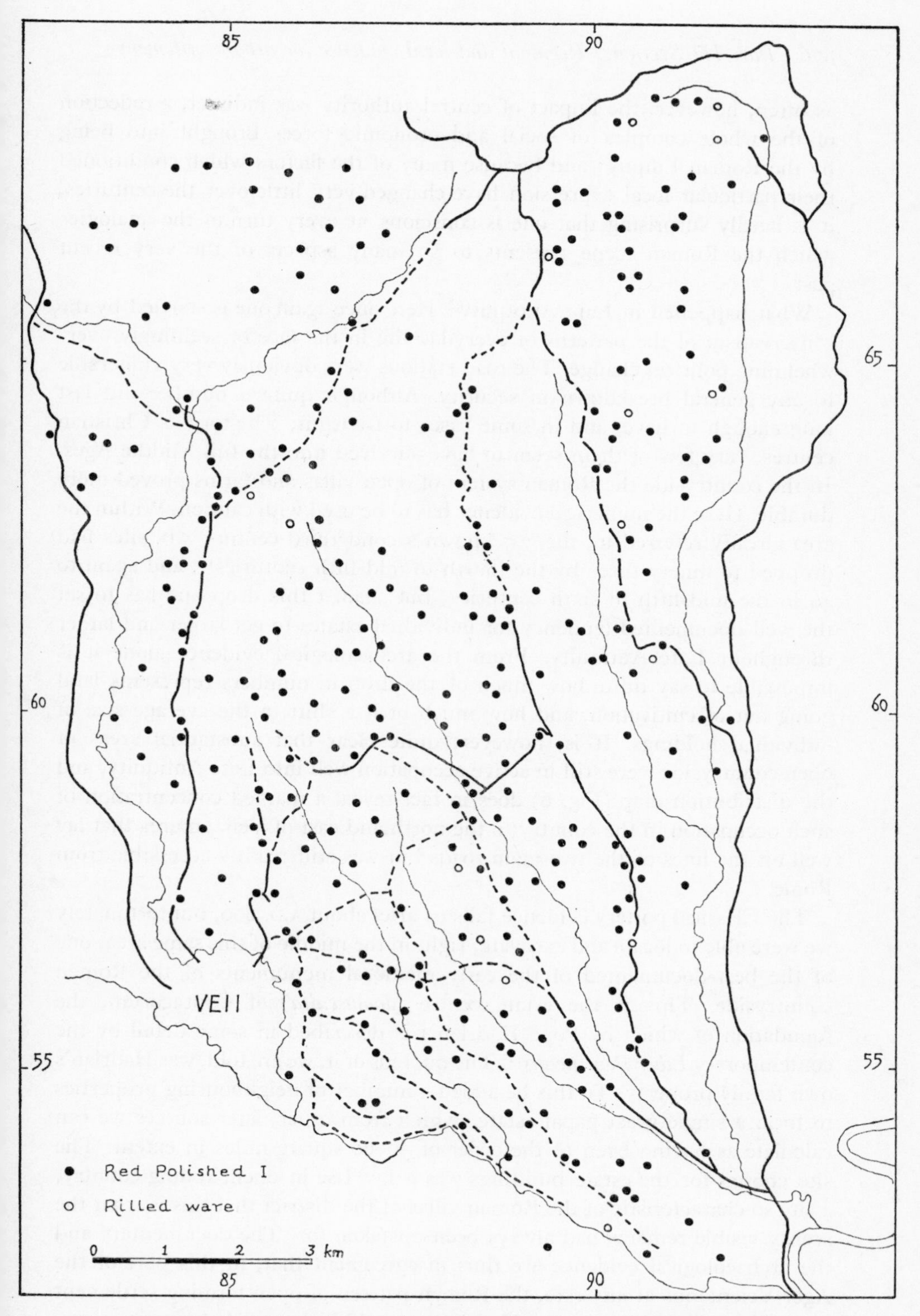

Fig 5 Distribution of settlement in the north-eastern Ager Veientanus, *c.* A.D. 200.

as often, however, the impact of central authority was indirect, a reflection of the whole complex of social and economic forces brought into being by the Roman Empire; and because many of the factors which conditioned their particular local expression have changed very little over the centuries, it is hardly surprising that one is conscious at every turn of the analogies which the Roman scene presents to so many aspects of the very recent past.

What happened in Late Antiquity? Here once again one is startled by the conservatism of the patterns of everyday life in the face of seemingly overwhelming political change. The road stations were obviously very vulnerable to any general breakdown in security. Although quite a number did last long enough to have, and in some cases to transmit, a history as Christian centres, very few of them seem to have survived into the full Middle Ages. In the countryside the Roman system of open villas and farms proved more durable. Here the numerical evidence has to be used with caution. Within the area already referred to, the 307 known second/third century A.D. sites had dropped to ninety-three by the fourth to mid-fifth centuries[16], and again to 46 in the mid-fifth to sixth centuries; but against this drop one has to set the well-documented tendency for individual estates to get larger and larger throughout Late Antiquity. From the archaeological evidence alone it is impossible to say quite how much of the drop in numbers represents land going out of cultivation, and how much of it a shift in the average size of individual holdings. It is, however, quite clear that substantial areas of open countryside were still in active occupation well into Late Antiquity, and the distribution map (Fig. 6) does in fact reveal a marked concentration of such occupation in the country to the north and east of Veii, an area that lay well off the lines of the two main roads but was still readily accessible from Rome.

The Classical pottery evidence fails us after about A.D. 600, but fortunately we were able to locate and excavate, right in the middle of this same area, one of the best-documented of the early medieval monuments of the Roman countryside. This is the estate centre (*domusculta*) of Capracorum, the foundation of which by Pope Hadrian I is described in some detail by the contemporary *Liber Pontificalis*[17]. The nucleus of it, we are told, was Hadrian's own family property. To this he added a number of neighbouring properties to form a single great papal estate, which from other, later sources we can calculate as having been of the order of 50–60 square miles in extent. The site chosen for the estate buildings was a low rise in open, rolling country, a site so characteristic of the Roman villas of the district that this is what the scanty visible remains had always been mistaken for. The documentary and the archaeological evidence are thus in agreement that, in this part of the Ager Veientanus at any rate, the Roman pattern of open farming settlement was still a going concern when Hadrian created his *domusculta*.

By the turn of the millennium there had been a drastic change in the

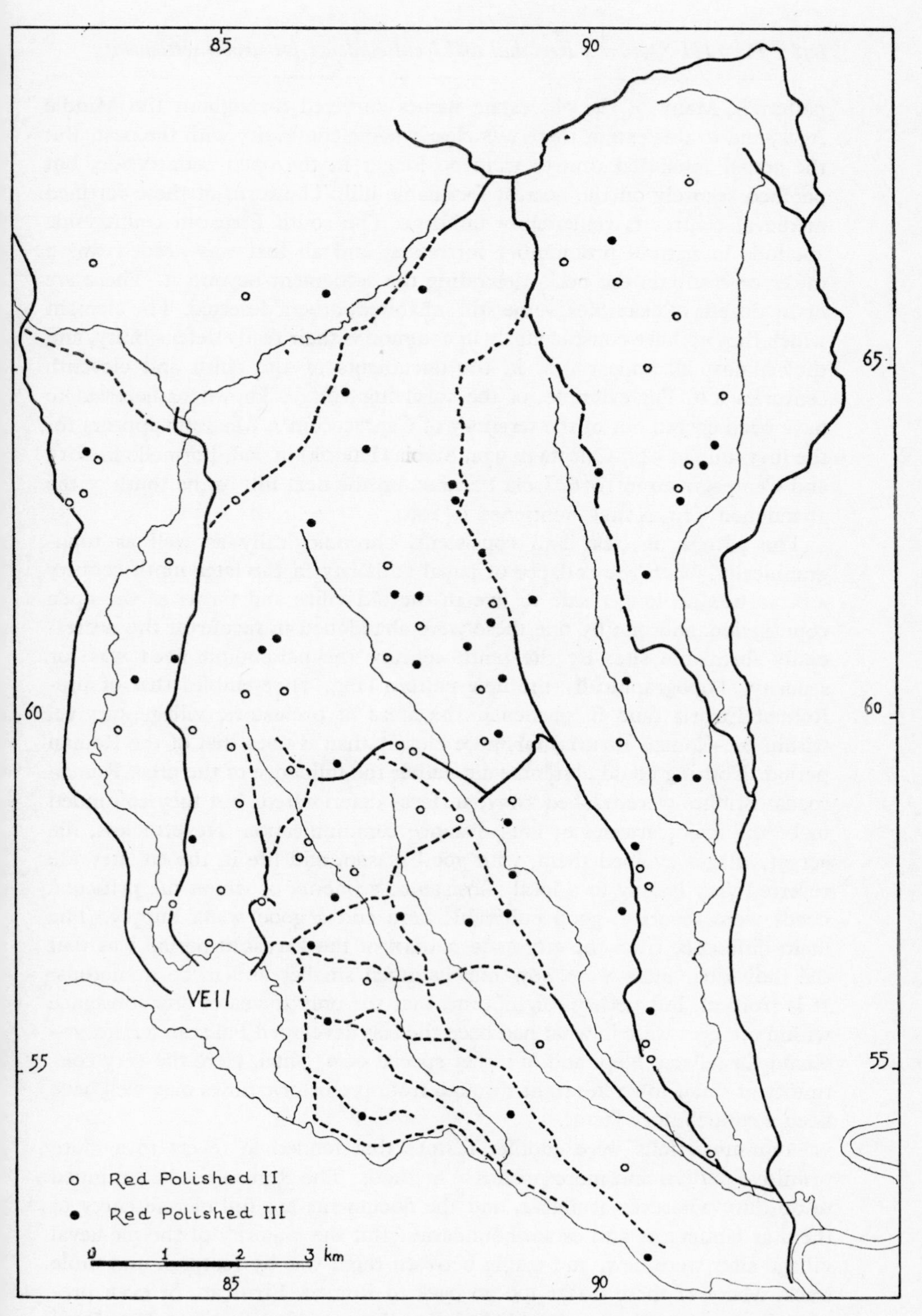

Fig 6 Distribution of Late Antique settlement in the north-eastern Ager Veientanus, *c.* A.D. 300–600.

pattern[18]. Many of the old estate names survived throughout the Middle Ages, and to this extent there was clearly some continuity with the past. But the actual inhabited centres were no longer in the open countryside, but perched securely on the nearest fortifiable hill. The form of these fortified medieval centres is remarkably uniform. The south Etruscan countryside abounds in natural promontory fortresses, and all that was needed was a tower or castle on the neck, defending the settlement beyond it. There are many dozens of examples, some still inhabited, others deserted. The element which they all have conspicuously in common is their ready defensibility, and they almost all appear first in the documents of the tenth and eleventh centuries A.D. For example, of the surviving villages known or believed to have been carved out of the territory of Capracorum[19], Mazzano appears for the first time in 945, Calcata in 974, Stabia (Faleria) in 998, Formello in 1013 and Campagnano in 1076. Isola Farnese, on the next hill to the south of the abandoned Veii, is first mentioned in 1003.

The pattern is clear and consistent, chronologically as well as topographically. With the collapse of papal authority in the later ninth century A.D. it was no longer safe to live in the old villas and farms of the open countryside, and one by one these were abandoned in favour of the nearest easily fortifiable site. By the tenth century the paramount need was for security. Topographically the new pattern (Fig. 7) resembles that of pre-Roman Etruria (and in particular the areas of prehistoric village survival within pre-Roman Etruria) far more closely than it does that of the Roman period. Nothing could eliminate altogether the influence of the great Roman roads. Bridges were washed away, surfaces deteriorated; but they continued to be the major arteries of long-distance communication. Nevertheless, the actual villages avoided them, with good reason, and life in the countryside reverted very largely to a local subsistence economy of which the principal needs were security, good cultivable land and a good water supply. The main difference from the city-state pattern of the Etruscan period was that the individual units were now mostly much smaller and more numerous. It is ironical, but perhaps significant, that the only towns of any substance within our area were in what had been the less developed Faliscan territory—Civita Castellana, Nepi and (a rather special case) Sutri. Here the very continuity of small-town life right through from prehistoric times may well have been a contributory factor.

Communications were another feature that tended to revert to a more primitive pattern and more primitive methods. The Roman roads continued to do duty wherever available, and the documents are full of references to them as landmarks and estate boundaries. But the majority of the medieval village sites were new, and traffic between them was by footpath and mule track. Many of these tracks too go back to Roman, Etruscan, or even prehistoric times; and once established they have tended to be as long-lived as the settlement pattern which they served. In the remoter countryside many

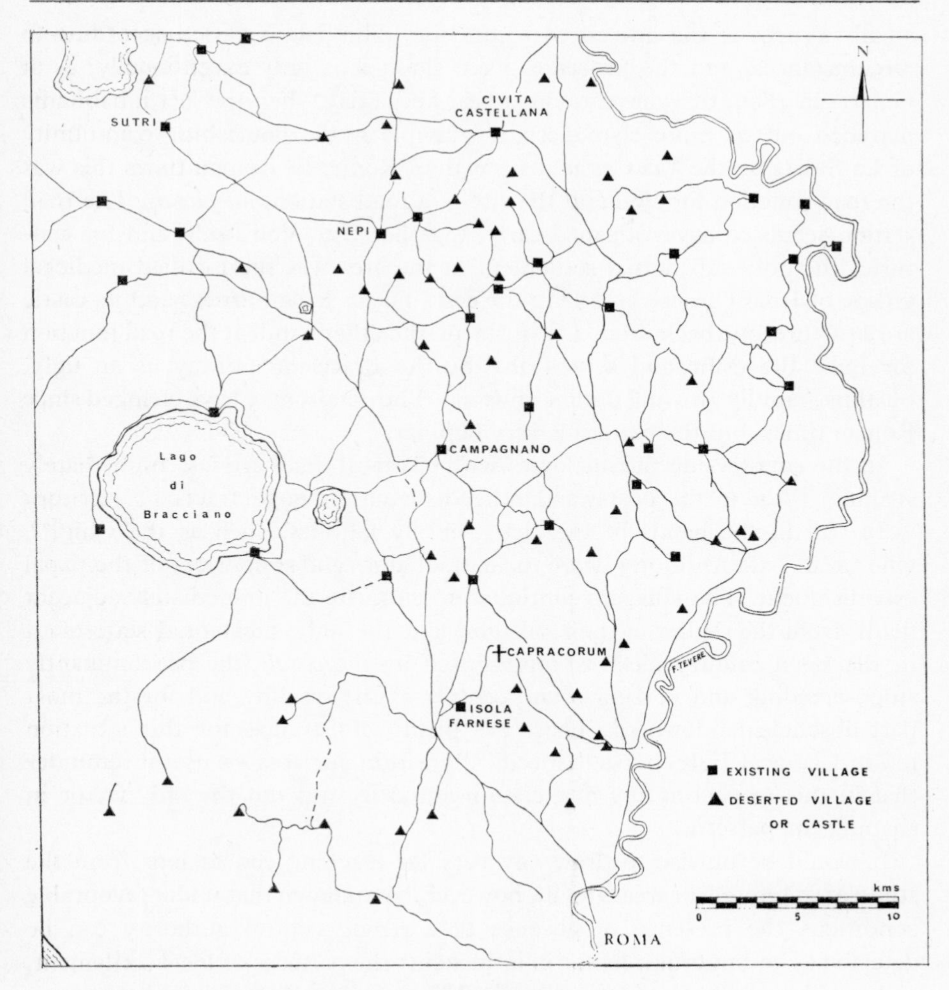

Fig 7 South-eastern Etruria in the Middle Ages.

of them still survive, and a great many more have been converted to wheeled traffic within living memory.

The final chapter of the story may be very briefly summarized. It was in the fifteenth century that the tide began to turn once more. Already in the previous century, in Italy as in many other parts of Europe, a large number of villages had been abandoned. Now, as a succession of strong Popes succeeding in re-establishing papal authority at home and in achieving a measure of security and orderly government within the papal territories, so inevitably the forces that had driven settlement away from the main roads and from the open countryside were relaxed. Many of the remoter villages were abandoned altogether; others survived as isolated monasteries or estate centres; others again began to develop towards the main roads or to establish small suburbs at the nearest road junction. The details varied according to circumstances, and the processes were slow; it is only exceptionally, as at Galeria in 1809, that one can cite a year and a day when the last inhabitants marched out. A more characteristic example is the flourishing community of La Storta on the Via Cassia just north of Rome. In Roman times this was the road junction for Veii and the site of a road station, *ad Nonas*. The road station seems to have vanished early, together with Veii itself, and for centuries the only substantial settlement in the area was the fortified medieval village of Isola Farnese, a mile to the east. Today Isola Farnese and its castle are a picturesque backwater. La Storta, on the other hand, at the road junction for Lake Bracciano and served also by the Bracciano railway, is an ugly, bustling, rapidly growing outlier of Rome. The details may have changed since Roman times, but the pattern is very familiar.

In the countryside the analogies with Classical times are less immediately striking. Prior to the post-war land redistribution huge tracts of territory were still in the hands of the rich princely families, such as the Chigi[20], who (as in Late Antiquity) were the natural allies and supporters of the papal establishment. The villagers continued to cultivate the immediately adjacent fields, from the shelter of their villages, and the only substantial settlement in the open countryside was represented by the *casali*, the predominantly stock-breeding and grazing estate centres of the wealthy and for the most part absentee landowners. There are plenty of parallels for this situation too in Classical Italy, but not specifically within our area—a useful reminder that in this respect at any rate central authority was not the only factor in shaping the pattern.

It would be unwise to draw any very far-reaching conclusions from the study of so limited an area. It will, however, have shown that under favourable conditions the presence or absence of a strong central authority can be detected as an important factor in influencing the patterns of local settlement, and it will have illustrated something of the ways in which it did so.

This influence might on occasion take the form of direct and deliberate intervention upon the shape of the pattern itself, as when (if our interpretation is correct) the Villanovan founders of Veii swept away the Bronze Age farmsteads of the Ager Veientanus to create the sort of city-state unit to which they were accustomed; or again in Roman and modern times when existing properties were broken up by the government of the day in order to settle small-holders on them. Pope Hadrian's creation of the *domusculta* of Capracorum, though undertaken within the framework of existing trends, would qualify as another instance of such direct central intervention. Such action is not always easy to detect archaeologically (the breaking-up of individual estates, for example) but it was unquestionably an important shaper of patterns in many parts of the Roman world, as it has since been once again in modern times.

Quite as often the influence was indirect, the result of policies undertaken with quite a different purpose. It is arguable that the most important single event in shaping the historical settlement patterns of central Italy was the laying out of the great military roads, the original builders of which cannot have had any conception of the forces that they were unleashing on the country-sides through which they happened to pass. Whether even at a later date the Romans ever built roads with this sort of object uppermost in mind may be questioned; but they can hardly have long remained unaware of the side-effects of their road-building programmes. Their indifference to the continued occupation of many of the old hill-top tribal centres of the newly-conquered provinces (Bibracte and Gergovia, for example, or the Magdalensberg in Noricum) may well have been coloured by their prior experience in Italy and by the knowledge that in a couple of generations economic opportunity and convenience would have done their work for them. The roads everywhere have been a powerful attractive force, and there can be little doubt that many of the roadside towns throughout the Empire were the product of circumstances comparable to those of Republican central Italy.

The Roman roads were not only a powerful instrument of social change: once established they were also remarkably durable. As we know very well in Britain, Roman roads could and did vanish from the map; but a great many more did manage to survive and, by the very fact of their survival, to add a new and enduring dimension to the topography of the areas through which they passed. London, like Rome itself, is a monument to this simple fact.

Finally, there is the indirect but all-pervading factor of public security. In a simple society, where this depends largely on the efforts of the individual community, the reactions to fresh circumstances are rapid. In a more complex society they depend ultimately on the effectiveness of central authority, and here the response to changing conditions could be surprisingly slow. To a contemporary observer of the dying Roman Empire it was the dramatic events of each succeeding year which most obviously affected the lives of himself and his friends and which are in consequence the stuff of nearly all contemporary history. And yet it was not the barbarian invasions, the movements of armies, or the clash of Empires which finally destroyed the Roman pattern of open farming in southern Etruria: it was the petty anarchy of the enfeebled post-Carolingian papacy. This was of course only the last of a long and complex chain of political situations and economic trends which over the years had been eroding the roots of the society that the open landed estates of Late Antiquity represented; and it is interesting to speculate which of these causes would have been picked up by either the documents or the archaeological record operating alone. Taken in conjunction, they make it very clear that in central Italy, both on this and other occasions, the authority of central government, though its mills ground slow and though its influence was quite as often indirect as direct, was a fundamental element in shaping the patterns of rural settlement.

Notes

1 For the principal published accounts of this survey, see: Frederiksen, M. W. and Ward-Perkins, J. B. (1957). The ancient road-systems of the central and northern Ager Faliscus, *Papers Brit. Sch. Rome*, **25**, pp. 68–208; Duncan, G. (1958). Sutri (Sutrium), *Papers Brit. Sch. Rome*, **26**, pp. 63–134; Ward-Perkins, J. B. (1961). Veii, *Papers Brit. Sch. Rome* **29**, pp. 1–123; Jones, G. D. B. (1962–3). Capena and the Ager Capenas, I and II, *Papers Brit. Sch. Rome*, **30**, pp. 116–207 and **31**, pp. 108–58; Kahane, A., Threipland, L. Murray and Ward-Perkins, J. B. (1968). The Ager Veientanus, north and east of Veii, *Papers Brit. Sch. Rome*, **36**, pp. 1–218. See also Ward-Perkins, J. B. (1957). Etruscan and Roman Roads in southern Etruria, *J. Roman Stud.*, **47**, pp. 139–43; Ward-Perkins, J. B. (1962). Etruscan towns, Roman roads and medieval villages, *Geogr. J.*, **218**, pp. 389–405; Ward-Perkins, J. B. (1964). *Landscape and History in Central Italy* (Second J. L. Myres Memorial Lecture). Oxford.

2 Mattias, P. P. and Ventriglia, U. (1970). La regione vulcanica dei Monti Sabatini e Cimini, *Memorie della Società Geologica Italiana*, **9**, pp. 331–84.

3 Hutchinson, G. E. (ed.) (1970). Ianula: an account of the history and development of the Lago di Monterosi, Latium, Italy, *Trans. Amer. Philosoph. Soc.*, **60**.

4 Kahane, A., Threipland, L. Murray and Ward-Perkins, J. B. (1968). *op. cit.* pp. 14–17; Cf. Rellini, U. (1920). Cavernette e ripari preistorici nell'Agro Falisco, *Monumenti Antichi*, **26**, cc. 5–180.

5 Kahane, A., Threipland, L. Murray and Ward-Perkins, J. B. (1968). *op. cit.*, pp. 18–67.

6 Kahane, A., Threipland, L. Murray and Ward-Perkins, J. B. (1968). *op. cit.*

7 Ward-Perkins, J. B. (1970). Città e Pagus, *in* Mansuelli, G. and Zangheri, R. (eds.) *Studi sulla Città antica*. Bologna. pp. 293–7.

8 VI.4.4, cf. 5.8. See also Taylor, L. R. (1960). *The Voting Districts of the Roman Republic*. Rome (American Academy). pp. 47–9.

9 Ward-Perkins, J. B. (1961). *op. cit.* pp. 52–7.

10 Kahane, A., Threipland, L. Murray and Ward-Perkins, J. B. (1968). *op. cit.* pp. 146–8.

11 Duncan, G. (1958). *op. cit.*

12 A few road stations, such as Forum Clodii on the Via Clodia west of Lake Bracciano, were artificial creations, but the majority were unquestionably spontaneous growths.

13 Kahane, A., Threipland, L. Murray and Ward-Perkins, J. B. (1968). *op. cit.* pp. 146–50.

14 Ward-Perkins, J. B. (1961). *op. cit.* pp. 57–9, 76–7.

15 *Ad Fam.* IX.17.2.

16 Kahane, A., Threipland, L. Murray and Ward-Perkins, J. B. (1968). *op. cit.* pp. 151–3.

17 Duchesne, L. (ed.) (1886-92). *Le Liber Pontificalis*. I. Paris. pp. 501–2, 506–7, cf. p. ccxxxiv. For the site: Kahane, A., Threipland, L. Murray and Ward-Perkins, J. B. (1968). *op. cit.* pp. 161–5. For the *domuscultae* in general: Partner, P. (1966). *Papers Brit. Sch. Rome*, **34**, pp. 67–78.

18 Kahane, A., Threipland, L. Murray and Ward-Perkins, J. B. (1968). *op. cit.* pp. 165–7.

19 Kahane, A., Threipland, L. Murray, and Ward-Perkins, J. B. (1968). *ibid.* Also Tomassetti, G. (1882). *Archivio della Società Romana di Storia Patria*, **5**, pp. 137–48; and Tomassetti, G. (1913). *La Campagna Romana*, **3**. Rome. pp. 109–12.

20 Who owned much of the northern Ager Veientanus, where they maintained two large country residences, the Villa Versaglia near Formello and Olgiata near Isola Farnese.

D: SUB-SAHARAN AFRICA

P. MORTON-WILLIAMS

Some factors in the location, growth and survival of towns in West Africa

Little can be said confidently about the circumstances governing the location and growth of the early towns in West Africa, but a fuller account of developments in recent centuries has been made possible by research in the last few years and there appear to be common factors and continuities to tempt speculation about the remote past. Social anthropology can draw attention to some of them and perhaps suggest lines of collaborative inquiry with archaeologists, historians and geographers. The anthropologists, who will be least sure of themselves in the question of what factors governed the location of different towns, can point up features of the social organization of urban communities that promote or limit growth and internal stability, and condition their relations with other communities.

It is interesting that for 1500 years or longer towns in West Africa have for the most part emerged in an economic environment of competitive trade[1]. Economic resources are patchily situated in this vast area and long distance trade, within the region and beyond (overland to the Mediterranean world, and overseas) is of great antiquity. Incidentally, trade provided the stimulus and opportunity for some of the earliest explorers, the Arab travellers and geographers, of the western Sudanic zone of West Africa. The traders distributed goods of high value: gold, copper, ivory, cornelian, spices, slaves, fine dress materials, and Saharan salt; and other commodities in which the bulk of the traffic rather than its materials made it valuable: kola, less costly local cotton materials, leather, iron tools and weapons, natron, dried fish (both fresh and salt water), and sea salt, to give only an incomplete list. Thus control of markets and organization for warfare are prominent features of the social systems of the towns. The centralized directing of those activities was vested in kingship characterized by attributes of fearsomeness and absolute ownership of territory and people, of sanctity and religious leadership, however constrained in reality by elaborate checks and balances the powers of a king might have been. The legitimacy of such ascribed powers seems directly related to the townspeople's love of orderly justice so often remarked upon by the early travellers[2].

Regular markets are held in the territories of many African peoples where there are no towns or even a nucleated settlement pattern, but in those circumstances problems of access and regulation are acute and the markets generally not very big. The market places are usually some distance away from dwelling places and often magical sanctions exist, which are important for the support of the local group attempting to keep the market peaceful. When all the men in a nucleated settlement are consanguines, the prospects for holding markets and for the development of specialized crafts and of commerce are poor. In such societies in which kinship determines lines of co-operation and hostility, and where there is absence of other forms of association it is difficult for outsiders to have access to the settlements. Hence, where markets do exist, they are again often small and may be held at crossroads near the edge of the territory of a particular local group[3].

The evidence that communities cannot always support a market in their midst, not because of economic but because of social conditions, helps to solve the problem of when a settlement can usefully be called a town. The contrast of urban and rural, town against village, is not simply one of size and nucleation of settlements. In West Africa social inventiveness created large settlements, of thousands of inhabitants, which strike the observer as villages rather than towns. (Umor, in south-east Nigeria, for instance, numbers more than 10,000 inhabitants[4].) Very big compact settlements, able to defend themselves effectively against powerful enemies and to survive for generations, were founded on the simple principles of descent, locality, and corporate association on the basis of cult affiliation supplemented sometimes by association into age sets. Kinship by unilineal descent was recognized for purposes of rights in land for farming or to build on; for claims to help in economic activities; for many kinds of jural rights and joint and individual obligations; for access to certain shrines; for qualification for membership in cult associations. Age grading sometimes qualified membership of other associations. Age and cult associations set up corporate ties, which transcended local grouping founded on bonds of kinship. Association and kinship thus offered complementary patterns and means of organization. The divisive effects of rival associations could be mitigated by the unifying effects of clan membership and clan ritual; while the rivalries of descent groups were limited before they resulted in the break-up of communities by the counter-claim of solidarity and privileges stemming from membership in cult and age associations. These were (and still are) adequate social devices to support large localized communities, but they keep only an internal equilibrium, without regulating external relationships. It is the lack of external relationships that makes us reluctant to call settlements of that sort towns rather than villages[5]. They were exclusive: strangers were enslaved or killed; admission rested upon kinship or slavery.

Within such villages there can be no big markets; most have none at all, since there is little specialization of occupation. The criterion of a truly urban society must, then, be accessibility, and in Africa the indication is commerce, made visible in markets. That implies for the archaeologist that evidence of a market place within a settlement is a better indication for an urban community, with the necessary centralized government, than the existence of cult centres which may indicate only competing cult associations within a closed community. If there is a market place, that is evidence for enough instituted authority to keep peace and to restrain predatory groups or bands from pillaging markets and roads.

Competition between neighbouring towns to attract merchant traffic and develop markets could lead to either of two results, and there are historical examples of both. Prolonged and indecisive turmoil caused long-distance traders to find alternative routes, so that the rival towns sank into political insignificance and reverted to agrarian economies. Or, through military prowess and skill in forming temporary alliances, one of the towns subdued others to the limits of its military power, thus founding one of the so-called empires of the western Sudan or (on the whole later, historically) of the forest lands. It is worth observing that the paramount town was not always best situated for the greatest market in its dependent territories and its rulers were then obliged to invent administrative procedures to exploit the market and keep it secure. Timbuktu (see Fig. 1), on a canoe-way to the Niger and on the edge of the Sahara, and the coastal market town of Ouidah are good examples of such market towns. Timbuktu[6] was for a time governed by a viceroy of the Songhai king, who reigned at Gao. It was safe while Songhai

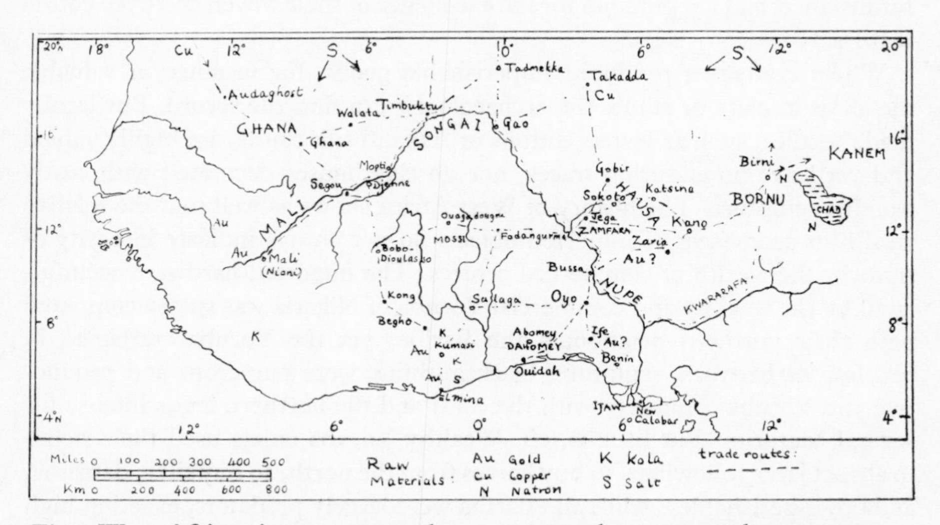

Fig 1 West Africa: important market towns, trade routes and resources.

kept firm control of traffic along the river in both directions and could intercept attacks from the south. Ouidah[7], formerly a port of the small kingdom of Allada, was seized by the inland kingdom of Dahomey and then showed analogies with the "desert-port" of Timbuktu. Alternative ports in both directions were suppressed and a Dahomey viceroy was installed. Much more rarely the king might have moved his capital to the market, as may have happened in ancient Ghana and when Gao became the Songhai capital. Once his hold on his new dependencies was firm, the sovereign of the new polity strengthened his economic position by suppressing, or reducing to merely local significance, all but a few select markets, and limiting the movements of traders to defined (but safe) routes.

That process of engrossing markets commonly resulted in a tendency for the population of most towns to remain constant or even dwindle, while the important *entrepôts*, especially if also capitals, grew big not simply through concentrations of resident traders and those providing them with domestic services, but also by becoming centres of crafts and industries. Throughout the region, and over a span of several centuries, the populations of the ordinary towns were estimated by travellers to lie between 1500 and 5000 inhabitants, with 2000 as a frequently given figure; while the commercial and political centres rarely were less than 10,000 strong and some may have exceeded 100,000.

The growth of local crafts sometimes brought a degree of prosperity to a wide area around the towns, the growing and sale of cotton or the mining and smelting of iron, for example, yielding incomes that enabled imported goods to be diffused from the markets. Some towns became celebrated as suppliers of certain manufactures of high quality, adding to the volume of trade to other places. Kano, Katsina, and the various successive Nupe capitals, for instance, had a reputation for the excellence of their woven and dyed cotton textiles.

Where craftsmen produced imperishable goods, for instance in valuable metal, or in glass or stone, the archaeologist can find the record. But locally made textiles, such as Hausa cottons or Ashanti silk cloths, are highly valued and yet leave no enduring traces, nor do clay houses decorated with costly wooden sculpture. The history of West Africa shows as well that the relative wealth or sparseness of imported objects do not always indicate intensity of trade or the wealth of commercial centres. The imported hardware accumulated by the wealthy and commercial Yoruba of Nigeria was sparse compared with their northern neighbours the Bariba; yet the Yoruba craftsmen in textiles, leatherwork, sculpture, blacksmithing, were numerous and productive and Yoruba commerce with the coast and the northern lands intense for several centuries. But little is left. Wealthy Yoruba chiefs used their riches to attract large followings, to buy horses from the north, to buy silks, damasks, and other perishables; while the Bariba were largely predators, exacting high tribute from the caravans carrying goods, including imported pottery, beads, and other enduring materials. Excavation, presumably, would show the much greater size of Yoruba towns and the absence of big market places in the Bariba settlements.

It has often been remarked that it was control over trade routes and markets rather than location among sources of raw materials that stimulated the growth of towns in West Africa. Many ancient cities sprang up along the great rivers, especially along the upper and middle Niger. These were points of trans-shipment, from camel to canoe, for distribution along the river systems, and for trans-shipment from canoe or camel to the beasts of burden of the savanna lands south of the Senegal-Niger-upper Benue systems. Gao, for long the capital of the Songhai *imperium*, was at a meeting of many routes, riverain, trans-Saharan, from the goldfields in the south-west, and on the route leading pilgrims through Bornu to Mecca and bearing natron from Bornu.

Further south, between latitudes 10°N and 12°N, lay another chain of towns along trade routes for gold and kola. These routes kept well to the north of the forest, going through the parkland savanna, near the southern limits of tsetse-free lands. Goods passing to or from the markets in these towns from places further south were generally headloaded. There were towns further south, on the edges of the forests, or within them. In the western block of forest the towns of the Akan peoples were founded, culminating in the seventeenth century in the Ashanti empire, centred on Kumasi. Even here the towns, though near the goldfields, owed their prominence as much to their control of the traffic in gold as to its production. Their inhabitants were, however, producers of the important export crop kola.

Another group of forest towns were those of the southern Yoruba kingdoms and Benin. Nothing is known with certainty of the circumstances of their foundation. The principal Yoruba forest town, Ife, which is at least 1000 years old, is on a small goldfield, but no proof that it was exploited in pre-colonial times has come to light. It was an important craft centre, producing glass, brass sculpture (but perhaps not for export) using imported ore, and ironwork. As iron implements figure in the cults of many of the old gods of Ife, and there are large old slag heaps near the town, it is tempting to surmise that its ascendency began as an important iron producing centre just within the forest, and other mineral resources were exploited in smelting and as fluxes.

Ports grew up on the coast following the establishment of the Atlantic trade after the Portuguese voyages of the late fifteenth century. They quickly became spoils in the competition to control trade, and most were taken over, destroyed, or engrossed by the kingdoms of the hinterland. On the coast of southern Nigeria, from the Niger Delta eastwards, trading states developed late with the growth in the nineteenth century of the export trade in palm oil, and recent studies have indicated how the transition from small fishing village to populous city-state was achieved. The older institutions of clusters of men of common descent bound together in their lineage ancestor cults, and of a few cult and recreational associations joining them in a common village culture, gave way to powerful "canoe-houses" (manning one or more big trading and fighting canoes) headed by powerful traders. The house heads joined in a common assembly and there followed a division into sets of big chiefs who had influence over related houses and of small chiefs whose influence was within their own houses only; and eventually one or more "big chiefs" came to dominate the state. Meanwhile the old associations proved themselves useful for the assimilation of slaves into the community[9].

Two further topics require brief mention in this context. The first concerns the stability of the growing towns and states, especially in the Sudan savanna region. Here, cavalry were important; and for a kingdom to maintain an adequate force wealth had to be dispersed among noblemen to obtain and equip horses and followings; but lengthening lines of controlled trade routes and the growth of a few important markets required effective administration. Various experiments were made to solve the problem of reconciling the needs of centralized administration and the dispersal of power. For instance, the Muslim Fulani, conquerers of the Hausa states of northern Nigeria in the early nineteenth century, made the loyalty of the new emirs of the new states to the Khalif of Sokoto a matter of their loyalty to Islam, so that the secession of one emir would call for the vengeance of all the others; and within the emirates appointed leaders to the tenures of fiefs which were tied to particular offices of state, so that when an incumbent was promoted to higher office he moved to a different fief. The northern Yoruba kingdom of Oyo offers another example; outlying territories were supervised by the king's slave officials (as in Bornu, too) while the great nobles in the capitals had limited, nearby, territorial jurisdiction over vassal chiefs and were rewarded mainly with booty which they had to disperse to maintain themselves[10].

The other topic is the large question of the role of Islam. The trans-Saharan traffic and the earliest towns of western Africa both antedate the first journeyings of Muslim Arabs into the region in the middle of the seventh century. Nevertheless in the following centuries most of the long-distance traders and merchants, notably the indigenous Malinke and Hausa, as well as Arabs and Berbers, were Muslims. Islam had obvious advantages for them, giving not only common cultural interests but a mutual code of conduct governing commercial transactions, the guarding of the property of deceased merchants, and hospitality. Trading communities were permitted by local rulers to establish their own communities, sometimes as separate townships, sometimes as sections of their towns, where they observed their own customs and laws, generally under the jurisdiction of the imam of the principal mosque. But Islam is a proselytizing faith and several important kings were converted to the Faith, and established it as the religion of the court and capital. At times, especially when a large populace in several neighbouring towns turned to Islam, the Faith served as a unifying force, at least for long periods. But kings were threatened by three dangers: first, in the early years after conversion there was a problem of finding new institutions to replace cult associations and the special ritual role of various noblemen in the different associations, in order to cement political loyalties; second, the surrounding rural areas commonly clung to the cults of the old gods and too rigorous an Islamic régime in the capital divided the kingdom; and third, the compromises made by Muslim rulers of such kingdoms provoked charges of apostasy from zealous Muslims, often members of the trading community, who led the reformist movements and holy wars that have punctuated the history of West Africa and brought about the downfall of many old royal dynasties.

In conclusion it is important to note that West African kings, like those of the ancient Near East and Egypt, had many attributes of divinity. They were the central figures in religion and all other expressions of power. Creation myths place the king in relation to gods and mankind, to powers of sky and earth and the world of civilized activity. They furnish cosmologies in which the organization of each kingdom epitomizes the whole cosmos. But the problem of how closely the cosmic symbolism reflects social reality is a difficult one. In the ancient world, we are told, the town was a microcosmos, its plan a deliberate allegory of the cosmos. In Yoruba towns in West Africa, an allegory can be discerned between the myths of a separate creation of kingship and ordinary, non-royal mankind, and the division of the town into royal and non-royal sectors; but there is not a thoroughgoing conscious creation of a microcosmos. Yet in Yoruba towns, and in the towns of most non-Islamic kingdoms, there is not so much an allegorical but a symbolic relationship between social organization and cosmic myths. Cosmology always symbolizes social organization fairly closely. Changes in cosmology, where it is founded on mythology rather than scientific observation, eventually follow changes in social structure; but changes in social structure, in the distribution of power between king and military chiefs or religious associations, can produce great change in the organization of religious activities and ceremonial without destroying the accepted picture of the cosmic order, of the organization of the different supra-human powers. Such changes can be documented for such African kingdoms as Benin, Dahomey, and Ashanti[11]. Ashanti is particularly interesting; in the nineteenth century the state ritual centred on the cult of the royal matrilineal ancestors who transmitted the powers of kingship. The cults of ancestors of the various Ashanti lineages were given their place in the public religion and they functioned to support the legal rights and military roles of the living heads of lineages, who were elders in the king's council. Cult associations of the various Ashanti gods had a minor place in state ritual and very little power. The Earth was honoured as a goddess, but had no cult association. Yet it is evident that two centuries earlier an earth cult and cult associations of the gods had been prominent. In the interval, the powers of the king had waxed, a more centralized polity had been evolved, and while the various orders of divinity in the cosmos continued to be represented in belief and myth, the pattern of ritual reflected a very different political reality.

Finally, it is plain that the needs of trade and the growth of important markets have a strong claim to being the main causal factors in the situation and growth of towns in West Africa, but at the same time economic rivalries, the demands of the slave trade, and even the spread by traders of Islam, have all contributed to the overthrow of kingdoms and the destruction of towns.

Notes

1 Useful general accounts are given by Bovill, E. W. (1958; 2nd ed. 1968). *The Golden Trade of the Moors.* London; Fage, J. D. (1969). *A History of West Africa.* London; and Mauny, R. (1961). *Tableau géographique de l'ouest africain au moyen âge.* Mém. IFAN, No. 61. Dakar.

2 Forde, C. D. and Kaberry, P. M. (1967). *West African Kingdoms in the Nineteenth Century.* London, present short descriptions of a representative variety of kingdoms.

3 See, for instance, Tait, D. (1961). *The Konkomba of Northern Ghana. Edited . . . by Jack Goody,* London; and Bohannan, L. (1958). Political aspects of Tiv social organization, *in* Middleton, J. and Tait, D. (eds.) *Tribes without Rulers.* London.

4 Forde, C. D. (1964). *Yakö studies.* London, gives a vivid account of the social structure and institutions of Umor.

5 Harris, R. (1965). *The Political Organization of Mbembe, Nigeria.* London, outlines the attempts of a number of such villages to adapt to the opportunities for trading provided by nearby trade routes.

6 See the discussions by Bovill, E. W. (1958). *op. cit.*; Fage, J. D. (1969). *op. cit.*; and Mauny, R. (1961). *op. cit.*

7 See Arnold, R. (1957). A port of trade: Whydah on the Guinea Coast, *in* Polanyi, K., Arensberg, C. M., and Pearson, H. W. *Trade and Markets in the Early Empires.* Illinois.

8 The general problems of the relationships of the differences in organization of the Ashanti capital Kumasi and the northern Yoruba metropolis Oyo are discussed by Morton-Williams, P. (1969). The influence of habitat and trade on the politics of Oyo and Ashanti, *in* Douglas, M. and Kaberry, P. M. (eds.) *Man in Africa.* London.

9 See Horton, R. (1969). From fishing village to city-state, in Douglas, M. and Kaberry, P. M. (eds.) *op. cit.*

10 See Last, D. M. (1967). *The Sokoto Caliphate.* London; and Smith, M. G. (1960). *Government in Zazzau.* London.

11 See, for instance, Bradbury, R. E. (1967). The kingdom of Benin, *in* Forde, C. D. and Kaberry, P. M. (eds.) *op. cit.*; Wilks, I. G. (1967). Ashanti government, *in* Forde, C. D. and Kaberry, P. M. (eds.) *op. cit.*; on Ashanti, see also Rattray, R. S. (1923). *Ashanti.* Oxford.

W. G. L. RANDLES

Pre-colonial urbanization in Africa south of the Equator

Can the capitals of the African states in the southern half of the continent, as described by the first European travellers, really be called "towns"? It is not proposed to review here in detail the justness of the criteria adopted by Childe to define his concept of the "urban revolution", although this undoubtedly requires a fresh appraisal[1]. We shall here limit ourselves to an examination of three pre-colonial capitals in the light of the following, admittedly limited and probably very insufficient criteria: (a) the size of the agglomeration; (b) the architectural features of the buildings and the layout of the capital; (c) the manner in which the inhabitants of the agglomeration obtained their food; (d) the evidence for craft specialization associated with residence in the agglomeration.

The first example examined is São Salvador (as it was later called), the capital of the Old Kingdom of the Congo, situated in northern Angola near the Congo Republic border, and first visited by the Portuguese in 1491. Attention will be concentrated on the specifically African features of the town, leaving aside all reference to the stone churches built by the missionaries and other Portuguese residents. Next, Musumba, capital of the Lunda Empire, in the western part of the present Katanga province of the Congo Republic, first described by a European in 1847, lastly, Zimbabwe, capital of an unnamed medieval state in present-day Rhodesia, whose stone ruins still exist.

Size of the agglomeration

Five sources give estimates of the population of São Salvador at different times in the sixteenth and seventeenth centuries:

Raeffaello Maffei da Volterra (1506)	100,000[2]
Filippo Pigafetta (1591)	100,000[3]
Anonymous source (1595)	10,000 families (=40,000 inhabitants ?)[4]
O. Dapper (1668)	40,000[5]
Gio. Ant. Cavazzi da Montecuccolo (1654–77)	60,000[6]

The first two estimates are probably exaggerated. It is known that the population of the Congo declined heavily in the middle of the seventeenth century (a source of 1656 says by half)[7], as a result of the slave trade and epidemics, probably smallpox. We can therefore hazard a fairly safe figure of at least 50,000 for the beginning of the sixteenth century.

The unreliability of rough estimates made by visiting travellers is brought out by Gutkind in his recent study of the royal capital of Buganda at the end of the nineteenth century. The number of its inhabitants was estimated by various observers at 77,000 in 1900, 60,000 in 1906, and 77,000 in 1910. The censuses of 1911 and 1948 however only gave figures of 32,441 and 34,337 respectively[8].

The prudent reserve which the above comparisons inspire may similarly be applied to the estimates which were made of the population of Musumba, the capital of the Lunda Empire:

Ladislaus Magyar (1855)	50,000[9]
Paul Pogge (1875)	8000–10,000[10]
Max Buchner (1882)	2000[11]
Lt Michaux (1897)	30,000[12]

The Empire had already started to decline with the end of the slave trade long before Pogge's visit and had completely collapsed with the Cokwe invasion of 1885[13]. Lt Michaux's figures seems therefore exaggerated. A figure of 20,000 does not seem unreasonable for the high period of the Lunda Empire in the first half of the nineteenth century.

No estimates have reached us for Zimbabwe, which was found abandoned by the first European visitors. However the archaeologist, R. Summers, who has spent over thirty years studying the archaeology of Rhodesia suggests with careful prudence that at the very most Zimbabwe might have had a population of about 10,000 inhabitants[14].

Architectural features and layout of the capitals

Most African capitals may be regarded as giant villages consisting of a collection of traditional African huts with walls of upright poles interlaced with withies and plastered over with *daga* (*adobe*) and having roofs thatched with grass or covered with palm leaves. Within the capital, the monarch inhabited what may be called a *compound*, usually surrounded by a substantial stockade of thick wooden stakes, though sometimes of plaited grass; in the case of Zimbabwe by a colossal thick massive stone wall. All observers use the word "labyrinth" to describe the tortuously arranged paths or corridors within the stockaded area leading to the royal presence[15].

At São Salvador the royal compound (*lumbu* in Kikongo) was inhabited by the monarch, his wives and private domestic servants[16]. At Musumba, the king's wives, slaves and certain members of the administrative staff lived within the compound (*kipanga* in Ruund), while the queen had a separate compound to herself[17]. The ruins of Zimbabwe are divided into two parts, the less massive and older Acropolis, and the more recent Great Enclosure presumably a royal residence, consisting of a massive stone wall elliptical in form, having three entrances, 800 ft long, 32 ft high at its maximum and a thickness varying from 17 ft 3 in to 6 ft 3 in at ground level, the whole having a volume of 182,000 cu ft[18]. The Great Enclosure has been dated to the fifteenth century[19].

The first Christian king of São Salvador, Dom Afonso (1506–43), probably under Portuguese influence, had a stone wall built round his compound[20]. It had, according to a seventeenth century source, a circumference of 1½ miles (Italian)[21]. A sixteenth century source declares that the entire "town", situated on a hill, had a circumference of 10 Italian miles[22].

The king's compound in Musumba was, according to Graça (1847), surrounded by a fence of thick poles[23], 10 ft high according to Pogge[24]. Lt Michaux speaks of an outer ditch 10 yds wide and 8 yds deep, the only access to the inner area being by means of a tree trunk thrown across the ditch[25]. Musumba, like certain other African capitals, was moved to a fresh site at the commencement of the reign of each new king and the three observers just mentioned describe different royal compounds. Pogge states that the Musumba he visited was half a German mile long[26].

The ruler's palace itself is never as a rule described as being other than an African hut slightly larger than normal. In 1648 the king of the Congo at São Salvador had, as a result of Portuguese influence, a double storeyed house of wood, his being the only such in the kingdom[27], but the king reigning in 1845 lived in a hut like the rest of his subjects[28].

Is it possible to speak of regular streets in these African capitals and of a more or less complete separation between the residential area and the garden plots of the inhabitants?

Instead of individual huts inhabited by nuclear families, each separated by streets, we find at São Salvador a series of compounds each surrounded by its own fence and containing several huts in which lived the members of an extended family. In seventeenth century São Salvador, according to a source of 1648, the houses were not aligned along streets but were arranged in compounds contiguous one with another[29]. Another source of 1668 states that the homes of the ordinary people were "built urbanwise in orderly fashion"[30]. Yet this same source speaks of the existence of cultivated gardens and enclosures to house domestic animals, within or beside the compounds and even within the royal compound[31]. In this respect, São Salvador does not seem to have differed greatly from the average medieval European town, where the same intermingling of urbanization and agriculture can be found[32].

Musumba, according to Graça, had "very long wide neat streets, which were swept every day. Each inhabitant had to sweep the part in front of his residence; defaulters were liable to a fine of a goat or an ivory tusk and each street had an inspector to watch over its cleanliness"[33].

Both São Salvador and Musumba had a large central open space situated outside, but next to, the royal compound. In this space the people gathered to hear the king speak and it was here that the king sat and gave judgement on cases brought before him[34]. But at Musumba, the space also served, according to Graça, as a market place where a daily market was held from ten in the morning to two in the afternoon[35]. There is no clear evidence of the layout of the huts and spaces in and round Zimbabwe, although Carl Mauch, the first European visitor to describe the site, suggested that the area between the Acropolis and the Great Enclosure (separated by a distance of about 600 yds) was once inhabited, thus implying a considerably larger total built-up area[36].

Manner in which the inhabitants were supplied with food

At São Salvador and at Zimbabwe there is no evidence that the inhabitants relied for their food supplies on anything other than their gardens situated in immediate proximity to the town area. São Salvador was situated on a hilltop; a source of 1607 states that the valley beneath was "well-cultivated"[37]. There is evidence of tribute from vassals, in the form of food, being delivered to the monarch in the case of São Salvador and of Musumba, but it appears to have been sent only once a year or at irregular intervals and there is nothing to prove that the ruler and his staff could have relied on it to feed themselves throughout the year[38]. The only direct evidence we have of food coming from a distance to the capital, but which, however, proves little, is that of two Portuguese mulatto traders known as the *Pombeiros*, who crossed the Lunda Empire in 1802–11 and who met at a distance of six hours march from Musumba a number of women carrying thither loads of manioc[39].

Association of the capital with craft specialization

The evidence on this subject is very meagre. Oral tradition relating to São Salvador, collected in 1930, declares that a certain quarter of the town was called *Mbangala Nzundu Tadi*, "the iron hammer", a probable allusion to *Ngangula ne Kongo* the blacksmith king of ancient fame[40], whose existence is referred to by the missionary Cavazzi (1654–77)[41], but there is no other clear evidence of crafts being practised at the capital. At Musumba, Carvalho (1890) tells of the existence of blacksmiths, weavers of cotton cloth, and artisans who carved gun stocks and who could even make gun locks[42]. There is no solid evidence to prove the existence of artisans at Zimbabwe, though there is every likelihood that they were to be found there. Of peculiar and negative significance is the evidence of the traveller Bent (1893), who found in the neighbourhood of Zimbabwe, then abandoned, a complete division of labour unaccompanied by urbanization. Whole villages devoted all their time and energies to iron working, "tilling no land and keeping no cattle, but exchanging their iron-headed assagais, barbed arrow heads and field tools for grain and such domestic commodities as they may require"[43].

No thorough investigation has so far been made of the ecological and strategic factors conditioning the siting of São Salvador. São Salvador and Zimbabwe may have been situated on hill-tops for defence reasons. A source of 1591 reports the occurrence of iron ore at São Salvador[44]. The site of Zimbabwe is known to be a favoured one in terms of rainfall and fertility[45] and further it was precisely situated to control the medieval gold trade between the Arab port of Sofala and the gold deposits round the present-day town of Bulawayo.

Conclusion

Can we now in conclusion justly describe these three basically similar agglomerations as "towns" ? Of Zimbabwe, Clark says that it had not reached the stage of "full urbanization" but was only "a first step toward it"[46]. Summers declares that it was "virtually a town"[47]. We would hesitantly suggest that all three may be called "towns", but more because they were centres of political control that attracted settlement through the advantages offered of personal security and hopes for political advancement, rather than as economic centres of specialization and of division of labour. Cavazzi states that the population of São Salvador only attained 60,000 in "time of peace"[48].

Notes

1 Childe, V. G. (1950). The urban revolution, *Town Planning Review*, Liverpool, **21**, pp. 3–17; Childe, V. G. (1957). Civilization, cities, towns, *Antiquity*, **31**, pp. 36–8.
2 Maffei, R. (da Volterra). (1506). *Commentariorum Urbanorum*. Rome. f° 138 (v°).
3 Pigafetta, F. (1591). *Relatione del Reame di Congo et delle circonvicine contrade* . . . Rome. facsimile ed. by Rosa Capeans (1951). Lisbon. Lib. II, Cap. I.
4 Anonymous text dated 1595 *in* Brásio, A. (1953). *Monumenta Missionária Africana*. Lisbon. Vol. III, pp. 500–4.
5 Dapper, O. (1668). *Naukeurige Beschrijvinge der Afrikaensche gewesten van*

Egypten, Barbaryen, Lybien . . . Guinea, Ethiopien Abyssinien. Amsterdam. p. 575.
6 Cavazzi, G. A. (da Montecuccolo). (1687). *Istorica Descrizione de' tre Regni Congo, Matamba, et Angola*. . . . 2nd ed. Milan (1690). Lib. I, para. 154 (1st ed. Bologna 1687). Cavazzi sojourned in Africa for two periods from 1654 to 1677.
7 Anonymous (1656). *Osservationi del Regno di Congo dell'anno* 1656. MS Bibliothèque Nationale, Paris. MS espagnol 324 (38), f° 150 (v°).
8 Gutkind, P. C. W. (1963). *The Royal Capital of Buganda*. The Hague. p. 15.
9 Magyar, L. (1858). Ladislaus Magyar's Erforschung von Inner Afrika, *Petermann's Mittheilungen* (1860). p. 232. The text is a letter from Magyar dated 1858 and describing his travels of 1850, 1851 and 1855.
10 Pogge, P. (1880). *Im Reich des Muata Jamwo*. Berlin. p. 231. Pogge visited Musumba in 1875.
11 Buchner, M. (1882). Vortrag uber seine Reise in's Lunda Reich, *Verhandl. der Gesellschaft f. Erdkunde zu Berlin*. p. 80.
12 Michaux, Lt. (1897). *in Mouvement Géographique*. Bruxelles. col. 470. Lt Michaux situated Musumba in Lat. 8° and Long. 23° 30′.
13 Carvalho, H. D. de (1890). *Expediçao Portuguesa ao Muataiânvua* (*Ethnographia e História Tradicional*). Lisbon. p. 553 and p. 652.
14 Letter to the writer of 8th July 1969.
15 The term "labyrinth" is used of the king's compound at São Salvador by an anonymous writer of the end of the sixteenth century, *in* Cuvelier, J. and Jadin, L. (1954). *L'Ancien Congo d'après les archives romaines*. Mémoire, Inst. Royal Colonial Belge, vol. 36, fas. 2. Bruxelles. p. 120; and by Bentley, W. H. (1900). *Pioneering on the Congo*. London. Vol. I, p. 123. It is used of Musumba by Pogge, P. (1880). *op. cit.* p. 229, and of Zimbabwe by Mauch, C. (1871). *The Journals of Carl Mauch* (1869–72). trans. by F. O. Bernhard, and ed. by E. E. Burke, Salisbury, National Archives of Rhodesia. (1969). p. 269.
16 Rome, J. F. de (1648). Brève Relation de la Fondation de la Mission des Frères Mineurs . . . au Royaume du Congo, *in* Bontinck, F. (1964). *La Fondation de la Mission des Capucins au Royaume du Congo* (1648). Publications de l'Université Lovanium de Leopoldville (=Kinshasa), Louvain/Paris. p. 115.
17 Graça, J. R. (1847). Expedicão ao Muataiânvua, *Boletim da Sociedade de Geografia de Lisboa*, 9ª Serie, No. 1 (1890). p. 454; and Pogge, P. (1880). *op. cit.* p. 230.
18 Summers, R., Robinson, K. R. and Whitty, A. (1962). *Zimbabwe Excavations* 1958. Occ. Papers of the Nat. Museums of southern Rhodesia, No. 23A, Salisbury. p. 319.
19 Garlake, P. S. (1968). Rhodesian Iron Age ceramics, *J. Afr. Hist.*, **9**, pp. 27–8.
20 Letter of the King of Portugal to Dom Afonso (1529), *in* Brásio, A. (1952). *Monumenta Missionária Africana*. Lisbon. vol. I, pp. 529–30.
21 Rome, J. F. de (1648). *op. cit.* p. 115.
22 Pigafetta, F. (1591). *op. cit.* Lib. II, Cap. I.
23 Graça, J. R. (1847). *op. cit.* p. 452.
24 Pogge, P. (1880). *op. cit.* p. 229.
25 Michaux, Lt. (1897). *ibid.*
26 Pogge, P. (1880). *op. cit.* p. 229.
27 Rome, J. F. de (1648). *op. cit.* p. 115.
28 Castro, A. J. (1845). Roteiro da Viagem ao Reino do Congo . . ., *Boletim da Sociedade de Geografia de Lisboa*, 2ª Serie (1880). pp. 53–67.
29 Rome, J. F. de (1648). *op. cit.* p. 113.
30 Dapper, O. (1668). *op. cit.* p. 562.

31 Dapper, O. (1668). *op. cit.* pp. 562–3. This is already corroborated by Pigafetta, F. (1591). *op. cit.* Lib. II, Cap. I.
32 Romano, R. and Le Goff, J. Paysages et Peuplement rural en Europe après le XIe siècle, *Etudes Rurales*, **17**, p. 10.
33 Graça, J. R. (1847). *op. cit.* p. 454.
34 For São Salvador: Cuvelier, J. (1930). Traditions Congolaises, *Congo*, **2**, p. 487. For Musumba: Graça, J. R. (1847). *op. cit.* p. 455.
35 Graça, J. R. (1847). *op. cit.* p. 454.
36 Mauch, C. (1871). *op. cit.* p. 147.
37 Brásio, A. (1955). *Monumenta Missionária African*, V. p. 385.
38 For São Salvador: Dapper, O. (1668). *op. cit.* p. 575; for Musumba: Pogge, P. (1880). *op. cit.* p. 226, and Graça, J. R. (1847). *op. cit.* p. 445.
39 Anonymous (1843). Explorações dos Portugueses no interior da Africa meridional, *Annaes Marítimos e Coloniaes*. Lisbon. p. 279.
40 Cuvelier, J. (1930). *ibid.*
41 Cavazzi, G. A. (da Montecuccolo). (1687). *op. cit.* Lib. I, para. 338.
42 Carvalho, H. D. de (1890). *op. cit.* p. 281 and p. 309.
43 Bent, Th. (1893). *The Ruined Cities of Mashonaland*. London. p. 145.
44 Pigafetta, F. (1591). *op. cit.* Lib. II, Cap. I.
45 Summers, R. (1963). Was Zimbabwe civilized ?, *in Conference of the History of the Central African Peoples*. Lusaka. (roneotyped).
46 Clark, J. D. (1962). African south of the Sahara, *in* Braidwood, R. J. and Willey, G. R. *Courses toward Urban Life*. Edinburgh. p. 29.
47 Summers, R. (1963). *ibid.*
48 Cavazzi, G. A. (da Montecuccolo). (1687). *op. cit.* Lib. I, para. 154.

ANTHONY WHITTY

Zimbabwe and Inyanga

The area under discussion, now known as Rhodesia, is about 150,000 sq. miles in extent and lies on the watershed between the Zambezi River to the north and the Limpopo River to the south. Westwards is the Kalahari Desert, while on the east mountainous country along the border with Mozambique drops away towards the coastal swamp lands lying along the the Indian Ocean. These natural barriers, and their great distance from centres of civilization, isolated this area from major outside influence until the nineteenth century A.D.

Bantu-speaking tribes were moving into the country from the north about 2000 years ago, gradually driving out the Late Stone Age inhabitants, and bringing with them a degenerated Iron Age culture based on a subsistence food economy. When Europeans, the British South Africa Company, occupied the country in 1890, they found the indigenous population organized on a tribal basis, the unit of settlement being the village. This varied greatly in size, from fifty to 100 inhabitants up to 1000 or more at the great kraal of a chief. The system of land-usage was that all workable land within easy distance of the village was progressively used until exhausted (perhaps 15–20 years), after which the whole community abandoned the site and set up again on new land. This way of life offered little incentive for the construction of permanent buildings and in fact villages consisted of lightly fabricated pole-and-mud huts placed in groups and often given some slight degree of privacy by screens between and around them.

Against this background there are, on sites spread over a large part of the country, the uninhabited and ruinous remains of several stone-building traditions. These at first sight seem to have little or no connection with the basic settlement pattern described, a fact which led some, but not all, early investigators to ascribe them to exotic sources. The most important among these ruins were Inyanga and Zimbabwe. Both have been the subject of considerable investigation[1]. There is no doubt that the builders of both Inyanga and Zimbabwe were Bantu-speaking people closely akin to those found in possession of the country at the end of the nineteenth century.

The Inyanga people flourished from about the sixteenth to the nineteenth centuries A.D. in the hilly country near the Mozambique border. They developed a building tradition of works especially concerned with agriculture and animal husbandry. An area of about 4000 sq. miles is more or less densely covered with agricultural terraces (mostly stone revetted), water furrows, small dams, large pits used as livestock pens, complex enclosures and platforms for domestic and other use, as well as walled villages and forts. The building material for all these was the local stone, granite or dolerite, usually scree-rock collected from the hillsides. Walls were always constructed in a crude dry-stone technique which in most cases consisted of rather roughly finished external faces with a rubble core. Notwithstanding these coarse methods and materials, the variety of types of building and the manner of their execution are evidence of the possession by the builders of a marked technical adaptability in meeting many different functional needs. But the layout of the settlements reflects no sign whatever of planning for more sophisticated ends than these immediate functional needs. Archaeological evidence suggests that the dense terracing over such a wide expanse was the result of repeated movement arising from soil exhaustion, and it is doubtful if a large population was involved. The repetitive patterning of building work over hundreds of square miles betrays a direct emergence from the subsistence economy basic to the whole area under discussion. It has been suggested, incidentally, that the use of terraces enabled these farmers to grow an early and a late crop each year on the upper and lower slopes respectively.

The Zimbabwe building tradition differed from that of Inyanga in every significant respect except in the use of a relatively crude stone-building technique. Building at Zimbabwe probably began in the twelfth century A.D. and continued until the nineteenth century. Zimbabwe itself is the largest and earliest of many sites dispersed over the greater part of Rhodesia. The characteristic building material was granite, together with its products of decomposition, and so close was the relationship between these materials and the method of construction that a Zimbabwe-type building made from other kinds of stone and earth would have been out of the question. Sites are therefore only to be found on granite soils, which cover approximately 40% of the country, mostly towards the east. The buildings are the prestige architecture of a dominant clan, probably owing its ascendancy in the first place to rain-making. In the initial siting of a building, preference was usually given to fitting the walls and huts among the heaped boulders on or near the top of a hill, but secondary groups were often placed on open ground. The wall-building technique was dry-stone, using tabular blocks of granite, roughly knapped and often laid in courses. Some of the walls are of impressive size, the greatest being 30 ft high and 17 ft wide at the base. Most are much smaller, varying around 8–10 ft high and 4–6 ft thick. They are almost invariably associated with circular huts made from a specially prepared earth containing, among other things, kaolin formed at the interface of the decomposing upper surface of bedrock granite and the residual soil above it. The prepared material was also used for pavings, benchings and hut furniture. On drying out it becomes extremely hard and resistant to weathering.

Apart from the actual selection of sites, the layout of the buildings and groups of buildings probably resulted merely from short-term needs; "planning" in any broader sense is not evident. No stone wall was to any significant extent built level, straight, vertical or regularly curved, nor were walls ever intended to carry roofs; their function appears to have been simply to form enclosures for huts. They were usually built after the huts had been completed, although this does not apply in the case of additions and alterations to existing structures. Except where they were used to retain earth (frequently due to the accumulation of occupation deposit rather than to deliberate design), they could effectively have been replaced by light timber screens as used in an ordinary village. The architecture was therefore very limited in scope, totally lacking the functional versatility found at Inyanga. Conceptually, the two are poles apart.

Nevertheless, at Zimbabwe itself (and to a lesser extent at many of the smaller sites) large numbers of gold objects were found by the early explorers, and these were almost certainly manufactured on the site, indicating a localized specialization in this technique. (There may well have been a direct connection between Zimbabwe and the primitive gold mining industry through which operated an extremely tenuous contact with the outside world.) Indeed, it can be argued that Zimbabwe offered a field of specialized activity for a ruling family, priests or rain-makers, peculiarly skilled builders in stone, potters capable of producing a finer ware than is found elsewhere, goldsmiths and sculptors in soapstone. It is therefore arguable that the concentration even of so small a range of specialists as this might be indicative of a progress towards urbanization. Other factors suggest that this was not so. In particular it is difficult to avoid the conclusion that Zimbabwe was little more than a great kraal which became crystallized into immobility by the dead weight of its stone walls.

Neither Zimbabwe nor Inyanga reached a stage of development approaching true urbanization. In a sense each represented, as far as it went, one side of the coin, Inyanga having a potential for food production above a subsistence level and Zimbabwe the beginnings of centralization and technical leadership. But although the two cultures were contemporary and contiguous they did not react on each other. The reasons are not far to see. Each in its own way was the over-development of a normal characteristic of the basic social and economic pattern which they shared in common with all the Iron Age people of Rhodesia. A village of the nineteenth century differed little from one of the ninth as regards its material level of culture. In this connection it is worth noting that there was little or no reaction to ideas which may have been presented from the outside world. The Portuguese and the Arabs both penetrated the country during the periods when Inyanga and Zimbabwe were flourishing, albeit in small numbers. But their mark on the local way of life was negligible. The use of the wheel and the plough, for instance, had to await the massive influence of European occupation. But some fifty years before that event the whole country was overrun by warlike tribes from the south, who sacked Zimbabwe and all known Zimbabwe-type sites beyond the point of recovery. At about the same time the Inyanga building tradition degenerated into the mere construction of hill-top refuges. All building in stone had virtually ceased by the middle of the nineteenth century.

Note

1 Summers, R. (1958). *Inyanga*. Cambridge; Caton-Thompson, G. (1931). *Zimbabwe Culture*. Oxford; Robinson, K. R. *et al.* (1961). Zimbabwe excavations 1958, *Occ. Papers Nat. Mus. S. Rhodesia*, **23** A, sections 1–6. Whitty, A. (1959). A classification of prehistoric store buildings in Mashonaland, *S. Afr. Archaeol. Bull.*, **14**.

E: CENTRAL AND SOUTH AMERICA

G. H. A. BANKES

Settlement patterns in the lower Moche Valley, north Peru, with special reference to the Early Horizon and Early Intermediate Period[1]

Environment

The Moche Valley is set on the arid part of the north coast of Peru and, like the other rivers of this coast, the Moche draws most of its water from seasonal sierra rains. The flow of water is sufficient for irrigation but not for navigation. The drainage basin of the Moche river covers about 1500 km². For most of its course the river valley is less than 3 km wide but in the lower 20 km or so it widens into a triangular area with a maximum width of about 15 km. This section slopes gently towards the sea, falling about 100 m in its 20 km and, known locally as the Santa Catalina Valley, it constitutes the flood plain of the Moche river and is marked with meanders, cut-off meanders and small ox-bow lakes.

The soil on the valley floor is alluvial while round the edge of the valley there are some large areas of wind-blown sand. Apart from the gravels on the floor of the Moche river and the dry bed which emerges at Huanchaco the water-laid silts in the centre of the valley provide very fertile soil, up to several metres thick in places. Considerable supplies of artesian water are available so that the modern city of Trujillo, with a population of over 200,000, is largely supplied by water from wells.

Ancient sites

The four broad categories of ancient sites suggested here are based upon field observations made by the author. Firstly there are the living sites which can range from a midden to a large group of house structures. Secondly there are the cemeteries which are set apart from the living sites—a phenomenon found in modern Peru as well.

Scattered around the valley are a number of ancient mounds which are known locally and throughout Peru as *huacas*, which are generally built of some kind of adobe brick. The exact purpose of these *huacas* is not fully understood. Occupation débris can be seen on top of and underneath some of them. None of those in the Santa Catalina valley contain any authenticated tomb. Some have a platform on top which could have served as a base for a religious or secular building and Kutscher illustrates a fine line drawing[2], the date of which is probably a century either side of A.D. 500, showing a Moche chief seated in a house on top of a mound receiving naked prisoners.

Finally there are the agricultural areas. In some places these consist of terracing, especially where the river narrows about 20 km from the sea. In the more open level areas small rectangular irrigated fields sometimes survive. Canals for irrigating both the terraces and the fields usually followed the contours along the valley sides. Terraces with structures and domestic pottery were noted by the author in the areas of agricultural terraces in the valley narrows. Ancient houses seem to have been built on terraces just as in modern Peruvian *barriadas* like El Porvenir just north of Trujillo where dwellings are built up on the hillside.

These four categories are not mutually exclusive. For example, it can be very difficult to distinguish agricultural terraces from those used for dwellings just by surface examination. *Huacas* can consist entirely of occupation débris or they can just be platforms built of adobe bricks.

Early horizon and early intermediate period

The term Early Horizon is generally taken, in Peruvian archaeology, to refer to the first period of cultural unification and it centres round the Chavin religion and art style. On the north coast the period of Chavin influence seems to have lasted from 1000 to 500 B.C. After this time (the dates are not absolute as there is considerable disagreement among Peruvianists about radiocarbon dates) each coast valley appears to have developed its own culture in the Early Intermediate Period. For the purpose of this paper the author will define the Early Intermediate Period as running from 500 B.C. to 700 A.D.

On the basis of the work of Max Uhle[3], Larco Hoyle[4], Strong and Evans[5], excavations carried out in 1969 by Donnan of the University of California, Los Angeles and the work of the Harvard Chan Chan-Moche Valley Project from June 1969 to June 1970, the Early Intermediate Period of the Santa Catalina Valley has been divided into three successive periods, Salinar, Gallinazo and Moche. The only available radiocarbon date in Santa Catalina for this period is 873 B.C. ± 500 from charcoal associated with early Moche-style sherds obtained by Kubler[6].

In the Santa Catalina Valley the Harvard Project located two definite Early Horizon sites and one very likely one up to June 1970. Two of these were *huacas* whilst the third consisted of a small enclosure. One of the *huacas* was located on the valley floor in irrigated land about 2½ km north of the Moche river and from this site both the author and Dr. Moseley collected Early Horizon and Moche style sherds. The other *huaca* was in the northern desert margins along the line of inland communication routes. The small enclosure lay about 30 m east of the line of an ancient road, still occasionally used by compesinos, connecting the Moche and Chicama valleys (Fig. 1). The external wall of the enclosure was U-shaped, about 1 m high and dome-shaped in cross-section, and it was bound on its north-western side by a line of upright granite slabs. There was relatively little occupation débris although a hearth was found alongside both the U-shaped and granite slab walls. A late Chavin style incised blackware sherd was found inside one section of the fill of the U-shaped wall. The structure possibly served as a mixture of a wayside shrine and a resting place for travellers. This same site was later occupied by the Moche people, probably about 500 years after it was originally built. The Moche tore down part of the U-shaped wall to make a rough and ready horseshoe-shaped wall inside the original. It is quite common to find that the Moche had used buildings which had originally been erected in the Chavin period.

No Early Horizon sites were found in hill-top positions. This and evidence about the Early Horizon occupation of Huaca Prieta in Chicama, immediately north of Santa Catalina, suggests that defence was not of primary importance. Junius Bird's work at the Huaca Prieta showed that the Early Horizon people settled there peacefully and there was no evidence of destruction or fortification[7].

With the end of the Early Horizon the Santa Catalina Valley appears to have undergone a gradual increase in population which reached its peak in the fourth phase of Larco's scheme for the Moche culture. Field surveys, surface collections and excavations by the Harvard Project have shown that the number of sites which can be securely dated to the Early Intermediate Period is about ten times that for the Early Horizon and the total area occupied is also about ten times greater in the Early Intermediate Period. Most of these sites were located along the northern and southern margins of the valley, outside the zone of modern cultivation. This picture may be distorted since there has been considerable destruction of sites on the valley floor by mechanized sugar haciendas. There are roughly twice the number of Moche living sites, cemeteries and *huacas*, compared to the total for Salinar and Gallinazo. There were about twice as many Gallinazo cemeteries as Salinar ones but the number of living sites is roughly equal. The figures given here are subject to modification by further field work.

The Moche period sites can be presented in Larco's classification scheme. The only site securely dated to Moche I and II is Uhle's site F cemetery, near

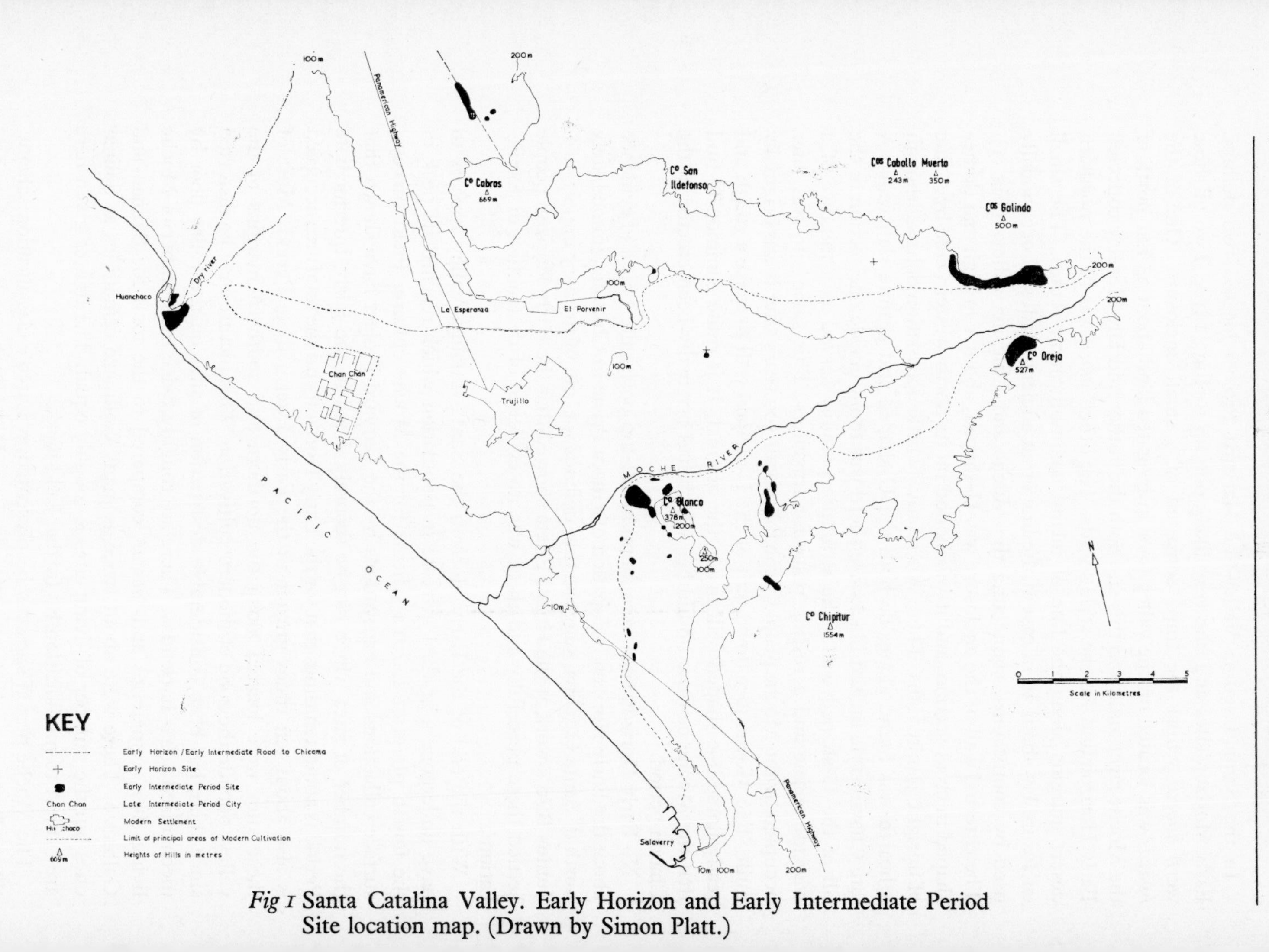

Fig 1 Santa Catalina Valley. Early Horizon and Early Intermediate Period Site location map. (Drawn by Simon Platt.)

the Huaca de La Luna. Ten sites were dated to Moche III, thirteen to Moche IV and only two, on the north side of the valley, to Moche V. This scheme is based upon pottery styles and is usually represented in terms of chronological stages. The finding of Moche V style sherds in a structure along and on the surface near the ancient road from Santa Catalina to Chicama is very likely connected with previous finds of large quantitites of Moche V style pottery in the Chicama valley.

The position of Early Intermediate sites along the valley margins, away from the most fertile land, could well reflect the increase in population and consequent pressure on irrigable agricultural land. In one now desert area on the south side of the valley Dr. Moseley located the line of an ancient canal associated with Gallinazo and Salinar sites and a Moche period sherd scatter which suggested that this area was irrigated to its maximum extent in Moche times as no later associated sites or sherds were found.

Nearly all of the Early Horizon and Early Intermediate Period sites were located away from the banks of the Moche river partly because irrigation water could be drawn off in canals or hauled up from wells. In addition the river banks were not an ideal environment as the river periodically flooded and changed its course. Perhaps most important was that transport was by land along roads for humans and llamas and not by river. The author has seen one such ancient road on the lower slopes of the Cerros Galindo on the north side of the valley connecting the ancient sites with the mountains to the east and the sea to the west. The earliest roads seem to date to Early Horizon times and later settlements grew up alongside them. The road past the U-shaped structure going from Moche to Chicama is an example of an Early Horizon route along which Moche houses were built at a later date. Just as modern settlements grew up alongside the Panamerican Highway in Peru so the Moche built their houses along their equivalent roadway.

Notes

1 This work was carried out while the author was working as a research assistant with the Harvard University Chan Chan-Moche Valley Project directed by Dr. M. E. Moseley. Financial support was received from the National Science Foundation (Grant No. GS-2472), the National Geographic Society, the William Waldorf Astor Foundation and the Central Research Fund of London University.

2 Kutscher, G. (1950). Iconographic studies as an aid in the reconstruction of the Early Chimu Civilization, in *Trans. of New York Acad. of Sciences, Ser. II*, **12** (6), Fig. 4.

3 Kroeber, A. L. (1925). The Uhle pottery collections from Moche, *Univ. Calif. Publ. Anthrop.*, **21**, pp. 191–234.

4 Larco Hoyle, R. (1948). *Cronologia Arqueologia del Norte del Peru*. Buenos Aires.

5 Strong, W. M. and Evans, C., Jr. (1952). *Cultural Stratigraphy in the Viru Valley, Northern Peru*. New York. p. 12.
6 Bird, J. (1951). South American radiocarbon dates, *Mem. Soc. Amer. Archaeol.*, 8, pp. 41–2.
7 Bird, J. (1948). A reappraisal of Peruvian archaeology, *Amer. Antiq.*, 8 (4), pp. 26–8.

WARWICK BRAY

Land-use, settlement pattern and politics in prehispanic Middle America: a review

Early colonial documentary sources and recent ethnological studies make a distinction between two basic Middle American patterns of land-use, one extensive, the other intensive.

Long-fallow agriculture (*roza*, slash-and-burn, or swidden cultivation) is characteristic of the *tierra caliente*, the lowland tropical forest zone where the wet season provides regular and abundant rainfall, but where the soil rapidly loses its fertility so that cultivated plots must be allowed to revert to forest after only a few years' cropping. Some techniques of intensive agriculture (e.g. terracing and irrigation) are irrelevant to lowland conditions, but the area is not ecologically uniform. Rainfall, vegetation cover, soil depth, and fallowing period vary considerably from one region to another, and any study of subsistence must take into consideration such micro-environments as lake shores, river levees and flood plains, savanna grassland, and *bajos* (seasonal swamps) as well as the forest itself. Even within the forest zone, the ancient Maya preferred certain types of soil[1], and modern farmers distinguish several qualities of land on the basis of altitude, vegetation cover, angle of slope, drainage, shelter from wind, and exposure to sun[2]. Nearly all of these ecological niches, including the *bajos* and savannas, can be made to yield a crop.

In extreme cases, shifting cultivation is not merely the most productive form of agriculture but the only possible one. The farmer can choose whether or not to cultivate a particular patch, but he cannot choose *how* to cultivate it. Areas of this kind may, however, turn out to be unusual rather than typical, and should not be considered as representative of the lowlands as a whole. Coe's study of the district around the Olmec site of San Lorenzo Tenochtitlán, in the Gulf Coast plain, draws attention to the ecological diversity of a low-land riverine locality[3]. Here the cultivator is forced to decide how his labour can be most profitably invested: whether to plant a single crop on the high-yielding river levee, which is flooded for a part of every year, or to sow two less productive crops on the higher ground; whether or not to try for an extra crop between the main ones; which variety of maize to plant; how much time to spend on hunting or fishing, and so on. Each of the local micro-environments, and even each season of the year, has its own appropriate "micro-technology", or modified version of "basic" milpa cultivation.

In other favoured areas of the lowlands, cacao plantations were commercially important during prehistoric times, and it has recently been argued that the fruit of the breadnut tree (*Brosimum alicastrum*) was one of the staple foods of the Petén Maya during the Classic period[4]. It is becoming clear that the notion of a pan-lowland system of agriculture, based exclusively on long-fallow, shifting cultivation, must be abandoned, to be replaced by a series of case studies of individual localities and sites.

Under a system based primarily on *roza* agriculture, whether of root or seed crops, each family needs to exploit a large area of land, most of which lies fallow in any one year. The precise amount of land depends on such factors as the number of harvests per year and the period needed for regeneration of the soil. Long-fallow cultivation therefore imposes limits on both the size of the population and its overall density, though it does not by itself determine how the population shall be distributed. Other things being equal, extensive agriculture will tend to produce a dispersed settlement pattern with each family living close to its fields, either in individual houses or in small clusters scattered at intervals throughout the countryside. Some of these sites may be permanent, while others may be occupied for only a short time[5]. Socio-political factors pull in the opposite direction, tending to draw the peasantry close to the central community with its governmental offices, religious buildings and markets. The result in lowland Mesoamerica is usually, though not inevitably, a compromise: a pattern in which religious and administrative centres with small resident populations serve a wide area occupied by farmers who live scattered throughout the forest.

In the highlands and the semi-arid parts of the country this situation is reversed. Because of the broken nature of the terrain and the greater range of altitude, several ecological zones occur within a small area. On the soils of the highlands two or three years of cultivation can be followed by as little as one year under fallow, while even the poorer highland soils can stand one year of cropping followed by two or three of regeneration. The average ratio of cultivated to reserve land is approximately 1 : 3[6], and because the fallow period is so short the farmland does not usually revert to forest. With crop rotation, terracing, manuring, and careful weeding the fallow period can be almost eliminated.

These figures refer to simple dry farming. Yield can be increased by the use of the *calmil-barbecho* (infield-outfield) system in which each family exploits cornfields (*milpas*) and also a garden plot (the *calmil*) surrounding the house and fertilized by garbage, dung, and vegetable refuse. The size of the *calmil* is limited mainly by the availability of manure. In the modern Totonac village of Eloxochitlán, in the Puebla highlands, the average family cultivates 2 ha of *milpa* and 0·5 ha of *calmil* per year. The milpa is cropped for three years and rested for three years, whereas the garden is permanently under cultivation and has a yield double that of the *milpa*[7].

A comparison between the *calmil-barbecho* system of Eloxochitlán (with one harvest per year) and the long-fallow, slash-and-burn farming of the Tajín lowlands of Veracruz (with two crops per year) shows that a lowland family cultivates 1·5 ha per annum compared with 2·5 ha in the highlands. But the cultivation cycle of Tajín requires 12 ha of arable land per family while the figure for Eloxochitlán drops to 6·5 ha per family[8].

In other words, *calmil-barbecho* farming allows almost twice the population density of the lowland slash-and-burn agriculture, and at the same time permits stable settlement with a degree of nucleation. Because each house has its attached *calmil*, the villages are of the dispersed type with houses spaced wide apart. This pattern has been recorded from the Guatemalan highlands, among the sixteenth century Otomí in central Mexico, in the Teotihuacán Valley, and from most of the dry-farming regions of upland Mesoamerica[9]. Under special conditions it also occurred in the Maya lowlands before the Spanish Conquest, for example at the large town of Potonchan at the mouth of the Grijalva River[10].

In arid and semi-arid zones the most striking increases in both crop yield and population density occur when water control is added to dry farming technology. Methods of water control vary according to local circumstances, but include floodwater farming, pot or well irrigation, canals, *chinampas*, and artificial flooding induced by damming streams. These techniques, in combination with terracing and other measures against erosion, can bring about a 50–300% increase in yield over the dry farming figure. Often the land owned by a single community cuts across several micro-environments, each of which requires a different technique for optimum exploitation[11].

The most intensive agriculture of all is to be found in the *chinampas*. These were much more widespread in prehispanic times than they are today, and have been recorded in Michoacán and the Toluca basin as well as in the Valley of Mexico. On these artificial islands surrounded by canals, all the necessary moisture is provided by filtration and hand irrigation, while soil fertility can be renewed at any time by dredging up fresh muck from the lake bottom[12]. The *chinampa* system with its highly specialized technology and intensive year-round exploitation probably developed around A.D. 1000[13], and was essential to the economy of the lakeside towns in the Basin of Mexico during the centuries before the Conquest. Since as little as 0·5 ha of *chinampa* will support a family, this form of cultivation permits a high population density. *Chinampa* agriculture is practicable only in restricted areas with shallow, sweet water, and this further tends to concentrate the population. The present-day *chinampa* villages of the southern Basin of Mexico have 240–360 people per km^2 [14], and each village has approximately 3000–6000 inhabitants, nearly all of whom are cultivators. Xochimilco, part market town and part *chinampa* village, had 15,000 inhabitants in 1910, when it was still uninfluenced by the modern expansion of Mexico City[15].

The figures listed below illustrate some relative densities of population under different systems of cultivation:

(A) Long-fallow slash-and-burn	
Chan Kom (Yucatan)[16]	1400 ha cultivable land/100 families
Tajín Totonac[17]	1200 „ „ „ „ „
Chichén Itzá (Yucatan)[18]	2360 „ „ „ „ „
San José (Petén)[19]	3200 „ „ „ „ „
Tabasco[20]	1800 „ „ „ „ „
(B) Intensive agriculture	
calmil-barbecho (1 crop per year)[21]	650 „ „ „ „ „
dry farming with supplementary irrigation (2 crops per year on irrigated land)[22]	86 „ „ „ „ „
chinampas with continuous cropping[23]	
(subsistence farming plus cash crops)	70 „ „ „ „ „
(commercial cash crops only)	37 „ „ „ „ „

These figures demonstrate very clearly the degree to which intensive agriculture (in which irrigation is only one among many factors) encourages nucleated settlement and large populations. The technology was already fully developed when the Spaniards reached Mesoamerica, and the following section summarizes the archaeological evidence for the growth and differentiation of native agricultural systems and settlement patterns.

The development of prehispanic settlement types

A. The Preclassic period

The divergence between lowland and highland patterns of settlement can be recognized during the Preclassic period (2000–1500 B.C. to A.D. 300), by the end of which the two extreme forms, ceremonial centre and nucleated town, had both made their appearance in the archaeological record.

The development of the ceremonial centre can best be studied at the Olmec site of San Lorenzo, which was constructed in a relatively fertile area of the Gulf Coast lowlands, with access to forest land, seasonally flooded savanna, and rich soil along the river levees[24]. San Lorenzo was occupied as early as 1500 B.C., but the first monumental structures date from around 1350 to 1250 B.C. with the building of a great artificial plateau some 7 m high. During the climax period of Olmec culture (the San Lorenzo phase, *c.* 1150 to 900 B.C.), the site was a major ceremonial centre with mounds and plazas, monumental sculpture, artificial lagoons, and stone-lined drains. Only about 200 house mounds have been found at San Lorenzo, giving a resident population of some 800–1000 inhabitants. A population of this size would have been far too small to have either built or maintained the centre. Two other ceremonial centres, Tenochtitlán and Portrero Nuevo, are within 3 km of San Lorenzo but the three sites together could not have held more than approximately 2500 people.

La Venta, occupied *c.* 1000 to 600 B.C., was also an Olmec centre and shares many architectural and cultural traits with San Lorenzo. Its situation, however, is much less favourable. The site is on an island of 5·22 km², surrounded by swamp with no arable land. Under slash-and-burn cultivation no more than 150 persons could have gained a living from the island itself and, as at San Lorenzo, the resident population could not have built the centre without help[25]. It must therefore have served a widely scattered rural peasantry which may perhaps have been organized as a centralized political state for which La Venta provided religious and governmental services[26]. In the absence of documentary records, however, this hypothesis remains unproven.

In the semi-arid highlands, settlement pattern studies are available for the Teotihuacán Valley[27], the Valley of Oaxaca[28], the Nochixtlan Valley of the Mixteca Alta[29], the Tehuacán Valley of Puebla[30], and the Texcoco region of the Basin of Mexico[31]. The sequence of events varies from one area to another but, beginning during the Preclassic period, the following general tendencies can be observed: (1) a progressive increase in population; (2) the adoption of various techniques of intensive agriculture, so that each micro-environment is exploited in the most productive way; and (3) the development of large, nucleated settlements with residential as well as ceremonial functions.

Not all of the sophisticated farming techniques were employed in any one place, and the full development of the highland pattern was not achieved until the Classic period. By the Late Preclassic, however, towns already existed in the highlands. Amalucan covered about 10 km² and is just one of the many similar towns in the Puebla Basin[32]; Cuicuilco, in the Valley of Mexico, embraced at least 2·0 to 2·5 km²; the Texcoco area during the last two centuries B.C. had sites up to 118 ha in extent[33]; and Teotihuacán during the Tzacualli phase of the early centuries A.D. was a settlement of 17·5 km² with about 30,000 inhabitants[34]. Other large sites are known from Oaxaca and the Basin of Mexico. In central Mexico, at least, these large sites are surrounded by dependent hamlets, and this may perhaps represent the beginning of the "city-state" pattern which is well documented for later periods in this area.

B. The Classic period (*c.* A.D. 300–900)

The patterns which had emerged during the Preclassic period were intensified during the Classic. In the Gulf Coast region, ceremonial centres like El Tajín were in the same tradition as San Lorenzo and La Venta, serving a scattered, non-resident population. The same is true of the Maya lowlands, which have yielded house mounds wherever there is suitable land for cultivation, sometimes as much as 30 km from any important ceremonial centre. Minor ceremonial nuclei, or "provincial" centres are scattered around and between the major centres[35]. It is estimated that in the southern Maya lowlands, which excludes the peninsula of Yucatan, there are about eighty centres important enough to have hieroglyphic inscriptions, and three to four times as many lesser centres[36].

In the north-east Petén, Bullard employs a three-tier classification to describe the hierarchy of settlements[37], and a rather similar pyramidal pattern has been recognized in the Chontalpa region of Tabasco[38]. The smallest of Bullard's units is the *cluster*, consisting of five to ten house mounds occupying an area of 200 to 300 m^2. Several of these clusters combine to make up a *zone*, composed of fifty to 100 house lots and occupying a territory which averages about 1 km^2, though the actual extent is very variable. Each of these zones includes a minor ceremonial centre, usually without stelae, altars, or ball courts. Finally, several zones go to make up a *district* which is served by one of the major centres. Districts are not all of the same size, and vary according to the configuration of the terrain, but in the area surveyed by Bullard they seem to control about 100 km^2 of cultivable land, plus *bajo* swamp and other infertile territory.

House mounds tend to cluster most thickly around the core of ceremonial buildings and to decrease in density towards the periphery of each district. Usually, however, there is a thin but continuous scatter of houses between one concentration and the next, making it impossible to draw any very precise frontier between districts.

This typical arrangement may be modified to suit local conditions, as in the Belize and Mopan valleys of British Honduras where there is something approaching ribbon development along the bottomlands[39], or at Dzibilchaltun, in Yucatan, where a truly urban level of development has been claimed[40].

Inscriptions on stelae at Piedras Negras, Yaxchilan, and other lowland Maya sites refer to political and dynastic events[41], which suggest that the principal centres also acted as state capitals. The sustaining area of a major ceremonial centre, corresponding with one of Bullard's districts, can perhaps be equated with a political state, whose capital exerted control over, and possibly exacted tribute from, several lesser centres. This view is supported by the study of the "emblem glyphs" which appear in Classic Maya inscriptions and are probably the symbols of individual states[42]. The same glyph is often shared by a major centre and several neighbouring ones of the second rank. Seibal, for example, shares an emblem glyph with the minor centres of La Amelia (more than 60 km away), and with Aguateca, Dos Pilas, and Tamarindito. Similarly Palenque appears to be the capital of a triangular area with sides about 60 km in length, which includes the secondary centres of Tortugero, Jonuta, and Miraflores. It is possible that several of these major centres were grouped into even larger territorial states like the "*provincias*" of sixteenth century Yucatan, which covered anything from 1200–9000 km^2 and had 30,000–120,000 inhabitants[43].

An important Maya centre consists of groups of ceremonial buildings (temples, palaces, governmental buildings, priests' quarters, etc.) surrounded by widely dispersed clumps of house mounds, usually arranged in threes and fours around courtyards. Apart from ceremonial causeways, there is no pattern of streets or traffic routes, and even within the central area the location of buildings is determined more by local topography than by any attempt at town planning.

A Maya centre is therefore non-urban in configuration, but offers all the services provided by a true city[44]. The confusion of form with function has led some scholars to maintain that Classic Maya civilization was urban-based, but in almost every respect the layout of a great centre like Tikal contrasts with that of its highland equivalent.

The most intensively studied of the highland cities is Teotihuacán, which occupies a flat, valley floor site in a branch of the Valley of Mexico[45]. Instead of a haphazard arrangement of house lots, Teotihuacán has a regular grid of civic and religious structures covering 3 km^2. Around this is an area of 5 km^2 forming a residential zone laid out to a precise grid plan, with streets intersecting at 57 m intervals. Each block forms a residential unit, usually a large communal dwelling rather than an individual family house. Around this planned core are 3–4 km^2 of ungridded but concentrated housing, with a further ring of 12 km^2 of suburbs, some of which had their own ceremonial structures.

The Teotihuacán Valley in the sixth and seventh centuries A.D. had a population somewhere between 76,000 and 157,000 people, of whom about 95% were concentrated in urban Teotihuacán. Elsewhere in the plain there were no large settlements at all, except at Portezuelo where a thin scatter of Classic sherds covered an area 2·5 km long by 200–500 m. The few rural sites were well back from the edge of the plain, and the only towns of any size were outside the area altogether. Just at the time when Teotihuacán was reaching its maximum power and extent there was a sudden drop in the number of archaeological sites in the eastern part of the Valley of Mexico proper, and it seems that the metropolis of Teotihuacán was attracting people from outside its normal catchment area[46]. A similar tendency can be recognized elsewhere in the central highlands, where the rise to prominence of such sites as Cholula and Xochicalco was accompanied by a corresponding fall in rural population. By contrast, in marginal highland areas (e.g. the Tehuacán Valley), where great cities did not emerge, the natural pattern of growth was not disrupted in this way[47].

C. The Postclassic period (*c.* A.D. 900 *to the Spanish conquest*)

With the possible exception of Dzibilchaltun, all the known prehispanic towns of the Maya lowlands belong to the Postclassic period, when the native culture had been considerably influenced by that of highland Mexico. Tulum, in Quintana Roo, was a walled town a little over 1·2 km^2 in extent, with a planned layout but an estimated population of no more than 500–600[48]. Topoxté, spread over three islands in Lake Yaxha in the Petén, consisted of a ceremonial nucleus and about 200 house lots (say 1000–2000 people)[49]. Towns existed also around Lake Petén Itzá, where the Spaniards discovered twelve pueblos as well as rancherías in the surrounding hills[50].

The largest town in the Maya lowlands was Mayapan, which was the dominant city of Yucatan from the thirteenth to mid-fifteenth century A.D. Its walls enclosed an area of 4·2 km^2, with approximately 4000 structures[51]. Most of these are residential; religious and civic buildings make up less than 4% of the total. The resident population has been estimated at around 12,000 persons. This figure, unusually high for a Maya town, is smaller than that for any of the major cities in the zone of intensive agriculture, and corresponds roughly with that for the Classic period ceremonial nucleus of Tikal. At Tikal, however, the 10,000 inhabitants of the ceremonial nucleus were scattered over 16 km^2 [52].

Conquest accounts are unanimous in maintaining that urbanism and intensive agriculture were the general rule in semi-arid zones during the Postclassic. Cholula, Aztec Tenochtitlan, and the principal Valley of Mexico sites were conurbations surrounded by irrigated lands. Irrigation was a common feature of highland farming, and Palerm has collected some 382 documentary references to the use of this technique throughout arid Mexico during the prehispanic period[53]. There is a marked correlation between true nucleated settlements and areas of intensive agriculture. This is best exemplified in the Gulf Coast lowlands where Cempoala, located in a wedge of arid territory, was a large city with irrigated farmlands, whereas the surrounding forest belt was exploited by slash-and-burn cultivators and had no towns[54].

The Postclassic settlement pattern of the Maya zone demonstrates that it is possible to regroup a scattered population of slash-and-burn cultivators into towns, but this redistribution is not accompanied by any change in the subsistence basis nor by any increase in the total number of people which a given area can support. There was always a tendency for the old pattern to reassert itself, as in Yucatan where attempts by the Spaniards to establish large towns during the years 1549–79 all ended in failure[55].

If, as has often been suggested, cities were introduced into the lowlands by newcomers from highland Mexico, the result was not an overnight change to the highland pattern of life. The environmental limitations of the forest zone were not overcome. Most settlements remained small, and all of them (except for Tulum) lacked the planned layout which was characteristic of highland Mexico.

Haviland[56] considers that these Maya towns are basically smaller and contracted versions of the Classic Maya ceremonial centre of Tikal type, with the population of the outlying zones brought in close to the governmental-religious centre. The implication is that town life may have developed spontaneously in the Maya lowlands in response to socio-political needs. The most obvious of such needs is defence, and it is worth emphasizing that—in contrast to most highland towns—the settlements in the Maya lowlands were often fortified. Describing Yucatan on the eve of the Conquest, Diego de Landa wrote[57]

> Before the Spaniards had conquered that country, the natives lived together in towns in a very civilized fashion . . . and they lived thus close together for fear of their enemies who took them captive

The Itzá capital of Tayasal, visited by Cortés in 1524, was described by the Spaniards as a large, fortified island town surrounded by a palisade. Mayapan, Tulum, and Ichpaatun were fortified, as were Cimatan, Potonchan, and the towns of the Cechache province[58]. The location and planning of these towns seems rather to reflect unstable political conditions (as recorded in the chronicles) than to be the result of any direct Mexican influence.

Some support for this view comes from the Maya zone of upland Chiapas, a region where contact with highland Mexico was minimal, and where the archaeological evidence shows a progressive nucleation of settlement and a shift to defensible situations[59]. In the Mexicanized areas of highland Guatemala, towns were also located with an eye to defence, and some such towns were fortified[60].

Political factors

Haviland maintains that it is an oversimplification to force all Mesoamerican sites into the mutually exclusive categories of either ceremonial centre or town/city[61]. He prefers to think in terms of a "continuum, with heavily populated centres at one end (e.g. Tenochtitlán) and centres inhabited by a few people at the other (e.g. La Venta)". Late Classic Tikal in this scheme would be placed somewhere between the two extremes.

To those who believe that detailed classification is an end in itself this proposal may seem reactionary, but it has a number of practical advantages. It can, for instance, accommodate sites in the highland Maya zone where the distinction between town and ceremonial centre has often proved difficult to draw. Highland Chiapas during the Classic period is a case in point. It had dispersed sites and also nucleated villages, including at least one site (Yerba Buena) which approached urban proportions in size and density[62]. It is probably no coincidence that this is an area where terracing (i.e. intensive cultivation) is unusually prominent, and where the Postclassic development shows a trend away from small independent settlements towards the "Mexican" pattern of political "proto-states" with large sites controlling many subsidiary settlements.

The situation in highland Guatemala is equally complex[63]. Various types of settlement co-existed, and the overall pattern cannot be interpreted in terms of a simple dichotomy between town and ceremonial centre. At the time of the Conquest, some highland groups had created complex political states with a settlement pattern which was urban rather than dispersed. The towns of the sixteenth century Pokom Maya had 3000–10,000 inhabitants living in proximity to a civic-ceremonial centre, but at the same time a part of the population lived away from the town, although sharing fully in the privileges and duties associated with citizenship. This pattern, which represents a compromise between lowland and Mexican highland traditions, has been termed the "extended town pattern", and some of the largest Pokom sites come within the category of true cities[64].

The Mixteca Alta is an example of a highland area which was slow to adopt the highland pattern of urban life. The survey of the Nochixtlan Valley revealed only small nucleated centres during the local Early Classic (*c.* 200 B.C. to A.D. 300), then "quasi-urban" sites of more than 2 km^2 between A.D. 300 and 1100. Large urban sites up to 3 km^2 were not constructed until the final centuries before the Conquest[65].

The distinction between a Classic lowland Maya centre and a highland Mexican city diminishes as soon as political factors are taken into consideration, and when one begins to think in terms of *states* rather than *sites*. Documentary evidence attests the existence of city states in highland Mexico for the Postclassic period after about A.D. 900, and the archaeological evidence suggests that a similar political structure was prevalent during the Classic, and probably part of the Preclassic as well. The Maya lowlands were also divided among various states during the Classic and Postclassic but, for reasons already discussed, it is difficult to delimit their political frontiers on the ground and to calculate the sustaining area for any one centre.

New data provided by Haviland[66] allow this to be done for Tikal with a fair degree of certainty. The central zone of Tikal covers about 63 km^2 and shows a greater density of buildings than the peripheral region of 60 km^2, where there is a marked drop-off as the natural and artificial boundaries are approached. The buildings in the nuclear area were continuously occupied and altered until their final abandonment in the ninth century, but in the peripheral zone not all house mounds were in use at the same time. To the east and west of Tikal the drop-off in density of mounds coincides with the natural limits formed by seasonal *bajo* swamps. Northwards and southwards a similar fall in density occurs at 3·5 and 6 km from the centre, even though suitable agricultural terrain continues. The northern and southern boundaries of greater Tikal are marked by earthworks at 4·5 and 8·5 km respectively from the nucleus, and the nearest major centre (Uaxactún) lies no more than 18 km from the Great Plaza of Tikal.

If these boundaries are assumed to mark the frontiers of the Tikal sustaining area (i.e. the area politically controlled by Tikal) we have a *state* covering at least 123 km^2, with an estimated population of 45,000 inhabitants, of whom about 25% lived within the nuclear zone of about 16 km^2 which included all the major ceremonial and governmental buildings.

This compares very closely with the average-sized city-states of central Mexico, although, for reasons explained below, the "superpowers" like Teotihuacán and Tenochtitlán fall outside this discussion.

In the highlands, intensive survey of the eastern side of the Valley of Mexico has revealed the settlement pattern of the Aztec period in considerable detail[67]. The area was divided between five city-states, and includes 110 archaeological sites in all. No matter where the state boundaries actually lay, each city must have had several dependent settlements, nearly all of which fall into the categories of hamlet, small dispersed village, or large dispersed village.

Similar results were obtained from the Teotihuacán Valley where the Aztec settlement pattern has been studied on the basis of both documentary and field evidence[68]. On the eve of the Spanish Conquest the Teotihuacán Valley was culturally and politically marginal to the centres of power in the Valley of Mexico proper, and contained six towns (apart from Teotihuacán itself) all of which paid tribute to the Aztec Alliance. Each town consisted of a nucleated core (with ceremonial or élite architecture) surrounded by a peripheral zone of poorer and more scattered dwellings. The estimated population of individual towns ranged from 1000–10,000 inhabitants. Each of these towns was the capital of a semi-autonomous state with an average territory of some 200 km^2 and a total population of 5000–40,000. The settlement pattern of the dependent rural population varied according to the agricultural potential of the different ecological zones of the Valley, and Sanders distinguishes the following categories: (1) compact, high-density villages covering up to 2·5 km^2 within or on the edge of the alluvial plain; (2) strip development, sometimes several kilometres long, with clusters of ten to fourteen houses arranged like strings of beads across areas of gentle or medium slope in the piedmont zone; and (3) isolated clusters of houses on mountain flanks.

Although the configuration of these highland towns is quite unlike that of lowland centres, as *states* they are not dissimilar. The size of the total sustaining area and the number of its inhabitants are roughly comparable, and in both kinds of state only about 25% of the population is concentrated in the ceremonial-governmental zone. There are also many similarities between the political and economic structures of highland and lowland states. In both areas we can recognize a centralized political organization under a secular or priestly ruler, and a social system based on a stratified class structure with administrators, craftsmen, common people and slaves. In his discussion of the Aztec states, Sanders emphasizes that the system "was characterized by close economic interdependence between villages, and between villages and the town via the town market and tribute system (taxes in labour, goods, and military services)"[69], and this too can be recognized in the Maya zone.

The problem of the superpowers

The previous section offers some support for Haviland's "continuum" theory by demonstrating that certain lowland states have a lot in common with the average-to-large sized city-states of the highland zone. In terms of overall size and degree of nucleation, however, the greatest of the highland cities are in a class of their own. By comparison with the ordinary run of sites they stand out as "superpowers", much larger than their rivals, politically more powerful, and often the centres of expansionist empires. Archaeologically they are frequently linked with the great "horizon styles" in pottery and art which leave their traces over the greater part of Mesoamerica. By definition, superpowers are few in number. In this category I would include only Classic Teotihuacán, Cholula, the Toltec state with its capital at Tula (tenth to twelfth centuries A.D.), and Tenochtitlán during the Aztec climax (1428 to the Spanish Conquest). A case can also be made out for including Mayapan among the superpowers (see below).

The sheer size of these superpowers creates problems which lesser states do not have to face. The main problem is that the superpowers have outgrown their sustaining areas and are no longer self-sufficient in basic necessities. In a situation where colonization or emigration are not feasible, the deficit must be made good from outside, either by extortion or trade, and it is no accident that the three successive expansionist "Empires" (Teotihuacán, Tula-Toltec and the Aztec confederation) had their centres in the semi-arid region of Mexico. All three of these superstates exerted a powerful influence on parts of the lowland zone, without, however, being able to alter the subsistence basis of lowland culture or to overcome the limitations imposed by slash-and-burn systems of cultivation. Only at an early period (*c.* 1300 to 600 B.C.), before the arid zone pattern of intensive agriculture had reached its fullest development, did lowland (Olmec) civilization exert any significant influence over the highlands.

It has been calculated that by the third century A.D. Teotihuacán would have outgrown the available local resources, and that during its period of maximum development in the fourth to eighth centuries only about 50,000 people could have been fed from the city's lands alone[70]. The city must have obtained food from outlying rural areas by trade or taxation, although Sanders believes that Teotihuacán could have been sustained by an area only slightly larger than the Valley itself.

Information is not yet available from Tula, but the documentary sources on the Aztec capital of Tenochtitlán are detailed and unambiguous. The Aztec city covered at least 12 km^2 [71], and occupied an island site in a densely settled region where there was no spare land to be brought into cultivation. The number of inhabitants is not easy to assess. Early Spanish estimates, in fact no more than guesses, range from 60,000 households (Gómara) to 120,000 (Torquemada). Borah and Cook suggest about 360,000 people, but this figure comes from a series of elaborate and roundabout calculations based on the tribute list in *Codex Mendoza*, with arbitrary correction factors introduced at various stages of the argument[72]. The Borah-Cook estimate has been criticized as being much higher than that for any comparable Old World City, and a more reasoned figure (based on a combination of archival research with mapping on air photographs) would be *c.* 150,000 to 200,000[73].

Calnek's analysis of the *chinampa* economy of Tenochtitlán makes it very obvious that the city could not have provided enough food for its population[74], and from *Codex Mendoza* and the *Matricula de Tributos* it is clear that Tenochtitlán was importing vast quantities of foodstuffs (maize, beans, amaranth and chia) from all over central Mexico to the tune of about 9120 tons per year[75]. Of this amount, 2720 tons was maize. Using the figure of 8 lbs as the average daily consumption of maize by a family of five people, the maize tribute alone would have provided the annual ration for some 10,500 persons. The city of Texcoco received the same quantity of tribute, and Tlacopan (the third partner in the Alliance) received half as much.

Only about a quarter of this tributary food was provided by the other Valley states, and some came from as far away as the province of Coyolapan (Oaxaca), some 350 km distant from Tenochtitlán.

Figures based on tribute lists do not, of course, include foodstuffs obtained on the free market. Extortion and legitimate trade are two sides of the same economic coin, but the lack of quantitative data makes it impossible to assess the relative importance of trade and tribute in the Aztec system. Native and Spanish sources are in agreement, however, that long-distance trade carried out by specialist merchants (the *pochteca*) was of great economic and political significance. Under conditions of stress, when imports were vital for the survival of Tenochtitlán, it is perhaps not surprising to learn that the Aztec ruler took a personal interest in merchant activities, and that refusal by a foreign power to trade with Aztec *pochteca* was one of the most common pretexts for declaring war.

It has been claimed by certain cultural ecologists that "symbiotic" trade relationships between adjacent ecological zones, each with its own distinctive range of products, was a major factor in the growth of urban life, but symbiosis alone is not enough to explain the observed trade pattern once city life had been achieved. Most of the foodstuffs brought as tribute into

Tenochtitlán were not foreign luxuries but the staple crops of the Valley of Mexico itself. The main exotic imports were in fact raw materials: gold and copper tropical forest products, incense, rubber, feathers, and great quantities of cotton cloth and garments. Tenochtitlán was therefore taking in both foodstuffs of local type and raw materials not available in the Basin of Mexico. Some of these raw materials were re-exported by the *pochteca* in the form of manufactured articles such as copper bells, gold ornaments, gems, and made-up garments[76]. There is here a suggestion of incipient industrialization—the traditional expedient for nations with too many people and too little land.

It must be emphasized that the status of superpower is a relative one. Teotihuacán and Tenochtitlán had passed beyond the critical point of self-sufficiency in foodstuffs and basic raw materials, but this process must be viewed against the highland background of large nucleated populations practising intensive cultivation. The same phenomenon can be recognized in the lowlands, although here in a forest setting with slash-and-burn cultivation and a more scattered pattern of settlement, the critical point comes at a much lower level of population.

Classic Tikal may just have reached the point of no return[77], and Post-classic Mayapan qualifies as a superpower even though it was a city of only 12,000 people. In a lowland Maya context it was much bigger than any of its neighbours, was the chief city of a large territorial state (controlling a region some 300 by 175 km in northern Yucatan for more than two centuries), and received tribute from many subject towns[78]. The area round Mayapan is such poor agricultural land that it could hardly have supplied the needs of the city, and the documentary sources confirm that foodstuffs and other commodities were brought in from rural districts ruled over by lords who lived in the town.

Conclusions

One of the most stimulating approaches to archaeology in recent years has been the study of the interaction between man and his environment. Cultural ecologists have examined patterns of subsistence and have analysed the ways in which environmental factors have stimulated or retarded cultural development, but many of these studies ignore the fact that man is a political animal with some control over his destiny. Ecological changes may have social and political effects, but the converse is also true. A town or ceremonial centre is the product of a particular society functioning in a given environment, and the variety of aboriginal settlement pattern in Middle America cannot be explained in terms of a simple dichotomy between lowland and highland subsistence bases. As a reaction (perhaps an over-reaction) against such environmental determinism, this paper examines the social, economic, and political implications of certain types of settlement pattern.

Notes

1 Stevens, R. L. (1964). The soils of Middle America and their relation to Indian peoples and cultures, *in* Wauchope, R. (gen. ed.) *Handbook of Middle American Indians*, **1**. Austin. pp. 298–309.
2 Reina, R. E. (1967). Milpas and Milperos: implications for prehistoric times, *Amer. Anthrop.*, **69**.
3 Coe, M. D. (1969). Photogrammetry and the ecology of Olmec civilization. Unpublished manuscript read at Working Conference on Aerial Photography, Cambridge, Mass. May 1969.
4 Puleston, D. E. (1968). New data from Tikal on Classic Maya subsistence. Unpublished paper presented at 33rd Annual Meeting of the *Soc. for Amer. Archaeol.*, May 11, 1968. Santa Fe.
5 See Harris, D. R. and Smith, P. E. L., this volume, p. 249 and p. 415.
6 Palerm, A. (1967). Agricultural systems and food patterns, *in* Wauchope, R. (gen. ed.) *Handbook of Middle American Indians*, **6**. Austin.
7 Palerm, A. (1955). The agricultural basis of urban civilization in Mesoamerica, *in* Steward, J. H. (ed.) *Irrigation Civilizations: a comparative study*. Pan American Union Social Science Monog. 1. Washington.
8 Palerm, A. (1955). *ibid.* The Tajín cycle includes an episode of vanilla cultivation after two to three years of maize. Without this, the figure of 12 ha per family could be reduced.
9 Sanders, W. T. (1967). Settlement patterns, *in* Wauchope, R. (gen. ed.) *Handbook of Middle American Indians*, **6**. Austin.
10 Scholes, F. W. and Roys, R. L. (1968). *The Maya Chontal Indians of Acalan-Tixchel*. Norman. p. 37.
11 See the study of the Teotihuacán Valley from ancient to modern times: Sanders, W. T. (1965). *The Cultural Ecology of the Teotihuacán Valley*. Pennsylvania State Univ. (multilithed).
12 Coe, M. D. (1964). The Chinampas of Mexico, *Sci. Amer.*, **211**; Moriarty, J. R. (1968). Floating Gardens (Chinampas). Agriculture in the old lakes of Mexico, *America Indigena*, **28**.
13 West, R. C. (1970). Population densities and agricultural practices in pre-Columbian Mexico, with emphasis on semi-terracing, *Verhandlungen 38 Internat. Amerikanistenkongresses*, **2**. München. p. 363; Coe, M. D. (1964). *ibid.* and Sanders, W. T. (1965). *ibid.* argue for an earlier date.
14 Sanders, W. T. (1956). The Central Mexican Symbiotic Region, *in* Willey, G. R. (ed.) *Prehistoric Settlement Patterns in the New World*. Viking Fund Publ. in Anthrop. New York.
15 Sanders, W. T. (1968). Hydraulic agriculture, economic symbiosis and the evolution of states in Central Mexico, *in* Meggers, B. J. (ed.) *Anthropological Archaeology in the Americas*. Washington. p. 91.
16 Recalculated from Heizer, R. F. (1960). Agriculture and the theocratic state in lowland south-eastern Mexico, *Amer. Antiq.*, **26**, p. 216.
17 Heizer, R. F. (1960). *ibid.*
18 Heizer, R. F. (1960). *ibid.*
19 Reina, R. E. (1967). *op. cit.*
20 Sanders, W. T. (1962). Cultural ecology of the Maya lowlands, Pt. 1, *Est. de Cultura Maya*, **2**, p. 100.
21 Palerm, A. (1955). *op. cit.* p. 30.
22 Palerm, A. (1955). *ibid.*
23 Palerm, A. (1955). *ibid.*
24 Coe, M. D. (1968). *Map of San Lorenzo, an Olmec Site in Veracruz, Mexico*. New Haven; Coe, M. D. (1968). San Lorenzo and the Olmec

Civilization, *in* Benson, E. (ed.) *Dumbarton Oaks Conference on the Olmec*. Washington.

25 Drucker, P. and Heizer, R. F. (1960). A study of the Milpa system of La Venta Island and its archaeological implication, *S.W. J. Anthrop.*, **16**; Heizer, R. F. (1960). *op. cit.* New investigations have shown that the archaeological site is larger than was realized when the above reports were written, and the number of man-days given by Heizer for its construction must be revised upwards. See Heizer, R. F., Graham, J. A. and Napton, L. K. (1968). The 1968 investigations at La Venta, *Univ. of California Archaeol. Research Facility, Contrib.*, **5**. Berkeley.

26 There is some doubt about the size and location of the La Venta sustaining area. Heizer, R. F. (1960). *op. cit.* thought the centre served an agricultural area of some 900 km² which lies to the west of the site and could have supported about 18,000 people. Another suggestion is that the rural population was concentrated along the fertile levees of the Tonala, Blasillo and and S. Felipe rivers; see Sisson, E. B. (1970). Settlement pattern and land-use in the north-western Chontalpa, Tabasco, Mexico, *Cerámica de Cultura Maya*, **6**.

27 Sanders, W. T. (1965). *op. cit.*

28 Flannery, K. V., Kirkby, A. T. V., Kirkby, M. J. and Williams, A. W. (1967). Farming systems and political growth in ancient Oaxaca, *Science*, **158**.

29 Spores, R. (1969). Settlement, farming technology and environment in the Nochixtlan Valley, *Science*, **166**.

30 MacNeish, R. S., this volume, pp. 67-93.

31 Parsons, J. R. (1969). Patrones de asentamiento prehispánico en la region texcocana, *Bol. Inst. Nac. de Antrop. e Hist.*, **35**; Parsons, J. R. (1970). An archaeological evaluation of the Codice Xolotl, *Amer. Antiq.*, **35**.

32 Fowler, M. L. (1968). The Temple Town Community: Cahokia and Amalucan Compared, *Actas 37 Cong. Internac. de Americanistas*, **1**. Buenos Aires; Fowler, M. L. (1969). A Preclassic water distribution system at Amalucan, Mexico, *Archaeology*, **22**.

33 Parsons, J. R. (1969). *op. cit.* p. 32.

34 Parsons, J. R. (1968). Teotihuacán, Mexico, and its impact on regional demography, *Science*, **162**, p. 875. Slightly lower figures in Millon, R. (1964). Teotihuacan Mapping Project, *Amer. Antiq.*, **29**; Sanders, W. T. (1965). *op. cit.* pp. 98–9 and 138–9; Sanders, W. T. and Price, B. J. (1968). *Mesoamerica: The Evolution of a Civilization*. New York. p. 200.

35 The settlement evidence from the Maya zone is conveniently summarized, with bibliography, in Haviland, W. A. (1966). Maya settlement patterns: a critical review, *Publ. Mid. Amer. Research Inst.*, **26**.

36 Thompson, J. E. S. (1965). Archaeological synthesis of the southern Maya lowlands, *in* Wauchope, R. (ed.) *Handbook of Middle American Indians*, **2**. Austin. p. 336.

37 Bullard, W. R. (1960). Maya settlement patterns in north-eastern Petén, Guatemala, *Amer. Antiq.*, **25**.

38 Sanders, W. T. (1963). Cultural ecology of the Maya lowlands, Pt. 2, *Est. de Cultura Maya*, **3**. pp. 211–8.

39 Willey, G. R., Glass, J. B. and Gifford, J. C. (1965). Prehistoric Maya settlements in Belize Valley, *Papers Peabody Mus.*, **54**.

40 Andrews, E. W. (1965). Archaeology and prehistory in the northern Maya lowlands: an introduction, *in* Wauchope, R. (ed.) *Handbook of Middle American Indians*, **2** (1). p. 300.

41 Proskouriakoff, T. (1960). Historical implications of a pattern of dates at Piedras Negras, Guatemala, *Amer. Antiq.*, **25**; Proskouriakoff, T. (1963). Historical data in the inscriptions of Yaxchilan: Part I, The reign of Shield Jaguar, *Est. de Cultura Maya*, **3**.

42 Thompson, J. E. S. (1965). *op. cit.* pp. 253–6. 579-92.
43 Sanders, W. T. (1963). *op. cit.*
44 See Trigger, B. G., this volume, pp. 579–92.
45 Millon, R. (1964). *op. cit*; Sanders, W. T. (1965). *op. cit.* pp. 101–7; Millon, R. (1967). Teotihuacán, *Sci. Amer.* **216.** I have not been able to consult a more recent article in *Science*, **170,** which apparently gives a slightly different population figure from that quoted here.
46 Parsons, J. R. (1968). *op. cit.*
47 MacNeish, R. S., this volume, pp. 67-93. The same appears to be true to some extent in Tlaxcala; see Snow, D. R. (1969). Ceramic sequence and settlement location in prehispanic Tlaxcala, *Amer. Antiq.*, **34.**
48 Sanders, W. T. (1960). Prehistoric ceramics and settlement patterns in Quintana Roo, Mexico. *Contrib.* 60, *Amer. Anthrop. and Hist., Carnegie Inst. Publ.*, **606.** Washington. p. 175.
49 Bullard, W. R. (1960). *op. cit.* pp. 370–1.
50 Reina, R. E. (1967). *op. cit.* p. 9.
51 Pollock, H. E. D., Roys, R. L., Proskouriakoff, T. and Smith, A. L. (1962). Mayapan, Yucatan, Mexico. *Carnegie Inst. Publ.* **619.**
52 Haviland, W. A. (1970). Tikal, Guatemala and Mesoamerican urbanism, *World Archaeol.*, **2**, p. 193.
53 Palerm, A. (1955). *op. cit.* pp. 34–5.
54 Palerm, A. (1955). *op. cit.* pp. 32–3.
55 Sanders, W. T. (1963). *op. cit.* pp. 92–3.
56 Haviland, W. A. (1969). A new population estimate for Tikal, Guatemala, *Amer. Antiq.*, **34**, p. 431.
57 Tozzer, A. M. (ed.). Landa's Relación de las Cosas de Yucatán, *Papers Peabody Mus. of Archaeol. and Ethnol.*, **18**, p. 62.
58 Scholes, F. W. and Roys, R. L. (1968). *op. cit.*
59 Adams, R. M. (1961). Changing patterns of territorial organization in the Central Highlands of Chiapas, *Amer. Antiq.*, **26** (3).
60 e.g. Iximché; Guillemin, J. F. (1965). *Iximché, Capital del Antiguo Reino Cakchiquel.* Guatemala City.
61 Haviland, W. A. (1969). *op. cit.* p. 432.
62 Adams, R. M. (1961). *op. cit.*
63 Borhegyi, S. F. de (1956). Settlement patterns in the Guatemalan Highlands, *in* Willey, G. R. (ed.) Prehistoric settlement patterns in the New World, *Viking Fund Publ. in Anthrop.* **23**; Shook, E. M. and Proskouriakoff, T. (1956). Settlement patterns in Mesoamerica and the sequence in the Guatemalan Highlands, *in* Willey, G. R. (ed.) *op. cit.*
64 Miles, S. W. (1958). An urban type: extended boundary towns, *S.W.J. Anthrop.*, **14**; Miles, S. W. (1965). Summary of pre-Conquest ethnology of the Guatemala-Chiapas Highlands and Pacific Slopes, *in* Wauchope, R. (gen. ed.) *Handbook of Middle American Indians*, **2**, pp. 278–80.
65 Spores, R. (1969). *op. cit.*
66 Haviland, W. A. (1969). *op. cit.*; Haviland, W. A. (1970). *op. cit.* Where there are minor differences between these two articles I have followed the 1970 figures.
67 Parsons, J. R. (1969). *op. cit.*; Parsons, J. R. (1970). *op. cit.*
68 Sanders, W. T. (1965). *op. cit.*; Sanders, W. T. (1968). *op. cit.*
69 Sanders, W. T. (1968). *op. cit.* p. 158.
70 Sanders, W. T. (1965). *op. cit.* p. 158.
71 Calnek, E. E. (1971). Settlement pattern and Chinampa agriculture at Tenochtitlán, *Amer. Antiq.*
72 Borah, W. and Cook, S. F. (1963). The aboriginal population of central Mexico on the eve of the Spanish Conquest, *Ibero-Americana*, **45.**
73 Personal communication from E. E. Calnek, who emphasizes that this figure is highly tentative.

74 Calnek, E. E. (1971). *op. cit.* Contrast Sanders, W. T. (1956). *op. cit.* p. 112, who maintained that Tenochtitlán could have been supported by the produce of *chinampa* cultivation within the Basin of Mexico. This calculation was based on a figure of only 62,000 inhabitants for Tenochtitlán, which in turn derived from an underestimate of the size of the pre-Conquest city and an assumption that the prehispanic density of occupation was the same as that of Colonial Mexico City on the same site.

75 This figure is only approximate. The various tribute lists do not agree with each other in detail, and there is some doubt about the exact equivalent of an Aztec *troje* (storage bin). The analysis by Borah, W. and Cook, S. F. (1963). *op. cit.* gives a total of 22,800 tons of food as the tribute due to the Triple Alliance *as a whole*, and the Tenochtitlán share would therefore be two fifths of this. Molíns Fábrega gives a total of 19,000 metric tons, but considers that the tribute lists refer to Tenochtitlán alone: see Molíns Fábrega, N. (1954–5). El Codice Mendocino y la Economía de Tenochtitlán, *Rev. Mexicana de Est. Antropológicos*, **14.** I have followed the Borah-Cook interpretation.

76 Chapman, A. M. (1957). Port of trade enclaves in Aztec and Maya civilizations, *in* Polanyi, K., Arensberg, C. M., and Pearson, H. W. (eds.) *Trade and Market in the Early Empires.* Glencoe, Ill.

77 Haviland, W. A. (1970). *op. cit.* p. 196.

78 Pollock, H. E. D., Roys, R. L., Proskouriakoff, T. and Smith, A. L. (1962). *op. cit.*

KENT C. DAY

Urban planning at Chan Chan, Peru

Chan Chan is located along a low, desolate ridge on the west side of the Moche Valley about 550 km north of Lima, Peru. At present the site covers an area of about 10 km^2, but was probably somewhat larger during the time it was occupied. At that time—between about A.D. 900–1000 to 1450—Chan Chan was the largest Chimu centre of the Peruvian north coast. Because of its magnitude, relatively few intensive archaeological investigations of Chan Chan have been made in the century since Squier[1] visited the site. Rivero and Tschudi[2], Hutchinson[3], Middendorf[4], and Bandelier[5] explored the site in the nineteenth century as part of their travels in Peru. More recently Bennett and Bird[6], Schaedel[7], Willey[8], and Kosok[9] have discussed some social implications of the *ciudadelas* as clan or high status residential areas. West[10] has reported on part of the large area of dwellings and "shrines" located outside the *ciudadelas*. Horkheimer's[11] suggestion that the formal layout of the *ciudadelas* was for storage and controlled movement of people is the most germane to the topic of this paper. In the years since its abandonment the site has suffered damage from infrequent rainstorms and, to a lesser degree, earthquakes. The greatest amount of destruction, however, has been due to centuries of looting.

Within the urban area of Chan Chan are crowded dwellings, small compounds, isolated burial platforms, a large area of melted earth, wells and, on the seaward side of the site, a vast cemetery. Along the eastern and northern edges of the site are three large mounds that were originally truncated pyramids. In the central part of the site are ten large compounds called *ciudadelas*. The best preserved architecture and most sophisticated planning at Chan Chan is found within the *ciudadelas*. A complex system of passages inter-connects the *ciudadelas* and a system of roads leads from the centre of the site to other parts of the valley and beyond. The purpose of this paper is to examine certain aspects of the best preserved, formally organized areas at Chan Chan. Of immediate interest are the entry courts, storage areas, U-shaped *audiencias* and burial platforms that occupy the bulk of the space in the *ciudadelas*. These, along with the communication system, are considered expressions of the city's well-developed commercial prosperity attendant with the socio-economic position of those corporate groups who controlled such enterprises.

Materials

Nearly all the structures at Chan Chan were built of materials found locally. Massive walls and platforms of the *ciudadelas* and other enclosures were built of adobe bricks laid over a foundation of boulders and cobbles. Some walls and platforms were built of large blocks of *tapia* (rammed gravel and mud) laid upon boulder foundations. Smaller structures and walls were usually built of adobe bricks mortared and plastered with clay. Additionally, gravel, cobbles, adobe brick rubble, and garbage were frequently used as floor ballast or hearting for walls, ramps, and platforms. Wood was used extensively as posts, columns, and, to some extent, lintels. Canes, reeds, plaited mats, and grass thatch, often tied with twine, were used as roofing material.

Layout and planning

The great *ciudadelas* are the most formally arranged parts of the site. Usually laid out as great rectangles, most of them are divided into three sectors by transverse walls. The northernmost, or entry sector of each *ciudadela* contains an elaborate entry passage, a spacious entry court flanked by smaller courts, a kitchen, several *audiencias*, and scores of storerooms. A tortuous system of corridors interconnects these interior spaces. Central sectors repeat the organization of the entry sectors but the use of space is somewhat different. Here, entry courts are smaller, and there are fewer *audiencias*. Storerooms, however, are more numerous. Specialized burial platforms with their own forecourts occur only in the central sectors. Similar to the entry sectors, structures in the central sectors are linked together by a system of tortuous corridors. The southernmost sectors at the seaward ends of the *ciudadelas* contain remains of dwellings, domestic débris, simple burials and usually a well. These areas contain no major architectural features, and, except for a single, long corridor, are isolated from the rest of the *ciudadela*.

Controlled circulation of people and goods or people with goods was apparently the primary function of the layout of the *ciudadelas*. Access to them was limited to one principal entry near the centre of the north wall. This entry opens to a dog-leg passage that leads to a large entry court. A ramp and corridor on the opposite side of the entry court give access to two smaller courts. Narrow corridors from this pair of courts lead to courts containing *audiencias*. Storerooms beyond the *audiencias* are reached via more corridors. Much the same layout of structures and system of corridors is repeated in the central sectors. In both sectors *audiencias* dominate the access to most of the corridors leading to the storage areas. Although there are a few alternate passages to storerooms that avoid *audiencias*, traffic on either route was necessarily single file or possibly one way. Even so, the *audiencias* are situated such that their occupants had considerable control over whoever or whatever passed to or from the storerooms.

Except for spaciousness, colonnades, and low platforms, there is little in the entry courts to indicate their function. These courts, however, are among the few places within the *ciudadelas* where public gatherings could have taken place. If some part of each *ciudadela* were devoted to marketing, the entry courts possibly served as the market places. The smaller lateral courts may have served as accessory clearing houses.

Several *ciudadelas* share another feature in common: burial platforms. These are massive one- or two-tiered structures built of adobe bricks or tapia blocks. There were T-shaped and rectangular chambers sealed in the platforms and usually a small, U-shaped building on the summit. The summit was reached by a system of lateral and inset ramps on the north face of the platforms.

All the burial platforms have been heavily looted, two to the point that they have been practically razed. Human bone, exotic shells, fragments of elaborate textiles and fine ceramics commonly occur in the backdirt piles, around the platforms and within the looted chambers. Although no unlooted chambers have been found, it is apparent that most once contained richly stocked interments.

Interpretation

The most formally arranged structures at Chan Chan—*the ciudadelas*—can be interpreted as centres of Chimu economic organization. Much of the wealth once contained in the burial platform tombs was probably accumulated through the commercial activities of the occupants of the *ciudadelas*. Assuming that movement of goods is essential to a prosperous exchange system, it follows that residents in the *audiencias* were in the proper place to manage, administer, and profit from exchange. Further, it is possible that marketing could have taken place in the entry courts and some manufacturing accomplished in the storeroom courtyards. Since all of these courts are an integral part of the *ciudadelas*, it is likely that any sales or manufactures there were also under the auspices of the *audiencia* residents. Luxury goods from distant sources (i.e. metal from the highlands, feathers from the Amazon basin, warm-water shells from the north), elaborate textiles, lavish use of wood, fine pottery, plus the labour and materials to build a *ciudadela*, point out the extravagant wealth and symbolize the high status of the living and dead occupants of the *ciudadelas*.

It is suggested that the wealth sealed in the burial platforms was deposited with deceased members of the *ciudadela* administration. Moreover, the planning of *ciudadelas* indicates that the administrators were actively in the process of storing and controlling the exchange of more goods. Thus, in the case of Chan Chan, planning served to orient people, safeguard goods, provide markets and administrative facilities, protect the wealthy dead, connect the city to the countryside and distant parts, and, finally, to assure the continued commercial prosperity of the city.

Notes

1 Squier, E. G. (1877). *Peru. Incidents of Travel and Exploration in the Land of the Incas.* New York.
2 Rivero, M. and J. von Tschudi (1851). *Las Antigüedades Peruanas.* Vienna.
3 Hutchinson, T. (1873). *Two Years in Peru with Explorations of its Antiquities* London.
4 Middendorf, E. W. (1894). *Peru, Das Kustenland von Peru,* II. Berlin.
5 Bandelier—see Kosok, P. (1965). *Life, Land and Water in Ancient Peru.* New York.
6 Bennett, W. C. and Bird, J. B. (1949). Andean culture history, *Amer. Mus. of Nat. Hist. Handbook,* **15**. New York.
7 Schaedel, R. (1951). Major ceremonial and population centres in northern Peru, *in Civilizations of Ancient America; Selected Papers of the XXIX International Congress of Americanists.* Chicago. pp. 232–43.
8 Willey, G. R. (1953). Prehistoric settlement patterns in the Viru Valley, Peru, *Bureau of Amer. Ethnology,* **155**. Washington, D.C.
9 Kosok, P. (1965). *ibid.*
10 West, M. (1970). Community settlement patterns at Chan Chan, Peru, *Amer. Antiq.,* **35**. pp. 320–5.
11 Horkheimer, H. (1944). *Vistas Arqueologicas del Noreste del Peru.* Trujillo.

DAVID J. ROBINSON

Changing settlement patterns in colonial Hispanic America

The past decade has seen the establishment of what may be termed the urban basis of colonial Hispanic activity in the New World. Scholars in a variety of disciplines have described, and in part analysed, what they regard as the significant forms and functions of the settlements referred to as *villas* or *ciudades*[1]. The town is portrayed as a key agent in the process of hispanicization: symbol of the new culture, locus of economic power, centre of innovative change, and focus of authority and administrative controls. Before such an urban-orientated analysis is accepted, however, it is suggested that the corresponding role of the rural components of the settlement hierarchy be examined in much more detail, and that many of the so-called "urban" centres be re-examined to isolate their distinctively urban characteristics. It is time for an acceptable definition of what constituted an "urban" settlement in colonial Hispanic America.

Reasons for the relative neglect of rural settlement studies in the area are not difficult to come by. First there are the problems associated with definitions[2]. Unlike the town, a plethora of terms have evolved to describe rural settlements that ranged in size from the hut occupied by a single family to the village community of several thousand inhabitants[3]. For the most part students of the colonial period have avoided the key issues of variation in type and function of such settlements[4]. Outside the confines of the town (*extra-muros*) problems proliferate. Immediately the researcher faces the task of locating the rural settlement, which has probably shifted its position several times, may have changed its name, or have been incorporated within later phases of settlement expansion[5]. For the countryside (*campana*) of colonial Hispanic America cartographic sources are very poor. Where descriptions of estates are found they are usually described in terms of features that have long since been modified out of all recognition[6]. At the regional scale it is possible to map rural-urban population distributions (Figs. 1 and 2); at the local level it involves many months of work in the field as well as the archive (Fig. 3)[7]. As well as these operational difficulties involved in the study of rural settlements, it has also become clear from historiographical work that in

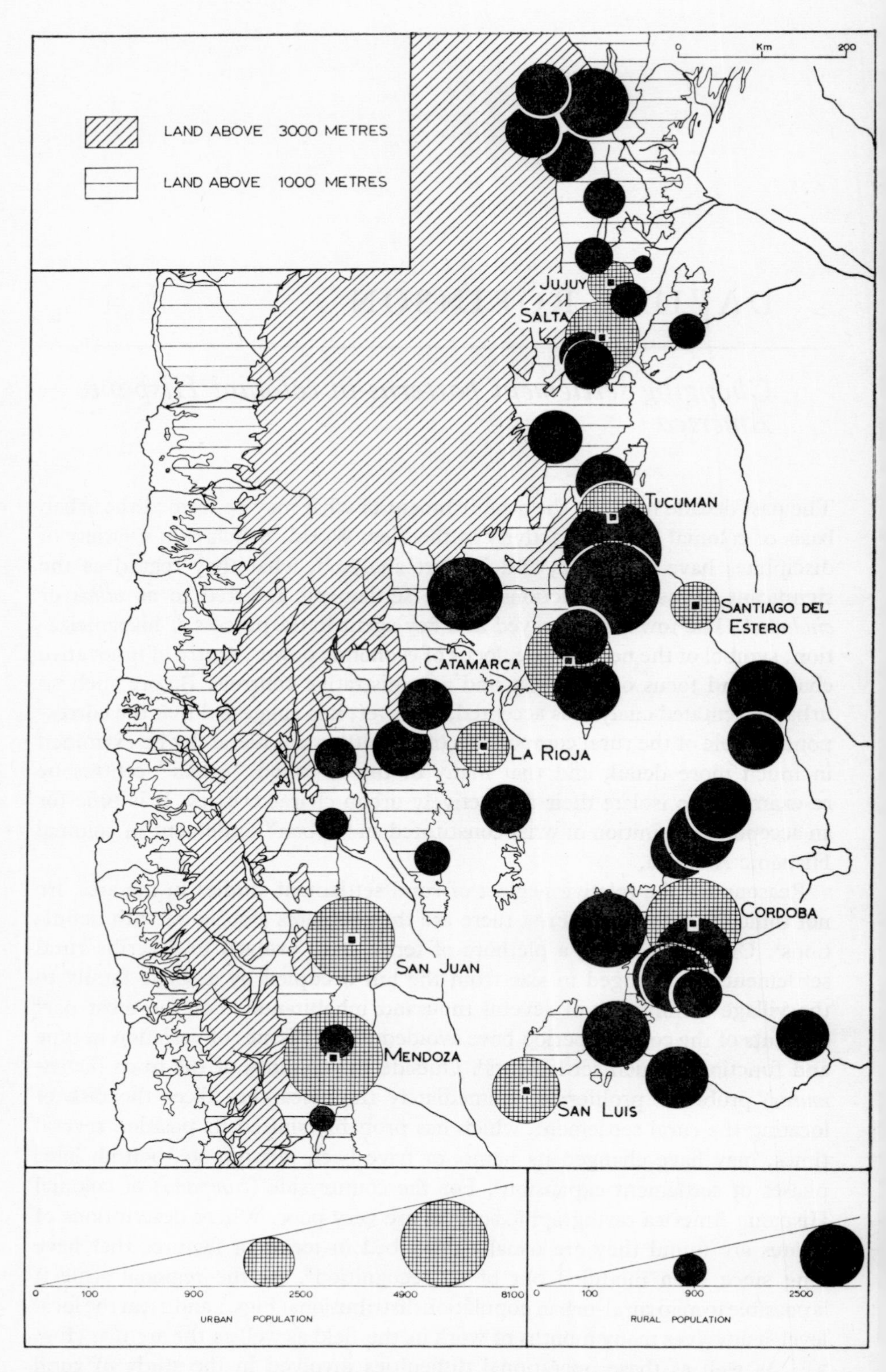

Fig 1 Distribution of rural-urban population in North-west Argentina in 1777.

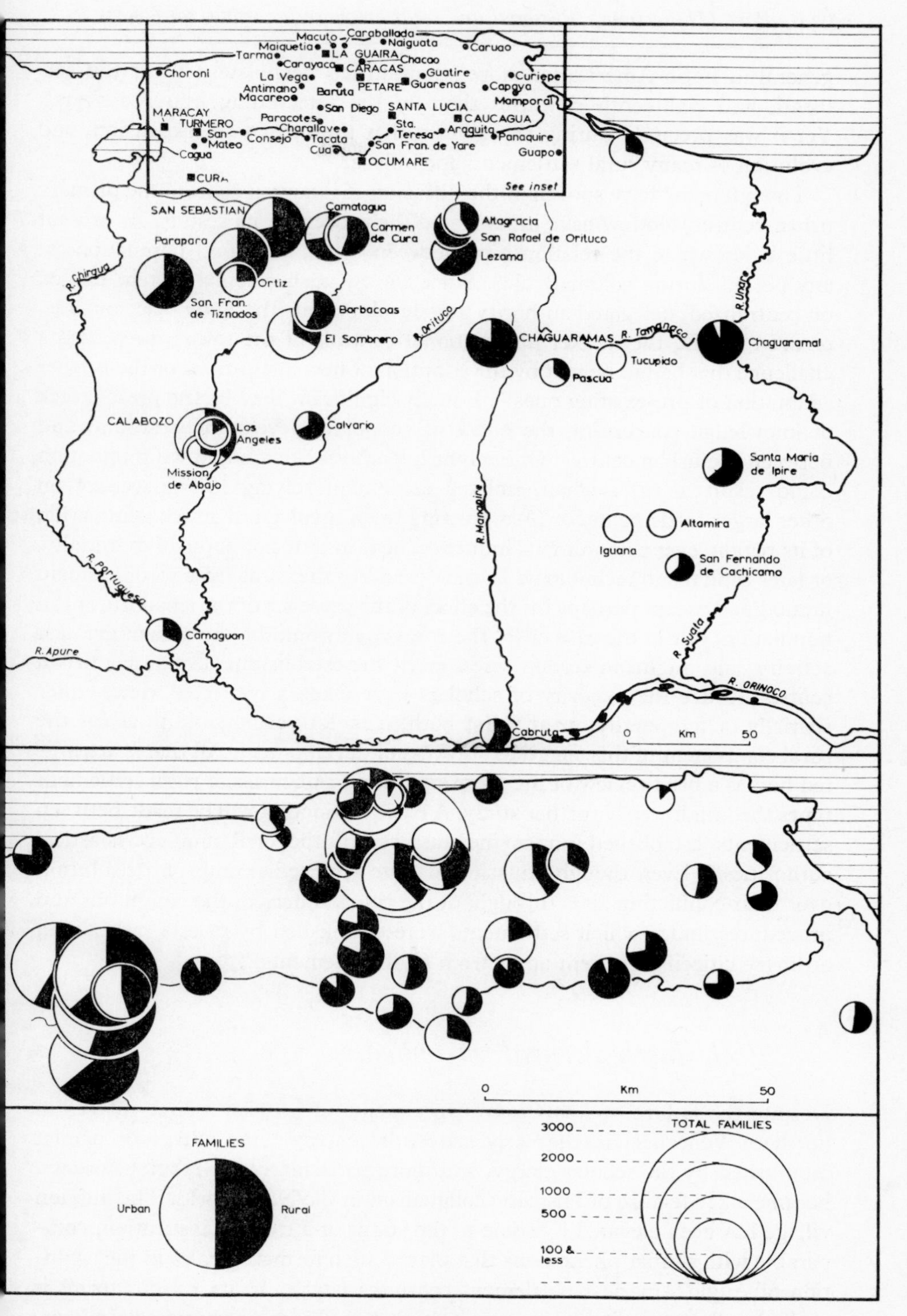

Fig 2 Distribution of rural-urban population in Central Venezuela around 1785.

general in Latin America far more attention has been lavished on the sixteenth and eighteenth centuries than on the intervening hundred years[8]. Yet it was precisely during those years that the formation, expansion, and evolution of many rural settlements took place.

Though many have spoken of the diffusion of innovations from the primary urban centres, too few have investigated the process or its results. At present little is known of the relationships between urban and rural populations at any period during colonial rule[9]. Since the Spanish colonial system hinged on centralized, delegated authority any developments that took place outside, or even within, the weaker jurisdictional controls of the towns, presented a challenge that had to be met by the adoption of new institutions or the further adaptation of pre-existing ones[10]. Equally significant may be the present lack of knowledge concerning the effect of economic development around and between the urban centres. One might hypothesize that any town foundation could result in (*a*) a small, isolated settlement relying for its welfare on other larger settlements, or (*b*) a growing town involved in active stimulation of its region, ever extending its influence, be it in terms of population initially, or later capital and technology. In the case of (*a*) the rural areas would remain unmodified except perhaps for the effect of the presence of the small European population, but in the case of (*b*) the rural zone would be the area of greatest activity, and might, in certain cases, merit the establishment of a new urban centre[11]. Since the majority of scholars have taken a restricted view, either spatially or temporally, or at worst both, it is not yet possible to assess the rural component of this suggested process of development. All that is attempted here is a brief review of the more significant categories of rural settlement types that might repay further study. A basic distinction will be made between settlements established under the auspices of the civil and ecclesiastical authorities[12]. Even though church and state achieved a compromise solution over their conflicting aims throughout the colonial period, the conditions and procedures under which settlements were established by priests rather than ordinary citizens set them apart from civil settlements[13].

Ecclesiastical rural settlements

Despite the failure of Las Casas' experiments in utopian model villages in northern Venezuela in the early sixteenth century[14], the village or hamlet established by the secular clergy, and more particularly the religious orders, became a key feature of Hispanic colonialism in the New World. The mission village has been elevated by some to the status of a frontier institution, comparable with similar phenomena that played such formative roles in the evolution of Anglo-American settlement centuries later[15]. In its various forms it undoubtedly transmitted Hispanic culture to millions of aboriginal Indians. It became the staging-post of culture contact, the strategic mode of land

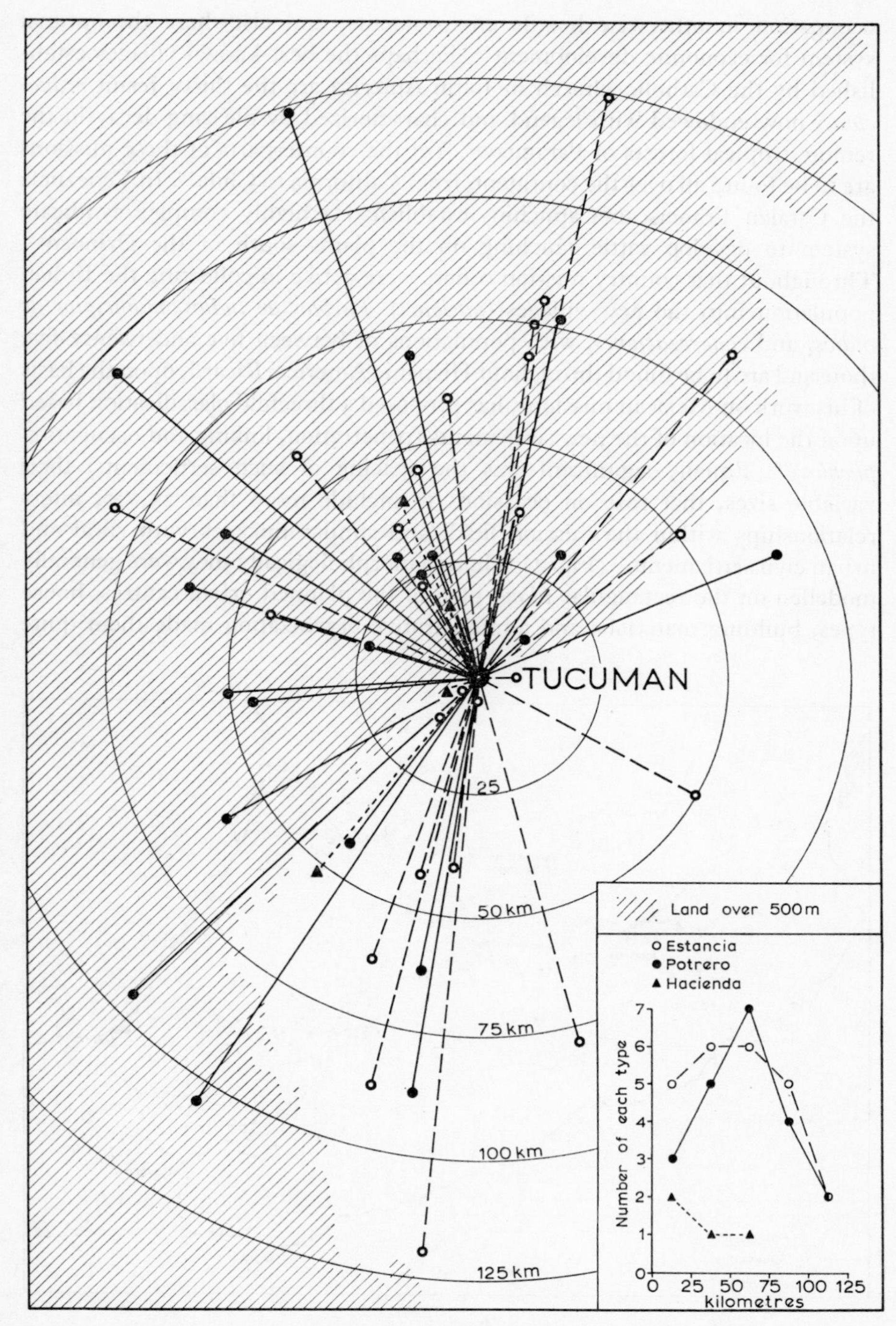

Fig 3 Distribution of agricultural estates with principal service links to Tucumán during the years 1670–1730.

occupation in otherwise hostile environments, and elsewhere the optimal system for economic development[16]. Perhaps the best-known villages established by the Catholic church in Latin America are the thirty Jesuit *reducciones* now endowed with legend and characterized by pleasing decay in the remote tropical forests of Paraguay[17]. Yet the counterparts of these missions are to be found all over the continent, from California to Chile[18]. In Venezuela the Catalan Franciscan Capuchins established a highly complex syndicate system to develop cattle ranching on the plains south of the Orinoco[19]. Throughout that country mission orders were active in collecting the Indian population into the new villages (*misiones*). *Pueblos de indios*, *municipios de indios*, and *congregaciones*, with populations of between five and twenty-five thousand are to be found throughout Hispanic America. Ricard, upon the basis of his work on Mexican missions, has proposed a threefold classification based upon the location of the new settlements (penetration, linking and occupation *pueblos*)[20]. Equally significant was the spacing of such settlements, their variable sizes, that rose on occasions to include more than 25,000, interrelationships within the mission field, and similar linkages with rural and urban civil settlements[21] (Fig. 4). Though morphologically they were generally modelled on the rectangular *plaza* plan of the planned urban centres, house types, building materials, and administrative organization made them quite

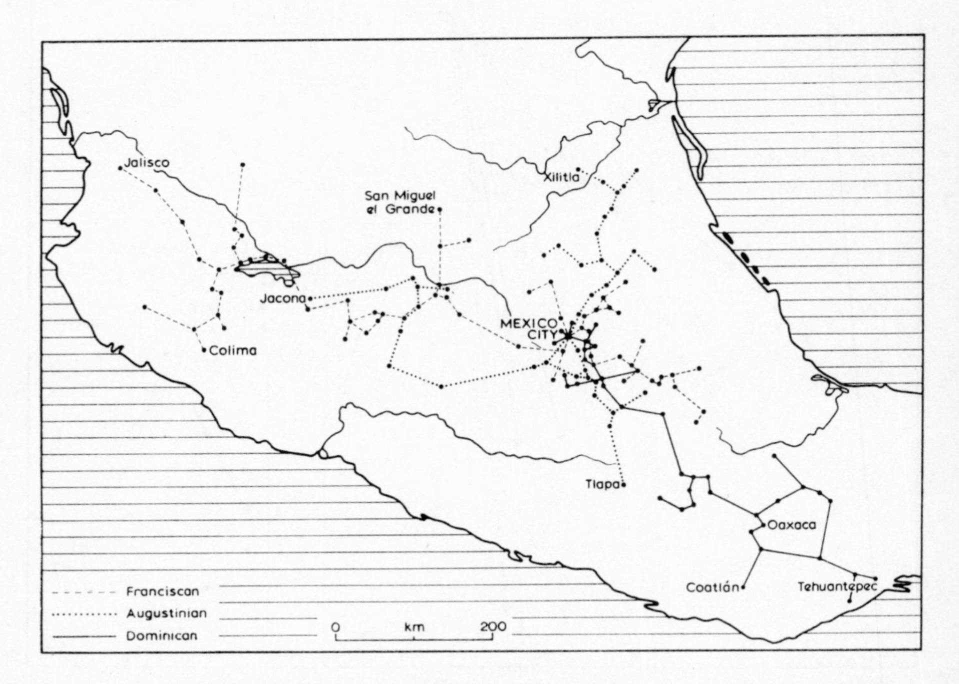

Fig 4 Distribution of mission establishments in Mexico around 1570 showing nearest neighbour links between the settlements of each mission order.

distinctive[22]. Through most of the colonial period, but especially during the late seventeenth century, the *corregidor*, *fraile*, and *doctrinero* in charge of these settlements became some of the most significant innovators amongst the Indian-mestizo population[23].

Civil rural settlements

Though the status of an urban settlement during the colonial period was generally measured by its legal title or corporate privileges, of long-term importance was the prosperity generated within the hinterland of the town. It was there that population rapidly increased after the seventeenth century. Since mining was an activity that achieved early and enduring significance in the New World, its associated rural settlements will be noted first. From northern Mexico to central Argentina the ephemeral camps, permanent hamlets and large villages have been described[24]. Containing sizable populations they were usually quite unlike their planned urban counterparts. Street plans, house types and the distribution and density of the dwellings reflected their origins and haphazard growth. Likewise they had somewhat different relationships with pre-existing settlements, and in some regions, such as Colombia, northern Mexico and northern Chile, formed the basic units in a complex network of communications[25]. The *real de minas* rapidly became a rural institution[26].

Much more important, however, in terms of number of rural settlements established were those that owed their origin to agricultural developments. It is a remarkable fact that in spite of the significance of the agricultural estate in colonial Hispanic America, few have studied it from the viewpoint of a settlement type. Such studies that have been completed on the plantation, cattle ranch and large arable farm had tended to concentrate on institutional or purely economic aspects. Few put these units of land occupation into the context of the overall settlement pattern. Even fewer mention the fact that one of the most significant changes in the settlement pattern in Hispanic America was the evolution of these new settlement forms[27]. Some of the blame for this lack of interest in the interaction of rural and urban components may be laid at the door of academic specialization. Scholars have tended to concentrate on one or other aspect of development, at the expense of seeing that both were, as they still are, interdependent[28]. Faced at the present time with the tide of migration from the countryside to the towns in Latin America, few would deny that the origins and evolution of such rural settlement zones are worthy of study. The changing pattern of settlement that altered the balance from urban to rural by the end of the colonial period was a temporary phenomenon; many would wish to repeat the process in the present century.

The most significant colonial agricultural settlements were the commercially orientated extensive pastoral ranch and agricultural estate producing a wide range of crops. Throughout the entire continent the nucleated agricultural settlement was to be found. While the medieval Iberian origins of these systems of economic organization are still debated[29], there is no doubt that some were merely adaptations of well-tried methods of land occupation, while others, such as the tropical plantation, necessitated much experimentation before their productivity could be ensured[30]. Characteristic features of these units were their large size, both in terms of labour force employed and land area within the holding, their almost "factory-like" methods of production, and centralized organization, and their linkages, by way of their owners' capital, to entrepreneurs in the towns[31]. The exact forms and functions of these settlements varied from region to region and from one century to the next. Some, such as those of northern Mexico, in form, at least, were reminiscent of the Andalucian *cortijo*, large fortified central farms (*casco*) housing up to 2000 people in isolated communities[32]. Elsewhere in Chile, Venezuela, Argentina and Colombia circumstances allowed the development of smaller estates containing between twenty and 200 persons[33]. Each of the major grassland zones of the New World enjoyed its regional type of cattle ranch called *hato*, *fundo*, *estancia*, *hacienda*, or *rancho*. Recent studies have shown that during the second half of the eighteenth century in several areas of Hispanic America hamlets and villages (*aldeas* and *pueblos*) were being established outside the control of the semi-feudal estates[34]. In many cases such developments made the creation of new administrative units necessary[35]. In other regions of central America the scatter of dispersed rural dwellings (*ranchos*) reflected the increasing significance of tenant farmers[36].

Mining and agricultural settlements were not the only ones established by the end of the colonial period. Also significant in certain regions were the frontier garrisons (*presidios*), set up as defensive posts to keep out marauding bands of hostile Indians. In Mexico and Argentina, Chichimec and Araucanian bands respectively harassed towns and farms, consequently lines of forts were established that led to further expansions of the rural settlement frontier[37]. In regions with relatively dense Indian population, yet other rural villages and hamlets were established in the reserves (*resguardos*) set aside for the protection of the Indians from ruthless landowners[38]. Elsewhere as trade developed between the larger urban centres it was found necessary to build small wayside stations for the mule and waggon trains. Many of these gradually grew into small villages[39].

The "reduction" of Indians into mission villages had its civil counterpart in the "congregated" villages (*congregaciones*) formed in their hundreds in Mexico and elsewhere in the late sixteenth and early seventeenth century[40]. However, although the population of many Indian hamlets (*rancherios*) was forcibly transferred to new planned village sites, often containing more than 500 inhabitants, it was not uncommon for these new villages to remain empty of population for extended periods of time, the settlement merely providing a new location for seasonal markets and other service facilities[41].

Conclusion

Even a brief outline such as this may demonstrate that in the colonial urbanization of Hispanic America, a good deal of "ruralization" was involved. Indeed it might be argued that the urbanization process itself depended upon the success of developments within the rural areas. What should not be overlooked in the rush to the easily accessible town records is the quantity of neglected problems associated with rural colonial America. Only when more detailed research has been undertaken on the settlement types as they evolved throughout the entire colonial period, will it be possible to assess the relative roles of the urban and rural components. Definitions have to be improved to take into account what were clearly recognized at the time as distinctive areas, *ciudad y campana*; only then will it be meaningful to speak of urban or rural phenomena.

Notes

1 Urban centres and towns are used in this paper as synonyms for both *villa* and *ciudad*. Notable contributions on urban developments in colonial Hispanic America have been made by Morse, R. M. (1962). Some characteristics of Latin American urban history, *Amer. Hist. Rev.*, **67**, pp. 317–38; Morse, R. M. (1961). Latin American cities: aspects of function and structure, *Comp. Studies in Society and Hist.*, **4**, pp. 473–93; Morse, R. M. (1968). Cities and society in nineteenth century Latin America, *Actas y Memorias del XXXVII Congreso de Americanistas*, **I**. Buenos Aires. pp. 306f. Also extremely useful are Hardoy, J. (1965). La influencia del urbanismo indígena en la localización y trazado de las ciudades coloniales, *Ciéncia y Investigación*, **21**, pp. 386–405; Hardoy, J. (1968). Escalas y funciones urbanas en América Hispánica hacia el año 1600, *Actas y Memorias del XXXVII Congreso de Americanistas*, I, (1968). pp. 171–208 and Gibson, C. (1968) Spanish-Indian institutions and colonial urbanism in New Spain, *Actas y Memorias del XXXVII Congreso de Americanistas*. pp. 225–39. More general accounts may be found in Davis, K. (1960). Colonial expansion and urban diffusion in the Americas, *Internat. J. Comp. Sociol.*, pp. 43–66; Kubler, G. A. (1964). Cities and culture in the colonial period in Latin America, *Diogenes*, **47**, pp. 53–62 and Houston, J. M. (1968). The foundation of colonial towns in Hispanic America, *in* Beckinsale, R. P. and Houston, J. M. (eds). *Urbanization and its Problems*. Oxford. pp. 352–90. For a comparative view of Brazilian rural colonial settlements see Buarque de Holanda, S. (1936). *Raizes do Brasil*. Rio de Janeiro.

2 Of inestimable value is Friederici, G. (1960). *Amerikanistisches Wörterbuch*. Hamburg, but even this omits many terms relating to settlements and one is best served by specialized glossaries appended to monographic studies such as West, R. C. (1942). *Colonial Placer Mining in Colombia*. Baton Rouge.

3 Almost unique in its coverage is Diégues, M. (1963). *Estabelecimentos rurais na América latina*. Geneva.

4 An exception is provided by Zavala, S. (1958). Las fronteras de Hispanoamerica, *Cuadernos Americanos*, **17**, pp. 374–84,—a description of rural life-styles.

5 This is especially true of the critical zone of "suburban" settlements that formerly girdled practically every sizable town of colonial America, and that have now been submerged in the urban expansion of the present century. This difficulty is often ignored in calculations of "urban" and "rural" population totals, for example, Martínez, P. S. (1961). *Historia Económica de Mendoza durante el Virreinato*. Madrid. Note is made of the problem in Rasini, B. (1965). Estructura demográfica de Jujuy, siglo XVIII, *in* Sanchez-Albornóz, N. and Torrado, S. (eds.) *América Colonial: Población y Economía*. Rosario. pp. 119–50.

6 For problems of locating rural colonial land units in Venezuela see Robinson, D. J. (in press). Problemas y soluciones acerca la localización de unidades agrícolas en Venezuela colonial, *Rev. Geogr.*, Mérida.

7 Two areas have been mapped, the first in central Venezuela using as a base the *visita* of Mariano Martí which listed urban and rural population separately; the second, north-west Argentina using a 1777 census. Detailed local surveys of colonial rural settlements have been completed for south-east Venezuela in Robinson, D. J. (unpublished). *Geographical Change in Venezuelan Guayana*, 1600–1880. Ph.D. thesis, London, 1967, and for the stock pasturing zone around the town of Tucumán in north-west Argentina (Fig. 2).

8 See Stein, S. J. (1964). Latin American historiography: status and opportunities, *in* Wagley, C. (ed.) *Social Science Research on Latin America*. New York. pp. 86–125; and Simpson, L. B. (1953). Mexico's forgotten century, *Pacific Hist. Rev.*, **22**, pp. 113–21.

9 Recent studies have paid some attention to this problem but have been primarily concerned with either one or the other population components. See Acevedo, E. O. (1956–7). La Gobernación del Tucumán: sus problemas y la divisóin de su territorio, *Rev. Hist. Amer. y Argentina*. **1**. pp. 119–57; Borde, J. and Góngora, M. (1956). *Evolución de la propiedad rural en el valle de Püangue*. Santiago; Brady, T. (1970). *The Application of computers to the analysis of census data. The Bishopric of Caracas, 1780–1830*. Manitoba; Comadran Ruíz, J. (1962). Nacimiento y desarrollo de los núcleos urbanos y del poblamiento de la compaña del País de Cuyo durante la época hispana (1550–1810), *Anuario de Estud. Amer.*, **19**, pp. 145–246; Comadran Ruíz, J. (1958). La población de la ciudad de Catamarca y su jurisdicción al crearse el Virreinato, *Rev. Hist. Amer. y Argentina*, **2**, pp. 125–44.

10 Relevant to this point are Altamira, R. (1945). *Autonomía y Decentralización legislativa en el régimen colonial espanol*. Coimbra; and Phelan, J. L. (1959). Authority and flexibility in the Spanish Imperial Bureaucracy. Mimeographed paper, Admin. Sci. Centre, Univ. of Pittsburgh; Phelan, J. L. (1967). *The Kingdom of Quito in the Seventeenth Century: Bureaucratic Politics in the Spanish Empire*. Madison.

11 See Pla, A. J. (1964). La propiedad rural en América Latina, *Anuario del Inst. de Invest. Hist.*, (Rosario), 7, pp. 31–69; García L. N. (1963). La sociedad rural de México en el siglo XVIII, *Anales de la Univ. Hispalense*, (Sevilla), **24**, pp. 19–53.

12 This basic division is described in general terms in Diffie, B. (1945). *Latin American Civilization: the Colonial Period*. Harrisburg; Zavala, S. (1943). *New Viewpoints on the Spanish Colonization of America*. Pennsylvania; and Viceres Vives, J. (1957). *Historia Social y Económica de España y América*. Barcelona.

13 On the mission methods see Specher, J. (1953). *Die Missionmethode in Spanisch-Amerika in 16 Jahrhundert mit besonderer Berücksichtung der Konzilien und Synoden*. Schönbeck. For civil practices useful information is given in Haring, H. (1952). *The Spanish Empire in America*. New York; and Zavala, S. (1944). *Ensayos sobre la colonización española en América*.

Buenos Aires. Early Brazilian missionary practices are described in Ricard, R. (1937). Les Jesuites au Brésil pendant la seconde moitié du XVIe siècle: Méthodes missionaires et conditions d'apolestat, *Rev. d'hist. des Missiones*, **15**, pp. 321–66; 435–70.

14 Hanke, L. (1935). *The First Social Experiments in America*. New York.

15 Bolton, H. E. (1917). The mission as a frontier institution in the Spanish American colonies, *Amer. Hist. Rev.*, **23**, pp. 42–61, began the long list of publications on this aspect of colonial rule. Other valuable contributions are Phelan, J. L. (1956). *The Millennial Kingdom of the Franciscans in the New World*. Berkeley. Useful as comparative studies are Sánchez-Albornóz, C. (1963). The frontier and Castilian liberties, *in* Lewis, A. C. and McGann, T. F. (eds.) *The New World looks at its History*. Austin. pp. 28–41, and Zavala, S. (1957). The frontiers of Hispanic America, *in* Wyman, W. D. and Kroeber, C. B. (eds.) *The Frontier in Perspective*. Madison. pp. 35–58.

16 Details of the colonization of the central Mexican area are to be found in Ricard, R. (1966). *The Spiritual Conquest of Mexico*. (Trans. by L. B. Simpson). Berkeley; a detailed account of the work of a Jesuit pioneer on the northern Mexican frontier is Shiel, W. E. (1965). *Gonzalo de Tapía, Jesuit Pioneer in New Spain*; the reduction procedure is explained in McShane, C. M. (1963). *Hernando de Santarén, Pioneer and Diplomat, 1565–1616*. pp. 149–52.

17 Mörner, M. (1953). *The Political and Economic Activities of the Jesuits in the La Plata Region. The Hapsburg Era*. Stockholm.

18 For California and Mexico see Dunne, P. M. (1944). *Pioneer Jesuits in Northern Mexico*. Berkeley; Dunne, P. M. (1940). *Pioneer Black Robes on the West Coast*. Berkeley; for Venezuela, Robinson, D. J. (1971). Venezuela and Colombia, *in* Blakemore, H. and Smith, C. T. (eds.). *Latin America: Geographical Perspectives*. Methuen; for Chile, Silva Vargos, F. (1962). *Tierras y pueblos de Indios en el reino de Chile*. Santiago. Also useful is Tibesar, A. (1953). *Franciscan Beginnings in Colonial Peru*.

19 Robinson, D. J. (in press). The syndicate system of the Catalan Capuchins in the Caroní mission field, south-east Venezuela, *Revista de Historia* (Mexico). This system may be compared with that described by Trentlein, T. E. (1939). Economic regime of Jesuit missions in eighteenth century Sonora, *Pacific Hist. Rev.*, **8**, pp. 12–22.

20 Ricard, R., *op. cit.* p. 77.

21 The inter-relationships between missions and civil settlements are discussed in Robinson, D. J. (1967)., *op. cit.* pp. 159–298; and Shiels, W. E. (1965). *op. cit.* p. 131.

22 Plans of mission villages are to be found in H[amilton], J. (1820). Journal of a trip from St. Thome de Angostura to the Capuchin Missions of the Caroni, *Quart. J. Sci. Lit. and Arts*, **8**, p. 250; and Mörner, M. (1953). *ibid.*

23 Lohmann, V. G. (1957). *El Corregidor de Indios en el Peru bajo los Austriacos*. Madrid; Rowe, J. H. (1957). The Incas under Spanish Colonial Institutions, *Hispanic Amer. Hist. Rev.*, **37**, pp. 155–99; Gibson, C. (1964). *The Aztecs under Spanish Rule*. Stanford. pp. 32–57; Fals-Borda, O. (1969). *Subversion and Social Change in Colombia*. New York.

24 West, R. C. (1942). *op. cit.*; West, R. C. (1949). *The Mining Community of Northern New Spain: the Parral Mining District*. Berkeley; West, R. C. (1958). The Mining Economy of Honduras during the Colonial Period, *33rd Congreso Internat. de Americanistas*, **2**, San José, pp. 767–77; Martínez, P. S. (1961). *ibid.*

25 For Chilean mineral developments see Pederson, L. R. (1966). The mining industry of the Norte Chico, *North-western Studies in Geogr.*, **11**.

26 Mecham, J. L. (1927). The *real de minas* as a political institution, *Hisp. Amer. Hist. Rev.*, **7**, pp. 45–83.

27 For surveys of plantations in the colonial period see Pan American Union

(1959). Plantation systems of the New World, *Soc. Sci. Res. Council Pub. no.* 7, Washington; Wolf, E. R. and Mintz, S. (1957). Haciendas and plantation in Middle America and the Antilles, *Soc. and Econ. Stud.*, pp. 380–412; Strickon, A. (1965). Hacienda and plantation in Yucatán: an historical-ecological consideration of the folk-urban continuum in Yucatán, *América Indígena*, **22**, pp. 32–51; Wagley, C. (1957). Plantation-America: a culture sphere, *in* Rubin, V. (ed.). *Caribbean Studies: a Symposium*, New York. pp. 3–13. For the evolution of the large estate in Hispanic America there is Chevalier, F. (1952). *La Formation des Grandes Domaines au Méxique*. Paris; Chevalier, F. (1960). La grande domaine au Méxique du XVIe au début du XIXe siècle, *1st Internat. Conference of Econ. Hist.*, Stockholm, pp. 399–407; Chevalier, F. (1963). The North Mexican hacienda: eighteenth and nineteenth centuries, *in* Lewis, A. R. and McGann, R. (eds.). *op. cit.* pp. 95–107. For the adjustments involved in adapting the *encomienda* system to land rather than labour-based agricultural units see: Faron, L. C. (1966). From encomienda to hacienda in Chancay valley Peru: 1533–1600, *Ethnohistory*, **13**, pp. 145–81; Zavala, S. (1948). De encomiendas y propiedad territorial en algunas regiones de la América española, *Estud. Indianos* (Mexico), **10**, pp. 2–12; Chamberlain, R. S. (1954). Simpson's 'The Encomienda in New Spain' and recent encomienda studies, *Hisp. Amer. Hist. Rev.*, **34**, pp. 238–40; Service, E. R. (1951). The encomienda in Paraguay, *Hisp. Amer. Hist. Rev.*, **31**, pp. 230–52, demonstrates the effect of regrouping Guaraní Indians into large *pueblezuelos*. For central Chile there exists the detailed study by Baraona, R. and Aranda, X. (1961). *Valle de Putaendo: estudio de estructura agraria*. Santiago; for Venezuela see Brito-Figueroa, F. (1963). *Estructura Económica de Venezuela Colonial*. Caracas.

28 Many of the historical works listed above are guilty of this, but it is a notable fact that relatively few geographers have likewise attached any significance to the colonial rural and urban settlement patterns and processes of change. See Parsons, J. J. (1964). The contribution of geography to Latin American Studies, *in* Wagley, C. *op cit.* pp. 33–86.

29 For example, Bishko, C. J. (1952). The peninsular background of Latin American cattle ranching, *Hisp. Amer. Hist. Rev.*, **32**, pp. 491–515.

30 For example Ratekin, M. (1954). The early sugar industry in Española, *Hisp. Amer. Hist. Rev.*, **34**, pp. 1–19.

31 Brito-Figueroa, F. (1963). *op. cit.* discusses such commercial connections as existed in colonial Venezuela.

32 Described by West, R. C. and Augelli, J. P. (1966). *Middle America: its Land and Peoples*. Englewood Cliffs.

33 Baraona, R. and Aranda X. (1961). *op. cit.*; Brito-Figueroa, F. (1963). *op. cit.*; McBride, G. M. (1936). *Chile: Land and Society*. New York.

34 Cattle ranching in the New World has been relatively well documented in Brand, D. (1961). The early history of the range cattle industry in Northern Mexico, *Agric. Hist.*, **35**, pp. 132–139; Mauro, F. (1961). Mexico y Brasil: dos economías coloniales comparadas, *Hist. Mexicana*, **10**, pp. 570–87; Morrisey, R. J. (1951). The northward expansion of cattle ranching in New Spain, *Agric. Hist.*, **25**, pp. 115–21; Dusenberry, W. H. (1963). *The Mexican Mesta*. Urbana; Crist, R. (1937). *Etude géographique des llanos de Venezuela*. Grenoble; Deffontaines, P. (1946). *Contribution à la géographie pastorale de l'Amérique latine*. Rio de Janiero.

35 Chevalier, F. (1963). *op. cit.* p. 98; also mentioned in Peru in Vargar Ugarte, R. (1956). *Historia del Peru: Virreinato (siglo XVIII)*. Lima. pp. 437–8.

36 West, R. C. and Augelli, J. P. (1966). *op. cit.* p. 292. Further details of the *rancho* are to be found in Vaquero, J. (1946). El Rancho centroamericano, *Revista de Indias*, **26**, pp. 905–12.

37 For presidio forts in Mexico see Powell, P. W. (1952). *Soldiers, Indians and Silver. The Northward Advance of New Spain, 1550-1600.* Berkeley. pp. 141f. Also useful is Bobb, E. B. (1962). *The Viceregency of Antonio Bucareli in New Spain, 1771-9.* Austin. The line of forts extending across the Argentine pampa in the late eighteenth century are to be seen in Robinson, D. J. (1970). Trade and trading links in western Argentina during the Viceroyalty, *Geogr. J.*, **136**, pp. 24–42.

38 Described for Colombia in Friede, J. (1944). *El Indio en lucha por la Tierra. Historia de los resguardos del macizo central Colombiano.* Bogotá

39 See Robinson, D. J. (1970). *op. cit.* p. 25.

40 Congregations in Colombia are described in Fals Borda, O. (1957). Indian congregations in the New Kingdom of Granada: land tenure aspects, *The Americas*, **13**, pp. 331–51. In Mexico the most important studies are Gibson, C. (1945). The transformation of the Indian community in New Spain, 1500–1800, *Cahiers d'Histoire Mondiale*, **11**, pp. 581–607; Cline, H. F. (1949). Civil congregations of the Indians in New Spain, 1598–1606, *Hisp. Amer. Hist. Rev.*, **29**, pp. 349–69.

Conclusion

STUART PIGGOTT

Conclusion

It is of course impossible for anybody, even the greatest polymath, adequately to summarize the results of our three days' discussions. I want to make some general points first, and then one or two detailed points which occurred to me in the course of the various communications and discussions. I think we must all recognize that this has been a momentous occasion. We have had a face-to-face encounter between social anthropologists and archaeologists, and one of the main themes that obviously emerged was that on both sides views were put forward that were sometimes compatible and sometimes incompatible as between the protagonists. This in itself was an extremely valuable thing to have achieved. I don't remember an occasion, with so many people qualified to speak all brought together, when this has happened before; and the problem as it presents itself to me is not only the obvious one of sorting out where our common ground lies, where in fact the two disciplines can assist one another in a further understanding of their own fields of study, but perhaps almost more important, the finding out where common ground can hardly be said to exist. I think this, although it is phrased as a negative, is something of positive value, because if we can eliminate the areas in which we feel we cannot help each other in practice and theory, then we shall not waste time and intellectual effort in pursuing lines of thought which will not be helpful to both sides.

I am an old and old-fashioned archaeologist, brought up in empirical and pragmatic ways of approach, and basically more concerned with a first-hand contact with material objects than having a great degree, not of interest, but of practice in theory. By temperament and accident I have been influenced by and attracted to historical disciplines, rather than those concerned with theories of human behaviour and social development and I hope, since this is perhaps not to be a wholly solemn occasion, you will not mind my saying that over the past three days I was reminded of what a friend of mine once wrote about the later novels of Henry James. James, he said, "takes us into a rich and varied world where no sound is heard save the sound of splitting hairs, and no more violent spectacle confronts us than that of a gentleman in the act of changing his own mind". We have heard the sound of splitting hairs, and, more important, we have seen gentlemen in the act of changing their own minds as a result of an interplay of different ideas; this has been all to the good, and those of us who have changed their minds are the better for it.

I find myself thinking there was perhaps one basic area of difficulty, that between the assumption that human behaviour by and large is, in all times and places, susceptible to study mainly, and perhaps only, in terms of a natural science, and on the other side, the view, largely held by historians, that the variables are too great for this approach to be applied to human history. I say human history advisedly, to avoid splitting history from pre-history, in the wider sense of the study of mankind in the past, whether this is literate or non-literate, as the distinction between prehistory and history is really that between non-literate and literate societies in the past. The main field in which there appeared to be a coming together of the disciplines was in the matter of the hunting/gathering communities, who in the course of the seminar made frequent inroads or raids from their territory into almost every aspect of the situation. With such communities there appeared to be a possibility of a fruitful application of methods which would not be those normally used by historians.

A sort of unease set in, it seemed to me, when one came to the establishment of village communities and it becomes more acute when entering into historically documented fields. As various speakers brought out, the nature of the documents on which any historical interpretation could be put are very often themselves scanty and ambiguous, and of course very often to be taken not in the sense in which they were originally written, but that is a thing historians have been familiar with all their lives. But in the field of anthropological comparison I was very dubious about the validity of comparing some of the very disparate societies, disparate in time and space and in structure, that were instanced. The New Guinea Highlanders were frequently being quoted, but also modern industrialized cities, early agriculturalists, developed ancient civilizations, and so on, right from man's beginnings up to modern times. Here I find myself unconvinced by the underlying assumption that there are unchanging or relatively unchanging patterns built into the behaviour of the hominids, which express themselves in any conditions of time and space.

Professor Wheatley made a point, as others had before, that we must avoid semantic confusion when we use certain words and names for things. We use the word "town" or "city", and in the classical world this was *polis* or *urbs*, and what we have to consider is whether we are falling into that well-known trap of confusing names with actual things, and while using a name embodying modern concepts, we forget that these concepts were not those of literate antiquity, and therefore by reasonable assumption not of non-literate antiquity. Consider for instance the Latin use of *urbs* in relation to the Celtic population of barbarian Europe. What did a Latin writer really mean when he called a hill-fort, *urbs*, as indeed on occasion they did? It did not mean it was

a

b

OATES: *a* View looking south-east from Choga Mami towards Mandali showing spoil banks of an ancient canal and, in the foreground, a modern irrigation channel identical with those of the 6th millennium.

b Looking north-east from Choga Mami towards the first line of the Zagros foothills, at the edge of the alluvial plain.

KEMP: Limestone libation tank from Memphis, apparently depicting the symbolically fortified enclosure wall of the temple of Ptah. Nineteenth Dynasty, thirteenth century B.C. Length 55 cm, height 31·5 cm. (*Photograph by courtesy of The University Museum, University of Pennsylvania.*)

PLATE 2

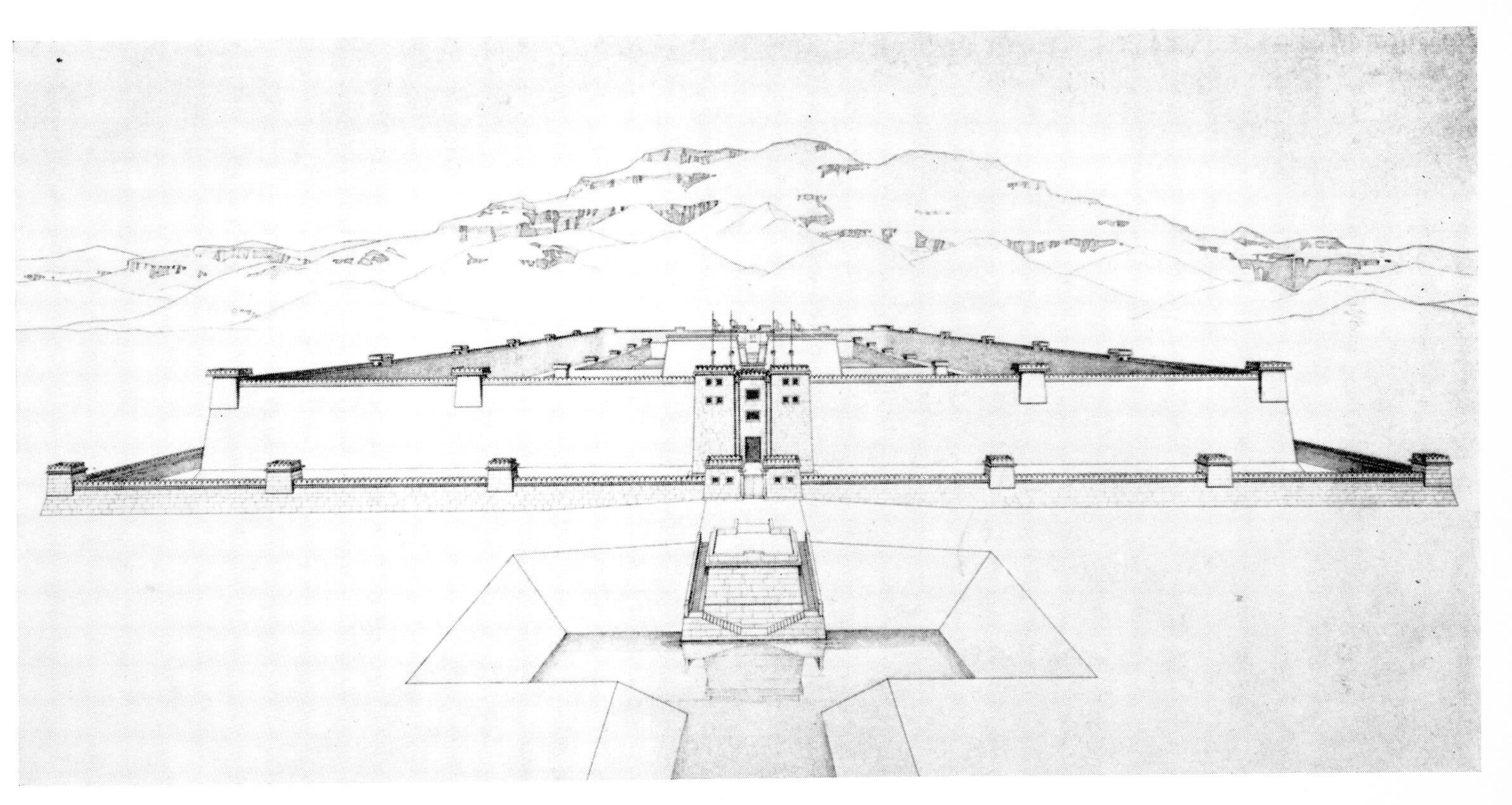

UPHILL: Approach from the canal. (After Hölscher, U. *Medinet Habu*, Vol. IV, Part II. University of Chicago Press, 1941.)

PLATE 3

a

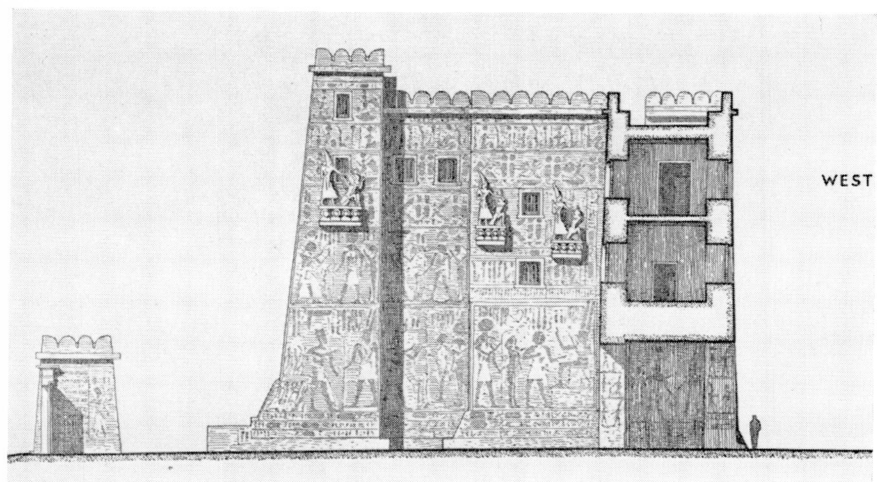

b

UPHILL: *a* Gate tower.

b Detail of gateway with statues. (After Hölscher, *op. cit.*)

PLATE 4

Printed in Germany by Ganymed, Berlin

UPHILL: Window of Appearances. (After Hölscher, *op. cit.*)

PLATE 5

CRANSTONE: "Summer and winter habitations in Kamtschatka", engraved by S. Smith, from the drawing made by J. Webber during Captain Cook's third voyage. In the foreground is the semi-subterranean winter house, entered by an opening in the roof; behind are the summer dwellings on piles. The two sorts of habitation would not normally be seen together at one site. (*By permission of the Trustees of the British Museum.*)

PLATE 6

a

b

CRANSTONE: *a* The skin covered *tipi* of the Dakota Indians, drawn by R. O. Sweeney in 1852. The shaped cover of bison skins and the adjustable smoke-flap are clearly seen. (*By permission of the Trustees of the British Museum.*)

b A bark-covered shelter with a pole frame of the *tipi* type, of the Dakota Indians, drawn by R. O. Sweeney in 1852. (*By permission of the Trustees of the British Museum.*)

PLATE 7

CRANSTONE: A Tongan assembly-house of tie-beam and king-post construction, engraved by W. Sharp from the drawing made by J. Webber during Captain Cook's third voyage. (*By permission of the Trustees of the British Museum.*)

PLATE 8

like Rome, although he used the same word for the city, the Imperial City, as he would for this barbarian earthwork enclosure, the functions of which, or the functions of any hill-fort, we very imperfectly understand. Let us avoid the ancient belief in the magic power of words, which can make us turn names into real things, and so fulfil a primitive conviction that when you have given a thing a name you have a command over it, like knowing somebody's secret name. It is possible to persuade oneself that having named a concept, therefore, it actually exists and can be dealt with accordingly. There is another point relating to the wisdom of using modern industrialized communities as parallels when we are studying very unindustrialized communities in the remote past. This is the danger of importing modern concepts into past cultures or civilizations, a point which occupied the attention of both Henri Frankfort and Ian Richmond and both of them pointed out the dangers that happened if we did import, particularly into literate civilizations where we could check to some extent on the thought of the period, ideas which were our own and therefore inapplicable.

We were reminded at the outset of the seminar that I was the first person to adapt the term "model", taken from the sciences, in reference to the interpretation of archaeological evidence. The inevitable trite comparison with Frankenstein is perhaps rather more applicable in this case as his monster was made by assembling dismembered fragments of humanity, and mine of fragments of evidence for humanity. Now that "model" has become such a vogue word in archaeology I can only apologize. I think however that the concept is potentially useful, but that we must recognize that there are not only explicit models which we know we are using, but implicit models which we do not always realize we are using. These are models which have turned into built-in assumptions on our part, as a result of the social conditioning which makes us and our models as much social artefacts as a house or a handaxe. It is impossible to take a wholly detached view: we are all children of our age and we think in terms of that age, so that our models, explicit and implicit, are equally conditioned and cannot be thought to embody immutable truths. But we do need some kind of a framework or a working hypothesis, or a group of these, and if it is to be a working hypothesis, its utility depends on whether it works (or appears to us to work). As we all know, the idea of conceptual models is primarily a scientist's idea, and it is interesting but not surprising that Charles Darwin, writing in the 1860's of what geology was like 30 years before, should have gone straight to the point. The geologists of the time, he wrote, were like people going into a gravel pit, and there counting and describing the colours of pebbles as they lay. "How odd it is," said Darwin, "that anyone should not see that all observation must be for or against some view if it is to be of any service"—in other words within a conceptual model, some framework of ideas. Darwin was writing of one of the earth sciences, but the question really is, can we use this approach to man and his activities, wholly and completely, and sometimes uncritically?

I must say that when I look at the archaeological and historical past I tend to see more muddles than models. But another geologist said a year or two ago in a discussion similar to ours, must our philosophical thinking be as clean as our laboratory glassware? In other words, have we got to think strictly in terms of one or more of possible models, and does it perhaps not matter if we mix them? When I originally wrote about models in archaeology I did say that you could use several models at once, and that all of them, or some of them, or none of them could be true. Here I feel we may really find our greatest common ground and as an archaeologist I believe we should not simply take one particular type of model, but should in fact appreciate that there are alternative ways of attempting to understand the past, provided we can utilize these various approaches within a supra-model, or whatever you like to call it, within which they function satisfactorily. Above all, models must be servants and not masters.

These assumptions or supra-models must themselves be defined, and the most obvious controlling factor in their formulation must always be the nature of archaeological evidence, its limitations and the legitimate inferences that can be drawn from it. As a pessimist by nature in all things, including archaeology, I become increasingly impressed by the imperfect nature of archaeological evidence, and by what appears to be a fact but perhaps is an assumption too, that there are certain built-in limits to the evidence which we cannot extend however much we advance our techniques. Over the last generation the techniques of field archaeology, and those of the subsequent examination of archaeological material, have been enormously enhanced and improved, very largely in the past by interaction with the natural and physical sciences; now we may hope not only with these, but with the social sciences and allied disciplines as well.

Now many years ago in America, Professor Christopher Hawkes gave a paper discussing the limitations of archaeological evidence and in it he drew up a scale of ascending difficulty in the use of what he called text-free archaeological evidence, as opposed to text-aided. Technology is something that can be illuminated by direct archaeological evidence because archaeology is based to a large extent on the products of technology, and from that, with an interdisciplinary approach with natural scientists, we can move towards inferring something of subsistence-economics and of man as a part of his ecological surroundings. But when to try to infer such things as social structure or in the broadest sense religious practices archaeological evidence becomes almost wholly ambiguous. In a settlement site a building larger and detached from the rest might be interpreted as the house of headman of the settlement or a club house, or a temple, or none of these things. It is only a building with a differentiated function and to determine that function is not inference, but guesswork. In a cemetery, if some of the graves are large and richly furnished, it might be thought a reasonable inference to see an indication of social stratification, but a cemetery in which all the graves are alike inhibits inferences of any kind of social terms. Such limitations in archaeological evidence will always be there. We cannot infer language; we cannot infer literature, tradition, dance, song; we cannot infer beliefs. We can at most see the evidence of activities presumed religious activities, because they are not utilitarian in our own society. Social structure again is not susceptible to direct archaeological inference and we should not try to persuade ourselves that it is, just because we would like it to be. Archaeologists should be more modest about the capabilities of their craft, and admit that it can only inform us of a limited range of man's activities. Here history comes to our aid, and here we should take advantage of the existence of historical approaches to the past, because they are approaches to the past, different from those of the sciences. It is right that the historical model should be another model, because history is another discipline, with rules of its own.

We can profitably utilize co-operation with historians and historical thinking if we are working in the prehistoric field, and if we are working in the historical field we should be foolish to think that we should not think historically. I hope we shall also be able to use the other approaches, the other models in the past, which have been put forward by the social anthropologists, and I do hope this is not going to be one-way traffic. What I would like to feel has come out of our seminar is that traffic in ideas about the human past should be multi-directional, between anthropologists, archaeologists, historians, social scientists and ecologists. One of the gravest errors of method, to my mind, is that arising from a belief that any one approach has a monopoly of "truth": there is no truth about the past, as Collingwood saw, only a series of pasts-as-known by means of the varied approaches available to the student.

I would now like to take up one or two points of detail, and the first is one I also brought up at the seminar on the domestication of animals and plants conference, the question of the circumstances of rejection or acceptance by communities of new ideas which may involve technological changes carrying with them social changes. It was latent in the denunciation of the phrase "incipient agriculture", which implied that hunter-gatherers inevitably evolved into agriculturalists, analogy that assumes that developments in technology follow a pattern of natural evolution irrespective of human references. A point more than once brought out was the way people can do things for irrational reasons, the breaking up and moving of hunting communities, and moving of other groups of people just for the hell of it. There are very good historically documented instances of this which were not mentioned, including not only the post-Roman Migration Period in Europe, but the archaeologically and historically attested movements of the pre-Roman Celtic peoples. While the Celtic tribes doubtless may have had population pressure and other contributory economic factors affecting them this in itself can hardly explain in all instances such sudden moves of people as that of the Helvetii, and of so many other Celtic tribes we know by name. Some, like those who went into Cisalpine Gaul, doubtless moved because there were better pickings south of the Alps, but unless you have a certain footloose quality, which was perhaps more common in antiquity in barbarian communities than we usually credit, such movement is less likely.

Malaria was mentioned as a factor influencing human settlement, but animal diseases that do not necessarily affect man may also be significant. Here one thinks of liver fluke in its relationship to sheep populations: Higham has recently studied this in connection with the distribution pattern of the middle third millennium Neolithic settlements in Denmark where sheep-pasturing people are not found along rivers and near lakes where the fresh-water snail essential to the liver-fluke life-cycle breeds, but they took every advantage of the good pasture along the coast where salinity prevented this. Another point not always appreciated in connection with the mobility of the Celtic and other ancient barbarian peoples, is the concept of war as fun and a Good Thing. We tend to reprobate war for good and sufficient reasons, but in many societies, especially those with a social stratification with a warrior-aristocracy warfare is the only way in which the members of this group can demonstrate the socially approved values for which they exist: this is often bound up with concepts of honour and the maintenance of feuds, from the Iliad to the sagas and beyond. Transhumance again is a factor we ought sometimes to bear in mind when dealing with ancient settlements, particularly in north Europe, where on uplands which in any climatic and floral conditions could carry nothing more than rough pasture, settlements occur which on excavation appear to show little or no signs of occupation. This situation might be explained in terms of transhumance and of temporary camping, perhaps in some sort of cycle. This was brought home to me many years ago when I once observed modern groups of similar structures on one of the Himalayan passes over the Rohtang-la from India to Ladakh, where they formed a series of camping sites, near a spring and pasture of what appeared to be hut foundations with nothing left behind except perhaps a few rags; if they had been prehistoric they would have been archaeologically interpreted as representing a very poverty-stricken part of the community. They were in fact the tent-stances which were used by the intinerant wool and salt merchants who were by far the richest section of the population, and drove large herds of goats on this trade route. The question of circular and square houses was discussed more than once: as far as prehistoric Europe is concerned there is a very curious and interesting geographical boundary which is very roughly the River Rhine, to the east of which the rectangular house is normal throughout prehistory, and to the west of which is a province, including the British Isles, Gaul, the Iberian peninsula and the west Mediterranean, where the prehistoric house-type is similarly consistent, but circular. If the circular house is related to a tent-type, this can hardly provide an explanation of such a curious dual distribution, though perhaps it might apply to the circular houses from earlier than the third millennium B.C. in east Turkey and the Caucasus region. The Kura-Araxes culture and its predecessors employed circular houses, which did in fact at a later stage turn into square houses simply by straightening their sides, but leaving rounded corners, and here is a curious house development over a very large area.

The question of markets and fairs deserves final mention: their potential existence has perhaps been unappreciated by prehistorians. They must have played an important part in many economies and we know they did in the early Celtic society, and particularly early Ireland. Here in late prehistoric times, we can infer from archaeology and from the later vernacular texts the existence of more or less sedentary, mainly pastoral economy based on cattle, with settlements of discrete farms and homesteads rather than nucleated villages or towns and periodic fairs, which went right on into the Middle Ages, were held. These were a social necessity quite apart from the question of interchange of goods and all the things one associates with a market, and the fun and games, which also go with a fair. But one sees from the texts what an important role they filled in a non-literate society for the promulgation of edicts and of regulations and laws, and the administration of justice by the chieftain of that particular group of small tribes that attended one of these central fairs. These fairs were held normally at the sites of ancient burial mounds, reminding one of the occurrence of market places and fairs together with shrines in many places and times.

I must apologize for these disjointed remarks and I would like to end by going back to what I said at the beginning. This has been a momentous occasion of scholarly exchange which we are unlikely to forget and I would like to think that we shall all go away, as I shall, feeling we have gained much from the last three days. We may be exhausted, but we have at least been exhausted in a profitable pursuit. When we recover and recollect in tranquility our impressions I hope on all sides we will think that the weekend has been worthwhile, stimulating, full of new ideas, and of promise for the future.

INDEX

General (with tribes in bold)

Index

Sites (in italics) and localities

INDEX

Authors